SECOND EDITION

Introductory Electronic Devices and Circuits

ELECTRON FLOW VERSION

Robert T. Paynter

St. Louis Community College at Forest Park
St. Louis, Missouri

Regents/Prentice Hall,
Englewood Cliffs, New Jersey 07632

Library of Congress Cataloging-in-Publication Data

Paynter, Robert T.
    Introductory electronic devices and circuits : electron flow
version / Robert T. Paynter.—2nd ed.
        p.   cm.
    Includes bibliographical references and index.
    ISBN 0–13–482985–9
    1. Solid state electronics.   2. Operational amplifiers.
I. Title.
TK7871.85.P39   1991
621.381—dc20   90–47582
                                              CIP

Editorial/production supervision: *Eileen M. O'Sullivan*
Cover design: *Lorraine Mullaney*
Cover photo: *Image Bank/Michel Tcherevkoff*
Manufacturing buyer: *Mary McCartney* and *Ed O'Dougherty*
Page layout: *Lorraine Mullaney*
Photo Editor: *Lorinda Morris-Nantz*
Photo Research: *Tobi Zausner*
Acquisitions Editor: *Holly Hodder*

 © 1991, 1989 by Prentice-Hall, Inc.
A Division of Simon & Schuster
Englewood Cliffs, New Jersey 07632

Printed in the United States of America

10 9 8 7 6 5 4 3

ISBN   0-13-482985-9

Prentice-Hall International (UK) Limited, *London*
Prentice-Hall of Australia Pty. Limited, *Sydney*
Prentice-Hall Canada Inc., *Toronto*
Prentice-Hall Hispanoamericana, S.A., *Mexico*
Prentice-Hall of India Private Limited, *New Delhi*
Prentice-Hall of Japan, Inc., *Tokyo*
Simon & Schuster Asia Pte. Ltd., *Singapore*
Editora Prentice-Hall do Brasil, Ltda., *Rio de Janeiro*

This book is dedicated to my parents, Bill and Mary, and to all parents who, like mine, cared enough to stay actively involved in the education of their children.

Contents

1

Fundamental Solid-State Principles   1

2

Diodes   2

3
Common Diode Applications: The Basic Power Supply 76

4
Common Diode Applications: Clippers, Clampers, Multipliers, and Displays 137

5

Bipolar Junction Transistors 175

6

DC Biasing Circuits 214

7

Common-Emitter Amplifiers 262

8

Other Transistor Configurations 315

9

Power Amplifiers 354

10

Field-Effect Transistors 417

11

MOSFETs   482

12

Operational Amplifiers   513

13

Amplifier Frequency Response

564

14

Negative Feedback

634

15

Tuned Amplifiers

677

16
Additional Op-Amp Applications

17
Oscillators

21
Special Applications Diodes

☐Appendix A
Specification Sheet Guide

☐Appendix B
Approximating Circuit Values

☐Appendix C
H-Parameter Equations and Derivations

☐Appendix D
Selected Derivations

☐Appendix E
Glossary

☐Appendix F
Answers to Selected Odd-Numbered Practice Problems

Index

The National Association of Radio and Telecommunications Engineers (NARTE) recommends this text as a reference for its Class III through Class I technicians' license examinations

Preface
To The Instructor

Since the release of the first edition of *Introductory Electronic Devices and Circuits* in the fall of 1988, I have received a lot of input from a wide variety of sources. While most of this input has been very positive, it has almost always included suggestions for improvement. It is in response to these suggestions that the second edition of the text has been written.

All too often, revisions of textbooks are written in such a way that they end up being little more than reproductions of previous editions. While many of the characteristics of the first edition have been retained, a review of this text and its supplements will show that this is much more than a "new-cover" revision.

New Features

- The use of *performance-based* objectives has been greatly expanded. The chapter objectives have been narrowly defined and written in checklist form to allow the students to track their progress.
- Objective identifiers have been included in the margins of the text and at the end of each review question and practice problem. This helps the students tie the objectives, text material, review questions, and practice problems together.
- *Summary illustrations* have been developed and included throughout the text. These illustrations list key points and provide comparisons of the various components and circuits.
- Margin notes have been added that stress the differences between theory and practice.
- Key terms listings have been added to each chapter.

- The number of practice problems and review questions has increased significantly. A section of challenging practice problems has been added to most chapters.
- New emphasis has been placed on the use of device specification sheets, the factors that affect component substitution, and the use of *worst-case* circuit analysis as a troubleshooting tool.
- Coverage of *discrete and integrated voltage regulators* and *optoelectronic devices* has been added.
- Coverage of the following areas has been expanded and improved: *diode ratings*, *clippers*, *clampers*, *voltage multipliers*, *power supplies*, *BJT ratings*, *FET ratings*, *negative feedback*, *thyristors*, *special applications diodes*, and op-amp applications, such as *integrators and differentiators*, *precision diodes*, *summing amplifiers*, and *active filters*.

In-Text Learning Aids

- Functional use of full-color to highlight key concepts, sections, and topics.
- Over 1000 illustrations and figures.
- Over 200 examples that demonstrate practical problem solving techniques.
- Over 2000 combined review questions and practice problems.
- Margin notes that define key terms, discuss practical considerations, and help your students to locate specific topic discussions and problem solving examples.
- Practice problems immediately follow most examples.

Extensive Supplements Package

For the Instructor:

- An Annotated Instructor's Edition with Supplements Sampler
- An Instructor's *Solutions Manual* that contains the step-by-step solutions to *every* practice problem in the text (with the exception of the Suggested Computer Applications Problems).
- An *Instructor's Resource Manual* that contains chapter outlines, a brief discussion of each exercise in the lab manual, and lab exercise solutions.
- A *Test Item File* that contains over 500 questions and problems.
- A set of *acetates* that can be used to enhance in-class discussions.
- A *Transparency Masters Manual* that contains enlarged versions of over 100 of the textbook illustrations.
- A Transparency Package
- Computerized Testing Service
- Video Cassette (Six, 10–15 minute lectures on key electronics topics)

For the students:

- An *Interactive Software Diskette* and *Student Study Guide* combine with the textbook to provide your students with over 3000 review questions and practice problems. Both the diskette and study guide contain chapter summaries.
- An accompanying *Lab Manual* is available that contains over 60 class tested exercises.

Review, Development, and Class Testing

The text and supplements have been reviewed continually during the entire development process. The most important input provided during this time has been the feedback from my students. Their suggestions, criticisms, and demonstrated learning have had a profound effect on the final form of the book.

My personal thanks is extended to the following individuals for their technical comments, suggestions, and/or reviews of the revised manuscript:

Bruce Bothwell, Chemeketa Community College
Ernest Arney, ITT Technical Institute, Indianapolis, IN
Donald R. Montgomery, ITT Technical Institute, Houston, Texas
Alan H. Czarapata, Montgomery Community College, Rockville, MD
Nasser Rashidi, Shawnee State University
Bruce Bosler, ITT Technical Institute—Nashville
Tom Bingham, St. Louis Community College at Florissant Valley
Scott G. Bisland, Equipment Engineer, Sematech
Dan Landiss, St. Louis Community College at Forest Park
Roger White, Chemeketa Community College
Brian Wirthlin, Systems Engineer, Beaumont Graphics
Walter R. Miller, St. Louis Community College at Forest Park
Neil F. Casey, Tampa Technical Institute (NEC)
William Muckler, St. Louis Community College at Forest Park
Clive Menezes, Microcomputer Technology Institute
Bernard Rudin, Philadelphia Community College
Firoze Gandhi, DeVry Institute of Technology—L.A.
Jim Deeter, NEC, Dallas, TX
Larry Leno, NEC, Houston, TX
Bob Leonard, NEC, Irvine, CA
Jerry Terrebrood, NEC, West Des Moines, IA
Kevin Price, NEC, San Bernardino, CA
John Hamilton, NEC, Irvine, CA

I would also like to thank the following individuals for participating in the developmental meetings that determined the direction of the text revision:

Jerry Reeder, IVV Tech, Indianapolis, IN
William Flowers, Kilgore College, Kilgore, TX
Gary Cardinale, DeVry Institute of Technology, Woodbridge, NJ
Robert Svarrer, Technical Career Institutes, New York, NY
Fred Nabatian, Technical Career Institutes, New York, NY
Robert Jones, Gulf Coast Community College, Panama City, FL
Bernard Rudin, Philadelphia Community College, Philadelphia, PA
William Burget, ITT Technical Institute, West Covina, CA
Anthony Baratta, NEC, NJ
George Haversang, NEC, NJ
Thomas Bingham, St. Louis Community College at Florissant Valley, MO
Chester Howard, American River College, Sacramento, CA
Neil F. Casey, Jr., NEC, Tampa, FL
Randall Youmen, Brevard Community College, Orlando, FL
Don Markel, DeVry Institute of Technology, Woodbridge, NJ
Bill Barnes, Middlesex Community College, NJ

Chapter Organization

Chapter 1 introduces the reader to the principles of atomic theory that relate to the operation of solid-state components. The coverage of the atomic structure of semiconductors is complete and detailed.

Chapter 2 introduces the reader to pn junction diodes, zener diodes, and LEDs. Other diodes, such as varactors and Schottky diodes are covered in a later chapter (Chapter 21). Chapter 2 is limited to the most common types of diodes to spare the reader from having to suffer through a diode overdose.

Chapter 3 covers the use of pn junction diodes and zener diodes in basic power supply circuits. Rectifiers, filters, and zener voltage regulators are covered in detail. The reader is first introduced to practical troubleshooting in this chapter, with the *worst-case analysis* and *troubleshooting* of a simple (but complete) dc power supply.

Chapter 4 covers the operation of clippers, clampers, voltage multipliers, and multisegment displays. The coverage of each of these circuit groups is complete and detailed.

Chapter 5 introduces the reader to the basic operating theory of bipolar junction transistors. Included in this chapter are sections on transistor ratings and specification sheets, as well as special-applications BJTs and practical BJT testing.

Chapter 6 introduces the reader to BJT biasing circuits. The operation, analysis techniques, troubleshooting, and relative strengths and weaknesses of each are addressed.

Chapter 7 introduces the reader to the ac operating principles of common-emitter amplifiers. The chapter includes the principles that are used in practical ac circuit analysis and troubleshooting.

Chapter 8 covers the operation, applications and troubleshooting of emitter followers, common-base amplifiers, and Darlington amplifiers.

Chapter 9 covers power amplifiers. The emphasis is placed on amplifier compliance, power calculations, and efficiency. Class A, class B, and class AB amplifiers are covered. The detailed coverage of class C amplifiers is reserved for Chapter 15.

Chapter 10 and 11 cover JFETs, MOSFETs, and their related circuits. The biasing circuits for these components are covered, with the emphasis placed on bias stability. The applications and troubleshooting of FET circuits are also covered in these chapters.

Chapter 12 covers op-amps and basic op-amp circuits. The foundation is laid for the coverage of op-amp circuits in the applications chapters that make up the bulk of the remaining text.

Chapter 13 discusses the frequency considerations that are involved in discrete and integrated circuit operation. The reader is introduced to frequency response curves, Bode plots, decibels (dB), and the effects of input frequency on all aspects of amplifier operation.

Chapter 14 introduces the reader to negative feedback and its effects on integrated and discrete amplifiers.

Chapter 15 covers active filters, discrete tuned circuits, and class C amplifiers. This chapter lays the foundation for continuing into communications electronics.

Chapter 16 covers several types of op-amp circuits that do not easily fit into the other chapters. Included are comparators, integrators, differentiators, summing amplifiers (including adders and subtractors), and precision diodes.

Chapter 17 covers the operation, applications, and troubleshooting of common oscillators.

Chapter 18 covers the operation of solid-state switching circuits. This chapter lays the foundation for continuing into the area of digital electronics.

Chapter 19 provides complete coverage of thyristors and optoelectronic devices. Included are the commonly used thyristors and thyristor triggering devices, along with their practical circuit applications. Photo-detectors, optoisolators, and optointerruptors are also covered in detail.

Chapter 20 covers discrete and integrated voltage regulators. Series and shunt regulators are discussed in detail with emphasis being placed on the strengths and limitations of each. The commonly used IC voltage regulator parameters are also discussed in detail.

Chapter 21 provides an introduction to many of the special applications diodes, such as varactors, transient suppressors, constant-current diodes, and Schottky diodes, etc. Emphasis is placed on the common applications of each of the components covered.

The appendices are included to provide the reader with additional information that will help in the reading of the text.

Appendix A includes an index of all of the specification sheets that are referred to throughout the text.

Appendix B has been included to show the reader when and how a circuit may be simplified into an equivalent circuit.

Appendix C covers h-parameter equations and their derivations. This material expands on the coverage provided in Chapter 7.

Appendix D contains the derivations of many of the equations used throughout the text. Although the validity of these equations is assumed in the text, the derivations are included for those readers who wish to know where they come from. In some cases, following the derivations requires an understanding of basic calculus.

Appendix E is a glossary of all the new terminology introduced throughout the text.

Appendix F includes the answers to selected odd-numbered practice problems.

Acknowledgments

A project of this size could not be completed without help from a wide variety of capable and concerned individuals. Some of the following people provided technical support, while others provided well-timed moral support:

Eileen O'Sullivan, Barb Cassel, Mary Carnis, Ray Mullaney, Lorraine Mullaney, Alice Barr, Susan Willig, Holly Hodder, Judy Casillo, Mark Hartman, Jeanne Bauer, Maureen ''Moe'' Lopez, John Hendron, Dan Miller, Audrey Hellman, Carol Bollati Jacoby, Mary McCartney, Ed O'Dougherty, Bunnie Neuman, and Barbara Palumbo.

I would also like to extend my thanks to my developmental editor, Andrea Rutherfoord, for her efforts in the production of this revision.

Finally, a special thanks goes out to my family and students for their constant support and patience, and to Susan for keeping everything from falling apart.

Bob Paynter
November, 1990

To The Student

"Why am I learning this?"

Have you ever found yourself asking this question? Most students have; especially those who are studying electronics. The funny part is that many electronics textbooks don't take the time to even consider such a fundamental question, much less answer it.

I believe that any subject is easier to learn if you know *why* you are learning it. For this reason, we are going to take a moment to discuss several things:

1. What *electronic devices* are
2. Why the study of electronic devices is important
3. How this area of study relates to the other areas of electronics
4. How you can get the most out of your study of electronic devices

One of the components that you have already studied is the *resistor*. In your basic dc electronics course, you were taught just about everything there is to know about resistors. Why? Because resistors are used in virtually every type of electronic circuit and/or system. If you take a moment to flip through this book, you'll see that there are very few circuits that do not contain at least one resistor. It would make sense, then, that you must thoroughly understand resistors if you are going to be successful in understanding the area of electronic devices. The point I am making here is simple: *Each area of electronics is learned because it contributes to understanding the next.* If you want to understand electronic devices, you must first understand basic dc electronics. And, if you want to understand the areas of electronics that are studied after devices, you must first come to understand the material in this book. Just as the knowledge of resistors is fundamental to this course, so the knowledge of devices is fundamental to the courses that are to follow. This is *why* you are studying devices at this point in your education.

What *are* electronic devices? They are components with *dynamic* resistance characteristics. That is, they are components whose resistance is determined by the voltage applied to them or by the current being drawn through them. Thus, some are *current-controlled resistances* while others are *voltage-controlled resistances*.

Electronic devices are somewhat complex components that are used in virtually every type of electronic system. They are used extensively in *communications systems* (such as televisions, radios, and VCRs), *digital systems* (such as PCs, calculators, etc.), and *industrial systems* (such as robotic and process control systems). Therefore, if you want to work in any of these areas, you must become familiar with the various types of electronic devices and their circuits and applications.

As you can see, the study of electronic devices is critical if you are going to advance beyond the point where you are now. The next question is

"What can I do to get the most out of this course?"

There are several steps that you can take to ensure that you will be successful in learning this area of electronics (or any other for that matter). The first is to realize that *learning electronics requires active participation on the part of you, the learner*. If you are going to learn electronics, you must take an active role in your education. It's like learning how to ride a bicycle. If you want to learn how to ride a bike, you have to hop on and take a few spills. You can't learn how to ride a bike just by "reading the book." The same can be said about learning electronics. You must become actively involved in the learning process.

How do you get involved in the learning process? If you haven't done so, you need to develop several habits:

1. *You must attend class on a regular basis.*
2. *You must take part in classroom problem solving sessions.* This means getting out your calculator and solving the problems along with the rest of the class . . . not waiting for someone else to come up with the answers.
3. *You must do all the assigned homework.* Circuit analysis is a skill. As with any skill, you gain competency only through practice.
4. *You must take part in classroom discussions.* More often than not, classroom discussion can serve to clarify points that may be confusing otherwise.
5. *You must be an active participant in the textbook discussions.*

Being an *active participant* in the textbook discussions means that you must do more than simply "read the book." When you are studying new material, there are several things that you need to do:

1. *Learn the terminology.* You are taught new terms because you need to know what they mean and how and when to use them. When you come across a new term in the text, take time to commit the new term to memory. How do you know when a new term is being introduced? Throughout this text, new terms are identified in the margins. When you see a new term and its definition in the margin, stop and learn the term before going on to the next section.
2. *Use your calculator to work through the examples.* When you come across an example, get your calculator out and work the example for yourself. When

you do this, you develop the skill necessary to solve the problems on your own.

3. *Solve the example practice problems.* Most of the examples in this book are followed by a practice problem that is identical in nature to the example. When you see these problems, solve them. Then you can check your solutions by looking up the answers at the end of the chapter.

4. *Use the chapter objectives to measure your learning.* Each chapter begins with an extensive list of performance-based objectives. *These objectives tell you what you should be able to do as a result of learning the material.* When you have mastered an objective, check that objective. That way, you'll know which of the intended skills you have developed and which ones you still need to work on.

Incidently, this book contains *objective identifiers* that are located in the margins of the text. For example, if you look at page 61, you'll see the number 19 contained in a box in Example 2.16. This identifier tells you that this is the point where you are taught the skill mentioned in objective 18 (see the objective list on page 22). These objective identifiers can be used to help you with your studies in many ways:

1. If you don't know how to perform the action called for in a specific objective, just flip through the chapter until you see the appropriate identifier. At that point in the chapter, you will find the information you need.

2. The review questions that follow each section are also tied into the objective identifiers. For example, look at the group of review questions on page 7. Each question is followed by an objective number. Let's say that you have studied the section, but can't remember the answer to question 7. The question is shown to be tied into objective 5. Therefore, all you have to do to find the answer is to locate the objective identifier that contains the number 5 (see page 5), and you will find the answer somewhere between that identifier and the identifier for objective 6. This will save you a lot of time and trouble when you are trying to locate the answer to a given review question.

3. The end-of-chapter practice problems are also tied into the objective identifiers. This will help you find the information you need in order to solve a given practice problem.

As you can see, the objective identifiers not only provide a guide to learning, they can also be used as a built-in cross reference that can save you a lot of time and trouble when you are working on review questions and practice problems.

One Final Note

There is a lot of work involved in being an active learner. However, the extra effort will pay off in the end. Your understanding of electronic devices will be better as a result of your efforts. Not only will this understanding improve your grades, it will also make learning the remaining areas of electronics much easier. I wish you the best of success.

Bob Paynter
November, 1990

1
Fundamental Solid-State Principles

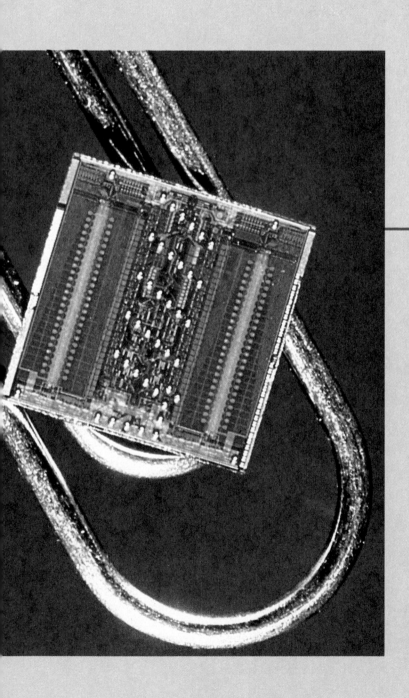

The circuit shown, if constructed with vacuum tubes, would have been approximately 365 cubic meters in size.

OBJECTIVES

☐ 1. Describe the makeup of the atom. (§1.1)

☐ 2. State the relationship between the number of valence electrons for a given element and its conductivity. (§1.1)

☐ 3. List the three semiconductor elements that are most commonly used in electronics. (§1.1)

☐ 4. List the rules that govern the relationship between electrons and orbital shells. (§1.1)

☐ 5. Explain what happens when an electron falls from the conduction band to the valence band. (§1.1)

☐ 6. Briefly discuss *covalent bonding* and list the resulting element properties. (§1.1)

☐ 7. Discuss conduction, from the forming of an electron–hole pair to recombination. (§1.1)

☐ 8. Discuss the relationship between conduction and temperature. (§1.1)

☐ 9. State the purpose served by *doping*. (§1.2)

☐ 10. State the difference between *trivalent* and *pentavalent* elements. (§1.2)

☐ 11. Explain why doped semiconductors are still electrically neutral. (§1.2)

☐ 12. List the similarities and differences between *n-type* and *p-type* semiconductors. (§1.2)

☐ 13. Discuss the results of *diffusion current* through a *pn* junction. (§1.3)

☐ 14. Discuss the forming of the *depletion layer* around a *pn* junction. (§1.3)

☐ 15. List the barrier potential values for silicon and germanium. (§1.3)

☐ 16. Discuss the relationship between depletion layer width, junction resistance, and junction current. (§1.4)

☐ 17. State the purpose served by *bias*. (§1.4)

☐ 18. State the effects of *forward bias* on depletion layer width, junction resistance, and junction current. (§1.4)

☐ 19. State the approximate values of V_F for silicon and germanium *pn* junctions. (§1.4)

☐ 20. List the commonly used methods of forward biasing a *pn* junction. (§1.4)

☐ 21. State the effects of *reverse bias* on depletion layer width, junction resistance, and junction current. (§1.4)

☐ 22. List the commonly used methods of reverse biasing a *pn* junction. (§1.4)

☐ 23. Explain how the voltage across a reverse-biased *pn* junction is determined. (§1.4)

☐ 24. Explain why silicon is used more commonly than germanium in the production of solid-state devices. (§1.4)

As you master each of the objectives listed, place a check mark (✔) in the appropriate box.

Electronic systems, such as radios, televisions, and computers, were originally constructed using *vacuum tubes*. Vacuum tubes were generally used to increase the strength of ac signals (*amplify*) and to convert ac energy to dc energy (*rectify*). Although they were able to perform these critical operations very well, vacuum tubes had several characteristic problems. They were large, fragile, and they wasted a tremendous amount of power through heat loss.

In the 1940s, a team of scientists working for Bell Labs developed the *transistor*, the first *solid-state* device capable of amplifying an ac signal. The term *solid state* was coined because the transistor was solid, rather than hollow like the vacuum tube it was designed to replace. The transistor was also smaller, more rugged, and wasted much less power than did its vacuum-tube counterpart. Since the development of the transistor, solid-state components have replaced vacuum tubes in almost every application.

Semiconductor. Neither an insulator nor a conductor.

Solid-state components are made from elements that are classified as *semiconductor* elements. A semiconductor element is one that is neither a conductor nor an insulator, but, rather, lies halfway between the two. Under certain circumstances, the resistive properties of a semiconductor can be varied between those of a conductor and those of an insulator. As you will see later, it is *this* characteristic of semiconductor elements that makes them useful as amplifiers and rectifiers.

1.1

Atomic Theory

Before we get started, let's take a look at the reason for reviewing atomic theory: *Some* coverage of atomic theory is essential to understanding the characteristics of semiconductors. However, all you need to be concerned with is *what* happens on the atomic level, not *why* it happens. As you read through this section, realize that the material is being presented to give you a *basic* understanding of what happens on the atomic level. If you simply *accept* these principles to be true, you will be much more comfortable with the principles of semiconductors.

Valence shell. The outermost shell that determines the conductivity of an atom.

The atom has been shown to contain three basic particles: the *protons* and *neutrons* that make up the nucleus (core) of the atom, and the *electrons* that orbit about the nucleus. The basic model of the atom, called the *Bohr model* is illustrated in Figure 1.1. The orbital paths, or *shells*, are identified using the letters K through Q. The innermost shell is the K shell, followed by the L shell, and so on. The

FIGURE 1.1

Bohr model of the atom.

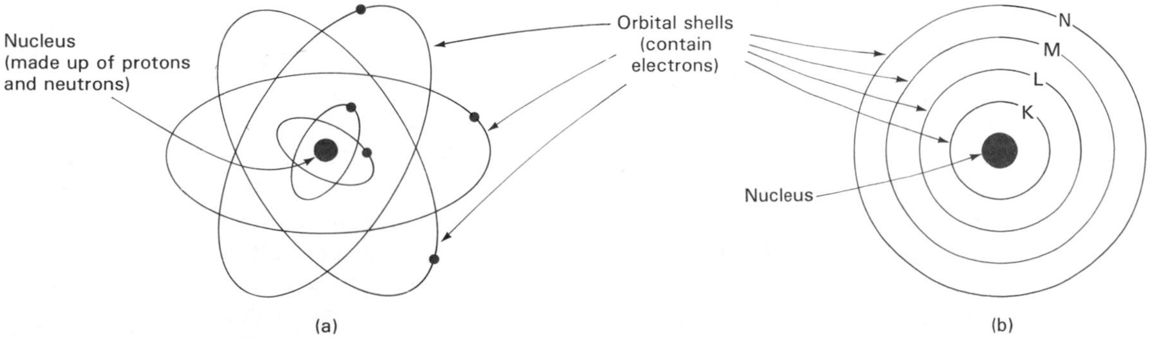

Nucleus (made up of protons and neutrons)

Orbital shells (contain electrons)

Nucleus

N
M
L
K

(a)

(b)

FIGURE 1.2

Semiconductor atoms.

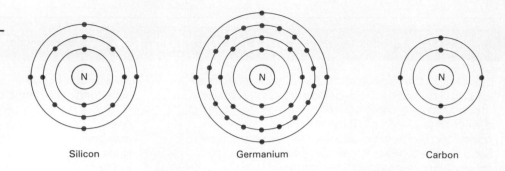

Silicon   Germanium   Carbon

outermost shell for a given atom is called the *valence shell*. The valence shell of an atom is critical because it determines the conductivity of the atom.

The valence shell of an atom can contain up to eight electrons. The conductivity of the atom depends on the number of electrons that are in the valence shell. When an atom has one valence electron, it is a nearly perfect conductor. When an atom has eight valence electrons, the valence shell is said to be *complete*, and the atom is an insulator. Therefore, *conductivity decreases with an increase in the number of valence electrons.*

Semiconductors are atoms that contain four valence electrons. Because the number of valence electrons in a semiconductor is halfway between one (for a conductor) and eight (for an insulator), a semiconductor atom is neither a good conductor nor a good insulator.

Three of the most commonly used semiconductor materials are *silicon* (Si), *germanium* (Ge), and *carbon* (C). These atoms are all represented in Figure 1.2. Note that all these elements contain four valence electrons. Of the semiconductors shown, silicon and germanium are used in the manufacture of solid-state components. Carbon is used mainly in the production of standard resistors.

Charge and Conduction

When no outside force causes conduction, the number of electrons in a given atom will equal the number of protons. Since the charges of electrons (−) and protons (+) are equal and opposite, the *net charge* on the atom is zero. If an atom *loses* one valence electron, the atom will then contain less electrons than protons, and the net charge on the atom will be positive. If an atom with an incomplete valence shell *gains* one valence electron, the atom will have more electrons than protons, and the net charge on the atom will be *negative*.

There are some fundamental laws regarding the relationship between electrons and orbital shells that have been proven over time. These are:

1. *Electrons must travel in one orbital shell or another. They cannot orbit the nucleus in the space that exists between any two orbital shells.*
2. *Each orbital shell relates to a specific energy level. Thus, all the electrons traveling in a given orbital shell will contain the same relative amount of energy.* Note that the energy levels for the shells increase as you move away from the nucleus of the atom. Thus, the valence electrons will always have the highest energy level in a given atom.
3. *For an electron to "jump" from one shell to another, it must absorb enough energy to make up the difference between its initial energy level and the energy level of the shell to which it is jumping.*
4. *If an electron absorbs enough energy to jump from one shell to another, it will eventually give up the energy it absorbed and return to a lower-energy shell.*

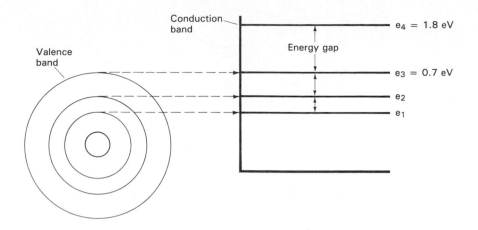

FIGURE 1.3
Silicon energy gaps and levels.

The first three of these principles are illustrated in Figure 1.3. The space between any two orbital shells is referred to as the *energy gap*. The electrons will travel through the energy gap when going from one shell to another, but they cannot orbit the nucleus of the atom in one of the energy gaps.

As Figure 1.3 shows, each orbital shell is related to a specific energy level. For an electron to jump from one orbital shell to another, it must absorb enough energy to make up the difference between the shells. For example, in Figure 1.3, the valence shell, or *band*, is shown to have an energy level of approximately 0.7 electron volt (eV). The *conduction band* is shown to have an energy level of 1.8 eV. Thus, for an electron to jump from the valence band to the conduction band, it would have to absorb an amount of energy equal to

$$1.8 \, \text{eV} - 0.7 \, \text{eV} = 1.1 \, \text{eV}$$

For conductors, semiconductors, and insulators, the valence-to-conduction band energy gaps are approximately 0.4, 1.1, and 1.8 eV, respectively. As you can see, the higher this energy gap, the harder it is to have conduction. This is due to the fact that more energy must be absorbed for an electron to enter the conduction band.

When an electron absorbs enough energy to jump from the valence band to the conduction band, the electron is said to be in the *excited* state. An excited electron will eventually give up the energy it absorbed and return to its original energy level. The energy given up by the electron is in the form of *light* or *heat*.

Energy gap. The space between orbital shells.

Band. Another name for an orbital shell.

Conduction band. The band outside of the valence shell.

eV (electron-volt). The energy absorbed by an electron when it is subjected to a 1-V difference of potential.

Covalent Bonding

Covalent bonding is the method by which atoms complete their valence shells by "sharing" valence electrons with other atoms. Remember, a valence shell is complete when it contains eight electrons. The covalent bonding of a group of silicon atoms is represented in Figure 1.4. To help illustrate the bonding process, each atom has been represented by an octagon (eight-sided figure) containing a square. The octagon represents the valence shell of the atom, and the square is used to identify the electrons that belong to the particular atom. As you can see, the center atom has eight valence electrons, four that belong to the center atom, plus one that belongs to each of the four surrounding atoms. These atoms, in turn, share electrons with four surrounding atoms, and so on. The results of this bonding are:

Covalent bonding holds atoms together.

1. The atoms are held together, forming a solid substance.
2. The atoms are all electrically stable because their valence shells are complete.

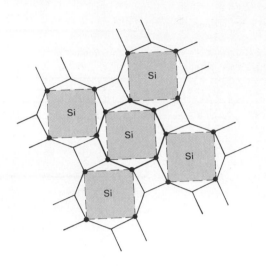

FIGURE 1.4

Silicon covalent bonding.

Intrinsic. Another way of saying *pure*.

3. The completed valence shells cause the silicon to act as an *insulator*. Thus pure (*intrinsic*) silicon is a very poor conductor. The same principle holds true for intrinsic germanium.

When semiconductor atoms bond together in a set pattern like the one shown in Figure 1.4, the resulting material is called a *crystal*. A crystal is a smooth, glassy solid. You have probably seen quartz crystals at some time. Silicon and germanium commonly crystallize in the same manner.

Carbon can crystallize in the same fashion as silicon and germanium, but carbon crystals are too expensive to use in solid-state component production. This is because carbon crystals are *diamonds*.

Conduction

Hole. A gap in a covalent bond.

Electron–hole pair. An electron and its matching valence band hole.

Recombination. When a free electron returns to the valence shell.

Lifetime. The time between electron-hole pair generation and recombination.

When sufficient energy is added to a valence electron, it will jump from the valence band to the conduction band. The result of this action is that a *gap* is left in the covalent bond. This gap is referred to as a *hole*. It would follow that *for every conduction band electron, there must exist a valence band hole*. The term used to describe this condition is *electron–hole pair*. The basic concept of the electron–hole pair is illustrated in Figure 1.5.

Within a few microseconds of becoming a free electron, an electron will give up its energy and fall into one of the holes in the covalent bond. This process is known as *recombination*. The time from when an electron jumps to the conduction band (becomes a free electron) until recombination occurs is called the *lifetime* of the electron–hole pair.

FIGURE 1.5

Generation of an electron-hole pair.

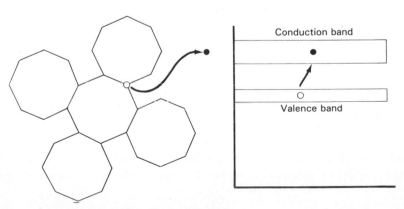

Conduction Versus Temperature

At room temperature, *thermal energy* (heat) causes the constant creation of electron–hole pairs, with their subsequent recombination. Thus, a semiconductor will have some number of free electrons, even when *no voltage is applied* to the element. As you increase the temperature, more electrons in the element will absorb enough energy to break free of their covalent bonds, and the number of free electrons will increase.

On the other hand, if you *decrease* the temperature, there is less thermal energy to release electrons from their covalent bonds, and the number of free electrons decreases. This situation will continue until the temperature reaches *absolute zero*. Absolute zero is, by definition, the temperature at which there is no thermal energy and occurs at $-273.16°C$ ($-459.69°F$). Since there is no thermal energy at this temperature, there are no free electrons. There is no energy for the electrons to absorb, so they cannot jump to the conduction band.

The important relationship here is the fact that *conductivity in a semiconductor is directly proportional to temperature*. This is why current increases in a circuit as the circuit warms up.

SECTION REVIEW

1. What are the three particles that make up the atom? [Objective 1]
2. What is the relationship between the number of valence electrons and the conductivity of a given element? [Objective 2]
3. How many valence electrons are there in a conductor? An insulator? A semiconductor? [Objective 2]
4. What three semiconductor elements are most commonly used in electronics? [Objective 3]
5. What are the relationships between electrons and orbital shells that were listed in this section? [Objective 4]
6. What is an *energy gap*? What are the energy gap values for insulators, semiconductors, and conductors? [Objective 4]
7. What forms of energy are given off by an electron that is falling into the valence band from the conduction band? [Objective 5]
8. What is *covalent bonding*? [Objective 6]
9. What are the effects of covalent bonding on intrinsic semiconductor materials? [Objective 6]
10. What is an *electron–hole pair*? [Objective 7]
11. What is *recombination*? [Objective 7]
12. What is the typical lifetime of an electron–hole pair? [Objective 7]
13. What is the relationship between temperature and conductivity? [Objective 8]
14. Why does current stop at absolute zero in a semiconductor? [Objective 8]

1.2

Doping

As you have been shown, *intrinsic* (pure) silicon and germanium are poor conductors. This is due partially to the number of valence electrons, the covalent bonding, and the relatively large energy gap. Because of their poor conductivity, intrinsic silicon and germanium are of little use.

Doping is the process of *adding impurity atoms to intrinsic silicon or germanium to improve the conductivity of the semiconductor*. The term *impurity* is used to describe the doping elements because the silicon or germanium is no longer pure once the doping has occurred. Since a doped semiconductor is no longer pure, it is called an *extrinsic* semiconductor.

Two element types are used for doping: *trivalent* and *pentavalent*. A trivalent element is one that has three valence electrons. A pentavalent element is one that has five valence electrons. When trivalent atoms are added to intrinsic semiconductors, the resulting material is called a *p-type* material. When pentavalent impurity atoms are used, the resulting material is called an *n-type* material. The commonly used doping elements are listed in Table 1.1.

TABLE 1.1
Commonly Used Doping Elements

Trivalent Impurities	Pentavalent Impurities
Aluminum (Al)	Phosphorus (P)
Gallium (Ga)	Arsenic (As)
Boron (B)	Antimony (Sb)
Indium (In)	Bismuth (Bi)

n-Type Materials

When pentavalent impurities are added to silicon or germanium, the result is an excess of electrons *in the covalent bonds*. This point is illustrated in Figure 1.6. As you can see, the pentavalent arsenic atom is surrounded by four silicon atoms. The silicon atoms will bond with the arsenic atom, each sharing one arsenic electron. However, the fifth arsenic electron is not bound to any of the surrounding silicon atoms. Because of this, the fifth arsenic electron requires little energy to break free and enter the conduction band. If literally millions of arsenic atoms are added to pure silicon or germanium, there will be millions of electrons that are not a part of the covalent bonding. All of these electrons could then be made to flow through the material with little difficulty.

One point that needs to be made and understood is this: Even though there are millions of electrons that are not a part of the covalent bonding, the material is still electrically neutral. This is because each arsenic atom has the same number of protons as electrons, just like the silicon or germanium atoms. Since the overall

FIGURE 1.6

N-type material.

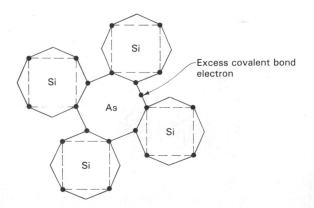

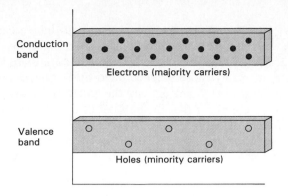

FIGURE 1.7

Energy diagram (*n*-type material).

number of protons and electrons in the material is equal, the net charge on the material is zero.

Since there are more conduction band electrons than valence band holes in an *n*-type material, the electrons are called *majority carriers*, and the valence band holes are called *minority carriers*. The relationship between majority and minority carriers in an *n*-type material can be better understood by taking a look at Figure 1.7. The valence band is shown to contain *some* holes. These holes are caused by thermal energy excitation of electrons, as was discussed earlier. With the excess of conduction band electrons, however, the lifetime of an electron–hole pair is shortened significantly. This is due to the fact that the hole is filled almost immediately by one of the excess electrons in the conduction band.

Since the only holes that exist in the covalent bonding are those caused by thermal energy, the number of holes is far less than the number of conduction band electrons. This is where the terms *majority* and *minority* come from.

Note that the term *n-type* implies an excess of electrons. As you will see, a *p-type* material is one with an excess of holes and a relative lack of free electrons.

p-Type Materials

When intrinsic silicon or germanium is doped with a trivalent element, the resulting material is called a *p-type* material. The use of a trivalent element causes the existence of a hole in the covalent bonding structure. This point is illustrated in Figure 1.8. As you can see, the aluminum atom is shown to be surrounded by four silicon atoms. This time, however, there is a gap in the covalent bond caused by the lack of a fourth valence electron in the aluminum atom. Now, instead of an excess of electrons, we have an excess of holes. This situation is represented

p-Type material. Silicon or germanium that has trivalent impurities.

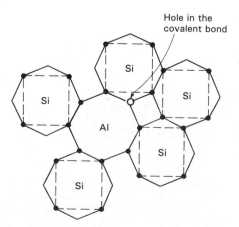

Hole in the covalent bond

FIGURE 1.8

P-type material.

further in Figure 1.9. The *p*-type material is shown to have an excess of holes in the valence band. At the same time, there are some electrons in the conduction band. Again, the free electrons in the conduction band are there because of thermal energy. Since there are many more valence band holes than conduction band electrons, the holes are the majority carriers and the electrons are the minority carriers.

Again, you must remember that the number of electrons in the *p*-type material is equal to the number of protons. Even though there are gaps in the covalent bonds, the proton–electron balance still exists. Therefore, the net charge on the *p*-type material is zero.

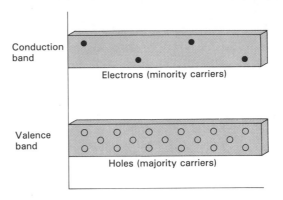

FIGURE 1.9

Energy diagram (*p*-type material).

Summary

Figure 1.10 is what we call a *summary illustration*. The figure is provided to give you a quick summary form of the material presented in this section. As you will see, summary illustrations will be used extensively throughout the text.

Whenever you come to a summary illustration, take time to study the illustration and the relationships it contains. This will help you to remember the points that were made in the section.

FIGURE 1.10

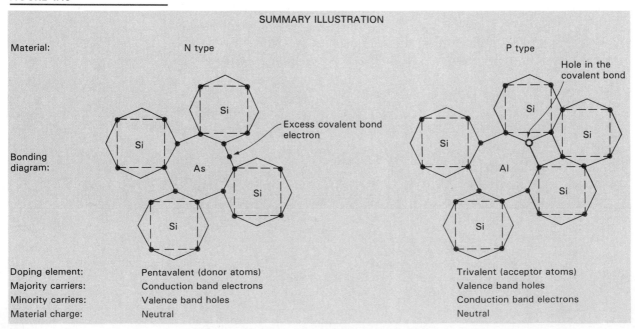

CHAP. 1   Fundamental Solid-State Principles

1. What is doping? Why is it necessary? [Objective 9]
2. What is an *impurity element*? [Objective 9]
3. What are *trivalent* and *pentavalent* elements? [Objective 10]
4. Despite their respective characteristics, *n-type* and *p-type* materials are still electrically neutral. Why? [Objective 11]
5. In what ways are *n-type* and *p-type* materials similar? In what ways are they different? [Objective 12]

1.3
The *pn* Junction

Alone, there is little use for either an *n*-type material or a *p*-type material. However, these materials become useful when *joined* together to form a *pn junction*. Figure 1.11 illustrates the relative conditions of the *n* and *p* materials prior to joining and at the moment they are joined together.

Figure 1.11a shows the individual materials. As you can see, the *n*-type material is shown (on top) as containing an excess of electrons (solid circles), and the *p*-type material is shown as having an excess of holes (open circles). The energy diagrams illustrate the relationship between the energy levels of the two materials. Note that the valence bands of the two materials are at slightly different energy levels, as are the conduction bands. This is due to differences in the atomic makeup of the two materials.

When the two materials are joined together (Figure 1.11b), the conduction and valence bands of the materials overlap. This allows free electrons from the *n*-type material to *diffuse* (wander) to the *p*-type material. This action and its results are illustrated in Figure 1.12. When a free electron wanders from the *n*-type material across the junction, it will become trapped in one of the valence band holes in the *p*-type material. As a result, there is one net *positive* charge in the *n*-type

13

FIGURE 1.11

pn-junction initial energy levels.

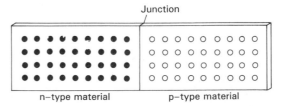
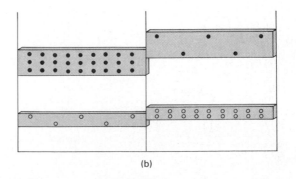
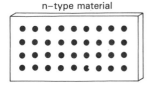
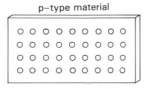
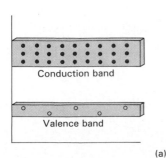
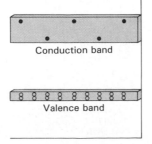

(a)　　　　　(b)

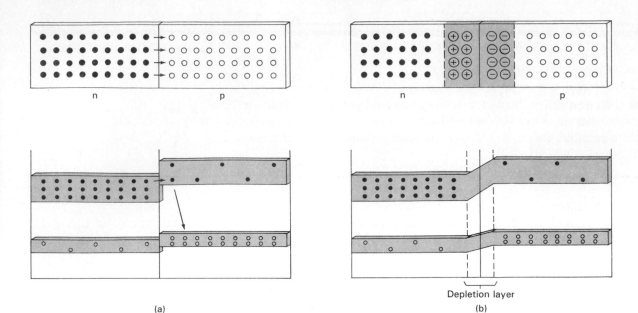

FIGURE 1.12

The forming of the depletion layer.

material and one net *negative* charge in the *p*-type material. This may seem confusing at first, but it really isn't that strange. Take a look at Figure 1.13. Under normal circumstances, the covalent bond shown in the *n*-type material would have a total of 21 valence electrons (4 per silicon atom + 5 for the pentavalent atom = 21) and 21 protons that offset the negative valence charges.* This balance between valence electrons and protons results in a net charge of zero. When an electron wanders into the *p*-type material as shown, there are two results:

* To simplify things, we consider only *valence* electrons and their "matching" protons. The actual number of total electrons and protons is not important because the electrical activity is determined by the valence shell.

FIGURE 1.13

Depletion layer charges.

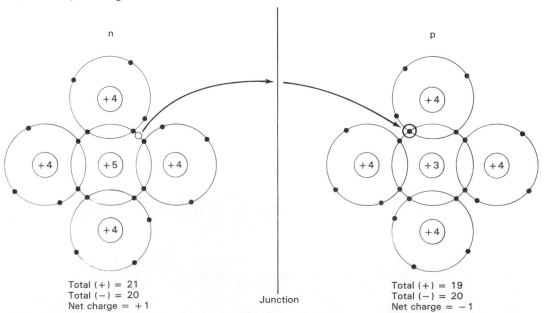

1. The bond shown in the *n*-type material has lost an electron. Now the bond has only 20 electrons, but *still* has 21 offsetting protons. As a result, the overall charge is positive (+1).

2. The covalent bond shown in the *p*-type material originally had 19 valence electrons that were offset by 19 protons, for a net charge of zero. However, when the electron from the *n*-type material falls into the hole, the bond has 20 electrons that are offset by only 19 protons. The net charge on the bond is therefore negative (−1).

Now it is time to consider what happens on a larger scale. When looking at the large-scale picture, there are several things to remember:

1. Each electron that diffuses across the junction leaves one positively charged bond in the *n*-type material and causes one bond in the *p*-type material to have a negative charge.

2. Both conduction band electrons and valence shell holes are needed for conduction through the materials. When an electron diffuses across the junction, the *n*-type material has lost a conduction band electron. When the electron falls into a hole in the *p*-type material, that material has lost a valence band hole. At this point, both bonds have been *depleted* of charge carriers.

Since the action just described happens on a large scale, the junction ends up with a layer (on both sides) that is depleted of carriers. This layer is called the *depletion layer*. The depletion layer is represented in Figure 1.12b. Note that the overall charge of the layer is shown to be positive on the *n* side of the junction and negative on the *p* side of the junction.

With the buildup of (−) charges on the *p* side of the junction and the buildup of (+) charges on the *n* side of the junction, you have a natural *difference of potential* between the two sides of the junction. This potential is referred to as the *barrier potential*. Recall that "difference of potential" is measured in volts. The barrier potential for silicon is approximately *0.6 V* and the barrier potential for germanium is approximately *0.2 V*.

Depletion layer. The area around a *pn* junction that is depleted of free carriers. Note that the depletion layer is also referred to as the *depletion region*.

Barrier potential. The natural potential across a *pn* junction.

Summary

When an *n*-type material is joined with a *p*-type material:

1. A small amount of *diffusion* occurs across the junction. The amount of diffusion is limited by the difference between the conduction band energy levels of the two materials.

2. When electrons diffuse into the *p* region, they give up their energy and "fall" into the holes in the valence band covalent bonds.

3. Since the pentavalent atoms (near the junction) in the *n* region have lost an electron, they have an overall *positive charge*.

4. Since the trivalent atoms (near the junction) in the *p* region have gained an electron, they have an overall *negative charge*.

5. The difference in charges on the two sides of the junction is called the *barrier potential*. The barrier potential is approximately equal to *0.6 V for silicon* and *0.2 V for germanium*.

Note that, since the pentavalent atoms in the *n* material are giving up electrons, they are often referred to as *donor atoms*. The trivalent atoms, on the other hand, are accepting electrons from the pentavalent atoms and thus are referred to as *acceptor atoms*.

Donor atoms. Another name for pentavalent atoms.

Acceptor atoms. Another name for trivalent atoms.

1. What is the overall charge on an *n-type* covalent bond that has just given up a conduction band electron? [Objective 13]
2. What is the overall charge on a *p-type* covalent bond that has just accepted an extra valence band electron? [Objective 13]
3. Discuss the forming of the *depletion layer*. [Objective 14]
4. What is *barrier potential*? What causes it? [Objective 15]
5. What is the barrier potential for silicon? For germanium? [Objective 15]

1.4
Bias

Depletion layer width and device current are inversely proportional.

A *pn* junction becomes useful when we are able to control the width of the depletion layer. By controlling the width of the depletion layer, we are able to control the resistance of the *pn* junction and thus the amount of current that can pass through the device. The relationship between the width of the depletion layer and the junction current is summarized as follows:

Depletion Layer Width	Junction Resistance	Junction Current
Minimum	Minimum	Maximum
Maximum	Maximum	Minimum

Bias. A potential used to establish depletion layer width.

A potential used to control the width of the depletion layer is referred to as *bias*. The two types of bias are *forward bias* and *reverse bias*. A forward-biased *pn* junction will have a minimum depletion layer width and the indicated resistance–current characteristics. A reverse-biased *pn* junction will have a maximum depletion layer width and the indicated resistance–current characteristics. In this section we will take a look at the two bias types.

Forward bias. A potential used to reduce the resistance of a *pn* junction.

Forward Bias

A *pn* junction is forward biased when the applied potential causes the *n*-type material to be more *negative* than the *p*-type material. When forward biased, a *pn* junction will allow current to pass with little opposition. The effects of forward bias are illustrated in Figure 1.14. Figure 1.14a shows the *pn* junction connected to a voltage source (V) and an open switch (SW$_1$). The energy diagram is included to show the initial energy states of the junction materials.

When SW$_1$ is closed, a negative potential is applied to the *n*-type material and a positive potential is applied to the *p*-type material. These potentials cause the following to occur:

The effects of *forward* bias on a *pn* junction.

1. The conduction band electrons in the *n*-type material are pushed toward the junction by the negative terminal potential.
2. The valence band holes in the *p*-type material are pushed toward the junction by the positive terminal potential.

Assuming that V is greater than the barrier potential of the junction, the electrons in the *n*-type material will gain enough energy to break through the depletion layer. When this happens, the electrons will be free to recombine with the holes in the *p*-type material and conduction will occur. Conduction through a *pn* junction is illustrated in Figure 1.14c.

When a forward-biased *pn* junction begins to conduct, the *forward voltage* (V_F) across the junction will be slightly greater than the barrier potential for the device. The values of V_F are approximated as

19

$$V_F \cong 0.7 \text{ V} \qquad \text{(for silicon)}$$

and

$$V_F \cong 0.3 \text{ V} \qquad \text{(for germanium)}$$

The difference between the barrier potential and the approximated value of V_F is a result of the current through the natural resistance of the device. This point is

Forward voltage. The voltage across a forward-biased *pn* junction.

FIGURE 1.14

The effects of forward bias.

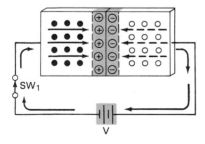

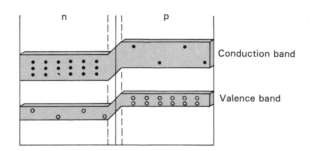

(a)

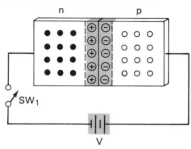

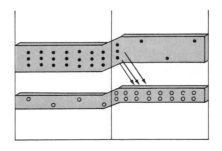

(b)

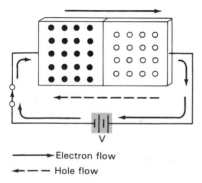

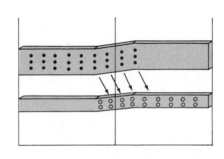

→ Electron flow

← — — Hole flow

(c)

discussed further in Chapter 2. The importance of the value of V_F will become evident in our discussions on diode circuits in Chapter 3.

It should be noted that there are two ways a *pn* junction can be forward biased. A junction can be forward biased by:

How a *pn* junction is forward biased.

1. Applying a potential to the *n*-type material that is more negative than the *p*-type material potential
2. Applying a potential to the *p*-type material that is more positive than the *n*-type material potential

FIGURE 1.15

Forward biasing a *pn* junction.

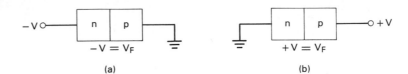

These two biasing methods are illustrated in Figure 1.15. In both cases, the fixed potential is shown as ground. Figure 1.15a shows the *n*-type material of the junction being driven negative by an amount that is sufficient to cause conduction. Assuming that the junction is made of silicon and $V \geq -0.7$ V, the junction will conduct. This is due to the fact that the *n*-type material is 0.7 V more negative than the *p*-type material. The junction shown in Figure 1.15b will conduct for the same reason. If the *p*-type material is based at +0.7 V with respect to the *n*-type material, the diode will conduct. Thus we can cause a junction to conduct by driving the *n*-type material more negative than the *p*-type material, or by driving the *p*-type material more positive than the *n*-type material. Both of these methods of forward biasing a *pn* junction are used in practice.

All the junctions shown in Figure 1.16 are forward biased. See if you can relate the components shown to the forward-bias conditions discussed.

Reverse Bias

Reverse bias. A potential that causes a *pn* junction to have a high resistance.

The effects of *reverse* bias on a *pn* junction.

A *pn* junction is reverse biased when the applied potential causes the *n*-type material to be more *positive* than the *p*-type material. When a *pn* junction is reverse biased, the depletion layer becomes wider, and junction current is reduced to *almost* zero. Reverse bias and its effects are illustrated in Figure 1.17. Figure 1.17a shows the junction in the forward-bias condition. As you have already been shown, the junction is allowing current to pass with little opposition. If the forward-biasing potential

FIGURE 1.16

Some forward-biased *pn* junctions.

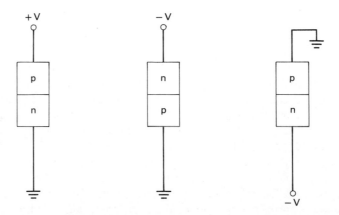

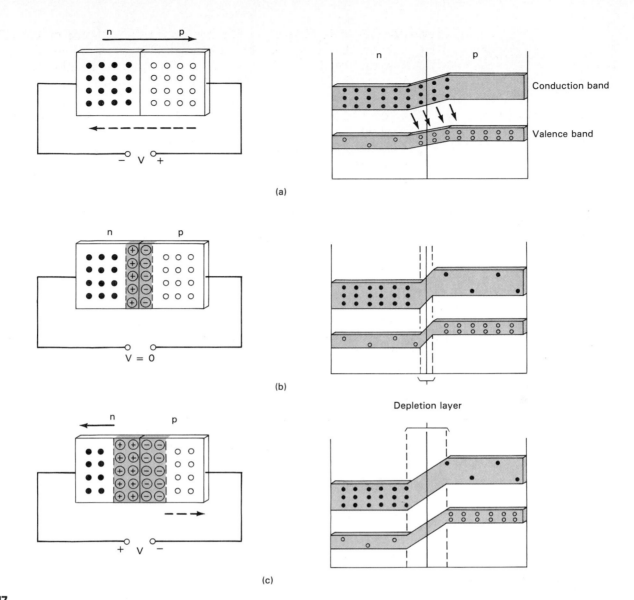

FIGURE 1.17

The effects of reverse bias.

returns to zero, the depletion layer re-forms as electrons diffuse across the junction. This is illustrated in Figure 1.17b.

If we apply a voltage with the polarity shown in Figure 1.17c, the electrons in the *n*-type material will head toward the positive terminal of the source. At the same time, the holes in the *p*-type material will move toward the negative source terminal. The electrons moving away from the *n* side of the junction will further deplete the material of free carriers. The depletion layer has effectively been widened. The same principle holds true for the *p* side of the material. Since there are fewer holes near the junction, the depletion region has grown. *The overall effect of the widening of the depletion layer is that the resistance of the junction has been drastically increased, and conduction drops to near zero.*

Figure 1.17 helps to demonstrate another point about reverse bias. During the time that the depletion layer is forming and growing, there is *still* majority carrier current in both materials. This current, called *diffusion current*, lasts only as long as it takes the depletion layer to reach its maximum width.

21

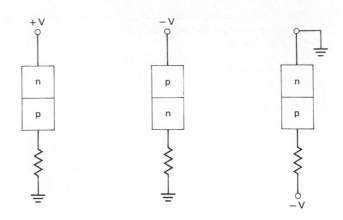

FIGURE 1.18

Some reverse-biased *pn* junctions.

How a pn junction is reverse biased.

Just as there are two ways to forward bias a junction, there are two ways to reverse bias a junction:

1. Applying a potential to the *n*-type material that is more positive than the *p*-type material potential
2. Applying a potential to the *p*-type material that is more negative than the *n*-type material potential

All the junctions shown in Figure 1.18 are reverse biased. Again, try to relate the diagrams to the statements made above.

The Bottom Line

Bias is a potential applied to a *pn* junction that determines the operating characteristics of the device. Bias polarities and effects are summarized as follows:

Bias Type	Junction Polarities	Junction Resistance
Forward	*n*-type material is more (−) than *p*-type material	Extremely *low*
Reverse	*p*-type material is more (−) than *n*-type material	Extremely *high*

The voltage drop across a forward-biased *pn* junction will be approximately equal to 0.7 V for silicon and 0.3 V for germanium. When a *pn* junction is reverse biased, it acts essentially as an *open* circuit. Therefore, the voltage across a reverse-biased *pn* junction will equal the applied voltage. These points are summarized in Figure 1.19.

A Final Note

Up to this point, we have been mentioning both silicon and germanium in our discussions of semiconductor materials and junctions. Of the two, silicon is almost always used because it has a higher tolerance of heat. For this reason we will limit future discussions to silicon. Just remember that germanium-based circuits work the same way as silicon-based circuits. The primary difference is in the value of V_F for germanium versus silicon.

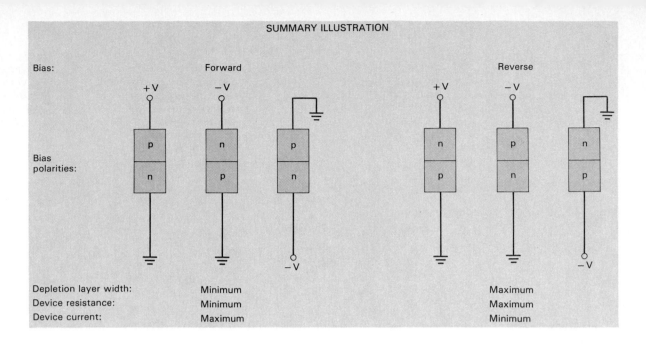

FIGURE 1.19

1. What are the resistance and current characteristics of a *pn* junction when its depletion layer is at its *maximum* width? *Minimum* width? [Objective 16]
2. What purpose is served by the use of *bias*? [Objective 17]
3. What is the depletion layer width of a forward-biased *pn* junction? [Objective 18]
4. What is the junction resistance of a forward-biased *pn* junction? [Objective 18]
5. What are the approximate values of V_F for forward-biased silicon and germanium *pn* junctions? [Objective 19]
6. List the commonly used methods for forward biasing a *pn* junction. [Objective 20]
7. What is the depletion layer width of a reverse-biased *pn* junction? [Objective 21]
8. What is the junction resistance of a reverse-biased *pn* junction? [Objective 21]
9. List the commonly used methods for reverse biasing a *pn* junction. [Objective 22]
10. What does the voltage across a reverse-biased *pn* junction equal? [Objective 23]
11. Why is silicon used more commonly than germanium in the production of solid-state components? [Objective 24]

KEY TERMS

The following terms were introduced and defined in this chapter:

acceptor atom	bias	depletion layer
band	conduction band	diffusion current
barrier potential	covalent bonding	donor atom

doping
electron-hole pair
energy gap
extrinsic
forward bias
forward voltage
germanium
hole

intrinsic
lifetime
majority carrier
minority carrier
n-type material
p-type material
pentavalent
recombination

reverse bias
reverse voltage
semiconductor
silicon
transient current
trivalent
valence shell

2
Diodes

Light-emitting diodes (LEDs) have been developed that will emit one color when forward biased, and another color when reverse biased.

OBJECTIVES

☐ 1. Identify the terminals of a *pn*-junction diode, given the schematic symbol of the component. (§2.1)

☐ 2. Analyze the schematic diagram of a simple diode circuit and determine:
a. Whether or not the diode is conducting. (§2.1)
b. The direction of current through a conducting diode. (§2.1)

☐ 3. List the three diode models and the applications for each. (§2.1)

☐ 4. Discuss the characteristics and applications of the *ideal diode model*. (§2.2)

☐ 5. Discuss the characteristics and applications of the *practical diode model*. (§2.3)

☐ 6. Perform a basic mathematical analysis of a diode circuit, using the ideal diode model and the practical diode model, and compare the results. (§2.3)

☐ 7. List the main parameters of the *pn*-junction diode and explain how each limits the use of the component. (§2.4)

☐ 8. Discuss the characteristics and applications of the *complete diode model*. (§2.5)

☐ 9. Explain the effects of bulk resistance and reverse current on circuit voltage measurements. (§2.5)

☐ 10. Discuss the effects of temperature on diode operation and circuit measurements. (§2.5)

☐ 11. Determine the value of any diode parameter or electrical characteristic using the spec sheet of the component. (§2.6)

☐ 12. Determine the suitability of a given diode for a given application, using diode spec sheets and/or selector guides. (§2.6)

☐ 13. Discuss the basic operating principles of the zener diode. (§2.7)

☐ 14. Identify the schematic symbol of the zener diode and determine the direction of current through the device. (§2.7)

☐ 15. List the main zener diode parameters and explain how each limits the use of the component. (§2.8)

☐ 16. Determine the value of any zener diode electrical characteristic, given the spec sheet for the component. (§2.8)

☐ 17. Determine the suitability of a given zener diode for a specific application, using zener diode spec sheets and/or selector guides. (§2.8)

☐ 18. Find a suitable replacement for a faulty zener diode. (§2.8)

☐ 19. Discuss the basic operating principles of the light-emitting diode (LED). (§2.9)

☐ 20. Calculate the value of the *current-limiting* resistor that is needed for an LED in a given circuit. (§2.9)

☐ 21. Determine whether a given *pn*-junction diode, zener diode, or LED is good or faulty. (§2.10)

The diode is the most basic of the solid-state components. There are many diode types, each with its own operating characteristics and applications. The various diode types are easily identified by name, circuit application, and schematic symbol. It should be noted that the term *diode*, used by itself, refers to the basic *pn*-junction diode. All other diode types have other identifying names, such as *zener diode*, *light-emitting diode*, and so on.

A *diode* is a *two-electrode* (two-terminal) device that acts as a *one-way conductor*. The most basic type of diode is the *pn-junction diode*, which is nothing more than a *pn* junction with a lead connected to each of the semiconductor materials. When forward biased, this type of diode will conduct. When reverse biased, diode conduction will drop to nearly zero.

In this chapter, we are going to take a look at the three most commonly used types of diodes: the *pn-junction diode*, the *zener diode*, and the *light-emitting diode*, or *LED*. Many of the other types of diodes are covered in Chapter 21.

Diode. A one-way conductor.

2.1
Introduction to the *pn* Junction Diode

The schematic symbol for the *pn*-junction diode is shown in Figure 2.1. The *n*-type material is called the *cathode*, and the *p*-type material is called the *anode*.

FIGURE 2.1

Pn junction diode schematic symbol.

Cathode. The *n*-type terminal of a diode.

Anode. The *p*-type terminal of a diode.

Recall that a *pn* junction will conduct when the *n*-type material (cathode) is more negative than the *p*-type material (anode). Relating this characteristic to the schematic symbol for the diode, we can make the following statement: *A diode will conduct when the arrow points to the more negative of the diode potentials.* This point is illustrated in Figure 2.2, which shows several forward biased (conducting) diodes. Note that the arrow in the diode schematic symbol points to the more negative potential in each case. Since the arrow in the schematic symbol points to the more negative potential when the diode is conducting, *diode forward current will be against the arrow*, as is shown in Figure 2.2.

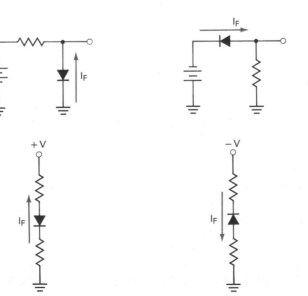

FIGURE 2.2

Forward-biased diodes.

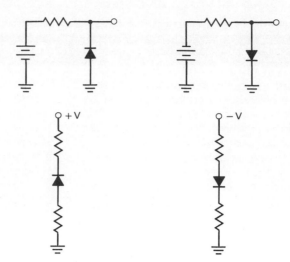

FIGURE 2.3
Reverse-biased diodes.

A *pn* junction diode is reverse biased when the *n*-type material (cathode) is more positive than the *p*-type material (anode). Relating this characteristic to the schematic symbol for the diode, we can make the following statement: *A diode will not conduct when the arrow points to the more positive of the diode potentials.* This point is illustrated in Figure 2.3, which shows several reverse biased (nonconducting) diodes. Note that the arrow in the diode schematic symbol points to the more positive potential in each case.

Diode Models

Model. A representation of a component or circuit.

Troubleshooting. The process of locating faults in electronic equipment.

In this chapter, you are going to be introduced to three diode *models*. A *model* is a representation of a component or circuit that contains one or more of the characteristics of that component or circuit. For example, the dc model of a capacitor may represent the component as an open circuit, since a capacitor blocks dc. At the same time, the ac model of a capacitor may represent the component as a variable impedance, since the impedance of a capacitor varies inversely with frequency.

Component models are usually used to represent the component under specific circumstances or in specific applications. As you will see, the diode model that you will use will depend on what you are trying to do.

The first diode model that we will cover is called the *ideal diode model*. This diode model represents the diode as a simple switch that is either *closed* (conducting) or *open* (nonconducting). This model is used only in the initial stages of *troubleshooting*, as will be explained in Section 2.2.

The *practical diode model* is a bit more complex than the ideal diode model. The practical model includes the diode characteristics that must be considered when mathematically analyzing a diode circuit and when determining whether or not a given diode can be used in a given circuit. The practical diode model will be covered in Sections 2.3 and 2.4.

The *complete diode model* is the most complex of the diode models. It includes the diode characteristics that are considered only under specific conditions such as *circuit development* (or *engineering*), high-frequency analysis, and so on. As you will see, the characteristics that are included in the complete diode model are not usually considered on a daily basis by the average technician. We will look at the complete diode model in Section 2.5.

The three diode models and their applications are summarized in Table 2.1.

TABLE 2.1
Diode Model Summary

Diode Model	Application(s)
Ideal	• Circuit troubleshooting
Practical	• Mathematical circuit analysis
	• Determining whether or not a given diode can be used in a given circuit
Complete	• Circuit development (engineering)
	• Uncommon special-condition circumstances

SECTION REVIEW

1. What is a diode? [Introduction]
2. What type of bias causes a diode to conduct? [Introduction]
3. What type of bias causes a diode to act as an insulator? [Introduction]
4. Draw the symbol for a diode and label the terminals. [Objective 1]
5. When analyzing a schematic diagram, how do you know whether or not a diode is conducting? [Objective 2]
6. When analyzing a schematic diagram, how do you determine the direction of diode current? Explain your answer. [Objective 2]
7. When is the *ideal diode model* used? [Objective 3]
8. When is the *practical diode model* used? [Objective 3]
9. When is the *complete diode model* used? [Objective 3]

2.2
The Ideal Diode

The *ideal* diode has the characteristics of an open switch when it is reverse biased and those of a closed switch when forward biased. You may recall that a switch has the following characteristics:

The ideal diode acts as a switch.

Condition	Characteristics
Open	• Infinite resistance, and thus no current
	• Full applied voltage dropped across the component terminals
Closed	• No resistance, and thus no control over switch current
	• No voltage dropped across the component terminals

Based on the characteristics of a switch, we can make the following statements about the ideal diode:

1. When *reverse biased* (open switch):
   a. The diode will have infinite resistance.
   b. The diode will not allow conduction.
   c. The diode will drop the entire applied voltage across its terminals.

Characteristics of the ideal reverse-biased diode.

2. When *forward biased* (closed switch):
   a. The diode will have no resistance.

Characteristics of the ideal forward-biased diode.

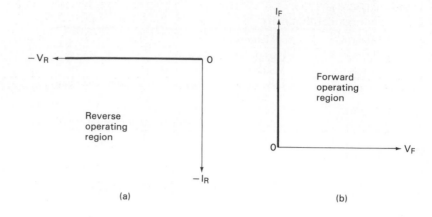

FIGURE 2.4

(a)   (b)

    b.   The diode will have no control over the current through it.
    c.   The diode will have no voltage drop across its terminals.

Reverse voltage. The voltage across a reverse-biased diode.

Reverse current. The current through a reverse-biased diode.

What is the voltage across a reverse-biased diode?

These characteristics of the ideal diode are illustrated in Figure 2.4. Figure 2.4a shows the reverse-bias characteristics of the ideal diode. Note that as reverse voltage (V_R) increases reverse current (I_R) remains at zero. This implies that the reverse-biased diode is an *open circuit* (just like an open switch), since there is no current through the device regardless of the value of the applied voltage. *Since the reverse-biased diode is acting as an open, the full applied voltage is dropped across the terminals of the device.* This point is illustrated in Example 2.1.

EXAMPLE 2.1

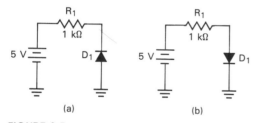

(a)   (b)

FIGURE 2.5

Determine the values of V_{D1}, V_{R1}, and I_T for the circuit shown in Figure 2.5a.

Solution: Because the arrow in the schematic symbol is pointing toward the positive terminal of the source, we know that the diode is reverse biased. Therefore:

    a.   The full applied voltage is dropped across D_1. By formula,

$$V_{D1} = V_S = 5 \text{ V}$$

    b. D_1 will not allow conduction. Therefore, $I_T = 0$ A.
    c.   Since there is no current through R_1, there is no voltage drop across the component ($V_{R1} = 0$ V).

PRACTICE PROBLEM 2–1

A series circuit consists of a 12-V source, a 470-Ω resistor (R_1), a 330-Ω resistor (R_2), and a diode. If the diode is reverse biased, what is the value of V_{R1}? What is the value of V_{R2}?

What determines the value of diode forward current?

Figure 2.4b shows the forward-bias characteristics of the ideal diode. Note that the forward voltage (V_F) across the diode is assumed to be 0 V for this diode model, while forward current (I_F) is shown to be at some measurable value. When the ideal diode is forward biased, I_F *is limited by the voltage and resistance values that are external to the component*. This point is illustrated in Example 2.2.

EXAMPLE 2.2

Determine the values of V_{D1}, V_{R1}, and I_T for the circuit shown in Figure 2.5b.

Solution: Because the arrow in the schematic symbol is pointing toward the negative terminal of the source, we know that the diode is forward biased. Therefore:

a. $V_{D1} = 0$ V, leaving the total applied voltage to be dropped across R_1.
b. $V_{R1} = V_S = 5$ V.
c. I_T is determined by the source voltage and R_1. By formula,

$$I_T = \frac{V_{R1}}{R_1} = \frac{5 \text{ V}}{1 \text{ k}\Omega} = 5 \text{ mA}$$

PRACTICE PROBLEM 2–2

A series circuit consists of a 12 V source, a 470 Ω resistor, a 330 Ω resistor, and a diode. If the diode is forward biased, what is the value of I_T for the circuit?

If you compare the values found in the last two examples to the switch characteristics described earlier, you will see that the diode acted as an ideal switch in both cases. The forward and reverse characteristics of the ideal diode are summarized in Figure 2.6.

When Do We Use the Ideal Diode Model?

Normally, the ideal model of the diode is used in the initial stages of circuit troubleshooting. When troubleshooting most diode circuits, your initial concern is only whether or not a given diode is acting as a one-way conductor. If it is, the component is assumed to be good. If not, it is faulty and must be replaced. This point will be illustrated further in this chapter.

One Final Note

It was stated earlier in the chapter that the ideal model of the diode is not normally used in the mathematical analysis of a diode circuit. Then, we used this model to mathematically analyze the circuits in Figure 2.5. Why? We did this only to illustrate how the ideal model of the diode is treated as a perfect switch: either *on* (closed) with no resistance, or *off* (open) with infinite resistance.

In the next section, you will be shown how the *practical diode model* is used to calculate current and voltage values.

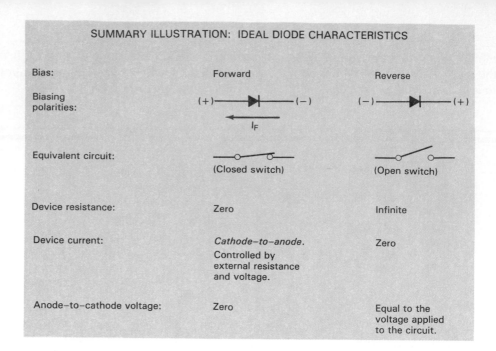

FIGURE 2.6

SUMMARY ILLUSTRATION: IDEAL DIODE CHARACTERISTICS

Bias:	Forward	Reverse
Biasing polarities:	$(+)$ ──▶│── $(-)$ I_F	$(-)$ ──▶│── $(+)$
Equivalent circuit:	(Closed switch)	(Open switch)
Device resistance:	Zero	Infinite
Device current:	*Cathode–to–anode.* Controlled by external resistance and voltage.	Zero
Anode–to–cathode voltage:	Zero	Equal to the voltage applied to the circuit.

SECTION REVIEW

1. What are the forward characteristics of the ideal diode model? [Objective 4]
2. What are the reverse characteristics of the ideal diode model? [Objective 4]
3. How can we use the ideal diode model? [Objective 4]

2.3
The Practical Diode Model

In our discussion of the ideal diode, we did not consider many of the diode characteristics that must be dealt with by working technicians on a regular basis. One of these characteristics, *forward voltage*, is normally considered in the mathematical analysis of a diode circuit. Many other practical diode characteristics are used when determining whether or not one diode may be used in place of another or in a specific circuit. These characteristics include *peak reverse voltage*, *average forward current*, and *forward power dissipation*.

In this section, we will take a look at *forward voltage* (V_F) and the effect that it has on the mathematical analysis of basic diode circuits. The other characteristics listed are covered in detail in Section 2.4.

Forward Voltage (V_F)

In Chapter 1, we established the fact that there is a slight voltage developed across a forward-biased *pn* junction. The effect of this V_F on the diode characteristic curve is illustrated in Figure 2.7.

Figure 2.7a is a composite of the ideal diode characteristic graphs that were shown in Figure 2.4. Note that the point where I_F suddenly increases is labeled V_k in the figure. This label is commonly used to identify what is called the *knee voltage* in a voltage versus current graph. The term *knee voltage* is often used to describe the point in a voltage versus current graph where current suddenly increases

Knee voltage. The voltage at which device current suddenly increases or decreases.

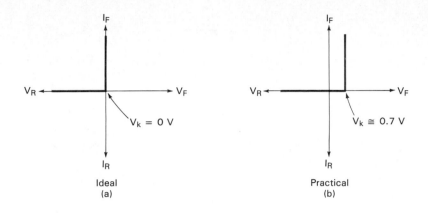

FIGURE 2.7

Diode characteristic curves.

or decreases. As you can see, the ideal diode model assumes a knee voltage of 0 V.

In Figure 2.7b, we see the characteristic curve for the practical diode model. The only difference between this curve and the one shown for the ideal diode model is the value of V_k. *In the practical diode model, the value of V_k is shown to be equal to the approximated value of V_F for a pn junction, 0.7 V.* (This value assumes that we are dealing with a silicon diode.)

Using the curve shown in Figure 2.7b, we can make the following statements about the forward operating characteristics of the practical diode:

Characteristics of the practical forward-biased diode.

1. Diode current will remain at zero until the knee voltage is reached.
2. Once the applied voltage reaches the value of V_k, the diode turns *on* and forward conduction occurs.
3. As long as the diode is conducting, the value of V_F will be approximately equal to V_k. In other words, V_F will be assumed to be approximately 0.7 V, regardless of the value of I_F.

Figure 2.8 will help you to see the differences between the forward operating characteristics of the ideal diode model and the practical diode model. Note the addition of the battery in the equivalent circuit for the practical diode model. This battery is used to represent the 0.7 V value of V_F for the component.

The Effect of V_F on Circuit Analysis

So, how does including the value of V_F change the analysis of a diode circuit? To answer this question, let's take a look at the circuit shown in Figure 2.9.

SUMMARY ILLUSTRATION

Diode model:	Ideal	Practical
Equivalent circuit:	A ○—○ K	A ○—○ —‖—‖— K (V_F)
Knee voltage:	0 V	0.7 V (Silicon) 0.3 V (Germanium)
Assumed V_F:	Same as V_k	Same as V_k

FIGURE 2.8

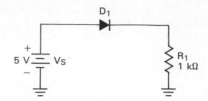

FIGURE 2.9

According to Kirchhoff's voltage law, the sum of the component voltages in the circuit must equal the applied voltage. By formula,

$$V_S = V_F + V_R$$

If we substitute the value of V_k (0.7 V) for V_F and rearrange the equation to solve for V_R, we get

$$\boxed{V_R = V_S - 0.7 \text{ V}} \tag{2.1}$$

According to Ohm's law,

$$I_T = \frac{V_R}{R_1}$$

Substituting equation (2.1) in place of V_R in the above equation, we get

$$\boxed{I_T = \frac{V_S - 0.7 \text{ V}}{R_1}} \tag{2.2}$$

Thus, for the circuit shown,

$$V_R = V_S - 0.7 \text{ V} = 4.3 \text{ V}$$

and

$$I_T = \frac{V_S - 0.7 \text{ V}}{R_1} = \frac{5 \text{ V} - 0.7 \text{ V}}{1 \text{ k}\Omega} = 4.3 \text{ mA}$$

If you compare these values with those obtained for the same circuit in Example 2.2 (page 27), you will see how including the value of V_F in the circuit analysis changes the results. The two sets of values are summarized as follows:

Value	Ideal	Practical
V_F	0 V	0.7 V
V_R	5 V	4.3 V
I_T	5 mA	4.3 mA

Percent of Error

How accurate must circuit calculation be?

In most circuit analysis problems, a calculated value is considered to be accurate enough if it is within ±10% of the actual measured value. The percent of error of a given calculation is determined using

CHAP. 2   Diodes

$$\boxed{\% \text{ of error} = \frac{|X - X'|}{X} \times 100} \qquad \textbf{(2.3)}$$

where X = the actual measured value
X' = the calculated value

Using the ideal diode model in circuit calculations can introduce a percent of error in the results that is not acceptable (that is, not within $\pm 10\%$). For example, we used the ideal and practical diode models to determine the value of V_R for the circuit shown in Figure 2.9. The percent of error introduced by using the ideal diode model in this case would be found as

$$\% \text{ of error} = \frac{|4.3 \text{ V} - 5 \text{ V}|}{4.3 \text{ V}} \times 100$$

$$= 16.28\%$$

As you can see, the percent of error is greater than 10% and therefore is not acceptable. This is why we use the practical diode model in circuit analysis problems. Now, let's go through a series of examples that will demonstrate the points that have been made so far in this section.

EXAMPLE 2.3

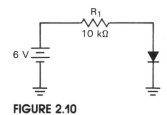

FIGURE 2.10

Determine the voltage across R_1 in Figure 2.10.

Solution: The voltage across the diode is assumed to be 0.7 V. Thus, the voltage across the resistor will be equal to the difference between the source voltage and the value of V_F. By formula,

$$V_R = V_S - 0.7 \text{ V} = 5.3 \text{ V}$$

EXAMPLE 2.4

Determine the total circuit current in the circuit shown in Figure 2.10.

Solution: The total circuit current is found as

$$I_T = \frac{V_S - 0.7 \text{ V}}{R_1}$$
$$= \frac{6 \text{ V} - 0.7 \text{ V}}{10 \text{ k}\Omega}$$
$$= 530 \text{ }\mu\text{A}$$

PRACTICE PROBLEM 2–4

A circuit like the one shown in Figure 2.10 has a 5-V source and a 510-Ω resistor. Determine the value of I_T for the circuit.

If there is more than one resistor in a simple diode circuit, the *total* resistance (R_T) must be used in determining the value of I_T, as was the case in all the circuits you studied in basic electronics. This point is illustrated in Example 2.5.

EXAMPLE 2.5

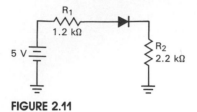

FIGURE 2.11

Determine the value of I_T for the circuit shown in Figure 2.11.

Solution: For the circuit shown, we can calculate the total circuit current using

$$I_T = \frac{V_S - 0.7 \text{ V}}{R_T}$$
$$= \frac{5 \text{ V} - 0.7 \text{ V}}{3.4 \text{ k}\Omega}$$
$$= 1.26 \text{ mA}$$

Note that we used R_T in place of R_1 in equation (2.2) to solve the circuit in Example 2.5.

Just as you must cŏnsider the total resistance in the analysis of a diode circuit, you must consider the *sum of the diode voltage drops* if there are several *series-connected* diodes in a circuit. This point is illustrated in Example 2.6.

EXAMPLE 2.6

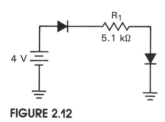

FIGURE 2.12

Determine the value of I_T for the circuit shown in Figure 2.12.

Solution: With two diodes in the circuit, the total value of V_F is assumed to be 1.4 V. Using this value in the place of 0.7 V in equation (2.2) allows us to accurately determine the value of I_T as follows:

$$I_T = \frac{V_S - 1.4 \text{ V}}{R_1}$$
$$= \frac{4 \text{ V} - 1.4 \text{ V}}{5.1 \text{ k}\Omega}$$
$$= 509.8 \text{ } \mu\text{A}$$

PRACTICE PROBLEM 2–6

A series circuit consists of two forward-biased diodes, a 470 Ω resistor, a 330 Ω resistor, and a 6 V source. What is the value of I_T for the circuit?

Now, let's do one more example to tie everything together.

EXAMPLE 2.7

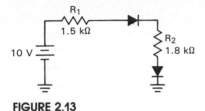

FIGURE 2.13

Determine the value of I_T for the circuit shown in Figure 2.13 *using the ideal diode model*. Then recalculate the value *using the practical diode model*. What is the percent of error introduced by using the ideal diode model?

Solution: The ideal diode model assumes that $V_F = 0$ V. Therefore, the total applied voltage is dropped across the two resistors, and I_T is found as

$$I_T = \frac{V_S}{R_T}$$
$$= \frac{10 \text{ V}}{3.3 \text{ k}\Omega}$$
$$= 3.03 \text{ mA}$$

The practical diode model assumes that $V_F = 0.7$ V (for each diode). Therefore, the value of I_T is found as

$$I_T = \frac{V_S - 1.4 \text{ V}}{R_T}$$
$$= \frac{10 \text{ V} - 1.4 \text{ V}}{3.3 \text{ k}\Omega}$$
$$= 2.61 \text{ mA}$$

The percent of error between the two calculations is found as

$$\% \text{ of error} = \frac{|\, 2.61 \text{ mA} - 3.03 \text{ mA} \,|}{2.61 \text{ mA}} \times 100$$
$$= 16.1\%$$

Note that the value of I_T found using the *practical diode model* was used in the denominator of the fraction in the percent of error calculation. This value was used because the practical value is always assumed to be closer than the ideal to the actual value of I_T.

PRACTICE PROBLEM 2–7

Refer back to Practice Problem 2–6. Recalculate the value of total circuit current using the *ideal diode model*. Then, determine the percent of error introduced by using this diode model.

One Final Note

It can be argued that it isn't always necessary to include the 0.7 V drop across a diode in the mathematical analysis of a diode circuit. For example, if we had a diode and resistor in series with a 100 V source, ignoring the 0.7 V diode drop would cause a percent of error of less than 1%, which is well within the acceptable limits for accuracy.

While this argument is valid, there are two other considerations. First, many diode circuits contain more than one diode. When this is the case, the percent of error introduced by using the ideal diode model becomes much larger. Second, we are always interested in getting the most accurate results (within reason) possible in the mathematical analysis of any circuit. For these reasons, we will always use the practical diode model in the mathematical analysis of any diode circuit. Remember, the only difference in using this model for analysis purposes is that you must take the value of V_F for each diode into account. Otherwise, the component is assumed to work just like the ideal model.

In the next section, we're going to take a look at the factors that must be considered when determining whether or not a specific diode can be used in a given situation. While these factors are not normally considered in voltage and current calculations, they are very important when it comes to actually working on a circuit.

SECTION REVIEW	1. What diode characteristic must be considered in the mathematical analysis of a diode circuit? [Objective 5]
	2. Which diode characteristics are normally considered when replacing one diode with another? [Objective 5]
	3. What is the assumed value of knee voltage for a silicon diode? [Objective 5]
	4. List the characteristics of the practical forward-biased diode. [Objective 5]
	5. How does including the value of V_F affect the accuracy of circuit calculations? [Objective 6]
	6. How accurate must a calculation be in order to be considered acceptable? [Objective 6]
	7. A circuit voltage is calculated to be 10 V. The actual value of this voltage is 12.2 V. What is the percent of error in the calculation? [Objective 6]

2.4
Other Practical Considerations

Assume that you have just finished troubleshooting a circuit. While troubleshooting, you found that the diode in the circuit is faulty and must be replaced. Now you discover that you have a wide variety of diodes in stock, but none of them has the same part number as the one that needs replacing. How can you determine which diode can be used in place of the faulty one?

There are several diode characteristics that must be considered when determining whether or not a specific diode can be used in a given circuit. These characteristics are *peak reverse voltage*, *average forward current*, and *forward power dissipation*.

Peak reverse voltage (V_{RRM}). The *maximum* reverse voltage allowable for a diode.

Peak Reverse Voltage (V_{RRM})

Any insulator will conduct if the applied voltage is high enough to cause the insulator to break down. For a reverse-biased diode, the *maximum* reverse voltage that won't force the diode to conduct is called the *peak reverse voltage* (V_{RRM}). When V_{RRM} is exceeded, the depletion layer may break down, and the diode may

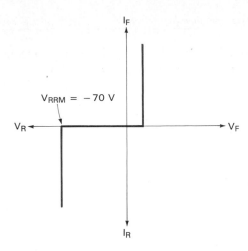

$V_{RRM} = -70$ V

FIGURE 2.14

conduct in the reverse direction. Typical values of V_{RRM} range from a few volts (for zener diodes) to thousands of volts.

The effect that V_{RRM} has on the diode characteristic curve is illustrated in Figure 2.14. Note that the value of reverse current (I_R) is shown to be zero until the value of V_{RRM} (-70 V in this case) is exceeded. When $V_R > V_{RRM}$, the value of I_R increases rapidly as the depletion layer breaks down. Normally, when a *pn* junction is forced to conduct in the reverse direction, the device is destroyed. A *zener diode*, on the other hand, is designed to work in the reverse direction without harming the diode. This point is discussed in detail later in this chapter.

The current that occurs when $V_R > V_{RRM}$ is called *avalanche current*. This name comes from the fact that one free electron bumps other electrons in the diode, causing them to break free from their covalent bonds, which then rapidly cause even more electrons to be broken free, and so on. The result is that the diode is destroyed by the excessive current and the heat that it produces.

Peak reverse voltage is a very important *parameter* (limit). When you are considering whether or not to use a given diode for some application, you must be sure that the reverse bias in the circuit will never meet or exceed the value of V_{RRM} for the diode. For example, consider the circuit shown in Figure 2.15. The ac signal being applied to the circuit has a peak value of 50 V. When the ac signal source has the polarity shown, the diode will be reverse biased and the full applied voltage will be dropped across its terminals. Thus, D_1 would have to have a V_{RRM} rating that is *greater than* 50 V. As long as this condition is met, the reverse bias on the diode will never meet or exceed V_{RRM} (under normal circumstances), and the diode will not be damaged.

One point should be made at this time: As long as the V_{RRM} rating is *greater than* the maximum reverse bias in the circuit, we do not really care what its value is. For example, let's say that the diode in Figure 2.15 needed to be replaced. We could use a diode with a V_{RRM} rating of 100 V, 200 V, or even 1000 V. This point is illustrated in the following example.

Avalanche current. The current that occurs when V_{BR} is reached. Avalanche current can destroy a *pn* junction diode.

A **parameter** is a *limit*.

A Practical Consideration: Most reverse breakdown voltage ratings for *pn* junction diodes are multiples of 50 V or 100 V. Common values are 50 V, 100 V, 150 V, 200 V, and so on.

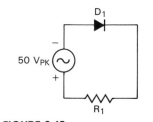

FIGURE 2.15

EXAMPLE 2.8

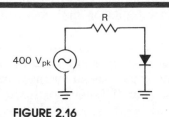

FIGURE 2.16

The diode in the circuit shown in Figure 2.16 is faulty and must be replaced. When checking the parts bin, you see that you have the following diodes in stock:

Diode part number	Peak reverse voltage
1N4001	50
1N4002	100
1N4003	200
1N4004	400
1N4005	600
1N4006	800
1N4007	1000

Which of these diodes could be used to replace the faulty diode?

Solution: The V_{RRM} rating of a diode *must be greater than* the maximum value of reverse bias that occurs in a circuit. Since the peak value of the source is 400 V, a full 400 V will be dropped across the diode terminals whenever the device is reverse biased. Therefore, the V_{RRM} rating of the replacement part must be *greater than* 400 V, and the only diodes we can use are the 1N4005, 1N4006, or 1N4007.

PRACTICE PROBLEM 2–8

A circuit like the one in Figure 2.16 has a 180 V signal source. Which of the diodes listed in Example 2.8 could be used in the circuit?

Specification sheet. A listing of all the important parameters and operating characteristics of a device or circuit.

The value of V_{RRM} for a given *pn*-junction diode can be obtained from the *specification sheet* (or data sheet) for the component. A *specification sheet* (spec sheet) lists all the important parameters and operating characteristics of a device or circuit. The specification sheet for a given component can be obtained from the manufacturer or from most electronics parts stores (usually free of charge). You will be shown in Section 2.6 how to find the value of V_{RRM} (and many other important parameters) on a specification sheet.

Peak reverse voltage is only one of several parameters that must be considered before trying to use a specific diode in a given application. Another parameter that must be considered is the *average forward current* rating of the diode.

Average Forward Current (I_o)

Average forward current. The maximum allowable value of dc forward current for a diode.

The *average forward current* rating of a diode is the *maximum allowable value of dc forward current*. For example, the 1N4001 diode has an average forward current rating of 1 A. This means that the dc forward current through the diode must never be allowed to exceed 1 A. If the dc forward current through the diode is allowed to exceed 1 A, the diode will be destroyed from excessive heat. The average forward current rating is another parameter found on the spec sheet for the component.

When you are considering whether or not to use a diode in a specific circuit, you must determine what the value of forward current (I_F) in the circuit will be. Then you must make sure that the diode you want to use has an average forward current rating that is *greater than or equal to* the value of I_T for the circuit. This is illustrated in the following example.

EXAMPLE 2.9

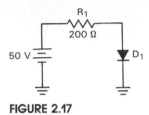

FIGURE 2.17

Determine the *minimum* average forward current rating that would be required for the diode in Figure 2.17.

Solution: The forward current in the circuit is calculated (as always) using the practical model of the diode. Thus,

$$I_T = \frac{V_S - 0.7 \text{ V}}{R_1}$$
$$= \frac{50 \text{ V} - 0.7 \text{ V}}{200 \text{ }\Omega}$$
$$= 246.5 \text{ mA}$$

A Practical Consideration: Let's say that you calculate the dc forward current in a circuit to be 1 A. If you use a diode that has an average forward current rating of 1 A in the circuit, you will be pushing the diode to its limit. This will shorten the lifetime of the diode. If, on the other hand, you use a diode with an average forward current rating of 10 A, the diode will not be pushed to its limit and will last much longer.

Thus, any diode used in the circuit would have to have an average forward current rating that is *greater than or equal to* 246.5. In practice, you would use a diode whose average forward current rating is *greater than* the calculated value of I_T.

PRACTICE PROBLEM 2–9

A circuit like the one shown in Figure 2.17 has a 100-V_{dc} source. If the resistor has a value of 51 Ω, what is the minimum allowable average forward current rating for the diode in the circuit?

In Chapter 3, you will be shown how to determine the *dc equivalent current* for an ac source. When dealing with ac circuits, you will need to determine the *average* (or *dc equivalent*) current for the circuit. Then this dc equivalent current value must be compared to the average forward current rating of the diode to see if it can be used in the circuit.

Forward Power Dissipation ($P_{D(MAX)}$)

Many diodes have a *forward power dissipation* rating. This rating indicates the *maximum possible power dissipation of the device when it is forward biased.*

Recall from your study of basic electronics that power is found as

Forward power dissipation ($P_{D(MAX)}$). A diode rating that indicates the *maximum* possible power dissipation of the forward-biased diode.

$$\boxed{P = IV}$$

where P = the power dissipated by a component
$\quad\quad I$ = the device current
$\quad\quad V$ = the voltage across the device

Using the basic power equation and the values of V_F and I_T for a diode, you can determine the required forward power dissipation rating for a replacement component, as is illustrated in the following example.

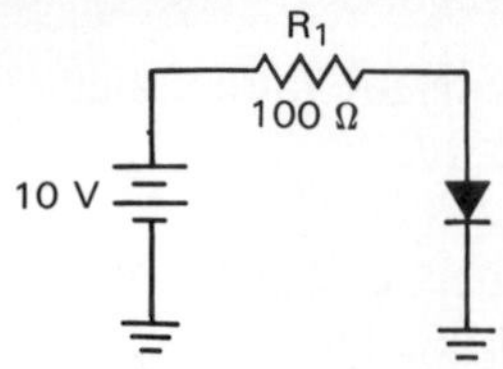

FIGURE 2.18

Calculate the *minimum* forward power dissipation rating for any diode that would be used in the circuit shown in Figure 2.18.

Solution: First, we have to calculate the total circuit current. This current is found as

$$I_T = \frac{10 \text{ V} - 0.7 \text{ V}}{100} = 93 \text{ mA}$$

Using $I_T = 93$ mA and $V_F = 0.7$ V, the power dissipation of the diode is found as

$$P = I_T V_F = (93 \text{ mA})(0.7 \text{ V}) = 65.1 \text{ mW}$$

Thus, any diode used in the circuit would have to have a forward power dissipation rating that is greater than 65.1 mW.

PRACTICE PROBLEM 2–10

A circuit like the one shown in Figure 2.18 has a 20 V_{dc} source and a 68 Ω series resistor. What is the minimum required forward power dissipation rating for any diode that is used in the circuit?

Some diode specification sheets will contain a forward power dissipation rating *instead* of an average forward current rating. This is especially true for *zener diode* specification sheets, as you will see in Section 2.5. When this is the case, the maximum forward current rating for the diode can be found using

$$\boxed{I_{F(\text{MAX})} = \frac{P_{D(\text{MAX})}}{V_F}} \qquad (2.4)$$

where $I_{F(\text{MAX})}$ = the maximum allowable forward current
$P_{D(\text{MAX})}$ = the forward power dissipation rating of the diode
V_F = the forward voltage across the diode, assumed to be 0.7 V for a silicon *pn*-junction diode

This equation is simply a variation on the standard power dissipation equation. Its use is illustrated in the following example.

A diode has a forward power dissipation rating of 500 mW. What is the maximum allowable value of forward current for the device?

Solution: The value of $I_{F(MAX)}$ is found as

$$I_{F(MAX)} = \frac{P_{D(MAX)}}{V_F}$$
$$= \frac{500 \text{ mW}}{0.7 \text{ V}}$$
$$= 714.29 \text{ mA}$$

As long as I_F is *less than* 714.29 mA, the forward power dissipation rating of the device will not be exceeded.

PRACTICE PROBLEM 2–11

Show that the power dissipation rating of 500 mW in Example 2.11 will be exceeded if forward current is 750 mA.

Summary

When you are trying to replace one diode with another, there are three main parameters that must be considered. Before substituting one diode for another, ask yourself:

1. Is the V_{RRM} rating of the replacement diode greater than the maximum reverse bias in the circuit?
2. Is the *average forward current* rating of the replacement diode greater than the average (dc) value of I_F in the circuit?
3. Is the *forward power dissipation* rating of the replacement diode high enough?

If the answer to any of these questions is *no*, then you cannot use the diode in the circuit.

One Final Note

In this section, we have considered the more practical aspects of diode operation that you will have to deal with on a regular basis. There are more diode characteristics, many of which are considered in *circuit development*, in *high-frequency circuit analysis*, and under some other special circumstances. These characteristics are included in the *complete diode model*. We will take a look at this diode model in the next section.

SECTION REVIEW

1. What is *peak reverse voltage*? [Objective 7]
2. Why do you need to consider the value of V_{RRM} for a diode before attempting to use the diode in a specific circuit? [Objective 7]
3. A diode circuit has a 290 V_{pk} ac source. Which of the diodes listed in Example 2.8 could be used in the circuit without having reverse breakdown problems? Explain your answer. [Objective 7]
4. What is *average forward current*? [Objective 7]
5. How do you determine whether or not the average forward current rating of a diode will be exceeded in a given circuit? [Objective 7]

6. What is the *forward power dissipation* rating of a diode? [Objective 7]

7. How do you determine whether or not a given diode has a *forward power dissipation* rating that is high enough to allow the device to be used in a specific circuit? [Objective 7]

8. When you know the *forward power dissipation* rating of a diode, how can you determine the maximum allowable forward current for the device? [Objective 7]

2.5

The Complete Diode Model

The complete diode model is the most accurate.

The complete diode model most accurately represents the true operating characteristics of the diode. Some of the factors that make this model so accurate are *bulk resistance* and *reverse current*. When these factors are taken into account, we get the diode characteristic curve shown in Figure 2.19. This illustration will be referred to throughout our discussion on the complete diode model.

Bulk Resistance (R_B)

Bulk resistance. The natural resistance of a forward-biased diode.

Bulk-resistance is the natural resistance of the diode *p*-type and *n*-type materials. The effect of bulk resistance on diode operation can be seen in the *forward operation region* of the curve. As Figure 2.19 illustrates, V_F is *not* constant, but, rather, varies with the value of I_F. The change in V_F (ΔV_F) is caused by the diode current

FIGURE 2.19

"Complete model" diode curve.

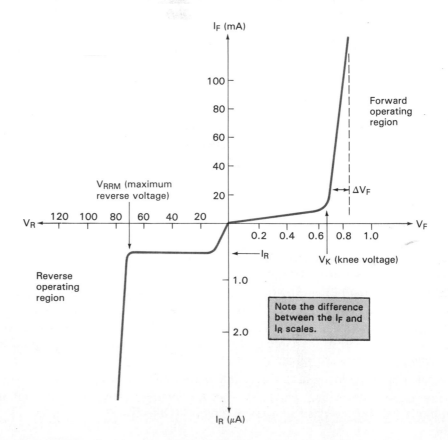

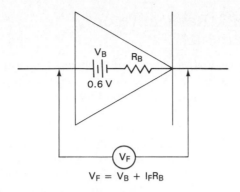

$$V_F = V_B + I_F R_B$$

FIGURE 2.20

Diode equivalent circuit.

passing through the bulk resistance of the diode. This concept is illustrated by the diode equivalent circuit shown in Figure 2.20. The 0.6 V source shown in Figure 2.20 represents the barrier potential of the diode. The bulk resistance of the diode is represented by the series resistor, R_B. The diode current, I_F, passing through the bulk resistance will develop a voltage equal to $I_F R_B$. The total voltage across the diode is the sum of $I_F R_B$ and the barrier potential. By formula,

$$\boxed{V_F = 0.6\ \text{V} + I_F R_B} \qquad \text{(for silicon)} \qquad \textbf{(2.5)}$$

and

$$\boxed{V_F = 0.2\ \text{V} + I_F R_B} \qquad \text{(for germanium)} \qquad \textbf{(2.6)}$$

Note that as I_F increases so does $I_F R_B$. Therefore, the total voltage across a diode is *directly proportional* to the value of I_F.

EXAMPLE 2.12

Determine the voltage across the diode in Figure 2.20 for values of $I_F = 1$ mA and $I_F = 5$ mA. Assume that the bulk resistance of the diode is 5 Ω and that the diode is silicon.

Solution: Because the diode is silicon, we use equation (2.5) to solve for the total voltage drop across the diode. For $I_F = 1$ mA, $V_F = 0.6 + I_F R_B$ and

$$\begin{aligned} V_F &= 0.6 + (1\ \text{mA})(5\ \Omega) \\ &= 0.6 + 0.005\ \text{V} \\ &= 0.605\ \text{V} \end{aligned}$$

For $I_F = 5$ mA,

$$\begin{aligned} V_F &= 0.6 + (5\ \text{mA})(5\ \Omega) \\ &= 0.6 + 0.025\ \text{V} \\ &= 0.625\ \text{V} \end{aligned}$$

PRACTICE PROBLEM 2–12

A silicon diode has a bulk resistance of 8 Ω and a forward current of 12 mA. What is the actual value of V_F for the device?

The Effect of Bulk Resistance on Circuit Measurements

The effect of bulk resistance on V_F readings.

In our discussion on practical diode circuit analysis, we assumed that the value of V_F for a silicon diode is 0.7 V. While this assumed value of V_F will work very well for the analysis of a circuit, you will find that measured values of V_F will generally vary between 0.6 and 0.8 V.

A diode that is used in a low-current circuit will have a very small amount of voltage developed across its bulk resistance. Because of this, the voltage across such a diode will tend to be closer to 0.6 V. At the same time, a diode that is being used in a high-current circuit will have a relatively large amount of voltage developed across its bulk resistance. This will cause it to have a value of V_F that is closer to 0.8 V.

For routine circuit measurements, using the approximate value of 0.7 V is acceptable because the *exact* value of V_F isn't critical. However, in *circuit development* (or *engineering*) applications, the circuit designer may need to very accurately predict the value of V_F in a circuit. This is when equations (2.5) and (2.6) and other factors of the complete diode model are most commonly used.

Reverse Current (I_R)

Reverse current (I_R). The minority carrier current through a reverse-biased diode.

Ideally, when a diode is reverse biased, the depletion layer reaches its maximum width, and conduction through the diode stops. In reality, this is not the case. There is a *very small* amount of minority carrier current that flows through the diode when it is reverse biased. This current is referred to as *reverse current* and is illustrated in the *reverse operating region* of the diode curve (Figure 2.19). Reverse current is made up of two independent currents: *reverse saturation current*, I_S, and *surface-leakage current*, I_{SL}. By formula,

$$\boxed{I_R = I_S + I_{SL}}$$ (2.7)

where I_R = the diode reverse current
I_S = the reverse saturation current
I_{SL} = the surface-leakage current

Reverse saturation current (I_S). A current that is caused by thermal activity in a reverse-biased diode. I_S is temperature dependent.

Reverse saturation current is the current that is caused by thermal activity in the two diode materials. *This current is effected by temperature*, but not by the amount of reverse bias applied to the diode. I_S accounts for the major portion of reverse current. Since I_S is not related to the amount of reverse bias, I_R remains relatively constant across a range of reverse voltages. This can be seen by referring back to Figure 2.19.

Surface leakage current (I_{SL}). A current along the surface of a reverse-biased diode. I_{SL} is V_R dependent.

Surface-leakage current is a current that flows along the surface of the diode. This current will increase with an increase in reverse bias. However, since its value is much lower than that of I_S, there is no noticeable change in I_R when I_{SL} changes.

The range of I_R values.

The total value of I_R for a diode is typically in the μA range or less. However, as temperature increases, I_S, and consequently I_R, increase. This point is discussed in detail later in this section.

The Effect of Reverse Current on Circuit Measurements

Until now we have assumed that the value of I_R is zero. As you now know, there is a small amount of reverse current that occurs in a diode circuit. This reverse current will cause a slight voltage to be developed across any resistance that is in the circuit. For example, consider the circuit shown in Figure 2.21. If the reverse current through the diode (under normal circumstances) is 20 μA, the voltage across the resistor will be

$$V_R = (20 \ \mu A)(10 \ k\Omega) = 200 \ mV$$

when the diode is reverse biased. While this voltage would not be significant in most cases, it can become important in any circuit where the normal value of I_F is also extremely small.

FIGURE 2.21

Diode Capacitance

Refer to Figure 2.22. An insulator placed between to closely spaced conductors forms a capacitor. When a diode is *reverse biased*, it forms a depletion layer (insulator) between the two semiconductor materials (conductors). Therefore, a reverse-biased diode forms a small capacitor. Under normal circumstances, the capacitance associated with a reverse-biased diode can be ignored. However, when the *high-frequency* operation of a diode is being analyzed, the junction capacitance can become extremely important. The effects of junction capacitance on high-frequency operation are discussed in detail in Chapter 13.

A reverse-biased *pn* junction has capacitance.

Incidentally, there is a type of diode that is designed to make use of its junction capacitance. This diode, called a *varactor*, is discussed in detail in Chapter 21.

Diffusion Current

Refer to the forward operating region in Figure 2.19. The 0.7 V point on the diode curve is labeled *knee voltage*, V_k. Below the knee voltage, you will notice that I_F does *not* instantly drop to the zero point. The reason for this is simple. When V_F goes below the barrier potential of the diode, the device starts to form a depletion layer. However, until the diode is actually reverse biased, the depletion layer will not reach its maximum width, and thus will not reach its maximum resistance. In other words, as long as there is *some* forward voltage the depletion layer will not be at its maximum width, and some amount of I_F will occur. This small amount of I_F is called *diffusion current*. Note that when a diode is switching from forward bias to reverse bias, diffusion current will last only as long as it takes the depletion layer to form, generally a few milliseconds or less.

Diffusion current. The I_F below the knee voltage.

Temperature Effects on Diode Operation

Temperature has a significant effect on most of the diode characteristics discussed in this section. The reason for this is simple: Increased temperature means increased

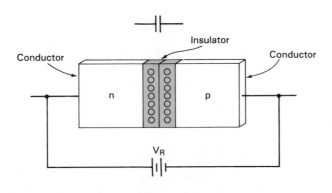

FIGURE 2.22
Diode capacitance.

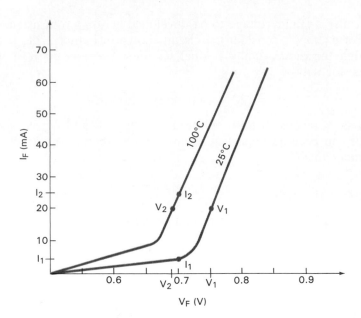

FIGURE 2.23

Temperature effects on
forward operation.

thermal activity and decreased diode resistance. This holds true for both forward
and reverse diode operation.

The effects of increased temperature on forward diode operation are illustrated
in Figure 2.23. As you can see, there are two forward operation curves. One
represents the diode operation at 25°C, and the other represents the diode operation
at 100°C. From the graph, we can draw two conclusions regarding forward diode
operation and temperature:

1. *As temperature increases, I_F will increase for a specified value of V_F.* This is
    illustrated in Figure 2.23 by the two points labeled I_1 and I_2. As you can
    see, both of these points fall on the $V_F = 0.7$ V line. However, I_2 is greater
    than I_1. This is due to the increased thermal activity in the diode. As temperature
    has increased from 25° to 100°C, the value of I_F has increased from 5 mA
    (I_1) to 25 mA (I_2).
2. *As temperature increases, V_F will decrease for a specified value of I_F.* This
    is illustrated by the V_1 and V_2 points on the curve. Note that both of these
    points correspond to a value of $I_F = 20$ mA. As temperature has increased,
    V_F has decreased from 0.75 V (V_1) to 0.68 V (V_2).

*In practice, a rise in temperature will usually be followed by both a slight
increase in I_F and a slight decrease in V_F.*

The effect of temperature on the reverse operation of a diode is basically the
same as the effect on I_F. The effect of temperature on I_R is illustrated in Figure
2.24. You may recall that I_R is equal to the sum of reverse saturation current (I_S)
and surface-leakage current (I_{SL}). Since I_S is normally much greater than I_{SL}, it is
safe to assume that I_R is approximately equal to I_S.

I_S will *double* with every 10°C rise in temperature. Since I_R is approximately
equal to I_S, we can say that I_R will double for every 10°C rise in temperature.
You can see this in Figure 2.24, where I_R doubled from 5 to 10 μA when the
temperature increased to 35°C, and then doubled again to 20 μA when temperature
increased to 45°C. A handy formula for determining the value of I_R at a given
temperature is

$$\boxed{I_R' = I_R(2^x)}$$

(2.8)

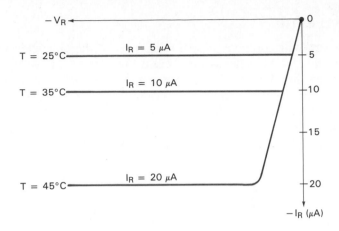

FIGURE 2.24

Temperature effects on reverse operation.

where I'_R = the value of I_R at the specified temperature
I_R = the value of I_R at 25°C
$x = (T - 25°C)/10$ (T is the temperature of interest)

===== EXAMPLE 2.13 =====

A given silicon diode is rated at a reverse current of 0.5 μA at 25°C. What is the value of I_R at 50°C? 75°C? 100°C?

Solution: For 50°C,

$$x = \frac{50 - 25}{10} = 2.5$$

and

$$I'_R = (0.5 \ \mu A)(2^{2.5}) = (0.5 \ \mu A)(5.66) = 2.83 \ \mu A$$

For 75°C,

$$x = \frac{75 - 25}{10} = 5$$

and

$$I'_R = (0.5 \ \mu A)(2^5) = (0.5 \ \mu A)(32) = 16 \ \mu A$$

For 100°C,

$$x = \frac{100 - 25}{10} = 7.5$$

and

$$I'_R = (0.5 \ \mu A)(2^{7.5}) = (0.5 \ \mu A)(181) = 90.5 \ \mu A$$

PRACTICE PROBLEM 2–13

A diode has a reverse current rating of 2 μA at 25°C. What is the value of I_R for the device at $T = 60$°C?

The Bottom Line

When the *complete* diode model is used.

In most cases, you will not have to deal with most of the characteristics discussed in this section unless you are designing a circuit with very low variation tolerances or very high operating frequencies. Even in most practical circuit designs, such as rectifiers and amplifiers, knowing the diode's parameters is more important than knowing the *exact* values of diode voltage, current, resistance, and capacitance.

When you are involved in practical circuit analysis and troubleshooting, you need only consider the more practical aspects of diode operation. These aspects are summarized as follows:

A summary of *practical* diode characteristics.

1. When the forward voltage across a diode reaches the barrier potential, the diode will start to conduct. From that point on, the value of V_F will be approximately equal to the barrier potential.
2. As V_F decreases below the barrier potential, the diode will start to turn off and begin to act like an open switch. This "open-switch" characteristic will continue for all reverse voltage values up to V_{RRM}.
   a. As long as $V_R \leq V_{RRM}$, the total applied voltage will be dropped across the diode.
   b. When $V_R > V_{RRM}$, the diode may break down and conduct in the reverse direction. When this happens, you will usually end up replacing the diode.
3. The forward current through a diode is limited by the applied voltage and the resistance values that are external to the diode. For all practical purposes, reverse current through a diode is zero.

SECTION REVIEW

1. What is *bulk resistance*? [Objective 8a]
2. What effect does bulk resistance have on circuit voltage measurements? [Objective 9a]
3. What is the relationship between I_F and V_F? [Objective 9a]
4. What is reverse current? [Objective 8b]
5. Which component of reverse current is affected by temperature? [Objective 8b]
6. Which component of reverse current is affected by the amount of reverse bias? [Objective 8b]
7. Which component of reverse current makes up a majority of that reverse current? [Objective 8b]
8. What effect will I_R have on the measured value of voltage across any series resistance? [Objective 9b]
9. Describe the reverse-biased diode as a capacitor. [Objective 8c]
10. What is diffusion current? [Objective 8d]
11. What effect will an increase in temperature have on I_F and V_F? [Objective 10]
12. What effect will an increase in temperature have on I_R? [Objective 10]

CHAP. 2 Diodes

2.6
Diode Specification Sheets

The *specification sheet*, or *spec sheet*, for any component lists the *parameters* and *operating characteristics* of the device. Because parameters are limits, they are always given as *minimum* or *maximum* values.

Parameters are listed as *minimum* or *maximum* values.

Diode operating characteristics and parameters are important for several reasons:

Why parameters are important.

1. They indicate whether or not a given diode can be used for a specific application.
2. They establish the operating limits of any circuit that is designed to use the diode.

You have already been introduced to most of the commonly used diode characteristics and parameters. In this section, we will take a look at how these characteristics and parameters may be found on the spec sheet of a given diode.

Spec Sheet Organization

Diode spec sheets are commonly divided into two sections: *maximum ratings* and *electrical characteristics*. This can be seen in Figure 2.25, which contains the spec sheet for the 1N4001–1N4007 (or 1N400X) series diodes.

The *maximum ratings* table contains the diode parameters that must not be exceeded under any circumstances. If any of these parameters are violated, you will more than likely have to replace the diode.

Maximum ratings. Diode parameters that must not be exceeded under any circumstances.
Electrical characteristics. The guaranteed operating characteristics of the device.

The *electrical characteristics* table contains the guaranteed operating characteristics of the device. As long as the *maximum ratings* limits are observed, the diode is guaranteed to work within the limits shown in the *electrical characteristics* table.

Confused? Let's take a look at the *maximum reverse current* rating in the *electrical characteristics* table. The 1N400X series of diodes is guaranteed to have a maximum reverse current (I_R) of 10 μA at 25°C. This assumes that you do not exceed the *reverse breakdown voltage* rating, which is listed on the spec sheet of the device. If you exceed any diode parameter, the *electrical characteristics* values of the device cannot be guaranteed.

The difference between *maximum ratings* and *electrical characteristics*.

Diode Maximum Ratings

As was stated earlier, many of the parameters listed in the *maximum ratings* table have already been covered in this chapter. Table 2.2 summarizes the parameters listed in Figure 2.25 and their meanings.

A few notes:

1. *Peak repetitive reverse voltage* (V_{RRM}) indicates the maximum allowable reverse voltage that can be applied to the device. The 1N4001 has a V_{RRM} rating of 50 V. If you apply a reverse voltage to the device that is *greater than* 50 V, the device *may* go into reverse breakdown. Note that this rating is also called *working peak reverse voltage* or *dc blocking voltage*.

Peak repetitive reverse voltage (V_{RRM}). The maximum allowable reverse voltage that can be applied to a diode.

2. The reverse voltage ratings are the only ratings that distinguish the diodes in the 1N400X series from each other. In other words, the diodes listed have the same maximum ratings and electrical characteristics *outside of their reverse voltage characteristics*. This is usually the case for a given series of diodes.

MOTOROLA Semiconductors
BOX 20912 • PHOENIX, ARIZONA 85036

Designers'Data Sheet

1N4001 thru 1N4007

"SURMETIC"▲ RECTIFIERS

. . . subminiature size, axial lead mounted rectifiers for general-purpose low-power applications.

Designers Data for "Worst Case" Conditions

The Designers▲ Data Sheets permit the design of most circuits entirely from the information presented. Limit curves — representing boundaries on device characteristics — are given to facilitate "worst case" design.

**LEAD MOUNTED
SILICON RECTIFIERS**

**50-1000 VOLTS
DIFFUSED JUNCTION**

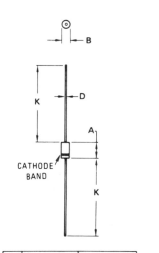

*MAXIMUM RATINGS

Rating	Symbol	1N4001	1N4002	1N4003	1N4004	1N4005	1N4006	1N4007	Unit
Peak Repetitive Reverse Voltage Working Peak Reverse Voltage DC Blocking Voltage	V_{RRM} V_{RWM} V_R	50	100	200	400	600	800	1000	Volts
Non-Repetitive Peak Reverse Voltage (halfwave, single phase, 60 Hz)	V_{RSM}	60	120	240	480	720	1000	1200	Volts
RMS Reverse Voltage	$V_{R(RMS)}$	35	70	140	280	420	560	700	Volts
Average Rectified Forward Current (single phase, resistive load, 60 Hz, see Figure 8, $T_A = 75^{\circ}C$)	I_O	1.0							Amp
Non-Repetitive Peak Surge Current (surge applied at rated load conditions, see Figure 2)	I_{FSM}	30 (for 1 cycle)							Amp
Operating and Storage Junction Temperature Range	T_J, T_{stg}	-65 to +175							°C

*ELECTRICAL CHARACTERISTICS

Characteristic and Conditions	Symbol	Typ	Max	Unit
Maximum Instantaneous Forward Voltage Drop ($i_F = 1.0$ Amp, $T_J = 25^{\circ}C$) Figure 1	v_F	0.93	1.1	Volts
Maximum Full-Cycle Average Forward Voltage Drop ($I_O = 1.0$ Amp, $T_L = 75^{\circ}C$, 1 inch leads)	$V_{F(AV)}$	–	0.8	Volts
Maximum Reverse Current (rated dc voltage) $T_J = 25^{\circ}C$ $T_J = 100^{\circ}C$	I_R	0.05 1.0	10 50	µA
Maximum Full-Cycle Average Reverse Current ($I_O = 1.0$ Amp, $T_L = 75^{\circ}C$, 1 inch leads)	$I_{R(AV)}$	–	30	µA

*Indicates JEDEC Registered Data.

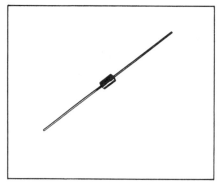

CATHODE BAND

DIM	MILLIMETERS		INCHES	
	MIN	MAX	MIN	MAX
A	5.97	6.60	0.235	0.260
B	2.79	3.05	0.110	0.120
D	0.76	0.86	0.030	0.034
K	27.94	–	1.100	–

CASE 59-04
Does Not Conform to DO-41 Outline.

MECHANICAL CHARACTERISTICS

CASE: Void free, Transfer Molded
MAXIMUM LEAD TEMPERATURE FOR SOLDERING PURPOSES: 350°C, 3/8" from case for 10 seconds at 5 lbs. tension
FINISH: All external surfaces are corrosion-resistant, leads are readily solderable
POLARITY: Cathode indicated by color band
WEIGHT: 0.40 Grams (approximately)

▲Trademark of Motorola Inc.

© MOTOROLA INC., 1975   DS 6015 R3

FIGURE 2.25

The Motorola 1N4001 specification   (Courtesy of Motorola, Inc.)

TABLE 2.2
Diode Parameters

Rating	*Discussion*
Peak repetitive reverse voltage, V_{RRM}	This is the maximum allowable V_R for the diode. This holds true for both dc and *peak* ac voltages.
Rms reverse voltage, $V_{R(rms)}$	This rating is found by converting the peak repetitive reverse voltage rating to an rms value. You may recall that V_{rms} = $0.707V_{pk}$. Using this formula, you can convert any peak maximum rating to an rms maximum rating.
Average half-wave rectified forward current, I_o	This rating tells you the maximum *average* forward current the diode can handle. For the 1N4001, this value is 1000 mA, or 1 A, when the temperature is 75°C. If the temperature increases to 100°C, the maximum allowable value drops to 750 mA. *This does not mean that diode forward current will decrease when temperature increases!* It means that at higher temperatures, the *limit* on forward current decreases.
Peak surge current, I_{FSM}	This is the maximum *surge* value of I_F that the diode can handle. Surge current is nonrepetitive, meaning that it does not happen at regular intervals. With the average diode, this rating will be exceeded *once*. After that, you will have to replace the diode.
Operating and storage temperature range, T_J, T_{stg}	This one is self-explanatory. The diode can be used and stored at any temperature between −65° and 175°C.

The meaning of the term *half-wave rectified* will be explained in Chapter 3.

3. The *average forward current* (I_0) rating shows that the maximum allowable average forward current *decreases* as temperature *increases*. The basis for this relationship is simple. The current through a diode generates *heat*. The maximum allowable current through a diode depends on *how much heat the component can dissipate*. When the air that surrounds a diode is hot, the diode cannot dissipate as much heat. Therefore, the limit on device current decreases.

4. The *peak surge current* (I_{FSM}) indicates the diode's ability to handle a *surge* (short duration, extremely high value) of current. We will cover surge current and one of its primary sources in Chapter 3.

Diode Parameters and Device Substitution

It was stated earlier in the chapter that *average forward current* and *peak reverse voltage* ratings are two of the primary concerns when you are substituting one diode for another. To make component substitutions easier, diode data books sometimes contain *selector guides*. These guides group diodes by *average forward current* (or *forward power dissipation*) and *peak reverse voltage*. An example of a selector guide is shown in Figure 2.26.

The selector guide allows you to select a diode based on circuit requirements. For example, let's say that you need a diode with an average forward current rating of 1.5 A and a peak reverse voltage of 100 V. You would locate the column that corresponds to 1.5 A and cross-match it with the row that has a V_{RRM} of 100 V. The block that corresponds to both of these values contains the number 1N5392. Thus, the 1N5392 would be suitable for the desired application.

A Practical Consideration: Any diode in the selector guide that is to the right and/or below the 1N5392 could be used in the application. As long as the forward current and reverse voltage ratings are *at least* 1.5 A and 100 V, the diode can be used.

Electrical Characteristics

As stated earlier, the values provided under *electrical characteristics* indicate the guaranteed operating characteristics of the diode. Table 2.3 gives a brief explanation of the electrical characteristics listed in Figure 2.25.

	I_O, AVERAGE RECTIFIED FORWARD CURRENT (Amperes)							
	12	**20**	**24**	**25**	**30**		**40**	**50**
V_{RRM} (Volts)	245A-02 (DO-203AA) Metal 		339-02 Plastic Note 1 	193-04 Plastic Note 2 	43-02 (DO-21) Metal 		42A-01 (DO-203AB) Metal 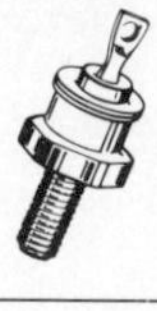	43-04 Metal 
50	MR1120 1N1199,A,B	MR2000	MR2400	MR2500	1N3491	1N3659	1N1183A	MR5005
100	MR1121 1N1200,A,B	MR2001	MR2401	MR2501	1N3492	1N3660	1N1184A	MR5010
200	MR1122 1N1202,A,B	MR2002	MR2402	MR2502	1N3493	1N3661	1N1186A	MR5020
400	MR1124 1N1204,A,B	MR2004	MR2404	MR2504	1N3495	1N3663	1N1188A	MR5040
600	MR1126 1N1206,A,B	MR2006	MR2406	MR2506		Note 3	1N1190A	Note 3
800	MR1128	MR2008		MR2508		Note 3	Note 3	Note 3
1000	MR1130	MR2010		MR2510		Note 3	Note 3	Note 3
I_{FSM} (Amps)	300	400	400	400	300	400	800	600
T_A @ Rated I_O (°C)								
T_C @ Rated I_O (°C)	150	150	125	150	130	100	150	150
T_J (Max) (°C)	190	175	175	175	175	175	190	195

Note 1. Meets mounting configuration of TO-220 outline.
Note 2. Request Data Sheet for Mounting Information.
Note 3. Available on special order.

	I_O, AVERAGE RECTIFIED FORWARD CURRENT (Amperes)					
	1.0	**1.5**	**3.0**			**6.0**
V_{RRM} (Volts)	59-03 (DO-41) Plastic 	59-04 Plastic 	60-01 Metal 	267-03 Plastic 	267-02 Plastic 	194-04 Plastic 
50	†1N4001	**1N5391	1N4719	**MR500	1N5400	MR750
100	†1N4002	**1N5392	1N4720	**MR501	1N5401	MR751
200	†1N4003	1N5393 *MR5059	1N4721	**MR502	1N5402	MR752
400	†1N4004	1N5395 *MR5060	1N4722	**MR504	1N5404	MR754
600	†1N4005	1N5397 *MR5061	1N4723	**MR506	1N5406	MR756
800	†1N4006	1N5398	1N4724	MR508		MR758
1000	†1N4007	1N5399	1N4725	MR510		MR760
I_{FSM} (Amps)	30	50	300	100	200	400
T_A @ Rated I_O (°C)	75	T_L = 70	75	95	T_L = 105	60
T_C @ Rated I_O (°C)						
T_J (Max) (°C)	175	175	175	175	175	175

FIGURE 2.26

(Courtesy of Motorola, Inc.)

† Package Size: 0.120" Max Diameter by 0.260" Max Length.
* 1N5059 series equivalent Avalanche Rectifiers.
** Avalanche versions available, consult factory.

TABLE 2.3

Maximum forward voltage drop, V_F	This is the maximum value that V_F will ever reach. All V_F values are guaranteed to be at 1.1 V or less. Note that this value was determined at a temperature of 25°C. On the average, V_F will decrease by about 1.8 mV for every 1°C rise in temperature.
Maximum full-cycle average forward voltage drop, $V_{F(av)}$	This is the maximum *average* forward voltage (V_F). For the 1N400X, this value is 0.8 V. Note that this parameter is also temperature dependent and was measured at 75°C.
Maximum reverse current, I_R	These ratings, 10 μA and 50 μA, were given for 25° and 100°C, respectively. The spec sheet lists this parameter "*at rated dc voltages.*" This means that the rating is valid for all dc values of V_R at or below the 50-V peak repetitive V_R rating.
Maximum full-cycle average reverse current, $I_{R(av)}$	This is the maximum average value of I_R. Note that this rating is at a temperature of 75°C. At all temperatures below 75°C, the average value of I_R will be less than 0.03 mA.

A few notes:

1. The *maximum reverse current* rating of the 1N400X series is 10 μA when T_J = 25°C. If we were to use these values in equation (2.8), we would calculate the maximum reverse current at 100°C to be

$$I_R = (10 \ \mu A)(2^{7.5}) = 1.81 \ mA$$

T_J stands for *junction temperature.*

However, the rated value of $I_{R(MAX)}$ at 100°C is 50 μA. You might wonder *which one is correct*? The fact is, equation (2.8) gives you a *worst-case* value, that is, the highest that it would be under any circumstances. When the *rated* value of I_R at a given temperature is less than the one calculated with equation (2.8), you assume that the rated value is the maximum.

2. Many of the characteristics list both *typical* and *maximum* values. When analyzing the operation of the component in a circuit, the value listed as *typical* would be used. When determining circuit tolerances for circuit development purposes, the *maximum* values would be used. When only one value is given, it is used for *all* circuit analyses.

Finding Data on Spec Sheets

When you are looking for data on a spec sheet, you follow a three-step procedure:

1. Determine whether the data you are looking for is a *maximum rating* (parameter) or an *electrical characteristic* (guaranteed minimum performance).
2. Look in the appropriate table for the desired data. If it isn't there, try the alternative-search technique covered in step 3. If you still can't find the data, contact the manufacturers of the component.
3. Sometimes it is difficult to locate a particular parameter or characteristic because of the wording used in the spec sheet. When you can't locate a particular bit of information, try this:
   a. Determine the unit that the desired parameter or characteristic would be measured in. For example, *average forward current* would typically be measured in *mA* or *A*, while *reverse current* would typically be measured in μ*A*.
   b. Search the *unit* column in the spec sheet tables to find the appropriate unit of measure.
   c. If you find the unit of measure, look at the data name to see if it is the value you are looking for.

The use of the third step can be illustrated by referring back to Figure 2.26. Let's say that we are interested in the temperature range of the 1N400X series of diodes. We know that temperature is measured in °C. Looking under the *unit* column, we see that only one parameter is measured in °C. Looking to the left side of that measurement, we see the parameter that is shown is *operating and storage junction temperature*. This is the parameter that we are looking for.

While the spec sheet for the 1N400X series of diodes is relatively short and simple, the spec sheets for many other types of devices are relatively complex. When dealing with these more complex spec sheets, you will find the steps listed under step 3 very useful for quickly finding the information that you need.

One Final Note

Not all operating parameters and characteristics used are listed on the 1N400X spec sheet. However, the most critical of the operating parameters and characteristics have been covered. Some other parameters you will often see that were not included in this sheet are *junction capacitance* (rated in *pF*), *forward power dissipation* (rated in *mW* or *W*), and *maximum switching frequency* (rated in *kHz* or *MHz*). Also, as mentioned earlier, not all spec sheets use the same terminology to describe the various parameters and characteristics. However, the wording is usually close enough that you will be able to find the information you need.

SECTION REVIEW

1. Why are parameters important?
2. What is a *maximum rating*? Give an example of one. [Objective 11a]
3. What is an *electrical characteristic*? Give an example of one. [Objective 11b]
4. What is the difference between a *maximum rating* and an *electrical characteristic*? [Objective 11a]
5. Why does the limit on forward current decrease when temperature increases [Objective 11b]
6. What is the procedure for locating information on a diode specification sheet? [Objective 11b]

2.7
Zener Diodes

Zener diodes. Diodes that are designed to work in the reverse operating region.

Reverse breakdown voltage (V_{BR}). The V_R that causes a diode to conduct in the reverse direction.

The *zener diode* is a special type of diode that is designed to work in the reverse breakdown region of the diode characteristic curve. Recall that a normal diode operated in this region will usually be destroyed by excessive current and the heat it produces. This is not the case for the zener diode.

The value of a component designed to operate in the reverse breakdown region can be seen by taking a look at Figure 2.27. This figure shows the reverse breakdown region of the diode characteristic curve. As the curve illustrates, two things happen when V_{BR} is reached:

1. The diode current increases drastically.
2. The reverse voltage across the diode, V_R, remains relatively constant.

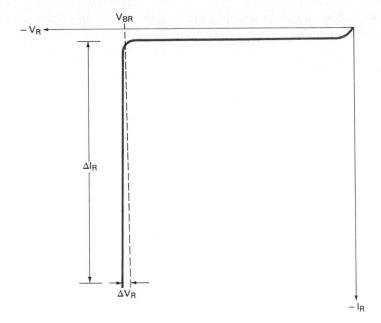

FIGURE 2.27

Reverse breakdown
characteristics.

In other words, *a diode operated in this region will have a relatively constant voltage across it, regardless of the value of current through the device.* The zener diode is used specifically for this purpose, to maintain a relatively constant voltage, regardless of variations in diode current. This makes the zener diode a good *voltage regulator*. A voltage regulator is a circuit designed to maintain a constant voltage regardless of variation in load current. Voltage regulators are discussed in detail in Chapter 3.

Since the zener diode is used for its reverse operating characteristics, it should not surprise you that electron flow through a zener is normally in the direction of the arrow, as shown in Figure 2.28. Since the diode is normally reverse biased, the cathode will be more *positive* than the anode when the diode is conducting. Thus the arrow will point in the direction of electron flow.

Will a zener ever conduct in the same direction as a conventional diode? Is it ever operated in the forward operating region? Rarely. Even though a zener diode has the same forward operating characteristics as a *pn*-junction diode, the value of the zener lies in its application as a voltage regulator. This application is lost in a forward-biased zener, so they are rarely used in the forward operating region.

When a zener is operating in the reverse operating region, the voltage across the device will be *nearly* constant and will be equal to the *zener voltage* (V_Z) rating of the device. Zener diodes have a range of V_Z ratings from about 1.8 V to several hundreds of volts. *Note that the zener rating always tells you the approximate voltage across the device when it is operating in the reverse breakdown region.*

Diode Breakdown

There are two types of diode breakdown. We have already briefly discussed the first type, *avalanche breakdown*. The other type of breakdown is called *zener breakdown*.

Zener breakdown occurs at much lower values of V_R than does avalanche breakdown. The heavy doping of the zener diode causes the device to have a much narrower depletion layer. As a result, it takes very little V_R to cause the diode to go into breakdown, typically 5 V or less. Note that the zener diodes

The zener diode as a voltage regulator.

Voltage regulator. A circuit designed to maintain a constant voltage regardless of variations in load current.

Zener voltage (V_z). The approximate voltage across a zener when operated in reverse breakdown.

Zener breakdown. A type of reverse breakdown that occurs at low values of V_R.

FIGURE 2.28

Zener current.

with low V_Z ratings experience zener breakdown, while the ones with higher V_Z ratings usually experience avalanche breakdown.

Zener Operating Characteristics

A zener diode will maintain a near-constant reverse voltage for a *range* of reverse current values. These values are identified in Figure 2.29. The minimum current required to maintain voltage regulation (constant voltage) is the *zener knee current*, I_{ZK}. When the zener is being used as a voltage regulator, the current through the diode must never be allowed to drop below this value.

The *maximum zener current*, I_{ZM}, is the maximum amount of current the diode can handle without being damaged or destroyed. The *zener test current*, I_{ZT}, is the current level at which the V_Z rating of the diode is determined. For example, if a given zener diode has values of $V_Z = 9.1$ V and $I_{ZT} = 20$ mA, this means that the diode has a reverse voltage of 9.1 V when the test current is 20 mA. At other current values, the value of V_Z will vary *slightly* above or below the rated value.

I_{ZT} is a *test* current value and thus is not a critical value for circuit analysis. However, it is an important value in circuit design. *If you need to have a zener voltage that is as close to V_Z as possible, you need to design the circuit to have a current value equal to I_{ZT}*. The further your circuit current is from I_{ZT}, the further V_R will be from the ideal V_Z value. I_{ZK} and I_{ZM} are very important for both design and circuit analysis. This point will be demonstrated in our discussion on voltage regulators in Chapter 3.

Zener impedance, Z_Z, is the zener diode's opposition to a *change in current*. This is evidenced by the fact that Z_Z is measured at a specific change in zener current. Figure 2.30 helps to illustrate this point. The spec sheet for the 1N746–759 series zener diodes lists the following test conditions for measuring Z_Z:

$$I_{ZT} = 20 \text{ mA}, \qquad I_{zt} = 2 \text{ mA}$$

This means that Z_Z is measured while *varying* zener current by 2 mA around the value of I_{ZT} (20 mA). This variation in zener current is illustrated in Figure 2.30. Note that the 2 mA variation in I_Z causes a 56 mV variation in V_Z. From this information, Z_Z is determined using

FIGURE 2.29

Zener reverse current values.

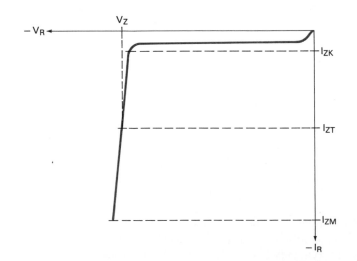

CHAP. 2   Diodes

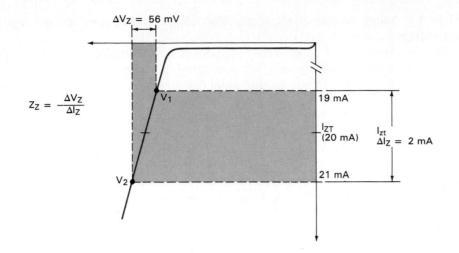

$$Z_Z = \frac{\Delta V_Z}{\Delta I_Z}$$

FIGURE 2.30

Determining zener
impedance.

$$\boxed{Z_Z = \frac{\Delta V_Z}{\Delta I_Z} \bigg| \begin{array}{l} \Delta V_Z = \text{the } \textit{change in } V_Z \\ \quad\quad = V_2 - V_1 \end{array}}$$

(2.9)

as follows:

$$Z_z = \frac{56 \text{ mV}}{2 \text{ mA}}$$
$$= 28 \ \Omega$$

Static reverse current, I_R, is the reverse current through the diode when V_R is less than V_Z. In other words, this is the reverse leakage current through the diode when it is off. For the 1N746, this value is 10 μA at 25°C and 30 μA at 150°C.

Static reverse current (I_R).
The reverse current through the
diode when $V_R < V_Z$.

Zener Equivalent Circuits

There are basically two equivalent circuits for the zener diode. Both are shown in Figure 2.31. The *ideal* model simply considers the zener to be a voltage source

The *ideal* and *practical* zener
models.

FIGURE 2.31

Zener equivalent circuits.

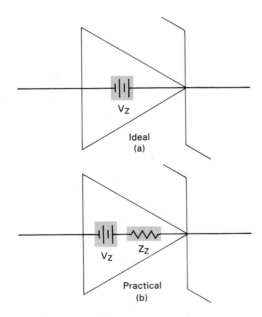

V_Z

Ideal
(a)

V_Z Z_Z

Practical
(b)

equal to V_Z. When placed in a circuit, this voltage source *opposes* the applied circuit voltage.

The *practical* model of the zener includes a series resistor, labeled Z_Z. This model of the zener is used mainly for predicting the response of the diode to a change in circuit current. This is another point that is demonstrated in our discussion on voltage regulators.

SECTION REVIEW

1. What is the primary difference between zener diodes and *pn*-junction diodes? [Objective 13a]
2. Why can the zener diode be used as a voltage regulator? [Objective 13a]
3. How do you determine the direction of zener current in a schematic diagram? [Objective 14]
4. What is the significance of the zener voltage rating? [Objective 14b]
5. Name and define the following symbols: V_Z, Z_Z, I_{ZK}, I_{ZM}, I_{ZT}, I_{zt}, and I_R. [Objective 13b]
6. Which rated zener currents would limit the total current in a zener diode circuit? [Objective 13b]

2.8
Zener Diode Specification Sheets

Zener diode spec sheets are somewhat different than *pn*-junction diode spec sheets. This can be seen by looking at Figure 2.32, which shows the spec sheet for the 1N746–1N759 series zener diodes. We will refer to this figure throughout this section.

Maximum Ratings

✔ 15

The primary parameters for the zener diode are *dc power dissipation* and the *power derating factor*. While *operating and storage junction temperature range* is also listed in the maximum ratings section, the average technician would not be concerned with this parameter under normal circumstances.

Dc power dissipation rating. The maximum allowable value of P_D for a zener diode that is operating in the reverse breakdown.

The *dc power dissipation rating* is extremely important for zener diodes. You see, most zener diode spec sheets do not list the values of I_{ZM} for the various diodes. When this is the case, you must calculate the value of I_{ZM} using the zener voltage rating (listed under "Electrical Characteristics") and the dc power dissipation rating (listed under "Maximum Ratings"). When you know the zener voltage and dc power dissipation ratings for a given diode, I_{ZM} can be determined using:

$$I_{ZM} = \frac{P_{D(MAX)}}{V_Z}$$

(2.10)

The following example illustrates the use of this equation.

ELECTRICAL CHARACTERISTICS ($T_A = 25°C$, $V_F = 1.5$ V max at 200 mA for all types)

Type Number (Note 1)	Nominal Zener Voltage V_Z @ I_{ZT} (Note 2) Volts	Test Current I_{ZT} mA	Maximum Zener Impedance Z_{ZT} @ I_{ZT} (Note 3) Ohms	*Maximum DC Zener Current I_{ZM} (Note 4) mA		Maximum Reverse Leakage Current $T_A = 25°C$ I_R @ $V_R=1$ V μA	$T_A = 150°C$ I_R @ $V_R=1$ V μA
1N4370	2.4	20	30	150	190	100	200
1N4371	2.7	20	30	135	165	75	150
1N4372	3.0	20	29	120	150	50	100
1N746	3.3	20	28	110	135	10	30
1N747	3.6	20	24	100	125	10	30
1N748	3.9	20	23	95	115	10	30
1N749	4.3	20	22	85	105	2	30
1N750	4.7	20	19	75	95	2	30
1N751	5.1	20	17	70	85	1	30
1N752	5.6	20	11	65	80	0.1	20
1N753	6.2	20	7	60	70	0.1	20
1N754	6.8	20	5	55	65	0.1	20
1N755	7.5	20	6	50	60	0.1	20
1N756	8.2	20	8	45	55	0.1	20
1N757	9.1	20	10	40	50	0.1	20
1N758	10	20	17	35	45	0.1	20
1N759	12	20	30	30	35	0.1	20

Type Number (Note 1)	Nominal Zener Voltage V_Z Volts	Test Current I_{ZT} mA	Maximum Zener Impedance Z_{ZT} @ I_{ZT} Ohms	Z_{ZK} @ I_{ZK} Ohms	I_{ZK} mA	*Maximum DC Zener Current I_{ZM} (Note 4) mA		Maximum Reverse Current I_R Maximum μA	Test Voltage V_{dc} 5% V_R	10% V_R
1N957A	6.8	18.5	4.5	700	1.0	47	61	150	5.2	4.9
1N958A	7.5	16.5	5.5	700	0.5	42	55	75	5.7	5.4
1N959A	8.2	15	6.5	700	0.5	38	50	50	6.2	5.9
1N960A	9.1	14	7.5	700	0.5	35	45	25	6.9	6.6
1N961A	10	12.5	8.5	700	0.5	32	41	10	7.6	7.2
1N962A	11	11.5	9.5	700	0.25	28	37	5	8.4	8.0
1N963A	12	10.5	11.5	700	0.25	26	34	5	9.1	8.6
1N964A	13	9.5	13	700	0.25	24	32	5	9.9	9.4
1N965A	15	8.5	16	700	0.25	21	27	5	11.4	10.8
1N966A	16	7.8	17	700	0.25	19	27	5	12.2	11.5
1N967A	18	7.0	21	750	0.25	17	23	5	13.7	13.0
1N968A	20	6.2	25	750	0.25	15	20	5	15.2	14.4
1N969A	22	5.6	29	750	0.25	14	18	5	16.7	15.8
1N970A	24	5.2	33	750	0.25	13	17	5	18.2	17.3
1N971A	27	4.6	41	750	0.25	11	15	5	20.6	19.4
1N972A	30	4.2	49	1000	0.25	10	13	5	22.8	21.6
1N973A	33	3.8	58	1000	0.25	9.2	12	5	25.1	23.8
1N974A	36	3.4	70	1000	0.25	8.5	11	5	27.4	25.9
1N975A	39	3.2	80	1000	0.25	7.8	10	5	29.7	28.1
1N976A	43	3.0	93	1500	0.25	7.0	9.6	5	32.7	31.0
1N977A	47	2.7	105	1500	0.25	6.4	8.8	5	35.8	33.8
1N978A	51	2.5	125	1500	0.25	5.9	8.1	5	38.8	36.7
1N979A	56	2.2	150	2000	0.25	5.4	7.4	5	42.6	40.3
1N980A	62	2.0	185	2000	0.25	4.9	6.7	5	47.1	44.6
1N981A	68	1.8	230	2000	0.25	4.5	6.1	5	51.7	49.0
1N982A	75	1.7	270	2000	0.25	4.0	5.5	5	56.0	54.0
1N983A	82	1.5	330	3000	0.25	3.7	5.0	5	62.2	59.0
1N984A	91	1.4	400	3000	0.25	3.3	4.5	5	69.2	65.5
1N985A	100	1.3	500	3000	0.25	3.0	4.5	5	76	72
1N986A	110	1.1	750	4000	0.25	2.7	4.1	5	83.6	79.2

FIGURE 2.32
(Courtesy of Motorola, Inc.)

MOTOROLA
SEMICONDUCTOR
TECHNICAL DATA

1.5KE6.8, A thru 1.5KE250, A
See Page 4-59

1N746 thru 1N759
1N957A thru 1N986A
1N4370 thru 1N4372

500-MILLIWATT HERMETICALLY SEALED
GLASS SILICON ZENER DIODES

Designer's Data Sheet

- Complete Voltage Range — 2.4 to 110 Volts
- DO-35 Package — Smaller than Conventional DO-7 Package
- Double Slug Type Construction
- Metallurgically Bonded Construction
- Oxide Passivated Die

Designer's Data for "Worst Case" Conditions

The Designer's Data sheets permit the design of most circuits entirely from the information presented. Limit curves — representing boundaries on device characteristics — are given to facilitate "worst case" design.

GLASS ZENER DIODES
500 MILLIWATTS
2.4–110 VOLTS

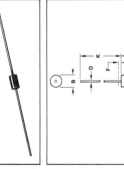

MAXIMUM RATINGS

Rating	Symbol	Value	Unit
DC Power Dissipation @ $T_L \leq 50°C$, Lead Length = 3/8″	P_D	400	mW
*Derate above $T_L = 50°C$		3.2	mW/°C
•JEDEC Registration			
Motorola Device Ratings		500	mW
*Derate above $T_L = 50°C$		3.33	mW/°C
Operating and Storage Junction Temperature Range	T_J, T_{stg}		°C
•JEDEC Registration		−65 to +175	
Motorola Device Ratings		−65 to +200	

*Indicates JEDEC Registered Data.

MECHANICAL CHARACTERISTICS

MAXIMUM LEAD TEMPERATURE FOR SOLDERING PURPOSES: 230°C, 1/16″ from case for 10 seconds

FINISH: All external surfaces are corrosion resistant with readily solderable leads.

POLARITY: Cathode indicated by color band. When operated in zener mode, cathode will be positive with respect to anode.

MOUNTING POSITION: Any

STEADY STATE POWER DERATING

NOTES:
1. PACKAGE CONTOUR OPTIONAL WITHIN A AND B. HEAT SLUGS, IF ANY, SHALL BE INCLUDED WITHIN THIS CYLINDER, BUT NOT SUBJECT TO THE MINIMUM LIMIT OF B.
2. LEAD DIAMETER NOT CONTROLLED IN ZONE F TO ALLOW FOR FLASH, LEAD FINISH BUILDUP AND MINOR IRREGULARITIES OTHER THAN HEAT SLUGS.
3. POLARITY DENOTED BY CATHODE BAND.
4. DIMENSIONING AND TOLERANCING PER ANSI Y14.5, 1973.

DIM	MILLIMETERS MIN	MAX	INCHES MIN	MAX
A	3.05	5.08	0.120	0.200
B	1.52	2.29	0.060	0.090
D	0.46	0.56	0.018	0.022
F	–	1.27	–	0.050
K	25.40	38.10	1.000	1.500

All JEDEC dimensions and notes apply.

CASE 299-02
DO-204AH
GLASS

EXAMPLE 2.14

A 1N754 zener diode has a dc power dissipation rating of 500 mW and a zener voltage rating of 6.8 V. What is the value of I_{ZM} for the device?

Solution: Using equation (2.10), the value of I_{ZM} is found as:

$$
\begin{aligned}
I_{ZM} &= \frac{P_{D(MAX)}}{V_Z} \\
&= \frac{500 \text{ mW}}{6.8 \text{ V}} \\
&= 73.5 \text{ mA}
\end{aligned}
$$

Remember, the value of I_{ZM} is important because it determines the maximum current the diode can tolerate! If the total current through this diode exceeds 73.5 mA, you'll end up replacing the diode.

PRACTICE PROBLEM 2–14

A zener diode has a dc power dissipation rating of 1 W and a zener voltage rating of 27 V. What is the value of I_{ZM} for the device?

Why are there two sets of power ratings on the spec sheet? The JEDEC (Joint Electronic Device Engineering Council) registered values are for military applications. Military specs call for large tolerances. By rating the device at 100 mW below its actual capability, the large tolerance required by the military is built into the specifications. For our purposes, we will use the manufacturer's parameter values.

The *power derating factor* (listed under "Maximum Ratings") tells you how much the dc power dissipation rating *decreases* when the operating temperature increases above a specified value. For example, the spec sheet shown in Figure 2.32 shows a derating factor of *3.33 mW/°C for temperatures above 50°C.* This means that you must *decrease* the dc power dissipation rating by 3.33 mW for every 1°C rise in operating temperature above 50°C. The following example illustrates the use of the derating factor.

EXAMPLE 2.15

A 1N746 is being operated at a temperature of 75°C. What is the maximum dc power dissipation for the device?

Solution: The first step is to determine the total derating value. This value is found as

$$(3.33 \text{ mW})(75°C - 50°C) = 83.25 \text{ mW}$$

We subtract this derating value from the dc power dissipation rating to get the dc power dissipation limit:

$$P_D = 500 \text{ mW} - 83.25 \text{ mW} = 416.75 \text{ mW}$$

Thus, at 75°C the 1N746 has a power dissipation limit of 416.75 mW.

Determine the power dissipation rating of the 1N746 at 125°C. Assume that the dc power dissipation rating of the device is 500 mW and the derating factor is 3.33 mW/°C above 50°C.

Electrical Characteristics

The electrical characteristics portion of the zener spec sheet shows the zener voltage, current, and impedance ratings for the entire group of diodes.

Take a look at the component numbers. You'll see that many of the device numbers end with the letter *A*. When a letter follows the part number, the letter indicates the *tolerance of the ratings*. In this case, the letter *A* indicates that there is a ±5% tolerance in the ratings. Thus, the 1N980A (rated $V_Z = 62$ V) could actually have a zener voltage that falls between

$$62 \text{ V} - 3.1 \text{ V} = 58.9 \text{ V} \quad (minimum)$$

and

$$62 \text{ V} + 3.1 \text{ V} = 65.1 \text{ V} \quad (maximum)$$

When no letter follows the part number, the tolerance of the component is ±10%.

If you look at the list of *nominal zener voltage* (rated zener voltages), you will notice something interesting. That is, standard zener voltage values follow the same basic progression as standard resistor values. The nominal zener voltages, listed were measured at the zener test current (I_{ZT}).

Nominal zener voltage. The *rated* value of V_Z for a given zener diode.

The *maximum zener impedance* (Z_{ZM}) rating is self-explanatory. Note that the listed maximum value of Z_Z would be used in any calculation that involves zener impedance. This point will be illustrated in Chapter 3 when we cover zener voltage regulators.

The *maximum dc zener current* (I_{ZM}) column needs a bit of explaining. First, you will notice that there are two sets of numbers. Those on the left correspond to the JEDEC registered value of 400 mW for dc power dissipation. Those on the right correspond to the Motorola device rating of 500 mW for dc power dissipation. Therefore, we would be interested in the values that appear in the right-hand column. Second, if you use equation (2.10) to calculate the values of I_{ZM}, you'll find that you get answers that are around 10% higher than the values shown. For example, using the 500 mW Motorola dc power dissipation rating, the value of I_{ZM} for the 1N758 is found as

$$
\begin{aligned}
I_{ZM} &= \frac{P_{D(\text{MAX})}}{V_Z} \\
&= \frac{500 \text{ mW}}{10 \text{ V}} \\
&= 50 \text{ mA}
\end{aligned}
$$

This value is approximately 10% higher than the $I_{ZM} = 45$ mA rating shown. The rated values of I_{ZM} are generally reduced by about 10% to keep you from driving the components to their absolute limits. This is the same situation as the one we encountered with the *average forward current* rating of the *pn*-junction diode.

As stated earlier, the *maximum reverse leakage current* (also known as *static reverse current*) is the reverse leakage current when the value of V_Z is less than the nominal zener voltage. As you can see, this reverse current rating is temperature dependent, just like the reverse current rating of the *pn*-junction diode.

Zener Diode Selector Guides

The selector guides for zener diodes are very similar to those used for the *pn*-junction diodes. A zener diode selector guide is shown in Figure 2.33.

As you can see, the critical parameters for device substitution are *nominal zener voltage* (vertical listing) and *dc power dissipation* (horizontal listing). The following series of examples shows how the selector guide is used.

FIGURE 2.33

(Courtesy of Motorola, Inc.)

Nominal Zener Voltage (*Note 1)	500 mW Cathode = Polarity Mark (*Notes 4,11) Glass Case 362-01	500 mW (*Notes 9,11)	1 Watt Cathode = Polarity Mark (*Note 6) Glass Case 59-04 (DO-41)	1 Watt (*Notes 6,12) Glass Case 362B-01	1 Watt Cathode to Case (*Note 7) Metal Case 52-03 (DO-13)	1.5 Watt Cathode = Polarity Mark (*Note 8) Sumetic 30 Case 59-03 (DO-41)	5 Watt Cathode = Polarity Mark (*Note 8) Sumetic 40 Case 17-02
1.8							
2.0							
2.2							
2.4	MLL4370	MLL5221A					
2.5		MLL5222A					
2.7	MLL4371	MLL5223A					
2.8		MLL5224A					
3.0	MLL4372	MLL4225A					
3.3	MLL746	MLL5226A	1N4728	MLL4728	1N3821	1N5913A	1N5333A
3.6	MLL747	MLL5227A	1N4729	MLL4729	1N3822	1N5914A	1N5334A
3.9	MLL748	MLL5228A	1N4730	MLL4730	1N3823	1N5915A	1N5335A
4.3	MLL749	MLL5229A	1N4731	MLL4731	1N3824	1N5916A	1N5336A
4.7	MLL750	MLL5230A	1N4732	MLL4732	1N3825	1N5917A	1N5337A
5.1	MLL751	MLL5231A	1N4733	MLL4733	1N3826	1N5918A	1N5338A
5.6	MLL752	MLL5232A	1N4734	MLL4734	1N3827	1N5919A	1N5339A
6.0		MLL5233A					
6.2	MLL753	MLL5234A	1N4735	MLL4735	1N3828	1N5920A	1N5341A
6.8	MLL754 / MLL957A	MLL5235A	1N4736	MLL4736	1N3829 / 1N3016A	1N5921A	1N5342A
7.5	MLL755 / MLL958A	MLL5236A	1N4737	MLL4737	1N3830 / 1N3017A	1N5922A	1N5343A
8.2	MLL756 / MLL959A	MLL5237A	1N4738	MLL4738	1N3018A	1N5923A	1N5344A
8.7		MLL5238A					1N5345A
9.1	MLL757 / MLL960A	MLL5239A	1N4739	MLL4739	1N3019A	1N5924A	1N5346A
10	MLL758 / MLL961A	MLL5240A	1N4740	MLL4740	1N3020A	1N5925A	1N5347A
11	MLL962A	MLL5241A	1N4741	MLL4741	1N3021A	1N5926A	1N5348A
12	MLL759 / MLL963A	MLL5242A	1N4742	MLL4742	1N3022A	1N5927A	1N5349A
13	MLL964A	MLL5243A	1N4743	MLL4743	1N3023A	1N5928A	1N5350A
14		MLL5244A					1N5351A
15	MLL965A	MLL5245A	1N4744	MLL4744	1N3024A	1N5929A	1N5352A
16	MLL966A	MLL5246A	1N4745	MLL4745	1N3025A	1N5930A	1N5353A
17		MLL5247A					1N5354A
18	MLL967A	MLL5248A	1N4746	MLL4746	1N3026A	1N5931A	1N5355A
19		MLL5249A					1N5356A
20	MLL968A	MLL5250A	1N4747	MLL4747	1N3027A	1N5932A	1N5357A
22	MLL969A	MLL5251A	1N4748	MLL4748	1N3028A	1N5933A	1N5358A
24	MLL970A	MLL5252A	1N4749	MLL4749	1N3029A	1N5934A	1N5359A
25		MLL5253A					1N5360A
27	MLL971A	MLL5254A	1N4750	MLL4750	1N3030A	1N5935A	1N5361A
28		MLL5255A					1N5362A
30	MLL972A	MLL5256A	1N4751	MLL4751	1N3031A	1N5936A	1N5363A
33	MLL973A	MLL5257A	1N4752	MLL4752	1N3032A	1N5937A	1N5364A
36	MLL974A	MLL5258A	1N4753	MLL4753	1N3033A	1N5938A	1N5365A
39	MLL975A	MLL5259A	1N4754	MLL4754	1N3034A	1N5939A	1N5366A
43	MLL976A	MLL5260A	1N4755	MLL4755	1N3035A	1N5940A	1N5367A
47	MLL977A	MLL5261A	1N4756	MLL4756	1N3036A	1N5941A	1N5368A
51	MLL978A	MLL5262A	1N4757	MLL4751	1N3037A	1N5942A	1N5369A
56	MLL979A	MLL5263A	1N4758	MLL4758	1N3038A	1N5943A	1N5370A
60		MLL5264A					1N5371A
62	MLL980A	MLL5265A	1N4759	MLL4759	1N3039A	1N5944A	1N5372A
68	MLL981A	MLL5266A	1N4760	MLL4760	1N3040A	1N5945A	1N5373A
75	MLL982A	MLL5267A	1N4761	MLL4761	1N3041A	1N5946A	1N5374A
82	MLL983A	MLL5268A	1N4762	MLL4762	1N3042A	1N5947A	1N5375A
87		MLL5269A					1N5376A
91	MLL984A	MLL5270A	1N4763	MLL4763	1N3043A	1N5958A	1N5377A
100	MLL985A		1N4764	MLL4764	1N3044A	1N5949A	1N5378A
110	MLL986A				1N3045A	1N5950A	1N5379A
120					1N3046A	1N5951A	1N5380A
130					1N3047A	1N5952A	1N5381A
150					1N3048A	1N5953A	1N5383A
160					1N3049A	1N5954A	1N5384A
170							1N5385A
175							
180					1N3050A	1N5955A	1N5386A
200					1N3051A	1N5956A	1N5388A

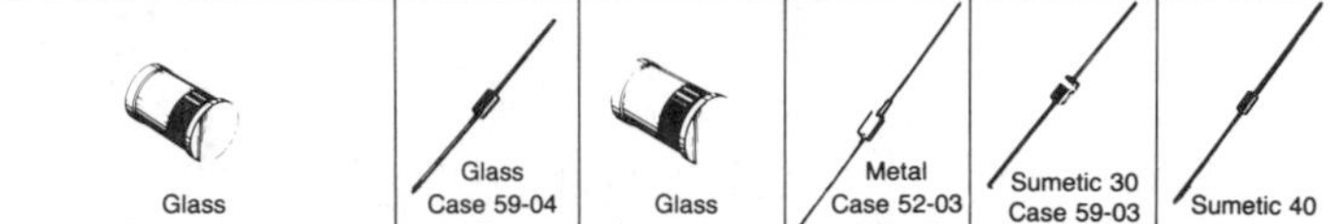

EXAMPLE 2.16

A Practical Consideration: When we were dealing with substituting one *pn* junction diode for another, we were only concerned with whether or not the current and peak reverse voltage ratings were high enough to survive in the circuit. When dealing with zener diodes, the power rating of the substitute diode may be higher than needed, but the V_Z rating of the substitute must *equal* that of the component it is replacing. We *cannot* use a diode with a higher (or lower) V_Z rating than the original component.

We need a substitute component for a zener diode that is faulty. The substitute component must have a nominal zener voltage of 12 V and must be capable of dissipating 1.2 W. Which of the zeners listed in Figure 2.33 can be used?

Solution: First, the zener diode must have a rating of 12 V. Therefore, our diode is listed in the row that corresponds to $V_Z = 12$ V.

There are two diodes listed in the 12-V row that have power dissipation ratings that are greater than 1.2 W. These are the 1N5927A (1.5 W) and the 1N5349A (5 W). Either of these diodes could be used as a substitute component.

PRACTICE PROBLEM 2–16

We need a zener diode with a nominal zener voltage of 75 V and a power dissipation capability of 4 W. Which zener diode(s) shown in Figure 2.33 can be used in this case?

When we know the zener voltage and current requirements of a component and need a substitute, we have to do a little calculating to find the right substitute part. This point is illustrated in the following example.

EXAMPLE 2.17

We need to replace a faulty zener diode. The substitute component must have a nominal zener voltage of 20 V and must be able to handle the power that is generated by a maximum current of 150 mA. Which diode(s) listed in Figure 2.33 can be used in this application?

Solution: First, we need to determine the power dissipation requirements of the substitute diode. The dc power requirement is found as

$$P_{D(MAX)} = I_{ZM}V_Z$$
$$= (150 \text{ mA})(20 \text{ V})$$
$$= 3 \text{ W}$$

Therefore, the substitute component must be able to dissipate *at least* 3 W. Checking the 20-V/5-W location on the selector guide, we see that the only component we can use is the 1N5357A.

PRACTICE PROBLEM 2–17

We need a zener diode with a nominal zener voltage of 6.8 V that can handle a maximum zener current of 175 mA. Which diode(s) listed in Figure 2.33 can be used for this application?

1. When a spec sheet does not list the value of I_{ZM}, how do you determine its value? [Objective 15]
2. What is a *power derating factor*? How is it used? [Objective 15]
3. Explain how you would determine whether or not one zener diode can be used in place of another. [Objective 17]

2.9
Light-Emitting Diodes (LEDs)

LEDs are diodes that will provide a limited amount of light when biased properly. The schematic symbol for the LED is shown in Figure 2.34. Although LEDs are available in several colors (red, green, white, yellow, and orange are the most common), the schematic symbol is the same for all LEDs. There is nothing in the symbol to indicate the color of a particular LED.

Since LEDs have clear (or semiclear) cases, there is normally no label on the case to identify the leads. The two leads of an LED are identified using one of the several schemes:

1. The leads may have different lengths, as shown in Figure 2.35a. When this scheme is used, the longer of the two leads is usually the *anode*.
2. One of the leads may be flattened, as shown in Figure 2.35b. The flattened lead is usually the *anode*.
3. One side of the case may be flattened, as shown in Figure 2.35c. The lead closest to the flattened side is usually the *anode*.

FIGURE 2.34

The LED schematic symbol.

LED Characteristics

A Practical Consideration: Some manufacturers use the schemes listed above to identify the *cathode*. When you are using an LED that you have never used before, check the spec sheet of the LED to determine which lead is being identified.

LEDs have characteristic curves that are very similar to those for *pn*-junction diodes. However, they tend to have *higher* forward voltage (V_F) values and *lower* reverse breakdown voltage (V_{BR}) ratings. The typical ranges for these values are as follows:

- Forward voltage: $+1$ V to $+3$ V (typical)
- Reverse breakdown voltage: -3 V to -10 V (typical)

The V_F and V_{BR} ratings of the LED mean that it typically drops more voltage than the *pn*-junction diode when forward biased and will tend to break down at lower reverse voltages.

The forward current ratings of LEDs typically range between 20 and 100 mA. Because they tend to have low forward current ratings, they usually require the use of a series *current-limiting resistor*.

LEDs need current-limiting resistors.

FIGURE 2.35

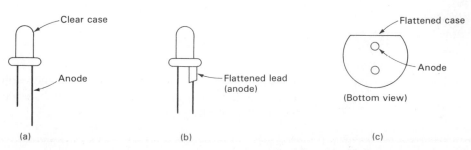

(a)

(b)

(c)

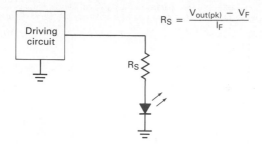

$$R_S = \frac{V_{out(pk)} - V_F}{I_F}$$

FIGURE 2.36

An LED needs a current-limiting resistor.

Current-Limiting Resistors (R_S)

When used in any practical application, the LED will normally have a series *current-limiting* resistor, as is shown in Figure 2.36. The resistor ensures that the maximum current rating of the LED will not be exceeded by the circuit. As the illustration shows, the value of the limiting resistor, R_S, is determined using the following equation:

Current-limiting resistor (R_S). A resistor in series with an LED to ensure a low I_F through the LED.

$$\boxed{R_S = \frac{V_{out(pk)} - V_F}{I_F}} \tag{2.11}$$

where $V_{out(pk)}$ = the peak output voltage of the driving circuit
V_F = the *minimum* value of V_F for the LED
I_F = the *maximum* allowable value of I_F for the LED

The following example illustrates the use of equation (2.11) in determining the needed value of a current-limiting resistor.

EXAMPLE 2.18

A Practical Consideration. The brightness of an LED is determined by the value of forward current. The more forward current, the brighter the light is. While we can use *any* standard-value resistor with a value higher than the calculated value of R_S, we want to keep the value as small as possible to ensure that the LED is as bright as possible.

The driving circuit shown in Figure 2.36 has a peak output voltage of 8 V. The LED has ratings of $V_F = 1.8$ to 2.0 V and $I_{F(MAX)} = 16$ mA. What value of current-limiting resistor is needed in the circuit?

Solution: Using the peak voltage, *minimum* V_F, and I_F, the value of R_S is found as

$$R_S = \frac{V_{out(pk)} - V_F}{I_F}$$
$$= \frac{8 \text{ V} - 1.8 \text{ V}}{16 \text{ mA}}$$
$$= 387.5$$

The smallest standard-value resistor that has a value *greater than* 387.5 Ω is the 390-Ω resistor. This is the component we would use in this circuit.

PRACTICE PROBLEM 2–18

An LED with a forward voltage rating of 1.4 to 1.8 V and a maximum forward current rating of 12 mA is being driven by a source with a peak voltage of 14 V. What is the smallest standard resistor value that can be used as R_S? (*Note*: A listing of the standard resistor values appears in Appendix A.)

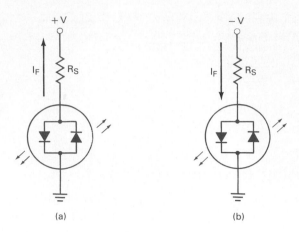

FIGURE 2.37

(a)   (b)

Multicolor LEDs

LEDs have been developed that will emit one color of light when forward biased and another when reverse biased. One commonly used schematic symbol for these multicolor LEDs is shown in Figure 2.37.

Multicolor LEDs actually contain two *pn* junctions that are connected in *reverse-parallel*, meaning that they are in parallel, with the anode of one being connected to the cathode of the other. If a positive potential is applied to the LED (as shown in Figure 2.37a), the *pn* junction on the *left* will light. Note that the device current passes through the left *pn* junction. If the polarity of the voltage source is reversed (as shown in Figure 2.37b), the *pn* junction on the right will light. Note that the direction of diode current has reversed and is now passing through the right *pn* junction.

Multicolor LEDs are typically *red* when biased in one direction and *green* when biased in the other. Incidently, if a multicolor LED is switched fast enough between the two polarities, the LED will produce a *third* color. A red/green LED will produce a *yellow* light when rapidly switched back and forth between biasing polarities.

One Final Note

LED applications (along with *pn*-junction and zener diode applications) will be covered in Chapter 4. At this point, we are going to take a look at diode testing.

Multicolor LED. An LED that emits one color when forward biased and another color when reverse biased.

SECTION REVIEW

1. How are the leads on an LED usually identified? [Objective 19]
2. How do the electrical characteristics of LEDs differ from those of *pn*-junction diodes? [Objective 19]
3. Why do LEDs need series current-limiting resistors? [Objective 20]

2.10
Diode Testing

We have seen the characteristics of the *pn*-junction diode, the zener diode, and the LED. Now we will cover the testing procedure for each of these devices.

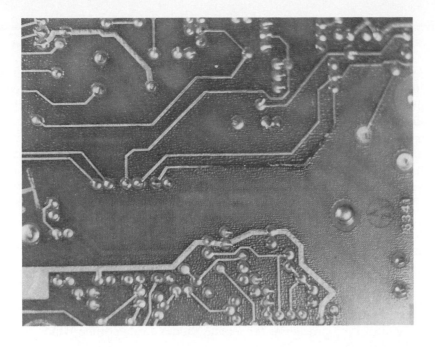

FIGURE 2.38

Most diode failures are caused by excessive current, component age, or surpassing the V_{RRM} rating. When a diode is damaged or destroyed by excessive current, the problem is easy to diagnose. In most cases, the diode will crack or fall apart completely. Burned connection points on a printed circuit board are also symptoms of excessive current through a diode. Figure 2.38 shows a printed circuit board after a diode failed from excessive current. Note the damage to the copper trace on the board.

Why diodes fail.

Not all diodes will show physical signs of failure. However, some simple tests will tell you whether or not a given diode is faulty.

Testing *PN* Junction Diodes

The common diode can be tested using an ohmmeter. When connected as shown in Figure 2.39a, the diode should have a very low resistance reading, typically less than 1 kΩ. When connected as shown in Figure 2.39b, the diode should have a very high resistance reading, typically in the megohm range. It is not difficult to understand why you get the readings described if you look at the ohmmeter as being a biasing source. In Figure 2.39a, the connection to the meter will forward bias the diode, causing a low resistance reading. In Figure 2.39b, the meter will reverse bias the diode, causing the high resistance reading.

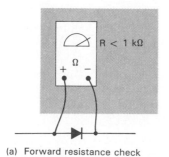

(a) Forward resistance check

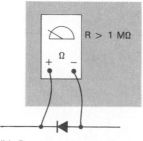

(b) Reverse resistance check

FIGURE 2.39
Diode testing.

Several notes of caution need to be made regarding this testing procedure. These are as follows:

A Practical Consideration: Some meters provide an output voltage that is less than V_F for resistance measurements. These meters will give an open indication on a good diode. You need to check your meter documentation before using it to test diodes.

1. Some ohmmeters on low resistance settings can supply enough current to destroy a low-current diode. To ensure that this does not occur, use only the high resistance scales on the ohmmeter. Also, check the current rating of the diode under test and compare it to the current capability of the meter.

2. Not all VOMs have a negative common lead when set for resistance measurements. Some meters will supply current from the common lead, while others will supply current from the "ohm" lead. Check the documentation on your meter to be sure that you know which lead is negative before testing any diodes. Otherwise, you may think that you have a faulty diode when you do not.

When you test a given diode, if the forward resistance is very high or the reverse resistance is very low, the diode must be replaced.

Incidentally, the procedure described in this section can serve another useful purpose. Diodes are usually marked to indicate which end of the component is the cathode. However, there are times when this marking will wear off. In this case, you can use the procedure for testing diodes to determine which end of the diode is the cathode.

Testing Zener Diodes

A zener diode cannot be tested in the same manner as a *pn*-junction diode simply because the zener is *designed* to conduct in *both* directions. Because of its design, you cannot test a zener diode with an ohmmeter.

The simplest test of a zener diode is to check the voltage across its terminals while it is in the circuit. If the voltage across the zener is within tolerance, the zener diode is good. If the voltage across the zener is out of tolerance, then there is a strong possibility that the zener is faulty.

FIGURE 2.40

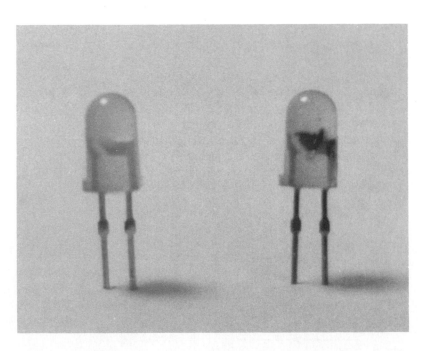

When you suspect that a zener diode is faulty, the simplest method of dealing with it is to go ahead and replace the diode.

Testing LEDs

You will rarely need to test an LED to see if it is faulty. When an LED goes bad, the physical signs are very obvious, as is shown in Figure 2.40.

1. What are the common causes of diode failure? [Objective 21]
2. What steps are involved in testing a *pn*-junction diode with an ohmmeter? [Objective 21]
3. What precautions should be taken before using an ohmmeter to test a *pn*-junction diode? [Objective 21]
4. How do you test a zener diode? [Objective 21]

Chapter Summary

You have been shown quite a lot in this chapter.* The summary illustration shown in Figure 2.41 will help you to remember the primary points that you need to remember about each of the diodes covered. If you have difficulty remembering any of the points listed, review the appropriate section of the chapter. Also, review the chapter summary that is contained in your student study guide.

* As if you really needed to be told that, right?

FIGURE 2.41

SUMMARY ILLUSTRATION

Diode Type:	Pn Junction	Zener	Light–Emitting
Schematic Symbol:			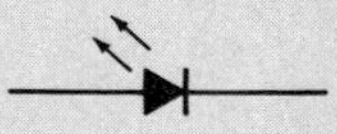
Bias for normal operation:	Switched back and forth between forward and reverse.	Reverse	Forward
*Normal V_F:	Approximately 0.7 V	Approximately 0.7 V (not normally operated in forward region.)	Varies between approximately 1 V and 3 V
Normal V_R:	Equal to applied voltage at values of $V_R < V_{BR}$.	Equal to the nominal zener rating, V_Z.	Equal to applied voltage at values of $V_R < V_{BR}$. Beware of low V_{BR} ratings.
Primary factors to consider for device substitution:	I_{AVE} and V_{RRM} ratings	$P_{D(Max)}$ and V_Z ratings. Beware of power derating factor.	$V_{F(min)}$, $I_{F(Max)}$, and V_{BR}

*Assumes that the device is Silicon.

KEY TERMS

The following terms were introduced and defined in this chapter:

anode
avalanche current
average forward current
bulk resistance
cathode
complete diode model
current-limiting resistor
dc blocking voltage
dc power dissipation
    rating
diffusion current
diode
electrical characteristics
forward power
    dissipation
ideal diode model

junction capacitance
knee voltage
light-emitting diode
    (LED)
maximum ratings
maximum zener current
model
multicolor LED
nominal zener voltage
parameter
peak repetitive reverse
    voltage
peak surge current
power derating factor
reverse breakdown
    voltage

reverse current
reverse saturation current
reverse voltage
selector guides
specification (spec) sheet
static reverse current
surface leakage current
troubleshooting
voltage regulator
zener breakdown
zener diode
zener impedance
zener knee current
zener test current
zener voltage

PRACTICE PROBLEMS

§2.1

1. Draw a circuit containing a dc voltage source, a resistor, and a forward-biased diode. [2]
2. Add an arrow to the circuit you drew in problem 1 to indicate the direction of diode current. [2]
3. Draw a circuit containing a dc voltage source, a resistor, and a reverse-biased diode. [2]
4. For each of the circuits shown in Figure 2.42, determine the direction (if any) of diode forward current. [2]
5. For each of the circuits shown in Figure 2.43, determine the direction (if any) of diode forward current. [2]

FIGURE 2.42

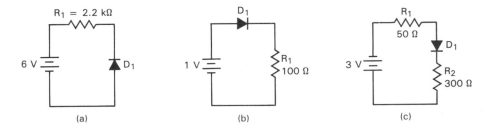

FIGURE 2.43

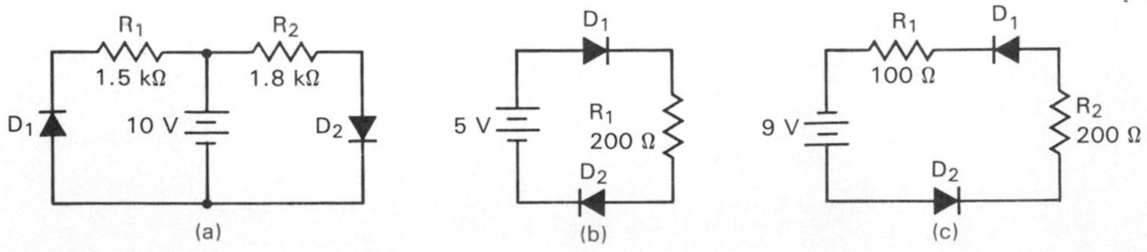

CHAP. 2   Diodes

6. Using the *ideal diode model*, determine the voltage drop across each of the diodes in Figure 2.42. [4]

7. Using the *ideal diode model*, determine the voltage drop across each of the components in Figure 2.43a. [4]

§2.3

8. Using the *practical diode model*, determine the values of V_{D1}, V_{R1}, and I_T for the circuit shown in Figure 2.42a. [6]

9. Using the *practical diode model*, determine the values of V_{D1}, V_R, and I_T for the circuit shown in Figure 2.42b. [6]

10. Using the *practical diode model*, determine the values of V_{D1}, V_{R1}, V_{R2}, and I_T for the circuit shown in Figure 2.42c. [6]

11. * Determine the values of V_{D1}, V_{R1}, I_1, V_{D2}, V_{R2}, and I_2 for the circuit shown in Figure 2.43a. [6]

12. Determine the values of V_{D1}, V_{D2}, V_{R1}, and I_T for the circuit shown in Figure 2.43b. [6]

13. Determine the values of V_{D1}, V_{D2}, V_{R1}, V_{R2}, and I_T for the circuit shown in Figure 2.43c. [6]

14. A voltage is calculated to be 12.8 V. The measured voltage is 13.2 V. What is the percent of error in the calculation? [6]

15. A current is calculated to be 750 μA. The measured current is 880 μA. Determine whether or not the percent of error in the calculation is acceptable. [6]

16. A voltage is calculated to be 144 mV. The measured voltage is 160 mV. Determine whether or not the percent of error is the calculation is acceptable. [6]

17. Calculate the value of V_{R2} for the circuit shown in Figure 2.44. When measured, V_{R2} is found to be 2.54 V. What is the percent of error between your calculated value of V_{R2} and the measured value? [6]

18. Calculate the value of V_{R2} for the circuit shown in Figure 2.45. When measured, V_{R2} is found to be 970 mV. What is the percent of error between your calculated value of V_{R2} and the measured value? [6]

§2.4

19. What is the minimum required peak reverse voltage rating for the diode in Figure 2.46a? [7]

* From now on, the practical diode model will be assumed unless indicated otherwise.

FIGURE 2.44

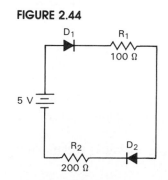

FIGURE 2.45

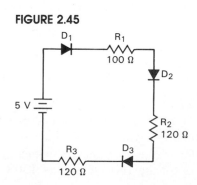

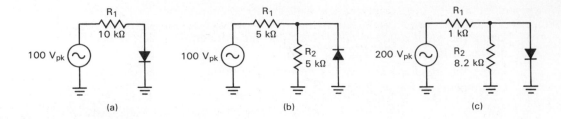

FIGURE 2.46

(a) (b) (c)

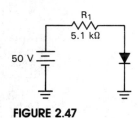

FIGURE 2.47

20. What is the minimum required peak reverse voltage rating for the diode in Figure 2.46b? (*Hint*: Don't forget how to work with voltage dividers!) [7]

21. It was stated in this section (page 35) that practical V_{RRM} ratings are usually multiples of 50 or 100 V. With this in mind, what would be the minimum acceptable V_{RRM} rating for the diode in Figure 2.46c? [7]

22. What is the minimum acceptable average forward current rating for the diode shown in Figure 2.47? [7]

23. What is the minimum acceptable forward power dissipation rating for the diode shown in Figure 2.47? [7]

24. A diode has a $P_{D(MAX)}$ rating of 1.2 W. What is the maximum allowable value of forward current for the device? [7]

25. A diode has a $P_{D(MAX)}$ rating of 750 mW. What is the maximum allowable value of forward current for the device? [7]

§2.5

26. A small-signal diode (silicon) has a forward current of 10 mA and a bulk resistance (R_B) of 5 Ω. What is the actual value of V_F for the device? [8a]

27. A small signal diode (silicon) has a forward current of 8.2 mA and a bulk resistance of 12 Ω. What is the actual value of V_F for the device? [8a]

28. A diode (silicon) has a bulk resistance of 20 Ω. At what value of I_F will the value of V_F actually *equal* 0.7 V? [8a]

29. Refer to Figure 2.46a. The diode in the circuit has a maximum rated value of $I_R = 10$ μA at 25°C. Assuming that I_R reaches its maximum value at each negative peak of the input cycle, what value of voltage will be measured across R_1 when I_R peaks? (Assume $T = 25$°C.) [9a]

30. The circuit described in problem 29 is operated at a temperature of 75°C. What is the voltage across R_1 when I_R peaks? [9b, 10]

31. A 150 Ω resistor is in series with a diode that has a rating of $I_R = 20$ μA at 25°C. What reverse voltage will be measured across the resistor if the diode is reversed biased and $T = 120$°C? (Assume that I_R reaches its maximum possible value.) [9b, 10]

§2.6

32. Refer to the diode spec sheet shown in Figure 2.48. In terms of maximum reverse voltage ratings, which of the diodes listed could be used in the circuit shown in Figure 2.46a? [11]

33. Refer to the spec sheet shown in Figure 2.48. What is the maximum value of I_R for the 1N5398 at $T = 150$°C? [11]

MOTOROLA
■ SEMICONDUCTOR ■
TECHNICAL DATA

**1N5391
thru
1N5399**

Designers Data Sheet

"SURMETIC" RECTIFIERS

... subminiature size, axial lead-mounted rectifiers for general-purpose, low-power applications.

Designers Data for "Worst Case" Conditions

The Designers Data Sheets permit the design of most circuits entirely from the information presented. Limits curves—representing boundaries on device characteristics—are given to facilitate "worst-case" design.

**LEAD-MOUNTED
SILICON RECTIFIERS**

**50–1000 VOLTS
DIFFUSED JUNCTION**

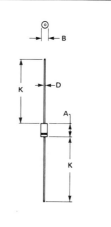

*MAXIMUM RATINGS

Rating	Symbol	1N5391	1N5392	1N5393	1N5395	1N5397	1N5398	1N5399	Unit
Peak Repetitive Reverse Voltage Working Peak Reverse Voltage DC Blocking Voltage	V_{RRM} V_{RWM} V_R	50	100	200	400	600	800	1000	Volts
Nonrepetitive Peak Reverse Voltage (Halfwave, Single Phase, 60 Hz)	V_{RSM}	100	200	300	525	800	1000	1200	Volts
RMS Reverse Voltage	$V_{R(RMS)}$	35	70	140	280	420	560	700	Volts
Average Rectified Forward Current (Single Phase, Resistive Load, 60 Hz, T_L = 70°C, 1/2" From Body)	I_O				1.5				Amp
Nonrepetitive Peak Surge Current (Surge Applied at Rated Load Conditions, See Figure 2)	I_{FSM}				50 (for 1 cycle)				Amp
Storage Temperature Range	T_{stg}				–65 to +175				°C
Operating Temperature Range	T_L				–65 to +170				°C
DC Blocking Voltage Temperature	T_L				150				°C

*ELECTRICAL CHARACTERISTICS

Characteristic and Conditions	Symbol	Typ	Max	Unit
Maximum Instantaneous Forward Voltage Drop (i_F = 4.7 Amp Peak, T_L = 170°C, 1/2 Inch Leads)	v_F	—	1.4	Volts
Maximum Reverse Current (Rated dc Voltage) (T_L = 150°C)	I_R	250	300	μA
Maximum Full-Cycle Average Reverse Current (1) (I_O = 1.5 Amp, T_L = 70°C, 1/2 Inch Leads)	$I_{R(AV)}$	—	300	μA

*Indicates JEDEC Registered Data.

NOTE 1: Measured in a single-phase, halfwave circuit such as shown in Figure 6.25 of EIA RS-282, November 1963. Operated at rated load conditions I_O = 1.5 A, V_r = V_{RWM}, T_L = 70°C.

MECHANICAL CHARACTERISTICS

CASE: Transfer molded plastic

MAXIMUM LEAD TEMPERATURE FOR SOLDERING PURPOSES: 240°C, 1/8" from case for 10 seconds at 5 lbs. tension

FINISH: All external surfaces are corrosion-resistant, leads are readily solderable

POLARITY: Cathode indicated by color band

WEIGHT: 0.40 grams (approximately)

NOTES:
1. ALL RULES AND NOTES ASSOCIATED WITH JEDEC DO-41 OUTLINE SHALL APPLY.
2. POLARITY DENOTED BY CATHODE BAND.
3. LEAD DIAMETER NOT CONTROLLED WITHIN "F" DIMENSION.

DIM	MILLIMETERS		INCHES	
	MIN	MAX	MIN	MAX
A	5.97	6.60	0.235	0.260
B	2.79	3.05	0.110	0.120
D	0.76	0.86	0.030	0.034
K	27.94	—	1.100	—

**CASE 59-04
PLASTIC**

FIGURE 2.48

(Courtesy of Motorola, Inc.)

34. Refer to the spec sheet shown in Figure 2.48. What is the surge current rating for the 1N5391? [11]

35. Refer to Figure 2.26. A circuit has an average forward current of 24.5 A and a 225 V_{pk} source. Which of the diodes listed has the *minimum* acceptable ratings for use in this circuit? [12]

36. Refer to Figure 2.26. A circuit has an average forward current of 3.6 A and a 170 V_{pk} source. Which diode has the *minimum* acceptable ratings for use in this circuit? [12]

37. Refer to Figure 2.26. A circuit has an average forward power dissipation (for the diode) of 2.8 W and a 470 V_{pk} source. Which diode has the *minimum* acceptable ratings for use in this circuit? [7, 12]

§2.7

38. A zener diode spec sheet lists values of $I_{ZT} = 20$ mA and $I_{zt} = 1$ mA. If the measured change in V_Z (at I_{zt}) is 25 mV, what is the value of zener impedance for the device? [13b]

39. Refer to Figure 2.49. In each circuit, determine whether or not the biasing voltage has the correct polarity for *normal* zener operation. [13a]

40. For each of the properly biased zener diodes in Figure 2.49, draw an arrow indicating the direction of zener current. [13a]

§2.8

41. A 6.8-V zener diode has a $P_{D(MAX)}$ rating of 1 W. What is the value of I_{ZM} for the device? [15]

42. A 24-V zener diode has a $P_{D(MAX)}$ rating of 10 W. What is the value of I_{ZM} for the device? [15]

43. A zener diode with a $P_{D(MAX)}$ rating of 5 W has a derating factor of 8 mW/°C above 50°C. What is the maximum allowable value of P_D for the device if it is operating at 120°C? [15]

44. The MLL4678 zener diode has a $P_{D(MAX)}$ rating of 250 mW and a derating factor of 1.67 mW/°C above 50°C. What is the maximum allowable value of P_D for the device if it is operated at 150°C? [15]

45. Refer to Figure 2.33. Which of the diodes could be used in place of a 28-V zener diode that has a maximum power dissipation of 1.8 W? [18]

FIGURE 2.49

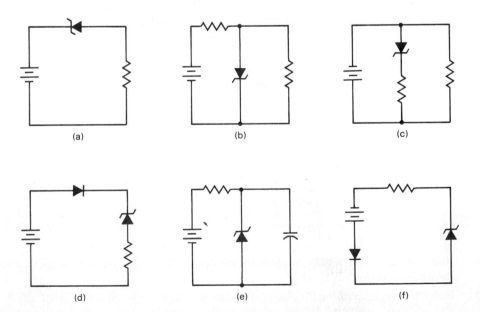

(a) (b) (c)

(d) (e) (f)

46. Refer to Figure 2.33. Which of the diodes listed could be used in place of a 6.8-V zener diode that has a maximum power dissipation of 1.2 W? [18]

47. Refer to Figure 2.33. Which of the diodes listed could be used in place of a 12-V zener diode that has a maximum operating current of 150 mA? [17, 18]

§2.9

48. An LED has a range of V_F = 1.5 to 1.8 V and $I_{F(MAX)}$ = 18 mA. If the LED is being driven by a 20-V_{pk} source, what standard value of current-limiting resistor is needed to protect the LED? [20]

49. An LED has a range of V_F = 1.6 to 2.0 V and $I_{F(MAX)}$ = 20 mA. Determine the minimum standard resistor value that could be used as a current-limiting resistor if the LED is being driven by a 32-V_{pk} source. [20]

TROUBLESHOOTING PRACTICE PROBLEMS

50. The following table lists the results of testing several diodes. In each case, determine whether the diode is good, open, or shorted. [21]

	Forward Resistance	Reverse Resistance
a.	1200 MΩ	1200 MΩ
b.	15 Ω	3500 MΩ
c.	75 Ω	175 MΩ
d.	30 Ω	50 Ω

51. Refer to Figure 2.46a. When the output from the source is positive, the peak voltage across R_1 is approximately 100 V. When the output from the source is negative, the peak voltage across R_1 is approximately 0 V. Is the diode good, open, or shorted? Explain your answer. [4, 21]

52. Refer to Figure 2.46a. The peak voltage across R_1 is always equal to the peak source voltage. Is the diode good, open, or shorted? Explain your answer. [4, 21]

53. Refer to Figure 2.46a. The voltage across R_1 is 0 V, regardless of the value of source voltage. Is the diode good, open, or shorted? Explain your answer. [4, 21]

THE BRAIN DRAIN

54. The spec sheet for the MLL755 zener diode is shown in Figure 2.50. What is the value of I_{ZM} for the device when it is being operated at 150°C?

55. The MLL756 zener diode *cannot* be used in the circuit shown in Figure 2.51. Why not? (*Note*: The temperature range shown is the normal operating temperature range for the circuit.)

500 MILLIWATT HERMETICALLY SEALED GLASS SILICON ZENER DIODES

- Complete Voltage Range — 2.4 to 110 Volts
- Leadless Package for Surface Mount Technology
- Double Slug Type Construction
- Metallurgically Bonded Construction
- Nitride Passivated Die
- Available in 8 mm Tape and Reel
  T1 Cathode Facing Sprocket Holes
  T2 Anode Facing Sprocket Holes

MLL746 thru MLL759

MLL957A thru MLL986A

MLL4370 thru MLL4372

LEADLESS GLASS ZENER DIODES
500 MILLIWATTS
2.4–110 VOLTS

MAXIMUM RATINGS

Rating	Symbol	Value	Unit
DC Power Dissipation @ $T_A \leq 50°C$ Derate above $T_A = 50°C$	P_D	500 3.3	mW mW/°C
Operating and Storage Junction Temperature Range	T_J, T_{stg}	−65 to +200	°C

MECHANICAL CHARACTERISTICS

CASE: Double slug type, hermetically sealed glass

MAXIMUM LEAD TEMPERATURE FOR SOLDERING PURPOSES: 230°C, for 10 seconds

FINISH: All external surfaces are corrosion resistant and readily solderable

POLARITY: Cathode indicated by color band. When operated in zener mode, cathode will be positive with respect to anode

MOUNTING POSITION: Any

CASE 362-01 GLASS

DIM	MILLIMETERS		INCHES	
	MIN	MAX	MIN	MAX
A	3.30	3.70	0.130	0.145
B	1.60	—	0.063	—
U	2.49	2.59	0.098	0.102
R	0.41	0.55	0.016	0.022

STEADY STATE POWER DERATING

P_D vs T_A

P_D vs T_C

ELECTRICAL CHARACTERISTICS (T_A = 25°C, V_F = 1.5 V Max @ 200 mA for all types)

Type Number (Note 1)	Nominal Zener Voltage V_Z @ I_{ZT} (Notes 1,2,3) Volts	Test Current I_{ZT} (Note 2) mA	Maximum Zener Impedance Z_{ZT} @ I_{ZT} (Note 4) Ohms	Maximum DC Zener Current I_{ZM} mA		Maximum Reverse Leakage Current	
						T_A = 25°C I_R @ V_R = 1 V μA	T_A = 150°C I_R @ V_R = 1 V μA
MLL4370	2.4	20	30	150	190	100	200
MLL4371	2.7	20	30	135	165	75	150
MLL4372	3.0	20	29	120	150	50	100
MLL746	3.3	20	28	110	135	10	30
MLL747	3.6	20	24	100	125	10	30
MLL748	3.9	20	23	95	115	10	30
MLL749	4.3	20	22	85	105	2	30
MLL750	4.7	20	19	75	95	2	30
MLL751	5.1	20	17	70	85	1	20
MLL752	5.6	20	11	65	80	1	20
MLL753	6.2	20	7	60	70	0.1	20
MLL754	6.8	20	5	55	65	0.1	20
MLL755	7.5	20	6	50	60	0.1	20
MLL756	8.2	20	8	45	55	0.1	20
MLL757	9.1	20	10	40	50	0.1	20
MLL758	10	20	17	35	45	0.1	20
MLL759	12	20	30	30	35	0.1	20

FIGURE 2.50

(Courtesy of Motorola, Inc.)

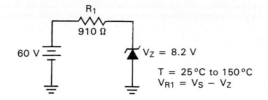

R_1
910 Ω

60 V

$V_Z = 8.2$ V

T = 25 °C to 150 °C
$V_{R1} = V_S - V_Z$

FIGURE 2.51

SUGGESTED COMPUTER APPLICATIONS PROBLEMS

56. Write a program to determine the total voltage drop across a forward-biased *pn*-junction diode. The program must take into account the type of semiconductor material being used, the value of V_k, the value of I_F, and the value of R_B.

57. Write a program to determine the total voltage drop across a zener diode when V_Z, Z_Z, and I_Z are known.

58. Write a program that will determine whether a given *pn*-junction diode is good, open, or shorted when provided with forward and reverse resistance readings.

59. Write a program that will determine the value of I_{ZM} at specified values of V_Z, $P_{D(MAX)}$, power derating factor, and temperature.

ANSWERS TO THE EXAMPLE PRACTICE PROBLEMS

2.1. The diode acts as an open, and an open drops the full applied voltage. Therefore, $V_{R1} = V_{R2} = 0$ V.

2.2. Using the ideal diode model, $I_T = 15$ mA.

2.4. $I_T = 8.43$ mA.

2.6. $I_T = 5.75$ mA.

2.7. I_T (ideal) = 7.5 mA; % of error = 30.4%.

2.8. Any diode with $V_{RRM} > 180$ V. Therefore, you could use any diode from 1N4003 to 7.

2.9. The value of I_T for the circuit is 1.95 A. The rating for the diode would have to be greater than this value.

2.10. $P_{D(min)} = 198.7$ mW.

2.11. $P_D = (750$ mA) $(0.7$ V$) = 525$ mW. This exceeds the 500 mW limit.

2.12. 0.696 V (or 696 mV)

2.13. 22.63 μA

2.14. $I_{ZM} = 37$ mA.

2.15. P_D (at 125°C) = 250.25 mW.

2.16. The 1N5374A.

2.17. P_D for the device is 1.19 W. Therefore, any diode in the 6.8 V row with a P_D of 1.5 W or higher could be used.

2.18. $R_S = 1.1$ kΩ (minimum standard value; the calculated value is 1050 Ω).

3

Common Diode Applications: The Basic Power Supply

". . . From that point on, research centered around the use of silicon rather than germanium."

OBJECTIVES

After studying the material in this chapter, you should be able to:

☐ 1. Briefly describe the purpose served by a power supply and the function of each circuit it contains. (Introduction)

☐ 2. Explain the forward and reverse operations of the half-wave rectifier. (§3.1)

☐ 3. Describe the output waveform of a given half-wave rectifier. (§3.1)

☐ 4. Determine whether a given half-wave rectifier is a *positive* or *negative* half-wave rectifier using a schematic diagram. (§3.1)

☐ 5. Calculate the *peak* and dc (*average*) output voltage and current values for a *positive* half-wave rectifier. (§3.1)

☐ 6. Calculate the *peak* and dc (*average*) output voltage and current values for a *negative* half-wave rectifier. (§3.1)

☐ 7. Determine the PIV for a diode in a half-wave rectifier. (§3.1)

☐ 8. Describe the operation of the full-wave rectifier. (§3.2)

☐ 9. Calculate the values of *peak* and dc (*average*) output voltage and current for any full-wave rectifier. (§3.2)

☐ 10. Determine the PIV for a diode in a full-wave rectifier. (§3.2)

☐ 11. Compare and contrast the half-wave rectifier and the full-wave rectifier. (§3.2)

☐ 12. Describe the operation of the bridge rectifier. (§3.3)

☐ 13. Calculate the *peak* and dc (*average*) output voltage and current values for a bridge rectifier. (§3.3)

☐ 14. Compare and contrast the bridge rectifier and the full-wave rectifier. (§3.3)

☐ 15. Determine the PIV for a diode in a bridge rectifier. (§3.3)

☐ 16. Discuss the effects that bulk resistance and reverse current can have on rectifier voltage measurements. (§3.4)

☐ 17. Discuss the relationship between transformer rating tolerances, measured rectifier voltages, and voltage regulators. (§3.4)

☐ 18. Discuss the differences between *power rectifiers* and small-signal diodes. (§3.4)

☐ 19. List the advantages of using IC (integrated-circuit) rectifiers. (§3.4)

☐ 20. Discuss the effects that *filtering* has on the output of a rectifier. (§3.5)

☐ 21. Describe the operation of the basic capacitive filter. (§3.5)

☐ 22. Discuss the primary cause and limiting of surge current in a power supply. (§3.5)

☐ 23. Describe the relationship between V_r and V_{dc} for a filtered rectifier. (§3.5)

☐ 24. Discuss the reason that full-wave rectifiers are preferred over half-wave rectifiers. (§3.5)

☐ 25. Calculate the values of V_r and V_{dc} for a filtered rectifier. (§3.5)

☐ 26. Discuss the effects of filtering on diode PIV. (§3.5)

☐ 27. Discuss the zener current requirements for proper zener voltage regulator operation. (§3.6)

☐ 28. Calculate the values of I_z, I_L, and/or I_T for a zener voltage regulator, given the needed circuit values and diode ratings. (§3.6)

☐ 29. Discuss and calculate the minimum allowable value of load resistance for a given zener voltage regulator. (§3.6)

☐ 30. Discuss and calculate the effect of zener impedance on regulator input ripple voltage. (§3.6)

☐ 31. Calculate the limits of R_S for a zener voltage regulator. (§3.6, optional)

☐ 32. Calculate the values of V_{dc}, I_L, and $V_{r(out)}$ for a basic power supply. (§3.7)

☐ 33. Discuss the purpose served by *worst-case analysis*. (§3.7)

☐ 34. Perform a worst-case analysis on a basic power supply. (§3.7)

☐ 35. List the faults that commonly occur in a basic power supply and the symptoms of each. (§3.8)

As you master each of the objectives listed, place a check mark (✔) in the appropriate box.

I t would take several volumes to discuss every diode application in modern electronics. For this reason, we are going to concentrate in this chapter only on the most common diode application, the *power supply*.

The power supply of an electronic system is used to convert the ac energy provided by the wall outlet to dc energy. The power cord of any electronic system

✔ 1

Introduction

A **power supply** converts *ac* to *dc*.

The circuits used in a basic power supply.

Rectifier. A circuit that converts ac to pulsating dc.

Filter. A circuit that reduces the variations in the output of a rectifier.

provides the line power to the system power supply, which then provides all internal dc voltages needed for proper circuit operation.

The basic power supply can be broken down into three circuit groups, as shown in Figure 3.1. The 60 Hz (or 50 Hz) line input is applied to a *rectifier*. A rectifier is a diode circuit that converts the ac to what is called *pulsating dc*. This pulsating dc is then applied to a *filter*, which reduces the variations in dc voltage. The filter is usually made up of passive components: resistors, capacitors, and sometimes inductors. The final stage is the *voltage regulator*. As you know, a voltage regulator is used to maintain a constant output voltage.

At one time, voltage regulators were designed using zener diodes as the regulating element. However, the development of the IC (integrated-circuit) voltage

FIGURE 3.1

Basic power supply block diagram and waveforms.

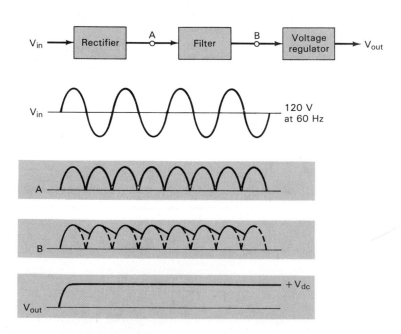

regulator has led to the replacement of zener diodes as regulating elements. IC voltage regulators are far more efficient than zener diodes, so power supply design currently emphasizes their use. At the same time, zener regulators are easier to understand and serve as valuable educational circuits. We will therefore concentrate on the zener regulator in this chapter. The IC voltage regulator is covered in detail in Chapter 20.

Voltage regulator. A circuit used to maintain a constant output voltage.

In this chapter, we're going to take a look at the operation and troubleshooting of basic power supplies. Then, in Chapter 4, we will take a look at several other common applications for *pn*-junction diodes, zener diodes, and LEDs. We'll begin our study of power supplies by looking at the first circuit in Figure 3.1, the rectifier.

3.1
Half-Wave Rectifiers

There are three basic types of rectifier circuits: the *half-wave*, *full-wave*, and *bridge* rectifiers. Of the three, the bridge rectifier is the most commonly used, followed by the full-wave rectifier. Our discussion on rectifiers begins with the half-wave rectifier, simply because it is the easiest to understand.

The half-wave rectifier is made up of a *diode* and a *resistor*, as shown in Figure 3.2. Since the half-wave rectifier in a power supply will be connected to a transformer, the transformer has also been included in the circuit. The half-wave rectifier is used to eliminate either the negative alternation of the input or the positive alternation of the input. As you will see, the diode direction determines which half-cycle will be eliminated.

What does a half-wave rectifier do?

Basic Circuit Operation

The negative half-cycle of the input to Figure 3.2 will be eliminated by the one-way conduction of the diode. Figure 3.3 details the operation of the circuit for one complete cycle of the input signal. During the positive half-cycle of the input, D_1 will be forward biased and will conduct. This allows a voltage to be developed across R_L that is approximately equal to the voltage across the secondary of the transformer (V_2).

◢ 2

When the polarity of the input signal reverses, D_1 will be reverse biased, preventing conduction in the circuit. With no current through R_L, no voltage is developed across the load. In this case, the output voltage remains at approximately 0 V, and the voltage across the diode (V_D) will be approximately equal to V_2.

If all this seems confusing, remember the *ideal diode model*. This diode model was shown in Chapter 2 to represent the component as either an *open* (reverse-biased) or *closed* (forward-biased) switch. When forward biased, this ideal switch will drop no voltage. When reverse biased, this ideal switch will drop all the applied voltage.

If we view the diode in Figure 3.3a to be a closed switch, it is easy to see that

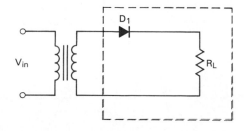

FIGURE 3.2

Half-wave rectifier.

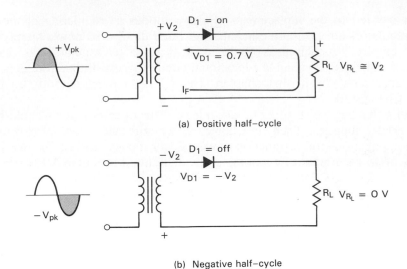

(a) Positive half-cycle

(b) Negative half-cycle

FIGURE 3.3

Ideal half-wave rectifier operation.

$$\boxed{V_{RL} \cong V_2} \qquad \text{(forward operation)} \qquad \textbf{(3.1)}$$

This is due to the fact that no voltage is dropped across a closed switch. By the same token, if we view the diode in Figure 3.3b as being an *open* switch, it is easy to see that

$$\boxed{V_{D1} = V_2} \qquad \text{(reverse operation)} \qquad \textbf{(3.2)}$$

This is due to the fact that an open switch drops all the applied voltage. This would leave no voltage to be dropped across R_L, and V_{RL} would equal 0 V, as is shown in the figure. The *ideal* circuit operating characteristics are summarized as follows:

Diode condition	V_{D1}	V_{RL}
Forward biased	0 V	Equal to V_2
Reverse biased	Equal to V_2	0 V

Using these relationships, it is easy to understand the input/output waveforms shown in Figure 3.4. During T_1, the diode is forward biased, and the output (V_{RL}) is approximately equal to the input (V_2). During T_2, the diode is reverse biased, and the output drops to 0 V. This is the way half-wave rectifiers work.

FIGURE 3.4

Input and output waveforms.

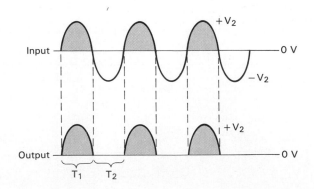

CHAP. 3   Common Diode Applications: The Basic Power Supply

FIGURE 3.5

Negative half-wave rectifier.

Negative Half-wave Rectifiers (Ideal)

Figure 3.5 shows a half-wave rectifier with the diode pointing the other way. In this circuit, the diode will conduct on the *negative* half-cycle of the input, and equation (3.1) will apply. The diode will be reverse biased on the *positive* half-cycle of the input, and equation (3.2) will apply. As a result, the positive half-cycle of the input is eliminated. Note that the operating principles for this circuit are *exactly* the same as those for the *positive* half-wave rectifier shown in Figure 3.2. The only difference is that the polarity of the output has been reversed.

As you can see, the *direction* of the diode determines whether the output from the rectifier is positive or negative. For circuit recognition, the following statements will generally hold true:

1. When the diode points toward the load (R_L), the output from the rectifier will be *positive*.
2. When the diode points toward the transformer, the output from the rectifier will be *negative*.

These two statements will also hold true for the full-wave rectifier. The points made so far about half-wave rectifiers are summarized in Figure 3.6.

Calculating Load Voltage and Current Values

In our discussion of the *ideal* half-wave rectifier, we ignored the value of V_F for the diode. When we take this value into account, the *peak load voltage*, $V_{out(pk)}$, is found as

$$\boxed{V_{out(pk)} = V_{2(pk)} - 0.7 \text{ V}}$$ (3.3)

A Practical Consideration: When we talk about the *positive* and *negative* half-cycles of the input, we are referring to the polarity of the transformer secondary voltage, measured from the top of the secondary to the bottom of the secondary.

Determining the output polarity of a half-wave rectifier.

FIGURE 3.6

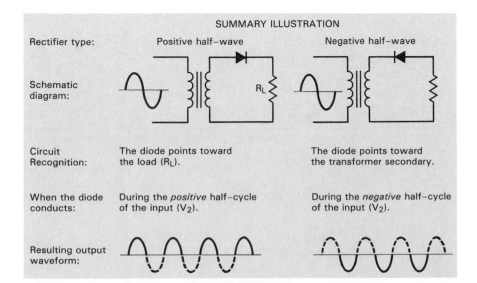

Equation 2.1: $V_R = V_S - 0.7$ V

You shouldn't have any difficulty with this equation. It is simply a variation on equation (2.1). $V_{2(pk)}$ is the *peak secondary voltage* of the transformer. $V_{2(pk)}$ is found as

$$V_{2(pk)} = \frac{N_2}{N_1} V_{1(pk)}$$

(3.4)

where $\dfrac{N_2}{N_1}$ = the ratio of transformer secondary turns to primary turns

$V_{1(pk)}$ = the peak transformer primary voltage

Note: You will be shown later in this section how to calculate the output values for a *negative* half-wave rectifier.

The following example illustrates the procedure for calculating the peak output voltage from a positive half-wave rectifier.

EXAMPLE 3.1

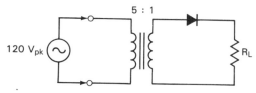

FIGURE 3.7

Determine the peak output voltage for the circuit shown in Figure 3.7.

Solution: The peak secondary voltage is found as

$$V_{2(pk)} = \frac{N_2}{N_1} V_{1(pk)}$$
$$= \frac{1}{5}(120 \text{ V}_{pk})$$
$$= 24 \text{ V}_{pk}$$

and

$$V_{out(pk)} = V_{2(pk)} - 0.7 \text{ V}$$
$$= 24 \text{ V}_{pk} - 0.7 \text{ V}$$
$$= 23.3 \text{ V}_{pk}$$

PRACTICE PROBLEM 3–1

A half-wave rectifier has values of $N_1 = 10$, $N_2 = 1$, and $V_{1(pk)} = 110$ V$_{pk}$. What is the peak output voltage from the circuit?

AC voltages are not always given in *peak* form.

A word of caution. In Figure 3.7, the primary voltage was identified as being 120 V *peak*. If a source voltage is *not identified as being a peak value, you must assume that it is an rms value.* In this case, V_1 must be converted to a peak value using

$$V_{1(pk)} = \frac{V_{1(rms)}}{0.707}$$

(3.5)

CHAP. 3   Common Diode Applications: The Basic Power Supply

EXAMPLE 3.2

Assume the rectifier circuit shown in Figure 3.7 is being fed by a 120 V wall outlet. Calculate the peak output from the rectifier.

Solution: Line voltage is always rated in rms. Therefore,

$$V_{1(pk)} = \frac{V_{1(rms)}}{0.707}$$
$$= \frac{120 \text{ Vac}}{0.707}$$
$$= 169.73 \text{ V}_{pk}$$

and

$$V_{2(pk)} = \frac{N_2}{N_1} V_{1(pk)}$$
$$= \frac{1}{5} (169.73 \text{ V}_{pk})$$
$$= 33.95 \text{ V}_{pk}$$

Now $V_{out(pk)}$ is found as

$$V_{out(pk)} = V_{2(pk)} - 0.7 \text{ V}$$
$$= 33.95 \text{ V}_{pk} - 0.7 \text{ V}$$
$$= 33.25 \text{ V}_{pk}$$

PRACTICE PROBLEM 3–2

A positive half-wave rectifier has values of $N_1 = 5$ and $N_2 = 1$. If the rectifier is being fed with a 115 Vac signal, what is the peak output voltage?

A Practical Consideration: Most transformers have rms output voltage ratings instead of turns ratio ratings.

Here is another practical situation: Most transformers are rated for a specific rms output voltage. For example, a 30 Vac transformer would have an rms output of 30 V when fed from a 120 V wall outlet. When a transformer has an output voltage rating, simply divide the rated output voltage by *0.707* to obtain the value of $V_{2(pk)}$.

EXAMPLE 3.3

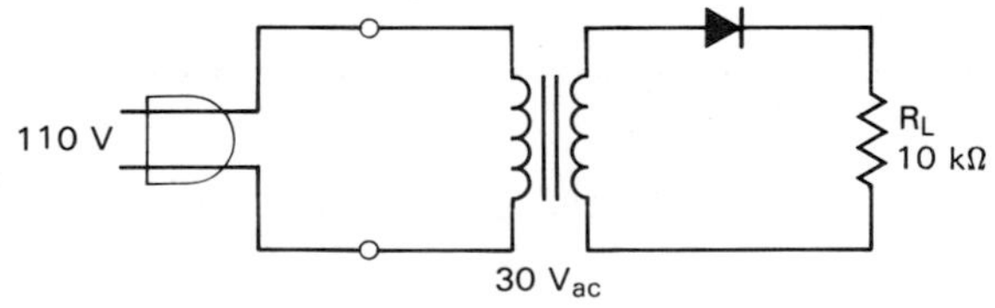

FIGURE 3.8

Determine the peak output voltage for the circuit shown in Figure 3.8.

Solution: The transformer is shown to have a 30 Vac rating. This value of V_2 is converted to peak form as follows:

$$V_{2(pk)} = \frac{V_{2(rms)}}{0.707}$$
$$= \frac{30 \text{ Vac}}{0.707}$$
$$= 42.43 \text{ V}_{pk}$$

Now the value of $V_{out(pk)}$ is found as

$$V_{out(pk)} = V_{2(pk)} - 0.7 \text{ V}$$
$$= 42.43 \text{ V}_{pk} - 0.7 \text{ V}$$
$$= 41.73 \text{ V}_{pk}$$

PRACTICE PROBLEM 3–3

A 24 Vac transformer is being used in a positive half-wave rectifier. What is the peak output voltage from the circuit?

Once the peak output voltage is determined, the peak output current is found as

$$I_{out(pk)} = \frac{V_{out(pk)}}{R_L} \qquad \textbf{(3.6)}$$

EXAMPLE 3.4

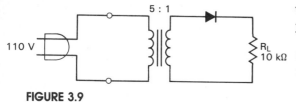

FIGURE 3.9

What is the peak output current for the circuit shown in Figure 3.9?

Solution: The input voltage is, again, an rms value. So $V_{1(pk)}$ is found as

$$V_{1(pk)} = \frac{V_{1(rms)}}{0.707}$$
$$= \frac{110 \text{ Vac}}{0.707}$$
$$= 155.59 \text{ V}_{pk}$$

and

$$V_{2(pk)} = \frac{N_2}{N_1} V_{1(pk)}$$
$$= \frac{1}{5}(155.59 \text{ V}_{pk})$$
$$= 31.12 \text{ V}_{pk}$$

Now

$$V_{out(pk)} = V_{2(pk)} - 0.7 \text{ V}$$
$$= 31.12 \text{ V} - 0.7 \text{ V}$$
$$= 30.42 \text{ V}_{pk}$$

and

$$I_{out(pk)} = \frac{V_{out(pk)}}{R_L}$$
$$= \frac{30.42 \text{ V}}{10 \text{ k}\Omega}$$
$$= 3.04 \text{ mA}$$

Average Load Voltage and Current

The *average* voltage, V_{ave}, from an ac circuit indicates *the reading you would get if the voltage was measured with a dc voltmeter*. In other words, V_{ave} is the *dc equivalent* of an ac signal. Since rectifiers are used to convert ac to dc, V_{ave} is a very important value. For a half-wave rectifier, V_{ave} is found as

Average load voltage. The dc equivalent of an ac signal.

V_{ave} is measured with a dc voltmeter.

$$\boxed{V_{ave} = \frac{V_{pk}}{\pi}} \quad \text{(half-wave rectified)} \tag{3.7}$$

Another form of this equation is

$$\boxed{V_{ave} = 0.318 \, (V_{pk})} \quad \text{(half-wave rectified)} \tag{3.8}$$

where $0.318 \cong 1/\pi$. Either of these equations can be used to determine the *dc equivalent* load voltage for a *half-wave* rectifier. The following example illustrates the process for determining the value of V_{ave} for a half-wave rectifier.

Question and Answer: Where did equations (3.7) and (3.8) come from? These are the equations that are taught in an ac circuits course for converting *peak* values to *average* values. You may want to review your basic electronics text for a complete explanation of the equations.

═══════ **EXAMPLE 3.5** ═══════

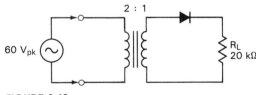

FIGURE 3.10

Determine the value of V_{ave} for the circuit shown in Figure 3.10.

Solution: The input signal to the transformer primary is given as a peak value, so we do not need to convert it. $V_{2(pk)}$ is found as

$$V_{2(pk)} = \frac{N_2}{N_1} V_{1(pk)}$$
$$= \frac{1}{2} (60 \text{ V}_{pk})$$
$$= 30 \text{ V}_{pk}$$

$V_{out(pk)}$ is found as

$$V_{out(pk)} = V_{2(pk)} - 0.7 \text{ V}$$
$$= 29.3 \text{ V}_{pk}$$

Now the average (dc) output voltage is found as

$$V_{ave} = \frac{V_{pk}}{\pi}$$
$$= \frac{29.3 \text{ V}_{pk}}{\pi}$$
$$= 9.33 \text{ Vdc}$$

PRACTICE PROBLEM 3–5

A half-wave rectifier like the one in Figure 3.10 has values of $N_1 = 14$, $N_2 = 1$, and $V_1 = 110$ Vac. What is the dc load voltage for the circuit?

Average load current. The dc equivalent current of an ac signal.

I_{ave} is measured with a dc ammeter.

Just as we can convert a *peak* voltage to *average* form, we can also convert a *peak* current to an *average* form. The value of *average current* for an ac waveform is the value that would be measured with a *dc ammeter*. Thus, the value of I_{ave} for a waveform gives us *equivalent dc current*. This is another very important value that we need to know for any circuit that is used to convert ac to dc. The value of I_{ave} can be calculated in one of two ways:

1. We can determine the value of V_{ave} and then use Ohm's law as follows:

$$I_{ave} = \frac{V_{ave}}{R}$$

2. We can convert I_{pk} to average form using the same basic equations (3.7 and 3.8) that we used to convert V_{pk} to V_{ave}. The current forms of these equations are

$$\boxed{I_{ave} = \frac{I_{pk}}{\pi}} \quad \text{(half-wave rectified)} \tag{3.9}$$

and

$$\boxed{I_{ave} = 0.318(I_{pk})} \quad \text{(half-wave rectified)} \tag{3.10}$$

The following example illustrates the first method for determining the value of I_{ave}.

═══ **EXAMPLE 3.6** ═══

Determine the value of I_{ave} for the circuit shown in Figure 3.10.

Solution: In Example 3.5, we determined the value of V_{ave} to be 9.33 V. Using this value and the value of R_L, the value of I_{ave} is found to be

$$
\begin{aligned}
I_{ave} &= \frac{V_{ave}}{R_L} \\
&= \frac{9.33 \text{ V}}{20 \text{ k}\Omega} \\
&= 466.5 \text{ }\mu\text{A}
\end{aligned}
$$

PRACTICE PROBLEM 3–6

A half-wave rectifier has an average output voltage that is equal to 24 Vdc. The load resistance is 2.2 kΩ. What is the value of dc load current for the circuit?

CHAP. 3   Common Diode Applications: The Basic Power Supply

The next example illustrates the second method of determining the value of I_{ave}.

EXAMPLE 3.7

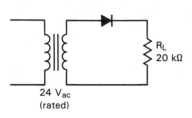

FIGURE 3.11

24 V$_{ac}$
(rated)

Determine the dc load current for the rectifier shown in Figure 3.11.

Solution: The transformer has a 24-Vac rating. Thus, the peak secondary voltage is found as

$$V_{2(pk)} = \frac{24 \text{ Vac}}{0.707}$$
$$= 33.9 \text{ V}_{pk}$$

The peak output voltage is now found as

$$V_{out(pk)} = V_{2(pk)} - 0.7 \text{ V}$$
$$= 33.2 \text{ V}_{pk}$$

The peak load current is found as

$$I_{out(pk)} = \frac{V_{out(pk)}}{R_L}$$
$$= \frac{33.2 \text{ V}}{20 \text{ k}\Omega}$$
$$= 1.66 \text{ mA}$$

Finally,

$$I_{ave} = \frac{I_{pk}}{\pi}$$
$$= \frac{1.66 \text{ mA}}{\pi}$$
$$= 528.4 \text{ }\mu\text{A}$$

PRACTICE PROBLEM 3–7

A half-wave rectifier is fed by a 30 Vac transformer. If the load resistance for the circuit is 12 kΩ, what is the dc load current for the circuit?

Component Substitution

Why knowing the value of I_{ave} is important.

The value of I_{ave} is important for another reason. You may recall from Chapter 2 that the *maximum dc forward current* that can be drawn through a diode is called the *average forward current* (I_0) rating of the device. When working with rectifiers, you may need at some point to substitute one diode for another. When this is the case, you must make sure that the value of I_{ave} for the diode in the circuit is less than the I_0 rating of the substitute component. This point will be discussed further in Section 3.4.

Negative Half-wave Rectifiers

The analysis of a *negative* half-wave rectifier is nearly identical to that for a positive half-wave rectifier. The only difference is that all the voltage polarities will be reversed.

You can use a simple method for performing the mathematical analysis of a negative half-wave rectifier. This method is as follows:

1. Analyze the circuit as if it was a positive half-wave rectifier.
2. After completing your calculations, change all of your voltage polarity signs from positive to negative.

This method for analyzing a negative half-wave rectifier is illustrated in Example 3.8.

EXAMPLE 3.8

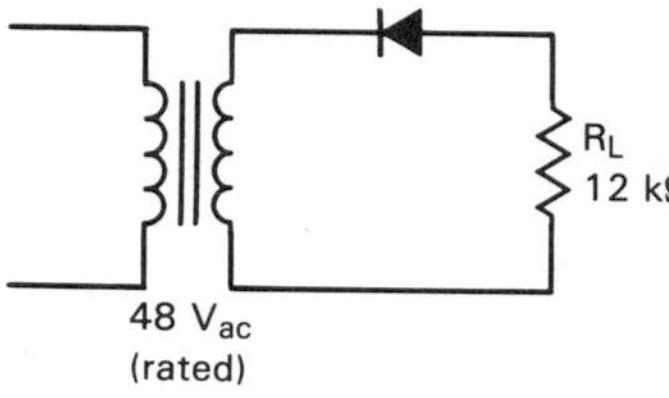

48 V$_{ac}$
(rated)

FIGURE 3.12

Determine the dc output voltage for the circuit shown in Figure 3.12.

Solution: We'll start by solving the circuit as if it was a *positive* half-wave rectifier. First,

$$V_{2(pk)} = \frac{48 \text{ Vac}}{0.707}$$
$$= 67.9 \text{ V}_{pk}$$

and

$$V_{out(pk)} = V_{2(pk)} - 0.7 \text{ V}$$
$$= 67.2 \text{ V}_{pk}$$

Finally,

$$V_{ave} = \frac{V_{pk}}{\pi}$$
$$= 21.39 \text{ Vdc}$$

Now we simply convert all the positive voltage values to negative voltage values. Thus, for the circuit shown in Figure 3.12,

$$V_{2(pk)} = -67.9 \text{ V}, \qquad V_{out(pk)} = -67.2 \text{ V},$$
$$V_{ave} = -21.39 \text{ V}$$

PRACTICE PROBLEM 3–8

A negative half-wave rectifier is fed by a 36 Vac transformer. Using the method illustrated in Example 3.8, calculate the values of $V_{2(pk)}$, $V_{out(pk)}$, and dc output voltage.

As you can see, there isn't really a whole lot of difference between the analysis of a negative half-wave rectifier and that of a positive half-wave rectifier.

Peak Inverse Voltage (PIV)

The maximum amount of reverse bias that a diode will be exposed to in a rectifier is called the *peak inverse voltage*, or *PIV*, of the rectifier. For the half-wave rectifier, the value of PIV is found as

$$\boxed{PIV = V_2} \qquad \text{(half-wave rectifier)} \qquad (3.11)$$

The basis for this equation can be seen by referring to Figure 3.3. When the diode is reverse biased (Figure 3.3b), there is no voltage dropped across the load. Therefore, all of V_2 is dropped across the diode in the rectifier.

The PIV of a given rectifier is important because it determines the *minimum allowable value of* V_{RRM} for any diode that is used in the circuit. This point is discussed further in Section 3.4.

Peak inverse voltage (PIV)
The maximum reverse bias that a diode will be exposed to in a rectifier.

Why PIV is important.

---

1. What does a power supply do? [Objective 1]
2. What are the circuits that make up a power supply? What does each do? [Objective 1]
3. Briefly explain the forward and reverse operation of a half-wave rectifier. [Objective 2]
4. Describe the difference between the output waveform of a positive and a negative half-wave rectifier. [Objective 3]
5. How can you tell the output polarity of a half-wave rectifier? [Objective 4]
6. List, in order, the steps you would take to calculate the dc output voltage from a rectifier if you were given the turns ratio of the transformer and the rms primary voltage. [Objective 5]
7. List, in order, the steps you would take to calculate the dc output voltage from a rectifier if you knew the rated rms output voltage. [Objective 5]
8. You have calculated the value of V_{ave} for a half-wave rectifier. What piece of test equipment would you use to measure it? [Objective 5]
9. You have calculated the value of I_{ave} for a half-wave rectifier. What piece of test equipment would you use to measure it? [Objective 5]
10. Describe the two methods for analyzing a *negative* half-wave rectifier. [Objective 6]
11. Why is PIV important? [Objective 7]
12. How do you determine the value of PIV for a half-wave rectifier? [Objective 7]

3.2
Full-wave Rectifiers

The full-wave rectifier consist of *two diodes* and a *resistor*, as shown in Figure 3.13a. The result of this change in circuit construction is illustrated in Figure 3.13b. Here the output from the full-wave rectifier is compared with that of a half-wave rectifier. Note that the full-wave rectifier has two positive half-cycles out for every one produced by the half-wave rectifier.

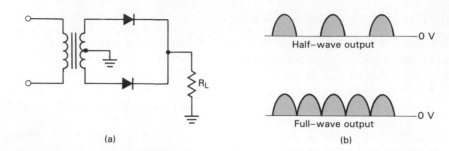

FIGURE 3.13

Full-wave rectifier.

(a)   (b)

Basic Circuit Operation

8

Figure 3.14 shows the operation of the full-wave rectifier during one complete cycle of the input signal. During the positive half-cycle of the input, D_1 is forward biased and D_2 is reverse biased. Note the direction of current through the load (R_L). Using the ideal operating characteristics of the diode, V_{RL} can be found as

$$V_{RL} = \frac{V_2}{2} \qquad (3.12)$$

Why load voltage is only half the value of transformer secondary voltage.

V_{RL} is equal to half of the secondary voltage because the transformer is center tapped. The voltage from one end of a center-tapped transformer to the center tap is *always* one-half of the total secondary voltage.

When the polarity of the input reverses, D_2 will be forward biased, and D_1 will be reverse biased. Note that the direction of current through the load *has not changed*, even though the polarity of the transformer secondary has. Thus another positive half-cycle is produced across the load. This gives the output waveform illustrated in Figure 3.13b.

Calculating Load Voltage and Current Values

9a

Using the *practical diode model*, the peak output voltage for a full-wave rectifier is found as

FIGURE 3.14

Full-wave operation.

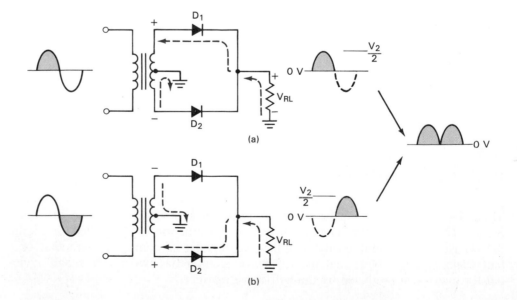

(a)

(b)

$$V_{out(pk)} = \frac{V_{2(pk)}}{2} - 0.7 \text{ V}$$ (3.13)

The full-wave rectifier will produce twice as many output pulses (per input cycle) as the half-wave rectifier. In other words, for every output pulse that is produced by a half-wave rectifier, two will be produced by a full-wave rectifier. For this reason, the *average output voltage* for the full-wave rectifier is found as

The full-wave rectifier has twice the output frequency of the half-wave rectifier.

$$V_{ave} = \frac{2V_{out(pk)}}{\pi}$$ (3.14)

or

$$V_{ave} = 0.636 \, V_{out(pk)}$$ (3.15)

where $0.636 \cong 2/\pi$. Note that the value *0.636* used in equation (3.15) is twice the value of *0.318* that was used in the V_{ave} equation, equation (3.8), for the half-wave rectifier. The procedure for determining the dc output voltage for a full-wave rectifier is illustrated in the following example.

EXAMPLE 3.9

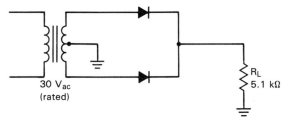

30 V$_{ac}$
(rated)

FIGURE 3.15

Determine the dc output voltage for the circuit shown in Figure 3.15.

Solution: The transformer is rated at 30 Vac. Therefore, the value of $V_{2(pk)}$ is found as

$$V_{2(pk)} = \frac{30 \text{ Vac}}{0.707}$$
$$= 42.4 \text{ V}_{pk}$$

The peak output voltage is now found as

$$V_{out(pk)} = \frac{V_{2(pk)}}{2} - 0.7 \text{ V}$$
$$= 21.2 \text{ V} - 0.7 \text{ V}$$
$$= 20.5 \text{ V}_{pk}$$

Finally, the dc output voltage is found as

$$V_{ave} = \frac{2V_{out(pk)}}{\pi}$$
$$= \frac{41 \text{ V}_{pk}}{\pi}$$
$$= 13.1 \text{ Vdc}$$

PRACTICE PROBLEM 3–9

A full-wave rectifier is fed by a 24 Vac rated center-tapped transformer. What is the dc output voltage for the circuit?

Once the peak and average load voltage values are known, it is easy to determine the values of $I_{out(pk)}$ and I_{ave}. Just use your known values and Ohm's law, as is illustrated in Example 3.10.

═══ **EXAMPLE 3.10** ═══

Determine the values of $I_{out(pk)}$ and I_{ave} for the circuit shown in Figure 3.15.

Solution: In Example 3.9, we calculated the peak and average output voltages for the circuit. Using these calculated values and the value of R_L shown in the circuit, we determine the circuit current values as follows:

$$I_{out(pk)} = \frac{V_{out(pk)}}{R_L}$$
$$= \frac{20.5 \text{ V}_{pk}}{5.1 \text{ k}\Omega}$$
$$= 4.02 \text{ mA}$$

and

$$I_{ave} = \frac{V_{ave}}{R_L}$$
$$= \frac{13.1 \text{ V dc}}{5.1 \text{ k}\Omega}$$
$$= 2.57 \text{ mA}$$

PRACTICE PROBLEM 3–10

The circuit described in practice problem 3–9 has a 2.2 kΩ load. What are the peak and dc output current values for the circuit?

Peak Inverse Voltage

✔ 10

When one of the diodes in a full-wave rectifier is reverse biased, the voltage across that diode will be approximately equal to V_2. This point is illustrated in Figure 3.16. The 24 V_{pk} across the primary develops voltages of $+12$ V and -12 V across the secondary (when measured from end to center tap). Note that V_2 equals the difference between these two voltages: 24 V. With the polarities shown, D_1 is conducting, and D_2 is reverse biased. If we assume D_1 to be ideal, the voltage drop across the component will equal 0 V. Thus the cathode of D_1 will

FIGURE 3.16

Full-wave rectifier PIV.

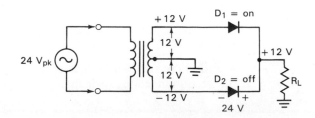

CHAP. 3   Common Diode Applications: The Basic Power Supply

also be at $+12$ V. Since this point is connected directly to the cathode of D_2, its cathode is also at $+12$ V. With -12 V being applied to the anode of D_2, and $+12$ V being applied to its cathode, the total voltage across the diode is 24 V.

The peak output voltage from the full-wave rectifier is equal to one-half the secondary voltage, V_2. Therefore, the reverse voltage across either diode will be twice the peak output voltage. By formula.

$$\boxed{\text{PIV} = 2\,V_{\text{out(pk)}}} \tag{3.16}$$

since the peak output voltage is half the secondary voltage, we can also find the value of PIV as

$$\boxed{\text{PIV} = V_2} \tag{3.17}$$

You may recall that equation (3.17) is the same as the equation we used for the half-wave rectifier, equation (3.11).

Full-wave Versus Half-wave Rectifiers

11

There are quite a few similarities between the full-wave rectifier and the half-wave rectifier. Figure 3.17 summarizes the relationships that you have been shown for the half-wave and full-wave rectifiers.

It would seem (at first) that the only similarity between the half-wave and the full-wave rectifiers is the method of finding the PIV values for the two circuits. However, there is another similarity that may not be too obvious. Let's assume for a moment that both of the rectifiers shown in Figure 3.17 are being fed by 24-V ac transformers. If you were to calculate the dc output voltages for the two circuits, you would get the following values:

$$V_{\text{ave}} = 10.58 \text{ Vdc}, \qquad \text{(for the half-wave rectifier)}$$
$$V_{\text{ave}} = 10.36 \text{ Vdc}, \qquad \text{(for the full-wave rectifier)}$$

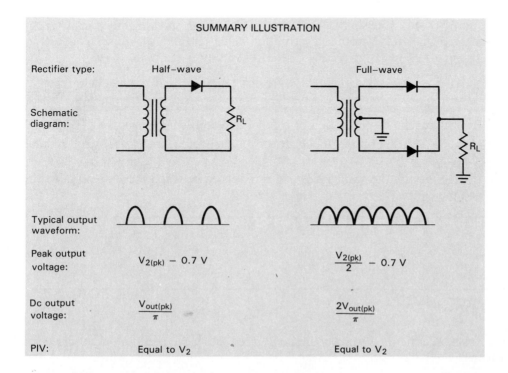

SUMMARY ILLUSTRATION

Rectifier type:	Half–wave	Full–wave
Schematic diagram:		
Typical output waveform:		
Peak output voltage:	$V_{2(\text{pk})} - 0.7$ V	$\dfrac{V_{2(\text{pk})}}{2} - 0.7$ V
Dc output voltage:	$\dfrac{V_{\text{out(pk)}}}{\pi}$	$\dfrac{2V_{\text{out(pk)}}}{\pi}$
PIV:	Equal to V_2	Equal to V_2

FIGURE 3.17

SEC. 3.2 Full-wave Rectifiers

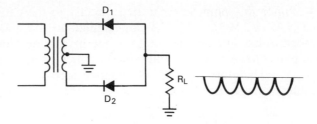

FIGURE 3.18

As you can see, the two circuits will produce nearly identical dc output voltages *for identical values of transformer secondary voltage*. In fact, if the values of R_L for the two circuits are equal, the dc output current values for the circuits will also be nearly identical.

So why do we bother with the full-wave rectifier when we can get the same dc output values with the half-wave rectifier? There are a couple of reasons. First, *if the peak output voltages for the two circuits are equal, the full-wave rectifier will have twice the dc output voltage that the half-wave rectifier has*. The second reason deals with operation of filters. As you will be shown in Section 3.5, the fact that the full-wave rectifier has twice the output frequency of the half-wave rectifier has an impact on the filtering of the rectifier output. When we add a filter to the half-wave and full-wave rectifiers, the advantages of the full-wave rectifier will become clear.

Negative Full-wave Rectifiers

9b

If we reverse the directions of the diodes in the positive full-wave rectifier, we will have a *negative* full-wave rectifier. The negative full-wave rectifier and its output waveform are shown in Figure 3.18. As you can see, the only difference between the two circuits is the direction that the diodes are pointing in and the polarity of the output voltage.

The method you were shown for analyzing a negative half-wave rectifier will also work for analyzing a negative full-wave rectifier: simply change the voltage polarity signs from positive to negative. While we will not go through these analysis procedures again, you should have no problem with making the transition from the negative half-wave to the negative full-wave rectifier. If you can't remember how to analyze a negative rectifier, refer to Section 3.1.

SECTION REVIEW

1. Briefly explain the operation of a full-wave rectifier. [Objective 8]
2. How do you determine the PIV across each diode in a full-wave rectifier? [Objective 10]
3. Briefly discuss the similarities and differences between the half-wave rectifier and the full-wave rectifier. [Objective 11]

3.3
Full-Wave Bridge Rectifiers

The bridge rectifier is the most commonly used full-wave rectifier circuit for several reasons:

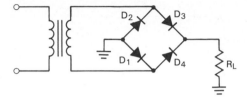

FIGURE 3.19

Bridge rectifier.

1. It does not require the use of a center-tapped transformer.
2. It provides a higher peak output voltage than the full-wave rectifier does. This ultimately provides a higher dc output voltage from the power supply.

Why bridge rectifiers are preferred.

The bridge rectifier is shown in Figure 3.19. As you can see, the circuit consists of *four diodes* and a *resistor*.

Basic Circuit Operation

The full-wave rectifier produces its output by alternating circuit conduction between the two diodes. When one is *on* (conducting), the other is *off* (not conducting), and vice versa. The bridge rectifier works basically the same way. The main difference is that the bridge rectifier alternates conduction between two diode *pairs*. When D_1 and D_3 (Figure 3.19) are on, D_2 and D_4 are off, and vice versa. This circuit operation is illustrated in Figure 3.20. During the positive half-cycle of the input, V_2 will have the polarities shown, causing D_1 and D_3 to conduct. Note the direction of current through the load resistor and the polarity of the resulting load voltage. During the negative half-cycle of the input, V_2 will reverse polarity. D_2 and D_4 will now conduct rather than D_1 and D_3. However, the current direction through the load has not changed, nor has the resulting polarity of the load voltage.

✔ 12

Calculating Load Voltage and Current Values

Recall that the full-wave rectifier has a peak voltage equal to one-half the secondary voltage (assuming that the conducting diode is ideal). This relationship was expressed in equation (3.13). As you know, the output voltage in a full-wave rectifier is reduced to one-half the secondary voltage by the center tap on the transformer secondary. The center-tapped transformer is essential for the full-wave rectifier to work, but it cuts the output voltage in half.

The bridge rectifier does not require the use of a center-tapped transformer. Assuming the diodes in the bridge to be ideal, the rectifier will have a peak output voltage of

✔ 13

$$V_{out(pk)} = V_{2(pk)} \quad \text{(ideal)}$$

(3.18)

Refer to Figure 3.20. If you consider the diodes to be ideal, the cathode and anode voltage will be equal for each diode. If you view the conducting diodes as being shorted connections to the transformer secondary, you can see that the voltage across the load resistor is equal to the voltage across the secondary.

When calculating circuit output values, you will get more accurate results if you take the voltage drops across the two conducting diodes into account. To include these values, use the following equation when calculating peak output voltage:

$$V_{out(pk)} = V_{2(pk)} - 1.4 \text{ V}$$

(3.19)

The 1.4-V value represents the sum of the diode voltage drops.

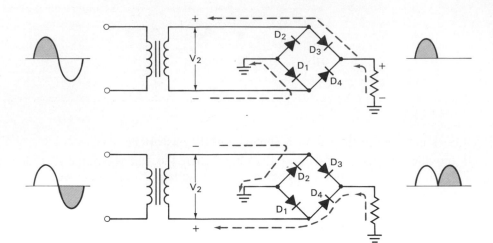

FIGURE 3.20

Bridge rectifier operation.

The rest of the load voltage and current values are found using the same equations as those used for the full-wave rectifier. This is illustrated in the following example.

═══ **EXAMPLE 3.11** ═══

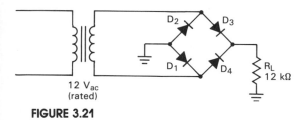

FIGURE 3.21

Determine the dc output voltage and current values for the circuit shown in Figure 3.21.

Solution: With the 12 Vac rated transformer, the peak secondary voltage is found as

$$V_{2(pk)} = \frac{12 \text{ Vac}}{0.707}$$
$$= 16.97 \text{ V}_{pk}$$

The peak output voltage is now found as

$$V_{out(pk)} = V_{2(pk)} - 1.4 \text{ V}$$
$$= 15.57 \text{ V}_{pk}$$

The dc output voltage is found as

$$V_{ave} = \frac{2V_{out(pk)}}{\pi}$$
$$= \frac{31.14 \text{ V}_{pk}}{\pi}$$
$$= 9.91 \text{ V dc}$$

Finally, the dc output current is found as

$$I_{ave} = \frac{V_{ave}}{R_L}$$
$$= \frac{9.91 \text{ V dc}}{12 \text{ k}\Omega}$$
$$= 825.8 \text{ } \mu A$$

PRACTICE PROBLEM 3–11

A bridge rectifier is fed by an 18 Vac transformer. Determine the dc output voltage and current for the circuit when it has a 1.2 kΩ load.

Bridge versus Full-Wave Rectifiers

Let's analyze one more full-wave rectifier. This will give us some values for comparing the outputs from the full-wave and bridge rectifiers.

EXAMPLE 3.12

A full-wave rectifier has a 12 Vac transformer and a 12-kΩ load. Determine the dc output voltage and current values for the circuit.

Solution: The peak secondary voltage for the circuit is found as

$$V_{2(pk)} = \frac{12 \text{ Vac}}{0.707}$$
$$= 16.97 \text{ V}_{pk}$$

The peak output voltage is now found as

$$V_{out(pk)} = \frac{V_{2(pk)}}{2} - 0.7 \text{ V}$$
$$= 7.79 \text{ V}_{pk}$$

The dc load voltage is found as

$$V_{ave} = \frac{2V_{out(pk)}}{\pi}$$
$$= 4.96 \text{ Vdc}$$

Finally, the dc output current is found as

$$I_{ave} = \frac{V_{ave}}{R_L}$$
$$= 413.3 \text{ } \mu A$$

Now let's compare the results from Examples 3.11 and 3.12. The only difference between these two circuits is that one is a full-wave rectifier and one is a bridge rectifier. For convenience, the results are summarized as follows:

Value	Bridge rectifier	Full-wave rectifier
Peak output voltage	15.57 V$_{pk}$	7.79 V$_{pk}$
DC output voltage	9.91 Vdc	4.96 Vdc
DC output current	825.8 μA	413.3 μA

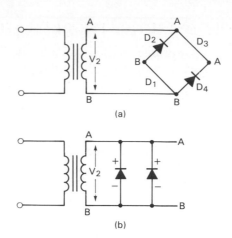

FIGURE 3.22

Bridge rectifier PIV.

The primary advantage of using a bridge rectifier instead of a full-wave rectifier.

As you can see, the bridge rectifier has output values that are twice as high as a comparable full-wave rectifier. This higher output is the primary advantage of using a bridge rectifier instead of a full-wave rectifier.

Peak Inverse Voltage

 15

The PIV across each diode in the bridge rectifier is equal to V_2, just like the full-wave rectifier. Figure 3.22 helps to illustrate this point. In Figure 3.22a, two things have been done:

1. The conducting diodes (D_1 and D_3) have been replaced by straight wires. Assuming that the diodes are ideal, they will have the same resistance as wire; therefore, the replacement is valid.
2. The positive side of the secondary has been labeled (A) and the negative side has been labeled (B).

FIGURE 3.23

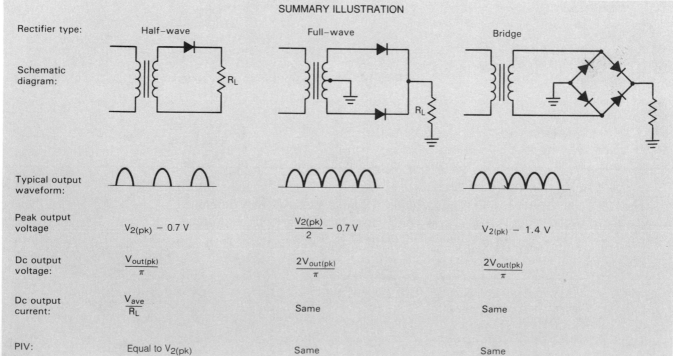

Rectifier type:	Half–wave	Full-wave	Bridge
Schematic diagram:			
Typical output waveform:			
Peak output voltage	$V_{2(pk)} - 0.7\text{ V}$	$\dfrac{V_{2(pk)}}{2} - 0.7\text{ V}$	$V_{2(pk)} - 1.4\text{ V}$
Dc output voltage:	$\dfrac{V_{out(pk)}}{\pi}$	$\dfrac{2V_{out(pk)}}{\pi}$	$\dfrac{2V_{out(pk)}}{\pi}$
Dc output current:	$\dfrac{V_{ave}}{R_L}$	Same	Same
PIV:	Equal to $V_{2(pk)}$	Same	Same

SUMMARY ILLUSTRATION

Connecting the common (A) points along a straight line, and doing the same with the (B) points, gives us the circuit shown in Figure 3.22b. With this equivalent circuit, it is easy to see that the two reverse-biased diodes and the secondary of the transformer are all in *parallel*. Since parallel voltages are equal, the PIV across each diode is equal to V_2. The same situation will exist for D_1 and D_3 when they are reverse biased.

Rectifier Summary

Figure 3.23 summarizes the operation of the rectifiers we have covered. Whenever you need to quickly review the relationships listed, you can refer to this illustration. If you need more detailed information, you should refer back to the appropriate discussion.

SECTION REVIEW

1. Describe the operation of the bridge rectifier. [Objective 12]
2. List the advantages that the bridge rectifier has over the full-wave rectifier. [Objective 14]
3. Describe the method for determining the PIV for a diode in a bridge rectifer. [Objective 15]

3.4
Working With Rectifiers

There are several practical factors involving rectifiers that we have ignored up to this point. Most of these factors explain the difference between *theory* and *practice*.

The Effects of Bulk Resistance and Reverse Current on Circuit Measurements

In Chapter 2, we discussed the effects of *bulk resistance* and *reverse current* on circuit voltage measurements. In that chapter, the following points were made:

1. The current through the bulk resistance of a diode can affect the actual value of V_F for the device.
2. When a diode is in series with a resistor, reverse current can cause a voltage to be developed across the resistor when the diode is biased *off*.

While the effects of bulk resistance and reverse current are not normally very drastic, they do serve to explain a few things:

1. Rectifiers tend to be very high current circuits. Thus, the effects of $I_F R_B$ on the value of V_F can be fairly significant. Many rectifier diodes will have a value of V_F than is closer to 1 V than to 0.7 V.

   A Practical Consideration: Forward voltage values in high-current rectifiers are typically as high as 1.4 V.

2. It is fairly common for *rectifier diodes* (those with high current capabilities) to have high reverse current ratings. For example, the MBR320–360 series of high-power rectifier diodes has a reverse current rating of *20 mA* (maximum) at $T = 100°C$. This value of I_R could have a major impact on the operation

   Reverse current is one of the reasons that many power supplies are fan cooled.

of a power supply. To keep the amount of reverse current to a minimum, power supplies must be kept as cool as possible. This is one of the reasons that power supplies are often fan cooled.

Transformer Rating Tolerance

The tolerance of a transformer's output rating can have a major effect on circuit voltage measurements. Depending on the quality of the transformer, the output voltage may (at times) be above or below the rated value by as much as 20%! This can have a major impact on any voltage measured in the circuit.

Line regulation. The ability of a voltage regulator to maintain a stable dc output voltage despite variations in rectifier output voltage.

Later in this chapter you will learn about *zener voltage regulators*. As you will be shown, voltage regulators provide varying degrees of *line regulation*. That is, *they are capable (up to a point) of maintaining a constant load voltage despite changes in the rectifier output voltage*. The bottom line here is simple: Voltage regulators make it possible to have a stable dc output voltage despite the variation in output voltage of the components in the rectifier. Thus, transformer rating tolerances may cause a measured rectifier voltage to be off by a considerable margin, but they will have little effect on dc load voltage when a voltage regulator is used. Again, the effect of voltage regulation on the operation of a power supply will be covered in detail later in this chapter.

Power Rectifiers

Most power supplies require the use of rectifier diodes that have extremely high forward current and/or power dissipation ratings. The characteristics of these *power rectifiers* are illustrated in Figure 3.24, which contains the spec sheet for the MUR50 series of power rectifiers.

Power rectifiers. Diodes with extremely high forward current and/or power dissipation ratings.

Question and Answer. The spec sheet for the MUR50 series calls these components *ultrafast* power rectifiers. What does *ultrafast* mean? The term *ultrafast* is used to describe diodes that can be switched between *on* and *off* at a very high frequency. We will discuss ultrafast components further in Chapter 21.

The spec sheet shown illustrates the primary differences between power rectifiers and small-signal diodes. These differences are as follows:

1. Power rectifiers have extremely high forward current ratings. The MUR50 series diodes are capable of handling an average forward current of up to 50 A!
2. Power rectifiers (as stated earlier) tend to have values of V_F that are greater than 0.7 V. The typical values of V_F for the MUR50 diodes are between 0.8 and 1.1 V.
3. Power diodes tend to have relatively high reverse current ratings. For the MUR50 series, I_R is typically 1 mA when $T = 100°C$.

These ratings agree with the statements that were made earlier about the effects of bulk resistance and reverse current on circuit measurements.

Integrated Rectifiers

Advances in semiconductor device manufacturing have made it possible to construct entire circuits on a single piece of semiconductor material. A circuit that is constructed entirely on a single piece of semiconductor material is called an *integrated circuit*.

Integrated circuit. An entire circuit that is constructed on a single piece of semiconductor material.

There are a number of advantages to using integrated circuits. Among them are:

1. Reduced cost (it takes fewer components to construct a circuit).
2. Troubleshooting is made easier.

CHAP. 3   Common Diode Applications: The Basic Power Supply

MOTOROLA
■ SEMICONDUCTOR ■
TECHNICAL DATA

MUR5005
MUR5010
MUR5015
MUR5020

SWITCHMODE POWER RECTIFIERS

. . . designed for use in switching power supplies, inverters and as free wheeling diodes, these state-of-the-art devices have the following features:

● Ultrafast 50 Nanosecond Recovery Time

● Low Forward Voltage Drop

● Hermetically Sealed Metal DO-203AB Package

ULTRAFAST RECTIFIERS

50 AMPERES
50 to 200 VOLTS

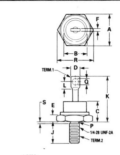

MAXIMUM RATINGS

Rating	Symbol	MUR 5005	MUR 5010	MUR 5015	MUR 5020	Unit
Peak Repetitive Reverse Voltage Working Peak Reverse Voltage DC Blocking Voltage	V_{RRM} V_{RWM} V_R	50	100	150	200	Volts
Nonrepetitive Peak Reverse Voltage	V_{RSM}	55	110	165	220	Volts
Average Forward Current T_C = 125°C	$I_{F(AV)}$	50				Amps
Nonrepetitive Peak Surge Forward Current (half cycle, 60 Hz, Sinusoidal Waveform)	I_{FSM}	600				Amps
Operating Junction and Storage Temperature	T_J, T_{stg}	–55 to +175				°C

THERMAL CHARACTERISTICS

Rating	Symbol	All Devices	Unit
Thermal Resistance, Junction to Case	$R_{\theta JC}$	1.0	°C/W

ELECTRICAL CHARACTERISTICS

Maximum Instantaneous Forward Voltage Drop (i_F = 50 Amp, T_J = 25°C) (i_F = 50 Amp, T_J = 125°C) (i_F = 100 Amp, T_J = 125°C)	v_F	1.15 0.95 1.10	Volts
Maximum Reverse Current @ DC Voltage (T_J = 25°C) (T_J = 125°C)	I_R	10 1.0	µA mA
Maximum Reverse Recovery Time (I_F = 1.0 Amp, di/dt = 50 Amp/µs, V_R = 30 V, T_J = 25°C)	t_{rr}	50	ns

NOTES:
1. DIM "P" IS DIA.
2. CHAMFER OR UNDERCUT ON ONE OR BOTH ENDS OF HEXAGONAL BASE IS OPTIONAL.
3. ANGULAR ORIENTATION AND CONTOUR OF TERMINAL ONE IS OPTIONAL.
4. THREADS ARE PLATED.
5. DIMENSIONING AND TOLERANCING PER ANSI Y14.5, 1973.

DIM	MILLIMETERS MIN	MILLIMETERS MAX	INCHES MIN	INCHES MAX
A	16.94	17.45	0.669	0.687
B	—	16.94	—	0.667
C	—	11.43	—	0.450
D	—	9.53	—	0.375
E	2.92	5.08	0.115	0.200
F	—	2.03	—	0.080
J	10.72	11.51	0.422	0.453
K	—	25.40	—	1.000
L	3.86	—	0.156	—
P	5.59	6.32	0.220	0.249
Q	3.56	4.45	0.140	0.175
R	—	20.16	—	0.794
S	—	2.26	—	0.089

CASE 257-01
DO-203AB
METAL

MECHANICAL CHARACTERISTICS
CASE: Welded, hermetically sealed
FINISH: All external surface corrosion resistant and terminal leads are readily solderable
POLARITY: Cathode to Case
MOUNTING POSITIONS: Any
MOUNTING TORQUE: 25 in-lb max

FIGURE 3.24

(Courtesy of Motorola, Inc.)

MOTOROLA
■ SEMICONDUCTOR ■
TECHNICAL DATA

MDA2500 Series

RECTIFIER ASSEMBLY

. . . utilizing individual void-free molded rectifiers, interconnected and mounted on an electrically isolated aluminum heat sink by a high thermal-conductive epoxy resin.

- 400 Ampere Surge Capability
- Electrically Isolated Base
- UL Recognized
- 1800 Volt Heat Sink Isolation

SINGLE-PHASE FULL-WAVE BRIDGE

25 AMPERES
50-600 VOLTS

MAXIMUM RATINGS

Rating (Per Diode)	Symbol	MDA							Unit
		2500	2501	2502	2504	2506	2508	2510	
Peak Repetitive Reverse Voltage Working Peak Reverse Voltage DC Blocking Voltage	V_{RRM} V_{RWM} V_R	50	100	200	400	600	800	1000	Volts

DC Output Voltage
  Resistive Load
  Capacitive Load
	Vdc	30
		50
			62
			100
				124
				200
					250
					400
						380
						600
							500
							600
								620
								1000
									Volts

Sine Wave RMS Input Voltage	V_R(RMS)	35	70	140	280	420	560	700	Volts
Average Rectified Forward Current (Single phase bridge resistive load, 60 Hz, T_C = 55°C)	I_O				25				Amp
Nonrepetitive Peak Surge Current (Surge applied at rated load conditions)	I_{FSM}				400				Amp
Operating and Storage Junction Temperature Range	T_J, T_{stg}				−65 to +175				°C

THERMAL CHARACTERISTICS

Characteristic	Symbol	Typ	Max	Unit

Thermal Resistance, Junction to Case
  Each Die
  Total Bridge
	$R_{\theta JC}$	
		4.5
		2.0
			
			6.0
			2.8
				°C/W

ELECTRICAL CHARACTERISTICS (T_C = 25°C unless otherwise noted)

Characteristic	Symbol	Min	Typ	Max	Unit
Instantaneous Forward Voltage (Per Diode) (i_F = 40 A)*	v_F	—	0.95	1.05	Volts
Reverse Current (Per Diode) (Rated V_R)	I_R	—	—	10	μA

MECHANICAL CHARACTERISTICS

CASE: Plastic case with an electrically isolated aluminum base.
POLARITY: Terminal designation embossed on case:
  + DC output
  − DC output
  AC not marked
MOUNTING POSITION: Bolt down. Highest heat transfer efficiency accomplished through the surface opposite the terminals. Use silicone heat sink compound on mounting surface for maximum heat transfer.
WEIGHT: 25 grams (approx.)
TERMINALS: Suitable for fast-on connections. Readily solderable, corrosion resistant. Soldering recommended for applications greater than 15 amperes.
MOUNTING TORQUE: 20 in-lb max

*Pulse Width = 100 ms, Duty Cycle ≤ 2%.

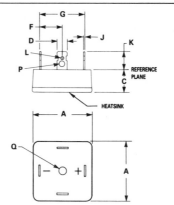

NOTES:
1. DIMENSION "Q" SHALL BE MEASURED ON HEATSINK SIDE OF PACKAGE.
2. DIMENSIONS "F" AND "G" SHALL BE MEASURED AT THE REFERENCE PLANE.

DIM	MILLIMETERS		INCHES	
	MIN	MAX	MIN	MAX
A	25.65	26.16	1.010	1.030
C	12.44	13.97	0.490	0.550
D	6.10	6.60	0.240	0.260
F	10.01	10.49	0.394	0.413
G	19.99	21.01	0.787	0.827
J	0.71	0.86	0.028	0.034
K	9.52	11.43	0.375	0.450
L	1.52	2.06	0.060	0.081
P	2.79	2.92	0.110	0.115
Q	4.42	4.67	0.174	0.184

CASE 309A-03

FIGURE 3.25

(Courtesy of Motorola, Inc.)

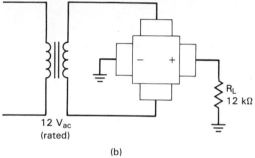

(b)

Figure 3.25 (Continued)

Integrated rectifier circuits are currently available. These integrated circuits (ICs) can be used as the only active component in a full-wave or bridge rectifier. The spec sheet for the MDA2500 series IC bridge rectifiers is shown in Figure 3.25a.

The single casing shown in the spec sheet contains the circuit shown above it: a bridge rectifier. This bridge rectifier is capable of handling an average forward current of 25 A and can handle surges that are as high as 400 A. The V_{RRM} ratings for the device are comparable to those for any rectifier diodes, and the reverse current rating is as low as most small-signal diodes.

If you look below the casing illustration, you will see a drawing that identifies the component leads. Figure 3.25b uses this symbol to show you how the single component would be wired to replace the circuit shown in Figure 3.21 (page 96).

Does the use of an IC bridge rectifier change the mathematical analysis of a rectifier circuit? Not really. If you look at the V_F rating for the MDA2500, you'll see that values of 0.95 to 1.05 V *per diode* are typical. Thus, you must still consider the effect of *two* diode voltage drops in the mathematical analysis of any circuit that uses the component.

SECTION REVIEW

1. What effect does bulk resistance have on the measured value of V_F for a rectifier diode? [Objective 16]
2. What effect can reverse current have on measured load voltages? [Objective 16]
3. How does cooling a power supply reduce the effects of reverse current? [Objective 16]
4. What is the typical tolerance range for transformers? [Objective 17]
5. What is *line regulation*? [Objective 17]
6. What are the primary differences between power rectifiers and small-signal diodes? [Objective 18]
7. What advantages of using ICs were listed in this section? [Objective 19]

3.5

Filters

The second circuit in a power supply is the *filter* (see Figure 3.1, p. 78). Filters are used in power supplies to *reduce the variations in the rectifier output signal.* Since our goal is to produce a *constant* dc output voltage, it is necessary to remove as much of the rectifier output variation as possible.

Filters reduce the variations in the rectifier output.

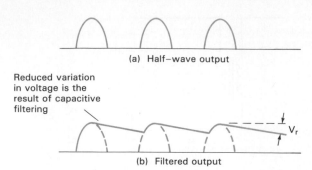

(a) Half-wave output

Reduced variation
in voltage is the
result of capacitive
filtering

(b) Filtered output

FIGURE 3.26

The effects of filtering on the
output of a half-wave rectifier.

The overall result of using a filter is illustrated in Figure 3.26. Here we see the output from the half-wave rectifier, both before and after filtering. Note that there are still voltage variations after filtering; however, the *amount* of variation has been greatly reduced.

Ripple voltage. The remaining variation in the output from a filter.

The remaining voltage variation in the output of the filter is called *ripple voltage*, V_r. As you will see, the amount of ripple voltage left by a given filter depends on the rectifier used, the filter component values, and the load resistance.

The Basic Capacitor Filter

☑ 21

The capacitor filter is the most basic filter type and the most commonly used. This filter is simply a capacitor connected in parallel with the load resistance, as shown in Figure 3.27. The filtering action is based on the *charge/discharge* action of the capacitor. During the positive half-cycle of the input, D_1 will conduct, and the capacitor will charge *rapidly* (Figure 3.27a). As the input starts to go negative, D_1 will turn off, and the capacitor will *slowly* discharge through the load resistance (Figure 3.27b). *It is the difference between the charge and discharge times of the capacitor that reduces the variations in the rectifier output voltage.*

The difference between the charge and discharge times of the capacitor is due to the fact that there are two distinct *RC time constants* in the circuit. You may recall from your study of basic electronics that a capacitor will charge (or discharge) in *five* time constants. One time constant (represented by the Greek letter *tau*) is found as

$$\tau = RC$$

(3.20)

where R and C are the total circuit resistance and capacitance, respectively. Since it takes five time constants for a capacitor to charge or discharge fully, this time period can be found as

$$T = 5(RC)$$

(3.21)

Now refer to Figure 3.27. The capacitor charges through the diode. For the sake of discussion, let's assume that D_1 has a forward resistance of 5 Ω. The time

FIGURE 3.27

The basic capacitive filter.

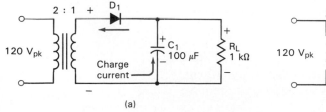

(a)

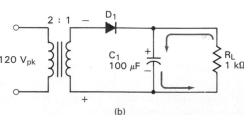

(b)

constant for the circuit would be found as

$$\tau = RC$$
$$= (5\ \Omega)(100\ \mu F)$$
$$= 500\ \mu s$$

and the total capacitor charge time would be found as

$$T = 5(RC)$$
$$= 5(500\ \mu s)$$
$$= 2.5\ ms$$

Thus the capacitor would charge to the peak input voltage in 2.5 ms. The discharge path for the capacitor is through the resistor (Figure 3.27b). For this circuit, the time constant would be found as

$$\tau = RC$$
$$= (1\ k\Omega)(100\ \mu F)$$
$$= 100\ ms$$

and the total capacitor discharge time would be found as

$$T = 5(RC)$$
$$= 5(100\ ms)$$
$$= 500\ ms$$

Therefore, the capacitor in Figure 3.27 would have a charge time of 2.5 ms and a discharge time of 500 ms. This is why it charges almost instantly, yet barely has time to discharge.

You have seen that the values of load resistance and capacitance in the filter determine the time required for the capacitor to discharge. When either of these two values is increased, the discharge time of the capacitor is increased. The effects of increased discharge time are illustrated in Figure 3.28. As the discharge time increases, the amount of ripple voltage (V_r) decreases. It would seem that the key to having no V_r would be to have an extremely long discharge time for the capacitor, which could be accomplished by using very high values of load resistance and/or capacitance. However, these two values are limited by other considerations. Since the load resistance limits the amount of output current, its value must be limited. If the load resistance is too high, output current from the rectifier will be reduced to the point where the circuit is useless. The value of C is limited by factors:

What limits the value of a filter capacitor?

1. The maximum allowable charge time for the component

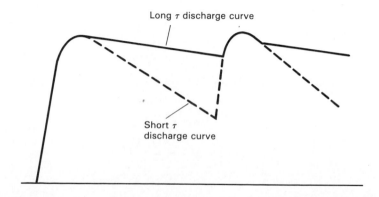

Long τ discharge curve

Short τ discharge curve

FIGURE 3.28

The effects of τ on filtering.

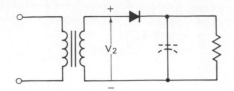

FIGURE 3.29

2. The amount of *surge current*, I_{surge}, that the rectifier diodes can withstand

3. The cost of "larger than needed" filter capacitors.

The capacitor is not only involved in the discharge action; it is involved in the charging action as well. If you make the value of C too high, your discharge time will be greatly increased, but so will the charge time. A charge time that is too high will reduce the average (dc) output voltage from the filter. As a general guideline, *C should be limited to 1000 μF or less*. This will prevent the output voltage from the filter from dropping. The other factor that limits the value of C is *surge current*. We'll take a look at surge current, its causes, and its effects now.

Surge Current

Surge current. The high initial current in a power supply.

When you first turn a power supply on, the filter capacitor has no accumulated charge to oppose V_2. For the first instant, the discharged capacitor acts as a short circuit, as shown in Figure 3.29. As you can see, the diode current is initially limited only by the resistance of the transformer secondary and the bulk resistance of the diode. Since these resistances are usually very low, the initial current will be extremely high. This high initial current is referred to as *surge current*, and it is calculated as follows:

$$I_{surge} = \frac{V_{2(pk)}}{R_W + R_B} \tag{3.22}$$

where V_2 = the *peak* secondary voltage
R_w = the resistance of the secondary windings
R_B = the total diode bulk resistance

EXAMPLE 3.13

If the circuit shown in Figure 3.27 has values of $R_w = 0.8$ Ω and $R_B = 5$ Ω, what is the initial surge current for the circuit?

Solution: The peak secondary voltage (V_2) is found as

$$V_{2(pk)} = \frac{N_2}{N_1} V_{1(pk)}$$
$$= 60 \text{ V}$$

Now the surge current is found as

$$I_{surge} = \frac{V_{2(pk)}}{R_w + R_B}$$
$$= \frac{60 \text{ V}_{pk}}{5.8 \text{ Ω}}$$
$$= 10.34 \text{ A}$$

If the circuit in Figure 3.27 has values of $R_w = 0.5\ \Omega$ and $R_B = 8\ \Omega$, what is the initial surge current for the circuit?

The surge current value found in Example 3.13 may seem to be extremely high, but it probably won't cause any problem. As you remember from Chapter 2, most rectifier diodes have extremely high surge current ratings. For example, the 1N4001 can withstand a surge current of 30 A for one cycle. For this diode, the surge current value found in Example 3.13 (10.34 A) would not be a problem.

When the amount of surge current produced by a circuit is more than the rectifier diodes can handle, the problem can be resolved by using a series *current-limiting* resistor, shown in Figure 3.30 as R_{surge}. The current-limiting resistor will usually be a small-value resistor. The resistor does help lower the surge current, but also lowers the output voltage from the circuit. This is because the output voltage from the rectifier is divided between the series resistor and the load resistance.

Limiting surge current.

Surge current can also be limited by using a smaller value of filter capacitor. Smaller-value capacitors have higher reactance values and will charge in a shorter period of time. The relationship between capacitance, current, and time is as follows:

$$C = \frac{I(t)}{V} \qquad \textbf{(3.23)}$$

where C = the capacitance, in farads
$\quad I$ = the dc (average) charge/discharge current
$\quad t$ = the charge/discharge time
$\quad V$ = the change in capacitor voltage during charge/discharge

Equation (3.23) can be rearranged to produce the following:

$$t = \frac{CV}{I} \qquad \textbf{(3.24)}$$

As equation (3.24) shows, the time required for a capacitor to charge to a specified value is directly proportional to the value of the capacitor. Thus a lower-value capacitor will charge faster and eliminate the surge current faster. Generally, if you stick to the 1000-μF guideline for filter capacitors, the surge current will not last long enough to damage the diodes.

Filter Output Voltages

23

Ideally, the filter capacitor will charge to the peak output voltage from the rectifier and will not discharge at all. However, as previous illustrations have shown, this

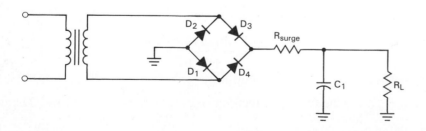

FIGURE 3.30

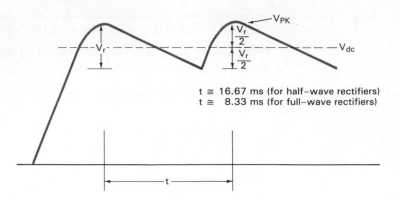

$$t \cong 16.67 \text{ ms (for half–wave rectifiers)}$$
$$t \cong 8.33 \text{ ms (for full–wave rectifiers)}$$

FIGURE 3.31

is not the case. Figure 3.31 relates the various voltage values contained in the filter output. The dc output voltage (Vdc) is shown to equal the peak voltage (V_{pk}) minus one-half of the peak-to-peak value of the ripple voltage. By formula,

$$Vdc = V_{pk} - \frac{V_r}{2} \qquad (3.25)$$

where V_{pk} = the peak rectifier output voltage
V_r = the peak-to-peak value of ripple voltage

The ripple voltage from the filter can be found using a variation of equation (3.23), as follows:

$$V_r = \frac{I_L t}{C} \qquad (3.26)$$

where I_L = dc load current
t = time between charging peaks
C = capacitance, in farads

The value of t is shown in Figure 3.31. Since this value is much lower for full-wave and bridge rectifiers than it is for half-wave rectifiers, the half-wave rectifier will have much higher ripple voltages than the other two.

━━━━━━━━━━━━━━━━━━ **EXAMPLE 3.14** ━━━━━━━━━━━━━━━━━━

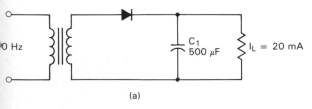

(a)

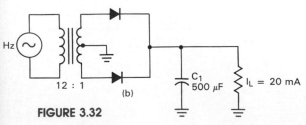

12 : 1

(b)

FIGURE 3.32

Determine the ripple output from each of the circuits shown in Figure 3.32. Assume that the load current is 20 mA for each circuit.

Solution: The half-wave rectifier (Figure 3.32) has a time of 16.67 ms between charging pulses. Therefore,

$$V_r = \frac{I_L t}{C}$$
$$= \frac{(20 \text{ mA})(16.67 \text{ ms})}{500 \text{ μF}} \qquad \Big| \quad t = \frac{1}{f}$$
$$= 666.8 \text{ mV}_{pp}$$

For the full-wave rectifier (Figure 3.32b), $t = 8.33$ ms. Therefore,

$$V_r = \frac{I_L t}{C}$$

$$= \frac{(20 \text{ mA})(8.33 \text{ ms})}{500 \text{ }\mu\text{F}}$$

$$= 333.2 \text{ mV}_{pp}$$

As Example 3.14 shows, the full-wave rectifier has exactly one-half the ripple output produced by the half-wave rectifier. This is due to the shortened time period between capacitor charging pulses.

You were told earlier in the chapter that the primary advantage of using full-wave rectifiers in place of half-wave rectifiers would be seen when we discussed filtering. As Example 3.14 has shown, a full-wave rectifier will have half the ripple output voltage of a comparable half-wave rectifier. Since our goal is to have a steady dc voltage that has as little ripple voltage as possible, the full-wave rectifier gets us much closer to our goal than does the half-wave rectifier.

Why full-wave rectifiers are preferred over half-wave rectifiers.

The Effects of Filtering on Rectifier Analysis

Refer to Figure 3.31. As the figure shows, you can find the value of Vdc for a filtered rectifier by subtracting $V_r/2$ from the peak rectifier output voltage. There is only one problem. To determine the value of V_r, we have to know the value of dc load current (I_L). And, to determine the value of I_L, we have to know the value of Vdc.

25

What we have here is referred to as a *loop*. We have to know the value of Vdc to find the value of V_r, but we have to know the value of V_r to find the value of Vdc.

So, what is the solution? If you look closely at Figure 3.31, you'll see that the final value of Vdc is *very close* to the value of V_{pk}. Thus, we can start our mathematical analysis of the circuit by making the following assumption:

$$\boxed{Vdc \cong V_{out(pk)}}$$

Then, using the *assumed* value of Vdc, we can calculate the approximate value of I_L. From there, we can find the approximate value of V_r and an even closer value of Vdc. As the following example illustrates, the final value of Vdc will prove to be very close to the value that was originally assumed.

EXAMPLE 3.15

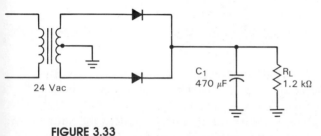

FIGURE 3.33

Determine the value of Vdc for the circuit shown in Figure 3.33.

Solution: The transformer is rated at 24 Vac, so the peak secondary voltage ($V_{2,pk}$) is 33.95 V. Using this value of $V_{2(pk)}$, the peak output voltage is found as

$$V_{out(pk)} = \frac{V_{2(pk)}}{2} - 0.7 \text{ V}$$

$$= 16.28 \text{ V}_{pk}$$

Now we will *assume* that

$$V_{dc} = V_{out(pk)} = 16.28 \text{ V}$$

Using this assumed value of Vdc, the dc load current is found as

$$I_L = \frac{Vdc}{R_L}$$
$$= 13.6 \text{ mA}$$

Now the value of the ripple voltage is found as

$$V_r = \frac{I_L t}{C}$$
$$= \frac{(13.6 \text{ mA})(8.33 \text{ ms})}{470 \text{ μF}}$$
$$= 241 \text{ mV}_{pp}$$

Finally, the calculated value of V dc is found as

$$V_{dc} = V_{pk} - \frac{V_r}{2}$$
$$= 16.3 \text{ V}_{pk} - 120.5 \text{ mV}$$
$$= 16.18 \text{ V dc}$$

PRACTICE PROBLEM 3–15

A full-wave rectifier has a 24 Vac transformer, a 330 μF filter capacitor, and a 1.5 kΩ load. Calculate the values of V_r and Vdc for the circuit.

Hey . . . that's not bad! We only had a percent of error between our assumed value of Vdc and our calculated value of Vdc of 0.62%. This shows that the initial assumption of $Vdc = V_{out(pk)}$ for the circuit is valid.

Filter Effects on Diode PIV

✔ 26

For the full-wave and bridge rectifiers, the filter will not really have any significant effect on the peak inverse voltage across each diode. However, the half-wave rectifier diode will have a peak inverse voltage equal to twice the secondary voltage. By formula,

$$\boxed{\text{PIV} = 2V_2} \qquad \text{(half-wave, filtered)} \qquad \textbf{(3.27)}$$

This point is illustrated in Figure 3.34. When the diode is conducting, C_1 will charge to V_2 (ignoring the voltage drop across the diode). This is shown in Figure 3.34a. When the polarity of the secondary voltage reverses (Figure 3.34b), D_1 turns off. For an instant, the reverse voltage across the diode is equal to the sum of the secondary voltage and the capacitor voltage. Since these two voltages are equal, the peak inverse voltage across the diode equals $2V_2$. Note that this value

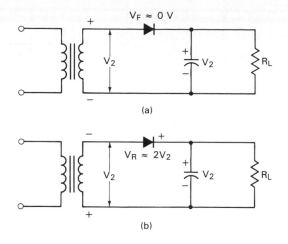

(a)

(b)

FIGURE 3.34

The effects of filtering on diode PIV.

will be reduced as the capacitor discharges through the load resistance. However, the diode must have a breakdown voltage rating that is greater than $2V_2$.

Other Filter Types

There are several other types of filters, as shown in Figure 3.35. Each filter type shown makes use of the reactance properties of the capacitors and/or inductors. In each filter, the series impedance is designed to be very high at the ripple frequency, while the shunt impedance is designed to be very low. Therefore, whatever ripple is not dropped by the series component is greatly reduced by the shunt component.

   The inductive filters provide an added benefit. Since an inductor opposes a rapid change in current, inductive filters do not have the surge current problems that the capacitive filters do. At the same time, capacitive filters are more commonly used because of cost and size factors.

Inductive filters provide protection against surge current problems.

One Final Note

The use of a filter will greatly reduce the variations in the output from a rectifier. At the same time, the *ideal* power supply would provide a stable dc output voltage that had no ripple voltage at all. While there will always be *some* ripple voltage at the output of a power supply, the use of a voltage regulator will reduce the filter output ripple even further. This point will be demonstrated in Section 3.6.

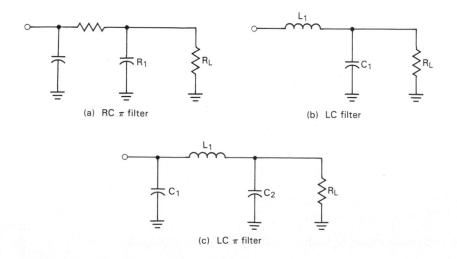

(a) RC π filter

(b) LC filter

(c) LC π filter

FIGURE 3.35

Some other filter circuits.

1. What are filters used for? [Objective 20]

2. What is *ripple voltage*? [Objective 20]

3. Describe the operation of the basic capacitive filter in terms of *charge/discharge time constants*. [Objective 21]

4. What limits the value of a filter capacitor? [Objective 21]

5. What causes *surge current* in a power supply? [Objective 22]

6. What are the *two* devices shown in this section that limit surge current? How do they limit I_{surge}? [Objective 22]

7. Describe the relationship between V_r and V_{dc} for a filtered rectifier. [Objective 23]

8. Why are full-wave rectifiers preferred over half-wave rectifiers? [Objective 24]

9. Describe the process used to calculate the values of V_r and V_{dc} for a filtered rectifier. [Objective 25]

10. Explain the effect that filtering has on the PIV rating of a half-wave rectifier. [Objective 26]

3.6
Zener Voltage Regulators

Voltage regulator. A circuit that maintains a constant load voltage despite variations in load current demand.

 27

The final circuit in the basic power supply is the *voltage regulator* (see Figure 3.1). There are many types of voltage regulators. Many of these circuits contain a number of transistors or an integrated-circuit (IC) voltage regulator. Some of these circuits are discussed later in the text. At this point, we are going to concentrate on the simple *zener regulator*. This regulator is shown in Figure 3.36a. You should recall that a zener diode operated in the reverse breakdown region will have a constant voltage across it as long as zener current remains between the knee current (I_{ZK}) and the maximum current rating (I_{ZM}). These operating characteristics are illustrated in Figure 3.36b.

Since the load resistance is in parallel with the zener diode, load voltage will remain constant as long as the zener voltage remains constant. If the zener current leaves the allowable range, zener voltage will change and so will the load voltage. Therefore, *the key to keeping the load voltage constant is to keep the zener current within its specified range.*

Total Circuit Current

28

For a zener regulator like the one shown in Figure 3.36a, the total current that is drawn from the source is found as

$$I_T = \frac{V_S - V_Z}{R_S} \tag{3.28}$$

where I_T = the total current drawn from the filtered rectifier
V_S = the source voltage
V_Z = the nominal zener voltage
R_S = the series resistor

The following example illustrates the use of this equation.

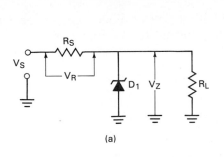

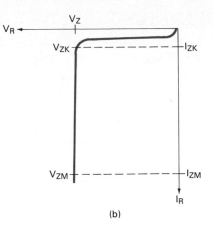

FIGURE 3.36
The basic zener voltage regulator.

(a)

(b)

════════════════ **EXAMPLE 3.16** ════════════════

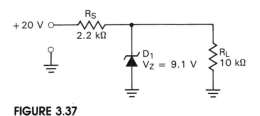

FIGURE 3.37

Determine the total circuit current for Figure 3.37.

Solution: With an applied voltage of 20 V and a zener voltage of 9.1 V, the total circuit current is found as

$$I_T = \frac{V_S - V_Z}{R_S}$$

$$= \frac{10.9 \text{ V}}{2.2 \text{ k}\Omega}$$

$$= 4.95 \text{ mA}$$

PRACTICE PROBLEM 3–16

A circuit like the one shown in Figure 3.37 has a 15 Vdc source, a 1.5 kΩ series resistor, and a 12 V zener diode. What is the value of I_T for the circuit?

The Basis of Equation (3.28)

In the regulator shown in Figure 3.36a, the total circuit current flows through the series resistor, R_S. Therefore,

$$I_T = I_S$$

According to Ohm's law, we can find the total circuit current as

$$I_T = \frac{V_R}{R_S}$$

Since V_R is equal to the difference between the source voltage (V_S) and the zener voltage (V_Z), the above equation can be written as

$$I_T = \frac{V_S - V_Z}{R_S} \tag{3.28}$$

Load Current

Since the load resistance (R_L) is in parallel with the zener diode, the voltage across the load will always equal V_Z. Thus, the load current can be found as

$$I_L = \frac{V_Z}{R_L} \qquad (3.29)$$

The following example illustrates the use of this equation.

EXAMPLE 3.17

Determine the load current for the circuit shown in Figure 3.37.

Solution: With a 9.1 V zener and a 10 kΩ resistor, the total load current is found as

$$\begin{aligned}
I_L &= \frac{V_Z}{R_L} \\
&= \frac{9.1 \text{ V}}{10 \text{ k}\Omega} \\
&= 910 \text{ } \mu\text{A}
\end{aligned}$$

PRACTICE PROBLEM 3–17

The zener voltage regulator described in Practice Problem 3–16 has a 12 kΩ load resistance. What is the value of load current in the circuit?

Zener Current

Since the zener diode and the load resistance are in parallel, the sum of I_Z and I_L will equal the total circuit current. Thus,

$$I_Z = I_T - I_L \qquad (3.30)$$

where I_Z = the total zener current
I_T = the total circuit current, found using equation (3.28)
I_L = the load current, found using equation (3.29)

The following example illustrates the procedure for determining the value of I_Z.

EXAMPLE 3.18

Determine the load current and zener current for the circuit shown in Figure 3.37.

Solution: The load current can be found using the value of V_Z for load voltage since the load and the zener diode are in parallel. Therefore,

$$I_L = \frac{V_Z}{R_L}$$
$$= \frac{9.1 \text{ V}}{10 \text{ k}\Omega}$$
$$= 910 \text{ } \mu\text{A}$$

In Example 3.16, we determined I_T to be 4.95 mA. Therefore,

$$I_Z = I_T - I_L$$
$$= 4.95 \text{ mA} - 910 \text{ } \mu\text{A}$$
$$= 4.04 \text{ mA}$$

PRACTICE PROBLEM 3–18

Determine the value of zener current for the circuit described in Practice Problems 3–16 and 3–17.

Load Variations

It was stated earlier that the zener regulator will maintain a constant output voltage as long as zener current stays between I_{ZK} and I_{ZM}. The possible effects of load resistance variations on zener current are illustrated in Figure 3.38. If load resistance is reduced to 0 Ω (Figure 3.38a), all circuit current will pass through the load. In this case, zener current will drop below I_{ZK}, and the diode will stop regulating the output voltage. If the load resistance is increased to ∞ Ω (Figure 3.38b), total circuit current will pass through the zener diode. In this case, the diode will be destroyed unless R_S is large enough to prevent I_{ZM} from being reached.

What are the practical limits on the value of R_L? The minimum value of R_L is determined by the zener voltage and the value of I_{ZK}. To maintain zener regulation, the minimum zener current must be equal to I_{ZK}. Therefore,

$$\boxed{I_{L(\text{max})} = I_T - I_{ZK}}$$

Since $I_{L(\text{max})}$ occurs when the load resistance is at a minimum,

$$\boxed{R_{L(\text{min})} = \frac{V_Z}{I_{L(\text{max})}}} \qquad \textbf{(3.31)}$$

29

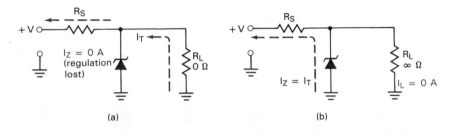

(a)   (b)

FIGURE 3.38

The effects of load variation on the operation of a zener voltage regulator.

The zener diode in Figure 3.39 has values of $I_{ZK} = 3$ mA and $I_{ZM} = 100$ mA. What is the minimum allowable value of R_L?

Solution: First, the total circuit current is found as

$$I_T = \frac{V_S - V_Z}{R_S}$$
$$= \frac{20 \text{ V} - 3.3 \text{ V}}{1 \text{ k}\Omega}$$
$$= 16.7 \text{ mA}$$

Now $I_{L(\text{max})}$ is found as

$$I_{L(\text{max})} = I_T - I_{ZK}$$
$$= 16.7 \text{ mA} - 3 \text{ mA}$$
$$= 13.7 \text{ mA}$$

and

$$R_{L(\text{min})} = \frac{V_Z}{I_{L(\text{max})}}$$
$$= \frac{3.3 \text{ V}}{13.7 \text{ mA}}$$
$$= 241 \ \Omega$$

PRACTICE PROBLEM 3–19

If the diode shown in Figure 3.39 has values of $V_Z = 5.1$ V and $I_{ZK} = 5$ mA, what is the minimum allowable value of R_L for the circuit?

A minimum load resistance that draws maximum current is called a **full load.**

For the circuit shown in Figure 3.39 the zener will maintain a constant output voltage as long as R_L does not go below 241 Ω. If R_L goes below this value, I_L will increase above its maximum allowable value, and I_Z will go below I_{ZK}. The diode will then stop regulating the output voltage. Note that a load that draws *maximum* current is referred to as a *full load*.

Load Regulation

Load regulation. The ability of a regulator to maintain a constant output voltage despite changes in load current demand.

You have been shown that the zener regulator can maintain a constant load voltage for a range of load current values. The ability of a regulator to maintain a constant load voltage under varying load current demands is called *load regulation*. Load regulation will be discussed in detail in Chapter 20.

FIGURE 3.39

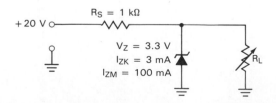

CHAP. 3   Common Diode Applications: The Basic Power Supply

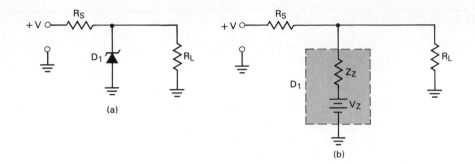

FIGURE 3.40

Zener Reduction of Ripple Voltage

The zener regulator provides an added bonus: It reduces the amount of ripple voltage that is present at the filter output. The effect that the regulator has on ripple voltage is easy to understand when you consider the equivalent circuit of the diode. The basic zener regulator and its equivalent circuit are shown in Figure 3.40. You may recall from Chapter 2 that the zener impedance, Z_Z, is an ac value. In other words, it must be considered in any analysis involving a *change* in current or voltage. Since ripple voltage is a changing quantity, it will be affected by Z_Z.

To the ripple waveform, there is a voltage divider present in the regulator. This voltage divider is made up of the series resistance (R_S) and Z_Z. The ripple output from the regulator can be found as

$$V_{r(out)} = \frac{Z_Z}{Z_Z + R_S} V_r \qquad (3.32)$$

where $V_{r(out)}$ = the ripple present at the regulator output
$\quad Z_Z$ = the zener impedance, obtained from the diode spec sheet
$\quad R_S$ = the regulator series resistance
$\quad V_r$ = the peak-to-peak ripple voltage present at the regulator input

════ EXAMPLE 3.20 ════

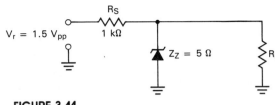

FIGURE 3.41

The filtered output from a full-wave rectifier has a peak-to-peak ripple voltage of 1.5 V. If this signal is applied to the circuit shown in Figure 3.41 what will the ripple at the load equal?

Solution: The zener has a Z_Z of 5 Ω, the circuit R_S is 1 kΩ, and the input ripple is 1.5 V_{PP}. Therefore,

$$
\begin{aligned}
V_{r(out)} &= \frac{Z_Z}{Z_Z + R_S} V_r \\
&= \frac{5\ \Omega}{5\ \Omega + 1\ k\Omega} (1.5\ V_{PP}) \\
&= (0.0049)(1.5\ V) \\
&= 7.46\ mV_{PP}
\end{aligned}
$$

A zener regulator has a 910 Ω series resistance and a zener impedance that equals 25 Ω. If the input ripple to the circuit is 1.2 V_{PP}, what is the amount of load ripple?

As the example shows, a voltage regulator will almost completely eliminate any ripple that is present at the output of a power supply filter.

*Circuit Development

When a zener regulator is designed, a value of series resistance must be chosen to fulfill the following requirements:

1. The value of R_S must be *small enough* to ensure that the value of zener current will not drop below I_{ZK} under normal circumstances.
2. The value of R_S must be *large enough* to ensure that the value of zener current will *never* exceed I_{ZM}.

Refer to Figure 3.38b. In this illustration, you were shown that the total regulator current will flow through the zener diode if the load resistance goes open circuit (infinite ohms). Since an open load is always a possibility, the value of R_S must be made large enough to prevent I_T from exceeding I_{ZM} if the load opens. This requirement is fulfilled by limiting the minimum value of R_S as follows:

$$R_{S(min)} = \frac{V_S - V_Z}{I_{ZM}} \qquad (3.33)$$

where $R_{S(min)}$ = the minimum allowable value of R_S
V_S = the source voltage
V_Z = the zener voltage
I_{ZM} = the maximum allowable zener current, as provided on the diode's spec sheet

The following example illustrates the use of this equation.

* Optional.

═══ EXAMPLE 3.21 ═══

Determine the minimum acceptable value of R_S for the circuit shown in Figure 3.42.

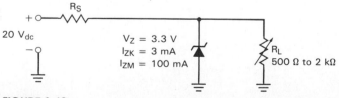

$V_Z = 3.3$ V
$I_{ZK} = 3$ mA
$I_{ZM} = 100$ mA

20 V_{dc}

R_L
500 Ω to 2 kΩ

FIGURE 3.42

Solution: The diode has an I_{ZM} rating of 100 mA. Therefore,

$$R_{S(min)} = \frac{V_S - V_Z}{I_{ZM}}$$

$$= \frac{20\ V - 3.3\ V}{100\ mA}$$

$$= 167\ \Omega$$

PRACTICE PROBLEM 3–21

You need to design a zener regulator that has a 12 V zener with an I_{ZM} rating of 250 mA. If the source voltage is 48 Vdc, what is the minimum allowable value of R_S?

As long as R_S in Example 3.21 is greater than 167 Ω, I_{ZM} cannot be reached in the circuit, even if the load goes open circuit.

The maximum value of R_S is limited by the total current requirements in the circuit. R_S must be small enough to allow I_{ZK} to flow through the zener diode when the load is drawing *maximum* current. For design purposes, the maximum allowable value of R_S is found as

$$R_{S(max)} = \frac{V_S - V_Z}{I_{L(max)} + I_{ZK}} \qquad (3.34)$$

where $I_{L(max)}$ = the *maximum* current that will be drawn by the load under normal circumstances.

The following example illustrates the process for determining the maximum value of R_S.

EXAMPLE 3.22

Determine the maximum allowable value of R_S for the circuit in Figure 3.42.

Solution: First, we need to determine the *maximum* value of load current. This value is found as

$$I_{L(max)} = \frac{V_Z}{R_{L(min)}}$$

$$= \frac{3.3\ V}{500\ \Omega}$$

$$= 6.6\ mA$$

Now the value of $R_{S(\max)}$ is found as

$$R_{S(\max)} = \frac{V_S - V_Z}{I_{L(\max)} + I_{ZK}}$$

$$= \frac{20 \text{ V} - 3.3 \text{ V}}{6.6 \text{ mA} + 3 \text{ mA}}$$

$$= \frac{16.7 \text{ V}}{9.3 \text{ mA}}$$

$$= 1796 \text{ }\Omega$$

PRACTICE PROBLEM 3–22

The regulator described in Practice Problem 3–21 has ratings of $I_{ZK} = 5$ mA and $R_L = 330$ Ω (minimum). What is the maximum allowable value of R_S?

In Example 3.21, we calculated a value of $R_{S(\min)} = 167$ Ω. Then, in Example 3.22, we calculated a value of $R_{S(\max)} = 1796$ Ω. Thus, we could use any standard-value resistor that falls between 167 and 1796 Ω as our series resistance. For optimum operation, the standard-value resistor chosen would have a value that falls approximately halfway between the calculated limits. In this case, a 1-kΩ resistor would be nearly perfect.

SECTION REVIEW

1. Why must zener current in a voltage regulator be kept within its specified limits? [Objective 27]
2. What is a *full load*? [Objective 29]
3. How does a zener voltage regulator reduce the ripple voltage from a filter? [Objective 30]

3.7
Putting It All Together

We have discussed the operation of rectifiers, filters, and zener regulators in detail. Now it is time to put them all together into a basic working power supply. In this section, we will analyze the basic power supply shown in Figure 3.43.

The power supply shown contains a bridge rectifier, a capacitive filter, and a zener voltage regulator. The bridge rectifier converts the transformer secondary ac voltage into a *positive* pulsating dc voltage. This pulsating dc voltage is applied to the capacitive filter, which reduces the variations in the rectifier dc output voltage. Finally, the zener voltage regulator performs two functions:

1. It reduces the ripple (variations) in the filter output voltage.
2. It ensures that the dc output voltage from the power supply (Vdc) will remain relatively constant despite variations in load current demand. Thus, the combina-

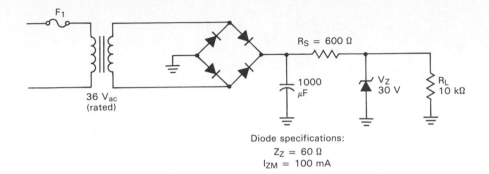

Diode specifications:
$Z_Z = 60 \ \Omega$
$I_{ZM} = 100 \ \text{mA}$

FIGURE 3.43

tion of the three circuits has converted an ac line voltage to a steady dc supply voltage and has ensured that this dc supply voltage will remain constant when load current demands change.

Our goal in analyzing a basic power supply is to determine the values of dc output voltage (Vdc), ripple voltage (V_r), and load current (I_L). The procedure for determining these values is as follows:

1. Using the required steps, determine the value of $V_{2(pk)}$.
2. Determine the value of V_{pk} at the rectifier output.
3. Assuming that Vdc is equal to the value obtained in step 2, determine the total current through the series resistor. This current value will be used as I_L when calculating the value of ripple voltage.
4. Determine the value of ripple voltage from the filter.
5. Find Vdc at the output. This value will equal the V_Z rating of the zener diode under normal circumstances.
6. Using the rated value of Z_Z, approximate the final ripple output voltage.
7. Using V_Z and R_L, determine the value of load current.

The following example illustrates the process for analyzing the schematic of a basic power supply.

===== **EXAMPLE 3.23** =====

Determine the values of Vdc, $V_{r(out)}$, and I_L for the power supply shown in Figure 3.43.

Solution: First, we must convert the Vac rating of the transformer to a peak value, as follows:

$$V_{2(pk)} = \frac{36 \ V\text{ac}}{0.707}$$
$$= 51 \ V_{pk}$$

Now, we determine the value of peak voltage at the filter.

$$V_{out(pk)} = V_{2(pk)} - 1.4 \ V$$
$$= 49.6 \ V_{pk}$$

Next, we assume that we have a dc source voltage of 49.6 V. Using 49.6 V as our value for V_S, we determine the value of the current through the series resistor as follows:

$$I_R = \frac{V_S - V_Z}{R_S}$$

$$= \frac{49.6 \text{ V} - 30 \text{ V}}{600 \text{ } \Omega}$$

$$= 32.66 \text{ mA}$$

We now use the value of $I_R = 32.66$ mA to determine the value of V_r.

$$V_r = \frac{I_R t}{C}$$

$$= \frac{(32.66 \text{ mA})(8.33 \text{ ms})}{1000 \text{ } \mu\text{F}}$$

$$= 272.14 \text{ mV}_{\text{PP}}$$

As was stated earlier, the dc output voltage from the power supply will equal the value of V_Z. By formula,

$$V_{\text{dc}} = V_Z = 30 \text{ V}_{\text{dc}}$$

and

$$I_L = \frac{V_Z}{R_L}$$

$$= \frac{30 \text{ V}}{10 \text{ k}\Omega}$$

$$= 3 \text{ mA}$$

Finally, the value of $V_{r(\text{out})}$ is found as

$$V_{r(\text{out})} = \frac{Z_Z}{Z_Z + R_S} V_r$$

$$= \frac{60 \text{ } \Omega}{660 \text{ } \Omega} (272.14 \text{ mV}_{\text{PP}})$$

$$= 24.74 \text{ mV}_{\text{PP}}$$

PRACTICE PROBLEM 3–23

A power supply like the one in Figure 3.43 has the following values: $V_2 = 24$ Vac (rated), $C = 470$ μF, $R_S = 500$ Ω, $V_Z = 10$ V, $Z_Z = 20$ Ω, and $R_L = 5.1$ kΩ. Determine the values of Vdc, I_L, and $V_{r(\text{out})}$ for the circuit.

Worst-case Analysis: A Practical Consideration

✔ 33

Worst-case analysis. A mathematical analysis that gives you the absolute limits of normal circuit operation.

The analysis in Example 3.23 is based on the assumption that every component in the power supply is exactly equal to its rated value. It assumes, for example, that the output from the transformer is *exactly* 36 Vac. It also assumes that the voltage drop across each diode in the rectifier is *exactly* 0.7 V, and so on.

Since power supply components *do* have tolerances, we are usually concerned with *worst-case* voltage and current values, rather than *exact* values. *Worst-case* circuit values are *limits*. They tell us the *maximum* or *minimum* value that a measured voltage or current will have under normal circumstances. For example, let's say that a given power supply had worst-case values of Vdc = 50 V, $I_L = 30$ mA (maximum), and $V_{r(\text{out})} = 550 \text{ mV}_{\text{PP}}$ (maximum). These worst-case values would tell us:

1. The dc output voltage should always be 50 V.
2. The dc load current should always be 30 mA or less.
3. The ripple voltage at the output should always be 550 mV$_{\text{PP}}$ or less.

The actual measured values of I_L and $V_{r(\text{out})}$ may be *less than* their worst-case values, but *they would never be greater than their worst-case values under normal circumstances*. The same holds true for the value of V_{dc}, which would always equal its rated value under normal circumstances.

Knowing the worst-case voltage and current values for a power supply is important because they can be used to determine whether or not the circuit is faulty. For example, let's say that the power supply described above has a measured output ripple voltage that is 1.2 V$_{\text{PP}}$. Since this measured voltage is *twice* its worst-case value, we know that there is a fault in the power supply.

The worst-case analysis of a basic power supply is based on three assumptions:

1. The transformer output voltage is at its *highest* limit (within tolerance).
2. The value of V_r is equal to the difference between $V_{\text{out(pk)}}$ and V_Z.
3. The power supply works; therefore, $V_{\text{dc}} = V_Z$.

The first assumption gives us the *maximum possible* value of $V_{\text{out(pk)}}$, that is, the maximum peak voltage that will appear across the filter capacitor.

The second assumption is really tied in with the third. Since we are assuming that the power supply works, it is safe to assume that the zener diode is *always* conducting (as it should be). The zener will conduct only if the voltage on the capacitor side of R_S is *more positive* than V_Z. Thus, the capacitor can never discharge to any value that is less than V_Z. Since the maximum charge on the capacitor is equal to $V_{\text{out(pk)}}$, the change in capacitor voltage (ripple) can never be greater than the difference between $V_{\text{out(pk)}}$ and V_Z. By formula,

$$\boxed{V_r = V_{\text{out(pk)}} - V_Z} \qquad \text{(worst case)} \qquad \textbf{(3.35)}$$

Knowing the worst-case value of V_r allows us to calculate the worst-case value of $V_{r(\text{out})}$. Any *measured* value of $V_{r(\text{out})}$ would (of course) be *lower than or equal to* this calculated value under normal operating conditions.

In the following example, we will determine the worst-case value of $V_{r(\text{out})}$ for the power supply we analyzed in Example 3.23. As you will see, the worst-case value will be significantly higher than the value we calculated earlier.

EXAMPLE 3.24

Determine the worst-case value of $V_{r(\text{out})}$ for the power supply in Figure 3.43. Assume that the transformer output tolerance is $\pm 20\%$.

Solution: Assuming that the transformer in the power supply has an output that is actually 20% above its rated value, the value of V_2 is found as

$$V_2 = 36 \text{ V} + 7.2 \text{ V} = 43.2 \text{ V}$$

The 7.2 V is 20% of 36 V. The value of $V_{2(pk)}$ is now found as

$$V_{2(pk)} = \frac{43.2 \text{ V}}{0.707}$$
$$= 61 \text{ V}_{pk}$$

The value of $V_{out(pk)}$ is now found as

$$V_{out(pk)} = V_{2(pk)} - 1.4 \text{ V}$$
$$= 59.6 \text{ V}_{pk}$$

The worst-case value of V_r is now found as

$$V_r = V_{out(pk)} - V_Z$$
$$= 59.6 \text{ V}_{pk} - 30 \text{ V}$$
$$= 29.6 \text{ V}_{PP} \quad \text{(worst case)}$$

Finally, the worst-case value of $V_{r(out)}$ is found as

$$V_{r(out)} = \frac{Z_Z}{Z_Z + R_S} V_r$$
$$= \frac{60 \text{ }\Omega}{660 \text{ }\Omega} (29.6 \text{ V}_{PP})$$
$$= 2.69 \text{ V}_{PP}$$

Thus, *under normal circumstances*, the ripple voltage at the output of the power supply could never be greater than 2.69 V_{PP}. If the value of $V_{r(out)}$ is greater than this value, a problem exists in the circuit.

PRACTICE PROBLEM 3–24

Determine the worst-case value of $V_{r(out)}$ for the circuit described in Practice Problem 3–23. Assume that the transformer tolerance is ±10%.

The worst-case value of V_{dc} is always assumed to be equal to V_Z. The worst-case value of I_L is calculated using the *minimum* value of load resistance, as was the case in Example 3.22.

The worst-case value of $V_{r(out)}$ in Example 3.24 may seem to be extremely high, but you have to remember two things:

1. We assumed the *maximum possible* output from the transformer.
2. We assumed that the value of V_r was extremely high.

In practice, the value of $V_{r(out)}$ from this supply would probably be closer to the value from Example 3.23 (24.74 mV$_{PP}$) than it would to 2.75 V$_{PP}$. However, in worst-case analysis, we are only trying to determine the absolute maximum (or minimum) values of voltage and current. These values can then be used to determine whether or not a circuit is operating properly.

CHAP. 3   Common Diode Applications: The Basic Power Supply

1. What is the goal of analyzing a power supply? In other words, what specific values are we looking for? [Objective 32]
2. What are *worst-case* values? [Objective 34]
3. Why are worst-case analyses important? [Objective 34]
4. What assumptions are made in the worst-case analysis of a basic power supply? [Objective 34]

3.8
Power Supply Troubleshooting

Power supply faults may occur in the transformer, the rectifier diodes, the filter, or the voltage regulator. When a fault develops in the power supply, the type of symptom and a few simple tests will tell you where the fault is located. In this section we will take a look at the common power supply fault symptoms and the tests used to isolate the faulty component.

The Primary Fuse

Every power supply contains a fuse that is located in the primary circuit of the supply. This fuse is normally placed in series with the primary of the transformer, as is shown in Figure 3.44.

If a fault in the power supply causes an extremely high amount of current to be drawn from the transformer, the primary fuse will *blow* (open). When this happens, the ac line voltage from the wall outlet is prevented from reaching the power supply. This will stop the excessive current, protecting the power supply and the technician who is working on it.

Many power supplies will use *slow-blow* fuses. These fuses are designed to handle the normal surge current that can occur in a power supply without opening. At the same time, if the current demand on a slow-blow fuse is too high for a long enough period of time (usually around 1 second), the slow-blow fuse will also open to protect the circuit.

When a power supply fuse needs to be replaced, *you must use the same value of fuse.* Never, under any circumstances, use a fuse that has a *higher* current rating than the one you are replacing. Using a higher-value fuse will defeat the purpose of the fuse (protecting the circuit from excessive current) and may even create a fire hazard.

As you will see throughout this section, many power supply faults will cause the primary fuse to blow, while many others will not. When a fault causes the primary fuse to blow, simply replace the fuse after the fault is diagnosed and corrected.

A Word of Caution: When you begin to work in the field, you may meet some technicians who will use a wire to "defeat" (bypass) the primary fuse in a faulty power supply. This allows the power supply to continue to operate under circumstances that would normally cause a system shutdown. *Never defeat the fuse in a power supply.* You not only risk starting a fire, but you could also be injured or killed if you should accidentally come into contact with any of the current-carrying components in the circuit!

Transformer Faults

The transformer in a power supply can develop one of several possible faults:

1. A shorted primary or secondary winding
2. An open primary or secondary winding
3. A short between the primary or secondary winding and the transformer frame

Transformer fault symptoms.

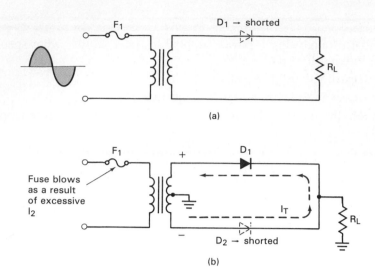

FIGURE 3.44

The effects of a shorted
rectifier diode.

In most cases a shorted primary or secondary winding will cause the fuse to blow. If the fuse does not blow, the dc output from the power supply will be extremely low and the transformer itself will be extremely hot.

When the primary or the secondary winding of the transformer opens, the output from the power supply will drop to zero. In this case the primary fuse will not blow. If you believe that either transformer winding is open, a simple resistance check of the winding will verify your suspicions. If either winding reads a very high resistance, the winding is open.

If either winding shorts to the transformer casing, the initial result will be a blown fuse. This fault is isolated by checking the resistance from the winding leads to the transformer casing. A low resistance measurement indicates that a winding-to-case short circuit exists.

With any of the problems above, the repair procedure is simple. You must replace the transformer.

Rectifier Faults

A Practical Consideration: If one diode in a rectifier goes bad, it will usually destroy one or more of the other rectifier diodes in the process. If you determine the fault to be in the rectifier of a power supply, you should go ahead and replace *all* the diodes.

The *half-wave rectifier* is the easiest rectifier to troubleshoot. If the diode in the rectifier *shorts*, the output from the rectifier will be a sine wave that is identical to V_2. This is illustrated in Figure 3.44a. Since the diode is shorted, it acts as a straight piece of wire. Therefore, neither half-cycle of the input (V_2) is eliminated. The output will be a replica of V_2.

If the diode in the half-wave rectifier opens, the output from the circuit will drop to zero. If the rectifier diode is either shorted or open, simply replace the diode.

In a *full-wave rectifier*, a shorted diode will cause the power supply fuse to blow. Figure 3.44b shows the effects of a shorted D_2 on circuit operation. Note that the diode has been replaced by a straight wire. When D_1 is forward biased by V_2, the transformer secondary will be shorted through D_1. This will cause excessive current to flow in the secondary, causing the primary fuse to blow.

If you suspect that either diode in a full-wave rectifier has shorted, simply measure the forward and reverse resistance of both diodes. In each case, the diode should have a very large reverse resistance and a very small forward resistance. If the reverse resistance of either diode is extremely low, replace the diode.

If a diode in a full-wave rectifier opens, the output from the rectifier will resemble the output from a half-wave rectifier. In this case, measure the diode resistances and look for a large value of forward resistance. When you have such a reading, replace that diode.

The symptoms for shorted and open diodes in the *bridge rectifier* are the same as those for full-wave rectifiers. In the case of the bridge rectifier, you simply have more diodes that need to be tested.

Filter Faults

When the *filter capacitor shorts*, the primary fuse will blow. The reason for this is illustrated in Figure 3.45. When the filter capacitor shorts, it shorts out the load resistance. This has the same effect as wiring the two sides of the bridge together (Figure 3.45a). If you trace from the high side of the bridge to the low side, you will see that the only resistance across the secondary of the transformer is the forward resistance of the two *on* diodes. This effectively shorts out the transformer secondary, causing excessive secondary current and a blown fuse in the primary.

When checking for a shorted filter capacitor, simply measure the resistance of the power supply output, as shown in Figure 3.45b. If the capacitor is shorted, you will measure a very low resistance. Be sure that you disconnect the bridge from the line before measuring the resistance. This will prevent you from forward biasing a rectifier diode with the meter, which will give you a faulty reading.

An *open filter capacitor* will cause the ripple in the power supply output to increase drastically. At the same time, the dc output voltage will show a significant drop. Since an open filter capacitor is the only fault that will cause *both* of these symptoms, no further testing is necessary. If both symptoms appear, replace the filter capacitor.

Caution: Be very careful when replacing electrolytic capacitors. If you connect an electrolytic capacitor backward in a high-current circuit (such as a power supply), it will either explode or become hot enough to cause third-degree burns.

Zener Regulator Faults

If the *zener diode shorts*, the symptoms will be the same as those for a shorted filter capacitor. When the output symptoms indicate a short, simply connect an ohmmeter to check for a shorted filter capacitor. With the ohmmeter connected, disconnect one end of the zener from the circuit. If the problem remains, the

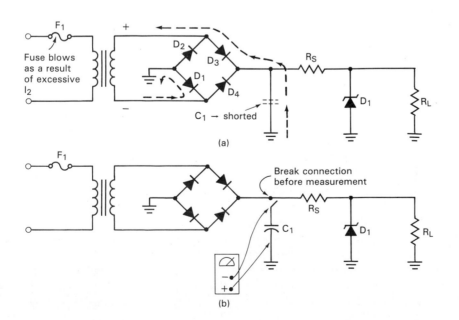

FIGURE 3.45
Shorted C effects and testing.

SEC. 3.8   Power Supply Troubleshooting

filter capacitor is the cause. If the problem disappears, the zener diode is the cause.

If the *zener diode opens*, the peak output voltage and ripple will both increase. At the same time, voltage regulation will be lost. Thus you will see a significant amount of variation in the output when the load current demands change.

Troubleshooting Applications

It is easy to tell you about common power supply faults and their causes. However, it is not always easy to initially apply these fault/cause relationships to actual circuit problems. In this section we are going to go through several example troubleshooting cases to show how a little thought and some simple testing will make power supply troubleshooting easy.

APPLICATION 1

Symptoms:
When the power supply in Figure 3.46 is first turned on, the output is okay. After a few minutes, the dc output voltage will drop to almost zero and then the primary fuse will blow.

Observation:
Some components will operate as they should when cool and will develop the symptoms of a short when they get hot. Electrolytic capacitors are extremely susceptible to this type of problem. For this reason the capacitor is suspected of being the cause of the problem.

Testing:
To diagnose the possible capacitor problem, the capacitor must be heated in one way or another. This is due to the fact that the problem is heat related. A soldering iron connected to one of the capacitor leads will heat the component enough for testing purposes.

An ohmmeter is connected in the circuit as shown in Figure 3.45b. When first connected, the ohmmeter reads a relatively high resistance. When the hot soldering iron is connected to the capacitor lead, the component starts to get hot. After a moment, the resistance reading on the meter drops to a very low value.

Conclusion:
Since the capacitor resistance dropped when temperature increased, the component is *leaky* (partially shorted) and must be replaced.

There is an alternative method of testing a capacitor that is suspected of being leaky. Replace the primary fuse and turn on the power supply. When the output voltage first drops, spray the capacitor with a specially made aerosol coolant. Such coolants are available at most electronic parts store. If the output voltage goes back to normal when the capacitor is cooled, the capacitor is leaky and must be replaced.

FIGURE 3.46

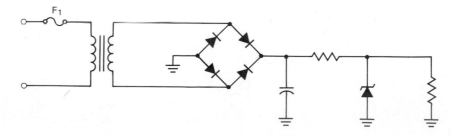

CHAP. 3 Common Diode Applications: The Basic Power Supply

Symptoms:
When the power supply in Figure 3.46 is turned on, the transformer becomes extremely hot and the dc output voltage is nearly zero. After a moment, the primary fuse blows.

Observation:
The symptoms listed are classic for a shorted transformer primary or secondary. For this reason, the transformer is tested immediately.

Testing:
The typical power supply transformer will have a primary resistance of nearly 50 Ω and a secondary resistance of 10 Ω or less. When tested, the transformer gives resistance readings of $R_P = 5\ \Omega$ and $R_S = 8\ \Omega$.

Conclusion:
Since the primary resistance is extremely low, the transformer primary must be shorted. Replacing the transformer corrects the problem.

Some transformers will have the value of secondary resistance printed on the side of the component. To determine the approximate value of primary resistance, multiply the secondary resistance by the turns ratio. For example, a transformer with a secondary resistance of 1.5 Ω and a turns ratio of 20:1 will have a primary resistance of approximately 1.5 $\Omega \times 20 = 30\ \Omega$.

Symptoms:
The power supply shown in Figure 3.46 has a dc output voltage that is approximately one-half of its rated value and a large amount of ripple voltage.

Observation:
These symptoms can be caused only by an open filter capacitor.

Testing:
None is required.

Conclusion:
The capacitor filter must be replaced.

What the Primary Fuse Tells You

If you look back through this section, you'll notice that there is a common thread to troubleshooting the circuits in the power supply: the fuse in the transformer primary.

The condition of the primary fuse indicates whether the problem in the power supply is a *short* or an *open*. If the fuse is blown, a short exists somewhere in the power supply. If it is not blown, an open exists somewhere in the supply. This, of course, assumes that there is a problem in the supply to begin with.

One Final Note

Most commercially used power supplies are far more complex than the ones we have seen in this chapter. The main difference lies in the voltage regulator circuitry, which usually contains one or more transistors and/or IC voltage regulators. However, the basic principles covered in this chapter apply to all power supplies. While the exact circuitry used may vary from one power supply to another, the basic principles of operation do not.

You will need to learn a great deal about transistors and IC voltage regulators before you will be ready to take on some of the more complex power supply circuits. However, by the time you get to Chapter 20 (Discrete and Integrated Voltage Regulators) you will be more than ready to deal with the complex circuits covered.

SECTION REVIEW

1. List the common transformer faults and their symptoms. [Objective 35]
2. What should you do when you diagnose a rectifier diode fault? [Objective 35]
3. List the common filter faults and their symptoms. [Objective 35]
4. List the common voltage regulator faults and their symptoms. [Objective 35]
5. What does the condition of the primary fuse tell you? [Objective 35]

Key Terms

The following terms were introduced and defined in this chapter:

average load current
average load voltage
bridge rectifier
capacitive filter
filter
full load
full-wave rectifier
half-wave rectifier
inductive filter

integrated circuit (IC)
integrated rectifier
line regulation
load regulation
peak inverse voltage (PIV)
peak load current
peak load voltage
peak secondary voltage
power rectifier

power supply
rectifier
ripple voltage
surge current
transformer secondary
rating
voltage regulator
worst-case analysis

PRACTICE PROBLEMS

§3.1

1. Determine the peak output voltage for the circuit shown in Figure 3.47. [5]
2. Determine the peak output voltage for the circuit shown in Figure 3.48. Assume that the line voltage is 115 V_{ac}. [5]
3. Determine the peak output voltage for the circuit shown in Figure 3.49. [5]
4. Determine the peak output current for the circuit shown in Figure 3.49. [5]

CHAP. 3   Common Diode Applications: The Basic Power Supply

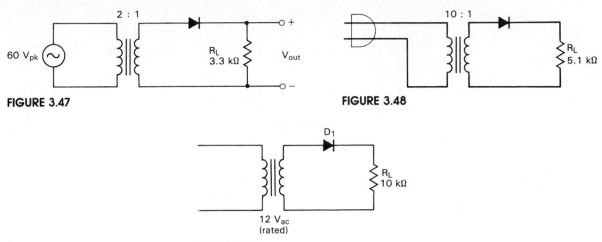

FIGURE 3.47

FIGURE 3.48

FIGURE 3.49

5. Determine the average (dc) output voltage for the circuit shown in Figure 3.47. [5]

6. Determine the average (dc) output voltage for the circuit shown in Figure 3.48. [5]

7. Determine the average (dc) output voltage for the circuit shown in Figure 3.49. [5]

8. Assume that the diode in Figure 3.47 is reversed. Determine the new values of $V_{out(pk)}$, V_{ave}, and I_{ave} for the circuit. [6]

9. Assume that the diode in Figure 3.48 is reversed. Determine the new values of $V_{out(pk)}$, V_{AVE}, and I_{AVE} for the circuit. [6]

10. Assume that the diode in Figure 3.49 is reversed. Determine the new values of $V_{out(pk)}$, V_{ave}, and I_{ave} for the circuit. [6]

11. Determine the PIV for the diode in Figure 3.48. [7]

12. A negative half-wave rectifier with a 12-kΩ load is being driven by a 30-Vac transformer. Draw the schematic for the circuit and determine the following values: PIV, $V_{out(pk)}$, V_{ave}, and I_{ave}. [6, 7]

§3.2

13. Determine the values of $V_{out(pk)}$, V_{ave}, and I_{ave} for the circuit shown in Figure 3.50. [9]

14. Determine the values of $V_{out(pk)}$, V_{ave}, and I_{ave} for the circuit shown in Figure 3.51. [9]

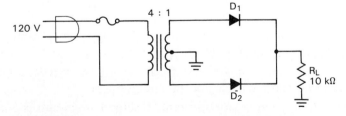

FIGURE 3.50

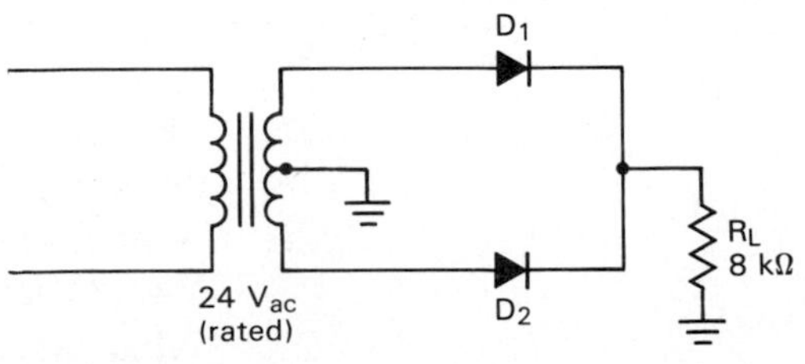

FIGURE 3.51

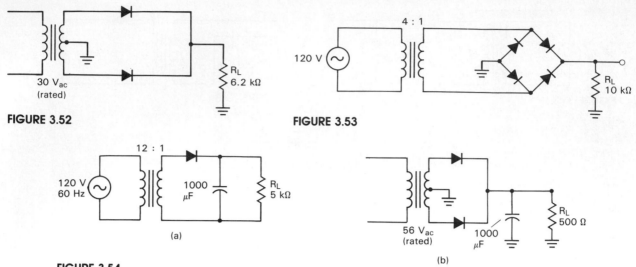

FIGURE 3.52

FIGURE 3.53

FIGURE 3.54

15. Determine the values of $V_{out(pk)}$, V_{ave}, and I_{ave} for the circuit shown in Figure 3.52. [9]

16. Determine the PIV of the circuit shown in Figure 3.50. [10]

∨ 17. Determine the PIV of the circuit shown in Figure 3.51. [10]

18. Determine the PIV of the circuit shown in Figure 3.52. [10]

19. Assume that the diodes in Figure 3.50 are both reversed. Determine the new values of $V_{out(pk)}$, V_{ave}, and I_{ave} for the circuit. [9]

20. A negative full-wave rectifier with a 910-Ω load is being driven by a 16-Vac transformer. Draw the schematic diagram of the circuit and determine the following values: $V_{out(pk)}$, PIV, V_{ave}, and I_{ave}. [9, 10]

§3.3

21. Determine the peak and average output voltage and current values for the circuit shown in Figure 3.53. [13]

22. Repeat problem 21. Assume that the transformer is a 16-V_{ac} transformer. [13]

23. What is the minimum allowable value of V_{RRM} for each of the diodes in problem 22? [15]

24. A bridge rectifier with a 1.2-kΩ load is being driven by a 48-V_{ac} transformer. Draw the schematic diagram for the circuit and calculate the dc load voltage and current values. Also, determine the PIV for each diode in the circuit. [13, 15]

§3.5

25. The circuit in Figure 3.54a has values of $R_w = 1\ \Omega$ and $R_B = 6\ \Omega$. What is the value of surge current for the circuit? [22]

26. A half-wave rectifier has values of $R_w = 2\ \Omega$, $R_B = 12\ \Omega$, and $V_2 = 36$ Vac. What is the value of surge current for the circuit? [22]

27. What are the values of V_{dc} and V_r for the circuit shown in Figure 3.54a? [25]

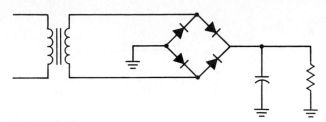

FIGURE 3.55

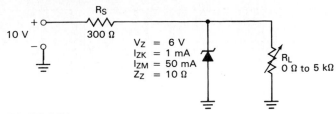

FIGURE 3.56

28. What are the values of V_{dc} and V_r for the circuit shown in Figure 3.54b? [25]
29. The circuit shown in Figure 3.55 has the following values: $V_2 = 18$ V$_{ac}$ (rated), $C = 470$ μF, and $R_L = 820$ Ω. Determine the values of V_r, V_{dc}, and I_L for the circuit. [25]
30. The circuit shown in Figure 3.55 has the following values: $V_2 = 24$ V$_{ac}$ (rated), $C = 1200$ μF, and $R_L = 200$ Ω. Determine the values of V_r, V_{dc}, and I_L for the circuit. [25]
31. What is the PIV for the circuit described in problem 30? [26]
32. What is the PIV for the circuit shown in Figure 3.54a? [26]

§3.6

33. For the circuit shown in Figure 3.56, determine the total circuit current. [28]
34. For the circuit shown in Figure 3.56, determine the value of I_L for $R_L = 2$ kΩ. [28]
35. For the circuit shown in Figure 3.56, determine the value of I_Z for $R_L = 2$ kΩ. [28]
36. For the circuit shown in Figure 3.56, determine the value of I_Z for $R_L = 3$ kΩ. [28]
37. For the circuit shown in Figure 3.56, determine the minimum allowable value of R_L. [29]
38. For the circuit shown in Figure 3.57, determine the total circuit current. [28]
39. For the circuit shown in Figure 3.57, determine the value of I_Z for $R_L = 5$ kΩ. [28]
40. For the circuit shown in Figure 3.57, determine the minimum allowable value of R_L. [29]
41. Determine the values of I_T, I_L, and I_Z for the circuit shown in Figure 3.58. [28]
42. The 12-V input to Figure 3.58 has 770 mV$_{PP}$ of ripple voltage. What is the output ripple voltage for the circuit? [30]

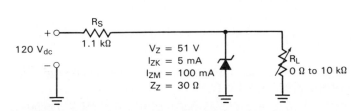

FIGURE 3.57

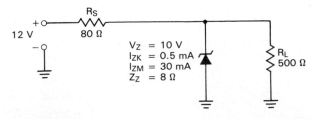

FIGURE 3.58

Practice Problems

133

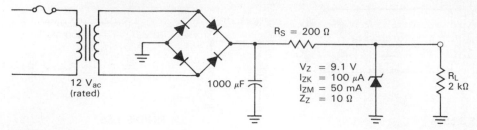

FIGURE 3.59

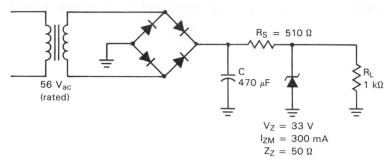

FIGURE 3.60

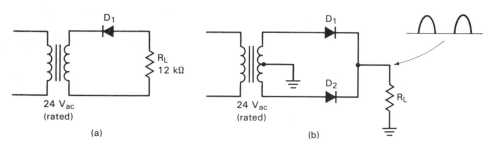

FIGURE 3.61

§3.7

43. Calculate the dc output voltage and current values for the circuit shown in Figure 3.59. [32]

44. Calculate the output ripple voltage for the circuit shown in Figure 3.59. [32]

45. Calculate the dc output values and output ripple voltage for the circuit shown in Figure 3.60. [32]

46. Calculate the worst-case values for the circuit shown in Figure 3.59. Assume that the tolerance rating of the transformer is ±10%. [33]

47. Calculate the worst-case values for the circuit shown in Figure 3.60. Assume that the tolerance rating of the transformer is ±10%. [33]

TROUBLESHOOTING PRACTICE PROBLEMS

48. The circuit shown in Figure 3.61a has an output signal that is identical to V_2. What is the cause of the problem? [35]

49. The circuit shown in Figure 3.61b has the output waveform shown. Discuss the possible cause(s) of the problem. [35]

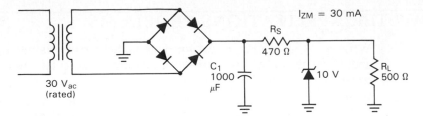

FIGURE 3.62

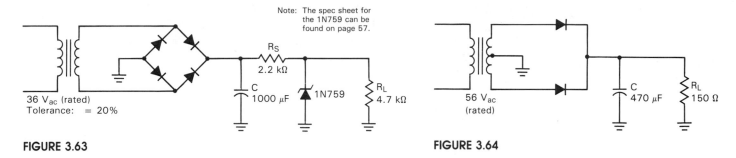

Note: The spec sheet for the 1N759 can be found on page 57.

FIGURE 3.63

FIGURE 3.64

50. The circuit in Figure 3.62 has an output that equals 0 V_{dc}. The primary fuse is not blown. Discuss the possible cause(s) of the problem. [35]

51. The load resistance in Figure 3.62 opens. Will this cause the zener diode to be destroyed? Use circuit calculations to explain your answer. [35]

THE BRAIN DRAIN

52. Refer back to Figure 2.33 (page 60). Assume that the zener diode in Figure 3.62 has opened. Which of the diodes in Figure 2.33 could be used as a replacement component?

53. The circuit shown in Figure 3.63 has measured output values of $V_{dc} = 12$ V and $V_r = 1.22$ V_{PP}. Determine whether or not there is a problem in the circuit.

54. The rectifier diodes in Figure 3.64 must be replaced. Using Figure 2.26 (page 50), determine which, if any, of the 1N4001–7 series diodes can be used as a replacement component in the circuit.

OPTIONAL PRACTICE PROBLEMS

§3.6 Zener Regulator Circuit Development

55. A zener voltage regulator is being designed using a 12-V zener that has an I_{ZM} rating of 250 mA. If the input voltage to the circuit is 18 V dc, what is the minimum allowable value of R_S? [30]

56. A zener regulator is being designed to the following specifications: (a) The load varies between 250 Ω and 5 kΩ, and $V_S = 20$ V dc. (b) The zener diode has the following ratings: $V_Z = 10$ V, $I_{zk} = 2$ mA, $I_{ZM} = 150$ mA. Determine the limits on the value of R_S. [30]

Answers to Example Practice Problems

SUGGESTED COMPUTER APPLICATIONS PROBLEMS

57. Write a program that is designed to calculate the values of $V_{out(pk)}$, V_{ave}, and I_{ave} when provided with the transformer secondary rating and the load resistance for a half-wave rectifier.

58. Write a program that will determine the values of $V_{out(pk)}$, V_{ave}, and I_{ave} for a full-wave rectifier when provided with the transformer secondary rating and the load resistance.

59. Modify the program in problem 58 to take into account the effects of a filter capacitor. The program should request the value of C (along with the values from problem 58) and should provide the values of dc output voltage and ripple voltage.

60. Write a program that will perform a worst-case analysis of a basic power supply when provided with the needed information.

ANSWERS TO EXAMPLE PRACTICE PROBLEMS

3–1. $V_{out(pk)} = 10.3\ V_{pk}$

3–2. $V_{out(pk)} = 31.83\ V_{pk}$

3–3. $V_{out(pk)} = 33.25\ V_{pk}$

3–4. $I_{out(pk)} = 1.57\ mA$

3–5. $V_{ave} = 3.31\ V_{dc}$

3–6. $I_{ave} = 10.91\ mA$

3–7. $I_{ave} = 1.11\ mA$

3–8. $V_{2(pk)} = -50.92\ V$, $V_{out(pk)} = -50.22\ V_{pk}$, $V_{ave} = -15.99\ Vdc$

3–9. $V_{ave} = 10.36\ V\ dc$

3–10. $I_{out(pk)} = 7.4\ mA$, $I_{ave} = 4.71\ mA$

3–11. $V_{ave} = 15.32\ V\ dc$, $I_{ave} = 12.77\ mA$

3–13. $I_{surge} = 7.06\ A$

3–15. $V_r = 274\ mV_{PP}$, $Vdc = 16.14\ V$

3–16. $I_T = 2\ mA$

3–17. $I_L = 1\ mA$

3–18. $I_Z = 1\ mA$

3–19. $R_{L(min)} = 515.2\ \Omega$

3–20. $V_{r(out)} = 32.09\ mV_{PP}$

3–21. $R_{S(min)} = 144\ \Omega$

3–22. $R_{S(max)} = 870\ \Omega$

3–23. $Vdc = 10\ V$, $I_L = 1.96\ mA$, $V_{r(out)} = 30.74\ mV_{PP}$

3–24. $V_{r(out)} = 1.13\ V_{PP}$

Common Diode Applications: Clippers, Clampers, Multipliers, and Displays

"Multisegment displays are used to display alpha-numeric symbols; that is, they are used to display numbers, letters, and punctuation marks."

OBJECTIVES

After studying the material in this chapter, you should be able to:

1. State the purposes served by clippers, clampers, and voltage multipliers. (Introduction)
2. Discuss the operation of *series clippers*. (§4.1)
3. Discuss and analyze the operation of *shunt clippers*. (§4.1)
4. Discuss and analyze the operation of *zener shunt clippers*. (§4.1)
5. Discuss and analyze the operation of *symmetrical zener shunt clippers*. (§4.1)
6. Discuss and analyze the operation of *biased shunt clippers*. (§4.1)
7. Discuss the use of clippers as transient protection circuits. (§4.2)
8. Describe the operation of the AM detector. (§4.2)
9. Describe the effects of *negative clampers* and *positive clampers* on an input waveform. (§4.3)
10. Describe the effects that clampers have on the value of V_{ave} for a given input waveform. (§4.3)
11. Describe the circuit operation of a clamper. (§4.3)
12. Determine whether a given clamper is a *positive* or *negative* clamper, given the schematic diagram of the circuit. (§4.3)
13. Determine the dc reference voltage for any biased clamper. (§4.3)
14. Describe and analyze the operation of the *half-wave voltage doubler*. (§4.4)
15. Describe and analyze the operation of the *full-wave voltage doubler*. (§4.4)
16. Describe and analyze the operation of the *voltage tripler*. (§4.4)
17. Describe and analyze the operation of the *voltage quadrupler*. (§4.4]
18. Describe the operation of a dual-polarity dc power supply. (§4.4)
19. Discuss the use of the LED as a power/level indicator. (§4.5)
20. Discuss the use of the LED in multisegment displays. (§4.5)
21. Describe the common fault symptoms that occur in clippers. (§4.6)
22. Describe the common fault symptoms that occur in clampers. (§4.6)
23. Describe the procedure for troubleshooting voltage multipliers. (§4.6)
24. Describe the procedure for troubleshooting a driver-display circuit. (§4.6)

As you master each of the objectives listed, place a check mark (✔) in the appropriate box.

Limiters (or **clippers**) are circuits that are used to eliminate a portion of an ac signal.

Dc restorers (or **clampers**) are circuits that are used to change the dc reference of an ac signal.

Voltage multiplers. Circuits used to produce a dc output voltage that is some multiple of an ac peak input voltage.

Multisegment display. An LED circuit used to display **alphanumeric symbols** (numbers, letters, and punctuation marks).

D iodes are among the primary components in power supplies, as we saw in Chapter 3, but they have many other common applications as well. In this chapter, we are going to take a look at several diode circuits and applications.

The first type of circuit that we will cover is the *limiter*, or *clipper*. Clippers are used to clip off or eliminate a portion of an ac signal. As you will see, there is a wide variety of clipper circuits.

The second circuit we will cover is the *dc restorer*, or *clamper*. Clampers are used to restore or change the dc reference of an ac signal. As you will see, it is very common to have an ac signal that is referenced at some dc voltage. For example, you may have a 12-V_{PP} ac signal that varies equally above and below 2 V dc. The clamper would be used to change the dc reference voltage.

The third group of circuits we will discuss are the *voltage multipliers*. A voltage multiplier will produce a dc voltage that is some multiple of an ac peak voltage.

Finally, we will look at the most common LED application: the *multisegment display*. Multisegment displays are used to display *alphanumeric symbols*; that is, they are used to display numbers, letters, and punctuation marks.

4.1
Clippers

You have already been introduced to the basic operating principles of the clipper. The half-wave rectifier is basically a clipper that eliminates one of the alternations of an ac signal.

There are two types of clippers: the *series* clipper and the *shunt* clipper. Each of these can be *positive* or *negative*. All of these clipper types are shown in Figure 4.1. The *series clipper* contains a diode that is *in series with the load*. The *shunt clipper*, on the other hand, contains a diode that is *in parallel with the load*.

Clippers are classified as being either *series clippers* or *shunt clippers*.

Series Clippers

The *series clipper* (shown in Figure 4.2) has the same circuit operating characteristics as the half-wave rectifier. In fact, if you compare the circuit shown in Figure 4.2 to those shown in Figure 3.6 (page 81), you will see that the half-wave rectifier is nothing more than a series clipper.

Since the operation of the series clipper is identical to that of the half-wave rectifier, we will not go into any great detail on series clipper operation at this point. As a review, the following points are made about series clippers:

A series clipper has an output when the diode is forward biased (conduction).

1. When the diode in a negative series clipper is *forward biased* by the input signal, it will conduct, and the output voltage is found as

$$V_{\text{out}} = V_{\text{in}} - 0.7 \text{ V} \qquad (4.1)$$

2. When the diode in the negative series clipper is *reverse biased* by the input signal, it will not conduct. Therefore,

$$V_{D1} = V_{\text{in}} \qquad (4.2)$$

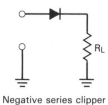

Negative series clipper

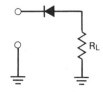

Positive series clipper

(a)

FIGURE 4.1

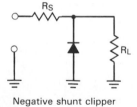

Negative shunt clipper

Positive shunt clipper

(b)

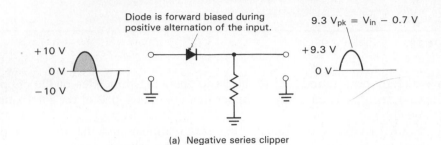

(a) Negative series clipper

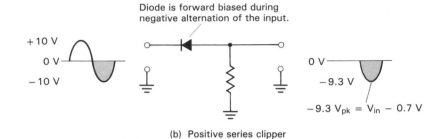

(b) Positive series clipper

FIGURE 4.2

Series clipper operation.

and

$$\boxed{V_{out} = 0 \text{ V}} \tag{4.3}$$

3. The positive series clipper operates in the same fashion. The only differences are:
    a. The output voltage polarities are reversed.
    b. The current directions through the circuit are reversed.

Reference:

For a complete review of series clippers, refer to the coverage of half-wave rectifiers in Chapter 3.

FIGURE 4.3

Shunt clipper operation.

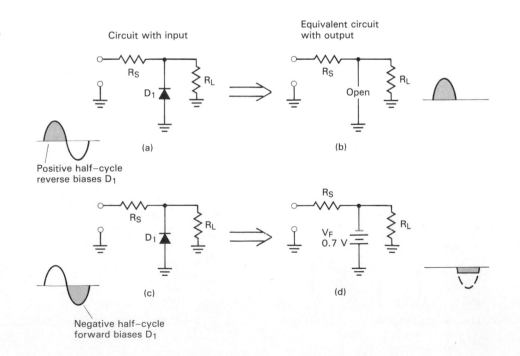

Shunt Clippers

3

The operation of the *shunt clipper* is exactly the opposite of that of the series clipper. The series clipper has an output when the diode is forward biased and no output when the diode is reverse biased. The *shunt clipper* has an output when the diode is *reverse biased* and will short the input signal to ground when the diode is forward biased. This operation is illustrated in Figure 4.3.

A **shunt clipper** has an output when the diode is reverse biased (not conducting).

The circuit shown in Figure 4.3 is a *negative shunt clipper*. When the diode in the negative shunt clipper is reverse biased, it is effectively removed from the circuit. This is shown in Figure 4.3b. With the diode removed from the circuit, the output voltage is found as

Note: For the time being, we will concentrate on the *negative shunt clipper*. The *positive shunt clipper* will be discussed later in this section.

$$V_{out} = \frac{R_L}{R_L + R_S} V_{in} \qquad (4.4)$$

since R_S and R_L form a voltage divider. While the output signal will resemble the positive alternation of the input, the peak output voltage will be somewhat less than the peak input voltage. This point is illustrated in the following example.

EXAMPLE 4.1

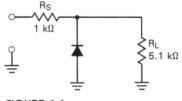

FIGURE 4.4

The negative shunt clipper shown in Figure 4.4 has a peak input voltage of $+12\ V_{pk}$. What is the peak output voltage from the circuit?

Solution: When the input is *positive*, the diode is reverse biased and does not conduct. Therefore, the peak output voltage is found as

$$V_{out} = \frac{R_L}{R_L + R_S} V_{in}$$

$$= \frac{5.1\ k\Omega}{6.1\ k\Omega} (+12\ V_{pk})$$

$$= 10\ V_{pk}$$

PRACTICE PROBLEM 4–1

A negative shunt clipper has values of $R_L = 510\ \Omega$ and $R_S = 100\ \Omega$. If the input voltage is $+15\ V_{pk}$, what is the peak output voltage?

During the negative alternation of the input signal, the diode in the negative shunt clipper is *forward biased* and the voltage across the component will equal V_F. Since the load is in parallel with the diode, load voltage will also equal the forward voltage across the diode. By formula

Just a Reminder: A diode won't conduct until its cathode is 0.7 V *more negative* than its anode. Thus, the diode in the negative shunt clipper won't conduct until the cathode is at −0.7 V. That is why V_{out} is given as V_F in equation (4.5).

$$V_{out} = -V_F \qquad (4.5)$$

Since the output side of R_S is held to approximately −0.7 V when the diode is forward biased, the voltage across R_S will be equal to the *difference between* V_{in} and the value V_F. By formula,

$$\boxed{V_{R_S} = -V_{in} + 0.7 \text{ V}}$$ (4.6)

when the diode is forward biased. The following example illustrates the circuit conditions that exist when the diode in a negative shunt clipper is forward biased.

=== EXAMPLE 4.2 ===

The circuit described in Example 4.1 has a -12 V_{pk} input voltage. Determine the values of V_{out} and V_{R_S} for the circuit.

Solution: Since the diode is forward biased, the output voltage is equal to the value of V_F for the diode. Thus,

$$V_{out} = -0.7 \text{ V}$$

The voltage across the series resistor (R_S) is now found as

$$V_{R_S} = -V_{in} + 0.7 \text{ V}$$
$$= -12 \text{ V}_{pk} + 0.7 \text{ V}$$
$$= -11.3 \text{ V}_{pk}$$

PRACTICE PROBLEM 4–2

The circuit described in Practice Problem 4–1 has a -15 V_{pk} input. Determine the values of V_{out} and V_{R_S} for the circuit.

So far we have worked with the *negative shunt clipper*. As the following example illustrates, the *positive shunt clipper* works according to the same basic principles and is analyzed in the same fashion.

=== EXAMPLE 4.3 ===

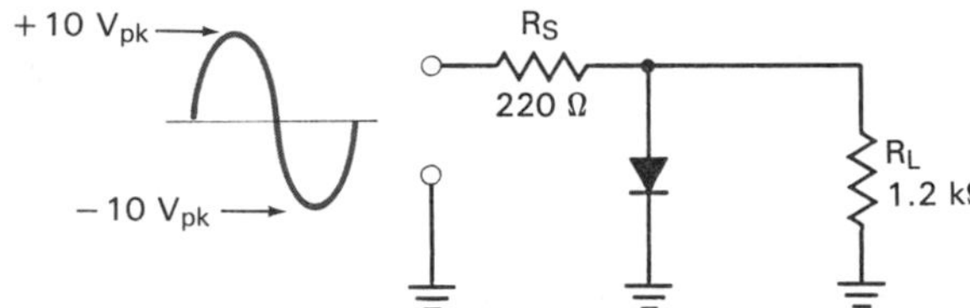

+10 V_{pk}

−10 V_{pk}

R_S
220 Ω

R_L
1.2 kΩ

FIGURE 4.5

The diode in a *positive* shunt clipper conducts when the input is *positive*. Thus, for the positive shunt clipper, equations (4.5) and (4.6) are modified as follows:

$$V_{out} = V_F$$
$$V_{R_S} = V_{in} - 0.7 \text{ V}$$

The *positive shunt clipper* shown in Figure 4.5 has the input waveform shown. Determine the value of V_{out} for each of the input alternations.

Solution: When the input is *positive*, the diode is forward biased. Thus,

and

$$V_{R_S} = V_{in} - 0.7 \text{ V}$$
$$= 10 \text{ V}_{pk} - 0.7 \text{ V}$$
$$= 9.3 \text{ V}_{pk}$$

When the input to the circuit is *negative*, the diode is reverse biased and effectively removed from the circuit. Thus,

$$V_{out} = \frac{R_L}{R_L + R_S} V_{in}$$
$$= \frac{1.2 \text{ k}\Omega}{1.42 \text{ k}\Omega} (-10 \text{ V}_{pk})$$
$$= -8.45 \text{ V}_{pk}$$

CHAP. 4 Common Diode Applications: Clippers, Clampers, Multipliers, and Displays

PRACTICE PROBLEM 4–3

A *positive shunt clipper* with values of $R_S = 100\Omega$ and $R_L = 1.1$ kΩ has a ±12-V_{pk} input signal. Determine the value of V_{out} for each alternation of the input. Also, determine the value of V_{R_S} when the diode is forward biased.

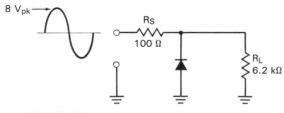

$V_{out} = 0.7$ V

$V_{out} = -8.45$ V_{pk}

FIGURE 4.6

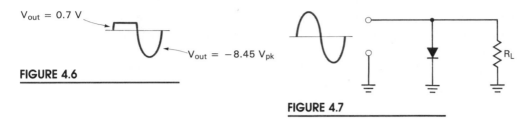

FIGURE 4.7

Using the values calculated in Example 4.3, we are able to draw the output waveform for the circuit. The waveform (shown in Figure 4.6) will be clipped at $V_{out} = 0.7$ V and will peak at $V_{out} = -8.45$ V_{pk}.

The Purpose Served by R_S

R_S is included in the shunt clipper as a *current-limiting* resistor. Consider what would happen if the input signal forward biased the diode and R_S was not in the circuit. This situation is illustrated in Figure 4.7.

Without R_S in the circuit, the diode will short the signal source to ground during the positive alternation of the input signal. This will result in either:

1. A "toasted" diode
2. A "toasted" component (or blown fuse) in the signal source

For example, assume that the circuit input signal in Figure 4.7 has a value of $+V_{pk} = +12$ V. When the input signal is at its positive peak, we have the diode shorting the $+12$ V_{pk} to ground. The excessive current produced by this situation will damage (or destroy) either the diode or the signal source.

In any practical situation, the value of R_S will be *much lower than* the value of R_L. When this is the case, the output voltage (when the diode is reverse biased) will be approximately equal to the value of V_{in}. This point is illustrated in the following example.

EXAMPLE 4.4

8 V_{pk}

R_S

100 Ω

R_L
6.2 kΩ

FIGURE 4.8

Determine the peak output voltage for the circuit shown in Figure 4.8. Assume that the diode is reverse biased.

Solution: The peak output voltage is found using equation (4.4) as follows:

$$V_{out} = \frac{R_L}{R_L + R_S} V_{in}$$
$$= \frac{6.2 \text{ k}\Omega}{6.3 \text{ k}\Omega} (8 \text{ V}_{pk})$$
$$= 7.87 \text{ V}_{pk}$$

With the relationship of $R_S + R_L$ in this circuit, we could have assumed that V_{in} and V_{out} were approximately equal in any practical analysis situation.

4

A **zener shunt clipper** uses a zener diode instead of a *pn*-junction diode.

Zener Shunt Clippers

The *zener shunt clipper* uses a zener diode in place of a *pn*-junction diode. A zener shunt clipper, which is used to clip *both* alternations of the input signal, is shown in Figure 4.9.

The zener shunt clipper uses both the forward and reverse operating characteristics of the zener diode. You may recall from Chapter 2 that the zener diode has the same forward characteristics as the *pn*-junction diode. Thus, when the zener diode is forward biased, this clipper acts just like the standard shunt clipper.

When the input signal in Figure 4.9 goes positive, the zener diode will remain off until the value of V_{in} reaches the value of V_Z. At that time, the zener diode will turn on, clipping the input signal. As long as V_{in} is greater than V_Z, the diode will remain on and the output voltage will be found as

$$\boxed{V_{out} = V_Z}$$

(4.7)

If we reverse the zener in the zener shunt regulator, we get the output waveform shown in Figure 4.10. When the input signal is *positive*, the zener diode is forward biased and the signal is clipped off at 0.7 V. When the input goes negative, the zener diode remains off until the value of V_Z is reached. When $V_{in} = V_Z$, the diode turns on and clips the signal as shown.

Incidentally, you may have noticed the similarity between the zener shunt clipper and the zener voltage regulator from Chapter 3. The only difference between these two circuits is the input. The zener voltage has a dc input, while the zener shunt clipper has an ac input.

5

The **symmetrical zener shunt clipper** uses two zener diodes that are connected back to back.

The Symmetrical Zener Shunt Clipper

The *symmetrical zener shunt clipper* uses two zener diodes that are connected as shown in Figure 4.11. The result of this connection can be seen by looking at the clipped symmetrical output waveform shown in the figure.

When the input is *positive*, D_1 is *forward biased* and D_2 is *reverse biased* (assuming that the value of V_{in} is high enough to turn both diodes on). When

FIGURE 4.9

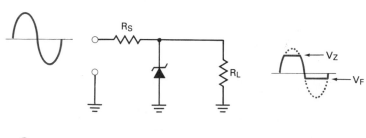

FIGURE 4.10

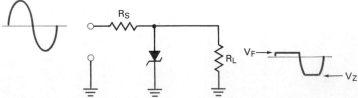

CHAP. 4 Common Diode Applications: Clippers, Clampers, Multipliers, and Displays

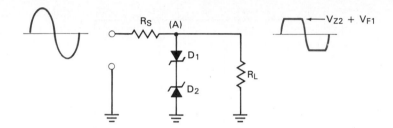

FIGURE 4.11

both diodes are conducting, the voltage from point (A) to ground will equal the sum of V_{Z2} and the forward voltage drop of D_1 (0.7 V).

When the input is *negative*, the opposite condition will exist. D_1 will be reverse biased and D_2 will be forward biased. The output voltage in this case is found as

$$V_{\text{out}} = -V_Z - 0.7 \text{ V} \qquad \textbf{(4.8)}$$

Even though the values in equation (4.8) are negative, we are still dealing with the *sum* of the V_Z and V_F ratings of the diodes.

In most symmetrical zener shunt clippers, the V_Z ratings of the two diodes are equal. You will rarely run into a situation where V_{Z1} is not equal to V_{Z2}.

The symmetrical zener shunt clipper is used primarily for *circuit protection*. This point will be discussed further in the next section.

Biased Clippers

Shunt clippers are sometimes designed so that the diode is biased by a dc voltage source. Such a clipper is called a *biased clipper*. Two examples of biased clippers are shown in Figure 4.12.

The *positive biased clipper* (Figure 4.12a) will clip the input signal at V_B + 0.7 V. The actual value at which the circuit will clip the input signal depends on the value of the biasing voltage (V_B). If V_B is 2 V, the input signal will be clipped at 2.7 V. If the value of V_B is 5 V, the input signal will be clipped at 5.7 V, and so on.

The *negative biased clipper* works in the same fashion, but it will clip the input signal at $-V_B$ - 0.7 V. If V_B is -2 V, the input signal will be clipped at

Biased clipper. A shunt clipper that uses a dc voltage source to bias the diode.

The **positive biased clipper** clips the input signal at V_B + 0.7 V.

The **negative biased clipper** clips the input signal at V_B − 0.7 V.

FIGURE 4.12

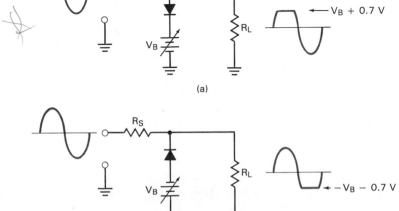

(a)

(b)

FIGURE 4.13

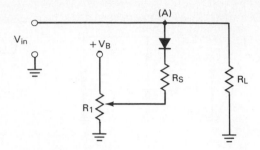

−2.7 V. If the value of V_B is −5 V, the input signal will be clipped at −5.7 V, and so on.

In practice, a potentiometer is used to provide an adjustable value of V_B, as shown in Figure 4.13. In this circuit, the biasing voltage ($+V_B$) is connected to the diode via the potentiometer (R_1) and the current-limiting resistor (R_S). R_1 is adjusted in this circuit to provide the desired clipping limit at point A in the circuit. Note that the clipping voltage will never be as high as V_B or as low as ground, due to the fact that there will be *some* voltage dropped across R_S whenever the diode is conducting. By reversing the direction of the diode and the polarity of V_B, the circuit in Figure 4.13 can be modified to work as a biased negative clipper.

One Final Note

As you can see, there are many different clipper configurations. These configurations, which are used for a variety of applications, are summarized in Figure 4.14.

What are they used for? This is what we will be discussing in the next section.

SECTION REVIEW

1. What purpose is served by a *clipper*? [Objective 1]
2. What is the other name for a clipper? [Objective 1]
3. What purpose is served by a *clamper*? [Objective 1]
4. What is the other name for a clamper? [Objective 1]
5. What purpose is served by the voltage multiplier? [Objective 1]
6. What purpose is served by a multisegment display? [Objective 1]
7. Discuss the differences between series and shunt clippers. [Objective 3]
8. What purpose is served by R_S in the shunt clipper? [Objective 3]
9. Describe the operation of the zener shunt clipper. [Objective 4]
10. What is the difference between the output of a zener shunt clipper and a comparable *pn*-junction diode shunt clipper? [Objective 4]
11. What is the circuit recognition feature of the symmetrical zener shunt clipper? [Objective 5]
12. What is the difference between the output of the symmetrical zener shunt clipper and that of a standard zener shunt regulator? [Objective 5]
13. Describe the operation of a biased clipper. [Objective 6]

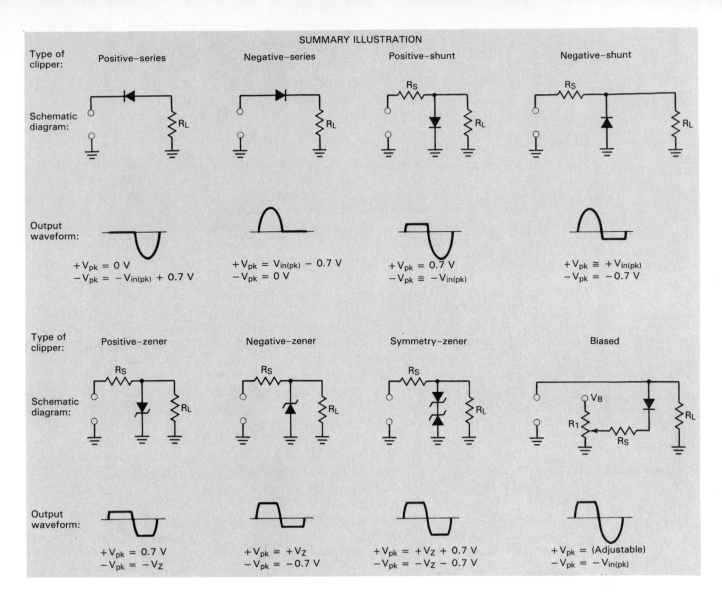

FIGURE 4.14

4.2
Clipper Applications

Clippers are used in a wide variety of electronic systems. They are generally used to perform one of two functions:

1. Altering the shape of a waveform
2. Circuit transient protection

You have already seen one example of the first function in the half-wave rectifier. This circuit alters the shape of an ac signal, changing it to pulsating dc. You will see another application of this type in this section, along with several transient protection circuits.

Transient Protection

A *transient* is an abrupt current or voltage spike that has an extremely short duration. A *current surge* would be one type of transient.

Transients can do a lot of damage to circuits whose inputs must stay within certain voltage or current limits. For example, many digital circuits have inputs that can only handle voltages that fall within a specified range. For these circuits, a voltage transient that goes outside the acceptable input voltage range cannot be allowed to reach the circuit. A clipper diode could be used in this case to prevent such a transient from reaching the digital circuit. This point is illustrated in Figure 4.15a.

The block labeled A represents a digital circuit whose input voltage must not be allowed to go outside of the range of 0 to +5 V. The clipper will protect the circuit from any voltage transients outside this range. For example, let's assume that the transient shown in the figure occurs on the input line. The transient will forward bias D_2, causing the diode to conduct. With D_2 conducting, the transient will be shorted to ground, protecting the input to circuit A.

The circuit in Figure 4.15b includes a clipper that is designed to protect the *output* of circuit B from transients that are produced when a square wave is used to drive a speaker (a common practice with computer sound effects). A speaker is essentially a big coil (inductor). When a square wave is used to drive the coil in a speaker, that coil produces a *counter emf*. This counter emf is produced each time that the output voltage makes the transition from +5 to 0 V. The counter emf will be equal in magnitude to the original voltage, but opposite in polarity. Thus, each time circuit B tries to drive the output line to 0 V, the speaker will try to force the line to continue to −5 V. This −5 V counter emf could destroy the original driving circuit. However, the clipping diode (D_1) will short the counter emf produced by the speaker coil to ground before it can harm the driving circuit.

The final circuit (Figure 4.15c) is used to protect a power supply from the transients that can occur in the primary winding of the supply's transformer. If a transient occurs in the wall outlet, that surge will normally be coupled to the power supply by the transformer. The symmetrical zener clipper in the primary

A Practical Consideration: Let's say that circuit B in Figure 4.15b goes bad. When it is replaced, the new circuit works fine until it tries to drive the speaker. After driving the speaker for a few seconds, the new circuit B also goes bad. The problem in this case is an open diode. If the diode is open, the counter emf produced by the speaker isn't shorted to ground and destroys each new circuit that is used to replace the previous one.

FIGURE 4.15

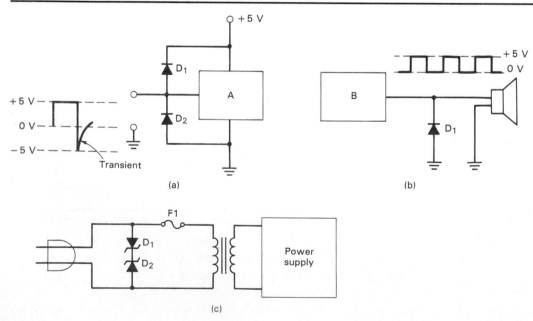

CHAP. 4 Common Diode Applications: Clippers, Clampers, Multipliers, and Displays

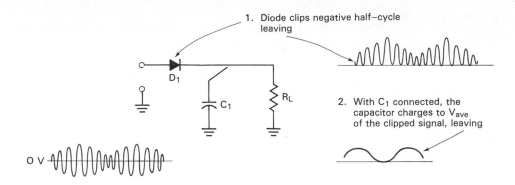

1. Diode clips negative half-cycle leaving

2. With C_1 connected, the capacitor charges to V_{ave} of the clipped signal, leaving

FIGURE 4.16

circuit prevents this from happening. Normally, 200-V zeners are used in this application.

You may be wondering where the load resistor is in each of the circuits shown. In Figure 4.15a, the load is the input impedance of circuit A. In Figure 4.15b, the load is the output impedance of circuit B. Finally, in Figure 4.15c, the load impedance is the reflected impedance of the power supply itself.

Incidentally, there is a special type of diode that is produced for just this type of application. This diode, called a *transient suppressor*, is covered in Chapter 21.

The AM Detector

◄ 8

Another common application for the diode clipper can be found in an AM radio. In this case, the clipper is part of a circuit called a *detector*. A simple detector is shown in Figure 4.16. The purpose of the detector is to provide a sine-wave output that is representative of the peak variations in the input signal. This function is accomplished by the capacitor, which is designed to charge to the average peak value of the input. However, with the input shown, the average voltage is 0 V. This is due to the fact that each positive peak has an equal and opposite negative peak.

The diode takes care of the problem by eliminating (clipping) the negative portion of the input signal. With the negative portion eliminated, the capacitor will charge and discharge at the rate of input variation. This provides an output signal that resembles the peak variations of the input.

One Final Note

The applications covered in this section were intended only to give you an idea of clipper applications. To cover every possible circuit would require several more chapters. Even though we have only covered a few clipper circuits, you should now have a good idea of the purposes they serve.

SECTION REVIEW

1. What is a *transient*? [Objective 7]
2. Why must digital circuit inputs be protected from voltage transients? [Objective 7]
3. Describe the operation of the AM detector. [Objective 8]

Clampers (DC Restorers)

A clamper is a circuit that is designed to shift a waveform either above or below a given reference voltage without distorting the waveform. There are two types of clampers: the *positive clamper* and the *negative clamper*. The input/output characteristics of these two circuits are illustrated in Figure 4.17.

Figure 4.17a shows a typical ac sine wave. As you can see, the sine wave has peak values of ± 10 V for an overall value of 20 V_{PP}. We will use this waveform to help illustrate the operation of the two clamper circuits.

A *positive clamper* will shift its input waveform so that the *negative* peak voltage of the waveform will be approximately equal to the dc reference voltage of the clamper. For example, let's say that we have a positive clamper with a dc reference voltage of 0 V. (You will be shown later where this *dc reference voltage* comes from.) If we were to apply the waveform in Figure 4.17a to this clamper, we would get the waveform shown in Figure 4.17b. As you can see, the waveform now has peak values of $+20$ V and 0 V (the negative peak) for an overall value of $20V_{PP}$. In this case, the positive clamper has shifted the entire waveform so that the negative peak voltage is approximately equal to the clamper's dc reference voltage.

The *negative clamper* will shift its input waveform so that the *positive* peak voltage of the waveform will be approximately equal to the dc reference voltage of the clamper. Now let's assume that we have a *negative* clamper with a dc reference voltage of 0 V. If we apply the original waveform to this circuit, we get the output waveform shown in Figure 4.17c. In this case, the positive peak voltage of the input has been shifted so that it is approximately equal to the clamper's dc reference voltage.

V_{ave} Shift

You were shown in Chapter 3 how to determine the dc average (V_{ave}) value of a *rectified* sine wave. When we are dealing with a sine wave that is *not* rectified,

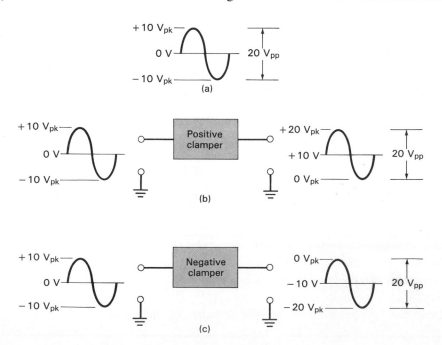

FIGURE 4.17

CHAP. 4 Common Diode Applications: Clippers, Clampers, Multipliers, and Displays

---

Positive clamper. A circuit that shifts an entire input signal *above* a dc reference voltage.

Negative clamper. A circuit that shifts an entire input signal *below* a dc reference voltage.

the value of V_{ave} falls halfway between the positive and negative peak voltage values. For example, if we were to measure the waveform in Figure 4.17a with a dc voltmeter, we would get a reading of 0 V. This is because the dc average of the waveform (0 V) falls halfway between +10 and −10 V_{pk}.

The effect of clamper operation on the value of V_{ave} for a waveform.

When a waveform is shifted by a clamper (or any other circuit for that matter), the value of V_{ave} for the waveform may not equal 0 V anymore. For example, consider the output waveform shown in Figure 4.17b. This waveform has peak values of +20 and 0 V_{pk}. The dc average, which falls halfway between the peak values, is +10 V. Thus, if we were to measure this output waveform with a dc voltmeter, we could get a reading of 10 Vdc. By the same token, if we were to measure the output waveform shown in Figure 4.17c with a dc voltmeter, we would get a reading of −10 Vdc.

There is an interesting point that can be made at this time. While the value of V_{ave} for a waveform will normally change when the waveform goes through the clamper, the *ac values* of a waveform will not. For example, all three waveforms in Figure 4.17 have a peak-to-peak voltage of 20 V_{PP}. If we were to convert this value to rms, we would get a value of 7.07 V_{ac}. If we were to measure the three waveforms with an ac voltmeter, they would all get a reading of approximately 7.07 Vac. Thus, while the clamper changes the peak and average (dc) values of a waveform, it does not change the peak-to-peak or rms value of the waveform.

Clampers do not affect the peak-to-peak and rms values of a waveform.

Clamper Operation

The schematic diagram of the positive clamper is shown in Figure 4.18. As you can see, this clamper is very similar to the shunt clipper with the exception of the added capacitor.

☑ 11

The clamper works on the basis of *switching time constants*. When the diode is forward biased, it provides a charging path for the capacitor. For this circuit, the charging time constant is found as

Clamper operation is based on switching time constants.

$$\tau = R_{D1}C$$

and the total charge time is found as

Switching time constants. A term used to describe a condition where a capacitor has different charge and discharge times.

$$\boxed{T_C = 5(R_{D1}C)} \tag{4.9}$$

where R_{D1} = the bulk resistance of the diode.

When the diode is reverse biased, the capacitor will start to discharge through the resistor. Thus, the discharge time constant for the circuit is found as

$$\tau = R_L C$$

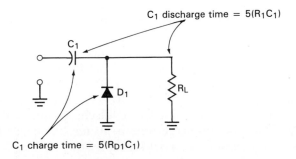

C$_1$ discharge time = $5(R_1C_1)$

C$_1$ charge time = $5(R_{D1}C_1)$

FIGURE 4.18

Clamper *charge* and *discharge* time constants.

and the total discharge time is found as

$$T_D = 5(R_L C)$$ (4.10)

As the following example shows, the difference between the capacitor charge and discharge times is significant.

EXAMPLE 4.5

Determine the capacitor charge and discharge times for the circuit shown in Figure 4.18. Assume that the forward resistance of the diode is 10 Ω, the value of R_L is 10 kΩ, and $C = 1$ μF.

Solution: The capacitor charges through the diode. Therefore, the charge time is found as

$$\begin{aligned} T_C &= 5(R_{D1} C) \\ &= 5(10\ \Omega \times 1\ \mu F) \\ &= 50\ \mu s \end{aligned}$$

The capacitor discharges through the resistor. Thus, the discharge time is found as

$$\begin{aligned} T_D &= 5(R_L C) \\ &= 5(10\ k\Omega \times 1\ \mu F) \\ &= 50\ ms \end{aligned}$$

As you can see, it takes a lot longer for the capacitor to discharge than it does for it to charge. This is the basis of clamper circuit operation.

PRACTICE PROBLEM 4–5

A clamper has values of $R_{D1} = 8\ \Omega$, $C = 4.7\ \mu F$, and $R_L = 1.2\ k\Omega$. Determine the charge and discharge times for the capacitor.

Using the values from Example 4.5, we will discuss the overall operation of the clamper. For ease of discussion, we are going to take a look at the circuit response to a square-wave input. This operation is illustrated in Figure 4.19. At T_0, the capacitor has a 0-V plate-to-plate charge. At T_1, the input to the circuit goes to +10 V, forward biasing D_1. Since the diode is forward biased, the time constant in the circuit is extremely short, and the capacitor almost instantly charges to the input voltage value, 10 V. Thus, 50 μs after T_1, the capacitor has a 10-V plate-to-plate charge, and the voltage across the output resistor has returned to zero.

At T_2, the input signal returns to 0 V. Since the capacitor cannot instantly change its plate-to-plate charge, the output side of the capacitor follows the input side. Thus, when the input goes negative by 10 V (to 0 V), the output side of the capacitor also goes negative by 10 V (to −10 V). This reverse biases the diode.

When the diode is reverse biased, the discharge time of the capacitor is determined by R_L. Since the discharge time is 50 ms, the capacitor has little time to lose any of its charge before the input signal returns to +10 V.

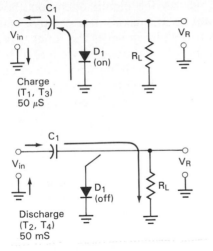

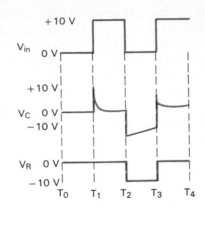

FIGURE 4.19

At T_3, the input returns to $+10$ V and D_1 turns on again. In a short period of time, the capacitor regains the charge it lost during the negative half-cycle of the input.

The result of this charge/discharge cycle is that a square wave is produced at the clamper output. However, this square wave varies between 0 and -10 V. Thus, the input waveform has been shifted so that its most positive value is approximately equal to the dc reference voltage of the clamper.

The dc reference voltage of a clamper is the potential to which the diode is returned. In the circuit shown in Figure 4.19, the diode is connected between the circuit and ground. Thus, ground (0 V) is the dc reference of the circuit. Had the diode been returned to a potential other than ground, the dc reference would have equaled that potential. This point will be discussed in detail later in this section.

The sine-wave operation of the clamper is the same as the square-wave operation. The charge/discharge times of the capacitor will shift a sine wave to some specified dc reference voltage, just as they shifted the square wave.

The diode return potential determines the dc reference voltage of a clamper.

Negative Clampers versus Positive Clampers

As was the case with clippers, *the difference between the negative clamper and the positive clamper is simply the direction of the diode*. This point is illustrated in Figure 4.20. Figure 4.20a shows a negative clamper and its effect on a 40-V_{PP}

 12

FIGURE 4.20

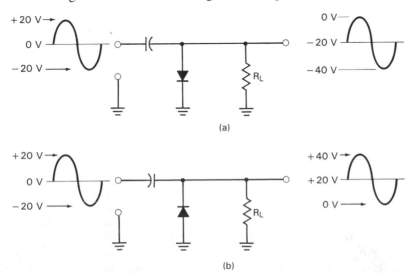

(a)

(b)

sine-wave input. By reversing the diode, we get the circuit operation shown in Figure 4.20b. Here the diode direction is reversed, making it a positive clamper. Since the diode is returned to ground (0 V), the circuit shifts the input waveform until the negative peak voltage of the waveform is approximately equal to 0 V.

There is a quick and easy memory trick for determining what type of clamper you are dealing with: *If the diode is pointing up (away from ground), the circuit is a positive clamper. If the diode is pointing down (toward ground), the circuit is a negative clamper.*

✔ 13

Biased clamper. Allows a waveform to be shifted above or below a dc reference other than 0 V.

Biased Clampers

Biased clampers allow us to shift a waveform so that it falls above or below a dc reference other than 0 V. Several biased clampers are shown in Figure 4.21.

The circuit in Figure 4.21a is easy to analyze using the procedure we have just established. The diode is pointing down, so the circuit is a negative clamper. What is the dc reference voltage for the circuit? That depends on the value of the dc voltage source (V) and the setting of R_1. If V is negative, the diode potential will fall somewhere between $-V$ and approximately 0 V. For example, let's say that V is -20 V, and R_1 is set so that the diode potential is -10V. With this setting, the clamper will shift the input waveform so that its positive peak voltage is approximately -10 V.

The clamper shown in Figure 4.21b works the same way, outside of the fact that it is a *positive* clamper. Thus, it will shift its input waveform so that its negative peak voltage is approximately equal to the potential set at the diode anode.

Zener clampers use zener diodes to determine the dc reference voltages.

The circuits in Figures 4.21c and 4.21d require a bit more explaining. The zener diodes are used to determine the dc reference voltages of the circuits. For example, if a 10-V zener is used, the dc reference voltage of Figure 4.21c will be approximately $+10$ V. If a 10-V zener is used in Figure 4.21d, the dc reference voltage of the circuit will be approximately -10 V. If 20-V zeners are used, the dc references of the circuits will be ±20 V, and so on.

Why are the *pn*-junction diodes included? The operation of the clamper depends on the *one-way* action of the diode. As you know, a zener will conduct in both the zener breakdown region *and* in the forward operating region. For example, if point A in Figure 4.21d is more positive than the value of V_Z, the zener will conduct, as will the forward-biased *pn*-junction diode. However, if point A is negative, the zener diode will try to conduct in its forward operating region. However,

FIGURE 4.21

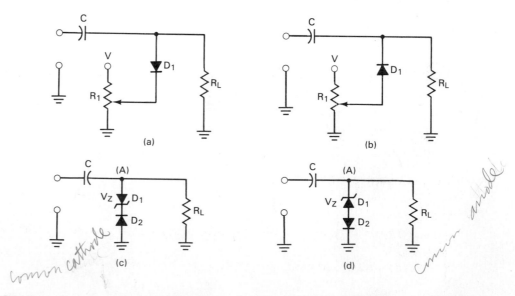

(a) (b) (c) (d)

CHAP. 4   Common Diode Applications: Clippers, Clampers, Multipliers, and Displays

this same negative potential will reverse bias the *pn*-junction diode, blocking conduction. Thus, the *pn*-junction diode ensures that the zener diode will conduct only in its reverse breakdown operating region.

Since the *pn*-junction diodes (D_2) and the zener diodes (D_1) are pointing in opposite directions, you may be wondering how you can tell whether the zener clampers are positive or negative clampers. *With the zener clamper, you still look at the direction of the pn-junction diode.*

One Final Note

Zener clampers are somewhat limited. That is, they come in one of only two varieties:

1. *Negative* clampers with *positive* dc reference voltages
2. *Positive* clampers with *negative* dc reference voltages

You cannot have a *positive* clamper with a *positive* dc reference voltage, for example. This is due to the fact that the *pn*-junction diode and the zener diode must be connected either as common-cathode (Figure 4.21c) or common-anode (Figure 4.21d) components. If D_2 in Figure 4.21c is turned around and point A is positive, nothing will stop D_1 from going into forward conduction. This will defeat the whole purpose of including D_2 in the circuit. Thus, D_2 in Figure 4.21c cannot be reversed to produce a negative clamper. By the same token, D_2 in Figure 4.21d cannot be reversed to produce a positive clamper.

Summary

The clamper shifts a waveform to a predetermined dc reference voltage. The dc reference voltage for a clamper is determined by the potential to which the diode in the circuit is returned.

When the diode in a clamper is pointing up (away from ground), the circuit is a positive clamper. The positive clamper will shift the input waveform so that its negative peak voltage is approximately equal to the dc reference voltage of the circuit. When the diode in a clamper is pointing down (toward ground), the circuit is a negative clamper. The negative clamper will shift the input waveform so that its positive peak voltage is approximately equal to the dc reference voltage of the circuit.

1. Describe the difference between *positive clampers* and *negative clampers* in terms of input/output relationships. [Objective 9]
2. What effect does a clamper have on the value of V_{ave} for a given input waveform? [Objective 10]
3. What piece of test equipment would you use to measure the value of V_{ave} for a clamper output signal? [Objective 10]
4. What effect does a clamper have on the rms voltage of a sine-wave input? [Objective 10]
5. Explain the operation of a clamper in terms of switching time constants. [Objective 11]
6. What determines the dc reference voltage of a clamper? [Objective 11]

7. How can you tell whether a given clamper is a *positive* or *negative* clamper? [Objective 12]
8. Why can't you have a negative zener clamper with a positive dc reference voltage? [Objective 13]
9. Describe the purpose served by biased clampers. [Objective 13]

4.4
Voltage Multipliers

Voltage multipliers are circuits that will provide a *dc output* that is a multiple of the *peak input voltage*. For example, a *voltage doubler* will provide a dc output voltage that is *twice* the peak input voltage, a *voltage tripler* will provide a dc output that is *three* times the peak input voltage; and so on.

While voltage multipliers will provide an output voltage that is much greater than the peak input voltage, they are not power generators. When a voltage multiplier increases the peak input voltage by a given factor, n, the peak input current is *decreased* by approximately the same factor. Thus, the actual output power from a voltage multiplier will *never* be greater than the input power.

In this section, we will take a look at several voltage multipliers. As you will see, these multipliers work on the same basic principle as the clamper: *switching time constants*.

A Practical Consideration. Since the diodes in a voltage multiplier will dissipate some power, the output power for the circuit will actually be *less than* the input power.

Half-Wave Voltage Doublers

The **half-wave voltage doubler** produces an output like that of a filtered half-wave rectifier.

The half-wave voltage doubler is made up of two diodes and two capacitors. The circuit is shown in Figure 4.22, along with its voltage source and load resistance.

The operation of the half-wave doubler is easier to understand if we go ahead and assume that the diodes are *ideal* components. During the negative alternation of the input (shown in Figure 4.23a), D_1 is forward biased and D_2 is reverse biased by the input signal polarity. If we represent D_1 as a *short* and D_2 as an *open*, we get the equivalent circuit shown in the figure. As you can see, C_1 will charge until its plate-to-plate voltage is equal to the source voltage. At the same time, C_2 will be in the process of discharging through the load resistance. (The source of this charge on C_2 will be explained in a moment.)

When the input polarity reverses, we have the circuit conditions shown in Figure 4.23b. Since D_1 is *off*, it is represented as an *open* in the equivalent circuit. Also, D_2 (which is *on*) is represented as a *short*. Using the equivalent circuit, it is easy to see that C_1 (which is charged to the peak value of V_S) and the source

FIGURE 4.22

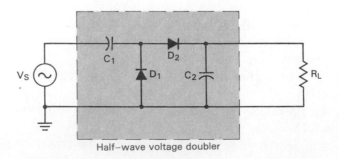

Half—wave voltage doubler

CHAP. 4   Common Diode Applications: Clippers, Clampers, Multipliers, and Displays

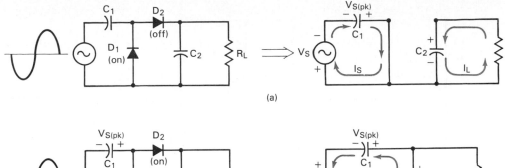

(a)

(b)

FIGURE 4.23

voltage (V_S) now act as *series-aiding* voltage sources. Thus, C_2 will charge to the sum of the series peak voltages, $2V_{S(pk)}$.

When V_S returns to its original polarity, D_2 is again turned off. With D_2 being off, the only charge path for C_2 is through the load resistance. Normally, the time constant of this circuit will be such that C_2 has little time to lose any of its charge before the input reverses polarity again. In other words, during the negative alternation of the input, C_2 will discharge *slightly*. Then, during the positive alternation, D_2 is turned on and C_2 recharges until its plate-to-plate voltage again equals $2V_{S(pk)}$.

Since C_2 barely discharges between input cycles, the output waveform of the half-wave voltage doubler closely resembles that of a filtered half-wave rectifier. Typical input and output waveforms for a half-wave voltage doubler are shown in Figure 4.24. As the figure shows, the circuit will have a dc output voltage and ripple voltage that closely resemble the output from a filtered rectifier. The dc output voltage is approximated as

> The time constant formed by the output capacitor and the load resistance prevents the capacitor from rapidly discharging.

$$\boxed{V_{dc} \cong 2V_{S(pk)}}$$

$V_r = \dfrac{I_L t}{C}$ page 108

(4.11)

The output ripple voltage is calculated using the same process that was used for the filtered half-wave rectifier.

Incidentally, if you reverse the directions of D_1 and D_2, you will have a *negative* half-wave voltage doubler. The proof that this is true is left as a "brain drain" practice problem at the end of the chapter (problem 49, page 173).

> Reversing the directions of the diodes reverses the output polarity from the half-wave doubler.

Full-Wave Voltage Doublers

The full-wave voltage doubler closely resembles the half-wave doubler. It contains two diodes and two capacitors (C_1 and C_2), as shown in Figure 4.25a. In the figure, the voltage source and load resistance are both shown, along with an added

15

FIGURE 4.24

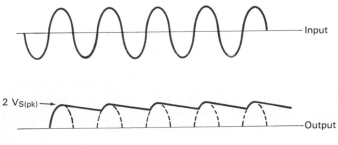

2 $V_{S(pk)}$ →

Input

Output

$V_{dc} = V_{pk} - \dfrac{V_r}{2}$

$V_{dc} - V_{pk} = \dfrac{-V_r}{2}$

$-2(V_{dc} - V_{pk}) = V_r$

$2(V_{pk} - V_{dc}) = -V_r$

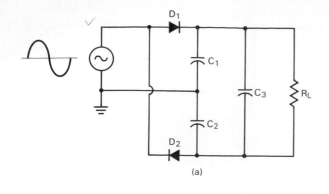

(a)

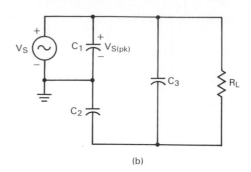

(b)

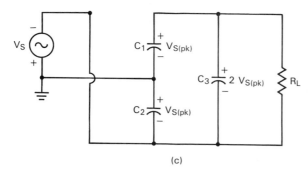

(c)

FIGURE 4.25

A Practical Consideration.
Sometimes a voltage multiplier will have a resistor in parallel with the output capacitor. This resistor, called a *bleeder resistor,* is used to provide a discharge path for the capacitor when the input voltage (V_S) is removed from the circuit.

16

Voltage tripler. Produces a dc output voltage that is three times the peak input voltage.

filter capacitor, C_3. This filter capacitor is used to reduce the ripple output from the voltage doubler.

During the positive half-cycle of the input, D_1 is forward biased and D_2 is reverse biased by the voltage source. This gives us the equivalent circuit shown in Figure 4.25b. Again, we have idealized the diodes to simplify the circuit. Using the equivalent circuit, it is easy to see that C_1 will charge to the value of $V_{S(pk)}$.

When the input polarity is reversed, D_1 will be reverse biased and D_2 will be forward biased. This gives us the equivalent circuit shown in Figure 4.25c. As this circuit shows, C_2 will now charge to the value of $V_{S(pk)}$. Since C_1 and C_2 are in series, the total voltage across the two components will be found as

$$\boxed{V_{dc} = 2V_{S(pk)}}$$

With the added filter capacitor (C_3), there will be very little ripple voltage at the output of the full-wave voltage doubler under normal circumstances. This is one of the advantages of using the full-wave voltage doubler in place of the half-wave voltage doubler. Another advantage of using this circuit is that it can be used to produce a *dual-polarity power supply*. This point will be shown later in this section.

Voltage Triplers and Quadruplers

Voltage triplers and *voltage quadruplers* are both variations on the basic half-wave voltage doubler. The schematic diagram for the voltage tripler is shown in Figure 4.26.

As you can see, the voltage tripler is very similar to the half-wave voltage doubler shown in Figure 4.23. In fact, the D_1, D_2, C_1, and C_2 circuitry forms a half-wave doubler. The added components (D_3, C_3, and C_4) form the rest of the voltage tripler.

CHAP. 4   Common Diode Applications: Clippers, Clampers, Multipliers, and Displays

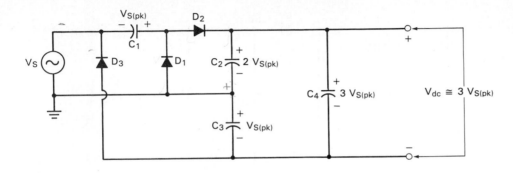

FIGURE 4.26

There are two keys to the operation of this circuit. The first is the fact that the half-wave voltage doubler works *exactly* as described earlier in this section. The second is the fact that D_3 will conduct whenever D_1 conducts.

D_1 and D_3 will conduct when V_S has the polarity shown in Figure 4.26. When D_3 is conducting, C_3 is connected in parallel to the voltage source. If we assume that D_3 is an *ideal* component, we can see that C_3 will charge until its plate-to-plate voltage is equal to $V_{S(pk)}$. As with the half-wave doubler, C_2 will have charged until its plate-to-plate voltage is equal to $2V_{S(pk)}$. Since C_2 and C_3 are in series, the total voltage across the two will now equal $3V_{S(pk)}$.

C_4 is a filter capacitor that is added to reduce the ripple in the dc output voltage. Since this capacitor is in parallel with the C_2/C_3 series circuit, it will charge to a plate-to-plate voltage that is equal to $3V_{S(pk)}$. Note that the value of C_4 is chosen so that the time constant formed with the load resistance will be extremely long. Thus, C_4 will barely discharge between charging cycles.

The *voltage quadrupler* will provide a dc output that is four times the peak input voltage. As Figure 4.27 shows, the voltage quadrupler is simply made up of parallel half-wave voltage doublers. The half-wave doubler that is made up of D_1, D_2, C_1, and C_2 will charge C_2 until its plate-to-plate voltage is equal to $2V_{S(pk)}$. The half-wave doubler formed by the other components will charge C_4 until its plate-to-plate voltage is also equal to $2V_{S(pk)}$. The series combination of C_2 and C_4 will charge the filter capacitor (C_5) until its plate-to-plate voltage is equal to $4V_{S(pk)}$. Again, the value of C_5 is chosen so that it will not lose very much of its charge between charging cycles.

 17

Voltage quadrupler.
Produces a dc output voltage that is four times the peak input voltage.

A Dual-polarity Power Supply

One application for the voltage multiplier can be seen in a basic dual-polarity dc power supply. A *dual-polarity supply* is one that provides both positive and negative dc output voltages. One such supply is shown in Figure 4.28.

The transformer in the figure is connected to a full-wave voltage doubler. The common point for the capacitors $(C_1$ and $C_2)$ is used as the power supply

 18

Dual-polarity power supply.
A dc supply that provides both positive and negative dc output voltages.

FIGURE 4.27

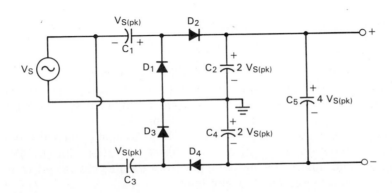

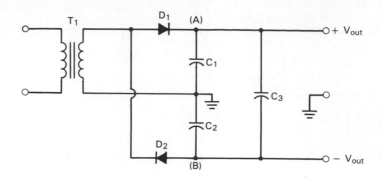

FIGURE 4.28

ground, meaning that all circuit voltages are measured with respect to this point. Thus, point A will be *positive* with respect to ground and point B will be *negative* with respect to ground. Note that the two dc output voltages will be approximately equal in magnitude to $V_{2(pk)}$. For example, if $V_{2(pk)}$ is 12 V_{pk}, the power supply will have outputs that are approximately equal to $+12$ V_{dc} and -12 V_{dc}. The filter capacitor (C_3) is added (as always) to reduce the ripple in the dc output voltages.

Practical dual-polarity power supplies are more complicated than the one shown in Figure 4.28. However, coverage of these more complex circuits is reserved for Chapter 21.

SECTION REVIEW

1. Describe the operation of the half-wave voltage doubler. [Objective 14]
2. Describe the operation of the full-wave voltage doubler. [Objective 15]
3. Why are full-wave voltage doublers preferred over half-wave voltage doublers? [Objective 15]
4. What is the similarity between the voltage tripler and the half-wave voltage doubler? [Objective 16]
5. What is the similarity between the voltage quadrupler and the half-wave voltage doubler? [Objective 17]
6. Describe the operation of the circuit shown in Figure 4.28. [Objective 18]

4.5
LED Applications

The most common application for the LED is as a *power indicator*. The light that appears on the front of your stereo set whenever the system is on is probably an LED.

Another common application for the LED is as a *level indicator* in switching circuits. Given a circuit whose output will always be at a *high dc voltage level* or a *low dc voltage level*, the LED can be used to indicate the output *state* at a given time. This application is illustrated in Figure 4.29.

The circuit in Figure 4.29a uses an LED to indicate when the output from the driving circuit is at the $+10$-V level. With a $+10$-V output, the diode is forward biased and lights. When the output from the driving circuit is at 0 V, the LED is not forward biased and will not light.

CHAP. 4 *Common Diode Applications: Clippers, Clampers, Multipliers, and Displays*

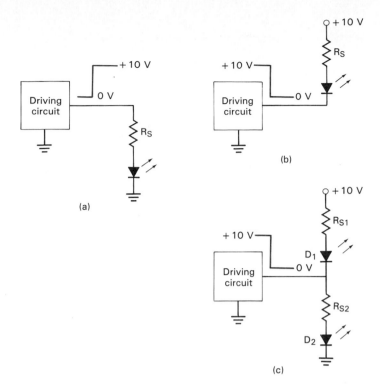

FIGURE 4.29

An LED as a level indicator.

In Figure 4.29b the LED is being used to indicate when the output from the driving circuit is 0 V. With a 0-V output, the diode is forward biased and will light. When the output from the driving circuit is +10 V, the diode has + 10 V on both sides of it and is not forward biased.

The circuit in Figure 4.29c is a combination of the other two. D_2 will light when the output from the driving circuit is at +10 V, and D_1 will light when the output from the driving circuit is 0 V.

Multisegment Displays

LEDs are used as the primary component in *multisegment displays*. Multisegment displays are used to display alphnumeric characters. The most common multisegment display is the *seven-segment display* shown in Figure 4.30. The seven-segment display uses seven LEDs shaped in a figure 8. By lighting individual LEDs, any number from 0 through 9 can be formed. For example, if LEDs b, c, f, and g are lit, the display will show the number 4.

The seven-segment display shown in Figure 4.30 is called a *common-cathode* display. This term means that the ground sides of the LEDs are tied together. Thus there is a single ground pin on this type of display. The individual LEDs are lit by applying a *high* voltage at the appropriate pins. Normally, this high voltage is within the range of +5 to +10 V. Note that each LED in the seven-segment display must have its own series current-limiting resistor.

Another type of display is the *common-anode* display. This type of display has a single +V input that goes to all the LEDs. Then individual LEDs are lit by applying a *ground* to the appropriate cathode.

Variations on the seven-segment display include displays with an added LED, as shown in Figure 4.31a. The additional LED in this display may be used as a decimal point. The display in Figure 4.31b has a segment pattern that allows you to produce any number or letter. The advantage of this display is obvious when you try to figure out a way for the seven-segment display to display the letter M.

✔ 20

Seven-segment display. A device made up of seven LEDs shaped in a figure 8 that is used to display numbers.

Common-cathode display. A display in which all LED cathodes are tied to a single pin.

more after use

Common-anode display. A display in which all LED anodes are tied to a single pin.

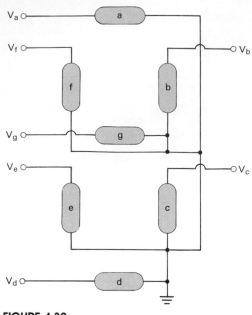

FIGURE 4.30

A seven-segment display.

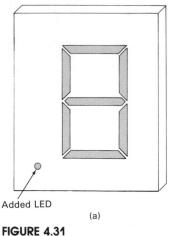 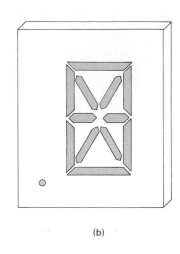

Added LED

(a)   (b)

FIGURE 4.31

Other multisegment displays.

SECTION REVIEW

1. Discuss the use of an LED as a power/level indicator. [Objective 19]
2. Discuss the use of LEDs in multisegment displays. [Objective 20]

4.6
Diode Circuit Troubleshooting

Now that we have covered the operating principles of clippers, clampers, voltage multipliers, and multisegment displays, we are going to take a look at the fault symptoms that commonly occur in these circuits.

FIGURE 4.32

◢ 21

A Practical Consideration. Carbon resistors do not internally short. If a short exists, the short is normally caused by a solder bridge or by accidentally using the wrong value of resistor. A visual inspection will find either of these two problems.

Clipper Faults

The faults that normally occur in series clippers are the same as those of the half-wave rectifier that were discussed in Chapter 3. The fault/symptom relationships shown in Table 4.1 are those for the basic shunt clipper (shown in Figure 4.32), the biased clipper, and the zener shunt clipper.

TABLE 4.1
Shunt Clipper Faults

Fault	Symptom(s)
R_S open	Since the source is isolated from the load, load voltage and current will both drop to zero.
D_1 open	Clipping action will be lost; so the output waveform will be nearly identical in shape to the input waveform.
D_1 shorted	The symptoms will be the same as those for R_S open because the full applied voltage will be dropped across R_S. The shorted diode will short out the load.

     CHAP. 4   Common Diode Applications: Clippers, Clampers, Multipliers, and Displays

Clamper Faults

The fault/symptom relationships shown in Table 4.2 are those for the basic clamper shown in Figure 4.33a.

TABLE 4.2
Clamper Faults

Fault	Symptom(s)
C_1 open	The voltage source is isolated from the rest of the circuit, so load and current will drop to zero.
C_1 shorted	The circuit will act as a *shunt clipper*. If the current drawn through the diode is sufficient, the diode will eventually open.
C_1 leaky	The capacitor will attempt to charge, but will not be able to hold the charge for any period of time. This will cause the dc reference of the output signal to constantly change. The waveform itself may also be extremely distorted.
D_1 open	All clamping action will be lost and the output waveform will be centered around 0 V.
D_1 shorted	The voltage source will be shorted to ground by the short *RC* time constant of the capacitor and the shorted diode. The output from the voltage source will be loaded down, and the voltage source itself may be damaged.

shorted

Faults that are common to most clampers.

The *biased clamper* (Figure 4.33b) may develop any of the fault/symptom combinations listed in Table 4.2. In addition, it can develop any of the fault/symptom combinations given in Table 4.3.

TABLE 4.3
Additional Biased-clamper Faults

Fault	Symptom(s)
V out	If the dc voltage source goes *out* (drops to 0 V), the clamper will work like an unbiased clamper. If the dc voltage source in an electronic system goes out, many other circuits in the system (including the clamper signal source) may also stop working.
R_1 open	If R_1 opens, the results will be the same as D_1 opening. All clamping action will be lost, and the output waveform will be centered around 0 V.

Additional faults that may develop in *biased clampers*.

FIGURE 4.33

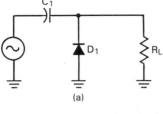

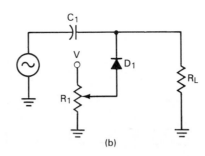

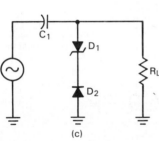

The *zener clamper* (Figure 4.33c) may develop any of the capacitor fault/symptom combinations listed in Table 4.2. Also, if either diode *opens*, the results will be the same as those shown in Table 4.2 for an open diode. If either diode in the zener clamper *shorts*, the results are slightly different than those listed in Table 4.2. The results of either diode shorting in the zener clamper are shown in Table 4.4.

Additional faults that may develop in the *zener clamper*.

TABLE 4.4
Additional Zener Clamper Faults

Fault	Symptom(s)
D_1 shorted	If D_1 shorts, the clamper will act as an unbiased clamper. As you can see, shorting D_1 in Figure 4.33c gives us the exact same circuit as the one shown in Figure 4.33a.
D_2 shorted	The zener diode will turn on during both alternations of the input waveform. The result will be an extemely distorted output waveform.

As always, there is a strong possibility that both diodes in the zener clamper will be destroyed if either diode shorts. You may recall from our discussion on rectifier circuits that it is not uncommon for a shorting diode to destroy any series diodes in the process. The best thing to do whenever you find a shorted diode is to go ahead and replace all diodes that are in series with the component.

Voltage Multiplier Faults

Of all the circuits covered in this chapter, voltage multipliers are by far the most difficult to troubleshoot. The maze of diodes and capacitors that make up the voltage multiplier can develop many faults, each having symptoms that may closely resemble those of some other fault.

The best approach to troubleshooting the voltage multiplier is to start by reading the voltage across the output filter capacitor. For example, if we were going to troubleshoot the voltage tripler shown in Figure 4.34, we would start by reading the voltage across C_4.

The value of voltage across the filter capacitor may help to isolate the area where the fault is located. For example, let's say that we are testing the voltage tripler shown in Figure 4.34, and the voltage across C_4 is close to $2V_{S(pk)}$. This

A Practical Consideration: Don't expect the capacitor voltages to be exactly equal to the peak values we have assumed in our discussions. The diodes will drop some voltage, and the capacitors will actually be charging and discharging continually. However, the average capacitor voltages should be close to the appropriate multiples of $V_{S(pk)}$ that you were shown earlier.

FIGURE 4.34

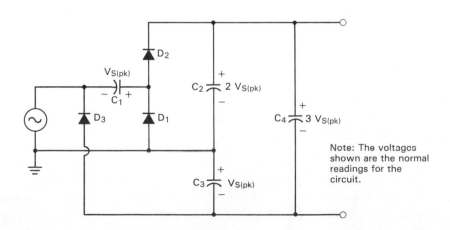

Note: The voltages shown are the normal readings for the circuit.

would indicate that C_3 is not charging, since the voltage across this component is usually equal to V_S. We would then test D_3 and C_3, the two components that lie in the C_4 charging path.

If the voltage across C_4 was approximately equal to $V_{S(pk)}$, we would then check the voltage across C_2, since the voltage across this component is normally equal to $2V_{S(pk)}$. If the voltage across the component is not correct, we would check the voltage across C_1. If the voltage across C_1 is correct, we would test D_2 and C_2. If the voltage across C_1 is not correct, we would test D_1 and C_1.

When testing voltage multipliers, it helps if you have a quick reference showing you the normal capacitor voltage readings for the various multiplier circuits. Such a reference is supplied in Table 4.5.

When testing a voltage multiplier, use the listing shown in Table 4.5 and the procedure outlined next:

1. *Measure the voltage across the output capacitor. The reading may direct you to test one of the parallel charging capacitors.* For example, if the voltage across C_4 (Figure 4.34) is short by V_S, you would check C_3. If it is short by $2V_{S(pk)}$, you would check C_2.

2. Measure the voltage across any capacitor that lies in the charge path of an uncharged capacitor. For example, if C_2 (Figure 4.34) is not charged, you would measure the voltage across C_1.

3. Continue the process until you find the uncharged capacitor that is closest to the source. Then test all the components in the charge path for that capacitor. For example, if C_2 (Figure 4.34) is not charged and the voltage across C_1 is normal, check D_2 and C_2. If the voltages across C_1 and C_2 are both wrong, check D_1 and C_1.

TABLE 4.5

Capacitor	Half-wave doubler	Full-wave doubler	Tripler	Quadrupler
C_1	$V_{S(pk)}$	$V_{S(pk)}$	$V_{S(pk)}$	$V_{S(pk)}$
C_2	$2V_{S(pk)}$	$V_{S(pk)}$	$2V_{S(pk)}$	$2V_{S(pk)}$
C_3		$2V_{S(pk)}$	$V_{S(pk)}$	$V_{S(pk)}$
C_4			$3V_{S(pk)}$	$2V_{S(pk)}$
C_5				$4V_{S(pk)}$

If none of the capacitor charges are correct, and there is no problem with either C_1 or D_1, the voltage source is the problem.

In most cases, the procedure outlined here will do the trick. When it doesn't, you may as well get comfortable . . . you're going to be working on the multiplier for quite a while.

Display Faults

Multisegment displays are driven by ICs called *decoder–drivers*. The most common symptom of a fault in the driver-display circuit is the failure of one or more segments to light.

The testing of a driver-display circuit is simple. When a segment should be *on*, check the driving pin for that segment. If the voltage there is correct (a positive voltage for a *common-anode* display and 0 V for a *common-cathode* display), the problem is the display. If the voltage is not there, the problem is likely the decoder–driver. This assumes that the inputs to the decoder–driver are correct.

Decoder–driver. An IC that is used to drive a multisegment display.

1. Describe the diode faults that can occur in the zener clamper and the symptoms of each. [Objective 22]
2. Why should you replace both diodes in a zener clamper if one of them shorts? [Objective 22]
3. Describe the process used to troubleshoot a voltage multiplier. [Objective 23]

KEY TERMS

The following terms were introduced and defined in this chapter:

alphanumeric symbols	half-wave voltage doubler	symmetrical zener shunt
am detector	limiter	clipper
biased clipper	multisegment display	transient
biased clamper	negative biased clipper	voltage multiplier
clamper	negative clamper	voltage quadrupler
clipper	negative shunt clipper	voltage tripler
common-anode display	positive biased clipper	zener clamper
common-cathode display	positive clamper	zener shunt clipper
dc reference voltage	positive shunt clipper	
dc restorer	series clipper	
decoder–driver	seven-segment display	
dual-polarity power	shunt clipper	
supply	switching time constants	
full-wave voltage doubler		

PRACTICE PROBLEMS

§4.1

1. Determine the positive peak output voltage from the circuit in Figure 4.35a. [3]
2. Draw the output waveform for the circuit in Figure 4.35a. Label the peak voltage values on the waveform. [3]
3. Determine the negative peak output voltage for the circuit shown in Figure 4.35b. [3]
4. Draw the output waveform for the circuit in Figure 4.35b. Label the peak voltage values on the waveform. [3]
5. Determine the peak voltage values for the output from Figure 4.35c. Then draw the waveform and include the voltage values in the drawing. [3]
6. Repeat problem 5 for the circuit shown in Figure 4.35d. [3]
7. The circuit in Figure 4.36a has values of $V_S = 12$ V_{pk} and $V_Z = 10$ V. Determine the peak output voltages and draw the output waveform. [4]
8. The circuit in Figure 4.36a has values of $V_S = 24$ V_{pk} and $V_Z = 12$ V. Determine the peak output voltages and draw the output waveform. [4]

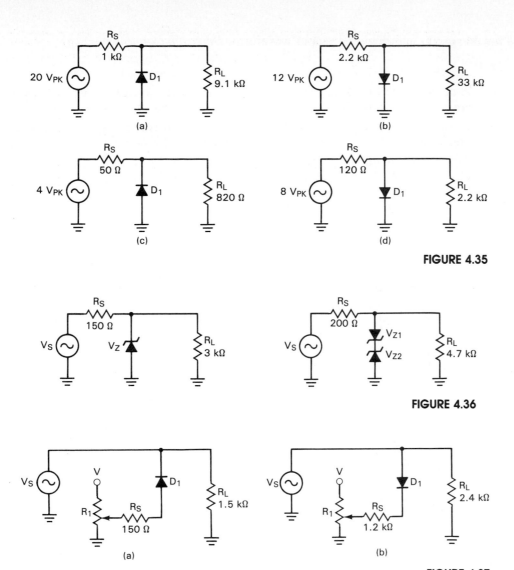

FIGURE 4.35

FIGURE 4.36

FIGURE 4.37

9. Reverse the direction of the diode and repeat problem 8. [4]

10. The circuit in Figure 4.36b has values of $V_S = 24$ V$_{PP}$ and $V_{Z1} = V_{Z2} = 6.2$ V. Determine the peak output voltages and draw the output waveform. [5]

11. The circuit in Figure 4.36b has values of $V_S = 36$ V$_{PP}$ and $V_{Z1} = V_{Z2} = 10$ V. Determine the peak output voltages and draw the output waveform. [5]

12. The potentiometer in Figure 4.37a is set so that the anode of D_1 is at -2 V. If the value of V_S is 14 V$_{PP}$, what are the peak output voltages for the circuit? [6]

13. The potentiometer in Figure 4.37a is set so that the anode voltage of D_1 is -8 V. Assuming that $V_S = 24$ V$_{PP}$, determine the peak output voltages for the circuit and draw the output waveform. [6]

14. The potentiometer in Figure 4.37b is set so that the cathode voltage of D_1 is $+4$ V. Assuming that $V_S = 22$ V$_{PP}$, determine the peak output voltages for the circuit and draw the output waveform. [6]

15. The potentiometer in Figure 4.37b is set so that the cathode voltage of D_1 is $+2$ V. Assuming that $V_S = 4$ V$_{PP}$, determine the peak output voltages

for the circuit and draw the output waveform. *Be careful on this one!* [6]

§4.3

16. A clamper with a 24-V_{PP} input shifts the waveform so that its peak voltages are 0 and -24 V. Determine the dc average of the output waveform. [10]

17. A clamper with a 14-V_{PP} input shifts the waveform so that its peak voltages are 0 and $+14$ V. Determine the dc average of the output waveform. [10]

18. The circuit in Figure 4.38a has values of $R_{D1} = 24\ \Omega$, $R_L = 2.2\ k\Omega$, and $C_1 = 4.7\ \mu F$. Determine the charge and discharge times for the capacitor. [11]

19. The circuit in Figure 4.38a has a 14-V_{PP} input signal. Draw the output waveform and determine its peak voltage values. [11]

20. The circuit shown in Figure 4.38b has values of $R_{D1} = 8\ \Omega$, $R_L = 1.2\ k\Omega$, and $C = 33\ \mu F$. Determine the charge and discharge times for the capacitor. [11]

21. The circuit shown in Figure 4.38b has values of $R_{D1} = 14\ \Omega$, $R_L = 1.5\ k\Omega$, and $C = 1\ \mu F$. Determine the charge and discharge times for the capacitor. [11]

22. The input to the circuit in Figure 4.38b is a 12-V_{PP} signal. Determine the peak output voltages for the circuit and draw the output waveform. [11]

23. The potentiometer in Figure 4.39a is set so that the voltage at the anode of D_1 is $+3$ V. The value of V_S is 9 V_{PP}. Draw the output waveform and determine its peak voltage values. [13]

24. The potentiometer in Figure 4.39b is set so that the cathode voltage of D_1 is $+6$ V. The value of V_S is 30 V_{PP}. Draw the output waveform for the circuit and determine its peak voltage values. [13]

25. Draw the output waveform for the circuit in Figure 4.40a and determine its peak voltage values. [13]

26. Draw the output waveform for the circuit in Figure 4.40b and determine its peak voltage values. [13]

§4.4

27. The circuit in Figure 4.41 has a 15-V_{pk} input signal. Determine the values of V_{C1} and V_{C2} for the circuit. Assume that the diodes are ideal components. [14]

28. The circuit in Figure 4.41 has a 48-V_{pk} input signal. Determine the values of V_{C1} and V_{C2} for the circuit. Assume that the diodes are ideal components. [14]

29. The circuit in Figure 4.42 has a 24-V_{pk} input signal. Determine the values of V_{C1}, V_{C2}, and V_{C3} for the circuit. Assume that the diodes are ideal components. [15]

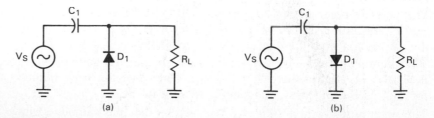

FIGURE 4.38

CHAP. 4   Common Diode Applications: Clippers, Clampers, Multipliers, and Displays

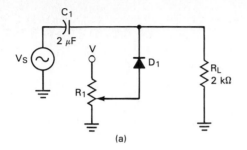

(a)

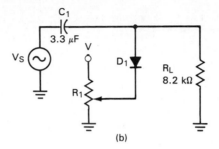

(b)

FIGURE 4.39

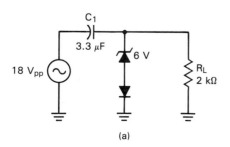

(a)

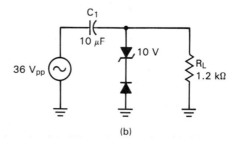

(b)

FIGURE 4.40

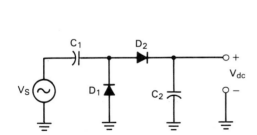

FIGURE 4.41

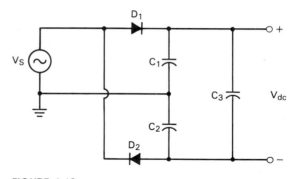

FIGURE 4.42

30. The circuit in Figure 4.42 has a 25-V_{ac} input signal. Determine the values of V_{C1}, V_{C2}, and V_{C3} for the circuit. Assume that the diodes are ideal components. *Be careful on this one!* [15]

31. Determine the dc output voltage for the circuit shown in Figure 4.43. Assume that the diodes are ideal components. [15]

32. Determine the dc output voltage for the circuit shown in Figure 4.44. Assume that the diodes are ideal components. [15]

33. The circuit shown in Figure 4.45 has a 20-V_{pk} input signal. Determine the values of V_{C1}, V_{C2}, V_{C3}, and V_{C4} for the circuit. Assume that the diodes are ideal components. [16]

34. The circuit shown in Figure 4.45 has a 15-V_{ac} input signal. Determine the values of V_{C1}, V_{C2}, V_{C3}, and V_{C4} for the circuit. Assume that the diodes are ideal components. [16]

Practice Problems

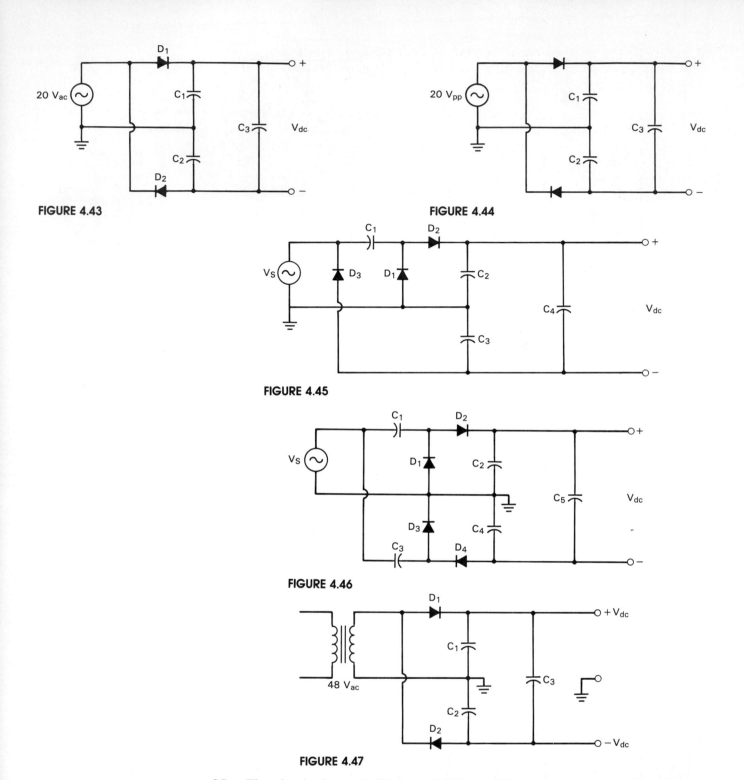

FIGURE 4.43

FIGURE 4.44

FIGURE 4.45

FIGURE 4.46

FIGURE 4.47

35. The circuit shown in Figure 4.46 has an 18-V_{pk} input signal. Determine the values of V_{C1}, V_{C2}, V_{C3}, V_{C4}, and V_{C5} for the circuit. Assume that the diodes are ideal components. [16]

36. The circuit shown in Figure 4.46 has a 36-V_{ac} input signal. Determine the values of V_{C1}, V_{C2}, V_{C3}, V_{C4}, and V_{C5} for the circuit. Assume that the diodes are ideal components. [16]

37. Determine the dc output voltage values for the dual-polarity power supply shown in Figure 4.47. Assume that the diodes are ideal components. [18]

38. Assume that the transformer in Figure 4.47 is a 36-V_{ac} transformer and that the diodes in the circuit are ideal components. Determine the output dc voltage values for the circuit. [18]

TROUBLESHOOTING PRACTICE PROBLEMS

39. The circuit in Figure 4.48a has the input/output waveforms shown. Determine whether or not there is a problem in the circuit. If there is, discuss the possible cause(s) of the problem. [21]

40. The circuit in Figure 4.48b has the input/output waveforms shown. Determine whether or not there is a problem in the circuit. If there is, discuss the possible cause(s) of the problem. [21]

41. The circuit in Figure 4.49a has the input/output waveforms shown. Determine whether or not there is a problem in the circuit. If there is, discuss the possible cause(s) of the problem. [22]

42. The circuit in Figure 4.49b has the input/output waveforms shown. Determine whether or not there is a problem in the circuit. If there is, discuss the possible cause(s) of the problem. [22]

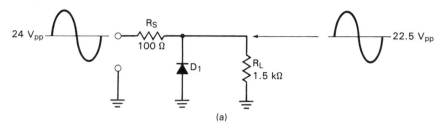

(a)

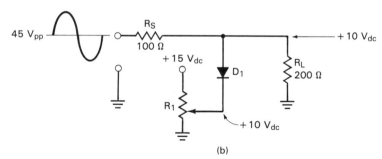

(b)

FIGURE 4.48

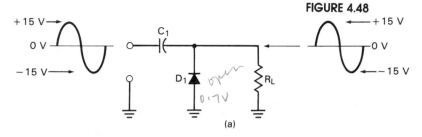

(a)

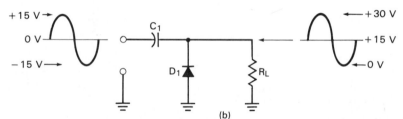

(b)

FIGURE 4.49

43. The circuit in Figure 4.50a has the input/output waveforms shown. Determine whether or not there is a problem in the circuit. If there is, discuss the possible cause(s) of the problem. [22]

44. The circuit in Figure 4.50b has the input/output waveforms shown. Determine whether or not there is a problem in the circuit. If there is, discuss the possible cause(s) of the problem. [22]

45. The circuit in Figure 4.51 has the capacitor voltages shown. Determine whether or not there is a problem in the circuit. If there is, list the steps you would take to find the faulty component. [23]

46. The circuit in Figure 4.52 has the capacitor voltages shown. Determine whether or not there is a problem in the circuit. If there is, list the steps you would take to find the faulty component. [23]

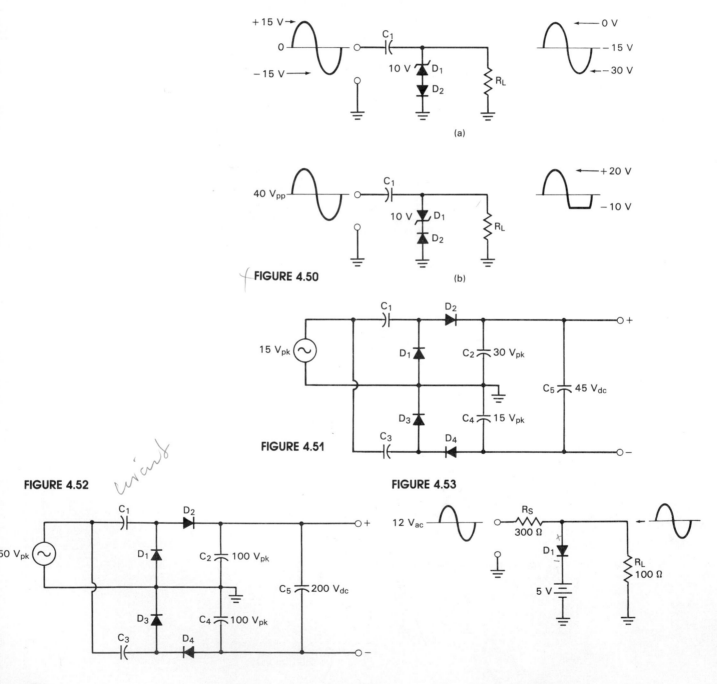

FIGURE 4.50

FIGURE 4.51

FIGURE 4.52

FIGURE 4.53

CHAP. 4 Common Diode Applications: Clippers, Clampers, Multipliers, and Displays

THE BRAIN DRAIN

47. Refer to Figure 4.36b (page 167). Assume that the circuit has values of $V_S = 24$ V_{ac}, $V_{Z1} = 6$ V, and $V_{Z2} = 8$ V. Draw the output waveform for the circuit and determine (a) the peak output voltage values and (b) the dc average of the waveform.

48. Explain the output waveform shown in Figure 4.53. (*Note:* There are no faulty components in the circuit.)

49. Figure 4.54 shows the basic half-wave voltage doubler with the diodes and capacitors reversed. Analyze the circuit and show that it will provide a *negative* dc output that is approximately twice the peak input voltage.

50. Schematic diagrams can be drawn in many different ways. For example, the circuit shown in Figure 4.55 is actually a voltage quadrupler. However, one component in the circuit has accidentally been placed in backward. Which one is it? Explain your answer.

51. If you look closely at the voltage multipliers discussed in this chapter, you may notice that there is a relationship between the circuit components and the value of the circuit multiplier. If you can determine the relationship, you will be able to determine the dc output voltage for the multipler shown in Figure 4.56.

SUGGESTED COMPUTER APPLICATIONS PROBLEMS

52. Write a program that will determine the peak output voltages for a biased shunt clipper. The program should allow the user to input the value of $V_{S(pk)}$ and the value of the diode biasing voltage. It should also be able to solve both negative and positive clippers.

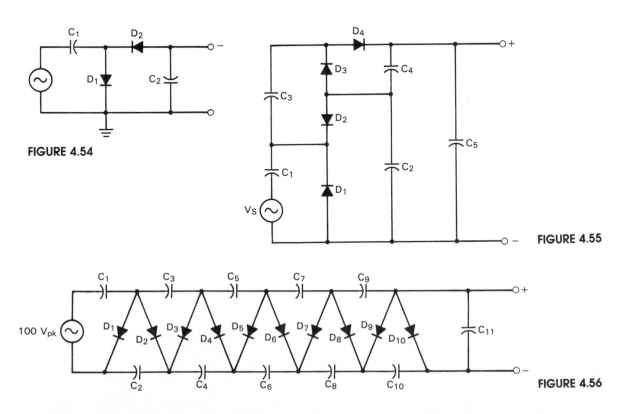

FIGURE 4.54

FIGURE 4.55

FIGURE 4.56

53. Modify the program in problem 52 to accept the value of V_S in peak, peak-to-peak, or rms form.

54. Write a program that will determine the peak output voltages and dc output voltage for either a positive or negative unbiased clamper. The program should provide for V_S inputs in any standard voltage form (peak, rms, and so on).

ANSWERS TO THE EXAMPLE PRACTICE PROBLEMS

4.1.   12.54 V_{pk}

4.2. $V_{out} = -0.7$ V, $-V_{R_S} = -14.3 V_{pk}$

4.3.   Positive alternation: $V_{out} = 0.7$ V, $V_{R_S} = 11.3$ V_{pk}; negative alternation: $V_{out} = -11$ V_{pk}

4.5. $T_C = 188$ μs, $T_D = 28.2$ ms

Bipolar Junction Transistors

In the 1940s, a team of scientists working at Bell Labs developed the transistor, the first solid-state device capable of amplifying an ac signal. The term solid-state was coined because the transistor was solid, rather than hollow like the vacuum tube it was designed to replace.

OBJECTIVES

☐ 1. Define *transistor* and state its primary application. (Introduction)

☐ 2. Name the three terminals of the bipolar junction transistor. (§5.1)

☐ 3. Explain the difference between the construction of the *npn* transistor and the construction of the *pnp* transistor. (§5.1)

☐ 4. Identify all the following, given the schematic symbol for a transistor: (§5.1)
   a. The type of transistor
   b. The base, collector, and emitter terminals
   c. The direction of the emitter current

☐ 5. Briefly discuss the transistor as a *current-controlled* device. (§5.1)

☐ 6. Define the following terms: *amplitude* and *current gain*. (§5.1)

☐ 7. Calculate the value of collector current (I_C), given the values of base current (I_B) and current gain (β). (§5.1)

☐ 8. State the relationship between the three transistor terminal currents. (§5.1)

☐ 9. List and describe the nine transistor circuit voltages. (§5.1)

☐ 10. List the three operating regions of the transistor and the junction biasing for each. (§5.2)

☐ 11. Describe the characteristics of a transistor that is in *cutoff*. (§5.2)

☐ 12. Describe the characteristics of a *saturated* transistor. (§5.2)

☐ 13. Describe the characteristics of a transistor that is being operated in the *active* region. (§5.2)

☐ 14. Define *beta* and perform transistor current calculations using the *dc beta* rating of the device. (§5.3)

☐ 15. Define *alpha* and state its value limit. (§5.3)

☐ 16. Perform transistor current calculations using the *dc alpha* rating of the device. (§5.3)

☐ 17. Discuss the relationship between beta and alpha. (§5.3)

☐ 18. Calculate the value of alpha for a given transistor, given the beta rating of the device. (§5.3)

☐ 19. Calculate the maximum allowable base current for a transistor, given the maximum allowable value of collector current and the maximum beta rating of the device. (§5.3)

☐ 20. List and discuss the importance of the transistor voltage ratings. (§5.3)

☐ 21. Describe the *common-emitter* configuration and list its characteristics. (§5.4)

☐ 22. Describe the *common-collector* configuration and list its characteristics. (§5.4)

☐ 23. Describe the *common-base* configuration and list its characteristics and applications. (§5.4)

☐ 24. Determine the configuration of a given transistor circuit. (§5.4)

☐ 25. Explain the operation of the transistor using the collector curve of the device. (§5.5)

☐ 26. Describe the base curve of a transistor. (§5.5)

☐ 27. Describe the relationship between *beta, temperature,* and *dc collector current*. (§5.5)

☐ 28. List and define the commonly used transistor maximum ratings and dc electrical characteristics as shown on the device specification sheet. (§5.6)

☐ 29. Describe the four checks used to test a faulty BJT. (§5.7)

☐ 30. Describe the differences between *npn* transistor circuits and *pnp* transistor circuits. (§5.8)

☐ 31. Explain how most electronic systems provide transistor supply voltages. (§5.8)

☐ 32. Explain the difference between *integrated* and *discrete* transistors. (§5.8)

☐ 33. Describe the characteristics of *high-voltage transistors*. (§5.8)

☐ 34. Describe the characteristics of *high-current transistors*. (§5.8)

☐ 35. Describe the characteristics of *high-power transistors*. (§5.8)

As you master each of the objectives listed, place a check mark (✔) in the appropriate box.

The main building block of modern electronic systems is the *transistor*. The transistor is *a three-terminal device whose output current, voltage, and/or power are controlled by its input current*. In communications systems, it is used as the primary component in the amplifier, a circuit that is used to increase the strength of an ac signal. In digital computer electronics, the transistor is used as a *high-speed electronic switch* that is capable of switching between two operating states (open and closed) at a rate of several billions of times per second.

There are two basic transistor types: the *bipolar junction transistor* (BJT) and the *field-effect transistor* (FET). As you will see, these two transistor types differ in both their operating characteristics and their internal construction. You should note that the single term *transistor* is used to identify the BJT. The field-effect transistor is referred to simply as a FET.

In this chapter we will take a look at the bipolar junction transistor and its basic operating principles. FETs are discussed in detail in Chapter 10.

Transistor. A three-terminal device whose output current, voltage, and/or power are controlled by its input current.

Amplifier. A circuit used to increase the strength of an ac signal.

5.1
Transistor Operation: An Overview

The bipolar junction transistor is a *three-terminal* component. The three terminals are called the *emitter*, *collector*, and *base*. The emitter and collector terminals are made up of the same type of semiconductor material, either *p-type* or *n-type*, while the base is made of the other. This point is illustrated in Figure 5.1.

As you can see, there are two types of bipolar junction transistors. The first type, called the *npn transistor*, has *n*-type emitter and collector materials and a *p*-type base. The *pnp transistor* is made in the opposite manner. This transistor has *p*-type emitter and collector materials and an *n*-type base.

***npn* Transistor.** A BJT with *n*-type emitter and collector materials and a *p*-type base.

***pnp* Transistor.** A BJT with *p*-type emitter and collector materials and an *n*-type base.

FIGURE 5.1

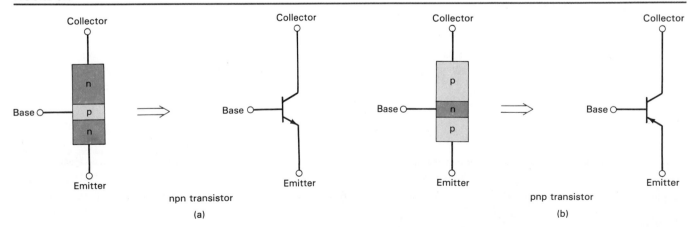

(a) npn transistor

(b) pnp transistor

Figure 5.1 also shows the schematic symbols for the *npn* and *pnp* transistors. The arrow on the schematic symbol is important for three reasons:

1. It identifies the component terminals. *The arrow is always drawn on the emitter terminal.* The terminal on the opposite side is the collector, and the center terminal is the base.

2. The arrow always points toward the *n*-type material. If the arrow points toward the base, the transistor is a *pnp* type. If the arrow points toward the emitter, the component is an *npn* type.

3. The arrow indicates the direction of current through the device. As with the *pn* junction diode, electron flow will be *against the arrow*. As you will see later in this chapter, knowing the direction of current in the emitter terminal will tell you the direction of the other two terminal currents (base current and collector current).

Transistor Currents

Base Current (I_B) can be varied to control the emitter current (I_E) and collector current (I_C).

Amplitude. The maximum value of a changing voltage or current

Current gain (β). The factor by which current increases from the base to collector.

The transistor is a *current-controlled* device. The base current (I_B) can be varied to control the amount of emitter current (I_E) and collector current (I_C). Because of the construction of the component, *a small change in base current causes a large change in both the emitter and collector currents.* This point is illustrated in Figure 5.2. As you can see, the ac emitter and collector currents are approximately equal, while the ac base current is at a much smaller value. If an ac input is applied to the base terminal of a transistor, and the output is taken from the collector terminal, the result is an increase in the *amplitude* of the ac signal.

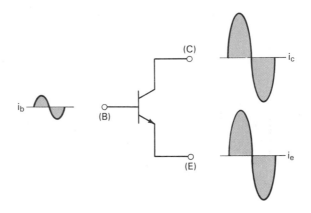

FIGURE 5.2

BJT current relationships.

The *factor* by which current increases from the base to the collector of the transistor is referred to as the *current gain* of the component. Current gain is often represented by the Greek letter *beta* (β). If you want to determine the amount of collector current in a transistor, simply multiply the base current by the beta rating of the component. This point is illustrated in the following example.

═══════════ **EXAMPLE 5.1** ═══════════

Determine the amount of ac collector current for the transistor shown in Figure 5.3.

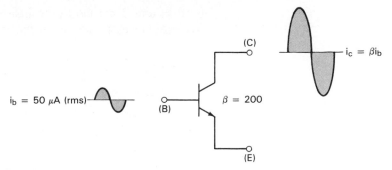

FIGURE 5.3

Solution: The ac base current in Figure 5.3 is shown as being 50 μA (rms). The beta rating of the transistor is 200. The ac collector current is found by multiplying I_B by β. By formula,

$$I_C = \beta I_B$$
$$= (200)(50 \text{ μA})$$
$$= 10 \text{ mA}$$

Thus 50 μA of ac base current produces 10 mA of ac collector current. This is definitely an *increase* (gain) in ac current from the transistor input to transistor output.

PRACTICE PROBLEM 5–1

A transistor has values of $I_B = 50$ μA and β $= 350$. Determine the value of I_C for the device.

The emitter current in a transistor is equal to the sum of the base and collector currents. By formula,

 8

$$\boxed{I_E = I_B + I_C} \tag{5.1}$$

Since I_B is normally *much less* than I_C, the collector and emitter currents are approximately equal. By formula,

$$\boxed{I_C \cong I_E} \tag{5.2}$$

A Practical Consideration. The higher the value of β is the closer I_C will be to the I_E. This can be seen by comparing your results from Practice Problem 5–1 with the results from Example 5.1.

The current relationships shown in equations (5.1) and (5.2) are correct for both the *npn* and *pnp* transistors.

The validity of equation (5.2) can be seen by looking at the results in Example 5.1. For the transistor, I_B was given as 50 μA and I_C was determined to be 10 mA. According to equation (5.1), I_E for the device would be found as

$$I_E = I_B + I_C$$
$$= 50 \text{ μA} + 10 \text{ mA}$$
$$= 10.05 \text{ mA}$$

As you can see, the values of I_E and I_C are approximately equal for the transistor in Example 5.1.

We will discuss the transistor currents in greater detail later in this chapter. Before we do, you need to be familiar with the voltage labels that are commonly used in any discussion on transistor circuit operation. We will look at the various transistor circuit voltages at this point.

Transistor Voltages

Several voltages are normally involved in any discussion of transistor operation. These voltages are listed in Table 5.1. The voltages listed are identified in Figure 5.4. V_{CC}, V_{EE}, and V_{BB} are identified as dc power supplies. V_C, V_B, and V_E are

TABLE 5.1
Transistor Voltages

Voltage Abbreviation	Definition
V_{CC}	*Collector biasing voltage*. This is a power supply voltage that is directly or indirectly applied to the collector of the transistor.
V_{BB}	*Base biasing voltage*. This is a dc voltage that is used to bias the base of the transistor. It may come directly from a dc voltage supply or may be applied indirectly to the base by a resistive circuit.
V_{EE}	*Emitter biasing voltage*. This, again, is a dc biasing voltage. In many cases, V_{EE} will be nothing more than a *ground* connection.
V_C	This is the dc voltage *measured from the collector terminal of the component to ground*.
V_B	This is the dc voltage *measured from the base terminal to ground*.
V_E	This is the dc voltage *measured from the emitter terminal to ground*.
V_{CE}	This is the dc voltage *measured between the collector and emitter terminals of the transistor*.
V_{BE}	This is the dc voltage *measured between the base and emitter terminals of the transistor*.
V_{CB}	This is the dc voltage *measured between the collector and base terminals of the component*.

A Memory Trick: There is a relatively simple way to remember the voltages listed here. When the voltage has a double subscript (such as CC, BB, or EE), it is a supply voltage. When two different subscripts are shown (such as CE, BE, or CB), the voltage is measured between the two terminals. When only one subscript is shown, the voltage is measured from that terminal to ground.

FIGURE 5.4

BJT amplifier voltages.

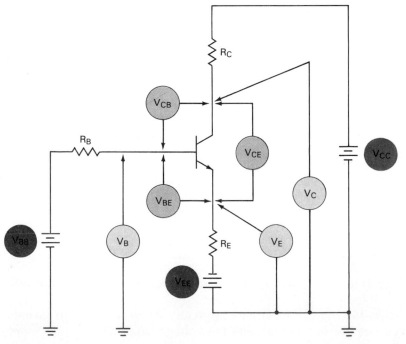

CHAP. 5   Bipolar Junction Transistors

all shown as voltages that are measured from their respective terminals *to ground*. V_{CE}, V_{BE}, and V_{CB} are all shown as being measured between the terminals indicated by the subscripts. Be sure that you are able to distinguish between these voltages, as it will make future discussions of transistor operation easier to follow.

SECTION REVIEW

1. What is the primary application for the transistor in communications electronics? [Objective 1]
2. What is the primary application for the transistor in digital electronics? [Objective 1]
3. What are the two basic types of transistors? [Objective 1]
4. What are the terminals of a bipolar junction transistor (BJT) called? [Objective 2]
5. What are the two types of BJTs? How are they different? [Objective 3]
6. What is indicated by the arrow on the BJT schematic symbol? [Objective 4]
7. Draw and label the schematic symbols of an *npn* and a *pnp* transistor. [Objective 4]
8. What is meant by the term *current-controlled device*? [Objective 5]
9. What is meant by the term *amplitude*? [Objective 6]
10. What is meant by the term *current gain*? [Objective 6]
11. What symbol is commonly used to represent current gain? [Objective 6]
12. Under normal circumstances, what is the relationship between: [Objective 8]
    a. Base current and collector current?   b. Collector current and emitter current?
13. Define the following transistor voltages: V_{CC}, V_{BB}, V_{EE}, V_C, V_B, V_E, V_{CE}, V_{BE}, and V_{CB}. [Objective 9]

5.2
Transistor Construction and Operation

The transistor is made up of three separate semiconductor materials. The three materials are joined together in such a way as to form two *pn* junctions. This is illustrated in Figure 5.5.

The two junctions in a BJT are referred to as the **emitter–base junction** and the **collector–base junction**.

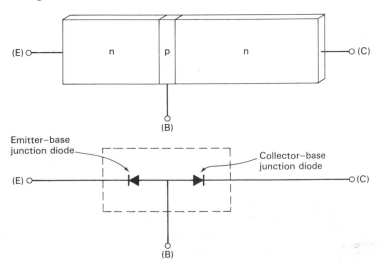

FIGURE 5.5

BJT construction.

The point at which the emitter and base are joined forms a single *pn* junction called the *emitter–base* junction. The *collector–base* junction is the point where the base and collector meet. The two junctions are normally operated in one of three biasing combinations, as follows:

✔ 10

Emitter–Base Junction	Collector–Base Junction	Operating Region
Reverse biased	Reverse biased	Cutoff
Forward biased	Reverse biased	Active
Forward biased	Forward biased	Saturation

The transistor operating regions are called *cutoff, active,* and *saturation.*

When both junctions are reverse biased, the transistor is said to be in *cutoff*. When the emitter–base junction is forward biased and the collector–base junction is reverse biased, the transistor is said to be operating in the *active* region. When both junctions are forward biased, the transistor is said to be operating in the *saturation* region. These "regions" refer to areas on a characteristic curve that will be discussed later in this chapter.

In our discussion of the transistor operating regions, we will be concentrating on the *npn* transistor. Note that all the principles covered apply equally to the *pnp* transistor. The differences between these two transistor types are found in the current directions and the biasing voltage polarities and are discussed later in this chapter.

Zero Biasing

Zero bias describes the biasing of the BJT at room temperature with no potentials applied.

Figure 5.6 shows the *npn* transistor at room temperature with no biasing potential applied. You may recall from Chapter 1 that an unbiased *pn* junction will form a depletion layer at room temperature due to recombination of free carriers that are produced by thermal energy. The depletion layers that form at room temperature are shown in Figure 5.6.

Since both junctions have a depletion layer, they are both *reverse biased* at room temperature. Note that the depletion layers extend farther into the base region than either of the other two. This is due to the fact that the base region has a lower doping level. With a lower doping level, there are fewer free carriers for recombination, so the depletion layer will extend farther into the region.

FIGURE 5.6

Zero biasing.

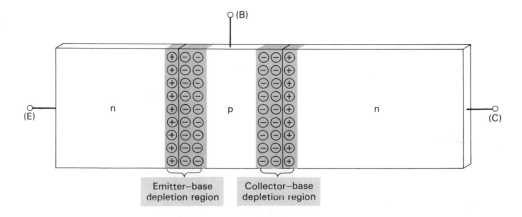

Cutoff

✔ 11

Cutoff. A BJT operating state where I_C is nearly zero.

Normally, two biasing potentials are applied to the transistor. These potentials are shown in Figure 5.7. Note that in this figure both biasing sources have polarities that will reverse bias their respective junctions.

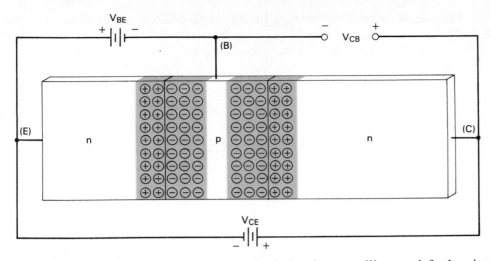

FIGURE 5.7

Cutoff.

With the polarities shown, the two depletion layers will extend farther into the emitter, base, and collector regions. With the larger depletion layers, only an extremely small amount of reverse current will flow from the emitter to the collector. For example, a 2N3904 transistor with a V_{CE} of 40 V and a reverse V_{BE} of only 3 V will have only 50 nA of collector current (I_C). This is extremely small when compared to the 200-mA value of I_C that the component is capable of handling.

Saturation

The opposite of cutoff is *saturation*. Saturation is the condition where *further increases in I_B will not cause increases in I_C*. When a transistor is saturated, I_C has reached its maximum possible value, as determined by V_{CC} and the total resistance in the emitter–collector circuit. This point is illustrated in Figure 5.8.

Saturation. A BJT operating region where I_C reaches its maximum value.

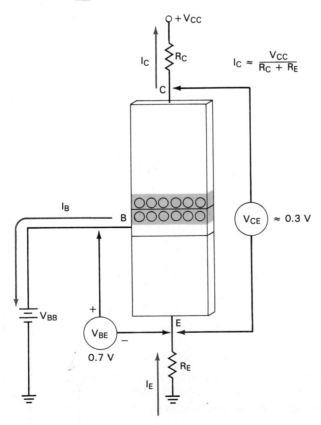

$$I_C \approx \frac{V_{CC}}{R_C + R_E}$$

FIGURE 5.8

Saturation circuit conditions.

Assume for a moment that V_{CE} for the transistor is 0 V (an *ideal* situation). If this is the case, I_C will depend completely on the values of V_{CC}, R_C, and R_E. By Ohm's law, the maximum value of I_C would be

$$I_C = \frac{V_{CC}}{R_C + R_E}$$

Now let's say that I_B is increased to the point where I_C approaches the value obtained in the formula above. If this is the case, the value of I_C will reach some maximum value and will be able to increase no further. Then, further increases in I_B will *not* increase I_C and the relationship $I_C = \beta I_B$ will no longer hold true.

If I_B is increased beyond the point where I_C can increase, both of the transistor junctions may become forward biased. This point is illustrated in Figure 5.9. V_{CE} for the transistor is shown to be approximately 0.3 V, which is typical for a saturated transistor. With the 0.7-V value of V_{BE}, the collector–base junction is biased to the difference between the two, 0.4 V. Note that this voltage indicates that the collector–base junction of the transistor is *forward* biased. When this occurs, the emitter, collector, and base currents are in the direction shown in the illustration.

FIGURE 5.9

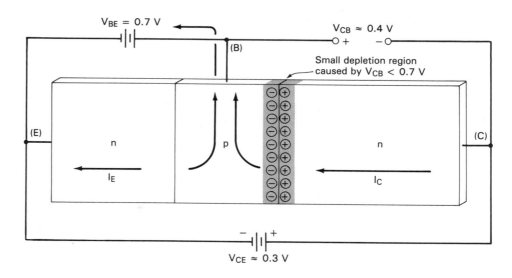

Active Operation *amplification takes place at this region*

Active region. The BJT operating region between saturation (maximum I_C) and cutoff (minimum I_C).

A transistor is said to be operating in the *active* region when the emitter–base junction is forward biased and the collector–base junction is reverse biased. Generally, the transistor is said to be in active operation when it is not in cutoff or saturation. This is the most commonly used biasing condition. The biasing for active operation is illustrated in Figure 5.10.

The operation of the transistor in this region is easiest to understand by considering just the emitter–base junction, shown in Figure 5.10b. When V_{BE} is great enough to surpass the barrier potential of the diode, current will start to flow in the emitter. You may recall that the barrier potential for silicon is approximately 0.7 V, and 0.3 V for germanium.

If we were considering a normal diode, the emitter current would all flow out of the base of the component. However, the base is a very small region that is lightly doped. Since the base is lightly doped, there is very little chance of a given electron being able to recombine with a valence band hole in the base region.

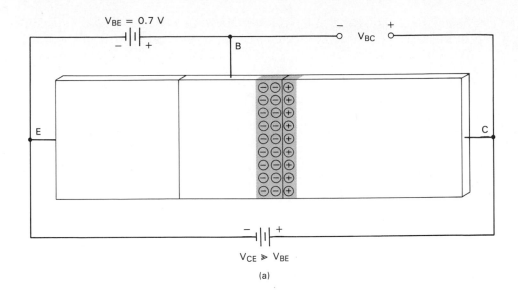

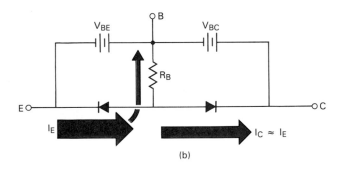

FIGURE 5.10

Active operation.

Thus, the vast majority of the electrons entering the base region pass on to the collector through the reverse-biased collector–base junction.

The idea of current through a reverse-biased *pn* junction should not seem that strange to you. After all, the zener diode is designed to allow current through the reverse-biased junction. The collector–base junction of the transistor is designed in the same way. It can handle a large amount of current when reverse biased without doing damage to the junction.

The Bottom Line

When a transistor is in *cutoff*, both junctions are reverse biased, and the current through all three terminals will be nearly zero. When *saturated*, I_C is at its maximum possible value. In this case, both transistor junctions may be forward biased, depending on the value of I_B. In either case, I_C is limited by V_{CC} and the resistance in the collector–emitter circuit.

The most commonly used operating region is the active region. When a transistor is operating in this region, *its emitter–base junction is forward biased and its collector–base junction is reverse biased*. Because of the light doping of the base, very little recombination occurs in the base region, and most of I_E passes on to the collector circuit. The base–emitter voltage (V_{BE}) will be approximately equal to the barrier potential of the junction: 0.7 V for silicon and 0.3 V for germanium. The collector–base voltage (V_{CB}) and the collector–emitter coltage (V_{CE}) will depend on the amount of current through the transistor and on the values of the external circuit components.

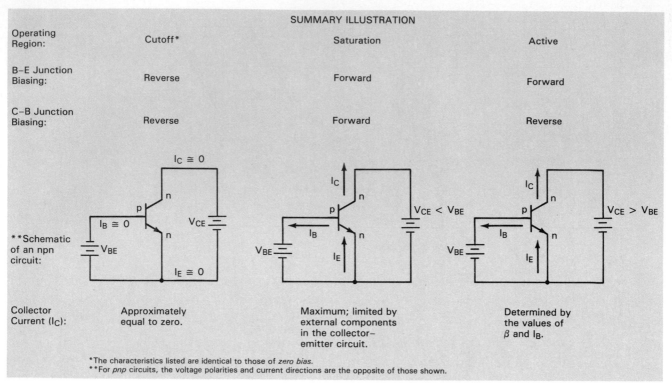

SUMMARY ILLUSTRATION

Operating Region:	Cutoff*	Saturation	Active
B–E Junction Biasing:	Reverse	Forward	Forward
C–B Junction Biasing:	Reverse	Forward	Reverse
**Schematic of an npn circuit:			
Collector Current (I_C):	Approximately equal to zero.	Maximum; limited by external components in the collector–emitter circuit.	Determined by the values of β and I_B.

*The characteristics listed are identical to those of *zero bias*.
**For *pnp* circuits, the voltage polarities and current directions are the opposite of those shown.

FIGURE 5.11

As a reference, the characteristics of the three transistor operating regions are summarized in Figure 5.11.

SECTION REVIEW

1. How are the two junctions of a transistor biased when the component is in: [Objective 10]
   a. Cutoff?
   b. The active region?
   c. Saturation?
2. What is the value of I_C when a transistor is in cutoff? [Objective 11]
3. What controls the value of I_C when a transistor is saturated? [Objective 12]
4. Describe the basic operation of a transistor biased for active-region operation. [Objective 13]

5.3
Transistor Current and Voltage Ratings

There are several transistor current and voltage ratings. Some of these ratings are parameters and some of them are typical electrical characteristics. In this section, we will take a look at several transistor current and voltage ratings and what they mean in everyday transistor applications.

DC Beta

The dc *beta* rating of a transistor is the *ratio of collector current to base current*. By formula,

$$\beta = \frac{I_C}{I_B}$$

(5.3)

dc Beta. The ratio of dc collector current to dc base current.

This is an extremely important rating, because *the most common transistor circuits have the input signal applied to the base and the output signal taken from the collector*. Thus, when the transistor is used in these circuits, the beta rating of the transistor represents the overall *dc current gain* of the transistor.

We can use equations (5.1) and (5.3) to define the other terminal currents as follows:

$$I_C = \beta I_B$$

(5.4)

and

$$I_E = I_B + I_C$$
$$= I_B + \beta I_B$$

or

$$I_E = I_B(1 + \beta)$$

(5.5)

As the following examples illustrate, you can use beta and any one terminal current to find the other two terminal currents.

━━━━━━━━━━━━━━━ **EXAMPLE 5.2** ━━━━━━━━━━━━━━━

Determine the values of I_C and I_E for the circuit shown in Figure 5.12a.

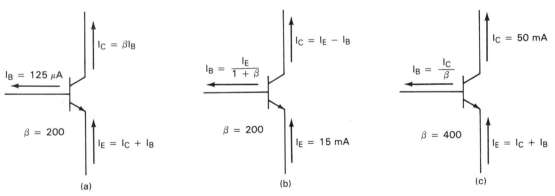

FIGURE 5.12

Solution: The value of I_C can be found as

$$I_C = \beta I_B$$
$$= (200)(125 \ \mu A)$$
$$= 25 \ mA$$

and I_E is found as

$$I_E = I_C + I_B$$
$$= 25\ mA + 125\ \mu A$$
$$= 25.125\ mA$$

PRACTICE PROBLEM 5–2

The transistor in Figure 5.12a has values of $I_B = 50\ \mu A$ and $\beta = 400$. Determine the values of I_C and I_E for the device.

EXAMPLE 5.3

Determine the values of I_C and I_B for the circuit shown in Figure 5.12b.

Solution: Equation (5.5) can be rearranged to give us

$$I_B = \frac{I_E}{1 + \beta}$$
$$= \frac{15\ mA}{1 + 200}$$
$$= 74.6\ \mu A$$

Now I_C can be found as

$$I_C = I_E - I_B$$
$$= 15\ mA - 74.6\ \mu A$$
$$= 14.93\ mA$$

PRACTICE PROBLEM 5–3

The transistor in Figure 5.12b has values of $I_E = 12\ mA$ and $\beta = 140$. Determine the values of I_B and I_C for the device.

EXAMPLE 5.4

Determine the values of I_B and I_E for the circuit shown in Figure 5.12c.

Solution: The base current can be found as

$$I_B = \frac{I_C}{\beta}$$
$$= \frac{50\ mA}{400}$$
$$= 125\ \mu A$$

Now the emitter current can be found as

$$I_E = I_C + I_B$$
$$= 50\ mA + 125\ \mu A$$
$$= 50.125\ mA$$

The transistor in Figure 5.12c has values of $I_C = 80$ mA and $\beta = 170$. Determine the values of I_B and I_E for the device.

Because the beta rating of a transistor is a *ratio* of current values, it has no unit of measure. The typical range for transistor beta ratings is from 50 to 400. This means that the typical transistor will have a dc collector current that is from 50 to 400 times as high as the dc base current when operated in the active region.

There is one point that should be made at this time: *Transistors have both dc beta ratings and ac beta ratings. Dc beta* is the ratio of *dc* collector current to *dc* base current, while *ac beta* is the ratio of *ac* collector current to *ac* base current. We will discuss ac beta and its applications in Chapter 7.

Transistors have both dc and ac beta ratings.

DC Alpha

The *alpha* rating of a transistor is the *ratio of collector current to emitter current.* By formula,

$$\alpha = \frac{I_C}{I_E} \tag{5.6}$$

dc Alpha. The ratio of dc collector current to dc emitter current.

The alpha rating of a given transistor *will always be less than unity (1).* The reason for this is illustrated in Figure 5.13.

Kirchhoff's current law states that the current leaving a point (or component) must equal the current entering the point (or component). The relationship between the three transistor terminal components was stated in equation (5.1) as

$$I_E = I_B + I_C$$

Therefore,

$$I_C = I_E - I_B$$

Since I_C is always less than I_E (by an amount equal to I_B), the fraction I_C/I_E must always work out to be less than 1.

The alpha rating of a transistor will usually be 0.9 or higher. Note that alpha (like beta) has no units, since it is a ratio between two current values. Equation (5.6) can be rearranged to give us the following useful relationships:

$$I_C = \alpha I_E \tag{5.7}$$

and

$$I_E = \frac{I_C}{\alpha} \tag{5.8}$$

Using these relationships, we can calculate base current (I_B) as

$$I_B = I_E - I_C$$
$$= I_E - \alpha I_E$$

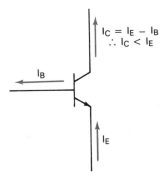

FIGURE 5.13

Why alpha is always less than unity.

or

$$I_B = I_E(1 - \alpha)$$ (5.9)

The following examples show how knowing the value of alpha and any single transistor current will allow you to determine the two unknown terminal currents.

EXAMPLE 5.5

Determine the values of I_B and I_C for the circuit shown in Figure 5.14a.

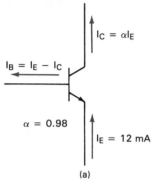

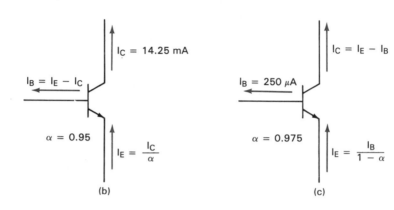

FIGURE 5.14

Solution: The collector current can be found as

$$I_C = \alpha I_E$$
$$= (0.98)(12 \text{ mA})$$
$$= 11.76 \text{ mA}$$

Now the base current is found as

$$I_B = I_E - I_C$$
$$= 12 \text{ mA} - 11.76 \text{ mA}$$
$$= 240 \text{ } \mu\text{A}$$

PRACTICE PROBLEM 5–5

The transistor in Figure 5.14a has values of $I_E = 10$ mA and $\alpha = 0.987$. Determine the values of I_C and I_B for the device.

EXAMPLE 5.6

Determine the values of I_E and I_B for the circuit shown in Figure 5.14b.

Solution: The emitter current can be found as

$$I_E = \frac{I_C}{\alpha}$$
$$= \frac{14.25 \text{ mA}}{0.95}$$
$$= 15 \text{ mA}$$

Now the base current is found as

$$I_B = I_E + I_C$$
$$= 15 \text{ mA} - 14.25 \text{ mA}$$
$$= 750 \text{ μA}$$

PRACTICE PROBLEM 5–6

The transistor in Figure 5.14b has values of $I_C = 15$ mA and $\alpha = 0.972$. Determine the values of I_B and I_E for the device.

EXAMPLE 5.7

Determine the values of I_C and I_E for the circuit shown in Figure 5.14c.

Solution: Rearranging equation (5.9) gives us

$$I_E = \frac{I_B}{1 - \alpha}$$
$$= \frac{250 \text{ μA}}{1 - 0.975}$$
$$= \frac{250 \text{ μA}}{0.025}$$
$$= 10 \text{ mA}$$

and now

$$I_C = I_E - I_B$$
$$= 10 \text{ mA} - 250 \text{ μA}$$
$$= 9.75 \text{ mA}$$

PRACTICE PROBLEM 5–7

The transistor in Figure 5.14c has values of $I_B = 300$ μA and $\alpha = 0.99$. Determine the values of I_C and I_E for the device.

Examples 5.5 through 5.7 showed you how to solve for all transistor currents when one current and the alpha rating are given. At the same time, they also reaffirmed the relationships between transistor terminal currents stated earlier in the chapter. Consider the results in the three examples:

Example	Alpha	I_E (mA)	I_C (mA)	I_B (μA)
5.5	0.98	12	11.76	240
5.6	0.95	15	14.25	750
5.7	0.975	10	9.75	250

In each case, I_E and I_C were very close to being equal. And in each case the value of I_B was much less than either of the other terminal currents. Note that the higher alpha is, the closer I_E and I_C are to being equal. This can be seen by comparing the results of Examples 5.5 and 5.6.

The Relationship between Alpha and Beta

✔ 17

As you will see later in this chapter, the spec sheet for a given transistor will list the value of beta for the device, but will not list the value of alpha. This is because beta is used far more commonly in transistor circuit calculations than is alpha. This fact will become very evident when we cover the dc and ac analyses of transistor circuits.

Since alpha is rarely listed on transistor spec sheets, you need to be able to determine its value using the value of beta. You can determine the value of alpha from the value of beta with the following equation:

$$\alpha = \frac{\beta}{1 + \beta}$$

h_{FE}

(5.10)

✔ 18

The following example illustrates the process of finding alpha when a beta value is known. It also demonstrates the validity of equation (5.10).

EXAMPLE 5.8

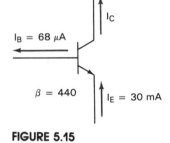

$I_B = 68\ \mu A$

$\beta = 440$ $I_E = 30$ mA

FIGURE 5.15

Determine the alpha rating for the transistor shown in Figure 5.15. Then determine the value of I_C using both the alpha rating and the beta rating of the transistor.

Solution: The beta rating is given as 440. Therefore, the alpha rating is found as

$$\alpha = \frac{\beta}{1 + \beta}$$
$$= \frac{440}{441}$$
$$= 0.9977$$

The value of I_C can be found as

$$I_C = \alpha I_E$$
$$= (0.9977)(30 \text{ mA})$$
$$= 29.93 \text{ mA}$$

or I_C can be found as

$$I_C = \beta I_B$$
$$= (440)(68\ \mu A)$$
$$= 29.92 \text{ mA}$$

The two values of I_C are close enough to validate both methods of calculating I_C.

PRACTICE PROBLEM 5–8

A transistor has the following values: $\beta = 349$, $I_E = 350$ mA, and $I_B = 1$ mA. Determine the value of α for the device. Then calculate the value of I_C using both α and β. How do the two values of I_C compare?

You may be wondering why we would bother with alpha at all when beta is listed on the spec sheets and will give us the same values of I_C in our calculations. You will be shown in Chapter 8 that some circuits require the use of alpha. When you are dealing with these circuits, you will need to be able to convert the value of beta to a value of alpha.

The Basis for Equation (5.10)

We can define alpha in terms of beta by starting with the equation

$$\alpha = \frac{I_C}{I_E}$$

If we rewrite the equation by using equation (5.4) in place of I_C and equation (5.5) in place of I_E, we get

$$\alpha = \frac{\beta I_B}{I_B(1 + \beta)}$$

Since I_B is common to both the numerator and denominator, it can be dropped, leaving the equation:

$$\boxed{\alpha = \frac{\beta}{1 + \beta}} \qquad \textbf{(5.10)}$$

Maximum Current Ratings

Most transistor specifications sheets will list a maximum collector current rating for both saturation and cutoff. When the transistor is saturated, the collector current can go as high as several hundred milliamperes. High-power transistors can have current ratings as high as several amperes.

The maximum allowable base current value can be found by dividing the maximum I_C value by the *maximum* dc beta rating. By formula,

$$\boxed{I_{B(max)} = \frac{I_{C(max)}}{\beta_{max}}} \qquad \textbf{(5.11)}$$

☛ 19

A Practical Consideration: The beta rating of a transistor is usually listed as a *range* of values on the device spec sheet. Dealing with beta *ranges* is addressed later in this chapter.

━━━━━━ **EXAMPLE 5.9** ━━━━━━

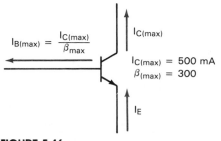

$$I_{B(max)} = \frac{I_{C(max)}}{\beta_{max}}$$

$I_{C(max)}$

$I_{C(max)} = 500$ mA
$\beta_{(max)} = 300$

I_E

FIGURE 5.16

The transistor shown in Figure 5.16 has the following ratings: $I_{C(max)} = 500$ mA and $\beta_{max} = 300$. Determine the maximum allowable value of I_B for the device.

Solution: Using the ratings given, the value of $I_{B(max)}$ is found as

$$\begin{aligned} I_{B(max)} &= \frac{I_{C(max)}}{\beta_{max}} \\ &= \frac{500 \text{ mA}}{300} \\ &= 1.67 \text{ mA} \end{aligned}$$

For the transistor shown, if the base current is allowed to exceed 1.67 mA, the collector current will exceed its maximum rating of 500 mA, and the transistor will probably be destroyed.

PRACTICE PROBLEM 5–9

A transistor has ratings of $I_{C(max)} = 1$ A and $\beta_{max} = 120$. Determine the values of $I_{B(max)}$ for the device.

Transistors also have maximum *cutoff* current ratings. These ratings are usually in the low nanoampere range and are specified for exact values of V_{CE} and reverse V_{BE}. It was stated earlier that the 2N3904 has a maximum cutoff current rating of 50 nA when the reverse value of V_{BE} is 3 V and the value of V_{CE} is 40 V. These values are illustrated in Figure 5.17.

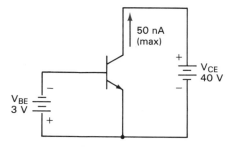

FIGURE 5.17

Transistor Voltage Ratings

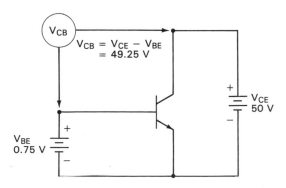

Most transistor specification sheets will list a maximum value of *collector–base voltage*, V_{CB}. This rating indicates the maximum amount of reverse bias that can be applied to the collector–base junction without damaging the transistor. This rating is important because the collector–base junction is reverse biased for active region operation. This is illustrated in Figure 5.18. In this circuit, V_{CE} is 50 V and V_{BE} is 0.75 V. The value of V_{CB} is equal to the difference between the other two voltages: 49.25 V. If this voltage is greater than the V_{CB} rating of the transistor, it will probably be destroyed.

FIGURE 5.18

Collector—base junction biasing.

Every transistor has three breakdown voltage ratings. These ratings indicate the *maximum reverse voltages* that the transistor can withstand. For the 2N3904, these voltage ratings are as follows:

Reverse voltage rating

Rating	Value (V_{dc})
BV_{CBO}	60
BV_{CEO}	40
BV_{EBO}	6

The voltages listed are illustrated in Figure 5.19. If any of the reverse voltage ratings are exceeded, the transistor may not survive the experience.

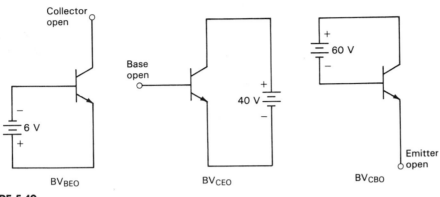

FIGURE 5.19

BJT breakdown voltage ratings.

There are many transistor ratings that have not been covered in this section. These ratings include *junction capacitance*, *maximum power dissipation*, *frequency limitations*, *operating temperature ranges*, and so on. All these ratings will be covered as they are needed in future chapters.

1. What is *dc beta*? [Objective 14]
2. Why doesn't beta have any units of measure? [Objective 14]
3. What are the two types of beta ratings? [Objective 14]
4. What is *dc alpha*? [Objective 15]
5. What is the limit on the value of alpha? [Objective 15]
6. Between beta and alpha, which raing is used more commonly? [Objective 17]
7. Why do you need to be able to determine the value of alpha using the value of beta? How is this determination made? [Objective 17]
8. What are the commonly used transistor voltage ratings? [Objective 20]
9. What will happen if any of the reverse voltage ratings of a transistor are exceeded? [Objective 20]

5.4
Basic Transistor Configurations

There are basically three ways in which a transistor can be connected for most applications. These three circuit configurations are called *common emitter*, *common*

SEC. 5.4 Basic Transistor Configurations

collector, and *common base*. In this section, we will take a brief look at these three circuit configurations.

✔ 21

Common emitter. A circuit configuration where the emitter is common to both supply voltages.

The capacitors shown in the figure are called *coupling capacitors*. We will discuss coupling capacitors in detail in Chapter 7.

Common Emitter

The *common-emitter* circuit is by far the most commonly used. This transistor configuration is illustrated in Figure 5.20. Note that the ac input to the circuit is applied to the base of the transistor, while the output ac voltage is taken from the collector.

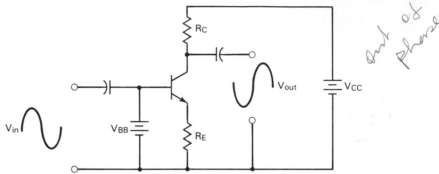

FIGURE 5.20

The name *common emitter* implies that the supply voltages are connected at the common point of the emitter. In other words, the emitter terminal is common to both supply voltages.

Because of the input/output connections, *beta* is a critical rating for this transistor circuit configuration. As you will see in Chapter 7, *the common-emitter circuit is characterized by high current gain, high voltage gain, and a 180° ac voltage phase shift from input to output*.

✔ 22

Common collector or **emitter follower.** A circuit configuration where the collector is common to both supply voltages.

Common Collector (Emitter Follower)

The second most commonly used circuit is the *common-collector* configuration. The common-collector circuit, which is often referred to as the *emitter follower*, *is characterized by high current gain, low voltage gain, and input/output ac voltages that are in phase*. The common-collector circuit is shown in Figure 5.21. Note that the input is applied to the base, and the output is taken from the emitter. As with the common-emitter circuit, beta is used extensively in analyses of the emitter follower.

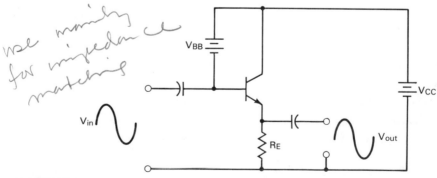

FIGURE 5.21

✔ 23

Common base. A circuit configuration where the base is common to both supply voltages.

Common Base

The *common-base* configuration is the least often used of the three transistor configurations. This circuit, which is shown in Figure 5.22, has the ac input applied to

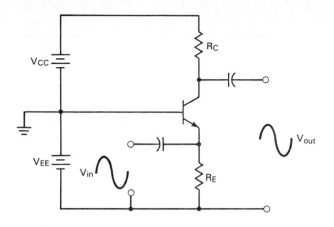

current in phase

FIGURE 5.22

the emitter and the ac output taken from the collector. As the name implies, the base is the common point for the two supply voltages.

The common-base circuit is used mainly in high-frequency applications. The circuit has *low current gain, high voltage gain, and input/output ac voltages that are in phase*.

Determining the Configuration of a Transistor Circuit

✔ 24

In practical circuits, you can quickly tell whether a given transistor is connected as a common emitter, common base, or common collector. Just locate the terminals where the input ac signal is applied to the transistor and where the ac output signal is taken from the transistor. The "odd terminal out" is the common one. If the ac input is applied to the base and the ac output is taken from the collector, the circuit is a *common emitter*. If the ac input is applied to the base and the ac output is taken from the emitter, the circuit is a *common collector*, and so on.

In later chapters we will be looking at the various circuit configurations in detail. As you will see, each has its advantages, disadvantages, analysis procedures, and applications.

SECTION REVIEW

1. Which of the transistor configurations is the most commonly used? [Objective 21]
2. What are the characteristics of the common-emitter configuration? [Objective 21]
3. What are the characteristics of the common-collector configuration? [Objective 22]
4. What name is commonly used for the common-collector circuit? [Objective 22]
5. What are the characteristics of the common-base configuration? [Objective 23]
6. What is the primary application for the common-base circuit? [Objective 23]
7. How do you determine whether a given circuit is a common-emitter, common-collector, or common-base circuit? [Objective 24]

Transistor Characteristic Curves

In this section, we are going to take a look at three characteristics curves that describe the operation of the transistor. We will start by looking at the collector and base curves. (There is no need for an emitter curve, since its current characteristics are the same as those of the collector.) Then we will take a look at the beta curve, which shows the relationship between beta, I_C, and temperature.

Collector curve. Relates the values of I_C, I_B, and V_{CE}.

Collector Curves

The collector characteristic curve relates the values of I_C, I_B, and V_{CE}. Such a curve is shown in Figure 5.23. As you can see, the collector curve is divided

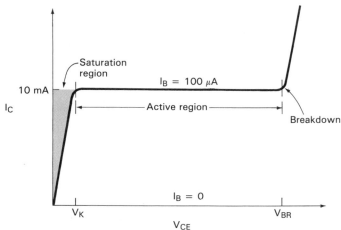

Note: Cutoff is the region of the graph that lies below the $I_B = 0$ line.

FIGURE 5.23

A collector characteristic curve.

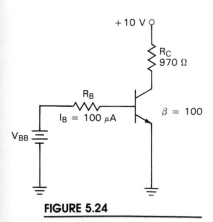

FIGURE 5.24

into three parts. The *saturation* region of the curve is the part of the curve where V_{CE} is less than the *knee voltage*, V_K. From the knee voltage to the breakdown region is the *active* region. Above the active region is the *breakdown* region. To help you understand these regions of the characteristic curve, we will be discussing the operation of the transistor shown in Figure 5.24. We will assume that the characteristic curve shown in Figure 5.23 applies to the transistor in this circuit.

The *saturation* region of the characteristic curve represents the transistor when both junctions are forward biased. In previous discussions of this operating region, it was stated that the collector current would be limited by the components external to the transistor. Now consider the circuit in Figure 5.24. The value of V_{CE} for this circuit can be found as

$$\boxed{V_{CE} = V_{CC} - I_C R_C}$$

(5.12)

Assuming that the input to the circuit is $I_B = 100$ μA and the value of R_C is 970 Ω, we have the following circuit conditions:

$$I_C = \beta I_B$$
$$= (100)(100 \text{ μA})$$
$$= 10 \text{ mA}$$

and

$$V_{CE} = V_{CC} - I_C R_C$$
$$= 10 \text{ V} - (10 \text{ mA})(970 \text{ } \Omega)$$
$$= 10 \text{ V} - 9.7 \text{ V}$$
$$= 0.3 \text{ V}$$

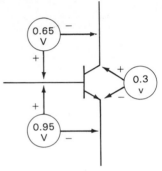

FIGURE 5.25

Now let's relate the values found to the transistor shown in Figure 5.25. As you can see, the value of V_{BE} is 0.95 V and the value of V_{CE} is 0.3 V. This leaves a V_{CB} of 0.65 V. In this case, the collector–base junction is forward biased, as is the emitter–base junction. The transistor is operating in the *saturation* region.

The *active* region of operation is the most commonly used. As Figure 5.23 indicates, this region is characterized by the fact that I_C is completely dependent on the value of I_B. As the curve illustrates, the value of I_C will remain constant for a range of values of V_{CE}.

Consider the operation of the circuit in Figure 5.24 when the input is 100 µA and the value of R_C is 400 Ω. For these conditions,

$$I_C = \beta I_B$$
$$= (100)(100 \text{ µA})$$
$$= 10 \text{ mA}$$

and

$$V_{CE} = V_{CC} - I_C R_C$$
$$= 10 \text{ V} - (10 \text{ mA})(400 \text{ } \Omega)$$
$$= 10 \text{ V} - 4 \text{ V}$$
$$= 6 \text{ V}$$

If the value of R_C is increased to 500 Ω, the value of I_C will not change. However, equation (5.12) shows that V_{CE} will decrease to 5 V. Thus, in the active region of operation, V_{CE} *will have no effect on the value of I_C*. I_C is determined solely by the values of I_B and beta for the transistor.

Figure 5.26 shows what happens when we increase the value of I_B. The only value that has changed is the value of I_C. Collector current has increased, but it is still independent of the value of V_{CE}.

FIGURE 5.26

The effect of changing I_B on a collector characteristic curve.

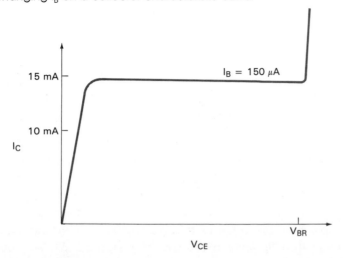

The *breakdown* region of the curve shows what happens when the breakdown voltage rating of V_{CE} is surpassed. The value of I_C increases drastically until the transistor burns up from excessive heat.

Most collector characteristic curves are composite curves.

Figures 5.23 and 5.26 show the characteristic curves for two separate values of I_B. If we plot several I_B versus I_C curves for a given transistor, we get a composite graph similar to the one shown in Figure 5.27. This graph is useful in that it shows you the collector currents produced by fixed increments of base current. Note that each curve in Figure 5.27 shows the same flat response in the active region of operation, the same relative knee voltage, and the same breakdown voltage.

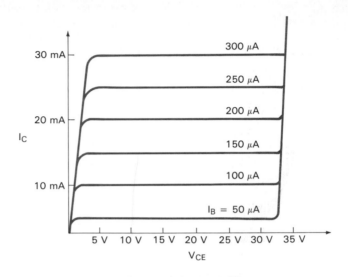

FIGURE 5.27

A composite of collector characteristic curves for a BJT.

Base Curves

Base curve. A curve illustrating the relationship between I_B and V_{BE}.

The *base curve* of a transistor plots I_B as a function of V_{BE}, as shown in Figure 5.28. Note that this curve closely resembles the forward operating curve of a typical *pn* junction diode.

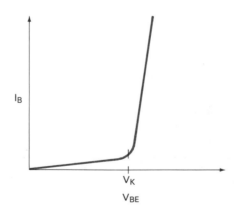

FIGURE 5.28

A base characteristic curve.

Beta curve. A curve that shows the relationship between beta and temperature and/or collector current.

Beta Curves

Beta curves show how the value of dc beta will vary with both *temperature* and *dc collector current*. This point is illustrated in Figure 5.29. As you can see, the

CHAP. 5 Bipolar Junction Transistors

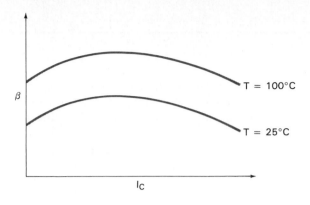

FIGURE 5.29

The relationship between beta, I_C, and temperature.

value of beta will be greater at $T = 100°C$ than it is at 25°C. Also, beta will increase (up to a point) for increases in the dc value of I_C. However, when I_C increases beyond a certain value, beta will start to decrease.

The spec sheet for the 2N3904 transistor lists the following minimum beta values, each measured at the indicated value of I_C:

Beta	Condition
40	$I_C = 0.1\text{mAdc}$
70	$I_C = 1.0\text{ mAdc}$
100	$I_C = 10\text{ mAdc}$
60	$I_C = 50\text{ mAdc}$
30	$I_C = 100\text{ mAdc}$

As you can see, the value of beta increases as I_C is increased to 10 mAdc. However, as I_C increases above 10 mAdc, the value of beta begins to decrease. This goes along with the curves shown in Figure 5.29.

SECTION REVIEW

1. What does the saturation region represent on the transistor collector curve? [Objective 25]
2. What does the active region represent on the transistor collector curve? [Objective 25]
3. Does the value of I_C depend on the value of V_{CE}? Explain your answer. [Objective 25]
4. What happens when a transistor is operated in its breakdown region? [Objective 25]
5. The base curve in Figure 5.28 indicates that the emitter–base junction of a transistor acts like what component? Explain your answer. [Objective 26)
6. What is the relationship between beta and temperature? [Objective 27]
7. What is the relationship between beta and dc collector current? [Objective 27]

5.6

Transistor Specification Sheets

The spec sheet for a given transistor contains a wide variety of dc and ac operating characteristics. In this section, we are going to look at some of the commonly used *dc* maximum ratings and electrical characteristics that are used. In Chapter 7, we will look at the *ac* characteristics and operating curves that are typically used. For our discussion on the transistor *dc* ratings, we will use the spec sheet for the Motorola 2N3904, which is shown in Figure 5.30.

Maximum Ratings

A Practical Consideration. The "O" in the V_{CEO}, V_{CBO}, and V_{EBO} ratings indicates that the third terminal is *open* when the rating is measured. For example, V_{CEO} is measured with the base terminal open. This ensures that the BJT is in cutoff when the parameter is measured.

You have already been introduced to many of the maximum ratings that are listed in Figure 5.30. The V_{CEO}, V_{CBO}, and V_{EBO} ratings are the maximum reverse voltage ratings that we discussed in the last section. The I_C rating is the maximum allowable continuous value of I_C, 200 mAdc.

The *total device dissipation* (P_D) rating of the transistor is the same type of rating that is used for the zener diode and *pn* junction diode. As you may recall, many devices have both *JEDEC ratings* and *manufacturer's ratings*. The JEDEC ratings are essentially military ratings that contain a built-in tolerance. This is the case with the 2N3904. While the JEDEC rating indicates that the device has a P_D rating of 625 mW, the manufacturer's rating indicates that the device can dissipate up to 1.5 W. Note that both of these ratings must be derated above 25°C, as is always the case.

Thermal Characteristics

The thermal ratings of the transistor are used primarily in circuit development applications and will not be discussed here. We will take a brief look at these ratings in Chapter 9 when we discuss power amplifiers.

Off Characteristics

Off characteristics. The characteristics of a transistor that is in *cutoff*.

The *off characteristics* describe the operation of the transistor when it is being operated in *cutoff*. Recall that cutoff means that I_C is nearly zero. The first three ratings, $V_{(BR)CEO}$, $V_{(BR)CBO}$, and $V_{(BR)EBO}$, are the same ratings that appeared in the maximum ratings listing. They are repeated here for the convenience of the technician or engineer who must rapidly find the ratings.

Base cutoff current (I_{BL}). The maximum amount of current through a reverse-biased emitter–base junction.

Collector cutoff current (I_{CEX}). The maximum value of I_C through a cutoff transistor.

The *base cutoff current* (I_{BL}) rating indicates the maximum amount of base current that will be present when the emitter–base junction is in cutoff. For the 2N3904, I_{BL} is a maximum of 50 nA.

The *collector cutoff current* (I_{CEX}) rating indicates the maximum value of I_C when the device is in cutoff. For the 2N3904, I_{CEX} is 50 nA.

As the I_{BL} and I_{CEX} ratings indicate, the terminal currents of a cutoff transistor are extremely low. In the case of the 2N3904, I_B and I_C will be no greater than 50 nA. The value of I_E will therefore be no greater than the sum of the two, or 100 nA.

On Characteristics

On characteristics. The characteristics of a transistor that is in either the active or saturation regions of operation.

The *on characteristics* describe the dc operating characteristics for both the active and saturation regions of operation.

The *dc current gain* (h_{FE}) rating of the transistor is the value of *dc beta*. Note that the label h_{FE} is normally used to represent dc beta, so we will be using

MAXIMUM RATINGS

Rating	Symbol	Value	Unit
*Collector-Base Voltage	V_{CB}	60	Vdc
*Collector-Emitter Voltage	V_{CEO}	40	Vdc
*Emitter-Base Voltage	V_{EB}	6.0	Vdc
*Collector Current	I_C	200	mAdc
Total Power Dissipation @ $T_A = 60^\circ$C	P_D	250	mW
**Total Power Dissipation @ $T_A = 25^\circ$C Derate above 25°C	P_D	350 2.8	mW mW/$^\circ$C
**Total Power Dissipation @ $T_C = 25^\circ$C Derate above 25°C	P_D	1.0 8.0	Watts mW/$^\circ$C
**Junction Operating Temperature	T_J	150	$^\circ$C
**Storage Temperature Range	T_{stg}	-55 to +150	$^\circ$C

NPN SILICON
SWITCHING & AMPLIFIER
TRANSISTORS

THERMAL CHARACTERISTICS

Characteristic	Symbol	Max	Unit
Thermal Resistance, Junction to Ambient	$R_{\theta JA}$	357	$^\circ$C/W
Thermal Resistance, Junction to Case	$R_{\theta JC}$	125	$^\circ$C/W

OFF CHARACTERISTICS

($T_A = 25^\circ$C unless otherwise noted)

Characteristic	Fig. No.	Symbol	Min	Max	Unit
Collector-Base Breakdown Voltage ($I_C = 10~\mu$Adc, $I_E = 0$)		BV_{CBO}	60	-	Vdc
Collector-Emitter Breakdown Voltage (1) ($I_C = 1.0$ mAdc, $I_B = 0$)		BV_{CEO}	40	-	Vdc
Emitter-Base Breakdown Voltage ($I_E = 10~\mu$Adc, $I_C = 0$)		BV_{EBO}	6.0	-	Vdc
Collector Cutoff Current ($V_{CE} = 30$ Vdc, $V_{EB(off)} = 3.0$ Vdc)		I_{CEX}	-	50	nAdc
Base Cutoff Current ($V_{CE} = 30$ Vdc, $V_{EB(off)} = 3.0$ Vdc)		I_{BL}	-	50	nAdc

ON CHARACTERISTICS

Characteristic		Fig. No.	Symbol	Min	Max	Unit
DC Current Gain (1) ($I_C = 0.1$ mAdc, $V_{CE} = 1.0$ Vdc)	2N3903 2N3904	15	h_{FE}	20 40	- -	-
($I_C = 1.0$ mAdc, $V_{CE} = 1.0$ Vdc)	2N3903 2N3904			35 70	- -	
($I_C = 10$ mAdc, $V_{CE} = 1.0$ Vdc)	2N3903 2N3904			50 100	150 300	
($I_C = 50$ mAdc, $V_{CE} = 1.0$ Vdc)	2N3903 2N3904			30 60	- -	
($I_C = 100$ mAdc, $V_{CE} = 1.0$ Vdc)	2N3903 2N3904			15 30	- -	
Collector-Emitter Saturation Voltage (1) ($I_C = 10$ mAdc, $I_B = 1.0$ mAdc) ($I_C = 50$ mAdc, $I_B = 5.0$ mAdc)		16, 17	$V_{CE(sat)}$	- -	0.2 0.3	Vdc
Base-Emitter Saturation Voltage (1) ($I_C = 10$ mAdc, $I_B = 1.0$ mAdc) ($I_C = 50$ mAdc, $I_B = 5.0$ mAdc)		17	$V_{BE(sat)}$	0.65 -	0.85 0.95	Vdc

FIGURE 5.30

(Courtesy of Motorola, Inc.)

this label in all our discussions from now on. Where did this label come from? The answer to that question is discussed in Chapter 7. For now, just remember that h_{FE} is the label that is commonly used to represent dc beta.

As you can see, the values of h_{FE} are measured at different values of I_C. This goes along with the discussion we had earlier on *beta versus collector current*.

The *collector–emitter saturation voltage* [$V_{CE(sat)}$] rating indicates the maximum value of V_{CE} when the device is in saturation. Recall that saturation means that I_C is maximum. For the 2N3904, the value of $V_{CE(sat)}$ is 0.2 V when $I_C = 10$ mAdc

h_{FE}. The label that is commonly used to represent dc beta.

Collector–emitter saturation voltage [$V_{CE(sat)}$]. The maximum value of V_{CE} when the transistor is in saturation.

and 0.3 V when $I_C = 50$ mA$_{dc}$. Thus, at the rated values of I_C, V_{CE} will be no greater than 0.2 or 0.3 V.

The *base–emitter saturation voltage* [$V_{BE(sat)}$] rating indicates the maximum value of V_{BE} when the device is in saturation. As you can see, this value is shown to be 0.85 or 0.95, depending on the rated value of I_B.

There are many other commonly used transistor specifications. However, these specifications deal mainly with the *ac* operation of the device. We will therefore cover these specs when we cover transistor ac operation in Chapter 7.

Base–emitter saturation voltage [($V_{BE(sat)}$)]. The maximum value of V_{BE} when the transistor is in saturation.

SECTION REVIEW

1. What do the *off characteristics* of a transistor indicate? [Objective 28]
2. What is *base cutoff current*? [Objective 28]
3. What is *collector cutoff current*? [Objective 28]
4. What do the *on characteristics* of a transistor indicate? [Objective 28]
5. What label is commonly used to represent dc beta? [Objective 28]
6. What is *collector–emitter saturation voltage*? [Objective 28]
7. What is *base–emitter saturation voltage*? [Objective 28]

5.7

Transistor Testing

29

You may recall that a diode can be tested with an ohmmeter. The transistor can be checked in the same basic manner. However, several more measurements are needed to verify that a transistor is good. The process for testing a transistor is illustrated in Figure 5.31. The emitter–base junction is tested for a low forward resistance and a high reverse resistance. When connected to the ohmmeter as in

FIGURE 5.31

Transistor resistance checks.

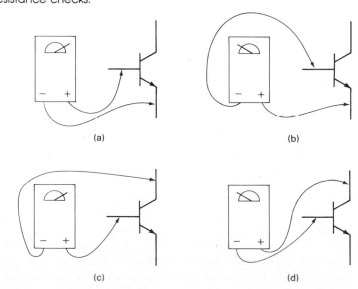

(a)

(b)

(c)

(d)

Figure 5.31a, the meter should read a low resistance, typically 1 kΩ or less. When connected as shown in Figure 5.31b, the resistance should be high, typically 100 kΩ or higher. If both of these tests show the emitter–base junction to be good, the collector–base junction should be tested in the same manner. These tests, illustrated in Figure 5.31c, should yield the same results. The forward resistance should be approximately 1 kΩ or less, and the reverse resistance should be approximately 100 kΩ or higher.

The last test involves checking the collector–emitter terminals. These terminals should have an extremely high resistance regardless of the polarity of the meter. If the transistor passes all the resistance checks, it is a good transistor. If it fails *any one check*, it is faulty and must be replaced.

A few final points: First, the polarities shown in Figure 5.31 are for an *npn* transistor. When testing a *pnp* transistor, the ohmmeter leads must be reversed. The results of the tests, however, will be the same. Also, before testing a transistor with a VOM, be sure to review the points made in Chapter 2 (page 66) regarding the use of a VOM for testing *pn* junctions.

Transistor Checkers

A *transistor checker* is a specialized piece of test equipment that is designed to test most *npn* and *pnp* transistors. In addition to determining whether or not a transistor is faulty, a transistor checker can be used to determine the actual value of h_{FE} (dc beta) for a given transistor. A typical transistor checker is shown in Figure 5.32.

FIGURE 5.32

(Courtesy of Sencore, Inc.)

SECTION REVIEW

1. Describe the method used to test a transistor with an ohmmeter. [Objective 29]
2. List the precautions that must be taken when checking a transistor with an ohmmeter. (These precautions can be found on page 66.) [Objective 29]
3. What test can be made with a transistor checker that cannot be made with a standard ohmmeter? [Objective 29]

In this section, we're going to wrap up the chapter by briefly discussing some topics that do not really fit into any of the previous sections.

pnp Versus *npn* Transistors

✔ 30

pnp and *npn* transistors differ only in voltage polarities and current directions.

In all our previous discussions about BJT operation, we concentrated on the *npn* transistor. The *pnp* transistor has the same basic operating characteristics as the *npn* transistor. The primary difference is the polarity of the biasing voltages and the direction of the terminal currents. These differences are illustrated in Figure 5.29. As you can see, V_{CC}, V_{CE}, and V_{BE} are all *negative* voltages for the *pnp* transistor. Also, collector current is shown as flowing *into* the collector terminal, base current is shown as flowing *into* the base terminal, and the sum of the two is shown as flowing *out of* the emitter.

All the ratings that have been discussed apply to the *pnp* transistor as well as to the *npn* transistor. Also, all the equations used in this chapter apply to the *pnp* transistor.

✔ 31

Supply Voltages

Throughout the chapter, V_{CC} and V_{BB} have been represented as batteries in all the circuit diagrams. In practice, the biasing voltages for transistor circuits are usually derived from the dc power supply of an electronic system. This point is illustrated in Figure 5.34.

In practice, the transistor biasing voltages are normally provided by the system's dc power supply.

In the circuit shown, the $(+)$ and $(-)$ outputs from the dc power supply are being used as V_{CC} and ground, respectively. Note that the output from the dc power supply would be used to bias a variety of other circuits as well. Figure 5.34 has been limited to one transistor circuit for the sake of simplicity.

There are a few more points that should be made: First, the value of V_{BB} in Figure 5.34 is developed by the combination of V_{CC}, R_B, and the transistor itself. *In most practical circuits, V_{BB} is actually drawn from the collector power supply, V_{CC}.* This point will be made clear throughout the next chapter when we cover *dc biasing circuits*. Second, *pnp transistors would be used in systems that have negative dc power supplies*. There are some cases where *pnp* transistors are used in systems with positive dc power supplies, but they are used mainly in negative-supply systems.

FIGURE 5.33

pnp voltage polarities and current directions.

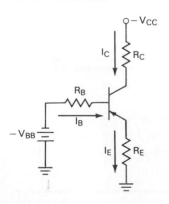

FIGURE 5.34

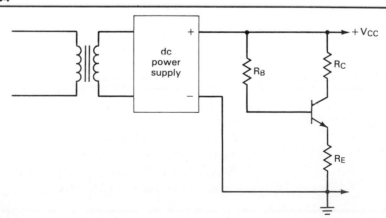

Integrated Transistors

✔ 32

Integrated transistors are single devices that house more than one transistor in integrated form. You may recall that integrated circuits contain more than one device in a single casing. Figure 5.35 shows a typical IC transistor casing. Note that the package contains four individual transistors, as is indicated by the *pin* listing.

   Each of the transistors in an IC like the one shown has the same types of maximum ratings and electrical characteristics as *individual* (or *discrete*) transistors. The only real difference is the small amount of space that is taken up by the four transistors in the IC package.

Integrated transistor. An IC that contains more than one transistor.

Discrete. A term used to describe devices that are packaged in individual casings.

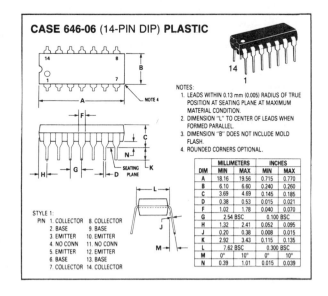

CASE 646-06 (14-PIN DIP) PLASTIC

NOTES:
1. LEADS WITHIN 0.13 mm (0.005) RADIUS OF TRUE POSITION AT SEATING PLANE AT MAXIMUM MATERIAL CONDITION.
2. DIMENSION "L" TO CENTER OF LEADS WHEN FORMED PARALLEL.
3. DIMENSION "B" DOES NOT INCLUDE MOLD FLASH.
4. ROUNDED CORNERS OPTIONAL.

DIM	MILLIMETERS MIN	MILLIMETERS MAX	INCHES MIN	INCHES MAX
A	18.16	19.56	0.715	0.770
B	6.10	6.60	0.240	0.260
C	3.69	4.69	0.145	0.185
D	0.38	0.53	0.015	0.021
F	1.02	1.78	0.040	0.070
G	2.54 BSC		0.100 BSC	
H	1.32	2.41	0.052	0.095
J	0.20	0.38	0.008	0.015
K	2.92	3.43	0.115	0.135
L	7.62 BSC		0.300 BSC	
M	0°	10°	0°	10°
N	0.39	1.01	0.015	0.039

STYLE 1:
PIN 1. COLLECTOR   8. COLLECTOR
2. BASE   9. BASE
3. EMITTER   10. EMITTER
4. NO CONN   11. NO CONN
5. EMITTER   12. EMITTER
6. BASE   13. BASE
7. COLLECTOR   14. COLLECTOR

FIGURE 5.35

(Courtesy of Motorola, Inc.)

✔ 33

High-voltage Transistors

High-voltage transistors have unusually high reverse breakdown voltage ratings. For example, the maximum ratings for the Motorola BFW43 transistor are shown in Figure 5.36. Note that the V_{CEO} and V_{CBO} ratings are much higher than those

High-voltage transistors. BJTs with high reverse breakdown voltage ratings.

FIGURE 5.36

(Courtesy of Motorola, Inc.)

MAXIMUM RATINGS

Rating	Symbol	Value	Unit
Collector-Emitter Voltage	V_{CEO}	150	Vdc
Collector-Base Voltage	V_{CBO}	150	Vdc
Emitter-Base Voltage	V_{EBO}	6.0	Vdc
Collector Current — Continuous	I_C	0.1	Adc
Total Device Dissipation @ $T_A = 25°C$ Derate above 25°C	P_D	0.4 2.66	Watt mW/°C
Total Device Dissipation @ $T_C = 25°C$ Derate above 25°C	P_D	1.4 8.0	Watt mW/°C
Operating and Storage Junction Temperature Range	T_J, T_{stg}	−65 to +200	°C

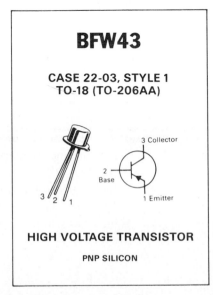

BFW43

CASE 22-03, STYLE 1 TO-18 (TO-206AA)

3 Collector

2 Base

1 Emitter

HIGH VOLTAGE TRANSISTOR

PNP SILICON

for the 2N3904 (see Figure 5.30). High-voltage transistors are used in circuits that have extremely high supply voltages, such as television CRT (cathode ray tube) control circuits.

High-current Transistors

High-current transistors. BJTs with high maximum I_C ratings.

High-current transistors have very maximum I_C ratings. For example, the maximum ratings for the Motorola 2N4237 transistor are shown in Figure 5.37. Note that the I_C rating of the component is 1 Adc. This is significantly higher than the I_C rating of the 2N3904. Obviously, these devices would be used in applications that have high current demands, such as current regulator circuits.

FIGURE 5.37

(Courtesy of Motorola, Inc.)

MAXIMUM RATINGS

Rating	Symbol	2N4237	2N4238	2N4239	Unit
Collector-Emitter Voltage	V_{CEO}	40	60	80	Vdc
Collector-Base Voltage	V_{CBO}	50	80	100	Vdc
Emitter-Base Voltage	V_{EBO}		6.0		Vdc
Base Current	I_B		500		mA
Collector Current — Continuous	I_C		1.0		Adc
Total Device Dissipation @ T_A = 25°C Derate above 25°C	P_D		1.0 5.3		Watt mW/°C
Total Device Dissipation @ T_C = 25°C Derate above 25°C	P_D		6.0 34		Watts mW/°C
Operating and Storage Junction Temperature Range	T_J, T_{stg}		−65 to +200		°C

**2N4237
thru
2N4239**

**CASE 79-04, STYLE 1
TO-39 (TO-205AD)**

**GENERAL PURPOSE
TRANSISTORS**

NPN SILICON

High-power Transistors

High-power transistors. BJTs with high power dissipation ratings.

High-power transistors are designed for use in high-power circuits, such as regulated and switching power supplies. High-power transistors, as you may have guessed, are devices that have extremely high power dissipation ratings. For example, the maximum ratings for the Motorola 2N3902 power transistor are shown in Figure 5.38.

As you can see, the P_D rating of the 2N3902 is 100 W. Power dissipation ratings of this magnitude are needed in many regulated power supply applications. (*Note*: Regulated power supplies are covered in Chapter 20.)

SECTION REVIEW

1. What are the primary differences between *pnp* and *npn* transistor circuits? [Objective 30]
2. Which transistor is used primarily in systems with *positive* dc supply voltages? [Objective 31]
3. Which transistor is used primarily in systems with *negative* dc supply voltages? [Objective 31]

Raster

MOTOROLA
SEMICONDUCTOR
TECHNICAL DATA

NPN
2N3902

HIGH VOLTAGE NPN SILICON TRANSISTORS

. . . designed for use in high-voltage inverters, converters, switching regulators and line operated amplifiers.

- High Collector-Emitter Voltage — V_{CEX} = 700 Vdc
- Excellent DC Current Gain —
  h_{FE} = 10 (Min) @ I_C = 2.5 Adc
- Low Collector-Emitter Saturation Voltage —
  $V_{CE(sat)}$ = 0.8 Vdc (Max) @ I_C = 1.0 Adc

3.5 AMPERE
POWER TRANSISTORS
NPN SILICON
400 VOLTS
100 WATTS

*MAXIMUM RATINGS

Rating	Symbol	2N3902	Unit
Collector-Emitter Voltage	V_{CEO}	400	Vdc
Collector-Emitter Voltage	V_{CEX}	700	Vdc
Emitter-Base Voltage	V_{EB}	5.0	Vdc
Collector Current — Continuous	I_C	3.5	Adc
Base Current	I_B	2.0	Adc
Total Device Dissipation @ T_C = 75°C Derate above 75°C	P_D	100 1.33	Watts W/°C
Operating Junction Temperature Range	T_J	–65 to +150	°C
Storage Temperature Range	T_{stg}	–65 to +200	°C

THERMAL CHARACTERISTICS

Characteristic	Symbol	Max	Unit
Thermal Resistance, Junction to Case	θ_{JC}	0.75	°C/W

*Indicates JEDEC Registered Data

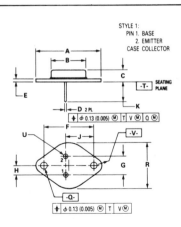

STYLE 1:
PIN 1. BASE
2. EMITTER
CASE COLLECTOR

NOTES:
1. DIMENSIONING AND TOLERANCING PER ANSI Y14.5M, 1982.
2. CONTROLLING DIMENSION: INCH.
3. ALL RULES AND NOTES ASSOCIATED WITH REFERENCED TO-204AA OUTLINE SHALL APPLY.

DIM	MILLIMETERS MIN	MILLIMETERS MAX	INCHES MIN	INCHES MAX
A	—	39.37	—	1.550
B	—	21.08	—	0.830
C	6.35	8.25	0.250	0.325
D	0.97	1.09	0.038	0.043
E	1.40	1.77	0.055	0.070
F	30.15 BSC		1.187 BSC	
G	10.92 BSC		0.430 BSC	
H	5.46 BSC		0.215 BSC	
J	16.89 BSC		0.665 BSC	
K	11.18	12.19	0.440	0.480
Q	3.84	4.19	0.151	0.165
R	—	26.67	—	1.050
U	4.83	5.33	0.190	0.210
V	3.84	4.19	0.151	0.165

CASE 1-06
TO-204AA
(TO-3)

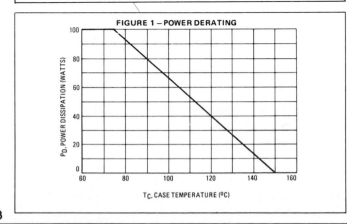

FIGURE 1 — POWER DERATING

P_D, POWER DISSIPATION (WATTS)

T_C, CASE TEMPERATURE (°C)

FIGURE 5.38

(Courtesy of Motorola, Inc.)

3-60

4. What is the difference between *integrated* and *discrete* transistors? [Objective 32]

5. What are *high-voltage transistors*? When are they typically used? [Objective 33]

6. What are *high-current transistors*? When are they typically used? [Objective 34]

7. What are *high-power transistors*? When are they typically used? [Objective 35]

KEY TERMS

The following terms were introduced and defined in this chapter:

active region
amplifier
amplitude
base
base current (I_B)
base curve
base cutoff current (I_{BL})
base–emitter saturation
  voltage ($V_{BE(\text{sat})}$)
beta (β)
beta curve
bipolar junction transistor
  (BJT)
collector
collector–base junction
collector biasing voltage
  (V_{CC})

collector current (I_C)
collector curve
collector cutoff current
  (I_{CEX})
collector–emitter
  saturation voltage
  ($V_{CE(\text{SAT})}$)
common-base
common-collector
common-emitter
current gain (β)
cutoff
dc alpha (α)
dc beta (β or h_{FE})
discrete
emitter

emitter–base junction
emitter current (I_E)
emitter follower
high-current transistor
high-power transistor
high-voltage transistor
integrated transistor
npn transistor
off characteristics
on characteristics
pnp transistor
saturation
transistor
transistor checker
zero bias

PRACTICE PROBLEMS

§5.1

1. A BJT has values of $\beta = 320$ and $I_B = 12$ μA. Determine the value of I_C for the device [7]

2. A BJT has values of $\beta = 400$ and $I_B = 30$ μA. Determine the value of I_C for the device. [7]

√3. A BJT has values of $\beta = 254$ and $I_B = 1.01$ mA. Determine the value of I_C for the device. [7]

4. A BJT has values of $\beta = 144$ and $I_B = 82$ μA. Determine the value of I_C for the device. [7]

√ 5. A BJT has values of $I_B = 20$ μA and $I_C = 1.1$ mA. Determine the value of I_E for the device. [8]

6. A BJT has values of $I_B = 1.1$ mA and $I_C = 344$ mA. Determine the value of I_E for the device. [8]

§5.3

7. Complete the following table. [14]

     CHAP. 5   Bipolar Junction Transistors

Beta	I_B	I_C
a. 150	25 μA	——
b. ——	75 μA	1.5 mA
c. 240	100 μA	——
d. 325	——	20 mA

8. Complete the following table. [14]

Beta	I_B	I_C
a. ——	50 μA	12 mA
b. 440	——	35 mA
c. 175	45 μA	——
d. ——	120 μA	84 mA

9. Complete the following table. [14]

I_B	I_C	I_E
a. 25 μA	1 mA	——
b. ——	1.8 mA	1.98 mA
c. 120 μA	——	3 mA
d. ——	7.5 mA	8 mA
e. 50 μA	——	20 mA
f. 175 μA	9.825 mA	——

10. A BJT has values of $I_B = 35$ μA and $\beta = 100$. Determine the values of I_C and I_E for the device. [14]

11. A BJT has values of $I_B = 150$ μA and $\beta = 400$. Determine the values of I_C and I_E for the device. [14]

12. A BJT has values of $I_B = 48$ μA and $\beta = 120$. Determine the values of I_C and I_E for the device. [14]

13. A BJT has values of $I_C = 12$ mA and $\beta = 440$. Determine the values of I_B and I_E for the device. [14]

14. A BJT has values of $I_C = 50$ mA and $\beta = 400$. Determine the values of I_B and I_E for the device. [14]

15. A BJT has values of $I_E = 65$ mA and $\beta = 380$. Determine the values of I_B and I_C for the device. [14]

16. A BJT has values of $I_E = 120$ mA and $\beta = 60$. Determine the values of I_B and I_C for the device. [14]

17. A BJT has values of $I_E = 28$ mA and $\alpha = 0.998$. Determine the values of I_C and I_B for the device. [16]

18. A BJT has values of $I_E = 34$ mA and $\alpha = 0.965$. Determine the values of I_B and I_C for the device. [16]

19. A BJT has values of $I_C = 12$ mA and $\alpha = 0.922$. Determine the values of I_B and I_E for the device. [16]

20. A BJT has values of $I_C = 42$ mA and $\alpha = 0.999$. Determine the values of I_E and I_B for the device. [16]

21. A BJT has a value of $\beta = 426$. Determine the value of α for the device. [18]

22. A BJT has a value of $\beta = 350$. Determine the value of α for the device. [18]

23. A BJT has the following parameters: $I_{C(max)} = 120$ mA and $\beta = 50$ to 120. Determine the maximum allowable value of I_B for the device. [19]

Key Terms

24. A BJT has the following parameters: $I_{C(max)} = 250$ mA and $\beta = 35$ to 100. Determine the maximum allowable value of I_B for the device. [19]

§5.5

25. Refer to Figure 5.39. What is the breakdown voltage of the transistor represented by the collector curve? [25]
26. Refer to Figure 5.39. What is the value of I_C when I_B is 40 μA? [25]
√27. Refer to Figure 5.39. What is the maximum value of V_{CE} when the device is saturated? [25]

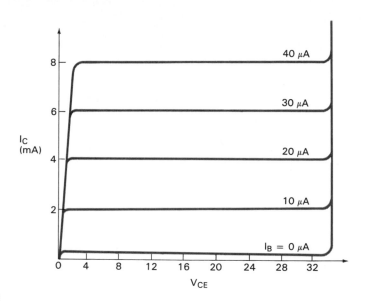

FIGURE 5.39

§5.8

28. For each circuit shown in Figure 5.40, identify the type of transistor and indicate the directions of the terminal currents. [4, 30]

FIGURE 5.40

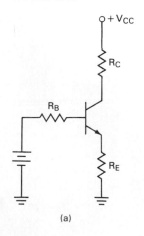

(a)

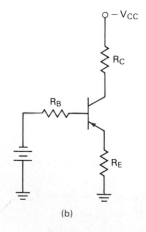

(b)

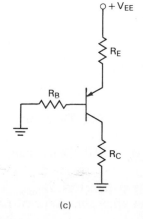

(c)

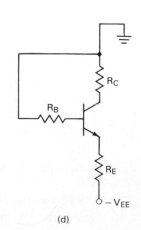

(d)

CHAP. 5   Bipolar Junction Transistors

TROUBLESHOOTING PRACTICE PROBLEMS

29. Following are the results of several transistor tests. In each case, determine whether or not the transistor is good. [29]

	Base–emitter		Collector–base		Emitter-to-collector
	Forward	Reverse	Forward	Reverse	resistance
a.	250 Ω	300 kΩ	150 kΩ	140 kΩ	1200 kΩ
b.	800 Ω	115 kΩ	800 Ω	115 kΩ	14 kΩ
c.	377 Ω	152 kΩ	900 Ω	180 kΩ	1500 kΩ
d.	100 kΩ	100 kΩ	190 Ω	144 kΩ	3500 kΩ

30. Draw a series of diagrams to show how you would connect a VOM with a negative ground lead to test a *pnp* transistor. In each case, indicate the reading you would obtain for a good transistor. [29]

31. Repeat problem 30, showing how you would connect a VOM with a positive ground lead to test a *pnp* transistor. [29]

SUGGESTED COMPUTER APPLICATIONS PROBLEMS

32. Write a program that will solve for I_C, I_E, and α, given the values of I_B and h_{FE} (dc beta).

33. Write a program that will solve the table shown in problem 9, given any two of the values.

34. Write a program that will accept information like that shown in problem 29 and tell you whether or not a given transistor is good. Assume a good transistor would have a minimum reverse-to-forward resistance ratio of 100:1.

ANSWERS TO THE EXAMPLE PRACTICE PROBLEMS

5.1. $I_C = 17.5$ mA
5.2. $I_C = 20$ mA, $I_E = 20.05$ mA
5.3. $I_B = 85.11$ μA, $I_C = 11.91$ mA
5.4. $I_B = 470.59$ μA, $I_E = 80.47$ mA
5.5. $I_C = 9.87$ mA, $I_B = 130$ μA
5.6. $I_E = 15.43$ mA, $I_B = 430$ μA
5.7. $I_E = 30$ mA, $I_C = 29.7$ mA
5.8. $\alpha = 0.997$, $I_C = 349$ mA (using β), $I_C = 348.95$ mA (using α)
5.9. $I_{B(max)} = 8.33$ mA

6
DC Biasing Circuits

At this point, manufacturing technology has developed to the point where literally hundreds of thousands of components can be produced on a single silicon wafer.

OBJECTIVES

☐ 1. State the purpose served by dc biasing circuits. (Introduction)

☐ 2. Describe the *dc load line* of an amplifier and explain what the line represents. (§6.1)

☐ 3. Plot the dc load line for an amplifier, given the values of V_{CC} and total collector–emitter circuit resistance. (§6.1)

☐ 4. Describe the *Q-point* of an amplifier and explain what the point represents. (§6.1)

☐ 5. Explain how the collector curves of a transistor can be used to determine the Q-point values of an amplifier. (§6.1)

☐ 6. Describe the output characteristics of a *midpoint-biased* amplifier. (§6.1)

☐ 7. Describe the construction and circuit currents of a base-bias circuit. (§6.2)

☐ 8. Describe the process used to perform the dc analysis of a base-bias circuit. (§6.2)

☐ 9. Perform the complete dc analysis of any base-bias circuit. (§6.2)

☐ 10. Determine whether or not a circuit is midpoint biased, given the values of I_C, V_{CE}, and V_{CC} for the circuit. (§6.2)

☐ 11. Explain *Q-point shift* and its causes. (§6.2)

☐ 12. Describe the construction and circuit currents of an emitter-bias circuit. (§6.3)

☐ 13. Explain why the output values of an emitter-bias circuit are beta independent. (§6.3)

☐ 14. Calculate the following values for an emitter-bias circuit: I_{CQ}, V_{CEQ}, $I_{C(sat)}$, $V_{CE(off)}$, $R_{IN(base)}$, and V_B. (§6.3)

☐ 15. Troubleshoot emitter-bias circuits. (§6.3)

☐ 16. Calculate the values of I_{CQ}, V_{CEQ}, and I_B for a voltage-divider bias circuit. (§6.4)

☐ 17. Determine the proper value of h_{FE} for the analysis of a biasing circuit, given the spec sheet of the transistor. (§6.4)

☐ 18. Calculate the values of $I_{C(sat)}$ and $V_{CE(off)}$ for a voltage-divider bias circuit. (§6.4)

☐ 19. Determine whether or not the input resistance of the base must be considered in the calculation of base voltage and adjust the analysis procedure accordingly. (§6.4)

☐ 20. Estimate the value of I_{CQ} for an amplifier without using the standard procedure for calculating its value and justify the use of this estimate. (§6.4)

☐ 21. Describe the troubleshooting procedure for a voltage-divider bias circuit. (§6.4)

☐ 22. Define *feedback* and state its purpose. (§6.5)

☐ 23. Describe the construction and current characteristics of the *collector-feedback bias*. (§6.5)

☐ 24. Calculate the values of I_{CQ} and V_{CEQ} for a collector-feedback bias circuit. (§6.5)

☐ 25. Explain how the collector-feedback bias circuit achieves its Q-point stability. (§6.5)

☐ 26. Describe the construction and current characteristics of the *emitter-feedback bias* circuit. (§6.5)

☐ 27. Calculate the values of I_{CQ} and V_{CEQ} for an emitter-feedback bias circuit. (§6.5)

☐ 28. Explain how emitter-feedback bias achieves its Q-point stability. (§6.5)

☐ 29. List the circumstances in which each of the feedback bias circuits is preferred. (§6.5)

☐ 30. Troubleshoot emitter-feedback bias circuits. (§6.5)

As you master each of the objectives listed, place a check mark (✓) in the appropriate box.

THE WORLD OF SILICON MINIATURE COMPONENTS

Not long after the development of the transistor, the race was on to make miniature versions of virtually every type of component and circuit. It was found that almost any type of circuit or component could be made using silicon. Resistors and capacitors, for example, could be produced on silicon. This made it possible for entire amplifiers (including the biasing components) to be made on a single silicon wafer.

At this point, manufacturing technology has developed to the point where literally hundreds of thousands of components can be produced on a single silicon wafer that is much smaller than a penny. Yet, with all of the advances in manufacturing miniature silicon components, they still have not been able to successfully produce the simplest electronic component on silicon: the *inductor*. While inductors can be made much smaller than was previously possible, they still cannot be produced in micro-miniature form.

1

The purpose of the *dc biasing circuit* is to set up the initial dc values of I_B, I_C, and V_{CE}.

The ac operation of an amplifier depends on the initial dc values of I_B, I_C, and V_{CE}. By varying I_B around an initial dc value, I_C and V_{CE} are made to vary around their initial dc values. This operation is illustrated in Figure 6.1. The purpose of the dc biasing circuit is to set up the initial dc values of I_B, I_C, and V_{CE}. In this chapter, we will take a look at the operation and troubleshooting of basic dc biasing circuits.

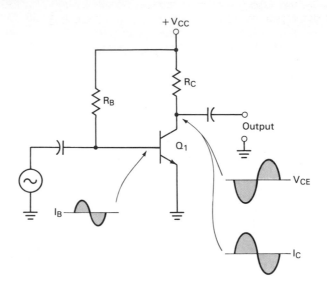

FIGURE 6.1

Typical amplifier operation.

6.1

Introduction to DC Biasing: The DC Load Line

The **dc load line** is a graph of all V_{CE} versus I_C combinations.

The *dc load line* is a graph that *represents all the possible combinations of I_C and V_{CE} for a given amplifier*. For every possible value of I_C, an amplifier will have a corresponding value of V_{CE}. The dc load line represents all the I_C/V_{CE} combinations for the circuit. A generic dc load line is shown in Figure 6.2.

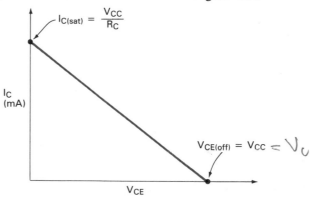

FIGURE 6.2

2

Equation (5.12):

$V_{CE} = V_{CC} - I_C R_C$

The ends of the load line are labeled $I_{C(sat)}$ and $V_{CE(off)}$. The value of the $I_{C(sat)}$ point represents the *ideal* value of saturation current for the circuit. If we look at the saturated transistor as being a short circuit from emitter to collector, then V_{CE} is zero, and equation (5.12) becomes

$$\boxed{V_{CC} = I_C R_C} \text{ (saturation)}$$

(6.1)

Rearranging the formula for I_C, we get

$$\boxed{I_{C(\text{sat})} = \frac{V_{CC}}{R_C}} \quad \text{(ideal)} \qquad\qquad \textbf{(6.2)}$$

Conversely, when the transistor is in cutoff, we can look at the component as being an open circuit form emitter to collector. Because of the open transistor, there is no current flowing in the collector circuit. Thus $I_C R_C = 0$, and equation (5.12) becomes

$$\boxed{V_{CE(\text{off})} = V_{CC}} \qquad\qquad \textbf{(6.3)}$$

As the following example shows, equations (6.2) and (6.3) are used to plot the end points of the dc load line.

✔ 3

EXAMPLE 6.1

Plot the dc load line for the circuit shown in Figure 6.3a.

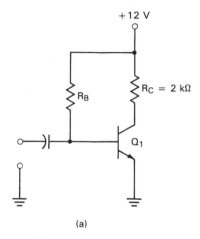

(a)

FIGURE 6.3

(b)

Solution: With the circuit values shown,

$$V_{CE(\text{off})} = 12 \text{ V}$$

and

$$\begin{aligned}
I_{C(\text{sat})} &= \frac{V_{CC}}{R_C} \\
&= \frac{12 \text{ V}}{2 \text{ k}\Omega} \\
&= 6 \text{ mA}
\end{aligned}$$

The load line therefore had end points of $V_{CE(\text{off})} = 12$ V and $I_{C(\text{sat})} = 6$ mA. The plotted line is shown in Figure 6.3b.

PRACTICE PROBLEM 6–1

A circuit like the one in Figure 6.3a has values of $V_{CC} = +8$ V and $R_C = 1.1$ kΩ. Plot the dc load line for the circuit.

As was stated earlier, the dc load line represents all the possible combinations of I_C and V_{CE} for a given amplifier. This point is illustrated in the following example.

EXAMPLE 6.2

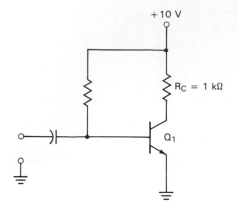

FIGURE 6.4

Plot the dc load line for the circuit shown in Figure 6.4. Then use equation (5.12) to verify the load line V_{CE} values for $I_C = 1$ mA, $I_C = 2$ mA, and $I_C = 5$ mA.

Solution: With the circuit values shown, $I_{C(sat)}$ is found as

$$
\begin{aligned}
I_{C(sat)} &= \frac{V_{CC}}{R_C} \\
&= \frac{10 \text{ V}}{1 \text{ k}\Omega} \\
&= 10 \text{ mA}
\end{aligned}
$$

Now $V_{CE(off)}$ is found as

$$
\begin{aligned}
V_{CE(off)} &= V_{CC} \\
&= 10 \text{ V}
\end{aligned}
$$

Using $I_{C(sat)}$ and $V_{CE(off)}$ as the end points, the dc load line for the circuit is plotted as shown in Figure 6.5a.

As Figure 6.5b shows, the following I_C versus V_{CE} combinations should be possible in this circuit:

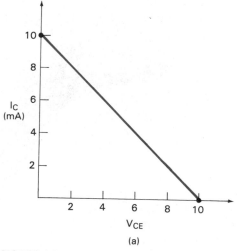

(a)

FIGURE 6.5

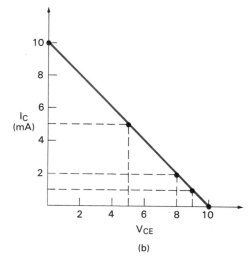

(b)

I_C (mA)	V_{CE} (V)
1	9
2	8
5	5

These combinations are verified by equation (5.12) as follows:
For $I_C = 1$ mA,

$$
\begin{aligned}
V_{CE} &= V_{CC} - I_C R_C \\
&= 10 \text{ V} - (1 \text{ mA})(1 \text{ k}\Omega) \\
&= 10 \text{ V} - 1 \text{ V} \\
&= 9 \text{ V}
\end{aligned}
$$

CHAP. 6 DC Biasing Circuits

For $I_C = 2$ mA,

$$V_{CE} = V_{CC} - I_C R_C$$
$$= 10 \text{ V} - (2 \text{ mA})(1 \text{ k}\Omega)$$
$$= 10 \text{ V} - 2 \text{ V}$$
$$= 8 \text{ V}$$

For $I_C = 5$ mA,

$$V_{CE} = V_{CC} - I_C R_C$$
$$= 10 \text{ V} - (5 \text{ mA})(1 \text{ k}\Omega)$$
$$= 10 \text{ V} - 5 \text{ V}$$
$$= 5 \text{ V}$$

These values verify the values obtained from the dc load line.

PRACTICE PROBLEM 6–2

A circuit like the one in Figure 6.4 has values of $V_{CC} = +16$ V and $R_C = 2$ kΩ. Plot the dc load line for the circuit and determine the values of V_{CE} for $I_C = 2$ mA, 4 mA, and 6 mA. Then verify your values using equation (5.12).

The Q-Point

When a transistor does not have an ac input, it will have specific dc values of I_C and V_{CE}. As you have seen, these values will correspond to a specific point on the dc load line. This point is called the *Q-point*. The letter *Q* comes from the word *quiescent*, meaning *at rest*. A quiescent amplifer is one that has no ac signal applied and therefore has constant dc values of I_C and V_{CE}.

Q-point. Indicates the values of V_{CE} and I_C for an amplifier at rest.

Quiescent. At rest.

When the dc load line of an amplifier is superimposed on the collector curves for the transistor, the *Q*-point value can easily be determined. This point is illustrated in Figure 6.6.

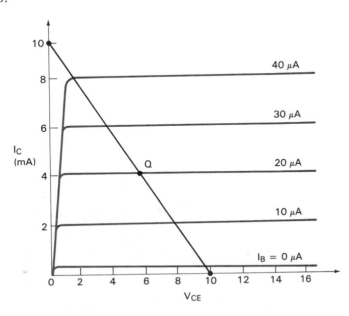

FIGURE 6.6

Assume that the collector curves shown in Figure 6.6 are the curves for the transistor in Figure 6.4. The load line found in Example 6.2 has been superimposed over the collector curves. The Q-point would be the point where the load line intersects the appropriate collector curve. For example, if the amplifier were being operated at $I_B = 20 \mu A$, the Q-point would be located at the point where the dc load line intersects the $I_B = 20 \mu A$ curve, as is shown in the illustration. From the curve we could then determine that the circuit had Q-point values of $I_C = 4$ mA and $V_{CE} = 6$ V, the coordinates that correspond to the Q-point location.

In most cases it is desirable to have the Q-point centered on the load line. When you have a centered Q-point, V_{CE} will be one-half of V_{CC}, and I_C will be one-half of $I_{C(sat)}$. This is illustrated in Figure 6.7. As you can see, the centered Q-point provides values of I_C and V_{CE} that are one-half of their maximum possible values. When a circuit is designed to have a centered Q-point, the amplifier is said to be *midpoint biased*.

 6

Midpoint bias. Having a Q-point that is centered on the load line.

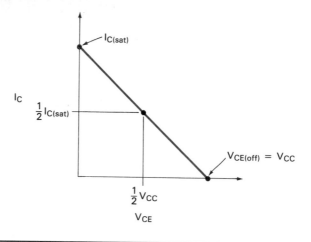

FIGURE 6.7

The Q-point for a midpoint biased circuit.

Midpoint biasing allows optimum ac operation of the amplifier. This point is illustrated in Figure 6.8. When an ac signal is applied to the base of the transistor, I_C and V_{CE} will both vary around their Q-point values. When the Q-point is centered, I_C and V_{CE} can both make the maximum possible transitions above and below

FIGURE 6.8

Optimum amplifier operation.

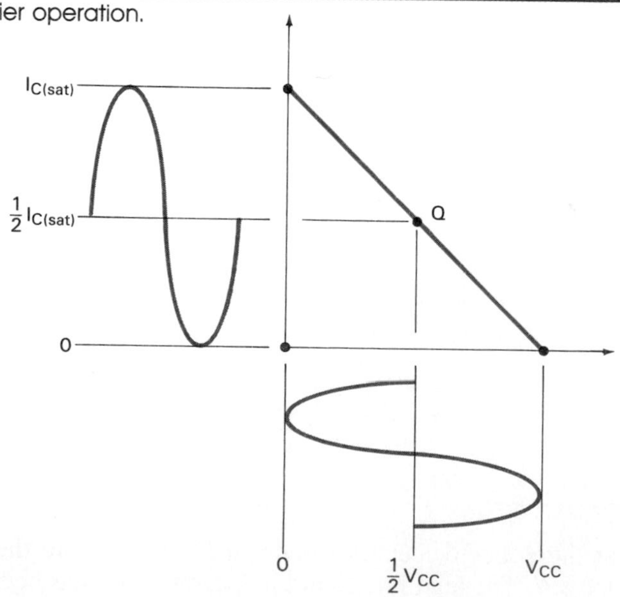

CHAP. 6   DC Biasing Circuits

their initial dc values. When the Q-point is above center on the load line, the input may cause the transistor to saturate. When this happens, a part of the output sine wave will be *clipped* off. If the Q-point is below midpoint on the load line, the input may cause the transistor to go into cutoff. This can also cause a portion of the output sine wave to be clipped.

Clipping can be caused by not designing for midpoint bias.

The ac operation of the amplifier is covered in detail in later chapters. For now, it is important that you understand the need for midpoint biasing. Having a midpoint-biased circuit will allow the best possible ac operation of the circuit.

1. What purpose is served by the dc biasing circuit? [Objective 1]
2. What is a *dc load line*? [Objective 2]
3. What is represented by the point where the load line meets the I_C axis on a graph? [Objective 2]
4. What is represented by the point where the load line meets the V_{CE} axis on a graph? [Objective 2]
5. What does the Q-*point* of an amplifier represent? [Objective 4]
6. What is meant by the term *quiescent*? [Objective 4]
7. When you say that an amplifier is in its quiescent state, what does that mean? [Objective 4]
8. How can the Q-point values for an amplifier be obtained from the collector curves of the circuit's transistor? [Objective 5]
9. What is *midpoint bias*? [Objective 6]
10. Why is midpoint biasing desirable? [Objective 6]

6.2
Base Bias

The simplest type of transistor biasing is *base bias*, or *fixed bias*. A base-bias circuit is shown in Figure 6.9. Electron flow in the circuit enters the emitter from ground and splits, with the larger portion of the current going out of the collector and the smaller portion going out of the base. The base current then goes through the base resistor (R_B) to V_{CC}. The collector current goes through the collector

Base bias. Consists of a single base resistor between the base terminal and V_{CC} and no emitter resistor. Also known as **fixed bias.**

FIGURE 6.9

Base bias.

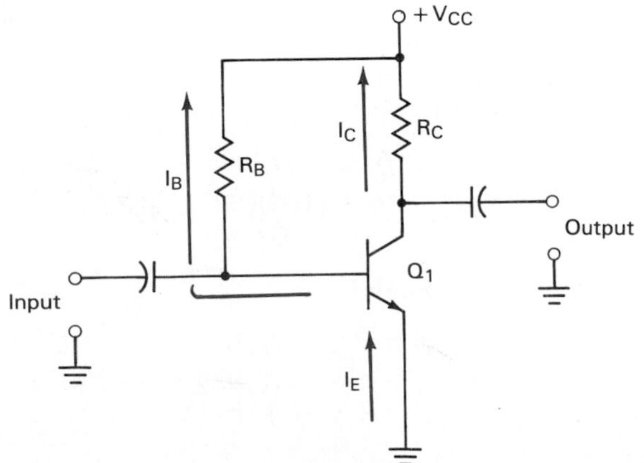

resistor (R_C) to V_{CC}. As long as power is applied to the circuit, the terminal currents will be in the direction indicated in Figure 6.9.

✔ 8

Circuit Analysis

The goal of the dc analysis of a transistor amplifier is to determine the Q-point values of I_C and V_{CE}. For the base-bias circuit, this analysis requires three steps. First, the value of I_B is found using

$$I_B = \frac{V_{CC} - V_{BE}}{R_B}$$ (6.4)

Don't Forget: h_{FE} is the spec sheet label for *dc beta*.

Once the value of I_B is known, the Q-point value of I_C is found as

$$I_C = h_{FE}I_B$$ (6.5)

where h_{FE} = the dc current gain (beta) of the device, as listed on its specification sheet

Then V_{CE} can be found as

$$V_{CE} = V_{CC} - I_C R_C$$

✔ 9

This is equation (5.12), which will give you the Q-point value of V_{CE}. Once the Q-point values of I_C and V_{CE} are known, the dc analysis of the circuit is complete. A complete analysis of a base-bias circuit is performed as shown in the following example.

EXAMPLE 6.3

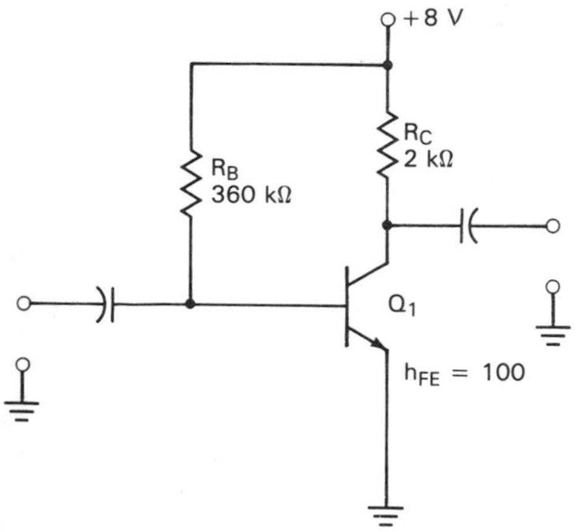

FIGURE 6.10

Determine the Q-point values of I_C and V_{CE} for the circuit shown in Figure 6.10.

Solution: First, I_B is found as

$$
\begin{aligned}
I_B &= \frac{V_{CC} - V_{BE}}{R_B} \\
&= \frac{8 \text{ V} - 0.7 \text{ V}}{360 \text{ k}\Omega} \\
&= \frac{7.3 \text{ V}}{360 \text{ k}\Omega} \\
&= 20.28 \text{ } \mu\text{A}
\end{aligned}
$$

Then I_C is found as

$$
\begin{aligned}
I_C &= h_{FE}I_B \\
&= (100)(20.28 \text{ } \mu\text{A}) \\
&= 2.028 \text{ mA}
\end{aligned}
$$

Finally, V_{CE} is found as

$$
\begin{aligned}
V_{CE} &= V_{CC} - I_C R_C \\
&= 8 \text{ V} - (2.028 \text{ mA})(2 \text{ k}\Omega) \\
&= 8 \text{ V} - 4.056 \text{ V} \\
&= 3.94 \text{ V}
\end{aligned}
$$

PRACTICE PROBLEM 6–3

A base-bias circuit like the one in Figure 6.10 has the following values: $V_{CC} = +14$ V, $R_C = 720$ Ω, $h_{FE} = 200$, and $R_B = 270$ kΩ. Determine the Q-point values of I_C and V_{CE} for the circuit.

Once the Q-point values of I_C and V_{CE} are known, you can determine whether or not the circuit is midpoint biased. As Examples 6.4 and 6.5 will show, there are two ways to determine whether or not a circuit is midpoint biased.

EXAMPLE 6.4

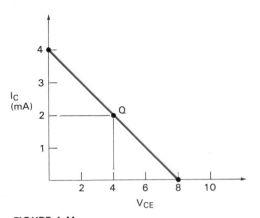

FIGURE 6.11

Construct the dc load line for the circuit in Figure 6.10, and plot the Q-point from the values obtained in Example 6.3. Determine whether or not the circuit is midpoint biased.

Solution: The end points of the load line are

$$I_{C(\text{sat})} = \frac{V_{CC}}{R_C}$$
$$= \frac{8 \text{ V}}{2 \text{ k}\Omega}$$
$$= 4 \text{ mA}$$

and

$$V_{CE(\text{off})} = V_{CC}$$
$$= 8 \text{ V}$$

Using these end points, the dc load line shown in Figure 6.11 is plotted. Plotting the point that corresponds to $I_C = 2.028$ mA and $V_{CE} = 3.94$ V (from Example 6.3), we get the Q-point shown in Figure 6.11. As you can see, the amplifier is midpoint biased.

EXAMPLE 6.5

Determine whether or not the circuit in Figure 6.10 is midpoint biased without drawing a dc load line for the circuit.

Solution: By definition, a circuit is midpoint biased when the Q-point value of V_{CE} is one-half of V_{CC}. This point was illustrated in Figure 6.7.

In Example 6.3, we determined that the amplifier in Figure 6.10 has the following values:

$$V_{CC} = +8 \text{ V}$$
$$V_{CE} = 3.94 \text{ V}$$

Since the Q-point value of V_{CE} is approximately one-half of V_{CC}, we can conclude that the circuit is midpoint biased without the use of a dc load line.

Determine whether or not the circuit described in Practice Problem 6–3 is midpoint biased. Use both methods to show your conclusion to be correct.

The process for determining the Q-point values of I_C and V_{CE} started with finding the value of I_B, as shown in equation (6.4). Let's take a look at where the I_B equation comes from.

The Basis of Equation (6.4)

The base circuit consists of the emitter–base junction of the transistor, the base resistor (R_B), and the collector supply voltage (V_{CC}). According to Kirchhoff's voltage law,

$$V_{CC} = V_{R_B} + V_{BE}$$

Thus, the voltage across R_B can be found as

$$V_{R_B} = V_{CC} - V_{BE}$$

According to Ohm's law,

$$I_B = \frac{V_{R_B}}{R_B}$$

Finally, if we substitute the V_{R_B} equation into the numerator of the equation for I_B, we get equation (6.4), as follows:

$$I_B = \frac{V_{CC} - V_{BE}}{R_B} \qquad\qquad \textbf{(6.4)}$$

Base-bias Applications

Base-bias circuits are used primarily in *switching* applications. When a transistor is being used as a switch, it is constantly going back and forth between saturation and cutoff. We will discuss this application for base-bias circuits in detail in Chapter 18.

Q-point shift. A condition where a change in operating temperature indirectly causes a change in I_C and V_{CE}.

A Practical Consideration: Cooling an amplifier (decreasing its temperature) will have the opposite effect: h_{FE} and I_C will *decrease*, and V_{CE} will *increase*.

Q-Point Shift

Even though they are easy to build and analyze, base-bias circuits are rarely used in any applications that require midpoint-biased amplifiers. The reason is the fact that base-bias circuits are extremely susceptible to a problem called *Q-point shift*. The term *Q-point shift* describes a condition where a change in operating temperature indirectly causes a change in the Q-point values of I_C and V_{CE}.

If you look at the equation for V_{CE}, you can see that V_{CE} will change if I_C changes. If you look at the equation for I_C, you'll see that the value of I_C will change if either I_B or h_{FE} changes. As you were shown in Chapter 5, dc beta (h_{FE}) varies with temperature. If temperature *increases*, h_{FE} will also increase (to a point). When h_{FE} increases, I_C will *increase*. This increase in I_C will cause a *decrease* in V_{CE}. Thus, a change in temperature will indirectly cause an increase in I_C and a decrease in V_{CE}. When this occurs, the circuit will no longer be midpoint biased. This point is illustrated in the following example.

EXAMPLE 6.6

A Practical Consideration: In practice, the value of base current will also increase slightly when temperature increases. Thus, the Q-point shift that would occur in this circuit is actually worse than is indicated in the example.

The transistor in Figure 6.12 has values of $h_{FE} = 100$ when $T = 25°C$ and $h_{FE} = 150$ when $T = 100 °C$. Determine the Q-point values of I_C and V_{CE} at both of these temperatures.

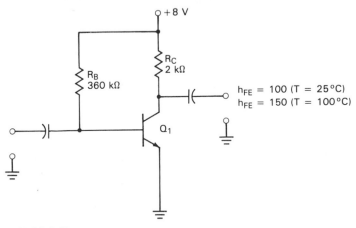

FIGURE 6.12

Solution: This is the same circuit that we analyzed in Example 6.3. At that time, we calculated the following values (when $h_{FE} = 100$):

$$I_B = 20.28 \ \mu A$$
$$I_C = 2.028 \ mA$$
$$V_{CE} = 3.94 \ V$$

When $T = 100°C$, the value of h_{FE} will change to 150. Assuming that we have the same initial value of I_B, the new value of I_C is found as

$$I_C = h_{FE}I_B$$
$$= (150)(20.28 \ \mu A)$$
$$= 3.04 \ mA$$

Finally, the new value of V_{CE} is found as

$$V_{CE} = V_{CC} - I_C R_C$$
$$= 8 \ V - (3.04 \ mA)(2 \ k\Omega)$$
$$= 1.92 \ V$$

As you can see, the amplifier is nowhere near midpoint biased when the temperature increases to 100°C. This is not acceptable for an amplifier.

PRACTICE PROBLEM 6–6

Refer back to Practice Problem 6–3. The value of h_{FE} shown in this practice problem was measured at 25°C. Determine the Q-point values for the circuit at 100°C if the value of h_{FE} at this temperature is 380.

Beta-dependent circuit. One whose Q-point values are affected by changes in h_{FE}.

Beta-independent circuit. One whose Q-point values are independent of changes in h_{FE}.

Since the Q-point values of I_C and V_{CE} for the base-bias circuit are affected by changes in h_{FE}, the circuit is referred to as a *beta-dependent* circuit. As we continue our discussion on dc biasing circuits, we will concentrate on several *beta-independent* circuits, that is, circuits whose Q-point values are independent of h_{FE} and, therefore, relatively independent of changes in temperature.

Summary

The base-bias circuit is the simplest of the dc biasing circuits. The characteristics and applications of this circuit are summarized in Figure 6.13.

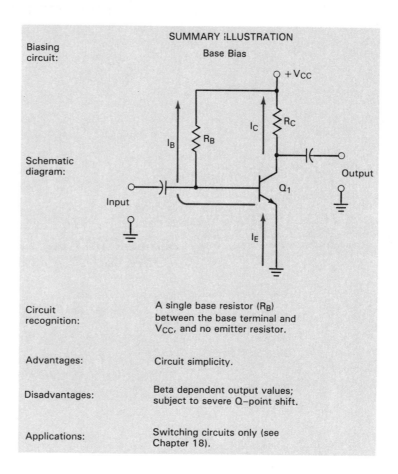

SUMMARY iLLUSTRATION

Biasing circuit: Base Bias

Schematic diagram:

Circuit recognition: A single base resistor (R_B) between the base terminal and V_{CC}, and no emitter resistor.

Advantages: Circuit simplicity.

Disadvantages: Beta dependent output values; subject to severe Q–point shift.

Applications: Switching circuits only (see Chapter 18).

One Final Note

I_{CQ}. The Q-point value of I_C.

V_{CEQ}. The Q-point value of V_{CE}.

Throughout this section, we have referred to *the Q-point values of I_C and V_{CE}*. To simplify further discussions, we will use the labels I_{CQ} and V_{CEQ} to represent the *Q-point values of I_C and V_{CE}*.

SECTION REVIEW

1. Describe the construction and current paths in a base-bias circuit. [Objective 7]
2. What is the goal of the dc analysis of an amplifier? [Objective 8]

3. List the steps that are taken, in order, to determine the Q-point values of I_C (I_{CQ}) and V_{CE} (V_{CEQ}). [Objective 8]
4. Once you have calculated the values of I_{CQ} and V_{CEQ} for an amplifier, how can you determine whether or not the circuit is midpoint biased? [Objective 10]
5. What is *Q-point shift*? [Objective 11]
6. Describe the Q-point shift process. [Objective 11]
7. What is a *beta-dependent circuit*? [Objective 11]
8. What is a *beta-independent circuit*? [Objective 11]

6.3
Emitter Bias

An emitter-bias circuit consists of several resistors and a *split power supply*, as shown in Figure 6.14. The current action in this circuit is basically the same as that of the base-bias circuit. Electron flow originates at the negative emitter power supply ($-V_{EE}$), goes through the emitter resistor (R_E), and enters the transistor. A small portion of the emitter current will leave the transistor through the base terminal, while the majority of the emitter current will go on to the collector. The base current will pass through R_B to ground, while the collector current will pass through R_C to V_{CC}.

Emitter bias. A bias circuit that consists of a split power supply and a grounded base resistor.

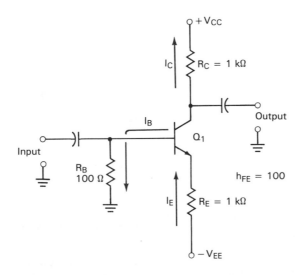

FIGURE 6.14

Emitter bias.

Circuit Analysis

We are going to start our analysis of the emitter-bias circuit by making an assumption. For reasons that will be explained later, we are going to assume that the transistor base is at ground. With this in mind, we can redraw Figure 6.14 as shown in Figure 6.15a.

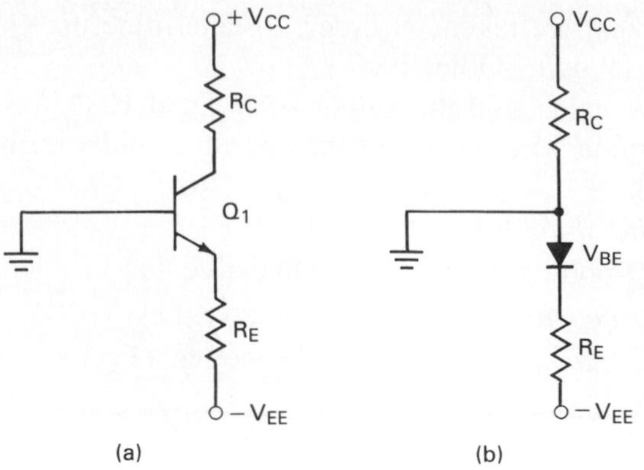

(a)   (b)

FIGURE 6.15

To analyze what is happening in the emitter circuit, the base–emitter junction is redrawn as a simple diode, as shown in Figure 6.15b. The emitter current for this circuit would be found using

$$I_E \cong \frac{V_{EE} - V_{BE}}{R_E} \tag{6.6}$$

Since $I_{CQ} \cong I_E$, I_{CQ} can be found as

$$I_{CQ} \cong \frac{V_{EE} - V_{BE}}{R_E} \tag{6.7}$$

Now, compare equation (6.7) with all our previous equations for I_C. You will notice that beta is not involved in the collector current equation for the emitter-bias circuit. This means that the output characteristics of this amplifier type are relatively independent of beta. A change in beta for the transistor will not have a major effect on the dc operation of this amplifier.

Since I_{CQ} is independent of the value of I_B, the practical analysis of the emitter-bias circuit involves only two steps. First, we calculate the approximate value of I_{CQ} using equation (6.7). Then we calculate the approximate value of V_{CEQ} using

$$V_{CEQ} = V_{CC} - I_{CQ}R_C \tag{6.8}$$

The base current and voltage will both be negligible. This point will be explained later in this section.

━━━━━━━━━━━━━━ **EXAMPLE 6.7** ━━━━━━━━━━━━━━

Determine the values of I_{CQ} and V_{CEQ} for the circuit shown in Figure 6.16.

Solution: First, the value of I_{CQ} is approximated as

$$
\begin{aligned}
I_{CQ} &= \frac{V_{EE} - V_{BE}}{R_E} \\
&= \frac{12 \text{ V} - 0.7 \text{ V}}{1.5 \text{ k}\Omega} \\
&= 7.53 \text{ mA}
\end{aligned}
$$

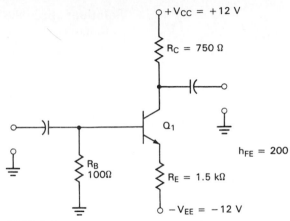

FIGURE 6.16

Now, since $I_E = I_C$, we can solve for V_{CEQ} as

$$
\begin{aligned}
V_{CEQ} &= V_{CC} - I_{CQ}R_C \\
&= 12\text{ V} - (7.53\text{ mA})(750\ \Omega) \\
&= 12\text{ V} - 5.65\text{ V} \\
&= 6.35\text{ V}
\end{aligned}
$$

PRACTICE PROBLEM 6–7

A circuit like the one shown in Figure 6.16 has the following values: $R_C = 1.5$ kΩ, $R_E = 3$ kΩ, $V_{CC} = +15$ V, and $V_{EE} = -15$ V. Determine the values of I_{CQ} and V_{CEQ} for the circuit.

Saturation and Cutoff

The presence of two power supplies in the emitter-bias circuit has an effect on the basic equations for $I_{C(\text{sat})}$ and $V_{CE(\text{off})}$. Normally, the two supply voltages will be *equal in value*. For example, if V_{CC} is $+15$ V$_{\text{dc}}$, $-V_{EE}$ will be -15 V$_{\text{dc}}$. When the two supply voltages are equal in value, we can use the following equations to find $I_{C(\text{sat})}$ and $V_{CE(\text{off})}$:

$$
I_{C(\text{sat})} = \frac{2V_{CC}}{R_C + R_E} \tag{6.9}
$$

and

$$
V_{CE(\text{off})} = 2V_{CC} \tag{6.10}
$$

Remember, these equations are valid *only* when the two supply voltages are equal in value. The following example illustrates the process used to find $I_{C(\text{sat})}$ and $V_{CE(\text{off})}$ in this case.

EXAMPLE 6.8

Determine the values of $I_{C(\text{sat})}$ and $V_{CE(\text{off})}$ for the circuit shown in Figure 6.16.

Solution: Since the supply voltages are equal in value, $I_{C(\text{sat})}$ is found as

$$I_{C(\text{sat})} = \frac{2V_{CC}}{R_C + R_E}$$
$$= \frac{24 \text{ V}_{dc}}{2250 \text{ }\Omega}$$
$$= 10.67 \text{ mA}$$

And the value of $V_{CE(\text{off})}$ is found as

$$V_{CE(\text{off})} = 2V_{CC}$$
$$= 24 \text{ V}_{dc}$$

PRACTICE PROBLEM 6–8

Refer to Practice Problem 6–7. Determine the values of $I_{C(\text{sat})}$ and $V_{CE(\text{off})}$ for the circuit described in that practice problem.

If you ever come across an emitter-bias circuit that has unequal supply voltages, simply use the *difference between the two supply voltages* in place of $2V_{CC}$ in equations (6.9) and (6.10).

The Basis of Equations (6.9) and (6.10)

The value of $I_{C(\text{sat})}$ is determined by two things:

1. *Total* voltage across the collector–emitter circuit
2. *Total* resistance in the collector–emitter circuit

The total voltage across the collector–emitter circuit in the emitter-bias amplifier is equal to the difference between the two supply voltages. Since these two supply voltages are normally equal, the difference between them will be twice the value of V_{CC}. Thus, we can normally use $2V_{CC}$ in the calculation of $I_{C(\text{sat})}$. The total resistance in the collector–emitter circuit is the sum of R_C and R_E. Using the values of $2V_{CC}$ and $(R_C + R_E)$, we obtain equation (6.9)

The value of $V_{CE(\text{off})}$ is equal to the total voltage applied to the collector–emitter circuit. In the base-bias circuit, the total voltage across the circuit was equal to V_{CC}. In the case of emitter bias, the total voltage is (again) equal to the difference between the supply voltages, or $2V_{CC}$ under normal circumstances, and thus we arrive at equation (6.10).

Base Input Resistance

Before we can discuss the base characteristics of the emitter-bias circuit, we need to take a look at the dc input resistance of the base terminal of the transistor. As you will see, this resistance value weighs heavily into our initial assumption that V_B is approximately equal to 0 V_{dc}.

The dc input resistance of the base of a transistor is found as

$$\boxed{R_{\text{IN(base)}} = h_{FE}R_E} \tag{6.11}$$

This is the equation we will use whenever we need to determine the dc input resistance of the base terminal, as is illustrated in the following example.

EXAMPLE 6.9

Determine the base input resistance for the circuit shown in Figure 6.16.

Solution: The value of h_{FE} is shown to be 200, and the value of R_E is shown to be 1.5 kΩ. Using these values, the value of $R_{\text{IN(base)}}$ is found as

$$R_{\text{IN(base)}} = h_{FE}R_E$$
$$= (200)(1.5 \text{ k}\Omega)$$
$$= 300 \text{ k}\Omega$$

PRACTICE PROBLEM 6–9

Determine the base input resistance for the circuit described in Practice Problem 6–7. Assume that the value of h_{FE} for the device is 180.

The Basis of Equation (6.11)

The basis of this equation is easy to understand if we idealize the transistor and some of its voltage and current relationships. First, equation (5.2) states that the emitter current is *approximately* equal to the collector current. The *ideal* form of this relationship is

$$I_C = I_E$$

Since $I_C = h_{FE}I_B$, we can rewrite the above equation as

$$I_E = h_{FE}I_B$$

Now, let's assume that V_E and V_B are equal. We can do this by assuming that the emitter–base junction diode is an ideal component. Based on our assumptions, let's represent the base and emitter circuits as two separate resistances connected to the same voltage. This is shown in Figure 6.17. Since the voltage across the two circuits is equal, we can use Ohm's law to write

$$I_E R_E = I_B R_{\text{IN(base)}}$$

FIGURE 6.17

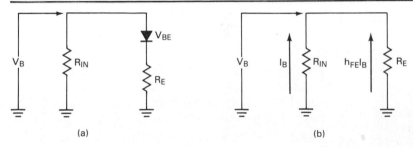

(a)   (b)

Substituting $I_E = h_{FE}I_B$ into the above equation, we have

$$h_{FE}I_BR_E = I_BR_{IN(base)}$$

Dividing both sides by I_B, we get

$$\boxed{h_{FE}R_E = R_{IN(base)}}$$

This is simply the reverse form of equation (6.11).

Base Characteristics

It was stated earlier that the base voltage in an emitter-bias circuit is approximately equal to ground. This is based on two assumptions:

1. The emitter potential will be very small, typically -1 V or less.
2. The base resistor will be much lower in value than $R_{IN(base)}$.

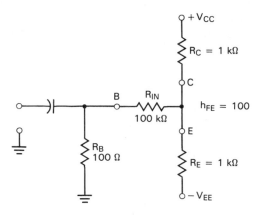

FIGURE 6.18

Now, consider the circuit shown in Figure 6.18. This is the equivalent of the emitter-bias base circuit shown in Figure 6.14. For simplicity, the emitter–base diode has been ignored. Using equation (6.11), $R_{IN(base)}$ was found as

$$\begin{aligned} R_{IN(base)} &= h_{FE}R_E \\ &= (100)(1 \text{ k}\Omega) \\ &= 100 \text{ k}\Omega \end{aligned}$$

Using the voltage-divider formula, V_B can be found as

$$\begin{aligned} V_B &= \frac{R_B}{R_T} V_E \\ &= \frac{100 \ \Omega}{100.1 \text{ k}\Omega}(-1 \text{ V}) \\ &= -999 \ \mu V \end{aligned}$$

With the voltage at the base terminal being so low, we can say that it is approximately equal to 0 V. Note that the base voltage in an emitter-bias circuit will always be very low as long as the base resistor is much lower in value than $R_{IN(base)}$. When this resistance relationship exists, most of V_E will be dropped across $R_{IN(base)}$, leaving very little to be dropped across R_B.

CHAP. 6   DC Biasing Circuits

Circuit Troubleshooting

When an emitter-bias circuit is operating properly (as shown in Figure 6.19a), it will have the following circuit conditions:

1. V_B will be approximately 0 V.
2. V_E will be approximately 0.7 V less than V_B.
3. V_C will be approximately one-half of V_{CC}.

When one or more of the terminal voltages shown are way out of range, there is probably something wrong with the circuit. In this section we will take a look at some of the problems that can develop in the dc emitter-bias circuit and how these problems will affect the overall operation of the circuit. In our discussion, we will refer to Figure 6.19.

R_C **Open.** When the collector resistor opens, there is no path for dc current in the collector circuit. Because of this, the transistor will be in cutoff, and only a small amount of current will flow in the emitter circuit. This emitter current will be caused by the forward-biased emitter–base junction of the transistor. The results of this circuit operation are illustrated in Figure 6.19b. Note that the following circuit conditions exist when R_C is open:

1. V_B is negative.
2. V_E is more negative than V_B.
3. V_C is negative, somewhere between the values of V_B and V_E.

FIGURE 6.19

Emitter bias fault symptoms.

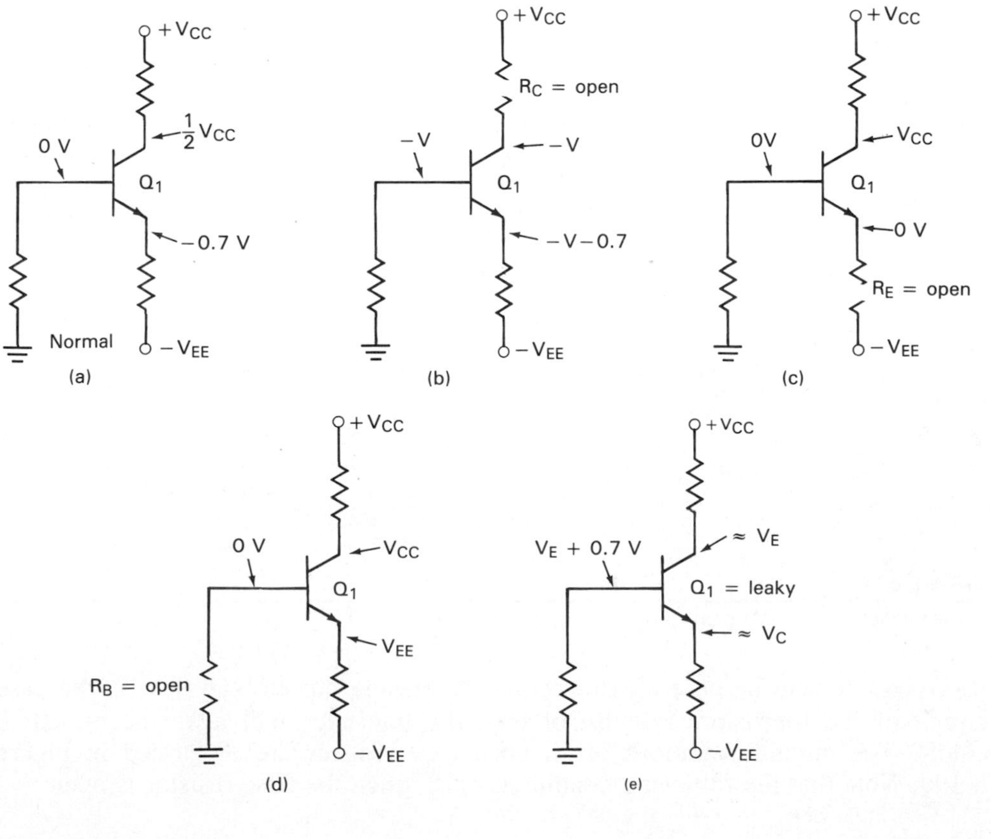

SEC. 6.3   Emitter Bias

233

The collector voltage reading (V_C) is caused by several things. First, with the collector circuit essentially open, the voltages at the other two terminals will be "reflected" in the collector circuit. Second, when you connect a voltmeter to the collector, the meter will allow a small amount of current to flow through its internal resistance to ground. This small amount of I_C will cause a negative reading to appear at the collector terminal.

R_E Open. When the emitter resistor opens, no current enters into the emitter terminal of the transistor and, again, the component will be in cutoff. However, since it is the *negative* supply that has been isolated from the circuit, the voltage readings will be different than those for an open collector resistor. The circuit conditions that correspond to an open emitter resistor are illustrated in Figure 6.19c. Note that the following conditions exist when the emitter resistor is open:

1. V_B is approximately 0 V.
2. V_E will be between 0 V and V_{CC}.
3. V_C is approximately equal to V_{CC}.

Since there is no current in the collector circuit, no voltage is dropped across R_C. This causes V_C to be approximately equal to V_{CC}. Since there is no base current, V_B will still be approximately equal to 0 V.

When measuring the voltage at the emitter terminal of the transistor, you will get one of two readings, depending on the point to which you are referencing the reading. If you read from the emitter terminal to ground, you will read approximately 0 V. However, if you read across the emitter resistor to $-V_{EE}$, you will get a positive voltage reading. The reason for this can be seen by taking a look at Figure 6.20. When the meter is connected to $-V_{EE}$, the open-emitter circuit is completed through the internal resistance of the meter. In other words, the meter becomes the new R_E. The current flowing through the meter will develop a voltage at the emitter terminal. This voltage will be read on the meter and may well affect other circuit voltages.

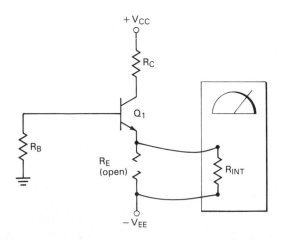

FIGURE 6.20

Meter input resistance can complete a circuit.

R_B Open. When the base resistor opens, no biasing current can flow in the base circuit of the transistor. Because of this, the transistor will again be biased to cutoff. The circuit conditions of an open base resistor are illustrated in Figure 6.19d. Note that the following conditions exist when the base resistor is open:

1. V_B is approximately 0 V.
2. V_E is approximately equal to $-V_{EE}$.
3. V_C is approximately equal to V_{CC}.

Since the transistor is the source of the open circuit between V_{EE} and V_{CC}, the difference between these two potentials will be dropped across the transistor. No current is present in either the emitter or collector circuits, so their respective resistors will not drop any voltage. Thus, the readings listed above.

Note that when reading the base voltage you will provide a path for base current with the meter. Thus you may read a slight negative voltage at the base of the transistor.

Q_1 **Open.** This condition will give you the same readings as having an open base resistor; that is, the transistor will be in cutoff. A simple check of the transistor (as outlined in Chapter 5) or a resistance reading of R_B will tell you which of the two components is faulty.

Q_1 **Leaky.** The term *leaky* is used to describe a transistor that is partially shorted from emitter to collector (Figure 6.19e). When a transistor is leaky, it may appear to be saturated, even when it should not be. The collector and emitter voltages may be nearly equal, and the 0.7-V drop from emitter to base will still be there. When a transistor is leaky, it must be replaced.

Leaky. A term used to describe a component that is partially shorted.

Practical Troubleshooting Considerations

When you are troubleshooting *any* dc biasing circuit, there are a few practical considerations that you should keep in mind. These points, which can save you a great deal of time and trouble, are as follows:

Causes of an apparent shorted carbon resistor.

1. *Standard carbon resistors do not internally short.* When a resistor appears to be short circuited, the short is caused by one of two conditions:
   a. Another component, wire, copper run (on a printed circuit board), or a solder bridge has shorted the component. In this case, the problem is solved by removing the cause of the short.
   b. The wrong resistor value was used in the circuit. It is not uncommon for people to misread the color code on a resistor. If a circuit calls for a 10-kΩ resistor, and you accidentally use a 100-Ω resistor, the resistor will appear as a short circuit.
   When either R_C or R_E is shorted, the terminal going to the shorted resistor will read the supply voltage. If R_C is shorted, the collector terminal will be at V_{CC}. If R_E is shorted, the emitter terminal will be at $-V_{EE}$. In either of these cases, the transistor will not last long.
   If R_B shorts, the dc biasing circuit will not be greatly affected. However, the ac operation of the transistor will stop.
2. *An open resistor can normally be checked without any test equipment.* When a resistor goes open circuit, it usually *burns* open. Because of this, the component will normally have a single color band: *black*. In most cases, you can troubleshoot an open resistor simply by looking at it. If it has turned black, or if the wire or copper run connected to the component shows signs of burning, the resistor probably needs to be replaced.

As a final note, do not overlook the obvious. If either of the power supplies is out, an emitter-bias circuit will not work. If a transistor is placed in a circuit using the wrong connections, the circuit will not work. The bottom line is this: When you are troubleshooting, look for the obvious first. Then, if you need to,

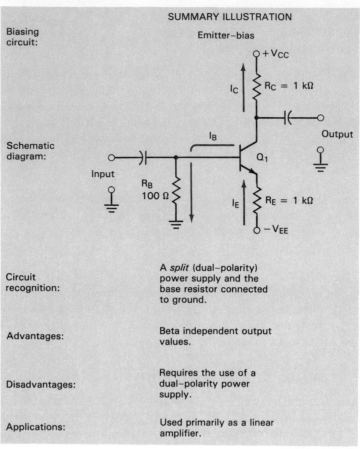

SUMMARY ILLUSTRATION

Biasing circuit: Emitter–bias

Schematic diagram:

Input

R_B 100 Ω

I_B

I_C $R_C = 1$ kΩ

$+V_{CC}$

Q_1

Output

I_E $R_E = 1$ kΩ

$-V_{EE}$

Circuit recognition: A *split* (dual–polarity) power supply and the base resistor connected to ground.

Advantages: Beta independent output values.

Disadvantages: Requires the use of a dual–polarity power supply.

Applications: Used primarily as a linear amplifier.

FIGURE 6.21

use your test equipment to troubleshoot the circuit. See what the circuit is doing as opposed to what it *should* be doing. Then, based on the comparison, check the suspected components.

Summary

The emitter-bias circuit is a beta-independent circuit in that its output characteristics (V_{CEQ} and I_{CQ}) are not affected by variations in beta. The characteristics of the emitter-bias circuit are summarized in Figure 6.21.

SECTION REVIEW

1. Draw an emitter-bias circuit and show the directions of the circuit currents. [Objective 12]
2. Explain why the emitter-bias circuit is a beta-independent circuit. [Objective 13]
3. List, in order, the steps required to determine the Q-point output values of an emitter-bias circuit. [Objective 14]
4. Why do we use the value $2V_{CC}$ when determining the values of $I_{C(sat)}$ and $V_{CE(off)}$ for an emitter-bias circuit? [Objective 14]
5. What adjustment must be made in equations (6.9) and (6.10) when V_{CC} and V_{EE} are *not* equal in value? [Objective 14]

6. What are the normal terminal voltage characteristics of the emitter-bias circuit? [Objective 15]
7. List the characteristics of an emitter-bias circuit with an open collector resistor. [Objective 15]
8. List the characteristics of an emitter-bias circuit with an open emitter resistor. [Objective 15]
9. List the characteristics of an emitter-bias circuit with an open base resistor. [Objective 15]
10. What are the symptoms of a *leaky* transistor? [Objective 15]
11. What are the most common causes of a ''shorted'' resistor [Objective 15]

6.4
Voltage-divider Bias

The voltage-divider bias circuit is by far the most commonly used. It has the stability of an emitter-bias circuit, yet does not require a dual-polarity power supply for its operation. The basic voltage-divider bias circuit is shown in Figure 6.22.

Voltage-divider bias. A biasing circuit that contains a voltage-divider in its base circuit.

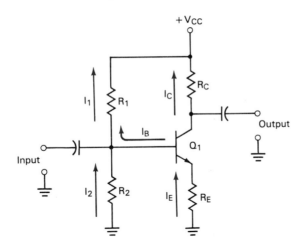

FIGURE 6.22

Voltage—divider bias.

The voltage-divider bias circuit uses a simple voltage divider to set the value of V_B. The voltage divider is made up of R_1 and R_2, and the base voltage of the transistor will equal V_2. Thus, V_B is found in this amplifier as

$$V_B = \frac{R_2}{R_1 + R_2} V_{CC} \qquad (6.12)$$

The emitter voltage is then found as

$$V_E = V_B - 0.7 \text{ V} \qquad (6.13)$$

Once the emitter voltage has been determined, Ohm's law can be used to find the value of I_E, as follows:

$$\boxed{I_E = \frac{V_E}{R_E}}$$ (6.14)

Now, assuming that $I_{CQ} = I_E$, V_{CEQ} can be found as

$$V_{CEQ} = V_{CC} - I_{CQ}R_C - I_E R_E$$

or

$$\boxed{V_{CEQ} = V_{CC} - I_E(R_C + R_E)}$$ (6.15)

The following example illustrates the process used to determine the values of I_{CQ} and V_{CEQ} for the voltage-divider bias circuit.

EXAMPLE 6.10

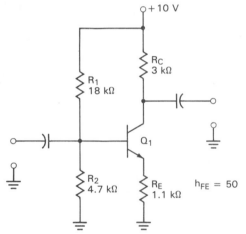

FIGURE 6.23

Determine the values of I_{CQ} and V_{CEQ} for the circuit shown in Figure 6.23.

Solution: R_1 and R_2 set up the base voltage for the amplifier. This voltage is found as

$$\begin{aligned}
V_B &= \frac{R_2}{R_1 + R_2} V_{CC} \\
&= \frac{4.7 \text{ k}\Omega}{22.7 \text{ k}\Omega} (10 \text{ V}) \\
&= (0.207)(10 \text{ V}) \\
&= 2.07 \text{ V}
\end{aligned}$$

V_E is found as

$$\begin{aligned}
V_E &= V_B - 0.7 \text{ V} \\
&= 2.07 \text{ V} - 0.7 \text{ V} \\
&= 1.37 \text{ V}
\end{aligned}$$

I_{CQ} is then found as

$$\begin{aligned}
I_{CQ} &\cong \frac{V_E}{R_E} \\
&= \frac{1.37 \text{ V}}{1.1 \text{ k}\Omega} \\
&= 1.25 \text{ mA}
\end{aligned}$$

Finally, V_{CEQ} is found as

$$\begin{aligned}
V_{CEQ} &= V_{CC} - I_E(R_C + R_E) \\
&= 10 \text{ V} - (1.25 \text{ mA})(3 \text{ k}\Omega + 1.1 \text{ k}\Omega) \\
&= 10 \text{ V} - (1.25 \text{ mA})(4.1 \text{ k}\Omega) \\
&= 10 \text{ V} - 5.13 \text{ V} \\
&= 4.87 \text{ V}
\end{aligned}$$

A circuit like the one in Figure 6.23 has the following values: R_C = 620 Ω, R_E = 180 Ω, R_1 = 12 kΩ, R_2 = 2.7 kΩ, and V_{CC} = 10 V. Determine the values of I_{CQ} and V_{CEQ} for the circuit.

Equation (5.5):
$I_E = I_B(1 + h_{FE})$

Once the value of I_E is known, the value of I_B can simply be found by using equation (5.5), which can be rewritten as

$$I_B = \frac{I_E}{1 + h_{FE}}$$

(6.16)

EXAMPLE 6.11

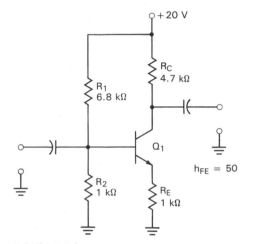

FIGURE 6.24

Determine the value of I_B for the circuit shown in Figure 6.24.

Solution: First, V_B is found as

$$V_B = \frac{R_2}{R_1 + R_2} V_{CC}$$
$$= \frac{1 \text{ kΩ}}{7.8 \text{ kΩ}} (20 \text{ V})$$
$$= 2.56 \text{ V}$$

and V_E is found as

$$V_E = V_B - 0.7 \text{ V}$$
$$= 1.86 \text{ V}$$

Now I_E is found as

$$I_E = \frac{V_E}{R_E}$$
$$= \frac{1.86 \text{ V}}{1 \text{ kΩ}}$$
$$= 1.86 \text{ mA}$$

Finally, I_B is found as

$$I_B = \frac{I_E}{1 + h_{FE}}$$
$$= \frac{1.86 \text{ mA}}{51}$$
$$= 36.5 \text{ μA}$$

✓ **PRACTICE PROBLEM 6–11**

Determine the value of I_B for the circuit described in Practice Problem 6–10. Assume that the value of h_{FE} for the transistor is 200.

Which Value of h_{FE} Do I Use?

17

When you are analyzing a voltage-divider bias circuit (or any other dc biasing circuit for that matter), you will often have to refer to the spec sheet for the transistor to obtain its value of h_{FE}. The only problem is that the spec sheet may not list a single value of h_{FE}. Normally, a transistor spec sheet, as we saw in Figure 5.30 (page 203), will list any combination of the following values:

1. A *maximum* value of h_{FE}
2. A *minimum* value of h_{FE}
3. A *typical* value of h_{FE}

When only one value of h_{FE} is listed on the spec sheet, you must use that value in any circuit analysis.

When two or more values of h_{FE} are listed, you should first look to see if one of them is the *typical* value of h_{FE}. If a *typical* value is listed, use that value.

When the spec sheet lists a *minimum* value and a *maximum* value of h_{FE}, you must use the *geometric average of the two values*. The geometic average of h_{FE} is found as

$$h_{FE(\text{ave})} = \sqrt{h_{FE(\text{min})} \times h_{FE(\text{max})}} \qquad \textbf{(6.17)}$$

The following example illustrates the use of $h_{FE(\text{ave})}$ in the analysis of a transistor amplifier.

EXAMPLE 6.12

Determine the value of I_B for the circuit shown in Figure 6.25a.

Solution: First, V_B is found as

$$V_B = \frac{R_2}{R_1 + R_2} V_{CC}$$
$$= 3.12 \text{ V}$$

and V_E is found as

$$V_E = V_B - 0.7 \text{ V}$$
$$= 2.42 \text{ V}$$

Now I_E is found as

$$I_E = \frac{V_E}{R_E}$$
$$\cong 10 \text{ mA}$$

Assuming that $I_{CQ} = 10$ mA, we check the specification sheet shown in Figure 6.25b. As shown, beta (h_{FE}) has a range of 100 to 300 when $I_C = 10$ mA. Thus $h_{FE(\text{ave})}$ is found as

$$h_{FE}(\text{ave}) = \sqrt{h_{FE\text{min}} \times h_{FE\text{max}}}$$
$$= \sqrt{100 \times 300}$$
$$= 173$$

ELECTRICAL CHARACTERISTICS

Characteristic		Symbol	Min	Max	Unit
ON CHARACTERISTICS					
DC Current Gain*		h_{FE}*			—
(I_C = 0.1 mAdc, V_{CE} = 1 Vdc)	2N3903		20	—	
	2N3904		40	—	
(I_C = 1.0 mAdc, V_{CE} = 1 Vdc)	2N3903		35	—	
	2N3904		70	—	
(I_C = 10 mAdc, V_{CE} = 1 Vdc)	2N3903		50	150	
	2N3904		100	300	
(I_C = 50 mAdc, V_{CE} = 1 Vdc)	2N3903		30	—	
	2N3904		60	—	
(I_C = 100 mAdc, V_{CE} = 1 Vdc)	2N3903		15	—	
	2N3904		30	—	
Collector-Emitter Saturation Voltage*		$V_{CE(sat)}$*			Vdc
(I_C = 10 mAdc, I_B = 1 mAdc)			—	0.2	
(I_C = 50 mAdc, I_B = 5 mAdc)			—	0.3	
Base-Emitter Saturation Voltage*		$V_{BE(sat)}$*			Vdc
(I_C = 10 mAdc, I_B = 1 mAdc)			0.65	0.85	
(I_C = 50 mAdc, I_B = 5 mAdc)			—	0.95	
SMALL SIGNAL CHARACTERISTICS					
High Frequency Current Gain	2N3903	$\lvert h_{fe} \rvert$	2.5	—	—
(I_C = 10 mA, V_{CE} = 20 V, f = 100 mc)	2N3904		3.0	—	
Current-Gain—Bandwidth Product	2N3903	f_T	250	—	mc
(I_C = 10 mA, V_{CE} = 20 V, f = 100 mc)	2N3904		300	—	
Output Capacitance		C_{ob}	—	4	pf
(V_{CB} = 5 Vdc, I_E = 0, f = 100 kc)					
Input Capacitance		C_{ib}	—	8	pf
(V_{OB} = 0.5 Vdc, I_C = 0, f = 100 kc)					
Small Signal Current Gain	2N3903	h_{fe}	50	200	—
(I_C = 1.0 mA, V_{CE} = 10 V, f = 1 kc)	2N3904		100	400	
Voltage Feedback Ratio	2N3903	h_{re}	0.1	5.0	X10^{-4}
(I_C = 1.0 mA, V_{CE} = 10 V, f = 1 kc)	2N3904		0.5	8.0	
Input Impedance	2N3903	h_{ie}	0.5	8	Kohms
(I_C = 1.0 mA, V_{CE} = 10 V, f = 1 kc)	2N3904		1.0	10	
Output Admittance		h_{oe}	1.0	40	µmhos
(I_C = 1.0 mA, V_{CE} = 10 V, f = 1 kc)	Both Types				
Noise Figure	2N3903	NF	—	6	db
(I_C = 100 µA, V_{CE} = 5 V, R_g = 1 Kohms, Noise Bandwidth = 10 cps to 15.7 kc)	2N3904		—	5	
SWITCHING CHARACTERISTICS					
Delay Time	V_{CC} = 3 Vdc, V_{OB} = 0.5 Vdc, I_C = 10 mAdc, I_{B1} = 1 mA	t_d	—	35	nsec
Rise Time		t_r	—	35	nsec
Storage Time	V_{CC} = 3 Vdc, I_C = 10 mAdc, I_{B1} = I_{B2} = 1 mAdc	2N3903	t_s	—	175	nsec
	2N3904		—	200	
Fall Time		t_f	—	50	nsec

*Pulse Test: Pulse Width = 300 µsec, Duty Cycle = 2% V_{OB} = Base Emitter Reverse Bias

(b)

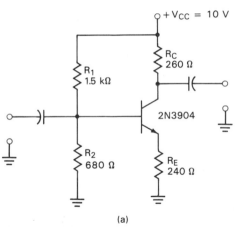

(a)

FIGURE 6.25

(Courtesy of Motorola)

Finally, I_B is found as

$$I_B = \frac{I_E}{h_{FE(ave)} + 1}$$
$$= \frac{10 \text{ mA}}{174}$$
$$= 57.47 \text{ μA}$$

Saturation and Cutoff

The dc load line for the voltage-divider bias circuit is plotted using the end points of $I_{C(sat)}$ and $V_{CE(off)}$. When the transistor is saturated, V_{CE} is approximately equal

18

SEC. 6.4   Voltage-divider Bias

241

to 0 V. Thus the collector current will equal the supply voltage divided by the total reistance between V_{CC} and gound. By formula,

$$I_{C(sat)} = \frac{V_{CC}}{R_C + R_E}$$ (6.18)

When the transistor is in cutoff, all the supply voltage is dropped across the transistor. Thus

$$V_{CE(off)} = V_{CC}$$ (6.19)

Using these equations, the dc load line for Figure 6.25 is plotted as shown in Figure 6.26.

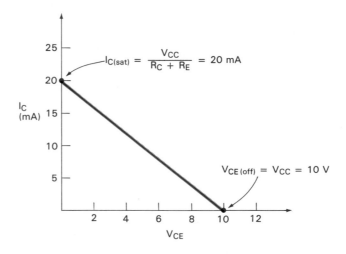

FIGURE 6.26

Voltage-divider bias DC load line.

Base Voltage: A Practical Consideration

In all our voltage-divider bias circuit analyses, we have solved for V_B using the equation

$$V_B = V_{CC} \frac{R_2}{R_1 + R_2}$$ (6.12)

Rule of Thumb: If $h_{FE}R_E \geq 10R_2$, you can ignore the value of $h_{FE}R_E$.

This equation is valid for V_B as long as one condition is fulfilled. For equation (6.12) to be valid, R_2 must be *less than or equal to one-tenth the base input resistance*. By formula,

$$R_2 \leq \frac{h_{FE}R_E}{10}$$ (6.20)

Note that cquation (6.12) is valid only when the condition shown in equation (6.20) is fulfilled. The reason for this can be seen by taking a look at Figure 6.27.

Figure 6.27a shows the base circuit and emitter circuit for a voltage-divider biased amplifier. Note the current arrows, which indicate that R_2 is *in parallel* with the input resistance of the amplifier. The value of $R_{IN(base)}$ is shown as $h_{FE}R_E$ in Figure 6.27b.

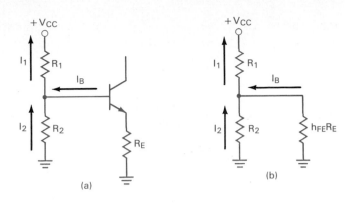

FIGURE 6.27

As long as $h_{FE}R_E$ is at least 10 times the value of R_2, the value of $h_{FE}R_E$ can be ignored in determining the value of V_B. However, if $h_{FE}R_E$ is *less than 10 times the value of R_2, it must be considered in the determination of V_B.*

When $h_{FE}R_E < 10R_2$, the following procedure must be used to determine the value of V_B:

1. Calculate the parallel equivalent resistance from the base of the transistor to ground. This resistance is found as

$$R_{eq} = R_2 \parallel h_{FE}R_E \qquad \textbf{(6.21)}$$

Don't Forget: The ∥ symbol means "in parallel with." Thus, R_{eq} would be found by using the values of R_2 and $h_{FE}R_E$ in a parallel equivalent resistance equation.

2. Solve for the base voltage as follows:

$$V_B = V_{CC}\frac{R_{eq}}{R_1 + R_{eq}} \qquad \textbf{(6.22)}$$

After solving for V_B, continue the normal sequence of calculations to find the values of I_{CQ} and V_{CEQ}. The entire process for determining the values of I_{CQ} and V_{CEQ} when $h_{FE}R_E < 10R_2$ is illustrated in the following example.

EXAMPLE 6.13

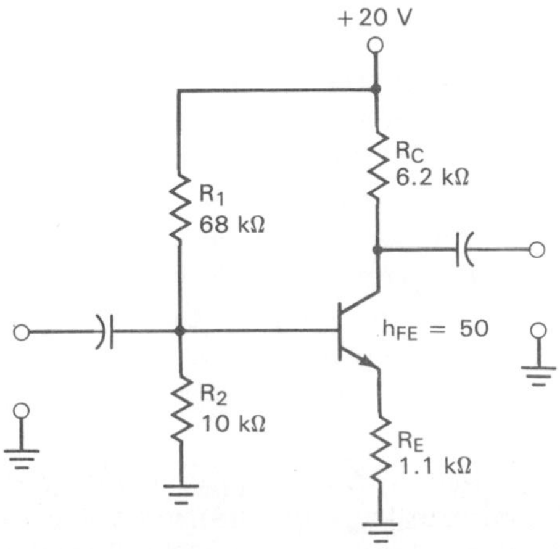

FIGURE 6.28

Determine the values of I_{CQ} and V_{CEQ} for the amplifier shown in Figure 6.28.

Solution: The value of $h_{FE}R_E$ for the amplifier is found as

$$h_{FE}R_E = (50)(1.1\ k\Omega)$$
$$= 55\ k\Omega$$

Since this value is less than 10 times the value of R_2, we must use equations (6.21) and (6.22) to determine the value of V_B, as follows:

$$R_{eq} = R_2 \parallel h_{FE}R_E$$
$$= (10\ k\Omega) \parallel (55\ k\Omega)$$
$$= 8.46\ k\Omega$$

and

$$V_B = V_{CC} \frac{R_{eq}}{R_1 + R_{eq}}$$
$$= (20 \text{ V}) \frac{8.46 \text{ k}\Omega}{68 \text{ k}\Omega + 8.46 \text{k}\Omega}$$
$$= 2.21 \text{ V}$$

The rest of the problem is solved in the usual fashion, as follows:

$$V_E = V_B - 0.7 \text{ V}$$
$$= 1.51 \text{ V}$$

and

$$I_{CQ} = I_E = \frac{V_E}{R_E}$$
$$= \frac{1.51}{1.1 \text{ k}\Omega}$$
$$= 1.37 \text{ mA}$$

Finally,

$$V_{CEQ} = V_{CC} - I_E(R_C + R_E)$$
$$= 20 \text{ V} - (1.37 \text{ mA})(7.3 \text{ k}\Omega)$$
$$= 9.999 \text{ V}$$

✓ **PRACTICE PROBLEM 6–13**

Assume that the transistor in Figure 6.28 has a value of $h_{FE} = 80$. Recalculate the values of I_{CQ} and V_{CEQ} for the circuit.

Estimating the Value of I_{CQ}

You may have noticed that we have a bit of a problem now. In order to determine whether or not $R_{IN(base)}$ must be considered in the calculation of V_B, we have to know the value of h_{FE}. However, as you may recall, we need to know the value of I_{CQ} to find the value of h_{FE} on any spec sheet. And we cannot determine the value of I_{CQ} without having first determined the value of V_B.

To determine the value of h_{FE} when the value of V_B isn't certain, we need a way of estimating the value of I_{CQ} without using V_B. We can approximate the value of I_{CQ} by making an assumption: *The circuit was designed for midpoint bias.* Assuming that the circuit was designed for midpoint bias, the value of I_{CQ} can be approximated as being one-half of $I_{C(sat)}$. So, the process for *estimating* the value of I_{CQ} is:

1. Calculate the value of $I_{C(sat)}$.
2. Approximate I_{CQ} as being one-half the calculate value of $I_{C(sat)}$. Once you have approximated the value of I_{CQ}, you can use that approximated value to look up the value of h_{FE} on the spec sheet of the transistor. Then use the value of h_{FE} from the spec sheet to proceed with your analysis.

The validity of this method can be seen by referring back to Example 6.13. Using equation (6.18), the value of $I_{C(sat)}$ for the circuit is calculated to be 2.74 mA. Half of $I_{C(sat)}$ would then be 1.37 mA, the same value as the calculated value of I_{CQ} in the example.

You may be wondering why we can assume that every voltage-divider bias circuit is designed for midpoint bias. The reason is simple: Voltage-divider bias circuits are used primarily in linear amplifiers. As was stated earlier in the chapter, these amplifiers are almost always designed for midpoint bias to provide for the largest possible output.

Circuit Troubleshooting

There really is not very much difference between troubleshooting a voltage-divider bias circuit and troubleshooting an emitter-bias circuit. In this section we will cover some of the problems that may arise in a voltage-divider bias circuit. As you will see, the symptoms are essentially the same as those found in the emitter-bias circuit.

✔ 21

R_1 **Open.** When R_1 is open, there is no path for current in the base circuit. Therefore, no voltage is developed across R_2, and V_B is zero. Because of the loss of V_B, the transistor will be biased off. Thus no current will flow in the emitter circuit, and V_E will equal 0 V.

Without any current flowing in the collector circuit, $I_C R_C$ will equal 0 V, and V_C will equal V_{CC}. Thus the following circuit conditions will exist when R_1 is open:

1. V_B will be 0 V.
2. V_E will be 0 V.
3. V_C will equal V_{CC}.

These circuit conditions are illustrated in Figure 6.29a.

R_2 **Open.** If R_2 goes open circuit, the current in the base circuit will equal I_B. Since this current normally equals $(I_2 + I_B)$, the current through R_1 will *decrease* significantly. This will cause the voltage drop across R_1 to decrease, and V_B will increase. The increase in V_B will cause the transistor to saturate. Thus V_C and V_E

FIGURE 6.29

Voltage-divider bias fault symptoms.

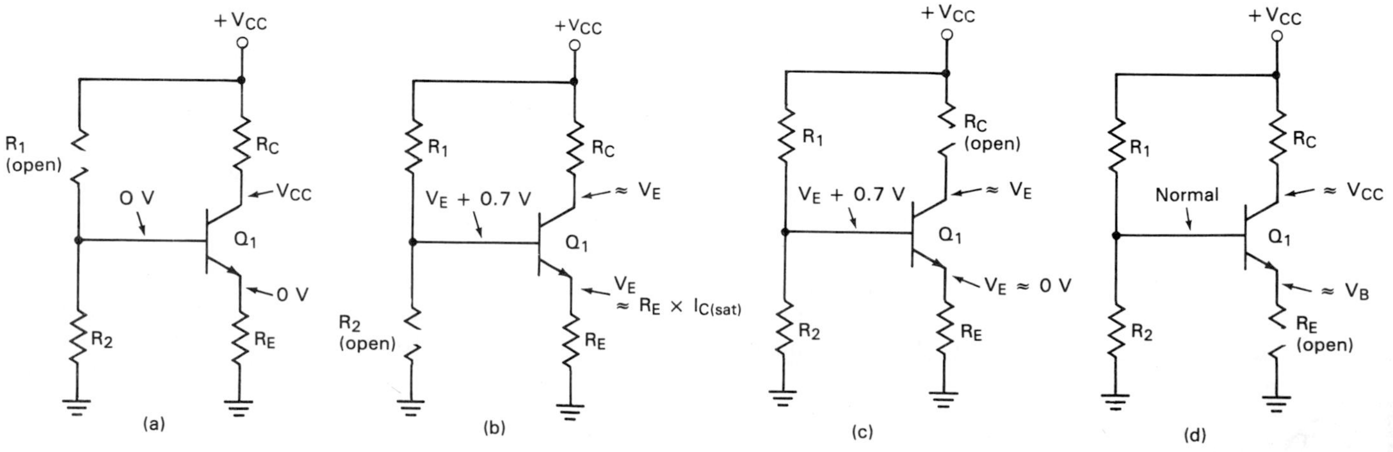

SEC. 6.4 Voltage-divider Bias

Feedback-bias Circuits

Feedback bias. A term used to describe a circuit that "feeds" a portion of the output voltage or current back to the input to control the circuit operation.

Collector-feedback bias. A bias circuit constructed so that V_C will directly affect V_B.

Emitter-feedback bias. A bias circuit constructed so that V_E will directly affect V_B.

In this section, we are going to take a look at two *feedback-bias* circuits. The term *feedback* is used to describe a circuit that "feeds" a portion of the output voltage or current back to the input. For example, the *collector-feedback bias* circuit is constructed so that the collector voltage (V_C) has a direct effect on the base voltage (V_B). The *emitter-feedback bias* circuit is constructed so that the emitter voltage (V_E) has a direct effect on the base voltage. The reason for wiring these circuits in this way is to help reduce the effects of beta variations on the Q-point values of the circuits. This point will be made clear in our discussions on the two circuits.

Collector-feedback Bias

The collector-feedback bias circuit obtains its stability by connecting the base resistor directly to the collector of the transistor. As you can see in Figure 6.31a, this circuit configuration looks quite similar to the base-bias circuit. The electron flow in this circuit enters the emitter from ground. The main part of I_E flows out of the collector terminal and through R_C to V_{CC}. The base current flow through R_B and rejoins the original current at the collector of the transistor. From there, this current flows to V_{CC}. The circuit currents are shown in Figure 6.31b.

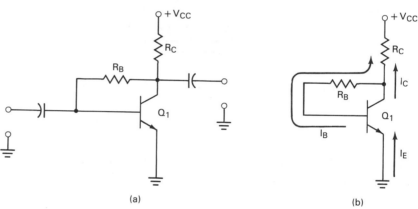

(a)   (b)

FIGURE 6.31

Collector-feedback bias.

The analysis of the collector-feedback bias circuit begins with finding the value of I_B. This value is found as

$$I_B = \frac{V_{CC} - V_{BE}}{R_B + h_{FE}R_C} \quad\quad \textbf{(6.23)}$$

After you have found the value of I_B, you proceed just as you did with the base-bias circuit, as follows:

$$I_{CQ} = h_{FE}I_B$$

and

$$V_{CEQ} = V_{CC} - I_{CQ}R_C$$

CHAP. 6   DC Biasing Circuits

The following example illustrates the Q-point analysis of a collector-feedback bias circuit.

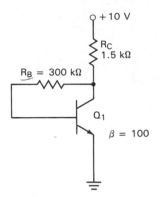

FIGURE 6.32

Determine the values of I_{CQ} and V_{CEQ} for the amplifier shown in Figure 6.32.

Solution: First, the value of I_B is found as

$$I_B = \frac{V_{CC} - V_{BE}}{R_B + h_{FE}R_C}$$

$$= \frac{10 \text{ V} - 0.7 \text{ V}}{(300 \text{ k}\Omega) + (100)(1.5 \text{ k}\Omega)}$$

$$= \frac{9.3 \text{ V}}{450 \text{ k}\Omega}$$

$$= 20.7 \text{ }\mu\text{A}$$

Next, I_{CQ} is found as

$$I_{CQ} = h_{FE}I_B$$
$$= 2.07 \text{ mA}$$

Finally, V_{CEQ} is found as

$$V_{CEQ} = V_{CC} - I_C R_C$$
$$= 10 \text{ V} - (2.07 \text{ mA})(1.5 \text{ k}\Omega)$$
$$= 6.895 \text{ V}$$

PRACTICE PROBLEM 6–14

A circuit like the one shown in Figure 6.32 has values of $R_C = 2$ kΩ, $R_B = 240$ kΩ, $h_{FE} = 120$, and $V_{CC} = +12$ V. Determine the values of I_{CQ} and V_{CEQ} for the circuit.

The Basis for Equation (6.23)

The base-emitter circuit consists of the base-emitter junction of the transistor, R_B, R_C, and the dc power supply. The Kirchhoff's voltage equation for this circuit is

$$V_{CC} = V_{BE} + I_B R_B + (I_B + I_C)R_C$$

If we approximate $(I_B + I_C)$ as $h_{FE}I_B$, the equation can be rewritten as

$$V_{CC} = V_{BE} + I_B R_B + h_{FE}I_B R_C$$

Subtracting V_{BE} from both sides of the equation gives us

$$V_{CC} - V_{BE} = I_B R_B + h_{FE}I_B R_C$$

or

$$V_{CC} - V_{BE} = I_B(R_B + h_{FE}R_C)$$

Finally, solving for I_B, we get

$$I_B = \frac{V_{CC} - V_{BE}}{R_B + h_{FE}R_C}$$

This is equation (6.23).

Circuit Stability

25

The collector-feedback bias circuit is relatively stable against changes in beta. The key to this stability is the fact that I_B and beta are *inversely related* in this circuit. As beta increases, I_B decreases, and vice versa. This relationship can be seen in equation (6.23), which shows I_B to vary inversely with the product of ($h_{FE}R_C$).

Now consider the response of the amplifer in Figure 6.32 to a change in temperature. If temperature increases, I_C increases. This causes beta to increase. In response to the increase in beta, I_B decreases, bringing the value of I_C down. The decrease in I_B has offset the initial increase in I_C.

The relationship just described also explains the fact that a collector-feedback bias circuit *cannot saturate*. No matter how large beta becomes, the transistor will not go into saturation because the increase in beta will be offset by a decrease in I_B.

Circuit Troubleshooting

If either resistor in the collector-feedback bias circuit opens, the transistor will go into cutoff. In the case of this circuit, the quickest method of troubleshooting is simply to replace both resistors. If the circuit continues to malfunction, replace the transistor.

Emitter-feedback Bias

26

The collector-feedback bias circuit is designed so that the collector circuit weighs into the base current calculations. Since I_B is controlled by the value of $h_{FE}R_C$, feedback has been obtained.

The emitter-feedback bias circuit works in basically the same fashion. However, in this case it is the *emitter circuit* that affects the value of I_B. The basic emitter-feedback bias circuit is shown in Figure 6.33. Note that the circuit is almost

FIGURE 6.33

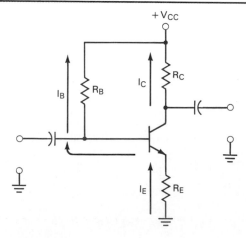

identical to the base-bias circuit, with the exception of the added emitter resistor. Also note the directions of the circuit currents.

The Q-point analysis of this circuit requires three steps. First, the base current is found as

✔ 27

$$I_B = \frac{V_{CC} - V_{BE}}{R_B + (h_{FE} + 1)R_E} \qquad (6.24)$$

After the value of I_B is found, the value of I_{CQ} is determined as

$$I_{CQ} = h_{FE}I_B$$

Finally, V_{CEQ} is found as

$$V_{CEQ} = V_{CC} - I_{CQ}(R_C + R_E)$$

Note that this is the same equation that we used to determine the value of V_{CEQ} for the voltage-divider bias circuit. It would make sense that these two circuits would have the same V_{CEQ} equation, since their collector–emitter circuits are identical. The following example illustrates the process of determining the Q-point values for an emitter-feedback bias circuit.

EXAMPLE 6.15

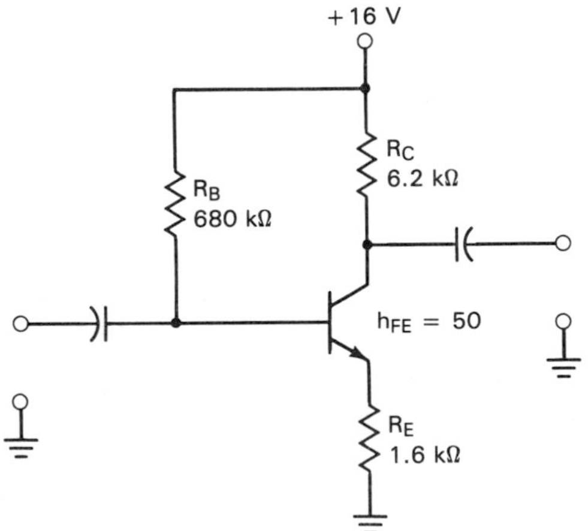

Determine the values of I_{CQ} and V_{CEQ} for the amplifier shown in Figure 6.34.

Solution: First, the value of I_B is found as

$$\begin{aligned} I_B &= \frac{V_{CC} - V_{BE}}{R_B + (h_{FE} + 1)R_E} \\ &= \frac{16 \text{ V} - 0.7 \text{ V}}{680 \text{ k}\Omega + (51)(1.6 \text{ k}\Omega)} \\ &= 20.09 \text{ }\mu\text{A} \end{aligned}$$

I_{CQ} is found as

$$\begin{aligned} I_{CQ} &= h_{FE}I_B \\ &= (50)(20.09 \text{ }\mu\text{A}) \\ &= 1 \text{ mA} \end{aligned}$$

Finally, the value of V_{CEQ} is found as

$$\begin{aligned} V_{CEQ} &= V_{CC} - I_{CQ}(R_C + R_E) \\ &= 16 \text{ V} - (1 \text{ mA})(7.8 \text{ k}\Omega) \\ &= 8.2 \text{ V} \end{aligned}$$

FIGURE 6.34

+16 V

R_C 6.2 kΩ

R_B 680 kΩ

$h_{FE} = 50$

R_E 1.6 kΩ

PRACTICE PROBLEM 6–15

A circuit like the one in Figure 6.34 has values of $V_{CC} = 16$ V, $R_B = 470$ kΩ, $R_C = 1.8$ kΩ, $R_E = 910$ Ω, and $h_{FE} = 100$. Determine the values of I_{CQ} and V_{CEQ} for the circuit.

The Basis of Equation (6.24)

The base circuit consists of R_E, the base–emitter junction of the transistor, R_B, and V_{CC}. The Kirchhoff's voltage equation of this circuit is

$$V_{CC} = I_B R_B + V_{BE} + I_E R_E$$

Since $I_E = (h_{FE} + 1)I_B$, we can rewrite the above equation as

$$V_{CC} = I_B R_B + V_{BE} + I_B(h_{FE} + 1)R_E$$

Subtracting V_{BE} from both sides of the equation, we get

$$V_{CC} - V_{BE} = I_B R_B + I_B(h_{FE} + 1)R_E$$

or

$$V_{CC} - V_{BE} = I_B[R_B + (h_{FE} + 1)R_E]$$

Finally, solving for I_B gives us

$$I_B = \frac{V_{CC} - V_{BE}}{R_B + (h_{FE} + 1)R_E}$$

This is equation (6.24).

Circuit Stability

Q-point stability in the emitter-feedback bias circuit is nearly identical to that of the collector-feedback bias circuit. If beta increases, the value of $(h_{FE} + 1)R_E$ also increases. This causes I_B to decrease, as can be seen in equation (6.24). Also, like the collector-feedback bias circuit, the emitter-feedback bias circuit cannot saturate, since I_B and h_{FE} vary inversely.

You may be wondering why the emitter-feedback bias circuit would be preferred over the collector-feedback bias circuit, or vice versa. As you will see in Chapter 8, the emitter-feedback bias circuit has a higher dynamic input impedance than does the collector-feedback bias circuit. For this reason, the emitter-feedback bias circuit has better ac operating characteristics than does the collector-feedback bias circuit. At the same time, the collector-feedback bias circuit requires fewer components. Thus, when the dynamic input impedance of the amplifier is not a concern, the collector-feedback bias circuit would be preferred.

A Practical Consideration: In practice, neither of the feedback-bias circuits will completely eliminate the effects of a change in beta. In fact, the average feedback-bias circuit will still have a ΔI_C that is nearly 50% when h_{FE} doubles.

Circuit Troubleshooting

The resistor faults that can develop in the emitter-feedback bias circuit are nearly identical to those of the voltage-divider bias circuit. The voltage-divider bias faults are listed on pages 245–246. The fault symptoms for R_C and R_E are identical between the two circuits. The R_B fault symptoms are identical to those for R_1 in the voltage-divider bias circuit.

Summary

The feedback bias circuits are designed so that the values of I_B and h_{FE} vary inversely. Thus, when h_{FE} increases, I_B decreases. This offsets the effect that the h_{FE} increase has on the value of I_C.

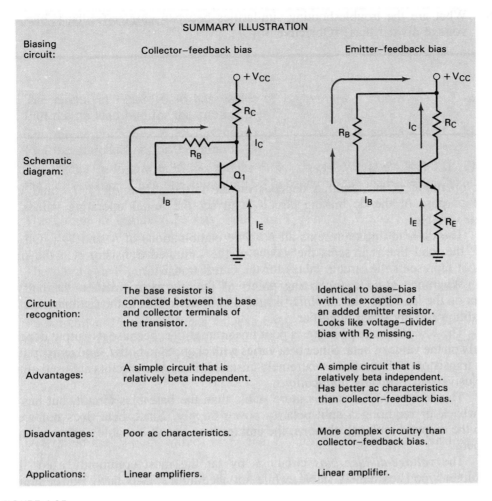

SUMMARY ILLUSTRATION

Biasing circuit:

Collector–feedback bias	Emitter–feedback bias

Schematic diagram:

Circuit recognition:	The base resistor is connected between the base and collector terminals of the transistor.	Identical to base–bias with the exception of an added emitter resistor. Looks like voltage–divider bias with R_2 missing.
Advantages:	A simple circuit that is relatively beta independent.	A simple circuit that is relatively beta independent. Has better ac characteristics than collector–feedback bias.
Disadvantages:	Poor ac characteristics.	More complex circuitry than collector–feedback bias.
Applications:	Linear amplifiers.	Linear amplifier.

FIGURE 6.35

While the *collector-feedback bias circuit* is the simpler of the two, the *emitter-feedback bias circuit* has ac characteristics that make it the more commonly used. Neither of the two circuits can be driven into saturation. The characteristics of the two feedback bias circuits are summarized in Figure 6.35.

SECTION REVIEW

1. What is *feedback*? What is it used for? [Objective 22]
2. Describe the construction and current characteristics of the *collector-feedback bias* circuit. [Objective 23]
3. Explain why collector-feedback bias circuits are stable against changes in beta. [Objective 25]
4. Describe the construction and current characteristics of the *emitter-feedback bias* circuit. [Objective 26]
5. List, in order, the steps required to determine the *Q*-point values of the emitter-bias circuit. [Objective 27]
6. Explain why emitter-feedback bias circuits are stable against changes in beta. [Objective 28]
7. Under what circumstances are each of the feedback-bias circuits preferred over the other? [Objective 29]

11. Plot the dc load line for the circuit in Figure 6.37a. Then, using your answers from problem 7, determine whether or not the circuit is midpoint biased. [10]

12. Plot the dc load line for the circuit in Figure 6.37b. Then, using your answers from problem 8, determine whether or not the circuit is midpoint biased. [10]

13. Without the use of the dc load line, determine whether or not the circuit in Figure 6.37c is midpoint biased. Use your answers from problem 9. [10]

14. Without the use of a dc load line, determine whether or not the circuit in Figure 6.37d is midpoint biased. Use your answers from problem 10. [10]

15. The value of h_{FE} shown in Figure 6.37a is measured at 25°C. At 100°C, the value of h_{FE} for the transistor is 120. Calculate the values of I_C and V_{CE} for the circuit when it is operated at 100°C. [11]

16. The value of h_{FE} shown in Figure 6.37b is measured at 25°C. at 100°C, the value of h_{FE} for the transistor is 120. Calculate the values of I_C and V_{CE} for the circuit when it is operated at 100°C. [11]

§6.3

17. Determine the values of I_{CQ} and V_{CEQ} for the circuit shown in Figure 6.38. [14]

18. For the circuit shown in Figure 6.38, *double* the values of R_C and R_E. Now recalculate the output values for the circuit. How do these values compare with those obtained in problem 17? [14]

19. Calculate the values of $I_{C(sat)}$ and $V_{CE(off)}$ for the circuit shown in Figure 6.38. Use the original component values shown. [14]

20. For the circuit shown in Figure 6.38, *double* the values of R_C and R_E. Now recalculate the values of $I_{C(sat)}$ and $V_{CE(off)}$. How do these values compare with the ones obtained in problem 19? [14]

21. Determine the values of I_{CQ}, V_{CEQ}, $I_{C(sat)}$, and $V_{CE(off)}$ for the circuit shown in Figure 6.39a. [14]

FIGURE 6.39

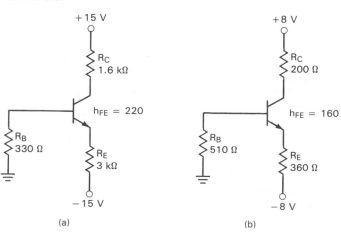

(a)   (b)

FIGURE 6.38

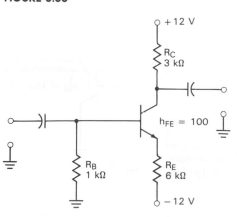

CHAP. 6   DC Biasing Circuits

22. Determine the values of I_{CQ}, V_{CEQ}, $I_{C(sat)}$, and $V_{CE(off)}$ for the circuit shown in Figure 6.39b. [14]
23. Determine the value of $R_{IN(base)}$ for the circuit shown in Figure 6.39a. [14]
24. Determine the value of $R_{IN(base)}$ for the circuit shown in Figure 6.39b. [14]
25. Determine the values of I_{CQ}, V_{CEQ}, and I_B for the circuit in Figure 6.40a. [16]
26. Determine the values of I_{CQ}, V_{CEQ}, and I_B for the circuit in Figure 6.40b. [16]
27. Determine the values of I_{CQ}, V_{CEQ}, and I_B for the circuit in Figure 6.40c. [16]
28. Determine the values of I_{CQ}, V_{CEQ}, and I_B for the circuit in Figure 6.40d. [16]
29. A transistor spec sheet lists the following values of h_{FE} at $I_C = 1$ mA: $h_{FE} = 50$ (min), $h_{FE} = 75$ (typical), and $h_{Fe} = 120$ (max). Determine which value would be used in any amplifier analysis. [16]
30. A transistor spec sheet lists the following values of h_{FE} at $I_C = 1$ mA: $h_{FE} = 80$ (min) and $h_{FE} = 220$ (max). Determine the value of h_{FE} that would be used in any circuit analysis. [16]
31. Determine the values of $I_{C(sat)}$ and $V_{CE(off)}$ for the circuit in Figure 6.40a. Then, using your answers from problem 25, determine whether or not the circuit is midpoint biased. [18]
32. Determine the values of $I_{C(sat)}$ and $V_{CE(off)}$ for the circuit in Figure 6.40b. Then, using your answers from problem 26, determine whether or not the circuit is midpoint biased. [18]
33. Determine the values of $I_{C(sat)}$ and $V_{CE(off)}$ for the circuit in Figure 6.40c. Then, using your answers from problem 27, determine whether or not the circuit is midpoint biased. [18]
34. Determine the values of $I_{C(sat)}$ and $V_{CE(off)}$ for the circuit in Figure 6.40d. Then, using your answers from problem 28, determine whether or not the circuit is midpoint biased. [18]
35. Determine whether or not the circuit in Figure 6.41a is midpoint biased. [16, 18, 19]

FIGURE 6.40

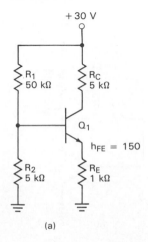 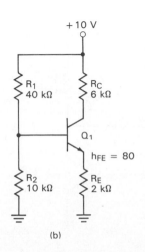 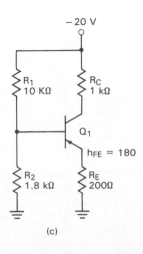 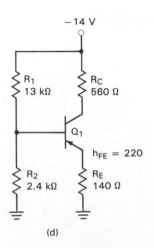

(a) (b) (c) (d)

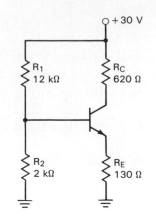

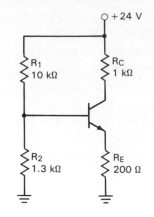

		hFE data		
	min	typ	max	@ I_C =
	40	85	180	1 mA
	60	—	240	10 mA
	100	—	400	20 mA

(a)

		hFE data		
	min	typ	max	@ I_C =
	35	60	100	1 mA
	50	—	400	10 mA
	70	—	500	20 mA

(b)

FIGURE 6.41

36. Determine whether or not the circuit in Figure 6.41b is midpoint biased. [16, 18, 19]

§6.5

37. Determine the values of I_{CQ} and V_{CEQ} for the circuit shown in Figure 6.42a. [24]

38. Determine the values of I_{CQ} and V_{CEQ} for the circuit shown in Figure 6.42b. [24]

√ 39. Determine the values of I_{CQ} and V_{CEQ} for the circuit shown in Figure 6.42c. [24]

40. Determine the values of I_{CQ} and V_{CEQ} for the circuit shown in Figure 6.42d. [24]

41. Determine the values of I_{CQ} and V_{CEQ} for the circuit shown in Figure 6.43a. [27]

FIGURE 6.42

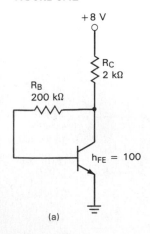

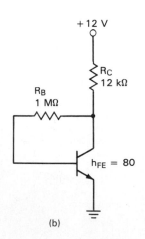

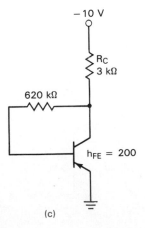

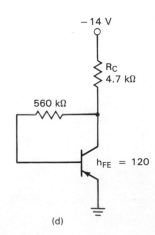

(a)   (b)   (c)   (d)

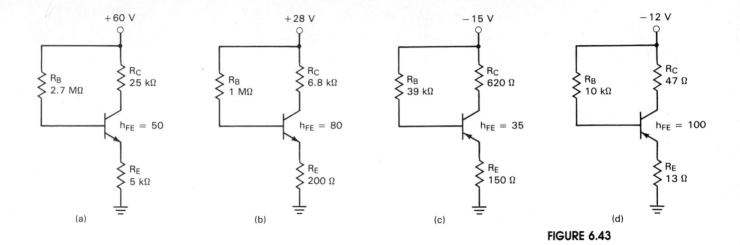

FIGURE 6.43

42. Determine the values of I_{CQ} and V_{CEQ} for the circuit shown in Figure 6.43b. [27]

43. Determine the values of I_{CQ} and V_{CEQ} for the circuit shown in Figure 6.43c. [27]

44. Determine the values of I_{CQ} and V_{CEQ} for the circuit shown in Figure 6.43d. [27]

TROUBLESHOOTING PRACTICE PROBLEMS

45. What fault is indicated in Figure 6.44a? Explain your answer. [15]
46. What fault is indicated in Figure 6.44b? Explain your answer. [15]
47. What fault is indicated in Figure 6.44c? Explain your answer. [15]
48. What fault is indicated in Figure 6.45a? Explain your answer. [21]
49. What fault is indicated in Figure 6.45b? Explain your answer. [21]

FIGURE 6.44

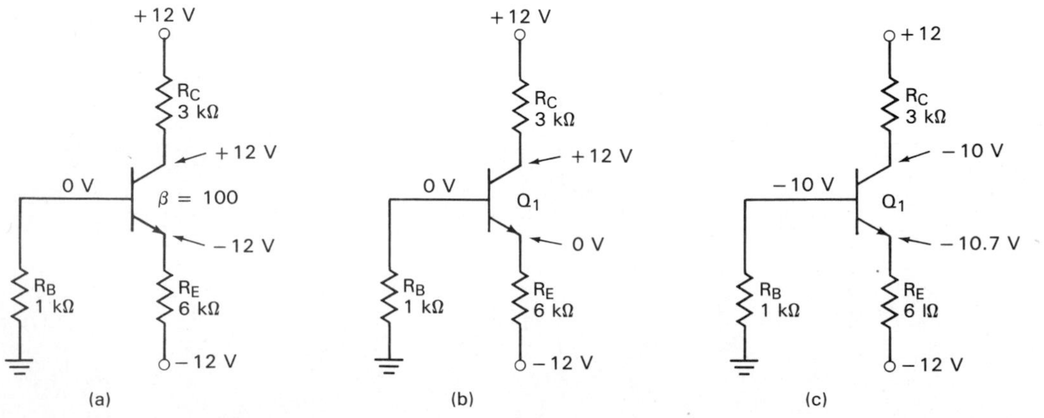

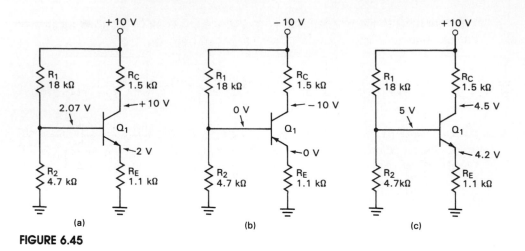

FIGURE 6.45

50. What fault is indicated in Figure 6.45c? Explain your answer. [21]

THE BRAIN DRAIN

51. The transistor in Figure 6.46 has a value of $h_{FE} = 200$ for $I_C = 1$ to 8 mA. The value of $V_{CE(sat)}$ for the component is 0.3 V. Derive the collector curves for the device for $I_C = 1$ to 5 mA and $V_{CE} = 0$ V to $V_{CE(off)}$. Then plot the dc load line for the circuit on the collector curves to determine the value of I_B required for midpoint-bias operation.

52. The spec sheet for the transistor in Figure 6.47 is located in Appendix A. Determine the values of I_{CQ}, V_{CEQ}, I_B, V_B, $I_{C(sat)}$, and $V_{CE(off)}$ for the circuit.

53. *A collector-feedback bias circuit will be midpoint biased if $R_B = h_{FE}R_C$. Using equation (6.23), prove this statement to be true. (Hint: Idealize the emitter–base junction diode in the equation.)*

54. The 2N3904 transistor cannot be used in the circuit shown in Figure 6.48.

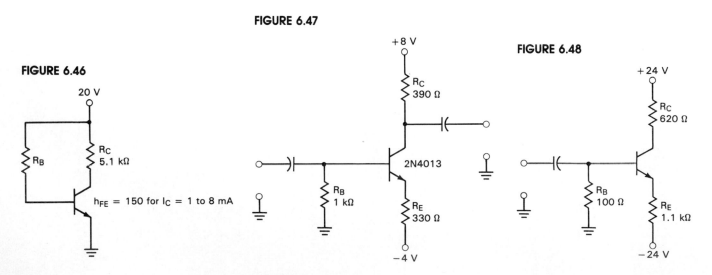

FIGURE 6.46

FIGURE 6.47

FIGURE 6.48

Why not? (*Hint*: Consider the dc load line for the amplifer and the *maximum ratings* shown in the 2N3904 spec sheet on page 203).

SUGGESTED COMPUTER APPLICATIONS PROBLEMS

55. Write a program that will determine whether or not a given base-bias circuit is midpoint biased, given the values of R_C, R_B, V_{CC}, and h_{FE} for the transistor. Set up the program so that it will provide you with the values of I_{CQ} and V_{CEQ}.

56. Write a program that will determine whether or not a given voltage-divider bias circuit is midpoint biased, given the values of R_1, R_2, R_C, R_E, and V_{CC}. Set up the program so that it will provide you with the values of I_{CQ} and V_{CEQ}.

ANSWERS TO THE EXAMPLE PRACTICE PROBLEMS

6.1. $V_{CE(\text{off})} = 8$ V, $I_{C(\text{sat})} = 7.27$ mA

6.2. The end points of the load line are $V_{CE(\text{off})} = 16$ V and $I_{C(\text{sat})} = 8$ mA. $V_{CE} = 12$ V at $I_C = 2$ mA, $V_{CE} = 8$ V at $I_C = 4$ mA, and $V_{CE} = 4$ V at $I_C = 6$ mA.

6.3. $I_C = 9.85$ mA, $V_{CE} = 6.91$ V

6.5. The circuit *is* midpoint biased.

6.6. $I_C = 18.72$ mA, $V_{CE} = 523$ mV

6.7. $I_{CQ} = 4.77$ mA, $V_{CEQ} = 7.85$ V

6.8. $I_{C(\text{sat})} = 6.67$ mA, $V_{CE(\text{off})} = 30$ V

6.9. $R_{\text{IN(base)}} = 540$ kΩ

6.10. $I_{CQ} = 6.33$ mA, $V_{CEQ} = 4.94$ V

6.11. $I_B = 31.49$ μA

6.13. $I_{CQ} = 1.48$ mA, $V_{CEQ} = 9.2$ V

6.14. $I_{CQ} = 2.83$ mA, $V_{CEQ} = 6.34$ V

6.15. $I_{CQ} = 2.72$ mA, $V_{CEQ} = 8.63$ V

7
Common-Emitter Amplifiers

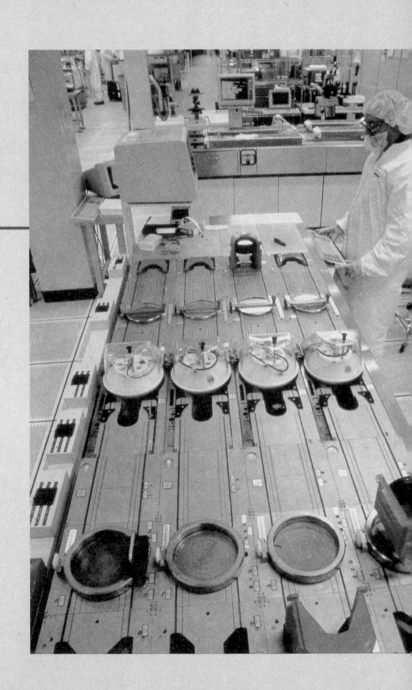

The vignette in this chapter describes "Mr. Meticulous," the first automated transistor manufacturing machine. As you can see, we've come a long way from the days of Mr. Meticulous.

OBJECTIVES

After studying the material in this chapter, you should be able to:

1. Describe *amplification*. (§7.1)
2. Describe *gain* and list the types of gain that are associated with each of the three transistor configurations. (§7.1)
3. Describe the input/output voltage and current phase relationships of the three transistor amplifier configurations. (§7.1)
4. Calculate the ac emitter resistance of a transistor in a given amplifier. (§7.1)
5. Compare and contrast *dc beta* (h_{FE}) and *ac beta* (h_{fe}). (§7.1)
6. List the two primary roles that capacitors serve in amplifiers. (§7.2)
7. Discuss the effects that *coupling capacitors* have on the dc and ac operation of amplifiers. (§7.2)
8. Discuss the effects that *bypass capacitors* have on the dc and ac operation of amplifiers. (§7.2)
9. Describe the effects that coupling and bypass capacitors have on the typical ac signals in a multistage amplifier. (§7.2)
10. List the steps involved in deriving the ac equivalent circuit for a given amplifier. (§7.3)
11. Derive the ac equivalent circuit for a given amplifier. (§7.3)
12. State the effect that the ac equivalent circuit has on the dc analysis and troubleshooting of an amplifier. (§7.3)
13. State the relationship between the *current gain* of an amplifier and the ac beta rating of the transistor. (§7.4)
14. Define *voltage gain* (A_v) and determine its value for a given common-emitter amplifier. (§7.4)
15. Calculate the value of v_{out} for a common-emitter amplifer, given the values of v_{in} and A_v, or given the values of R_C and e'_e. (§7.4)
16. Define *power gain* and determine its value for a given common-emitter amplifier. (§7.4)
17. Calculate the output power for a common-emitter amplifier, given the amplifier values of A_v, h_{fe}, and P_{in}. (§7.4)
18. Explain why the value of A_v for a given amplifier may be unstable. (§7.4)
19. Discuss the effects of *loading* on the voltage gain of a common-emitter amplifier. (§7.5)
20. Calculate the loaded voltage gain (A_{vL}) of a common-emitter amplifier. (§7.5)
21. Explain how the input impedance of a load stage affects the voltage gain of a source stage. (§7.5)
22. Calculate the value of Z_{in} for a common-emitter amplifier. (§7.5)
23. Calculate the values of A_{vT}, A_{iT}, and/or A_{pT} for a multistage amplifier. (§7.5)
24. State the means by which *amplifier swamping* improves gain stability. (§7.6)
25. Calculate the value of A_v for a swamped amplifier. (§7.6)
26. Explain the effect that swamping has on Z_{in}. (§7.6)
27. Calculate the values of $Z_{in(base)}$ and Z_{in} of a swamped amplifier. (§7.6)
28. Explain the effect that amplifier swamping has on the voltage gain of a source amplifier. (§7.6)
29. Discuss the disadvantage of swamping. (§7.6)
30. List and describe the four ac *h-parameters*. (§7.7)
31. Calculate the values of A_i, $Z_{in(base)}$, r'_e, and A_{vL} using h-parameters. (§7.7)
32. Obtain the values of h_{ie} and h_{fe} at a given value of I_{CQ} from the spec sheet of a given transistor, and use those values in the ac analysis of the circuit. (§7.7)
33. Explain the difference between the procedures for determining an *approximate* value of A_{vL} and determining the *worst-case* values of A_{vL}. (§7.7)
34. Troubleshoot a multistage common-emitter amplifier to determine which amplifier stage is faulty. (§7.8)

As you master each of the objectives listed, place a check mark (✔) in the appropriate box.

Now that we have analyzed the operation of most of the transistor dc biasing circuits, it is time to take a look at the basic principles of amplifier ac operation. In this chapter, we will concentrate on the characteristics of the *common-emitter amplifier*. Then, in Chapter 8, we will look at both the *common-collector amplifier* and the *common-base amplifier*.

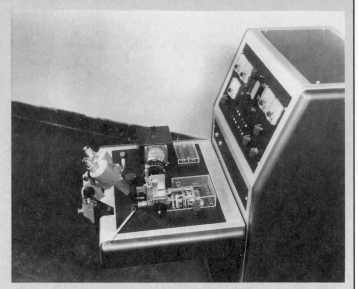
7.1

AC Concepts

Amplification. Increasing the power of an ac signal.

Gain. A multiplier that exists between the input and output of an amplifier.

There are three types of gain: *current* gain, *voltage* gain, and *power* gain.

The choice of amplifier configuration often depends on the type(s) of gain desired.

When you cannot hear a stereo system, you turn up the volume. When the picture on your television is too dark, you turn up the brightness control. In both of these cases, you are taking a relatively weak ac signal and making it stronger (increasing its power). The process of increasing the power of an ac signal is called *amplification*. The circuits used to perform this function are called *amplifiers*.

Gain

Amplifiers are circuits that exhibit a property called *gain*. The gain of an amplifier is a *multiplier* that exists between the input and output. For example, if a given amplifier has a current gain of 100, the output current will be 100 times the input current under normal operating conditions. If the current gain of an amplifier is 250, the output current will be 250 times the input current. The gain of an amplifier is simply a multiplier that is determined by the circuit component values.

There are three types of gain: *current gain*, *voltage gain*, and *power gain*. Common-emitter amplifiers exhibit all three types of gain. From input to output, current will increase, voltage will increase, and power will increase. Common-collector amplifiers have current gain and power gain, but no voltage gain. Common-base amplifiers have voltage gain and power gain, but no current gain. Each of these amplifier types and their gain characteristics will be covered in detail. For now, it is important that you understand that the type of gain an amplifier has depends on the transistor configuration. As a result, the choice of amplifier for a given application often depends on the type of gain that is desired. We will discuss gain in greater detail later in this chapter.

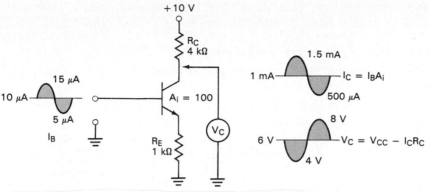

FIGURE 7.1

Common-emitter phase relationships.

Input/Output Phase Relationships

For every amplifier type, *the input and output currents are in phase*. When the input current increases, the output current increases. When the input current decreases, the output current decreases.

The common-emitter amplifier is unique in that the input and output voltages are 180° out of phase, even though the input and output currents are in phase. The cause of this input/output phase relationship is illustrated in Figure 7.1. The amplifier shown has a current gain (A_i) of 100. This means that the ac output current will be 100 times the ac input current. The input current varies by 5 μA both above and below a 10-μA dc level. In other words, the input current is initially at 10 μA, increases to 15 μA, and then decreases to 5 μA at the negative peak. Multiplying the input current values by the A_i of the amplifier gives us the output values shown. When the input current is at 10 μA, the output current is at (10 μA)(100) = 1 mA. When the input current is at 15 μA, the output current is at (15 μA)(100) = 1.5 mA, and so on. *Note that the input current and the output current are in phase*.

To understand the voltage relationship, there are two points that need to be considered:

✔ 3

The common-emitter amplifier is the only BJT amplifier with a 180° voltage phase shift from input to output.

1. The output of the circuit is taken from the collector and is measured with respect to ground. Therefore,

$$\boxed{V_{\text{out}} = V_C} \qquad (7.1)$$

2. The collector voltage is equal to the supply voltage minus the drop across the collector resistor. This relationship has been expressed already as

$$\boxed{V_C = V_{CC} - I_C R_C}$$

If you calculate the value of V_C for each of the values of I_C indicated, you will see that V_C *decreases as* I_C *increases, and vice versa*. When I_C increases to 1.5 mA, V_C *decreases* from 6 V to 4 V. As I_C decreases from 1.5 mA to 500 μA, V_C increases from 4 V to 8 V. The phase relationship is due strictly to the fact that I_C and V_C are inversely proportional.

Now that you have seen the relationship between I_C and V_C, the basis of the input/output voltage phase relationship can be given as follows:

1. Input voltage and current are in phase.
2. Input current and output current are in phase. Therefore, input voltage and output current are in phase.
3. Output current is 180° out of phase with output voltage (V_C). Therefore, *input voltage and output voltage are 180° out of phase*.

Remember that the common-emitter amplifier is the only configuration that has this input/output voltage phase relationship. For both the common-base and common-collector amplifiers, the input and output voltages are in phase. For all three amplifier configurations, input and output currents are in phase.

AC Emitter Resistance

AC emitter resistance. The dynamic resistance of the transistor emitter, used in voltage gain calculations.

You may recall that the zener diode has a *dynamic* resistance value that is considered only in ac calculations. The same holds true for the emitter–base junction diode of a transistor. This junction has a dynamic resistance that is used in ac current and voltage gain calculations. This resistance can be approximated by using the equation

$$r'_e = \frac{25 \text{ mV}}{I_E} \qquad (7.2)$$

where r'_e = the ac emitter resistance
I_E = the dc emitter current, found as V_E/R_E

The derivation of equation (7.2) is lengthy and involves calculus, and thus is not covered here. For those readers who are interested in seeing the derivation of equation (7.2), it is included in Appendix D of the text. However, for those not interested in the derivation, Appendix D need not be studied. In either case, the equation should be committed to memory since it will be used in many ac circuit analyses.

The following example shows how you go about determining the ac resistance of the emitter diode.

EXAMPLE 7.1

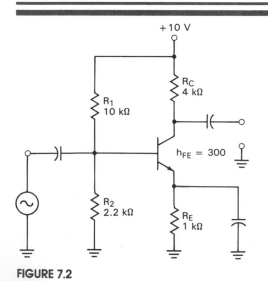

FIGURE 7.2

Determine the ac emitter resistance for the transistor in Figure 7.2.

Solution: First, the value of V_B is found as

$$V_B = \frac{R_2}{R_1 + R_2} V_{CC}$$
$$= 1.8 \text{ V}$$

V_E is found as

$$V_E = V_B - 0.7 \text{ V}$$
$$= 1.1 \text{ V}$$

and I_E is found as

$$I_E = \frac{V_E}{R_E}$$
$$= 1.1 \text{ mA}$$

CHAP. 7 Common-emitter Amplifiers

Finally, the ac emitter resistance is found as

$$r'_e = \frac{25\ mV}{I_E}$$

$$= \frac{25\ mV}{1.1\ mA}$$

$$= 22.73\ \Omega$$

PRACTICE PROBLEM 7–1

A voltage-divider biased amplifier has values of $R_1 = 40\ k\Omega$, $R_2 = 10\ k\Omega$, $R_C = 6\ k\Omega$, $R_E = 2\ k\Omega$, $V_{CC} = +10\ V$, and $h_{FE} = 80$. Determine the ac emitter resistance of the transistor.

AC Beta

5

The ac current gain of an amplifier is different than the dc current gain. This has to do with the fact that dc current gain is measured with I_B and I_C being constant, while ac current gain is measured with *changing* (ac) current values. Figure 7.3 helps to illustrate this point. Figure 7.3a shows the method by which dc beta is determined. A set value of I_C is divided by a set value of I_B. Since the curve of I_B versus I_C is not linear, the value of dc beta changes from one value of I_C to another.

Ac beta. The ratio of ac collector current to ac base current.

Figure 7.3b shows the measurement of ac beta. I_B is varied to cause a variation of I_C *around the Q-point*. The change in I_C is then divided by the change in I_B to obtain the value of ac beta. By formula,

$$\boxed{\beta_{ac} = \frac{\Delta I_C}{\Delta I_B}} \qquad (7.3)$$

or

$$\boxed{\beta_{ac} = \frac{i_c}{i_b}} \qquad (7.4)$$

FIGURE 7.3A&B

The determination of ac beta.

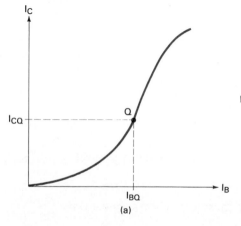

(a)

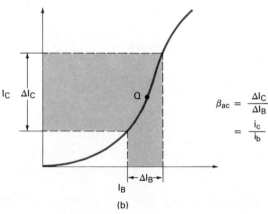

(b)

Note that the symbol Δ means *a change in* the indicated variable. This change is illustrated in Figure 7.3b. Since the input and output currents for a transistor change at the same rate, you can simply divide the ac output current by the ac input current to get the value of ac beta, as indicated in equation (7.4).

Transistor specification sheets list dc beta as h_{FE}. To distinguish the ac value of current gain, ac beta is listed as h_{fe}. As is usually the case, the lowercase subscript is being used to tell you that the current gain is an ac value, not a dc value.

For a common-emitter amplifier, the value of h_{fe} is the current gain (A_i) for the amplifier. We will look at current gain again later in this chapter. By formula,

$$\boxed{A_i = h_{fe}}\qquad \text{(for CE amplifiers)}\qquad\qquad\textbf{(7.5)}$$

Thus the overall current gain for a common-emitter amplifier is found on the spec sheet for the transistor. For practical applications, you do not need to go through any lengthy calculations to determine the overall amplifier current gain.

The concepts and equations introduced in this section will be used constantly throughout our discussions on transistor ac operation. As you use them more and more, you will become more comfortable with them.

h_{FE} = dc beta

h_{fe} = ac beta

SECTION REVIEW

1. What is *amplification*? [Objective 1]
2. What is *gain*? [Objective 2]
3. Given the input to an amplifier and the gain of the circuit, how do you determine the output? [Objective 2]
4. What are the three types of gain? [Objective 2]
5. What types of gain are associated with each amplifier configuration? [Objective 2]
6. Which amplifier type has an input/output voltage phase shift of 180°? [Objective 3]
7. What is the phase relationship between the input and output voltage and current of each transistor amplifier configuration? [Objective 3]
8. List, in order, the steps you need to take in order to determine the ac emitter resistance of a transistor. [Objective 4]
9. What is the difference between h_{FE} and h_{fe}? [Objective 5]
10. Discuss the means by which h_{FE} and h_{fe} are measured. [Objective 5]

7.2
The Role of Capacitors in Amplifiers

Capacitors serve two primary roles in transistor amplifiers.

1. *Coupling* capacitors pass an ac signal from one amplifier to another, while providing dc isolation between the two.
2. *Bypass* capacitors are used to "short circuit" ac signals to ground, while not affecting the dc operation of a circuit.

In this section we will take a look at both of these capacitor applications.

Coupling Capacitors

In most practical applications, you will not see a single transistor amplifier. Rather, there will be a series of *cascaded* amplifiers. The term *cascaded* means *connected in series*.

When amplifiers are cascaded, the original signal is increased by the gain of each individual stage. In the end, the original signal has been provided with more gain than could have been provided by a single amplifier. Each amplifier in the series string is referred to as a *stage*. The overall series of amplifier stages is referred to as a *multistage amplifier*. Multistage amplifiers are covered in detail throughout this chapter. What we are interested in at this point is the fact that *capacitors are commonly used to connect one amplifier stage to another*. A capacitor that is used in this application is called a *coupling capacitor*.

Coupling capacitors are easy to spot in a schematic diagram. They are always seen between the output of one amplifier and the input to the next. For example, take a look at the multistage amplifier shown in Figure 7.4. You will notice that a series capacitor is placed between the collector of each transistor and the base of the next. These capacitors are all coupling capacitors. Any time that you see a capacitor in this position, it is a coupling capacitor. The term *coupling* means *to connect two circuits together electronically so that a signal will pass from one to the other with little or no distortion*. Coupling capacitors:

Cascaded. A term meaning connected in series.

Stage. A single amplifier in a cascaded group of amplifiers.

Multistage amplifier. An amplifier with a series of stages.

Coupling capacitor. A capacitor connected between amplifier stages to provide dc isolation between the stages while allowing the ac signal to pass without distortion.

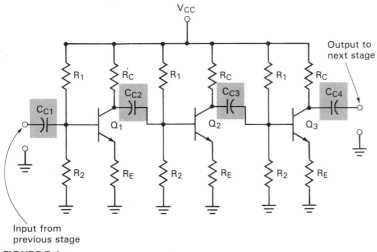

FIGURE 7.4

Coupling capacitors in a multistage amplifier.

1. Pass the ac signal from one stage to the next with little or no distortion
2. Provide dc isolation between the two stages

The purposes served by coupling capacitors.

These two functions are performed easily by the capacitor, which provides little opposition to an ac signal, while completely blocking a dc voltage or current. You may recall from your study of basic electronics that the reactance of a capacitor is found as

$$X_C = \frac{1}{2\pi f C}$$

(7.6)

where X_C = the opposition to current provided by the capacitor
f = the frequency of the signal being applied to the capacitor
C = the capacitance, in farads

From this formula it is easy to see that the *opposition of a capacitor to current is inversely proportional to the frequency of the applied signal.* As frequency goes up, the capacitor acts more and more like a short circuit. As frequency goes down, the capacitor acts more and more like an open circuit. At dc, the frequency of the applied signal is 0 Hz. The reactance at this frequency would be infinite. Thus the capacitor will block dc.

Ac equivalent circuit. A representation of a circuit that shows how the circuit appears to an ac source.

The effects of the coupling capacitor on cascaded amplifiers are illustrated in Figure 7.5. Figure 7.5a is the *ac equivalent* of the circuit in Figure 7.4. Note that the capacitors have been replaced by a wire connection. This represents the

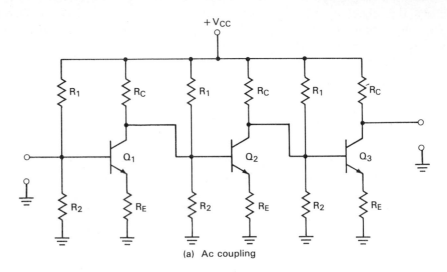

(a) Ac coupling

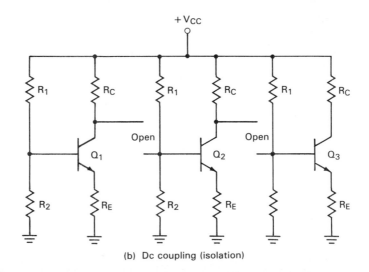

(b) Dc coupling (isolation)

FIGURE 7.5

The effects of coupling capacitors.

low opposition that the coupling capacitor presents to the ac signal. Figure 7.5b is the *dc equivalent* of Figure 7.4. In this circuit, the capacitors have been replaced by breaks in the circuit connections. This represents the infinite opposition the capacitor presents to the dc voltage and current levels of the two stages.

The need for dc isolation (blocking) between amplifier stages can be seen in Figure 7.6. In this circuit, two individual amplifiers are shown, with an open

CHAP. 7 **Common-emitter Amplifiers**

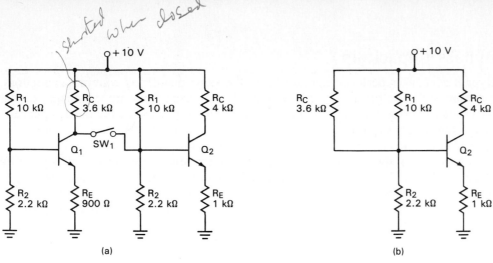

shorted when closed

(a)

(b)

FIGURE 7.6

Why dc isolation is important.

switch between the two. For the second stage, V_B is found as

$$V_B = \frac{R_2}{R_1 + R_2} V_{CC}$$
$$= 1.8 \text{ V}$$

This is the biasing potential for Q_2. Now, assume that the switch is closed. This would directly couple the output of the first stage to the input of the second. This coupling changes the picture, because R_C of the first stage is now in parallel with R_1 of the second stage. The equivalent circuit for this condition is shown in Figure 7.6b. The voltage divider at the input of stage two consists of the parallel combination of R_C and R_1, in series with R_2. To see what this does to V_B of the second stage, we first determine the total resistance of R_C in parallel with R_1. This resistance is found as

Note: The discussion here ignores the effects of I_{CQ} for the first stage in order to keep things as simple as possible.

$$R_{eq} = \frac{R_1 R_c}{R_1 + R_c}$$
$$= 2.65 \text{ k}\Omega$$

Now V_B of the second stage is found as

$$V_B = \frac{R_2}{R_{eq} + R_2} V_{CC}$$
$$= 4.54 \text{ V}$$

With this new value of V_B, the biasing for the second stage has been completely lost. The 4.54 V at the base of Q_2 would undoubtedly cause the transistor to saturate, and the linear amplification of the overall amplifier would be lost.

The use of the coupling capacitor allows each transistor amplifier stage to maintain its independent biasing characteristics, while allowing the ac output from one stage to pass on to the next stage. *The exact value of the coupling capacitor is not critical, provided that its reactance is extremely low at the lowest frequency used by the circuit,* which causes the capacitor to act as a short to the ac signal.

The bottom line.

SEC. 7.2   The Role of Capacitors in Amplifiers

271

Bypass capacitor. A capacitor that is used to establish an ac ground at a specific point in a circuit.

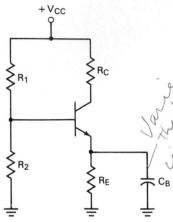

FIGURE 7.7

When the bypass is open RE goes down very slightly

Bypass Capacitors

Bypass capacitors perform functionally in the same manner as coupling capacitors, but they are used for a different reason. Bypass capacitors are connected across the emitter resistor of an amplifier to keep the resistor from affecting the ac operation of the circuit. A bypass capacitor is shown in the circuit in Figure 7.7. Note that the capacitor effectively *bypasses* the emitter resistor, thus the name.

Figure 7.8 shows how the bypass capacitor affects the ac and dc operation of the amplifier. Figure 7.8 shows the *ac equivalent* of the bypass capacitor. Note that the capacitor has been replaced by a wire, just as it is in the coupling application. In this application, the capacitor establishes an *ac ground* at the emitter of the transistor. Thus, for ac purposes, R_E does not exist.

What does this have to do with anything? Well, as you will be shown later in this chapter, the *voltage gain* of a common-emitter amplifier depends on the *total ac emitter resistance*. The lower this ac resistance, the higher the voltage gain of the amplifier. By shorting out R_E, the bypass capacitor ensures that the value of R_E will not weigh into the voltage gain calculations for the amplifier, and thus increases the voltage gain of the circuit. This point will be examined more thoroughly later in this chapter.

Figure 7.8b shows the *dc equivalent* of the bypass capacitor. As before, the capacitor has been replaced with an open wire. This represents the infinite opposition that the capacitor presents to any dc voltage or current. Since the capacitor acts as an open to dc voltages and currents, it has no effect on the dc level of I_E, and therefore no effect on the value of r_e'.

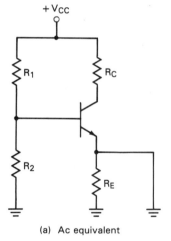
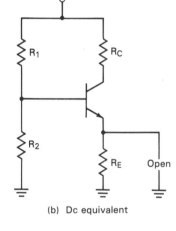

(a) Ac equivalent   (b) Dc equivalent

FIGURE 7.8
The effects of bypass capacitors.

Amplifier Signals

The effects of coupling and bypass capacitors on the ac signals in a multistage amplifier can be seen by looking at Figure 7.9. The signal at the collector of Q_1 is shown to be a sine wave that is riding on a 5.6-V dc voltage. This dc voltage is the quiescent value of V_C for the first stage. Note that this signal is connected to the coupling capacitor, C_{C2}. On the other side of the capacitor you see the same ac sine wave. The only difference is that the sine wave is now riding on the dc value of V_B for the second stage. Thus the ac signal has been passed from the

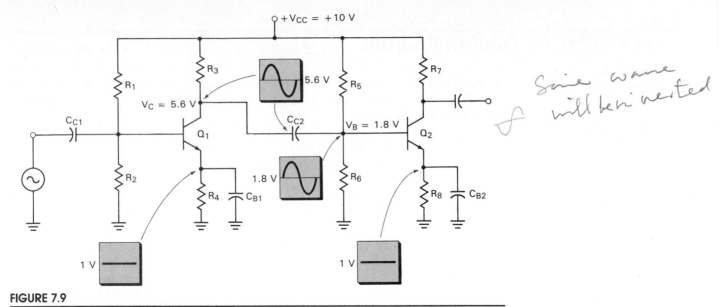

FIGURE 7.9

Typical common-emitter amplifier signals.

first stage to the next, while the dc reference of the signal has been changed from V_C of the first stage to V_B of the second.

Based on what we have seen, we can sum up the characteristics of the coupling capacitor as follows:

1. The ac signal on the input side of the capacitor will equal the ac signal on the output side.
2. The dc voltage on the input side of the capacitor will equal V_C of the source stage.
3. The dc voltage on the output side of the capacitor will equal the value of V_B for the second stage.

The emitters of both amplifiers show no change in V_E. This is caused by the bypass capacitors. The bypass capacitors short the ac component of the emitter signal to ground. Thus the signal present at each emitter is the pure dc value of V_E. Since V_E remains constant, I_E and r'_e remain constant. This is important, since r'_e will weigh into all voltage gain calculations.

The effects of bypass capacitors on ac operation.

---

1. What two purposes are served by capacitors in multistage amplifiers? [Objective 6]
2. What is meant by the term *coupling*? [Objective 7]
3. Why is dc isolation between amplifier stages important? [Objective 7]
4. What effect does a bypass capacitor have on the ac operation of an amplifier? [Objective 8]
5. What effect does a coupling capacitor have on the ac signal that is coupled from one amplifier stage to the next? [Objective 9]
6. Why don't you see an ac voltage at the emitter terminal of an amplifier with a bypass capacitor? [Objective 9]

SECTION REVIEW

7.3

The Amplifier AC Equivalent Circuit

How to derive an ac equivalent circuit.

At this point, we need to derive the ac equivalent circuit for the common-emitter amplifier. This circuit, which is used in many ac calculations, is easily derived using this two-step process:

1. *Short circuit all capacitors.*
2. *Replace all dc sources with a ground symbol.*

The basis for the first step is easy to understand. Since the capacitors act as short circuits to the ac signals, they are replaced with wires, as they were in Figures 7.5a and 7.8a.

The second step is based on the fact that *dc sources have extremely low ac resistance values.* This point is illustrated in Figure 7.10. In the circuit shown, V_{CC} is represented as a simple dc battery. You may recall from your study of basic electronics that the internal resistance of a dc source (battery) is *ideally* 0 Ω. Thus we can replace the dc source with a simple wire, as shown in Figure 7.10b. Note that this wire is replaced by the ground (reference) symbol in Figure 7.10c.

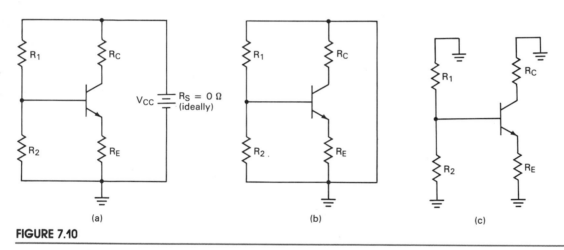

FIGURE 7.10

The process for finding the ac equivalent of a common-emitter amplifier is demonstrated in the following example.

═══════════════ **EXAMPLE 7.2** ═══════════════

Derive the ac equivalent circuit for the amplifier shown in Figure 7.11a.

Solution: The first step is to short circuit all the capacitors (C_C and C_B) and replace V_{CC}' with a ground symbol. This gives us the circuit shown in Figure 7.11b. As you can see, shorting the bypass capacitor leaves us with a straight wire across R_E. Thus, for the ac equivalent circuit, R_E is shorted and is replaced by a wire. Figure 7.11c shows the equivalent circuit without R_E. Note also that R_1 and R_C have been drawn

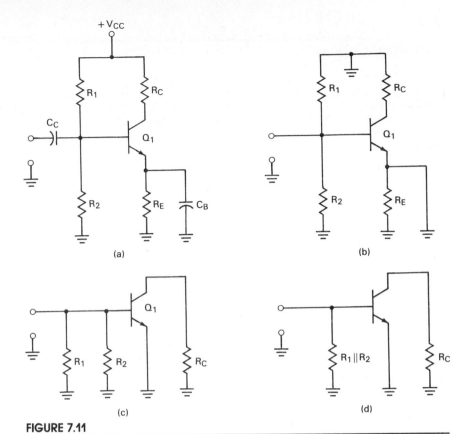

+V_CC

(a)

(b)

(c)

(d)

FIGURE 7.11

Deriving the common-emitter ac equivalent circuit.

as going to ground. (Be sure that you can see what is happening at this point before you go on.)

As a final step, R_1 and R_2 can be combined into a single equivalent resistance. Since the ac equivalent circuit shows these two resistors as both being connected to ground, they are in parallel. Figure 7.11d shows the final ac equivalent circuit for Figure 7.11a.

PRACTICE PROBLEM 7–2

Derive the ac equivalent circuit for the amplifier shown in Figure 7.2 (page 266). Include the component values in the equivalent circuit.

You need to be absolutely clear on one point: *The ac equivalent circuit is based on the ac characteristics of the amplifier and does not affect the dc analysis or troubleshooting of the amplifier.* The ac equivalent circuit is used for ac operation analysis only.

 12

1. What steps are taken to derive the ac equivalent circuit for a given amplifier? **SECTION REVIEW** [Objective 10]
2. What effects do the ac characteristics of an amplifier have on the dc characteristics of the circuit? [Objective 12]

As was stated earlier, there are three types of gain: *current*, *voltage*, and *power* gain. In this section, we will take a closer look at all three. The main emphasis in this section is on the common-emitter circuit. The gain characteristics of the common-base and common-collector circuits are covered in detail in Chapter 8.

Current gain (A_i). The factor by which ac current increases from the input of an amplifier to the output.

Current Gain

Current gain (A_i) is the factor by which ac current increases from the input of an amplifier to the output. The value of A_i for a given amplifier can be found by dividing the ac output current by the ac input current. By formula,

$$A_i = \frac{i_{\text{out}}}{i_{\text{in}}} \tag{7.7}$$

where A_i = the current gain of the amplifier
  i_{out} = the ac output current from the amplifier
  i_{in} = the ac input current to the amplifier

Since the input to a common-emitter amplifier is always applied to the *base* and the output is always taken from the *collector*,

$$\frac{i_{\text{out}}}{i_{\text{in}}} = \frac{i_c}{i_b}$$

Since $h_{fe} = i_c/i_b$ [equation (7.4)],

$$A_i = h_{fe} \tag{7.8}$$

In other words, the current gain of a common-emitter amplifier is equal to the value of ac beta (h_{fe}) for the transistor, as was stated earlier.

Voltage gain (A_v). The factor by which ac signal voltage increases from the amplifier input to the amplifier output.

Voltage Gain

The *voltage gain* (A_v) of an amplifier is the factor by which the ac signal voltage increases from the input of the amplifier to the output. The value of A_v for a common-emitter amplifier can be found by dividing the ac output voltage by the ac input voltage, as follows:

$$A_v = \frac{v_{\text{out}}}{v_{\text{in}}} \tag{7.9}$$

where A_v = the voltage gain of the amplifier
  v_{out} = the ac output voltage from the amplifier
  v_{in} = the ac input voltage to the amplifier

Equation (7.9) is useful for determining the voltage gain of an amplifier when v_{out} and v_{in} are known, as is illustrated in the following example.

EXAMPLE 7.3

An amplifier has values of v_{out} = 12 V and v_{in} = 60 mV. What is the voltage gain of the circuit?

A Practical Consideration: The value of A_v can be determined using peak, peak-to-peak, or rms values. As long as v_{out} and v_{in} are measured in the same fashion (that is, both peak, both rms, etc.), you will obtain the same value of A_v.

Solution: The voltage gain for the circuit is found as follows:

$$A_v = \frac{v_{out}}{v_{in}}$$
$$= \frac{12 \text{ V}}{60 \text{ mV}}$$
$$= 200$$

as with current gain (A_i), voltage gain is a ratio and thus has no units. Thus, the value of A_v for the amplifier described above is simply 200.

PRACTICE PROBLEM 7–3

An amplifier has measured values of v_{out} = 14.2 V and v_{in} = 120 mV. Determine the voltage gain of the circuit.

Equation (7.9) works well enough when the values of v_{out} and v_{in} are known, but there is a problem. The values of v_{out} and v_{in} are not always shown on the schematic diagram for an amplifier. When this is the case, you need to be able to determine the value of A_v without knowing the values of v_{out} and v_{in}.

The value of A_v for a common-emitter amplifier will equal the ratio of *total ac collector resistance to total ac emitter resistance*. For the circuit shown in Figure 7.12, this ratio is found as

$$\boxed{A_v = \frac{R_C}{r'_e}} \qquad (7.10)$$

where R_c = the value of the collector resistor
r'_e = the ac resistance of the transistor emitter

FIGURE 7.12

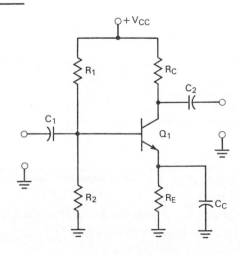

The following series of examples will show you how to determine the voltage gain of an amplifier by analyzing the schematic diagram.

EXAMPLE 7.4

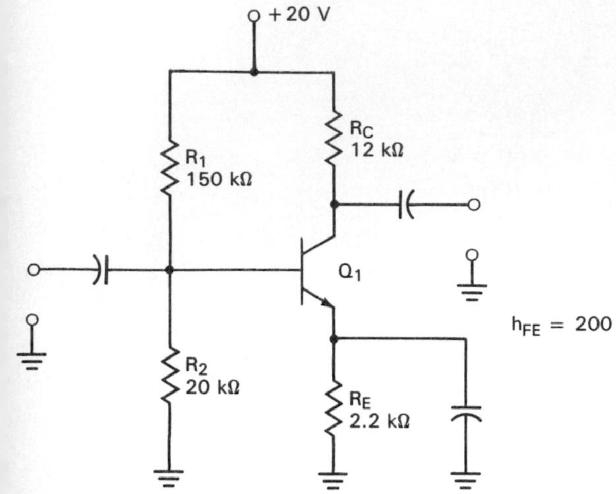

FIGURE 7.13

Determine the voltage gain (A_v) of the circuit in Figure 7.13.

Solution: The value of A_v is found by using the following step:

Step 1: V_B is found as

$$V_B = \frac{R_2}{R_1 + R_2} V_{CC}$$
$$= 2.35 \text{ V}$$

Step 2: V_E is found as

$$V_E = V_B - 0.7 \text{ V}$$
$$= 1.65 \text{ V}$$

Step 3: I_E is found as

$$I_E = \frac{V_E}{R_E}$$
$$= 750 \text{ } \mu\text{A}$$

Step 4: r_e' is found as

$$r_e' = \frac{25 \text{ mV}}{I_E}$$
$$= 33.3 \text{ } \Omega$$

Step 5: A_v is found as

$$A_v = \frac{R_C}{r_e'}$$
$$= 360$$

Therefore, the output ac voltage will be 360 times the input ac voltage level for this circuit.

PRACTICE PROBLEM 7–4

An amplifier like the one in Figure 7.13 has the following values: $V_{CC} = 30$ V, $R_1 = 51$ kΩ, $R_2 = 5.1$ kΩ, $R_C = 5.1$ kΩ, $R_E = 910$ Ω, and $h_{FE} = 250$. Determine the voltage gain of the amplifier.

EXAMPLE 7.5

Determine the value of A_v for the circuit shown in Figure 7.14.

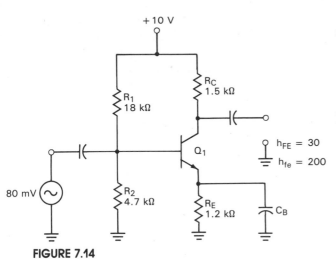

FIGURE 7.14

Solution: Since $h_{FE}R_E < 10R_2$, we have to consider the base input resistance in our V_B calculation, as follows:

$$R_{eq} = R_2 \| h_{FE}R_E$$
$$= 4.16 \text{ k}\Omega$$

and

$$V_B = \frac{R_{eq}}{R_1 + R_{eq}} V_{CC}$$
$$= 1.88 \text{ V}$$

Now

$$V_E = V_B - 0.7 \text{ V}$$
$$= 1.18 \text{ V}$$

and

$$I_E = \frac{V_E}{R_E}$$
$$= 983 \text{ } \mu\text{A}$$

The value of r'_e is now found as

$$r'_e = \frac{25 \text{ mV}}{I_E}$$
$$= 25.4 \text{ } \Omega$$

Finally

$$A_v = \frac{R_C}{r'_e}$$
$$= 59.1$$

PRACTICE PROBLEM 7–5

An amplifier like the one shown in Figure 7.14 has the following values: $V_{CC} =$ +10 V, $R_1 = 39$ kΩ, $R_2 = 10$ kΩ, $R_C = 6.2$ kΩ, $R_E = 2.2$ kΩ, and $h_{FE} = 40$. Determine the value of A_v for the circuit.

The Basis for Equation (7.10)

To understand where equation (7.10) comes from, take a look at Figure 7.15. Figure 7.15a shows a common-emitter amplifier. The ac equivalent of this circuit is shown in Figure 7.15b. In Figure 7.15c, the transistor has been replaced with its own equivalent circuit. The equivalent circuit for the transistor contains three components:

1. A resistor (r'_e) which represents the ac resistance of the emitter
2. A diode, which represents the emitter–base junction of the transistor
3. A current source, which represents the current being supplied to R_C from the collector of the transistor

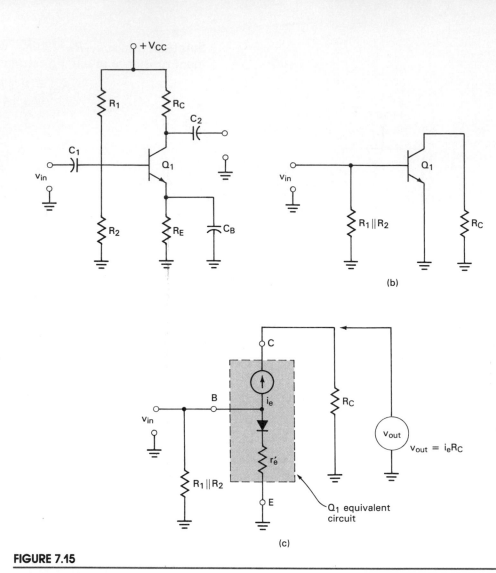

FIGURE 7.15

Note that the input voltage (v_{in}) is applied across the diode and r'_e. Assuming the diode to be ideal, the ac emitter current can be found as

$$\boxed{i_e = \frac{v_{in}}{r'_e}} \qquad (7.11)$$

Thus,

$$\boxed{v_{in} = i_e r'_e} \qquad (7.12)$$

Assuming that $i_c = i_e$, v_{out} is found as

$$\boxed{v_{out} = i_e R_c} \qquad (7.13)$$

This equation is valid because the ac output voltage will be developed across R_C by the ac emitter current flowing to ground from the current source.

Now we can substitute equations (7.12) and (7.13) into equation (7.9) as follows:

$$A_v = \frac{v_{\text{out}}}{v_{\text{in}}}$$

$$= \frac{i_e R_C}{i_e r_e'}$$

Since (i_e) appears in both the numerator and the denominator, it can be dropped completely, leaving

$$\boxed{A_v = \frac{R_C}{r_e'}} \qquad \textbf{(7.10)}$$

Voltage Gain Instability

One important consideration for the common-emitter amplifier is the *stability* of its voltage gain. An amplifier should have gain values that are stable so that the output of the circuit is predictable under all normal circumstances.

Why A_v may be unstable.

The voltage gain of a common-emitter amplifier can tend to be somewhat unstable. The reason for this can be seen in Figure 7.16. The voltage gain of an amplifier is equal to the ratio of collector resistance to the emitter resistance. For the amplifier shown in Figure 7.16, the emitter resistance is equal to r_e'. Like beta, r_e' is somewhat dependent on temperature. Because of this, the voltage gain of the amplifier tends to be unstable. To reduce the effects of temperature on voltage gain, the *swamped* amplifier is used. Swamped amplifiers are discussed in detail in Section 7.6.

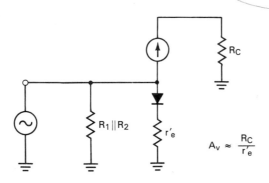

$$A_v \approx \frac{R_C}{r_e'}$$

FIGURE 7.16

Calculating v_{out}

When you know the voltage gain of an amplifier, determining the ac output voltage is simple. Simply use a rewritten form of equation (7.9), as follows:

$$\boxed{v_{\text{out}} = A_v(v_{\text{in}})} \qquad \textbf{(7.14)}$$

===== **EXAMPLE 7.6** =====

The circuit used in Example 7.5 has an ac input signal of 80 mV. Determine the ac output voltage from the circuit.

Solution: The output voltage is found as

$$v_{\text{out}} = A_v(v_{\text{in}})$$

Using the value of A_v from Example 7.5, we obtain

$$v_{\text{out}} = 59.1 \ (80 \ \text{mV})$$
$$= 4.73 \ \text{V}$$

PRACTICE PROBLEM 7–6

The amplifier described in Practice Problem 7.5 has a value of $v_{\text{in}} = 20$ mV. Determine the value of v_{out} for the amplifier.

Power Gain

Power gain (A_p). The factor by which ac signal power increases from the input of an amplifier to the output.

The power gain (A_p) of an amplifier is the factor by which the ac signal power increases from the input of the amplifier to the output. Power gain is found by multiplying current gain by voltage gain. By formula,

$$\boxed{A_p = A_i A_v} \tag{7.15}$$

Since the current gain for a common-emitter amplifier is equal to h_{fe}, equation (7.15) can be rewritten as

$$\boxed{A_p = h_{fe} A_v} \tag{7.16}$$

Once the power gain of an amplifier has been determined, you can calculate the output power of an amplifier as

$$\boxed{P_{\text{out}} = A_p P_{\text{in}}} \tag{7.17}$$

where P_{out} = the amplifier output power
A_p = the power gain of the amplifier
P_{in} = the amplifier input power

▬▬▬ EXAMPLE 7.7 ▬▬▬

The amplifier shown in Figure 7.14 has values of $A_v = 59.1$ and $h_{fe} = 200$. Determine the power gain (A_p) of the amplifier and the output power when $P_{\text{in}} = 80 \ \mu$W.

Solution: The power gain of the amplifier is found as

$$A_p = h_{fe} A_v$$
$$= (200)(59.1)$$
$$= (11{,}820)$$

The output power is therefore found as

$$P_{\text{out}} = A_p P_{\text{in}}$$
$$= (11{,}820)(80 \ \mu\text{W})$$
$$= 945.6 \ \text{mW}$$

PRACTICE PROBLEM 7–7

The circuit described in Practice Problem 7–5 has the following values: $h_{fe} = 200$ and $P_{\text{in}} = 60 \ \mu$W. Determine the output power for the circuit.

---

1. What is *current gain*? [Objective 13]
2. What is the relationship between current gain and the ac beta rating of the amplifier transistor? [Objective 13]
3. What is *voltage gain*? [Objective 14]
4. How can you determine the value of A_v when both v_{in} and v_{out} are known? [Objective 14]
5. How can you determine the value of A_v when the values of v_{in} and/or v_{out} are *not* known? [Objective 14]
6. What is *power gain*? [Objective 16]
7. How do you go about determining the value of P_{out} for an amplifier? [Objective 17]
8. Why is the voltage gain of the standard common-emitter amplifier unstable? [Objective 18]

7.5
The Effects of Loading

Amplifiers are used to provide ac power to a load. Since amplifiers are capacitively coupled to the load, the load will affect the ac gain calculations. For example, take a look at Figure 7.17. The amplifier shown in Figure 7.17a has a 5-kΩ load

19

Amplifier loading *reduces* A_v.

FIGURE 7.17

The effects of loading on the ac equivalent circuit.

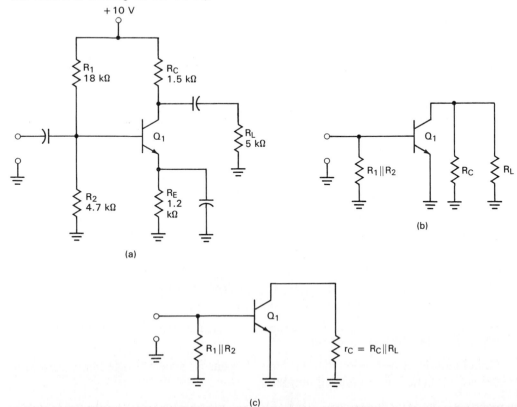

(a)

(b)

(c)

resistor connected to the output. This resistor is capacitively coupled to the collector terminal of the transistor. Therefore, the ac equivalent circuit for the amplifier would place this resistor in parallel with R_C, as shown in Figure 7.17b. Thus the total ac resistance of the collector circuit is found as

$$r_C = R_C \| R_L \qquad (7.18)$$

Since the voltage gain of the amplifier depends partially on the ac resistance of the collector circuit, equation (7.10) must be modified as follows:

$$A_{vL} = \frac{r_C}{r'_e} \qquad (7.19)$$

where A_{vL} = the loaded voltage gain of the amplifier
      r_C = the total ac resistance of the collector circuit, as determined by equation (7.18)
      r'_e = the ac resistance of the emitter circuit

✔ 20

The following example shows how connecting a load *reduces* the overall voltage gain of an amplifier.

══════════════════════ EXAMPLE 7.8 ══════════════════════

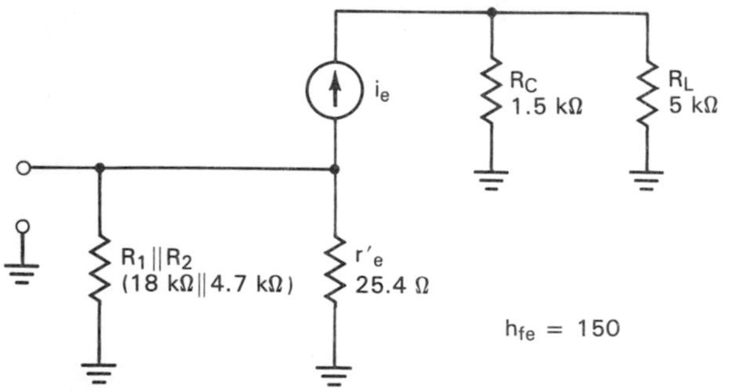

FIGURE 7.18
The ac equivalent circuit for Figure 7.17.

A Practical Consideration: This example illustrates the fact that amplifier voltage gain is reduced when a load is connected. Conversely, if the load is removed from an amplifier, the voltage gain of the amplifier will *increase*.
    If a load goes open circuit, the effects will be the same as removing the load entirely. Thus, the primary symptom of an open load in an amplifier is an *increase* in the voltage gain of the circuit.

Determine the loaded voltage gain for the amplifier shown in Figure 7.17a.

Solution: This circuit was analyzed in Example 7.5. From this example, r'_e was found to be 25.4 Ω. The unloaded voltage gain was found to be 59.1.
    With the load connected, the ac equivalent circuit appears as shown in Figure 7.18. For the circuit, the total collector resistance is found as

$$r_C = R_C \| R_L$$
$$= 1.15 \text{ k}\Omega$$

The loaded voltage gain for the circuit is found as

$$A_{vL} = \frac{r_C}{r'_e}$$
$$= \frac{1.15 \text{ k}\Omega}{25.4 \text{ }\Omega}$$
$$= 45.3$$

Thus, the voltage gain was reduced from 59.1 to 45.3 when the load was connected to the circuit.

───

PRACTICE PROBLEM 7–8

The circuit described in Practice Problem 7–5 has an added 10-kΩ load. Determine the value of A_{vL} for the loaded amplifier.

When one common-emitter amplifier is being used to drive another, the input impedance of the second amplifier will serve as the load resistance of the first. Thus, to calculate the A_{vL} of the first amplifier stage correctly, you must be able to calculate the input impedance of the second stage.

Calculating Amplifier Input Impedance

The input impedance of an amplifier is determined using the ac equivalent circuit for the amplifier. As Figure 7.19 shows, the input impedance of an amplifier is the parallel impedance formed by R_1, R_2, and the base of the transistor. By formula,

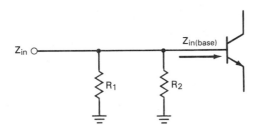

FIGURE 7.19

$$\boxed{Z_{in} = R_1 \| R_2 \| Z_{in(base)}} \tag{7.20}$$

(whole amplifier)

where Z_{in} = the input impedance to the amplifier
$Z_{in(base)}$ = the input impedance of the transistor base

You may recall that the dc input resistance of the base was found as

$$R_{IN(base)} = h_{FE}R_E$$

This relationship was derived in Chapter 6. The input impedance of the base is derived in the same manner. The only differences are:

1. The ac emitter resistance is used in the calculation.
2. The value of ac beta (h_{fe}) is used in place of dc beta (h_{FE}).

Therefore, $Z_{in(base)}$ is found as

$$\boxed{Z_{in(base)} = h_{fe}r'_e} \tag{7.21}$$

Once the input impedance of the transistor is found, it is considered to be in parallel with the base biasing resistors. This parallel circuit gives us the total input impedance to the amplifier, as illustrated in the following example.

═══════════ **EXAMPLE 7.9** ═══════════

Determine the input impedance to the amplifier represented by the equivalent circuit in Figure 7.18.

Solution: The value of $Z_{in(base)}$ is found as

$$Z_{in(base)} = h_{fe}r'_e$$
$$= (150)(25.4\ \Omega)$$
$$= 3.81\ k\Omega$$

Z_{in} for the amplifier is found as

$$Z_{in} = R_1 \| R_2 \| Z_{in(base)}$$
$$= 18 \text{ k}\Omega \| 4.7 \text{ k}\Omega \| 3.81 \text{ k}\Omega$$
$$= 1.88 \text{ k}\Omega$$

PRACTICE PROBLEM 7–9

Determine the value of Z_{in} for the amplifier described in Practice Problem 7–5. Assume that $h_{fe} = 200$ for the transistor.

Multistage Amplifier Gain Calculations

23

When you want to determine the overall value of A_v, A_i, and/or A_p for a multistage amplifier, *you must begin by determining the appropriate gain values for the individual stages*. Once the overall gain values for the individual stages are determined, you can determine the desired overall gain value by using any (or all) of the following equations:

$$A_{vT} = (A_{v1})(A_{v2})(A_{v3}) \ldots \tag{7.22}$$

$$A_{iT} = (h_{fe1})(h_{fe2})(h_{fe3}) \ldots \tag{7.23}$$

$$A_{pT} = (A_{vT})(A_{iT}) \tag{7.24}$$

Equations (7.22) and (7.23) indicate that the overall value of A_v or A_i is simply the product of the individual stage gain values. Equation (7.24) indicates that the overall power gain is found as the product of the overall values of A_v and A_i.

The exact sequence of equations used to determine the overall value of A_{vL} for a given amplifier is as follows:

1. Perform a basic dc analysis of the amplifier to determine the value of I_E.
2. Using equation (7.2), determine the value of r_e' for the amplifier.
3. Using equation (7.18), determine the value of r_C for the amplifier.
4. Using the values of r_e' and r_C found in steps 2 and 3 and equation (7.19), determine the value of A_{vL} for the stage.

A review of Examples 7.5 and 7.8 will show that this is the sequence of steps we have used up to this point.

When you are dealing with a two-stage amplifier, the value of Z_{in} for the second stage must be used in place of R_L for the r_C calculation of the first stage. This point is illustrated in the following example.

EXAMPLE 7.10

Determine the loaded voltage gain for the first stage in Figure 7.20.

Solution: Using the established procedure, r_e' for the first stage is found to be 19.8 Ω, and r_e' for the second stage is

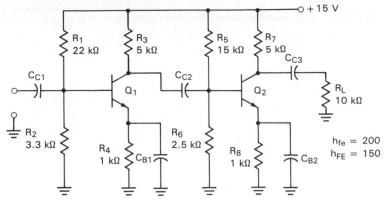

FIGURE 7.20

found to be 17.4 Ω. For the second stage, h_{fe} is 200; therefore,

$$Z_{in(base)} = h_{fe}r'_e$$
$$= (200)(17.4 \ \Omega)$$
$$= 3.48 \ k\Omega$$

The input impedance for the second stage is found as

$$Z_{in} = R_5 \| R_6 \| Z_{in(base)}$$
$$= 1.33 \ k\Omega$$

The input impedance of the second stage (1.33 kΩ) is the load for the first stage. Therefore, r_C for the first stage is found as

$$r_C = R_3 \| Z_{in}$$
$$= 1.05 \ k\Omega$$

Finally, the value of A_{vL} for the first stage is found as

$$A_{vL} = \frac{r_C}{r'_e}$$
$$= \frac{1.05 \ k\Omega}{19.8 \ \Omega}$$
$$= 53.03$$

PRACTICE PROBLEM 7–10

Assume for a moment that the value of h_{fe} for the second stage in Figure 7.20 has increased to 280. Determine the new value of A_{vL} for the first amplifier stage.

As Example 7.10 showed, you must determine the Z_{in} of a load stage before you can determine the actual A_{vL} of a source stage. After determining the Z_{in} of the load stage, this value is used in place of R_L in equation (7.18). Then, using r_C, the loaded voltage gain of the first stage is calculated.

Once the value of A_{vL} for the first stage is known, we can calculate the value of A_{vL} for the second stage and for the overall amplifier. This is illustrated in the following example.

EXAMPLE 7.11

Determine the value of A_{vT} for the amplifier in Figure 7.20.

Solution: We know (from Example 7.10) that the value of A_{vL} for the first stage is 53.03 and that the value of r'_e for the second stage is 17.4 Ω.

The first step in determining the value of A_{vL} for the second stage is to determine the value of r_C, as follows:

$$r_C = R_7 \| R_L$$
$$= 3.33 \text{ k}\Omega$$

A_{vL} for the second stage is found as

$$A_{vL} = \frac{r_C}{r'_e}$$
$$= \frac{3.33 \text{ k}\Omega}{17.4 \text{ }\Omega}$$
$$= 191.38$$

Now that we know the values of A_{vL} for both stages, the overall voltage gain for the two-stage amplifier is found as

$$A_{vT} = (A_{v1})(A_{v2})$$
$$= (53.03)(191.38)$$
$$= 10,149$$

√ PRACTICE PROBLEM 7.11

Assume that the amplifier in Figure 7.20 has the following values: $h_{fe} = 240$ and $R_L = 22$ kΩ. Determine the value of A_{vT} for the two-stage amplifier.

We mentioned earlier that *swamping* reduces the effects of changes in r'_e on voltage gain. It also reduces the loading of an amplifier on a previous stage. We will now look at swamped amplifiers and how they overcome variations in r'_e while reducing circuit loading.

SECTION REVIEW

1. How does the addition of a load affect the voltage gain of an amplifier? [Objective 19]
2. List, in order, the steps required to determine the value of A_{vL} for a common-emitter amplifier. [Objective 20]
3. Why is it important to be able to determine the value of Z_{in} for a load stage? [Objective 21]
4. List, in order, the steps required to determine the value of Z_{in} for a common-emitter amplifier. [Objective 22]
5. How do you determine the values of A_{vT} and A_{iT} for a multistage amplifier? [Objective 23]
6. How do you determine the value of A_{pT} for a multistage amplifier? [Objective 23]

7.6
Swamped Amplifiers

✔ 24a

A *swamped amplifier* reduces variations in voltage gain by *increasing the ac resistance of the emitter circuit.* By increasing the resistance, it also increases $Z_{\text{in(base)}}$, and thus reduces the amplifier's loading effects on a previous stage. A swamped amplifier is shown in Figure 7.21a.

Swamped amplifier. An amplifier that uses a partially bypassed emitter resistance to increase the ac emitter resistance.

    The swamped amplifier has a higher ac emitter resistance because *only part of the dc emitter resistance is bypassed.* The bypass capacitor eliminates only the value of R_E. The other emitter resistor, r_E, is part of the ac equivalent circuit, as shown in Figure 7.21b. Since the voltage gain of the amplifier is equal to the ratio of ac collector resistance to ac emitter resistance, the voltage gain for the amplifier is found as

$$A_v = \frac{R_C}{r_e' + r_E} \qquad (7.25)$$

Just a Reminder: Don't forget to use r_C in the numerator of equation (7.25) when the swamped amplifier has a load resistance.

FIGURE 7.21

The swamped common-emitter amplifier.

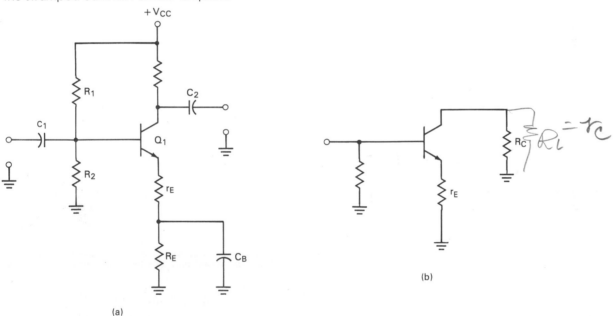

(a)

(b)

The following example illustrates the use of equation (7.25) in the calculation of A_v for a swamped amplifier.

═══════════════════ **EXAMPLE 7.12** ═══════════════════

✔ 25

Determine the value of A_v for the amplifier in Figure 7.22.

Solution: Using the established procedure, V_E is found to equal 1.37 V. The total dc resistance in the emitter circuit equals $(R_E + r_E)$. Therefore, I_E is found as

$$I_E = \frac{V_E}{R_E + r_E}$$
$$= 1.14 \text{ mA}$$

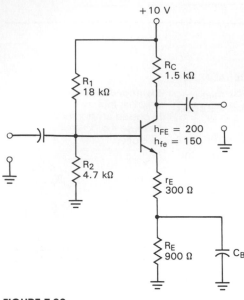

FIGURE 7.22

Now r_e' is found as

$$r_e' = \frac{25 \text{ mV}}{I_E}$$
$$= 21.9 \text{ }\Omega$$

Finally, A_v is found as

$$A_v = \frac{R_C}{r_e' + r_E}$$
$$= \frac{1.5 \text{ k}\Omega}{321.9 \text{ }\Omega}$$
$$= 4.66$$

PRACTICE PROBLEM 7–12

Assume that the amplifier in Figure 7.22 has values of $R_E = 820 \text{ }\Omega$ and $r_E = 330 \text{ }\Omega$. Determine the value of A_v for the circuit.

Why swamping improves gain stability.

Swamping improves the gain stability of an amplifier because $r_E \gg r_e'$. Since a majority of the ac emitter resistance is determined by the value of r_E, any change in r_e' will have little effect on the overall gain of the amplifier. This point is illustrated in the following example.

EXAMPLE 7.13

Determine the change in gain for the amplifier in Example 7.12 when r_e' doubles in value.

Solution: When r_e' doubles in value, A_v becomes

$$A_v = \frac{R_C}{r_e' + r_E}$$
$$= \frac{1.5 \text{ k}\Omega}{343.8 \text{ }\Omega}$$
$$= 4.36$$

The change in gain is therefore

$$4.66 - 4.36 = 0.30$$

this is a change of only 6.44% from the original value of A_v.

In an amplifier that is not swamped, doubling the value of r_e' would cause the value of A_v to decrease by 50%. Thus the gain of an amplifier is made to be much more stable by swamping the emitter circuit.

The Effect of Swamping on Z_{in}

✔ 26

The input impedance of a transistor base was shown in equation (7.21) to equal beta times the ac resistance of the emitter. This ac resistance is equal to $(r_e' + r_E)$ for the swamped amplifier. Thus,

$$Z_{in(base)} = h_{fe}(r_e' + r_E) \qquad \textbf{(7.26)}$$

The result is that the input impedance of the transistor base is increased by an amount equal to $h_{fe}r_E$, as shown in the following example.

EXAMPLE 7.14

Determine the value of $Z_{in(base)}$ for the circuits shown in Figure 7.23.

✔ 27

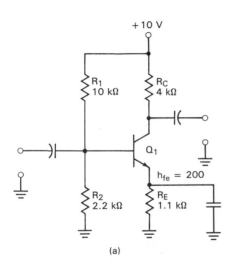

(a)

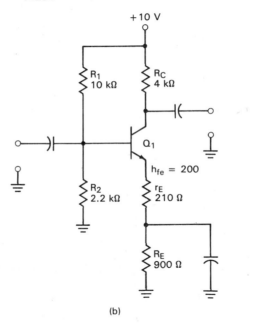

(b)

FIGURE 7.23

Solution: Both of the circuits have the same value of dc emitter resistance. Therefore, following the established procedure for finding r_e' gives us a value of 25 Ω for both circuits. Now, for Figure 7.23a,

$$Z_{in(base)} = h_{fe}r_e'$$
$$= 5 \text{ k}\Omega$$

For Figure 7.23b,

$$Z_{in(base)} = h_{fe}(r_e' + r_E)$$
$$= (200)(25 \text{ }\Omega + 210 \text{ }\Omega)$$
$$= 47 \text{ k}\Omega$$

Increasing the value of $Z_{in(base)}$ *increases* the overall value of Z_{in} for the amplifier, as shown in the following example.

EXAMPLE 7.15

Determine the value of Z_{in} for the amplifiers shown in Figure 7.23.

Solution: For Figure 7.23a,

$$Z_{in} = R_1 \parallel R_2 \parallel Z_{in(base)}$$
$$= 10 \text{ k}\Omega \parallel 2.2 \text{ k}\Omega \parallel 5 \text{ k}\Omega$$
$$= 1.33 \text{ k}\Omega$$

For Figure 7.23b,

$$Z_{in} = R_1 \parallel R_2 \parallel Z_{in(base)}$$
$$= 10 \text{ k}\Omega \parallel 2.2 \text{ k}\Omega \parallel 47 \text{ k}\Omega$$
$$= 1.74 \text{ k}\Omega$$

Thus, we can see that increasing the value of $Z_{in(base)}$ increases the overall value of Z_{in} for an amplifier.

✓ PRACTICE PROBLEM 7–15

Determine the value of Z_{in} for the amplifier shown in Figure 7.22.

↙ 28

The increased Z_{in} for Figure 7.23b means that this circuit will have less effect on the A_v of a source amplifier. This point is illustrated in the following example.

EXAMPLE 7.16

The amplifiers in Figure 7.23 are both being driven by a source amplifier with values of $r'_e = 25 \ \Omega$ and $R_C = 8 \text{ k}\Omega$. Determine the value of A_{vL} for the source amplifier when each of the circuits is connected as the load.

Solution: When the circuit in Figure 7.23a is connected to the source amplifier, the loaded voltage gain of the amplifier is found as

$$A_{vL} = \frac{r_C}{r'_e}$$
$$= \frac{8 \text{ k}\Omega \parallel 1.33 \text{ k}\Omega}{25 \ \Omega} \qquad \left| r_C = R_C \parallel Z_{in} \right.$$
$$= 45.6$$

When the circuit in Figure 7.23b is connected to the source amplifier, the loaded voltage gain of the amplifier is found as

$$A_{vL} = \frac{r_C}{r'_e}$$

$$= \frac{8 \text{ k}\Omega \parallel 1.74 \text{ k}\Omega}{25 \text{ }\Omega} \qquad \Big| \; r_C = R_C \parallel Z_{in}$$

$$= 57.2$$

Thus the reduced loading by the circuit in Figure 7.23b increased the gain of the driving amplifier.

The Disadvantage of Swamping

You have been shown that a swamped amplifier will be more stable against variations in r'_e. You have also been shown how this amplifier can increase the value of A_v for a source amplifier.

Swamping improves stability but reduces A_v.

The main disadvantage of swamping is that the overall voltage gain of this amplifier is lower than that of a comparable standard common-emitter amplifier. While the gain of a swamped amplifier is more stable, it is lower than the gain of an amplifier with no swamping. This can be seen by looking at the two amplifiers in Figure 7.23 again. Note that the two circuits are identical for dc analysis purposes. Both have a total of 1.1 kΩ in their emitter circuits.

The two amplifiers differ only in their ac characteristics. Note that the total ac resistance in the emitter circuit of Figure 7.23a is 25 Ω, the value of r'_e. This low resistance provides the circuit with a gain of 160. The total ac resistance in the emitter circuit of Figure 7.23b is 235 Ω. This higher emitter resistance reduces

FIGURE 7.24

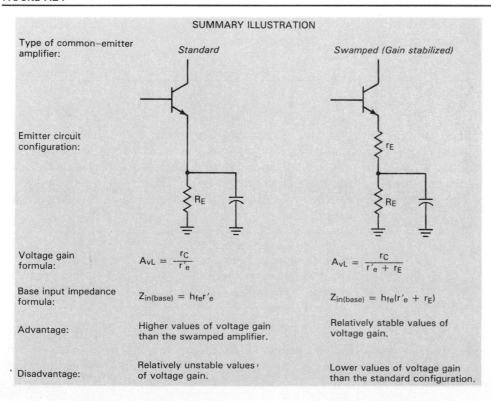

SUMMARY ILLUSTRATION

Type of common–emitter amplifier:	Standard	Swamped (Gain stabilized)
Emitter circuit configuration:	R_E	r_E , R_E
Voltage gain formula:	$A_{vL} = \dfrac{r_C}{r'_e}$	$A_{vL} = \dfrac{r_C}{r'_e + r_E}$
Base input impedance formula:	$Z_{in(base)} = h_{fe}r'_e$	$Z_{in(base)} = h_{fe}(r'_e + r_E)$
Advantage:	Higher values of voltage gain than the swamped amplifier.	Relatively stable values of voltage gain.
Disadvantage:	Relatively unstable values of voltage gain.	Lower values of voltage gain than the standard configuration.

the overall gain of the amplifier to 17. Thus swamping reduces the overall gain of the amplifier. At the same time, you must remember that the swamped amplifier will have a much more stable value of A_v than the standard common-emitter amplifier. Also, the increase in A_v experienced by a source amplifier that is driving a swamped amplifier will partially offset the reduction in gain.

The Emitter Circuit Is the Key

When you are dealing with common-emitter amplifers, you must be able to distinguish between *swamped* (or *gain-stabilized*) amplifiers and standard amplifiers. The key to distinguishing one common-emitter amplifier type from the other is to look at the emitter circuit. If the emitter resistance is *completely bypassed*, you are dealing with a standard common-emitter amplifier. If the emitter resistance is only partially bypassed, you are dealing with a swamped, or gain-stabilized, amplifier. The circuit recognition features of these two types of common-emitter amplifiers, along with the different equations for the two, are summarized in Figure 7.24.

SECTION REVIEW

1. How does swamping reduce the effects of variations in the value of r_e'? [Objective 24]

2. How does swamping affect the value of $Z_{in(base)}$ of a common-emitter amplifier? [Objective 26]

3. How does swamping affect the value of Z_{in} of a common-emitter amplifier? [Objective 26]

4. How does swamping affect the value of A_v for a source amplifier? Explain your answer. [Objective 28]

5. What is the primary disadvantage of using amplifier swamping? Explain your answer. [Objective 29]

7.7
h-Parameters

h-Parameters. Transistor specifications that describe the operation of the device under full-load or no-load conditions.

Hybrid parameters, or *h-parameters*, are transistor specifications that describe the component operating limits under specific circumstances. Each of the four *h*-parameters is measured under *no-load* or *full-load* conditions. These *h*-parameters are then used in circuit analysis applications.

The four *h*-parameters for a transistor in a common-emitter amplifier are as follows:

h_{ie} = the base input impedance
h_{fe} = the base-to-collector current gain
h_{oe} = the output admittance
h_{re} = the reverse voltage feedback ratio

It should be noted that all these values represent *ac characteristics* of the transistor, as measured under specific circumstances. Before discussing the applications of *h*-parameters, let's take a look at the parameters and the method of measurement used for each.

CHAP. 7   Common-emitter Amplifiers

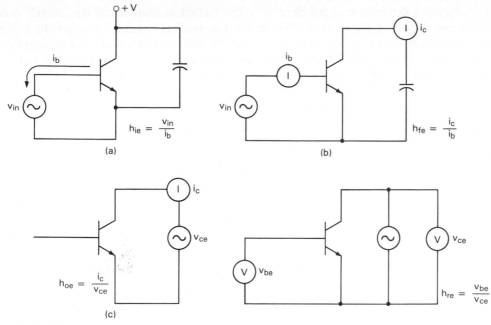

FIGURE 7.25

Input Impedance (h_{ie})

The input impedance parameter, h_{ie}, is measured when the *output is shorted*. A shorted output is a full load, so h_{ie} *represents the input impedance to the transistor under full-load conditions*. The measurement of h_{ie} is illustrated in Figure 7.25a. As you can see, the collector and emitter terminals are shorted. Then an ac signal is applied to the base–emitter junction of the transistor. With the input voltage applied, the base current is measured. Then h_{ie} is determined as

$$h_{ie} = \frac{v_{in}}{i_b}\qquad \text{(output shorted)} \qquad\qquad (7.27)$$

Why short the output? You may recall that any resistance in the emitter circuit is reflected back to the base. This condition was described in the equation

$$Z_{in(base)} = h_{fe}(r'_e + r_E)$$

By shorting the collector and emitter terminals, the measured value of h_{ie} does not reflect any external resistance in the circuit.

Current Gain (h_{fe})

The base-to-collector current gain, h_{fe}, is also measured when the *output is shorted*. Again, this represents a full load, so h_{fe} *represents the current gain of the transistor under full-load conditions*. The measurement of h_{fe} is illustrated in Figure 7.25b. With the output shorted and a signal applied to the base, both the base and collector currents are measured. Then h_{fe} is determined as

$$h_{fe} = \frac{i_c}{i_b}\qquad \text{(output shorted)} \qquad\qquad (7.28)$$

Input impedance (h_{ie}). The input impedance of the transistor, measured under full-load conditions.

minimum
maximum current

Why h_{ie} is measured under full-load conditions.

Current gain (h_{fe}). The ac beta of the component, measured under full-load conditions.

In this case it should be clear why the output is shorted. If the output was left open, i_c would equal zero. Shorting the output gives us a measurable value of i_c that can be reproduced in a practical test. In other words, i_c will be a measurable value, and anyone can achieve the same results simply by shorting the output terminals and applying the same signal to the transistor.

Output Admittance (h_{oe})

Output admittance (h_{oe}). The admittance of the collector-emitter circuit, measured under no-load conditions. This parameter is used mainly in amplifier design. Since admittance is the reciprocal of impedance, the unit of measure is listed as *mhos* (*siemens* is the preferred unit of measure).

The output admittance, h_{oe}, is measured with the *input open*. The measurement of h_{oe} is illustrated in Figure 7.25c. As you can see, a signal is applied across the collector–emitter terminals. Then, with this signal applied, the value of i_c is measured. The value of h_{oe} is then determined as

$$h_{oe} = \frac{i_c}{v_{ce}} \quad \text{(input open)} \tag{7.29}$$

Measuring h_{oe} with the input open makes sense when you consider the effect of shorting the input. If the input to the transistor were shorted, some base current (i_b) would flow. Since this current would come from the emitter, the value of i_c would not be at its maximum potential. In other words, by not allowing any i_b to flow, i_c is at its absolute maximum value. Therefore, h_{oe} is an accurate measurement of the *maximum* output admittance.

Reverse Voltage Feedback Ratio (h_{re})

Reverse voltage feedback ratio (h_{re}). The ratio of v_{be} to v_{ce}, measured under no-load conditions. This parameter is used mainly in amplifier design. Since h_{re} is a ratio of two voltages, it has no unit of measure.

The reverse voltage feedback ratio, h_{re}, is *the amount of output voltage that is reflected back to the input*. This value is measured with the *input open*. The measurement of h_{re} is illustrated in Figure 7.25d. A signal is applied to the collector–emitter terminals. Then, with the input open, the voltage that is fed back to the base–emitter junction is measured. The value of h_{re} is then determined as

$$h_{re} = \frac{v_{be}}{v_{ce}} \quad \text{(input open)} \tag{7.30}$$

Since the voltage at the base terminal will always be less than the voltage across the emitter–collector terminals, h_{re} will always be less than 1. By measuring h_{re} with the input open, you ensure that the voltage fed back to the base will always be at its maximum possible value, since maximum voltage is always developed across an open circuit.

Circuit Calculations Involving *h*-Parameters

Circuit calculations involving *h*-parameters can be very simple or very complex, depending on:

1. What you are trying to determine
2. How exact you want your calculations to be

For our purposes, we are interested in only four *h*-parameter circuit equations. These equations are as follows:

$$A_i = h_{fe} \tag{7.5}$$

$$Z_{\text{in(base)}} = h_{ie} \tag{7.31}$$

$$r'_e = \frac{h_{ie}}{h_{fe}} \qquad \text{(7.32)}$$

$$A_{vL} = \frac{h_{fe}r_C}{h_{ie}} \qquad \text{(7.33)}$$

Equation (7.5) needs no explanation, as it was introduced earlier in the chapter. Equation (7.31) is relatively easy to understand if you refer back to Figure 7.11d (page 275). Here we see the ac equivalent circuit for the amplifier shown in Figure 7.11a. Looking at the ac equivalent circuit, it is easy to see that the only impedance between the base of the transistor and the emitter-ground connection is the input impedance of the transistor. This input impedance is h_{ie}.

Equation (7.32) is derived using equations (7.21) and (7.31). You may recall from earlier discussions that

$$Z_{\text{in(base)}} = h_{fe}r'_e \qquad \text{[equation (7.21)]}$$

Substituting equation (7.31) for $Z_{\text{in(base)}}$, we obtain

$$h_{ie} = h_{fe}r'_e$$

Simply rearranging for r'_e gives us equation (7.32).

You may be wondering why we would go to all this trouble to find r'_e with h-parameters when we can simply use

$$r'_e = \frac{25 \text{ mV}}{I_E}$$

The fact of the matter is that the 25 mV/I_E equation really is not that accurate. By using the h-parameter equation, we are able to get a much more accurate value of r'_e and thus can more closely calculate the value of A_{vL} for a given common-emitter amplifier.

Equation 7.33 is derived as follows:

$$\begin{aligned} A_{vL} &= \frac{r_C}{r'_e} \\ &= \frac{1}{r'_e}r_C \\ &= \frac{h_{fe}}{h_{ie}}r_C \\ &= \frac{h_{fe}r_C}{h_{ie}} \end{aligned}$$

A Practical Consideration: Depending on the transistor being used, the given value of 25 mV in the r'_e equation can actually be any value between 25 mV and 52 mV. This gives us

$$\frac{25 \text{ mV}}{I_E} \le r'_e \le \frac{52 \text{ mV}}{I_E}$$

This is why we use the more accurate h-parameter equations whenever possible.

The equations that we have discussed in this section will be used throughout our discussion on ac amplifier analysis. It should be noted that the entire subject of h-parameters and h-parameter derivations is far more complex than has been presented here. The subject is covered thoroughly in Appendix C for those readers who are interested in going deeper into the subject. For those who do not wish to get involved in the h-parameter derivations, you may continue from this point without loss of continuity.

Determining *h*-parameter Values

A Practical Consideration:
h-parameter values are often
listed in an electrical
characteristics section called
small-signal characteristics.

The spec sheet for a given transistor will list the values of the device's *h*-parameters in the *electrical characteristics* portion of the sheet. This is illustrated in Figure 7.26, which shows the electrical characteristics portion of the spec sheet for the Motorola 2N4400–4401 series transistors.

2N4400, 2N4401

(continued) (T_A = 25°C unless otherwise noted.)

Characteristic		Symbol	Min	Max	Unit
Emitter-Base Capacitance (V_{BE} = 0.5 Vdc, I_C = 0, f = 100 kHz)		C_{eb}	—	30	pF
Input Impedance (I_C = 1.0 mAdc, V_{CE} = 10 Vdc, f = 1.0 kHz)	2N4400 2N4401	h_{ie}	0.5 1.0	7.5 15	k ohms
Voltage Feedback Ratio (I_C = 1.0 mAdc, V_{CE} = 10 Vdc, f = 1.0 kHz)		h_{re}	0.1	8.0	X 10^{-4}
Small-Signal Current Gain (I_C = 1.0 mAdc, V_{CE} = 10 Vdc, f = 1.0 kHz)	2N4400 2N4401	h_{fe}	20 40	250 500	—
Output Admittance (I_C = 1.0 mAdc, V_{CE} = 10 Vdc, f = 1.0 kHz)		h_{oe}	1.0	30	μmhos

FIGURE 7.26

(Courtesy of Motorola, Inc.)

When *minimum* and *maximum* *h*-parameter values are given, we must determine the *geometric average* of the two values, as was discussed in Chapter 6. Thus, the values of h_{ie} and h_{fe} that we would use in the analysis of a 2N4400 circuit would be found as

$$h_{ie} = \sqrt{h_{ie(\text{min})} \times h_{ie(\text{max})}}$$
$$= \sqrt{(500\ \Omega)(7.5\ \text{k}\Omega)}$$
$$= 1.94\ \text{k}\Omega$$

and

$$h_{fe} = \sqrt{h_{fe(\text{min})} \times h_{fe(\text{max})}}$$
$$= \sqrt{(20)(250)}$$
$$= 71$$

The following example shows how the above values would be used in the ac analysis of a 2N4400 amplifier.

=== **EXAMPLE 7.17** ===

Determine the values of Z_{in} and A_{vL}' for the circuit shown in Figure 7.27.

Solution: Using the established analysis procedure, the value of I_C for the circuit is found to be approximately 1 mA. Since the values of h_{ie} and h_{fe} for the 2N4400 are listed at I_C = 1 mA (see the spec sheet in Figure 7.26), we can use the geometric averages of h_{ie} = 1.94 kΩ and h_{fe} = 71 that were determined earlier. Using these values,

$$Z_{\text{in(base)}} = h_{ie}$$
$$= 1.94\ \text{k}\Omega$$

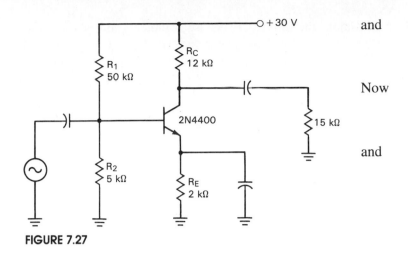

FIGURE 7.27

and

$$Z_{in} = R_1 \| R_2 \| Z_{in(base)}$$
$$= 1.35 \text{ k}\Omega$$

Now

$$r_C = R_C \| R_L$$
$$= 6.67 \text{ k}\Omega$$

and

$$A_{vL} = \frac{h_{fe}r_C}{h_{ie}}$$
$$= \frac{(71)(6.67 \text{ k}\Omega)}{1.94 \text{ k}\Omega}$$
$$= 244$$

PRACTICE PROBLEM 7–17

An amplifier has values of $R_C = 12$ kΩ, $R_L = 4.7$ kΩ, $R_1 = 33$ kΩ, $R_2 = 4.7$ kΩ, and $I_C = 1$ mA. At 1 mA, the transistor has h-parameter values of $h_{ie} = 1$ kΩ to 5 kΩ and $h_{fe} = 70$ to 350. Determine the values of Z_{in} and A_{vL} for the circuit.

As Example 7.17 pointed out, the h-parameter values listed on the spec sheet of the 2N4400 were measured at $I_C = 1$ mA. As was the case with h_{FE}, the ac h-parameters will all vary with the value of I_C. This point is illustrated in Figure 7.28, which shows the h-parameter curves for the 2N4400–4401 series transistors.

FIGURE 7.28

(Courtesy of Motorola, Inc.)

2N4400, 2N4401

h PARAMETERS
V_{CE} = 10 Vdc, f = 1.0 kHz, T_A = 25°C

This group of graphs illustrates the relationship between h_{fe} and other "h" parameters for this series of transistors. To obtain these curves, a high-gain and a low-gain unit were selected from both the 2N4400 and 2N4401 lines, and the same units were used to develop the correspondingly numbered curves on each graph.

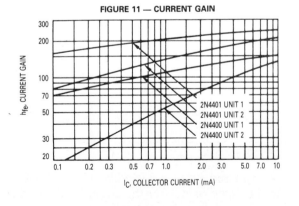

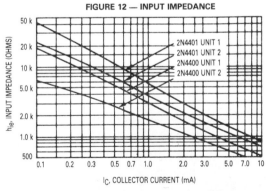

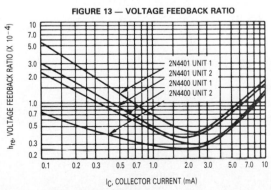

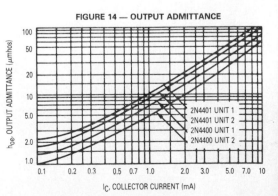

First, a word or two about the curves. Each *h*-parameter graph shows four curves. Two of the curves are identified as being for the 2N4400, and two of them are identified as being for the 2N4401. Just as we were given *minimum* and *maximum* *h*-parameter values at $I_{CQ} = 1$ mA, we have minimum (Unit 2) and maximum (Unit 1) curves for the transistors.

To determine the value of a given parameter at a given value of I_{CQ}, use the following procedure:

1. Using the graph, determine the minimum and maximum *h*-parameter values at the given value of I_{CQ}.
2. Use the geometric average of the two values obtained in the analysis of the amplifier.

The use of this procedure is illustrated in the following example.

EXAMPLE 7.18

An amplifier that uses a 2N4400 has values of $I_{CQ} = 5$ mA and $r_C = 460$ Ω. Determine the value of A_{vL} for the circuit.

Solution: From the h_{fe} curve, we can approximate the limits of h_{fe} to be 110 to 140 when $I_{CQ} = 5$ mA. The geometric average of these two values is found as

$$h_{fe} = \sqrt{h_{fe(\min)} \times h_{fe(\max)}}$$
$$= 124$$

The h_{ie} graph shows the limits of h_{ie} to be 600 and 800 Ω when $I_{CQ} = 5$ mA. The geometric average of these two values is found as

$$h_{ie} = \sqrt{h_{ie(\min)} \times h_{ie(\max)}}$$
$$= 693 \ \Omega$$

Using $h_{fe} = 124$ and $h_{ie} = 693$ Ω, the value of A_{vL} is determined to be

$$A_{vL} = \frac{h_{fe} r_C}{h_{ie}}$$
$$= \frac{(124)(460)}{693 \ \Omega}$$
$$= 82.3$$

PRACTICE PROBLEM 7–18

A 2N4401 is being used in a circuit with values of $I_{CQ} = 2$ mA and $r_C = 1.64$ kΩ. Determine the value of A_{vL} for the amplifier using the curves shown in Figure 7.28.

A Few More Points

33

In this section, we have concentrated on h_{ie} and h_{fe}. That's because these two parameters are the only ones required for "everyday" circuit analysis. The other two *h*-parameters, h_{oe} and h_{re}, are used primarily in circuit development applications.

CHAP. 7 Common-emitter Amplifiers

Whenever you need to determine the voltage gain of an amplifier, the required *h*-parameter values can easily be obtained from the spec sheet of the transistor. If a range of values is given, you can actually follow either of two procedures:

1. If you want an *approximate* value of A_{vL}, you can use the geometric average of the *h*-parameter values listed.

2. If you are interested in the *worst-case* values of A_{vL}, you can analyze the amplifier using *both* the minimum and maximum values of h_{fe} and h_{ie}. By doing this, you will obtain the *minimum* and *maximum* values of A_{vL}. These two values would be the worst-case values of voltage gain for the amplifier.

SECTION REVIEW

1. What is h_{ie}? How is the value of h_{ie} measured? [Objective 30]
2. What is h_{fe}? How is the value of h_{fe} measured? [Objective 30]
3. What is h_{oe}? How is the value of h_{oe} measured? [Objective 30]
4. What is h_{re}? How is the value of h_{re} measured? [Objective 30]
5. Which of the four *h*-parameters are commonly used in circuit analysis? [Objective 31]
6. Where are *h*-parameter values usually listed on a transistor spec sheet? [Objective 32]
7. When minimum and maximum *h*-parameter values are listed, what value do you usually use for circuit analyses? [Objective 32]
8. What is the procedure for determining the values of h_{ie} and h_{fe} at a specified value of I_{CQ}? [Objective 32]
9. How is the circuit analysis procedure changed for determining the *worst-case* values of A_{vL}? [Objective 33]

7.8
Amplifier Troubleshooting

 34

We have already discussed the troubleshooting procedure for an amplifier. However, that discussion assumed that you already knew there was a problem in the amplifier. How can you tell when one out of several amplifiers is the cause of a problem? For example, take a look at Figure 7.29. If the final output from this series of

FIGURE 7.29

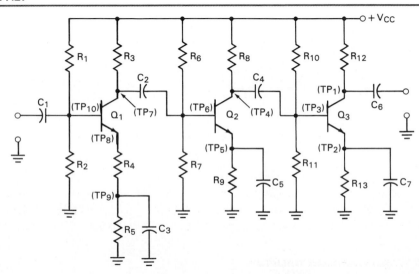

amplifiers is bad, how do you know which of the amplifiers is the source of the trouble?

When you go to troubleshoot a series of amplifiers, *you start at the final output stage*. When you verify that the output from the amplifier is bad, check the input. *If the amplifier's input is bad, go to the next amplifier output.* There is no reason to troubleshoot an amplifier with a bad input. It should be obvious that an amplifier cannot possibly have a good output if its input is not a good signal.

When you verify that the input to an amplifier is bad, check the output of the previous stage. If that output is good, there is a problem in the circuit that is coupling the two amplifiers. If the output from the previous stage is not good, check the input, and so on. The complete procedure for troubleshooting a three-stage amplifier is illustrated in the flow chart in Figure 7.30. A *flowchart* is a step-by-step illustration of a problem-solving procedure.

FIGURE 7.30

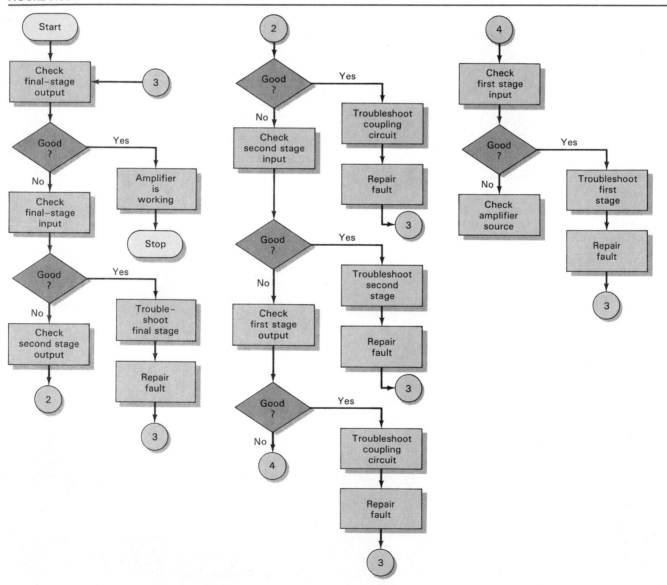

CHAP. 7   Common-emitter Amplifiers

Amplifier Input/Output Signals

Figure 7.31 shows the signals you should see at the inputs and outputs of each stage of Figure 7.29. Note that the standard common-emitter amplifiers (stages 2 and 3) show no ac signal at the emitter terminal of the transistors. This is due to the fact that the ac voltage is developed across r_e' in the transistors themselves. The swamped amplifier (stage 1) has a small ac signal present at the emitter terminal, but only a dc level at the bypass capacitor connection point. In all three cases, the collector signal is much larger in amplitude than the input signal at the base, and the two signals are 180° out of phase.

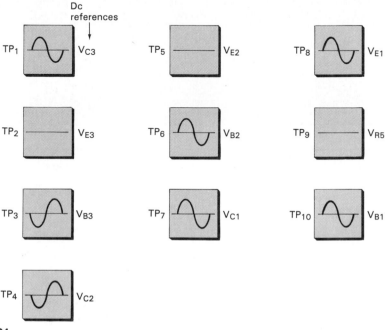

FIGURE 7.31

Signals for Figure 7.29.

When troubleshooting the amplifier, you should find the amplifier stage that has the bad input/output signal condition. In other words, find the stage that has a good input signal and a bad output signal. When you find that stage, troubleshoot the dc circuit as you were shown in Chapter 5.

Nonlinear Distortion

Distortion is *an unwanted change in the shape of an ac signal*. One problem that commonly occurs in common-emitter amplifiers is called *nonlinear distortion*. The output waveform from an amplifier experiencing nonlinear distortion is shown in Figure 7.32. Note the difference between the shape of the negative alternation of the signal (normal) and that of its positive alternation (distorted). The change in the shape of the positive alternation is a result of nonlinear distortion.

Nonlinear distortion is *a type of distortion that is caused by driving the base–emitter junction of the transistor into its nonlinear operating region*. This

Distortion. An unwanted change in the shape of an ac signal.

Nonlinear distortion. A type of distortion that is caused by driving the base–emitter junction of a transistor into its nonlinear operating region.

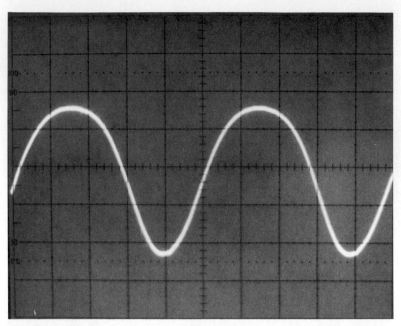

FIGURE 7.32

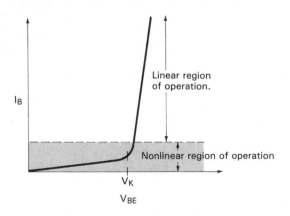

FIGURE 7.33

A base characteristic curve.

point is illustrated with the help of Figure 7.33, which shows the base characteristic curve of a transistor.

Normally, a transistor is operated so that the base–emitter junction stays in the linear region of operation. In this region, a change in V_{BE} causes a *linear* (constant rate) change in I_B. If the transistor is operated in the nonlinear operating region, a change in V_{BE} will cause a change in I_B that is not linear. For example, compare the slope of the curve at the points immediately above and below V_K. Since the slope of the curve changes, the rate of change in I_B also changes.

The base–emitter junction of a transistor may be driven into its nonlinear operating region in one of two sets of circumstances:

1. A transistor is normally biased so that I_B will have a value that is well within the linear region of operation. If an amplifier is poorly biased so that the Q-point value of I_B is near the nonlinear region of operation, a relatively small ac input signal will drive the amplifier into the nonlinear region of operation, and nonlinear distortion will result.

2. The amplitude of the amplifier input signal may be sufficient to drive even a well-biased amplifier into nonlinear operation, causing nonlinear distortion.

If the first of these two problems (poor biasing) is present, the only solution is to redesign the amplifier for a higher value of I_B. When the second problem (overdriving the amplifier) is present, the solution is to reduce the amplitude of the amplifier input signal.

Nonlinear distortion can cause a variety of problems in communications electronics. In stereos, it may cause the audio to sound "grainy" or may even cause the tones of various musical instruments to change. In televisions, crossover distortion in the video circuitry can cause distortion in the picture on the CRT. To avoid these types of problems, amplifier design normally emphasizes avoiding crossover distortion. Since amplifiers are normally designed to avoid nonlinear distortion problems, the most common cause of nonlinear distortion is overdriving an amplifier.

A Few Pointers

When you are troubleshooting amplifiers, remember that the input signal for a given stage will have a much lower amplitude than the output. For this reason, you should reduce the VOLTS/DIV setting on your oscilloscope when going from output to input. For example, consider a circuit with a voltage gain of 100. *Under normal circumstances, this amplifier will have an output signal that is 100 times as large as the input signal.* If you try to read these two signals using the same VOLTS/DIV setting on the oscilloscope, you will not be able to see the input signal. This could lead you to believe there is a problem where no problem exists.

Another point is this: Don't worry about seeing the 180° phase shift on your oscilloscope. If you want to, you can *external trigger* the oscilloscope on the amplifier input and will then be able to see the phase shift. However, viewing the phase shift is not really important. If the amplifier is working, the phase shift will be there. You will never run into a problem where the amplifier has the right signal but the phase shift is not there, so don't worry about it. Just look to see if the amplifier has a small-signal input and a large-signal output. If it does, it is all right. If neither signal is good, you need to move on to the preceding stage. If the input signal is good and the output signal is bad, you have found the bad stage.

Use a square wave to compensate scope probe.

SECTION REVIEW

1. What is the procedure for troubleshooting a multistage amplifier? [Objective 34]
2. When troubleshooting an amplifier, why do you need to pay close attention to the VOLTS/DIV setting of the oscilloscope? [Objective 34]

Chapter Summary

Amplification is the process of increasing the strength of an ac signal. Amplifiers are capable of performing this operation because they have inherent *gain*. Gain is a multiplier that exists between the input and output of an amplifier.

There are three types of gain: *current gain*, *voltage gain*, and *power gain*. The types of gain exhibited by each amplifier type are listed as follows:

Amplifier Type	Types of Gain
Common emitter	Current/voltage/power
Common collector	Current/power
Common base	Voltage/power

For every type of amplifier, the input and output currents are in phase. The common-emitter amplifier has a 180° voltage phase shift from input to output, while the common-base and common-collector input and output voltages are in phase.

The current gain (A_i) of a common-emitter amplifier is the ac beta (h_{fe}) rating of the transistor. The value of h_{fe} is dependent on the value of I_{CQ}, as is the value of dc beta (h_{FE}).

The voltage gain (A_v) of the common-emitter amplifier depends on the ac values of collector and emitter resistance. In an amplifier where the external emitter resistance is *completely* bypassed, the ac emitter resistance is equal to r_e', the ac resistance of the emitter–base junction. Because of this, the voltage gain of a standard common-emitter amplifier is not entirely stable, as r_e' varies with temperature and the value of I_{CQ}.

To overcome the instability of A_v, *swamping* is used. This involves bypassing only a small part of the emitter resistance. This results in a portion of the emitter resistance still weighing into the ac model of the amplifier. The overall effect is to reduce the effects of r_e', at a loss of overall voltage gain. The power gain of the common-emitter amplifier is equal to the product of h_{fe} and A_v.

The four *h*-parameters are current gain (h_{fe}), input impedance (h_{ie}), output admittance (h_{oe}), and reverse voltage transfer ratio (h_{re}). Of the four, only h_{ie} and h_{fe} are used in common circuit analyses. The values of h_{ie} and h_{fe} for a given transistor can be obtained from the spec sheet of the device. As with h_{FE}, you should use the geometric averages of h_{fe} and h_{ie} when minimum and maximum parameter values are listed on the spec sheet.

Troubleshooting amplifier stages involves analyzing the ac signals to determine which amplifier has a good input and a bad output. When this amplifier is found, the dc troubleshooting procedures covered in Chapter 6 are used to locate the faulty component.

KEY TERMS

The following new terms were introduced and defined in this chapter:

ac beta	current gain (A_i or h_{fe})	nonlinear distortion
ac emitter resistance (r_e')	distortion	output admittance (h_{oe})
ac equivalent circuit	gain	power gain (A_p)
amplification	gain-stabilized amplifier	reverse voltage feedback
bypass capacitor	*h*-parameters	ratio (h_{re})
cascaded	input impedance (h_{ie})	stage
common-emitter amplifier	linear	swamped amplifier
coupling capacitor	multistage amplifier	voltage gain (A_v)

PRACTICE PROBLEMS

§7.1

1. An amplifier has an emitter current of 12 mA. Determine the value of r_e' for the circuit. [4]

2. An amplifier has an emitter current of 10 mA. Determine the value of r_e' for the circuit. [4]

3. An amplifier has values of $V_E = 2.2$ V and $R_E = 910$ Ω. Determine the value of r_e' for the circuit. [4]

4. An amplifier has values of $V_E = 12$ V and $R_E = 4.7$ kΩ. Determine the value of r_e' for the circuit. [4]

5. An amplifier has values of $V_B = 3.2$ V and $R_E = 1.2$ kΩ. Determine the value of r_e' for the circuit. [4]

6. An amplifier has values of $V_B = 4.8$ V and $R_E = 3.9$ kΩ. Determine the value of r_e' for the circuit. [4]

7. Determine the value of r_e' for the circuit shown in Figure 7.34. [4]

8. Determine the value of r_e' for the circuit shown in Figure 7.35 [4]

§7.3

9. Derive the ac equivalent circuit for the amplifier shown in Figure 7.34. Include all component values. [11]

10. Derive the ac equivalent circuit for the amplifier shown in Figure 7.35. Include all component values. [11]

11. Derive the ac equivalent circuit for the amplifier shown in Figure 7.36. Include all component values. [11]

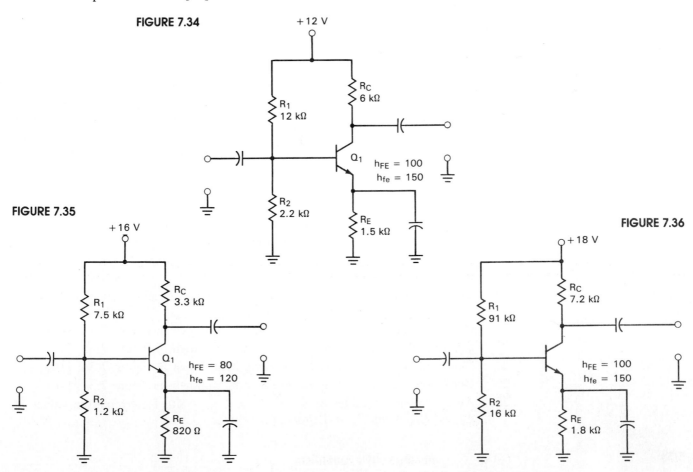

FIGURE 7.34

FIGURE 7.35

FIGURE 7.36

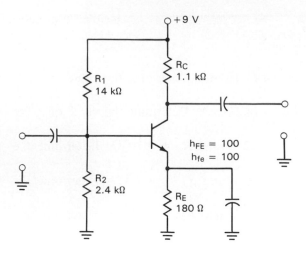

FIGURE 7.37

12. Derive the ac equivalent circuit for the amplifier shown in Figure 7.37. Include all component values. [11]

§7.4

13. An amplifier has values of v_{in} = 120 mV and V_{out} = 4 V. Determine the value of A_v for the circuit. [14]
14. An amplifier has values of v_{in} = 82 mV and v_{out} = 6.4 V. Determine the value of A_v for the circuit. [14]
15. An amplifier has values of R_C = 2.2 kΩ and r'_e = 22.8 Ω. Determine the value of A_v for the circuit. [14]
16. An amplifier has values of R_C = 4.7 kΩ and r'_e 32 Ω. Determine the value of A_v for the circuit. [14]
17. An amplifier has values of R_C = 2.7 kΩ and I_E = 1 mA. Determine the value of A_v for the circuit. [14]
18. An amplifier has values of R_C = 3.3 kΩ and I_E = 2 mA. Determine the value of A_v for the circuit. [14]
19. Determine the value of A_v for the amplifier shown in Figure 7.34. [14]
20. Determine the value of A_v for the amplifier shown in Figure 7.35. [14]
21. Determine the value of A_v for the amplifier shown in Figure 7.36. [14]
22. Determine the value of A_v for the amplifier shown in Figure 7.37. [14]
23. The amplifier in Figure 7.36 has a 12-mV input signal. Determine the value of v_{out} for the circuit. [15]
24. The amplifier in Figure 7.35 has a 22-mV input signal. Determine the value of v_{out} for the circuit. [15]
25. Determine the value of A_p for the amplifier shown in Figure 7.34. [16]
26. Determine the value of A_p for the amplifier shown in Figure 7.35. [16]
27. Determine the value of A_p for the amplifier shown in Figure 7.36. [16]
28. Determine the value of A_p for the amplifier shown in Figure 7.37. [16]
29. An amplifier has values of A_v = 110 and h_{fe} = 40. Determine the values of A_p and P_{out} when P_{in} = 10 mW. [17]

30. An amplifier has values of A_v = 68.8 and h_{fe} = 144. Determine the values of A_p and P_{out} when P_{in} = 240 μW. [17]

§7.5

31. An amplifier has values of R_C = 3.3 kΩ, R_L = 10 kΩ, and r'_e = 22.5 Ω. Determine the value of A_{vL} for the circuit. [20]
32. An amplifier has values of R_C = 4.7 kΩ, R_L = 300 Ω, and r'_e = 12 Ω. Determine the value of A_{vL} for the circuit. [20]
33. An amplifier has values of R_C = 6.2 kΩ, R_L = 1.2 kΩ, and r'_e = 16 Ω. Determine the value of A_{vL} for the circuit. [20]
34. An amplifier has values of R_C = 2.2 kΩ, R_L = 820 Ω, and r'_e = 8.8 Ω. Determine the value of A_{vL} for the circuit. [20]
35. Determine the values of $Z_{in(base)}$ and Z_{in} for the amplifier in Figure 7.34. [22]
36. Determine the values of $Z_{in(base)}$ and Z_{in} for the amplifier in Figure 7.35. [22]
37. Determine the values of $Z_{in(base)}$ and Z_{in} for the amplifier in Figure 7.36. [22]
38. Determine the values of $Z_{in(base)}$ and Z_{in} for the amplifier in Figure 7.37. [22]
39. Determine the value of A_{vL} for the third-stage amplifier in Figure 7.38. [23]
40. Determine the value of A_{vL} for the second-stage amplifier in Figure 7.38. [23]
41. Determine the value of A_{vL} for the first-stage amplifier in Figure 7.38. [23]
42. A two-stage amplifier has values of A_{v1} = 23.8, A_{v2} = 122, h_{fe1} = 240, and h_{fe2} = 38. Determine the values of A_{vT}, A_{iT}, and A_{pT} for the circuit. [23]
43. A two-stage amplifier has values of A_{v1} = 88.6, A_{v2} = 90.3, h_{fe1} = 110, and h_{fe2} = 210. Determine the values of A_{vT}, A_{iT}, and A_{pT} for the circuit. [23]
44. A two-stage amplifier has values of A_{v1} = 24.8, A_{v2} = 77.1, h_{fe1} = 300,

FIGURE 7.38

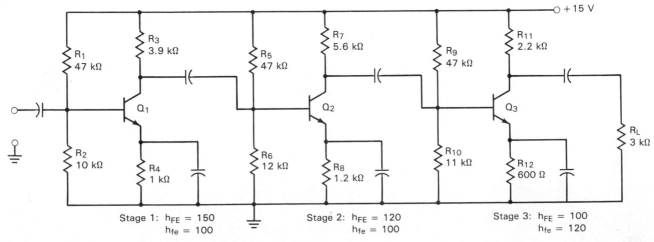

and $h_{fe2} = 90$. Determine the values of A_{vT}, A_{iT}, and A_{pT} for the circuit. [23]

45. Determine the values of A_{vT}, A_{iT}, and A_{pT} for the amplifier shown in Figure 8.36. [23]

§7.6

46. Determine the value of A_v for the amplifier shown in Figure 7.39. [25]

47. Determine the value of A_v for the amplifier shown in Figure 7.40. [25]

48. Determine the value of A_{vL} for the amplifier shown in Figure 7.41. [25]

49. Determine the values of $Z_{in(base)}$ and Z_{in} for the amplifier shown in Figure 7.39. [27]

FIGURE 7.39

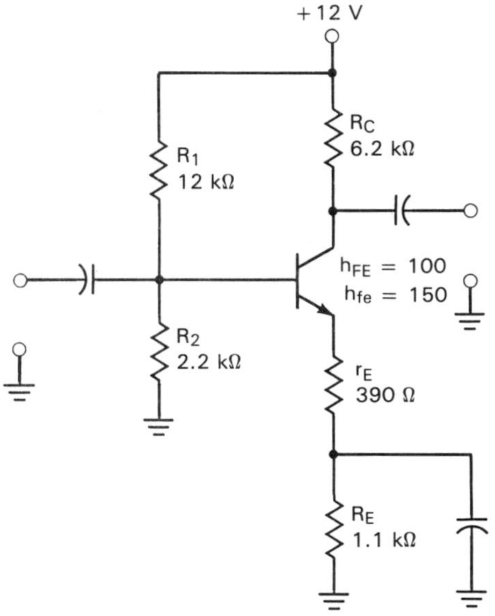

FIGURE 7.40

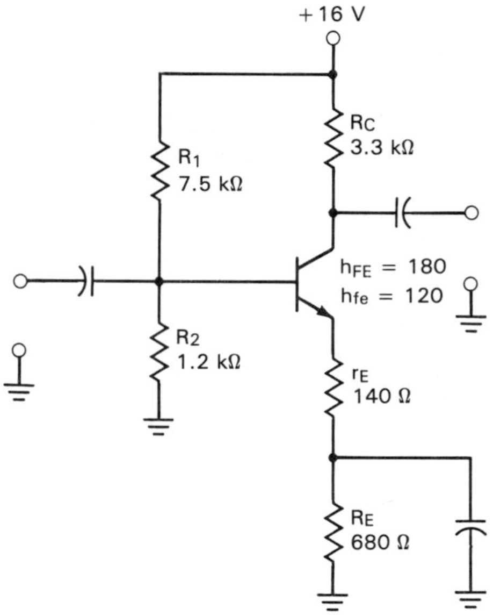

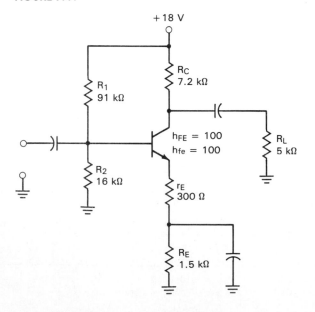

FIGURE 7.41

CHAP. 7   Common-emitter Amplifiers

50. Determine the values of $Z_{in(base)}$ and Z_{in} for the amplifier shown in Figure 7.40. [27]

§7.7

51. An amplifier has values of $h_{fe} = 100$, $h_{ie} = 5\ k\Omega$, and $r_C = 3.8\ k\Omega$. Determine the values of A_i, $Z_{in(base)}$, r'_e, and A_{vL} for the circuit. [31]

52. An amplifier has values of $h_{fe} = 120$, $h_{ie} = 4\ k\Omega$, and $r_C = 3.8\ k\Omega$. Determine the values of A_i, $Z_{in(base)}$, r'_e, and A_{vL} for the circuit. [31]

53. Refer to Figure 7.42. The transistor described is being used in an amplifier with values of $I_{CQ} = 1\ mA$ and $r_C = 2.48\ k\Omega$. Determine the values of A_i, $Z_{in(base)}$, and A_{vL} for the circuit. [32]

54. Refer to Figure 7.42. The transistor described is being used in an amplifier with values of $I_{CQ} = 2\ mA$ and $r_C = 1.18\ k\Omega$. Determine the values of A_i, $Z_{in(base)}$, and A_{vL} for the circuit. [32]

55. Refer to Figure 7.42. The transistor described is being used in an amplifier with values of $I_{CQ} = 5\ mA$ and $r_C = 878\ \Omega$. Determine the value of A_i, $Z_{in(base)}$, and A_{vL} for the circuit. [32]

56. Refer to Figure 7.42. The transistor described is being used in an amplifier with values of $I_{CQ} = 10\ mA$ and $r_C = 1.05\ k\Omega$. Determine the values of A_i, $Z_{in(base)}$, and A_{vL} for the circuit. [32]

FIGURE 7.42

Motorola 2N2222 curves.
(Courtesy of Motorola, Inc.)

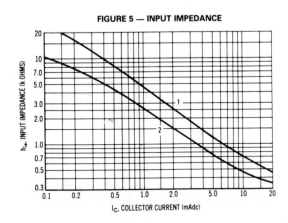

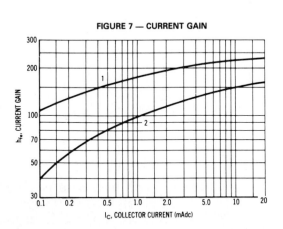

TROUBLESHOOTING PRACTICE PROBLEMS

57. In Figure 7.43, several sets of waveforms are shown. Each row represents a series of signal checks that were performed on the circuit shown in Figure 7.44. For each set of waveforms shown, answer the following series of questions:
   a. Do the waveforms indicate any problem in the circuit?

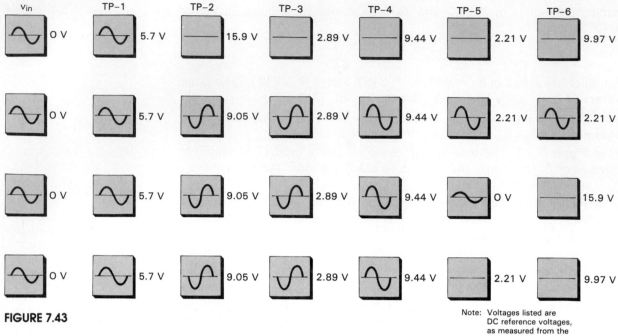

FIGURE 7.43

Note: Voltages listed are DC reference voltages, as measured from the designated test point to ground.

FIGURE 7.44

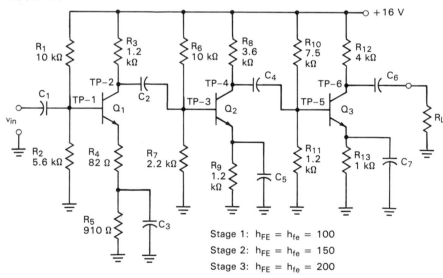

Stage 1: $h_{FE} = h_{fe} = 100$
Stage 2: $h_{FE} = h_{fe} = 150$
Stage 3: $h_{FE} = h_{fe} = 200$

    b. If your answer to part (a) is *yes*, where do you think the problem is probably located? (Simply give the amplifier stage, or indicate the coupling between two specific stages.)

58. (Review) For each circuit shown in Figure 7.45, indicate whether or not a problem is indicated by the dc voltages shown, and state the possible cause(s) of the problem, if any.

THE BRAIN DRAIN

59. Determine the values of Z_{in}, A_{vL}, A_i, and A_p for the amplifier shown in Figure 7.46.

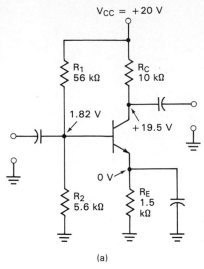

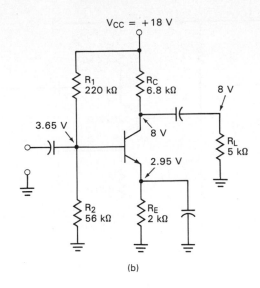

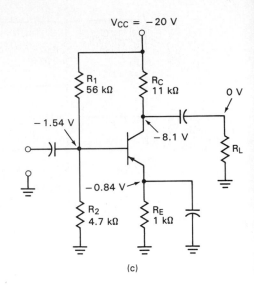

(a)   (b)   (c)

FIGURE 7.45

FIGURE 7.46

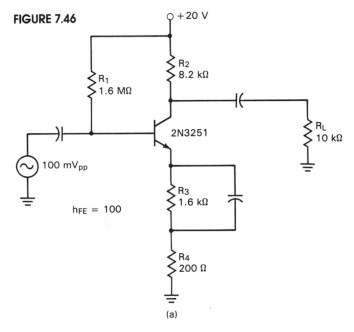

(a)

h PARAMETERS
$V_{CE} = 10$ V, f = 1 kc, $T_A = 25\,°C$

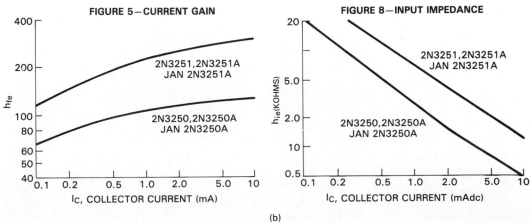

(b)

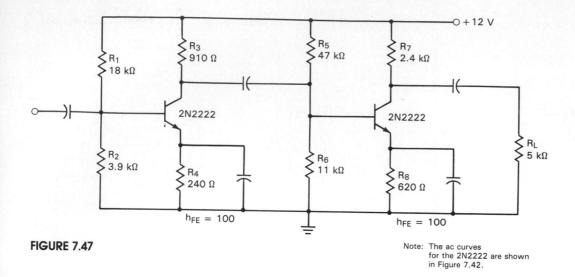

FIGURE 7.47

Note: The ac curves
for the 2N2222 are shown
in Figure 7.42.

60. The load resistor in Figure 7.46 opens. What is the resulting change in the peak-to-peak output voltage?

61. Determine the worst-case values of A_{vL} for the circuit shown in Figure 7.47.

SUGGESTED COMPUTER APPLICATIONS PROBLEMS

62. Write a program to determine the values of A_{vL}, $Z_{in(base)}$, and Z_{in} for a voltage-divider biased common-emitter amplifier, given the circuit resistor values, the load resistance, the values of h_{ie} and h_{fe}, and the value of V_{CC}.

63. If you wanted to write a program for determining the input impedance and gain of a swamped amplifier, what input information would the user have to provide?

64. Write the program described in Problem 63.

ANSWERS TO THE EXAMPLE PRACTICE PROBLEMS

7.1. 38.46Ω

7.2. See Figure 7.48.

7.3. 118.33

7.4. 455.4

7.5. 132

7.6. 2.64 Vac

7.7. 1.58 W

7.8. 81.49

7.9. 4.31 kΩ

7.10. 58.08

7.11. 12.89×10^3

7.12. 4.26

7.15. 3.46 kΩ

7.17. $Z_{in} = 1.45$ kΩ, $A_{vL} = 236.9$

7.18. 132.5

FIGURE 7.48

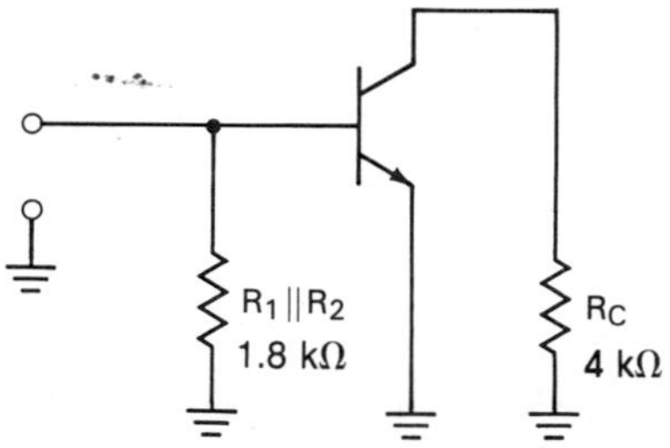

CHAP. 7   Common-emitter Amplifiers

Other Transistor Configurations

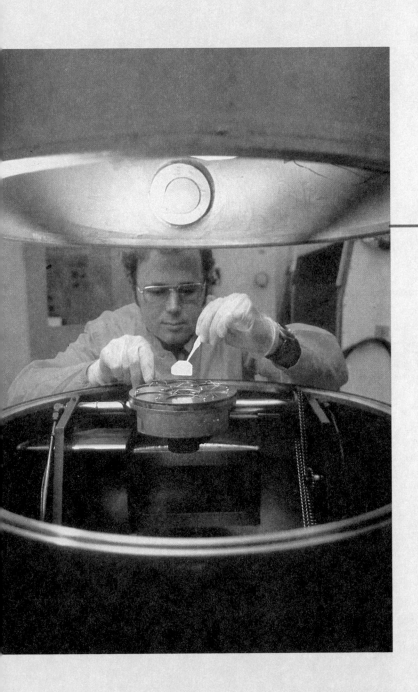

The process of producing solid-state components requires extremely high temperatures. The ovens shown here are used to heat the silicon wafers to a temperature of approximately 1400°C. Temperatures in this range are needed to remove any remaining impurities from the silicon wafers.

OBJECTIVES

After studying the material in this chapter, you should be able to:

☐ 1. Describe the gain characteristics and circuit recognition features of the emitter follower (common-collector amplifier). (§8.1)

☐ 2. Perform a dc analysis of an emitter follower. (§8.1)

☐ 3. Derive and draw the dc load line of an emitter follower. (§8.1)

☐ 4. State and explain the limit on the value of voltage gain (A_v) for an emitter follower. (§8.2)

☐ 5. Calculate the value of A_v for a given emitter follower. (§8.2)

☐ 6. Determine the value of A_i for a given emitter follower. (§8.2)

☐ 7. State the limit on the value of A_p for an emitter follower and calculate its value. (§8.2)

☐ 8. Describe the effects of loading on the value of A_v for an emitter follower. (§8.2)

☐ 9. Calculate the loaded voltage gain (A_vL) for a given emitter follower. (§8.2)

☐ 10. Calculate the amplifier input impedance of an emitter follower. (§8.2)

☐ 11. Define and calculate the value of output impedance (Z_{out}) of an emitter follower. (§8.2)

☐ 12. Describe the overall ac analysis procedure for an emitter follower. (§8.2)

☐ 13. Explain why emitter-feedback bias is commonly used in emitter followers. (§8.3)

☐ 14. Calculate the value of amplifier input impedance for an emitter follower using emitter-feedback bias. (§8.3)

☐ 15. Explain why emitter followers often require collector bypass capacitors. (§8.3)

☐ 16. List and discuss the two common emitter-follower applications. (§8.3)

☐ 17. Discuss the troubleshooting procedure for an emitter follower. (§8.3)

☐ 18. List the biasing circuits that are commonly used in common-base amplifiers. (§8.4)

☐ 19. Perform a complete ac analysis of a common-base amplifier. (§8.4)

☐ 20. List the gain and impedance characteristics of the common-base amplifier. (§8.4)

☐ 21. Discuss the two common applications for common-base amplifiers. (§8.5)

☐ 22. Compare and contrast the troubleshooting of common-base and common-emitter amplifiers. (§8.5)

☐ 23. Describe the Darlington amplifier. (§8.6)

☐ 24. Perform the dc analysis of a Darlington amplifier. (§8.6)

☐ 25. Perform the complete ac analysis of a Darlington amplifier. (§8.6)

☐ 26. List the major ac characteristics of the Darlington amplifier. (§8.6)

☐ 27. Perform the "quick" ac analysis of a Darlington amplifier. (§8.6)

As you master each of the objectives listed, place a check mark (✔) in the appropriate box.

Common-collector amplifier. A current amplifier that has no voltage gain. Also called an **emitter follower.**

Common-base amplifier. A voltage amplifier that has no current gain.

Darlington amplifier. A two-transistor circuit used to increase the overall amplifier current gain and input impedance.

In Chapter 7, we concentrated on the common-emitter amplifier. Now we are going to take a look at the *common-collector* amplifier and the *common-base* amplifier. We will also discuss the *Darlington amplifier*, a two-transistor configuration that is commonly used in common-emitter and common-collector applications.

The *common-collector* circuit is used to provide *current gain* and in *impedance-matching* applications. The input to this circuit is applied to the base while the output is taken from the emitter. The voltage gain of the common-collector circuit is typically less than 1, and the output voltage is in phase with the input voltage. Since the output signal ''follows'' the input signal, the common-collector amplifier is commonly referred to as the *emitter follower*.

The *common-base* circuit is used to provide *voltage gain* and for *high-frequency applications*. The common-base circuit has current gain that is typically less than 1 and, again, the output voltage is in phase with the input voltage.

The *Darlington* amplifier consists of two transistors that are connected in such a way as to provide dc and ac beta values that are equal to the product of

You have just finished a long chapter that focused on the operating principles of common-emitter (CE) amplifiers. Now you are starting the *single* chapter that covers the two other types of BJT amplifiers: the common-base (CB) and the common-collector (CC) amplifiers. At this point, you may be wondering why we spent so much more time and energy on CE amplifiers than we are on the other two types combined.

There are two reasons for the extended coverage of the CE amplifier. First, the CE amplifier is used much more extensively than either the CC or the CB amplifiers. While CB and CC amplifiers are used in a wide variety of applications and systems, they are not used nearly as often as the CE amplifier. Second, many of the principles that were covered in connection with the CE amplifier apply to the CB And CC amplifiers as well. Since these principles were covered earlier, they only need to be briefly reviewed in this chapter. In this chapter, we will spend most of our time discussing the ways in which CB and CC amplifiers differ from CE amplifiers.

the individual transistor beta values. The Darlington amplifier can be connected in any configuration. Most often, it is used in common-emitter and emitter-follower circuits.

8.1
The Emitter Follower (Common-Collector Amplifier)

The *emitter follower* is a *current amplifier* that has no voltage gain. The input is applied to the base of the transistor, while the output is taken from the emitter. The emitter follower is shown in Figure 8.1.

Emitter-follower circuit recognition.

As you can see, there is no collector resistor in the circuit. Notice also that there is no emitter bypass capacitor. These are the two circuit recognition features of the emitter follower.

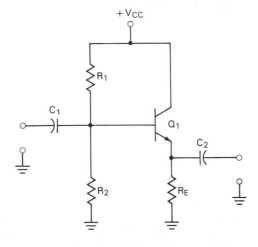

FIGURE 8.1

The emitter follower (common-collector) circuit.

DC Operation

For the circuit shown in Figure 8.1, the base bias is established by a voltage divider. This circuit works in the same way as the voltage-divider bias circuit of the common-emitter amplifier. Thus the base voltage is found as

$$V_B = \frac{R_2}{R_1 + R_2} V_{CC}$$

(8.1)

Once the base voltage is found, the emitter voltage is found as

$$V_E = V_B - 0.7 \text{ V}$$

(8.2)

and the emitter current is found as

$$I_E = \frac{V_E}{R_E}$$

(8.3)

Now V_{CEQ} is found as

$$V_{CEQ} = V_{CC} - V_E$$

(8.4)

If you refer back to Chapter 6, you will see that these formulas are nearly identical to those we used for the analysis of common-emitter bias circuits. The following example shows how the dc analysis of an emitter follower is performed.

EXAMPLE 8.1

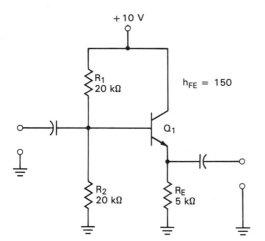

FIGURE 8.2

Determine the values of V_B, V_E, and I_E for the circuit shown in Figure 8.2.

Solution: The base voltage is found as

$$V_B = \frac{R_2}{R_1 + R_2} V_{CC}$$
$$= 5 \text{ V}$$

The emitter voltage is then found as

$$V_E = V_B - 0.7 \text{ V}$$
$$= 4.3 \text{ V}$$

Now I_E is found as

$$I_E = \frac{V_E}{R_E}$$
$$= 860 \text{ }\mu\text{A}$$

PRACTICE PROBLEM 8–1

A circuit like the one in Figure 8.2 has the following values: V_{CC} = +18 V, R_E = 910 Ω, R_1 = 16 kΩ, R_2 = 22 kΩ, and h_{FE} = 200. Determine the values of V_B, V_E, and I_E for the circuit.

The DC Load Line

3

As with the common-emitter circuit, the dc load line of the emitter follower is determined by the saturation and cutoff characteristics of the amplifier.

When the transistor is saturated, the *ideal* value of V_{CE} is 0 V. Since $V_E = I_E R_E$, the value of $I_{C(sat)}$ is found as

$$\boxed{I_{C(sat)} \cong \frac{V_{CC}}{R_E}}$$

(8.5)

This equation is derived by substituting $V_{CE} = 0$ V and $V_E = I_E R_E$ into equation (8.4).

As with the common-emitter circuit, the total applied voltage will be dropped across the transistor when it is in cutoff. Thus $V_{CE(off)}$ is found as

$$\boxed{V_{CE(off)} = V_{CC}}$$

(8.6)

These two equations are used to derive the dc load line of an emitter-follower amplifier. This is illustrated in the following example.

EXAMPLE 8.2

Derive and draw the dc load line for the circuit shown in Figure 8.2.

Solution: The value of $I_{C(sat)}$ is found as

$$I_{C(sat)} = \frac{V_{CC}}{R_E}$$
$$= 2 \text{ mA}$$

and now the value of $V_{CE(off)}$ is found as

$$V_{CE(off)} = V_{CC}$$
$$= 10 \text{ V}$$

Thus the dc load line for Figure 8.2 is drawn as shown in Figure 8.3.

FIGURE 8.3
The dc load line for Figure 8.2.

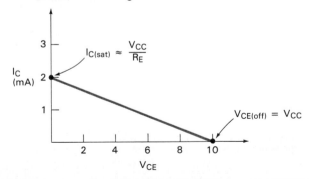

PRACTICE PROBLEM 8–2

Derive and draw the dc load line for the emitter follower described in Practice Problem 8–1.

That's all there is to the dc analysis of the emitter follower. As you can see, the dc analysis of the emitter follower is nearly identical to that of the common-

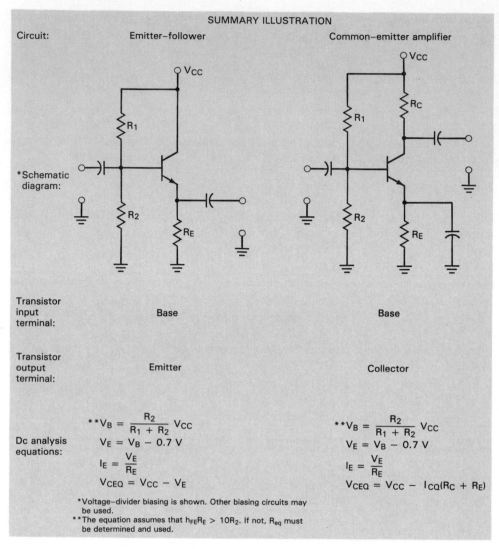

SUMMARY ILLUSTRATION

Circuit:	Emitter–follower	Common–emitter amplifier

*Schematic diagram:

Transistor input terminal: Base — Base

Transistor output terminal: Emitter — Collector

Dc analysis equations:

Emitter–follower:
$$**V_B = \frac{R_2}{R_1 + R_2} V_{CC}$$
$$V_E = V_B - 0.7 \text{ V}$$
$$I_E = \frac{V_E}{R_E}$$
$$V_{CEQ} = V_{CC} - V_E$$

Common–emitter amplifier:
$$**V_B = \frac{R_2}{R_1 + R_2} V_{CC}$$
$$V_E = V_B - 0.7 \text{ V}$$
$$I_E = \frac{V_E}{R_E}$$
$$V_{CEQ} = V_{CC} - I_{CQ}(R_C + R_E)$$

*Voltage–divider biasing is shown. Other biasing circuits may be used.
**The equation assumes that $h_{FE}R_E > 10R_2$. If not, R_{eq} must be determined and used.

FIGURE 8.4

emitter amplifier. The similarities and differences between the dc characteristics of the emitter-follower and the common-emitter amplifier are illustrated in Figure 8.4.

SECTION REVIEW

1. What is an emitter follower? [Objective 1]
2. What are the circuit recognition features of the emitter follower? [Objective 1]
3. List, in order, the steps that are involved in the dc analysis of an emitter follower. [Objective 2]
4. How do you determine the end points of the emitter follower dc load line? [Objective 3]

8.2
AC Analysis of the Emitter Follower

The key to the operation of the emitter follower is the fact that the emitter resistor is not bypassed. Because of this, the emitter follower has the ac equivalent circuit shown in Figure 8.5. Without the emitter resistor being bypassed, the ac resistance of the emitter circuit is found as

$$\boxed{r_E = r'_e + R_E}$$

(8.7)

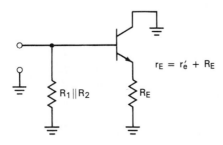

$$r_E = r'_e + R_E$$

FIGURE 8.5

The basic emitter follower ac equivalent circuit.

Voltage Gain (A_v)

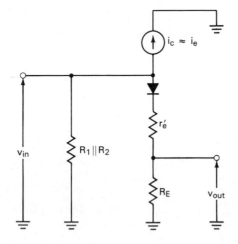

The derivation of the voltage gain equation for the emitter follower can be better understood if you refer to the circuit shown in Figure 8.6. As this figure shows, the input voltage is applied across the ac resistance of the emitter circuit. This is due to the fact that this resistance is in parallel with the base resistance, as shown in Figure 8.5.

FIGURE 8.6

The complete ac equivalent circuit.

With the output being taken across R_E, there is only a small voltage difference between the input and the output. By formula,

$$A_v = \frac{v_{\text{out}}}{v_{\text{in}}}$$

and

$$A_v = \frac{i_e R_E}{i_e(r'_e + R_E)}$$

Eliminating (i_e) from both the numerator and denominator of the equation above gives us

$$\boxed{A_v = \frac{R_E}{r'_e + R_E}} \qquad \text{(8.8)}$$

In most practical applications, R_E will be much greater than ($>>$) r'_e. When $R_E >> r'_e$, the following approximation can be used for voltage gain:

$$\boxed{A_v \cong 1 \quad |R_E >> r'_e} \qquad \text{(8.9)}$$

A_v is always less than 1.

Note that the value of $A_v = 1$ is the *ideal* voltage gain of the emitter follower, since R_E can never *completely* swamp out r'_e. However, the greater the difference between R_E and r'_e, the closer the amplifier voltage gain will come to 1. In practice, the voltage gain of the emitter follower is usually between 0.8 and 0.999. This is illustrated in the following example.

✔ 5

EXAMPLE 8.3

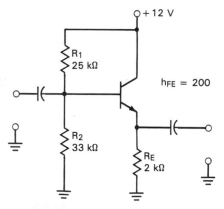

FIGURE 8.7

Don't forget. When the values of h_{fe} and h_{ie} for a transistor are known, they should be used to calculate the value of r'_e.

Determine the value of A_v for the emitter follower shown in Figure 8.7.

Solution: Using the established procedure, the value of I_E is found to be 3.06 mA. We do not know the h-parameter values of the transistor, so r'_e is approximated as

$$r'_e = \frac{25\ \text{mV}}{I_E}$$
$$= 8.17\ \Omega$$

and

$$A_v = \frac{R_E}{r'_e + R_E}$$
$$= \frac{2\ \text{k}\Omega}{(8.17\ \Omega + 2\ \text{k}\Omega)}$$
$$= 0.9959$$

PRACTICE PROBLEM 8–3

Determine the voltage gain of the amplifier described in Practice Problem 8–1.

CHAP. 8 Other Transistor Configurations

Current Gain

In Chapter 7 you were shown that the current gain of a common-emitter amplifier is approximately equal to the h-parameter value of h_{fe}. The same relationship holds true for the emitter follower. By formula,

$$\boxed{A_i = h_{fc}} \tag{8.10}$$

where $h_{fc} \cong h_{fe}$. Note that the subscript c merely indicates that the parameter applies to an emitter follower (common-collector) amplifier rather than to a common-emitter amplifier. Although there is a more exact h-parameter derivation for h_{fc}, and thus A_i, equation (8.10) will always do the trick for circuit analysis calculations.

✔ 6

A Practical Consideration: If you look in Appendix C you'll see that the exact equation for the value of h_{fc} is

$$h_{fc} = h_{fe} + 1$$

This is shown in the appendix, along with the equation derivation. Since h_{fe} is normally much greater than 1, we can simply assume that $h_{fc} \cong h_{fe}$.

Power Gain

As was the case with the common-emitter amplifier, the power gain of an emitter follower is found as the product of A_i and A_v. By formula,

$$\boxed{A_p = A_i A_v} \tag{8.11}$$

Since the value of A_v is always slightly less than 1, the power gain (A_p) of an emitter follower will always be slightly less than the current gain (A_i) of the circuit. This point is illustrated in the following example.

EXAMPLE 8.4

The amplifier in Figure 8.7 (Example 8.3) has a value of $h_{fe} = 220$. Determine the power gain (A_p) of the amplifier.

Solution: In Example 8.3, we determined the voltage gain (A_v) of the amplifier to be 0.9959. The transistor has a current gain of $h_{fe} = 220$. Since $h_{fc} \cong h_{fe}$, the current gain of the amplifier is found as

$$A_i = h_{fc}$$
$$= 220$$

Now the power gain of the circuit is found as

$$A_p = A_i A_v$$
$$= (220)(0.9959)$$
$$= 219.1$$

As you can see, the value of A_p for the emitter follower is slightly less than the value of current gain, A_i.

PRACTICE PROBLEM 8–4

The amplifier described in Practice Problem 8–1 has a current gain of $h_{fe} = 240$. Determine the power gain of the amplifier.

Amplifier Loading

8

When a load is added to an emitter follower, it is connected via a coupling capacitor to the emitter terminal of the circuit. This point is illustrated in Figure 8.8a, which shows the emitter follower from Figure 8.7 with an added 5-kΩ load resistance.

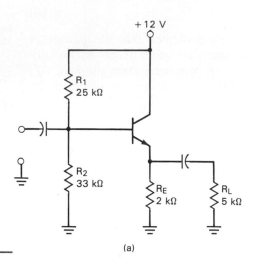

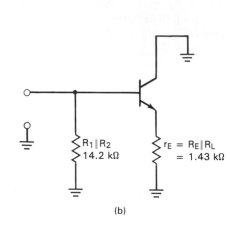

FIGURE 8.8

(a)

(b)

When a load resistance is added to an emitter follower, it forms an ac parallel equivalent resistance with R_E. You may recall that a capacitor acts as a short circuit to ac signals. If you view the output coupling capacitor (C_2) in Figure 8.8a as a short circuit, it is easy to see that R_E and R_L form an ac parallel circuit. Thus, R_E and R_L are combined in the ac equivalent circuit (shown in Figure 8.8b) as follows:

$$r_E = R_E \| R_L$$

(8.12)

9

When a load resistance is present, r_E is used in equation (8.8) in place of R_E. The result is an equation for the *loaded voltage gain* (A_{vL}) of the emitter follower, as follows:

$$A_{vL} = \frac{r_E}{r'_e + r_E}$$

(8.13)

The effect of loading on the voltage gain of an emitter follower is illustrated with the help of the following example.

EXAMPLE 8.5

Determine the value of A_{vL} for the circuit shown in Figure 8.8a.

Solution: As shown in Figure 8.8b, the value of R_E is found as

$$r_E = R_E \| R_L$$
$$= 1.43 \text{ k}\Omega$$

The value of r'_e was found in Example 8.3 to be 8.17 Ω.

Using this value of r'_e and the value of r_E, the loaded voltage gain of the amplifier is found as

$$A_{vL} = \frac{r_E}{r'_e + R_E}$$

$$= \frac{1.43 \text{ k}\Omega}{8.17 \ \Omega + 1.43 \text{ k}\Omega}$$

$$= 0.9943$$

PRACTICE PROBLEM 8–5

The amplifier described in Practice Problem 8–1 has an added 4-kΩ load resistance. Determine the value of A_{vL} for the circuit.

As was stated earlier, the amplifier in Figure 8.8 is identical to the one in Figure 8.7, with the exception of the added load resistance. If you compare the results of Example 8.5 with those of Example 8.3, you'll see that the addition of the load caused the voltage gain of the amplifier to drop from 0.9959 to 0.9934.

If you turn back to Chapter 7, you'll see that the addition of a load resistance to the common-emitter amplifier can have a drastic effect on the voltage gain of the amplifier. The emitter follower, on the other hand, will show a much smaller drop in voltage gain when a load resistance is added. This is another difference between the emitter follower and the common-emitter amplifier.

Input Impedance (Z_{in})

As with the common-emitter amplifier, the input impedance of the emitter follower is found as the parallel equivalent resistance of the base resistors and the transistor input impedance. By formula,

$$\boxed{Z_{in} = R_1 \| R_2 \| Z_{in(base)}} \qquad \textbf{(8.14)}$$

The base input impedance is equal to the product of the device current gain and the total ac emitter resistance. By formula,

$$\boxed{Z_{in(base)} = h_{fc}(r'_e + r_E)} \qquad \textbf{(8.15)}$$

where h_{fc} = the transistor current gain
  r'_e = the ac resistance of the transistor emitter
  r_E = the parallel combination of R_E and R_L

The following example illustrates the procedure used to determine the input impedance of an emitter follower.

EXAMPLE 8.6

Determine the input impedance of the amplifier shown in Figure 8.8a. Assume that the value of h_{fe} for the transistor is listed on the spec sheet as 220.

Solution: In Example 8.3 the value of r_e' for the circuit was found to be 8.17 Ω. The figure shows the value of r_E to be 1.43 kΩ. Using these values, the base input impedance is found as

$$h_{fc} \cong h_{fe} = 220$$

and

$$\begin{aligned} Z_{in(base)} &= h_{fc}(r_e' + r_E) \\ &= (220)(8.17\ \Omega + 1.43\ k\Omega) \\ &= 316.4\ k\Omega \end{aligned}$$

Now the input impedance of the amplifier is found as

$$\begin{aligned} Z_{in} &= R_1 \parallel R_2 \parallel Z_{in(base)} \\ &= 25\ k\Omega \parallel 33\ k\Omega \parallel 316.4\ k\Omega \\ &= 13.6\ k\Omega \end{aligned}$$

PRACTICE PROBLEM 8–6

The amplifier described in Practice Problem 8–1 has a current gain of $h_{fe} = 240$ and a load resistance of $R_L = 2$ kΩ. Determine the input impedance of the circuit.

When the value of r_E is much greater than r_e', as was the case in Example 8.6, equation (8.15) can be simplified as follows:

$$\boxed{Z_{in(base)} = h_{fc} r_E} \qquad (r_E >> r_e') \tag{8.16}$$

The validity of this approximation can be seen by recalculating the values of $Z_{in(base)}$ and Z_{in} for the circuit in Figure 8.8a. If you ignore the value of r_e' in the calculations, you will end up with the same value of Z_{in} that was found in Example 8.6. (This is left as an exercise for the reader.)

✔ 11

Output Impedance (Z_{out})

Output impedance. The impedance that a circuit presents to its load. The output impedance of a circuit is effectively the source impedance of its load.

Just as an amplifier has input impedance, it also has *output impedance*. Output impedance is *the impedance that a circuit presents to its load*. When a load is connected to a circuit, the output impedance of the circuit acts as the source impedance for that load.

We have not been concerned with amplifier output impedance before this point. However, since the emitter follower is often used in *impedance-matching* applications, Z_{out} is an important consideration. One of the main functions of the emitter follower is to "match" a source impedance to a load impedance so that maximum power is transferred from the source to the load. This point will be made clear in the following section.

The output impedance of an emitter follower is found as

$$\boxed{Z_{out} = R_E \parallel \left(r_e' + \frac{R_{in}'}{h_{fc}} \right)} \tag{8.17}$$

where Z_{out} = the output impedance of the amplifier
$R_{in}' = R_1 \parallel R_2 \parallel R_S$
R_S = the output resistance of the amplifier voltage source

The following example illustrates the procedure used to determine the output impedance of an emitter follower.

━━━━━━━━━━━━━━━━━━━━ **EXAMPLE 8.7** ━━━━━━━━━━━━━━━━━━━━

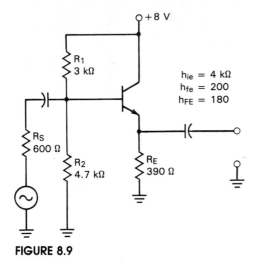

$h_{ie} = 4\ k\Omega$
$h_{fe} = 200$
$h_{FE} = 180$

FIGURE 8.9

Determine the output impedance (Z_{out}) of the amplifier shown in Figure 8.9.

Solution: Using the *h*-parameter method of analysis, the value of r_e' is found as

$$r_e' = \frac{h_{ie}}{h_{fe}}$$

$$= \frac{4\ k\Omega}{200}$$

$$= 20\ \Omega$$

The value of R_{in}' is now found as

$$R_{in}' = R_1 \| R_2 \| R_S$$
$$= 3\ k\Omega \| 4.7\ k\Omega \| 600\ \Omega$$
$$= 452\ \Omega$$

Finally, the output impedance of the amplifier is found as

$$Z_{out} = R_E \| \left(r_e' + \frac{R_{in}'}{h_{fc}} \right)$$

$$= 390 \| \left(20 + \frac{452\ \Omega}{200} \right)$$

$$= 390\ \Omega \| 22.26\ \Omega$$

$$= 21.06\ \Omega$$

───

PRACTICE PROBLEM 8–7

An emitter follower has the following values: $V_{CC} = +12$ V, $R_E = 200\ \Omega$, $R_1 = 2.5\ k\Omega$, $R_2 = 3.3\ k\Omega$, $R_S = 500\ \Omega$, $h_{FE} = 180$, $h_{fe} = 200$, and $h_{ie} = 3\ k\Omega$. Determine the value of Z_{out} for the amplifier.

Calculating the output impedance of an amplifier is not an everyday practice for most technicians. However, when we discuss the emitter follower as an impedance-matching circuit in the next section, you will see that being able to determine the output impedance of an emitter follower provides us with a useful circuit analysis tool.

The Basis for Equation (8.17)

The derivation of Z_{out} for the emitter follower is best understood by looking at the circuits shown in Figure 8.10. Figure 8.10a shows the ac equivalent of the emitter follower. Note that the input circuit consists of the source resistance (R_s)

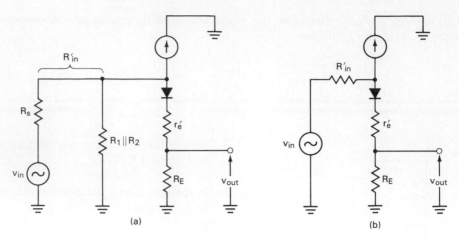

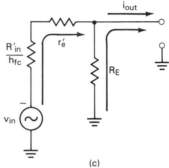

FIGURE 8.10

The derivation of Z_{out} for the emitter follower.

in parallel with the combination of R_1 and R_2. Thus the total resistance in the base circuit is found as

$$R'_{in} = R_1 \| R_2 \| R_s$$

This resistance is shown as a series resistor in Figure 8.10b.

To determine the value of Z_{out}, we start by writing the Kirchhoff's voltage equation for the circuit and solving that equation for i_e. The voltage equation is

$$v_{in} = i_b R'_{in} + i_e(r'_e + i_e R_E)$$

or

$$v_{in} = i_b R'_{in} + i_e(r'_e + R_E)$$

Since $i_b = i_e/h_{fc}$, the voltage equation can be rewritten as

$$v_{in} = \frac{i_e R'_{in}}{h_{fc}} + i_e(r'_e + R_E)$$

Solving for i_e yields

$$i_e = \frac{v_{in}}{r'_e + R_E + R'_{in}/h_{fc}}$$

This equation indicates that the emitter current "sees" a three-resistor circuit, as shown in Figure 8.10c. If we Thevenize the circuit, we find that the load sees

CHAP. 8   Other Transistor Configurations

R_E in parallel with $(r'_e + R'_{in}/h_{fc})$. Thus Z_{out} is found as

$$Z_{out} = R_E \,\|\, \left(r'_e + \frac{R'_{in}}{h_{fc}} \right)$$

(8.17)

where $R_{in} = R_1 \,\|\, R_2 \,\|\, R_s$.

Putting It All Together

The ac analysis of an emitter follower typically involves determining the values of A_i, A_{vL}, A_p, and Z_{in} for the circuit. Assuming that h-parameters for the transistor are available, the current, voltage, and power gain values are found as

$$
\begin{aligned}
A_i &= h_{fc} \cong h_{fe} \\
A_{vL} &= \frac{r_E}{r'_e + r_E} \quad \bigg| \quad r_E = R_E \,\|\, R_L \\
A_p &= A_{vL} A_i
\end{aligned}
$$

and the input impedance is found as

$$Z_{in} = R_1 \,\|\, R_2 \,\|\, Z_{in(base)}$$

where

$$Z_{in(base)} = h_{fc}(r'_e + r_E)$$

The value of $Z_{in(base)}$ for an emitter follower is typically higher than that of a common-emitter amplifier. Because of this, the value of Z_{in} will typically be much higher for an emitter follower than for a comparable common-emitter amplifier. This assumes, of course, that the common-emitter amplifier is not swamped. When compared to a swamped common-emitter amplifier, the emitter follower typically has the same range of $Z_{in(base)}$ and Z_{in} values.

The current gain of an emitter follower is equal to the h_{fc} rating of the transistor. Since $h_{fc} \cong h_{fe}$, the emitter follower and the common-emitter amplifier will typically have A_i values that are very nearly equal, all other factors being equal.

The voltage gain (A_v) of an emitter follower is always less than 1. If you have trouble seeing this in the A_v equation for the circuit, consider this: The emitter voltage of a transistor must always be slightly less than the base voltage for the transistor to be biased properly. Since V_E is always slightly less than V_B, it would make sense that any ac emitter voltage would have to be slightly less than any ac base voltage.

The output impedance of an emitter follower is considered only in *impedance-matching* applications. When the circuit is being used as an impedance-matching circuit, you need to be able to calculate the value of Z_{out} for the amplifier. In any other application, this calculation would not be necessary.

---

1. What is the limit on the value of voltage gain (A_v) for an emitter follower? [Objective 4]
2. List, in order, the steps that you would take to determine the value of A_v for an emitter follower. [Objective 5]
3. How is the value of current gain (A_i) for an emitter follower determined? [Objective 6]

SECTION REVIEW

SEC. 8.2   AC Analysis of the Emitter Follower

329

4. What is the limit on the value of power gain (A_p) for an emitter follower? [Objective 7]

5. What effect does loading have on the voltage gain of an emitter follower? [Objective 8]

6. List, in order, the steps you would take to determine the value of loaded voltage gain (A_{vL}) for an emitter follower. [Objective 9]

7. List, in order, the steps you would take to determine the value of amplifier input impedance (Z_{in}) for an emitter follower. [Objective 10]

8. What is output impedance (Z_{out})? [Objective 11]

9. List, in order, the steps you would take to determine the value of Z_{out} for an emitter follower. [Objective 11]

10. Describe the procedure used to perform a complete ac analysis of an emitter follower. [Objective 12]

11. When is the value of Z_{out} for an emitter follower an important consideration? [Objective 12]

8.3
Practical Considerations, Applications, and Troubleshooting

There are two practical considerations regarding emitter followers that we have ignored up to this point. These considerations are the use of emitter-feedback bias and the need for a *collector bypass capacitor*. Each of these practical considerations will be discussed in detail in this section.

We will also discuss the most common applications for the emitter follower: as a *current amplifier* and as an *impedance-matching circuit*. Finally, we will take a brief look at the troubleshooting of the emitter follower.

Using Emitter-Feedback Bias to Increase Amplifier Input Impedance

Why Z_{in} is higher when emitter-feedback bias is used.

In many applications, such as impedance matching, one goal of emitter-follower design is to have as high an input impedance value as possible. When high input impedance is desirable, emitter-feedback bias is often used. An emitter follower with emitter-feedback bias is shown in Figure 8.11a. A comparable emitter follower with voltage-divider bias is shown in Figure 8.11b.

An emitter follower with emitter-feedback bias will have a much higher input impedance than a comparable circuit with voltage-divider bias. This point is easy to understand if you consider the equation for the input impedance of a voltage-divider biased emitter follower. As you know, this equation is

$$Z_{in} = R_1 \| R_2 \| Z_{in(base)}$$

Since the emitter-feedback bias circuit has only one base resistor, the input impedance is found as

$$\boxed{Z_{in} = R_1 \| Z_{in(base)}} \tag{8.18}$$

As you know, the total resistance in a parallel circuit will be less than the lowest individual resistance value. Since R_1 in Figure 8.11b is much lower than R_1 in Figure 8.11a, it would make sense that the value of Z_{in} for the voltage-divider bias circuit would have to be lower than the value of Z_{in} for the emitter-feedback bias circuit. This point is illustrated further in the following example.

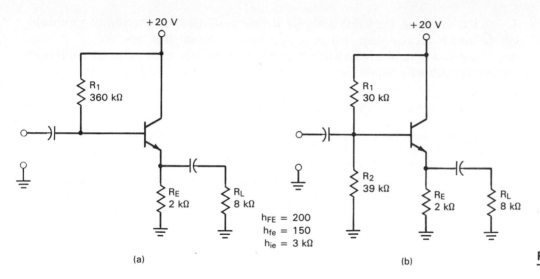

$h_{FE} = 200$
$h_{fe} = 150$
$h_{ie} = 3\ k\Omega$

(a)

(b)

FIGURE 8.11

EXAMPLE 8.8

The amplifiers in Figure 8.11 were designed to have nearly identical output characteristics, using identical transistors with the specifications shown. Determine the input impedance values for the two circuits.

Solution: Since the circuits are using identical transistors, emitter resistance values, and load resistance values, the value of $Z_{in(base)}$ will be the same for the two circuits. This value of $Z_{in(base)}$ is found using the following procedure:

$$r'_e = \frac{h_{ic}}{h_{fc}}$$
$$= 20\ \Omega$$
$$r_E = R_E \| R_L$$
$$= 1.6\ k\Omega$$

and

$$Z_{in(base)} = h_{fc}(r'_e + r_E)$$
$$= (150)(20\ \Omega + 1.6\ k\Omega)$$
$$= 243\ k\Omega$$

For the emitter-feedback bias circuit, the value of Z_{in} is found as

$$Z_{in} = R_1 \| Z_{in(base)}$$
$$= 360\ k\Omega \| 243\ k\Omega$$
$$= 145\ k\Omega$$

For the voltage-divider biased circuit, the value of Z_{in} is found as

$$Z_{in} = R_1 \| R_2 \| Z_{in(base)}$$
$$= 30\ k\Omega \| 39\ k\Omega \| 243\ k\Omega$$
$$= 15.9\ k\Omega$$

As you can see, the value of Z_{in} for an emitter-feedback biased emitter follower will be much greater than that of a comparable circuit that has voltage-divider bias. This is why emitter-feedback biasing is commonly used in emitter-follower impedance-matching applications. This point will become clear later in this section.

The Collector Bypass Capacitor

✔ 15

Why a collector bypass capacitor is used.

There is a problem that can occur as a result of the normal operation of an emitter follower. This problem can be seen in the oscilloscope display shown in Figure 8.12. The upper trace in Figure 8.12 is the input signal to an emitter follower. As you can see, the output waveform (lower trace) is very distorted. The cause of this distortion can be explained with the help of Figure 8.13.

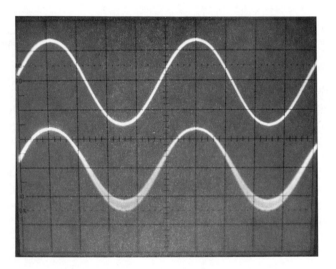

FIGURE 8.12

In Figure 8.13, V_{CC} is represented as a simple battery with some value of internal resistance (R_{int}). As the color highlight shows, the dc power supply is part of a loop that also contains the transistor (Q_1), C_{C2}, and R_L.

FIGURE 8.13

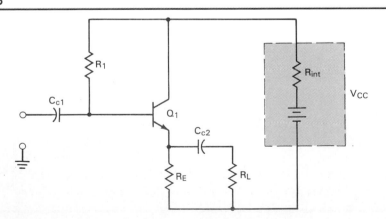

If the value of R_L is low enough, the capacitor may try to charge and discharge when an ac signal is applied to the circuit. As the capacitor charges and discharges through the dc power supply, the value of V_{CC} will actually vary slightly, as shown in the following equation:

$$\Delta V = (\Delta I)(R_{int})$$

where ΔV = the change in V_{CC}
  ΔI = the change in loop current, caused by the capacitor charge/discharge action
  R_{int} = the internal resistance of the power supply

Thus, the charge/discharge action of the capacitor will cause a change in the value of V_{CC}, resulting in the distorted output waveform shown in Figure 8.12.

There are two ways to prevent the output distortion shown in Figure 8.12. The first is to use a *regulated dc power supply*. A regulated dc power supply is one that has *extremely low* internal resistance, typically less than 10 Ω. When a regulated dc power supply is used, the value of ΔV will be extremely small. In most cases, it will be too small to see in an oscilloscope display.

Another practical solution to the problem is the use of a *collector bypass capacitor* or *decoupling capacitor*. An emitter follower with a collector bypass capacitor (C_B) is shown in Figure 8.14.

Regulated dc power supply.
A dc power supply with extremely low internal resistance.

Collector bypass capacitor. A capacitor connected between V_{CC} and ground, parallel with the dc power supply. Often referred to as a **decoupling capacitor.**

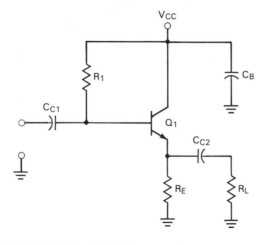

FIGURE 8.14

The collector bypass capacitor is connected between V_{CC} and ground and thus is in parallel with the dc power supply. When C_{C2} tries to charge or discharge through the dc power supply, the change in current (ΔI) is shorted around the supply by C_B. Thus, V_{CC} does not change, and the distortion shown in Figure 8.12 is eliminated.

For C_B to be effective, its value must be extremely small. A small-value C_B will ensure that the dc power supply is completely bypassed. Typically, C_B should be 0.01 μF or lower in value.

Typically, a collector bypass capacitor will be 0.01 μF or less.

Emitter-follower Applications

The emitter follower is used when you need *current amplification with no voltage gain*. There are many instances (especially in digital electronics) when an increase in current is required but an increase in voltage is not. Because the emitter follower ideally has a voltage gain of 1 and a high current gain, it is especially suited to these applications.

Emitter followers are used when A_i is needed and A_v is not needed.

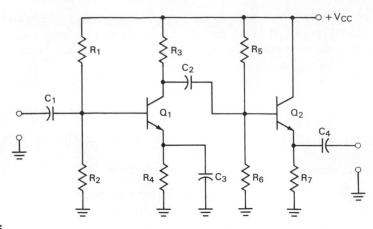

FIGURE 8.15

Consider the circuit shown in Figure 8.15. The first stage of the amplifier has specific values of A_i and A_v. What if A_v is high enough to provide the final required output voltage, but A_i is not high enough? The emitter follower would be used to provide the additional current gain without increasing A_v beyond its desired value.

Impedance matching. A circuit design technique that ensures maximum power transfer between source and load.

Another application for the emitter follower is as an *impedance-matching* circuit. Such an application is illustrated in Figure 8.16. The input and output impedance values shown were calculated for the circuit in Figure 8.11a.*

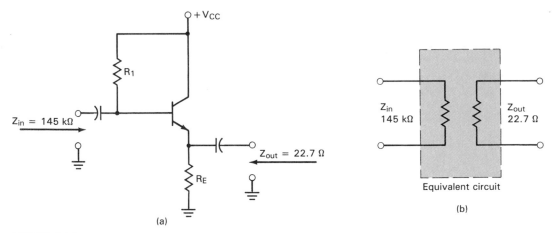

FIGURE 8.16

You may recall from your study of basic electronics that *maximum power is transferred to a load when the source and load impedances are equal*. When the source impedance equals the load impedance, the two impedances are said to be *matched*. When the impedances are matched, maximum power is delivered from the source to the load. The closer you get to a perfect match, the more power you will get at the load.

Emitter followers may be used to provide maximum power transfer to low Z loads.

Note that the emitter follower shown in Figure 8.16 has a *relatively high input impedance* and a *relatively low output impedance*. This makes it a perfect circuit for *connecting a low-impedance load to a high-impedance source*. For example, look at the circuits shown in Figure 8.17.

---

* The value of $Z_{out} = 22.7 \, \Omega$ was calculated using an assumed value of $R_S = 600 \, \Omega$. The value $Z_{in} = 145 \, k\Omega$ was calculated in Example 8.8.

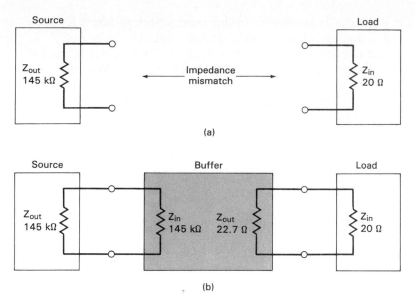

FIGURE 8.17

In Figure 8.17a, a source with 145 kΩ of internal resistance is being connected to a 20-Ω load. In this case, very little of the source power would actually be delivered to the load. In fact, most of the source power would be used by its own internal resistance. To match the source to the load, the emitter follower from Figure 8.16 could be placed between the original source and load. This connection is shown in Figure 8.17b. Note that the source impedance very closely matches the input impedance of the amplifier. Because of this, there is a maximum power transfer from the source to the amplifier input. The low output impedance of the amplifier matches the load resistance very closely. Therefore, there is a maximum power transfer from the amplifier to the load. The end result is that maximum power has been transferred from the original source to the original load. An emitter follower that is used for this purpose is usually called a *buffer*.

Buffer. An emitter follower that is used for impedance matching.

The specific purpose served by a given emitter follower depends on the circuit in which it is being used. However, you will find that it is always used for one of the two purposes mentioned in this section: providing current gain and/or impedance matching.

Troubleshooting

Troubleshooting an emitter follower is not very different from troubleshooting a common-emitter amplifier. However, there is one important fact that you must keep in mind: *The emitter follower does not provide any voltage gain.*

When you troubleshoot a common-emitter amplifier, you expect to see a much higher output voltage than input voltage. This is because the common-emitter amplifier has a relatively high amount of voltage gain. If you checked out a common-emitter circuit and saw that its ac output voltage and ac input voltage were equal, you would have good reason to believe that the circuit was faulty. However, for the emitter follower, this condition is *normal*. In fact, the output voltage will always be slightly less than the input voltage for the emitter follower. *Do not let the near-equal input and output ac voltages lead you to believe that an emitter follower is not working.*

When any component in an emitter follower goes bad, you will generally have one of two conditions: Either the transistor will saturate or it will cut off. When the output is very nearly equal to V_{CC} or ground, there is reason to believe that the emitter follower is faulty. The output voltage (V_E) will be close to V_{CC}

only if the transistor is *saturated*. In this case, check to see if R_2 (base circuit) is open. If the output is close to ground, the transistor is in *cutoff*. In this case, check R_1 (base circuit) to see if it is open. If not, replace the transistor.

If no ac signal appears at the output (emitter terminal) and V_E is at some value between ground and V_{CC}, check the input signal to the base. If a signal is there, check R_E to see if it is open. If it is not, the transistor is probably faulty and should be replaced.

Review

The major characteristics of the emitter follower are as follows:

1. Low voltage gain (less than 1)
2. Relatively high current and power gain
3. High input impedance and low output impedance
4. Input and output ac voltages are in phase

SECTION REVIEW

1. Why is emitter-feedback bias commonly used in emitter followers? [Objective 13]
2. List, in order, the steps you would take to determine the value of amplifier input impedance for an emitter follower with emitter-feedback bias. [Objective 14]
3. What problem is solved by the use of a collector bypass capacitor? [Objective 15]
4. Why are collector bypass capacitors preferred over the use of regulated dc power supplies? [Objective 15]
5. Describe the two common applications of the emitter follower. [Objective 16]
6. What is the purpose of a collector bypass capacitor? [Objective 16]
7. What is the purpose of a *buffer*? [Objective 16]
8. How does the buffer solve the problem referenced in question 6? [Objective 16]
9. What must be kept in mind when troubleshooting an emitter follower? [Objective 17]

8.4
The Common-Base Amplifier

Common-base applications.

The *common-base* amplifier is used to provide *voltage gain with no current gain* and for *high-frequency applications*. The input to the common-base amplifier is applied to the emitter, while the output is taken from the collector. The schematic of the common-base amplifier is shown in Figure 8.18.

As you can see, there are two ways in which the common-base amplifier is commonly biased. Figure 8.18a shows the transistor in an *emitter-bias* configuration. Figure 8.18b shows the transistor in a *voltage-divider bias* configuration. The dc analysis of these two circuits is the same as for the equivalent common-emitter configurations.

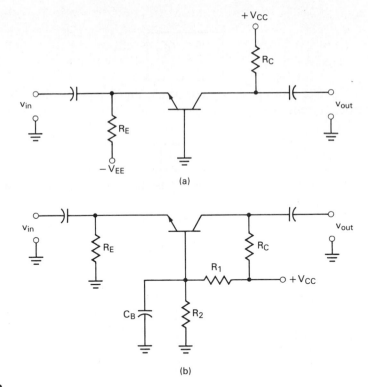

FIGURE 8.18

The common-base amplifier.

Note that the input to each amplifier is applied to the emitter, while the output is taken from the collector. The base of Figure 8.18a is coupled directly to ground. The base of Figure 8.18b is grounded (in terms of ac operation) by the coupling capacitor across R_2.

AC Analysis

The ac equivalent circuit for the common-base amplifier is shown in Figure 8.19. We will use this circuit for our discussion on the ac characteristics of the common-base amplifier.

The input signal is applied to the emitter, so the input signal is developed across r'_e. The output signal is developed across the ac collector resistance. Therefore, voltage gain is found as

$$A_{vL} = \frac{v_{out}}{v_{in}}$$
$$= \frac{i_e r_C}{i_e r'_e}$$

A Practical Consideration: The conversion of h_{ie} and h_{fe} to h_{ib} and h_{fb} is fairly complex (see Appendix C.). In this case, you're better off using

$$r'_e = \frac{25\,mV}{I_E}$$

FIGURE 8.19

The common-base ac equivalent circuit.

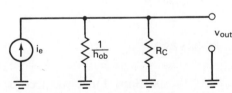

and

$$A_{vL} = \frac{r_C}{r_e'} \tag{8.19}$$

Since the emitter current is approximately equal to the collector current, the input and output currents are approximately equal. Therefore, current gain is found as

$$A_i \cong 1 \tag{8.20}$$

The input impedance to the amplifier is the parallel combination of r_e' and R_E. By formula,

$$Z_{\text{in}} = r_e' \, \| \, R_E \tag{8.21}$$

Since R_E is usually much greater than r_e', the input impedance to the amplifier can be approximated as

$$Z_{\text{in}} \cong r_e' \tag{8.22}$$

h_{ob}. The common-base value of output admittance. You can assume that $1/h_{ob}$ is much greater than any practical value of R_C.

The output impedance of the amplifier is the parallel combination of R_C and the collector resistance of the transistor $(1/h_{ob})$. By formula,

$$Z_{\text{out}} = R_C \, \| \, \frac{1}{h_{ob}} \tag{8.23}$$

Since $1/h_{ob}$ is typically very high (in the hundreds of kilohms), the output impedance is approximated as

$$Z_{\text{out}} \cong R_C \tag{8.24}$$

The following example illustrates the ac analysis of a common-base amplifier.

━━━━━ **EXAMPLE 8.9** ━━━━━

Determine the gain and impedance values for the circuit shown in Figure 8.20.

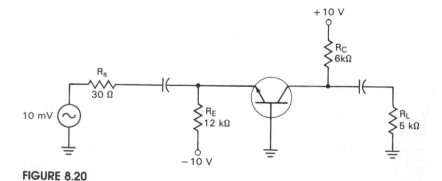

FIGURE 8.20

Solution: First, we need to determine the value of I_E. Since the circuit is an emitter-bias circuit, I_E is found as

$$I_E = \frac{V_{EE} - V_{BE}}{R_E}$$
$$= 775 \ \mu A$$

The value of r_e' can now be approximated as

$$r_e' = \frac{25 \ mV}{I_E}$$
$$= 32.3 \ \Omega$$

Since $R_E \gg r_e'$, we can approximate the value of Z_{in} as

$$Z_{in} \cong r_e'$$
$$= 32.3 \ \Omega$$

The output impedance is found as

$$Z_{out} \cong R_C$$
$$= 6 \ k\Omega$$

The ac resistance of the collector circuit is found as

$$r_C = R_C \| R_L$$
$$= 2.73 \ k\Omega$$

and the voltage gain of the amplifier is found as

$$A_{vL} = \frac{r_C}{r_e'}$$
$$= 84.5$$

Finally, since the output current is approximately equal to the input current, the current gain is found as

$$A_i \cong 1$$

PRACTICE PROBLEM 8–9

A base-bias circuit like the one in Figure 8.20 has values of $V_{CC} = +15$ V, $V_{EE} = -15$ V, $R_E = 30$ kΩ, $R_C = 15$ kΩ, and $R_L = 3$ kΩ. Determine the approximate values of A_{vL}, A_i, Z_{in}, and Z_{out} for the amplifier.

Example 8.9 demonstrated the ac analysis of a common-base amplifier. The results found in the example also serve to illustrate the ac characteristics of this type of amplifier. These are:

1. Relatively high voltage gain
2. Current gain that is less than 1
3. Low input impedance
4. High output impedance

As you will see in the next section, the characteristics listed above make the common-base amplifier well suited for several applications.

1. Which biasing circuits are commonly used for common-base amplifiers? [Objective 18]
2. List, in order, the steps you would take to perform a complete ac analysis of a common-base amplifier. [Objective 19]
3. What are the gain characteristics of a common-base amplifier? [Objective 20]
4. What are the input/output characteristics of a common-base amplifier? [Objective 20]

8.5

Common-Base Applications and Troubleshooting

As was stated earlier, the common-base amplifier is used to provide *voltage gain* *without current gain* and for *high-frequency* applications. Of the two, the high frequency applications are far more common.

FIGURE 8.21

The common-base amplifier as an impedance-matching circuit.

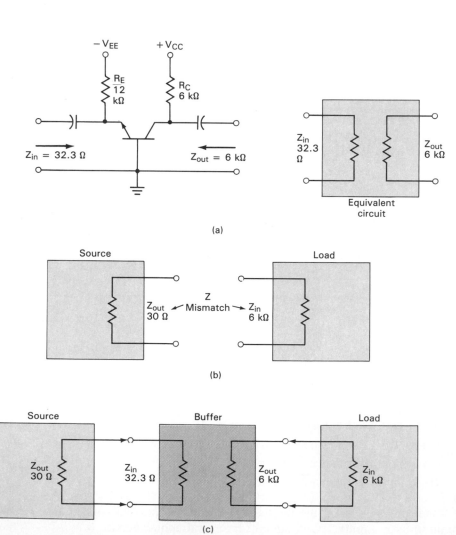

Since the common-base circuit has a high voltage gain and a value of $A_i \cong$ 1, the circuit can be used to provide voltage gain without increasing the value of circuit current. Consider the case where the output current from an amplifier is more than enough for the application, but the voltage needs to be boosted. The common-base amplifier would serve well in this situation because it would increase the voltage without increasing the current.

The high-frequency application for common-base amplifiers ties in closely with the impedance-matching application of the emitter follower. The common-base amplifier has the *opposite* impedance characteristics of the emitter follower, as shown in Figure 8.21a. Note that the common-base circuit typically has *low input impedance* and *high output impedance*. The emitter follower has just the opposite impedance characteristics.

How does this relate to high-frequency operation? Most high-frequency voltage sources have *very low output impedance*. When you want to connect a high-frequency, low-Z_{out} source to a high-impedance load, you need a circuit to match the source impedance to the load impedance. This function is served very well by the common-base circuit, as shown in Figure 8.21b and c.

Troubleshooting

The troubleshooting of the common-base amplifier is very nearly the same as troubleshooting the common-emitter circuit. The output voltage should show gain when compared to the input voltage. However, *there is no phase shift from input to output*.

When you troubleshoot a common-base circuit, check the input and output signals. If both signals are present and there is voltage gain from emitter to collector, the amplifier is good. If there is a valid input signal but no output signal, check the amplifier using the techniques discussed in Chapter 6.

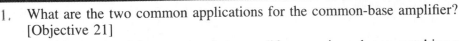

SECTION REVIEW

1. What are the two common applications for the common-base amplifier? [Objective 21]
2. Explain the use of the common-base amplifier as an impedance-matching circuit. [Objective 21]
3. What is the difference between troubleshooting a common-base amplifier and troubleshooting a common-emitter amplifier? [Objective 22]

8.6
The Darlington Amplifier

✔ 23

Darlington amplifier. A special-case emitter follower that uses two transistors to increase the overall current gain and amplifier input impedance.

The Darlington amplifier is a special-case emitter follower that uses two transistors to increase the overall values of circuit current gain (A_i) and input impedance (Z_{in}). The two transistors are connected as shown in Figure 8.22. The emitter of

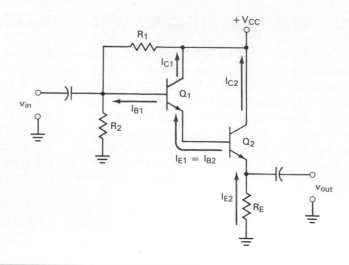

FIGURE 8.22

The Darlington amplifier.

the first is tied to the base of the second, and the transistor collector terminals are tied together. The transistor connections result in the current paths shown in the figure.

DC Analysis

24

Don't forget:

$h_{FC} \cong h_{FE}$
$h_{fc} \cong h_{fe}$

The circuit consists of two transistors, each with a base-to-emitter voltage, V_{BE}. Thus I_{E2} will be found by taking $2V_{BE}$ from the base voltage and dividing the differences by R_E. By formula,

$$I_{E2} = \frac{V_B - 2V_{BE}}{R_E} \tag{8.25}$$

Since the transistors are directly coupled,

$$I_{E1} = I_{B2} \tag{8.26}$$

and since $I_{B2} \cong I_{E2}/h_{FC2}$,

$$I_{E1} \cong \frac{I_{E2}}{h_{FC2}} \tag{8.27}$$

You may recall that the dc input resistance of a transistor is found as

$$R_{IN} = h_{FC}R_E$$

For the input transistor, the emitter resistance is the input resistance of the second transistor. By formula,

$$R_{IN1} = h_{FC1}R_{IN2}$$

and since $R_{IN2} = h_{FC2}R_E$, the equation above becomes

$$R_{IN1} = h_{FC1}h_{FC2}R_E \tag{8.28}$$

CHAP. 8   Other Transistor Configurations

EXAMPLE 8.10

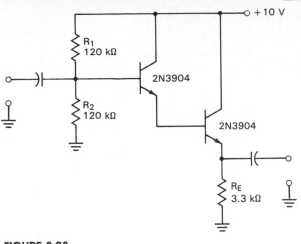

FIGURE 8.23

Determine the input resistance to the *Darlington pair* shown in Figure 8.23. Use the spec sheet on page 241 of the text.

Solution: The dc value of I_E is found as

$$I_E = \frac{V_B - 2V_{BE}}{R_E}$$
$$= 1.09 \text{ mA}$$

The spec sheet for the 2N3904 lists a minimum h_{FE} of 70 at $I_{CQ} = 1$ mA. Using this value as h_{FC}, the input resistance of the Darlington pair is found as

$$R_{IN1} = h_{FC1}h_{FC2}R_E$$
$$= (70)(70)(3.3 \text{ k}\Omega)$$
$$= 16.17 \text{ M}\Omega$$

PRACTICE PROBLEM 8–10

When two transistors are connected as shown in Figure 8.23, they are referred to as a **Darlington pair**.

A Darlington amplifier like the one in Figure 8.23 has the following values: $h_{FC} = 240$ (each transistor) and $R_E = 39 \ \Omega$. Determine the input resistance to the Darlington pair.

As Example 8.10 illustrated, the Darlington pair has a much higher input resistance than the typical transistor.

AC Analysis

For the Darlington amplifier, the input impedance is found using the same equations as the standard emitter follower. When voltage-divider bias is used, Z_{in} is found as

$$Z_{in} = R_1 \| R_2 \| Z_{in(base)}$$

When emitter-feedback bias is used,

$$Z_{in} = R_1 \| Z_{in(base)}$$

The value of $Z_{in(base)}$ for the Darlington pair is found as

$$Z_{in(base)} = h_{ic1} + h_{fc1}(h_{ic2} + h_{fc2}r_E)$$

where h_{ic} = the input impedance of the identified transistor
  h_{fc} = the ac current gain of the identified transistor
  r_E = the parallel combination of R_E and R_L

The derivation of equation (8.29) is covered in Appendix D.

The output impedance of the amplifier is found in the same way as the output impedance of a standard emitter follower. However, the gain of the second transistor must be considered. Thus, Z_{out} is found as

$$Z_{out} = R_E \parallel \left[r'_{e2} + \frac{r'_{e1} + (R_{in}/h_{fc1})}{h_{fc2}} \right]$$

(8.30)

Since R_E is usually much greater than the rest of the equation, Z_{out} can be approximated as

$$Z_{out} \cong r'_{e2} + \frac{r'_{e1} + (R'_{in}/h_{fc1})}{h_{fc2}}$$

(8.31)

where $R'_{in} = R_1 \parallel R_2 \parallel R_S$
r'_e = the ac emitter resistance of the identified transistor
h_{fc} = the ac current gain of the identified transistor

The ac current gain of the transistors is equal to the product of the individual gains. This is due to the fact that input current is increased by the gain of the first transistor and, after being applied to the base of the second transistor, is increased again by the gain of that component. By formula,

$$A_i = h_{fc1}h_{fc2}$$

(8.32)

As before, the voltage gain of the amplifier (A_v) is slightly less than 1.

EXAMPLE 8.11

Determine the circuit gain and impedance values for the amplifier in Figure 8.23. Assume that the circuit has values of R_S = 3.3 kΩ, h_{fc1} = 120, h_{fc2} = 150, h_{ic1} = 40 kΩ, and h_{ic2} = 3 kΩ.

Solution: The overall current gain of the amplifier is found as

$$\begin{aligned} A_i &= h_{fc1}h_{fc2} \\ &= (120)(150) \\ &= 18,000 \end{aligned}$$

This extremely high current gain is typical for a Darlington amplifier. As with any emitter follower, the overall voltage gain will be slightly less than 1.

The input impedance to the Darlington pair is found as

$$\begin{aligned} Z_{in(base)} &= h_{ic1} + h_{fc1}(h_{ic2} + h_{fc2}r_E) \\ &= 40 \text{ k}\Omega + (120)[3 \text{ k}\Omega + (150)(3.3 \text{ k}\Omega)] \\ &= 40 \text{ k}\Omega + (120)(498 \text{ k}\Omega) \\ &= 59.8 \text{ M}\Omega \end{aligned}$$

The amplifier input impedance is now found as

$$\begin{aligned} Z_{in} &= R_1 \parallel R_2 \parallel Z_{in(base)} \\ &= 59.9 \text{ k}\Omega \end{aligned}$$

Again, this high input impedance is characteristic of a Darlington amplifier.

To determine the value of Z_{out}, we need to start by determining the values of r_e' for the two transistors. These values are found as

$$r_{e1}' = \frac{h_{ic1}}{h_{fc1}}$$
$$= 333 \; \Omega$$

and

$$r_{e2}' = \frac{h_{ic2}}{h_{fc2}}$$
$$= 20 \; \Omega$$

The input resistance of the circuit (R_{in}') is found as

$$R_{in}' = R_1 \| R_2 \| R_S$$
$$= 3.13 \; \text{k}\Omega$$

Finally, the value of Z_{out} is found as

$$Z_{out} \cong r_{e2}' + \frac{r_{e1}' + (R_{in}'/h_{fc1})}{h_{fc2}}$$
$$= 20 \; \Omega + \frac{333 \; \Omega + (3.13 \; \text{k}\Omega/120)}{150}$$
$$= 20 \; \Omega + 2.4 \; \Omega$$
$$= 22.4 \; \Omega$$

The low output impedance is expected in this amplifier.

PRACTICE PROBLEM 8–11

Determine the circuit gain and impedance values for the circuit in Figure 8.23. Assume that the circuit has the following values: $R_1 = R_2 = 240 \; \text{k}\Omega$, $R_E = 510 \; \Omega$, $R_S = 6 \; \text{k}\Omega$, $h_{fc1} = 80$, $h_{fc2} = 180$, $h_{ic1} = 38 \; \text{k}\Omega$, and $h_{ic2} = 4 \; \text{k}\Omega$.

Example 8.8 demonstrates the major characteristics of the Darlington amplifier. These characteristics are as follows:

1. *Extremely high base input impedance.* From the example, you can see that $Z_{in(base)}$ values in the MΩ region are possible.
2. *Extremely high current gain.* Current gain values in the thousands are typical for Darlington amplifiers.
3. *Extremely low output impedance.* For the circuit in the example, Z_{out} was 22.4 Ω. This is a very typical value of Z_{out}.

Darlington amplifier characteristics.

Because the characteristics of the Darlington amplifier are basically the same as those of the emitter follower, the two circuits are used for similar applications. When you need higher input impedance and current gain and/or lower output impedance than the standard emitter follower can provide, you use a Darlington amplifier.

The "Quick" Analysis

 27

As you can see the detailed analysis of the Darlington amplifier is long and involved. When you are only interested in quickly determining approximate circuit values, the following approximations can be used:

$$A_i = h_{fc1}h_{fc2}$$
$$A_v \cong 1$$
$$Z_{in(base)} \cong A_i r_E$$

(8.33)

$$Z_{out} \cong r'_{e2} + \frac{r'_{e1}}{h_{fc2}}$$

(8.34)

The first two equations are the same ones we used in our earlier analysis. Equation (8.33) ignores the values of r'_e for the two transistors. Had we used this equation in Example 8.11, we would have obtained a value of $Z_{in(base)} = 3.17$ MΩ. Using this value of $Z_{in(base)}$, we would have obtained a value of $Z_{in} = 58.9$ kΩ, a difference of only 1%. The final approximation ignores the value of (R'_{in}/h_{fc1}) in the calculation of Z_{out}. Had we used this equation in Example 8.11, we would have obtained a value of $Z_{out} = 22.2$ Ω, rather than the 22.4 Ω value found in the example.

One Final Note

Darlington transistors are commonly available. Like standard transistors, they have only three terminals, but they have much higher values of h_{fe}, h_{ie}, and h_{FE}. When you work with one of these transistors, all the relationships discussed in this section will apply.

TRANSISTOR AMPLIFIERS: A SUMMARY

In Chapters 5 through 8, you were introduced to a wide variety of transistor amplifiers. While the dc characteristics of the amplifiers do not vary significantly from one configuration to another, the ac characteristics of the circuits do have significant differences. As a review, the ac characteristics of the common-emitter, emitter-follower, common-base, and Darlington amplifiers are compared in Table 8.1.

TABLE 8.1
Amplifier Configurations and Their AC Characteristics

	Common emitter	Emitter follower	Common base	Darlington
A_v	High	Less than 1	High	Less than 1
A_i	High	High	Less than 1	Extremely high
A_p	High	High[a]	High[b]	High[a]
Z_{in}	Low	High	Extremely low	Extremely high
Z_{out}	High	Very low	High	Extremely low

[a] Slightly less than the value of A_i.
[b] Slightly less than the value of A_v.

SECTION REVIEW

1. What is a *Darlington amplifier*? [Objective 24]
2. How are the two transistors in a Darlington amplifier interconnected? [Objective 24]

3. Why is the current gain of a Darlington amplifier higher than that of a standard emitter follower? [Objective 25]
4. List the major characteristics of the Darlington amplifier. [Objective 26]

KEY TERMS

The following terms were introduced and defined in this chapter:

buffer
collector-bypass capacitor
common-base amplifier
common-collector
   amplifier

Darlington amplifier
Darlington pair
emitter follower

impedance matching
output impedance
regulated dc power supply

PRACTICE PROBLEMS

§8.1

1. Determine the values of V_B, V_E, and I_E for the circuit in Figure 8.24. [2]
2. Determine the values of V_B, V_E, and I_E for the circuit in Figure 8.25. [2]
3. Plot the dc load line for the circuit in Figure 8.24 [3]
4. Plot the dc load line for the circuit in Figure 8.25. [3]
5. Determine the values of A_{vL}, A_i, and A_p for the circuit in Figure 8.24. [6, 7, 9]
6. Determine the values of A_{vL}, A_i, and A_p for the circuit in Figure 8.25. [6, 7, 9]
7. Determine the value of Z_{in} for the circuit in Figure 8.24. [10]
8. Determine the value of Z_{in} for the circuit in Figure 8.25. [10]
9. Determine the value of Z_{out} for the circuit in Figure 8.24. [11]

FIGURE 8.24

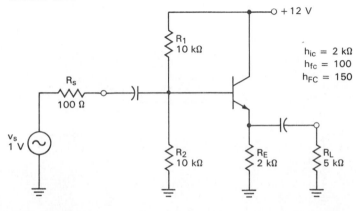

$h_{ic} = 2$ kΩ
$h_{fc} = 100$
$h_{FC} = 150$

FIGURE 8.25

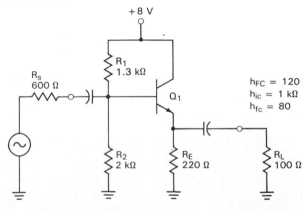

$h_{FC} = 120$
$h_{ic} = 1$ kΩ
$h_{fc} = 80$

10. Determine the value of Z_{out} for the circuit in Figure 8.25. [11]
11. Perform a basic ac analysis on the amplifier in Figure 8.26. [12]
12. Perform a basic ac analysis on the amplifier in Figure 8.27. [12]

§8.2

13. Determine the value of Z_{in} for the circuit in Figure 8.28. [14]
14. Determine the value of Z_{in} in the circuit in Figure 8.29. [14]

§8.3

15. Determine the value of A_{vL} for the amplifier in Figure 8.30. [19]
16. Determine the value of A_{vL} for the amplifier in Figure 8.31. [19]
17. Determine the values of Z_{in} and Z_{out} for the amplifier in Figure 8.30. [19]
18. Determine the values of Z_{in} and Z_{out} for the amplifier in Figure 8.31. [19]

§8.6

19. Determine the values of I_E and R_{IN1} for the amplifier in Figure 8.32. [24]

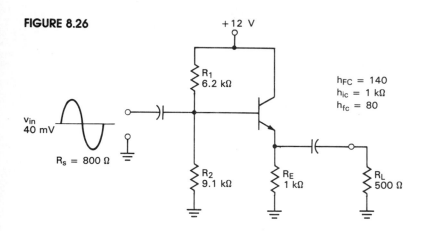

FIGURE 8.26

$+12$ V

R_1 6.2 kΩ

$h_{FC} = 140$
$h_{ic} = 1$ kΩ
$h_{fc} = 80$

v_{in} 40 mV

$R_s = 800$ Ω

R_2 9.1 kΩ

R_E 1 kΩ

R_L 500 Ω

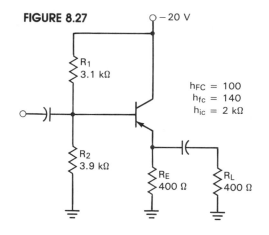

FIGURE 8.27

-20 V

R_1 3.1 kΩ

$h_{FC} = 100$
$h_{fc} = 140$
$h_{ic} = 2$ kΩ

R_2 3.9 kΩ

R_E 400 Ω

R_L 400 Ω

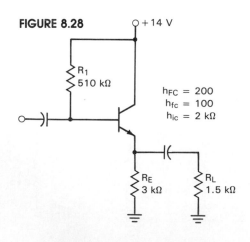

FIGURE 8.28

$+14$ V

R_1 510 kΩ

$h_{FC} = 200$
$h_{fc} = 100$
$h_{ic} = 2$ kΩ

R_E 3 kΩ

R_L 1.5 kΩ

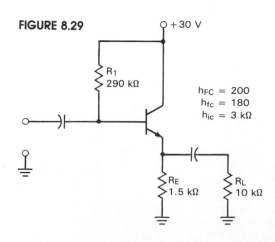

FIGURE 8.29

$+30$ V

R_1 290 kΩ

$h_{FC} = 200$
$h_{fc} = 180$
$h_{ic} = 3$ kΩ

R_E 1.5 kΩ

R_L 10 kΩ

20. Determine the values of I_E and R_{IN1} for the amplifier in Figure 8.33. [24]
21. Determine the value of A_i for the circuit in Figure 8.32. [25]
22. Determine the value of A_i for the circuit in Figure 8.33. [25]
23. Determine the values of Z_{in} and Z_{out} for the circuit in Figure 8.32. [25]
24. Determine the values of Z_{in} and Z_{out} for the circuit in Figure 8.33. [25]
25. Perform a "quick" analysis on the amplifier in Figure 8.32. [27]
26. Perform a "quick" analysis on the amplifier in Figure 8.33. [27]

FIGURE 8.30

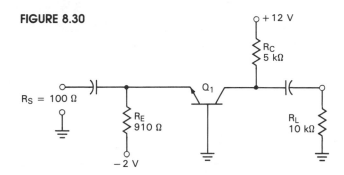

FIGURE 8.31

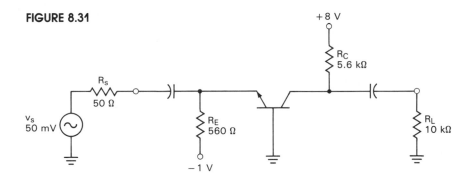

FIGURE 8.32

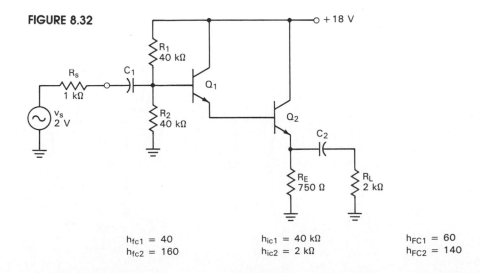

| $h_{fc1} = 40$ | $h_{ic1} = 40\ k\Omega$ | $h_{FC1} = 60$ |
| $h_{fc2} = 160$ | $h_{ic2} = 2\ k\Omega$ | $h_{FC2} = 140$ |

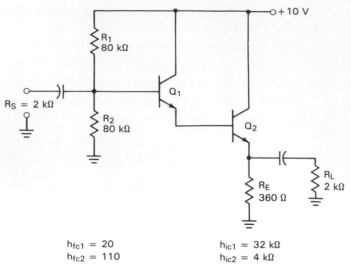

| $h_{fc1} = 20$ | $h_{ic1} = 32\ k\Omega$ | $h_{FC1} = 50$ |
| $h_{fc2} = 110$ | $h_{ic2} = 4\ k\Omega$ | $h_{FC2} = 110$ |

FIGURE 8.33

TROUBLESHOOTING PRACTICE PROBLEMS

27. Refer to Figure 8.25. The circuit has a constant dc emitter voltage of 0 V. Discuss the possible causes of the problem.

28. Refer to Figure 8.26. The dc emitter voltage for this circuit is approximately 1.7 V. Discuss the possible causes for this problem.

29. The circuit in Figure 8.34 has the waveforms shown in Figure 8.35. Discuss the possible causes of the problem.

30. The circuit shown in Figure 8.30 has a constant output of $+12\ V_{dc}$. Discuss the possible causes of the problem.

FIGURE 8.34

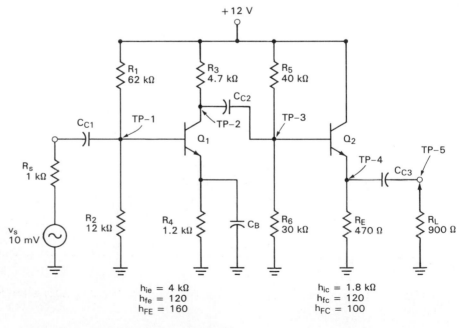

$h_{ie} = 4\ k\Omega$	$h_{ic} = 1.8\ k\Omega$
$h_{fe} = 120$	$h_{fc} = 120$
$h_{FE} = 160$	$h_{FC} = 100$

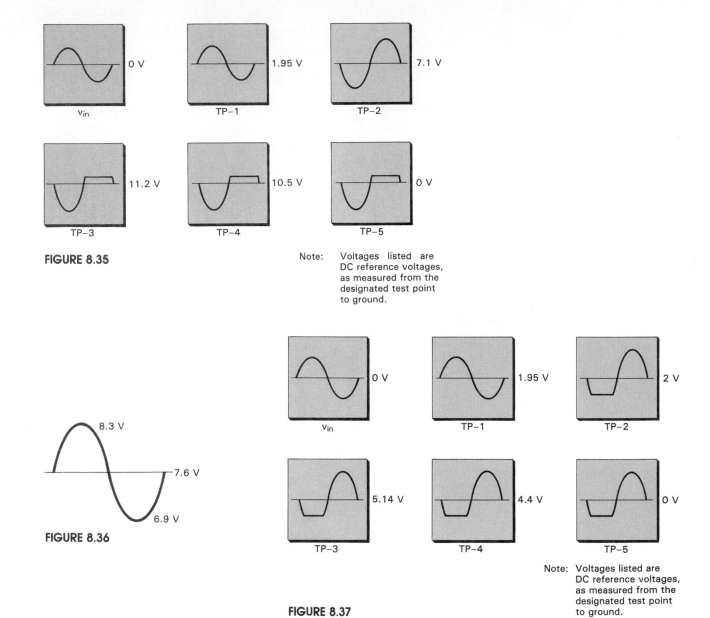

FIGURE 8.35

Note: Voltages listed are DC reference voltages, as measured from the designated test point to ground.

FIGURE 8.36

FIGURE 8.37

Note: Voltages listed are DC reference voltages, as measured from the designated test point to ground.

31. The circuit shown in Figure 8.33 has a dc emitter voltage (Q_2) of approximately 0 V. Discuss the possible causes of the problem.

32. The circuit shown in Figure 8.32 has the output waveform shown in Figure 8.36. Discuss the possible causes of the problem.

33. The circuit shown in Figure 8.34 has the signals shown in Figure 8.37. Determine the possible causes of the problem.

34. The circuit in Figure 8.34 has the output waveforms shown in Figure 8.38. Determine the possible causes of the problem.

THE BRAIN DRAIN

35. Determine the values of A_{vT}, A_{iT}, A_{pT}, and Z_{in} for the amplifier shown in Figure 8.34.

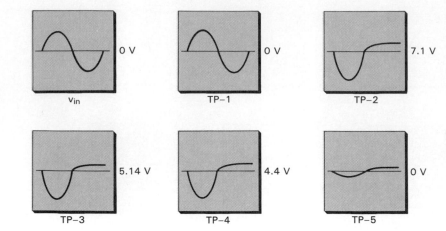

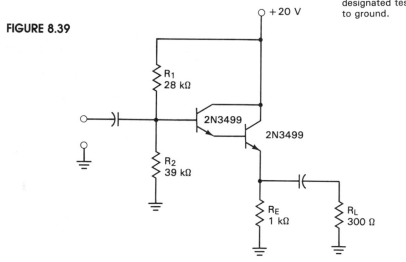

FIGURE 8.38

Note: Voltages listed are DC reference voltages, as measured from the designated test point to ground.

FIGURE 8.39

$h_{FC1} = h_{FC2} = 50$

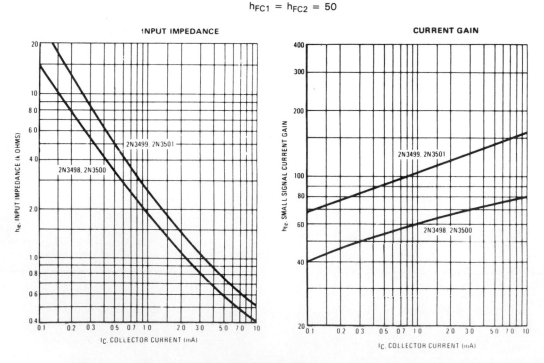

CHAP. 8   Other Transistor Configurations

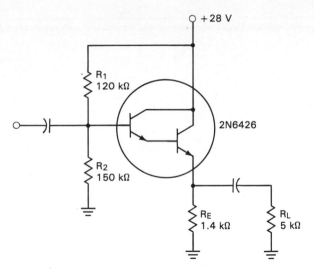

FIGURE 8.40

36. Determine the values of A_i, Z_{in}, and Z_{out} for the amplifier in Figure 8.39 using the "quick" analysis equations. The h_{ie} and h_{fe} curves for the two transistors are shown in the figure.

37. Determine the values of A_{vL}, A_i, A_p, Z_{in}, and Z_{out} for the amplifier in Figure 8.40. The spec sheet for the Darlington transistor is located in Appendix A. Use the "quick" analysis approximations in your analysis.

SUGGESTED COMPUTER APPLICATIONS PROBLEMS

38. Write a program that will perform the dc analysis of a standard emitter follower or a Darlington emitter follower, given the required inputs.

39. Write a program that will perform the ac analysis of a standard emitter follower, given the required input values.

40. Write a program that will perform the "quick" analysis of a Darlington emitter follower, given the required input values.

ANSWERS TO THE EXAMPLE PRACTICE PROBLEMS

8.1. $V_B = 9.92$ V, $V_E = 9.22$ V, $I_E = 10.13$ mA

8.2. $V_{CE(Off)} = 18$ V, $I_{C(Sat)} = 19.78$ mA

8.3. 0.9973

8.4. 239.4

8.5. 0.9967

8.6. 8.73 kΩ

8.7. 15.54 Ω

8.9. $A_{vL} = 47.71$, $A_i = 1$, $Z_{in} \cong 52.4$ Ω, $Z_{out} \cong 15$ kΩ

8.10. 2.25 MΩ

8.11. $A_i = 14,400$, $A_v = 0.9582$, $Z_{in} \cong 118.2$ kΩ, $Z_{out} = 25.2$ Ω

9
Power Amplifiers

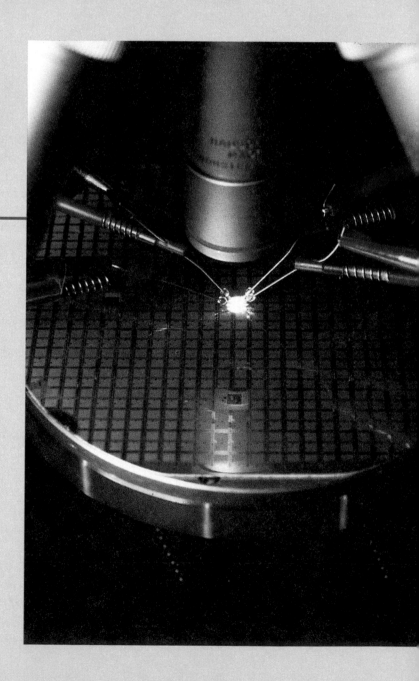

The final stage of device production is testing one of a group of devices for proper operation.

OBJECTIVES

After studying the material in this chapter, you should be able to:

☐ 1. Discuss the concept of amplifier *efficiency*. (Introduction)

☐ 2. List the three primary amplifier classifications and state the conduction characteristics of each. (§9.1)

☐ 3. State the maximum theoretical efficiency ratings for class A, class B, and class C amplifiers. (§9.1)

☐ 4. Compare and contrast the *ac load line* and *dc load line* of an amplifier. (§9.2)

☐ 5. Explain the concept of amplifier *compliance*. (§9.2)

☐ 6. Calculate the compliance of an *RC*-coupled class A amplifier. (§9.2)

☐ 7. Calculate the amount of power that a given *RC*-coupled A amplifier will draw from its dc power supply. (§9.3)

☐ 8. Calculate the value of ac load power for an *RC*-coupled class A amplifier. (§9.3)

☐ 9. Describe the relationship between maximum load power and amplifier compliance. (§9.3)

☐ 10. Explain why high amplifier efficiency ratings are desirable in power amplifiers. (§9.3)

☐ 11. Calculate the efficiency of an *RC*-coupled class A amplifier. (§9.3)

☐ 12. Describe the *transformer-coupled class A amplifier*. (§9.4)

☐ 13. List the characteristics of step-down and step-up transformers. (§9.4)

☐ 14. Describe *counter emf* and explain how it is produced. (§9.4)

☐ 15. Describe the dc operating characteristics of the transformer-coupled class A amplifier. (§9.4)

☐ 16. Plot the ac load line for a given transformer-coupled class A amplifier. (§9.4)

☐ 17. State the maximum theoretical efficiency of a transformer-coupled class A amplifier and explain why the actual efficiency rating is always lower than the theoretical rating. (§9.4)

☐ 18. Calculate the maximum load power and efficiency of a transformer-coupled class A amplifier. (§9.4)

☐ 19. List and discuss the two advantages that transformer-coupled class A amplifiers have over *RC*-coupled class A amplifiers. (§9.4)

☐ 20. List the advantages and disadvantages of class B amplifiers. (§9.5)

☐ 21. Compare and contrast *push–pull amplifiers* and *complementary-symmetry amplifiers*. (§9.5)

☐ 22. Discuss the relationship between class B operation and *crossover distortion*. (§9.5)

☐ 23. Describe the dc operating characteristics of the complementary-symmetry amplifier. (§9.5)

☐ 24. Describe the ac operating characteristics of the complementary-symmetry amplifier. (§9.5)

☐ 25. Plot the dc and ac load lines for a given complementary-symmetry amplifier. (§9.5)

☐ 26. Perform all the calculations necessary to determine the values of source power and load power for a complementary-symmetry amplifier. (§9.5)

☐ 27. Calculate the efficiency of a complementary-symmetry amplifier and explain why it is less than the maximum theoretical value. (§9.5)

☐ 28. List the problems that can occur in standard class B amplifiers and identify the biasing circuit that is used to eliminate those problems. (§9.6)

☐ 29. Explain how diode bias produces the required values of V_B for the amplifier transistors. (§9.6)

☐ 30. Define and describe *class AB* operation. (§9.6)

☐ 31. Explain how the class AB amplifier eliminates crossover distortion. (§9.6)

☐ 32. Explain how the class AB amplifier eliminates thermal runaway. (§9.6)

☐ 33. Perform the complete analysis of a class AB amplifier. (§9.6)

☐ 34. Discuss the procedure used to troubleshoot class AB amplifiers. (§9.7)

☐ 35. Describe the construction of, and purpose served by, each of the following amplifiers: the *Darlington complementary-symmetry amplifier*, the *transistor-biased amplifer*, and the *split-supply class AB amplifier*. (§9.7)

☐ 36. Calculate the transistor power dissipation requirements for a given class A, class B, or class AB amplifier. (§9.8)

☐ 37. State the purpose of *heat sinks* and *heat-sink compound*, and describe the procedure used for the proper replacement of transistors that are connected to *heat sinks*. (§9.8)

As you master each of the objectives listed, place a check mark (✔) in the appropriate box.

P ower amplifiers are used to deliver a relatively high amount of power, usually to a low resistance load. Typical load values range from 300 Ω (for transmission antennas) to 8 Ω (for audio speakers). Although these load values do not cover every possibility, they do illustrate the fact that

In this chapter, you are going to be studying a group of amplifiers referred to as *power amplifiers*. Most of the amplifier circuits and principles covered in this chapter are used in *communications electronics*. Communications electronics is the study of the circuits and systems that are used to transmit information from one point to another. Television and radio transmitters and receivers, two-way transceivers, and radar systems are all examples of communications systems.

There are three basic types of power amplifiers: class A, class B, and class C power amplifiers. In this chapter, we will concentrate on the first two, the class A and class B amplifiers. These two types of circuits are used primarily in *audio amplifiers*. Audio amplifiers are circuits that are used to amplify signals whose frequencies lie in the range of approximately 2 to 20 kHz.

power amplifiers usually drive low-resistance loads. The typical output power rating of a power amplifier will be 1 W or higher.

1

The ideal power amplifier will deliver 100% of the power it draws from the supply to the load. In practice, however, this can never occur. The reason for this is the fact that the components in the amplifier will all dissipate *some* of the power that is being drawn from the supply. For example, consider the circuit shown in Figure 9.1. Assuming that the amplifier is operating normally, current will constantly be flowing through both base resistors, the emitter resistor, the collector resistor, and the transistor itself. Since all these components have current flowing through them and some amount of voltage across them, they all dissipate a measurable amount of power. If you add all these power values together, you will get the total amount of power being dissipated by the amplifier. The difference between this total value and the total power being drawn from the supply is the power that actually goes to the load.

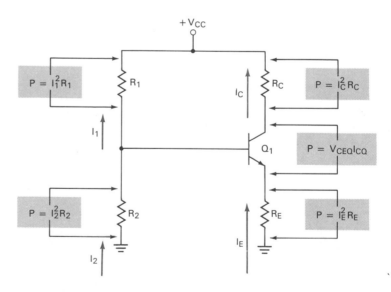

FIGURE 9.1

Amplifier power dissipation.

Efficiency (η). The ratio of *ac* output power to *dc* input power, in percent.

A *figure of merit* for the power amplifier is its *efficiency*, η. The efficiency of an amplifier is the *amount of power drawn from the supply that is actually delivered to the load*, given as a percent. By formula,

CHAP. 9   Power Amplifiers

$$\eta = \frac{\text{ac output power}}{\text{dc input power}} \times 100\% \qquad \textbf{(9.1)}$$

As you will see, certain amplifier configurations have much higher efficiency ratings than others. This is a primary consideration when deciding which type of power amplifier to use for a specific application.

9.1
Amplifier Classifications

The common-emitter, emitter-follower, and common-base amplifier covered in Chapters 6 and 7 all have one thing in common: the transistor in each amplifier conducts during the entire 360° of the ac input cycle. In other words, during the entire cycle of the ac input signal, the transistor never goes into cutoff.

Any amplifer having a transistor that conducts during the entire 360° of the input cycle is referred to as a *class A amplifier*. In contrast, some amplifiers are designed so that their transistors conduct for less than 360° of the ac input cycle. Most of these amplifiers are classified as being either *class B* or *class C amplifiers*.

The *class B* amplifier contains *two* transistors that are typically connected as shown in Figure 9.2a. Each transistor in this amplifier conducts for approximately 180° of the ac input cycle. With one transistor conducting during the negative alternation of the ac input and the other conducting during the positive alternation, a complete 360° output waveform is produced. The *class AB* amplifier, a variation on the basic class B amplifier, produces an output waveform in a similar fashion.

Class A amplifier. An amplifier containing a transistor that conducts during the entire cycle of the ac input.

Class B amplifier. An amplifier containing two transistors that each conduct for 180° of the ac input cycle.

Class C amplifier. An amplifier containing a transistor that conducts for less than 180° of the ac input cycle.

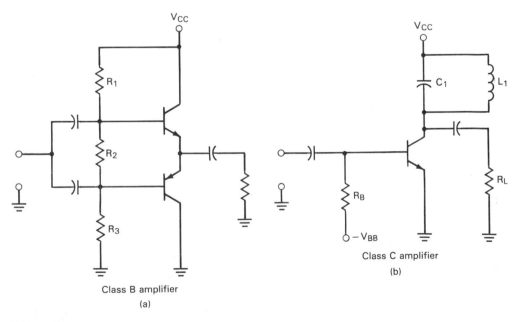

Class B amplifier
(a)

Class C amplifier
(b)

FIGURE 9.2

The *class C* amplifier, shown in Figure 9.2b, contains a single transistor that conducts for less than 180° of the ac input cycle. The rest of the output waveform is produced by the reactive components in the transistor collector circuit. The class C amplifier is easily recognized by the *LC* collector circuit.

Efficiency Ratings

The maximum theoretical efficiency ratings of class A, B, and C amplifiers are summarized as follows:

Amplifier	Maximum Theoretical Efficiency (%)
Class A	
RC coupled	25
Transformer coupled	50
Class B, AB	78.5
Class C	99 (approximate)

As you can see, the class C amplifier has the highest efficiency rating, while the class A amplifier has the poorest. Also, the maximum efficiency rating of the class A amplifier depends on the type of coupling used between the amplifier and the load. This point is discussed further in Section 9.3.

It would seem at first that a class C amplifier would be the natural choice for every power amplifier application. However, the class C amplifier has operating limitations that make it useless in many applications. This point is covered in Chapter 15.

SECTION REVIEW

1. For the *ideal* amplifier, how much of the power being drawn from the source is actually delivered to the load? [Objective 1]
2. Why can't the ideal value in question 1 be obtained? [Objective 1]
3. What is *efficiency*? How is it calculated? [Objective 1]
4. What are the three common amplifier classifications? [Objective 2]
5. What are the conduction characteristics of the *class A* amplifier? [Objective 2]
6. What are the conduction characteristics of the *class B* amplifier? [Objective 2]
7. What are the conduction characteristics of the *class C* amplifier? [Objective 2]
8. What are the ideal efficiency ratings of each of the amplifier classifications? [Objective 3]

9.2
The AC Load Line

AC load line. A graph that represents all possible combinations of i_c and v_{ce}.

In Chapter 6, you were introduced to the dc load line. This line is used to represent all possible combinations of V_{CE} and I_C for a given amplifier. The *ac load line* is used for the same basic purpose: It represents all possible ac combinations of i_c and v_{ce}.

Up to this point, the ac load line has not been an important consideration. However, the ac load line serves as an effective tool for teaching many of the principles of class A and class B amplifier operation. For this reason, we will take a brief look at the ac load line before covering any of the common power amplifier circuits.

The ac load line for a given amplifier will *not* follow the plot of the dc load line. In other words, the two load lines will have different values of $I_{C(sat)}$ and $V_{CE(off)}$. The reason for this lies in the fact that the dc load of an amplifier is different from the ac load. For example, consider the circuit shown in Figure 9.3. In terms of dc operation, the load on the transistor is equal to R_C. The value of R_L does not weigh into the picture because the coupling capacitor provides dc isolation between the transistor and R_L. The ac equivalent circuit (Figure 9.3b) shows the ac load to consist of R_C in parallel with R_L. As you will see, this change in the load resistance has a considerable effect on the ac load line.

Why dc and ac load lines are different.

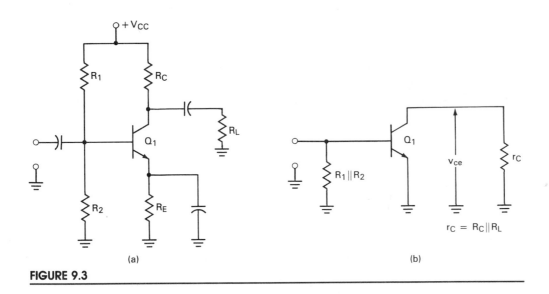

FIGURE 9.3

AC Saturation Current

If you take a close look at the ac equivalent circuit in Figure 9.3b, you will notice that the transistor is in parallel with the ac load resistance, r_C. Since the ac collector current (i_c) flows through this resistance, the following Ohm's law relationship must hold true:

$$i_c = \frac{v_{ce}}{r_C} \tag{9.2}$$

Now take a look at Figure 9.4a. As illustrated, the value of i_c is equal to a given change in I_C. This is expressed as

$$i_c = I_C - I_{CQ}$$

where i_c = the *change in* collector current
I_C = the peak value of ac collector current
I_{CQ} = the *quiescent* value of I_C

Figure 9.4b shows that v_{ce} is equal to the change in V_{CE}. By formula,

$$v_{ce} = V_{CEQ} - V_{CE}$$

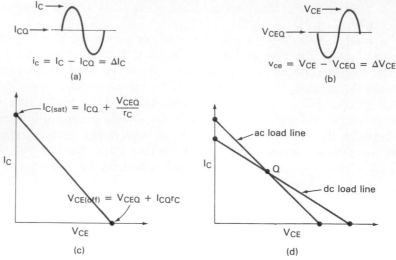

FIGURE 9.4

The ac load line.

where v_{ce} = a *change in* collector–emitter voltage

V_{CE} = the minimum value of collector–emitter voltage

V_{CEQ} = the *quiescent* value of V_{CE}

If we substitute these two equations in place of i_c and v_{ce} in equation (9.2), we get

$$I_C - I_{CQ} = \frac{V_{CEQ} - V_{CE}}{r_c}$$

or

$$\boxed{I_C - I_{CQ} = \frac{V_{CEQ}}{r_C} - \frac{V_{CE}}{r_C}} \qquad (9.3)$$

At saturation, V_{CE} is ideally equal to 0 V. Substituting this into equation (9.3) gives us

$$\boxed{I_{C(\text{sat})} = I_{CQ} + \frac{V_{CEQ}}{r_C}} \qquad (9.4)$$

As Figure 9.4c shows, this is the saturation point for the ac load line.

$V_{CE(\text{off})}$

The equation for $V_{CE(\text{off})}$ is derived from equation (9.3). At cutoff, the value of I_C is ideally zero. Substituting this value into equation (9.3) gives us

$$\boxed{V_{CE(\text{off})} = V_{CEQ} + I_{CQ}r_c} \qquad (9.5)$$

As Figure 9.4c shows, this is the cutoff point for the ac load line. Note that the ac load line crosses the dc load line at the Q-point, as is shown in Figure 9.4d. This will always be the case for a given amplifier.

What Does the AC Load Line Tell You?

The ac load line is used to tell you the *maximum possible output voltate swing for a given common-emitter amplifier*. In other words, the ac load line will tell you the maximum possible peak-to-peak output voltage from a given amplifier. This maximum V_{PP} is referred to as the *compliance* of the amplifier.

Compliance. The maximum V_{PP} output of an amplifier.

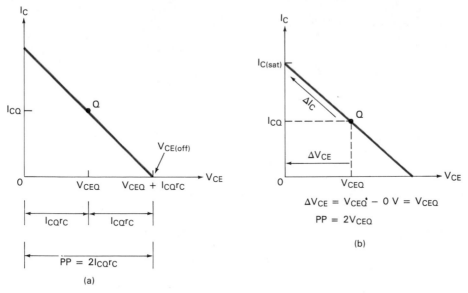

FIGURE 9.5

Amplifier compliance.

The compliance of an amplifier is found by determining the maximum possible of I_C and V_{CE} from their respective values of I_{CQ} and V_{CEQ}. The maximum possible transitions are illustrated in Figure 9.5. As Figure 9.5a shows, the maximum possible transition for V_{CE} is equal to the difference between $V_{CE(off)}$ and V_{CEQ}. Since this transition is equal to $I_{CQ}r_C$, the maximum peak output voltage from the amplifier is equal to $I_{CQ}r_C$. Two times this value will give us the maximum peak-to-peak transition of the output voltage, as follows:

$$PP = 2I_{CQ}r_C \qquad (9.6)$$

where PP = the output compliance, in peak-to-peak voltage
    I_{CQ} = the *quiescent* value of I_C
    r_C = the ac resistance in the emitter circuit

When $I_C = I_{C(sat)}$, V_{CE} is ideally equal to 0 V. When $I_C = I_{CQ}$, V_{CE} is at V_{CEQ}. These two points are illustrated in Figure 9.5b. Note that when I_C makes its maximum possible transition (from I_{CQ} to $I_{C(sat)}$), the output voltage changes by an amount equal to V_{CEQ}. Thus, the maximum peak-to-peak transition would be equal to twice this value, as follows:

$$PP = 2V_{CEQ} \qquad (9.7)$$

where PP = the output compliance, in peak-to-peak voltage
    V_{CEQ} = the *quiescent* value of V_{CE}

Having two different equations for compliance may seem a bit confusing at first, but it really isn't so strange. Equation (9.6) sets the limit in terms of $V_{CE(off)}$. In other words, if you exceed the value obtained by the equation, the output

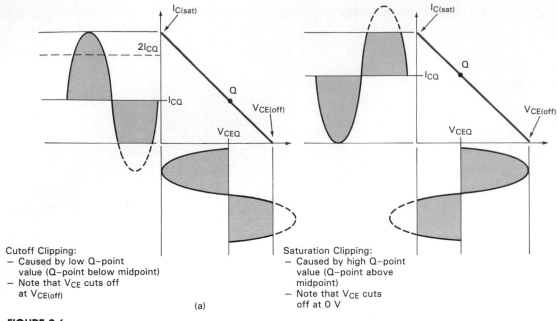

Cutoff Clipping:
– Caused by low Q–point
  value (Q–point below midpoint)
– Note that V_{CE} cuts off
  at $V_{CE(off)}$

Saturation Clipping:
– Caused by high Q–point
  value (Q–point above
  midpoint)
– Note that V_{CE} cuts
  off at 0 V

(a)

FIGURE 9.6

Cutoff and saturation clipping.

Cutoff clipping. A type of distortion caused by driving a transistor into cutoff.

Saturation clipping. A type of distortion caused by driving a transistor into saturation.

voltage will try to exceed $V_{CE(off)}$, which it cannot do. The result of exceeding the value obtained by equation (9.6) is shown in Figure 9.6a. This is called *cutoff clipping*, because the output voltage is clipped off at the value of $V_{CE(off)}$.

Equation (9.7) sets the limits in terms of $I_{C(sat)}$. If you exceed the value obtained by this equation, the output will experience *saturation clipping*. Saturation clipping is illustrated in Figure 9.6b.

When determining the output compliance for a given amplifier, solve both equations (9.6) and (9.7). *The lower of the two results is the compliance of the amplifier.*

═══════ **EXAMPLE 9.1** ═══════

Determine the maximum output compliance for the amplifier shown in Figure 9.7.

Solution: Using the established procedure, I_{CQ} and V_{CEQ} are found to be 950 µA and 5.45 V, respectively. (*Refer to Chapter 6 if you have trouble solving for these values.*) The value of r_C is equal to the parallel combination of R_C and R_L: 3.2 kΩ. Now the amplifier is solved for *both* compliance values, as follows:

$$PP = 2V_{CEQ}$$
$$= 2(5.45 \text{ V})$$
$$= 10.9 \text{ V}_{PP}$$

and

$$PP = 2I_{CQ}r_C$$
$$= 2(950 \text{ µA})(3.2 \text{ kΩ})$$
$$= 6.08 \text{ V}_{PP}$$

$h_{fe} = h_{FE} = 200$

FIGURE 9.7

(Circuit: +12 V, R_1 33 kΩ, R_C 4.7 kΩ, R_2 10 kΩ, R_E 2.2 kΩ, R_L 10 kΩ, Q_1)

Since the overall compliance is always equal to the *smaller* of the two values obtained, the maximum PP for this amplifier is equal to 6.08 V_{PP}. This means that the peak-to-peak output voltage can be no more than 6.08 V. If this value is exceeded, cutoff clipping will occur.

PRACTICE PROBLEM 9–1

A common-emitter amplifier has the following values: $V_{CC} = 20$ V, $R_1 = 10$ kΩ, $R_2 = 1.8$ kΩ, $R_C = 620$ Ω, $R_E = 200$ Ω, $R_L = 1.2$ kΩ, and $h_{FE} = 180$. Determine the compliance (PP) of the amplifier.

A Practical Consideration

In Chapter 7, you were introduced to the concept of *nonlinear distortion*. In many case, nonlinear distortion will occur long before cutoff clipping will occur. In other words, if the amplifer input amplitude is sufficient to cause cutoff clipping to occur, it is more than sufficient to cause nonlinear distortion in the amplifier output.

When the compliance of an amplifier is limited by cutoff clipping ($2I_{CQ}r_C$), you need to be aware that nonlinear distortion may occur before the compliance of the amplifier is reached. At the same time, if saturation clipping ($2V_{CEQ}$) is the limiting factor, nonlinear distortion will not be a problem.

Determining the compliance of an amplifier is not an everyday task for the average technician. At the same time, you need to be aware that the *ac equivalent circuit* of an amplifier will regulate the output from that amplifier. This will become an important consideration as we continue our discussion on power amplifiers.

---

1. What is the *ac load line*? [Objective 4]
2. Why is the ac load line of an amplifier different from the dc load line? [Objective 4]
3. What is *compliance*? [Objective 5]
4. How is the compliance of an amplifier determined? [Objective 5]

9.3
RC-coupled Class A Amplifiers

We have already performed most of the analyses that are commonly required for *RC*-coupled class A amplifiers. The ac and dc analyses of these amplifiers were covered in Chapters 6 through 8.

The relationships listed on the formula card hold true for the class A power amplifiers, and we will not repeat their derivations or applications here. If you have trouble with any of the standard dc or ac relationships for the common-emitter or emitter-follower amplifiers, refer to the appropriate discussion(s) to review the material. In this section, we will concentrate on the power relationships in the *RC*-coupled class A amplifier.

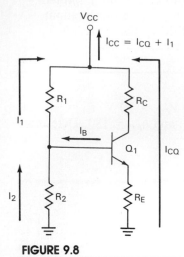

FIGURE 9.8

Total supply current.

Amplifier DC Power

The total dc power that an amplifier draws from the power supply is found as

$$\boxed{P_S = V_{CC}I_{CC}}$$

(9.8)

As Figure 9.8 illustrates, I_{CC} is equal to the sum of I_{CQ} and the current flowing through the transistor base circuit, I_1. The calculation of total dc power is illustrated in the following example.

EXAMPLE 9.2

7

Determine the total dc power being drawn from the supply by the amplifier shown in Figure 9.9.

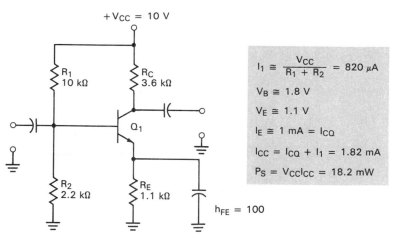

FIGURE 9.9

Solution: The value of I_1 is found as

$$I_1 = \frac{V_{CC}}{R_1 + R_2}$$
$$= 820 \ \mu A$$

Using the established procedure, the value of I_{CQ} is found as

$$I_{CQ} \cong 1 \ mA$$

Now the value of I_{CC} is found as

$$I_{CC} = I_1 + I_{CQ}$$
$$= 1.82 \ mA$$

CHAP. 9 Power Amplifiers

Finally, P_S is found as

$$P_S = V_{CC}I_{CC}$$
$$= 18.2 \text{ mW}$$

PRACTICE PROBLEM 9–2

Determine how much power is being drawn from the dc power supply by the amplifier described in Practice Problem 9–1.

AC Load Power

AC load power can be calculated using the standard V^2/R equation. Specifically,

$$\boxed{P_L = \frac{V_L^2}{R_L}}$$

(9.9)

A Practical Consideration: This equation is used when V_{out} is being measured with an ac voltmeter.

where P_L = the ac load power
V_L = the *rms* load voltage

The application of this equation is illustrated in the following example.

EXAMPLE 9.3

Determine the ac load power for the circuit shown in Figure 9.10.

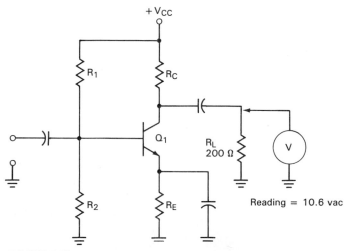

FIGURE 9.10

Solution: The voltmeter reads 10.6 Vac, as indicated. Since ac voltmeters read rms voltage, the value shown can be plugged directly into equation (9.9). Thus

$$P_L = \frac{(10.6 \text{ Vac})^2}{200 \ \Omega}$$
$$= 561.8 \text{ mW}$$

PRACTICE PROBLEM 9–3

An amplifier with a 1.2-kΩ load resistance has an output voltage value of 4.62 Vac. Determine the value of load power for the amplifier.

A Practical Consideration: This equation can be used when V_{out} is being measured with an oscilloscope.

A convenient form of equation (9.9) that can be used when peak output voltage is known is as follows:

$$P_L = \frac{(0.707V_{pk})^2}{R_L} \qquad (9.10)$$

This equation simply converts V_{pk} to an rms value in the numerator. The use of this equation is illustrated in the following example.

EXAMPLE 9.4

Determine the ac load power for the circuit shown in Figure 9.11.

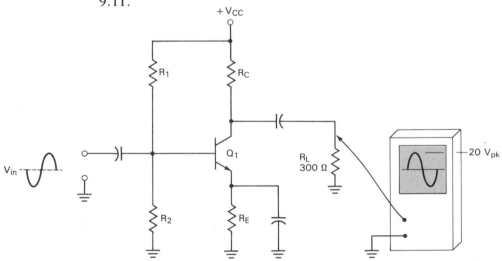

FIGURE 9.11

Solution: The oscilloscope is shown to be measuring 20 V_{pk}. Using equation (9.10), the ac load power is found as

$$
\begin{aligned}
P_L &= \frac{(0.707V_{pk})^2}{R_L} \\
&= \frac{(14.14 \ V_{pk})^2}{300 \ \Omega} \\
&= 666.5 \ \text{mW}
\end{aligned}
$$

PRACTICE PROBLEM 9–4

An amplifier has a 670-Ω load resistance. Using an oscilloscope, the peak output voltage is measured at $+8V_{pk}$. Determine the amount of power that the amplifier is supplying to the load.

When the peak-to-peak output voltage is known, the ac load power can be found as

A Practical Consideration: This equation can be used when V_{out} is measured with an oscilloscope.

$$P_L = \frac{V_{PP}^2}{8R_L}$$ **(9.11)**

This equation is derived by substituting $V_{PP}/2$ in place of V_{pk} in equation (9.10), as follows:

$$P_L = \frac{(0.707V_{pk})^2}{R_L}$$
$$= \frac{(0.3535V_{PP})^2}{R_L}$$
$$= \frac{0.125V_{PP}^2}{R_L}$$

Now, since $0.125 = \frac{1}{8}$, this equation is rewritten as

$$P_L = \frac{V_{PP}^2}{8R_L}$$

Equation (9.11) is important because it allows us to calculate the *maximum possible value of ac load power*. You may recall that the maximum possible output voltage from an amplifier is called *compliance*, PP. Since compliance is a maximum peak-to-peak output value, the maximum possible ac load power can be found as

$$P_{L(max)} = \frac{PP^2}{8R_L}$$ **(9.12)**

The following example illustrates the use of this equation.

EXAMPLE 9.5

A given amplifier has a compliance of 18 V_{PP}. If the load value is 100 Ω, what is the maximum possible ac load power?

Solution: Using the values given, the maximum possible load power is found as

$$P_{L(max)} = \frac{PP^2}{8R_L}$$
$$= \frac{(18 \ V_{PP})^2}{800 \ \Omega}$$
$$= 405 \ mW$$

PRACTICE PROBLEM 9–5

An amplifier has a compliance of 20 V_{PP} and a load resistance of 140 Ω. What is the maximum possible load power for the circuit?

Amplifier Efficiency

When the values of P_S and P_L are known, we can determine the *efficiency rating* of an amplifier. You may recall that the efficiency rating of an amplifier tells us the percentage of the power drawn from the dc power supply that is actually delivered to the load.

The higher the efficiency of an amplifier, the better. Why? Because a high efficiency rating indicates that a very small percentage of the power drawn from the dc power supply is being wasted by the amplifier itself. For example, if an amplifier has an efficiency rating of 20%, then 80% of the power being drawn from the dc power supply is being used by the amplifier itself. On the other hand, if an amplifier has an efficiency rating of 90%, then only 10% of the power being drawn from the source is being used by the amplifier.

As was stated earlier, the maximum theoretical efficiency of an *RC*-coupled class A amplifier is 25%. In practice, the efficiency of this type of amplifier will generally be much lower than the maximum theoretical value of 25%. This point is illustrated in the following example.

■ EXAMPLE 9.6 ■

Determine the maximum efficiency of the amplifier described in Example 9.1

Solution: In the example, we determined the compliance of the amplifier to be PP = 6.08 V. With a 10-kΩ load resistance (shown in Figure 9.7), we can find the maximum value of load power as

$$P_{L(MAX)} = \frac{PP^2}{8R_L}$$
$$= \frac{(6.08\ V_{PP})^2}{80\ k\Omega}$$
$$= 462\ \mu W$$

Using the established procedures, the following values are determined for the amplifier:

$$I_1 = 279.1\ \mu A$$
$$I_{CQ} = 950.3\ \mu A$$

and

$$I_{CC} = I_{CQ} + I_1$$
$$= 1.23\ mA$$

Now P_S is found as

$$P_S = V_{CC}I_{CC}$$
$$= (12\ V)(1.23\ mA)$$
$$= 14.76\ mW$$

Finally,

$$\eta = \frac{P_L}{P_S} \times 100$$
$$= \frac{463\ \mu W}{14.76\ mW} \times 100$$
$$= 3.13\%$$

As you can see, even when the amplifier is driven to compliance its efficiency rating will be only 3.13%. This is considerably less than the maximum theoretical value of 25%.

PRACTICE PROBLEM 9–6

The amplifier described in Practice Problem 9–1 is driven to compliance. Determine the maximum efficiency of the amplifier.

The maximum theoretical efficiency value of 25% for the *RC*-coupled class A amplifier is derived in Appendix D for those who wish to review the derivation. Those who are not interested in knowing where the value of 25% comes from may continue from this point without loss of continuity.

As was stated earlier, the maximum efficiency of the *transformer-coupled* class A amplifier is 50%. The reason for the higher efficiency rating of this amplifier is due to its operating characteristics. We will take a look at this type of amplifier in the next section.

---

1. List, in order, the steps required to determine the value of I_{CC} for an *RC*-coupled class A amplifier. [Objective 7]   **SECTION REVIEW**
2. How do you calculate the value of P_L when V_{out} is measured with an ac voltmeter? [Objective 8]
3. How do you calculate the value of P_L when V_{out} is measured with an oscilloscope? [Objective 8]
4. When is load power (P_L) at its maximum value? [Objective 9]
5. What is the maximum theoretical efficiency of an *RC*-coupled class A amplifier? [Objective 10]
6. Why is a high efficiency rating desirable for a power amplifier? [Objective 10]

9.4
Transformer-coupled Class A Amplifiers

A *transformer-coupled* class A amplifier uses a transformer to couple the output signal from the amplifier to the load. A typical transformer-coupled amplifier is shown in Figure 9.12. The ac characteristics of the transformer shown play an important role in the overall operation of the amplifier. For this reason, we will start our discussion on the amplifier with a summary of the basic transformer operating characteristics.

Transformer-coupled class A amplifier. A class A amplifier that uses a transformer to couple the output signal to the load.

Transformers

You may recall that the *turns ratio* of a transformer determines the relationship between primary and secondary values of voltage, current, and impedance. These

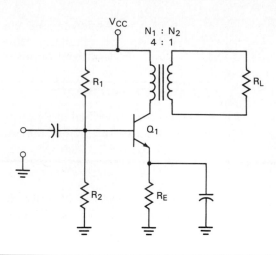

FIGURE 9.12

A transformer-coupled class A amplifier.

Just a Note: The relationships shown in this discussion are covered in detail in your basic electronics textbook. You may want to review the appropriate material from that book. The relationships shown will be reinforced in the circuit analysis examples later in this section.

relationships are summarized as follows:

$$\frac{N_1}{N_2} = \frac{V_1}{V_2} = \frac{I_2}{I_1}$$

(9.13)

and

$$\left(\frac{N_1}{N_2}\right)^2 = \frac{Z_1}{Z_2}$$

(9.14)

where N_1, N_2 = the number of turns in the primary and secondary, respectively
V_1, V_2 = the primary and secondary voltages
I_1, I_2 = the primary and secondary currents
Z_1, Z_2 = the primary and secondary impedances

Step-down transformer. One with a secondary voltage that is less than the primary voltage.

Most transformers are classified as being either a *step-down transformer* or a *step-up* transformer. A step-down transformer is one that has a secondary voltage that is less than the primary voltage. A step-down transformer will always have the following characteristics:

Step-down transformer characteristics.

$$N_1 > N_2$$
$$V_1 > V_2$$
$$I_1 < I_2$$
$$Z_1 > Z_2$$

Step-up transformer. One that has a secondary voltage that is greater than the primary voltage.

A step-up transformer is one that has a secondary voltage that is greater than the primary voltage. A step-up transformer will always have the following characteristics:

Step-up transformer characteristics.

$$N_1 < N_2$$
$$V_1 < V_2$$
$$I_1 > I_2$$
$$Z_1 < Z_2$$

CHAP. 9   Power Amplifiers

As you will see, these relationships play an important role in the ac operation of the transformer-coupled class A amplifier.

Another important characteristic of the transformer is one that is shared by all inductors, that is, the ability to produce a *counter emf*, or *"kick" emf*. When an indicator experiences a rapid change in supply voltage, it will produce a voltage *with a polarity that is opposite to the original voltage polarity*. For example, take a look at the circuit shown in Figure 9.13. Figure 9.13a shows an inductor with 10 V applied to it. Note that point A is shown as being *positive* and point B is shown as being *negative*. Figure 9.13b shows the same inductor with the voltage source removed. The inductor is shown as still having 10 V across the component. However, now point A is *negative* and point B is *positive*. This is the counter *emf* that is produced when there is a rapid change in the inductor supply voltage. Note that the amplitude of the counter emf may be much greater than that of the original applied voltage.

The counter emf shown in Figure 9.13b is caused by the electromagnetic field that surrounds the inductor. When this field collapses, it induces a voltage that is equal to (or greater than) the original supply voltage in value and opposite to the original voltage in polarity. *Note that this counter emf will be present only for an instant.* As the field collapses into the inductor, the voltage *decreases* in value until it eventually reaches 0 V. As you will see, this characteristic also plays an important role in the operation of the transformer-coupled class A amplifier.

We have reviewed all the transformer principles needed to understand the operation of transformer-coupled amplifiers. Now we will take a look at the dc and ac operating characteristics of these circuits.

DC Operating Characteristics

The dc biasing of a transformer-coupled class A amplifier is very similar to any other class A amplifier with one important exception: *The value of V_{CEQ} is designed to be as close as possible to V_{CC}.* Having a V_{CEQ} that is approximately equal to V_{CC} is the only acceptable situation in this amplifier type. The reason for this will become clear in our discussion of the ac characteristics of the amplifier.

Figure 9.14 shows a basic transformer-coupled amplifier and its dc load line. Note that the load line shown for the amplifier is very close to being a verticial line, indicating that V_{CEQ} will be approximately equal to V_{CC} for all the values of I_C shown. The nearly vertical load line of the transformer-coupled amplifier is

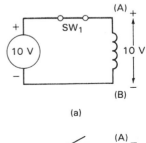

Counter emf. A voltage produced by a transformer that has the opposite polarity of the voltage that caused it. Often referred to as *kick emf*.

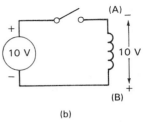

FIGURE 9.13

Counter emf.

$V_{CEQ} \cong V_{CC}$

Why the dc load line for a transformer-coupled class A amplifier is nearly vertical.

FIGURE 9.14

The dc load line of a transformer-coupled circuit.

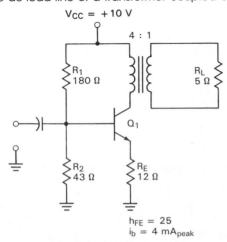

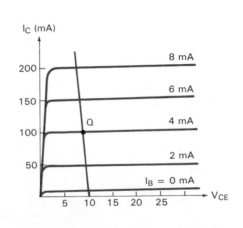

caused by the extremely low dc resistance of the transformer primary. You may recall that V_{CEQ} for an amplifier is found as

$$V_{CEQ} = V_{CC} - I_{CQ}(R_C + R_E)$$

In Figure 9.14, R_E is shown to be 12 Ω. Let's assume for a moment that the primary winding resistance (R_w) of the transformer is 2 Ω. If this is the case, V_{CEQ} at $I_{CQ} = 50$ mA is found as

$$V_{CEQ} = 10 \text{ V} - (50 \text{ mA})(14 \text{ Ω})$$
$$= 9.3 \text{ V}$$

V_{CEQ} at $I_{CQ} = 200$ mA would be found as

$$V_{CEQ} = 10 \text{ V} - (200 \text{ mA})(14 \text{ Ω})$$
$$= 7.2 \text{ V}$$

Thus, the change in V_{CEQ} was only 2.1 V over a range of $I_{CQ} = 50$ to 200 mA. This slight change in V_{CEQ} would produce the nearly vertical dc load line shown.

You should note that the value of R_L would be ignored in the dc analysis of the transformer-coupled class A amplifier. The reason for this is the fact that transformers provide dc isolation between the primary and secondary. Since the load resistance is in the secondary of the transformer, it does not affect the dc analysis of the primary circuitry.

AC Operating Characteristics

16

The first step in analyzing the ac operation of the transformer-coupled amplifer is to plot the ac load line of the circuit. The process for plotting this load line is as follows:

Plotting the ac load line.

1. Determine the maximum possible change in V_{CE}.
2. Determine the corresponding change in I_C.
3. Plot a line that passes through the Q-point and the value of $I_{C(max)}$.
4. Locate the two points where the load line passes through the lines representing the minimum and maximum values of I_B. These two points are then used to find the maximum and minimum values of I_C and V_{CE}.

The first step is simple. Since V_{CE} cannot change by an amount greater than $(V_{CEQ} - 0$ V), the maximum value of ΔV_{CE} is approximately equal to V_{CEQ}. This is illustrated in Figure 9.15. Finding the maximum corresponding change in I_C takes several calculations. First, you must use equation (9.14) to find the value of Z_1 for the transformer. For the circuit represented in Figure 9.14 and 9.15, Z_1 is found as

$$Z_1 = \left(\frac{N_1}{N_2}\right)^2 Z_2$$
$$= 4^2(5 \text{ Ω})$$
$$= 80 \text{ Ω}$$

Now, using the values of ΔV_{CE} and Z_1, ΔI_C is found as

$$\boxed{\Delta I_C = \frac{\Delta V_{CE}}{Z_1}} \tag{9.15}$$

CHAP. 9   Power Amplifiers

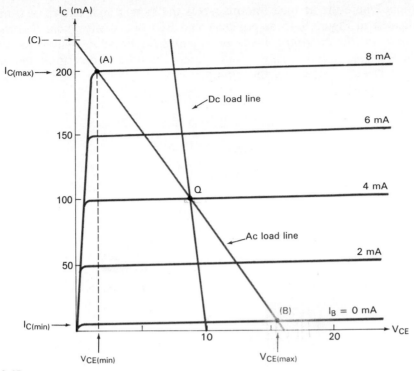

FIGURE 9.15

AC operating characteristics.

For the circuit in Figure 9.14,

$$\Delta I_C = \frac{8.8 \text{ V}}{80 \text{ } \Omega}$$
$$= 110 \text{ mA}$$

The value of $I_{C(\text{max})}$ is now found using

$$\boxed{I_{C(\text{max})} = I_{CQ} + \Delta I_C} \qquad (9.16)$$

For the circuit shown,

$$I_{C(\text{max})} = 100 \text{ mA} + 110 \text{ mA}$$
$$= 210 \text{ mA}$$

The ac load line is now drawn from the point on the y-axis that corresponds to 210 mA (indicated by the dashed arrow, C, in Figure 9.15) through the Q-point to the x-axis. The completed line is shown in Figure 9.15.

Now the minimum and maximum values of I_B are determined. Figure 9.14 indicates a peak ac value of I_B equal to 4 mA. Thus I_B can change by 4 mA both above and below the initial value of I_B. With an initial value of $I_B = 4$ mA, we get the following values:

$$I_{B(\text{max})} = 4 \text{ mA} + 4 \text{ mA} = 8 \text{ mA}$$

and

$$I_{B(\text{min})} = 4 \text{ mA} - 4 \text{ mA} = 0 \text{ mA}$$

The points where the ac load line intersects the $I_B = 8$ mA and $I_B = 0$ mA lines are indicated in Figure 9.15 as points A and B, respectively. *Note that these two points indicate the operating limits of V_{CE} and I_C.* Point A is used to determine the values of $I_{C(max)}$ and $V_{CE(min)}$, while point B is used to determine the values of $I_{C(min)}$ and $V_{CE(max)}$. Thus, for this circuit,

$$I_{C(max)} = 200 \text{ mA}$$

$$I_{C(min)} = 5 \text{ mA}$$

$$V_{CE(max)} = 16.5 \text{ V}$$

$$V_{CE(min)} = 1.5 \text{ V}$$

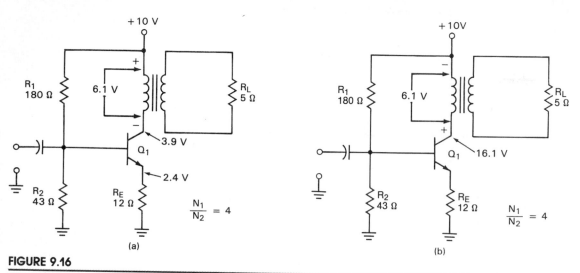

FIGURE 9.16

The effects of counter emf.

At this point, you may be wondering how we managed to get a value of $V_{CE(max)}$ that is greater than the value of V_{CC} for the circuit. The high value of $V_{CE(max)}$ is caused by the *counter emf* produced by the transformer primary. This point is illustrated by the circuits in Figure 9.16, which show the circuit conditions that correspond to points A and B in Figure 9.15. When the circuit is operating at point A, the value of I_C is approximately 200 mA. With 200 mA flowing through R_E, V_E is equal to 2.4 V. As indicated above, the value of $V_{CE(min)}$ is 1.5 V. This means that there is a total of 3.9 V from the collector of the transistor to ground, leaving 6.1 V across the primary of the transformer. Figure 9.16b shows what happens when I_B and I_C drop to a minimum. At the instant that $I_{C(min)}$ (point B) is reached, the 6.1 V is still across the primary of the transformer, but the voltage polarity has reserved. This means that the bottom side of the transformer is now 6.1 V *more positive* than V_{CC}. Thus the voltage from the transistor collector to ground is equal to 16.1 V. This is very close to the 16.5 V derived from the characteristic curves.

As you can see, a transformer-coupled amplifier will have two very important characteristics:

1. V_{CEQ} will be very close to the value of V_{CC}.
2. The maximum output voltage will be very close to $2V_{CEQ}$. Thus it can approach the value of $2V_{CC}$.

CHAP. 9   Power Amplifiers

As you will see, these characteristics are the bases for the higher efficiency rating for the transformer-coupled amplifier.

Amplifier Efficiency

✔ 17

As was stated earlier, the maximum theoretical efficiency rating of the transformer-coupled class A amplifier is 50%. The derivation of this value is shown in Appendix D.

The actual efficiency rating of a transformer-coupled class A amplifier will generally be less than 40%. There are several reasons for the difference between the practical and theoretical efficiency ratings for the amplifier:

1. The derivation of the $\eta = 50\%$ value assumes that $V_{CEQ} = V_{CC}$. In practice, V_{CEQ} will always be some value that is less than V_{CC}.
2. The transformer is subject to various power losses, as you were taught in your study of ac electronics. Among these losses are *copper loss* and *hysteresis loss*. These transformer power losses are not considered in the derivation of the $\eta = 50\%$ value.

Calculating Maximum Load Power and Efficiency

✔ 18

To calculate the maximum load power for a transformer-coupled class A amplifier, you start by determining the value of I_{CQ} for the circuit, using the established procedure for the common-emitter circuit. Once the value of I_{CQ} is known, we can approximate the value of source power as

$$P_S = V_{CC}I_{CQ} \qquad (9.17)$$

This approximation is valid because $I_{CQ} \gg I_1$ in a typical transformer-coupled class A amplifier. Once the value of P_S is known, we have to calculate the maximum load power for the amplifier.

A transformer-coupled amplifier will have a maximum peak-to-peak output that is approximately equal to the difference between the values of $V_{CE(\max)}$ and $V_{CE(\min)}$ that are found using the ac load line for the circuit. Once these values have been found,

$$PP = V_{CE(\max)} - V_{CE(\min)} \qquad (9.18)$$

A Practical Consideration: The value of PP for a transformer-coupled class A amplifier will be approximately equal to the $V_{CE(\text{off})}$ value found with the ac load line.

The value of compliance (PP) found in equation (9.18) indicates the maximum possible peak-to-peak voltage *across the primary of the transformer*. Using this value and the turns ratio of the transformer, the maximum possible peak-to-peak load voltage is found as

$$V_{PP} = \frac{N_2}{N_1} PP \qquad (9.19)$$

Once the value of peak-to-peak load voltage (V_{PP}) is known, the values of load power and efficiency are calculated as shown in Section 9.3. The entire process for calculating the maximum efficiency of a transformer-coupled class A amplifier is illustrated in the following example.

▬ EXAMPLE 9.7 ▬

Determine the maximum efficiency of the amplifier in Figure 9.16.

Don't Forget: The value of I_{CQ} is found by finding V_B, V_E, and then I_E. The value of I_{CQ} is then assumed to be equal to I_E.

Solution: Using the established procedure, the value of I_{CQ} is found to be 102 mA. If you refer to Figure 9.15, you'll see that the calculated value of $I_{CQ} = 102$ mA is very close to the load line value of $I_{CQ} = 100$ mA.

The value of P_S is now found as

$$P_S = V_{CC}I_{CQ}$$
$$= (10 \text{ V})(102 \text{ mA})$$
$$= 1.02 \text{ W}$$

The compliance of the amplifier is now found as

$$PP = V_{CE(\text{max})} - V_{CE(\text{min})}$$
$$= 16.5 \text{ V} - 1.5 \text{ V}$$
$$= 15 \text{ V}_{PP}$$

Note that the values of $V_{CE(\text{max})}$ and $V_{CE(\text{min})}$ were found using the ac load line in Figure 9.15.

Once the compliance of the amplifier is known, the maximum peak-to-peak load voltage is found as

$$V_{PP} = \frac{N_2}{N_1} PP$$
$$= \frac{1}{4}(15 \text{ V}_{PP})$$
$$= 3.75 \text{ V}_{PP}$$

Now the maximum load power is found using equation (9.11) as follows:

$$P_{L(\text{max})} = \frac{V_{PP}^2}{8R_L}$$
$$= \frac{(3.75 \text{ V}_{PP})^2}{40}$$
$$= 351.56 \text{ mW}$$

Finally, the efficiency rating of the amplifier is found as

$$\eta = \frac{P_L}{P_S} \times 100$$
$$= \frac{351.56 \text{ mW}}{1.02 \text{ W}} \times 100$$
$$= 36.46\%$$

The maximum efficiency found here would be slightly higher than the actual efficiency of the amplifier, since the calculation does not take transformer losses into account.

PRACTICE PROBLEM 9–7

A transformer-coupled class A amplifier has the following values:
$V_{CE(max)} = 22.4$ V, $V_{CE(min)} = 2.4$ V, $N_1 = 5$, $N_2 = 1$, $R_L = 4$ Ω, $V_{CC} = 12$ V, and $I_{CQ} = 120$ mA. Calculate the maximum efficiency of the circuit.

One Final Note

One of the primary advantages of using the transformer-coupled class A amplifer is the increased efficiency over the *RC*-coupled class A circuit. Another advantage is the fact that the transformer-coupled amplifier is easily converted into a type of amplifier that is used extensively in communications: the *tuned amplifer*. A tuned amplifier is a circuit that is designed to have a specific value of power gain over a specific range of frequencies. Tuned amplifiers are discussed in Chapter 15.

In the next section, we're going to move on to the class B amplifier. A summary of the two class A amplifiers is shown as a review in Figure 9.17.

Tuned amplifier. A circuit that is designed to have a specific value of power gain over a specified range of frequencies.

FIGURE 9.17

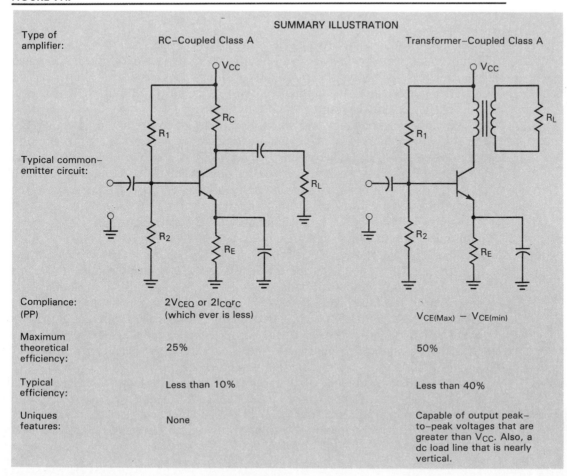

SUMMARY ILLUSTRATION

Type of amplifier:	RC–Coupled Class A	Transformer–Coupled Class A
Compliance: (PP)	2V_CEQ or 2I_CQr_C (which ever is less)	V_CE(Max) − V_CE(min)
Maximum theoretical efficiency:	25%	50%
Typical efficiency:	Less than 10%	Less than 40%
Uniques features:	None	Capable of output peak–to–peak voltages that are greater than V_CC. Also, a dc load line that is nearly vertical.

1. What is a *transformer-coupled class A amplifier*? [Objective 12]
2. List the characteristics of the step-down transformer. [Objective 13]
3. List the characteristics of the step-up transformer. [Objective 13]
4. Describe *counter emf* and its causes. [Objective 14]
5. Describe the dc load line of a transformer-coupled amplifier. [Objective 15]
6. What is the process for plotting the ac load line of a transformer-coupled class A amplifier? [Objective 16]
7. What is the maximum theoretical efficiency of the transformer-coupled class A amplifier? Why is the actual efficiency alway less than this value? [Objective 17]
8. List, in order, the steps required to calculate the maximum load power and efficiency rating of a transformer-coupled class A amplifier. [Objective 18]
9. What are the advantages of using the transformer-coupled class A amplifier? [Objective 19]
10. What is a tuned amplifier? [Objective 19]

9.5
Class B Amplifiers

Efficiency is improved by using a class B amplifier.

Complementary-symmetry amplifier (push–pull emitter follower). A class B circuit configuration using **complementary transistors** (a pair, one *npn* and one *pnp*, with matched characteristics).

Standard push–pull amplifier. A class B circuit that uses two identical transistors and a center-tapped transformer.

Why complementary-symmetry amplifiers are preferred.

The primary disadvantage of using class A power amplifiers is the fact that their efficiency ratings are so low. As you have been shown, a majority of the power that is drawn from the supply by a class A amplifier is used up by the amplifier itself. This goes against the primary purpose of a power amplifier, which is to deliver the power drawn from the supply to the load.

The class B amplifier was developed to improve on the low efficiency rating of the class A amplifier. The maximum *ideal* efficiency rating of a class B amplifier is approximately 78.5%. This means that up to 78.5% of the power drawn from the supply will ideally be delivered to the load. This is a major improvement over the class A amplifier. The main disadvantage of class B operation is that each amplifier requires *two* transistors to operate.

Figure 9.18 shows the most commonly used type of class B configuration. This circuit configuration is referred to as a *complementary-symmetry amplifier*, or *push–pull emitter follower*. The circuit recognition feature is the use of *complementary* transistors (that is, one of the transistors is an *npn* and the other is a *pnp*). The biasing circuit components may change from one amplifier to another, but complementary-symmetry amplifiers will always contain complementary transistors.

The standard push–pull amplifier contains two transistors of the same type with the emitters tied together. It also uses a *center-tapped transformer*. This amplifier type is shown in Figure 9.19. Note the transistor types and the transformer. This is the standard push–pull amplifier configuration.

Why is the complementary-symmetry configuration preferred over the standard push–pull? The center-tapped transformer makes the standard push–pull circuit much more expensive to construct than the complementary-symmetry amplifier. Since the complementary-symmetry amplifier is by far the most commonly used, we will concentrate on this circuit configuration in our discussion on class B operation.

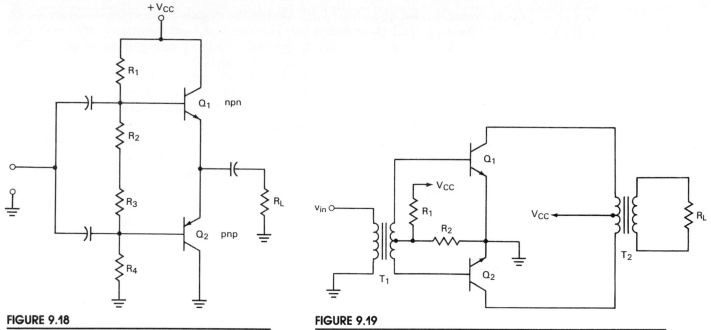

FIGURE 9.18

The class B complementary-symmetry amplifier.

FIGURE 9.19

The class B push-pull amplifier.

Operation Overview

The term *push–pull* comes from the fact that the two transistors in a class B amplifier conduct on alternating half-cycles of the input. For example, consider the circuit shown in Figure 9.20. During the positive half-cycle of the input, Q_1

FIGURE 9.20

Class B operation.

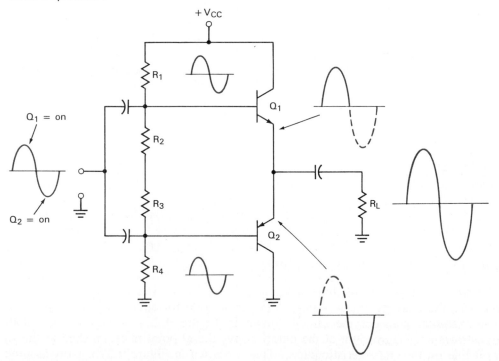

is biased *on* and Q_2 is biased *off*. During the negative half-cycle of the input, Q_1 is biased *off*, and Q_2 is biased *on*. The fact that *both transistors are never fully on at the same time* is the key to the high efficiency rating of the amplifier. This point will be discussed in detail later in this section.

The biasing of the two transistors is the key to its operation. When the transistor is in its quiescent state (no input), both transistors are biased at *cutoff*. When the input goes positive, Q_1 is biased above cutoff, and conduction results through the transistor. During this time, Q_2 is still biased at cutoff. When the input goes into the negative half-cycle, Q_1 is returned to the cutoff state, and Q_2 is biased above cutoff. As a result, conduction through Q_2 starts to build, while Q_1 remains off.

Because of the biasing arrangement, class B amplifiers are subject to a type of distortion called *crossover distortion*. Crossover distortion is illustrated in Figure 9.21. Note the flat lines between the half-cycles of the output signal. This is the crossover distortion. Crossover distortion is caused by having a short period of time when *both* transistors are *off*. When both transistors are off, the output drops to zero. To prevent crossover distortion, both transistors will normally be biased at a level that is slightly *above* cutoff. As you will be shown, biasing both transistors slightly above cutoff will allow the amplifier to provide a linear output that contains no distortion.

Crossover distortion.
Distortion caused by class B transistor biasing. Crossover distortion occurs when both transistors are in cutoff.

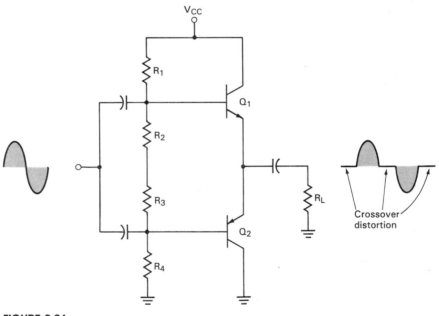

FIGURE 9.21
Crossover distortion.

DC Operating Characteristics

Class B amplifiers have vertical load lines.

The class B amplifier will have a vertical dc load line. The reason for this is the fact that there are no resistors in the emitter or collector circuits of the transistors. For example, consider the circuit shown in Figure 9.22a. Assuming that both transistors are biased right at the cutoff point, the Q-point is established at the $I_B = 0$ line on the characteristic curve. This is shown in Figure 9.22b. Now assume

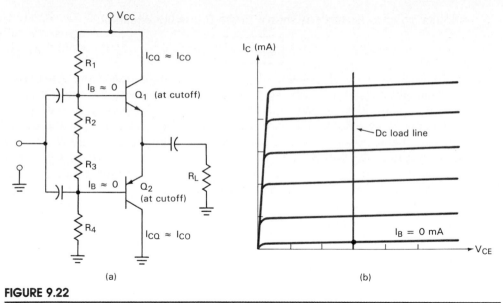

FIGURE 9.22

The dc load line for a class B amplifier.

that we could turn both transistors on. If they were both on, the following conditions would exist:

What would happen if both transistors were biased *on*.

1. The voltage drops across the two transistors (from emitter to collector) would remain the same because the resistance *ratio* of the two components would not change.
2. There would be no limit to the value of I_C, because there are no resistors present in the emitter or collector circuits to restrict the flow of charge.

Thus the voltage across the transistors would remain fairly constant, and the current through the collector and emitter circuits would not be restricted. This gives us the vertical dc load line. Note that this line shows little change in V_{CE}, but *does* show a large possible increase in the value of I_C.

The dc load line shows two other points about the dc operation of the class B amplifier. First, note that

$$V_{CEQ} = \frac{V_{CC}}{2}$$

(9.20)

This relationship is due to the fact that the amplifier is built using *matched* transistors. Matched transistors have the same operating characteristics, outside of the fact that one is an *npn* and the other is a *pnp*. For example, the 2N3904 and the 2N3906 are matched transistors. They have the same operating parameters and specifications, except for the fact that the 2N3904 is an *npn* transistor and the 2N3906 is a *pnp* transistor. When matched transistors are used, the value of V_{CE} for the two will be equal when I_C of one transistor is approximately equal to I_C of the other. Now refer to the circuit shown in Figure 9.22. Since the transistors are wired in series, I_{C1} will approximately equal I_{C2}. Thus V_{CE1} will equal V_{CE2}. Since the two voltages are equal and must add up to V_{CC}, each will equal one-half of V_{CC}.

SEC. 9.5   Class B Amplifiers

Another point to be made is shown in the following equation:

$$\boxed{I_{CQ} \cong 0} \tag{9.21}$$

This approximation is valid because of the fact that both transistors are biased just inside the cutoff region. If they were both biased farther into the cutoff region, the value of I_{CQ} would approach the ideal value, 0 A. To help you understand this point better, take a look at Figure 9.23. This is a "close-up" of the cutoff region of the characteristic curve for the transistors shown in Figure 9.22. For discussion purposes, we will refer to the two biasing points shown as *soft cutoff* and *hard cutoff*. When the transistors are biased at the soft-cutoff point, I_{CQ} will be at a higher level than that for the hard-cutoff point. As the transistor biasing is adjusted nearer the hard-cutoff point, the value of I_{CQ} approaches zero. (It can never reach the ideal value of zero because there will always be *some* amount of leakage current through the transistors.)

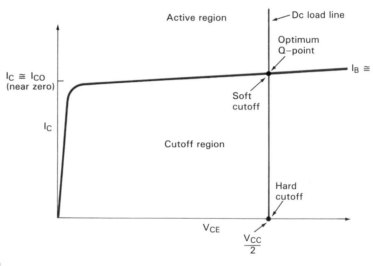

FIGURE 9.23

The cutoff region of operation.

Hard-cutoff causes crossover distortion.

It would seem that hard cutoff would be the ideal type of biasing to use on a class B amplifier. However, this is not the case. Biasing a class B amplifier at the hard-cutoff point causes crossover distortion. This is due to the transition time required for the transistor to come out of cutoff into the active region of operation. Biasing a transistor at soft cutoff eliminates this transition time, thus eliminating crossover distortion. Later in this section we will take a look at the methods used to bias a class B amplifier.

AC Operating Characteristics

Before we can cover the methods used to bias class B amplifiers, it is important for you to understand the ac characteristics of this amplifier type. The ac characteristics of the class B amplifier are illustrated in Figure 9.24.

Figure 9.24b shows the ac equivalent circuit for the class B amplifier. You may recall that this circuit is obtained by grounding the dc voltage source (V_{CC}) and shorting the capacitors in the circuit. With this equivalent circuit it is easy to see that the voltage across R_L is equal to V_{CE} of the transistors. Earlier we stated

CHAP. 9 Power Amplifiers

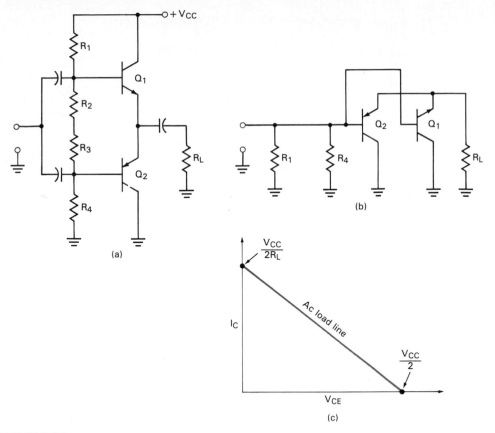

FIGURE 9.24

Class B ac characteristics.

that this voltage would equal one-half of V_{CC}. We can therefore find the value of $I_{C(\text{sat})}$ as

$$I_{C(\text{sat})} = \frac{V_{CC}}{2R_L} \tag{9.22}$$

We stated earlier that the transistors in a class B amplifier are normally biased at cutoff. Thus the normal value of $V_{CE(\text{off})}$ is found as

$$V_{CE(\text{off})} = \frac{V_{CC}}{2} \tag{9.23}$$

These two points are used to derive the ac load line of the class B amplifier. This load line is illustrated in Figure 9.24c.

═══════ **EXAMPLE 9.8** ═══════

 25

Determine the end-point values of the ac load line for the circuit shown in Figure 9.25.

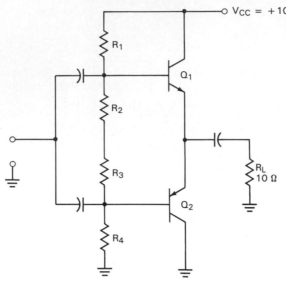

FIGURE 9.25

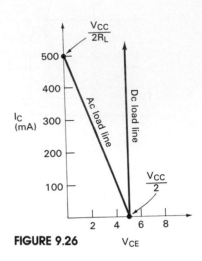

FIGURE 9.26

Solution: The value of $I_{C(\text{sat})}$ is found as

$$I_{C(\text{sat})} = \frac{V_{CC}}{2R_L}$$
$$= \frac{10 \text{ V}}{2(10 \text{ }\Omega)}$$
$$= 500 \text{ mA}$$

Now the value of $V_{CE(\text{off})}$ is found as

$$V_{CE(\text{off})} = \frac{V_{CC}}{2}$$
$$= \frac{10 \text{ V}}{2}$$
$$= 5 \text{ V}$$

The ac/dc load lines for the circuit are shown in Figure 9.26.

PRACTICE PROBLEM 9–8

A class B amplifier has values of $V_{CC} = +12$ V and $R_L = 2.2$ kΩ. Plot the dc and ac load lines for the circuit.

Amplifier Impedance

You may recall that the input impedance to the base of an emitter follower is found as

$$Z_{\text{in(base)}} = h_{fc}(r'_e + r_E)$$

If you take a look at the class B amplifier, you will notice that the load resistor is connected to the emitters of the two transistors. Since the load resistor is not bypassed, its value must be considered in the calculation of $Z_{\text{in(base)}}$ for the amplifier. The transistor input impedance is therefore found as

$$Z_{\text{in(base)}} = h_{fc}(r'_e + R_L) \qquad \textbf{(9.24)}$$

Dont forget: r'_e is the ac resistance of the transistor emitter, found as

$$r'_e = \frac{h_{ic}}{h_{fc}}$$

The output of the class B amplifier is taken from the emitters of the transistors, so the amplifier output impedance is equal to the ac resistance of the emitter circuit. As you may recall from Chapter 8, this impedance is found as

$$\boxed{Z_{\text{out}} = r'_e + \frac{R'_{\text{in}}}{h_{fc}}} \qquad \textbf{(9.25)}$$

where

$$\boxed{R'_{\text{in}} = R_1 \parallel R_4 \parallel R_S}$$

Note that the values of R_2 and R_3 in the class B amplifier are not used in equation (9.25). This is due to the fact that they are shorted in the ac equivalent circuit of the amplifier.

Amplifier Gain

Since the complementary-symmetry amplifer is basically an *emitter follower*, the current gain is found as with any emitter follower. By formula,

$$\boxed{A_i = h_{fc}}$$

The voltage gain of the class B amplifier is found in the same manner as a standard emitter follower, as follows:

$$\boxed{A_v = \frac{R_L}{R_L + r'_e}}$$

As with any amplifier, the power gain is the product of A_v and A_i. By formula,

$$\boxed{A_p = A_v A_i}$$

Power Calculations

The class B amplifer has the same output power characteristic as the class A amplifier. By formula,

$$\boxed{P_L = \frac{V_{\text{PP}}^2}{8R_L}} \qquad \textbf{(9.11)}$$

The maximum load power is also found in the same manner as it is for the class A amplifier. By formula,

$$\boxed{P_{L(\text{max})} = \frac{\text{PP}^2}{8R_L}} \qquad \textbf{(9.12)}$$

✔ 26

To calculate the maximum possible load power for a class B amplifer, we need to be able to determine its *compliance*. The compliance of a class B amplifier is found as

$$PP = 2V_{CEQ} \tag{9.26}$$

Since $V_{CEQ} \cong V_{CC}/2$, equation (9.26) can be rewritten as

$$PP \cong V_{CC} \tag{9.27}$$

The compliance of a class B amplifer is illustrated in Figure 9.27. Each transistor is capable of making the full transition from V_{CEQ} to approximately 0 V. This means that each transistor has a ΔV_{CE} that is approximately equal to V_{CEQ}. Since there are two transistors that conduct on alternate half-cycles of the input, the total transition of the output is approximately equal to $2V_{CEQ}$.

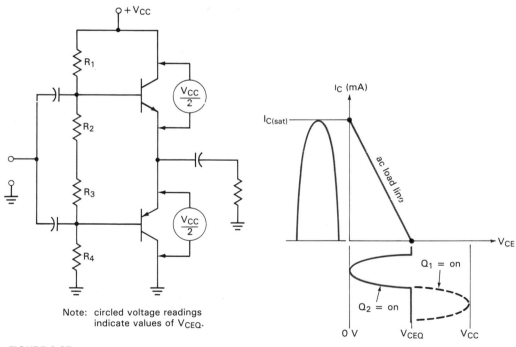

Note: circled voltage readings indicate values of V_{CEQ}.

FIGURE 9.27

Class B amplifier compliance.

=== EXAMPLE 9.9 ===

Determine the maximum load power for the circuit shown in Figure 9.28.

Solution: The compliance of the amplifier is found as

$$PP = V_{CC}$$
$$= 12 \text{ V}$$

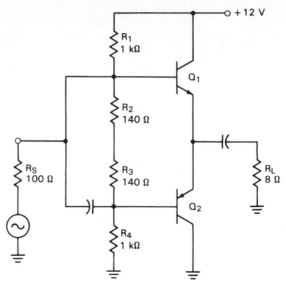

FIGURE 9.28

Now the maximum load power is found as

$$P_{L(\text{max})} = \frac{PP^2}{8R_L}$$

$$= \frac{(12\ V_{PP})^2}{8(8\ \Omega)}$$

$$= \frac{144\ V}{64\ \Omega}$$

$$= 2.25\ W$$

PRACTICE PROBLEM 9–9

Determine the maximum load power for the circuit described in Practice Problem 9–8.

The total power that the amplifier draws from the supply is found as

$$\boxed{P_S = V_{CC}I_{CC}} \tag{9.8}$$

where

$$\boxed{I_{CC} = I_{C1(\text{ave})} + I_1} \tag{9.28}$$

The equation for P_S is the same one that is used for the class A amplifier. However, equation (9.28) needs some explaining.

The total current drawn from the supply is the sum of the *average* Q_1 collector current and the current flowing through the amplifier base circuit, as is shown in Figure 9.29. The average value of the current flowing in the collector of Q_1 is given as

$$\boxed{I_{C(\text{ave})} = \frac{I_{Pk}}{\pi}}$$

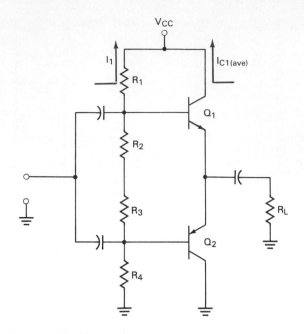

FIGURE 9.29

Class B amplifier supply current.

or

$$I_{C(ave)} \cong 0.318 I_{Pk} \quad \left| \quad \left(0.318 \cong \frac{1}{\pi}\right) \right.$$

where I_{Pk} is the peak current through the transistor. Note that this is the standard I_{ave} equation for the half-wave rectifier. Since the transistor is on for alternating half cycles, it effectively acts as a half-wave rectifier.

If you refer to Figure 9.27, you'll see that the peak current through each transistor is equal to $I_{C(sat)}$ for the transistor, given as

$$I_{C(sat)} = \frac{V_{CC}}{2R_L}$$

Combining these two equations,

$$I_{C1(ave)} = \frac{0.159 V_{CC}}{R_L} \tag{9.29}$$

The following example illustrates the process of determining the total power drawn from the supply.

EXAMPLE 9.10

Determine the value of P_S for the circuit shown in Figure 9.30.

Solution: Neglecting the base current from the two transistors,

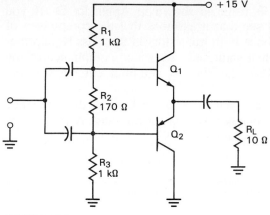

FIGURE 9.30

I_1 is found as

$$I_1 = \frac{V_{CC}}{R_T}$$
$$= \frac{15 \text{ V}}{2.17 \text{ k}\Omega}$$
$$= 6.91 \text{ mA}$$

Now $I_{C1(ave)}$ is found as

$$I_{C(ave)} = \frac{0.159 \, V_{CC}}{R_L}$$
$$= \frac{(0.159)(15 \text{ V})}{10 \, \Omega}$$
$$= 238.5 \text{ mA}$$

Using these two values, I_{CC} is found as

$$I_{CC} = I_{C1(ave)} + I_1$$
$$= 245.4 \text{ mA}$$

Finally, the total power demand on the source is determined as

$$P_S = V_{CC}I_{CC}$$
$$= (15 \text{ V})(245.4 \text{ mA})$$
$$= 3.68 \text{ W}$$

PRACTICE PROBLEM 9–10

Refer to Figure 9.28. Determine the value of P_S for the circuit.

EXAMPLE 9.11

Determine the total power being delivered to the load in Figure 9.30.

Solution: The maximum compliance of the amplifier is equal to V_{CC}, or 15 V. Using this value, the maximum load power is calculated as

$$P_{L(max)} = \frac{PP^2}{8R_L}$$
$$= \frac{(15 \text{ V}_{PP})^2}{80 \, \Omega}$$
$$= 2.81 \text{ W}$$

PRACTICE PROBLEM 9–11

Refer to Figure 9.28. Determine the maximum value of P_L for the circuit.

Class B Amplifier Efficiency

✔ 27

It was stated earlier in the chapter that the maximum theoretical efficiency of a class B amplifier is 78.5%. As is the case with the class A amplifiers, any practical efficiency rating must be less than the maximum theoretical value.

The derivation of the $\eta = 78.5\%$ value is shown in Appendix D. If you take a look at the derivation, you will see that it assumes that the compliance of the class B amplifier is equal to V_{CC}. Since both transistors in the class B amplifier will still have a slight value of V_{CE} when saturated, the actual compliance of the amplifier is slightly less than V_{CC}. Thus, the class B amplifier efficiency rating never reaches the value of 78.5%.

Once the values of P_S and P_L for a given class B are known, the efficiency of the circuit is calculated in the same manner as it is for the class A amplifiers. This point is illustrated in the following example.

EXAMPLE 9.12

Determine the efficiency of the amplifier used in Examples 9.10 and 9.11 (Figure 9.30).

Solution: The values of load power and source power were calculated in Examples 9.10 and 9.11 as

$$P_L = 2.81 \text{ W} \qquad P_S = 3.68 \text{ W}$$

Using these two values, the maximum efficiency of the amplifier is found as

$$\begin{aligned} \eta &= \frac{P_L}{P_S} \times 100\% \\ &= \frac{2.81 \text{ W}}{3.68 \text{ W}} \times 100\% \\ &= 76.36\% \end{aligned}$$

PRACTICE PROBLEM 9–12

Determine the efficiency of the amplifier described in Practice Problems 9–10 and 9–11.

Section Summary

The class B amplifier is a two-transistor circuit that has a higher maximum efficiency rating than either of the common class A amplifiers. There are two types of class B amplifiers: the *push–pull amplifier* and the *complementary-symmetry amplifier*. Of the two, the complementary-symmetry amplifier is more commonly used for two reasons:

1. The complementary-symmetry amplifier does not require the use of a transformer and thus is cheaper to produce.
2. Since the complementary-symmetry doesn't have a transformer, it is not subject to transformer losses. Thus, it will have a higher efficiency than a comparable push–pull amplifier.

The two transistors in a class B amplifier are biased at cutoff. When an ac signal is applied to the amplifier, the positive alternation of the ac signal will turn one transistor on, and the negative alternation will turn the other transistor on.

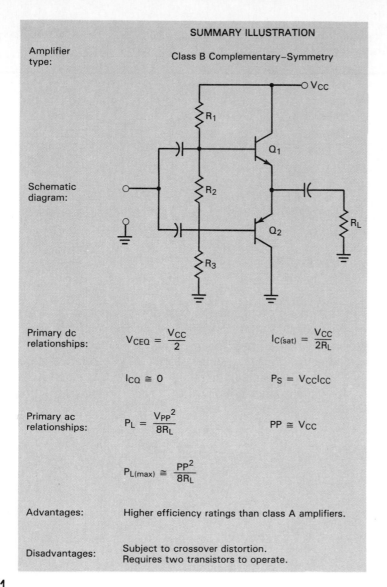

SUMMARY ILLUSTRATION

Amplifier type:

Class B Complementary–Symmetry

Schematic diagram:

Primary dc relationships:

$$V_{CEQ} = \frac{V_{CC}}{2}$$

$$I_{CQ} \cong 0$$

$$I_{C(sat)} = \frac{V_{CC}}{2R_L}$$

$$P_S = V_{CC}I_{CC}$$

Primary ac relationships:

$$P_L = \frac{V_{PP}^2}{8R_L}$$

$$PP \cong V_{CC}$$

$$P_{L(max)} \cong \frac{PP^2}{8R_L}$$

Advantages: Higher efficiency ratings than class A amplifiers.

Disadvantages: Subject to crossover distortion.
Requires two transistors to operate.

FIGURE 9.31

The resulting amplifier output is an ac signal that has a peak-to-peak value that is approximately equal to V_{CC}. The dc and ac characteristics of the complementary-symmetry amplifier are summarized in Figure 9.31.

SECTION REVIEW

1. What advantage does the class B amplifier have over the class A amplifier? [Objective 20]
2. What are the two common types of class B amplifiers? What are the advantages and disadvantages of each? [Objective 21]
3. What are *complementary transistors*? What kind of amplifier requires the use of complementary transistors? [Objective 21]
4. Describe the operating characteristics of the class B amplifier. [Objective 22]
5. What are the typical values of I_{CQ} and V_{CEQ} for a class B amplifier? [Objective 23]
6. What is the typical compliance of a class B amplifier? [Objective 24b]

7. Why is the practical efficiency rating of a class B amplifier less than the maximum theoretical value of 78.5%? [Objective 27]

9.6
Diode Bias

Up to this point, we have used voltage-divider bias for all our class B amplifiers. There are several problems that can develop when voltage-divider bias is used in class B amplifiers:

1. *Crossover distortion* can occur.
2. *Thermal runaway* can occur.

You may recall that crossover distortion occurs when both transistors in the class B amplifier are biased into cutoff during the input ac cycle. Thermal runaway is a problem that occurs when an increase in temperature causes an increase in beta, causing an increase in I_C. The increase in I_C then causes further increases in temperature, and so on. When voltage-divider bias is used in a class B amplifier, it is relatively easy for the amplifier Q-point to run right up the dc load line.

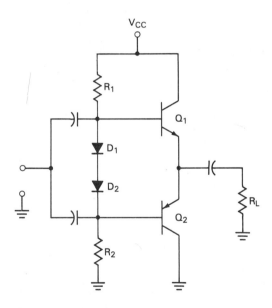

FIGURE 9.32

Diode biasing.

Diode bias. A biasing circuit that uses two diodes in place of the resistor(s) between the bases of the two transistors.

A biasing circuit that can be used to eliminate the problems of crossover distortion and thermal runaway is shown in Figure 9.32. The circuit shown, called *diode bias*, uses two diodes in place of the resistor(s) between the base of Q_1 and the base of Q_2. The diodes in the bias circuit are chosen to match the characteristic values of V_{BE} for the two transistors. As you will see, the diodes will eliminate both crossover distortion and thermal runaway when they are properly matched to the amplifier transistors.

CHAP. 9 Power Amplifiers

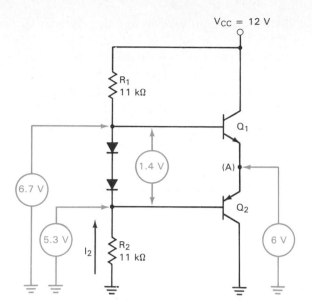

FIGURE 9.33

✔ 29

Diode Bias DC Characteristics

You may recall that the transistors in a class B amplifier are biased at cutoff, causing the value of I_{CQ} for the amplifier to be approximately equal to zero. When diode bias is used, the transistors are actually biased just above cutoff. That is, there will be some measurable amount of I_{CQ} when diode bias is used. This point is illustrated with the help of Figure 9.33.

In order to understand the operation of the circuit shown, we must start with a few assumptions:

1. V_{CEQ} is approximately one-half the value of V_{CC}, as is shown in the figure.
2. The current through R_2 causes 5.3 V to be developed across the resistor (R_2).

Assuming the above conditions are met, the base of Q_2 will be at 5.3 V and V_E of Q_2 will be 6 V. Since V_B for this *pnp* transistor is 0.7 V more negative than V_E, Q_2 will conduct.

With 1.4 V being developed across the biasing diodes, V_B of Q_1 will be 5.3 V + 1.4 V = 6.7 V. The value of V_E for Q_1 is 6 V, as is shown in the figure. Since V_B for this *npn* transistor is 0.7 V more positive than V_E, Q_1 will also conduct. Thus, both transistors in the diode-biased amplifier will conduct, and some measurable amount of I_{CQ} will be present, as is shown in Figure 9.34.

Now let's take a look at the assumptions we made (above) and where they came from. Even though we have changed the transistor biasing circuit, the output transistors are in the same configuration as they were in the standard class B amplifier. Thus, the dc load line characteristics of the amplifier haven't changed, and V_{CEQ} will always be approximately one-half of V_{CC}.

To understand where the assumed value of $V_{R2} = 5.3$ V came from, we have to take a look at the operation of the biasing circuit. The total voltage across the two resistors in the biasing circuit is equal to the difference between V_{CC} and the 1.4-V drop across the diodes. By formula,

$$V_{R1} + V_{R2} = V_{CC} - 1.4 \text{ V}$$

Using the voltage-divider equation, we can find the value of V_{R2} as

$$V_{R2} = \frac{R_2}{R_1 + R_2}(V_{CC} - 1.4 \text{ V})$$

Since $V_{B(Q2)} = V_{R2}$, the above equation can be rewritten as

$$V_{B(Q2)} = \frac{R_2}{R_1 + R_2}(V_{CC} - 1.4 \text{ V}) \tag{9.30}$$

When $R_1 = R_2$ (as was the case with the circuit in Figure 9.33), the fraction in equation (9.30) will equal ½, and the equation can be rewritten as

$$V_{B(Q2)} = \frac{V_{CC}}{2} - 0.7 \text{ V}$$

Don't forget:

$V_{CEQ} \cong \dfrac{V_{CC}}{2}$

or

$$V_{B(Q2)} = V_{CEQ} - 0.7 \text{ V} \quad | \quad (R_1 = R_2) \tag{9.31}$$

Equation (9.31) is very important for several reasons: First, it points out the fact that Q_2 (and thus Q_1) will automatically be properly biased when $R_1 = R_2$. The actual values of the resistors are relatively unimportant, as long as they are equal. When $R_1 = R_2$, you can assume that the base voltages are equal to $V_{CEQ} \pm 0.7$ V. When $R_1 \neq R_2$, you must use equation (9.30) to find the value of $V_{B(Q2)}$ and then add 1.4 V to this value to find $V_{B(Q1)}$. The second reason that equation (9.31) is so important is because almost all diode-biased circuits are designed with equal-value resistors. Thus, equation (9.30) rarely needs to be used.

A Practical Consideration. The values of the biasing resistors must be low enough to allow the diodes and the transistor base–emitter junctions to conduct. Normally, the biasing resistors will be in the low-kΩ range.

You may be wondering at this point why we didn't consider the values of I_{B1} and I_{B2} in our analysis of the biasing circuit. The reason is simple: When the transistors are properly matched, I_{B1} and I_{B2} will be equal in value. Therefore, the resistor currents (I_1 and I_2) will also be equal in value.

When diode bias is used, the value of I_1 (which is needed in the calculation of I_{CC}) is found as

$$I_1 = \frac{V_{CC} - 1.4 \text{ V}}{R_1 + R_2} \tag{9.32}$$

Again, we do not need to consider the effects of I_{B1} and I_{B2}, since they do not affect the resistor current values.

Class AB Operation

✔ 30

You have been shown that I_{CQ} will have some measurable value when diode bias is used. For this reason, the amplifier can no longer technically be called a *class B* amplifier. This point is illustrated with the help of Figure 9.35, which shows a typical input waveform for the circuit in Figure 9.34.

To simplify our discussion, we are going to make an assumption: A transistor will conduct until its base and emitter voltages are equal, at which time it will turn off. With this in mind, let's look at the circuit's response to the waveform shown in Figure 9.35.

Q_1 will conduct as long as its base voltage is more positive than 6 V. The value of V_B drops to 6 V at t_1 and t_3, since the -0.7-V value of v_{in} subtracts

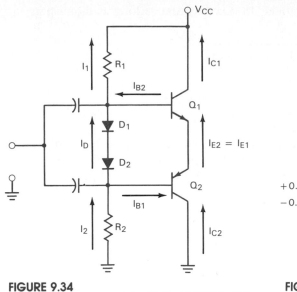

FIGURE 9.34

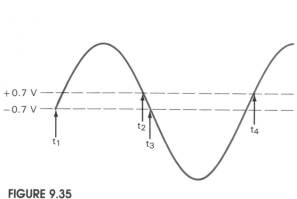

FIGURE 9.35

from the 6.7-V value of $V_{B(Q1)}$. Thus, we can assume that Q_1 conducts for the entire time between t_1 and t_3. The same principle applies to Q_2. At t_2 and t_4, Q_2 will turn off, since the +0.7-V value of v_{in} adds to the value of $V_{B(Q2)}$, causing V_B and V_E to be equal. Thus, Q_2 conducts for the entire time between t_2 and t_4.

As you can see, the transistors in the diode bias circuit will conduct for *slightly more than* 180°. Because of this, the circuit is classified as a *class AB amplifier*. In class AB operation, the transistors conduct for a portion of the input cycle that is between 180° and 360°.

While the diode bias circuit is technically classified as a class AB amplifier, most technicians simply refer to it as a class B amplifier. This is because of the strong similarities in operation between the two types of amplifiers.

Eliminating Crossover Distortion

Since both transistors in the class AB amplifier are conducting when the input signal is at zero volts, the amplifier does not have the crossover distortion problems that the class B amplifier may have. Crossover distortion only occurs when *both* transistors are in cutoff, so the problem will not normally occur in class AB amplifiers.

◢ 31

Eliminating Thermal Runaway

For diode bias to eliminate the problem of thermal runaway, two conditions must be met:

◢ 32

1. The diodes and the transistor base–emitter junctions must be very nearly *perfectly* matched.
2. The diodes and the transistors must be in *thermal contact*. This means that they must be in physical contact with each other so that the operating temperature of the diodes will always equal the operating temperature of the transistors.

When the diodes and transistors are matched, the forward voltage drops across the diodes (V_{F1} and V_{F2}) will be approximately equal to the V_{BE} drops of the transistors, all other factors being equal. In other words, as long as their operating temperatures are the same,

$$\boxed{V_{F1} = V_{BE(Q1)}}$$

and

$$\boxed{V_{F2} = V_{BE(Q2)}}$$

When the transistors are in thermal contact with the diodes, the components will all experience the same operating temperature. Thus, if the operating temperature of the transistors increases by a given amount, the operating temperature of the diodes will increase by the same amount, and the values of V_F and V_{BE} will remain approximately equal. With this in mind, let's take a look at the circuit response to an increase in temperature.

You may recall from our discussion on *temperature versus diode conduction* that an increase in diode temperature will cause a slight increase in diode current and a slight decrease in the value of V_F. Assuming that the transistor and diode temperature increases are equal, here's what happens:

1. The increase in diode temperature causes the 1.4-V drop across D_1 and D_2 to *decrease*. At the same time, I_D *increases* (see Figure 9.34).
2. With the decrease in diode voltage, I_1 and I_2 *increase*. This causes the voltage drops across R_1 and R_2 to *increase*.
3. With the change in resistor voltage drops, the value of $V_{B(Q1)}$ *decreases* [$V_{B(Q1)} = V_{CC} - V_1$], and the value of $V_{B(Q2)}$ *increases*.
4. The base voltage changes bring the values of V_B closer to the value of V_E for the two transistors, reducing the forward bias on the transistors.
5. The reduction in forward bias decreases I_B, causing a decrease in I_C. The decrease in I_C prevents thermal runaway from occurring.

Thus, the class AB amplifier reduces the chance of thermal runaway occurring.

Class AB Amplifier Analysis

► 33

We can now use the information on dc biasing to approach the process for analyzing a class AB amplifier. For this discussion, we will use the example circuit shown in Figure 9.36.

We will start our analyses by determining the values of $I_{C(sat)}$ and $V_{CE(off)}$. For this circuit, $I_{C(sat)}$ is found as

$$I_{C(sat)} = \frac{V_{CC}}{2R_L}$$
$$= 750 \text{ mA}$$

and $V_{CE(off)}$ is found as

$$V_{CE(off)} = \frac{V_{CC}}{2}$$
$$= 6 \text{ V}$$

Using these two values, the dc and ac load lines in Figure 9.36 are plotted.

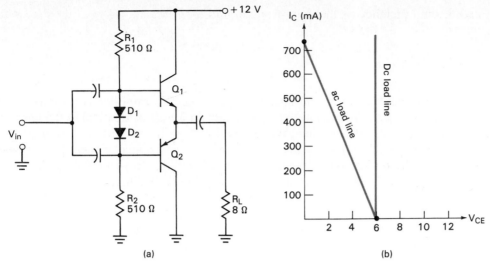

FIGURE 9.36

A class AB amplifier and its load lines.

Next, we determine the value of I_1. The value of I_1 is found as

$$I_1 = \frac{V_{CC} - 1.4 \text{ V}}{R_1 + R_2}$$
$$= \frac{12 \text{ V} - 1.4 \text{ V}}{1020}$$
$$= 10.4 \text{ mA}$$

The average current in the collector circuit of the amplifier is found as

$$I_{C1(\text{ave})} = \frac{0.159 V_{CC}}{R_L}$$
$$= 238.5 \text{ mA}$$

Using this value and the value of I_1 calculated earlier, we can find the value of I_{CC} as

$$I_{CC} = I_{C1(\text{ave})} + I_1$$
$$= 248.9 \text{ mA}$$

Now we can calculate the total power being drawn from the supply as

$$P_S = V_{CC} I_{CC}$$
$$= 2.99 \text{ W}$$

Assuming that the compliance of the amplifier is at its maximum value (V_{CC}), we can calculate the load power as

$$P_L \cong \frac{V_{CC}^2}{8R_L}$$
$$= 2.25 \text{ W}$$

SEC. 9.6   Diode Bias

397

Finally, the efficiency of the amplifier is found as

$$\eta = \frac{P_L}{P_S} \times 100\%$$
$$= 75.25\%$$

As you can see, the basic analysis of a class AB amplifier is actually fairly simple. In fact, with the exception of the I_1 calculation, it is identical to the analysis of the standard class B amplifier. When you deal with a standard class B amplifier, I_1 is found by dividing V_{CC} by the total series base resistance. When dealing with the class AB amplifier, you must take the diode voltage drops into account.

The class AB amplifier is used far more commonly than the standard class B amplifier. For this reason, we will concentrate on the class AB amplifier from this point on. Just remember, outside of the biasing circuit and the value of I_{CQ}, the class AB and the class B amplifiers are nearly identical. Thus, the term *class B* is commonly used to describe both amplifier types.

SECTION REVIEW

1. What are the two primary problems that can occur in standard class B amplifiers? What type of circuit is used to eliminate these problems? [Objective 28]
2. Explain how the diode bias circuit develops the proper values of V_B for the amplifier transistors. [Objective 29]
3. How do you find the values of V_B in a diode bias circuit when $R_1 = R_2$? [Objective 29]
4. How do you find the values of V_B in a diode bias circuit when $R_1 \neq R_2$? [Objective 29]
5. What is *class AB operation*? [Objective 30]
6. How does the class AB amplifier eliminate crossover distortion? [Objective 31]
7. How does the class AB amplifier eliminate thermal runaway? [Objective 32]
8. List, in order, the steps you would take to perform the complete analysis of a class AB amplifier. [Objective 33]
9. Why is the class AB amplifier often referred to as a class B amplifier? [Objective 30]

9.7
Troubleshooting Class AB Amplifiers

In this section we are going to concentrate on the faults that may develop in the diode-biased complementary-symmetry push–pull amplifier. We will also discuss some techniques you can use to isolate those faults. At the end of the section, you will be introduced to some other biasing circuits for class AB amplifiers. Although they are a bit more complex than the diode-biasing circuit, they each serve their own purpose. Do not let the more complex circuits confuse you; they may contain more components than the amplifiers discussed previously, but they work in the same basic fashion.

Troubleshooting

Class AB amplifiers can be more difficult to troubleshoot than class A amplifiers because of the dual-transistor configuration. Any of the standard transistor faults can develop in *either* of the two transistors. However, there are some techniques you can use that will make troubleshooting these amplifiers relatively simple.

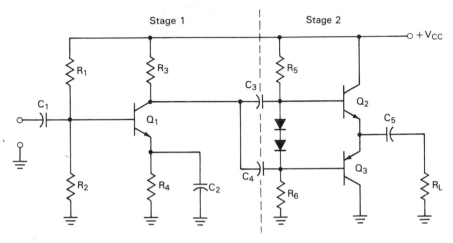

FIGURE 9.37

The first step in troubleshooting a class AB amplifier is the same as with any amplifier: *You must make sure that the amplifier is the source of the trouble.* For example, assume that the two-stage amplifier shown in Figure 9.37 is not working (there is no output signal). For this circuit, there are *three* checks that must be made before you can assume there is a problem with the push–pull amplifier:

Initial tests.

1. You must disconnect the load to make sure that it is not loading down the amplifier and preventing it from working.
2. You must be sure that the push–pull is getting the proper input from the first stage.
3. You must verify that the V_{CC} and ground connections in the push–pull amplifier are good.

Picture this: You spend a couple of hours troubleshooting the push–pull amplifier, only to discover that the amplifier never had an input signal to begin with! This may sound a bit farfetched, but 99 out of 100 technicians would admit to having done something like this at one time or another. (And I would question the 100th!)

Always start by making sure that the load is not the source of the trouble. A shorted load will always prevent a source amplifier from having an output. If you disconnect the load and the amplifier starts to work properly, then the load is the cause of the problem. Once the load has been verified as being okay, make sure that the amplifier is getting an input. If the input is bad, you need to continue on to the first stage (that is, there is no reason to suspect that the push–pull is the cause of the problem). If the input is good, check the V_{CC} and ground connections in the amplifier. If all these tests check out, there is a problem with the amplifier.

Once you have concluded that the push–pull amplifier is the source of the problem, the first step is to *disconnect the ac signal source.* For the circuit in Figure 9.37, this means disconnecting the output of the first stage from the input

DC troubleshooting starts with disconnecting the signal source.

to the push–pull amplifier. There are two ways to do this, depending on the type of circuit you are dealing with:

1. If the input signal is being fed to the push–pull by a wire, you can desolder the wire to isolate the stage.
2. If the input arrives by a copper run on a printed circuit board (which is usually the case), desolder and remove the coupling component that connects the two stages. For Figure 9.37, this would mean removing the coupling capacitors between the two stages, C_3 and C_4.

After disconnecting the signal source, perform the dc voltage checks.

When the amplifier has been isolated from the signal source, you can perform some relatively simple voltage checks. Figure 9.38 shows the types of readings you should get in Figure 9.37 when the amplifier has been isolated and all the components are working properly. For example, assume that the circuit has been biased properly and that all components are good. If V_{CC} is +20 V, you should get the *approximate* readings shown in Figure 9.39.

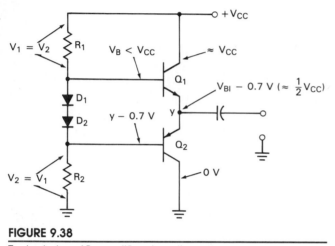

FIGURE 9.38

Typical class AB amplifier dc voltages.

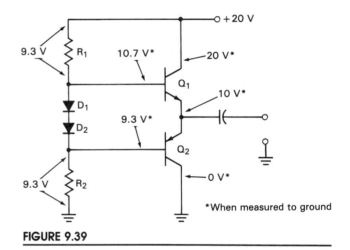

FIGURE 9.39

With most class A amplifiers, the troubleshooting procedure begins with analyzing the emitter and collector voltages of the transistor and then working your way back to the input. The procedure is essentially the same for the class AB amplifier. We check the voltage at the point where the two emitters are connected, and then we check the base voltages. Table 9.1 summarizes some common faults and their symptoms. The test points referenced in the table are shown in Figure 9.40.

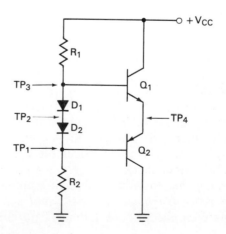

FIGURE 9.40

Commonly used dc test points.

CHAP. 9   Power Amplifiers

TABLE 9.1.
Open Component Troubleshooting

Open Component	Symptoms
R_2	The voltages at TP_1 and TP_3 will be higher than normal (near V_{CC}). The voltage at TP_4 will be higher than normal.
D_1 or D_2	The voltages at TP_1, TP_3, and TP_4 will all be slightly lower than normal.
R_1	The voltages at TP_1 and TP_3 will be at or near 0 V. The voltage at TP_4 will be slightly lower than normal.
Q_1	All base circuit voltages will be normal. The voltage at TP_4 will be very low.
Q_2	All base circuit voltages will be normal. The voltage at TP_4 will be very high.

If the biasing circuit proves to be okay and the amplifier does not work, you should replace both transistors. For one thing, it saves time (as stated earlier). Another reason is this: If one of the two transistors has gone bad, odds are that it has damaged the other in the process. Transistors are not as durable as resistors and other passive components. It is very easy for the bad transistor to damage the other. So even if the second transistor has not been destroyed, odds are that it will soon develop trouble and will have to be replaced anyway. Save yourself some trouble from the start. Replace both transistors when you suspect that one of them is bad.

What to do if you suspect the transistors to be faulty.

Now, to help you get more comfortable with the information in this section, let's go through some troubleshooting applications.

APPLICATION 1

The circuit shown in Figure 9.41 has no output. After checking the load, input signal, and V_{CC}/ground connections, the input to the circuit is disconnected. Then a voltmeter is used to get the readings shown. In this case, the problem is easy to spot. Since the readings on both sides of D_2 are equal, the diode is shorted. Note that the reading at TP_1 is 5.65 V, up from the 5.3 V that should be there. In this case, replacing the diode solves the problem.

FIGURE 9.41

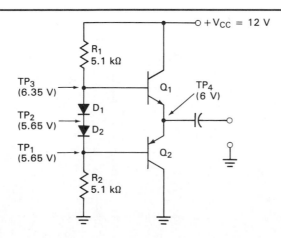

APPLICATION 2

The circuit in Figure 9.42a is tested. The results are shown in the diagram. The readings in Figure 9.42a indicate that there is an open diode in the biasing circuit. A resistance check on D_2 indicates that it is the open diode.

When D_2 is replaced, the circuit readings change to those shown in Figure 9.42b. These readings indicate that D_1 has shorted. In this circuit, D_1 apparently shorted, causing D_2 to burn open. Replacing D_1 causes the circuit to operate properly again.

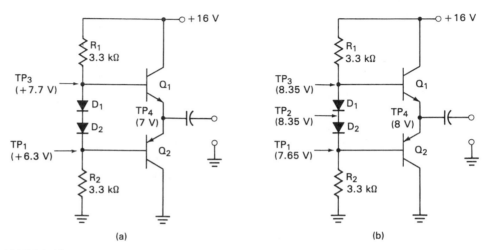

(a)   (b)

FIGURE 9.42

APPLICATION 3

What problem is indicated by the readings in Figure 9.43? What would you do to repair the problem? (Assume that the amplifier inputs all check out and that the load is not the problem.) The answer can be found at the end of Application 4.

FIGURE 9.43

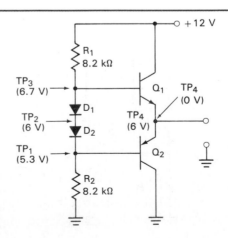

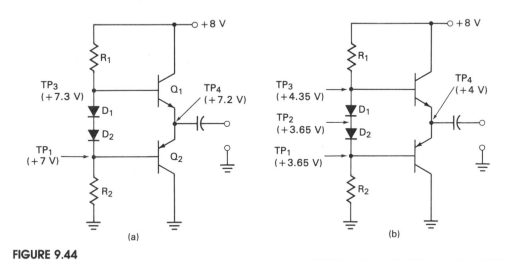

FIGURE 9.44

In Application 3, all the biasing voltages are correct. The only logical assumption at this point would be that there is a problem with one of the transistors. Therefore, the next step would be to remove/replace both transistors.

Other Class AB Amplifiers

Next we will look at a few other types of class AB amplifiers and biasing circuits. The purpose of this discussion is *not* to teach you *everything* about these circuits, but, rather, to introduce you to alternative circuit configurations.

The first circuit is the *Darlington complementary-symmetry amplifer*. This circuit is shown in Figure 9.45. In this amplifier, the two transistors have been replaced by Darlington pairs. You may recall from earlier discussions that the Darlington pair has a very high characteristic current gain. This circuit has a much higher current gain than the standard complementary-symmetry amplifier. Because of this, it is used in applications where a high amount of load power is required. Recall that load power is calculated as

$$A_P = A_v A_i$$

Since the value of A_i is much higher for a Darlington pair than it is for a single transistor, the overall gain of the Darlington class AB amplifier will be much higher than for a standard push–pull.

Note the four biasing diodes between the bases of Q_1 and Q_3. Four diodes are needed to compensate for the 1.4-V value of V_{BE} for *each* Darlington pair.

Darlington complementary-symmetry amplifier.

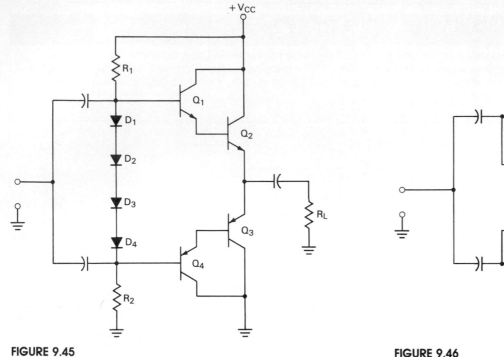

FIGURE 9.45

The Darlington class B amplifier.

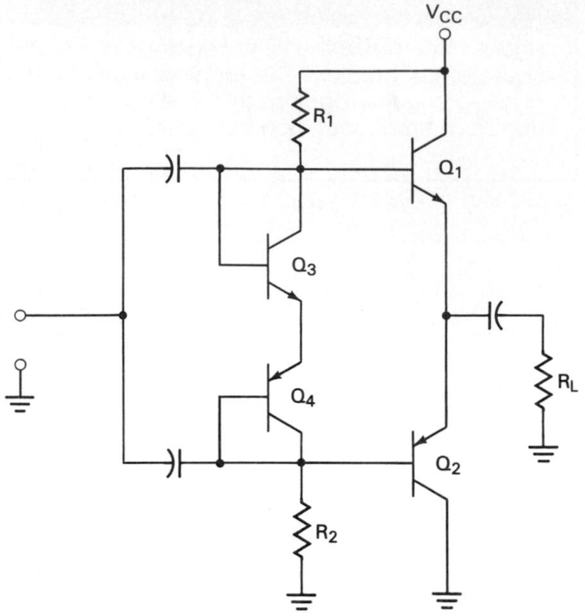

FIGURE 9.46

The transistor-biased amplifier.

Since each pair has a V_{BE} of 1.4 V, there is a 2.8-V difference of potential between the bases of Q_1 and Q_3. The four diodes, when properly matched, will maintain the 2.8 V difference.

Transistor-biased complementary-symmetry amplifier.

The *transistor-biased complementary symmetry amplifier* uses transistors instead of diodes in the biasing circuit. This amplifier is shown in Figure 9.46. The amplifier is used primarily in circuits that contain integrated (IC) transistors. Note that the biasing transistors are wired in such a way as to cause them to act as diodes. The collector-base junction of each biasing transistor is shorted, leaving on an emitter-base junction in the circuit. This junction acts as a simple diode. Why go to the trouble of using transistors in the biasing circuit? Most transistor

FIGURE 9.47

The dual-polarity class AB amplifier.

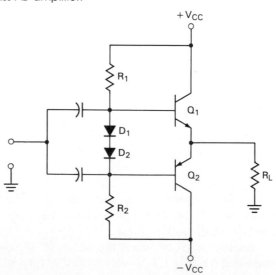

ICs contain four transistors. By using the extra transistors in an IC as the biasing diodes, you ensure that the diodes are perfectly matched to the circuit transistors (Q_1 and Q_2).

When the output from a class AB amplifier must be centered around 0 V (instead of around the value of $V_{CC}/2$), the *split-supply class AB amplifier* may be used. This amplifier is shown in Figure 9.47. The two power supply connections for this circuit will be equal and opposite in polarity. For example, voltage supplies of ± 10 V may be used, but voltage supplies of $+10$ V and -5 V would not be used. With matched power supplies, each transistor will drop its own supply voltage, and the output signal will be centered at 0 V. This point is illustrated in Figure 9.47.

Split-supply class AB amplifier.

There are many different biasing configurations for class AB amplifiers, many more than could be reasonably covered here. However, the ones you have been shown are fairly common. When you come up against a biasing circuit you have never seen before, just remember the basic principles of class B and class AB operation. Within reason, the operation of any class B or class AB amplifier will follow these principles.

SECTION REVIEW

1. What is the first step in troubleshooting a class AB amplifier (or any other amplifier, for that matter)? [Objective 34]
2. Why are class AB amplifiers more difficult to troubleshoot than class A amplifiers? [Objective 34]
3. What is the general approach to troubleshooting a class AB amplifier? [Objective 34]
4. Your are troubleshooting a class AB amplifier and come to the conclusion that the trouble is one of the two transistors. Briefly discuss the reasons for replacing both transistors rather than one or the other of them. [Objective 34]
5. When is the *Darlington complementary-symmetry amplifier* used? [Objective 35]
6. Why are four biasing diodes needed in the Darlington complementary-symmetry amplifier? [Objective 35]
7. When is transistor biasing used in place of diode biasing? [Objective 35]
8. When is a *split-supply class AB amplifier* used? [Objective 35]

9.8
Related Topics

In this section we will take a look at a few topics that relate to power amplifiers, specifically, maximum power ratings and calculations and the use of heat sinks.

Maximum Power Ratings

 36

You may recall that transistors have a *maximum power dissipation rating*. When a transistor is being used for a specific power amplifier application, you must make sure that the power dissipated by the transistor in the circuit does not exceed the rating of the transistor you are trying to use. For example, the $P_{D(max)}$ rating for the 2N3904 is 310 mW (assuming that no power derating is required). Thus, you could not use the 2N3904 in any power amplifier that would require its transistor to dissipate more than 310 mW.

How do you determine the amount of power a transistor will have to handle in a specific circuit? For class A amplifiers, use the equation

$$P_D = V_{CEQ} I_{CQ} \qquad \text{(9.33)}$$

For class B and class AB amplifiers, use the equation

$$P_D = \frac{V_{PP}^2}{40 \, R_L} \qquad \text{(9.34)}$$

where V_{PP} = the *peak-to-peak* load voltage. The derivations of these equations are fairly involved, and thus are reserved for Appendix D. The following examples show how the equations are used to determine the transistor power requirements of class A and class AB amplifiers.

EXAMPLE 9.13

What is the value of P_D for the transistor in Figure 9.48?

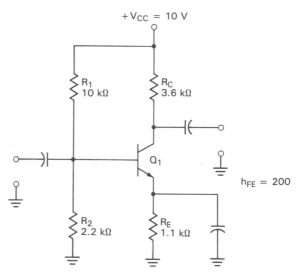

FIGURE 9.48

Solution: Using the established procedure, I_{CQ} and V_{CEQ} are found to be:

$$I_{CQ} = 1 \text{ mA}, \qquad V_{CEQ} = 5.3 \text{ V}$$

Using these two values, P_D is found as

$$\begin{aligned} P_D &= V_{CEQ} I_{CQ} \\ &= (5.3 \text{ V})(1 \text{ mA}) \\ &= 5.3 \text{ mW} \end{aligned}$$

A class A amplifier has the following values: $R_E = 1.2$ kΩ, $R_C = 2.7$ kΩ, $V_{CC} = 16$ V, and $V_E = 2.4$ V. Determine the required value of P_D for the transistor in the amplifier.

EXAMPLE 9.14

What is the value of P_D for each transistor in Figure 9.49? (Assume that the output is equal to the compliance of the amplifier.)

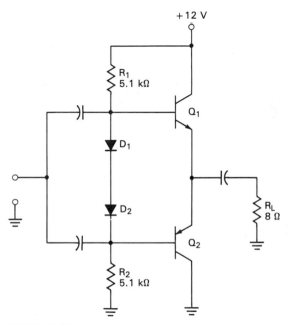

+12 V

R_1
5.1 kΩ

Q_1

D_1

D_2

R_L
8 Ω

Q_2

R_2
5.1 kΩ

FIGURE 9.49

Solution: The compliance of the amplifier is 12 V. Thus $V_{PP} = 12$ V, and

$$P_D = \frac{V_{PP}^2}{40R_L}$$
$$= \frac{144 \text{ V}_{PP}}{320 \text{ Ω}}$$
$$= 450 \text{ mW}$$

A class AB amplifier with values of $V_{CC} = 15$ V and $R_L = 12$ Ω is driven to compliance. Determine the required value of P_D for each transistor in the circuit.

Component Cooling

✔ 37

When large numbers of electronic components are used in an enclosed system, the power dissipated by the components can cause a significant rise in the temperature of the components. In this situation, the derating factor of the transistors becomes important.

To help decrease the loss in P_D values caused by a closed system, fans are often used to reduce the temperature within the system. If you have ever worked with a computer terminal, you may have noticed that it is usually fan cooled. Fan cooling decreases component temperature and thus helps to restore the transistors and ICs to their original P_D values.

Another device that is used to reduce the case temperature of a given transistor is the heat sink. Heat sinks are large metallic objects that help cool transistors by increasing their surface area. Several heat sinks are shown in Figure 9.50.

Heat sink. A large metallic object that helps to cool transistors by increasing their surface area.

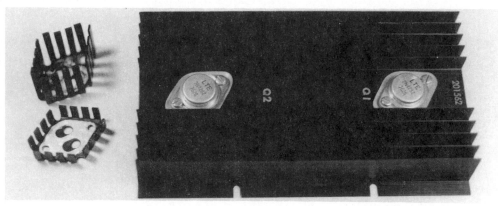

FIGURE 9.50

When a heat sink is *properly* connected to a transistor, the case heat from the transistor is transferred to the heat sink. Because the heat sink has more surface area than the transistor casing, it is able to dissipate a much greater amount of heat. Thus the transistor case temperature is held at a lower value, and the P_D derating *amount* is reduced.

Heat-sink compound. A compound used to aid in the transfer of heat from a transistor to a heat sink.

When a heat sink is connected to a transistor, a chemical compound called *heat-sink compound* is used to aid in the transfer of heat to the sink. This compound can be purchased at any electronics parts store. The heat sink compound is applied to the transistor casing, and then the transistor is connected (attached) to the heat sink itself. The transistor must be attached to the heat sink mechanically (that is, heat-sink compound is *not* glue; it does not act as an adhesive between the transistor and the heat sink).

A Practical Consideration. Most power transistors are bolted to the heat sink (see Q_1 and Q_2 in Figure 9.50). When properly mounted, the transistor leads will *not* be able to touch the heat sink. If the leads *are* touching the heat sink, you have inserted the transistor backward.

When a transistor is connected to a heat sink, care must be taken to ensure that the leads of the transistor do not touch the heat sink. This causes a short circuit from the transistor input to its output. The bottom line is this: You must follow the proper procedure for replacing a transistor that is connected to a heat sink. The proper procedure is as follows:

1. Remove the bad transistor from the heat sink and wipe off the old heat-sink compound (from the heat sink).

2. *Lightly* coat the new transistor with heat-sink compound. Do not use more than is necessary to build a thin coat on the component.
3. Connect the transistor to the heat sink. If there were any mechanical connectors between the old transistor and the heat sink, such as screws, be sure to replace them in the new circuit.
4. *Be sure that the transistor leads are not touching the heat sink.*

SECTION REVIEW

1. What is a *heat sink*? [Objective 37]
2. What is *heat-sink compound*? [Objective 37]
3. What is the proper procedure for replacing a transistor that is connected to a heat sink? [Objective 37]
4. What precautions must be taken when performing the procedure in question 3? [Objective 37]

KEY TERMS

The following terms were introduced and defined in this chapter:

ac load line
class A amplifier
class AB operation
class B amplifier
class C amplifier
complementary-symmetry
  amplifier
complementary
  transistors
compliance

counter emf
crossover distortion
cutoff clipping
diode bias
efficiency
heat sink
heat-sink compound
RC-coupled class A
  amplifier
saturation clipping

standard push–pull
  amplifier
step-down transformer
step-up transformer
thermal contact
transformer-coupled class
  A amplifier
tuned amplifier

PRACTICE PROBLEMS

§9.2

1. Calculate the compliance of the amplifier in Figure 9.51. [6]
2. Calculate the compliance of the amplifier in Figure 9.52. [6]
3. Calculate the compliance of the amplifier in Figure 9.53. Determine the type of clipping that the circuit would be most likely to experience. [6]
4. Calculate the compliance of the amplifier in Figure 9.54. Determine the type of clipping that the circuit would be most likely to experience. [6]

§9.3

5. Calculate the value of P_S for the amplifier in Figure 9.51. [7]
6. Calculate the value of P_S for the amplifier in Figure 9.52. [7]
7. Calculate the value of P_L for the amplifier in Figure 9.51. [8]
8. Calculate the value of P_L for the amplifier in Figure 9.52. [8]

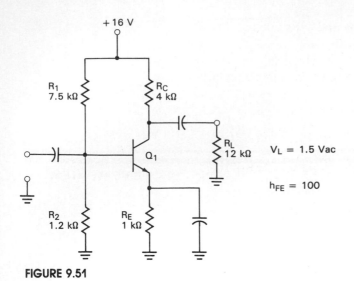

FIGURE 9.51

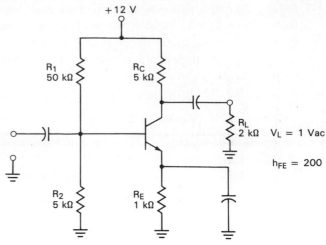

FIGURE 9.52

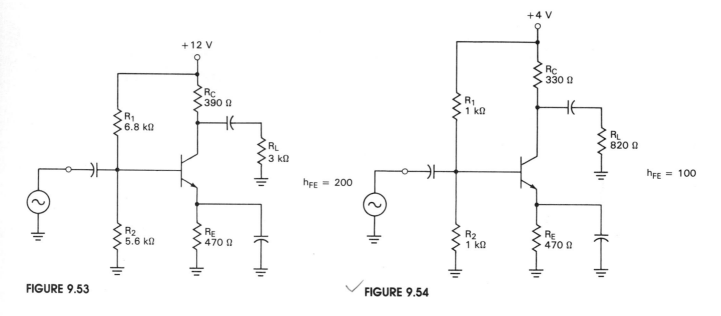

FIGURE 9.53

FIGURE 9.54

9. Calculate the maximum possible load power for the circuit in Figure 9.51. [9]

10. Calculate the maximum possible load power for the circuit in Figure 9.52. [9]

11. Assuming that the amplifier in Figure 9.51 is driven to compliance, calculate the efficiency of the circuit. [10]

12. Assuming that the amplifier in Figure 9.52 is driven to compliance, calculate the efficiency of the circuit. [10]

13. Determine the values of P_S, P_L, and the efficiency for the amplifier in Figure 9.53. Assume that the circuit is driven to compliance. [7, 8, 10]

14. Determine the values of P_S, P_L, and the amplifier in Figure 9.54. Assume that the circuit is driven to compliance. [7, 8, 10]

15. Calculate the maximum efficiency for the amplifier in Figure 9.55. [6, 11]

16. Calculate the maximum efficiency for the amplifier in Figure 9.56.

§9.4

17. Derive the dc and ac load lines for the circuit in Figure 9.57. [16]
18. Derive the dc and ac load lines for the circuit in Figure 9.58. [16]
19. Determine the maximum load power for the circuit in Figure 9.57. [18]
20. Determine the maximum load power for the circuit in Figure 9.58. [18]
21. Calculate the maximum efficiency of the amplifier in Figure 9.57. [18]
22. Calculate the maximum efficiency of the amplifier in Figure 9.58. [18]

FIGURE 9.56

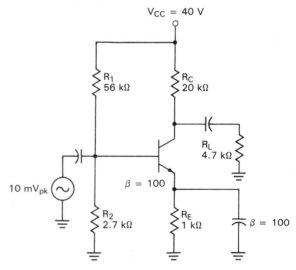

FIGURE 9.55

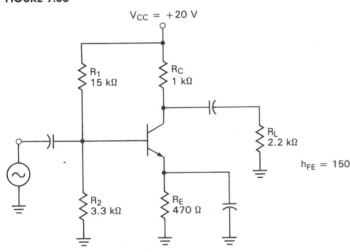

FIGURE 9.57

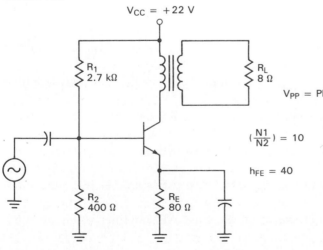

FIGURE 9.58

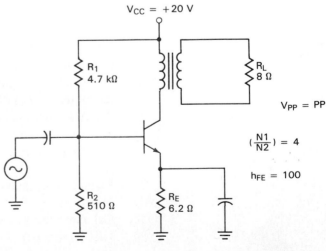

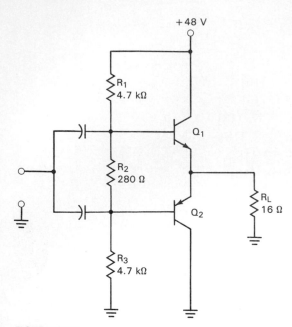

FIGURE 9.59

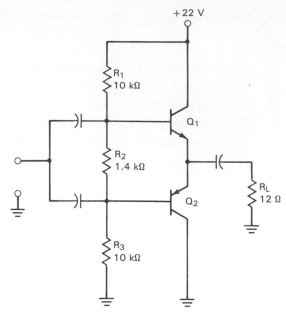

FIGURE 9.60

§9.5

23. A class B amplifier has values of $V_{CC} = +18$ V and $R_L = 3$ kΩ. Plot the ac and dc load lines for the circuit. [25]

24. A class B amplifier has values of $V_{CC} = +24$ V and $R_L = 200$ Ω. Plot the ac and dc load lines for the circuit. [25]

25. Calculate the maximum load power for the amplifier described in problem 23. [26]

26. Calculate the maximum load power for the amplifier described in problem 24. [26]

27. Calculate the value of P_S for the amplifier in Figure 9.59. [26]

28. Calculate the value of P_S for the amplifier in Figure 9.60. [26]

29. Calculate the value of $P_{L(max)}$ for the amplifier in Figure 9.59 [26]

30. Calculate the value of $P_{L(max)}$ for the amplifier in Figure 9.60. [26]

31. Calculate the maximum efficiency for the amplifier in Figure 9.59. [27]

32. Calculate the maximum efficiency for the amplifier in Figure 9.60. [27]

§9.6

33. Determine the values of V_{CEQ}, $V_{B(Q1)}$, and $V_{B(Q2)}$ for the class AB amplifier in Figure 9.61. [29]

34. Determine the values of V_{CEQ}, $V_{B(Q1)}$, and $V_{B(Q2)}$ for the class AB amplifier in Figure 9.62. [29]

35. Calculate the maximum efficiency of the class AB amplifier in Figure 9.61. [33]

36. Calculate the maximum efficiency of the class AB amplifier in Figure 9.62. [33]

37. Calculate the maximum efficiency of the class AB amplifier in Figure 9.63. [33]

38. Calculate the maximum efficiency of the class AB amplifier in Figure 9.64. [33]

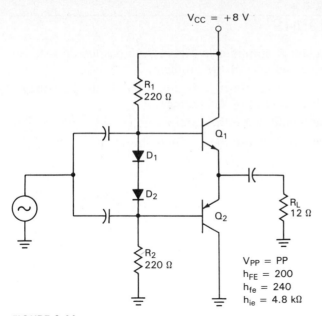

FIGURE 9.61

$V_{CC} = +8$ V

R₁ 220 Ω

D₁

D₂

Q₁

Q₂

R_L 12 Ω

R₂ 220 Ω

$V_{PP} = PP$
$h_{FE} = 200$
$h_{fe} = 240$
$h_{ie} = 4.8$ kΩ

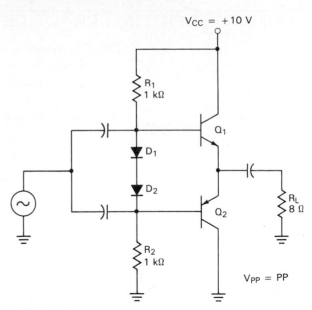

FIGURE 9.62

$V_{CC} = +10$ V

R₁ 1 kΩ

D₁

D₂

Q₁

Q₂

R_L 8 Ω

R₂ 1 kΩ

$V_{PP} = PP$

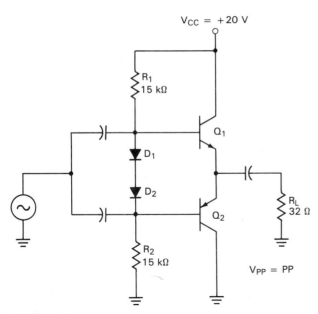

FIGURE 9.63

$V_{CC} = +20$ V

R₁ 15 kΩ

D₁

D₂

Q₁

Q₂

R_L 32 Ω

R₂ 15 kΩ

$V_{PP} = PP$

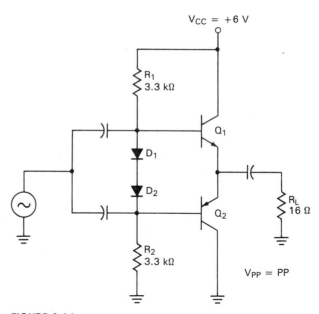

FIGURE 9.64

$V_{CC} = +6$ V

R₁ 3.3 kΩ

D₁

D₂

Q₁

Q₂

R_L 16 Ω

R₂ 3.3 kΩ

$V_{PP} = PP$

§9.8

39. Calculate the value of P_D for the transistor in Figure 9.53. [36]
40. Calculate the value of P_D for the transistor in Figure 9.54. [36]
41. Calculate the value of P_D for the transistor in Figure 9.57. [36]
42. Calculate the value of P_D for the transistor in Figure 9.58. [36]
43. Calculate the value of P_D for each transistor in Figure 9.61. [36]
44. Calculate the value of P_D for each transistor in Figure 9.62. [36]
45. Calculate the value of P_D for each transistor in Figure 9.63. [36]
46. Calculate the value of P_D for each transistor in Figure 9.64. [36]

TROUBLESHOOTING PRACTICE PROBLEMS

47. The transformer-coupled class A amplifier in Figure 9.65 has the dc voltages shown. Discuss the possible cause(s) of the problem.

48. The transformer-coupled class A amplifier in Figure 9.66 has the dc voltages shown. Discuss the possible cause(s) of the problem.

49. Determine the fault(s) that would cause the readings shown in Figure 9.67a.

50. Determine the fault(s) that would cause the readings shown in Figure 9.67b.

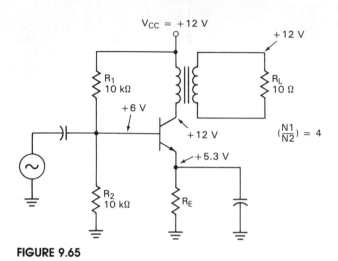

FIGURE 9.65

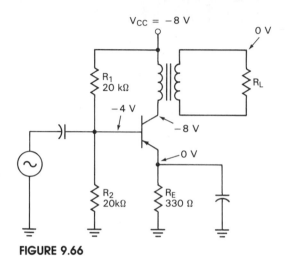

FIGURE 9.66

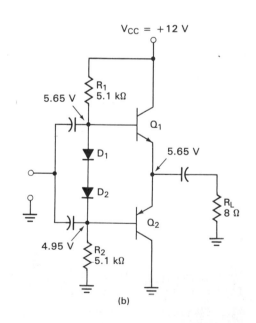

FIGURE 9.67

THE BRAIN DRAIN

51. Calculate the maximum allowable input power for the amplifier in Figure 9.61.

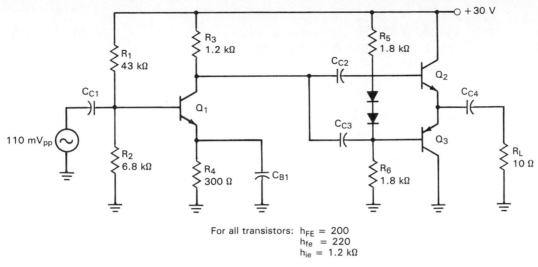

For all transistors: $h_{FE} = 200$
$h_{fe} = 220$
$h_{ie} = 1.2 \text{ k}\Omega$

FIGURE 9.68

52. Calculate the values of A_{vT}, T_{iT}, and A_{pT} for the two-stage amplifier in Figure 9.68.

53. Answer the following questions for the circuit in Figure 9.68. Use circuit calculations to explain your answers.
    a. Is the amplifier being driven to compliance?
    b. What type of clipping would the amplifier be most likely to experience?

54. Transistor Q_2 in Figure 9.68 is faulty. Can the 2N3904 be used in place of this transistor? Explain your answer using circuit calculations. The spec sheet for the 2N3904 is located on page 203. (*Hint*: The power-handling requirement of the circuit is the primary consideration in this problem.)

SUGGESTED COMPUTER APPLICATIONS PROBLEMS

55. Write a program that will determine the efficiency of a class AB amplifier when provided with the needed input values.

56. Write a program that will perform the complete dc analysis of a transformer-coupled class A amplifier, given the needed circuit values.

57. Write a program that will perform the complete ac analysis of a transformer-coupled class A amplifier, given the needed circuit values.

58. Write a program that will perform the complete dc and ac analysis of a class AB amplifier, given the needed values.

ANSWERS TO THE EXAMPLE PRACTICE PROBLEMS

9.1. 9.61 V_{PP}
9.2. 269 mW
9.3. 17.79 mW
9.4. 47.75 mW
9.5. 357.14 mW
9.6. 3.58%

9.7. 34.72%

9.8. The dc load line is vertical at $V_{CEQ} = 6$ V. The ac load line has end points of $V_{CE(\text{off})} = 6$ V and $I_{C(\text{sat})} = 2.73$ mA.

9.9. 8.18 mW

9.10. 2.93 W

9.11. 2.25 W

9.12. 76.8%

9.13. 16.4 mW

9.14. 468.75 mW

10
Field-Effect Transistors

The first field-effect transistor (shown here) was developed by the Bell Labs team of Ross and Dacey in 1955.

OBJECTIVES

☐ 1. List the two types of field-effect transistors (FETs). (Introduction)

☐ 2. Describe the physical construction of the *junction FET* (JFET). (§10.1)

☐ 3. Compare and contrast the physical construction, schematic symbols, and supply voltage polarities of the *n-channel* JFET and the *p-channel* JFET. (§10.1)

☐ 4. Explain the relationship between JFET channel width and drain current (I_D). (§10.1)

☐ 5. State the relationship between gate–source voltage (V_{GS}) and I_D. (§10.1)

☐ 6. Discuss the relationship between drain–source voltage (V_{DS}) and I_D. (§10.1)

☐ 7. State the significance of the V_P and I_{DSS} parameters. (§10.1)

☐ 8. State the significance of the $V_{GS(off)}$ parameter. (§10.1)

☐ 9. State the restriction on the biasing of the JFET gate–source junction. (§10.1)

☐ 10. Describe the gate input impedance characteristics of the JFET. (§10.1)

☐ 11. Calculate the value of I_D, given the values of V_{GS}, $V_{GS(off)}$, and I_{DSS}. (§10.1)

☐ 12. Plot the *transconductance curve(s)* of a JFET, given the device values of $V_{GS(off)}$ and I_{DSS}. (§10.1)

☐ 13. Calculate the Q-point values of I_D and V_{DS} for a *gate-bias* circuit. (§10.2)

☐ 14. State and explain the primary disadvantage of gate bias. (§10.2)

☐ 15. Explain how the *self-bias* circuit produces a negative value of V_{GS}. (§10.2)

☐ 16. Plot the dc bias line for a *self-bias* circuit and determine the Q-point values of I_D and V_{DS} using the bias line and the circuit schematic. (§10.2)

☐ 17. Plot the dc bias line for a *voltage-divider bias* circuit and determine the Q-point values of I_D and V_{DS} using the bias line and the circuit schematic. (§10.2)

☐ 18. Explain the common-source amplifier response to an ac input voltage. (§10.3)

☐ 19. Define *transconductance* and discuss its relationship with the position of a circuit Q-point on the transconductance curve. (§10.3)

☐ 20. Calculate the value of transconductance (g_m), given the values of V_{GS}, $V_{GS(off)}$, and g_{m0}. (§10.3)

☐ 21. Explain the relationship between JFET amplifier voltage gain and V_{GS}. (§10.3)

☐ 22. Calculate the minimum and maximum values of A_{vL} for a common-source amplifier. (§10.3)

☐ 23. State the purpose of swamping a JFET amplifier. (§10.3)

☐ 24. Calculate the minimum and maximum values of A_{vL} for a swamped common-source amplifier. (§10.3)

☐ 25. Explain why the input impedance of a JFET amplifier tends to be higher than that of a comparable BJT amplifier. (§10.3)

☐ 26. List the gain and impedance characteristics of the source follower. (§10.4)

☐ 27. Calculate the values of A_v, Z_{in}, and Z_{out} for a source follower. (§10.4)

☐ 28. List the gain and impedance characteristics of the common-gate amplifier. (§10.4)

☐ 29. Calculate the values of A_v, Z_{in}, and Z_{out} for a common-gate amplifier. (§10.4)

☐ 30. Compare and contrast the troubleshooting of a JFET amplifier to that of a comparable BJT amplifier. (§10.5)

☐ 31. List the characteristic symptoms of a *shorted-gate junction* and those of an *open-gate junction*. (§10.5)

☐ 32. Describe the procedure used to troubleshoot a JFET amplifier. (§10.5)

☐ 33. List and define the commonly used JFET parameters and ratings. (§10.6)

☐ 34. Discuss the use of a JFET amplifier as a *buffer*. (§10.6)

☐ 35. Discuss the difference between FET meters and standard VOMs. (§10.6)

☐ 36. Discuss the use of the JFET in an *RF amplifier*. (§10.6)

As you master each of the objectives listed, place a check mark (✔) in the appropriate box.

✔ 1

Field effect transistors (FETs).
Voltage-controlled devices.

You may recall that the *bipolar junction transistor* is a *current-controlled device*. That is, the output characteristics of the device are controlled by the base *current*, not the base voltage. Another type of transistor, called the *field-effect transistor*, or FET, is a *voltage-controlled device*. The output characteristics of the FET are controlled by the input voltage, not by the input current.

There are two basic types of FETs. The *junction field-effect transistor*, or JFET, and the *metal oxide semiconductor FET*, or MOSFET. As you will be shown, the operation of these two components vary, as do their applications and limitations. In this chapter, we will discuss JFETs and their circuits. MOSFETs and their circuits are covered in Chapter 11.

There are two types of FETs.

10.1
Introduction to JFETs

The physical construction of the JFET is significantly different from that of the bipolar transistor. You may recall that the bipolar transistor has three separate materials: either two *n*-type materials and a single *p*-type material, or two *p*-type materials and a single *n*-type material. The JFET has only *two* materials, a single *n*-type material and a single *p*-type material. The construction of the JFET is illustrated in Figure 10.1.

As you can see, the device has three terminals, just like the bipolar transistor. However, the terminals of the JFET are labeled *source*, *drain*, and *gate*. The *source* can be viewed as being the equivalent of the BJT's *emitter*, the *drain* is the equivalent of the *collector*, and the *gate* is the equivalent of the *base*.

✔ 2

Source. The JFET equivalent of the BJT's emitter.

Drain. The JFET equivalent of the BJT's collector.

Gate. The JFET equivalent of the BJT's base.

FIGURE 10.1

JFET construction.

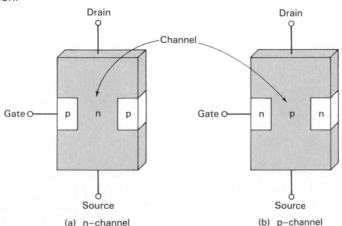

(a) n–channel   (b) p–channel

The material that connects the source to the drain is referred to as the *channel*. When this material is an *n*-type, the JFET is referred to as an *n-channel JFET*. Obviously, an FET with a *p*-type channel would be referred to as a *p*-channel *JFET*. Both of these JFET types are shown in Figure 10.1.

It would seem from the illustration that there are three material regions. However, this is not the case. The gate material *surrounds* the channel in the same manner as a belt surrounding your waist. The diagram shows a cross section of the component, so the gate material appears to be two separate materials.

The schematic symbols for the *p*-channel and *n*-channel JFETs are shown in Figure 10.2. Note that the arrow points "in" or "out" from the component. Just as with the BJT schematic symbol, the arrow points toward the *n*-type material. When the arrow is pointing *in*, it is pointing toward the channel. Thus Figure 10.2a is the schematic symbol for the *n*-channel JFET. When the arrow is pointing *out*, it can be viewed as pointing toward the gate. If the gate is an *n*-type material, the channel must be a *p*-type material. Thus Figure 10.2b is the schematic symbol for the *p*-channel JFET. If you think you will have trouble distinguishing the two symbols, just think "*n* . . . in." This will remind you that the *n*-channel JFET has the arrow pointing *in*.

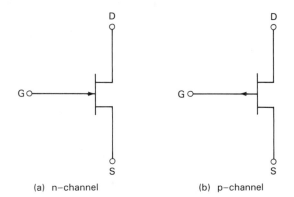

FIGURE 10.2

JFET schematic symbols.

(a) n–channel   (b) p–channel

JFET supply voltage polarities.

You may recall that the *pnp* transistor is normally used with *negative* supply voltages, while the *pnp* transistor is normally used with *negative* supply voltages. The same relationship holds true for the JFETs. The *n-channel* JFET is normally used with *positive* supply voltages, while the *p-channel* JFET is normally used with *negative* supply voltages. This point is illustrated in Figure 10.3. Note that the *n*-channel JFET circuit (Figure 10.3a) has a *positive* drain supply voltage (V_{DD}), while the *p*-channel JFET circuit (Figure 10.3b) has a *negative* supply voltage.

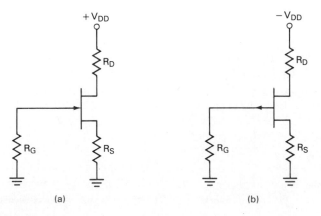

FIGURE 10.3

JFET supply voltages.

(a)   (b)

Operation Overview

In our discussions on BJTs, we concentrated on the *npn* transistor. At the same time, it was stated that all the relationships discussed also held true for the *pnp* transistor. The only differences were the voltage polarities and the current directions. In the same manner, our discussions on JFETs will concentrate on the *n-channel JFET*. Again, all the relationships covered will also apply to the *p*-channel JFET. The only difference will be the voltage polarities and current directions.

The overall operation of the JFET is based on *varying the width of the gate to control the drain current*. This point is illustrated in Figure 10.4. In Figure 10.4a, the drain source voltage (V_{DS}) is causing conduction through the JFET. You may recall that the amount of current through a conductor is *directly proportional* to the cross-sectional area (width) of the conductor. Thus, as the width of a conductor *decreases*, so does the amount of current through it at a set voltage. Now assuming that V_{DD} is a set voltage, we can decrease the amount of drain current by decreasing the width of the channel. This is shown in Figure 10.4b.

Drain current is controlled by the channel width.

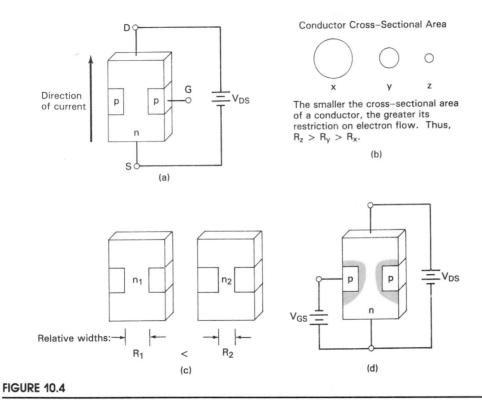

FIGURE 10.4

The relationship between channel width and drain current.

How do we decrease the width of the channel? By effectively *increasing* the size of the gate. For example, in Figure 10.4c you can see two JFETs with different gate widths. Since the JFET on the right has a wider gate, it has a narrower channel. Thus the resistance of this JFET (from source to drain) is greater than the resistance of the JFET on the left.

How do we increase the width of the gate? By applying a *reverse bias* to the gate–source junction. In Figure 10.4d, the biasing supply is shown as V_{GS}. Note that the negative terminal of the supply goes to the *p*-type gate, while the positive terminal is connected to the *n*-type source. This reverse biases the gate–source junction, causing a depletion layer to form. As you know, a depletion

Channel width is controlled by the gate.

layer will prevent current. Thus the area of the channel is effectively *decreased*, causing the drain current to decrease.

There are two ways to control channel width. First, by varying the value of V_{GS}, we can vary the width of the channel and, in turn, vary the amount of drain current. This point is illustrated in Figure 10.5. Note that, as V_{GS} increases, so does the size of the depletion layer. As the size of the depletion layer increases, the effective size of the channel decreases. As the effective size of the channel decreases, so does the amount of drain current. The relationship between V_{GS} and I_D can be summarized as follows: *As V_{GS} increases (becomes more negative), I_D decreases.*

The relationship between V_{GS} and I_D.

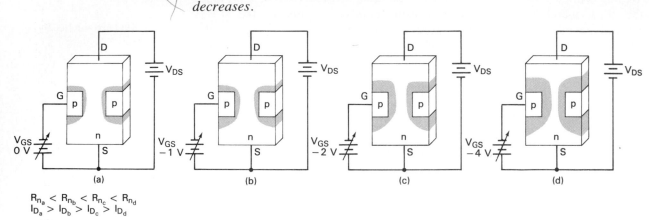

$R_{n_a} < R_{n_b} < R_{n_c} < R_{n_d}$
$I_{D_a} > I_{D_b} > I_{D_c} > I_{D_d}$

FIGURE 10.5

The relationship between V_{GS} and I_D.

The relationship between V_{DS} and I_D.

Figure 10.5 serves to illustrate another important point. In Figure 10.5a, V_{GS} is shown to be 0 V. At the same time, a small depletion layer is shown to be surrounding the gate. This small depletion layer is caused by the current through the channel. Increasing I_D will cause the size of the depletion layer to increase. Thus, the second method of increasing the size of the depletion layer is to hold V_{GS} constant and increase the value of V_{DS}. Increasing V_{DS} causes I_D to increase, which in turn causes the size of the depletion layer to increase. This is illustrated in Figure 10.6.

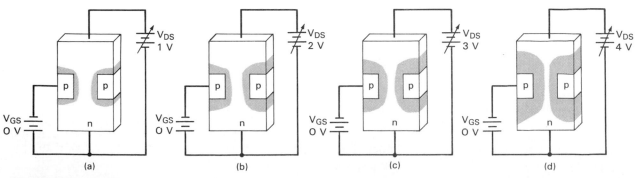

FIGURE 10.6

The effects of varying V_{DS} with constant V_{GS}.

At this point, there seems to be a contradiction in the theory of operation. It would seem that the increase in the size of the depletion layer would increase the resistance of the channel, preventing I_D from increasing. How is it that I_D is increasing if the conduction of the JFET channel is decreasing? While the channel of the component *is* becoming narrower, it is doing so at a very small rate. Because of this, the value of I_D continues to increase. However, a point will be reached where further increases in V_{DS} will be offset by proportional increases in the resistance of the channel. Thus, after a specified value of V_{DS}, further increases in V_{DS} will *not* cause an increase in I_D. The value of V_{DS} at which this occurs is called the *pinch-off voltage*. As V_{DS} is increased, I_D will increase until $V_{DS} = V_P$ (pinch-off voltage). If V_{DS} is increased above the value of V_P, the resistance of the channel will increase at the same rate and I_D will remain constant. This is shown in the JFET drain curve in Figure 10.7a.

Pinchoff voltage (V_P). With V_{GS} constant, the value of V_{DS} that causes maximum I_D. The pinchoff voltage rating of a JFET is measured at $V_{GS} = 0$ V.

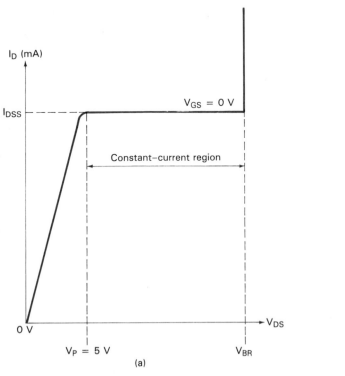
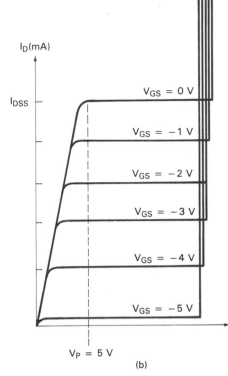

FIGURE 10.7

JFET drain curves.

Look at the portion of the curve below V_P. As V_{DS} rises from 0 V, I_D shows a constant increase. During this time, the resistance of the channel is also increasing, but at a lower rate than V_{DS}. By the time the value of V_P is reached, V_{DS} and the resistance of the channel are increasing at the same rate. Thus I_D remains fairly constant for further increases in V_{DS}. Note that the region of operation between V_P and V_{BR} (breakdown voltage) is called the *constant-current* region. As long as V_{DS} is kept within this range, I_D will remain constant for a constant value of V_{GS}.

The constant-current region.

When V_{GS} is equal to 0 V, you essentially have the gate and source terminals *shorted* together. When you have a shorted gate–source junction, the I_D will reach its maximum possible value, I_{DSS}. I_{DSS} is the *maximum* of current that can pass through the channel of a given JFET. This value, which is stated on the spec sheet for a given JFET, is measured under the following conditions:

Shorted-gate drain current (I_{DSS}). The maximum possible value of I_D.

$$\boxed{V_{GS} = 0 \text{ V}, \qquad V_{DS} = V_P}$$

Any current value for a given JFET will have to be less than or equal to the value of I_{DSS}. In this respect, it can be viewed as being the equivalent of $I_{C(sat)}$ for a given BJT circuit. The relationship between V_{GS}, V_{DS}, and I_{DSS} is also shown in Figure 10.7.

I_D is always less than or equal to I_{DSS}.

As V_{GS} is made more negative than 0 V, the JFET will pinch off at a lower voltage than the V_P rating for the component. Also, the constant-current value of I_D will decrease from I_{DSS}. This is shown in Figure 10.7b.

Let's focus on the decrease in I_D for a moment. As Figure 10.7b illustrates, continual increases in V_{GS} cause continual decreases in the constant-current value of I_D. If we continue to increase V_{GS}, we will eventually hit a point where minimum current will pass through the JFET. The value of V_{GS} that causes I_D to drop to approximately *zero* is called $V_{GS(off)}$. Note that the value of $V_{GS(off)}$ defines the limits on the value of V_{GS}. For conduction to occur from source to drain, the value of V_{GS} must be somewhere between 0 V and $V_{GS(off)}$.

Here's an interesting point: $V_{GS(off)}$ will always have the same magnitude value as V_P. For example, if V_P is 8 V, then $V_{GS(off)}$ will be -8 V. Since these two values are always equal and opposite, only one will be listed on the spec sheet for a given JFET.

You must be sure to remember that there is a definite difference between V_P and $V_{GS(off)}$. V_P is the value of V_{DS} that causes the JFET to become a constant-current component. It is measured at $V_{GS} = 0$ V and will have a constant drain current of $I_D = I_{DSS}$. $V_{GS(off)}$, on the other hand, is the value of V_{GS} that causes I_D to drop to nearly zero.

JFET Biasing

The gate–source junction of a JFET will *always* be reverse biased. Even when the gate and source terminals are shorted, I_D will cause the junction to be reverse biased, as shown in Figures 10.5 and 10.6. *The gate–source junction of a JFET must never be allowed to become forward biased* because the gate material is not constructed to handle any significant amount of current. If the junction is allowed to become forward biased, current will start to flow in the gate circuit. Under most circumstances, this current will destroy the component.

The fact that the gate is always reverse biased leads us to another important characteristic of the device. *JFETs have an extremely high characteristic gate impedance.* This impedance is typically in the mid to high MΩ range. For example, the 2N5457 lists the following input characteristics:

$$\text{gate reverse current } (I_{GSS}) = 1.0 \text{ nA maximum at } 25°C$$

when

$$V_{DS} = 0 \text{ V} \quad \text{and} \quad V_{GS} = -15 \text{ V}$$

If we use Ohm's law on the values of V_{GS} and I_{GSS} listed above, we can calculate a gate resistance of 15 GΩ!

The advantage of having this extremely high input impedance can be seen by taking a look at the circuit shown in Figure 10.8. Here the JFET amplifier is being driven by a high-impedance source. Because of the extremely high input impedance of the JFET gate, it draws no current from the source. Thus it is as if there were no load on the source at all. The high input impedance of the JFET has also led to its extensive use in microcomputer circuits. The low current requirements of the component make it perfect for use in ICs, where thousands of transistors must be etched onto a single piece of silicon. The low current draw helps the IC

Gate–source cutoff voltage ($V_{GS(off)}$). The value of V_{GS} that causes I_D to drop to zero.

V_P and $V_{GS(off)}$ are equal in *magnitude*.

Forward biasing a JFET gate–source junction may destroy the component.

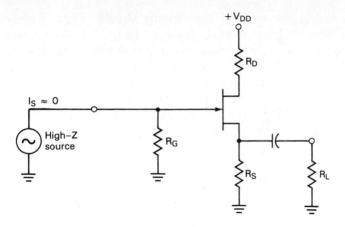

FIGURE 10.8

remain relatively cool, thus allowing more components to be placed in a smaller physical area.

Component Control

It was stated earlier that the JFET is a *voltage-controlled* device, while the BJT is a *current-controlled* device. We have now established the foundation necessary to look at this point in detail. We will start by reviewing the BJT as a current-

For both circuits: $I_B = \dfrac{V_{BB} - V_{BE}}{R_B}$

$I_C = h_{FE}I_B$
$V_{CE} = V_{CC} - I_C R_C$

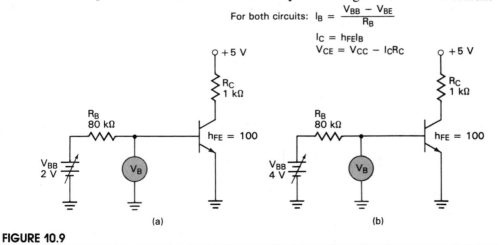

FIGURE 10.9

controlled component. Consider the circuit shown in Figure 10.9. Using the formulas listed, the circuit values for Figure 10.9a can be calculated as

$$I_B = 16.3 \ \mu A$$
$$I_C = 1.63 \ mA$$
$$V_{CE} = 3.38 \ V$$

Using the same formulas on the circuit in Figure 10.9b gives the following results:

$$I_B = 41.3 \ \mu A$$
$$I_C = 4.13 \ mA$$
$$V_{CE} = 0.88 \ V$$

Note that $V_B = V_{BE}$ for each of the circuits shown. Thus, from one circuit to another, the value of base voltage did not change. It stayed at approximately 0.7 V. Yet there were drastic changes in the output characteristics of the two amplifiers. These changes were caused by the change in I_B, not by any change in V_B. Therefore, it can be seen that the BJT is a current-controlled component. The output characteristics are determined by the input current, not the input voltage.

In contrast, the JFET has *no* gate current. It has been shown that the size of the channel is controlled by the amount of reverse bias applied to the gate–source junction. Since V_{GS} controls the JFET, it is easy to see that the JFET is *not* a current-controlled device, but rather a voltage-controlled device. This is the major difference between the JFET and the BJT.

Transconductance

Since the JFET has no gate current, there is no beta rating for the device. However, the output current (I_D) can be defined in terms of the circuit input as follows:

$$I_D = I_{DSS}\left(1 - \frac{V_{GS}}{V_{GS(\text{off})}}\right)^2 \qquad \textbf{(10.1)}$$

where I_{DSS} = the shorted-gate drain current rating of the device
V_{GS} = the gate–source voltage
$V_{GS(\text{off})}$ = the gate–source cutoff voltage

The following example illustrates the use of equation (10.1).

━━━━━━━━━━━━━━ **EXAMPLE 10.1** ━━━━━━━━━━━━━━

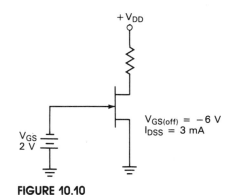

FIGURE 10.10

Determine the value of drain current for the circuit shown in Figure 10.10.

Solution: The drain current for the circuit is found as

$$I_D = I_{DSS}\left(1 - \frac{V_{GS}}{V_{GS(\text{off})}}\right)^2$$
$$= (3 \text{ mA})\left(1 - \frac{-2 \text{ V}}{-6 \text{ V}}\right)^2$$
$$= (3 \text{ mA})(0.444)$$
$$= 1.33 \text{ mA}$$

PRACTICE PROBLEM 10–1

A JFET with parameters of $I_{DSS} = 12$ mA and $V_{GS(\text{off})} = -6$ V is used in a circuit that provides a V_{GS} of -3 V. Determine the value of I_D for the circuit.

If you take a closer look at equation (10.1), you will notice that V_{GS} is the only value on the right side of the equation that will change for a specified JFET. In other words, for a specified JFET, I_{DSS} and $V_{GS(\text{off})}$ are parameters that will not change from one circuit to another or from one point to another. Based on the fact that I_{DSS} and $V_{GS(\text{off})}$ are constants for a given JFET, we can assume that I_D

is a function of the value of V_{GS}. As V_{GS} changes (which it will), I_D changes.

We can use a series of V_{GS} versus I_D values to plot what is called the *transconductance* curve for a specified JFET. A transconductance curve is a graph of all possible combinations of V_{GS} and I_D. The process for plotting the transconductance curve of a given JFET is as follows:

1. Plot a point on the x-axis that corresponds to the value of $V_{GS(\text{off})}$.
2. Plot a point on the y-axis that corresponds to the value of I_{DSS}.
3. Select two or three values of V_{GS} between 0 V and $V_{GS(\text{off})}$. For each value of V_{GS} selected, determine the corresponding values of I_D using equation (10.1).
4. Plot the points from step 3, and connect all the plotted points with a smooth curve.

Step 1 is based on the fact that $I_D = 0$ when $V_{GS} = V_{GS(\text{off})}$. Thus the point for $V_{GS(\text{off})}$ versus I_D will always fall on the x-axis of the graph. Step 2 is based on the fact that $I_D = I_{DSS}$ when $V_{GS} = 0$ V. Thus the point for $V_{GS} = 0$ versus I_D will always fall on the I_{DSS} value point on the y-axis. The following example demonstrates the use of the process stated for plotting the transconductance curve of a given JFET.

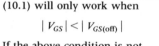

Transconductance curve. A plot of all possible combinations of V_{GS} and I_D.

How to plot a transconductance curve.

EXAMPLE 10.2

Plot the transconductance curve for the 2N5457. For this device, $V_{GS(\text{off})}$ is -6 V and I_{DSS} is 3 mA.

A Practical Consideration: Equation (10.1) will only work when

$$|V_{GS}| < |V_{GS(\text{off})}|$$

If the above condition is not fulfilled, the equation will give you a false value of I_D. Just remember, if V_{GS} is greater than (or equal to) $V_{GS(\text{off})}$, I_D will be zero.

Solution: With the values given, we know that our end points for the curve are $(-6$ V, 0 mA) and (0 V, 3 mA). We now use three values of V_{GS} (-1 V, -3 V, and -5 V) and calculate the corresponding value of I_D for each.

At $V_{GS} = -1$ V,

$$I_D = (3 \text{ mA})\left(1 - \frac{-1 \text{ V}}{-6 \text{ V}}\right)^2$$
$$= 2.08 \text{ mA}$$

At $V_{GS} = -3$ V,

$$I_D = (3 \text{ mA})\left(1 - \frac{-3 \text{ V}}{-6 \text{ V}}\right)^2$$
$$= 0.75 \text{ mA} \qquad (750 \text{ μA})$$

At $V_{GS} = -5$ V,

$$I_D = (3 \text{ mA})\left(1 - \frac{-5 \text{ V}}{-6 \text{ V}}\right)^2$$
$$= 0.083 \text{ mA} \qquad (83 \text{ μA})$$

We now have the following sets of V_{GS} versus I_D values:

V_{GS} (V)	I_D (mA)
-6	0
-5	0.083
-3	0.75
-1	2.08
0	3

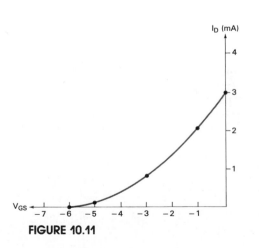

FIGURE 10.11

The points are now plotted and connected to form the curve shown in Figure 10.11.

PRACTICE PROBLEM 10–2

A JFET has parameters of $V_{GS(\text{off})} = -20$ V and $I_{DSS} = 12$ mA. Plot the transconductance curve for the device using V_{GS} values of 0 V, -5 V, -10 V, -15 V, and -20 V.

A Practical Consideration: Many JFET spec sheets will list only a *maximum* value of $V_{GS(\text{off})}$. When only a maximum value is listed:

a. Only one transconductance curve is plotted for the device.
b. It is generally expected (though not guaranteed) that the ratio of $V_{GS(\text{off})}$ (maximum) to $V_{GS(\text{off})}$ (minimum) will be approximately equal to the ratio of I_{DSS} (maximum) to I_{DSS} (minimum). For example, if the ratio of I_{DSS} values is approximately 5:1, the ratio of $V_{GS(\text{off})}$ values can be generally assumed to also be around 5:1. However, this ratio is neither guaranteed nor exact.

The transconductance curve for a given JFET will be used in both the dc and ac analyses of any amplifier using that JFET. For this reason it is important that you be able to plot the transconductance curve for any given JFET. When you need to plot the transconductance curve for a JFET, simply obtain the values of $V_{GS(\text{off})}$ and I_{DSS} from the spec sheet for the device, and then follow the procedure outlined in this section.

There is one important point that needs to be made at this time: Most JFET spec sheets will list more than one value of $V_{GS(\text{off})}$ and I_{DSS}. For example, the spec sheet for the 2N5457 lists the following:*

$V_{GS(\text{off}) \, (\text{minimum})}$	$= -0.5$ V,	I_{DSS} (minimum) $= 1$ mA
$V_{GS(\text{off}) \, (\text{maximum})}$	$= -6$ V,	I_{DSS} (maximum) $= 5$ mA

When a range of values is given, you must use the two *minimum* values to plot one curve and the two *maximum* values to plot another curve on the same graph. This procedure is illustrated in the following example.

EXAMPLE 10.3

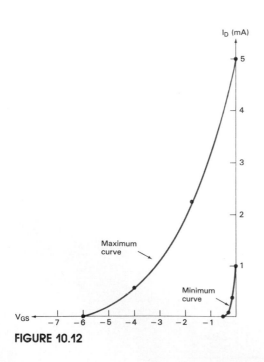

FIGURE 10.12

Using the minimum and maximum values of $V_{GS(\text{off})}$ and I_{DSS} for the 2N5457, plot the two transconductance curves for the device.

Solution: The maximum values of $V_{GS(\text{off})}$ and I_{DSS} are -6 V and 5 mA, respectively. Using these values and equation (10.1), we can solve for the following points:

V_{GS} (V)	I_D (mA)	
-6	0	
-4	0.556	(556 μA)
-2	2.222	
0	5	

Using the minimum values of $V_{GS(\text{off})} = -0.5$ V and $I_{DSS} = 1$ mA, we can solve for the following points with equation (10.1):

V_{GS} (V)	I_D (mA)	
-0.5	0	
-0.4	0.04	(40 μA)
-0.2	0.36	(360 μA)
0	1	

---

* We used the value of 3 mA for I_{DSS} in Examples 10.1 and 10.2. This value is listed as being "typical" on the 2N5457 spec sheet. (You can see this in Appendix A.)

The point sets listed above are used to plot two separate curves. These curves are shown in Figure 10.12.

PRACTICE PROBLEM 10–3

The 2N5486 JFET has values of $V_{GS(off)} = -2$ V to -6 V and $I_{DSS} = 8$ mA to 20 mA. Plot the minimum and maximum transconductance curves for the device.

There are a few more relationships that must be discussed involving the principle of transconductance. However, these principles relate to the ac operation of the JFET. Therefore, we will cover them in our discussion on the ac characteristics of JFETs. At this point, we have the foundation needed to cover JFET biasing circuits.

One Final Note

In this section, we have used n-channel JFETs in all the examples. This is due to the fact that n-channel JFETs are used far more commonly than p-channel JFETs. Just remember, all of the n-channel JFET principles apply equally to p-channel JFETs. The only differences are the voltage polarities and the direction of the drain and source currents. The n-channel and p-channel JFETs are contrasted in Figure 10.13.

FIGURE 10.13

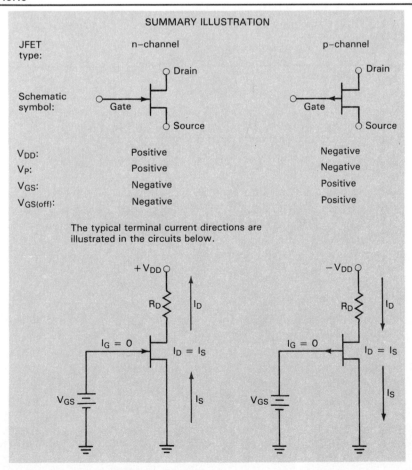

1. What is a *field-effect transistor*? What are the two types of JFETs? [Objective 1]
2. What do we call the three terminals of a JFET? [Objective 2]
3. Describe the physical relationship between the JFET gate and channel. [Objective 2]
4. What are the two types of JFETs? [Objective 2]
5. Draw the schematic symbol for each type of JFET. [Objective 3]
6. What supply voltage polarity is typically used for each type of JFET? [Objective 3]
7. What is the relationship between channel width and drain current? [Objective 4]
8. What is the relationship between V_{GS} and channel width? [Objective 4]
9. What is the relationship between V_{GS} and drain current? [Objective 5]
10. What is *pinchoff voltage*? [Objective 6]
11. What effect does an increase in V_{DS} have on I_D when $V_{DS} < V_P$? [Objective 6]
12. What effect does an increase in V_{DS} have on I_D when $V_{DS} > V_P$? [Objective 6]
13. What is the operating region above V_P called? [Objective 6]
14. What is I_{DSS}? [Objective 7]
15. What is the relationship between I_D and I_{DSS}? [Objective 7]
16. What is $V_{GS(\text{off})}$? [Objective 8]
17. What is the primary restriction on the value of V_{GS}? [Objective 9]
18. Why do JFETs typically have extremely high gate input impedance? [Objective 10]

10.2
JFET Biasing Circuits

JFET biasing circuits are very similar to BJT biasing circuits. The main difference between JFET circuits and BJT circuits is the operation of the active components themselves. In this section we cover those dc operating principles and relationships that differ from the BJT circuits already covered. We will not go into lengthy explanations of the circuit operation, as these explanations were made earlier.

Gate Bias

[13]

Reference:

Gate bias. The JFET equivalent of BJT base bias.

Do you remember *base bias*? The JFET equivalent of this circuit type is called *gate bias*. The gate-bias circuit is shown in Figure 10.14. The gate supply voltage $(-V_{GG})$ is used to ensure that the gate–source junction is reverse biased. Since there is no gate current, there is no voltage dropped across R_G, and the value of V_{GS} is found as

$$V_{GS} = V_{GG} \tag{10.2}$$

Using V_{GG} in equation (10.1) as V_{GS} allows us to calculate the value of I_D. Once

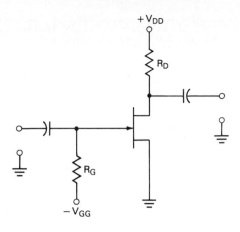

+V_{DD}

R_D

FIGURE 10.14

Gate bias.

I_D is known, V_{DS} for the JFET can be found as

$$\boxed{V_{DS} = V_{DD} - I_D R_D}$$

(10.3)

The following example illustrates the complete dc analysis of a simple gate–bias circuit.

EXAMPLE 10.4

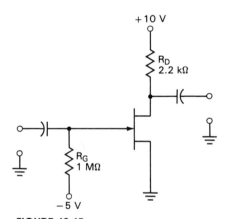

FIGURE 10.15

The JFET in Figure 10.15 has values of $V_{GS(off)} = -8$ V and $I_{DSS} = 16$ mA. Determine the values of V_{GS}, I_D, and V_{DS} for the circuit.

Solution: Since none of V_{GG} is dropped across the gate resistor, V_{GS} is found as

$$V_{GS} = V_{GG}$$
$$= -5 \text{ V}$$

Using this value of V_{GS} and the parameters listed above, the value of I_D is found as

$$I_D = I_{DSS}\left(1 - \frac{V_{GS}}{V_{GS(off)}}\right)^2$$

$$= 16 \text{ mA}\left(1 - \frac{-5 \text{ V}}{-8 \text{ V}}\right)^2$$

$$= 16 \text{ mA } (0.1406)$$

$$= 2.25 \text{ mA}$$

Now the value of V_{DS} is found as

$$V_{DS} = V_{DD} - I_D R_D$$
$$= 10 \text{ V} - (2.25 \text{ mA})(2.2 \text{ k}\Omega)$$
$$= 5.05 \text{ V}$$

PRACTICE PROBLEM 10–4

Assume that the JFET in Figure 10.15 has values of $V_{GS(off)} = -10$ V and $I_{DSS} = 12$ mA. Determine the values of V_{GS}, I_D, and V_{DS} for the circuit.

Example 10.4 illustrated the process for analyzing a gate-bias circuit. However, it assumed that $V_{GS(off)}$ and I_{DSS} have one specified value each. As you know, this is not usually the case.

The fact that a given type of JFET can have a range of values for $V_{GS(off)}$ and I_{DSS} leads us to a major problem with the gate bias circuit: *Gate bias will not provide a stable Q-point from one JFET to another*. This problem is illustrated in the following example.

EXAMPLE 10.5

A Practical Consideration: When only a maximum curve can be plotted, the Q-point can fall anywhere between the intersection of the bias line and the curve and the intersection of the bias line and either axis of the graph.

Determine the range of Q-point values for the circuit shown in Figure 10.16. Assume that the JFET has ranges of $V_{GS(off)}$ = -1 to -7 V and I_{DSS} = 2 to 9 mA.

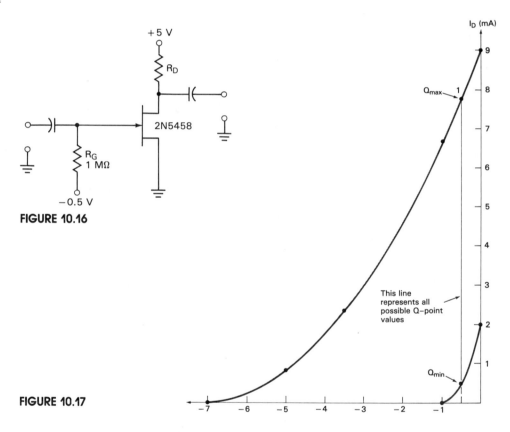

FIGURE 10.16

FIGURE 10.17

Solution: Using the established procedure, the two transconductance curves for the 2N5458 are plotted. These curves are shown in Figure 10.17. Since V_{GS} is set to -0.5 V, a vertical line, called a *bias line*, is drawn from this point through the two transconductance curves, as shown in Figure 10.17. The point where the line intersects the maximum curve is the value of Q_{max}. The point where the line intersects the minimum curve is the value of Q_{min}. The line between the two points represents all the possible Q-points for a 2N5458 used in the circuit in Figure 10.16.

As Example 10.5 shows, the circuit does not provide a stable Q-point. You may place one 2N5458 in the circuit and get an I_D of 5 mA. Another 2N5458 may give you an I_D of 2 mA, and so on. Because of the instability of the circuit, gate bias is rarely used for anything other than switching applications.

You may be wondering why the gate-bias circuit contains a gate resistor when there is no gate current. Since resistors are usually used to develop a voltage, it would seem that this resistor would be useless. After all, there is no voltage drop across the component and no current for it to limit. Actually, R_G is used for ac operation purposes. Take a look at the circuits shown in Figure 10.18. Figure 10.18a shows the results of eliminating R_G. The ac signal being produced by the first stage is coupled to the second, only to be shorted to the dc supply, $-V_{GG}$. The resistor is needed to prevent this from happening. As Figure 10.18b shows, placing a gate resistor in the second stage restores the ac signal. Therefore, while the gate resistor has little to do with the dc operation of the amplifier, it is a vital part of the ac operation of the circuit. As you will see, the *self-bias* circuit uses a gate resistor for the same reason. to prevent ac from going to Ground.

The purpose served by R_G.

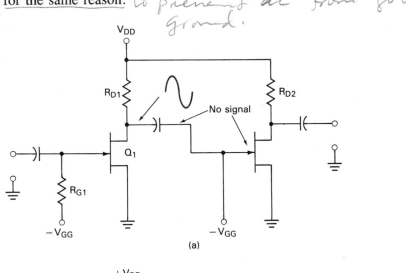

(a)

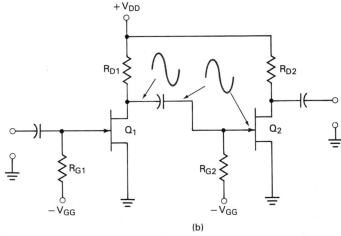

(b)

FIGURE 10.18

The purpose served by the gate resistor, R_G

✔ 15

Self-bias

Self-bias is a more viable type of JFET biasing. The self-bias circuit replaces the gate supply ($-V_{GG}$) with a source resistor (R_S). The self-bias circuit is shown in

Self-bias. A JFET biasing circuit that uses a source resistor to establish a negative V_{gs}.

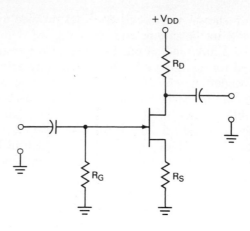

FIGURE 10.19

Self-bias.

Figure 10.19. Note that the gate is grounded through R_G, and a resistor has been added in the source circuit. This resistor is used to produce the $-V_{GS}$ needed for the operation of the JFET. Figure 10.20 helps to illustrate this point.

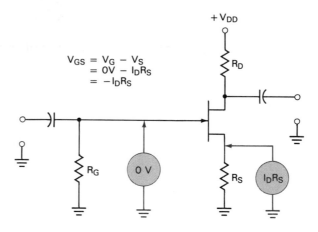

$$V_{GS} = V_G - V_S$$
$$= 0V - I_DR_S$$
$$= -I_DR_S$$

FIGURE 10.20

V_{GS} for the self-bias current.

In the JFET circuit, *all* the source current flows out of the drain. This is due to the fact that there is no gate current. Therefore, we can define source current as

$$\boxed{I_S = I_D} \tag{10.4}$$

With the full value of drain current flowing through the source resistor, the voltage across the resistor is found as

$$\boxed{V_S = I_DR_S} \tag{10.5}$$

Since no current flows in the gate circuit, no voltage is developed across the gate resistor. Thus the gate voltage (with respect to ground) is given as

$$\boxed{V_G = 0 \text{ V}} \tag{10.6}$$

Now we can use the relationships shown in equations (10.5) and (10.6) to show

how the source resistor is used to reverse bias the gate–source junction. The voltage *from gate to source* can be expressed as

$$V_{GS} = V_G - V_S \qquad \textbf{(10.7)}$$

Substituting the values in equations (10.5) and (10.6) in place of V_G and V_S gives us

$$V_{GS} = 0 \text{ V} - I_D R_S$$

which can be written as

$$V_{GS} = -I_D R_S \qquad \textbf{(10.8)}$$

Thus $-V_{GS}$ is developed by the drain current flowing through the source resistor.

If equation (10.8) is confusing, just remember that V_{GS} is the voltage measured from the gate terminal *to the source terminal*. There is no rule stating that the voltage at the gate terminal must be negative, only that the gate terminal must be more negative than the source terminal. If the source terminal is at some positive voltage and the gate is grounded, the gate terminal is *more negative* (less positive) than the source terminal. Figure 10.21 helps to illustrate this point. In all three circuits, the gate is 2 V more negative than the source. Thus the value of V_{GS} for each circuit is -2 V. Note that the value of V_G changes from one circuit to another, as does the value of V_S. However, the value of V_{GS} is constant for the three circuits. Make sure that you are comfortable with this point before continuing.

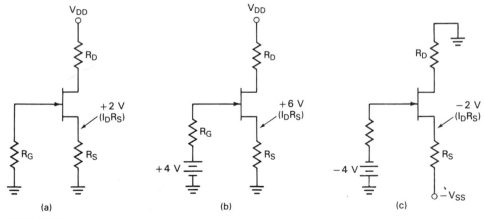

FIGURE 10.21

Three circuits with $V_{GS} = -2V$.

Equation (10.8) is the *bias line equation* for the self-bias circuit. To plot the dc bias line for a self-bias circuit, you simply follow this procedure:

1. Plot the minimum and maximum transconductance curves for the JFET being used in the circuit.

   How to plot the self-bias dc bias line.

2. Choose any value of V_{GS} and determine the corresponding value of I_D using

$$I_D = \frac{-V_{GS}}{R_S} \qquad \textbf{(10.9)}$$

3. Plot the point determined by equation (10.9), and draw a line from this point to the graph *origin* (the [0, 0] point).

4. The points where the line crosses the two transconductance curves define the limits of the Q-point operation of the circuit.

The procedure outlined here is illustrated in the following example.

EXAMPLE 10.6

Determine the range of Q-point values for the circuit shown in Figure 10.22a.

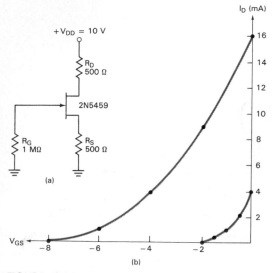

FIGURE 10.22

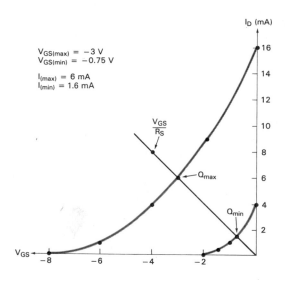

FIGURE 10.23

Solution: The 2N5459 spec sheet lists the following values for $V_{GS(off)}$ and I_{DSS}:

$$V_{GS(off)} = -2 \text{ to } -8 \text{ V}$$

$$I_{DSS} = 4 \text{ to } 16 \text{ mA}$$

Using these two sets of values and equation (10.1), the transconductance curves are plotted, as is shown in Figure 10.22b.

The value of $V_{GS} = -4$ V is chosen to calculate I_D. Using equation (10.9), I_D at $V_{GS} = -4$ V is found as

$$I_D = \frac{-V_{GS}}{R_S}$$
$$= 8 \text{ mA}$$

This point $(-4, 8)$ is plotted on the graph, and a line is drawn from the point to the graph origin. The plot of the line is shown in Figure 10.23. Note that the points where the bias line intersects the transconductance curves are labeled Q_{max} and Q_{min}. Using the graph coordinates of these two points, we obtain the maximum and minimum values of V_{GS} and I_D shown in Figure 10.23.

As Example 10.6 has shown, there is still some instability in the Q-point. In other words, we have not succeeded in obtaining a completely stable Q-point value by switching to self-bias. However, as Figure 10.24 indicates, the Q-point of a self-bias circuit is far more stable for a given JFET than the gate-bias circuit. In Figure 10.24, the bias line from Example 10.6 is shown as plotted in the example. A dashed line is plotted showing the same transistor in a gate-bias circuit. From these two dc bias lines, a few points can be made:

1. The gate-bias circuit is not subject to any change in V_{GS} and thus is more stable in this respect.
2. The possible change in I_D from one 2N5459 to another is much greater for the gate-bias circuit than for the self-bias circuit.

Thus the gate-bias circuit has the more stable value of V_{GS}, while the self-bias circuit has the more stable value of I_D. Which is more important? This question is best answered by considering the ac operation of amplifiers in general. AC characteristics such as compliance, amplifier power dissipation, and load power all depend on the output voltage and current characteristics of the amplifier. You may recall that PP, P_L, and P_S for the BJT amplifier are all defined in terms of V_{CEQ} and I_{CQ}. For the JFET amplifier, the ac output characteristics are defined in terms of V_{DS} and I_D. The relationship between V_{DS} and I_D is given as

$$\boxed{V_{DS} = V_{DD} - I_D(R_D + R_S)} \qquad (10.10)$$

If the ac output characteristics of the JFET amplifier are to be stable, I_D must be as stable as possible. Since self-bias provides a more stable value of I_D than gate bias, it is the preferred biasing method of the two.

The complete dc analysis of a self-bias circuit is actually pretty simple. The process starts with plotting the transconductance curves and the dc bias line, as was done in Example 10.6. Then, using the Q-point limits, the minimum and maximum values of V_{GS} and I_D are determined. These values can then be used to determine the minimum and maximum values of V_{DS}. The analysis process is illustrated in the following example.

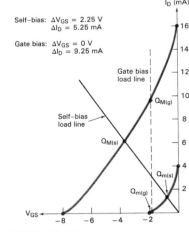

FIGURE 10.24

Self-bias stability versus gate-bias stability.

A Practical Consideration: Even though self-bias provides more stability, gate bias is still preferred for many switching applications. JFET switching applications are discussed in Chapter 18.

EXAMPLE 10.7

Determine the dc characteristics of the amplifier used in Example 10.6 (Figure 10.22).

Solution: We have already determined the limits of V_{GS} and I_D (Figure 10.23) as

$$V_{GS} = -0.75 \text{ to } -3 \text{ V}$$

$$I_D = 0.75 \text{ to } 6 \text{ mA}$$

Using the two values of I_D listed, we can determine the limits of V_{DS}. When $I_D = 0.75$ mA, V_{DS} is found as

$$\begin{aligned}V_{DS} &= V_{DD} - I_D(R_D + R_S)\\ &= 10 \text{ V} - (1.6 \text{ mA})(1 \text{ k}\Omega)\\ &= 10 \text{ V} - 1.6 \text{ V}\\ &= 8.4 \text{ V}\end{aligned}$$

When $I_D = 6$ mA, V_{DS} is found as

$$\begin{aligned}V_{DS} &= V_{DD} - I_D(R_D + R_S)\\ &= 10 \text{ V} - (6 \text{ mA})(1 \text{ k}\Omega)\\ &= 10 \text{ V} - 6 \text{ V}\\ &= 4 \text{ V}\end{aligned}$$

Thus the value of V_{DS} will fall between 4 and 8.4 V, depending on the particular 2N5459 used in the circuit.

PRACTICE PROBLEM 10–7

The circuit described in Practice Problem 10–6 has a 1-kΩ drain resistor and a 9-V_{dc} source. Determine the range of V_{DS} values for the circuit.

Example 10.7 illustrates one important point: Even though the self-bias circuit is more stable than gate bias, it still leaves a lot to be desired. A range of $V_{DS} = 4$ to 9.25 V would hardly be stable enough for most linear applications. We must therefore look for a circuit to provide a much greater amount of stability. *Voltage-divider bias* does the job extremely well.

Voltage-divider Bias

The voltage-divider bias JFET amplifier is very similar to its BJT counterpart. This biasing circuit is shown in Figure 10.25. The gate voltage for this amplifier is found in the same manner as V_B in the BJT circuit. By formula,

$$V_G = V_{DD}\frac{R_2}{R_1 + R_2} \tag{10.11}$$

Also, the difference between V_G and V_S is equal to V_{GS}. Therefore, the value of I_D can be found as

$$I_D = \frac{V_S}{R_S}$$

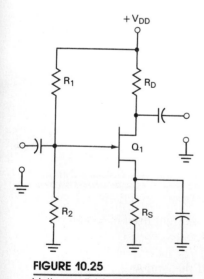

FIGURE 10.25

Voltage divider bias.

or

$$I_D = \frac{V_G - V_{GS}}{R_S} \tag{10.12}$$

Now we have a problem. If you refer to Examples 10.6 and 10.7 on the self-bias circuit, you will see that V_{GS} changes from one transistor to another. The same

CHAP. 10   Field-Effect Transistors

holds true for the voltage-divider biased circuit. This can make it difficult to determine the exact value of I_D. However, plotting the dc bias line for this circuit will show you that the problem is not really a major one.

The method used to plot the dc bias line for the voltage-divider bias circuit is a bit strange—easy, but strange. The exact process is as follows:

17

1. As usual, plot the transconductance curves for the specific JFET.
2. Calculate the value of V_G using equation (10.11).
3. *On the positive x-axis of the graph*, locate the voltage point equal to V_G.
4. Solve for I_D using

How to plot the voltage-divider bias dc bias line.

$$I_D = \frac{V_G}{R_S} \qquad (10.13)$$

5. Locate the point on the y-axis that corresponds to the value found in step 4.
6. Draw a line from the V_G point through the point found with equation (10.13), and continue the line to intersect *both* transconductance curves. The intersection points represent the Q_{max} and Q_{min} points.

The following example illustrates this procedure.

EXAMPLE 10.8

Plot the dc bias line for the circuit shown in Figure 10.26.

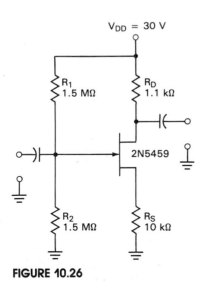

FIGURE 10.26

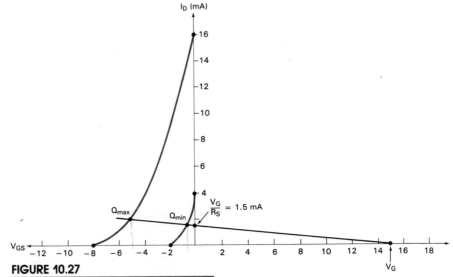

FIGURE 10.27

The voltage-divider bias dc bias line.

Solution: Figure 10.26 contains the 2N5459, the JFET whose transconductance curves were plotted in previous examples. Those transconductance curves can be seen in Figure 10.27. The value of V_G for the circuit is calculated as

$$V_G = V_{DD} \frac{R_2}{R_1 + R_2}$$
$$= 15 \text{ V}$$

This point is now plotted on the *positive* x-axis, as shown in Figure 10.27. The value of V_G is also used to find the y-axis intersect point, as follows:

$$I_D = \frac{V_G}{R_S}$$
$$= 1.5 \text{ mA}$$

Note that this point is marked on the y-axis of the graph. A line is then drawn through the two points and on through the two curves to mark the Q_{max} and Q_{min} points.

PRACTICE PROBLEM 10–8

In Practice Problem 10–3, you plotted the two transconductance curves for the 2N5486. This component is being used in a voltage-divider biased amplifier with values of $V_{DD} = 36$ V, $R_1 = 10$ MΩ, $R_2 = 3.3$ MΩ, $R_D = 1.8$ kΩ, and $R_S = 3$ kΩ. Plot the dc bias line for the amplifier.

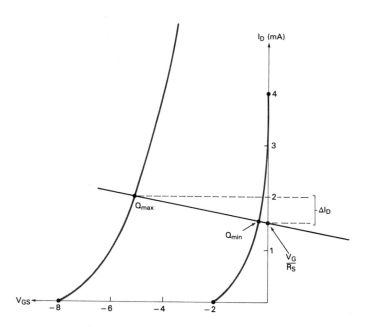

FIGURE 10.28

A close-up of Figure 10.27.

Figure 10.27 shows that we again have a relatively unstable value of V_{GS}. However, the stability of I_D has improved a great deal. Figure 10.28 shows a close-up of the lower portion of Figure 10.27 to help illustrate this point. While V_{GS} is varying between approximately −5 and −0.5 V, I_D varies only from 2 mA to approximately 1.5 mA. Thus, this amplifier provides a much more stable value of I_D than either gate bias or self-bias. For comparison, the results from Figures 10.24 and 10.28 are summarized as follows:

Circuit Type	Change in 2N5459 Drain Current Values (mA)
Gate bias	9.25
Self-bias	5.25
Voltage-divider bias	0.5

As you can see, the voltage-divider bias circuit is by far the most stable of the three.

It would seem that the stability of I_D would be impossible given the large possible variations in V_{GS} for the amplifier in Example 10.8. Since I_D is a function of V_{GS}, as was stated in equation (10.1), it would seem that a large change in V_{GS} would cause a large change in I_D. However, an analysis of the circuit would show the Q-point values obtained to be valid. This analysis is performed in the following example.

═══ **EXAMPLE 10.9** ═══

Example 10.8 gave minimum and maximum Q-point values for Figure 10.26 as

$$Q_{max} = -5 \text{ V at 2 mA}$$

$$Q_{min} = -0.5 \text{ V at 1.5 mA} \quad \text{(approximate)}$$

Verify these two Q-point values as being possible operating combinations of V_{GS} and I_D.

Solution: We will tackle the Q_{max} values of $V_{GS} = -5$ V and $I_D = 2$ mA first. Just as before, we start by determining the value of V_G, as follows:

$$V_G = V_{DD} \frac{R_2}{R_1 + R_2}$$
$$= 15 \text{ V}$$

Now recall that V_S is equal to the difference between V_G and V_{GS}. By formula,

$$V_S = V_G - V_{GS}$$

Since V_{GS} is a *negative* value, the value of V_S will end up being *higher* than the value of V_G, as follows:

$$\begin{aligned} V_S &= V_G - V_{GS} \\ &= 15 \text{ V} - (-5 \text{ V}) \\ &= 15 \text{ V} + 5 \text{ V} \\ &= 20 \text{ V} \end{aligned}$$

Therefore, I_D is found as

$$\begin{aligned} I_D &= \frac{V_G - V_{GS}}{R_S} \\ &= \frac{20 \text{ V}}{10 \text{ k}\Omega} \\ &= 2 \text{ mA} \end{aligned}$$

Thus the combination of $V_{GS} = -5$ V and $I_D = 2$ mA has been shown to be a possible combination of V_{GS} and I_D for the circuit. Next, we will verify the Q_{min} values of $V_{GS} =$

−0.5 V and $I_D \cong 1.5$ mA. At $V_{GS} = -0.5$ V and $V_G = 15$ V,

$$
\begin{aligned}
I_D &= \frac{V_G - V_{GS}}{R_S} \\
&= \frac{15\text{ V} - (-0.5\text{ V})}{10\text{ k}\Omega} \\
&= \frac{15\text{ V} + 0.5\text{ V}}{10\text{ k}\Omega} \\
&= \frac{15.5\text{ V}}{10\text{ k}\Omega} \\
&= 1.55\text{ mA}
\end{aligned}
$$

Again, the results obtained from plotting the bias line of Figure 10.24 have been proved to be accurate.

PRACTICE PROBLEM 10–9

Using the technique shown in Example 10.9, verify your Q-point values for Practice Problem 10–8.

When you are analyzing a voltage-divider biased JFET amplifier, the procedure is the same as that which we followed in plotting the transconductance curves and verifying the results. The only other value that must be determined is V_{DS}. V_{DS} is determined for this circuit as it is with all the others—with equation (10.10). In the following example, we calculate the range of possible V_{DS} values for the circuit we have covered in this section.

EXAMPLE 10.10

Determine the minimum and maximum values of V_{DS} for the circuit in Figure 10.26.

Solution: When $I_D = 2$ mA,

$$
\begin{aligned}
V_{DS} &= V_{DD} - I_D(R_D + R_S) \\
&= 30\text{ V} - (2\text{ mA})(11.1\text{ k}\Omega) \\
&= 30\text{ V} - 22.2\text{ V} \\
&= 7.8\text{ V}
\end{aligned}
$$

When $I_D = 1.55$ mA,

$$
\begin{aligned}
V_{DS} &= V_{DD} - I_D(R_D + R_S) \\
&= 30\text{ V} - (1.55\text{ mA})(11.1\text{ k}\Omega) \\
&= 30\text{ V} - 17.21\text{ V} \\
&= 12.79\text{ V}
\end{aligned}
$$

PRACTICE PROBLEM 10–10

Determine the minimum and maximum values of V_{DS} for the circuit described in Practice Problem 10–8. Use the calculated values of I_D from Practice Problem 10–9.

We now have a situation where the value of I_D is much more stable than it has been with previous circuits. However, the smaller variation in I_D *still* allows a rather drastic change in V_{DS}.

Summary

Gate bias, self-bias, and voltage-divider bias are the three most commonly used JFET biasing circuits. Each of these circuits has its own type of dc bias line and circuit advantages and disadvantages. The characteristics of these three dc biasing circuits are summarized in Figure 10.29.

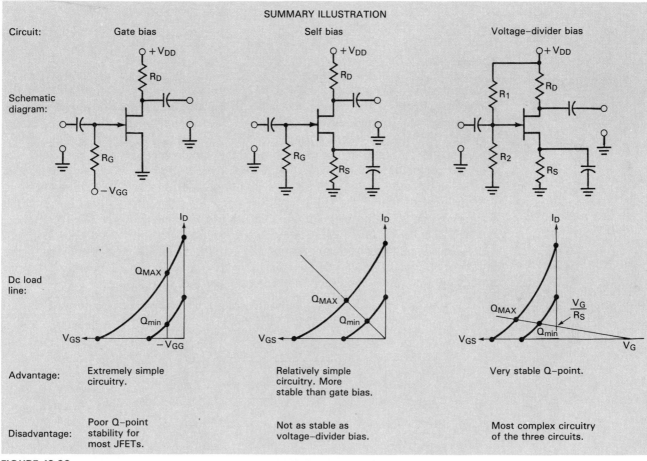

FIGURE 10.29

---

1. What is the primary disadvantage of gate bias? [Objective 14]
2. How does self-bias produce a negative value of V_{GS}? [Objective 15]
3. What is the process used to plot the dc bias line for the self-bias circuit? [Objective 16]
4. What is the process used to plot the dc bias line for the voltage-divider bias circuit? [Objective 17]

SECTION REVIEW

In many ways, the ac operation of a JFET amplifier is just like that of the BJT amplifier. However, there are some differences that need to be examined in detail, such as the effects of changes in JFET transconductance.

In the next two sections, we will examine the ac operation of JFET amplifiers. We will start with the *common-source* amplifier, simply because it is the most widely used of the JFET amplifiers. Then, in the next section, we will look at the operation of the *common-gate* and *source follower* (common-drain) amplifiers.

Operation Overview

✔ 18

There are many similarities between BJT amplifiers and JFET amplifiers. At the same time, there are differences that make JFET amplifiers more desirable in some applications and limit their use in others. One of the biggest differences is the fact that JFETs are voltage-controlled components and BJTs are current-controlled components. Another major difference is the input impedance of the JFET, as opposed to the input impedance of the BJT. As you will see, these differences play a major role in which transistor type is selected for a specific application.

Common-source amplifier.
The JFET equivalent of a common-emitter amplifier.

In general terms, the JFET amplifier responds to an input signal in much the same way as a BJT amplifier. Figure 10.30 shows a *common-source* JFET amplifier, which is the JFET equivalent of the *common-emitter* amplifier. Note that the input and output signals for the common-source amplifier closely resemble those of the common-emitter amplifier. Both amplifiers have a 180° signal phase shift from input to output. Both amplifiers provide *gain* for the input signal.

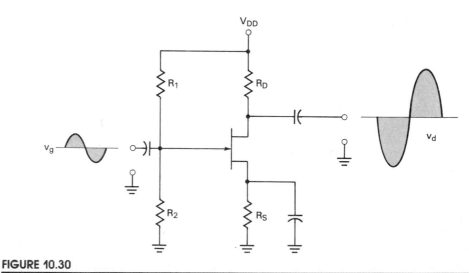

FIGURE 10.30

The common-source amplifier.

Even though the common-source and common-emitter amplifiers serve the same basic purpose, the means by which they operate are quite different. AC operation of the common-source amplifier is better explained with the help of

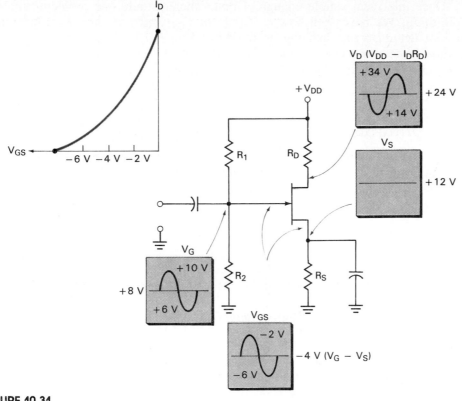

FIGURE 10.31

Common-source amplifier ac operation.

Figure 10.31. The initial dc operating values are represented as straight-line values in the blocks. These voltage values are assumed to be as follows:

$$V_G = +8 \text{ V}$$

$$V_S = +12 \text{ V}$$

$$V_D = +24 \text{ V}$$

$$V_{GS} = V_G - V_S = -4 \text{ V}$$

Here's what happens when V_G is caused to increase by the input signal:

1. V_G increases to $+10$ V.
2. V_{GS} becomes -2 V ($V_{GS} = V_G - V_S = 10$ V $- 12$ V $= -2$ V).
3. The *decrease* in V_{GS} causes an *increase* in I_D, as shown in the transconductance curve.
4. Increasing I_D causes V_D to decrease ($V_D = V_{DD} - I_D R_D$).

Thus, when the input voltage increases, the output voltage decreases. Here is what happens when V_G is caused to decrease by the input signal:

1. V_G decreases to $+6$ V.
2. V_{GS} becomes -6 V ($V_{GS} = V_G - V_S = 6$ V $- 12$ V $= -6$ V).
3. The *increase* in V_{GS} causes a *decrease* in I_D, as shown in the transconductance curve.
4. Decreasing I_D causes V_D to increase ($V_D = V_{DD} - I_D R_D$).

How the common-source amplifier responds to an ac input voltage.

Does this seem simple enough? Good. There is only one problem with the whole thing: We have seen how *a change in V_{GS} causes a change in I_D*. When you start dealing with changing quantities of V_{GS} and I_D, you begin to get into the whole issue of transconductance. Next, we will take a closer look at transconductance and its effects on ac operation.

Transconductance

We have been plotting transconductance curves throughout this chapter. Up to this point, we have considered transconductance to be a graph of V_{GS} versus I_D. *Transconductance* is actually *a ratio of a change in drain current to a change in gate–source voltage*. By formula,

$$g_m = \frac{\Delta I_D}{\Delta V_{GS}} \tag{10.14}$$

where g_m = the transconductance of the JFET at a given value of V_{GS}
  ΔI_D = the change in I_D
  ΔV_{GS} = the change in the initial value of V_{GS}

As a rating, transconductance is measured in *microsiemens* (μS) or *micromhos* (μmhos). You may recall that conductance is measured in *siemens* or *mhos*.

We have plotted several transconductance curves throughout the chapter. The maximum-value transconductance curve for the 2N5459 will help you see one of the problems with using transconductance in ac calculations. This curve is shown in Figure 10.32. Note that two sets of points were selected on the curve, each

FIGURE 10.32

The maximum value transconductance curve for the 2N5459.

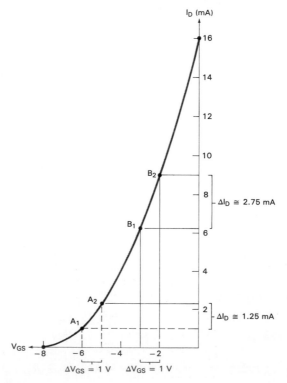

marking a ΔV_{GS} of 1 V. Points A_1 and A_2 correspond to a change in V_{GS} of 1 V, as do points B_1 and B_2. For each set of points, the corresponding change in I_D is shown. The values of ΔV_{GS} and their corresponding values of ΔI_D are summarized as follows:

ΔV_{GS} (V)	ΔI_D (mA)
-6 to $-5 = 1$	1.25
-3 to $-2 = 1$	2.75

Using equation (10.14), we can calculate the values of g_m that correspond to the changes above as

$$g_{m(A)} = \frac{\Delta I_D}{\Delta V_{GS}}$$
$$= \frac{1.25 \text{ mA}}{1 \text{ V}}$$
$$= 1250 \text{ } \mu S$$

and

$$g_{m(B)} = \frac{\Delta I_D}{\Delta V_{GS}}$$
$$= \frac{2.75 \text{ mA}}{1 \text{ V}}$$
$$= 2750 \text{ } \mu S$$

As you can see, the value of g_m is not constant for the entire transconductance curve. Figure 10.33 shows a single transconductance curve, with three possible

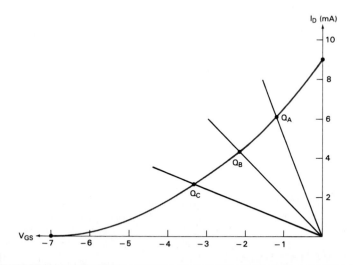

FIGURE 10.33

dc bias lines. A single JFET operated at these three points would have three separate values of g_m. Luckily, there is an equation that will tell you the value of g_m at a specific value of V_{GS}:

$$g_m = g_{m0}\left(1 - \frac{V_{GS}}{V_{GS(\text{off})}}\right) \qquad \textbf{(10.15)}$$

where g_m = the value of transconductance for the specific application
$g_{m(0)}$ = the maximum value of g_m, measured at $V_{GS} = 0$ V

The following example shows how the value of g_m is calculated for a specific application.

EXAMPLE 10.11

The 2N5459 has a value of $g_{m0} = 4000$ μS. Determine the values of g_m at $V_{GS} = -3$ V and $V_{GS} = -5$ V for this device. For the 2N5459, $V_{GS(off)} = -8$ V.

Solution: At $V_{GS} = -3$ V,

$$g_m = g_{m0}\left(1 - \frac{V_{GS}}{V_{GS(off)}}\right)$$
$$= 4000\ \mu S \left(1 - \frac{-3\ V}{-8\ V}\right)$$
$$= 4000\ \mu S\ (0.625)$$
$$= 2500\ \mu S$$

At $V_{GS} = -5$ V,

$$g_m = 4000\ \mu S \left(1 - \frac{-5\ V}{-8\ V}\right)$$
$$= 4000\ \mu S\ (0.375)$$
$$= 1500\ \mu S$$

PRACTICE PROBLEM 10–11

The 2N5486 has maximum values of $g_{m0} = 8000$ μS and $V_{GS(off)} = -6$ V. Using these values, determine the values of g_m at $V_{GS} = -2$ V and $V_{GS} = -4$ V.

On many specification sheets, the value of g_{m0} will be listed as y_{fs} or g_{fs}. When no value of maximum transconductance is given, you can approximate the value of g_{m0} as

$$\boxed{g_{m0} \cong \frac{2I_{DSS}}{V_{GS(off)}}} \tag{10.16}$$

While equation (10.16) will not give you an exact value of g_{m0}, the value obtained will usually be close enough to serve as a valid approximation.

Amplifier Voltage Gain

The ac operation of the JFET amplifier is closely related to the value of g_m at the dc value of V_{GS}. Consider the circuit shown in Figure 10.34. Figure 10.34a shows the basic common-source JFET amplifier. The circuit shown in Figure 10.34b shows the ac equivalent of the original circuit. Note that the biasing resistors (R_1 and R_2) have been replaced with a single parallel equivalent resistor, r_G. Also, the drain and load resistances (R_D and R_L) have been replaced by the single parallel equivalent resistor r_D. This is the standard ac equivalent circuit, as developed in our discussions on BJT amplifiers.

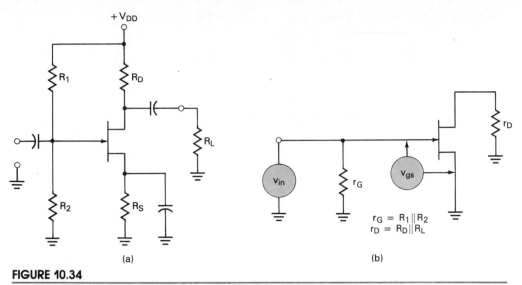

FIGURE 10.34

The common-source amplifier and its ac equivalent circuit.

For Figure 10.34b, the voltage gain would be found as

✔ 21

$$A_{vL} = g_m r_D \qquad (10.17)$$

Equation (10.17) points out another problem with the JFET amplifier. As you can see, the value of A_{vL} depends on the value of g_m. You have also been shown that:

1. The value of g_m depends on the value of V_{GS}.
2. The value of V_{GS} can vary from one JFET to another.

This leads to the conclusion that *the voltage gain of an amplifier can vary from one JFET to another*. This point is illustrated in the following example.

EXAMPLE 10.12

✔ 22

Determine the maximum and minimum values of A_v for the amplifier shown in Figure 10.35. Assume that the 2N5459 has values of $g_{m0} = 6000$ μS at $V_{GS(off)} = -8$ V and $g_{m0} = 2000$ μS at $V_{GS(off)} = -2$ V.

Solution: The transconductance curves and dc bias line for the amplifier are shown in Figure 10.35b. As shown, the maximum and minimum values of V_{GS} are -5 and -1 V, respectively. Using the maximum values of $V_{GS(off)}$ and V_{GS}, g_m is found as

$$g_m = g_{m0}\left(1 - \frac{V_{GS}}{V_{GS(off)}}\right)$$
$$= 6000 \ \mu S\left(1 - \frac{-5 \ V}{-8 \ V}\right)$$
$$= 2250 \ \mu S$$

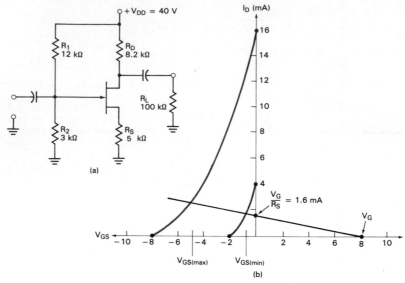

FIGURE 10.35

Using the minimum values of $V_{GS(off)}$ and V_{GS}, g_m is found as

$$g_m = 2000\ \mu S\left(1 - \frac{-1\ V}{-2\ V}\right)$$
$$= 1000\ \mu S$$

The value of r_D is found as

$$r_D = R_D \parallel R_L$$
$$= 7.58\ k\Omega$$

The maximum value of A_{vL} can now be found as

$$A_{vL} = r_D g_{m(max)}$$
$$= 7.58\ k\Omega(2250\ \mu S)$$
$$= 17.06$$

The minimum value of A_{vL} is found as

$$A_{vL} = r_D g_{m(min)}$$
$$= 7.58\ k\Omega(1000\ \mu S)$$
$$= 7.58$$

PRACTICE PROBLEM 10–12

The amplifier described in Practice Problem 10–8 has a load resistance of 4.7 kΩ. The 2N5486 has a g_{m0} range of 4000 to 8000 μS. Using the values of V_{GS} from Practice Problem 10–10, calculate the range of A_{vL} values for the amplifier.

The Basis for Equation (10.17)

The ac equivalent circuit shown in Figure 10.34b would have an output voltage found as

$$\boxed{v_{out} = i_d r_D} \tag{10.18}$$

where v_{out} = the ac output voltage
i_d = the change in I_D caused by the ac input signal

Now recall that g_m is defined in equation (10.14) as

$$g_m = \frac{\Delta I_D}{\Delta V_{GS}}$$

Another way of writing this equation is

$$g_m = \frac{i_d}{v_{gs}} \qquad \textbf{(10.19)}$$

where v_{gs} is the change in V_{GS} caused by the amplifier input signal, equal to that input signal. Equation (10.19) can be rewritten as

$$i_d = g_m v_{gs}$$

Substituting this equation into equation (10.19) gives us

$$v_{out} = g_m v_{gs} r_D \qquad \textbf{(10.20)}$$

Now recall that voltage gain (A_v) is the ratio of output voltage to input voltage. By formula,

$$A_{vL} = \frac{v_{out}}{v_{in}} \qquad (v_{in} = v_{gs})$$

We can rearrange equation (10.20) to get a valid voltage-gain formula for the common-source JFET amplifier:

$$\frac{v_{out}}{v_{in}} = g_m r_D$$

or

$$A_{vL} = g_m r_D$$

This is equation (10.17).

JFET Swamping

Just as we swamped the BJT amplifier to reduce the effects of variations in r'_e, we can swamp the JFET amplifier to reduce the effects of variations in g_m. The swamped JFET amplifier closely resembles the swamped BJT amplifier, as shown in Figure 10.36. As you can see, the source resistance is only partially bypassed, just as the emitter resistance is partially bypassed in the swamped. BJT amplifier.

The voltage gain of the swamped JFET amplifier is found as

$$A_{vL} = \frac{r_D}{r_S + (1/g_m)} \qquad \textbf{(10.21)}$$

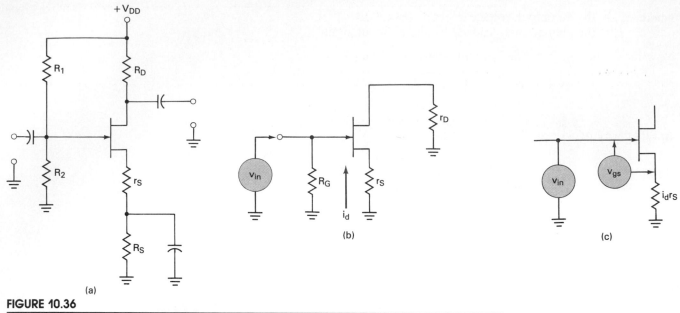

FIGURE 10.36

A swamped common-source amplifier.

If the value of $r_S \gg 1/g_m$, the resistor will swamp out the effects of variations in g_m. This point is illustrated in the following example.

EXAMPLE 10.13

✔ 24

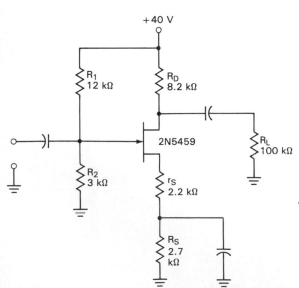

FIGURE 10.37

Determine the maximum and minimum gain values for the circuit in Figure 10.37.

Solution: Basically, the circuit in Figure 10.37 is the same as the one shown in Figure 10.35. The difference of 100 Ω in the source circuit is not enough to significantly change the values obtained in Example 10.12, so we will use those values here.

Using the values from Figure 10.37 and Example 10.12, the maximum value of A_{vL} is found as

$$A_{vL} = \frac{r_D}{r_S + (1/g_m)}$$
$$= \frac{7.58 \text{ k}\Omega}{2.2 \text{ k}\Omega + 444 \text{ }\Omega}$$
$$= 2.87$$

The minimum value of A_{vL} can be calculated as

$$A_{vL} = \frac{r_D}{r_S + (1/g_m)}$$
$$= \frac{7.58 \text{ k}\Omega}{2.2 \text{ k}\Omega + 1 \text{ k}\Omega}$$
$$= 2.37$$

A swamped JFET amplifier has values of $r_D = 6.6$ kΩ, $r_S = 1.5$ kΩ, and $g_m = 2000$ to 4000 μS. Determine the range of A_{vL} values for the circuit.

As you can see, the stability of A_v is much better for the swamped amplifier than it is for the standard common-source amplifier.

Example 10.13 demonstrates the one drawback to using swamping. If you compare the values of A_v for Examples 10.12 and 10.13, you will notice that you lose quite a bit of gain when you swamp the JFET amplifier. This is the same problem that we had with swamping BJT amplifiers. *Anytime you swamp an amplifier, you improve stability but you lose some voltage gain.* Your value of A_v may be much more stable, but it will also be much lower than it was without the swamping.

The Basis for Equation (10.21)

The input voltage to the circuit in Figure 10.36b is in parallel with the gate–source circuit. Because of this, v_{in} must equal the sum of v_{gs} and the voltage developed across r_S. By formula,

$$v_{in} = v_{gs} + i_d r_S$$

This point is illustrated by Figure 10.36c. Since $i_d = v_{gs}g_m$, the equation above can be rewritten as

$$v_{in} = v_{gs} + v_{gs}g_m r_S$$

or

$$v_{in} = v_{gs}(1 + g_m r_S)$$

The voltage gain of the amplifier is found as

$$A_{vL} = \frac{v_{out}}{v_{in}}$$
$$= \frac{g_m v_{gs} r_D}{v_{gs}(1 + g_m r_S)}$$

Eliminating v_{gs} from the equation above gives us

$$A_{vL} = \frac{g_m r_D}{1 + g_m r_S}$$

Finally, we can multiply the fraction above by 1 in the form of

$$\frac{1/g_m}{1/g_m}$$

to obtain

$$\boxed{A_{vL} = \frac{r_D}{r_S + (1/g_m)}}$$

This is equation (10.21).

Amplifier Input Impedance

Due to the extremely high input impedance of the JFET, the overall input impedance to a JFET amplifier will be higher than for a similar BJT amplifier. For the *gate-bias* and *self-bias* circuits, the amplifier input impedance is found as

$$\boxed{Z_{in} \cong R_G} \tag{10.22}$$

For the *voltage-divider biased* amplifier, Z_{in} is found as

$$\boxed{Z_{in} \cong R_1 \,\|\, R_2} \tag{10.23}$$

The following example serves to illustrate the fact that a JFET amplifier will have a higher value of Z_{in} than a similar BJT amplifier.

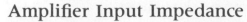

EXAMPLE 10.14

Determine the values of Z_{in} for the amplifiers shown in Figure 10.38.

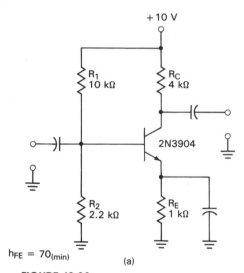

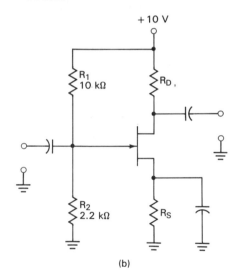

$$V_B = V_{CC}\, \frac{R_2}{R_1 + R_2} = 1.8 \text{ V}$$

$$V_E = V_B - 0.7 \text{ V} = 1.1 \text{ V}$$

$$I_E = \frac{V_E}{R_E} = 1.1 \text{ mA}$$

$$r_e' = \frac{25 \text{ mV}}{I_E} = 22.7 \ \Omega$$

FIGURE 10.38

Solution: Figure 10.38a shows the calculations for the BJT amplifier up to the determination of r_e'. The value of current gain shown is the minimum spec sheet value for $I_C = 1$ mA. Using the values shown, the input impedance to the BJT amplifier is found as

$$Z_{in} = R_1 \,\|\, R_2 \,\|\, h_{fe}(r_e')$$
$$= 844.7 \ \Omega$$

For the JFET amplifier, the input impedance is found as

$$Z_{in} \cong R_1 \,\|\, R_2$$
$$= 1.8 \text{ k}\Omega$$

Thus, using the same biasing resistors, the JFET amplifier will have a much higher overall value of Z_{in} than that of the BJT amplifier.

Because of the higher input impedance of the JFET amplifier, the circuit will present less of a load to the circuit driving the amplifier. Thus, in any multistage amplifier, the overall gain of an amplifier will be higher if the load is a JFET amplifier than it will be if the load is an equivalent BJT amplifier. The following example serves to illustrate this point.

EXAMPLE 10.15

The amplifier in Example 10.12 is being used to drive the amplifiers described in Example 10.14. Assume that g_m for the driving amplifier is fixed at 2000 μS. Determine the value of A_v for the driving amplifier when each load amplifier is connected to the output.

Solution: When the BJT amplifier is the load, the value of r_D for Figure 10.35 is found as

$$r_D = R_D \| Z_{\text{in}}$$
$$= 8.2 \text{ k}\Omega \| 844.7 \text{ }\Omega$$
$$= 765.8 \text{ }\Omega$$

With this load, the voltage gain of the amplifier is

$$A_{vL} = g_m r_D$$
$$= (2000 \text{ }\mu\text{S})(765.8 \text{ }\Omega)$$
$$= 1.53$$

When the JFET amplifier is the load, the value of r_D for Figure 10.35 is found as

$$r_D = R_D \| Z_{\text{in}}$$
$$= 8.2 \text{ k}\Omega \| 1.8 \text{ k}\Omega$$
$$= 1.476 \text{ k}\Omega$$

With this load, the voltage gain of the amplifier is

$$A_{vL} = g_m r_D$$
$$= (2000 \text{ }\mu\text{S})(1.476 \text{ k}\Omega)$$
$$= 2.952$$

Thus the gain of the circuit in Figure 10.35 is higher when driving the JFET amplifier than it is when driving the BJT amplifier.

One Final Note

You may have noticed that the bulk of material in this section was related only to the *voltage-divider biased* JFET amplifier. If you derive the ac equivalent circuits for the gate-bias and self-bias circuits, you will see that these ac equivalent circuits are the same as the voltage-divider bias equivalent circuit. For this reason, these other biasing circuits were not discussed in this section.

In this section, you have been shown that the JFET common-source amplifier has extremely high input impedance when compared to the common-emitter amplifier. You have also been shown that the voltage gain of the JFET amplifier (which tends to be relatively low) changes when the value of transconductance for the JFET changes. By swamping the JFET amplifier, we can reduce the variations in voltage gain that occur when transconductance varies. However, the increased stability in voltage gain is obtained at the cost of even lower voltage gain.

SECTION REVIEW

1. Explain the common-source amplifier circuit response to an ac input signal. [Objective 18]
2. What is *transconductance*? [Objective 19]
3. What are the units of transconductance? [Objective 19]
4. How does the value of g_m relate to the position of a circuit Q-point on the transconductance curve? [Objective 19]
5. What is the relationship between JFET amplifier voltage gain and the value of V_{GS} for the circuit [Objective 21]
6. What purpose is served by swamping a JFET amplifier? [Objective 23]
7. How does the JFET amplifier input impedance compare to that of a comparable BJT amplifier? Explain your answer. [Objective 25]

10.4
AC Operating Characteristics: Common-Drain and Common-Gate Amplifiers

The *common-drain* and *common-gate* configurations are the JFET equivalents for the *common-collector* and *common-base* BJT amplifiers. In this section we will take a brief look at both of these amplifier types.

The Common-Drain (Source-Follower) Amplifier

✔ 26

Source follower (common-drain amplifier). The JFET equivalent of the emitter follower.

The *source follower* accepts an input signal at its gate and provides an output signal at its source terminal. The input and output signals for this amplifier are *in phase*, thus the name *source follower*. The basic source follower is shown in Figure 10.39. The characteristics of the source follower are summarized as follows:

Input impedance: High
Output impedance: Low
Gain: $A_{vL} < 1$

Because of its input and output impedance characteristics, the source follower is used primarily as an *impedance-matching* circuit. That is, it is used to couple a high-impedance source to a low-impedance load. You may recall from Chapter 8 that the *emitter follower* is also used for this purpose.

The voltage gain (A_{vL}) of the source follower is found using the following equation:

$$A_{vL} = \frac{r_S}{r_S + (1/g_m)}$$

(10.24)

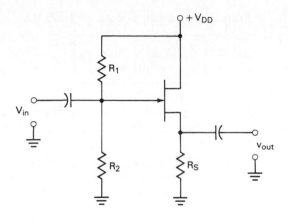

FIGURE 10.39

The source follower.

where $r_S = R_S \| R_L$. The basis for equation (10.24) is easy to see if you compare the source follower in Figure 10.39 with the swamped common-source amplifier shown in Figure 10.36. As you can see, both circuits have some value of unbypassed source resistance. Thus, the voltage gains of the two circuits are calculated in the same basic fashion. The only differences are:

1. The unbypassed source resistance in the source follower is the parallel combination of the source resistor (R_S) and the circuit load resistance (R_L).
2. The output voltage from the source follower appears across r_S. Therefore, r_S is used in place of r_D in the swamped amplifier equation.

Since r_S appears in both the numerator and the denominator of equation (10.24), it should be obvious that the value of A_{vL} for the source follower will always be less than 1.

The high input impedance of the amplifier is caused, again, by the high Z_{in} of the JFET. As with the common-source amplifier, the total input impedance to the amplifier will equal the Thevenin resistance of the input circuit. When voltage-divider bias is used (which is the most common), Z_{in} is found as

$$\boxed{Z_{in} = R_1 \| R_2} \qquad \textbf{(10.25)}$$

When gate bias or self-bias is used,

$$\boxed{Z_{in} = R_G} \qquad \textbf{(10.26)}$$

Again, the high input impedance of the gate eliminates it from any Z_{in} calculations.

The output impedance of the amplifier is determined with a little help from equation (10.24). If you look closely at the formula, you will notice that you essentially have a voltage-divider formula. Based on this, the circuit in Figure 10.40 is drawn. This is a resistive equivalent of the source-follower circuit. To the output (v_{out}) the circuit appears as two resistors connected *in parallel*. Thus the output impedance of the amplifier is given as

$$\boxed{Z_{out} = R_S \| \frac{1}{g_m}} \qquad \textbf{(10.27)}$$

Now, how does equation (10.27) show that the value of Z_{out} is always low? The

The voltage gain of the swamped common-source amplifier is found as

$$A_{vL} = \frac{r_D}{r_S + (1/g_m)}$$

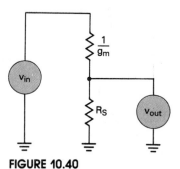

FIGURE 10.40

Don't Forget: Output impedance is measured with the load disconnected. Thus, r_S is replaced by R_S in Figure 10.40 and equation (10.27).

normal range of g_m is from 1000 μS up. For $g_m = 1000$ μS,

$$\frac{1}{g_m} = \frac{1}{0.001} = 1 \text{ k}\Omega$$

Since we used the lowest typical value of g_m in the calculation above, it is safe to say that $1/g_m$ *is typically lower than 1 kΩ*. Since the total resistance of a parallel circuit must be lower than the lowest individual resistance value, it is safe to say that Z_{out} *will typically be less than 1 kΩ*.

The following example will serve to illustrate the process of analyzing a source-follower circuit. It also illustrates the fact that the source follower has a high Z_{in}, a low Z_{out}, and an overall voltage gain of less than unity.

EXAMPLE 10.16

27

Determine the maximum and minimum values of A_{vL} and Z_{out} for the circuit shown in Figure 10.41. Also, determine the value of Z_{in} for the circuit.

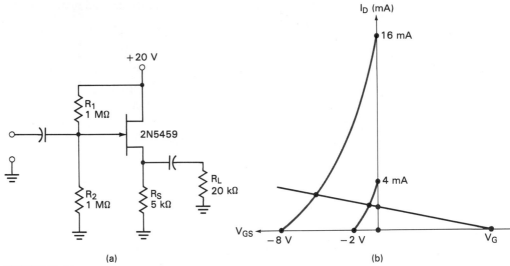

FIGURE 10.41

Solution: Using the dc bias line shown in Figure 10.41b, we determine the minimum and maximum values of V_{GS} to be −1 V and −6 V, respectively. Again, the 2N5459 has values stated on the spec sheet of $g_{m0} = 6000$ μS at $V_{GS(\text{off})} = -8$ V, and $g_{m0} = 2000$ μS at $V_{GS(\text{off})} = -2$ V. These values are used with our minimum and maximum values of V_{GS} to determine the minimum and maximum values of g_m, as follows:

$$g_{m(\text{max})} = g_{m0}\left(1 - \frac{V_{GS}}{V_{GS(\text{off})}}\right)$$

$$= 6000 \text{ μS}\left(1 - \frac{-6 \text{ V}}{-8\text{V}}\right)$$

$$= 1500 \text{ μS}$$

and

$$g_{m(min)} = 2000 \ \mu S \left(1 - \frac{-1.0 \ V}{-2 \ V}\right)$$
$$= 1000 \ \mu S$$

The value of r_S is now found as

$$r_S = R_S \,\|\, R_L$$
$$= 5 \ k\Omega \,\|\, 20 \ k\Omega$$
$$= 4 \ k\Omega$$

Using the maximum value of g_m, the value of A_{vL} is found as

$$A_{vL} = \frac{r_S}{r_S + (1/g_m)}$$
$$= \frac{4 \ k\Omega}{4 \ k\Omega + 667 \ \Omega}$$
$$= 0.8571 \quad \text{(maximum)}$$

Using the minimum value of g_m, the value of A_{vL} is found as

$$A_{vL} = \frac{r_S}{r_S + (1/g_m)}$$
$$= \frac{4 \ k\Omega}{4 \ k\Omega + 1 \ k\Omega}$$
$$= 0.8 \quad \text{(minimum)}$$

Using the maximum and minimum values of g_m, the corresponding values of Z_{out} are found as

$$Z_{out} = R_S \,\|\, \frac{1}{g_m}$$
$$= 5 \ k\Omega \,\|\, 1 \ k\Omega$$
$$= 833 \ \Omega \quad \text{(maximum)}$$

and

$$Z_{out} = R_S \,\|\, \frac{1}{g_m}$$
$$= 5 \ k\Omega \,\|\, 667 \ \Omega$$
$$= 588 \ \Omega \quad \text{(minimum)}$$

Finally, the value of Z_{in} is found as

$$Z_{in} = R_1 \,\|\, R_2$$
$$= 500 \ k\Omega$$

PRACTICE PROBLEM 10–16

The 2N5486 is used in the amplifier shown in Figure 10.41. Determine the range of A_{vL} and Z_{out} values for the circuit.

Reference: The data sheet for the 2N5486 JFET is shown on page 469.

Example 10.16 served to show us several things. First, the example demonstrated the fact that source followers have high input impedance and low output impedance values. Even when Z_{out} was at its maximum value, the ratio of Z_{in} to Z_{out} was close to 1000:1. The example also demonstrated that there is a relationship between g_m, A_v, and Z_{out}. When g_m is at its maximum value, A_v is at its maximum value and Z_{in} is at its minimum value. The reverse also holds true. Based on these observations, we can state the following relationships for the source-follower:

1. *A_v is directly proportional to g_m.*
2. *Z_{out} is inversely proportional to g_m.*

As always, the value of Z_{in} is not related to the value of g_m.

You have been shown that the source follower is a JFET amplifier that has high input impedance, low output impedance, and a voltage gain that is less than 1. At this point, we will move on to take a look at the common-gate amplifier.

The Common-gate Amplifier

☑ 28

The *common-gate* amplifer accepts an input signal at its source terminal and provides an output signal at its drain terminal. The basic common-gate amplifier is shown in Figure 10.42.

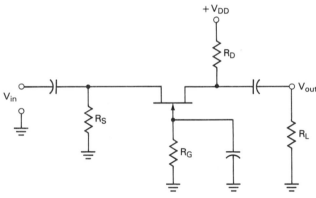

FIGURE 10.42

The common-gate amplifier.

Common-gate amplifier. The JFET equivalent of the common-base amplifier.

The characteristics of the common-gate circuit are summarized as follows:

Input impedance: Low
Output impedance: High (as compared to Z_{in})
Gain: $A_{vL} > 1$

As with the common-base circuit, the common-gate amplifier can be used to couple a low Z_{out} source to a high Z_{in} load. In this respect, the common-gate circuit (like the source follower) is being used as an *impedance-matching* circuit. Also, just as with the common-base amplifier, the common-gate amplifier will have a value of A_v that is greater than 1. Therefore, the overall gain of the amplifier is greater than 1.

☑ 29

Like the common-source amplifier, the common-gate amplifier has a voltage gain that is equal to the product of its transconductance and ac drain resistance. By formula,

$$\boxed{A_{vL} = g_m r_D}$$

The derivation of this equation is shown on pages 450–451.

CHAP. 10   Field-Effect Transistors

The input signal to the common-gate amplifier is applied to the source. You were shown on page 457 that the impedance of the source is found as R_S in parallel with $1/g_m$. Thus, the input impedance of the common-gate amplifier is found as

$$Z_{\text{in}} = R_S \parallel \frac{1}{g_m} \qquad (10.28)$$

As you can see, this is identical to the Z_{out} equation (10.27) for the source follower. This would make sense, since the input impedance of the common-gate amplifier is measured at the same terminal (the source) as the output impedance of the source follower.

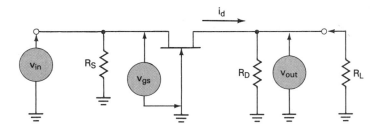

FIGURE 10.43

Figure 10.43 shows that the output impedance would be made up of the parallel combination of R_D and the resistance of the JFET drain. By formula,

$$Z_{\text{out}} = R_D \parallel r_d \qquad (10.29)$$

where r_d = the resistance of the JFET drain terminal. The resistance of the JFET drain can be calculated using another JFET parameter: *output admittance*, y_{os}. In your study of basic electronics, you were taught that admittance is the reciprocal of impedance. We can therefore determine the value of the JFET drain resistance by taking the reciprocal of y_{os}. By formula,

Output admittance (y_{os}). The JFET admittance parameter.

$$r_d = \frac{1}{y_{os}} \qquad (10.30)$$

The value of r_d is typically *much greater* than R_D. Because of this, the output impedance can normally be determined as

$$Z_{\text{out}} \cong R_D \qquad (10.31)$$

The following example illustrates this point.

EXAMPLE 10.17

Determine the value of r_d for the JFET in Figure 10.44. Also, determine the exact and approximated values of Z_{out}.

Solution: The spec sheet for the 2N5459 lists a maximum value of y_{os} = 50 μS. Using this value, the *minimum* value

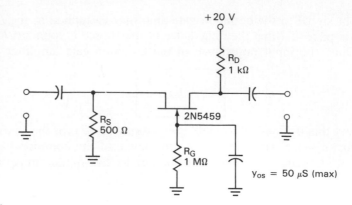

FIGURE 10.44

of r_d is found as

$$r_d = \frac{1}{y_{os}}$$
$$= \frac{1}{50 \ \mu S}$$
$$= 20 \ k\Omega$$

Using this value of r_d, the exact value of Z_{out} is found as

$$Z_{out} = R_D \| r_d$$
$$= 1 \ k\Omega \| 20 \ k\Omega$$
$$= 952 \ \Omega$$

The approximated value of Z_{out} is found as

$$Z_{out} = R_D$$
$$= 1 \ k\Omega$$

In Example 10.17 the *maximum* value of y_{os} was used to find the *minimum* value of r_d. This was done to illustrate a point. When the minimum value of r_d was used, the "exact" value of Z_{out} was still within 10% of the approximated value. If you were to use any higher value of r_d, the exact and approximated values of Z_{out} would be even closer to each other. For example, the spec sheet for the 2N5459 lists a *typical* value of $y_{os} = 10 \ \mu S$. Using this value, $r_d = 100$ $k\Omega$. Using this value for r_d, Z_{out} calculates to 990 Ω. This is even closer to the approximated value of Z_{out} than the value obtained in Example 10.17. We can therefore assume that equation (10.31) is valid in all cases.

In our discussion of the common-gate circuit, we have used only self-bias circuits. This is certainly not the only type of biasing used with common-gate circuits. Figure 10.45 shows how voltage-divider biasing can be used in common-gate applications.

At this point we have covered the most basic JFET circuits. Other circuits that use JFETs will appear off and on throughout the remainder of this book and will be discussed at the appropriate points. At this point, we will take a look at the methods used to test JFETs and JFET circuits.

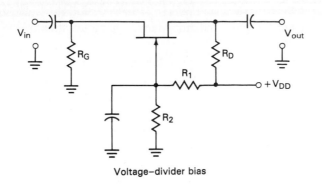

Voltage–divider bias

FIGURE 10.45

Summary

You have now been shown the gain and impedance characteristics of the common-source, source-follower, and common-gate amplifiers. These three amplifiers and their characteristics are summarized in Figure 10.46.

FIGURE 10.46

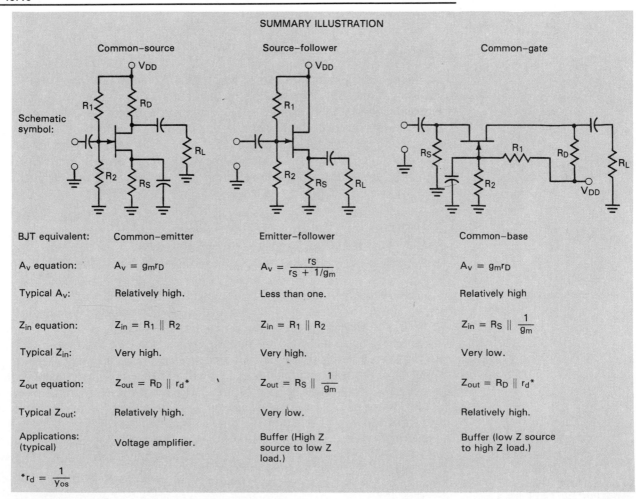

SUMMARY ILLUSTRATION

	Common-source	Source-follower	Common-gate
Schematic symbol:	(circuit)	(circuit)	(circuit)
BJT equivalent:	Common-emitter	Emitter-follower	Common-base
A_v equation:	$A_v = g_m r_D$	$A_v = \dfrac{r_S}{r_S + 1/g_m}$	$A_v = g_m r_D$
Typical A_v:	Relatively high.	Less than one.	Relatively high
Z_{in} equation:	$Z_{in} = R_1 \parallel R_2$	$Z_{in} = R_1 \parallel R_2$	$Z_{in} = R_S \parallel \dfrac{1}{g_m}$
Typical Z_{in}:	Very high.	Very high.	Very low.
Z_{out} equation:	$Z_{out} = R_D \parallel r_d^*$	$Z_{out} = R_S \parallel \dfrac{1}{g_m}$	$Z_{out} = R_D \parallel r_d^*$
Typical Z_{out}:	Relatively high.	Very low.	Relatively high.
Applications: (typical)	Voltage amplifier.	Buffer (High Z source to low Z load.)	Buffer (low Z source to high Z load.)

$^*r_d = \dfrac{1}{y_{os}}$

1. What are the gain and impedance characteristics of the source follower? [Objective 26]
2. List, in order, the steps taken to determine the range of A_{vL} values for a source follower. [Objective 27]
3. What are the gain and impedance characteristics of the common-gate amplifier? [Objective 28]
4. List, in order, the steps taken to determine the exact and approximated values of Z_{out}. [Objective 29]

10.5

Troubleshooting JFET Circuits

One of the big advantages of troubleshooting circuits that use JFETs is the fact that JFET faults are easy to diagnose. This is due to the fact that the JFET has only one junction that can go bad. As you will be shown, JFETs can be treated without removing them from their circuits. You will also see that the procedures used for troubleshooting JFET circuits are the same as those used for troubleshooting BJT amplifiers.

JFET Faults

JFET faults are about as difficult to detect as a flat tire. The reason for this is that there is *one* junction that can develop either a short or an open. To make things even easier, the shorted-junction and open-junction JFETs have distinct symptoms that can be observed in the circuit. Therefore, a minimum of effort is required to determine whether or not the JFET is the cause of an amplifier failure.

Normal JFET operating characteristics.

Figure 10.47a shows the normal operating characteristics in a voltage-divider biased JFET amplifier. Note that the circuit has the following operating characteristics:

1. The current through R_1 is equal to the current through R_2.
2. The source and drain terminal currents are equal.
3. There is no current through the gate terminal of the device.

Shorted-gate symptoms.

The *shorted-junction* JFET will not have the same current characteristics. This is shown in Figure 10.47b. Since the gate–source junction is shorted, V_{GS} will be 0 V. Because of this, the current flowing out of the drain circuit will be approximately equal to I_{DSS}. This current comes from the JFET, with part coming from the source terminal and part coming from the gate terminal. With part of I_D coming from the gate terminal, I_S and I_D will *not* be equal, as is usually the case. The current flowing into the gate terminal is part of the R_2 current. Therefore, I_1 will be less than I_2 (that is, the current flowing through R_2 will *not* equal the current through R_1). These are the classic symptoms of a shorted-gate junction. In a summary:

1. $I_1 \neq I_2$
2. $I_S \cong I_{DSS}$

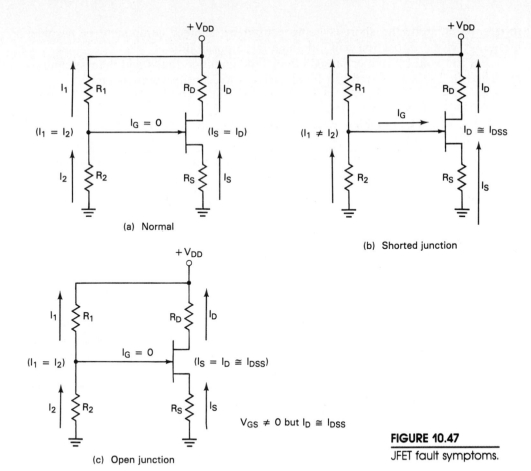

(a) Normal

(b) Shorted junction

(c) Open junction

$V_{GS} \neq 0$ but $I_D \cong I_{DSS}$

FIGURE 10.47

JFET fault symptoms.

3. $I_S \neq I_D$
4. $I_G > 0$

The *open-junction* JFET can be detected by observing two simple conditions: Open-gate symptoms.

1. $V_{GS} \neq 0$ V
2. $I_D \cong I_{DSS}$

You see, with the gate junction open, the depletion layer that normally restricts I_D cannot form. At the same time, V_{GS} will be at its normal value. If you have a measurable amount of V_{GS} and I_D is approximately equal to I_{DSS}, the gate–source junction of the JFET is open. The circuit conditions that relate to this fault are shown in Figure 10.47c.

JFET Circuit Troubleshooting

As with any other amplifier, you have to start the troubleshooting process by making sure that the amplifier is the source of the trouble. This point cannot be stressed often enough. When you find a JFET amplifier with no output, check to make sure that the amplifier has a good input signal, and check the V_{DD} and ground connections to the circuit. If all these tests have the proper results, you can assume that the amplifier is faulty.

We are going to restrict our discussion of amplifier faults to *open* components. Recall that resistors generally do not simply develop a *shorted* condition. If a resistor is shorted, the short is being caused by another component or a solder bridge. A visual inspection will eliminate these possibilities.

For the voltage-divider biased circuit, the common faults and their symptoms are listed in Table 10.1 (refer to Figure 10.47a to identify components).

TABLE 10.1
JFET Troubleshooting

Fault	Symptoms
R_1 open	V_{GS} increases; $I_2 = 0$; V_D increases; the circuit acts as a self-bias circuit.
R_2 open	JFET is destroyed by forward bias; if the JFET is replaced, the new component is also destroyed; $V_2 = V_{DD}$.
R_D open	$V_D = 0$ V; all terminal currents drop to 0.
R_S open	$V_S = V_{DD}$; $V_{RD} = 0$ V.

Of the problems listed, R_1 going open is the most difficult to detect. If R_1 opens, it is like removing the component from the circuit. If you do that, you simply have a self-bias circuit. Because of this, R_1 could open and the unit containing the amplifier could take a long time to fail. The main effect of R_1 opening on circuit operation is a loss of stability. If the amplifier develops a problem with output signal clipping, check the voltage across R_1. If this voltage is equal to V_{DD}, the component is open.

If R_2 opens, the gate circuit consists of R_1 and V_{DD}. At the instant R_2 opens, the current through R_1 will drop to zero, dropping the voltage across the component to zero. This places V_{DD} at the gate of the JFET, which will forward bias and destroy the component. If you replace the JFET, the same thing will happen to the new component.

When R_D opens, it effectively removes V_{DD} from the drain–source circuit. All the supply voltage will be dropped across the open component, so the voltage from drain to ground will be 0 V. Also, the lack of supply voltage will cause I_D and I_S to drop to zero. At this point, the JFET may become forward biased by V_G, destroying the component. If you test a circuit and the value of V_D indicates that R_D is open, be prepared to replace the JFET.

If R_S opens, source current will drop to zero. Since the resistor is open, the supply voltage will be dropped across the resistor, R_S. With V_{DD} being dropped across R_S, V_{GS} will be abnormally high, probably higher than $V_{GS(\text{off})}$. The voltages across the gate-biasing resistors will be normal.

Now let's put all this into perspective by going through a few troubleshooting applications.

APPLICATION 1

The circuit shown in Figure 10.48 has no output signal. After verifying that all inputs were valid, the input was disconnected from the circuit. Then the dc voltages shown were measured. The problem: R_2 is open. The first indication that R_2 is open is the fact that V_{DD} can be measured across this component. But where did the other readings come from? Is there another problem in the circuit? Yes, there is. When R_2 opened, it caused the JFET junction to become forward biased. This, in turn, caused the junction of the JFET to open.

When the junction of the JFET opened, there was no depletion region to restrict the flow of electrons through the channel. Thus the current in the source–drain circuit increased as much as R_S, R_D, and the resistance of the channel would allow, in this case to about 1.62 mA. This value of I_D caused 11 V to be developed across R_S. The difference between V_S and V_G is now approximately 9 V, as shown. *Note the polarity of V_{GS}.* Normally, V_{GS} is a negative voltage. *The positive value of V_{GS} is a good indication that the junction has opened.* For this circuit, both R_2 and the JFET would have to be replaced.

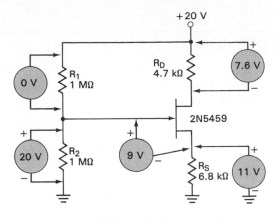

FIGURE 10.48

APPLICATION 2

The circuit in Figure 10.49 has been verified as being faulty. After disconnecting the input, the voltages shown were measured. This problem is a simple one. The fact that V_{DD} is being measured across the source resistor indicates that R_S is open. Note that the lack of I_D causes the reading shown across the drain resistor. Also, the gate circuit is not affected by the fault, so the voltage readings across R_1 and R_2 are normal.

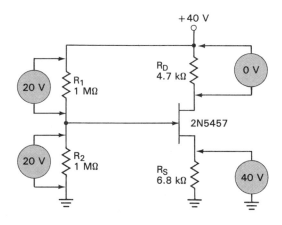

FIGURE 10.49

Again, we have not covered every problem that could possibly develop with A JFET amplifier. However, using a solid understanding of the component and a little common sense, you should have no difficulty in diagnosing JFET amplifier faults.

---

1. What is the primary difference between troubleshooting JFET amplifiers and **SECTION REVIEW** BJT amplifiers? [Objective 30]
2. List the common symptoms of a shorted-gate junction. [Objective 31]
3. List the common symptoms of an open-gate junction. [Objective 31]
4. Describe the procedure used to troubleshoot a JFET amplifier. [Objective 32]

JFET Specification Sheets and Applications

Up to this point, we have based our calculations on several JFET parameters. Now it is time to take a look at the common JFET parameters, what they mean, and when they are used. In this section, we will also take a look at some practical JFET applications.

JFET Specification Sheets

✔ 33

We have used the parameters of the 2N5486 in most of the example practice problems. In this section, we will use the spec sheet for this device. The 2N5485 through 2N5486 JFET spec sheet is shown in Figure 10.50.

Maximum Ratings

Most of the maximum ratings for these devices are self-explanatory at this point. In our BJT spec sheet discussions, we covered all the commonly used maximum voltage, current, and power ratings. The JFET maximum ratings are no different.

There is one maximum rating that may surprise you. In the section of maximum ratings in Figure 10.50, you will see the following:

Rating	Symbol	Value	Unit
Gate current	I_G	10	mAdc

Gate current (I_G). The maximum amount of current that can be drawn through the JFET gate without damaging the device.

This rating indicates that the gate can handle a maximum forward current of 10 mA. Does this mean that the device could be operated in the forward region? No. The control of drain current depends on the amount of *reverse* bias on the JFET gate. The maximum drain current (I_{DSS}) is reached when the reverse bias reaches 0 V. Any increase in V_{GS} above 0 V will not cause an increase in I_D. However, should the gate *accidentally* become forward biased, I_G must be greater than 10 mAdc for the device to be destroyed. As long as $I_G < 10$ mAdc, the device will be safe.

Off Characteristics

Gate–source breakdown voltage ($V_{BR(GSS)}$). The value of V_{GS} that will cause the gate–source junction to break down.

The *gate–source breakdown voltage* defines the limit on V_{GS}. If V_{GS} is allowed to exceed this value, the junction will break down and the JFET will have to be replaced. For the 2N5484–6 series JFETs, the gate–source breakdown voltage is shown to be -25 V.

Gate reverse current (I_{GSS}). The maximum amount of gate current that will occur when the gate–source junction is reverse biased.

The *gate reverse current* rating indicates the maximum value of gate current that will occur when the gate–source junction is reverse biased. For the 2N5484–6 series JFETs, this rating is 1 nAdc when the ambient temperature (T_A) is 25°C. Note that the negative current value is used to indicate the direction of the gate current. I_{GSS} is a *temperature-dependent* rating. As the spec sheet shows, the value of I_{GSS} increases as temperature increases. The relationship between junction reverse current and temperature was first discussed in Chapter 2.

Since we have already discussed $V_{GS(off)}$ and I_{DSS} (listed under ''On Characteristics'') in detail, we will not discuss them further here. You should note that there are wide ranges of $V_{GS(off)}$ and I_{DSS} values within the 2N5484–6 series.

MAXIMUM RATINGS

Rating	Symbol	Value	Unit
Drain-Gate Voltage	V_{DG}	25	Vdc
Reverse Gate-Source Voltage	V_{GSR}	25	Vdc
Drain Current	I_D	30	mAdc
Forward Gate Current	$I_{G(f)}$	10	mAdc
Total Device Dissipation @ T_C = 25°C Derate above 25°C	P_D	310 2.82	mW mW/°C
Operating and Storage Junction Temperature Range	T_J, T_{stg}	−65 to +150	°C

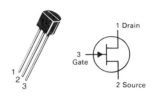

2N5484
thru
2N5486

CASE 29-04, STYLE 5
TO-92 (TO-226AA)

1 Drain

3 Gate

2 Source

JFET
VHF/UHF AMPLIFIERS

N-CHANNEL — DEPLETION

Refer to 2N4416 for graphs.

ELECTRICAL CHARACTERISTICS (T_A = 25°C unless otherwise noted.)

Characteristic		Symbol	Min	Typ	Max	Unit
OFF CHARACTERISTICS						
Gate-Source Breakdown Voltage (I_G = −1.0 μAdc, V_{DS} = 0)		$V_{(BR)GSS}$	−25	—	—	Vdc
Gate Reverse Current (V_{GS} = −20 Vdc, V_{DS} = 0) (V_{GS} = −20 Vdc, V_{DS} = 0, T_A = 100°C)		I_{GSS}	— —	— —	−1.0 −0.2	nAdc μAdc
Gate Source Cutoff Voltage (V_{DS} = 15 Vdc, I_D = 10 nAdc)	2N5484 2N5485 2N5486	$V_{GS(off)}$	−0.3 −0.5 −2.0	— — —	−3.0 −4.0 −6.0	Vdc
ON CHARACTERISTICS						
Zero-Gate-Voltage Drain Current (V_{DS} = 15 Vdc, V_{GS} = 0)	2N5484 2N5485 2N5486	I_{DSS}	1.0 4.0 8.0	— — —	5.0 10 20	mAdc
SMALL-SIGNAL CHARACTERISTICS						
Forward Transfer Admittance (V_{DS} = 15 Vdc, V_{GS} = 0, f = 1.0 kHz)	2N5484 2N5485 2N5486	$\|y_{fs}\|$	3000 3500 4000	— — —	6000 7000 8000	μmhos
Input Admittance (V_{DS} = 15 Vdc, V_{GS} = 0, f = 100 MHz) (V_{DS} = 15 Vdc, V_{GS} = 0, f = 400 MHz)	2N5484 2N5485, 2N5486	Re(y_{is})	— 	— 	100 1000	μmhos
Output Admittance (V_{DS} = 15 Vdc, V_{GS} = 0, f = 1.0 kHz)	2N5484 2N5485 2N5486	$\|y_{os}\|$	— — —	— — —	50 60 75	μmhos
Output Conductance (V_{DS} = 15 Vdc, V_{GS} = 0, f = 100 MHz) (V_{DS} = 15 Vdc, V_{GS} = 0, f = 400 MHz)	2N5484 2N5485, 2N5486	Re(y_{os})	— 	— 	75 100	μmhos
Forward Transconductance (V_{DS} = 15 Vdc, V_{GS} = 0, f = 100 MHz) (V_{DS} = 15 Vdc, V_{GS} = 0, f = 400 MHz)	2N5484 2N5485 2N5486	Re(y_{fs})	2500 3000 3500	— — —	— — —	μmhos

FIGURE 10.50

(Courtesy of Motorola, Inc.)

Small-signal Characteristics

At first glance, this section of the spec sheet can be somewhat confusing. It seems that we have several y_{fs} and y_{os} parameters listed. The questions are which ones do we use, and when? To answer these questions, we have to briefly review some basic ac principles.

Conductance is *the reciprocal of resistance*. *Susceptance* is *the reciprocal of reactance*. Combined, they make up *admittance*, which is *the reciprocal of impedance* (the combination of resistance and reactance). While resistance, reactance, and impedance are all measures of *opposition to current*, conductance, susceptance, and admittance are all measures of *the relative ease with which current will pass through a component*.

The *forward transfer admittance*, *input admittance*, and *output admittance* ratings all take the effects of susceptance into account. You see, the JFET has measurable input and output *capacitances*. These capacitances all have some amount of reactance, and thus some amount of susceptance. The admittance ratings include these values of susceptance. The *output conductance* and *forward transconductance* ratings, on the other hand, take only the component resistance into account.

The bottom line is this: Admittance values are based on both component resistance and component reactance. Conductance values are based solely on component resistance values. The question now is, which ratings should we use for circuit analysis?

If you look closely at the 2N5484–6 series spec sheet, you'll see that the following ratings were measured in the 100 to 400-MHz frequency range:

- Input admittance
- Output conductance
- Forward transconductance

If you are analyzing a circuit that is being operated at or near this frequency range, you should use the above ratings in your analysis.

The following ratings were all measured at a frequency of approximately 1 kHz:

- Forward transfer admittance
- Output admittance

If you are analyzing a circuit that is being operated at or near this frequency, you should use the above ratings in your analysis.

The problem we have encountered in this section of the JFET spec sheet brings up a very interesting point. On many spec sheets, you will find multiple ratings that use the same unit of measure. When this occurs, you will always find that one or more of the conditions under which the ratings were measured differ between the two ratings. (In this case, the frequency of operation differed between the various ratings.) When this occurs, you should check to see which ratings were measured under the conditions that most closely resemble those of the circuit you are analyzing. Then use those ratings in your circuit analysis. For example, if we were analyzing a circuit that has an operating frequency of 10 kHz, we should use the *forward transfer admittance* rating for our g_{m0} values and the *output admittance* to find the value of r_d. If we were analyzing a circuit that has an operating frequency of 70 MHz, we would use the *forward transconductance* rating to obtain our g_{m0} value and the *output conductance* rating to obtain our value of r_d.

JFET Capacitance Ratings

JFETs have three capacitance ratings that affect the high-frequency operation of the components. These ratings are discussed in detail in Chapter 13. At this point, we will move on to briefly discuss several JFET circuit applications.

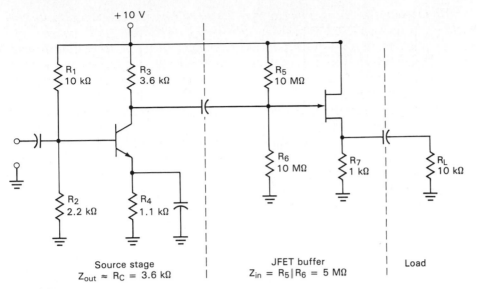

FIGURE 10.51

The JFET buffer.

JFET Applications

The high input impedance of the JFET amplifier makes it ideal for use in situations where source loading is a critical consideration. For example, consider the circuit shown in Figure 10.51. The JFET amplifier is being used as a *buffer* between the low Z_{out} source and the load. Because of the high input impedance of the JFET amplifier (5 MΩ), the amplifier presents virtually no load to the source amplifier. This has the effect of increasing the gain of the source stage. Recall that the gain of a BJT common-emitter amplifier is found as

$$A_v = \frac{R_C \| R_L}{r'_e}$$

If the load resistor had been connected directly to the source stage, the gain of the source stage would have been found as

$$A_v = \frac{2.65 \text{ k}\Omega}{r'_e}$$

where 2.65 kΩ is the parallel combination of R_C and R_L. With the JFET buffer, the gain of the first stage is found as

$$A_v = \frac{3.59 \text{ k}\Omega}{r'_e}$$

Regardless of the value of r'_e for the first stage, the gain of that stage is increased due to the use of the buffer amplifier. The low Z_{out} of the buffer would also aid in transferring voltage to the load. Thus the overall circuit works much more efficiently with the buffer amplifier than without it.

The high input impedance of the JFET also makes it useful in voltmeters. The FET voltmeter provides more reliable voltage readings than a standard VOM because the input to the meter is a high-impedance circuit that does not load the circuit under test. The input to an FET meter is represented (rather loosely) in

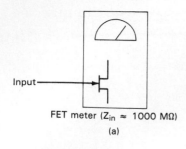

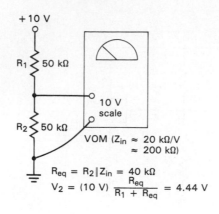

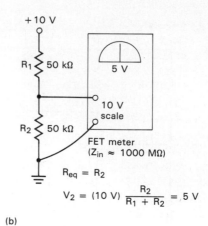

FIGURE 10.52

The FET meter.

$R_{eq} = R_2 \| Z_{in} = 40 \text{ k}\Omega$

$V_2 = (10 \text{ V}) \dfrac{R_{eq}}{R_1 + R_{eq}} = 4.44 \text{ V}$

$R_{eq} = R_2$

$V_2 = (10 \text{ V}) \dfrac{R_2}{R_1 + R_2} = 5 \text{ V}$

(b)

Figure 10.52. Figure 10.52a is used only to show you that the input to an FET meter is applied to the gate of a reverse-biased JFET. While the actual circuitry is more complicated, the diagram illustrates the fact that the FET meter has a very high input impedance. The effect of this high input impedance on circuit measurements is illustrated in Figure 10.52b. Note that the VOM has an input impedance (typically) of 20 kΩ/V. Thus, when the meter is on the 10-V scale, the input impedance is 20 kΩ × 10 V = 200 kΩ. When the 200-kΩ input impedance of the meter is across R_2, the parallel combination of resistances reduces the voltage across the resistor. As is shown, the meter gives a reading of 4.44 V. The FET meter, on the other hand, provides no loading on the circuit. Thus, the more accurate reading of 5 V is obtained.

In communications electronics, the JFET is used for a variety of functions. One of these is as the active component in an *RF amplifier* (radio-frequency amplifier). The RF amplifier is the first circuit to which the input signal to a receiver is applied. A JFET RF amplifier is shown in Figure 10.53. The gate and drain circuits of the amplifier are *tuned* circuits, meaning that they are designed to operate on and around specified frequencies. We will cover tuned amplifiers later in the text. The advantage of using the JFET in this circuit is the fact that JFETs are *low-*

☑ 36

RF amplifier. Radio-frequency amplifier; the input circuit of a radio receiver.

FIGURE 10.53

A JFET rf amplifier.

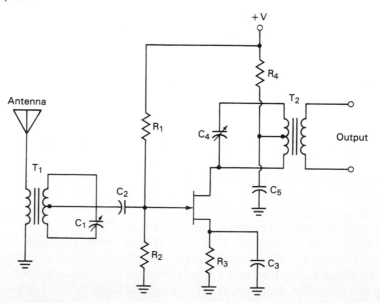

CHAP. 10   Field-Effect Transistors

noise components. *Noise* is any unwanted interference with a transmitted signal. Noise can be generated by either the transmitter or receiver or by factors outside of the communications system. The JFET will not generate any significant amount of noise, and thus is useful as an RF amplifier. Another advantage of using the JFET RF amplifier is the fact that it requires no input current to operate. The antenna of the receiver receives a very weak signal that has an extremely low amount of current. Since the JFET is a voltage-controlled component, it is well suited to respond to the low current signal provided by the antenna.

There are many other applications for JFET amplifiers. In this section we have looked at only a few of them. The main thing to remember is that a JFET amplifier is used to serve the same basic purpose as a BJT amplifier. In applications where it is advantageous to have a high Z_{in} amplifier, the JFET amplifier is preferred over the BJT amplifier.

SECTION REVIEW

1. Define each of the following JFET ratings: gate current, gate–source breakdown voltage, gate reverse current, gate–source cutoff voltage, and zero gate voltage drain current. [Objective 33]
2. What is the advantage of using a JFET buffer over a BJT buffer? [Objective 34]
3. Why are FET meter voltage readings more accurate than those made by other VOMs? [Objective 35]
4. What is an *RF amplifier*? [Objective 36]
5. What advantages does the JFET RF amplifier have over the BJT RF amplifier? [Objective 36]

KEY TERMS

The following terms were introduced and defined in this chapter:

channel	gate reverse current	self-bias
common-drain amplifier	gate–source cutoff voltage,	shorted-gate drain current
common-gate amplifier	$V_{GS(off)}$	(I_{DSS})
common-source amplifier	gate–source breakdown	source
drain	voltage	source follower
field-effect transistor	noise	transconductance (g_m, g_{fs},
gate	output admittance (y_{os})	or y_{fs})
gate bias	pinchoff voltage (V_P)	transconductance curve
gate current	RF amplifier	

PRACTICE PROBLEMS

§10.1

1. A JFET amplifier has values of $I_{DSS} = 8$ mA, $V_{GS(off)} = -12$ V, and $V_{GS} = -6$ V. Determine the value of I_D for the circuit. [11]
2. A JFET amplifier has values of $I_{DSS} = 16$ mA, $V_{GS(off)} = -5$ V, and $V_{GS} = -4$ V. Determine the value of I_D for the amplifier. [11]
3. A JFET amplifier has values of $I_{DSS} = 14$ mA, $V_{GS(off)} = -6$ V, and $V_{GS} = 0$ V. Determine the value of I_D for the circuit. [11]

4. A JFET amplifier has values of $I_{DSS} = 10$ mA, $V_{GS(off)} = -8$ V, and $V_{GS} = -5$ V. Determine the value of I_D for the circuit. [11]

5. A JFET amplifier has values of $I_{DSS} = 12$ mA, $V_{GS(off)} = -10$ V, and $V_{GS} = -4$ V. Determine the value of I_D for the circuit. [11]

6. The JFET amplifier described in problem 5 has an input of $V_{GS} = -14$ V. Determine the value of I_D for the circuit. [11]

7. The 2N5484 has values of $V_{GS(off)} = -0.3$ to -3 V and $I_{DSS} = 1.0$ to 5.0 mA. Plot the minimum and maximum transconductance curves for the device. [12]

8. The 2N5485 has values of $V_{GS(off)} = -0.5$ to -4.0 V and $I_{DSS} = 4.0$ to 10 mA. Plot the minimum and maximum transconductance curves for the device. [12]

9. The 2N5457 has values of $V_{GS(off)} = -0.5$ to -6.0 V and $I_{DSS} = 1.0$ to 5.0 mA. Plot the minimum and maximum transconductance curves for the device. [12]

10. The 2N5458 has values of $V_{GS(off)} = -1.0$ to -7.0 V and $I_{DSS} = 2.0$ to 9.0 mA. Plot the minimum and maximum transconductance curves for the device. [12]

11. The 2N3437 has parameters of $V_{GS(off)} = 5.0$ V (maximum) and $I_{DSS} = 0.8$ to 4.0 mA. Plot the maximum transconductance curve for the device. [12]

12. The 2N3438 has parameters of $V_{GS(off)} = -2.5$ V (maximum) and $I_{DSS} = 0.2$ to 1.0 mA. Plot the maximum transconductance curve for the device. [12]

§10.2

13. Determine the values of I_D and V_{DS} for the amplifier in Figure 10.54a. [13]

14. Determine the values of I_D and V_{DS} for the amplifier in Figure 10.54b. [13]

15. The 2N5486 (described in Practice Problem 10.3) is being used in the circuit in Figure 10.54a. Determine the range of I_D values for the circuit. [14]

16. A JFET with values of $V_{GS(Off)} = -5$ to -10 V and $I_{DSS} = 4$ to 8 mA is being used in the circuit in Figure 10.54b. Determine the range of I_D values for the circuit. [14]

17. The 2N3437 (described in problem 11) is being used in the circuit in Figure 10.55. Determine the range of I_D values for the circuit. (*Hint*: See the margin note under *A Practical Consideration* on page 432 of the text.) [14]

FIGURE 10.54   **FIGURE 10.55**

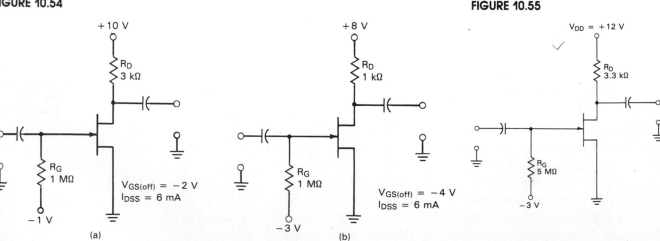

(a)   (b)

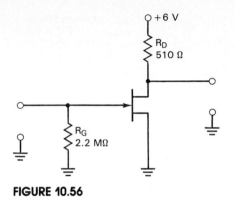

FIGURE 10.56

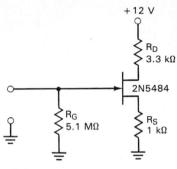

FIGURE 10.57

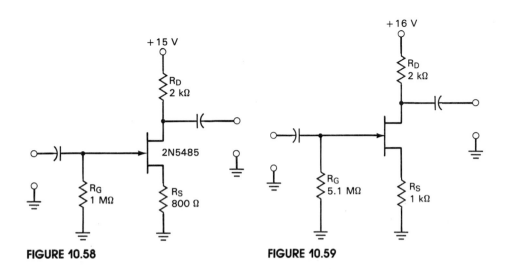

FIGURE 10.58   FIGURE 10.59

18. The 2N3438 (described in problem 12) is being used in the circuit in Figure 10.56. Determine the range of I_D values for the circuit. (*Hint*: See the margin note under *A Practical Consideration* on page 432 of the text.) [14]

19. Determine the ranges of V_{GS}, I_D, and V_{DS} for the circuit in Figure 10.57. The 2N5484 is described in problem 7. [16]

20. Determine the ranges of V_{GS}, I_D, and V_{DS} for the circuit in Figure 10.58. The 2N5485 is described in problem 8. [16]

21. Determine the ranges of V_{GS}, I_D, and V_{DS} for the circuit in Figure 10.59. The 2N5458 is described in problem 10. [16]

22. Determine the ranges of V_{GS}, I_D, and V_{DS} for the circuit in Figure 10.60. The 2N3437 is described in problem 11. [16]

23. Determine the ranges of I_D and V_{DS} for the circuit in Figure 10.61. [17]

24. Determine the ranges of I_D and V_{DS} for the circuit in Figure 10.62. [17]

25. Determine the ranges of I_D and V_{DS} for the circuit in Figure 10.63. [17]

26. Determine the ranges of I_D and V_{DS} for the circuit in Figure 10.64. [17]

§10.3

27. The 2N5484 has values of g_{m0} = 3000 to 6000 µS and $V_{GS(off)}$ = −0.3 to −3.0 V. Determine the range of g_m when V_{GS} = −0.2 V. [20]

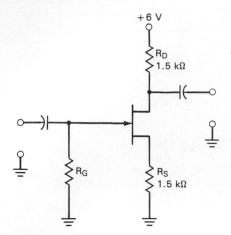

+6 V

R_D 1.5 kΩ

R_G

R_S 1.5 kΩ

FIGURE 10.60

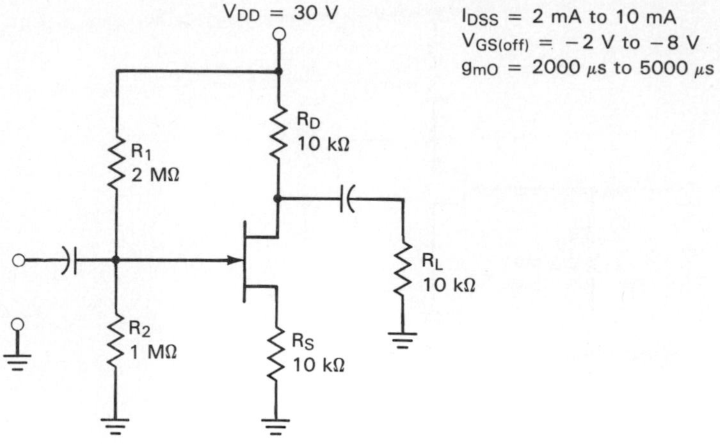

V_{DD} = 30 V

I_{DSS} = 2 mA to 10 mA
$V_{GS(off)}$ = −2 V to −8 V
g_{m0} = 2000 μs to 5000 μs

R_1 2 MΩ

R_D 10 kΩ

R_L 10 kΩ

R_2 1 MΩ

R_S 10 kΩ

FIGURE 10.61

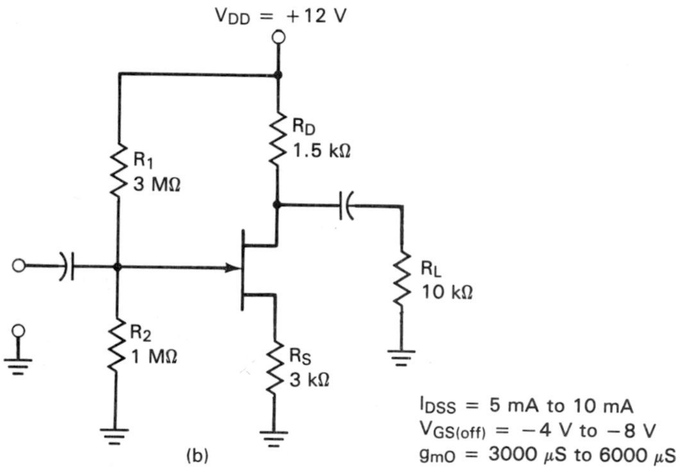

V_{DD} = +12 V

R_1 3 MΩ

R_D 1.5 kΩ

R_L 10 kΩ

R_2 1 MΩ

R_S 3 kΩ

(b)

FIGURE 10.62

I_{DSS} = 5 mA to 10 mA
$V_{GS(off)}$ = −4 V to −8 V
g_{m0} = 3000 μS to 6000 μS

I_{DSS} = 5 mA to 14 mA
$V_{GS(off)}$ = −2 V to −6 V
g_{m0} = 1000 μS to 2500 μS

V_{DD} = + 24 V

R_1 10 MΩ

R_D 2 kΩ

v_{in}

R_L 20 kΩ

R_2 2 MΩ

R_S 2 kΩ

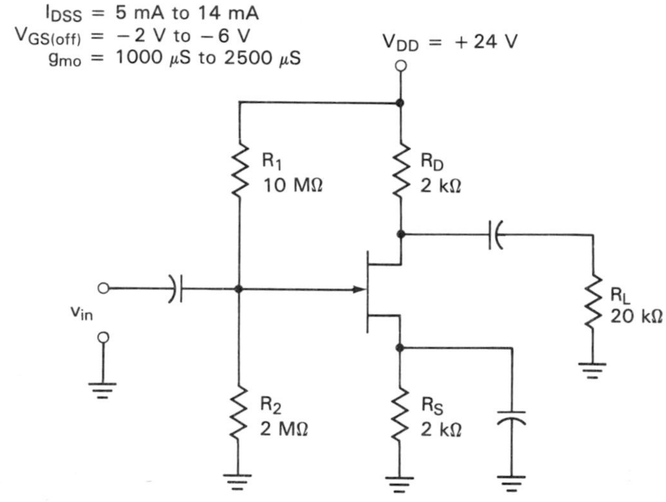

FIGURE 10.63

28. The 2N5485 has values of g_{m0} = 3500 to 7000 μS and $V_{GS(off)}$ = −0.5 to −4.0 V. Determine the range of g_m when V_{GS} = −0.4 V. [20]

29. The 2N3437 has maximum values of $V_{GS(off)}$ = 5.0 V and g_{m0} = 6000 μS. Determine the maximum values of g_m at V_{GS} = −1 V, V_{GS} = −2.5 V, and V_{GS} = −4 V. [20]

30. The 2N3438 has maximum values of $V_{GS(off)}$ = 2.5 V and g_{m0} = 4500 μS. Determine the maximum values of g_m at V_{GS} = −1 V, V_{GS} = −1.5 V, and V_{GS} = −2 V. [20]

31. Determine the range of A_{vL} values for the circuit in Figure 10.61. [22]

32. Determine the range of A_{vL} values for the circuit in Figure 10.62. [22]

33. Determine the range of A_{vL} values for the circuit in Figure 10.63. [22]

34. Determine the range of A_{vL} values for the circuit in Figure 10.64. [22]

35. Determine the range of A_{vL} values for the swamped amplifier shown in Figure 10.65. [24]

36. Determine the range of A_{vL} values for the swamped amplifier shown in Figure 10.66. [24]

37. Determine the value of Z_{in} for the amplifier in Figure 10.61. [25]

38. Determine the value of Z_{in} for the amplifier in Figure 10.62. [25]

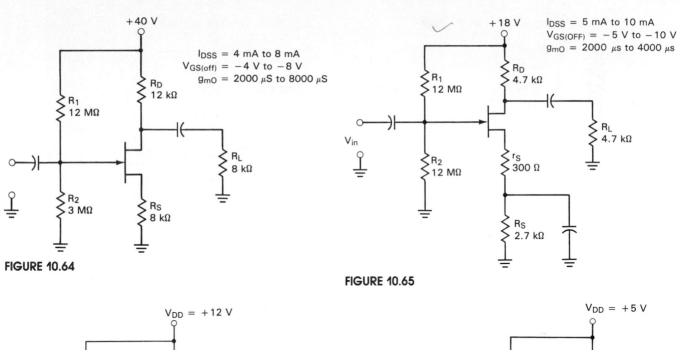

FIGURE 10.64

FIGURE 10.65

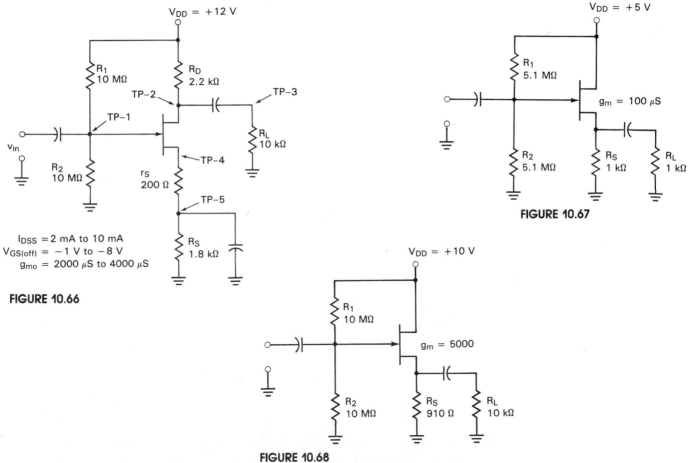

FIGURE 10.66

FIGURE 10.67

FIGURE 10.68

§10.4

39. Determine the values of A_{vL}, Z_{in}, and Z_{out} for the amplifier in Figure 10.67.
[27]

40. Determine the values of A_{vL}, Z_{in}, and Z_{out} for the amplifier in Figure 10.68.
[27]

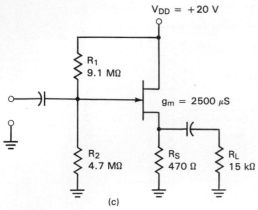

(c)

FIGURE 10.69

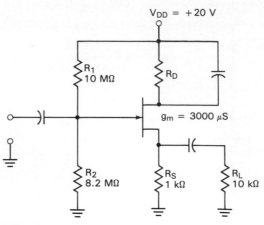

FIGURE 10.70

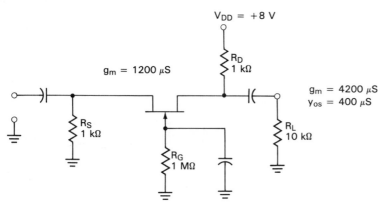

FIGURE 10.71

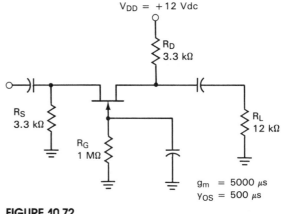

FIGURE 10.72

41. Determine the ranges of A_{vL}, Z_{in}, and Z_{out} for the amplifier in Figure 10.69. [27]

42. Determine the ranges of A_{vL}, Z_{in} and Z_{out} for the amplifier in Figure 10.70. [27]

43. Determine the ranges of A_{vL}, Z_{in}, and Z_{out} for the circuit in Figure 10.71. [29]

44. Determine the ranges of A_{vL}, Z_{in} and Z_{out} for the circuit in Figure 10.72. [29]

TROUBLESHOOTING PRACTICE PROBLEMS

45. The circuit in Figure 10.66 has the waveforms shown in Figure 10.73. Discuss the possible causes of the problem. [31]

46. A voltage-divider biased JFET amplifier suddenly starts to have the instability of a self-bias circuit. Changing the JFET does not solve the problem. What is wrong? [31]

47. The circuit in Figure 10.74 has the waveforms shown. Discuss the possible causes of the problem. [31]

48. The circuit in Figure 10.75 has the dc voltages shown. Discuss the possible causes of the problem.

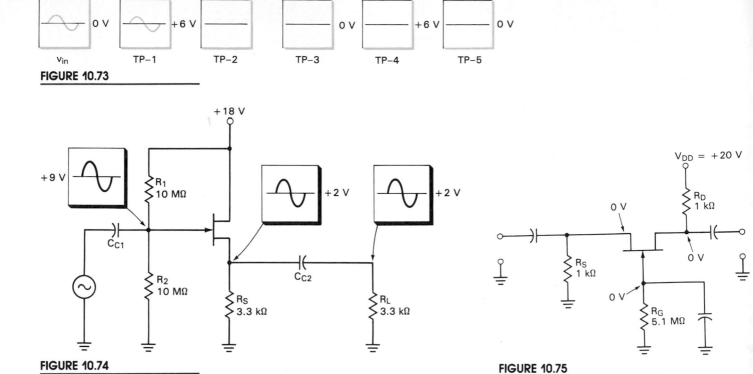

FIGURE 10.73

FIGURE 10.74

FIGURE 10.75

THE BRAIN DRAIN

49. The biasing circuit shown in Figure 10.76 is called a *current-source bias* circuit. Using your knowledge of BJT circuits, determine how this circuit obtains a stable value of I_D.

50. Can the 2N5486 be used safely in the circuit shown in Figure 10.77? Explain your answer using circuit calculations.

51. The 2N5486 cannot be substituted for the JFET in Figure 10.78. Why not?

FIGURE 10.76

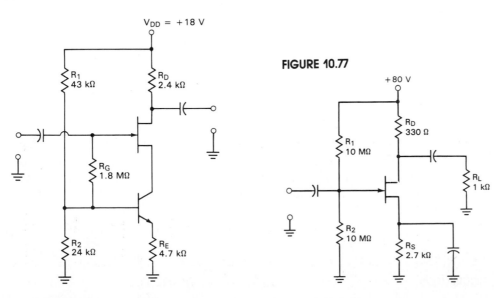

FIGURE 10.77

FIGURE 10.78

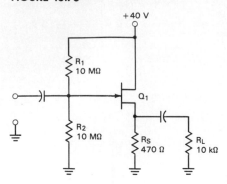

ANSWERS TO THE EXAMPLE PRACTICE PROBLEMS

10.1.   3 mA

10.2.   See Figure 10.79.

10.3.   See Figure 10.80.

10.4. $V_{GS} = -5$ V, $I_D = 3$ mA, $V_{DS} = 3.4$ V

10.6.   See Figure 10.81.

10.7.   1.5 V to 5.25 V

10.8.   See Figure 10.82.

10.10.   15.98 V (minimum), 20.4 V (maximum)

10.11. $g_m = 5333$ μS @ $V_{GS} = -2$ V, $g_m = 2667$ μS @ $V_{GS} = -4$ V

10.12. $A_{vL} = 4.71$ to 8.58

10.13. $A_{vL} = 3.3$ to 3.77

10.16. $A_{vL} = 0.8889$ to 0.9143, $Z_{out} = 349\Omega$ to 455Ω

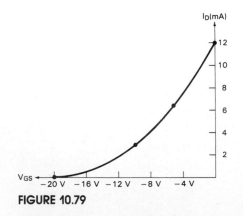

FIGURE 10.79

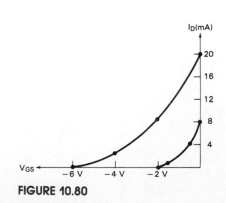

FIGURE 10.80

FIGURE 10.81

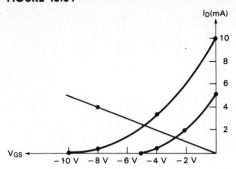

FIGURE 10.82

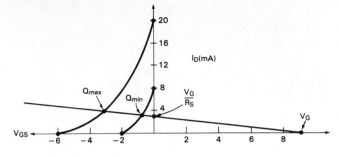

11
MOSFETs

Advances in MOSFET technology have lead to the development of VLSI (very large scale integrated) circuits. Many of these circuits (like the ones shown) are essentially entire electronic systems that are etched on a single piece of silicon.

OBJECTIVES

After studying the material in this chapter, you should be able to:

☐ 1. Define the following terms: *depletion-mode operation*, *enhancement-mode operation*, and *MOSFET*. (Introduction)

☐ 2. Discuss the similarities and differences between the operating modes, physical construction, and schematic symbols for *D-MOSFETs* and *E-MOSFETs*. (§11.1)

☐ 3. Describe the depletion-mode operation of the D-MOSFET. (§11.2)

☐ 4. Describe the enhancement-mode operation of the D-MOSFET. (§11.2)

☐ 5. Plot the transconductance curve(s) for a D-MOSFET. (§11.2)

☐ 6. Describe the relationship between g_m and g_{m0} for a D-MOSFET. (§11.2)

☐ 7. Describe *zero bias* and list the other biasing circuits that are commonly used for D-MOSFETs. (§11.2)

☐ 8. List the similarities and differences between JFETs and D-MOSFETs. (§11.2)

☐ 9. Describe the means by which a channel is generated in an E-MOSFET. (§11.3)

☐ 10. Define *threshold voltage* $V_{GS(th)}$ and discuss its significance. (§11.3)

☐ 11. Calculate the value of the transconductance constant (k) for a given E-MOSFET. (§11.3)

☐ 12. Calculate the value of I_D for an E-MOSFET at a given value of V_{GS}. (§11.3)

☐ 13. List the FET biasing circuits that can and cannot be used with E-MOSFETs. (§11.3)

☐ 14. Describe and analyze the operation of the *drain-feedback bias* circuit. (§11.3)

☐ 15. Describe the physical construction and capacitance characteristics of the *dual-gate MOSFET*. (§11.4)

☐ 16. Describe the physical construction and current-handling capacilities of the VMOS. (§11.5)

☐ 17. Describe the physical characteristics, channel resistance, and current handling capabilities of the LDMOS. (§11.5)

☐ 18. Define the terms *logic level*, *logic family*, and *inverter*. (§11.6)

☐ 19. Describe the construction and operation of the CMOS inverter. (§11.6)

☐ 20. List the advantages that CMOS logic circuits have over TTL logic circuits. (§11.6)

☐ 21. Describe the construction of the *cascode amplifier* and discuss the means by which it reduces overall amplifier input capacitance. (§11.7)

☐ 22. Discuss the use of the dual-gate MOSFET as a cascode amplifier. (§11.7)

☐ 23. Discuss the use of the dual-gate MOSFET as an RF amplifier with *automatic gain control* (*AGC*) capabilities. (§11.7)

☐ 24. Describe the purpose served by power MOSFET drivers in digital communictions. (§11.7)

☐ 25. Describe the types of applications where VMOS amplifiers are typically used. (§11.7)

☐ 26. Describe the devices and precautions that are used to protect MOSFETs. (§11.8)

As you master each of the objectives listed, place a check mark (✔) in the appropriate box.

The main drawback to JFET operation is the fact that the JFET gate *must* be reverse biased for the device to operate properly. Recall that the reverse bias of the gate is adjusted to deplete the channel of free carriers, thus controlling the effective size of the channel. This type of operation is referred to as *depletion-mode* operation. A depletion-type device would be one that uses an input voltage to *reduce* the channel size from its *zero-bias* size.

The *metal oxide semiconductor FET*, or *MOSFET*, is a device that can be operated in the *enhancement mode*. This means that the input signal can be used to *increase* the size of the channel. This also means that the device is not restricted to operating with its gate reverse biased. This is an improvement over the standard JFET. In this chapter we will look at the various types of MOSFETs, their construction and operation, as well as their circuit operation, applications, and troubleshooting techniques.

✔ 1

Depletion-mode operation. Using an input voltage to reduce the channel size.

Enhancement-mode operation. Using an input voltage to increase the channel size.

MOSFET. An FET that can be operated in the enhancement mode.

483

THE MOSFET: A MAJOR COMPONENT IN INTEGRATED CIRCUIT TECHNOLOGY

Odds are that you have never heard of MOS (metal-oxide semiconductor) technology before. At the same time, the odds are that you probably have one or more items with you that were made using MOSFET circuitry.

Most calculators, digital watches, and desktop computers are made using VLSI (very large scale integrated) circuits. These circuits contain up to hundreds of thousands of components that are all etched on a single piece of semiconductor material. The development of VLSI technology has made many items, such as those listed above, possible.

All electronic components dissipate heat. All electronic components must have some measurable size. These are two of the problems that have always faced the developers of integrated circuits, or *ICs*. The goal of IC technology has always been to produce the maximum number of components in the smallest amount of space possible. However, the need for heat-dissipation space and the actual space taken up by the components that could be etched into a single piece of silicon.

The development of MOS technology has provided one solution to the problem of component spacing. Since MOSFET circuits are extremely low input current circuits, the heat-dissipation problem has been greatly reduced in these circuits. At the same time, MOSFET circuits can be made much smaller than their BJT counterparts. For these reasons, MOSFET technology has come to dominate most of the VLSI circuit market.

11.1
Introductory Concepts

D-MOSFETs can be operated in both the depletion mode and the enhancement mode.

There are two basic types of MOSFETs: *depletion-type MOSFETs*, or *D-MOSFETs*, and *enhancement-type MOSFETs*, or *E-MOSFETs*. D-MOSFETs can be operated in both the depletion mode and the enhancement mode, whereas E-MOSFETs are restricted to operating in the enhancement mode. The primary difference between D-MOSFETs and E-MOSFETs is their physical construction. The construction difference between the two is illustrated in Figure 11.1. As you can see, the D-MOSFET has a physical channel (shaded) between the source and drain terminals.

FIGURE 11.1

MOSFET construction and schematic symbols.

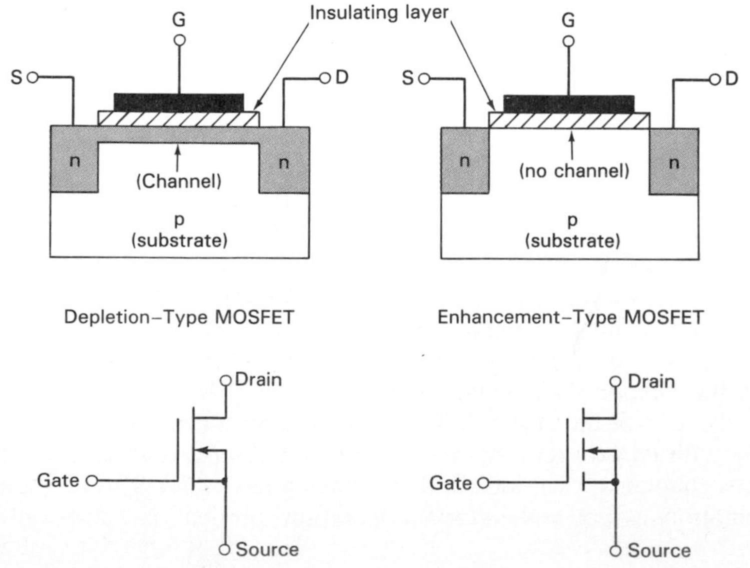

Depletion–Type MOSFET

Enhancement–Type MOSFET

The E-MOSFET, on the other hand, has no such channel. The E-MOSFET depends on the gate voltage to *form* a channel between the source and drain terminals. This point will be discussed in detail later in this chapter.

E-MOSFETs are restricted to enhancement-mode operation.

Note that both MOSFETs in Figure 11.1 show an *insulating layer* between the gate and the rest of the component. This insulating layer is made up of *silicon dioxide* (SiO_2), a glasslike material that is an insulator. The gate terminal is made of a metal conductor. Thus, going from gate to substrate, you have a *metal oxide semiconductor*, which is where the term *MOSFET* comes from. Since the gate is insulated from the rest of the component, the MOSFET is sometimes referred to as an *insulated-gate FET*, or *IGFET*. However, this term is rarely used in place of the term *MOSFET*.

The foundation of the MOSFET is called the *substrate*. This material is represented in the schematic symbol by the center line that is connected to the source terminal. The connection of these lines indicates that the substrate and the source are connected internally. Note that an *n*-channel MOSFET will have a *p*-material substrate, and a *p*-channel MOSFET will have an *n*-material substrate.

Substrate. The foundation of a MOSFET.

In the schematic symbol for the MOSFET, the arrow is placed on the substrate. As with the JFET, an arrow pointing *in* represents an *n-channel* device, while an arrow pointing *out* represents a *p-channel* device.

SECTION REVIEW

1. What is a *MOSFET*? [Objective 1]
2. What is *depletion-mode operation*? [Objective 1]
3. What is *enhancement-mode operation*? [Objective 1]
4. In terms of operating modes, what is the difference between D-MOSFETs and E-MOSFETs? [Objective 2]
5. In terms of physical construction, what is the difference between D-MOSFETs and E-MOSFETs? [Objective 2]
6. What is the difference between the schematic symbol of a D-MOSFET and that of an E-MOSFET? [Objective 2]

11.2
D-MOSFETs

As stated earlier, the D-MOSFET is capable of operating in both the depletion mode and the enhancement mode. When it is operating in the depletion mode, the characteristics of the D-MOSFET are very similar to those of the JFET. The overall operation of the D-MOSFET is illustrated in Figure 11.2.

The D-MOSFET can operate in both modes.

Figure 11.2a shows the D-MOSFET operating conditions when $V_{GS} = 0$ V (that is, when the gate and source terminals are *shorted*). As stated in Chapter 10, I_{DSS} is the value of I_D when the source and gate terminals are shorted together. Therefore, when $V_{GS} = 0$ V, $I_D = I_{DSS}$.

When the gate–source junction is *reverse* biased (V_{GS} is negative), the biasing voltage depletes the channel of free carriers. This *effectively* reduces the width of the channel, increasing its resistance. This is shown in Figure 11.2b. Note that this operating state is exactly the same as the normal JFET operating state. The white area below the insulating layer in Figure 11.2b represents the depleted carrier

3

Depletion-mode operation is very similar to JFET operation.

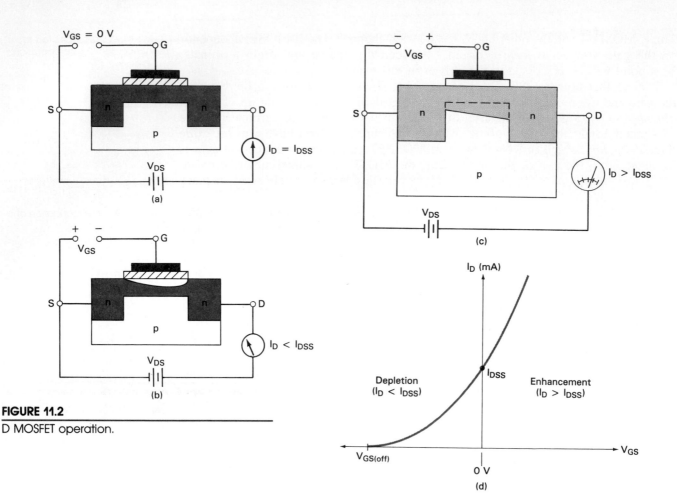

FIGURE 11.2

D MOSFET operation.

How enhancement-mode
operation works.

For enhancement-mode
operation, $I_D > I_{DSS}$.

region. The shaded area below this region is the effective channel width. Since this area is less than the area shown in Figure 11.2a, the resistance of the MOSFET channel is higher when V_{GS} is negative. Thus I_D will be at some value less than I_{DSS}. The exact value of I_D depends on the value of I_{DSS} and V_{GS}.

Figure 11.2c shows the *enhancement mode* of operation. In this circuit, V_{GS} is *positive*. When V_{GS} is positive, the channel is *effectively* widened. This reduces the resistance of the channel, and I_D increases above I_{DSS}.

How does a positive V_{GS} effectively *widen* the channel? You must remember that the majority carriers in a *p*-type material are valence band *holes*. The holes in the *p*-type substrate are *repelled* by the positive gate voltage. Left behind is a region that is depleted of valence band holes. At the same time, the conduction band electrons (minority carriers) in the *p*-type material are *attracted* toward the channel by the positive gate voltage. With the buildup of electrons near the channel, the area below the physical channel *effectively* becomes an *n*-type material. The extended *n*-type channel now allows more current to pass through the channel and $I_D > I_{DSS}$.

The combination of these three operating states is represented by the D-MOSFET transconductance curve in Figure 11.2d. Note that $I_D = I_{DSS}$ when $V_{GS} = 0$ V. When V_{GS} is *negative*, I_D decreases below the value of I_{DSS}. I_D reaches zero when $V_{GS} = V_{GS(off)}$, just as with the JFET. When V_{GS} is *positive*, I_D increases above the value of I_{DSS}. The maximum allowable value of I_D is given on the spec sheet of a given MOSFET.

You may have noticed that the transconductance curve for the D-MOSFET is very similar to the curve for a JFET. Because of this similarity, the JFET and the D-MOSFET have the same transconductance equation. This equation was given in Chapter 10 as

$$I_D = I_{DSS}\left(1 - \frac{V_{GS}}{V_{GS(off)}}\right)^2 \qquad \textbf{(11.1)}$$

The following example illustrates the use of equation (11.1) in plotting the transconductance curve for a D-MOSFET.

EXAMPLE 11.1

A D-MOSFET has parameters of $V_{GS(off)} = -6$ V and $I_{DSS} = 1$ mA. Plot the transconductance curve for the device.

Solution: From the parameters, we can determine two of the points on the curve as follows:

$$V_{GS} = -6 \text{ V}, \qquad I_D = 0$$
$$V_{GS} = \quad 0 \text{ V} \qquad I_D = 1 \text{ mA}$$

These points define the conditions at $V_{GS} = V_{GS(off)}$ and at $I_D = I_{DSS}$. Now we will use equation (11.1) to determine the coordinates of several more points. Remember that the only value we change is V_{GS}. When $V_{GS} = -3$ V,

$$I_D = 1 \text{ mA} \left(1 - \frac{-3 \text{ V}}{-6 \text{ V}}\right)^2$$
$$= 0.25 \text{ mA} \qquad (250 \text{ μA})$$

When $V_{GS} = -1$ V,

$$I_D = 1 \text{ mA} \left(1 - \frac{-1 \text{ V}}{-6 \text{ V}}\right)^2$$
$$= 0.694 \text{ mA} \qquad (694 \text{ μA})$$

When $V_{GS} = +1$ V,

$$I_D = 1 \text{ mA} \left(1 - \frac{-1 \text{ V}}{+6 \text{ V}}\right)^2$$
$$= 1.36 \text{ mA}$$

When $V_{GS} = +3$ V,

$$I_D = 1 \text{ mA} \left(1 - \frac{+3 \text{ V}}{-6 \text{ V}}\right)^2$$
$$= 2.25 \text{ mA}$$

We now have the following combinations of V_{GS} and I_D:

V_{GS} (V)	I_D (mA)
−6	0
−3	0.25
−1	0.69
0	1.00
+1	1.36
+3	2.25

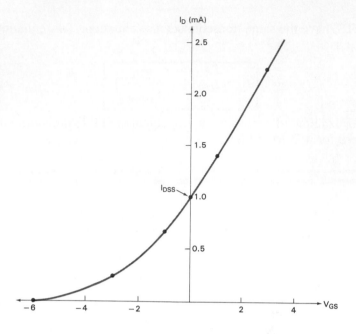

FIGURE 11.3

A *Practical Consideration.*
D-MOSFETs normally have minimum and maximum $V_{GS(\text{off})}$ and I_{DSS} ratings, just like JFETs. Thus, you normally need to plot both the minimum and maximum transconductance curves just as we did for the JFET in Chapter 10.

Using the values listed, we can plot the transconductance curve shown in Figure 11.3.

PRACTICE PROBLEM 11–1

A D-MOSFET has values of $V_{GS(\text{off})} = -4$ to -6 V and $I_{DSS} = 8$ to 12 mA. Plot the minimum and maximum transconductance curves for the device. Use V_{GS} values of -4 V, -2 V, 0 V, 2 V, and 4 V.

As you can see, the process for plotting a D-MOSFET transconductance curve is almost the same as plotting one for a JFET. The only differences are that you must use positive values of V_{GS} (as well as negative) and that you will get values of I_D that are greater than I_{DSS}.

D-MOSFET Drain Curves

The drain curves of a given D-MOSFET can be used to plot the dc load line of a given circuit, just as the collector curves can be used to plot the load line of a given BJT. The drain curves for a D-MOSFET are plotted using corresponding values of V_{GS} and I_D. Such a plot of drain curves is shown in Figure 11.4. The curves shown were plotted using the values from Example 11.1. Also, the curves for $V_{GS} = -2$ V and $V_{GS} = 2$ V were added. Note that above V_P the value of V_{DS} becomes constant for each combination of V_{GS} and I_D. Also note that the value of V_P is relative to the gate voltage, and thus changes from one curve to another. The value of V_P for $V_{GS} = +3$ V is highlighted in Figure 11.4 to illustrate this point. At higher values of V_{GS}, V_P is higher. Reducing the value of V_{GS} reduces the value of V_P.*

The actual dc load line for a D-MOSFET would depend on the circuit the device is being used in. The *ideal* saturation point on the load line is found as

* Note that the V_P rating of a MOSFET is measured at $V_{GS} = 0$ V.

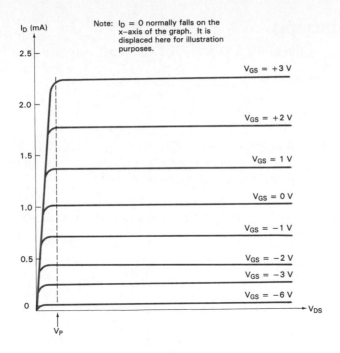

FIGURE 11.4

MOSFET drain curves.

$$I_{D(\text{sat})} = \frac{V_{DD}}{R_D} \qquad (11.2)$$

The *ideal* value of $V_{DS(\text{off})}$ would be found as

$$V_{DS(\text{off})} = V_{DD} \qquad (11.3)$$

Compare these equations to the load-line equations for the BJT amplifier. The load line for the D-MOSFET would be plotted as shown in Figure 11.5. Note the differences between the *real* and *ideal* values of $I_{D(\text{sat})}$ and $V_{DS(\text{off})}$.

FIGURE 11.5

Plotting a D-MOSFET load line.

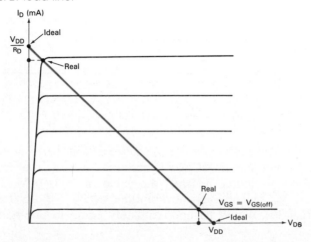

Transconductance

The value of g_m is found for a D-MOSFET the same way that it is for the JFET. By formula,

$$g_m = g_{m0}\left(1 - \frac{V_{GS}}{V_{GS(off)}}\right) \tag{11.4}$$

This equation leads to another characteristic of the D-MOSFET. Just as I_D can be greater than I_{DSS}, g_m can be greater than g_{m0}. Of course, this happens only when the device is being operated in the enhancement mode.

D-MOSFET Biasing Circuits

D-MOSFET biasing is very similar to JFET biasing.

Again, the similarities between D-MOSFETs and JFETs pay off. The primary D-MOSFET biasing circuits are exactly the same as those used for JFETs. The D-MOSFET will also have the same overall characteristics as the JFET when being used in the common-source, common-drain, and common-gate configurations. All the dc and ac relationships covered for the JFET will work for D-MOSFET circuits as well.

Zero bias. A D-MOSFET biasing circuit that has quiescent values of $V_{GS} = 0$ V and $I_D = I_{DSS}$.

The primary difference between D-MOSFETs and JFETs is the fact that the D-MOSFET does *not* require a negative value of V_{GS}. In fact, one common method of biasing a D-MOSFET is to set V_{GS} to 0 V. This biasing circuit configuration, called *zero bias*, is illustrated in Figure 11.6.

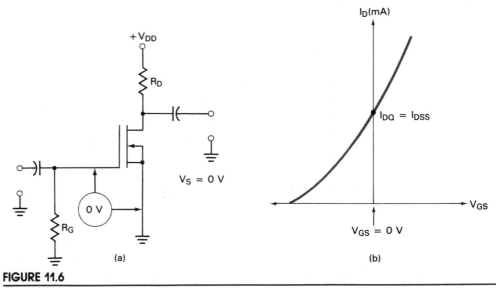

FIGURE 11.6

Zero biasing.

With the JFET self-bias circuit, I_{DQ} caused a voltage to develop across the source resistor, R_S. For the D-MOSFET zero-bias circuit, the source resistor is not necessary. With no source resistor, the value of V_S is 0 V. This gives us a value of $V_{GS} = 0$ V. This biases the amplifier at $I_D = I_{DSS}$. The value of R_D would be such that V_{DS} would be approximately equal to one-half of V_{DD}. It is just that simple.

D-MOSFET Input Impedance

The gate impedance of a D-MOSFET is extremely high. For example, one MOSFET has a maximum gate current of 10 pA when V_{GS} is 35 V. This would calculate to a gate impedance of 3.5×10^{12} Ω! With an input impedance in this range, the MOSFET would present virtually no load to a source circuit.

D-MOSFETs versus JFETs

Figure 11.7 summarizes many of the characteristics of JFETs and D-MOSFETs. As you can see, the components are very similar in a number of respects. The D-MOSFET has the advantages of higher input impedance and the ability to operate in the enhancement mode. It can also use zero bias, while the JFET cannot. At the same time, the D-MOSFET is more sensitive to increases in temperature, and there are precautions that must be taken when handling many MOSFETs. These two MOSFET disadvantages are discussed in detail later in this chapter.

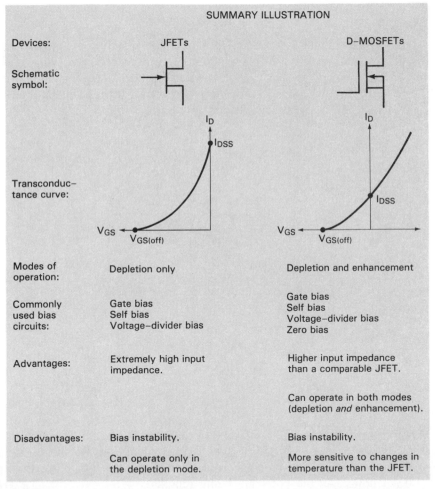

FIGURE 11.7

1. Describe the depletion-mode relationship between V_{GS} and I_D for a D-MOSFET. **SECTION REVIEW**
    [Objective 3]

2. Describe the enhancement-mode relationship between V_{GS} and I_D for a D-MOSFET. [Objective 4]
3. Describe the relationship between I_D and I_{DSS} for the D-MOSFET. [Objective 4]
4. Describe the relationship between g_m and g_{m0} for the D-MOSFET. [Objective 6]
5. Describe the quiescent conditions of the *zero-bias* circuit. [Objective 7]
6. List the similarities and differences between the JFET and the D-MOSFET. [Objective 8]

11.3
E-MOSFETs

It was stated earlier that the E-MOSFET is capable of operating only in the enhancement mode. In other words, the gate potential must be positive with respect to the source. The reason for this can be seen in Figure 11.8. When the value of V_{GS} is 0 V, there is no channel connecting the source and drain materials. As a result, there can be no significant amount of drain current. This is illustrated in Figure 11.8a.

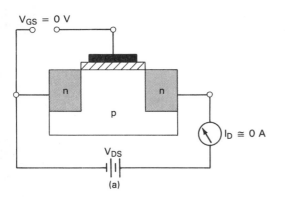
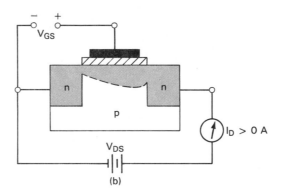

FIGURE 11.8

E-MOSFET operation.

When a positive potential is applied to the gate, the E-MOSFET responds in the same manner as the D-MOSFET. As Figure 11.8b shows, the positive gate voltage forms a channel between the source and drain by depleting the area of valence band holes. As the holes are being repelled by the positive gate voltage, that voltage is also attracting the minority carrier electrons in the p-type material toward the gate. This forms an *effective* n-type bridge between the source and drain, allowing some measurable value of I_D to flow.

If the value of V_{GS} is *increased*, the newly formed channel becomes wider, causing I_D to increase. If the value of V_{GS} *decreases*, the channel will become narrower, and I_D will decrease. This is illustrated by the E-MOSFET transconductance curve shown in Figure 11.9. As you can see, this transconductance curve is

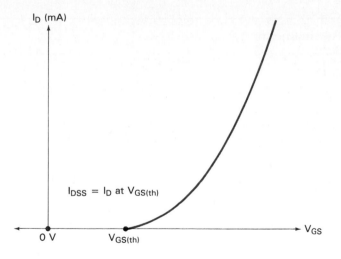

FIGURE 11.9

The E-MOSFET transconductance curve.

similar to those covered previously. The primary differences are:

1. All values of V_{GS} that cause the device to conduct are *positive*.
2. The point at which the device turns on (or off, depending on how you look at it) is called *threshold voltage*, $V_{GS(\text{th})}$.

Threshold voltage ($V_{GS(\text{th})}$).
The value of V_{GS} that turns the
E-MOSFET *on*.

Note that the value of I_{DSS} for the E-MOSFET is approximately 0 A. For example, the 3N169 E-MOSFET spec sheet lists an I_{DSS} of 10 nA when the ambient temperature is 25°C. For all practical purposes, this value of I_D is zero. Obviously, since the value of I_{DSS} for the E-MOSFET is near zero, the standard transconductance formula will not work for the E-MOSFET. To determine the value of I_D at a given value of V_{GS}, you must use the following formula:

$$I_D = k[V_{GS} - V_{GS(\text{th})}]^2 \qquad (11.5)$$

where k is a *constant* for the MOSFET, found as

$$k = \frac{I_{D(\text{on})}}{[V_{GS} - V_{GS(\text{th})}]^2} \qquad (11.6)$$

The values used in equation (11.6) to determine the value of k are obtained from the spec sheet of the E-MOSFET being used. For example, consider the spec sheet shown in Figure 11.10. This is the sheet for the 3N169, 70, 71 series E-MOSFETs. The 3N169 has the following parameters listed:

$$I_{D(\text{on})} = 10 \text{ mA dc} \quad \text{(minimum)}$$

$$V_{GS(\text{th})} = 0.5 \text{ V dc} \quad \text{(minimum)}$$

Now, if you look closely at the test conditions for the $I_{D(\text{on})}$ rating, you will see that this current was measured when $V_{GS} = 10$ V. Using this value and the two

ELECTRICAL CHARACTERISTICS (T_A = 25°C unless otherwise noted)
Substrate connected to source.

Characteristic	Figure No.	Symbol	Min	Max	Unit
OFF CHARACTERISTICS					
Drain-Source Breakdown Voltage (I_D = 10 µAdc, V_{GS} = 0)	–	$V_{(BR)DSS}$	25	–	Vdc
*Gate Leakage Current (V_{GS} = –35 Vdc, V_{DS} = 0) (V_{GS} = –35 Vdc, V_{DS} = 0, T_A = 125°C)	–	I_{GSS}		10 100	pAdc
*Zero-Gate-Voltage Drain Current (V_{DS} = 10 Vdc, V_{GS} = 0) (V_{DS} = 10 Vdc, V_{GS} = 0, T_A = 125°C)	–	I_{DSS}		10 1.0	nAdc µAdc
***ON CHARACTERISTICS**					
Gate-Source Threshold Voltage 3N169 (V_{DS} = 10 Vdc, I_D = 10 µAdc) 3N170 3N171	–	$V_{GS(th)}$	0.5 1.0 1.5	1.5 2.0 3.0	Vdc
"ON" Drain Current (V_{GS} = 10 Vdc, V_{DS} = 10 Vdc)	3	$I_{D(on)}$	10	–	mAdc
Drain-Source "ON" Voltage (I_D = 10 mAdc, V_{GS} = 10 Vdc)	–	$V_{DS(on)}$	–	2.0	Vdc
SMALL SIGNAL CHARACTERISTICS					
*Drain-Source Resistance (V_{GS} = 10 Vdc, I_D = 0, f = 1.0 kHz)	4	$r_{ds(on)}$	–	200	Ohms
Forward Transfer Admittance (V_{DS} = 10 Vdc, I_D = 2.0 mAdc, f = 1.0 kHz)	1	|Yfs|	1000	–	µmhos
*Reverse Transfer Capacitance (V_{DS} = 0, V_{GS} = 0, f = 1.0 MHz)	2	C_{rss}	–	1.3	pF
*Input Capacitance (V_{DS} = 10 Vdc, V_{GS} = 0, f = 1.0 MHz)	2	C_{iss}	–	5.0	pF
*Drain-Substrate Capacitance ($V_{D(SUB)}$ = 10 Vdc, f = 1.0 MHz)	–	$C_{d(sub)}$	–	5.0	pF
***SWITCHING CHARACTERISTICS**					
Turn-On Delay Time	6,10	$t_{d(on)}$	–	3.0	ns
Rise Time (V_{DD} = 10 Vdc, $I_{D(on)}$ = 10 mAdc,	7,10	t_r	–	10	ns
Turn-Off Delay Time $V_{GS(on)}$ = 10 Vdc, $V_{GS(off)}$ = 0,	8,10	$t_{d(off)}$	–	3.0	ns
Fall Time R_G' = 50 Ohms)	9,10	t_f	–	15	ns

*Indicates JEDEC Registered Data.

FIGURE 1   FORWARD TRANSFER ADMITTANCE

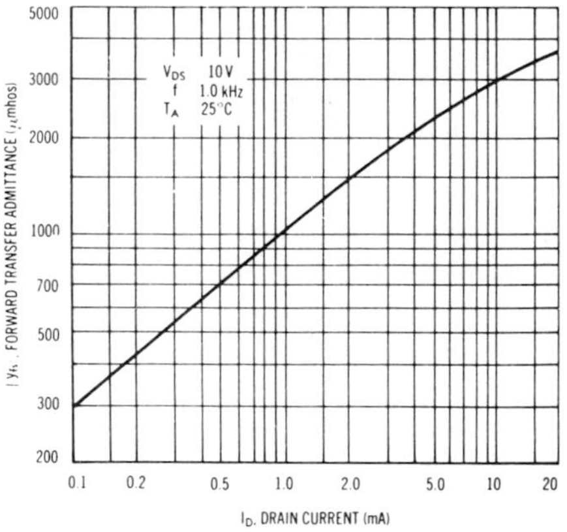

FIGURE 2 – CAPACITANCE

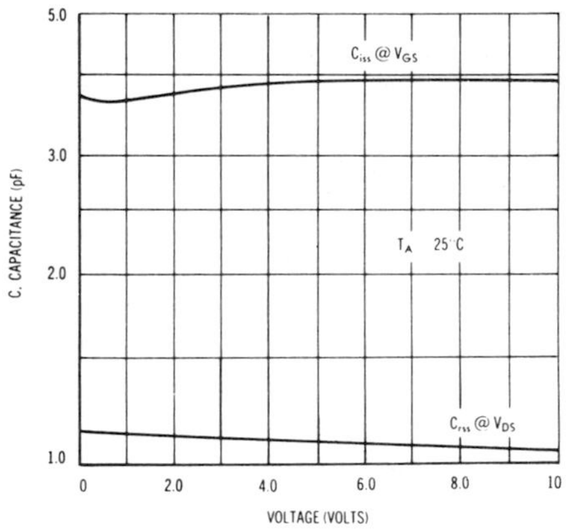

MOTOROLA Semiconductor Products Inc. Ⓜ

FIGURE 11.10

The 3N169, 70, 71 series specification sheet. (Courtesy of Motorola, Inc.)

parameters listed above, k would be found as

$$k = \frac{I_{D(on)}}{[V_{GS} - V_{GS(th)}]^2}$$
$$= \frac{10\ \text{mA}}{(10\ \text{V} - 0.5\ \text{V})^2}$$
$$= 1.11 \times 10^{-4}$$

The value of $k = 1.11 \times 10^{-4}$ would now be used in equation (11.5) to determine the value of I_D for a given value of V_{GS}. The entire process is illustrated in the following example.

■ EXAMPLE 11.2 ■

Determine the value of I_D for the circuit shown in Figure 11.11. Use the spec sheet in Figure 11.10 to obtain the values needed to calculate the k of the E-MOSFET.

Solution: The spec sheet for the 3N171 lists the following parameters:

$$I_{D(\text{on})} = 10 \text{ mA at } V_{GS} = 10 \text{ V}$$

$$V_{GS(\text{th})} = 1.5 \text{ V}$$

Using these *minimum* values, the value of k is found as

$$k = \frac{10 \text{ mA}}{(10 \text{ V} - 1.5 \text{ V})^2}$$
$$= 1.38 \times 10^{-4}$$

From the circuit, the source voltage is seen to be 0 V. Therefore, $V_{GS} = V_G$. The value of V_G is found as

$$V_G = V_{DD} \frac{R_2}{R_1 + R_2}$$
$$= 5 \text{ V}$$

Using the value of $V_{GS} = 5$ V in equation (11.5), the value of I_D for the circuit is found as

$$I_D = k[V_{GS} - V_{GS(\text{th})}]^2$$
$$= (1.38 \times 10^{-4})(5 \text{ V} - 1.5 \text{ V})^2$$
$$= 1.69 \text{ mA}$$

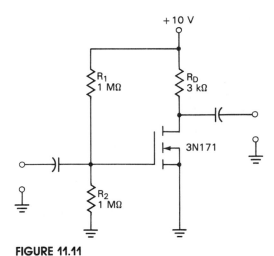

FIGURE 11.11

PRACTICE PROBLEM 11–2

An E-MOSFET has values of $I_{D(\text{on})} = 14$ mA at $V_{GS} = 12$ V and $V_{GS}(\text{th}) = 2$ V. Determine the value of I_D at $V_{GS} = 16$ V.

E-MOSFET Biasing Circuits

One of the problems with the E-MOSFET is the fact that many of the biasing circuits used for JFETs and D-MOSFETs cannot be used with this device. The reason for this is the fact that V_{GS} must be *positive*. Since self-bias and zero bias produce values of V_{GS} that are equal to zero or more negative than zero, neither of these circuits can be used. However, voltage-divider bias can be used, as can gate bias and *drain-feedback bias*.

In our study of BJTs, we covered a biasing circuit called *collector-feedback bias*. Drain-feedback bias is the MOSFET equivalent of this biasing method. Figure

13

14

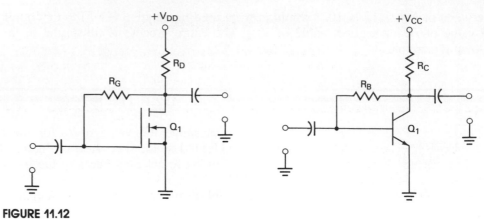

FIGURE 11.12

Drain-feedback bias.

Drain-feedback bias. The
MOSFET equivalent of collector-
feedback bias.

11.12a shows a drain-feedback bias circuit. For reference purposes, a collector-
feedback bias circuit is shown in Figure 11.12b.

The analysis of this circuit is relatively simple. First, refer back to the spec
sheet in Figure 11.10. If you look at the $I_{D(on)}$ parameter, you will notice that it
was measured under the following condition:

$$\boxed{V_{GS} = V_{DS}} \tag{11.7}$$

Now take a look at the drain-feedback circuit. With the superhigh impedance of
the gate, no current will flow in the gate circuit. Therefore, no voltage will be
dropped across the gate resistor, R_G. Since no voltage is dropped across R_G, the
gate will be at the same potential as the drain. Therefore, the circuit will fulfill
the condition specified in equation (11.7), and I_D will equal $I_{D(on)}$. Therefore, the
value of V_{DS} for a drain-feedback bias circuit can be found as

$$\boxed{V_{DS} = V_{DD} - R_D I_{D(on)}} \tag{11.8}$$

The following example illustrates the complete analysis of a drain-feedback bias
circuit.

EXAMPLE 11.3

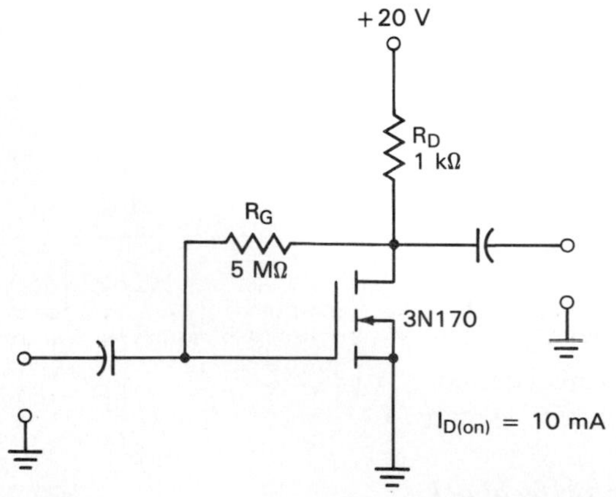

+20 V

R_D
1 kΩ

R_G
5 MΩ

3N170

$I_{D(on)} = 10$ mA

FIGURE 11.13

Determine the values of I_D and V_{DS} for the circuit shown in
Figure 11.12.

Solution: The spec sheet for the 3N170 lists an $I_{D(on)}$ of
10 mA when $V_{GS} = V_{DS}$. Since these voltages *are* equal in
the drain-feedback bias circuit, I_D will equal $I_{D(on)}$: 10 mA.
Using this value, V_{DS} (and thus V_{GS}) is found as

$$V_{DS} = V_{DD} - R_D I_{D(on)}$$
$$= 20 \text{ V} - (1 \text{ k}\Omega)(10 \text{ mA})$$
$$= 10 \text{ V}$$

PRACTICE PROBLEM 11–3

The E-MOSFET described in Practice Problem 11–2 is used in a drain-feedback bias circuit with values of $V_{DD} = 14$ V and $R_D = 510$ Ω. Determine the values of I_D and V_{DS} for the circuit.

That is all there is to analyzing a drain-feedback bias circuit. Don't forget: The drain-feedback bias circuit sets up a *positive* value of V_{GS}. Because of this, it cannot be used with JFETs or D-MOSFETs. D-MOSFETs can be operated in the enhancement mode, but they are never biased with a value of V_{GS} that is more positive than 0 V.

Summary

So far, you have been introduced to D-MOSFETs and E-MOSFETs. A comparison of the two can be seen in Figure 11.14, which summarizes many of the important characteristics of these devices.

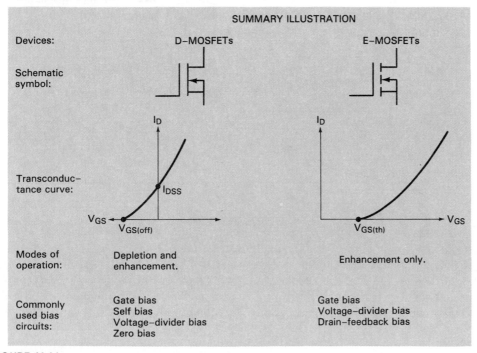

FIGURE 11.14

1. How does a positive value of V_{GS} develop a channel in an E-MOSFET? [Objective 9]
2. What is *threshold voltage*, $V_{GS(\text{th})}$? [Objective 10]
3. When $V_{GS} = V_{GS(\text{th})}$, what does the value of I_D for an E-MOSFET equal? [Objective 10]

4. List, in order, the steps required to determine the value of I_D for an E-MOSFET at a given value of V_{GS}. [Objective 12]

5. Which D-MOSFET biasing circuits can be used with E-MOSFETs? [Objective 13]

6. Which D-MOSFET biasing circuits cannot be used with E-MOSFETs? [Objective 13]

7. Describe the quiescent conditions for the *drain-feedback bias* circuit. [Objective 14]

11.4
Dual-gate MOSFETs

The operation of MOSFETs is limited at high frequencies because of their high gate-to-channel capacitance. The cause of this high capacitance can be seen in Figure 11.15. The metal plate that is used for the gate is a conductor. The silicon dioxide between the gate and channel is an insulator. The channel itself can be viewed as being a conductor when compared to the silicon dioxide layer. Thus the combination of the three forms a capacitor.

FIGURE 11.15

MOSFET input capacitance.

Dual-gate MOSFET. A MOSFET constructed with two gates to reduce gate input capacitance.

The *dual-gate MOSFET* uses two gate terminals to reduce the overall capacitance of the component. The physical construction and schematic symbols for the dual-gate MOSFET are shown in Figure 11.16. The reduced capacitance of the dual-gate MOSFET is a result of the way in which the component is used. Normally, the component is used in such a way as to act as two series-connected MOSFETs. This is shown in the applications section of this chapter. When the dual-gate MOSFET

FIGURE 11.16

Dual-gate MOSFET construction and schematic symbols.

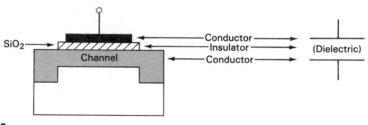

(a) n–channel   (b) p–channel

is used as series MOSFETs, the effect is similar to connecting two capacitors in series. You should recall that the total capacitance in a series connection is lower than either individual component value. Thus, by connecting the two gates in a series configuration, the overall capacitance is reduced.

SECTION REVIEW

1. Why does the MOSFET have high input capacitance? [Objective 15]
2. How does the dual-gate MOSFET reduce the MOSFET input capacitance? [Objective 15]

11.5
Power MOSFETs

Advances in engineering have produced a wide variety of MOSFETs that are designed specifically for high current, voltage, and/or power applications. In this section, we will take a look at two of these high-power MOSFETs. Some applications for these components are covered in section 11.7.

VMOS

The *vertical MOSFET*, or *VMOS*, is a component designed to handle much larger drain currents than the standard MOSFET. The current-handling capability of the VMOS is a result of its physical construction, which is illustrated in Figure 11.17. As you can see, the component has materials that are labeled p, $n+$, and $n-$. The n-material labels indicate differences in doping levels. Also, there is no physical channel connecting the source (at top) and the drain (at bottom). Thus, VMOS is an *enhancement-type* MOSFET.

Vertical MOS (VMOS). An E-MOSFET that is designed to handle high values of drain current.

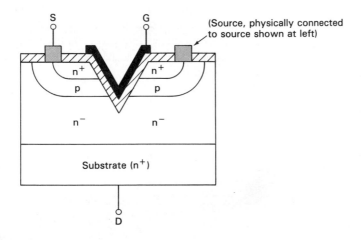

FIGURE 11.17

Vertical—MOS (VMOS) construction.

With the V-shaped gate, a larger channel is formed by a positive gate voltage. With a larger channel, the device is capable of handling a larger amount of drain current. The operation of the VMOS is illustrated in Figure 11.18.

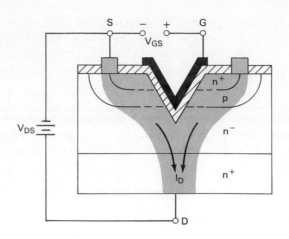

FIGURE 11.18

VMOS operation.

When a positive gate voltage is applied to the device, an *n*-type channel forms in the *p*-material region. This effective channel connects the source to the drain. As Figure 11.18 illustrates, the shape of the gate causes a wider channel to form than is created in the standard MOSFET. Because of this, the amount of drain current is much higher for this component.

Another advantage of using VMOS is the fact that it is not susceptible to thermal runaway. The VMOS has a *positive* temperature coefficient. This means that the resistance of the component *increases* when temperature increases. Thus an increase in temperature will cause a *decrease* in drain current.

Because of its higher drain current ratings and positive temperature coefficients, the VMOS device can be used in several applications where the standard MOSFET cannot be used. Some of these applications will be discussed briefly in section 11.7.

> VMOS is not susceptible to thermal runaway.

 17

> **LDMOS.** A high-power MOSFET that uses a narrow channel and a heavily doped *n*-type region to obtain high I_D and low $r_{d(on)}$.

LDMOS

Another type of power MOSFET is the *lateral double-diffused MOSFET*, or LDMOS. This type of MOSFET uses a very small channel region and a heavily doped *n*-type region ($n-$) to obtain a high drain current and low channel resistance [$r_{d(on)}$]. The basic construction of this enhancement-type MOSFET is shown in Figure 11.19.

FIGURE 11.19

LDMOS.

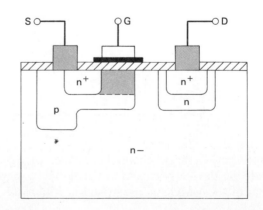

> *A Practical Consideration.* VMOS and LDMOS devices have g_{m0} values in the mho (or siemens) range. This means that they are capable of very high voltage gain values when used in common-source amplifiers.

The narrow channel (shaded) is made up of the *p*-type material that lies between the *n*− substrate and the *n*+ (lightly doped) source material. Since only *n*-type material lies between the channel and the drain, the effective length of the channel is extremely short. This, coupled with the *n* material in the channel-to-drain path, provides an extremely low typical value of $r_{d(on)}$. With a low channel resistance, the LDMOS device can handle very high amounts of current without generating any damaging amount of heat (power dissipation).

The LDMOS has typical values of $r_{d(on)}$ that are in the range of 2 Ω or less. With this low channel resistance, it is typically capable of handling currents as high as 20 A.

SECTION REVIEW

1. Describe the physical construction of the VMOS. [Objective 16]
2. How does the physical construction of the VMOS allow for high value of I_D? [Objective 16]
3. Describe the physical construction of the LDMOS. [Objective 17]
4. How does the physical construction of the LDMOS allow for high values of I_D? [Objective 17]

11.6
Complementary MOSFETs (CMOS): A MOSFET Application

The main contribution to electronics that is made by the MOSFET can be found in the area of *digital* (computer) electronics. In digital circuits, signals are made up of rapidly switching dc levels. An example of a digital signal is shown in Figure 11.20. This type of signal, referred to as a *square wave*, is made up of two dc levels, called *logic levels*. For the waveform shown in Figure 11.20, the logic levels are 0 V and +10 V.

A group of circuits that have similar operating characteristics is referred to as a *logic family*. All the circuits in a given logic family will respond to the same logic levels, have similar speed and power dissipation characteristics, and can be connected together directly. One such logic family is *complementary MOS*, or *CMOS* logic. This logic family is made up entirely of MOSFETs. The basic CMOS *inverter* is shown in Figure 11.21. An *inverter* is a digital circuit that converts one logic level to the other. When the input is at one logic level, the output will be at the other. For the circuit in Figure 11.21, a 0-V input will produce a +10-V output and a +10-V input will produce a 0-V output. The purpose served by such a circuit is beyond the scope of this book, but we will take a minute to look at its operation.

The operation of the CMOS inverter is easy to understand if we take a moment to look at each of the transistors as an individual *switch*. Figure 11.22 shows each of the MOSFETs as individual circuits. The *p*-channel MOSFET (Figure 11.22a) is an *enhancement*-type MOSFET, meaning that there is no physical channel from source to drain. When the value of V_{GS} for the circuit is 0 V, there is no current through the component. When V_{GS} is −10 V, a channel is formed, and current can flow through the device. Note that the source is connected to the

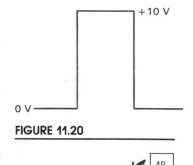

FIGURE 11.20

Logic families. Groups of digital circuits with nearly identical characteristics.

Inverter. A logic-level converter.

Complementary MOS (CMOS). A logic family made up of MOSFETs.

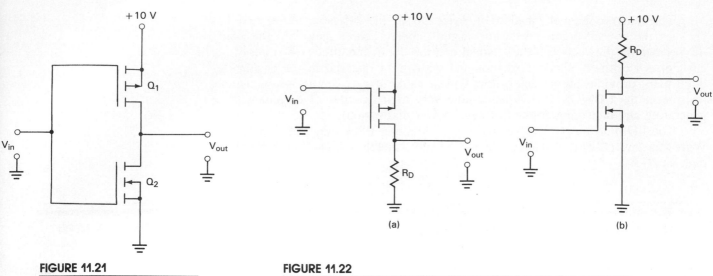

FIGURE 11.21

The CMOS inverter.

FIGURE 11.22

+10-V supply. Assuming that the input to the circuit can only be 0 or +10 V, we have the following voltage relationships for the circuit:

V_{in} (V)	V_{DD} (V)	$V_{GS} = V_{in} - V_{DD}$ (V)
+10	+10	0
0	+10	−10

Thus, when the input is +10 V, the MOSFET does not conduct, and when the input is 0 V, the MOSFET does conduct. Note that the MOSFET is said to be *on* when it is conducting and *off* when it is not. Now consider the *n*-channel MOSFET shown in Figure 11.22b. This is also an enhancement-type MOSFET. However, because it is an *n*-channel MOSFET, it has exactly the opposite characteristics of the *p*-channel MOSFET. In other words, it will conduct when $V_{GS} = +10$ V and will not conduct when the value of $V_{GS} = 0$ V. This is summarized as follows:

V_{in} (V)	V_S (V)	$V_{GS} = V_{in} - V_S$ (V)
+10	0	+10
0	0	0

Thus, when the input is +10 V, the MOSFET will conduct, and when the input is 0 V, the MOSFET will not conduct. Now let's put the two circuits together, as shown in Figure 11.21. Note that Q_1 serves as the drain resistor for Q_2, while Q_2 serves as the drain resistor for Q_1. When the input to the circuit is +10 V, Q_1 is off and Q_2 is on. Thus the conduction path is between the output and ground through Q_2. When the input is at 0 V, Q_1 is on and Q_2 is off. Thus the conduction path is between the output and V_{DD} through Q_1. This operation is summarized as follows:

V_{in} (V)	Q_1	Q_2	V_{out} (V)
0	On	Off	+10
+10	Off	On	0

The relationship between V_{in} and V_{out} is as it should be for an inverter.

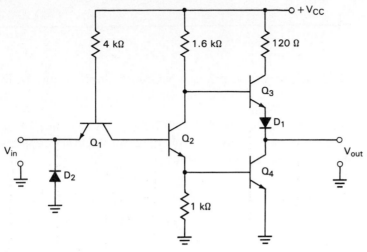

FIGURE 11.23

The TTL inverter.

So, why is the CMOS logic circuit so popular? This question is best answered by taking a quick look at the BJT counterpart of the CMOS inverter.

Why CMOS logic is so popular.

The basic TTL (a bipolar transistor logic family) inverter is shown in Figure 11.23. It is obvious that the CMOS circuit is far less complex than its TTL counterpart. This means that many more CMOS circuits can be developed in a given amount of space on an integrated circuit; that is, the CMOS circuits have a greater *packing density* than TTL circuits. In addition to the improved density, CMOS circuits have the following advantages over TTL circuits:

1. They draw little current from the supply and thus have better power dissipation ratings.
2. The low current draw also allows one CMOS output to be used to drive any number of inputs to other CMOS circuits. The TTL circuit is usually restricted to driving no more than 10 inputs to other TTL circuits.

These advantages will become clearer to you as your study of electronics progresses in the area of digital electronics. At this point, however, you should have a pretty good idea of the benefits of using CMOS logic.

SECTION REVIEW

1. What are *logic levels*? [Objective 18]
2. What is a *logic family*? [Objective 18]
3. What is an *inverter*? [Objective 18]
4. Describe the physical construction of the CMOS inverter. [Objective 19]
5. Describe the operation of the CMOS inverter. [Objective 19]
6. List the advantages that CMOS logic has over TTL logic. [Objective 20]

While CMOS logic is one of the primary MOSFET applications, there are many other applications as well. In this section, we will look at a representative sample of these MOSFET applications.

It is important to note that the D-MOSFET can be used in most applications where the JFET can be used. Thus, the full range of MOSFET applications goes well beyond those discussed in this section.

Cascode Amplifiers

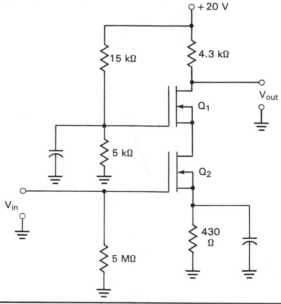

Cascode amplifier A low C_{in} amplifier used in high-frequency applications.

Both JFETs and MOSFETs have relatively high input capacitances, as stated earlier. These input capacitances can adversely affect the high-frequency operation of the components. To overcome the effects of high input capacitance, the *cascode amplifier* was developed. The cascode amplifier consists of a common-source amplifier in series with a common-gate amplifier, as shown in Figure 11.24.

Q_1 in Figure 11.24 is a common-gate amplifier that is voltage divider biased. Q_2 is the common-source circuit and is self-biased. The input to the circuit is applied to Q_2, and the output is taken from the drain of Q_1. Therefore, the two MOSFETs are in series.

FIGURE 11.24
The cascode amplifier.

Since each MOSFET has a given amount of input capacitance, the series connection of the components is the same as hooking two capacitors in series. The resulting total capacitance is therefore less than either individual capacitance value. For example, assume that each MOSFET has an input capacitance of 10 pF. The total circuit capacitance would be the series combination of the individual values, a total of 5 pF. The actual calculations are a little more complicated than shown here, but the principle is correct. (We will cover high-frequency operation in depth in Chapter 13.) The lower overall capacitance of the cascode amplifier makes it more suitable for high-frequency operation, where values of capacitive reactance can become critical.

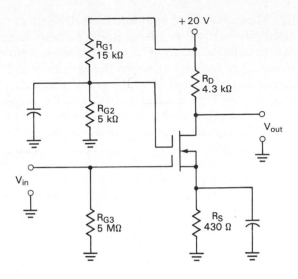

FIGURE 11.25

The dual-gate MOSFET cascode amplifier.

A simple cascade amplifier can be constructed using a dual-gate MOSFET. The dual-gate MOSFET equivalent of Figure 11.24 is shown in Figure 11.25. Gate 1 of the MOSFET is connected to act as a common-gate amplifier, while gate 2 is connected to act as the common-source amplifier. This amplifier has the same capacitance characteristics and high-frequency capabilities as those of the amplifier in Figure 11.24.

RF Amplifier

The dual-gate MOSFET can also be used in an RF amplifier. Recall that the RF amplifier is the "front-end" amplifier in an AM receiver. A dual-gate MOSFET RF amplifier is shown in Figure 11.26. Note the zener diodes drawn inside the

FIGURE 11.26

The dual-gate MOSFET RF amplifier.

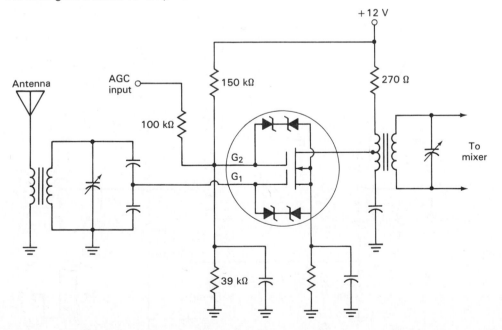

dual-gate MOSFET. These diodes are built into the MOSFET to protect the component from static electricity. As will be discussed in the next section, the MOSFET is extremely sensitive to static electricity. The zener diodes will protect the component inputs. Outside of that, they have no effect on the operation of the circuit.

The RF amplifier shown is similar to the one discussed in Chapter 10. The antenna is connected to gate 1 by a tuned circuit, and the output of the amplifier is coupled to the next circuit (the mixer) by another tuned circuit. When the dual-gate MOSFET is being used, virtually no current is required from the antenna to cause the circuit to operate properly.

The advantage of using the dual-gate MOSFET is that it allows for the use of an *automatic gain control* (AGC) of the RF amplifier. A dc voltage is applied to gate 2 from the AGC circuit. This voltage is directly proportional to the strength of the signal being received. When this AGC voltage is at a relatively high level (indicating a strong signal), the gain of the dual-gate MOSFET is decreased. When a weak signal is received, the voltage from the AGC circuitry decreases, and the gain of the dual-gate MOSFET increases. The overall effect of this operation is to prevent *fading*. If you have ever been in a car with an AM radio, you have experienced fading when going under a bridge. When the AM signal strength decreases, the radio fades out. The use of AGC circuitry makes this fading minimal.

✔ 24

Digital communications. A method of transmitting and receiving information in digital form.

Power MOSFET Drivers

At some point, you will probably study *digital communications*. In digital communications, information is converted (using one of several methods) into a series of digital signals. Those signals are then transmitted and received in digital form. Finally, the receiver converts the information back into its original form.

Many digital communications systems require the use of power amplifiers that can produce high-speed, high-current digital outputs. These signals can be produced using a power MOSFET driver circuit like the one shown in Figure 11.27.

FIGURE 11.27

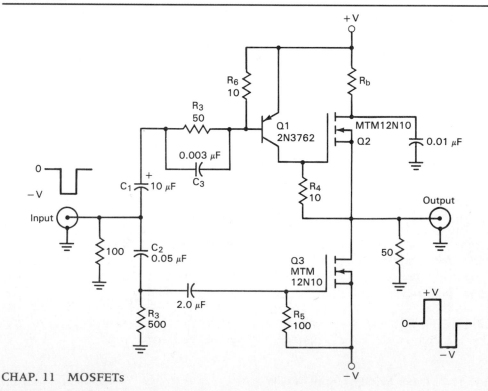

CHAP. 11 MOSFETs

The high-current capability of the driver is provided by the low channel resistance [$r_{d(\text{on})}$] of the power MOSFETs. Remember . . . power MOSFETs typically have $r_{d(\text{on})}$ values of 2 Ω or less. Thus, depending on the values of $+V$, $-V$, and R_b, the MOSFET pair can output currents as high as 20 to 35 A.

The high-speed quality of the circuit is provided by a number of factors. First, the resistance values in the driver circuit are extremely low. This allows the capacitances in the circuit to charge and discharge extremely rapidly. Also, note the parallel circuit made up of R_3 and C_3 in the base circuit of Q_1. The capacitor (C_3) is called a *speed-up capacitor*. This capacitor is used to improve the switching time (the time taken to go back and forth between saturation and cutoff) of Q_1. The operation of speed-up capacitors is discussed in detail in Chapter 18.

Speed-up capacitor. A capacitor used in the base circuit of a BJT to allow the device to switch rapidly between saturation and cutoff.

VMOS Applications

You may remember that the Darlington pair BJT configuration is used in applications where the standard BJT will not handle the current requirements. The same relationship exists between VMOS devices and standard MOSFETs. A VMOS device may be used in virtually any enhancement-type MOSFET circuit. However, it is usually used only when its high current capabilities are required.

SECTION REVIEW

1. What is a *cascode amplifier*? [Objective 21]
2. How does the cascode amplifier reduce input capacitance? [Objective 21]
3. Describe the dual-gate MOSFET cascode amplifier. [Objective 22]
4. Describe the dual-gate MOSFET RF amplifier. [Objective 23]
5. What is *automatic gain control* (*AGC*)? How is it achieved in the dual-gate MOSFET RF amplifier? [Objective 23]
6. What is *fading*? [Objective 23]
7. What is *digital communications*? [Objective 24]
8. What purpose does the power MOSFET driver serve in digital communications? [Objective 24]
9. When is VMOS typically used? [Objective 25]

11.8
Component Handling

The layer of SiO_2 that insulates the gate from the channel is extremely thin and can easily be destroyed by *static electricity*. The static electricity generated by the human body can be sufficient to ruin a MOSFET. Because of this, some precautions must be taken to protect the MOSFET.

Many MOSFET devices are now manufactured with protective diodes etched between the gate and the source, as shown in Figure 11.28. The diodes are configured so that they will conduct in either direction, provided that a predetermined voltage is reached. This voltage will be higher than any working voltage normally applied to the MOSFET. For example, if a MOSFET with a protected input is rated for a maximum V_{GS} of ±30 V, the zener diodes will be designed to conduct at any voltage outside the ±30-V range. This will protect the device from excessive static buildup, as well as accidentally high values of V_{GS}.

FIGURE 11.28

MOSFET static protection.

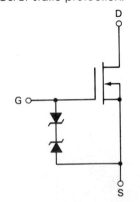

When MOSFETs without protected inputs are used, some basic handling precautions will keep the components from being damaged. These precautions are as follows:

1. Store the devices with the leads shorted together or in conductive foam. *Never store MOSFETs in styrofoam.* (Styrofoam is the best static electricity generator ever devised.)
2. Do not handle MOSFETs unless you need to. When you handle any MOSFET, hold the component by the case, not the leads.
3. Do not install or remove any MOSFET while power is applied to a circuit. Also, be sure that any signal source is removed from a MOSFET circuit before turning the supply voltage off or on.

If you follow the guidelines listed above, you will have no difficulty in handling MOSFET devices. Also, remember that these guidelines apply to the handling of CMOS integrated circuits.

SECTION REVIEW

1. Discuss the means by which MOSFET inputs are internally protected in some devices. [Objective 26]
2. What precautions should be observed when handling MOSFETs? [Objective 26]
3. Why are the precautions in question 1 necessary? [Objective 26]

KEY TERMS

The following terms were introduced and defined in this chapter:

cascode amplifier
complementary MOS (CMOS)
depletion-mode operation
Depletion MOSFET (D-MOSFET)
digital communications
drain-feedback bias
dual-gate MOSFET

enhancement-mode operation
enhancement MOSFET (E-MOSFET)
inverter
lateral double-diffused MOSFET (LDMOS)
logic families
logic levels

metal oxide semiconductor FET (MOSFET)
speed-up capacitor
substrate
threshold voltage, $V_{GS(\text{th})}$
vertical MOSFET (VMOS)
zero bias

PRACTICE PROBLEMS

§11.2

1. A D-MOSFET has values of $V_{GS(\text{off})} = -4$ V and $I_{DSS} = 8$ mA. Plot the transconductance curve for the device. [5]
2. A D-MOSFET has values of $V_{GS} = -8$ V and $I_{DSS} = 12$ mA. Plot the transconductance curve for the device. [5]

3. A D-MOSFET has values of $V_{GS(off)} = -4$ to -6 V and $I_{DSS} = 6$ to 9 mA. Plot the minimum and maximum transconductance curves for the device. [5]

4. A D-MOSFET has values of $V_{GS(off)} = -4$ to -10 V and $I_{DSS} = 5$ to 10 mA. Plot the minimum and maximum transconductance curves for the device. [5]

5. The D-MOSFET described in problem 1 has a value of $g_{m0} = 2000$ μS. Determine the values of g_m at $V_{Gs} = -2$ V, $V_{GS} = 0$ V, and $V_{GS} = +2$ V. [6]

6. The D-MOSFET described in problem 2 has a value of $g_{m0} = 3000$ μS. Determine the values of g_m at $V_{GS} = -5$, $V_{GS} = 0$ V, and $V_{GS} = +5$ V. [6]

§11.3

7. An E-MOSFET has ratings of $V_{GS(th)} = 4$ V and $I_{D(on)} = 12$ mA at $V_{GS} = 8$ V. Determine the value of I_D for the device when $V_{GS} = 6$ V. [12]

8. An E-MOSFET has ratings of $V_{GS(th)} = 2$ V and $I_{D(on)} = 10$ mA at $V_{GS} = 8$ V. Determine the value of I_D for the device when $V_{GS} = +12$ V. [12]

9. An E-MOSFET has ratings of $V_{GS(th)} = 1$ V and $I_{D(on)} = 8$ mA at $V_{GS} = 4$ V. Determine the values of I_D for the device when $V_{GS} = 1$ V, $V_{GS} = 4$ V, and $V_{GS} = 5$ V. [12]

10. An E-MOSFET has values of $V_{GS(th)} = 3$ V and $I_{D(on)} = 2$ mA at $V_{GS} = 5$ V. Determine the values of I_D for the device when $V_{GS} = 3$ V, $V_{GS} = 5$ V, and $V_{GS} = 8$ V. [12]

11. Calculate the values of I_D and V_{DS} for the amplifier in Figure 11.29. [14]

12. Calculate the values of I_D and V_{DS} for the amplifier in Figure 11.30. [14]

13. Calculate the values of I_D and V_{DS} for the amplifier in Figure 11.31. [14]

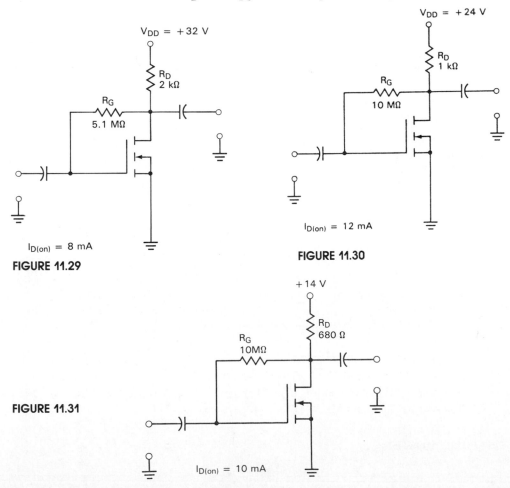

FIGURE 11.29

FIGURE 11.30

FIGURE 11.31

14. Calculate the values of I_D and V_{DS} for the amplifier in Figure 11.32. [14]
15. Calculate the values of I_D and V_{DS} values for the amplifier in Figure 11.33. [14]
16. Calculate the values of I_D and V_{DS} for the amplifier in Figure 11.34. [14]

TROUBLESHOOTING PRACTICE PROBLEMS

17. The circuit in Figure 11.35 has the waveforms shown. Discuss the possible cause(s) of the problem.
18. The circuit in Figure 11.36 has the dc voltages indicated. Discuss the possible cause(s) of the problem.

THE BRAIN DRAIN

19. Determine the range of A_v values for the amplifier in Figure 11.37.
20. Determine the range of A_v values for the amplifier in Figure 11.38.
21. Determine the range of I_D values for the amplifier in Figure 11.39.
22. The constant-current bias circuit in Figure 11.40 won't work. Why not?

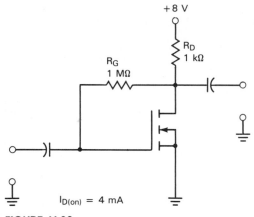

FIGURE 11.32

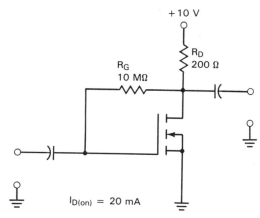

FIGURE 11.33

FIGURE 11.35

FIGURE 11.34

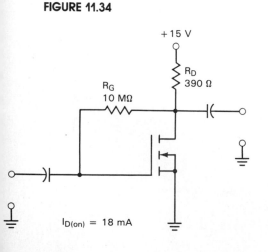

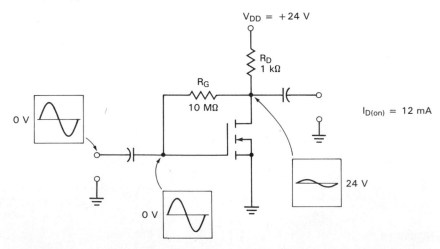

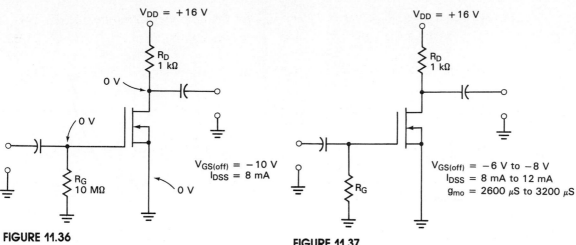

FIGURE 11.36

FIGURE 11.37

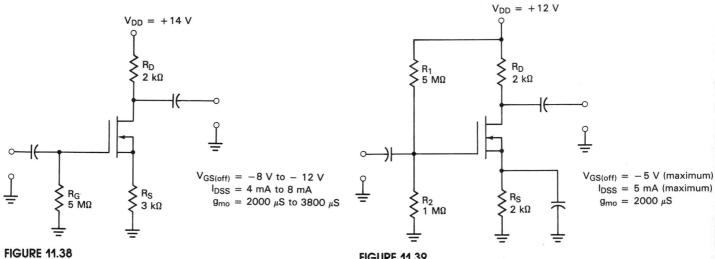

FIGURE 11.38

FIGURE 11.39

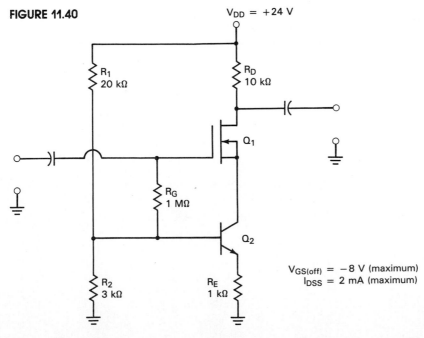

FIGURE 11.40

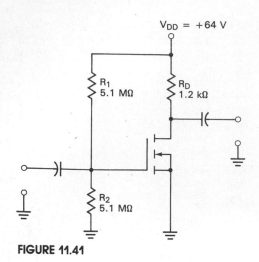

FIGURE 11.41

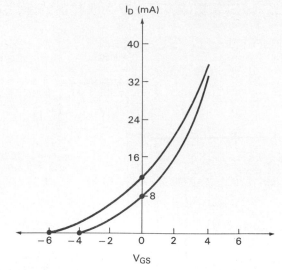

FIGURE 11.42

23. The MOSFET shown in Figure 11.41 has a maximum gate current of 800 pA when $V_{GS} = 32$ V. Determine the input impedance of the device and the circuit. Then determine the input impedance of the circuit, assuming that the MOSFET has infinite input impedance. What is the difference between the two values of Z_{in}?

SUGGESTED COMPUTER APPLICATIONS PROBLEMS

24. Write a program that will perform the complete dc analysis of a zero-bias circuit when provided with the proper information.

25. Write a program that will determine the value of I_D for an E-MOSFET at any given value of V_{GS}. The program should include the determination of the value of k for the device.

26. Write a program that will perform the complete dc analysis of a drain-feedback bias circuit when provided with the required information.

ANSWERS TO THE EXAMPLE PRACTICE PROBLEMS

11–1. See Figure 11.42

11–2. 27.44 mA

11–3. $I_D = 14$ mA, $V_{DS} = 6.86$ V

12

Operational Amplifiers

The production of integrated circuits starts with the design of the circuit. Here, several engineers check the final circuit design to ensure the accuracy of the design.

OBJECTIVES

After studying the material in this chapter, you should be able to:

☐ 1. Describe the difference between the physical construction of *discrete components* and that of *integrated circuits*. (Introduction)

☐ 2. Describe the *operational amplifier*, or *op-amp*. (§12.1)

☐ 3. Draw and identify the signal and supply voltage inputs to an op-amp on the component's schematic symbol. (§12.1)

☐ 4. List the three parts of the op-amp identification code and state the meaning of each. (§12.1)

☐ 5. List and identify (by physical shape) the commonly used op-amp packages. (§12.1)

☐ 6. Define the term *differential amplifier*. (§12.2)

☐ 7. List the factors that determine the output voltage for an op-amp. (§12.2)

☐ 8. Define *open-loop voltage gain* and state its typical range of values. (§12.2)

☐ 9. Predict the output voltage polarity for an op-amp with specified input voltages using both the *observation method* and the *mathematical method*. (§12.2)

☐ 10. Discuss the effects that supply voltage values and load resistance values have on the maximum possible peak-to-peak output voltage from an op-amp. (§12.2)

☐ 11. Calculate the maximum possible peak-to-peak output voltage and actual peak-to-peak output voltage for an op-amp. (§12.2)

☐ 12. Calculate the maximum allowable peak-to-peak input voltage for an op-amp circuit. (§12.2)

☐ 13. Describe the discrete differential amplifier and its response to input signals at its inverting and noninverting inputs. (§12.3)

☐ 14. Define *output offset voltage*. (§12.3)

☐ 15. Explain the relationship between *input offset voltage*, *output offset voltage*, and the *offset null inputs* of an op-amp. (§12.3)

☐ 16. Define *input offset current*. (§12.3)

☐ 17. Define *input bias current* and explain the restriction that it places on op-amp circuit wiring. (§12.3)

☐ 18. Define *common-mode rejection ratio* (*CMRR*) and explain why a high CMRR value is desirable for an op-amp. (§12.3)

☐ 19. Define *power supply rejection ratio*. (§12.3)

☐ 20. Define *output short-circuit current* and discuss its relationship to the limit that load resistance places on the output voltage swing of an op-amp. (§12.3)

☐ 21. Define *slew rate* and state its relationship to the maximum operating frequency of an op-amp. (§12.3)

☐ 22. Calculate the maximum operating frequency for an op-amp, given the values of slew rate and peak output voltage. (§12.3)

☐ 23. Explain what happens when you attempt to exceed the slew-rate limit of an op-amp. (§12.3)

☐ 24. Define each of the following op-amp electrical characteristics: *input voltage rating*, *large-signal voltage gain*, *supply current rating*, and *power consumption rating*. (§12.3)

☐ 25. Explain why the *closed-loop voltage gain* (A_{CL}) of an inverting amplifier is equal to the ratio of the feedback resistance (R_f) to the amplifier input resistance (R_i). (§12.4)

☐ 26. Explain why the input impedance of an inverting amplifier is always lower than the input impedance of the op-amp. (§12.4)

☐ 27. Explain why the value of feedback resistance (R_f) is not normally considered in the calculation of inverting amplifier output impedance. (§12.4)

☐ 28. Explain why the CMRR of an inverting amplifier is always lower than that of its op-amp. (§12.4)

☐ 29. Calculate the values of A_{CL}, Z_{in}, Z_{out}, CMRR, and f_{max} for an inverting amplifier. (§12.4)

☐ 30. List and explain the methods used to determine the gain and impedance values for a noninverting amplifier. (§12.5)

☐ 31. Calculate the values of A_{CL}, Z_{in}, Z_{out}, CMRR, and f_{max} for a noninverting amplifier. (§12.5)

☐ 32. Compare and contrast the operating characteristics of the inverting amplifier with the noninverting amplifier. (§12.5)

☐ 33. Describe and analyze the operation of the *voltage follower*. (§12.5)

☐ 34. List the common op-amp circuit faults and the symptoms of each. (§12.6)

As you master each of the objectives listed, place a check mark (✔) in the appropriate box.

✔ ☐ 1

Discrete components.
Components housed in individual packages, that is, one package—one component.

I ndividual components, such as the 2N3904 BJT and the 2N5459 FET, are classified as *discrete* components. The term *discrete* indicates that each physical package contains only one component. For example, when you purchase a 2N3904, you are buying a single component housed in its own casing.

Over the years, advances in manufacturing technology have made it possible to produce entire *circuits* on a single piece of semiconductor material. This type of circuit, which is housed in a single casing, is referred to as an *integrated circuit*, or *IC*. ICs range in complexity from simple circuits containing a few active and/or passive components to complex circuits containing hundreds of thousands of components. The more complex the internal circuitry of an IC, the more complex its function is.

The major impacts of ICs all relate to their internal operation and relatively low manufacturing cost. Circuit operations that once took hundreds of discrete components to perform can now be accomplished with a single IC. This has made circuits easier to design and troubleshoot. At the same time, the cost of an IC is generally lower than the cost of a comparable discrete-component circuit. This has made electronic systems less expensive to manufacture.

It would be impossible for a single book to cover *every* type of IC that is available. As your study of electronics continues, you will be introduced to more and more types of ICs. In this book we are going to concentrate on the most commonly used *linear* IC, the *operational amplifier*, or *op-amp*. We will discuss the operating principles and basic amplifier applications of op-amps, including op-amp circuit troubleshooting.

Integrated circuit (IC). A single package that contains any number of active and/or passive components, all constructed on a single piece of silicon.

12.1
Op-amps: An Overview

The op-amp is a *high gain* dc amplifier that has *high input impedance* and *low output impedance*. The internal circuitry, schematic symbol, and pin diagram for the 741 general-purpose op-amp are shown in Figure 12.1.

Check out that circuit! How would you like to have to troubleshoot *that* on your average Monday morning? Fortunately, the circuitry of the 741 is all contained in a single component. Since we're dealing with a single component, all we need to be concerned with are the input/output relationships and characteristics of the component. You cannot get into a 741 to repair the internal circuitry, so its complexity is of no consequence.

The signal inputs to the op-amp are labeled *inverting* and *noninverting*. Normally, the active input to the amplifier will be applied to one of these two inputs. The other is usually used to control the operating characteristics of the component. The particular application determines which input pin is used as the active input.

Operational amplifier (op-amp). A high gain dc amplifier that has high input impedance and low output impedance.

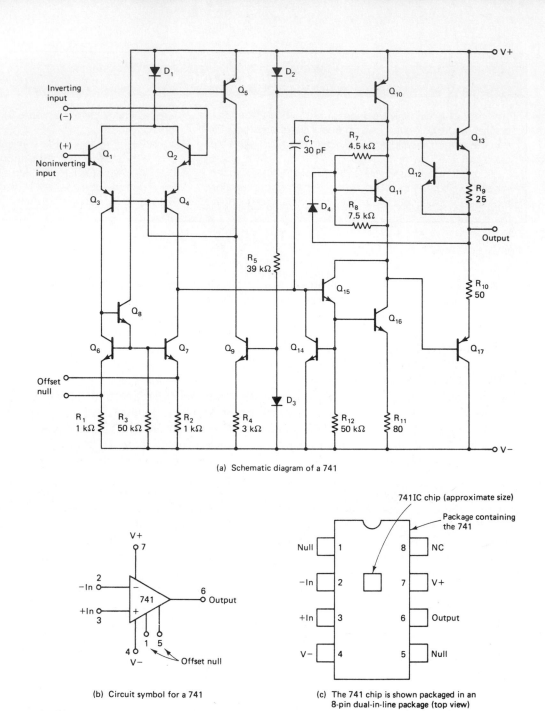

(a) Schematic diagram of a 741

(b) Circuit symbol for a 741

(c) The 741 chip is shown packaged in an 8-pin dual-in-line package (top view)

FIGURE 12.1

The A741 operational amplifier (op-amp). (Courtesy of Fairchild, a division of National Semiconductor.)

The op-amp has *two* supply voltage inputs, labeled $+V$ and $-V$. These two supply pins are normally connected in one of two ways, as illustrated in Figure 12.2. One way is to have $+V$ and $-V$ set to equal voltages that are of opposite polarity, as shown in Figure 12.2a. The other way is to provide a single supply voltage to one of the supply pins while grounding the other. This connection is shown in Figure 12.2b. Again, the specific wiring of these two pins depends on

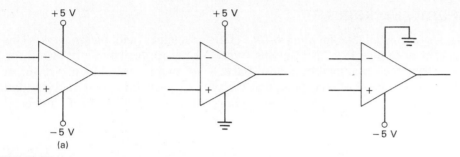

FIGURE 12.2

Op-amp supply voltages.

the particular application. The *offset null* pins (Figure 12.1) are discussed later in this chapter.

IC Identification

Hundreds of types of op-amps are produced by various manufacturers. These op-amps are all identified using a seven-character ID code. This code is shown in Figure 12.3. The *prefix* is used to identify the particular manufacturer. A listing of the most common prefixes is provided in Table 12.1. The *designator code* indicates two things:

1. The three-digit number indicates the specific type of op-amp.
2. The final letter indicates the operating temperature range.

The commonly used temperature codes are shown in Table 12.2. The designator code is used not only to determine the specific type of op-amp you are dealing with; it is also used to help you determine which op-amps can be substituted for each other. This point is discussed further in the section on op-amp circuit trouble-shooting. The *suffix* in the ID code indicates the type of package in which the op-amp is housed. The commonly used suffix codes are listed in Table 12.3.

Designator code. An IC code that indicates the type of circuit and its operating temperature range.

Prefix	Designator	Suffix
MC	741C	N

FIGURE 12.3

An op-amp ID code.

TABLE 12.1
Manufacturers' Prefixes

Prefix	Manufacturer
AD	Analog Devices
CA	RCA
LM	National Semiconductor
MC	Motorola
NE/SE	Signetics
OP	Precision Monolithics
RC/RM	Raytheon
SG	Silicon General
TL	Texas Instruments
UA	Fairchild[a]

[a] Fairchild is now a division of National Semiconductor.

TABLE 12.2
Temperature Codes

Code	Application	Temperature Range (°C)
C	Commercial	0 to 70
I	Industrial	−25 to 85
M	Military	−55 to 125

TABLE 12.3
Suffix Codes

Code	Package Type
D	Plastic dual-in-line (DIP)
J	Ceramic DIP
N, P	Plastic DIP with longer lead

Op-amp Packages

Op-amps are available in *dual-in-line* (DIP) packages and *metal cans*. This is illustrated in Figure 12.4. Metal cans (type TO-5) are available with 8, 10, or 12 leads. DIPs for op-amps commonly have 8 or 14 pins as shown. Of the two, DIPs are far more common. Note that the type of package is extremely important when considering whether or not one op-amp may be substituted for another.

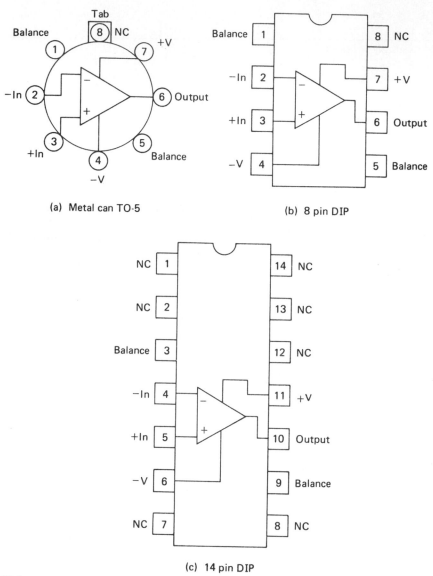

(a) Metal can TO-5

(b) 8 pin DIP

(c) 14 pin DIP

FIGURE 12.4

Op-amp packages. (Courtesy of Prentice Hall.)

SECTION REVIEW

1. What is a *discrete component*? [Objective 1]
2. What is an *integrated circuit*? [Objective 1]
3. What is an *operational amplifier*, or *op-amp*? [Objective 2]
4. Draw the schematic symbol for an op-amp and identify the signal and supply voltage inputs. [Objective 3]

5. What are the three parts of the op-amp identification code? What does each part of the code tell you? [Objective 4]

6. What are the two types of op-amp packages? [Objective 5]

12.2
Operation Overview

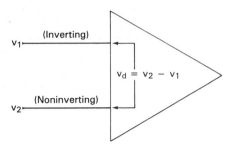

v_1 — (Inverting)

$v_d = v_2 - v_1$

v_2 — (Noninverting)

FIGURE 12.5

The input stage of the op-amp is a *differential amplifier*. This means that the component amplifies the *difference between* inputs v_1 and v_2. This point is illustrated in Figure 12.5. Note that the amplifier sees the input as being a *difference voltage* between the two input terminals. By formula,

$$v_{in} = v_2 - v_1 \qquad (12.1)$$

where v_{in} = the difference voltage that will be amplified
  v_1 = the voltage applied to the *inverting* input.
  v_2 = the voltage applied to the *noninverting* input.

It is important for you to remember that the op-amp is amplifying the difference between the input terminal voltages.

The output from the amplifier for a given pair of input voltages depends on several factors:

1. The *gain* of the amplifier.
2. The polarity relationship between v_1 and v_2.
3. The values of the supply voltage, $+V$ and $-V$.

We will now look at these three factors in detail.

Op-amp Gain

The *maximum* possible gain from a given op-amp is referred to as the *open-loop voltage gain*, A_{OL}. The value of A_{OL} is generally greater than 10,000. For example, the Fairchild μA741 op-amp has an open-loop voltage gain of 200,000 (typical).

The term *open-loop* indicates a circuit condition where there is *no feedback path from the output to the op-amp input*. You see, op-amps are usually operated so that a part of the output signal is fed back to the input. Such a circuit condition is illustrated in Figure 12.6. When part of the output signal is fed back to the input, the overall gain of the op-amp is reduced. As we cover specific circuits,

Differential amplifier. A circuit that amplifies the difference between two input voltages.

Open-loop voltage gain. The maximum possible gain of a given op-amp.

What the term *open loop* means.

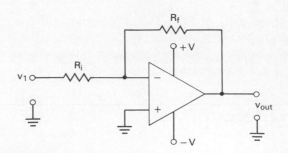

FIGURE 12.6

The op-amp feedback path.

you will be shown how to determine the exact gain of each. In the meantime, you should remember.

1. The maximum gain of a given op-amp is A_{OL}: typically, 10,000 or greater.
2. The actual gain of a specific op-amp circuit is reduced when a feedback path exists between the component output and input.

The high gain of the op-amp is another advantage that this component has over BJTs and FETs. Values of A_v near the 200,000 mark are virtually impossible for discrete components.

Input/Output Polarity

The polarity relationship between v_1 and v_2 will determine whether the op-amp output voltage swings toward $+V$ or $-V$. If v_1 is *more negative* than v_2, the op-amp output voltage will swing toward $+V$. This is illustrated in Figure 12.7a. If v_1 is *more positive* than v_2, the op-amp output voltage will swing toward $-V$. This is illustrated in Figure 12.7b. Note the relationships between the input voltage

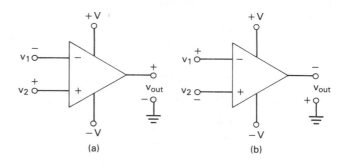

FIGURE 12.7

Op-amp input polarities.

polarities and the input signs in the schematic symbol. In Figure 12.7a, the polarity symbols of v_1 and v_2 match the polarity signs in the schematic symbol, and the output is positive. In Figure 12.7b, the polarity signs for v_1 and v_2 do not match the polarity signs in the schematic symbol, and the output is negative. This leads us to the following statement regarding the input/output polarity relationships for the op-amp:

When the input voltage polarities match the polarity signs in the schematic symbol, the output voltage is positive. When they do not match, the output is negative.

Op-amp input/output polarity relationship.

CHAP. 12 Operational Amplifiers

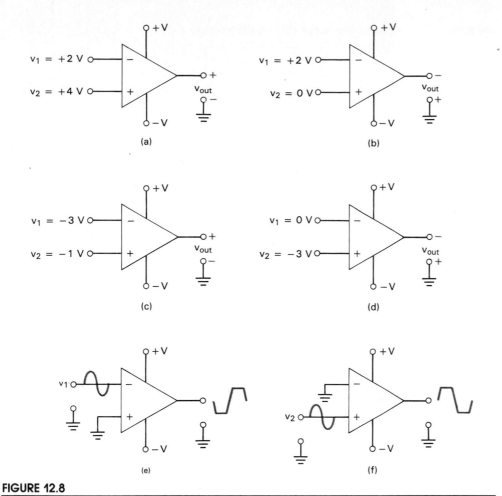

FIGURE 12.8

Input/output polarity relationships.

This point is illustrated further in Figure 12.8.

It is important to note that the output polarity is determined by the *relationship between* the polarities of v_1 and v_2, not by their polarities with respect to ground. For example, look at circuits (a) and (c) in Figure 12.8. In both of these cases, v_1 is *more negative* than v_2 (the voltage polarities match the polarity symbols), and the output is *positive*. It did not matter whether both inputs were positive or negative, only that v_1 was *negative with respect to* v_2. If you take a moment to look at circuits (b) and (d), you'll see that v_1 is *more positive* than v_2 (the voltage polarities do not match the polarity signs), and the output is *negative*. Again, the *relationship between* v_1 and v_2 has determined the output polarity.

Now take a look at circuit (e). Note that the noninverting input (v_2) is *grounded*. When v_1 goes more positive than ground, the voltage polarities do not match signs, and the output goes negative. When v_1 goes more negative than ground, the voltage polarities match the polarity signs, and the output goes positive. Note the 180° phase shift that occurs between the op-amp input and output signals. This is where the term *inverting input* comes from. If you apply the same reasoning to circuit (f), you will see why the (+) input is referred to as the *noninverting input*.

There is another method you can use to determine the polarity of the op-amp output voltage for a given set of input voltages. You may remember that

Inverting input. The op-amp input that produces an output 180° voltage phase shift.

Noninverting input. The op-amp input that does not produce an output voltage phase shift.

equation (12.1) defined the input difference voltage as

$$v_{in} = v_2 - v_1$$

When the result of this equation is *positive*, the op-amp output voltage will be *positive*. When the result of this equation is *negative*, the output voltage will be *negative*. This is illustrated in the following example.

EXAMPLE 12.1

Determine the output voltage polarity for circuits (a) and (b) in Figure 12.8 using equation (12.1)

Solution: For circuit (a),

$$v_{in} = v_2 - v_1$$
$$= 4 \text{ V} - 2 \text{ V}$$
$$= 2 \text{ V}$$

Since v_{in} is *positive*, the op-amp output voltage will be *positive*. For circuit (b),

$$v_{in} = v_2 - v_1$$
$$= 0 \text{ V} - 2 \text{ V}$$
$$= -2 \text{ V}$$

Since v_{in} is *negative*, the op-amp output voltage will be *negative*.

PRACTICE PROBLEM 12–1

Using equation (12.1), determine the output voltage polarities for circuits (c) and (d) in Figure 12.8.

If you compare the polarities obtained in the example with those determined in the discussion, you will see that we reached the same conclusions in both cases.

You may be wondering why we went to all the trouble of developing an *observation method* of analysis (comparing the voltage polarities to the polarity signs in the schematic symbol) when the *mathematical method* of analysis [using equation (12.1)] seems so much simpler. The reason for establishing two methods of analysis is because each is simpler to use under different circumstances. When you are analyzing the schematic diagram of an op-amp-based circuit or system, it is easier to use the mathematical method of analysis. When you are troubleshooting an op-amp circuit or system with an oscilloscope, the observation method is easier to use because you do not need to determine exact voltage values. You only need to quickly determine the polarity relationship between the inputs, and then you can predict and observe the output voltage polarity.

Up to this point, we have simplified matters by not considering the effects of the $+V$ and $-V$ values on the op-amp output. As you were shown earlier, these supply pins can be set to different values. We will now take a look at the effects of $+V$ and $-V$ on the output voltage from the op-amp.

Supply Voltages

The supply voltages ($+V$ and $-V$) determine the *limits* of the output voltage swing. No matter what the gain or input signal strength, the output cannot exceed some value *slightly less than* $+V$ or $-V$. For example, consider the circuits shown in Figure 12.9. Circuit (a) has supply voltages of ± 15 V. Assuming that the output can make the full transition between $+V$ and $-V$, the output cannot go any more positive than $+15$ V or any more negative than -15 V. For circuit (b), the output would be limited to $+5$ V on the positive transition and -5 V on the negative transition. The output from circuit (c) would have limits of $+10$ V and ground, while the output from circuit (d) would have limits of ground and -10 V.

The supply voltages limit the output V_{PP}.

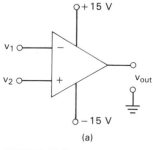
(a)

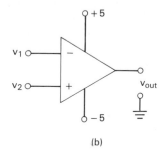
(b)

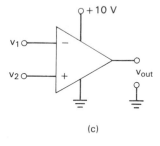
(c)

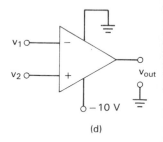
(d)

FIGURE 12.9

Op-amp supply voltages.

In practice, the output voltage will not reach either $+V$ or $-V$. The reason for this is the fact that some voltage is dropped across the components in the op-amp output circuit (refer to Figure 12.1a). The actual maximum output voltages depend on the op-amp being used and the value of the load resistance. For example, the spec sheet for the μA741 op-amp lists the following parameters:

Parameter	Condition	Typical Value[a]
Output voltage swing	$R_L \geqslant 10\text{ k}\Omega$	± 14 V
	$R_L \geqslant 2\text{ k}\Omega$	± 13 V

[a] $V_S = \pm 15$ V, $T_A = 25°C$.

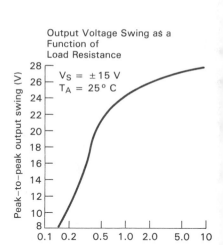

FIGURE 12.10

Output voltage as a function of load resistance.

For this op-amp, with source voltages of ± 15 V, the output will typically be limited to ± 14 V if the load is 10 kΩ or more, and to ± 13 V if the load is between 2 and 10 kΩ. Note that ± 14 V represents a peak-to-peak voltage of 28 V and ± 13 V represents a peak-to-peak voltage of 26 V.

What happens if the load resistance is less than 2 kΩ? The spec sheet for the μA741 contains a graph showing maximum output voltage as a function of load resistance. This graph is shown in Figure 12.10. Note that the output voltages are given as peak-to-peak values. Also note that the curve was derived using supply voltage of ± 15 V. You determine the maximum output voltage for a given load resistance by finding that resistance on the *x*-axis and determining the corresponding value of maximum output voltage. For example, a load resistance of 200 Ω would limit the output to a peak-to-peak voltage of 10 V, or ± 5 V.

Now another question arises. How do you determine the maximum output swing when the source voltages are at a value other than ± 15 V? The graph of output voltage swing versus supply voltage for the μA741 is shown in Figure

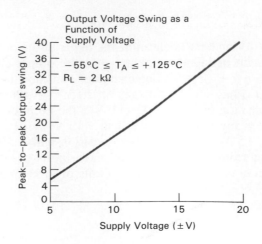

FIGURE 12.11

Output voltage as a function of supply voltage.

12.11. As you can see, the maximum output voltage increases at a linear rate as the supply voltage values increase. Generally, the following guidelines can be used when dealing with the 741:

Value of Load Resistance (kΩ)	Max (+) Output	Max (−) Output
Greater than 10	$(+V) - 1$ V	$(-V) + 1$ V
2 to 10	$(+V) - 2$ V	$(-V) + 2$ V

For example, consider the circuit shown in Figure 12.12. The supply voltages are shown to be +10 V and ground. If the value of R_L is 10 kΩ or more, the maximum output transition is from +9V to +1V. If the load resistance is between 2 kΩ and 10 kΩ, the maximum output voltage transition is from +8 to +2V.

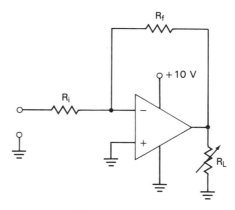

FIGURE 12.12

Putting It All Together

We have considered the factors of gain, $v_2 - v_1$ polarity relationships, load resistance, and supply voltages on the operation of an op-amp. In this section, we are going to go through a series of examples that put all these factors together. In each example, a value of gain will be assumed in order to simplify the problems. Later in the chapter you will be shown how to determine the gain of each of these circuits.

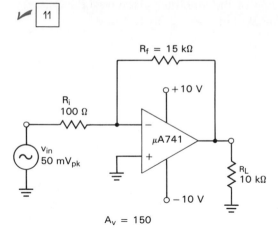

$A_v = 150$

FIGURE 12.13

In Example 12.2, we converted v_{in} to a peak-to-peak form. When you convert a negative peak voltage to peak-to-peak form, you can drop the minus sign, as was done in the example.

Determine the peak-to-peak output voltage value for the circuit shown in Figure 12.13. Also determine the maximum possible output values for the amplifier. Assume that the gain of the amplifier is 150.

Solution: Since the noninverting input to the amplifier is grounded, the value of v_2 is 0 V. The difference voltage is therefore found as

$$v_{in} = v_2 - v_1$$
$$= -v_1$$

Converting the peak input value of 50 mV$_{pk}$ to 100 mV$_{PP}$, the peak-to-peak output voltage is determined as

$$v_{out} = A_v v_{in}$$
$$= (150)(100 \text{ mV}_{pp})$$
$$= 15 \text{ V}_{pp}$$

Since the load resistance is 10 kΩ, we can determine the maximum possible output peak values using the general guidelines established for the 741 on page 524. Thus the maximum positive and negative peak values are found as

$$V_{pk(+)} = (+V) - 1 \text{ V} = +9 \text{ V} \quad \text{(maximum)}$$

and

$$V_{pk(-)} = (-V) + 1 \text{ V} = -9 \text{ V} \quad \text{(maximum)}$$

PRACTICE PROBLEM 12–2

An op-amp circuit has the following values: $+V = 12$ V, $-V = -12$ V, $v_1 = 20$ mV$_{pk}$, $v_2 = 0$ V (ground), and $A_v = 140$. Determine the value of v_{out} for the circuit.

━━━ **EXAMPLE 12.3** ━━━

Determine the maximum allowable value of v_{in} for the circuit shown in Figure 12.14. Assume that the gain of the amplifier is 200.

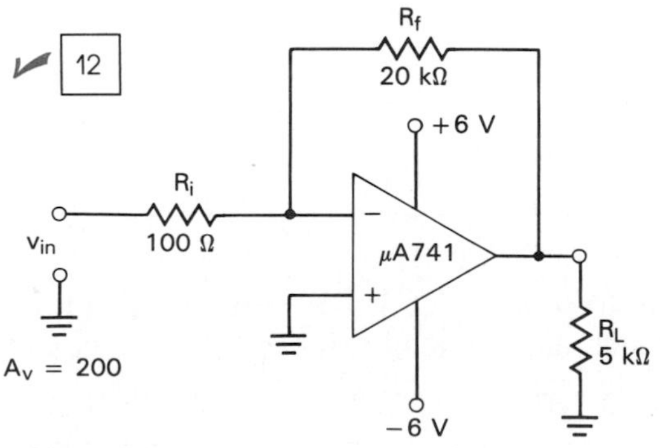

Solution: The first step is to determine the maximum allowable peak output voltage values. Since the load resistance is between 2 and 10 kΩ, the peak output values are found as

$$V_{pk(+)} = (+V) - 2 \text{ V} = +4 \quad \text{(maximum)}$$

and

$$V_{pk(-)} = (-V) + 2 \text{ V} = -4 \text{ V} \quad \text{(maximum)}$$

$A_v = 200$

FIGURE 12.14

The maximum value of peak-to-peak output voltage would be the difference between these two values, 8 V. Dividing this value by the gain of the amplifier gives us the maximum allowable peak-to-peak value of v_{in}, as follows:

$$v_{in,PP} = \frac{V_{PP(max)}}{A_v}$$
$$= \frac{8\ V}{200}$$
$$= 40\ mV_{PP}$$

PRACTICE PROBLEM 12–3

Determine the maximum allowable peak-to-peak input voltage for the amplifier described in Practice Problem 12–2. Assume that the load resistance for the circuit is 20 kΩ.

EXAMPLE 12.4

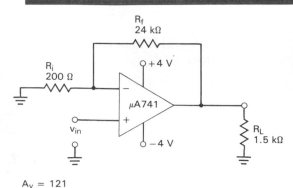

R_f
24 kΩ

R_i
200 Ω

+4 V

μA741

v_{in}

−4 V

R_L
1.5 kΩ

$A_v = 121$

FIGURE 12.15

Determine the maximum peak output values for the circuit in Figure 12.15. Also, determine the maximum allowable value of v_{in} for the circuit. Assume that the gain of the amplifier is 121.

Solution: Since the load resistance is less than 2 kΩ, we must make sure that the output voltage swing will not be limited by the load. Checking the graph in Figure 12.10, we see that the maximum possible output swing with a load of 1.5 kΩ is 26 V_{PP}, or ± 13 V_{pk}. Since the supply voltages are well below this value, we can safely assume that the load will not affect the output voltage swing. With this in mind, we can determine the maximum output values using the guidelines established for loads *under* 10 kΩ. Thus

$$V_{pk(+)} = (+V) - 2\ V = +2\ V \quad (maximum)$$

and

$$V_{pk(-)} = (-V) + 2\ V = -2\ V \quad (maximum)$$

and our maximum V_{PP} for the circuit is 4 V. Using this value, the maximum V_{PP} of the input is found as

$$v_{in,PP} = \frac{V_{PP(max)}}{A_v}$$
$$= 33.06\ mV_{PP}$$

PRACTICE PROBLEM 12–4

Determine the maximum allowable peak-to-peak input voltage for the amplifier described in Practice Problem 12–2. Assume that the load resistance for the circuit is 1 kΩ. Use the curve in Figure 12.10 to determine the effect of the load resistance on the op-amp output voltage limits.

Incidentally, the circuit in Figure 12.15 had the input applied to the *noninverting* terminal, while the circuits in Figures 12.13 and 12.14 had the inputs applied to the *inverting* terminals. The main difference between the circuit in Figure 12.15 and the other two is that there will be no phase shift from input to output. The circuits in Figures 12.13 and 12.14 will both have a 180° phase shift from input to output.

Another point can be made at this time. The circuit configuration shown in Figures 12.13 and 12.14 are the op-amp equivalents of the *common-emitter* amplifier. If you compare these circuits to the voltage-divider biased amplifier, you will see that the op-amp circuits contain fewer components. This is another advantage of using op-amp circuits rather than discrete component circuits. Op-amp circuits require fewer external components to establish the desired operation.

SECTION REVIEW

1. What is a *differential amplifier*? [Objective 6]
2. What determines the output from an op-amp? [Objective 7]
3. What is *open-loop voltage gain* (A_{OL})? [Objective 8]
4. What are the typical values of A_{OL} for op-amps? [Objective 8]
5. Describe the *observation method* for determining the output voltage polarity for an op-amp. [Objective 9]
6. Describe the *mathematical method* for determining the output voltage polarity for an op-amp. [Objective 9]
7. What effect do the supply voltages of an op-amp have on the maximum peak-to-peak output voltage? [Objective 10]
8. What effect does the load resistance of an op-amp have on the maximum peak-to-peak output voltage? [Objective 10]

12.3
The Differential Amplifier and Op-Amp Specifications

Diodes, BJTs, and FETs all have parameters and electrical characteristics that affect their operation. The op-amp is no exception. Up to this point, we have ignored the op-amp parameters and electrical characteristics to give you a chance to grasp the fundamental concepts of component operation without being bogged down with detail. Now it is time to bring the op-amp parameters and electrical characteristics into the picture.

To understand many of the op-amp electrical characteristics, you need to understand the operation of its input circuit. This input circuit, which is driven by the inverting and noninverting inputs, is called a *differential amplifier*. In this section, we will take a look at the operation of a discrete differential amplifier. We will then cover most of the op-amp electrical characteristics.

The Basic Differential Amplifier

A *differential amplifier* is a circuit that accepts two inputs and produces an output that is proportional to the *difference between* those inputs. The basic differential amplifier is shown in Figure 12.16. Note that the input to Q_2 is identified as

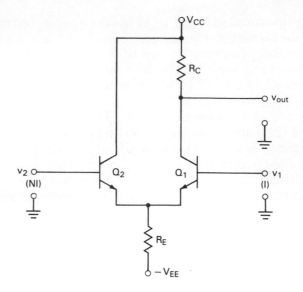

FIGURE 12.16

The basic differential amplifier.

being the *noninverting input (NI)*, and the input to Q_1 is identified as being the *inverting input (I)*.

Ideally, the transistor emitter currents will be equal. If the transistors are perfectly matched and their base voltages and currents are equal, their emitter currents will be equal to each other and equal to one-half of the current through R_E. This point is illustrated in Figure 12.17. Note that the two base voltages, v_1

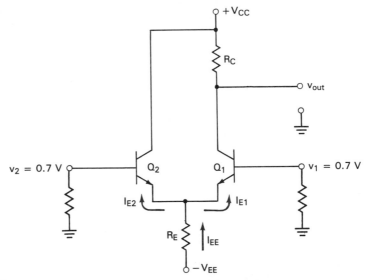

FIGURE 12.17

and v_2, are shown as being equal. This ensures that the current through each transistor is equal to the current through the other. The current through each transistor is equal to one-half of the current through R_E. By formula,

$$I_E = \frac{I_{EE}}{2}$$

(12.2)

With both transistor bases being at 0.7 V, the emitters are at approximately 0 V.

When this condition exists, the value of I_{EE} is found as

$$I_{EE} = \frac{V_{EE}}{R_E}$$ **(12.3)**

and I_E can be found as

$$I_E = \frac{1}{2}\frac{V_{EE}}{R_E}$$ **(12.4)**

or

$$I_E = \frac{V_{EE}}{2R_E}$$ **(12.4)**

Assuming that the transistors are perfectly matched, the value of I_C for the two circuits will be equal. The fact that R_C is placed in the collector circuit of Q_1 makes no difference in the values of I_C, because the collector currents are determined by the emitter circuit, not the collector circuit. As long as the transistor bases are biased to the same operating point, the emitter currents will be equal, and so will the collector currents.

The output voltage from the differential amplifier is equal to V_{CC} minus the voltage drop across R_C. By formula,

$$V_{out} = V_{CC} - I_{C1}R_C$$

where I_{C1} is the collector current of Q_1. The distinction of I_{C1} in the above equation is important because it explains the overall operation of the differential amplifier. If we assume that $I_C = I_E$, then

$$I_{C1} + I_{C2} = I_{EE}$$

Since I_{EE} is a fixed value, the sum of the transistor collector currents must also be fixed. With this in mind, let's take a look at what happens when I_{C2} changes. This situation is illustrated in Figure 12.18.

FIGURE 12.18

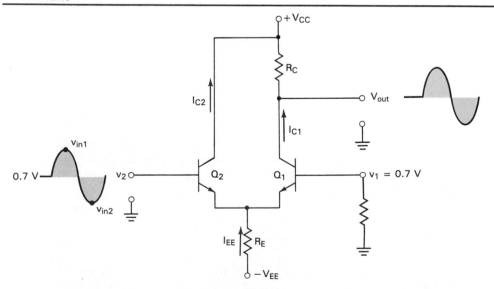

When the input signal to Q_2 causes I_{C2} to *increase*, I_{C1} must *decrease* by the same amount. This is due to the fact that the two currents must add up to I_{EE}. When I_{C1} decreases, the voltage drop across R_C decreases, and the output voltage *increases*. Thus an increase in v_2 causes an increase in v_{out}. The same logic explains what happens when v_2 decreases to $v_{in,2}$. The decrease in v_2 causes I_{C2} to decrease, causing an equal increase in I_{C1}. This increase in I_{C1} causes the drop across R_C to increase, decreasing the circuit output voltage. As Figure 12.18 illustrates, v_{out} follows v_2 when v_1 is constant. This verifies that v_2 is a *noninverting* (no phase shift) input.

Now let's assume that v_2 remains constant and v_1 changes. This situation is illustrated in Figure 12.19. Note that v_2 in this circuit is shown to be constant. When v_1 increases, I_{C1} increases, causing the voltage drop across R_C to increase. This causes v_{out} to decrease. When v_1 decreases, I_{C1} decreases, causing the voltage drop across R_C to decrease. This *increases* v_{out}. Thus the output signal will be 180° out of phase with any signal applied to Q_1. This is where the name *inverting input* comes from.

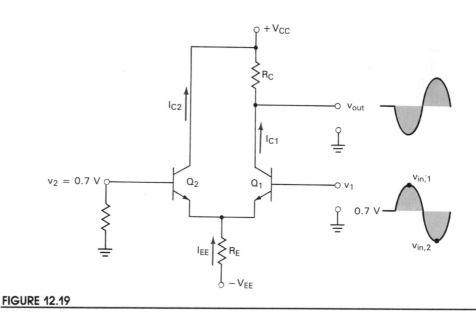

FIGURE 12.19

For any combination of v_1 and v_2, the value of v_{out} will depend on the *difference* between the two voltages. To help you understand this concept, picture what would happen if v_1 and v_2 were to increase to a given value at the same time. Since the two collector currents must add up to I_{EE}, neither collector current would increase. While each transistor would try to conduct harder, the attempted increase would be offset by the attempted increase in the other transistor. Thus the values of I_{C1} and I_{C2} would remain constant, and the output from the amplifier would also remain constant. Only when there is an *imbalance* between the two inputs will the output change.

Figure 12.20 summarizes the operation of the basic differential amplifier. When you need to review the operation of the basic differential amplifier, you can simply refer back to this illustration.

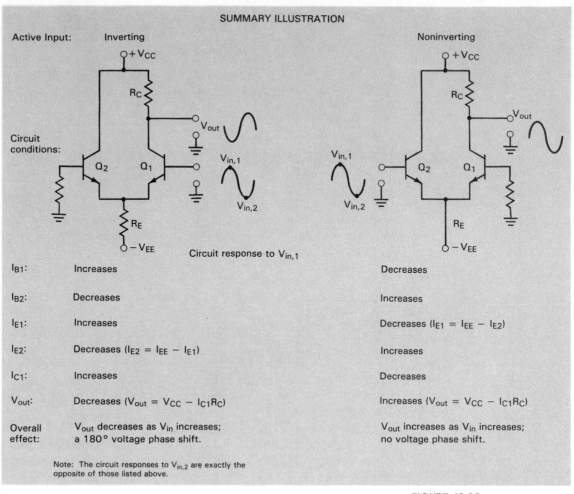

SUMMARY ILLUSTRATION

Active Input: Inverting / Noninverting

Circuit response to $V_{in,1}$

	Inverting	Noninverting
I_{B1}:	Increases	Decreases
I_{B2}:	Decreases	Increases
I_{E1}:	Increases	Decreases ($I_{E1} = I_{EE} - I_{E2}$)
I_{E2}:	Decreases ($I_{E2} = I_{EE} - I_{E1}$)	Increases
I_{C1}:	Increases	Decreases
V_{out}:	Decreases ($V_{out} = V_{CC} - I_{C1}R_C$)	Increases ($V_{out} = V_{CC} - I_{C1}R_C$)
Overall effect:	V_{out} decreases as V_{in} increases; a 180° voltage phase shift.	V_{out} increases as V_{in} increases; no voltage phase shift.

Note: The circuit responses to $V_{in,2}$ are exactly the opposite of those listed above.

FIGURE 12.20

Our discussion on the differential amplifier has been based on the assumption that the transistors are *perfectly matched*, meaning that they have the exact same electrical characteristics. In reality, this cannot happen. There will always be *some* difference between the characteristics of the two transistors. This leads us to the first of the op-amp parameters.

Output Offset Voltage

Even though the transistors in the differential amplifier are very closely matched, there are some differences in their electrical characteristics. One of these differences is found in the values of V_{BE} for the two transistors. When $V_{BE1} \neq V_{BE2}$, an imbalance is created in the differential amplifier which may show up as an *output offset voltage*. This point is illustrated in Figure 12.21. Note that with the op-amp inputs grounded, the output shows a measurable voltage. This voltage is a result of the imbalance in the differential amplifier, which causes one of the transistors to conduct harder than the other.

14

Output offset voltage. A voltage that may appear at the output of an op-amp, caused by an imbalance in the differential amplifier.

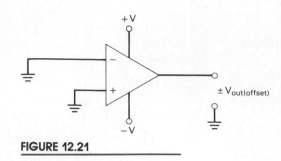

FIGURE 12.21

There are several methods that may be used to eliminate output offset voltage. One of these is to apply an *input offset voltage* between the input terminals of the op-amp, as shown in Figure 12.22. The value of V_{io} required to eliminate the output offset voltage is found as

$$V_{io} = \frac{V_{out(offset)}}{A_v} \qquad \textbf{(12.5)}$$

Another method of eliminating output offset voltage is to connect the op-amp's *offset null* pins as shown in Figure 12.23. When the offset null is being used, power is applied to the circuit and the potentiometer is adjusted to eliminate the output offset.

Figure 12.1 (page 516) shows that the offset null pins are connected (indirectly) to the input differential amplifier. When properly adjusted, the offset null circuitry corrects the imbalance in the differential amplifier, causing the output of the op-amp to go to 0 V.

Input Offset Current

Input offset current. A difference in the input currents, caused by differences in the transistor beta ratings.

When the output offset voltage of an op-amp is eliminated (as shown in either Figure 12.22 or 12.23), there will be a slight difference between the input currents to the noninverting and inverting inputs of the device (i_1 and i_2). This slight difference in input currents is called *input offset current*, and is caused by a beta mismatch

FIGURE 12.22

Applying an input offset voltage.

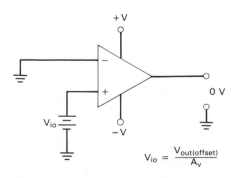

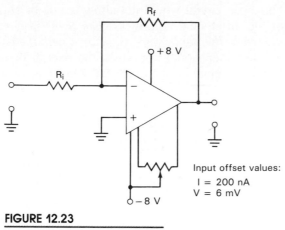

FIGURE 12.23

The offset null inputs.

Input offset values:

I = 200 nA
V = 6 mV

between the transistors in the differential amplifier. Note that there is no way of predicting which of the two input currents will be greater when the output offset voltage is eliminated.

Input Bias Current

The inputs to an op-amp require some amount of dc biasing current for the BJTs in the differential amplifier. The average quiescent value of dc biasing current that is drawn by the signal inputs of the op-amp is the *input bias current rating* of the amplifier. For the μA741, this rating is between 80 nA (typical) and 500 nA (maximum). This means that the op-amp signal inputs will draw between 80 and 500 nA from the external circuitry when no active signal is applied to the device.

The fact that both transistors in the differential amplifier require an input biasing current leads to the following operating restriction: *An op-amp will not work if either of its inputs is open*. For example, look at the circuit shown in Figure 12.24. The noninverting input is shown to have an open between the op-

Input bias current. The average quiescent value of dc biasing current drawn by the signal inputs of an op-amp.

Why both op-amp inputs must have a current path.

FIGURE 12.24

An open input prevents the op-amp from working.

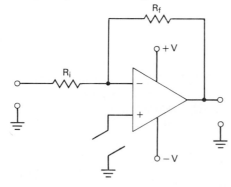

amp and ground. This open circuit would not allow the dc biasing current required for the operation of the differential amplifier. (*The transistor associated with the inverting input would work, but not the one associated with the noninverting input.*) Since the differential amplifier would not work, the overall op-amp circuit would not work either. Thus an input bias current path must always be provided for *both* op-amp inputs.

Common-mode signals. Signals that appear simultaneously at the two inputs of an op-amp.

Common-Mode Rejection Ratio

Common-mode signals are *signals that appear simultaneously at the two inputs of an op-amp*. For example, the two signals shown in Figure 12.25 are common-mode signals.

Assuming that the two signals shown *occur at the same time and have the same amplitude*, the *ideal* op-amp will not respond to their presence. Remember that the op-amp is designed to respond to *the difference between* two input signals. If there is no difference between the two inputs, the op-amp should not respond to those inputs.

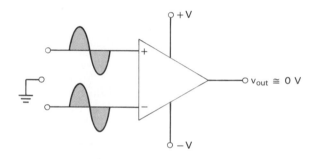

FIGURE 12.25

Common-mode signals.

Common-mode rejection ratio (CMRR). The ratio of differential gain to common-mode gain.

The ability of an op-amp to "ignore" common-mode signals is called the *common-mode rejection ratio*. Technically, it is *the ratio of differential gain to common-mode gain*. For example, let's assume for a moment that the op-amp in Figure 12.25 has a differential voltage gain of 1500. Let's also assume that the op-amp has a common-mode gain of 0.01. The common-mode rejection ratio for the op-amp would be found as

$$\begin{aligned} \text{CMRR} &= 1500\!:\!0.01 \\ &= 150{,}000\!:\!1 \end{aligned}$$

This number means that the output produced by a difference between the inputs would be 150,000 times as great as an output produced by a common-mode signal.

Common-mode signals are usually undesired signals.

The ability of the op-amp to reject common-mode signals is one of its main advantages. Common-mode signals are usually *undesired* signals, caused by external interference. For example, any RF signals picked up by the op-amp inputs would be considered undesirable. The common-mode rejection ratio indicates the op-amp's ability to reject such unwanted signals.

Decibels (dB). Logarithmic representations of numbers.

The common-mode rejection ratio for a given op-amp is usually measured in units called *decibels*, or *dB*. Decibels are logarithmic representations of numbers and are covered in detail in Chapter 13. The μA741 op-amp has a common-mode rejection ratio of 70 dB. This means that a differential input signal will have a gain that is at least 3163 times as great as the gain of a common-mode signal.

Power Supply Rejection Ratio

This ratio indicates *how much the output from an op-amp will change when the supply voltages change*. The μA741 has a power supply rejection ratio of 150 μV/V (maximum). This means that the dc output voltage from the op-amp will change by no more than 150 μV when the power supply voltage changes by 1 V. If the supply voltage for the op-amp was to change by 3 V, the dc output voltage would change by no more than 3×150 μV $= 450$ μV.

Power Supply Rejection Ratio. The ratio of a change in op-amp output voltage to a change in supply voltage.

Output Short-Circuit Current

Op-amps are protected internally from excessive current caused by a shorted load. The *output short-circuit current* rating of an op-amp is the maximum amount of current that will be supplied when the load is shorted. For the μA741, this rating is 25 mA. Thus, with a load of 0 Ω, the output current from this op-amp will be no more than 25 mA.

Output Short-Circuit Current. The maximum output current for an op-amp, measured under shorted load conditions.

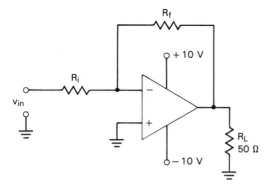

FIGURE 12.26

    The short-circuit current rating helps to explain why the output voltage from an op-amp drops when the load resistance decreases. For example, consider the circuit shown in Figure 12.26. The load resistance is shown to be 50 Ω. Assuming that the op-amp is producing the maximum output current of 25 mA, the load voltage (maximum) would be found as

$$V_L = I_{out}R_L$$
$$= 25 \text{ mA } (50 \text{ }\Omega)$$
$$= 1.25 \text{ V}$$

This is considerably less than the $\pm$ 10-V supply that normally limits the output voltage. Note that whenever you use a load resistance that is less than the output resistance of the op-amp, the value of the short-circuit current will drop the maximum output voltage to a very low value.

Slew Rate

The *slew rate* of an op-amp is a measure of *how fast the output voltage can change*. The slew rate of the μA741 is 0.5 V/μs (typical). This means that the output from this amplifier can change by 0.5 V every μs. Since frequency is a function of time, the *slew rate can be used to determine the maximum operating frequency of the op-amp*, as follows:

Slew rate. How fast the op-amp output voltage can change.

$$f_{max} = \frac{\text{slew rate}}{2\pi V_{pk}}$$

(12.6)

The following example illustrates the use of this formula.

━━━━━━━━━━ **EXAMPLE 12.5** ━━━━━━━━━━

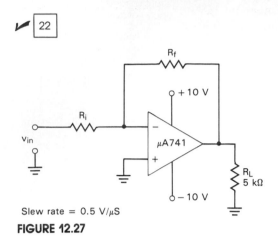

Slew rate = 0.5 V/μS

FIGURE 12.27

Determine the maximum operating frequency for the circuit shown in Figure 12.27.

Solution: The maximum peak output voltage for this circuit is approximately 8 V. Using this value as V_{pk} in equation (12.6), the maximum operating frequency for the amplifier is determined as

$$f_{max} = \frac{\text{slew rate}}{2\pi V_{pk}}$$
$$= \frac{0.5 \text{ V/μs}}{6.28\,(8)}$$
$$= \frac{500 \text{ kHz}}{50.24} \quad \left(500 \text{ kH}_z = \frac{0.5 \text{ V}}{1 \text{ μs}}\right)$$
$$= 9.95 \text{ kHz}$$

─────────────────────────────

PRACTICE PROBLEM 12–5

An op-amp with a slew rate of 0.4 V/μs has a 10-V_{pk} output. Determine the maximum operating frequency for the component.

━━━━━━━━━━━━━━━━━━━━━━━━━━━━━

While 9.95 kHz may not seem to be a very high output frequency, you must realize that the amplifier was assumed to be operating at its maximum output voltage. Let's see what happens when the peak output voltage is reduced.

━━━━━━━━━━ **EXAMPLE 12.6** ━━━━━━━━━━

The amplifier in Figure 12.27 is being used to amplify an input signal to a peak output voltage of 100 mV. What is the maximum operating frequency of the amplifier?

Solution: Using 100 mV as the peak output voltage, f_{max} is determined as

$$f_{max} = \frac{\text{slew rate}}{2\pi V_{pk}}$$
$$= \frac{0.5 \text{ V/μs}}{6.28(0.1)}$$
$$= \frac{500 \text{ kHz}}{0.628} \quad \left(500 \text{ kHz} = \frac{0.5 \text{ V}}{1 \text{ μs}}\right)$$
$$= 796 \text{ kHz}$$

PRACTICE PROBLEM 12–6

The output peak voltage for the op-amp in Practice Problem 12-5 is reduced to 2 V_{pk}. Determine the new maximum operating frequency for the device.

The previous examples have served to show that the op-amp can be operated at a much higher frequency when being used as a small-signal amplifier than when being used as a large-signal amplifier.

The effects of operating an op-amp circuit above the value of f_{max} can be seen in Figure 12.28. This figure shows the *distortion* caused to the input signal by the op-amp. In both cases shown, the input is changing at a rate that is higher than the op-amp slew rate can handle. When this happens, the output from the op-amp changes more slowly than does the input signal, and the distortion shown is the result. Note that the distortion can be eliminated by reducing the input frequency or by using an op-amp with a higher-frequency capability. The distortion can also be reduced by reducing the peak output voltage of the op-amp. If the gain of the component is reduced, the peak output voltage is also reduced for a given input signal. As you were shown in Example 12.8, the reduction in peak output voltage results in an increase in maximum operating frequency.

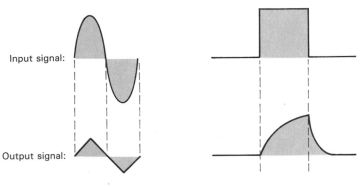

FIGURE 12.28

Signal distortion.

Input/Output Resistance

As was stated earlier in the chapter, op-amps typically have high input resistance and low output resistance ratings. These ratings for the μA741 are 2 MΩ and 75 Ω, respectively. Note that the high input resistance and low output resistance of the op-amp closely resemble the characteristics of the *ideal* amplifier, which has infinite input resistance and zero output resistance.

Putting It All Together

You have been exposed to many op-amp operating characteristics in this section. Now we are going to take a look at how they relate to each other. As the graphs in Figure 12.29 illustrate, the basic op-amp characteristics are not independent of each other. Table 12.4 has been derived using the graphs in Figure 12.29. As you read the comments on each characteristic, refer back to the indicated graph

FIGURE 12.29

Performance curves for the
μA741 op-amp. (Courtesy of
Fairchild, a division of National
Semiconductor.)

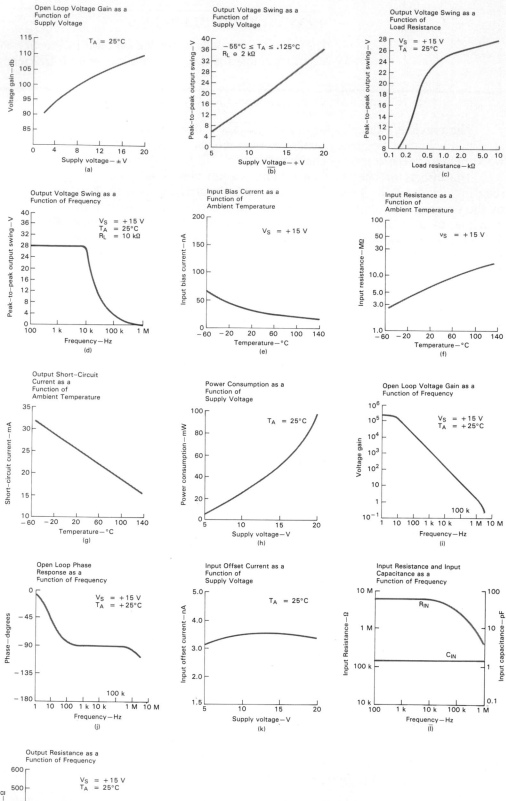

to verify the components for yourself. Graph references are given in parentheses after each statement.

TABLE 12.4
Op-amp Characteristics

Characteristic	Comments
Open-loop voltage gain	Directly proportional to supply voltage (a); decreases at a linear rate as frequency increases from 10 Hz to 1 MHz (i)
Output voltage swing	Directly proportional to R_L; drop rate increases significantly at values of $R_L < 500\ \Omega$ (c); directly proportional to supply voltage (b); constant for frequencies up to 10 kHz, then drops off rapidly (d)
Input resistance	Directly proportional to ambient temperature (f); constant for frequencies up to 10 kHz, then starts to drop off (1)
Input bias current	Inversely proportional to temperature (e); this follows the increase in input resistance
Output short-circuit current	Rated for temperature of 25°C; increases below this temperature, and decreases above this temperature (g)
Power consumption	The power used by the device is directly proportional to the supply voltage (h)
Output resistance	Remains constant for frequencies up to 100 kHz, then increases rapidly (m)

The following series of statements will provide a *cause-and-effect* viewpoint of the information contained in the chart:

1. As you *increase supply voltage*, you also:
   increase the open-loop voltage gain.
   increase the possible output voltage swing.
   increase the component power consumption.
   vary the input offset current very slightly.

2. As you *increase the operating frequency*, you also:
   decrease the output voltage swing ($f > 10$ kHz).
   decrease the open-loop voltage gain.
   decrease the input resistance ($f > 10$ kHz).
   increase the output resistance ($f > 100$ kHz).

3. As the *ambient temperature increases*:
   input bias current decreases.
   input resistance increases.
   output short-circuit current decreases.

As you will see, these cause-and-effect statements come in very handy when you are trying to build a circuit in lab or when troubleshooting a given op-amp circuit.

Other Op-amp Characteristics

Figure 12.30 shows the specification sheet for the μA741 op-amp. As you can see, the spec sheet contains all the electrical characteristics we have discussed in this section. Now we are going to take a brief look at the four electrical characteristics that are highlighted in the spec sheet.

The *input voltage range* indicates *the maximum value that the op-amp can accept without risking damage to the internal differential amplifier*. The minimum value for the μA741 is ± 12 V. If either input is driven beyond this range, the result could be damage to the amplifier.

Electrical Characteristics $V_S = \pm 15$ V, $T_A = 25°C$ unless otherwise specified

Typical (handwritten)

Characteristic	Condition	μA741			μA741C			Unit
		Min	Typ	Max	Min	Typ	Max	
Input Offset Voltage	$R_S \leq 10$ kΩ		1.0	5.0		2.0	6.0	mV
Input Offset Current			20	200		20	200	nA
Input Bias Current			80	500		80	500	nA
Power Supply Rejection Ratio	$V_S = +10, -20$ $V_S = +20, -10$ V, $R_S = 50$ Ω		30	150		30	150	μV/V
Input Resistance		.3	2.0		.3	2.0		MΩ
Input Capacitance			1.4			1.4		pF
Offset Voltage Adjustment Range			± 15			± 15		mV
Input Voltage Range					± 12	± 13		V
Common Mode Rejection Ratio	$R_S \leq 10$ kΩ				70	90		dB
Output Short Circuit Current			25			25		mA
Large Signal Voltage Gain	$R_L \geq 2$ kΩ, $V_{OUT} = \pm 10$ V	50k	200k		20k	200k		
Output Resistance			75			75		Ω
Output Voltage Swing	$R_L \geq 10$ kΩ				± 12	± 14		V
	$R_L \geq 2$ kΩ				± 10	± 13		V
Supply Current			1.7	2.8		1.7	2.8	mA
Power Consumption			50	85		50	85	mW
Transient Response (Unity Gain) — Rise Time	$V_{IN} = 20$ mV, $R_L = 2$ kΩ, $C_L \leq 100$ pF		.3			.3		μs
Transient Response (Unity Gain) — Overshoot			5.0			5.0		%
Bandwidth (Note 4)			1.0			1.0		MHz
Slew Rate	$R_L \geq 2$ kΩ		.5			.5		V/μs

Notes

4. Calculated value from $BW(MHz) = \dfrac{0.35}{\text{Rise Time } (\mu s)}$

5. All $V_{CC} = 15$ V for μA741 and μA741C.

6. Maximum supply current for all devices
   25°C = 2.8 mA
   125°C = 2.5 mA
   −55°C = 3.3 mA

FIGURE 12.30

The μA741 specification sheet. (Courtesy of Fairchild, a division of National Semiconductor.)

The *large-signal voltage gain* is the open-loop voltage gain for the op-amp. As you can see, the value of A_{OL} for the μA741 is typically around 200,000. This agrees with the earlier statement that the op-amp has extremely high voltage gain capabilities.

The *supply current* rating indicates *the quiescent current that is drawn from the power supply*. When the μA741 does not have an active input, it will draw a maximum of 2.8 mA from $+V$ and $-V$.

The *power consumption* rating indicates *the amount of power that the op-amp will dissipate when it is in its quiescent state*. As you can see, the power dissipation rating of the μA741 has a maximum value of 85 mW.

There are two more ratings that we have not covered. These are *bandwidth* and *transient response*. Bandwidth is covered in Chapter 13. Transient response is covered in Chapter 18. In the next section, we are going to discuss the operation of the two most basic op-amp circuits, the *inverting amplifier* and the *noninverting amplifier*.

SECTION REVIEW

1. Describe the differential amplifier's response to an input signal at the inverting input. [Objective 13]
2. Describe the differential amplifier's response to an input signal at the noninverting input. [Objective 13]
3. What is *input offset voltage*? [Objective 14]
4. What is *output offset voltage*? [Objective 15]
5. What effect does the *offset null* control have on the input offset voltage and output offset voltage? Explain your answer. [Objective 15]
6. What is *input offset current*? [Objective 16]
7. What is *input bias current*? What restriction does input bias current place on the wiring of an op-amp? [Objective 17]
8. What are *common-mode signals*? [Objective 18]
9. What is the *common-mode rejection ratio (CMRR)*? Why is a high CMRR considered to be desirable? [Objective 18]
10. What is the *power supply rejection ratio*? [Objective 19]
11. What is the *output short-circuit current* rating? [Objective 20]
12. Discuss the relationship between load resistance and maximum output voltage swing in terms of the output short-circuit current rating. [Objective 20]
13. What is *slew rate*? What circuit parameter is limited by the op-amp slew rate? [Objective 21]
14. What is the result of trying to operate an op-amp beyond its slew rate limit? [Objective 23]
15. Define each of the following op-amp characteristics: [Objective 24]
    a. Input voltage range
    b. Large-signal voltage range
    c. Supply current rating
    d. Power consumption rating

12.4
Inverting Amplifiers

In most of our examples, we have used an op-amp circuit that contains a single input resistor (R_i) and a single feedback resistor (R_f). This amplifier is the basic *inverting amplifier*, which is the op-amp equivalent of the common-emitter and common-source circuits. The operation of the inverting amplifier is illustrated in Figure 12.31.

The key to the operation of the inverting amplifier lies in the differential input circuit of the op-amp. For reference purposes, this input circuit is shown

Inverting amplifier. A basic op-amp circuit that produces a 180° signal phase shift. The op-amp equivalent of the common-emitter and common-source circuits.

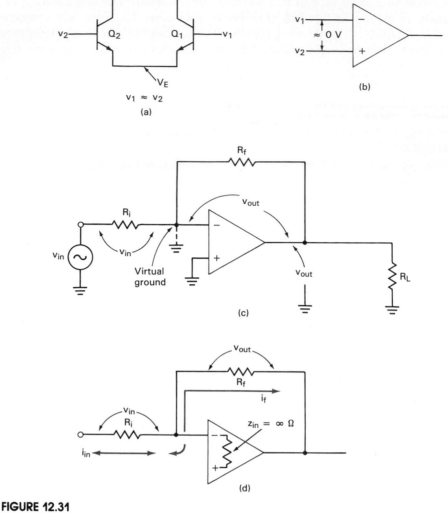

FIGURE 12.31

Inverting amplifier operation.

again in Figure 12.31a. Assuming that the two transistors are perfectly matched, the values of v_1 and v_2 will always be *approximately* equal to each other. They may differ by a few millivolts, but that is all.

Now, for a moment, let's idealize the differential circuit. For the sake of discussion, we will assume that v_1 and v_2 are *exactly* equal. With this in mind, look at the circuit in Figure 12.31c. If the voltages at the two inputs are equal, and the noninverting input is grounded, the inverting input to the op-amp is also at ground. Remember, this *virtual ground* is caused by the fact that v_1 and v_2 are approximately equal, and v_2 is grounded.

With the inverting input of the op-amp at virtual ground, the total input voltage can be measured across R_i. Also, the total output voltage can be measured across R_f. These points are illustrated in Figure 12.31c.

We need to make another point about the op-amp. The high input impedance of the op-amp greatly restricts the amount of current that can flow into the input terminals. If we assume that the op-amp is ideal (Z_{in} is infinite), we can say that the input current to the circuit must equal the current through the feedback resistor. This point makes sense when you think about it. If the input impedance of the

CHAP. 12   Operational Amplifiers

op-amp is infinite, the input current has only one place to go—through R_f. This current action is illustrated in Figure 12.31d.

Now let's tie all of this together. Since the input voltage to the circuit is measured across R_i, we can state the following:

$$v_{in} = i_{in}R_i \qquad (12.7)$$

Also, with v_{out} being measured across R_f, we can state the following:

$$v_{out} = i_f R_f \qquad (12.8)$$

We have already established the fact that i_{in} and i_f are equal, so equation (12.8) can be rewritten as

$$v_{out} = i_{in}R_f \qquad (12.9)$$

Recall that the voltage gain (A_v) of an amplifier is defined as the ratio of output voltage to input voltage. By formula,

$$A_v = \frac{v_{out}}{v_{in}}$$

Substituting equations (12.7) and (12.9) in place of v_{in} and v_{out}, we get

$$A_v = \frac{i_{in}R_f}{i_{in}R_i}$$

or

$$A_v = \frac{R_f}{R_i} \qquad (12.10)$$

Hey, that's not bad! For the inverting amplifier, you only have to divide the value of R_f by the value of R_i to determine the approximate gain of the amplifier. In fact, if you go back to look at the examples earlier in this chapter, you'll see where the "assumed" values of voltage gain came from. In every case, the value of A_v was approximately equal to the value of R_f/R_i.

Now that we have discussed a basic op-amp circuit, we can give practical meaning to a few terms. The *open-loop voltage gain*, A_{OL}, of an op-amp is the gain that is measured when there is no *feedback path* (physical connection) between the output and the input of the circuit. When a feedback path is present, such as the R_f connection in the inverting amplifier, the resulting circuit gain is referred to as the *closed-loop voltage gain*, A_{CL}.

A Practical Consideration: The gain of an inverting amplifier is determined by the ratio of R_f to R_i. It would follow that a low value of R_i would be desirable to provide a high voltage gain. However, lowering the value of R_i also lowers the amplifier input impedance. Thus, increasing voltage gain is normally done by increasing the value of R_f rather than decreasing the value of R_i.

Feedback path. A signal path from the output of an amplifier back to the input.

Closed-loop voltage gain (A_{CL}). The gain of an op-amp with a feedback path; always lower than the value of A_{OL}.

Amplifier Input Impedance

While the op-amp has an extremely high input impedance, the inverting amplifier does not. The reason for this can be seen by referring back to Figure 12.31c. As this figure shows, the voltage source "sees" an input resistance (R_i) that is going to (virtual) ground. Thus, the input impedance for the inverting amplifier is found as

$$Z_{in} \cong R_i \qquad (12.11)$$

The value of R_i will always be much less than the input impedance of the op-amp. Therefore, the overall input impedance of an inverting amplifier will also be much lower than the op-amp input impedance.

Amplifier Output Impedance

If you look at the circuit in Figure 12.32, you can see that the output impedance of the inverting amplifier is the parallel combination of R_f and the output impedance of the op-amp itself.

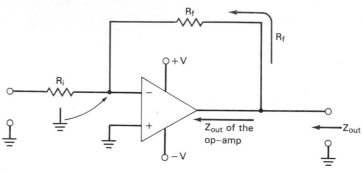

FIGURE 12.32

The presence of the feedback network reduces the output impedance of the amplifier to a value that is less than the output impedance of the op-amp. In Chapter 14, you will be shown how to calculate the output impedance of an inverting amplifier. In the mean time, we will simply assume that the circuit output impedance is less than the output impedance rating of the op-amp.

Amplifier CMRR

In the last section, the *common-mode rejection ratio* (CMRR) of an op-amp was defined as *the ratio of differential gain to common-mode gain*. Since the differential gain of the inverting amplifier (A_{CL}) is much less than the differential gain of the op-amp (A_{OL}), the CMRR of the inverting amplifier will be much lower than that of the op-amp.

The CMRR of the inverting amplifier is found as the ratio of closed-loop gain to common-mode gain. By formula,

$$CMRR = \frac{A_{CL}}{A_{CM}}$$

(12.12)

where A_{CL} = the closed loop voltage gain of the inverting amplifier
A_{CM} = the *common-mode* gain of the op-amp

Even though the CMRR of an inverting amplifier will be lower than that of the op-amp, it will still be extremely high in most cases. This point will be illustrated later in this section.

Inverting Amplifier Analysis

The complete analysis of an inverting amplifier involves determining the circuit values of A_{CL}, Z_{in}, Z_{out}, CMRR, and f_{max}. The complete analysis of an inverting amplifier is illustrated in the following example.

Common-mode gain (A_{CM}). The gain a differential amplifier provides to common-mode signals, typically less than 1.

A Practical Consideration. Most op-amp spec sheets do not provide a value of common-mode gain (A_{CM}). Rather, they list values of A_{OL} and CMRR. When this is the case, the value of A_{CM} can be obtained using

$$A_{CM} = \frac{A_{OL}}{CMRR}$$

If the CMRR is listed in decibel (dB) form, it must be converted to standard numeric form before using the above equation. To convert the CMRR to standard form, use

$$CMRR = \log^{-1}\left(\frac{CMRR (dB)}{20}\right)$$

The basis of this equation is discussed in detail in Chapter 13 (pages 573–581).

CHAP. 12   Operational Amplifiers

EXAMPLE 12.7

✔ 29

Perform the complete analysis of the circuit shown in Figure 12.33.

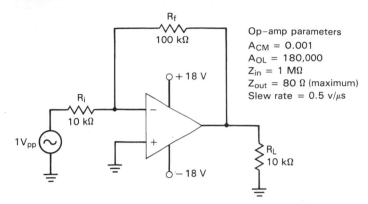

FIGURE 12.33

Solution: First, the closed-loop voltage gain (A_{CL}) of the circuit is found as

$$A_{CL} = \frac{R_f}{R_i}$$
$$= \frac{100 \text{ k}\Omega}{10 \text{ k}\Omega}$$
$$= 10$$

The input impedance of the circuit is found as

$$Z_{in} \cong R_i$$
$$= 10 \text{ k}\Omega$$

The output impedance is lower than the output impedance of the op-amp, 80 Ω (maximum).

The CMRR of the circuit is found as

$$\text{CMRR} = \frac{A_{CL}}{A_{CM}}$$
$$= \frac{10}{0.001}$$
$$= 10{,}000$$

To calculate the maximum operating frequency for the inverting amplifier, we need to determine its peak output voltage. With values of $V_{in} = 1 \text{ V}_{PP}$ and $A_{CL} = 10$, the peak-to-peak output voltage is found as

$$V_{out} = (1 \text{ V}_{PP})(10)$$
$$= 10 \text{ V}_{PP}$$

Thus, the peak output voltage is 5 V_{pk}, and the maximum operating frequency is found as

$$f_{max} = \frac{\text{slew rate}}{2\pi V_{pk}}$$
$$= \frac{0.5 \text{ V/}\mu\text{s}}{31.42}$$
$$= \frac{500 \text{ kHz}}{31.42}$$
$$= 15.9 \text{ kHz}$$

PRACTICE PROBLEM 12–7

An op-amp has the following parameters: A_{CM} = 0.02, A_{OL} = 150,000, Z_{in} = 1.5 MΩ, Z_{out} = 50 Ω (maximum), and slew rate = 0.75 V/μs. The op-amp is being used in an inverting amplifier with ±12-V_{dc} supply voltages and values of V_{in} = 50 mV$_{PP}$, R_f = 250 kΩ, and R_i = 1 kΩ. Perform the complete analysis of the circuit.

Other Inverting Amplifier Configurations

In this section, we have covered only the simplest of the inverting amplifiers. In Chapter 14, you will be introduced to a variety of inverting amplifier configurations, each with its own circuit analysis procedure. While the calculations of gain, input impedance, and output impedance will vary, the calculations of f_{max} and CMRR will not. You will also see that the various inverting amplifiers all have the characteristic 180° voltage phase shift from input to output.

SECTION REVIEW

1. Explain why the voltage gain of an inverting amplfiier (A_{CL}) is equal to the ratio of R_f to R_i. [Objective 25]
2. Why is the input impedance of an inverting amplifier lower than the input impedance of the op-amp? [Objective 26]
3. Why isn't the value of R_f normally considered in the output impedance calculation for an inverting amplifier? [Objective 27]
4. Why is the CMRR ratio of an inverting amplifier lower than that of its op-amp? [Objective 28]
5. List, in order, the steps involved in performing the complete analysis of an inverting amplifier. [Objective 29]

12.5
Noninverting Amplifiers

Noninverting amplifier. An op-amp circuit with no signal phase shift.

The basic noninverting amplifier has most of the characteristics of the inverting amplifier, with two exceptions:

1. The noninverting amplifier has much higher input impedance.

CHAP. 12   Operational Amplifiers

2. The noninverting amplifier does not produce a 180° voltage phase shift from input to output. Thus, the input and output signals are in phase.

The basic noninverting amplifier is shown in Figure 12.34. Note that the input is applied to the noninverting op-amp input, and the input resistance (R_i) is returned to ground.

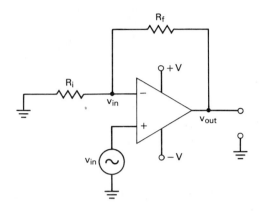

FIGURE 12.34

The noninverting amplifier.

Because the input signal is applied to the noninverting terminal, we must calculate the gain of this circuit differently. We will start our discussion of circuit gain by again assuming that $v_1 = v_2$ and that the currents through R_i and R_f are equal. Since $v_1 = v_2$ and $v_2 = v_{in}$, we can state the following relationship:

$$\boxed{v_1 = v_{in}}$$ **(12.13)**

30

Note that this is different from the relationship of $v_1 = 0$ V stated for the inverting amplifier with a grounded v_2. Once again, we have v_{in} being measured across R_i, and i_{in} can be found as

$$i_{in} = \frac{v_{in}}{R_i}$$

or

$$v_{in} = i_{in}R_i$$

Since the voltage across R_f is equal to the difference between v_{in} and v_{out}, we can state that the current through the resistor is found as

$$\boxed{i_f = \frac{v_{out} - v_{in}}{R_f}}$$ **(12.14)**

and

$$v_{out} = i_{in}R_f + v_{in}$$

Note that the value of i_{in} was used in place of i_f. This is valid, since the two currents are equal.

Now we can use the equations established to define A_{CL} as

$$A_{CL} = \frac{v_{out}}{v_{in}}$$

$$= \frac{i_{in}R_f + i_{in}R_i}{i_{in}R_i} \qquad (v_{in} = i_{in}R_i)$$

$$= \frac{i_{in}R_f}{i_{in}R_i} + \frac{i_{in}R_i}{i_{in}R_i}$$

or

$$\boxed{A_{CL} = \frac{R_f}{R_i} + 1} \qquad (12.15)$$

where A_{CL} is the *closed-loop* voltage gain of the amplifier. Thus the gain of a noninverting amplifier will always be greater than the gain of an equivalent inverting amplifier by a value of 1. If an inverting amplifier has a gain of 150, the equivalent noninverting amplifier will have a gain of 151.

Amplifier Input and Output Impedance

Z_{in} for the noninverting amplifier is greater than or equal to the rated value for the op-amp, typically in the megohm range.

Since the input signal is applied directly to the op-amp, the noninverting amplifier has extremely high input impedance. In fact, the presence of the feedback network causes the amplifier input impedance to be greater than the input impedance of the op-amp in most cases. In Chapter 14, you will be shown how to calculate the input impedance of a noninverting amplifier. In the mean time, we will simply assume that the circuit input impedance is greater than (or equal to) the rated input impedance of the op-amp.

Z_{out} is also lower than the rated Z_{out} of the op-amp.

The output impedance of the noninverting amplifier is lower than the output impedance of the op-amp, as was the case with the inverting amplifier.

The extremely high input impedance and extremely low output impedance of the noninverting amplifier make the circuit very useful as a *buffer*. You may recall that the *emitter follower* and the *source follower* are circuits that can be used to match a high-impedance source to a low-impedance load. The noninverting amplifier can be used for the same purpose. The primary difference is that the basic noninverting amplifier is capable of high voltage gain values, while the emitter follower and the source follower have A_v values that are less than 1.

Noninverting Amplifier Analysis

✔ 31

The complete analysis of the noninverting amplifier is almost identical to that of the inverting amplifier. The values of Z_{out}, CMRR, and f_{max} are found using the same equations we used for the inverting amplifier. The values of A_{CL} and Z_{in} are found using the equations and principles established in this section. The complete analysis of the noninverting amplifier is illustrated in the following example.

━━━━━━━━━━ **EXAMPLE 12.8** ━━━━━━━━━━

Perform the complete analysis of the noninverting amplifier shown in Figure 12.35.

Solution: The closed-loop voltage gain (A_{CL}) of the circuit

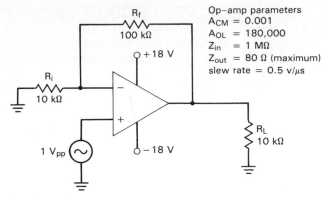

Op-amp parameters
$A_{CM} = 0.001$
$A_{OL} = 180,000$
$Z_{in} = 1\ M\Omega$
$Z_{out} = 80\ \Omega$ (maximum)
slew rate $= 0.5\ v/\mu s$

FIGURE 12.35

is found as

$$A_{CL} = \frac{R_f}{R_i} + 1$$
$$= \frac{100\ k\Omega}{10\ k\Omega} + 1$$
$$= 11$$

The input impedance of the amplifier is at least 1 MΩ and the output impedance of the circuit is less than 80 Ω.

The CMRR for the circuit is found as

$$CMRR = \frac{A_{CL}}{A_{CM}}$$
$$= \frac{11}{0.001}$$
$$= 11,000$$

To determine the value of f_{max}, we need to calculate the peak output voltage for the amplifier. The peak-to-peak output voltage is found as

$$V_{out} = A_{CL}V_{in}$$
$$= 11\ (1\ V_{PP})$$
$$= 11\ V_{PP}$$

The peak output voltage would be one-half of this value, 5.5 V_{pk}. Using this value and the slew rate shown in the figure, the value of f_{max} is calculated as

$$f_{max} = \frac{\text{slew rate}}{2\pi V_{pk}}$$
$$= \frac{0.5\ V/\mu s}{34.56}$$
$$= \frac{500\ kHz}{34.56}$$
$$= 14.47\ kHz$$

PRACTICE PROBLEM 12–8

The circuit described in Practice Problem 12–7 is wired as a noninverting amplifier. Perform a complete analysis on the new circuit.

The circuit values and op-amp parameters used in Example 12.10 were identical to those used for the inverting amplifier in Example 12.9. The results found in the two examples are summarized below.

Value	Inverting Amplifier	Noninverting Amplifier
A_{CL}	10	11
Z_{in}	1 kΩ	1 MΩ (minimum)
Z_{out}	80 Ω (maximum)	80 Ω (maximum)
CMRR	10,000	11,000
f_{max}	15.9 kHz	14.47 kHz

As you can see, the noninverting amplifier has slightly greater values of A_{CL} and CMRR. It also has a much greater value of input impedance. At the same time, the inverting amplifier has a slightly greater maximum operating frequency. This is due to the higher peak output voltage of the comparable noninverting amplifier. Figure 12.36 summarizes the operation of the noninverting and inverting amplifiers.

FIGURE 12.36

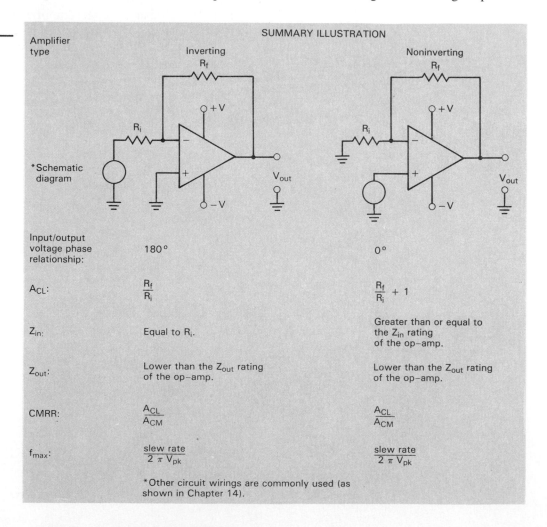

SUMMARY ILLUSTRATION

Amplifier type	Inverting	Noninverting
*Schematic diagram	R_f, R_i, +V, $-$, +, $-$V, V_{out}	R_f, R_i, +V, $-$, +, $-$V, V_{out}
Input/output voltage phase relationship:	180°	0°
A_{CL}:	$\dfrac{R_f}{R_i}$	$\dfrac{R_f}{R_i} + 1$
Z_{in}:	Equal to R_i.	Greater than or equal to the Z_{in} rating of the op–amp.
Z_{out}:	Lower than the Z_{out} rating of the op–amp.	Lower than the Z_{out} rating of the op–amp.
CMRR:	$\dfrac{A_{CL}}{A_{CM}}$	$\dfrac{A_{CL}}{A_{CM}}$
f_{max}:	$\dfrac{\text{slew rate}}{2\,\pi\,V_{pk}}$	$\dfrac{\text{slew rate}}{2\,\pi\,V_{pk}}$

*Other circuit wirings are commonly used (as shown in Chapter 14).

The Voltage Follower

If we remove R_i and R_f from the noninverting amplifier and short the output of the amplifier to the inverting input, we have the *voltage follower*. This circuit, which is the op-amp equivalent of the emitter follower and source follower, is shown in Figure 12.37. You may recall that the characteristics of the emitter and source followers are as follows:

1. High Z_{in} and low Z_{out}
2. A_v that is approximately equal to 1
3. Input and output signals that are in phase

Characteristics 1 and 3 are accomplished by using an op-amp in a noninverting circuit configuration. The voltage gain for the voltage follower is calculated as

$$A_{CL} = \frac{R_f}{R_i} + 1$$
$$= \frac{0\ \Omega}{R_i} + 1 \qquad (\text{since } R_f = 0\ \Omega)$$
$$= 1$$

Voltage follower. The op-amp equivalent of the emitter follower and the source follower.

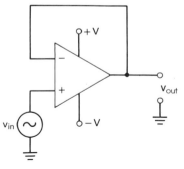

FIGURE 12.37

The voltage follower.

Simple enough? (*If you think that calculating A_{CL} for this circuit is easy, wait until you try troubleshooting it!*)

The values of Z_{in}, Z_{out}, and f_{max} for the voltage follower are calculated using the same equations that we used for the basic noninverting amplifier. Since $A_{CL} = 1$ for the voltage follower, the circuit CMRR is found using

$$\boxed{\text{CMRR} = \frac{1}{A_{CM}}} \qquad (12.16)$$

As the following example illustrates, the voltage follower is by far the easiest op-amp circuit to analyze.

═══════════════ **EXAMPLE 12.9** ═══════════════

Perform the complete analysis of the voltage follower in Figure 12.38.

Op–amp parameters
$A_{CM} = 0.001$
$A_{OL} = 180,000$
$Z_{in} = 1\ M\Omega$
$Z_{out} = 80\ \Omega$ (maximum)
slew rate = 0.5 v/μs

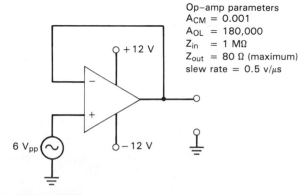

FIGURE 12.38

Solution: For the voltage follower,

$$A_{CL} = 1$$

The values of Z_{in} and Z_{out} are equal to the rated values for the op-amp, 1 MΩ and 80 Ω (maximum), respectively. The CMRR of the circuit is found as

$$CMRR = \frac{1}{A_{CM}}$$
$$= \frac{1}{0.001}$$
$$= 1000$$

Since $A_{CL} = 1$ for the circuit, $V_{out} = V_{in}$. Thus, the peak output voltage is one-half of 6 V_{PP}, or 3 V_{pk}. Now f_{max} is found as

$$f_{max} = \frac{\text{slew rate}}{2\pi V_{pk}}$$
$$= \frac{500 \text{ kHz}}{18.85}$$
$$= 26.53 \text{ kHz}$$

PRACTICE PROBLEM 12–9

The op-amp described in Practice Problem 12–7 is used in a voltage follower with ±14-V supply voltages and a 12-V_{PP} input. Perform the complete analysis of the amplifier.

As you can see, a voltage follower will have significantly lower A_{CL} and CMRR values and a higher maximum operating frequency than a comparable "standard" noninverting amplifier.

One Final Note

Just as there are a wide variety of ways to wire an inverting amplifier, there are many ways to wire a noninverting amplifier. Several of these noninverting amplifier wiring schemes are discussed in Chapter 14.

SECTION REVIEW

1. How do you calculate the value of A_{CL} for a noninverting amplifier? [Objective 30]
2. How do you determine the values of Z_{in} and Z_{out} for a noninverting amplifier? [Objective 30]
3. How do you determine the values of CMRR and f_{max} for a noninverting amplifier? [Objective 31]
4. Compare and contrast the inverting amplifier with the noninverting amplifier. [Objective 32]
5. Describe the gain and impedance characteristics of the voltage follower. [Objective 33]

12.6
Troubleshooting Basic Op-Amp Circuits

From the technician's viewpoint, op-amp circuits are a dream come true. The basic op-amp circuit has only three or four components that can become faulty, and each fault has distinct symptoms. For example, consider the circuit shown in Figure 12.39. Assuming that the load resistor, the input signal, and the supply voltages are all good, there are only four components that could cause a given fault: R_i, R_f, the offset resistor, and the op-amp itself. Let's take a look at what happens when one of the resistors goes bad.

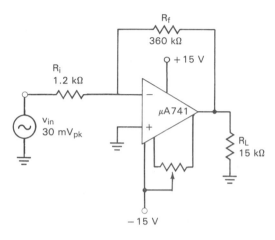

FIGURE 12.39

R_f Open

When the feedback resistor opens, the entire feedback loop is effectively removed from the circuit. This causes the gain of the amplifier to increase from A_{CL} to the open-loop voltage gain of the op-amp, A_{OL}. Now consider what this would do to the circuit in Figure 12.39. The value of A_{CL} for this circuit is 300. The value of A_{OL} for the μA741 is 200,000. With the amplifier gain going to A_{OL}, the output would try to go to ±6000 V, which is clearly impossible. The result of this circuit action is shown in Figure 12.40.

FIGURE 12.40

The effect of an open R_f on the inverting amplifier output.

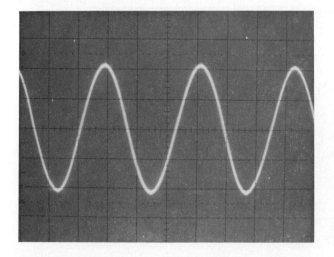

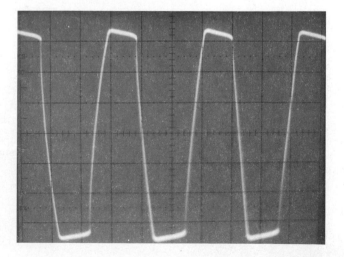

Figure 12.40a shows the normal output from the circuit in Figure 12.39. As you can see, the signal is a clear, unclipped sine wave. The waveform shown in Figure 12.40b is the result of R_f opening. The gain of the amplifier has become so great that the output waveform is clipped on both the positive and negative alternations. This output waveform is classic for an open feedback loop.

R_i Open

This fault is an interesting one. You would think that an open R_i would simply block the input signal, and that the output would therefore go to 0 V. However, this may not be the case. Let's take a look at the circuit you have when R_i opens. This circuit is shown in Figure 12.41.

Now we are going to assume for a moment that the output from this circuit is equal to $+V$ at the moment when R_i opens. Here's what may happen:

1. A *positive* signal is fed back to the inverting input from the output via R_f.
2. The positive inverting input causes the output from the amplifier to go negative toward $-V$.
3. A *negative* signal is now fed back to the inverting input from the output.
4. As the inverting input goes negative, the output again goes positive. This takes the amplifier back to step 1.

The process above repeats over and over, *causing the amplifier to produce an output signal with no input signal*. This signal will be in the low-millivolt range. Incidentally, there are circuits that are *designed* to work in the fashion described above. These circuits, called *oscillators*, are described later in the book.

R_{Offset} Open

If the offset adjust resistor opens, the output from the op-amp will be offset from its normal level by an amount equal to the offset voltage times the closed-loop

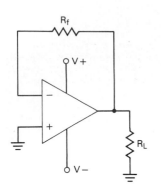

FIGURE 12.41

FIGURE 12.42

The effect of an open offset null resistor on the inverting amplifier output.

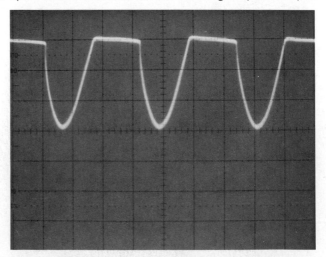

gain of the amplifier, A_{CL}. The effects of an open offset resistor are shown in the photograph in Figure 12.42.

Whether the op-amp offsets in the positive direction or the negative direction depends on the circuit. Just remember, if the output goes to a dc level that is above or below the proper output when there is no input signal, the offset resistor is either open or needs to be adjusted.

What Happens If the Op-Amp Is Bad?

The answer to this question depends on what goes wrong with the op-amp. If you refer back to Figure 12.1, you will see that there are a lot of components in the op-amp that could go bad.

The best way to determine that the op-amp is faulty is to determine that everything else is okay. If all the resistors in the amplifier are good and all supply voltages are as they should be, and the amplifier still does not work, the op-amp is the source of the problem and must be replaced.

Component substitution. **Most common op-amps are available at any electronics parts store. If an op-amp is bad, you can generally replace it with an equivalent from any manufacturer. Equivalent op-amps will have the same package, designator code, and suffix code (see Figure 12.3).**

Working With ICs

ICs are sometimes placed into circuits using IC sockets, such as the one shown in Figure 12.43. These sockets are used to allow you to easily remove and replace the ICs.

FIGURE 12.43

I.C. sockets.

When IC sockets are used, a problem can develop. Sometimes an IC will develop an *oxide* layer on its pins. If this oxidation becomes severe enough, it can cause the circuit to have an erratic output. The circuit may work one minute and not the next. When this happens, remove the IC and clean the pins and the socket with contact cleaner. This will eliminate the problem.

One more point. When you replace an IC, make sure that you put it into the circuit properly. It is very simple to put an IC in backward if you are not paying attention. If you *do* place an IC in a socket backward and apply power to the circuit, odds are that you are going to have to replace that IC with a new one.

1. What is the primary symptom of an open feedback resistor? [Objective 34]
2. What is the primary symptom of an open input resistor? [Objective 34]

SECTION REVIEW

3. What is the primary symptom of an open offset null resistor? [Objective 34]
4. How can you tell when an op-amp is faulty? [Objective 34]

KEY TERMS

The following terms were introduced and defined in this chapter:

closed-loop voltage gain (A_{CL})
common-mode gain (A_{CM})
common-mode rejection ratio (CMRR)
common-mode signals
decibels
designator code
differential amplifier
discrete component
feedback path
input bias current

input offset current
input offset voltage
integrated circuit (IC)
inverting amplifier
inverting input
noninverting amplifier
noninverting input
offset null
open-loop voltage gain (A_{OL})
operational amplifier (op-amp)

output offset voltage
output short-circuit current
power supply rejection ratio
slew rate

PRACTICE PROBLEMS

§12.2

1. Determine the output polarity for each of the op-amps in Figure 12.44. [9]
2. Determine the output polarity for each of the op-amps in Figure 12.45. [9]
3. Determine the maximum peak-to-peak output voltage for the amplifier in Figure 12.46. [11]
4. Determine the maximum peak-to-peak output voltage for the amplifier in Figure 12.47. [11]
5. The amplifier in problem 3 has a voltage gain of 120. Determine the maximum allowable peak-to-peak input voltage for the circuit. [12]
6. The amplifier in problem 4 has a voltage gain of 220. Determine the maximum allowable peak-to-peak input voltage for the circuit. [12]
7. Determine the maximum allowable peak-to-peak input voltage for the amplifier in Figure 12.48. [12]
8. Determine the maximum allowable peak-to-peak input voltage for the amplifier in Figure 12.49. [12]

§12.3

9. The amplifier in Figure 12.48 has an output offset voltage of 2.4 V. Determine the input offset voltage for the circuit. [14]
10. The amplifier in Figure 12.49 has an output offset voltage of 960 mV. Determine the input offset voltage for the circuit. [14]
11. Determine the maximum operating frequency for the amplifier in Figure 12.48. Assume that the circuit has a 10 mV$_{PP}$ input signal. [22]

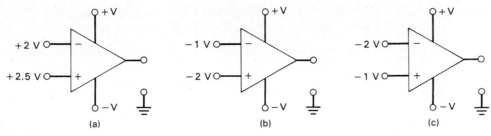

(a)   (b)   (c)

FIGURE 12.44

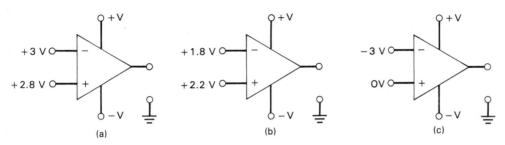

(a)   (b)   (c)

FIGURE 12.45

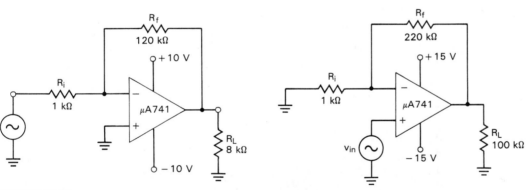

FIGURE 12.46

FIGURE 12.47

FIGURE 12.48   **FIGURE 12.49**

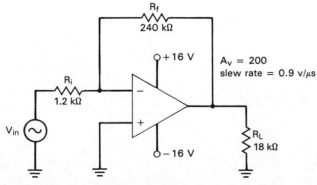

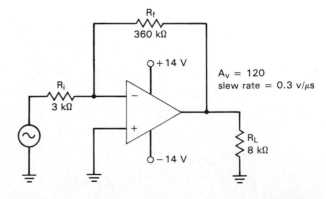

Practice Problems

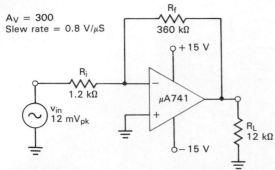

FIGURE 12.50

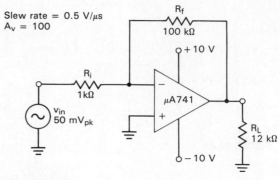

FIGURE 12.51

12. Determine the maximum operating frequency for the amplifier in Figure 12.49. Assume that the circuit has a 40 mV$_{PP}$ input signal. [22]

13. Determine the maximum operating frequency for the amplifier in Figure 12.50. [22]

14. Determine the maximum operating frequency for the amplifier in Figure 12.51. [22]

§12.4

15. Perform the complete analysis of the amplifier in Figure 12.52. [29]
16. Perform the complete analysis of the amplifier in Figure 12.53. [29]
17. Perform the complete analysis of the amplifier in Figure 12.54. [29]
18. Perform the complete analysis of the amplifier in Figure 12.55. [29]
19. Perform the complete analysis of the amplifier in Figure 12.56. [31]
20. Perform the complete analysis of the amplifier in Figure 12.57. [31]
21. Perform the complete analysis of the amplifier in Figure 12.58. [31]

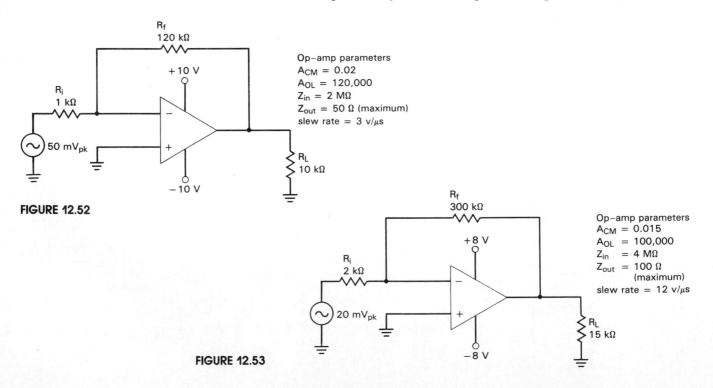

FIGURE 12.52

FIGURE 12.53

CHAP. 12   Operational Amplifiers

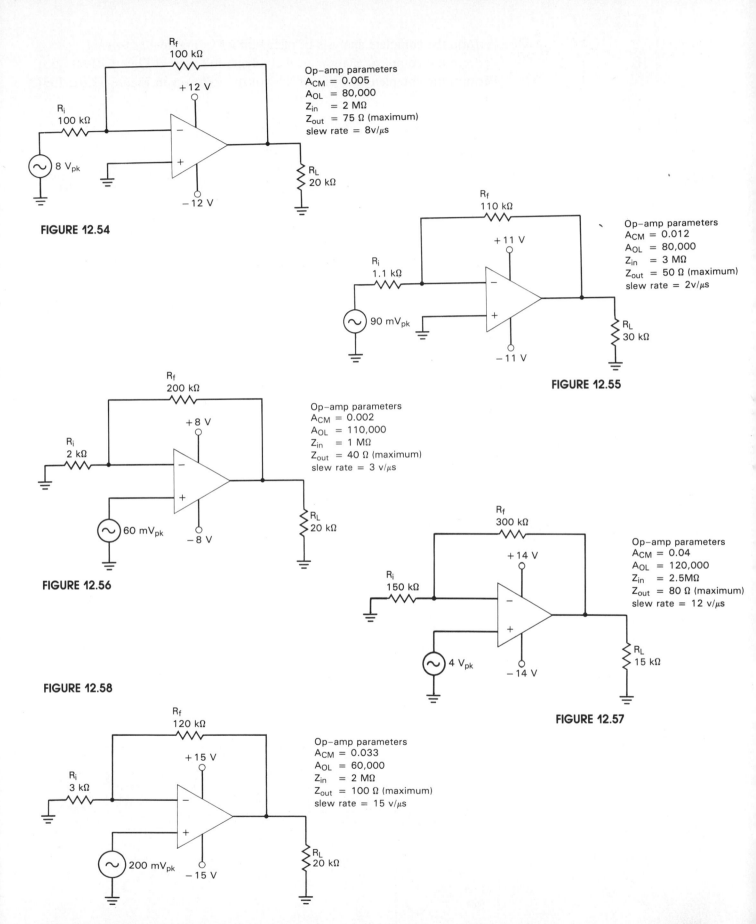

FIGURE 12.54

Op-amp parameters
$A_{CM} = 0.005$
$A_{OL} = 80,000$
$Z_{in} = 2\ M\Omega$
$Z_{out} = 75\ \Omega$ (maximum)
slew rate = 8v/μs

R_f 100 kΩ

+12 V

−12 V

R_i 100 kΩ

8 V$_{pk}$

R_L 20 kΩ

Op-amp parameters
$A_{CM} = 0.012$
$A_{OL} = 80,000$
$Z_{in} = 3\ M\Omega$
$Z_{out} = 50\ \Omega$ (maximum)
slew rate = 2v/μs

R_f 110 kΩ

+11 V

−11 V

R_i 1.1 kΩ

90 mV$_{pk}$

R_L 30 kΩ

FIGURE 12.55

Op-amp parameters
$A_{CM} = 0.002$
$A_{OL} = 110,000$
$Z_{in} = 1\ M\Omega$
$Z_{out} = 40\ \Omega$ (maximum)
slew rate = 3 v/μs

R_f 200 kΩ

+8 V

−8 V

R_i 2 kΩ

60 mV$_{pk}$

R_L 20 kΩ

FIGURE 12.56

Op-amp parameters
$A_{CM} = 0.04$
$A_{OL} = 120,000$
$Z_{in} = 2.5M\Omega$
$Z_{out} = 80\ \Omega$ (maximum)
slew rate = 12 v/μs

R_f 300 kΩ

+14 V

−14 V

R_i 150 kΩ

4 V$_{pk}$

R_L 15 kΩ

FIGURE 12.57

FIGURE 12.58

Op-amp parameters
$A_{CM} = 0.033$
$A_{OL} = 60,000$
$Z_{in} = 2\ M\Omega$
$Z_{out} = 100\ \Omega$ (maximum)
slew rate = 15 v/μs

R_f 120 kΩ

+15 V

−15 V

R_i 3 kΩ

200 mV$_{pk}$

R_L 20 kΩ

Practice Problems

22. Perform the complete analysis of the amplifier in Figure 12.59. [31]
23. Perform the complete analysis of the voltage follower in Figure 12.60. [33]
24. Perform the complete analysis of the voltage follower in Figure 12.61. [33]

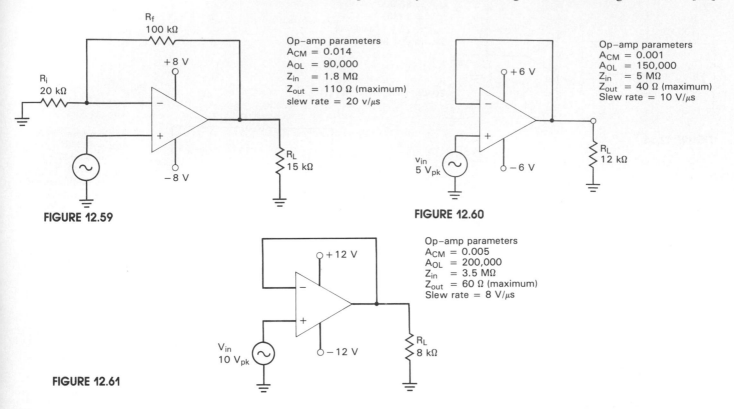

FIGURE 12.59

FIGURE 12.60

FIGURE 12.61

TROUBLESHOOTING PRACTICE PROBLEMS

25. The circuit in Figure 12.62 has the waveforms shown. Discuss the possible cause(s) of the problem.
26. The circuit in Figure 12.63 has the waveforms shown. Determine the cause of the problem. (*Hint*: The measured value of R_f is exactly as shown; that is, R_f is not the problem.)
27. The circuit in Figure 12.64 has the following readings: TP-1 = +10 mV, TP-2 = +20 mV, TP-3 = +120 mV, TP-4 = 0 V, and TP-5 = −240 mV. Determine the possible cause(s) of the problem.

FIGURE 12.62

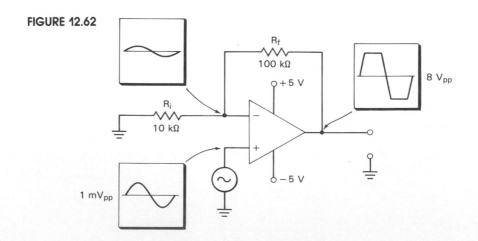

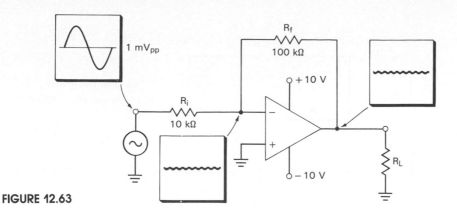

FIGURE 12.63

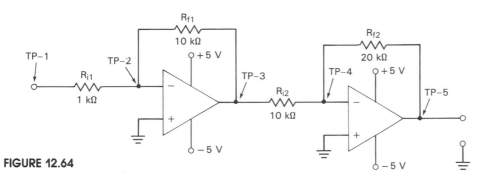

FIGURE 12.64

28. The circuit in Figure 12.64 has the following readings: TP-1 = −5 mV, TP-2 = 0 V, TP-3 = +4 V, TP-4 = 0 V, and TP-5 = −4 V. Determine the possible cause(s) of the problem.

29. The circuit in Figure 12.64 has the following readings: TP-1 = −2 mV, TP-2 = +1.8 mV, TP-3 = 0 V, TP-4 = 0 V, and TP-5 = 0 V. Determine the possible cause(s) of the problem.

30. The circuit in Figure 12.64 has the following readings: TP-1 = −1 V, TP-2 = 0 V, TP-3 = 0 V, TP-4 = 0 V, and TP-5 = −4 V. Determine the possible cause(s) of the problem.

THE BRAIN DRAIN

31. The op-amp in Figure 12.65 has the input and output voltages shown. Determine the CMRR of the device.

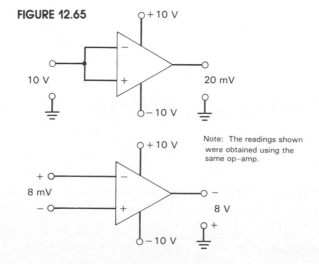

FIGURE 12.65

Note: The readings shown were obtained using the same op-amp.

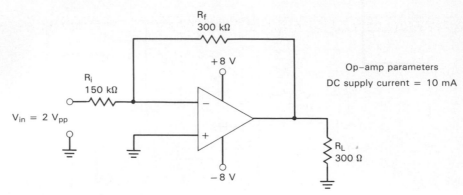

FIGURE 12.66

Op–amp parameters
DC supply current = 10 mA

R_f = 300 kΩ

R_i = 150 kΩ

+8 V

−8 V

V_{in} = 2 V_{pp}

R_L = 300 Ω

FIGURE 12.67

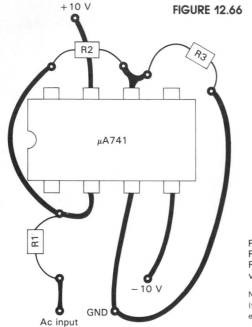

μA741

+10 V

R2

R3

R1

−10 V

GND

Ac input

R1 = 2 kΩ
R2 = 200 kΩ
R3 = 3 kΩ
V_{in} = 80 mV_{pk}

Note: The μA741 spec sheet lists a value of CMRR = 90 dB
(typical). In standard numeric form, 90 dB is approximately
equal to 31.6 K.

32. In Chapter 9, you were shown how to calculate the efficiency of an amplifier. Using those principles, calculate the efficiency of the amplifier in Figure 12.66. (*Hint*: Don't forget to consider the average output current from the op-amp in your current calculations.)

33. Using the μA741, design a noninverting amplifier that will deliver a 25-V_{PP} output to a 20-kΩ load resistance with a 100-mV_{pk} input signal. The available supply voltages are ±18 V_{dc}. Include the 8-pin DIP pin numbers in your schematic diagram.

34. Figure 12.67 is the parts placement diagram for an op-amp circuit. Analyze the circuit and draw the schematic diagram for the circuit. Then perform a complete analysis of the circuit to determine its values of A_{CL}, Z_{in}, Z_{out}, CMRR, and f_{max}.

SUGGESTED COMPUTER APPLICATIONS PROBLEMS

35. Write a program that will determine the maximum allowable input voltage for a given inverting amplifier when provided with the values of R_f, R_i, R_1, $+V$, and $-V$.

36. Write a program like the one described in problem 37 for the noninverting amplifier.

ANSWERS TO THE EXAMPLE PRACTICE PROBLEMS

12.1. 12.8c: positive. 12.8d: negative

12.2. 5.6 V_{pp}

12.3. 157 mV_{pp}

12.4. 128.6 mV_{pp}

12.5. 6.4 kHz

12.6. 31.7 kHz

12.7. $A_{CL} = 250$, $Z_{in} \cong 1\ k\Omega$, $Z_{out} < 50\Omega$, CMRR = 12,500 $f_{max} = 19.1$ kHz

12.8. $A_{CL} = 251$, $Z_{in} \geq 1.5\ M\Omega$, $Z_{out} \leq 50\Omega$, CMRR = 12,550, $f_{max} = 18.99$ kHz

12.9. $A_{CL} = 1$, $Z_{in} = 1.5\ M\Omega$, $Z_{out} = 50\Omega$, CMRR = 50, $f_{max} = 19.9$ kHz

13
Amplifier Frequency Response

Modern production technology is used to produce circuit boards that are far more complex than their predecessors.

OBJECTIVES

After studying the material in this chapter, you should be able to:

☐ 1. Define *bandwidth, cutoff frequency*, and *midpoint frequency* and identify them on a frequency-response curve. (§13.1)

☐ 2. Calculate the bandwidth (BW) and midpoint frequency (f_0) of an amplifier, given its cutoff frequencies. (§13.1)

☐ 3. Explain the relationship between the values of f_1, f_0, and f_2 for a given amplifier. (§13.1)

☐ 4. Calculate the value of f_1 or f_2 for an amplifier, given the midpoint frequency and the other cutoff frequency. (§13.1)

☐ 5. Describe the procedure used to measure the values of f_1 and f_2 with an oscilloscope. (§13.1)

☐ 6. State and define the units of measure that are commonly used on frequency-response curves. (§13.2)

☐ 7. Calculate any two of the following values, given the other two: dB power gain $[A_{p(dB)}]$, power gain ratio (A_p), input power (P_{in}), and output power (P_{out}). (§13.2)

☐ 8. Compare and contrast *positive dB values* with *negative dB values* (§13.2)

☐ 9. List the advantages of using dB gain values. (§13.2)

☐ 10. Explain why amplifier cutoff frequencies are usually referred to as *the upper 3-dB point* and *the lower 3-dB point*. (§13.2)

☐ 11. Define *dBm reference* and discuss the difference between dBm values and standard dB values. (§13.2)

☐ 12. Calculate the output power of an amplifier, given the dBm power gain of the circuit. (§13.2)

☐ 13. Calculate any two of the following values, given the other two: dB voltage gain $[A_{v(dB)}]$, voltage gain ratio (A_v), input voltage (v_{in}), and output voltage (v_{out}). (§13.2)

☐ 14. State the relationship between changes in dB power gain and changes in dB voltage gain. (§13.2)

☐ 15. Discuss the difference between *algebraic scales* and *geometric scales*. (§13.2)

☐ 16. Describe the use of the *decade* and *octave* frequency multipliers. (§13.2)

☐ 17. Compare and contrast the *Bode plot* with the frequency-response curve. (§13.2)

☐ 18. Describe the relationship between the values of R and C in an amplifier terminal circuit and the cutoff frequency of that circuit. (§13.3)

☐ 19. Calculate the values of f_{1B}, f_{1C}, and f_{1E} for a given BJT amplifier. (§13.1)

☐ 20. Define *roll-off rate*. (§13.3)

☐ 21. Calculate the change in dB gain that occurs when operating frequency drops to a specified value. (§13.3)

☐ 22. List the three characteristics of amplifier roll-off rates. (§13.3)

☐ 23. Describe the low-frequency Bode plot for a BJT amplifier. (§13.3)

☐ 24. Compare and contrast the Bode plot and the actual frequency-response curve for the low-frequency operation of a BJT amplifier. (§13.3)

☐ 25. List the differences between f_2 calculations and f_1 calculations for a BJT amplifier. (§13.4)

☐ 26. Calculate the value of C_{be} for a BJT amplifier, given the current–gain bandwidth product, h_{fe}, and h_{ie} values for the device. (§13.4)

☐ 27. Explain why BJT internal capacitances are not considered in low-frequency analyses. (§13.4)

☐ 28. Explain Miller's theorem and the purpose it serves. (§13.4)

☐ 29. Calculate the Miller input and output capacitance values for a BJT amplifier. (§13.4)

☐ 30. Perform the complete high-frequency analysis of a BJT amplifier. (§13.4)

☐ 31. Compare high-frequency roll-off rates to low-frequency roll-off rates. (§13.4)

☐ 32. Compare and contrast frequency-response calculations and measurements. (§13.4)

☐ 33. Perform the low-frequency response analysis of a FET amplifier. (§13.5)

☐ 34. Perform the high-frequency response analysis of a FET amplifier. (§13.5)

☐ 35. List the typical JFET capacitance ratings and convert those ratings into usable terminal capacitance values. (§13.5)

☐ 36. Summarize the relationship between slew rate, peak output voltage, and maximum operating frequency for an op-amp. (§13.6)

☐ 37. Discuss the relationship between operating frequency and voltage gain for an op-amp. (§13.6)

☐ 38. Define *gain–bandwidth product* and explain its significance. (§13.6)

☐ 39. Calculate any one of the following values, given the other two: f_{unity}, A_{CL}, and f_2. (§13.6)

☐ 40. Discuss the means by which the value of f_{unity} may be obtained from a spec sheet and measured. (§13.6)

☐ 41. Discuss and analyze the frequency response of a multistage amplifier. (§13.7)

As you master each of the objectives listed, place a check mark (✔) in the appropriate box.

In this chapter, we are going to take a look at the effects of operating frequency on BJT, FET, and op-amp circuit gain. As we progress through our discussions, you will see that there are several principles that apply to all types of amplifiers. In this section, we will take a look at each of these principles.

Bandwidth

Bandwidth (BW). The range of frequencies over which gain is relatively constant.

Most amplifiers have relatively constant gain over a certain range, or *band*, of frequencies. This band of frequencies is called the *bandwidth* of the amplifier. The bandwidth for a given amplifier depends on the circuit component values and the type of active component(s) used. Later in this chapter, you will be shown how to calculate the bandwidth for a given amplifier.

When an amplifier is operated within its bandwidth, the current, voltage, and power gain values for the amplifier are calculated as shown earlier in the text. For clarity, these values of A_i, A_v, and A_p are referred to as *midband gain* values. For example, the *midband power gain*, $A_{p(mid)}$, of an amplifier is the power gain of the circuit when it is operated within its bandwidth. Again, the exact value of $A_{p(mid)}$ (as well as the other gain values) will be calculated as shown earlier in the text and will vary from one circuit to another.

Frequency-response curve. A curve showing the relationship between amplifier gain and operating frequency.

A graphic representation of the relationship between amplifier gain and operating frequency is the *frequency-response curve*. A *simplified* frequency-response curve is shown in Figure 13.1.

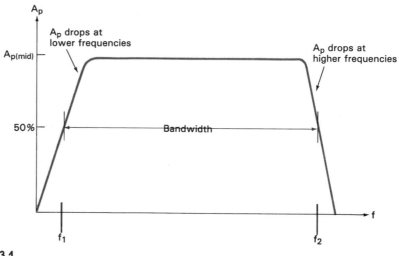

FIGURE 13.1

As the frequency-response curve shows, the power gain of an amplifier will remain relatively constant for a band of frequencies. When the operating frequency starts to go outside of this range, the gain of the amplifier begins to drop off rapidly.

A Practical Consideration. The gain of an amplifier will start to drop off well before the operating frequency reaches f_1 or f_2. However, by the time f_1 or f_2 is reached, the value of A_p will have dropped by approximately 50%.

Two frequencies of interest, f_1 and f_2, are marked on the frequency-response curve. These are the frequencies at which power gain drops to 50% of $A_{p(mid)}$. For example, let's assume that the amplifier represented by Figure 13.1 has the following values: $f_1 = 10$ kHz, $f_2 = 100$ kHz, and $A_{p(mid)} = 500$. As long as the amplifier is operated between f_1 and f_2, the power gain of the amplifier will be approximately 500. If the operating frequency increases to 100 kHz, or drops to 10 kHz, the power gain of the amplifier will drop to 250. If the input frequency goes far enough outside of the f_1 or f_2 limits, the gain of the amplifier will continue to drop until it eventually reaches *unity gain* ($A_p = 1$). At this point, the amplifier no longer serves any useful purpose.

F_1 and f_2 are called the *cutoff frequencies* of an amplifier. These two frequencies are considered to be the limits of the bandwidth of the amplifier. This point is illustrated in Figure 13.1. Since f_1 and f_2 are the bandwidth limits, the bandwidth of an amplifier can be found as the *difference* between f_1 and f_2, as follows:

Cutoff frequencies. The frequencies at which A_p drops to 50% of $A_{p(mid)}$. These frequencies are designated as f_1 and f_2.

$$\boxed{BW = f_2 - f_1} \tag{13.1}$$

where BW = the amplifier bandwidth, in hertz
f_2 = the *upper* cutoff frequency
f_1 = the *lower* cutoff frequency

The following chart shows some examples of f_1, f_2, and BW value combinations.

f_1 (kHz)	f_2 (kHz)	BW (kHz)
10	100	90
25	250	225
0 (dc)	400	400

Note that the bandwidth in each case is found as the *difference* between f_1 and f_2. As with any gain value, the specific values of f_1 and f_2 will vary from one amplifier to another. However, every amplifier has upper and lower cutoff frequencies, and these frequencies are used to determine the bandwidth of the amplifier. This point is illustrated further in the following example.

EXAMPLE 13.1

2

The frequency-response curve in Figure 13.2 represents the operation of a given amplifier. Determine the bandwidth of the circuit.

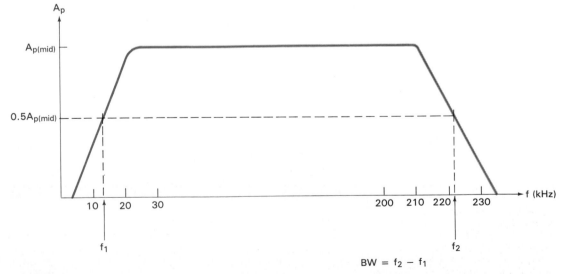

FIGURE 13.2

Solution: The cutoff frequencies correspond to the *half-power points* on the curve, as is shown. For this particular circuit, these values are approximately

$$f_1 = 13 \text{ kHz}$$

and

$$f_2 = 222 \text{ kHz}$$

The bandwidth is found as

$$\begin{aligned} BW &= f_2 - f_1 \\ &= 222 \text{ kHz} - 13 \text{ kHz} \\ &= 209 \text{ kHz} \end{aligned}$$

PRACTICE PROBLEM 13–1

Determine the bandwidth for the amplifier represented by the frequency-response curve in Figure 13.3.

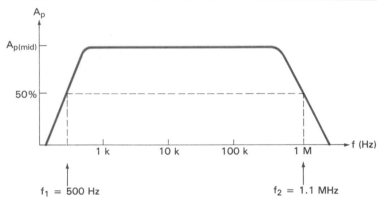

FIGURE 13.3

Center frequency (f_O). The frequency that equals the geometric average of f_1 and f_2.

A_p is at its maximum value when an amplifier is operated at f_O.

There are cases where it is desirable to know the *center frequency* of a given amplifier. The center frequency (f_O) is the frequency that equals *the geometric average of the cutoff frequencies*. By formula,

$$\boxed{f_O = \sqrt{f_1 f_2}} \tag{13.2}$$

When the operating frequency of an amplifier is equal to the value of f_O, the power gain of the amplifier will be at its maximum value. Then, as the operating frequency varies toward f_1 or f_2, the power gain begins to slowly drop off. By the time the operating frequency reaches either of the cutoff frequencies, power gain will have dropped to 50% of $A_{p(\text{mid})}$. The following example illustrates the calculation of f_O for an amplifier.

EXAMPLE 13.2

Determine the value of f_O for the circuit in Figure 13.2.

Solution: The values of f_1 and f_2 were found to be 13 kHz and 222 kHz, respectively. Using these two values in equation

(13.2), the value of the center frequency is found as

$$f_O = \sqrt{f_1 f_2}$$
$$= \sqrt{(13 \text{ kHz})(222 \text{ kHz})}$$
$$= 53.72 \text{ kHz}$$

PRACTICE PROBLEM 13–2

Determine the value of f_O for the amplifier represented in Figure 13.3.

Since f_O is the *geometric average* of the cutoff frequencies, the ratio of f_O to f_1 will equal the ratio of f_2 to f_O. By formula,

$$\boxed{\frac{f_O}{f_1} = \frac{f_2}{f_O}}$$ (13.3)

The following example illustrates the fact that these ratios are equal.

EXAMPLE 13.3

Show that the ratio of f_O to f_1 is equal to the ratio of f_2 to f_O for the amplifier described in Examples 13.1 and 13.2.

Solution: The following values were obtained in the two examples:

$$f_1 = 13 \text{ kHz}$$
$$f_2 = 222 \text{ kHz}$$
$$f_O = 53.72 \text{ kHz}$$

The ratio of f_O to f_1 is found as

$$\frac{f_O}{f_1} = \frac{53.72 \text{ kHz}}{13 \text{ kHz}}$$
$$= 4.13$$

The ratio of f_2 to f_O is found as

$$\frac{f_2}{f_O} = \frac{222 \text{ kHz}}{53.72 \text{ kHz}}$$
$$= 4.13$$

As you can see, the frequency ratios are equal.

PRACTICE PROBLEM 13–3

Show that equation (13.3) holds true for the circuit described in Practice Problems 13–1 and 13–2

Equation (13.3) is important because it provides us with some insight into the relationship between f_O, f_1, and f_2. When an amplifier is operated at its center frequency, the input frequency can vary by the same *factor* in each direction before the power gain of the amplifier drops to 50% of $A_{p(mid)}$. For example, let's say that an amplifier has the following values:

$$f_1 = 6 \text{ kHz}$$

$$f_2 = 150 \text{ kHz}$$

$$f_O = 30 \text{ kHz}$$

As you can see, f_O is 5 times the value of f_1. At the same time, f_2 is 5 times the value of f_O. Thus, if the amplifier is operated at f_O, the input frequency can decrease by a factor of 5 before hitting the lower cutoff frequency (f_1), or can increase by a factor of 5 before hitting the upper cutoff frequency (f_2). This is where the term *center frequency* comes from.

Equation (13.3) can be transposed to provide us with two other useful equations. Each of these equations can be used to predict the value of f_1 or f_2 when the other cutoff frequency and the center frequency are known, as follows:

$$\boxed{f_1 = \frac{f_O^2}{f_2}} \qquad \textbf{(13.4)}$$

and

$$\boxed{f_2 = \frac{f_O^2}{f_1}} \qquad \textbf{(13.5)}$$

Once the unknown cutoff freuqency has been determined, we can predict the bandwidth of the amplifier, as the next two examples will illustrate.

■ EXAMPLE 13.4 ■

The values of f_O and f_2 for an amplifier are measured at 60 kHz and 300 kHz, respectively. Determine the values of f_1 and BW for the amplifier.

Solution: The value of f_1 is found using equation (13.4) as follows:

$$\begin{aligned}
f_1 &= \frac{f_O^2}{f_2} \\
&= \frac{(60 \text{ kHz})^2}{300 \text{ kHz}} \\
&= 12 \text{ kHz}
\end{aligned}$$

The value of BW is now found as

$$\begin{aligned}
BW &= f_2 - f_1 \\
&= 300 \text{ kHz} - 12 \text{ kHz} \\
&= 288 \text{ kHz}
\end{aligned}$$

PRACTICE PROBLEM 13–4

An amplifier has values of $f_2 = 300$ kHz and $f_O = 30$ kHz. Determine the bandwidth of the amplifier.

CHAP. 13 Amplifier Frequency Response

The values of f_O and f_1 for an amplifier are measured at 40 kHz and 8 kHz, respectively. Determine the values of f_2 and BW for the amplifier.

Solution: The value of f_2 is found using equation (13.5), as follows:

$$f_2 = \frac{f_O^2}{f_1}$$
$$= \frac{(40 \text{ kHz})^2}{8 \text{ kHz}}$$
$$= 200 \text{ kHz}$$

The value of BW is now found as

$$BW = f_2 - f_1$$
$$= 200 \text{ kHz} - 8 \text{ kHz}$$
$$= 192 \text{ kHz}$$

PRACTICE PROBLEM 13–5

An amplifier has values of $f_O = 25$ kHz and $f_1 = 5$ kHz. Determine the bandwidth of the amplifier.

You have been shown the relationship between the cutoff frequencies (f_1 and f_2), the center frequency (f_O), and the bandwidth (BW) of an amplifier. In later sections, you will be shown the circuit analysis equations that are used to determine all these values. As you will see, the equations used for BJT, FET, and op-amp circuits will vary. However, the relationships between the various frequencies and amplifier bandwidth values will not vary significantly from the relationships covered in this section.

Measuring f_1 and f_2

When you are working with an amplifier, the f_1 and f_2 frequencies for the amplifier can easily be measured using an oscilloscope. The procedure used is as follows:

1. Set up the amplifier for the maximum undistorted output signal. This is done by varying the *amplitude* of v_{in} to the amplifier. As a first step, set the input frequency to approximately 100 kHz.
2. Establish that you are operating in the midband of the amplifier by varying the *frequency* of the input several kilohertz in both directions. If you are in the midband range of the amplifier, the *amplitude* of the amplifier output will *not* change significantly as you vary the input frequency.
3. If you are not in the midband frequency range, adjust f_{in} until you are.
4. To determine the value of f_1, decrease the input frequency until the peak value of v_{out} drops to 70.7% of its midband value. For example, if the peak value of v_{out} is normally $10V_{pk}$, decrease the input frequency until v_{out} drops to 7.07 V_{pk}. Measure f_1 at the frequency where this drop occurs.

A Practical Consideration. Most ac voltmeters do not have the frequency handling capability needed to measure bandwidth. For this reason, the method covered here emphasizes the use of the oscilloscope.

When using the oscilloscope to measure bandwidth, you should use a ×10 probe to reduce the effects of the scope's input capacitance on your measurements.

5. Increase the frequency until the same thing happens on the high end of operation. Measure f_2 at this frequency.

Where did the value of 0.707 V_{pk} come from? In basic electronics, you learned that *power varies with the square of voltage*. The same principle holds true for A_p and A_v. Thus, when A_p drops to 0.5 of $A_{p(mid)}$, the voltage gain will drop by the *square root* of the *0.5* factor, 0.707. As you will see, the fact that A_v equals $0.707A_{v(mid)}$ at the cutoff frequencies is important when deriving the equations for f_1 and f_2.

SECTION REVIEW

1. What is *bandwith*? [Objective 1]
2. Draw a frequency-response curve and identify the following: f_1, f_2, bandwidth, and midpoint frequency. [Objective 1]
3. What is the numeric relationship between the values of f_1, f_O, and f_2 for a given amplifier? [Objective 3]
4. What is the procedure for measuring f_1 and f_2 with an oscilloscope? [Objective 5]
5. Which oscilloscope probe should be used for high-frequency measurements? Why? [Objective 5]

13.2
Gain and Frequency Measurements

Up to this point, we have used *simplified* frequency-response curves in the various illustrations. A more practical frequency-response curve is shown in Figure 13.4.

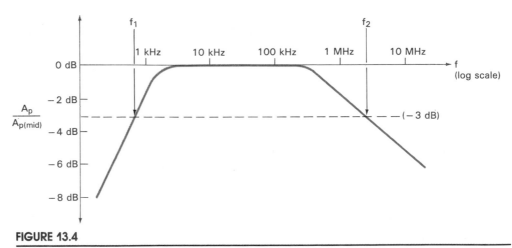

FIGURE 13.4

A more practical frequency response curve.

As you can see, this curve is a bit more complicated than any you have seen thus far. The primary differences are:

1. The *y*-axis of the graph represents a power gain *ratio* rather than specific power gain values.
2. The power gain ratios are measured in *decibels* (dB).

3. The frequency units are based on a *logarithmic* scale. That is, each major division is a whole number multiple of the previous major division.

We are going to look at all three of these differences so that you will have no problem with reading an actual frequency-response curve.

Decibel Power Gain

A decibel (dB) is a ratio of output power to input power, equal to 10 times the common logarithm of that ratio. One of the advantages of using decibels is that they allow us to represent very large power ratios as relatively small numbers. For example a power gain of 1,000,000 would equal 60 dB. In this section, you will be shown how to convert back and forth between standard numeric form and dB form. You will also be shown the advantages and limitations of using dB gain ratios.

Decibel (dB). A ratio of output power to input power, equal to 10 times the common logarithm of that ratio.

Frequency response curves and specification sheets often list gain values that are measured in decibels. The decibel power gain of an amplifier is found as

$$A_{p(\text{dB})} = 10 \log A_p \qquad (13.6)$$

or

$$A_{p(\text{dB})} = 10 \log \frac{P_{\text{out}}}{P_{\text{in}}} \qquad (13.7)$$

The following example illustrates the process for finding the power gain of an amplifier in decibels.

EXAMPLE 13.6

A given amplifier has values of $P_{\text{in}} = 100 \ \mu\text{W}$ and $P_{\text{out}} = 2 \ \text{W}$. What is the dB power gain of the amplifier?

Solution: The dB power gain of the amplifier is found as

$$A_{p(\text{dB})} = 10 \log \frac{P_{\text{out}}}{P_{\text{in}}}$$
$$= 10 \log \frac{2 \ \text{W}}{100 \ \mu\text{W}}$$
$$= 10 \ \log (20{,}000)$$
$$= 10(4.301)$$
$$= 43.01 \ \text{dB}$$

PRACTICE PROBLEM 13–6

An amplifier has values of $P_{\text{in}} = 420 \ \mu\text{W}$ and $P_{\text{out}} = 6 \ \text{W}$. What is the dB power gain of the amplifier?

If you want to convert a dB gain value to standard numerical value, use the following equation:

$$A_p = \log^{-1} \frac{A_{p(dB)}}{10}$$

(13.8)

where $\log^{-1}$ is the *inverse log* of the fraction. The following example illustrates the use of this equation.

EXAMPLE 13.7

A given amplifier has a power gain of 3 dB. What is the ratio of output power to input power?

Solution: The ratio of output power to input power is found as

$$A_p = \log^{-1} \frac{A_{p(dB)}}{10}$$
$$= \log^{-1} \frac{3\ dB}{10}$$
$$= \log^{-1}(0.3)$$
$$= 1.995$$

Thus a gain of 3 dB represents a ratio of output power to input power that is approximately equal to 2.

PRACTICE PROBLEM 13–7

An amplifier has a dB power gain of 20 dB. What is the ratio of output power to input power for the circuit?

Two other useful relationships are based on equation (13.7). These are

$$P_{out} = P_{in} \left(\log^{-1} \frac{A_{p(dB)}}{10} \right)$$

(13.9)

and

$$P_{in} = \frac{P_{out}}{\log^{-1} (A_{p(dB)}/10)}$$

(13.10)

The following examples help to illustrate the use of these two equations.

EXAMPLE 13.8

An amplifier with a gain of 3 dB has an input power of 50 mW. What is the output power of the amplifier?

Solution: Using equation (13.9), the output power is found as

$$P_{out} = P_{in}\left(\log^{-1}\frac{A_{p(dB)}}{10}\right)$$

$$= (50 \text{ mW})\left(\log^{-1}\frac{(3 \text{ dB})}{10}\right)$$

$$= (50 \text{ mW})[\log^{-1}(0.3)]$$

$$= (50 \text{ mW})(2)$$

$$= 100 \text{ mW}$$

PRACTICE PROBLEM 13–8

An amplifier with a power gain of 5.2 dB has an input power of 40 μW. Determine the output power of the circuit.

Note that the inverse log of 0.3 was rounded off to 2 in Example 13.8. As a standard, the value of *3dB* is rounded off to a gain value of 2. Now, let's reverse the problem.

EXAMPLE 13.9

A circuit has a power gain of −3 dB. If the output power is 50 mW, what is the input power?

Solution: Using equation (13.10), the input power of the circuit is found as

$$P_{in} = \frac{P_{out}}{\log^{-1}(A_{p(dB)}/10)}$$

$$= \frac{50 \text{ mW}}{\log^{-1}(-3 \text{ dB}/10)}$$

$$= \frac{50 \text{ mW}}{\log^{-1}(-0.3)}$$

$$= \frac{50 \text{ mW}}{0.5}$$

$$= 100 \text{ mW}$$

PRACTICE PROBLEM 13–9

An amplifier with a power gain of 12.4 dB has an output power of 2.2 W. Determine the input power to the circuit.

If we take a moment to compare the results of Examples 13.8 and 13.9, we can come to some very important conclusions regarding the use of dB values.

The results of the examples are as follows:

Example	dB power gain (dB)	P_{in} (mW)	P_{out} (mW)
13.8	3	50	100
13.9	−3	100	50

The conclusions we can draw from these results are as follows:

Positive versus negative decibel values.

1. *Positive dB values represent a power gain, while negative dB values actually represent a power loss.* When the gain was +3 dB, the output power was greater than the input power. When the gain was −3 dB, the output power was actually less than the input power.
2. *Positive and negative decibels of equal value represent gains/losses of equal value.* A positive 3-dB gain caused power to double while a negative 3-dB loss caused power to be cut in half.

The relationship described in conclusion 2 can be stated mathematically as follows:

- If x dB represents a power gain of y, then $-x$ dB represents a power loss equal to $1/y$.

The following table lists some examples of this relationship.

(+)dB value	Gain	(−)dB value	Loss
3	2	− 3	½
6	4	− 6	¼
12	16	−12	1/16
20	100	−20	1/100

The fact that equivalent (+) and (−) dB values represent gains and losses that are reciprocals is one of the reasons that decibels are commonly used. Granted, they are a bit intimidating when you first start to deal with them. However, once you get used to seeing gains represented as dB values, you will have no problem using them.

Another advantage of using dB gain values is as follows: In multistage amplifiers, the total gain of the circuit is equal to the *sum* of the individual amplifier dB gains. This point is illustrated in the following example.

EXAMPLE 13.10

For the multistage amplifier represented in Figure 13.5, prove that the total power gain is equal to $A_{p1(dB)} + A_{p2(dB)} + A_{p3(dB)}$.

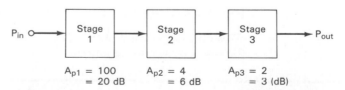

FIGURE 13.5

Solution: The gain of each stage is shown as a ratio value and as a dB value. As you recall, the total power gain of a

CHAP. 13 Amplifier Frequency Response

multistage amplifier is equal to the product of the individual power ratio values. By formula

$$A_{p(\text{total})} = (A_{p1})(A_{p2})(A_{p3})$$
$$= (100)(4)(2) \quad \text{standard numeric form}$$
$$= 800$$

If we convert the ratio value of (800) to decibels, we get the following:

$$A_{p(\text{dB})} = 10 \log A_p$$
$$= 10 \log 800$$
$$= 29 \text{ dB}$$

Now, if we simply add the dB gains of the three stages, we get the following:

$$A_{p(\text{total})} = A_{p1(\text{dB})} + A_{p2(\text{dB})} + A_{p3(\text{dB})}$$
$$= 20 \text{ dB} + 6 \text{ dB} + 3\text{dB}$$
$$= 29 \text{ dB}$$

Since both methods of determining total dB power gain yielded the same results, we have proven that you can determine total multistage gain by simply adding the individual dB power gain values.

PRACTICE PROBLEM 13–10

A three-stage amplifier has gain values of 100, 86, and 45. Show that the dB gain values can be added to determine the total amplifier gain.

Now, let's relate what you have learned about dB gain values to the frequency-response curve in Figure 13.4. Note that the y-axis is labeled as being the *ratio* of power gain at a given frequency to midband power gain. When the circuit being represented is operated in the midband range,

$$A_p = A_{p(\text{mid})}$$

and

$$\frac{A_p}{A_{p(\text{mid})}} = 1$$

If we convert a power ratio of 1 to decibels, we get the following:

$$A_{p(\text{dB})} = 10 \log 1$$
$$= 10(0)$$
$$= 0 \text{ dB}$$

Thus the midband power ratio shown is labeled as 0 dB. This means that the power gain of the amplifier is equal to $A_{p(\text{mid})}$ throughout the midband frequency range. This is what we have been saying all along.

When the value of $A_p/A_{p(\text{mid})}$ is equal to -3 dB, the output power of the amplifier has dropped to 50% of $A_{p(\text{mid})}$. This is proven as follows:

$$\frac{A_p}{A_{p(\text{mid})}} = -3 \text{ dB}$$

Therefore,

$$A_p = (-3 \text{ dB})A_{p(\text{mid})}$$

Now, since $-3 \text{ dB} = \frac{1}{2}$, or 0.5,

$$A_p = (0.5)A_{p(\text{mid})}$$

Based on this relationship, we can redefine f_1 and f_2 (the cutoff frequencies) as follows: *The cutoff frequencies of an amplifier are those frequencies at which the power gain of the amplifier drops by 3 dB.* For this reason the cutoff frequencies of an amplifier are often called the *upper* and *lower 3-dB points*.

The dBm Reference

dBm Reference. A dB value that references output power to 1 mW.

On some specification sheets, you will see power values that are listed as *dBm* values. For example, a stereo may be listed as having a maximum output power of 50 dBm. This rating tells you that the output power from the stereo is 50 dB *above 1 mW*. Power, measured in dBm, is found as

$$\boxed{P_{\text{dBm}} = 10 \log \frac{P}{1 \text{ mW}}} \tag{13.11}$$

Note that dBm values represent *actual power values*. In contrast, dB values represent *power ratios*. This point is illustrated by the following two examples.

EXAMPLE 13.11

An amplifier has a gain of 50 dB. Determine the output power of the amplifier.

Solution: The problem cannot be solved as given. The value of 50 dB represents the *ratio* of output power to input power. Since no input power value is given, we cannot determine the value of output power.

EXAMPLE 13.12

An amplifier has an output power of 50 dBm. What is the actual output power in watts?

Solution: Since the power value is given in dBm, we can determine the actual output power as

$$P = \left(\log^{-1} \frac{P_{\text{dBm}}}{10} \right) (1 \text{ mW})$$
$$= \left(\log^{-1} \frac{50 \text{ dBm}}{10} \right) (1 \text{ mW})$$
$$= (\log^{-1} 5)(1 \text{ mW})$$
$$= (100{,}000)(1 \text{ mW})$$
$$= 100 \text{ W}$$

An amplifier has an output power of 32 dBm. Determine the actual output power of the circuit in Watts.

As you can see, dBm values can be used to determine actual power values, while dB values simply indicate the *ratio* of output power to input power.

dB Voltage Gain

13

The dB voltage gain of an amplifier is found as

$$A_{v(\text{dB})} = 20 \log A_v \qquad \text{(13.12)}$$

or

$$A_{v(\text{dB})} = 20 \log \frac{v_{\text{out}}}{v_{\text{in}}} \qquad \text{(13.13)}$$

The following example illustrates the process for finding dB voltage gain.

EXAMPLE 13.13

An amplifier has values of $v_{\text{in}} = 10$ V and $v_{\text{out}} = 14.14$ V. Determine the dB voltage gain of the amplifier.

Solution: Using equation (13.13), the dB voltage gain of the amplifier is found as

$$
\begin{aligned}
A_{v(\text{dB})} &= 20 \log \frac{v_{\text{out}}}{v_{\text{in}}} \\
&= 20 \log \frac{14.14 \text{ V}}{10 \text{ V}} \\
&= 20 \log 1.414 \\
&= 20(0.15) \\
&= 3 \text{ dB}
\end{aligned}
$$

PRACTICE PROBLEM 13–13

An amplifier with a 25-mV input has a 7.8-V output. Determine the dB voltage gain of the amplifier.

When you want to convert a dB voltage gain value to a standard voltage ratio, use the following equation:

$$A_v = \log^{-1} \frac{A_{v(\text{dB})}}{20} \qquad \text{(13.14)}$$

The following example illustrates the use of this equation.

EXAMPLE 13.14

A given amplifier has a dB voltage gain of 6 dB. What is the ratio of output voltage to input voltage for the amplifier?

Solution: Using equation (13.14), the ratio of output voltage to input voltage is found as

$$A_v = \log^{-1} \frac{A_{v(dB)}}{20}$$
$$= \log^{-1} \frac{6\ dB}{20}$$
$$= \log^{-1}(0.3)$$
$$= 2 \quad \text{(again, rounded off)}$$

PRACTICE PROBLEM 13–14

An amplifier has a 12-dB voltage gain. What is the ratio of output voltage to input voltage for the circuit?

Just as with power dB values, negative voltage dB values indicate a voltage *loss*. This point is illustrated in the following example.

EXAMPLE 13.15

A given amplifier has a dB voltage gain of -6 dB. What is the ratio of output voltage to input voltage?

Solution: Using equation (13.14), the ratio of output voltage to input voltage for the amplifier is found as

$$A_v = \log^{-1} \frac{A_{v(dB)}}{20}$$
$$= \log^{-1} \frac{-6\ dB}{20}$$
$$= \log^{-1}(-0.3)$$
$$= 0.5$$

Thus, with a gain of -6 dB, the output voltage is approximately one-half of the input voltage.

PRACTICE PROBLEM 13–15

A given amplifier has a voltage gain of 4 dB. What is the ratio of output voltage to input voltage?

Do you see a pattern here? All the rules for dB power gains apply to dB voltage gains as well. The only difference is that dB voltage gain values are found using a multiplier of 20 instead of 10.

The difference between voltage and power *dB* calculations.

Changes in dB Gain

When the power gain of an amplifier changes by a given number of decibels, the voltage gain of the amplifier will change by the same number of decibels. By formula,

$$\Delta A_{p(\text{dB})} = \Delta A_{v(\text{dB})}$$ (13.15)

For example, if the power gain of an amplifier changes by -6 dB, the voltage gain will also change by -6 dB. The values of A_p and A_v may or may not be equal, but they will *change by* the same number of decibels.

Now, consider the case of the half-power points (the cutoff frequencies). As you know, the power gain of an amplifier will change by -3 dB when it reaches either of these points. This means that the voltage gain (in decibels) must also change by -3 dB by the time the half-power points are reached. From this information, we can derive a voltage multiplier for the half-power points, as follows:

$$A_v = \log^{-1}\left(\frac{-3 \text{ dB}}{20}\right)$$
$$= \log^{-1}(-0.15)$$
$$= 0.707$$

This means that A_v will be approximately equal to $0.707A_{v(\text{mid})}$ at the half-power points. Because of this, the peak output voltage from an amplifier will be approximately equal to $0.707 \text{ V}_{\text{pk}}$ at the half-power points. This verifies the values used in our discussion on measuring the f_1 and f_2 frequencies in Section 13.1.

Be sure you understand that the values of dB voltage and power gain for a given amplifier are not necessarily equal. However, if one gain value *changes* by a given number of decibels, the other will change by the same amount. This point is illustrated in the following example.

EXAMPLE 13.16

A given amplifier has values of $A_{p(\text{mid})} = 22$ dB and $A_{v(\text{mid})} = 14$ dB. What are the values of A_p and A_v at the half-power points?

Solution: At the half-power points, the power gain decreases by 3 dB. Thus the value of A_p is found as

$$A_p = A_{p(\text{mid})} - 3 \text{ dB}$$
$$= 22 \text{ dB} - 3 \text{ dB}$$
$$= 19 \text{ dB}$$

The dB voltage gain will change by the same amount. Therefore, at the half-power points,

$$A_v = A_{v(\text{mid})} - 3 \text{ dB}$$
$$= 14 \text{ dB} - 3 \text{ dB}$$
$$= 11 \text{ dB}$$

One Final Note on Decibels

Experience has shown that most people take a while to get used to working with dB gain values. Although the whole concept of decibels may seem a bit strange to you now, rest assured that everything will "click" for you if you stick with it. In the meantime, the following list of summary points on decibels will help. Whenever you are not sure about the meaning of a given set of dB values, refer back to this list:

1. Decibels are *logarithmic* representations of voltage and power gain ratios.
2. Decibel power gain is found as $10 \log A_p$.
3. Decibel voltage gain is found as $20 \log A_v$.
4. When A_p changes by a given number of decibels, A_v will change by the same number of decibels.
5. You *cannot* use dB voltage and power gain values as multipliers. For example, if you want to determine v_{out}, given v_{in} and the dB voltage gain, you must convert the dB voltage gain to standard numerical form before multiplying to find v_{out}.
6. For multistage amplifiers, the dB gain values of each stage are added to find the total dB gain of the circuit.

There is an extensive number of practice problems involving decibels at the end of this chapter.

Frequency Units and Terminology

Frequency-response curves use logarithmic scales.

The frequency-response curve in Figure 13.4 utilizes a *logarithmic scale* to measure frequency values. A logarithmic scale is one that uses a *geometric* progression of units. For example, take a look at Figure 13.6. Figure 13.6 shows the two types of scales that are used for measurement. Figure 13.6a uses an *algebraic* progression of values. That is, the *difference* between the values of the major divisions is fixed. For Figure 13.6a, this difference (*d*) is 100. Thus from any division to the next represents an increase of 100.

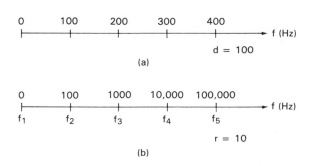

FIGURE 13.6

Algebraic and geometric frequency sales.

Figure 13.6b uses a *geometric* progression of values. That is, the value of each division is determined by multiplying the previous division value by a multiplier *rate* (*r*). For Figure 13.6b, $r = 10$. This means that each division has a value that is 10 times the previous division value. It also means that the *difference* between division values changes from one point to another. This is shown in the following table.

From:	To:	Difference
f_1	f_2	100 Hz
f_2	f_3	1000 Hz − 100 Hz = 900 Hz
f_3	f_4	10,000 Hz − 1000 Hz = 9000 Hz
f_x	f_y	$f_y − f_x$

The advantage in using a geometric scale is that you can represent a wide range of values with a relatively short scale. For example, we could represent the entire megahertz frequency range simply by adding four more divisions onto the line in Figure 6.6b. To do the same with the algebraic scale in Figure 13.6a, we would have to continue the line for several miles. For this reason, most frequency-response curves use logarithmic frequency scales.

When the multiplier rate for a frequency scale is 10, the scale is called a *decade scale*. Note that a *decade* is an increase *multiplier* equal to 10. Thus, each frequency division in Figure 13.6b represents a *decade*.

Another term that is usually used to express the relationship between two frequency values is *octave*. When you increase frequency by one octave, you *double* the original frequency. The following table shows a series of *fundamental* (starting) frequencies and the decade and octave frequencies for each.

Fundamental	Octave	Decade
100 Hz	200 Hz	1000 Hz
500 Hz	1000 Hz	5000 Hz
1.5 MHz	3 MHz	15 MHz
f_x	$2f_x$	$10f_x$

The terms *octave* and *decade* are used extensively in describing the rate at which gain varies with frequencies. For example, an amplifier may be said to have *a roll-off of 20 dB/decade*. This means that the gain of the amplifier decreases at a rate to 20 dB for each decade that the frequency goes beyond the f_1 and f_2 limits. Thus, if f_2 for the amplifier is 10 kHz, the gain of the amplifier will drop by 20 dB at 100 kHz, and by another 20 dB by 1 MHz, and by another 20 dB by 10 MHz, and so on. This point will be discussed further later in this chapter.

Bode Plots

The Bode plot (pronounced "bō-dē" plot) is a variation on the basic frequency-response curve. A Bode plot is shown in Figure 13.7. As you can see, the Bode plot differs from the frequency-response curves you have seen thus far in one respect: The $\Delta A_{p(mid)}$ is assumed to be zero until the cutoff frequencies are reached. Then the power gain of the amplifier is assumed to drop at a set rate, 20 dB/

FIGURE 13.7

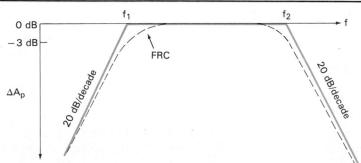

decade. For comparison, the standard frequency-response curve (labeled FRC) is included in the figure. Note that the power gain of the amplifier is shown to be down by 3 dB at each of the cutoff frequencies on the FRC, while the power gain is still shown to be at its midband value on the Bode plot.

Bode plots are typically used in place of other types of frequency-response curves because they are easier and faster to draw. Since Bode plots are the most commonly used frequency-response curves, we will use them throughout the remainder of this chapter. Just remember, while the Bode plot shows $\Delta A_{p(\text{dB})}$ to be zero at the cutoff frequencies, the actual value is -3 dB.

At this point, we have looked at the overall relationship between gain and frequency. We have also discussed the basic units used to measure gain and frequency. Now we are going to move on to specific types of circuits, their respective calculations, and the reasons for their response characteristics.

SECTION REVIEW

1. In what form are gain values typically measured? [Objective 6]
2. What type of frequency scale is normally used in frequency-response curves? [Objective 6]
3. In what way are positive and negative dB values similar? In what way are they different? [Objective 8]
4. What are the advantages of using dB gain values? [Objective 9]
5. Why are the cutoff frequencies referred to as the *upper* and *lower 3-dB points*? [Objective 10]
6. What is the *dBm reference*? [Objective 11]
7. How do dBm values differ from standard dB values? [Objective 11]
8. When dB power gain drops by a given value, what happens to dB voltage gain? [Objective 14]
9. What is the difference between an *algebraic* frequency scale and a *geometric* frequency scale? [Objective 15]
10. Describe the use of the *decade* and *octave* frequency multipliers. [Objective 16]
11. Compare and contrast the *Bode plot* with a frequency-response curve. [Objective 17]

13.3
BJT Amplifier Low-frequency Response

The cutoff frequencies of an amplifier are determined by the combination of amplifier capacitance and resistance values. In this section, you will be shown how the capacitance and resistance values in a BJT amplifier determine the low-frequency response characteristics of the circuit.

Low-frequency Response

We'll start our discussion on BJT amplifier low-frequency response by taking a look at the amplifier base circuit. For reference, this circuit is shown in Figure

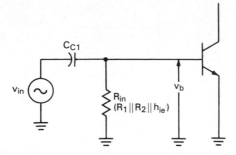

FIGURE 13.8

13.8. As the figure shows, the voltage that is applied to the base of the transistor is equal to the voltage across R_{in}. Since R_{in} is in series with the coupling capacitor, the two components form a voltage divider. For this divider circuit, the voltage across R_{in} is found as

$$v_b = \frac{R_{in}}{\sqrt{X_C^2 + R_{in}^2}} \qquad \textbf{(13.16)}$$

Now, let's take a look at what happens to the fraction in equation (13.16) when $X_C = R_{in}$. When this happens, the fraction simplifies as follows:

$$
\begin{aligned}
\frac{R_{in}}{\sqrt{X_C^2 + R_{in}^2}} &= \frac{R_{in}}{\sqrt{R_{in}^2 + R_{in}^2}} \\
&= \frac{R_{in}}{\sqrt{2R_{in}^2}} \\
&= \frac{R_{in}}{\sqrt{2}\,R_{in}} \\
&= \frac{1}{\sqrt{2}} \\
&= 0.707 \quad \text{(approximately)}
\end{aligned}
$$

If we replace the fraction in equation (13.16) with the 0.707 value obtained, we get the following relationship:

$$v_b = 0.707 v_{in} \quad | \quad X_C = R_{in} \qquad \textbf{(3.17)}$$

Thus, for the circuit in Figure 13.8, v_b will drop to $0.707 v_{in}$ when $X_C = R_{in}$. If this happens, the gain of the amplifier is *effectively* reduced to $0.707 A_{v(mid)}$. As you remember, $A_v = 0.707 A_{v(mid)}$ at the half-power points. Thus the lower half-power point (f_1) for Figure 13.8 will occur at the frequency that causes X_C to equal R_{in}.

At what frequency does this occur? If we swap the values of f and X_C in the basic capacitive reactance formula, we obtain the following equation:

$$f = \frac{1}{2\pi X_C C} \qquad \textbf{(13.18)}$$

Since X_C is equal to the total ac resistance in the base circuit (R) at f_1, we can rewrite equation (13.8) as

$$f_{1B} = \frac{1}{2\pi RC}$$

(13.19)

where f_{1B} = the lower cutoff frequency of the base circuit
R = the *total* ac resistance in the base circuit
C = the value of the base coupling capacitor

In equation (13.19), R is shown as being the *total* ac resistance in the base circuit. For Figure 13.8, the total ac resistance was shown to be R_{in}. However, if you want a more accurate value of f_{1B}, you must take the resistance of the source into

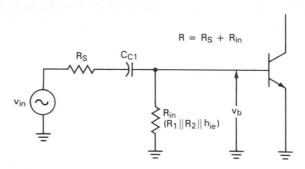

FIGURE 13.9

account. This resistance is shown in Figure 13.9. For the circuit shown, the half-power point will occur at the frequency where X_C is equal to the *series* total of R_S and R_{in}. By formula,

$$f_{1B} = \frac{1}{2\pi(R_S + R_{in})C_{C1}}$$

(13.20)

where R_S = the resistance of the voltage source
$R_{in} = R_1 \| R_2 \| h_{ie}$

19

This is the equation that is used to determine the lower cutoff frequency of the base circuit, as the following example illustrates.

EXAMPLE 13.17

Determine the value of f_{1B} for the circuit shown in Figure 13.10.

Solution: First, the value of R_{in} is found as

$$R_{in} = R_1 \| R_2 \| h_{ie}$$
$$= 18 \text{ k}\Omega \| 4.7 \text{ k}\Omega \| 4.4 \text{ k}\Omega$$
$$= 2018 \ \Omega$$

Now, using this value in equation (13.20), we obtain the

Do you remember how to determine the values of h_{ie} and h_{fe} for a BJT? You can review the procedure by referring back to page 300.

CHAP. 13 Amplifier Frequency Response

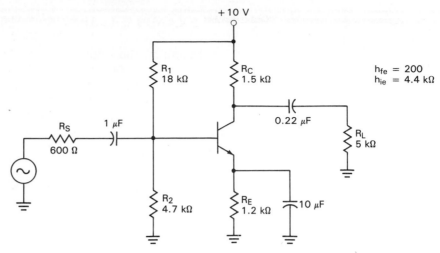

FIGURE 13.10

value of f_{1B} as follows:

$$f_{1B} = \frac{1}{2\pi(R_S + R_{in})C}$$

$$= \frac{1}{2\pi(2618\ \Omega)(1\ \mu F)}$$

$$= \frac{1}{16.45 \times 10^{-3}}$$

$$= 60.79\ \text{Hz}$$

Thus, at 60.79 Hz, the base circuit of Figure 13.15 would cause the power gain of the amplifier to drop by 3 dB.

PRACTICE PROBLEM 13–17

A BJT capacitor has values of $R_1 = 56\ \text{k}\Omega$, $R_2 = 5.6\ \text{k}\Omega$, $R_S = 1\ \text{k}\Omega$, $C_{C1} = 1\ \mu F$, and $h_{ie} = 5\ \text{k}\Omega$. Determine the value of f_{1B} for the circuit.

The collector circuit of the BJT amplifier works according to the same logic as the base circuit. When X_C of the output coupling capacitor is equal to the total ac resistance in the collector circuit $(R_C + R_L)$, the output voltage will drop to 0.707 times its midband value. Again, the value of A_v has then been effectively reduced to $0.707A_{v(mid)}$.

It should come as no surprise that the equation for the cutoff frequency of the collector circuit is very similar to that for the base circuit. The actual equation is as follows:

$$\boxed{f_{1C} = \frac{1}{2\pi(R_C + R_L)C}} \qquad \textbf{(13.21)}$$

where
f_{1C} = the lower cutoff frequency of the collector circuit
$(R_C + R_L)$ = the total ac resistance in the collector circuit
C = the value of the output coupling capacitor

The following example illustrates the use of this equation.

EXAMPLE 13.18

Determine the value of f_{1C} for the circuit in Figure 13.10.

Solution: Using the values shown in the figure, the value of f_{1C} is found as

$$f_{1C} = \frac{1}{2\pi(R_C + R_L)C}$$

$$= \frac{1}{2\pi(6.5 \text{ k}\Omega)(0.22 \text{ }\mu\text{F})}$$

$$= \frac{1}{8.984 \times 10^{-3}}$$

$$= 111.3 \text{ Hz}$$

PRACTICE PROBLEM 13–18

The circuit described in Practice Problem 13–17 has values of $R_C = 3.6 \text{ k}\Omega$, $R_L = 1.5 \text{ k}\Omega$, and $C_{C2} = 2.2 \text{ }\mu\text{F}$. Determine the value of f_{1C} for the circuit.

In Examples 13.17 and 13.18, we got values of $f_{1B} = 60.79$ Hz and $f_{1C} = 111.3$ Hz for the amplifier in Figure 13.10. Based on these two values, what would the cutoff frequency for the amplifier be? If you said 111.3 Hz, you were right. An amplifier will cut off at the frequency that is closest to f_0. Thus, for low-frequency response, the amplifier will cut off at the *highest* calculated value of f_1.

Before deriving the equation for the emitter cutoff frequency, we need to refer back to some relationships established in Chapter 8. These were

$$Z_{\text{out}} = R_E \,\|\, \left(r_e' + \frac{R_{\text{in}}'}{h_{fe}} \right) \tag{13.22}$$

where

$$R_{\text{in}}' = R_1 \,\|\, R_2 \,\|\, R_s$$

In equation (13.22), the output impedance of the amplifier is considered to be the parallel combination of R_E and $(r_e' + R_{\text{in}}'/h_{fe})$. Under normal circumstances, the value of R_E is much greater than the value of $(r_e' + R_{\text{in}}'/h_{fe})$. Therefore, we can approximate the total ac resistance in the emitter circuit as

$$R_{\text{out}} \cong r_e' + \frac{R_{\text{in}}'}{h_{fe}} \tag{13.23}$$

Using this value, the value of f_{1E} is found as

$$f_{1E} = \frac{1}{2\pi R_{\text{out}} C_B} \tag{13.24}$$

The following example illustrates the use of this equation.

EXAMPLE 13.19

Determine the value of f_{1E} for the circuit in Figure 13.10.

Solution: First, the value of R_{out} is found as

Remember:

$$r'_e = \frac{h_{ie}}{h_{fe}}$$

$$R_{out} = r'_e + \frac{R'_{in}}{h_{fe}}$$

$$= 22\ \Omega + \frac{4.7\ k\Omega \parallel 18\ k\Omega \parallel 600\ \Omega}{200}$$

$$= 22\ \Omega + 2.58\ \Omega$$

$$= 24.58\ \Omega$$

Now the value of f_{1E} can be found as

$$f_{1E} = \frac{1}{2\pi R_{out} C_B}$$

$$= \frac{1}{2\pi(24.58\ \Omega)(10\ \mu F)}$$

$$= \frac{1}{1.544 \times 10^{-3}}$$

$$= 647.5\ Hz$$

PRACTICE PROBLEM 13–19

The amplifier described in Practice Problems 13–17 and 13–18 has values of h_{fe} = 200, R_E = 1.3 kΩ, and C_B = 100 μF. Determine the value of f_{1E} for the circuit.

At this point, what would you say is the value of f_1 for the circuit in Figure 13.15? Would you say 647.5 Hz? Good!

Gain Roll-off

✔ 20

The term *roll-off rate* is used to describe the rate at which the gain of an amplifier drops off after f_1 or f_2 has been passed. The *low-frequency* roll-off for the common-emitter amplifier is calculated using the following equation:

Roll-off rate. The rate of gain reduction when operated beyond the cutoff frequencies.

$$\boxed{\Delta A_{v(dB)} = 20 \log \frac{1}{\sqrt{1 + (f_1/f)^2}}} \qquad \textbf{(13.25)}$$

where f = the frequency of operation
f_1 = the lower cutoff frequency of the amplifier

The derivation of equation (13.25) is shown in Appendix D.

Before we get involved in any calculations, one point needs to be made about equation (13.25): *Equation (13.25) is used to determine the change in voltage gain*. For example, if equation (13.25) yields a result of -3 dB at a given frequency, f, this means that $A_{v(dB)}$ is 3 dB lower than its midband value at that frequency. The following example serves to illustrate this point.

EXAMPLE 13.20

A given amplifier has values of $A_{v(mid)} = 45$ dB and $f_1 = 2$ kHz. What is the gain of the amplifier when it is operated at 500 Hz?

Solution: The *change in* A_v is found using equation (13.25), as follows:

$$\Delta A_{v(dB)} = 20 \log \frac{1}{\sqrt{1 + (f_1/f)^2}}$$

$$= 20 \log \frac{1}{\sqrt{1 + (2000/500)^2}}$$

$$= 20 \log \frac{1}{\sqrt{1 + 16}}$$

$$= 20 \log(0.2425)$$

$$= -12.3 \text{ dB}$$

Now the value of $A_{v(dB)}$ at 500 Hz is found as

$$A_{v(dB)} = A_{v(mid)} + \Delta A_{v(dB)}$$
$$= 45 \text{ dB} + (-12.3 \text{ dB})$$
$$= 32.7 \text{ dB}$$

PRACTICE PROBLEM 13–20

An amplifier with a gain of 23 dB has a lower cutoff frequency of 8 kHz. If the circuit is operated at 4 kHz, what is the gain of the amplifier?

Values of *R* and *C* do not affect roll-off rates.

Equation (13.25) tells us one very important thing about the low-frequency roll-off of an amplifier: *The values of R and C in the amplifier have nothing to do with the roll-off rate of a given terminal circuit.* For each terminal circuit (base versus emitter versus collector), the roll-off rate is strictly a function of the difference between f and f_c. The *RC* circuit determines *when* the roll-off begins. However, once it has begun, the values of R and C are no longer significant. We can therefore conclude that *all RC circuits will roll off at the same rate*. Confusing? It shouldn't be. Equation (13.25), which is used to determine the low-frequency roll-off rate of an *RC* circuit, does not take the values of R and C into consideration. The only conclusion that can be drawn from this is that the values of R and C do not affect the roll-off rate. Therefore, it must be consistent from one *RC* circuit to another.

So, what *is* the low-frequency roll-off rate? We can determine this by setting f to several multiples of f_1 and solving the equation. For example, when $f = f_1$, $\Delta A_{v(dB)}$ is found as

$$\Delta Av_{(dB)} = 20 \log \frac{1}{\sqrt{1 + 1^2}}, \qquad \frac{f_1}{f} = 1$$
$$= -3 \text{ dB}$$

When $f = 0.1f_1$,

$$\Delta A_{v(\text{dB})} = 20 \log \frac{1}{\sqrt{1 + 10^2}} \qquad \frac{f_1}{f} = 10$$
$$= -20 \text{ dB}$$

When $f = 0.01f_1$,

$$\Delta A_{v(\text{dB})} = 20 \log \frac{1}{\sqrt{1 + 100^2}} \qquad \frac{f_1}{f} = 100$$
$$= -40 \text{ dB}$$

When $f = 0.001f_1$,

$$\Delta A_{v(\text{dB})} = 20 \log \frac{1}{\sqrt{1 + 1000^2}} \qquad \frac{f_1}{f} = 1000$$
$$= -60 \text{ dB}$$

Do you see a pattern here? Every time that we decrease the value of f by a factor of 10 (a *decade*), the gain of the amplifier drops another 20 dB. Therefore, we can state that the low-frequency roll-off for voltage gain is *20 dB/decade*. The Bode plot that corresponds to this roll-off rate is shown in Figure 13.11.

What is the roll-off rate of an *RC* circuit?

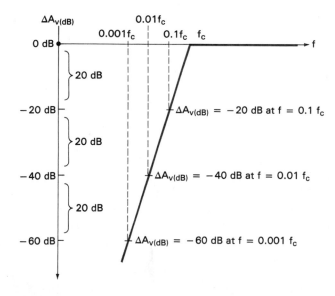

FIGURE 13.11

The Bode plot in Figure 13.11 is best explained in a series of steps, as follows:

1. At f_1, the change in voltage gain is shown to be 0 dB. (Don't forget, at f_1 the voltage gain of the amplifier is actually down by 3 dB.)
2. At $f = 0.1f_1$ (one decade down), the voltage gain of the amplifier has dropped by 20 dB to -20 dB.
3. At $f = 0.01f_1$ (next decade down), the voltage gain of the amplifier has dropped by another 20 dB. The total value of $\Delta A_{v(\text{dB})}$ is -40 dB.

SEC. 13.3   BJT Amplifier Low-frequency Response

4. At $f = 0.001f_1$ (next decade down), the voltage gain of the amplifier has dropped by another 20 dB. The total value of $\Delta A_{v(\text{dB})}$ is now -60 dB.

If the process were to continue, the next several decades would have total $\Delta A_{v(\text{dB})}$ values of -80 dB, -100 dB, and so on. At each decade interval, the voltage gain of the circuit would be 20 dB down from the previous decade interval.

Let's relate this to the circuit in Figure 13.10. We determined that the cutoff frequency for the emitter circuit (f_{1E}) was 647.49 Hz. For the sake of discussion, we are going to make two assumptions:

1. We are going to assume that the cutoff frequency is 600 Hz. This will simplify the numbers with which we are dealing.
2. We are going to assume that the emitter circuit is the *only* circuit that will cut off. In other words, we are going to neglect the effects of the base and collector circuits. (You will be shown very shortly what happens when these circuits are brought into the picture.)

Figure 13.12 shows the Bode plot for the emitter circuit in Figure 13.10. The cutoff frequency is 600 Hz. One decade down from this frequency is 60 Hz. At this frequency, the voltage gain of the amplifier has dropped by 20 dB, giving us a total $\Delta A_{v(\text{dB})}$ of -20 dB, and so on.

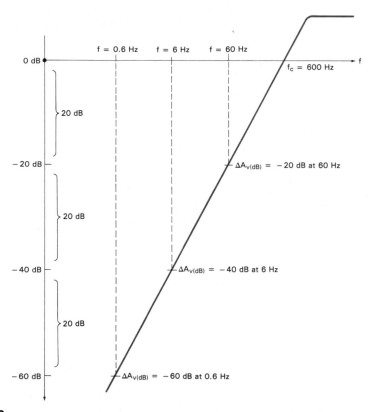

FIGURE 13.12

The Bode plot for the emitter circuit of Figure 13.14.

If we had wished to plot the roll-off curve for the base circuit in Figure 13.10, the roll-off rate of 20 dB per decade would not have been very useful. Since f_{1B} was about 60 Hz to begin with, one or two decades down does not

CHAP. 13 Amplifier Frequency Response

leave us with a whole lot of frequency! It helps, then, to be able to determine the voltage gain roll-off at *octave* frequency rates. The following table shows the $\Delta A_{v(dB)}$ values that are associated with octave frequency intervals. The values shown were determined by plugging the value of (f_1/f) into equation (13.25) and solving for $\Delta A_{v(dB)}$.

f	f_1/f	$A_{v(dB)}$ (approximate)
$\frac{1}{2}f_c$	2	-6
$\frac{1}{4}f_c$	4	-12
$\frac{1}{8}f_c$	8	-18

The pattern here is easy to see. The voltage gain of the circuit would roll off at a rate of *6 dB per octave*. Using the 6 dB per octave roll-off rate, we can easily derive the Bode plot for the base circuit of Figure 13.10. This plot is shown in Figure 13.13. Note that the $\Delta A_{v(dB)}$ intervals are now equal to 6 dB and that the frequency is measured in octaves instead of decades. Other than these two points, the plot shown is read in the same manner as the one in Figure 13.12.

Roll-off rates can also be measured at octave intervals.

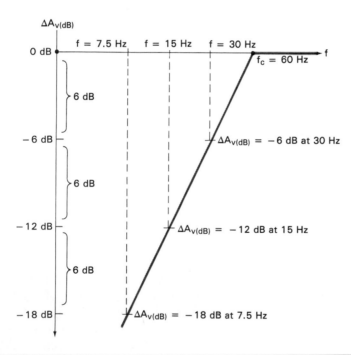

FIGURE 13.13

Roll-off measured in dB/octave.

Before we go on to look at the effect of having three cutoff frequencies in an amplifier (f_{1B}, f_{1C}, and f_{1E}), let's summarize the points that have been made in this section. These are:

1. The low frequency roll-off rate for an *RC* circuit is *20 dB per decade*. This is equal to a rate of *6 dB per octave*.

Summary of RC roll-off rates.

2. The roll-off rate is independent of the values of *R* and *C* in the terminal circuit. Thus, the rates listed in (1) apply to all *RC* terminal circuits.
3. Equation (13.25) can be used to determine the value of $A_{v(dB)}$ *relative to* $A_{v(mid)}$ at a given frequency.

The Combined Effects of f_{1B}, f_{1C}, and f_{1E}

You have seen that a circuit like the one in Figure 13.10 has three RC terminal circuits, each with its own cutoff frequency. Next, we are going to take a look at what happens when more than one cutoff frequency is passed. The overall concept is easy to understand with the help of Figure 13.14. Here we have three separate RC circuits, each having the indicated value of f_1.

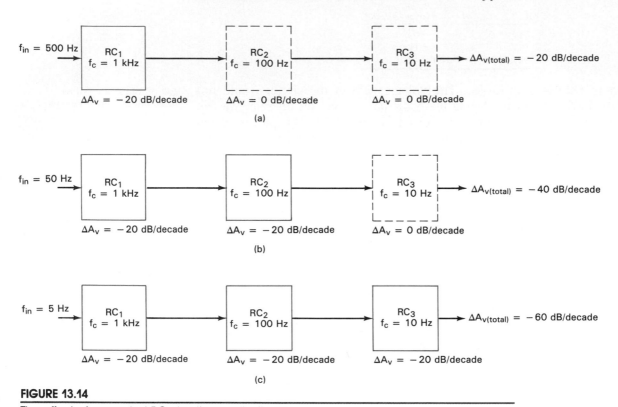

FIGURE 13.14

The effect of cascaded RC circuits on roll-off rates.

In Figure 13.14a, the input frequency is shown to be 500 Hz. Since this frequency is below the value of f_1 for RC_1, RC_1 has cut off and is causing a voltage gain roll-off of 20 dB/decade. At this point, the input frequency is still above RC_2 and RC_3, so these two circuits are still operating at their respective values of $A_{v(min)}$. Thus, at this point, the overall gain is dropping by 20 dB/decade, the drop rate of RC_1.

In Figure 13.14b, the input frequency has dropped to 50 Hz, and the value of f_c for RC_2 has also been passed. Now we have RC_1 introducing a 20-dB/decade drop, and RC_2 introducing a 20-dB/decade drop. *The overall roll-off is equal to the sum of these drops, 40 dB/decade.* This should not seem strange to you. As you recall, the total dB gain of a multistage circuit is equal to the *sum* of the individual dB gain values. The same principle holds true for roll-off rates.

When the input frequency decreases to 5 Hz (Figure 13.14c), all three of the RC circuits are being operated below their respective values of f_c. Thus all three circuits are introducing a 20-dB/decade roll-off. The sum of these roll-off values is 60 dB/decade.

The circuit action that was just described is represented by the Bode plot in Figure 13.15. As you can see, the roll-off rate is 20 dB/decade for input frequencies that are between 1 kHz and 100 Hz. When the input frequency drops past 100

CHAP. 13 Amplifier Frequency Response

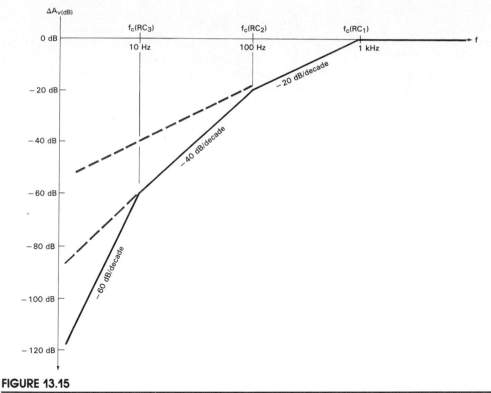

FIGURE 13.15

The effects of multiple cutoff frequencies.

Hz, the roll-off increases to 40 dB/decade. When the input frequency drops past 10 Hz, all three circuits are cut off, and the roll-off rate is shown to be 60 dB/ decade.

Now let's relate this principle to the circuit in Figure 13.10. The low-frequency Bode plot for this circuit is shown in Figure 13.16. Note that we are using the more convenient rate of *6 dB/octave* in this figure. This is acceptable, since a roll-off of 6 dB/octave is equal to a roll-off of 20 dB/decade. We have also rounded off the calculated values of f_{1B}, f_{1C}, and f_{1E} as shown in the illustration.

At 600 Hz, the *RC* circuit in the emitter cuts off. Thus, as frequency decreases, we get a roll-off of 6 dB/octave. This continues until the 120-Hz cutoff frequency of the collector is reached. Once this is reached, the roll-off rate increases to 12 dB/octave. This rate continues until the frequency reaches the 60-Hz cutoff frequency of the base. At this point, all three of the terminal circuits are introducing a 6-dB/octave drop, and the total roll-off rate is 18 dB/octave.

A Practical Consideration

The Bode plot represents the *ideal* amplifier response to a change in frequency. For example, the Bode plot in Figure 13.16 idealizes the low-frequency response of the circuit in Figure 13.10 in that:

1. It assumes that all cutoff frequencies are exactly as calculated. In other words, there are no stray capacitances or resistor tolerances that will affect the actual cutoff frequency values.

2. It assumes that each transistor terminal circuit will have a 0-dB value of $A_{v(dB)}$ until the cutoff frequency is reached.

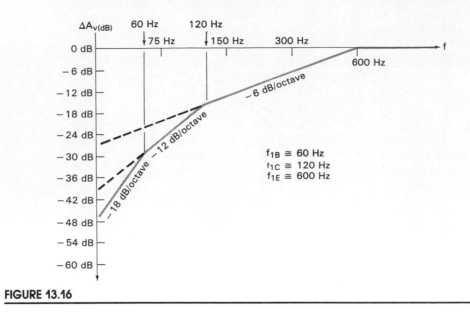

FIGURE 13.16

In practice, neither of the above assumptions is valid. Circuit component tolerances and stray capacitance values *will* affect the actual cutoff frequencies. You will find that, in practice, measured cutoff frequencies will vary from the calculated cutoff frequencies.

Another factor that affects measured frequency-response values is the fact that *RC* circuits will *begin* to reduce amplifier gain well before their lower cutoff frequencies are reached. For example, let's assume that a given *RC* circuit is being operated at *twice* its value of f_1. When this is the case, the value of $\Delta A_{v(dB)}$ is found as

$$\Delta A_{v(dB)} = 20 \log \frac{1}{\sqrt{1 + (0.5)^2}} \quad \Bigg| \quad f = 2f_1$$
$$\cong -1 \text{ dB}$$

This means that a given *RC* circuit will begin to affect the frequency response of an amplifier at twice its lower cutoff frequency. For example, refer back to Figure 13.16. The collector circuit ($f_{1C} = 120 \text{ Hz}$) would actually begin to cause a noticeable drop in gain at 240 Hz. The base circuit ($f_{1B} = 60 \text{ Hz}$) would actually begin to cause a noticeable drop in gain at 120 Hz, and so on. The result is a frequency-response curve that, while resembling the one shown in the figure, actually shows the amplifier gains to be dropping at a higher rate than predicted as each of the cutoff frequencies is approached.

What is the bottom line? There are two ways to view the low-frequency response of an amplifier. The first is the ideal approach that we have taken in this section. This approach assumes that a given BJT amplifier low-frequency response curve will be nearly identical to the Bode plot for the circuit. That is, *RC* circuits will cut off at specific frequencies and then will roll-off at an exact rate of 20 dB/decade (6 dB/octave). When multiple *RC* circuits are involved, the roll-off rates will remain constant between each cutoff frequency and the next. Then the roll-off rates will instantly change to reflect the added *RC* circuit effects.

A more practical approach to the whole situation is as follows:

1. An *RC* circuit will begin to reduce the gain of an amplifier at approximately *twice* its value of f_1. The gain of the amplifier will drop by approximately 3 dB by the time the lower cutoff frequency is reached.

　　CHAP. 13　Amplifier Frequency Response

2. The roll-off rate for a given *RC* circuit is approximately 20 dB/decade (6 dB/octave).

3. When multiple *RC* circuits are involved (as in the BJT amplifier), the effects of each will be felt well before their actual cutoff frequencies are reached.

How much will these considerations affect the frequency-response curve of a given amplifier? Figure 13.17 shows the original Bode plot for the amplifier in Figure 13.10 and the more realistic frequency-response curve. As you can see, the curve does resemble the Bode plot. However, the combined effects of the *RC* circuits cause the actual curve to have more rounded turning points that occur at higher frequencies than those calculated.

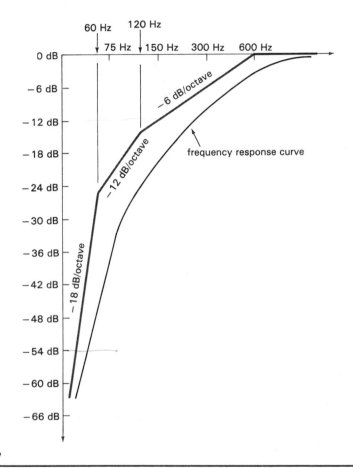

FIGURE 13.17

Does this mean that the Bode plot is a waste of time? No. The Bode plot provides us with a means of predicting the response of a circuit to a change in frequency. However, when we are making these predictions, we need to keep in mind the fact that they are *approximations* that are based on assuming a number of ideal conditions. Therefore, we should not be surprised when the percent of error between our predicted and measured frequency-response values approaches 10% or 20%.

One point of interest: The greater the frequency range from one cutoff frequency to the next, the closer the actual frequency-response curve will come to the Bode plot for an amplifier. For example, an amplifier with cutoff frequencies of 10 kHz, 1 kHz, and 10 Hz would have a frequency-response curve that is closer to

When using Bode plots, keep in mind that they provide approximate values. This is because they are plotted assuming a number of ideal conditions.

the ideal (the Bode plot) than an amplifier with cutoff frequencies of 10 kHz, 5 kHz, and 1 kHz. The proof of this concept is included as a Brain Drain problem at the end of the chapter.

SECTION REVIEW

1. Explain why the power gain of a given amplifier drops by 50% when the reactance of the input coupling capacitor equals the input resistance of the amplifier. [Objective 18]
2. Define the term *roll-off rate*. [Objective 20]
3. What is the relationship between the roll-off rate for a given RC circuit and the circuit values of R and C? Explain your answer. [Objective 20]
4. What are the standard low-frequency roll-off rates? [Objective 22]
5. Describe the BJT amplifier low-frequency Bode plot. [Objective 23]
6. Contrast the Bode plot with the actual low-frequency response curve of a BJT amplifier. [Objective 24]

13.4
BJT Amplifier High-frequency Response

In a lot of ways, the high-frequency operation of the BJT amplifier is similar to the low-frequency operation. You have a cutoff frequency for each terminal circuit. Once a given value of f_2 is passed, the circuit will cut off and will have a roll-off rate that is approximately equal to 20 dB/decade (6 dB/octave).

How f_2 calculations differ from f_1 calculations.

The primary differences between low- and high-frequency response are:

1. The values of capacitance used in f_2 calculations.
2. The methods used to determine the total resistance in each terminal circuit.

These differences are discussed in detail in this section.

BJT Internal Capacitance

You may recall from Chapter 2 that a pn junction will have some measurable amount of capacitance. Since the BJT has two internal pn junctions, it has two internal capacitances, as is shown in Figure 13.18. As you will be shown in this section, these internal capacitances control the high frequency response of the device.

FIGURE 13.18

BJT internal capacitance.

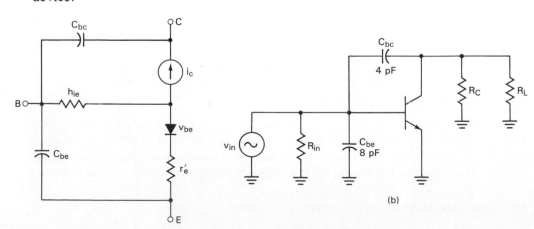

598

The capacitance of the collector–base junction (C_{bc}) is usually listed on the specification sheet of a given transistor. Note that it may be listed as C_{bc}, C_c, C_{ob}, or C_{obo}. If you refer back to the specification sheet for the 2N3904 (Figure 6.12), you will see that it lists a value of collector–base capacitance (C_{ob}) equal to 4 pF. The value of C_{be} may or may not be given on the specification sheet. When given, it is usually listed as C_{in}, C_{be}, C_{ib}, or C_b. The specification sheet for the 2N3904 lists this value as $C_{ib} = 8$ pF.

When the value of C_{be} is *not* listed on the spec sheet of a given transistor, it can be calculated as

$$C_{be} = \frac{1}{2\pi f_T r'_e} \qquad \textbf{(13.26)}$$

where f_T is the *current gain–bandwidth product* of the transistor as listed on the spec sheet. The current gain–bandwidth product of a transistor is the frequency where C_{be} will have a low enough reactance to cause the transistor current gain to drop to *unity*. The following example illustrates the use of equation (13.26) in determining the value of C_{be}.

Current–gain bandwidth product. The frequency at which BJT current gain drops to unity.

EXAMPLE 13.21

 26

The value of r'_e used in Example 13.21 was determined using

$$r'_e = \frac{25 \text{ mV}}{I_E}$$

As you know, this equation gives only an approximation of r'_e. A more accurate value can be obtained using

$$r'_e = \frac{h_{ie}}{h_{fe}}$$

While the *h*-parameter equation is more accurate, the 25 mV/I_E equation will give us a value of r'_e that is accurate enough for our purposes.

The spec sheet for the 2N3904 lists a value of $f_T = 300$ MHz at $I_c = 10$ mA. Determine the value of C_{be} at this current.

Solution: Assuming that $I_C = I_E$, we can determine the value of r'_e to equal 2.5 Ω. Using this value in equation (13.26), we get the value of C_{be} as

$$C_{be} = \frac{1}{2\pi f_T r'_e}$$
$$= \frac{1}{2\pi (300 \text{ MHz})(2.5 \ \Omega)}$$
$$= 212 \text{ pF}$$

PRACTICE PROBLEM 13–21

A transistor has a current gain–bandwidth product of $f_T = 200$ MHz at $I_C = 10$ mA. Determine the value of C_{be} for the device.

As you can see, the value of $C_{be} = 212$ pF in the example is significantly different from the spec sheet rating of 8 pF. The difference between the two values is due to the fact that they were determined for two different values of I_E. The spec sheet rating indicates that C_{be} will be no greater than 8 pF when $I_E = 0$ mA. However, as I_E changes, so does the value of r'_e and the width of the base–emitter junction, as shown in the following table.

I_E	r'_e	Junction Width
Increase	Decrease	Decrease
Decrease	Increase	Increase

Since the value of C_{be} is *inversely* proportional to the width of the emitter–base junction, it will vary directly with variations in I_E. In other words, as I_E increases, so will C_{be}. As I_E decreases, so will C_{be}.

So which value of C_{be} do you use for a given circuit, the calculated value or the spec sheet value? In most cases, you should use the *calculated* value of C_{be}, as it will provide much more accurate results in later calculations. However, remember that even the calculated value of C_{be} will give you only a *ballpark* figure. Since the calculation of r'_e is only an approximation, the value obtained by using equation (13.26) will only be an approximation. However, in most cases, this approximated value of C_{be} will be much closer to the actual emitter–base capacitance than the spec sheet rating.

Why we don't consider transistor capacitance in low-*f* analysis.

Why have we ignored these capacitance values until now? As you can see, C_{be} and C_{bc} are very small capacitance values. Because of this, their low-frequency reactance is extremely high. For example, the reactance of C_{bc} is approximately 3.98 MΩ at $f = 10$ kHz (calculate this for yourself). This reactance is so high that it can be considered to be an open for all practical purposes. But what happens when the operating frequency of the transistor increases? Since reactance is inversely proportional to frequency for a given capacitor, the value of X_C for C_{bc} will decrease. At 100 MHz, the reactance of C_{bc} is only 398 Ω. This reactance *cannot* be ignored because it will have a definite impact on the operation of the transistor.

Miller's Theorem

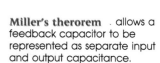

Before we can get on to some actual circuit calculations, we have to solve one more problem that involves the value of C_{bc}. As you saw in Figure 13.18, C_{bc} is the capacitance between the collector and base terminals. We can represent this capacitance as shown in Figure 13.18.

FIGURE 13.19

Miller equivalent circuit for a feedback capacitor.

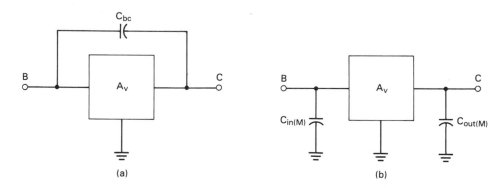

(a)   (b)

Miller's therorem . allows a feedback capacitor to be represented as separate input and output capacitance.

A Practical Consideration. Miller's theorem applies only to inverting amplifiers (those with a 180° voltage phase shift).

In Figure 13.19a, the transistor has been shown as a block having a certain value of voltage gain, A_v. Since C_{bc} exists between the collector and base terminals, it is shown as an external capacitor. The question now is this: Is C_{bc} part of the input (base) circuit or part of the output (collector) circuit? The answer is: *both*.

Miller's theorem allows C_{bc} to be represented as *two* capacitors, one in the base circuit and one in the collector circuit. These two capacitors are shown in Figure 13.24b. For this circuit the *Miller input capacitance* is found as

$$C_{in(M)} = C_{bc}(A_v + 1) \qquad \textbf{(13.27)}$$

and the *Miller output capacitance* is found as

$$C_{out(M)} = C_{bc}\frac{A_v + 1}{A_v} \qquad \textbf{(13.28)}$$

In most cases, $A_v \gg 1$. When this is the case, the above equations can be simplified as

$$\boxed{C_{\text{in(M)}} \cong A_v C_{bc}} \qquad (13.29)$$

and

$$\boxed{C_{\text{out(M)}} \cong C_{bc}} \qquad (13.30)$$

Note that the above equations can be used whenever the value of A_v is greater than 10.

So, why go to all this trouble? The *feedback* capacitor shown in Figure 13.18a can make circuit calculations extremely complex. A much simpler approach is to represent C_{bc} as two separate capacitors. This way, one of the capacitors is used strictly for input (base) circuit calculations, and the other is used strictly for output (collector) circuit calculations. The following example illustrates the method for determining the Miller input/output capacitance values for a given amplifier.

Why we use Miller's theorem.

═══════════════ **EXAMPLE 13.22** ═══════════════

A given amplifier has values of C_{bc} = 6 pF and A_v = 120. Determine the Miller input and output capacitance values for the circuit.

FIGURE 13.20

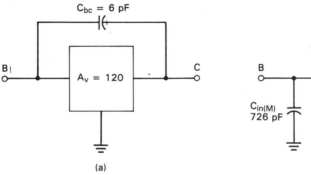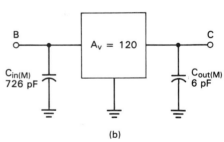

(a)   (b)

Solution: The Miller input capacitance is determined as

$$C_{\text{in(M)}} \cong A_v C_{bc}$$
$$= (120)(6\text{pF})$$
$$= 720 \text{ pF}$$

The Miller output capacitance is determined as

$$C_{\text{out(M)}} \cong C_{bc}$$
$$= 6 \text{ pF}$$

Using the values obtained here, we can derive the equivalent of the circuit in Figure 13.20a. This circuit is shown in Figure 13.20b.

─────────────────────────────

PRACTICE PROBLEM 13–22

A BJT with a value of C_{bc} = 4 pF is being used in an amplifier that has a value of A_v = 240. Determine the Miller input and output equivalent capacitances for C_{bc}.

─────────────────────────────

After the Miller capacitance values for an amplifier are determined, we are ready to determine the upper cutoff frequencies for a given BJT amplifier.

High-frequency Operation

✔ 30

When we discussed the low-frequency operation of the common-emitter amplifier, you were shown that the value of f_1 depended on the various values of circuit resistance and capacitance. The same holds true for the value of f_2. However, as you will be shown, the resistances are combined in a different fashion, and we use different values of circuit capacitance.

The high-frequency equivalent circuit for the common-emitter amplifier is shown in Figure 13.21a. We are going to reduce this circuit to its simplest *RC* equivalent circuit so that you will be able to see *why* the values of *R* and *C* are combined as they are. Before we do this, however, there are two values shown that should be explained. C_{in} is the *input capacitance of the next stage*. If this amplifier is driving another, C_{in} will equal the total *high-frequency* input capacitance of the next stage. (*This point is discussed further in our coverage of multistage amplifiers. At this point we will simply assume a value of C_{in} for any example problems.*) Also, the *base input impedance*, h_{ie}, is shown in the circuit. As you will see, this value is needed for the determination of f_2.

FIGURE 13.21

BJT high-frequency ac equivalent circuit.

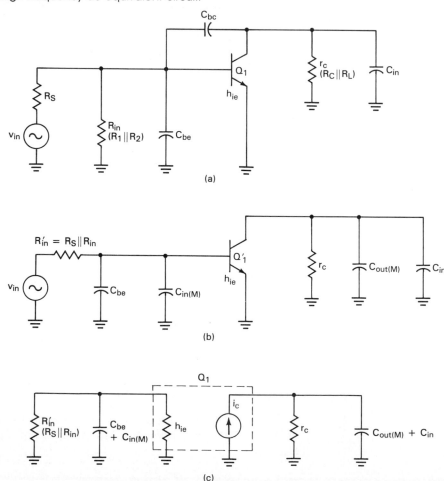

CHAP. 13   Amplifier Frequency Response

The equivalent circuit in Figure 13.21b was derived by performing *two* steps:

1. The collector–base capacitor, C_{bc}, was replaced by its equivalent Miller input and output capacitors.
2. The source resistance (R_S) and the amplifier input resistance (R_{in}) were combined into a single *parallel* equivalent resistance, R'_{in}.

Now, we are going to simplify the circuit further by taking three additional steps. As Figure 13.21c shows:

3. The parallel capacitors in the base circuit, C_{be} and $C_{in(M)}$, are combined into a single input capacitance value.
4. The two parallel capacitors in the collector circuit, C_{in} and $C_{out(M)}$, are combined into a single output capacitance value.
5. The transistor has been replaced to show it as a resistance, equal to h_{ie}, and a current source.

We can now use this equivalent circuit to analyze the high-frequency response of the original amplifier.

The high-frequency cutoff, f_2, is determined using the same basic equation as is used to determine f_1:

$$f_c = \frac{1}{2\pi RC}$$

For the base circuit, R is the parallel combination of R'_{in} and h_{ie}. Also, C is the combination of C_{be} and $C_{in(M)}$. Thus, for the base circuit,

$$f_{2B} = \frac{1}{2\pi(R'_{in} \| h_{ie})(C_{be} + C_{in(M)})} \tag{13.31}$$

where $R'_{in} = R_S \| R_1 \| R_2$
h_{ie} = the input impedance to the amplifier
C_{be} = the base–emitter junction capacitance
$C_{in(M)}$ = the Miller input capacitance

The following example illustrates the process for determining the value of f_2 for the base circuit.

A Practical Consideration. The value of C_{in} in the collector circuit of Figure 13.21b assumes that the amplifier load has some amount of capacitance. If the load is purely resistive, C_{in} for the load will be zero, and the total capacitance in the output circuit can be assumed to be equal to $C_{out(M)}$ for the transistor.

EXAMPLE 13.23

Determine the value of f_2 for the base circuit of the amplifier shown in Figure 13.22.

FIGURE 13.22

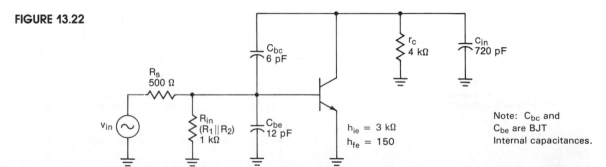

Note: C_{bc} and C_{be} are BJT Internal capacitances.

Solution: The first step is to determine the value of A_{vL} for the amplifier, as follows:

$$A_{vL} = \frac{h_{fe}r_C}{h_{ie}}$$
$$= 200$$

The value of A_v is now used (along with C_{bc}) to determine the value of $C_{in(M)}$:

$$C_{in(M)} = A_v C_{bc}$$
$$= (200)(6 \text{ pF})$$
$$= 1.2 \text{ nF}$$

The value of R'_{in} is found as

$$R'_{in} = R_S \| R_{in}$$
$$= 333 \ \Omega$$

Now that we have all the needed values of resistance and capacitance, we can find the value of f_2 for the base circuit as follows:

$$f_{2B} = \frac{1}{2\pi(R'_{in} \| h_{ie})(C_{be} + C_{in(M)})}$$
$$= \frac{1}{2\pi(333 \ \Omega \| 3 \text{ k}\Omega)(12 \text{ pF} + 1.2 \text{ nF})}$$
$$= \frac{1}{2\pi(300 \ \Omega)(1.212 \text{ nF})}$$
$$= \frac{1}{2.283 \times 10^{-6}}$$
$$= 437.94 \text{ kHz}$$

PRACTICE PROBLEM 13–23

The amplifier described in Practice Problem 13–17 has values of $h_{fe} = 150$, $f_t = 300$ MHz at $I_c = 10$ mA, and $C_{bc} = 4$ pF. Determine the value of f_{2B} for the circuit.

For the collector circuit, the value of R is equal to r_C, and C is the combination of C_{in} for the next stage and $C_{out(M)}$ for the amplifier. Thus, for the collector circuit,

$$\boxed{f_{2C} = \frac{1}{2\pi r_C[C_{out(M)} + C_{in}]}} \tag{13.32}$$

The following example illustrates the use of this equation.

EXAMPLE 13.24

Determine the value of f_{2C} for the circuit in Figure 13.22.

Solution: Since the value of $A_v \gg 1$, the value of $C_{out(M)}$ is equal to C_{bc}. Thus

$$C_{out(M)} = 6 \text{ pF}$$

Since the other circuit values are already known, we can proceed to find f_2 for the collector circuit, as follows:

$$
\begin{aligned}
f_{2C} &= \frac{1}{2\pi r_C[C_{out(M)} + C_{in}]} \\
&= \frac{1}{2\pi(4 \text{ k}\Omega)(6 \text{ pF} + 720 \text{ pF})} \\
&= \frac{1}{1.825 \times 10^{-5}} \\
&= 54.8 \text{ kHz}
\end{aligned}
$$

PRACTICE PROBLEM 13–24

What is the value of f_{2C} for the amplifier described in Practice Problems 13–23 and 13–17?

Gain Roll-off

✔ 31

The *high-frequency* gain roll-off is found in much the same manner as the low-frequency gain roll-off. The exact equation is

$$\Delta A_{v(dB)} = 20 \log \frac{1}{\sqrt{1 + (f/f_2)^2}} \qquad \textbf{(13.33)}$$

As you can see from equation (13.33), the gain roll-off for the high-frequency circuit is also independent of the values of R and C. We can use equation (13.33) to determine the high-frequency roll-off in *dB per decade*, as follows. When $f = f_2$,

$$
\begin{aligned}
\Delta A_{v(dB)} &= 20 \log \frac{1}{\sqrt{1 + 1^2}} \\
&= -3 \text{ dB}
\end{aligned}
$$

When $f = 10f_2$,

$$
\begin{aligned}
\Delta A_{v(dB)} &= 20 \log \frac{1}{\sqrt{1 + 10^2}} \\
&= -20 \text{ dB}
\end{aligned}
$$

When $f = 100f_2$,

$$
\begin{aligned}
\Delta A_{v(dB)} &= 20 \log \frac{1}{\sqrt{1 + 100^2}} \\
&= -40 \text{ dB}
\end{aligned}
$$

A Practical Consideration. When you connect an oscilloscope or frequency counter to the output of an amplifier, the value of f_{2C} will decrease. The reason is the added input capacitance of the particular piece of test equipment. For example, if an oscilloscope has an input capacitance of 30 pF, you are adding 30 pF when you connect the oscilloscope to the output of the circuit. This added 30 pF may significantly reduce the value of f_2 for the circuit.

One way of reducing the percent of error introduced by an oscilloscope is to use a ×10 probe when making high-frequency measurements. A ×10 probe decreases the input capacitance of the oscilloscope and thus reduces the error introduced by the device.

High-frequency roll-off rates are equal to the low-frequency roll-off rates.

And so on. As you can see, the high-frequency roll-off of the common emitter amplifier is equal to the low-frequency roll-off, *20 dB/decade*. Similarly, the high-frequency roll-off can also be computed as being equal to *6 dB/octave*.

As you can see, the high-frequency operation of a given common-emitter amplifier is exactly the same as the low-frequency operation. The only difference is in the method used to determine f_2.

One Final Note

✔ 32

A wide variety of factors can affect the frequency response of a given amplifier. Component tolerances, estimations of BJT internal capacitances, and the effects of test equipment input capacitance can cause the measured values of f_1 and f_2 to vary significantly from their predicted (calculated) values.

When measuring the values of f_1 and f_2 for a BJT amplifier, you will tend to find that the largest percent of error will occur in the f_2 measurements. There are two reasons for this:

Why f_2 measurements tend to have large percent of error values.

1. The BJT internal capacitance values are estimated.
2. The capacitance values that are used in f_2 calculations are in the picofarad (pF) range, as are the input capacitance ratings of most pieces of test equipment. Thus, by connecting the test equipment to the circuit, you are significantly altering the total capacitance that determines the values of f_2.

When you are predicting the values of f_1 and f_2 for your standard BJT amplifier, you need to keep in mind that you are dealing with approximated values and, therefore, will get results that are approximations. While this may seem frustrating, it shouldn't be. You see, the purpose of performing the frequency analysis of an amplifier is to ensure that the circuit gain will not be affected by the input frequency. For example, if you want to operate a given amplifier at a frequency of 50 kHz, you want to ensure that an input frequency of 50 kHz will not cause the amplifier gain to be reduced. Beyond this application, the exact values of f_1 and f_2 are rarely of any consequence for standard *RC*-coupled BJT amplifiers.

There is one common application where the exact values of f_1 and f_2 *are* important. This is the case where you are dealing with *tuned amplifiers*. You may recall that tuned amplifiers are designed for specific bandwidths. Since they are designed for specific values of f_1 and f_2, these values are important when you are dealing with tuned amplifiers. We will discuss tuned amplifiers in detail in Chapter 15.

SECTION REVIEW

1. What are the primary differences between BJT f_1 and f_2 calculations? [Objective 25]
2. Why aren't the BJT internal capacitances considered in the low-frequency analysis of a given amplifier? [Objective 27]
3. What does Miller's theorem state? Why is it used? [Objective 28]
4. List, in order, the steps taken to perform the high-frequency analysis of a BJT amplifier. [Objective 30]
5. Compare high-frequency roll-off rates to low-frequency roll-off rates. [Objective 31]
6. Explain why f_2 measurements tend to have larger percent of error values than f_1 measurements. [Objective 32]

FET Amplifier Frequency Response

The transition from BJT circuits to FET circuits is actually very simple. All we have to do is come up with the equations for f_1 and f_2 for a given FET amplifier. Everything else is identical to what we have been doing up to this point.

In this section we are going to concentrate on the voltage-divider biased common-source amplifier. This circuit, along with its low-frequency ac equivalent, is shown in Figure 13.23.

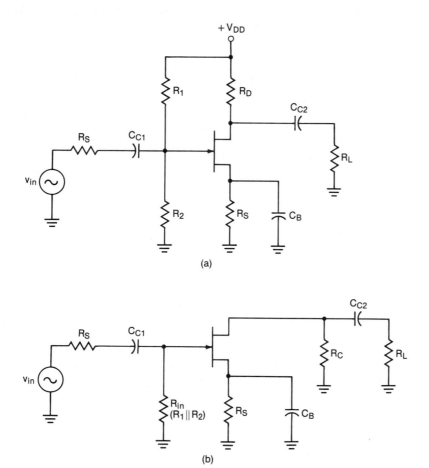

(a)

(b)

FIGURE 13.23

FET amplifier low-frequency ac equivalent circuit.

Low-frequency Response

33

As you can see, there is very little difference between the FET amplifier and the BJT amplifier as far as the low-frequency equivalent circuit is concerned. In fact, the gate and drain current equations are exactly the same as those used for the BJT amplifier. Thus

$$f_{1G} = \frac{1}{2\pi(R_S + R_{in})C_{C1}} \qquad (13.34)$$

and

$$f_{1D} = \frac{1}{2\pi(R_D + R_L)C_{C2}}$$ (13.35)

Because of the extremely high input impedance of the FET, the value of R_{in} for the amplifier is found as

$$R_{in} = R_1 \| R_2$$ (13.36)

As the following example shows, the high input impedance of the FET means that the FET amplifier will normally have a much lower input cutoff frequency than that of a similar BJT amplifier.

━━━ **EXAMPLE 13.25** ━━━

Determine which of the two circuits shown in Figure 13.24 has the higher input circuit cutoff frequency.

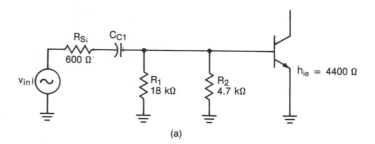

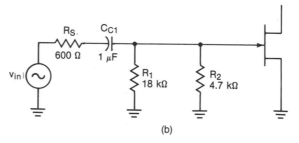

FIGURE 13.24

Solution: Figure 13.24a is the ac circuit that we analyzed in Example 13.17. At that point we determined the value of f_{1B} for the circuit to be 60.81 Hz (review Example 13.17 if necessary).

The circuit in Figure 13.24b is identical to the one in Figure 13.24a except for the fact that the BJT has been replaced by a FET. Because of the high input impedance of the FET,

$$R_{in} = R_1 \| R_2$$
$$= 18 \text{ k}\Omega \| 4.7 \text{ k}\Omega$$
$$= 3727 \ \Omega$$

Using this value for R_{in}, the value of f_{1G} is found as

$$f_{1G} = \frac{1}{2\pi(R_S + R_{in})C_{C1}}$$

$$= \frac{1}{2\pi(4327 \ \Omega)(1 \ \mu F)}$$

$$= 36.78 \ Hz$$

Thus, by replacing the BJT with a FET, we have decreased the value of f_1 for the input circuit by nearly 50%.

In practice, FET amplifiers normally have much higher input resistance values than do BJT amplifiers. As the following example serves to show, this has the effect of lowering the value of f_{1G} even more.

EXAMPLE 13.26

Determine the value of f_{1G} for the amplifier shown in Figure 13.25.

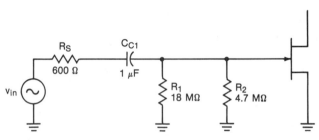

FIGURE 13.25

Solution: For this circuit,

$$R_{in} = R_1 \| R_2$$

$$= 18 \ M\Omega \| 4.7 \ M\Omega$$

$$= 3.727 \ M\Omega$$

Using this value of R_{in}, the gate cutoff frequency is found as

$$f_{1G} = \frac{1}{2\pi(R_S + R_{in})C_{C1}}$$

$$= \frac{1}{2\pi(3.727 \ M\Omega)(1 \ \mu F)}$$

$$= 0.043 \ Hz$$

PRACTICE PROBLEM 13–26

A FET amplifier has values of $R_1 = R_2 = 10 \ M\Omega$, $R_S = 100 \ \Omega$, and $C_{C1} = 3.3 \ \mu F$. Determine the value of f_{1G} for the circuit.

The low-frequency analysis of the drain circuit is just like that for the collector circuit of a BJT amplifier. For this reason, it needs no further explanation. The entire low-frequency analysis of a basic FET amplifier is demonstrated in the following example.

EXAMPLE 13.27

Determine the overall value of f_1 for the circuit shown in Figure 13.26.

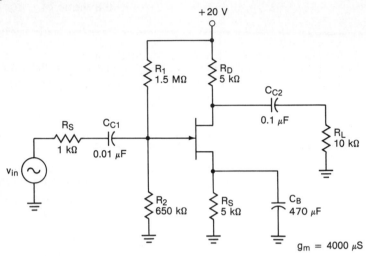

FIGURE 13.26

Solution: The input resistance to the amplifier is found as

$$R_{in} = R_1 \| R_2$$
$$= 453.5 \text{ k}\Omega$$

Now f_{1G} is found as

$$f_{1G} = \frac{1}{2\pi(R_S + R_{in})C_{C1}}$$
$$= \frac{1}{2\pi(454.5 \text{ k}\Omega)(0.01 \text{ } \mu\text{F})}$$
$$= 35 \text{ Hz}$$

The value of f_{1D} is now found as

$$f_{1D} = \frac{1}{2\pi(R_D + R_L)C_{C2}}$$
$$= \frac{1}{2\pi(15 \text{ k}\Omega)(0.1 \text{ } \mu\text{F})}$$
$$= 106 \text{ Hz}$$

PRACTICE PROBLEM 13–27

A FET amplifier has values of $R_1 = 10 \text{ M}\Omega$, $R_2 = 1 \text{ M}\Omega$, $R_D = 1.1 \text{ k}\Omega$, $R_S = 820 \text{ }\Omega$, $R_L = 5 \text{ k}\Omega$, $C_{C1} = 0.01 \text{ }\mu\text{F}$, $C_{C2} = 0.1 \text{ }\mu\text{F}$, $R_L = 10 \text{ k}\Omega$, and $g_m = 2200 \text{ }\mu\text{S}$. Determine the overall value of f_1 for the amplifier.

The value of f_1 for the amplifier would be assumed to be 106 Hz, the higher of the two cutoff frequencies. Just as with the BJT amplifier, the gain would drop by 3 dB by the time the frequency had dropped to 106 Hz. At this point, the gain would continue to drop (with a continual decrease in frequency) at a rate of *6 dB/octave* until 35 Hz is reached. Then the roll-off rate would increase to *12 dB/octave*; again, exactly the same with the BJT amplifier.

High-frequency Analysis

↙ 34

The high-frequency response of the FET is limited by values of *internal* capacitance, just like the BJT. These capacitances are shown in Figure 13.27. As you can see, we again have a situation that is very similar to that of the BJT. There is a measurable amount of capacitance between each terminal *pair* of the FET. These capacitances each have a reactance that decreases as frequency increases. As the reactance of a given terminal capacitance decreases, more and more of the signal at the terminal is shorted through the component. This is the same problem that we ran into with the BJT.

The high-frequency equivalent circuit of the basic FET amplifier is analyzed in the same fashion as the BJT amplifier. The high-frequency equivalent for the amplifier in Figure 13.23a is shown in Figure 13.28. As you can see, all the terminal capacitance values are included, with the exception of C_{gd}. This capacitor has been replaced with the Miller equivalent input and output capacitance values. This is the same thing that we did with C_{bc} in the BJT amplifier.

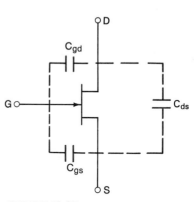

FIGURE 13.27
FET internal capacitances.

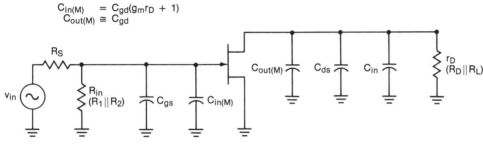

FIGURE 13.28
FET amplifier high-frequency ac equivalent circuit.

The value of $C_{in(M)}$ is found using a form of equation (13.27) as

$$C_{in(M)} = C_{gd}(A_v + 1)$$

Since the value of A_v for the FET amplifier is equal to $g_m r_D$, the equation above is rewritten as

$$C_{in(M)} = C_{gd}(g_m r_D + 1) \qquad (13.37)$$

The following example illustrates the use of this equation.

A Practical Consideration: When the value of $g_m r_D$ for the amplifier is greater than (or equal to) 10, equation (13.37) can be simplified as

$$C_{in(M)} \cong C_{gd} g_m r_D$$

═══════ **EXAMPLE 13.28** ═══════

A given FET amplifier has values of $C_{gd} = 4$ pF, $g_m = 2500\ \mu S$, and $r_D = 5.6$ kΩ. Determine the value of $C_{in(M)}$ for the amplifier.

Solution: The value of $C_{in(M)}$ is found as

$$C_{in(M)} \cong C_{gd}g_m r_D$$
$$= (4 \text{ pF})[2500 \text{ μS})(5.6 \text{ kΩ})$$
$$= (4 \text{ pF})(14)$$
$$= 56 \text{ pF}$$

PRACTICE PROBLEM 13–28

A given FET amplifier has values of $C_{gd} = 3$ pF, $g_m = 3200$ μS, and $r_D = 1.8$ kΩ. Determine the value of $C_{in(M)}$ for the circuit.

Equation (13.28) defined $C_{out(M)}$ as being

$$C_{out(M)} = C \frac{A_v + 1}{A_v}$$

If we replace C with C_{gd}, and A_v with $g_m r_D$, we obtain the following equation for the FET amplifier value of $C_{out(M)}$:

$$C_{out(M)} = C_{gd} \frac{g_m r_D + 1}{g_m r_D} \tag{13.38}$$

Again, when $g_m r_D \gg 1$, the equation for $C_{out(M)}$ can be simplified to

$$C_{out(M)} \cong C_{gd} \tag{13.39}$$

From the circuit in Figure 13.28, it is easy to see where the following equations come from:

$$C_G = C_{gs} + C_{in(M)} \tag{13.40}$$

and

$$C_D = C_{out(M)} + C_{ds} + C_{in} \tag{13.41}$$

where C_{in} is the input capacitance of the following stage. We can now use these values to define the values of f_2 for the gate and drain circuits. These equations, which will look very familiar at this point, are as follows:

$$f_{2G} = \frac{1}{2\pi R'_{in} C_G} \tag{13.42}$$

where

$$R'_{in} = R_S \| R_{in}$$

and

$$f_{2D} = \frac{1}{2\pi r_D C_D} \tag{13.43}$$

The following example illustrates the process for determining the upper cutoff frequency for a FET amplifier.

━━━━━━━━━━━━━━━━ **EXAMPLE 13.29** ━━━━━━━━━━━━━━━━

Determine the values of f_{2G} and f_{2D} for the amplifier in Figure 13.33. Assume that the FET has values of $C_{gd} = 4$ pF, $C_{gs} = 5$ pF, and $C_{ds} = 2$ pF. Also assume that the input capacitance to the load is 1 pF.

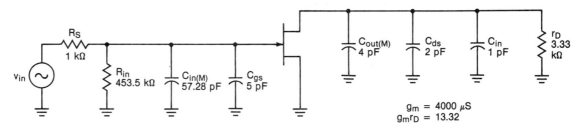

FIGURE 13.29

Solution: The first step is to draw the high-frequency equivalent of the circuit. This equivalent circuit is shown in Figure 13.29.

The value of R_{in} was found by taking R_1 in parallel with R_2. Combining this value with R_S, we get the total ac resistance in the gate circuit, as follows:

$$R'_{in} = R_S \,\|\, R_{in}$$
$$= 998 \ \Omega$$

Now the Miller input capacitance is found as

$$C_{in(M)} \cong C_{gd} g_m r_D$$
$$= (4 \text{ pF})(13.32)$$
$$= 53.28 \text{ pF}$$

Using this value of $C_{in(M)}$ and the value of C_{gs}, the total gate circuit capacitance is found as

$$C_G = C_{gs} + C_{in(M)}$$
$$= 5 \text{ pF} + 53.28 \text{ pF}$$
$$= 58.28 \text{ pF}$$

We can now use this value along with the value of R'_{in} to find f_{2G}, as follows:

$$f_{2G} = \frac{1}{2\pi R'_{in} C_G}$$
$$= \frac{1}{2\pi(998 \ \Omega)(58.28 \text{ pF})}$$
$$= 2.74 \text{ MHz}$$

Since $g_m r_D \gg 1$ for this circuit, the Miller output capacitance is equal to C_{gd}, 4 pF. Combining this value with C_{ds} and C_{in}, we get the total capacitance in the drain circuit, as follows:

$$C_D = C_{out(M)} + C_{ds} + C_{in}$$
$$= 4 \text{ pF} + 2 \text{ pF} + 1 \text{ pF}$$
$$= 7 \text{ pF}$$

Using the values of C_D and r_D, we can find the upper cutoff frequency of the drain circuit as follows:

$$f_{2D} = \frac{1}{2\pi r_D C_D}$$

$$= \frac{1}{2\pi(3.33 \text{ k}\Omega)(7 \text{ pF})}$$

$$= 6.83 \text{ MHz}$$

PRACTICE PROBLEM 13–29

The amplifier described in Practice Problem 13–27 (page 610) has values of C_{gd} = 4 pF, C_{gs} = 5 pF, and C_{ds} = 2 pF. Determine the values of f_{2G} and f_{2D} for the circuit. Assume a load capacitance of 0 F.

Since the value of f_{2G} is lower than f_{2D}, the gate circuit determines the overall value of f_2 for the amplifier. Thus the upper cutoff frequency is 2.74 MHz. If the input frequency reaches this value, the value of A_v will be 3 dB lower than $A_{v(mid)}$. Also, further increases in input frequency would cause the value of A_v to continue to drop at a rate of 6 dB/octave until 6.83 MHz is reached. If frequency continues to increase, the value of A_v will drop at a rate of 12 dB/octave, since both the gate and drain circuits will be cut off. The Bode plot representing this circuit action is shown in Figure 13.30.

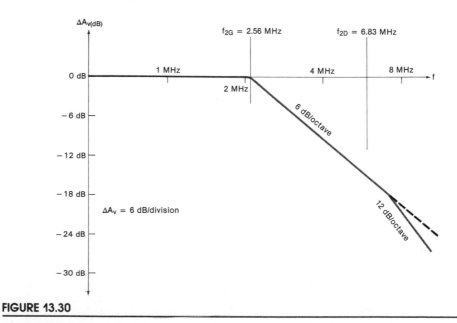

FIGURE 13.30

Capacitance Specifications

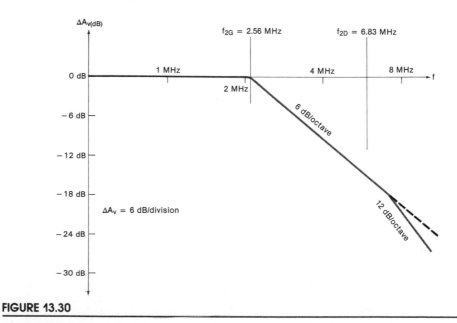35

The only catch with analyzing the high-frequency operation of a FET amplifier is the fact that the values of C_{ds}, C_{gd}, and C_{gs} are not usually listed on the specification

CHAP. 13 Amplifier Frequency Response

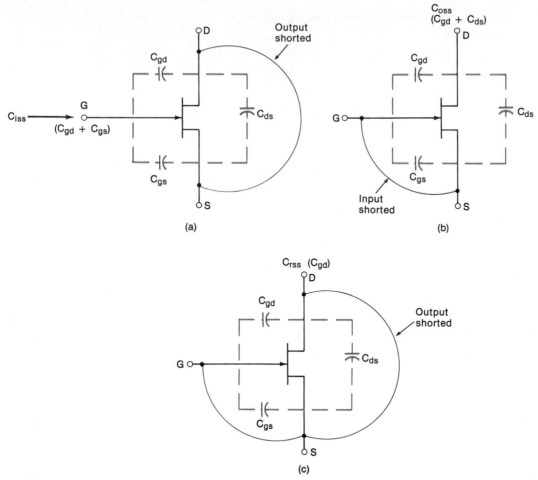

FIGURE 13.31

Measuring C_{iss}, C_{rss}, and C_{oss}.

---

sheet for a given FET. Rather, the spec sheet will list the following: C_{iss}, C_{oss}, and C_{rss}.

C_{iss} is the input capacitance to the FET, measured with the output shorted. The measurement of C_{iss} is illustrated in Figure 13.31a. When the output is shorted and the input capacitance is measured, the measurement is equal to the parallel combination of C_{gs} and C_{gd}. Thus

$$C_{iss} = C_{gs} + C_{gd} \qquad \textbf{(13.44)}$$

C_{oss} is the output capacitance, measured with the gate–source junction shorted. The measurement of C_{oss} is shown in Figure 13.31b. When the gate–source junction is shorted, the total capacitance is the parallel combination of C_{ds} and C_{gd}. Thus

$$C_{oss} = C_{ds} + C_{gd} \qquad \textbf{(13.45)}$$

Finally, C_{rss} is the capacitance measured with both the drain–source and gate–source junctions shorted. With these two junctions shorted, the only capacitance present is C_{gd}, as shown in Figure 13.31c. Thus

$$C_{rss} = C_{gd} \qquad \textbf{(13.46)}$$

Equations (13.43), (13.44), and (13.45) can be used to derive the following useful equations:

$$\boxed{C_{gd} = C_{rss}} \qquad (13.47)$$

$$\boxed{C_{ds} = C_{oss} - C_{rss}} \qquad (13.48)$$

and

$$\boxed{C_{gs} = C_{iss} - C_{rss}} \qquad (13.49)$$

The following example illustrates the use of these equations.

EXAMPLE 13.30

A given FET has values of $C_{oss} = 6$ pF, $C_{rss} = 2$ pF, and $C_{iss} = 10$ pF. Determine the values of C_{gs}, C_{ds}, and C_{gd}.

Solution: The gate–drain capacitance is found as

$$C_{gd} = C_{rss}$$
$$= 2 \text{ pF}$$

The gate–source capacitance is found as

$$C_{gs} = C_{iss} - C_{rss}$$
$$= 10 \text{ pF} - 2 \text{ pF}$$
$$= 8 \text{ pF}$$

Finally, the drain–source capacitance is found as

$$C_{ds} = C_{oss} - C_{rss}$$
$$= 6 \text{ pF} - 2 \text{ pF}$$
$$= 4 \text{ pF}$$

One Final Note

In our discussion on BJT amplifiers, you were shown that there tends to be a high percent of error in the circuit f_2 calculations. For the FET amplifier, the largest percent of error tends to show up in the f_{2G} calculation. The reason for this is the fact that g_m is involved in the calculation of $C_{in(M)}$.

In Chapter 10, you were shown that a JFET has two transconductance curves and, therefore, a range in values of g_m. The large possible range in g_m values can cause a wide range in the value of A_v for the amplifier. The range of A_v values for a FET amplifier will show up in the calculation of $C_{in(M)}$ and, therefore, in the calculation of f_{2G}.

With the wide possible variation in the value of f_{2G}, you should not be surprised when a measured value of f_{2G} is significantly lower (or higher) than the measured value. Again, the exact value of f_2 for an amplifier is only critical when you are dealing with tuned amplifiers. In any standard FET amplifier, you are only concerned with the approximate value of f_2 for the circuit.

1. List, in order, the steps taken to perform the low-frequency analysis of a FET amplifier. [Objective 33]
2. List, in order, the steps taken to perform the high-frequency analysis of a FET amplifier. [Objective 34]
3. List the typical FET capacitance ratings and the equations used to convert them into usable terminal capacitances. [Objective 35]

13.6
Op-Amp Frequency Response

You have already been introduced to some of the frequency considerations that are involved in dealing with op-amps. As a review, here are some of the major points that were made regarding the frequency response of an op-amp:

Reference: These points were introduced on pages 535 to 539.

1. The *slew rate* of an op-amp is a measure of *how fast the output voltage can change*, measured in volts per microsecond V/μs.
2. The maximum operating frequency of an op-amp is found as

$$f_{max} = \frac{\text{slew rate}}{2\pi\ v_{pk}}$$

   Thus, the *peak output voltage limits the maximum operating frequency*.
3. When the maximum output frequency of an op-amp is exceeded, the result is a *distorted* output waveform.
4. Increasing the operating frequency of an op-amp beyond a certain point will:
   a. Decrease the maximum output voltage swing.
   b. Decrease the open-loop voltage gain.
   c. Decrease the input resistance.
   d. Increase the output resistance.

In this section we are going to take a closer look at how frequency affects the operation of an op-amp.

Frequency versus Gain

The gain of an op-amp will remain stable from 0 Hz up to some upper cutoff frequency, f_2. Then the gain will drop at the standard rate of *20 dB/decade*. This operating characteristic is represented in the Bode plot in Figure 13.32. Since the op-amp is a *dc amplifier*, it will exhibit midband voltage gain at 0 Hz. As the frequency of operation increases from 0 Hz, a point will be reached where the gain starts to drop. This drop in gain is due to internal values of input and output capacitance, as will be discussed later in this section.

Don't forget. The Bode plot doesn't show the 3-dB drop at f_2.

    As with any other circuit, f_2 is the upper cutoff frequency for the op-amp. When this frequency is reached, the gain of the op-amp will have dropped by 3 dB. As frequency continues to increase, the gain of the op-amp will continue to drop at the standard 20-dB/decade rate. Thus, *increasing the operating frequency decreases the component gain.*

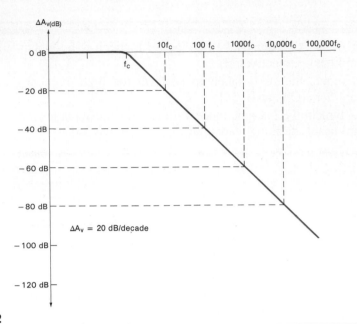

FIGURE 13.32

Op-amp frequency response.

There is another way that we can look at this frequency–gain relationship. That is, *decreasing the gain of an op-amp will increase the maximum operating frequency*. This point is easy to see by taking a look at the Bode plot shown in Figure 13.33. This plot represents the operating characteristics of the μA741 op-amp.

The maximum voltage gain on the Bode plot is shown to be equal to the open-loop voltage gain of the component, A_{OL}. For this op-amp, A_{OL} is shown to be approximately 103 dB (150,000). If you want to operate the μA741 so that $A_{v(dB)}$ is equal to the maximum possible value of A_{OL}, you are limited to a maximum

FIGURE 13.33

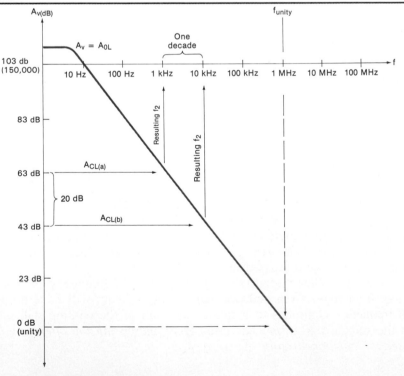

CHAP. 13   Amplifier Frequency Response

operating frequency of about 10 Hz. Above this frequency, the maximum gain of the op-amp drops by more than 3 dB, and the device is considered to be beyond the cutoff frequency.

Now, what if we were to use a *feedback path*, such that the closed-loop gain of the amplifier was equal to 63 dB? What would your maximum operating frequency be now? Figure 13.33 shows that a value of $A_{CL} = 63$ dB has a corresponding value of $f_2 = 1$ kHz. In other words, by intentionally decreasing the gain of the amplifier, we have increased the value of f_2, and thus the *bandwidth* of the device. If we further reduce the value of A_{CL} to 43 dB, f_2 increases to 10 kHz, as does the bandwidth of the amplifier. In fact, each time we decrease the value of A_{CL} by 20 dB, we increase the bandwidth of the amplifier by one decade. Eventually, a point is reached where the value of $A_{CL} = 0$ dB (unity). The frequency that corresponds to this A_{CL} is designated as f_{unity}. For the μA741 op-amp, f_{unity} is approximately equal to 1 MHz, as shown in Figure 13.33.

Based on the fact that gain and bandwidth are *inversely proportional*, we can make the following statements:

The relationship between op-amp gain and bandwidth.

1. *The higher the gain of an op-amp, the narrower its bandwidth.*
2. *The lower the gain of an op-amp, the wider its bandwidth.*

Thus we have a *gain–bandwidth trade-off*. If you want a wide bandwidth, you have to settle for less gain. If you want high gain, you have to settle for a narrower bandwidth.

Gain–Bandwidth Product

A *figure of merit* for a given op-amp is called the *gain–bandwidth product*. The gain–bandwidth product can be used to find:

1. The maximum value of A_{CL} at a given value of f_2
2. The value of f_2 for a given value of A_{CL}

Gain–bandwidth product. A constant, equal to the unity-gain frequency of an op-amp. The product of A_{CL} and bandwidth will always be approximately equal to this constant.

For example, let's say that you want to know the value of f_2 when $A_{CL} = 45$ dB or the value of A_{CL} that will allow an f_2 of 200 kHz. The gain–bandwidth product can be used to solve either of these problems.

The gain–bandwidth product is always equal to the value of f_{unity} for an op-amp. By formula,

$$\boxed{A_{CL}f_2 = f_{unity}} \tag{13.50}$$

At any frequency, the product of A_{CL} and f_2 must equal the unity-gain frequency of the op-amp. For example, refer back to Figure 13.33. As you were shown, $f_{unity} = 1$ MHz for the μA741 op-amp. At 10 Hz,

$$A_{CL}f_2 = (100{,}000)(10 \text{ Hz})$$
$$= 1 \text{ MHz}$$

At 100 Hz,

$$A_{CL}f_2 = (10{,}000)(100 \text{ Hz})$$
$$= 1 \text{ MHz}$$

At 1 kHz,

$$A_{CL} f_2 = (1000)(1 \text{ kHz})$$
$$= 1 \text{ MHz}$$

and so on. By rearranging equation (13.50), we can derive the following useful equations:

$$\boxed{A_{CL} = \frac{f_{unity}}{f_2}} \qquad \textbf{(13.51)}$$

and

$$\boxed{f_2 = \frac{f_{unity}}{A_{CL}}} \qquad \textbf{(13.52)}$$

The following example illustrates the usefulness of these two equations.

■ EXAMPLE 13.31 ■

The LM318 op-amp has a gain–bandwidth product of 15 MHz. Determine the bandwidth of the LM318 when $A_{CL} = 500$, and the maximum value of A_{CL} when $f_2 = 200$ kHz.

Solution: When $A_{CL} = 500$, the value of f_2 is found as

$$f_2 = \frac{f_{unity}}{A_{CL}}$$
$$= \frac{15 \text{ MHz}}{500}$$
$$= 30 \text{ kHz}$$

Since the op-amp is capable of operating as a dc amplifier,

$$BW = f_2$$
$$= 30 \text{ kHz}$$

When $f_2 = 200$ kHz, the maximum value of A_{CL} is found as

$$A_{CL} = \frac{f_{unity}}{f_2}$$
$$= \frac{15 \text{ MHz}}{200 \text{ kHz}}$$
$$= 75 \quad (37.5 \text{ dB})$$

PRACTICE PROBLEM 13–31

An op-amp has a gain–bandwidth product of 25 MHz. What is the bandwidth of the device when $A_{CL} = 200$?

The greatest thing about using the gain–bandwidth product is that it allows us to solve various gain–bandwidth problems without the use of a Bode plot. Here's another type of problem that can be solved using the gain–bandwidth product.

================ **EXAMPLE 13.32** ================

We need to construct an amplifier that has values of $A_{CL} = 500$ and BW = 80 kHz. Can the μA741 op-amp be used in this application?

Solution: The μA741 has a value of $f_{unity} = 1$ MHz. Therefore, any product of $A_{CL}f_2$ must be *less than or equal to* this value. In other words, if $A_{CL}f_2$ is *greater than* f_{unity}, the op-amp cannot be used. For our application,

$$A_{CL}f_2 = (500)(80 \text{ kHz})$$
$$= 40 \text{ MHz}$$

Since this value of $A_{CL}f_2$ is greater than the μA741 f_{unity} rating, the μA741 *cannot* be used for this application.

PRACTICE PROBLEM 13–32

We need to construct an amplifier with values of $A_{CL} = 52$ dB and BW = 10 kHz. We have an op-amp with a gain–bandwidth product of 5 MHz. Determine whether or not the op-amp can be used in this application.

In Chapter 14, we are going to discuss the effects of *feedback* on various types of amplifiers. At that point, you will be shown how to determine the cutoff frequencies, gain, and bandwidth of a given amplifier employing feedback.

Op-amp Internal Capacitance

If you refer back to the internal diagram of the μA741 op-amp (Figure 12.1), you'll see that the circuit contains an internal *compensating capacitor*, C_1. This capacitor, which is used to improve the internal frequency response of the device, limits the high-frequency operation of the component.

As frequency increases, the reactance of C_1 decreases. As this reactance decreases, the capacitor acts more and more like a short circuit. Eventually, a point is reached where a portion of the op-amp internal circuitry is short circuited, effectively reducing the gain of the amplifier to unity. The frequency at which this occurs is the unity-gain frequency of the device.

Determining the Value of f_{unity}

The unity-gain frequency of an op-amp can be determined in a number of ways. Some op-amp spec sheets will list an f_{unity} rating. Others will simply list a *bandwidth* rating. For example, the spec sheet for the μA741 op-amp (Figure 12.30) lists a bandwidth rating of 1 MHz. For this device, 1 MHz is the unity-gain frequency.

When the operating curves for an op-amp are available, the *voltage gain versus operating frequency* curve can be used to determine the value of f_{unity}. For example, refer to the curve for the μA741 op-amp shown in Figure 13.34. By

FIGURE 13.34

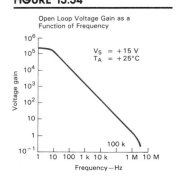

Open Loop Voltage Gain as a Function of Frequency

taking any frequency value and multiplying it by the corresponding limit on voltage gain, you can obtain the value of f_{unity}. From the curve in Figure 13.34, we can approximate the value of f_{unity} as

$$\boxed{\begin{aligned} f_{unity} &= (100 \text{ kHz})(10) \\ &= 1 \text{ MHz} \end{aligned}}$$

Note that the value of $f = 100$ kHz and the corresponding value of $A_{OL} = 10$ were obtained from the curve.

Measuring f_{unity}.

The value of f_{unity} for an op-amp can be measured using the following procedure:

1. Set up an inverting amplifier with a closed-loop gain of 100. (We use this value because it is easy to construct with resistor values of $R_f = 100$ kΩ and $R_i = 1$ kΩ.)
2. Apply an input signal to the amplifier and increase the operating frequency until f_2 is reached, that is, until the peak-to-peak output voltage drops to 0.707 times the midband value.
3. Take the measured value of f_2 and plug it into the following equation:

$$\boxed{f_{unity} = 100 f_2}$$

Note that this equation is simply a form of equation (13.50) that assumes a voltage gain of 100.

One Final Point

As you can see, the bandwidth calculations for the op-amp are much simpler than those for the BJT or FET amplifier. This is another of the many advantages that op-amp circuits have over discrete amplifiers.

Now that we have covered the basics of BJT, FET, and op-amp frequency response, we will conclude this chapter by taking a brief look at the frequency response of multistage amplifiers.

SECTION REVIEW

1. What is the relationship between slew rate and maximum operating frequency? [Objective 36]
2. What is the relationship between peak output voltage and maximum operating frequency? [Objective 36]
3. What is the relationship between op-amp operating frequency and voltage gain? [Objective 37]
4. What is *gain–bandwidth product*? [Objective 38]
5. How can the *voltage gain versus operating frequency* curve for an op-amp be used to determine the value of f_{unity} for the device? [Objective 40]
6. How can the value of f_{unity} for an op-amp be measured? [Objective 40]

CHAP. 13   Amplifier Frequency Response

13.7
Multistage Amplifiers

When you cascade op-amp circuits with identical values of f_2, the overall value of f_2 is found as

$$f_{2T} = f_2 \sqrt{2^{1/n} - 1}$$

(13.53)

where f_{2T} = the overall value of f_2 for the circuit
f_2 = the value of f_2 for one stage
n = the total number of stages

The derivation of equation (13.53) is included in Appendix D.

When you cascade several op-amp stages, the overall value of f_2 will be *lower* than the value of f_2 for a single stage. This point is illustrated in the following example.

===== EXAMPLE 13.33 =====

Two op-amp circuits are cascaded, each having a value of f_2 = 500 kHz. Determine the overall value of f_2 for the two-stage amplifier.

Solution: The overall value of f_2 is found as

$$f_{2T} = f_2 \sqrt{2^{1/n} - 1}$$
$$= (500 \text{ kHz}) \sqrt{2^{1/2} - 1}$$
$$= (500 \text{ kHz})(0.643)$$
$$= 321.8 \text{ kHz}$$

PRACTICE PROBLEM 13–33

Four op-amp stages are cascaded, each having an upper cutoff frequency of f_2 = 800 kHz. Determine the overall value of f_2 for the amplifier.

Most BJT and FET amplifiers have values of f_1, as you have seen in this chapter. When a value of f_1 exists for identical cascaded amplifiers, the overall value of f_1 changes as follows:

$$f_{1T} = \frac{f_1}{\sqrt{2^{1/n} - 1}}$$

increases the value of f1   narrow bandwidth

(13.54)

This equation is also derived in Appendix D.

When you cascade several amplifiers with values of f_1, the overall value of f_1 will be *higher* than the value of f_1 for a single stage. This point is illustrated in the following example.

EXAMPLE 13.34

The amplifiers with values of $f_1 = 5$ kHz are cascaded. Determine the overall value of f_1 for the two-stage amplifier.

Solution: The overall value of f_1 is found as

$$f_{1T} = \frac{f_1}{\sqrt{2^{1/n} - 1}}$$
$$= \frac{5 \text{ kHz}}{\sqrt{2^{1/2} - 1}}$$
$$= \frac{500 \text{ kHz}}{0.643}$$
$$= 7.78 \text{ kHz}$$

PRACTICE PROBLEM 13–34

Three amplifier stages, each having a value of $f_1 = 800$ Hz, are cascaded. Determine the overall value of f_1 for the three-stage amplifier.

For any multistage amplifier, the total bandwidth is found as

$$\boxed{\text{BW}_T = f_{2T} - f_{1T}} \tag{13.55}$$

Since op-amp circuits do not have values of f_1, the overall bandwidth for these circuits is found as

$$\boxed{\text{BW}_T = f_{2T}} \tag{13.56}$$

These two equations should need no further explanation at this point.

When you cascade two or more amplifiers that are *not* identical, the overall circuit is analyzed in the same fashion as we used to analyze the combined effects of f_{1B}, f_{1C}, and f_{1E} for the BJT amplifier. For example, consider the Bode plot shown in Figure 13.35. This plot represents the overall response of a two-stage amplifier with the following characteristics:

Stage	f_1	f_2
1	1 kHz	100 kHz
2	10 kHz	1 MHz

On the low end, the two-stage amplifier cuts off at 10 kHz, the value of f_1 for stage 2. The gain of the amplifier decreases at a *20-dB/decade* rate until the value of f_1 for stage 1 is reached, 1 kHz. At this point, the overall gain decreases at a rate of *40 dB/decade*. The same principles apply to the high-frequency operation. As you can see, the two-stage gain drops by 3 dB at the cutoff frequency of stage 1, 100 kHz. The *20-dB/decade* drop continues until f_2 of stage 2 (1 MHz) is reached. At that point, the roll-off rate increases to *40 dB/decade*.

Note that the bandwidth of the overall amplifier is determined by the *highest* f_1 value and the *lowest* f_2 value. Also, these principles apply to circuits with more

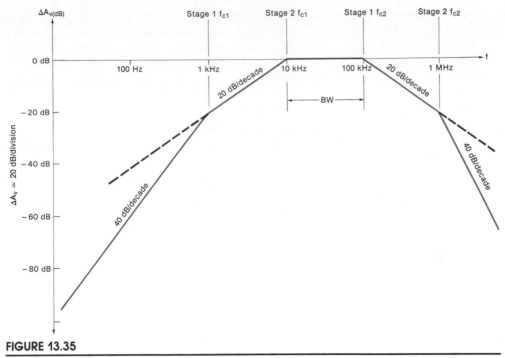

FIGURE 13.35

Overall amplifier frequency response.

than two stages. As each amplifier cuts off, the drop in gain will increase by another 20 dB/decade.

SECTION REVIEW

1. As you cascade identical amplifier stages, what happens to the overall value of f_2? [Objective 41]
2. As you cascade identical amplifer stages, what happens to the overall value of f_1? [Objective 41]
3. As you cascade identical amplifier stages, what happens to the overall value of BW? [Objective 41]
4. How do you plot the frequency response of a cascaded amplifier that does not contain identical stages? [Objective 41]

13.8
Chapter Summary

The *bandwidth* of an amplifier is the range of frequencies over which gain remains relatively constant. The bandwidth of a given amplifier depends on the value of the circuit components and on the type of active component(s) used. When an amplifier is operated within its bandwidth, the gain values for the amplifier are calculated as shown earlier in the text. For clarity, these values of A_i, A_v, and A_p are referred to as *midband* gain values.

The *Bode plot* is a graphic representation of the frequency response of a given amplifier. The bandwidth of the amplifier is bound by the lower and upper *cutoff frequencies*, f_1 and f_2. At these frequencies, the gain of the amplifier is 3 dB down from the value of midband gain. As the Bode plot shows, gain continues to drop at a rate of 20 dB/decade (6 dB/octave) when the frequency continues beyond f_1 or f_2.

The low-frequency response of an amplifier is a function of the circuit resistance values and the values of the bypass and coupling capacitors. The high-frequency response is a function of the circuit resistance values and the *internal* capacitance values of the active component.

When making high-frequency calculations for an inverting circuit, you must take into account the *Miller input/output capacitance values*. These values are used in place of the feedback capacitance in order to make circuit calculations easier. When a given amplifier cuts off, the *gain roll-off rate* is independent of the circuit values of R and C. In fact, these values are used only to determine the exact values of f_1 and f_2. Once these frequencies are passed, the decrease in gain is proportional to the actual frequency.

Since the power gain (A_p) of an amplifier drops by 3 dB (50%) at f_1 and f_2, these two points are commonly referred to as the *lower and upper 3-dB points*. At f_1 and f_2, A_v will equal $0.707A_{v(\text{mid})}$. This relationship gives us a convenient method of measuring the lower and upper 3-dB points with the oscilloscope.

When more than one circuit is in cutoff, the roll-off rate will equal 20 dB *for each cutoff circuit*. Thus, if three circuits are in cutoff, the roll-off rate will equal three times 20 dB/decade. This holds true for any number of cascaded stages.

KEY TERMS

The following terms were introduced and defined in this chapter:

bandwidth	dBm Reference	gain–bandwidth product
Bode plot	decade	Miller's theorem
center frequency	decibel (dB)	octave
current–gain bandwidth product	frequency-response curve	roll-off rate
cutoff frequencies		

PRACTICE PROBLEMS 13.1

1. An amplifier has cutoff frequencies of 1.2 kHz and 640 kHz. Calculate the bandwidth and center frequency of the circuit. [2]

2. An amplifier has cutoff frequencies of 3.4 kHz and 748 kHz. Calculate the bandwidth and center frequency for the circuit. [2]

3. An amplifier has cutoff frequencies of 5.2 kHz and 489.6 kHz. Calculate the bandwidth and center frequency for the circuit. [2]

4. An amplifier has cutoff frequencies of 1.4 kHz and 822.7 kHz. Calculate the bandwidth and center frequency for the circuit. [2]

5. An amplifier has cutoff frequencies of 2.6 kHz and 483.6 kHz. Show that the ratio of f_O to f_1 is equal to the ratio of f_2 to f_O for the circuit. [2]

6. An amplifier has cutoff frequencies of 1 kHz and 345 kHz. Show that the ratio of f_O to f_1 is equal to the ratio of f_2 to f_O for the circuit. [2]

7. The values of f_O and f_2 for an amplifier are measured at 72 kHz and 548 kHz, respectively. Calculate the values of f_1 and bandwidth for the amplifier. [4]

8. The values of f_O and f_2 for an amplifier are measured at 22.8 kHz and 321 kHz, respectively. Calculate the values of f_1 and bandwidth for the circuit. [4]

9. The values of f_O and f_1 for an amplifier are measured at 48 kHz and 2 kHz, respectively. Calculate the values of f_2 and bandwidth for the circuit. [4]

10. The values of f_O and f_1 for an amplifier are measured at 36 kHz and 4.3 kHz, respectively. Calculate the values of f_2 and bandwidth for the circuit. [4]

§13.2

11. Complete the following table. [7]

	P_{in}	P_{out}	$A_{p(dB)}$
a.	10 mW	5 W	_____
b.	100 mW	1 W	_____
c.	1.25 mW	4 W	_____
d.	12 W	6 W	_____

12. Complete the following table. [7]

	P_{in}	$P_{out}(W)$	$A_{p(dB)}$
a.	_____	12	3
b.	_____	6.2	12
c.	_____	300	50
d.	_____	25	6
e.	_____	8	−5

13. Complete the following table. [7]

	P_{in}	P_{out}	$A_{p(dB)}$
a.	14.7 mW	_____	12
b.	25 mW	_____	60
c.	5 W	_____	−8
d.	12 W	_____	−14

14. A three-stage amplifier has power gain values of $A_{p1} = 2600$, $A_{p2} = 950$, and $A_{p3} = 200$. Convert each power gain value to dB form, and then prove that the total power gain is equal to the sum of the individual dB power gain values. [9]

15. A three-stage amplifier has values of $A_{p1} = 1400$, $A_{p2} = 320$, and $A_{p3} = 400$. Convert each of the power gain values to dB form, and then prove that the total power gain is equal to the sum of the individual dB power gain values. [9]

16. An amplifier has an output power of 62 dBm. Calculate the actual output power of the circuit. [12]

17. An amplifier has an output power of 16 dBm. Calculate the actual output power of the circuit. [12]

18. An amplifier has an output power of 22 dBm. Calculate the actual output power of the circuit. [12]

19. Complete the following table. [13]

	v_{in}	v_{out}	$A_{v(dB)}$
a.	1 mV	10 V	_____
b.	120 μV	16 V	_____
c.	5 V	2.5 V	_____
d.	14.4 mV	14 mV	_____

20. Complete the following table. [13]

	v_{in}	v_{out}	$A_{v(dB)}$
a.	_____	12 V	6
b.	_____	10 V	40
c.	_____	8 V	-10
d.	_____	16 V	0

21. Complete the following table. [13]

	v_{in}	v_{out}	$A_{v(dB)}$
a.	12 mV	_____	16
b.	10 mV	_____	5
c.	100 mV	_____	-3
d.	80 V	_____	-22

22. An amplifier has values of $A_{p(mid)} = 31$ dB and $A_{v(mid)} = 12$ dB. Determine the values of A_p and A_v (in standard numeric form) at the cutoff frequencies. [14]

13.3

23. Calculate the value of f_{1B} for the amplifier in Figure 13.36. [19]
24. Calculate the value of f_{1B} for the amplifier in Figure 13.37. [19]
25. Calculate the value of f_{1C} for the amplifier in Figure 13.36. [19]
26. Calculate the value of f_{1C} for the amplifier in Figure 13.37. [19]
27. Calculate the value of f_{1E} for the amplifier in Figure 13.36. [19]
28. Calculate the value of f_{1E} for the amplifier in Figure 13.37. [19]

FIGURE 13.36

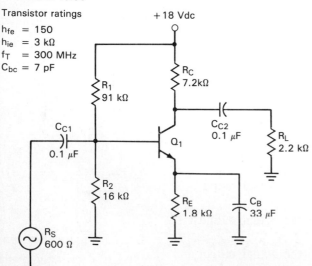

Transistor ratings
h_{fe} = 150
h_{ie} = 3 kΩ
f_T = 300 MHz
C_{bc} = 7 pF

FIGURE 13.37

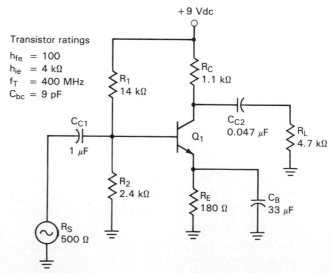

Transistor ratings
h_{fe} = 100
h_{ie} = 4 kΩ
f_T = 400 MHz
C_{bc} = 9 pF

CHAP. 13   Amplifier Frequency Response

29. A given amplifier has values of $A_{v(mid)}$ = 32 dB and f_1 = 8 kHz. Calculate the dB voltage gain of the amplifier at operating frequencies of 7, 5, 4, and 1 kHz. [21]

30. A given amplifier has values of $A_{v(mid)}$ = 16 dB and f_1 = 12 kHz. Calculate the dB voltage gain of the amplifier at operating frequencies of 18, 12, 10, and 2 kHz. [21]

31. Compare the results of problems 23, 25, and 27. What is the approximate value of f_1 for the amplifier described in these three problems? [23]

32. Compare the results of problems 24, 26, and 28. What is the approximate value of f_1 for the amplifier described in these three problems? [23]

§13.4

33. A transistor has a value of f_T = 200 MHz at I_C = 10 mA. Determine the value of C_{be} at this current. [26]

34. A transistor has a value of f_T = 400 MHz at I_C = 1 mA. Determine the value of C_{be} at this current. [26]

35. An inverting amplifier has values of C_{bc} = 7 pF and A_v = 100. Determine the Miller input and output capacitance values for the circuit. [29]

36. An inverting amplifier has values of C_{bc} = 3 pF and A_v = 4. Determine the Miller input and output capacitance values for the circuit. [29]

37. Determine the value of f_{2B} for the amplifier in Figure 13.36. [30]

38. Determine the value of f_{2B} for the amplifier in Figure 13.37. [30]

39. Determine the value of f_{2C} for the amplifier in Figure 13.36. [30]

40. Determine the value of f_{2C} for the amplifier in Figure 13.37. [30]

41. Calculate the overall values of f_1 and f_2 for the amplifier in Figure 13.38. [19, 30]

42. Calculate the overall values of f_1 and f_2 for the amplifier in Figure 13.39. [19, 30]

§13.5

43. Calculate the value of f_{1G} for the amplifier in Figure 13.40. [33]

44. Calculate the value of f_{1G} for the amplifier in Figure 13.41. [33]

FIGURE 13.38

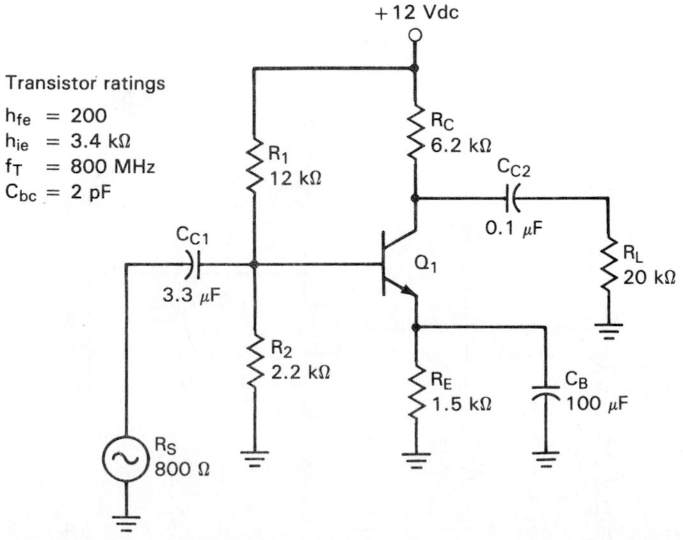

FIGURE 13.39

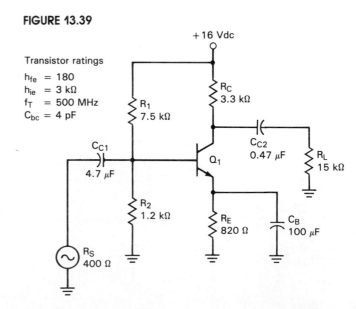

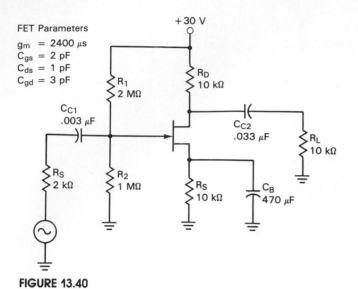

FIGURE 13.40

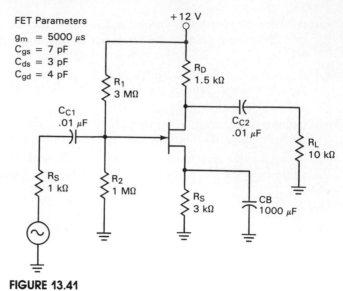

FIGURE 13.41

45. Calculate the value of f_{1D} for the amplifier in Figure 13.40. [33]

46. Calculate the value of f_{1D} for the amplifier in Figure 13.41. [33]

47. Calculate the values of f_{2G} and f_{2D} for the amplifier in Figure 13.40. [34]

48. Calculate the values of f_{2G} and f_{2D} for the amplifier in Figure 13.41. [34]

49. A FET has ratings of $C_{iss} = 14$ pF, $C_{rss} = 9$ pF, and $C_{oss} = 10$ pF. Determine the values of C_{gd}, C_{gs}, and C_{ds} for the device. [35]

50. A FET has ratings of $C_{oss} = 16$ pF, $C_{rss} = 9$ pF, and $C_{iss} = 12$ pF. Determine the values of C_{gd}, C_{gs}, and C_{ds} for the device. [35]

51. Calculate the overall values of f_1 and f_2 for the amplifier in Figure 13.42. [33, 34]

52. Calculate the overall values of f_1 and f_2 for the amplifier in Figure 13.43. [33, 34]

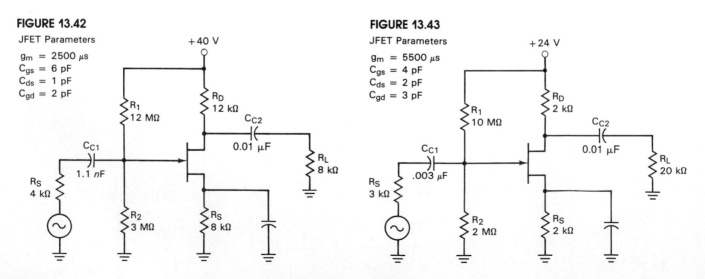

FIGURE 13.42

FIGURE 13.43

53. An op-amp has a gain-bandwidth product of 12 MHz. Determine the bandwidth of the device when $A_{CL} = 400$. [39]

54. An op-amp has a gain–bandwidth product of 14 MHz. Determine the bandwidth of the device when $A_{CL} = 320$. [39]

55. An op-amp has a gain–bandwidth product of 25 MHz. Determine the bandwidth of the device when $A_{CL} = 42$ dB. [13, 39]

56. An op-amp has a gain–bandwidth product of 1 MHz. Determine the bandwidth of the device when $A_{CL} = 20$ dB. [13, 39]

57. We need to construct an amplifier that has values of $A_{CL} = 220$ and $f_2 = 120$ kHz. Can we use an op-amp with $f_{unity} = 28$ MHz for this application? [39]

58. We need to construct an amplifier that has values of $A_{CL} = 24$ dB and $f_2 = 40$ kHz. Can we use an op-amp with $f_{unity} = 1$ MHz for this application? [39]

59. An op-amp circuit with $A_{CL} = 120$ has a measured value of $f_2 = 100$ kHz. What is the gain–bandwidth product of the op-amp? [40]

60. An op-amp circuit with $A_{CL} = 300$ has a measured value of $f_2 = 88$ kHz. What is the gain–bandwidth product of the op-amp? [40]

61. The circuit in Figure 13.44 has a measured f_2 of 250 kHz. What is the value of f_{unity} for the op-amp? [40]

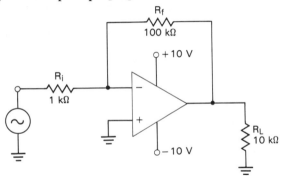

FIGURE 13.44

62. The circuit in Figure 13.45 has a measured f_2 of 100 kHz. What is the value of f_{unity} for the op-amp? [40]

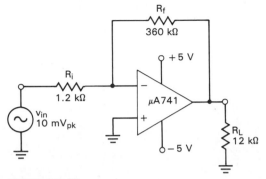

FIGURE 13.45

§13.7

63. Three op-amps are cascaded, each with a value of $f_2 = 200$ kHz. Determine the overall value of f_2 for the circuit. [41]

64. Two op-amps are cascaded, each having a value of $f_2 = 120$ kHz. Determine the overall value of f_2 for the circuit. [41]

65. Two amplifiers are cascaded, each having a value of $f_1 = 3$ kHz. Determine the overall value of f_1 for the circuit. [41]

66. Three amplifiers are cascaded, each having a value of $f_1 = 4$ kHz. Determine the overall value of f_1 for the circuit. [41]

67. Two amplifiers are cascaded, each having values of $f_1 = 2$ kHz and $f_2 = 840$ kHz. Determine the bandwidth of the circuit [41]

68. Four amplifiers are cascaded, each having values of $f_1 = 1.5$ kHz and $f_2 = 620$ kHz. Determine the bandwidth of the circuit. [41]

69. Three op-amps are cascaded, each having a value of $f_2 = 325$ kHz. Determine the bandwidth of the circuit. [41]

70. Two op-amps are cascaded, each having a value of $f_2 = 428$ kHz. Determine the bandwidth of the circuit. [41]

THE BRAIN DRAIN

71. We have two amplifiers. The first amplifier has values of $f_{1B} = 2$ kHz, $f_{1C} = 16$ kHz, and $f_{1E} = 32$ kHz. The second amplifier has values of $f_{1B} = 8$ kHz, $f_{1C} = 12$ kHz, and $f_{1E} = 4$ kHz. For each of these amplifiers:

   a. Draw the Bode plot, including the appropriate roll-off rates. The Bode plots should have an octave frequency scale that starts at 2 kHz and continues up to 32 kHz.

   b. Plot the frequency-response curve on the same graph as the Bode plot. The frequency-response curve must take into account the combined effects of f_{1B}, f_{1C}, and f_{1E} at each major division on the graph. In other words, at each major division, calculate the values of $\Delta A_{v(\text{dB})}$ for each of the terminal circuits, determine the total value of $\Delta A_{v(\text{dB})}$, and plot the point that corresponds to the total value of $\Delta_{v(\text{dB})}$.

   After completing the two graphs, compare the results to see which frequency-response curve more closely resembles its Bode plot.

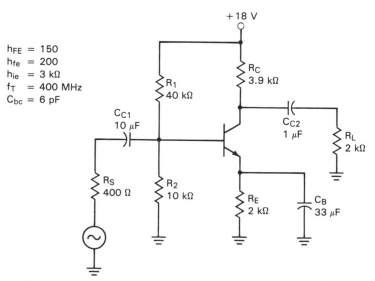

FIGURE 13.46

72. Calculate the change in f_0 that occurs in the circuit in Figure 13.46 if the load resistance opens.

SUGGESTED COMPUTER APPLICATIONS PROBLEMS

73. Write a program that will determine the lower cutoff frequencies of a BJT amplifier, given the proper input values.

74. Write a program that will determine the upper cutoff frequencies of a BJT amplifier, given the proper input values.

75. Combine the programs from problems 74 and 75. Then add program steps to determine the overall values of f_1, f_2, f_O, and bandwidth for the circuit.

76. Repeat problems 74 through 76 for a JFET amplifier.

77. Repeat problem 77 for an inverting amplifier.

ANSWERS TO EXAMPLE PRACTICE PROBLEMS

13.1.	1.0995 MHz	13.18.	14.2 Hz
13.2.	23.45 kHz	13.19.	54.5 Hz
13.3.	Both ratios equal 46.9	13.20.	16.01 dB
13.4.	297 kHz	13.21.	318 pF
13.5.	120 kHz	13.22.	4 pF
13.6.	41.55 dB	13.23.	1.17 MHz
13.7.	100	13.24.	37.54 MHz
13.8.	132.5 μW	13.26.	3.2 Hz
13.9.	126.6 mW	13.27.	260.9 Hz
13.10.	$A_{pT} = 55.88$ dB	13.28.	17.3 pF
13.12.	1.59 W	13.29.	$f_{2G} = 14.2$ MHz, $f_{2D} = 34.2$ MHz
13.13.	49.88 dB	13.31.	125 kHz
13.14.	3.98 dB	13.32.	Yes, it can be used.
13.15.	0.631	13.33.	348 kHz
13.17.	45.2 Hz	13.34.	1.57 kHz

14
Negative Feedback

Feedback can take a wide variety of forms. Here, sensors feed stress information back to a computer, allowing the robotic hand to hold the egg without crushing it.

OBJECTIVES

After studying the material in this chapter, you should be able to:

☐ 1. Define *feedback amplifier* and state the purpose served by the feedback network. (Introduction)

☐ 2. Describe *negative feedback* and its effects on circuit operation. (§14.1)

☐ 3. Describe *positive feedback* and its primary application. (§14.1)

☐ 4. Explain how *negative voltage feedback* reduces the input voltage to an amplifier. (§14.1)

☐ 5. Explain how *negative current feedback* reduces the input current to an amplifier. (§14.1)

☐ 6. Compare and contrast the effects that negative voltage and current feedback have on circuit values of gain and impedance. (§14.1)

☐ 7. Describe the effect that negative voltage feedback has on the effective voltage gain of an amplifier. (§14.2)

☐ 8. Define the terms *attenuation factor* and *feedback factor*. (§14.2)

☐ 9. Calculate the voltage gain of an amplifier with negative voltage feedback (A_{vf}), given the original value of voltage gain (A_v) and the attenuation factor (α_v). (§14.2)

☐ 10. Discuss and calculate the effects of negative voltage feedback on the input impedance of an amplifier. (§14.2)

☐ 11. Discuss and calculate the effects of negative voltage feedback on the bandwidth and cutoff frequencies of an amplifier. (§14.2)

☐ 12. Explain the basic approach that is taken in performing the mathematical analysis of an amplifier with negative feedback. (§14.2)

☐ 13. Describe and analyze the effects that negative voltage feedback has on the noninverting op-amp values of gain, input and output impedance, and bandwidth. (§14.3)

☐ 14. Describe and analyze the effects that negative voltage feedback has on the inverting op-amp values of gain, input and output impedance, and bandwidth. (§14.3)

☐ 15. Describe and analyze the effects that negative current feedback has on the open-loop values of gain, impedance, and bandwidth for a given amplifier. (§14.4)

☐ 16. Describe and analyze the effects that negative current feedback has on the open-loop gain, impedance, and bandwidth values of a noninverting op-amp. (§14.5)

☐ 17. Describe and analyze the effects that negative current feedback has on the open-loop gain, impedance, and bandwidth values of an inverting op-amp. (§14.5)

☐ 18. State the primary symptom of a feedback network fault. (§14.6)

☐ 19. Describe and analyze the effects of negative voltage feedback on the gain, impedance, and frequency characteristics of a discrete amplifier. (§14.7)

☐ 20. Describe and analyze the effects of negative current feedback on the gain, impedance, and frequency characteristics of a discrete amplifier. (§14.7)

As you master each of the objectives listed, place a check mark (✔) in the appropriate box.

A *feedback amplifier* is one that provides a path from the amplifier output back to the input. The purpose of this path is to "feed" a portion of the output signal back to the amplifier input. The basic feedback amplifier block diagram is shown in Figure 14.1. With the *feedback network*

Feedback amplifier. An amplifier that provides a signal path from the output back to the input.

FIGURE 14.1

Feedback.

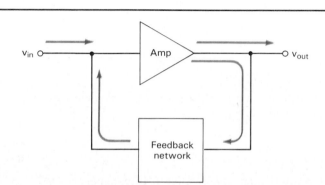

Feedback network. The feedback signal path.

(circuit or path) connecting the output of the amplifier back to the input, a small portion of the output signal is felt at the amplifier input. The effect that this feedback signal has on the amplifier depends on the *type* of feedback used.

14.1
Introductory Concepts

Negative **feedback.** The feedback signal is out of phase with the input signal.

There are two basic types of feedback: *negative feedback* and positive *feedback*. When *negative feedback* is used, the feedback signal is *out of phase* with the input signal. This process is illustrated in Figure 14.2. As you can see, the amplifier introduces a 180° phase shift into the circuit, while the feedback network does not. The result is that the *feedback voltage*, v_f, is 180° out of phase with v_{in}.

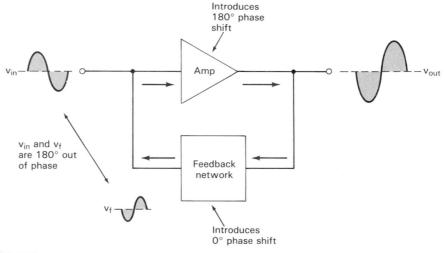

FIGURE 14.2

Negative feedback.

Negative feedback can be used to serve several useful purposes. As you will be shown, negative feedback can:

The effects of *negative* feedback.

1. Increase bandwidth
2. Increase amplifier stability
3. Change the amplifier input and/or output impedance

CHAP. 14   Negative Feedback

The primary disadvantage of using negative feedback is that is reduces the effective *gain* of the amplifier. However, you'll see that this is not a problem when we're dealing with op-amp circuits.

Positive feedback provides a feedback signal that is *in phase* with the amplifier input signal. This process is illustrated in Figure 14.3. As you can see, both the amplifier and feedback network introduce a 180° phase shift. The result is a 360° phase shift around the loop, causing v_f to be in phase with the original signal, v_{in}.

Positive feedback. The feedback signal is in phase with the input signal.

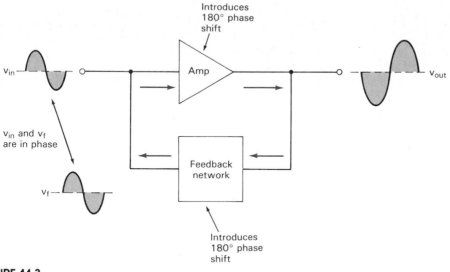

FIGURE 14.3

Positive feedback.

Positive feedback is used mostly in a special type of circuit called an *oscillator*. As you will be shown in Chapter 17, an oscillator is an amplifier that will convert a dc voltage to an ac signal. In this chapter, we will concern ourselves with negative feedback and its effects on circuit operation.

Oscillator. An amplifier that converts dc to ac.

Current versus Voltage Feedback

Negative feedback can be further divided into two types: *voltage feedback* and *current feedback*. Both of these feedback configurations are illustrated in Figure 14.4. When *voltage feedback* is used, the inputs of the feedback network are

Voltage feedback feeds a portion of the output voltage back to the amplifier input.

FIGURE 14.4

Voltage feedback versus current feedback.

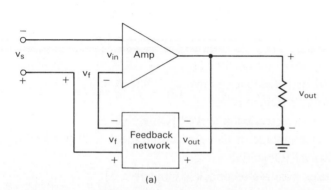

(a)

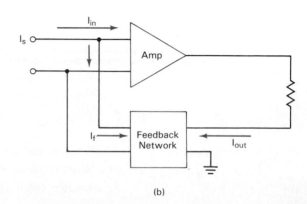

(b)

connected across the load. Thus the input voltage to the feedback network is equal to the output voltage of the amplifier. The feedback network then *attenuates* (reduces) this voltage and provides the feedback voltage at its output. As you can see, this feedback voltage is applied *in series* with the amplifier input. Since v_f is 180° out of phase with the source voltage, v_s, the input voltage to the amplifier is equal to the *difference* between the two. By formula,

$$v_{in} = v_s - v_f \qquad \textbf{(14.1)}$$

Thus the *voltage feedback amplifier* has reduced the *input voltage* to the amplifier.

✔ 5

Current feedback feeds a portion of the output current back to the amplifier input.

When *current feedback* is used, the amplifier output and the feedback network input are connected in *series* through the load. Thus the input current into the feedback network is equal to the output and load current. The feedback network is designed to draw a *proportional* amount of current from the amplifier input. Because of this, the input current to the amplifier is equal to the *difference* between the source current, I_s, and the feedback current, I_f. By formula,

$$I_{in} = I_s - I_f \qquad \textbf{(14.2)}$$

Again, the feedback network has reduced the amplifier input. However, in this case it is the input *current* that is reduced.

Why use two different types of feedback? The effects of voltage feedback are significantly different than those of current feedback. This is shown in the following table.

Feedback Type	Effect on:			
	A_v	A_i	Z_{in}	Z_{out}
Voltage	Decreases	None	Varies[a]	Decreases
Current	None	Decreases	Varies[a]	Increases

[a] Depending on the type of circuit.

✔ 6

As you can see, voltage feedback causes A_v and Z_{out} to decrease. It does not affect the current gain of the circuit, and will either increase or decrease the circuit input impedance, depending on the exact circuit configuration.

Current feedback decreases current gain and increases output impedance. It does not affect the voltage gain of the amplifier, and will either increase or decrease the circuit input impedance, depending on the exact circuit configuration.

Since A_p is the product of A_v and A_i, both types of feedback will decrease the power gain of an amplifier. However, *the drawback of reduced power gain is offset by the advantage of increased bandwidth*. This point will be demonstrated throughout this chapter.

SECTION REVIEW

1. What is a *feedback amplifier*? [Objective 1]
2. What purpose is served by a feedback network? [Objective 1]
3. What is *negative feedback*? [Objective 2]
4. What purposes does negative feedback serve? [Objective 2]
5. What is the primary disadvantage of using negative feedback? [Objective 2]
6. What is *positive feedback*? [Objective 3]
7. What type of circuit uses positive feedback? [Objective 3]

8. Describe the means by which the input voltage is reduced in the circuit in Figure 14.4a. [Objective 4]
9. Describe the means by which the input current is reduced in the circuit in Figure 14.4b. [Objective 5]
10. List the differences between voltage feedback and current feedback. [Objective 6]

14.2
Negative Voltage Feedback: An Overview

It was stated earlier that negative voltage feedback *reduces A_v, changes Z_{in}, and decreases Z_{out}*. To see how negative voltage feedback does all of this, refer to the block diagram in Figure 14.5. Here we see the feedback network being represented as an impedance, Z_f. (The basis for this representation will be shown later in this section.)

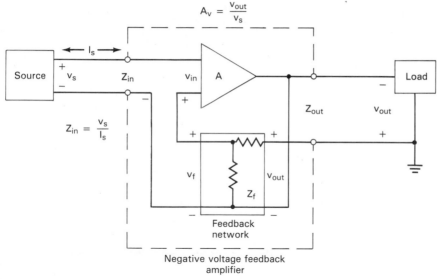

Negative voltage feedback amplifier

FIGURE 14.5

The effect of negative feedback on amplifier input impedance.

If you follow from the $(-)$ terminal of the source to the input of A, you'll see that Z_f is a *series* impedance in this circuit. Because we have added Z_f in series with the input circuit, the effective value of Z_{in} for the amplifier has been increased.

The same impedance is shown as being in *parallel* with the output impedance of A. Since the combined parallel impedance *must* be lower than the output impedance of A, the overall value of Z_{out} for the amplifier has been reduced.

The reduction of A_v for the amplifier is easiest to see if we mess around with some numbers. Let's assume for a moment that the circuit in Figure 14.5 has the following values: $A_{v(A)} = 100$, $v_s = 100$ mV, and $v_f = 20$ mV. Note that $A_{v(A)}$ is the voltage gain of A, as determined by the circuit component values. Now, using the values listed above, equation (14.1) would be used to determine

How voltage feedback affects gain.

the value of v_{in} as follows:

$$v_{in} = v_s - v_f$$
$$= 100 \text{ mV} - 20 \text{ mV}$$
$$= 80 \text{ mV}$$

Now the output voltage from A would be found as

$$v_{out} = A_{v(A)}v_{in}$$
$$= (100)(80 \text{ mV})$$
$$= 8 \text{ V}$$

However, the *effective* voltage gain of the amplifier is the ratio of output voltage to source voltage, found as

$$A_{v(eff)} = \frac{v_{out}}{v_s}$$
$$= \frac{8 \text{ V}}{100 \text{ mV}}$$
$$= 80$$

Thus, even though the active component, A, has a voltage gain of 100, the *effective* voltage gain of the amplifier is 80. In effect, the voltage gain of the amplifier has been reduced by using negative voltage feedback.

How voltage feedback affects amplifier bandwidth.
How does this feedback increase the bandwidth of the amplifier? Take a look at the frequency-response curve shown in Figure 14.6. The dashed portion of the Bode plot represents the bandwidth of Figure 14.5 *without feedback*. As you can see, this amplifier would have a bandwidth defined by the half-power points, f_{1A} and f_{2A}. Also, the standard roll-off rate of 20 dB/decade is shown.

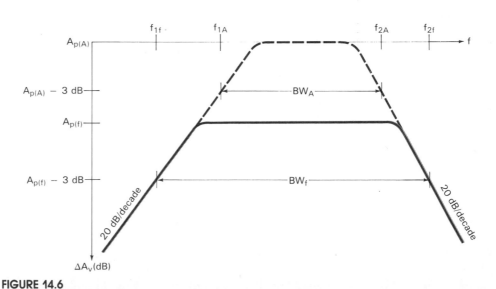

FIGURE 14.6

The effect of negative feedback on amplifier gain and bandwidth.

By adding the feedback network, the power gain of the amplifier is reduced to the value $A_{p(f)}$. The bandwidth for the new amplifier is defined by the half-power points, f_{1f} and f_{2f}. As you can see, the bandwidth of the amplifier has increased significantly. If you have trouble seeing this, just consider the concept

CHAP. 14   Negative Feedback

of *gain–bandwidth* product. Any time that you decrease the gain of an amplifier, the bandwidth *must* increase in order to maintain the constant gain–bandwidth product for the circuit.

Mathematical Analysis

Now that you have been introduced to the basic principles of negative feedback, it is time to get into the mathematical analysis of a feedback amplifier. For this discussion we will be referring to the block diagram shown in Figure 14.7. First, we are going to consider the feedback amplifier to be a pair of gain values. The amplifier itself is shown to be a voltage gain, defined as

$$A_v = \frac{v_{\text{out}}}{v_{\text{in}}}$$

Be sure that you do not confuse this gain with the *effective* gain of the overall amplifier. In Figure 14.7 we are referring only to the gain of the active component itself.

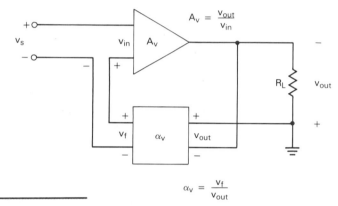

FIGURE 14.7

$$\alpha_v = \frac{v_f}{v_{\text{out}}}$$

The feedback network is represented as an *attenuation*, α. Note that the Greek letter α is commonly used to represent a voltage and/or power loss. Don't confuse this use of α with the collector-emitter current ratio discussed earlier in the text.

The *attenuation factor*, α_v, of the feedback network is found as

$$\alpha_v = \frac{v_f}{v_{\text{out}}} \tag{14.3}$$

Note that v_f will always be less than v_{out}. For this reason, α_v will always be less than 1.

Now, as stated earlier, the *effective gain* of the feedback amplifier is found as

$$A_{vf} = \frac{v_{\text{out}}}{v_s} \tag{14.4}$$

In Appendix D, we use equations (14.1), (14.3), and (14.4) to derive the following useful equation:

$$A_{vf} = \frac{A_v}{1 + \alpha_v A_v} \tag{14.5}$$

✔ 8

Attenuation factor (α_v). The ratio of feedback voltage to output voltage. The value of α_v will always be less than 1.

where

A_{vf} = the *effective* gain of the voltage feedback amplifier
A_v = the *open-loop* voltage gain of the amplifier (that is, the voltage gain that the amplifier would have if there was no feedback path present)

As you will see, the value that appears in the denominator of equation (14.5), $1 + \alpha_v A_v$, appears in almost every equation for a given negative feedback amplifier. To simplify things, we will call this value the *feedback factor* for a given amplifier. In most cases, the values of gain, impedance, cutoff frequencies, and bandwidth will either be increased or decreased by the feedback factor of the circuit. The following example illustrates the effect of the feedback factor on the voltage gain of an amplifier.

EXAMPLE 14.1

The amplifier in Figure 14.7 has values of $A_v = 500$ and $\alpha_v = 0.001$. What is the effective gain of the amplifier?

Solution: Using the given values, A_{vf} is found as

$$
\begin{aligned}
A_{vf} &= \frac{A_v}{1 + \alpha_v A_v} \\
&= \frac{500}{1 + (0.001)(500)} \\
&= 333.33
\end{aligned}
$$

PRACTICE PROBLEM 14-1

An amplifier has values of $A_v = 800$ and $\alpha_v = 0.03$. Calculate the value of A_{vf} for the circuit.

When the value of $\alpha_v A_v$ is *much greater than* 1, you can approximate the effective gain of the amplifier as

$$
\boxed{A_{vf} \cong \frac{1}{\alpha_v}} \tag{14.6}
$$

This is shown in the following example.

EXAMPLE 14.2

The amplifier in Figure 14.7 has values of $A_v = 1000$ and $\alpha_v = 0.01$. Determine the effective gain of the amplifier using *both* equations (14.5) and (14.6) and compare the results.

Solution: Using equation (14.5), the value of A_{vf} is found as

$$
\begin{aligned}
A_{vf} &= \frac{A_v}{1 + \alpha_v A_v} \\
&= \frac{1000}{1 + (0.01)(1000)} \\
&= 90.9
\end{aligned}
$$

CHAP. 14 Negative Feedback

Using equation (14.6), the value of A_{vf} is found as

$$A_{vf} = \frac{1}{\alpha_v}$$
$$= \frac{1}{0.01}$$
$$= 100$$

Since the margin of error is less than 10% for the approximation formula, it can be used in this case.

PRACTICE PROBLEM 14–2

An amplifier has values of $A_v = 420$ and $\alpha_v = 0.035$. Determine whether or not equation (14.6) could be used to approximate the value of A_{vf} for the circuit.

Note that the value of $\alpha_v A_v$ was 10 in Example 14.2. Since this value is much greater than 1, the approximation formula is valid. In Example 14.1, the value of $\alpha_v A_v$ is 0.5. In this case, the approximation formula would have yielded an answer of $A_{vf} = 1000$, which is way off from the actual value. Thus the approximation formula could *not* have been used for a situation like the one in Example 14.1.

The Effects of Negative Voltage Feedback on Circuit Impedance Values

For most amplifiers, the addition of negative voltage feedback will *increase* the input impedance of the amplifier. The exception to this rule is the *inverting amplifier*, as you will be shown in Section 14.4. In this discussion, we will concentrate on the means by which negative feedback can be used to *increase* the input impedance of a given noninverting or discrete amplifier.

The input impedance of a given amplifier can be found as

$$Z_{in} = \frac{v_s}{i_s} \qquad (14.7)$$

This was shown in Figure 14.5. Now v_s has been shown to be

$$v_s = v_{in} + v_f$$

and, since $v_f = \alpha_v v_{out}$,

$$v_s = v_{in} + \alpha_v v_{out}$$
$$= v_{in} + \alpha_v (A_v v_{in})$$
$$= v_{in}(1 + \alpha_v A_v)$$

Substituting this value into equation (14.7) yields the following:

$$Z_{in(f)} = Z_{in}(1 + \alpha_v A_v) \qquad (14.8)$$

where
$Z_{in(f)}$ = the input impedance to the feedback amplifier
Z_{in} = the input impedance to the amplifier without feedback, calculated as shown earlier in the text

The following example demonstrates the effect that adding a feedback network can have on amplifier input impedance.

EXAMPLE 14.3

The amplifier described in Example 14.2 has a value of $Z_{in} = 1$ kΩ when there is no feedback network. What is the value of Z_{in} *with* the feedback?

Solution: In Example 14.2 we were given values of $A_v = 1000$ and $\alpha_v = 0.01$. Using these values in equation (14.8), the value of $Z_{in(f)}$ is found as

$$
\begin{aligned}
Z_{in(f)} &= Z_{in}(1 + \alpha_v A_v) \\
&= (1 \text{ k}\Omega)[1 + (0.01)(1000)] \\
&= (1 \text{ k}\Omega)(11) \\
&= 11 \text{ k}\Omega
\end{aligned}
$$

PRACTICE PROBLEM 14–3

An amplifier has values of $Z_{in} = 4.2$ kΩ, $A_v = 220$, and $\alpha_v = 0.01$. Determine the value of $Z_{in(f)}$ for the circuit.

As you can see, the addition of a feedback network can significantly increase the input impedance of the amplifier. This is an advantage, since the amplifier will now present less of a load to its source circuit.

While the input impedance of the amplifier is increased by a factor of $(1 + \alpha_v A_v)$, the *output* impedance is *decreased* by the same factor. By formula,

$$
\boxed{Z_{out(f)} = \frac{Z_{out}}{1 + \alpha_v A_v}} \tag{14.9}
$$

where $Z_{out(f)}$ = the output impedance of the feedback amplifier
Z_{out} = the output impedance of the amplifier *without* feedback

The following example illustrates the use of this equation.

EXAMPLE 14.4

The amplifier in Example 14.2 has a Z_{out} of 420 Ω when no feedback is used. Determine the value of Z_{out} with the feedback circuit added.

Solution: In Example 14.3, we determined the value of $(1 + \alpha_v A_v)$ to be 11. Using this value in equation (14.9), we obtain the following value of $Z_{out(f)}$:

$$
\begin{aligned}
Z_{out(f)} &= \frac{Z_{out}}{1 + \alpha_v A_v} \\
&= \frac{420 \text{ }\Omega}{11} \\
&= 38.18 \text{ }\Omega
\end{aligned}
$$

As you can see, the feedback network has greatly reduced the output impedance of the amplifier. Again, this is an added benefit of using negative voltage feedback. With the lower value of Z_{out}, the amplifier is much better suited to driving low-impedance loads.

The Effects of Feedback on Bandwidth

You have seen the feedback factor, $1 + \alpha_v A_v$, used in several circuit calculations. As you will be shown, it is also used in the calculations of f_1 and f_2 for a given amplifier.

✔ 11

You were shown in Chapter 13 that the product of A_v and f_2 for a given amplifier is a constant value, called the gain–bandwidth product. You were also shown that a change in A_v causes an equal change in the values of f_1 and f_2. If we apply this principle to the negative feedback amplifier, we obtain the following equations:

$$f_{1(f)} = \frac{f_1}{1 + \alpha_v A_v}$$
(14.10)

and

$$f_{2(f)} = f_2(1 + \alpha_v A_v)$$
(14.11)

where

$f_{1(f)}, f_{2(f)}$ = the cutoff frequencies for the feedback amplifier
f_1, f_2 = the cutoff frequencies for the amplifier *without* a feedback path

The following example illustrates the use of these equations.

EXAMPLE 14.5

The amplifier in Example 14.3 had cutoff frequencies of $f_1 = 1.5$ kHz and $f_2 = 501.5$ kHz *before* the feedback path was added. What are the new cutoff frequencies for the circuit?

Solution: The new lower cutoff frequency is found as

$$f_{1(f)} = \frac{f_1}{1 + \alpha_v A_v}$$
$$= \frac{1.5\ \text{kHz}}{11}$$
$$= 136.4\ \text{Hz}$$

The new upper cutoff frequency is determined as

$$f_{2(f)} = f_2(1 + \alpha_v A_v)$$
$$= (501.5\ \text{kHz})(11)$$
$$= 5.52\ \text{MHz}$$

The amplifier described in Practice Problem 14–3 has values of $f_1 = 12$ kHz and $f_2 = 640$ kHz. Determine the values of $f_{1(f)}$ and $f_{2(f)}$ for the circuit.

The bandwidth for the amplifier described in Example 14.5 would be approximately 5.52 MHz. We could obtain the same approximation of the amplifier bandwidth using the following equation:

$$\boxed{BW_f \cong BW(1 + \alpha_v A_v)}$$

(14.12)

The following example illustrates the validity of this equation.

EXAMPLE 14.6

The amplifier described in Example 14.5 had the following values *before* feedback was added:

$$f_1 = 1.5 \text{ kHz}$$
$$f_2 = 501.5 \text{ kHz}$$
$$BW = 500 \text{ kHz}$$

Don't forget:
$BW = f_2 - f_1$

Determine the value of BW_f for the circuit.

Solution: The value of BW_f can be approximated as

$$BW_f = BW(1 + \alpha_v A_v)$$
$$= (500 \text{ kHz})(11)$$
$$= 5.5 \text{ MHz}$$

This is essentially the same value that we approximated using the values of f_1 and f_2 calculated in Example 14.5.

PRACTICE PROBLEM 14–6

Calculate the value of BW_f for the circuit in Practice Problem 14–5 using equation (14.2). Then calculate its value using the values of $f_{1(f)}$ and $f_{2(f)}$ obtained in the practice problem and compare the results.

There are two important points that should be made at this time:

1. Equation (14.12) is consistent with the idea of gain–bandwidth product. If the voltage gain of an amplifier changes by a given factor, the bandwidth must also change by the same approximate factor in order to fulfill the requirement of a relatively constant gain–bandwidth product.

2. Equation (14.12) gives only an *approximation* of BW_f. If we had determined the *exact* bandwidth for the amplifier in Example 14.5, we would have obtained

a value of

$$BW_f = f_{2(f)} - f_{1(f)}$$
$$= 5.5 \text{ MHz} - 136.4 \text{ Hz}$$
$$= 5.4998636 \text{ MHz}$$

Thus, using equation (14.12) will introduce an error into the BW_f calculation. However, in most cases, the percent of error will be 2% or less.

Analyzing a Negative Feedback Amplifier

✔ 12

In most of the examples in this section, we assumed a given gain, impedance, or frequency value for an amplifier *without* feedback. We then determined the effect of the feedback factor on this value. This is the method that is normally used for analyzing a negative feedback amplifier.

A Practical Consideration. The method used to determine the open-loop values of A_v, Z_{in}, Z_{out}, f_1, f_2, and BW will vary from one amplifier type to another, as you have been shown throughout this text. As we progress through this chapter, you will be shown how to calculate the feedback factor for the various types of amplifiers.

FIGURE 14.8

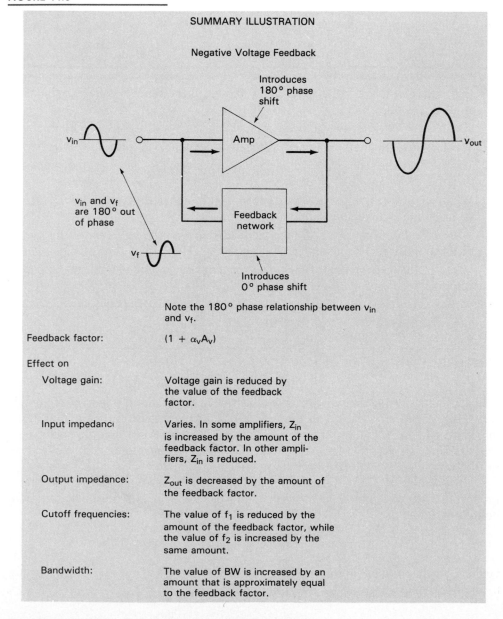

SUMMARY ILLUSTRATION

Negative Voltage Feedback

Introduces 180° phase shift

Amp

v_{in}

v_{out}

v_{in} and v_f are 180° out of phase

Feedback network

v_f

Introduces 0° phase shift

Note the 180° phase relationship between v_{in} and v_f.

Feedback factor:	$(1 + \alpha_v A_v)$
Effect on	
Voltage gain:	Voltage gain is reduced by the value of the feedback factor.
Input impedance	Varies. In some amplifiers, Z_{in} is increased by the amount of the feedback factor. In other amplifiers, Z_{in} is reduced.
Output impedance:	Z_{out} is decreased by the amount of the feedback factor.
Cutoff frequencies:	The value of f_1 is reduced by the amount of the feedback factor, while the value of f_2 is increased by the same amount.
Bandwidth:	The value of BW is increased by an amount that is approximately equal to the feedback factor.

When faced with the mathematical analysis of a negative feedback amplifier, you start by assuming that the feedback path isn't present. Based on this assumption, you perform the appropriate analysis of the amplifier, predicting the values of circuit gain, input impedance, output impedance, cutoff frequencies, and bandwidth. Then you determine the value of the feedback factor for the circuit and use this value to predict the changes in the original gain, impedance, and frequency values.

As we study some specific circuits in this chapter, you will be shown the equations that are used to analyze each. As you will see, while the equations will vary from one circuit to another, the basic progression of the analyses will remain constant. That is, we will always analyze the amplifier as if the feedback path wasn't present, determine the feedback factor of the circuit, and then calculate the change that the feedback factor causes in the original gain, impedance, and frequency calculations.

Section Summary

Figure 14.8 summarizes the principles on negative voltage feedback that were introduced in this section. Make sure that you are comfortable with these principles before continuing to study any specific circuits.

SECTION REVIEW

1. How does negative voltage feedback reduce the effective voltage gain of an amplifier? [Objective 7]
2. What is the *attenuation factor* of a feedback network? [Objective 8]
3. What is the *feedback factor* of a negative feedback amplifier? [Objective 8]
4. What effect does negative voltage feedback normally have on the input impedance of an amplifier? [Objective 10]
5. What effect does negative voltage feedback have on the output impedance of an amplifier? [Objective 10]
6. What effect does negative voltage feedback have on the upper cutoff frequency of an amplifier? The lower cutoff frequency? The bandwidth? [Objective 11]
7. What is the general approach to performing the mathematical analysis of a negative feedback amplifier? [Objective 12]

14.3
Noninverting and Inverting Voltage Feedback

One application for negative voltage feedback is in the standard noninverting amplifier. This amplifier is shown in Figure 14.9. If you compare the amplifier shown in Figure 14.9 with the one shown in Figure 12.34, you'll see that the two circuits are identical. In Figure 14.9, the circuit has simply been redrawn to make some of the principles we will be discussing easier to visualize.

In Chapter 12, you were shown that the op-amp provides an output that is proportional to the *difference* between two input voltages. By formula, this input voltage is found as

$$v_d = v_2 - v_1$$

where v_2 = the voltage at the *noninverting* input
v_1 = the voltage at the *inverting* input

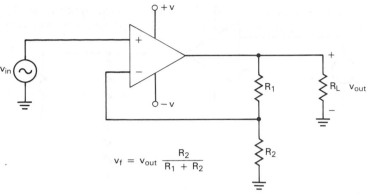

FIGURE 14.9

Noninverting voltage feedback.

Relating this equation to Figure 14.9, we can see that the difference voltage is found as

$$v_d = v_{in} - v_f$$

Note the similarity between the above equation and equation (14.1). The similarity between the two equations shows that the noninverting amplifier is an amplifier that uses negative voltage feedback.

The value of v_f for the noninverting amplifier is found as

$$v_f = v_{out} \frac{R_2}{R_1 + R_2}$$

If we rearrange this equation, we get

$$\frac{v_f}{v_{out}} = \frac{R_2}{R_1 + R_2}$$

and, since $\alpha_v = v_f/v_{out}$,

$$\alpha_v = \frac{R_2}{R_1 + R_2}$$

Using this value of α_v, the value of A_{CL} for the circuit can be found as

$$A_{CL} = \frac{A_{OL}}{1 + \alpha_v A_{OL}} \qquad \textbf{(14.14)}$$

where A_{OL} is the open-loop gain of the op-amp, as listed on the spec sheet of the component. Using these equations, the basic analysis of the noninverting voltage feedback amplifier is performed as shown in the following example.

━━━━━━━━━━━ **EXAMPLE 14.7** ━━━━━━━━━━━

Determine the closed-loop gain of the circuit shown in Figure 14.10.

formula,

$$Z_{out(CL)} = \frac{Z_{out}}{1 + \alpha_v A_{OL}} \qquad (14.17)$$

where $Z_{out(CL)}$ = the *closed-loop* output impedance of the amplifier
Z_{out} = the Z_{out} of the op-amp, as given on the component spec sheet

EXAMPLE 14.9

Assume that the op-amp in Figure 14.10 has a Z_{out} rating of 100 Ω. Determine the closed-loop output impedance of the amplifier.

Solution: Using the Z_{out} rating of the op-amp and the values obtained in Example 14.7, the value of $Z_{out(CL)}$ is found as

$$\begin{aligned}
Z_{out(CL)} &= \frac{Z_{out}}{1 + \alpha_v A_{OL}} \\
&= \frac{100 \ \Omega}{1 + (0.0099)(150,000)} \\
&= 0.067 \ \Omega
\end{aligned}$$

This value is too small to even consider.

PRACTICE PROBLEM 14–9

The circuit described in Practice Problem 14–7 has an op-amp output impedance of 180 Ω. Determine the value of $Z_{out(f)}$ for the circuit.

Now consider the impact of the results we have obtained from Examples 14.7 through 14.9. The circuit we have analyzed has the following characteristics:

1. Input and output signals that are in phase
2. A voltage gain of $A_{CL} = 101$
3. A virtually infinite input impedance
4. Virtually zero output impedance

These characteristics describe the *ideal buffer amplifier*. This circuit would provide virtually no load on its source circuit, while being able to drive any load, no matter how low its impedance. At the same time, we have a respectable amount of voltage gain, a situation that is not possible with either the emitter follower or the source follower. No wonder that op-amp circuits are rapidly replacing everything else!

You were shown in Chapter 13 how to determine the bandwidth for an op-amp with a given value of A_{CL}. As a review,

$$BW = \frac{f_{unity}}{A_{CL}} \qquad (14.18)$$

and since op-amps are dc amplifiers,

$$\boxed{f_2 = BW} \qquad\qquad (14.19)$$

EXAMPLE 14.10

The op-amp in Figure 14.10 has a unity-gain frequency of 30 MHz. Determine the bandwidth of the amplifier.

Solution: Using the unity-gain frequency rating and the value of A_{CL} obtained in Example 14.7, the bandwidth of the amplifier is found as

$$
\begin{aligned}
BW_f &= \frac{f_{unity}}{A_{CL}} \\
&= \frac{30\ \text{MHz}}{101} \\
&= 297\ \text{kHz}
\end{aligned}
$$

Since we are dealing with an op-amp, the value of f_1 is 0 Hz. Therefore, the value of f_2 is equal to the bandwidth, 297 kHz.

PRACTICE PROBLEM 14–10

The op-amp in Practice Problem 14–7 has a rating of $f_{unity} = 5$ MHz. Determine the bandwidth of the circuit.

Inverting Voltage Feedback

We are not really going to spend much time and effort on the *inverting voltage feedback* amplifier simply because we have already done so. The inverting voltage feedback amplifier is shown in Figure 14.11. The closed-loop gain of this amplifier is found as shown. Also, the relationship between bandwidth, the cutoff frequencies,

FIGURE 14.11

Inverting voltage feedback.

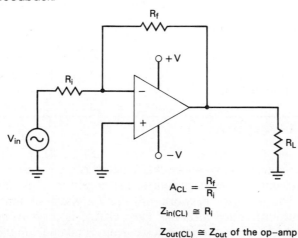

$$A_{CL} = \frac{R_f}{R_i}$$

$$Z_{in(CL)} \cong R_i$$

$$Z_{out(CL)} \cong Z_{out} \text{ of the op-amp}$$

and gain is the same as for the noninverting feedback amplifier. The primary difference comes in the determination of Z_{in} for the circuit.

The input signal to an inverting amplifier is applied to the input resistor, R_i. As you may recall from Chapter 12, the connection point between R_i and the op-amp is at *virtual ground*. Based on this fact, we can derive the equivalent circuit shown in Figure 14.12. Since the op-amp side of R_i is at virtual ground, the source sees only R_i as the input impedance of the amplifier. Thus, for the inverting amplifier,

$$\boxed{Z_{in} \cong R_i} \quad\quad (14.20)$$

Since R_i is normally much lower in value than the input impedance of the op-amp, *the addition of negative voltage feedback to the inverting op-amp reduces the input impedance of the circuit.*

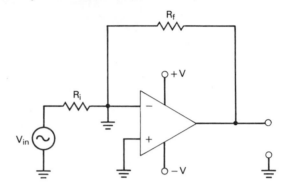

FIGURE 14.12

The reduction of Z_{in} is the primary difference between the inverting and the noninverting negative feedback circuits. Otherwise, the effects of negative voltage feedback are nearly identical for the two circuits. It reduces voltage gain, decreases output impedance, and increases the bandwidth of both circuit types.

SECTION REVIEW

1. Describe the effects that negative voltage feedback has on the noninverting op-amp values of gain, input impedance, output impedance, and bandwidth. [Objective 13]
2. Describe the effects that negative voltage feedback has on the inverting op-amp values of gain, input impedance, output impedance, and bandwidth. [Objective 14]

14.4
Negative Current Feedback: An Overview

As stated earlier, negative *current* feedback decreases the current gain (A_i) of an amplifier by decreasing the input current to the amplifier. This point is illustrated in Figure 14.13. The feedback network will draw a given amount of current from the input. This current, called *feedback current* (i_f), is proportional to the output current of the amplifier. Since the feedback current is taken from the source current,

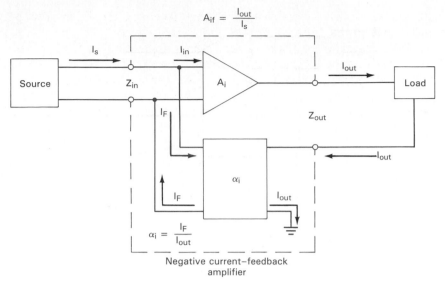

FIGURE 14.13

Negative current feedback.

I_s, the value of input current is found as

$$\boxed{I_{\text{in}} = I_s - I_f} \qquad (14.21)$$

The feedback current is always less than the value of I_{out}. Therefore, we have a *current attenuation*, α_i, for the circuit. As shown in Figure 14.13, the value of α_i is found as

$$\boxed{\alpha_i = \frac{I_f}{I_{\text{out}}}} \qquad (14.22)$$

As you can see, we have a familiar pattern starting to develop. We can take this pattern one step further, as follows:

$$\boxed{A_{if} = \frac{A_i}{1 + \alpha_i A_i}} \qquad (14.23)$$

where A_{if} = the *effective* gain of the feedback amplifier
A_i = the gain of the active component *without* feedback

This equation is exactly the same as that for the negative voltage feedback amplifier. The only difference is that we are dealing with current gain rather than voltage gain. The following example illustrates the determination of A_{if} for a given amplifier.

EXAMPLE 14.11

Determine the effective current gain of the amplifier shown in Figure 14.14.

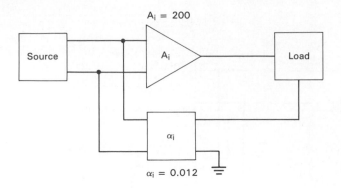

FIGURE 14.14

Solution: Using the circuit values shown, the value of A_{if} is found as

$$A_{if} = \frac{A_i}{1 + \alpha_i A_i}$$
$$= \frac{200}{1 + (0.012)(200)}$$
$$= 58.82$$

PRACTICE PROBLEM 14–11

An amplifier has values of $A_i = 240$ and $\alpha_i = 0.015$. Determine the value of A_{if} for the circuit.

Another similarity between current feedback and voltage feedback is the fact that bandwidth is *increased* by an amount that is approximately equal to the feedback factor. For the current feedback amplifier, this translates into the following equation:

$$\boxed{BW_f = BW(1 + \alpha_i A_i)} \qquad (14.24)$$

EXAMPLE 14.12

The amplifier in Figure 14.14 has a bandwidth of 125 kHz when no feedback path is present. Determine the approximate value of BW_f for the circuit.

Solution: Using the circuit values shown, the value of BW_f is found as

$$BW_f \cong (125 \text{ kHz})[1 + (0.012)(200)]$$
$$= (125 \text{ kHz})(3.4)$$
$$= 425 \text{ kHz}$$

The Effects of Negative Current Feedback on Circuit Impedance Values

Now we get to the primary difference between negative current feedback and negative voltage feedback. Negative current feedback will *reduce* the input impedance of most amplifiers by an amount that is equal to the feedback factor of the circuit. By formula,

$$Z_{in(f)} = \frac{Z_{in}}{1 + \alpha_i A_i} \qquad \textbf{(14.25)}$$

Thus, for most current-feedback amplifiers, the initial input impedance is *decreased* by the presence of the feedback network. This point is illustrated further in the following example.

EXAMPLE 14.13

The amplifier in Figure 14.14 has an open-loop input impedance of 15 kΩ. What is the effective value of Z_{in} when the feedback network is added to the circuit?

Solution: Using the gain and attenuation values shown, $Z_{in(f)}$ is found as

$$
\begin{aligned}
Z_{in(f)} &= \frac{Z_{in}}{1 + \alpha_i A_i} \\
&= \frac{15\ k\Omega}{1 + (0.012)(200)} \\
&= 4.41\ k\Omega
\end{aligned}
$$

If you refer again to Figure 14.14, you will see that the feedback network is in *series* with the amplifier output. Therefore (*you guessed it*), the output impedance is *increased* by the feedback factor of the amplifier. By formula,

$$Z_{out(f)} = Z_{out}(1 + \alpha_i A_i) \qquad \textbf{(14.26)}$$

EXAMPLE 14.14

The amplifier in Figure 14.14 has an open-loop output impedance of 2 kΩ. Determine the value of $Z_{out(f)}$ for the amplifier.

Solution: The value of $Z_{out(f)}$ is found as

$$Z_{out(f)} = Z_{out}(1 + \alpha_i A_i)$$
$$= (2 \text{ k}\Omega)[1 + (0.012)(200)]$$
$$= (2 \text{ k}\Omega)(3.4)$$
$$= 6.8 \text{ k}\Omega$$

PRACTICE PROBLEM 14–14

The circuit described in Practice Problem 14–11 has an open-loop output impedance of 120 Ω. Determine the value of $Z_{out(f)}$ for the circuit.

Summary

The characteristics of negative current feedback are summarized in Figure 14.15. As you can see, they are generally the opposite of those for the negative voltage feedback amplifier.

FIGURE 14.15

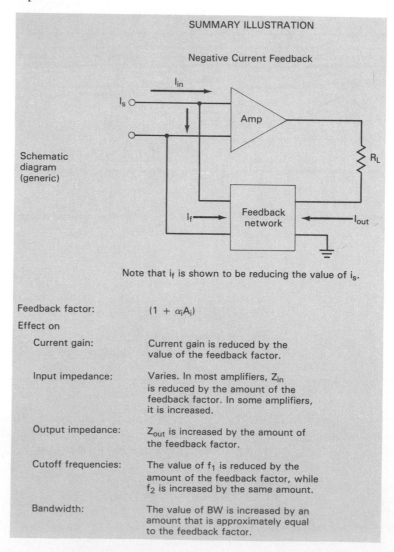

SUMMARY ILLUSTRATION

Negative Current Feedback

Schematic diagram (generic)

Note that i_f is shown to be reducing the value of i_s.

Feedback factor:	$(1 + \alpha_i A_i)$
Effect on	
Current gain:	Current gain is reduced by the value of the feedback factor.
Input impedance:	Varies. In most amplifiers, Z_{in} is reduced by the amount of the feedback factor. In some amplifiers, it is increased.
Output impedance:	Z_{out} is increased by the amount of the feedback factor.
Cutoff frequencies:	The value of f_1 is reduced by the amount of the feedback factor, while f_2 is increased by the same amount.
Bandwidth:	The value of BW is increased by an amount that is approximately equal to the feedback factor.

1. What effect does negative current feedback have on the open-loop current gain of an amplifier? [Objective 15]
2. What effect does negative current feedback have on the open-loop bandwidth of an amplifier? [Objective 15]
3. What effect does negative current feedback have on the open-loop input impedance of most amplifiers? [Objective 15]
4. What effect does negative current feedback have on the open-loop output impedance of an amplifier? [Objective 15]

14.5
Noninverting and Inverting Current Feedback Amplifiers

Noninverting current feedback is very similar to noninverting voltage feedback. The basic circuit is shown in Figure 14.16. As you can see, we again use a voltage-divider circuit that is connected to the inverting input of the op-amp. However, in this case, the *load* is one of the resistances in the voltage-divider circuit. In the voltage-feedback amplifier, the load was in parallel with the voltage-divider circuit, not a part of it.

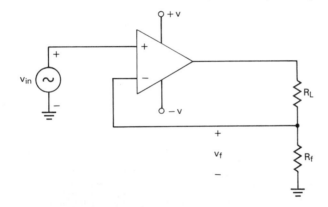

FIGURE 14.16

Noninverting current feedback.

The closed-loop gain for this circuit is determined as it is with the noninverting voltage-feedback circuit. That is,

$$A_{CL} = \frac{A_{OL}}{1 + \alpha_i A_{OL}} \qquad (14.27)$$

where

$$\alpha_i = \frac{R_f}{R_L + R_f} \qquad (14.28)$$

Both of these equations should seem very familiar to you by now. The following example illustrates their use in determining the value of A_{CL} for a noninverting current-feedback amplifier.

Determine the value of A_{CL} for the circuit shown in Figure 14.17.

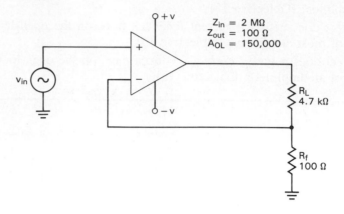

FIGURE 14.17

Solution: First, the value of α_i is found as

$$\alpha_i = \frac{R_f}{R_L + R_f}$$
$$= 0.021$$

Now this value is used to determine A_{CL}, as follows:

$$A_{CL} = \frac{A_{OL}}{1 + \alpha_i A_{OL}}$$
$$= \frac{150,000}{1 + (0.021)(150,000)}$$
$$= 47.6$$

PRACTICE PROBLEM 14–15

An amplifier like the one in Figure 14.17 has values of $R_f = 150\ \Omega$, $R_L = 8.2\ k\Omega$, and $A_{OL} = 120,000$. Determine the value of A_{if} for the circuit.

As with noninverting voltage feedback, current feedback increases the input impedance of the amplifier by a factor equal to the feedback factor of the amplifier. By formula,

$$\boxed{Z_{in(CL)} = Z_{in}(1 + \alpha_i A_{OL})}$$ (14.29)

Again, this is the same situation that we encountered with the use of noninverting voltage feedback.

■ EXAMPLE 14.16 ■

Determine the closed-loop input impedance of the amplifier in Figure 14.17.

Solution: Using the circuit values shown, the value of $Z_{in(CL)}$ is found as

$$Z_{in(CL)} = Z_{in}(1 + \alpha_i A_{OL})$$
$$= (2\ M\Omega)(3151)$$
$$= 6.302\ G\Omega!$$

Again, we have an extremely high amplifier input impedance.

PRACTICE PROBLEM 14–16

The circuit in Practice Problem 14–15 has an open-loop input impedance of 1.2 $M\Omega$. Determine the value of $Z_{in(f)}$ for the circuit.

Now we get to the primary difference between current feedback and voltage feedback. The noninverting current feedback amplifier has a very *high* output impedance. The reason for this is easy to see with the help of Figure 14.18.

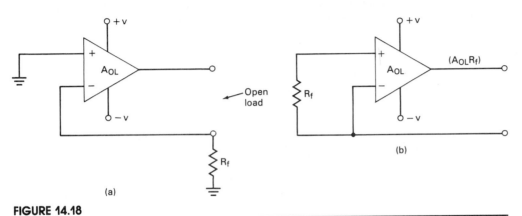

FIGURE 14.18

The effect of noninverting current feedback on output impedance.

If we Thevenize the circuit in Figure 14.16, we obtain the circuit shown in Figure 14.18a. Here the load has been opened. Now we can redraw the circuit as shown in Figure 14.18b. As you can see, the value of R_f appears across the input terminals. If we view the op-amp as being an impedance amplifier, it is easy to see that Z_{out} of the op-amp will equal the value of R_f times the open-loop gain of the component. Thus

$$\boxed{Z_{out} = A_{OL}R_f} \qquad \textbf{(14.30)}$$

EXAMPLE 14.17

Determine the value of $Z_{out(CL)}$ for the circuit shown in Figure 14.17.

Solution: Using the circuit values shown, $Z_{out(CL)}$ is found as

$$Z_{out(CL)} = A_{OL}R_f$$
$$= (150{,}000)(100\ \Omega)$$
$$= 15\ M\Omega$$

PRACTICE PROBLEM 14–17

The amplifier described in Practice Problem 14–15 has an open-loop output impedance of 90Ω. Determine the value of $Z_{out(CL)}$ for the circuit.

Why the noninverting current feedback amplifier makes a good current source.

The high input and output impedances of the noninverting current feedback amplifier make it well suited for use as a *current source*. You should recall that the ideal current source has an infinite output impedance. For all practical purposes, this requirement is fulfilled by the circuit discussed in this section.

The effects of noninverting current feedback on bandwidth and cutoff frequencies are exactly as described for every amplifier covered to this point. The closed-loop bandwidth for this circuit is found just as it always is for an op-amp circuit. Therefore, we will not take any time to discuss the effects of noninverting current feedback on bandwidth. Rather, we will move on to discuss *inverting* current feedback.

Inverting Current Feedback

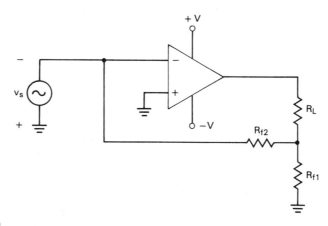

17

The effects of inverting current feedback.

You won't find any great surprises here. Inverting current feedback has the same effects on an op-amp as current feedback has on discrete circuits. In other words, inverting current feedback decreases gain, increases bandwidth, decreases input impedance, and increases output impedance. The basic inverting current feedback amplifier is shown in Figure 14.19.

FIGURE 14.19

Inverting current feedback.

The closed-loop current gain of the amplifier is found, as always, as

$$A_{CL}^{\downarrow} = \frac{A_{OL}}{1 + \alpha_i A_{OL}}$$

CHAP. 14   Negative Feedback

where

$$\alpha_i = \frac{R_{f1}}{R_{f1} + R_{f2}} \qquad (14.31)$$

If we were to Thevenize the circuit as we did in Figure 14.18, we would see that the equations for $Z_{in(CL)}$ and $Z_{out(CL)}$ are as follows:

$$Z_{in(CL)} = \frac{Z_{in}}{1 + \alpha_i A_{OL}} \qquad (14.32)$$

and

$$Z_{out(CL)} = A_{OL}(R_{f1} + R_{f2}) \qquad (14.33)$$

The effects of A_{CL} on bandwidth and the cutoff frequencies are the same as with all the other circuits we have covered.

SECTION REVIEW

1. What effect does negative current feedback have on the open-loop values of gain, input impedance, output impedance, and bandwidth for a noninverting op-amp? [Objective 16]
2. What effect does negative current feedback have on the open-loop values of gain, input impedance, output impedance, and bandwidth for an inverting op-amp? [Objective 17]

14.6
Feedback Network Troubleshooting

A failure of the feedback network in a given amplifier is one of the easiest problems in the world to diagnose. The most noticeable effect is that *gain of the amplifier will change drastically*. Sometimes the gain will increase; sometimes it will decrease. It all depends on which component goes bad. For example, consider the circuit shown in Figure 14.20. Under normal circumstances, the output from the amplifier

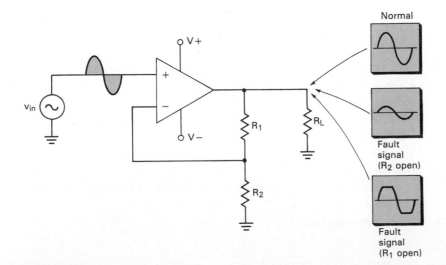

FIGURE 14.20

would be found as

$$v_{\text{out}} = A_{\text{CL}}v_{\text{in}}$$

If R_2 were to open, the feedback circuit would consist solely of R_1. In this case, the gain would *drop*. Why? Because the circuit would now act as a voltage follower. In other words, the circuit would now be a buffer with an output voltage that is equal to the input voltage. Thus we would have the output signal shown. The waveform would be correct, but we would have only *unity* gain.

If R_1 were to open, the entire feedback network would be effectively removed. This would cause the gain of the amplifier to increase to the value of A_{OL}. Obviously, this would cause the output voltage to clip at or near the values of $V+$ and $V-$. This results in the signal shown in Figure 14.20.

No matter what type of feedback you are dealing with, a fault in the network will cause a change in gain. Whether gain increases or decreases depends on the bad component. However, any time that you see either a drastic increase or decrease in gain, while the overall functioning of the amplifier is still there, you check out the feedback network.

SECTION REVIEW

1. What is the primary symptom of a fault in the feedback network of a given op-amp circuit? [Objective 18]

14.7

Discrete Feedback Amplifiers

In this section, we are going to apply the principles covered in this chapter to two discrete (BJT) amplifiers, one having negative voltage feedback and the other having negative current feedback. This will give you a good idea of how discrete amplifiers respond in general to negative feedback networks.

Negative Voltage Feedback

When negative voltage feedback is used in a discrete amplifier, the voltage gain of the circuit is reduced, the input impedance is increased, the output impedance is decreased, and the bandwidth is increased. In other words, the discrete amplifier generally reacts to the presence of a negative voltage feedback network in the same way that a noninverting amplifier reacts. In our discussion of the negative voltage feedback amplifier, we will be referring to the circuit in Figure 14.21.

First, a word or two about the circuit. The feedback path consists of C_5, $R9$, and $R10$. This path goes from the collector of the second stage back to the emitter of the first stage. The following series of statements shows how *negative* voltage feedback is accomplished with this circuit:

1. The emitter voltage and the collector voltage in the first stage are 180° out of phase.
2. The stage 1 collector voltage is in phase with the stage 2 base voltage. Therefore, the stage 1 emitter voltage is 180° out of phase with the stage two base voltage.

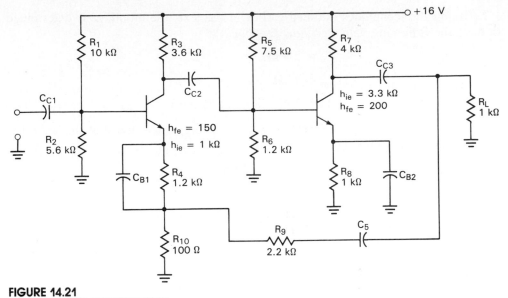

FIGURE 14.21

A two-stage negative voltage feedback amplifier.

3. The stage two base voltage is 180° out of phase with the stage 2 collector voltage. Therefore, the stage 2 collector voltage is in phase with the stage 1 emitter signal. This effectively reduces the difference between the base and emitter voltages in the first stage, which has the same effect as reducing the base voltage of the first stage.

The ac output signal from the second stage is coupled to Q_1 via C_5. The resistors, $R9$ and $R10$ form a *voltage divider*. Since the emitter of the first stage is connected between $R9$ and $R10$, the feedback voltage is going to equal the voltage across $R10$. Thus,

$$v_f = v_{out} \frac{R_{10}}{R_9 + R_{10}}$$

and

$$\frac{v_f}{v_{out}} = \frac{R_{10}}{R_9 + R_{10}}$$

or

$$\alpha_v = \frac{R_{10}}{R_9 + R_{10}}$$

For this circuit,

$$\alpha_v = \frac{100 \ \Omega}{2.2 \ k\Omega + 100 \ \Omega}$$
$$= 0.043$$

How's *that* for being simple? To find α_v, all you have to do is solve the feedback resistors as you would in any voltage divider.

Using the techniques outlined earlier in the text, the amplifier is determined to have the following values *without feedback*:

$$Z_{in} = 2.9 \text{ k}\Omega$$

$$Z_{out} = 4 \text{ k}\Omega$$

$$A_v = 294 \quad \text{(approximately)}$$

Also, just for the sake of discussion, we are going to assume that the amplifier has the following values:

$$f_1 = 5 \text{ kHz}$$

$$f_2 = 305 \text{ kHz}$$

$$\text{BW} = 300 \text{ kHz}$$

Next, we are going to save ourselves a lot of time and trouble by determining the voltage-feedback factor for this amplifier, as follows:

$$1 + \alpha_v A_v = 1 + (0.043)(5200)$$
$$= 224.6$$

Using this feedback factor, we can determine the value of A_{vf} to be

$$A_{vf} = \frac{A_v}{1 + \alpha_v A_v}$$
$$= \frac{294}{224.6}$$
$$= 1.31$$

The value of $Z_{in(f)}$ is now found as

$$Z_{in(f)} = Z_{in}(1 + \alpha_v A_v)$$
$$= (2.93 \text{ k}\Omega)(224.6)$$
$$= 658 \text{ k}\Omega$$

and the value of $A_{out(f)}$ is found as

$$Z_{out(f)} = \frac{Z_{out}}{1 + \alpha_v A_v}$$
$$= \frac{4 \text{ k}\Omega}{224.6}$$
$$= 17.8 \text{ }\Omega$$

The new bandwidth is approximated as

$$\text{BW}_f \cong \text{BW}(1 + \alpha_v A_v)$$
$$= (300 \text{ kHz})(224.6)$$
$$= 67.38 \text{ MHz}$$

The lower cutoff frequency is now

$$f_{1(f)} = \frac{f_1}{1 + \alpha_v A_v}$$
$$= \frac{5 \text{ kHz}}{224.6}$$
$$= 22.26 \text{ Hz}$$

CHAP. 14 Negative Feedback

Finally, the new upper cutoff frequency is found as

$$f_{2(f)} = f_2(1 + \alpha_v A_v)$$
$$= (305 \text{ kHz})(224.6)$$
$$= 68.503 \text{ MHz}$$

As you can see, the analysis of the discrete amplifier with negative voltage feedback is performed in three basic steps:

1. Determineethe open-loop gain, impedance, and frequency characteristics of the amplifier. That is, analyze the circuit neglecting the effects of the feedback.
2. Determine the voltage feedback factor.
3. Using the feedback factor, recalculate the values obtained in step 1.

The above procedure is the same as that used for the op-amp circuits.
That's all there is to it. Now we'll move on to look at the discrete negative current feedback amplifier.

Negative Current Feedback

The circuit in Figure 14.22 shows a discrete amplifier that uses negative current feedback. We'll start our discussion with a few points on the operation of this circuit. The feedback network consists of R_8 in the emitter of the second stage, C_f, and R_f. The reduction of I_{in} for this circuit is accomplished as follows:

✔ 20

1. The amplifier is a two-stage amplifier. Since the output voltage is taken from the collector of the second stage, v_{in} and v_{out} are in phase. Since v_{out} is 180° out of phase with the ac emitter voltage of the second stage, v_{in} and v_{R8} are also 180° out of phase.
2. The combined current gain of the two stages (open loop) is 30,000. It is therefore safe to assume that $v_{R8} >> v_{in}$.
3. When v_{in} goes positive, for example, v_{R8} will go negative.

FIGURE 14.22

A two-stage negative current feedback amplifier.

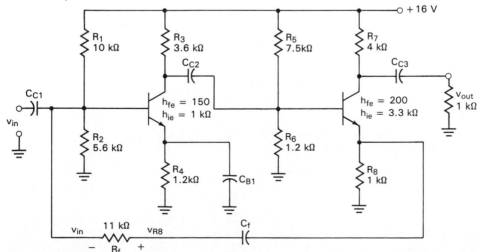

4. The difference between v_{in} and v_{R8} will be felt across R_f, assuming that the reactance of C_f is 0 Ω. Note the polarities shown on the two sides of R_f. These polarities are correct, based on the assumption that $V_{R8} \gg v_{in}$.

5. With the polarities shown across R_f, current through this component is *from the stage 1 base circuit toward the stage 2 emitter circuit*. Thus, I_{in} is made to be less than the current being supplied by the original voltage source.

The circuit shown in Figure 14.22 is basically the same as the one we analyzed earlier in this section. At that time, we established the following values for the circuit:

$$A_v = 294$$

$$Z_{in} = 2.9 \text{ k}\Omega$$

$$Z_{out} = 4 \text{ k}\Omega$$

$$f_1 = 5 \text{ kHz}$$

$$f_2 = 305 \text{ kHz}$$

$$\text{BW} = 300 \text{ kHz}$$

Remember. Swamping decreases A_v and increases Z_{in}. Thus, removing a swamping circuit will increase A_v and decrease Z_{in}.

Of the above values, two have been changed by the modifications made to the original circuit. In Figure 14.21, the first stage of the amplifier was swamped to provide a connection point for the feedback network. Since the first stage of the amplifier is no longer swamped, the open-loop voltage gain has increased, and the open-loop input impedance has decreased. The values of A_v, A_i, and Z_{in} for the circuit are

$$A_v = 4698$$

$$A_i = 30,000$$

$$Z_{in} = 782 \ \Omega$$

The values of A_i and Z_{in} will be used in our mathematical analysis of the amplifier.

The first step in analyzing the effects of the feedback network is to determine the value of α_i. The attenuation factor for this circuit is found as

$$\alpha_i = \frac{R_8}{R_8 + R_f} \tag{14.34}$$

This equation is based on the fact that R_8 and R_f form an ac current-divider circuit. Using the values shown in the circuit, α_i is found to be

$$\begin{aligned}
\alpha_i &= \frac{R_8}{R_8 + R_f} \\
&= \frac{1 \text{ k}\Omega}{1 \text{ k}\Omega + 11 \text{ k}\Omega} \\
&= 0.083
\end{aligned}$$

Now the effective current gain of the amplifier is found as

$$\begin{aligned}
A_{if} &= \frac{A_i}{1 + \alpha_i A_i} \\
&= \frac{30,000}{1 + (0.083)(30,000)} \\
&= 12.04
\end{aligned}$$

668   CHAP. 14   Negative Feedback

The bandwidth for the amplifier has been increased by the feedback factor as shown:

$$BW_f \cong BW(1 + \alpha_i A_i)$$
$$= (300 \text{ kHz})(2491)$$
$$= 747.3 \text{ MHz}$$

The lower and upper 3-dB frequencies are found for the current-feedback amplifier in nearly the same fashion as for the voltage-feedback amplifier, as follows:

$$f_{1(f)} = \frac{f_1}{1 + \alpha_i A_i}$$
$$= \frac{5 \text{ kHz}}{2491}$$
$$= 2.07 \text{ Hz}$$

(14.35)

and

$$f_{2(f)} = f_2(1 + \alpha_i A_i)$$
$$= (305 \text{ kHz}) (2491)$$
$$= 759.8 \text{ MHz}$$

(14.36)

Since the feedback network is in parallel with the amplifier input, the value of z_{in} is *decreased*, as follows:

$$Z_{in(f)} = \frac{Z_{in}}{1 + \alpha_i A_i}$$
$$= \frac{782 \text{ }\Omega}{2491}$$
$$= 0.314 \text{ }\Omega$$

As you can see, the operation of the current-feedback amplifier is very similar to that of the voltage-feedback amplifier.

SECTION REVIEW

1. What are the effects of negative voltage feedback on the gain, impedance, and frequency characteristics of a discrete amplifier? [Objective 19]
2. What are the effects of negative current feedback on the gain, impedance, and frequency characteristics of a discrete amplifier? [Objective 20]

KEY TERMS

The following terms were introduced and defined in this chapter:

attenuation factor	feedback factor	oscillator
current feedback	feedback network	positive feedback
feedback amplifier	negative feedback	voltage feedback

PRACTICE PROBLEMS

§14.2

1. An amplifier has values of $A_v = 1000$ and $\alpha_v = 0.22$. Determine the value of A_{vf} for the circuit. [9]

2. An amplifier has values of $A_v = 588$ and $\alpha_v = 0.092$. Determine the value of A_{vf} for the circuit. [9]

3. An amplifier has values of $A_v = 900$ and $\alpha_v = 0.012$. Determine the value of A_{vf} for the circuit. [9]

4. An amplifier has values of $A_v = 640$ and $\alpha_v = 0.158$. Determine the value of A_{vf} for the circuit. [9]

5. The amplifier described in problem 1 has values of $Z_{in} = 48$ kΩ and $Z_{out} = 220$ Ω. Determine the values of $Z_{in(f)}$ and $Z_{out(f)}$ for the circuit. [10]

6. The amplifier described in problem 2 has values of $Z_{in} = 8.8$ kΩ and $Z_{out} = 186$ Ω. Determine the values of $Z_{in(f)}$ and $Z_{out(f)}$ for the circuit. [10]

7. Determine the values of A_{vf}, $Z_{in(f)}$, and $Z_{out(f)}$ for the amplifier in Figure 14.23. [9, 10]

8. Determine the values of A_{vf}, $Z_{in(f)}$, and $Z_{out(f)}$ for the amplifier in Figure 14.24. [9, 10]

9. Determine the values of A_{vf}, $Z_{in(f)}$, and $Z_{out(f)}$ for the amplifier in Figure 14.25. [9, 10]

10. The value of A_v in Figure 14.25 changes to 600. Recalculate the value of A_{vf} for the circuit. [9]

11. The circuit in Figure 14.23 has open-loop values of $f_1 = 2$ kHz and $f_2 = 826$ kHz. Determine the values of $f_{1(f)}$, $f_{2(f)}$, and BW$_f$ for the circuit. [11]

12. The circuit in Figure 14.24 has open-loop values of $f_1 = 8$ kHz and $f_2 = 980$ kHz. Determine the values of $f_{1(f)}$, $f_{2(f)}$, and BW$_f$ for the circuit. [11]

13. The circuit in Figure 14.25 has open-loop values of $f_1 = 1$ kHz and $f_2 = 462$ kHz. Determine the values of $f_{1(f)}$, $f_{2(f)}$, and BW$_f$ for the circuit. [11]

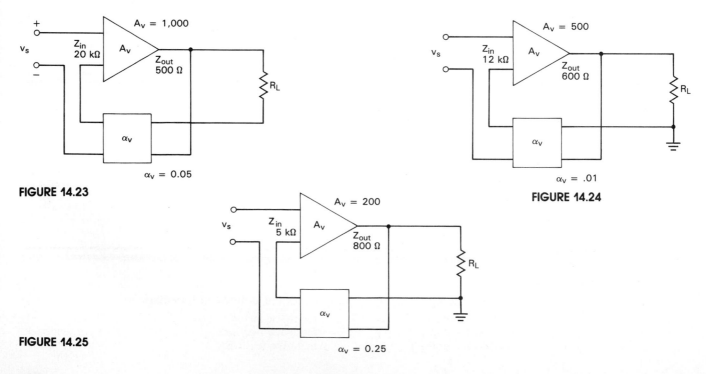

FIGURE 14.23

FIGURE 14.24

FIGURE 14.25

CHAP. 14   Negative Feedback

14. An amplifier has open-loop values of $A_v = 420$, $Z_{in} = 4.4$ kΩ, $Z_{out} = 900$ Ω, $f_1 = 4$ kHz, and $f_2 = 435$ kHz. Determine the gain, impedance, and frequency values for the amplifier when $\alpha_v = 0.036$. [9, 10, 11]

§14.3

15. Calculate the value of A_{CL} for the amplifier in Figure 14.26. [13]
16. Calculate the value of A_{CL} for the amplifier in Figure 14.27. [13]
17. Calculate the value of A_{CL} for the amplifier in Figure 14.28. [13]
18. Calculate the values of $Z_{in(CL)}$ and $Z_{out(CL)}$ for the amplifier in Figure 14.26. [13]
19. Calculate the values of $Z_{in(CL)}$ and $Z_{out(CL)}$ for the amplifier in Figure 14.27. [13]
20. Calculate the values of $Z_{in(CL)}$ and $Z_{out(CL)}$ for the amplifier in Figure 14.28. [13]
21. The op-amp in Figure 14.26 has a value of $f_{unity} = 10$ MHz. Calculate the value of BW for the circuit. [13]
22. The op-amp in Figure 14.27 has a value of $f_{unity} = 2$ MHz. Calculate the value of BW for the circuit. [13]
23. Calculate the values of A_{CL}, $Z_{in(CL)}$, $Z_{out(CL)}$, and BW for the amplifier in Figure 14.29. [14]

FIGURE 14.26

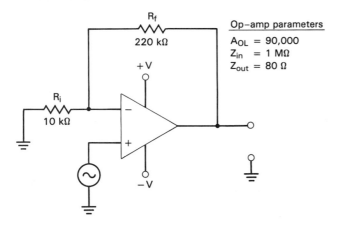

FIGURE 14.27

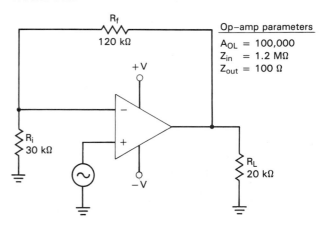

FIGURE 14.29

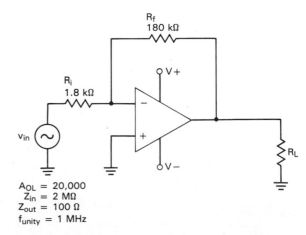

FIGURE 14.28

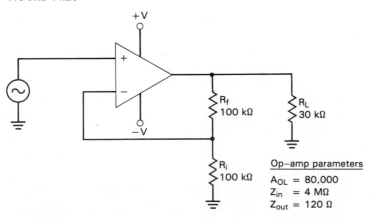

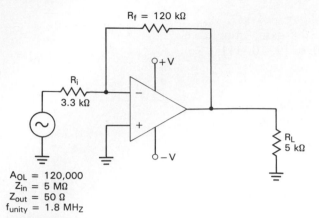

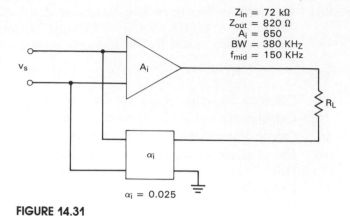

FIGURE 14.30

FIGURE 14.31

$R_f = 120$ kΩ

R_i
3.3 kΩ

+V

$-$

$+$

$-V$

R_L
5 kΩ

$A_{OL} = 120{,}000$
$Z_{in} = 5$ MΩ
$Z_{out} = 50$ Ω
$f_{unity} = 1.8$ MHz

$Z_{in} = 72$ kΩ
$Z_{out} = 820$ Ω
$A_i = 650$
BW $= 380$ KHz
$f_{mid} = 150$ KHz

v_s

A_i

α_i

$\alpha_i = 0.025$

R_L

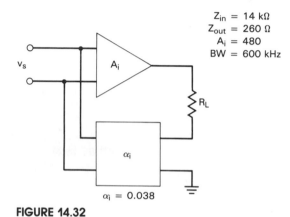

$Z_{in} = 14$ kΩ
$Z_{out} = 260$ Ω
$A_i = 480$
BW $= 600$ kHz

v_s

A_i

R_L

α_i

$\alpha_i = 0.038$

FIGURE 14.32

24. Calculate the values of A_{CL}, $Z_{in(CL)}$, $Z_{out(CL)}$, and BW for the amplifier in Figure 14.30. [14]

§14.4

25. An amplifier has values of $A_i = 240$ and $\alpha_i = 0.128$. Determine the value of A_{if} for the circuit. [15]
26. An amplifier has values of $A_i = 320$ and $\alpha_i = 0.022$. Determine the value of A_{if} for the circuit. [15]
27. The amplifier in problem 25 has an open-loop bandwidth of 488 kHz. Calculate the value of BW_f for the circuit. [15]
28. The amplifier in problem 26 has an open-loop bandwidth of 624 kHz. Calculate the value of BW_f for the circuit. [15]
29. The amplifier in problem 25 has open-loop values of $Z_{in} = 36$ kΩ and $Z_{out} = 40$ Ω. Calculate the values of $Z_{in(f)}$ and $Z_{out(f)}$ for the circuit. [15]
30. The amplifier in problem 26 has open-loop values of $Z_{in} = 12$ kΩ and $Z_{out} = 800$ Ω. Calculate the values of $Z_{in(f)}$ and $Z_{out(f)}$ for the circuit. [15]
31. Calculate the values of A_{if}, $Z_{in(f)}$, $Z_{out(f)}$, and BW_f for the amplifier in Figure 14.31. [15]
32. Calculate the values of A_{if}, $Z_{in(f)}$, $Z_{out(f)}$, and BW_f for the amplifier in Figure 14.32. [15]

CHAP. 14   Negative Feedback

33. Calculate the values of A_{CL}, $Z_{in(CL)}$, $Z_{out(CL)}$, and BW for the circuit in Figure 14.33. [16]

34. Calculate the values of A_{CL}, $Z_{in(CL)}$, $Z_{out(CL)}$, and BW for the circuit in Figure 14.34. [16]

35. Calculate the values of A_{CL}, $Z_{in(CL)}$, $Z_{out(CL)}$, and BW for the circuit in Figure 14.35. [16]

√36. The op-amp in Figure 14.35 is replaced with an op-amp having the following parameters: $A_{OL} = 150{,}000$, $Z_{in} = 5$ MΩ, $Z_{out} = 200$ Ω, and $f_{unity} = 2.5$ MHz. Recalculate the circuit gain, impedance, and BW values. [16]

√37. Calculate the gain, impedance, and BW values for the amplifier in Figure 14.36. [17]

38. Calculate the gain, impedance, and BW values for the amplifier in Figure 14.37. [17]

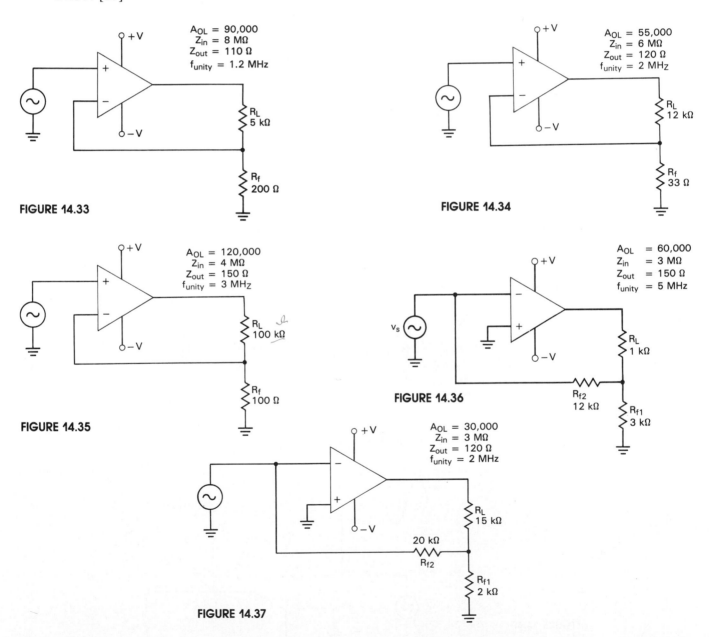

FIGURE 14.33

FIGURE 14.34

FIGURE 14.35

FIGURE 14.36

FIGURE 14.37

TROUBLESHOOTING PRACTICE PROBLEMS

39. The circuit in Figure 14.38 has the waveforms shown. Discuss the possible cause(s) of the problem.

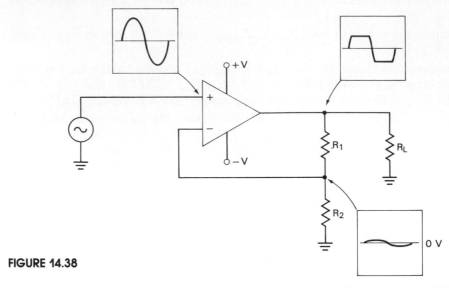

FIGURE 14.38

40. The circuit shown in Figure 14.39 has the waveforms shown. Discuss the possible cause(s) of the problem.

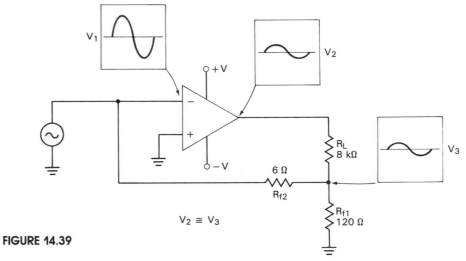

FIGURE 14.39

41. The circuit in Figure 14.40 has the waveforms shown. Discuss the possible cause(s) of the problem.

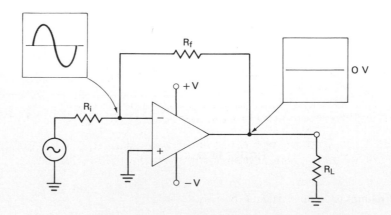

FIGURE 14.40

THE BRAIN DRAIN

42. Calculate the gain, input impedance and output impedance of the two-stage amplifier shown in Figure 14.41.

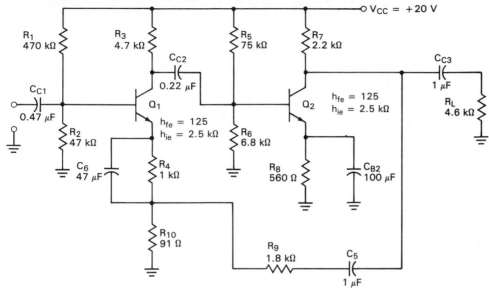

FIGURE 14.41

43. Calculate the current gain, input impedance and output impedance of the two-stage amplifier in Figure 14.42.

44. Refer to Figure 14.42. Transistor Q_2 is found to be faulty and is replaced by a transistor having values of $h_{fe} = 200$ and $h_{ie} = 5$ kΩ. Determine the change in A_{if} caused by the change in transistors.

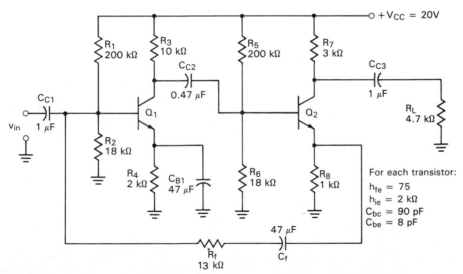

FIGURE 14.42

SUGGESTED COMPUTER APPLICATIONS PROBLEMS

45. Write a program that will determine the values of A_{CL}, $Z_{in(CL)}$, $Z_{out(CL)}$, and BW for a noninverting amplifier with negative voltage feedback.

46. Modify the program in problem 45 to calculate the same values for either a noninverting or inverting amplifier with negative voltage feedback.

ANSWERS TO THE EXAMPLE PRACTICE PROBLEMS

14.1. 32
14.2. The values of A_{vf} are 26.75 and 28.57.
14.3. 13.4 kΩ
14.4. 162.5Ω
14.5. $f_{1(f)} = 3.75$ kHz, $f_{2(f)} = 2.05$ MHz
14.6. The values of BW$_f$ are 2.01 MHz and 2.05MHz
14.7. 111
14.8. 865 MΩ
14.9. 0.25 Ω
14.10. 45.05 kHz
14.11. 52.17
14.12. 2.39 MHz
14.13. 113 kΩ
14.14. 522 Ω
14.15. 55.53
14.16. 2.59 GΩ
14.17. 10.8 MΩ

Tuned Amplifiers

The heart of any communications system is the tuned amplifier. Communications receivers, like the one shown, use tuned circuits to select one out of the many hundreds of signals they constantly receive.

OBJECTIVES

After studying the material in this chapter, you should be able to:

☐ 1. Compare and contrast the ideal and practical frequency-response characteristics of tuned amplifiers. (§15.1)

☐ 2. Discuss the relationship between the roll-off rate and the bandwidth for a tuned amplifier. (§15.1)

☐ 3. Describe the *quality* (Q) rating of a tuned amplifier. (§15.1)

☐ 4. Calculate any one of the following values, given the other two: Q, f_0, and BW. (§15.1)

☐ 5. State, in general terms, how the value of Q is determined for a given tuned amplifier. (§15.1)

☐ 6. Describe how the value of Q for a tuned amplifier affects the calculation of f_0 for the circuit. (§15.1)

☐ 7. List the four types of active filters and describe the frequency-response characteristics of each. (§15.2)

☐ 8. Define a *pole* and describe its frequency-response characteristics. (§15.2)

☐ 9. Compare and contrast the frequency-response characteristics of *Butterworth* and *Chebyshev* active filters. (§15.2)

☐ 10. Discuss and analyze the operation of single-pole and multipole Butterworth low-pass filters. (§15.2)

☐ 11. Discuss and analyze the operation of single-pole and multipole Butterworth high-pass filters. (§15.2)

☐ 12. Describe the dB gain and roll-off characteristics of low-pass and high-pass active filters. (§15.2)

☐ 13. Describe how a low-pass filter and a high-pass filter can be cascaded to form a band-pass filter. (§15.3)

☐ 14. Analyze the operation of a two-stage band-pass filter. (§15.3)

☐ 15. Compare and contrast the two-stage band-pass filter with the multiple-feedback band-pass filter. (§15.3)

☐ 16. Describe the operation of the multiple-feedback band-pass filter. (§15.3)

☐ 17. Calculate the values of f_0, f_1, f_2, Q, and bandwidth for a multiple-feedback band-pass filter. (§15.3)

☐ 18. Calculate the closed-loop voltage gain of a multiple-feedback band-pass filter. (§15.3)

☐ 19. Describe and analyze the operation of the multistage notch filter. (§15.3)

☐ 20. Describe the operation of the multiple-feedback notch filter. (§15.3)

☐ 21. Discuss the use of active filters in audio crossover networks. (§15.4)

☐ 22. Discuss the use of active filters in graphic equalizers. (§15.4)

☐ 23. Discuss the use of active filters in eliminating low-frequency noise in audio systems. (§15.4)

☐ 24. Determine the possible cause(s) for a given active filter fault symptom using the symptom–cause tables. (§15.4)

☐ 25. Troubleshoot active filters. (§15.4)

☐ 26. Describe the circuit recognition feature of a discrete tuned amplifier. (§15.5)

☐ 27. Explain why the net current in a parallel resonant tank circuit is approximately equal to zero. (§15.5)

☐ 28. Describe how the tank circuit in a tuned amplifier reacts to frequencies that are outside and within its bandwidth. (§15.5)

☐ 29. Calculate the values of Q, Q_L, f_0, bandwidth, f_1, and f_2 for a tuned BJT amplifier. (§15.5)

☐ 30. Explain the methods used to provide a variable center frequency and the reasons for using them. (§15.6)

☐ 31. Discuss the procedures used for adjusting the tuning of an amplifier. (§15.6)

☐ 32. List the faults that commonly occur in tuned amplifiers, along with their symptoms and solutions. (§15.6)

☐ 33. Describe and analyze the dc and ac operation of the class C amplifier. (§15.7)

As you master each of the objectives listed, place a check mark (✔) in the appropriate box.

The heart of any communications system is the *tuned amplifier*. A tuned amplifier is one that is designed to have a specific bandwidth. For example, a given tuned amplifier may be designed to amplify only those frequencies that are within ±20 kHz of 1000 kHz, that is, between 980 and 1020 kHz. As long as the input signal is within these frequencies, it will be amplified. If it goes outside of this frequency range, amplification will be drastically reduced.

In Chapter 13, you were introduced to the concept of bandwidth. You were also shown how amplifier resistance and capacitance values limit the frequency

Tuned amplifier. An amplifier that is designed to have a specific bandwidth.

response of a given amplifier. You may be wondering, then, what distinguishes a *tuned* amplifier from any other type of amplifier. After all, don't *all* amplifiers provide gain over a limited band of frequencies? The difference is that tuned amplifiers are designed to have specific, usually narrow, bandwidths. This point is illustrated in Figure 15.1.

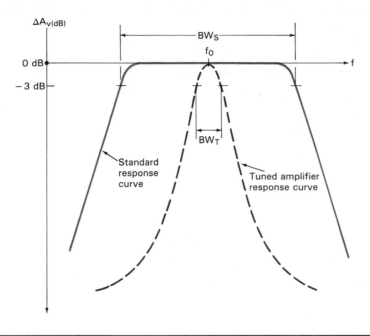

FIGURE 15.1

Tuned amplifier frequency response.

In this chapter, we will take a look at how discrete and op-amp circuits are tuned. We will also take a brief look at some of the typical applications for tuned amplifiers.

15.1
Tuned Amplifier Characteristics

◤ 1

Before we get into any specific circuits, let's take a moment to discuss some of the bandwidth characteristics of tuned amplifiers. The *ideal* characteristics of such an amplifier are illustrated in Figure 15.2.

Characteristics of the *ideal* tuned amplifier.

FIGURE 15.2

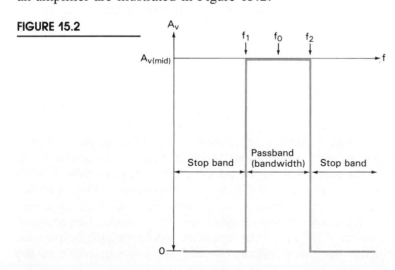

The ideal tuned amplifier would have zero ($-\infty$ dB) gain for all frequencies from 0 Hz up to the lower cutoff frequency, f_1. At that point, the gain would *instantly* jump to $A_{v(\text{mid})}$. The gain would stay at $A_{v(\text{mid})}$ until f_2 was reached. At that time, the gain would *instantly* drop back to zero. Thus, all the frequencies within the bandwidth of the amplifier would be *passed* by the circuit, while all others would be effectively *stopped*. This is where the terms *pass band* and *stop band* come from.

In practice, the ideal characteristics of the tuned amplifier cannot be achieved. Figure 15.3 compares the characteristics of the ideal tuned amplifier to those of a more practical circuit. Note that the gain roll-off of the practical circuit curve is not instantaneous. Rather, the gain rolls off from $A_{v(\text{mid})}$ gradually as frequency passes outside the pass band. As you will now be shown, the roll-off rate is an extremely important factor in the overall operation of a tuned amplifier.

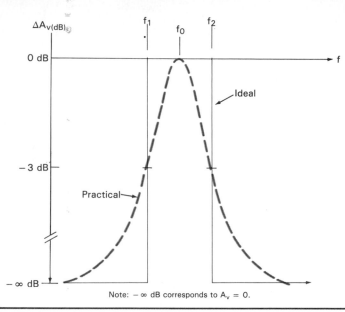

FIGURE 15.3

Ideal versus practical pass band characteristics.

Roll-off Rate versus Bandwidth

We need to start this discussion by establishing two "bottom line" principles:

1. The closer the frequency response curve of a given tuned amplifier to that of the ideal, the better.
2. *In many applications*, the narrower the bandwidth of a tuned amplifier, the better.

The first statement almost goes without saying. As with any circuit, the closer we come to the ideal characteristics of that circuit, the better off we are. This has been the case with every circuit and component we have discussed in this text. The second statement warrants discussion. In Chapter 13 we looked at the frequency response of an amplifier as being a *limitation*. In other words, we considered an *infinite* bandwidth to be the ideal. This is not the case with tuned amplifiers. *With tuned amplifiers, we want a specific, usually narrow, bandwidth.* For example, take a look at Figure 15.4. Here we see a *pass band* (shaded) with two *stop bands*, A and B. If a tuned amplifier was designed to have a center frequency of 1000 kHz and a bandwidth of 40 kHz, that amplifier would be able to pass all the frequencies within the pass band, while rejecting all those in either of the

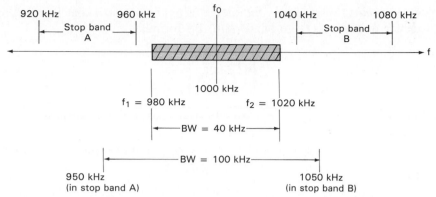

FIGURE 15.4

An Example: The frequency ranges that are shown in Figure 15.4 could represent the frequency ranges of three AM radio stations. If the tuned amplifier has a bandwidth of 40 kHz, it will only pick up the radio station to which it is tuned (the center station). If the tuned amplifier has a bandwidth of 100 kHz, it will not only pick up the center station, it will pick up portions of the other two station signals as well.

stop bands. However, if we used a tuned amplifier with the same center frequency and a bandwidth of 100 kHz, that amplifier would pass a portion of the frequencies in both stop bands. This is not an acceptable situation. A tuned amplifier must pass all the frequencies within the pass band, while stopping all others. Therefore, we want a given tuned amplifier to have a specific, usually narrow, bandwidth. With this in mind, let's take a look at the relationship between roll-off rate and bandwidth.

If we *decrease* the roll-off rate of a given amplifier, we *increase* its bandwidth. This point is illustrated in Figure 15.5. Here we see the frequency-response curves of four different circuits, (a), (b), (c), and (d). All the circuits are shown to have the same center frequency, f_0. Also, the roll-off rates of the circuits become longer as we go from circuit (a) to (b), from (b) to (c), and from (c) to (d). As you can see, the lower the roll-off rate, the wider the bandwidth of the circuit.

The relationship between roll-off rates and bandwidth.

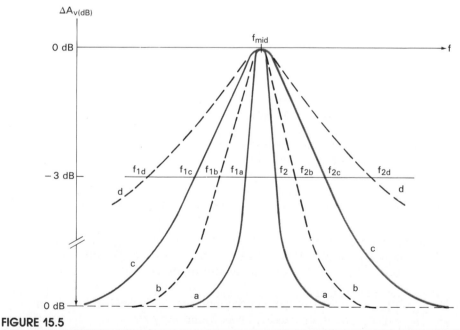

FIGURE 15.5

Bandwidth versus roll-off rate.

Now, apply the principle illustrated in Figure 15.5 to the situation shown in Figure 15.4. It is easy to see that we must limit the roll-off rate of a tuned amplifier if it is to be used for a specific application. The roll-off rate (and therefore, bandwidth) of an amplifier is controlled by what is called the Q of the circuit.

3

The Q (quality) of a tuned amplifier is a *figure of merit* that is equal to the ratio of center frequency (f_o) to bandwidth. By formula,

$$Q = \frac{f_0}{\text{BW}}$$

(15.1)

EXAMPLE 15.1

What is the required Q of a tuned amplifier that is used for the application shown in Figure 15.4?

Solution: The pass band in Figure 15.4 has a center frequency of 1000 kHz and a bandwidth of 40 kHz. Using these two values, the required Q is found as

$$Q = \frac{f_0}{\text{BW}}$$
$$= \frac{1000 \text{ kHz}}{40 \text{ KHz}}$$
$$= 25$$

Thus any tuned amplifier used in this application would have to have a Q of approximately 25.

PRACTICE PROBLEM 15–1

An amplifier with a center frequency of 1200 kHz has a bandwidth of 20 kHz. Determine the Q of the amplifier.

It should be noted that the value of Q for a tuned circuit determines its bandwidth. The Q of a given amplifier depends on component values that are within the amplifier. Once the Q is determined, the bandwidth is found as

$$\text{BW} = \frac{f_0}{Q}$$

(15.2)

The following example illustrates this point.

EXAMPLE 15.2

Using the proper circuit calculations, the Q of a given amplifier is found to be 60. If the center frequency for the amplifier is 1200 kHz, what is the bandwidth of the amplifier?

Solution: With a center frequency of 1200 kHz and a Q of 60, the amplifier bandwidth is found as

$$\text{BW} = \frac{f_0}{Q}$$
$$= \frac{1200 \text{ kHz}}{60}$$
$$= 20 \text{ kHz}$$

With a center frequency of 1200 kHz and a bandwidth of 20 kHz, the cutoff frequencies would be found as 1200 ± 10 kHz, that is, 1190 and 1210 kHz.

As we discuss the various tuned circuits, you will be shown how to determine the value of Q for each. Once the Q of an amplifier is known, determining the bandwidth of the amplifier is simple. For now, just remember that the Q of an amplifier is a figure of merit that depends on circuit component values. The bandwidth of the amplifier depends on the value of Q.

Center Frequency

As you may recall, center frequency is found as the geometric average of f_1 and f_2. By formula,

$$\boxed{f_0 = \sqrt{f_1 f_2}} \qquad\qquad (15.3)$$

When the Q of an amplifier is greater than (or equal to) 2, the value of f_0 approaches the *algebraic average* of f_1 and f_2 (designated f_{ave}). That is, f_0 approaches a value that is exactly halfway between f_1 and f_2. By formula,

$$\boxed{f_0 \cong f_{ave} \qquad Q \geqslant 2}$$

where f_{ave} is the algebraic average of f_1 and f_2, found as

$$\boxed{f_{ave} = \frac{f_1 + f_2}{2}} \qquad\qquad (15.4)$$

The following example illustrates the relationship between equations (15.3) and (15.4).

═══ **EXAMPLE 15.3** ═══

A Practical Consideration. The cutoff frequencies of a tuned amplifier are measured using the technique shown in Chapter 13.

The cutoff frequencies for a tuned amplifier are measured as $f_1 = 1160$ kHz and $f_2 = 1240$ kHz. Using equations (15.3) and (15.4) determine the values of f_0 and f_{ave} for the circuit. Then verify that the value of Q for the circuit is greater than 2.

Solution: Using equation (15.3), the value of f_0 is found as

$$f_0 = \sqrt{f_1 f_2}$$
$$= \sqrt{(1160 \text{ kHz})(1240 \text{ KHz})}$$
$$= 1199 \text{ kHz}$$

using equation (15.4), f_{ave} is found as

$$f_{ave} = \frac{f_1 + f_2}{2}$$

$$= \frac{1120 \text{ kHz} + 1240 \text{ kHz}}{2}$$

$$= 1200 \text{ kHz}$$

The value of Q for the circuit is now found as

$$Q = \frac{f_0}{\text{BW}}$$

$$= \frac{1199 \text{ kHz}}{80 \text{ kHz}}$$

$$= 14.99$$

As you can see, the value of f_0 approaches the value of f_{ave} when the Q of an amplifier is greater than (or equal to) 2.

PRACTICE PROBLEM 15–3

A tuned amplifier has measured cutoff frequencies of 980 kHz and 1080 kHz. Show that $f_0 \cong f_{mid}$ for the circuit and that the circuit value of Q is greater than 2.

You may be wondering how you can tell when the value of Q for an amplifier is greater than 2. You can assume that the value of Q is greater than (or equal to) 2 *when the value of f_0 for the circuit is at least two times its bandwidth*. As long as this condition is met, the Q of the amplifier will be greater than (or equal to) 2, and f_0 will be approximately equal to f_{ave}. As you will be shown later in this chapter, this fact simplifies the frequency analysis of many tuned amplifiers.

Integrated versus Discrete Tuned Circuits

Discrete circuits, such as BJT and FET amplifiers, are normally tuned using inductive–capacitive (LC) circuits. An example of a discrete tuned amplifier is shown in Figure 15.6a. The frequency response of the circuit is determined primarily by the values of C_T and L_T.

FIGURE 15.6

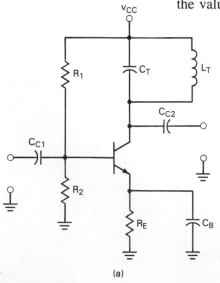

(a)

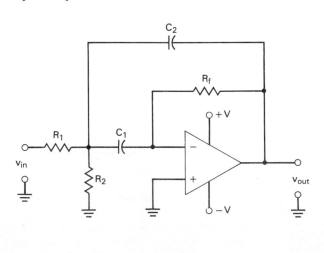

(b)

An op-amp (IC) equivalent of the discrete tuned amplifier is shown in Figure 15.6b. As you can see, the op-amp tuned circuit does not require the use of inductors. The frequency response of the circuit is determined by its resistor and capacitor values.

In the next sections, we will cover the operation of various op-amp tuned circuits. Then we will cover the operation of the basic discrete tuned amplifier. As you will see, the exact equations used to analyze the operation of these amplifiers will vary from one circuit to another. However, the relationships between bandwidth, Q, and f_0 discussed in this section will not vary significantly between the various circuits.

SECTION REVIEW

1. What are the characteristics of the *ideal* tuned amplifier? [Objective 1]
2. How does the frequency response of the practical tuned amplifier vary from that of the ideal tuned amplifier? [Objective 1]
3. What is the relationship between the roll-off rate and the bandwidth of a tuned amplifier? [Objective 2]
4. What is the *quality* (Q) rating of a tuned amplifier? [Objective 3]
5. What is the relationship between the values of Q and BW for a tuned amplifier? [Objective 3]
6. What determines the value of Q for a tuned amplifier? [Objective 5]
7. What is the relationship between the values of Q and f_0 for a tuned amplifier? [Objective 6]

15.2
Active Filters: Low-pass and High-pass Circuits

Tuned op-amp circuits are generally referred to as *active filters*. There are four basic types of active filters, each with its own circuit configuration and frequency response curve.

Active filter. A tuned op-amp circuit.

Overview

The frequency-response curves for the four types of active filters are shown in Figure 15.7. The *low-pass filter*, whose frequency-response curve is shown in Figure 15.7a, is designed to pass all frequencies from dc up to the upper cutoff frequency, f_2. The *high-pass filter*, represented in Figure 15.7b, is designed to pass all frequencies that are above its lower cutoff frequency, f_1. All frequencies that would normally be applied to this circuit that are above f_1 would be passed by the circuit. Of course, the op-amp does have a unity-gain rating. Therefore, it is impossible to design a high-pass filter that does not have a value of f_2. However, the value of f_{unity} for the op-amp will be well above the range of frequencies that the circuit is being used to pass.

The *band-pass filter*, represented by the curve in Figure 15.7c, is designed to pass only the frequencies that fall between its values of f_1 and f_2. The *band-stop filter*, or *notch filter*, is used to eliminate all signals within the stop band (between f_1 and f_2) while passing all frequencies outside this band. The frequency-response curve for the notch filter is shown in Figure 15.7d. Note that the function of the notch filter is the opposite of that of the band-pass filter.

Low-pass filter. A filter designed to pass all frequencies *below* a given cutoff frequency.

High-pass filter. A filter designed to pass all frequencies *above* a given cutoff frequency.

Band-pass filter. One designed to pass all frequencies that fall between f_1 and f_2.

Band-stop ("notch") filter. One designed to block all frequencies that fall between f_1 and f_2.

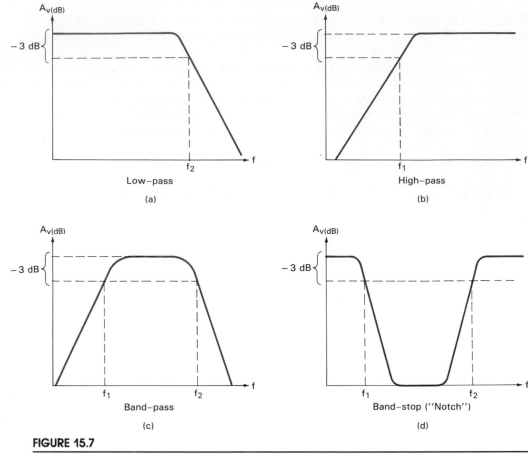

FIGURE 15.7

Active filter frequency response curves.

There is one critical point that needs to be made at this time. The concepts of Q, center frequency, and bandwidth are primarily related to the band-pass and notch filters. When we are dealing with band-pass and notch filters, we are generally concerned with the bandwidths of such amplifiers, along with their respective values of Q and f_0. However, when we are dealing with low-pass and high-pass filters, we are only concerned with one value, f_2 or f_1, respectively. The concepts of bandwidth and center frequency generally are not applied to these circuits.

General Terminology

8

Pole. A single *RC* circuit.

Remember: Any amplifier with more than one *RC* circuit will have roll-off rates that are additive. This point was discussed in detail in Chapter 13.

Before we get into any circuits, we need to establish some of the terms that are commonly used to describe active filters. The first of these is the term *pole*. A *pole* is nothing more than an *RC circuit*. Thus a one-pole filter is one that contains a single *RC* circuit. A two-pole filter would contain two *RC* circuits, and so on. As you will be shown, the most commonly used type of active filter, the *Butterworth filter*, will have a 20-dB/decade roll-off for each pole found in the circuit. This is shown in the following table.

Filter Type	Number of RC Circuits	Roll-off Rate (dB/Decade)
One pole	1	20
Two pole	2	40
Three pole	3	60

CHAP. 15   Tuned Amplifiers

The *Butterworth filter* is one that has constant gain across the pass band of the circuit. The term *flat response* is used to describe the constant-gain characteristic. Thus, when we talk of a flat response, we mean that the value of $A_{v(dB)}$ is relatively constant across the pass band of the circuit. Since Butterworth filters have the best flat-response characteristics, they are often referred to as *maximally flat*, or *flat-flat*, filters.

Another type of active filter, the *Chebyshev filter*, has a higher roll-off rate than the Butterworth filter containing the same number of poles. A Chebyshev filter has a roll-off rate of 40 dB/decade per pole. However, the frequency-response curve of the Chebyshev filter is *not* flat across the pass band. The typical frequency response of this filter type is illustrated in Figure 15.8. As you can see, there are variations in the pass band gain of the Chebyshev filter. For comparison sake, the response of the Butterworth filter is included in the graph (dashed line).

Butterworth filter. An active filter characterized by flat pass-band response and 20 dB/decade roll-off rates.

Chebyshev filter. An active filter characterized by irregular pass-band response and 40 dB/decade roll-off rates.

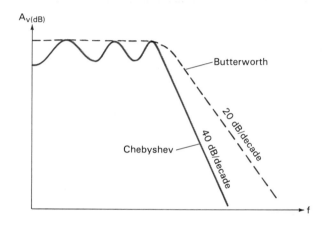

FIGURE 15.8

Butterworth and Chebyshev response.

Even though the Chebyshev filter has a roll-off rate that is closer to the ideal tuned circuit than the Butterworth filter, it is not used as often. There are two reasons for this:

Why Butterworth filters are preferred over Chebyshev filters.

1. The inconsistent gain of the Chebyshev filter is undesirable.
2. By using a two-pole Butterworth filter, the roll-off rate of the Chebyshev filter can be achieved, while still having a flat response.

Since the Butterworth filter is by far the most common active filter, we will concentrate on this circuit type in our discussions on active filters.

The Single-pole Low-pass Filter

The single-pole low-pass Butterworth filter is constructed in one of two ways, as either a *high-gain* circuit or a *voltage follower*. (Recall that a voltage follower has a voltage gain of 0 dB, or unity.) The high-gain circuit is shown in Figure 15.9. As you can see, this circuit is simply a noninverting amplifier with an added input shunt capacitor. Since the reactance of the capacitor will decrease as frequency increases, the high-frequency response is limited. The value of f_2 for this type of circuit is found as

$$f_2 = \frac{1}{2\pi RC}$$

(15.5)

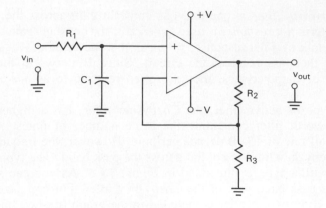

FIGURE 15.9

The single-pole low-pass active filter.

This is the same basic relationship that we used to find f_2 for the circuits we discussed in Chapter 13. The following example illustrates the process for determining the bandwidth of a single-pole low-pass filter.

EXAMPLE 15.4

Determine the bandwidth of the single-pole low-pass filter in Figure 15.10. Also, as a review, determine the value of A_{CL} for the circuit.

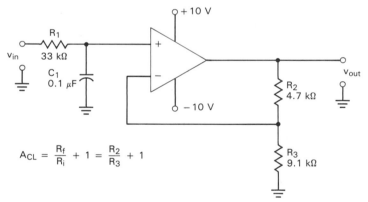

$$A_{CL} = \frac{R_f}{R_i} + 1 = \frac{R_2}{R_3} + 1$$

FIGURE 15.10

Solution: Using the circuit values of R_1 and C_1, the upper cutoff frequency, f_2, is found as

$$f_2 = \frac{1}{2\pi R_1 C_1}$$

$$= \frac{1}{2\pi(33 \text{ k}\Omega)(0.1 \text{ }\mu\text{F})}$$

$$= 48.23 \text{ Hz}$$

Since the amplifier is capable of working as a dc amplifier (0 Hz), there is no value of f_1 for the circuit. Thus, the bandwidth is equal to f_2, or 48.23 Hz.

CHAP. 15 Tuned Amplifiers

The circuit is a noninverting amplifier. Thus, its value of A_{CL} is found using

$$A_{CL} = \frac{R_f}{R_i} + 1$$

In this case, $R_f = R_2 = 4.7$ kΩ and $R_i = R_3 = 9.1$ kΩ. Therefore, the value of A_{CL} for the circuit is found as

$$A_{CL} = \frac{4.7 \text{ k}\Omega}{9.1 \text{ k}\Omega} + 1$$
$$= 1.52$$

PRACTICE PROBLEM 15–4

A single-pole low-pass filter like the one in Figure 15.10 has values of $R_1 = 47$ kΩ, $C_1 = 0.033$ μF, $R_2 = 10$ kΩ, and $R_3 = 10$ kΩ. Determine the values of f_2 and A_{CL} for the circuit.

The frequency-response curve for the circuit in Example 15.4 is shown in Figure 15.11. As you can see, the midband gain of the amplifier is 1.52, or 3.637 dB. At 48.23 Hz, the gain of the amplifier will be down to 1.08, or 0.637 dB. If the input frequency increases to one decade above the cutoff frequency (482.3 Hz), the gain of the circuit will drop to 0.637 dB − 20 dB = −19.36 dB. This means that the filter can actually *attenuate* the input signal. In other words, at some point, the output voltage will actually be less than the input voltage.

Don't forget.
$A_{CL(dB)} = 20 \log A_{CL}$

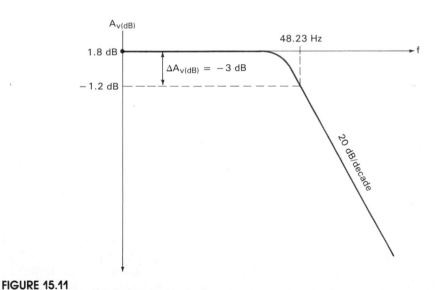

FIGURE 15.11

Although the value of A_{CL} in Example 15.4 was relatively low, it can be made much higher by changing the $R_3:R_2$ ratio. When unity (0 dB) gain is desired, the unity-gain circuit shown in Figure 15.12 may be used. Note that the bandwidth for this circuit is determined in the same manner as described previously. As you can see, this cirucit is nothing more than a *voltage follower*. Thus the voltage gain of the circuit will be approximately 0 dB. All circuit calculations for this amplifier are the same as described in previous chapters.

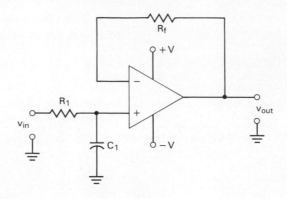

FIGURE 15.12

The Two-pole Low-pass Filter

As stated earlier, the two-pole low-pass filter will have a roll-off rate of 40 dB/decade. Several commonly used two-pole low-pass filter configurations are shown in Figure 15.13. As you can see, each circuit has two RC circuits, R_1–C_1 and R_2–C_2. As the operating frequency increases beyond f_2, each RC circuit will be dropping A_{CL} by 20 dB, giving a total roll-off rate of 40 dB/decade when operated above f_2.

The cutoff frequency for each of the circuits in Figure 15.13 is found as

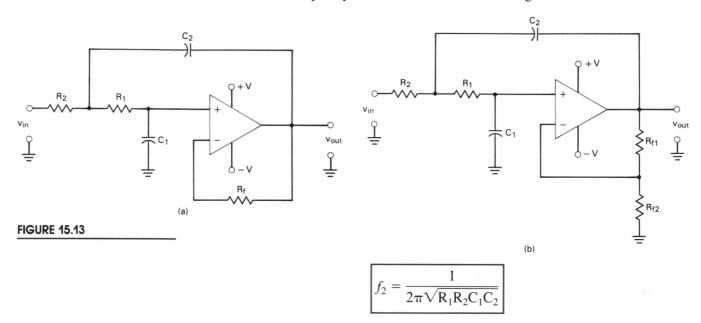

(a)

(b)

FIGURE 15.13

$$f_2 = \frac{1}{2\pi\sqrt{R_1 R_2 C_1 C_2}}$$

Here's where things get weird. In order for the two-pole low-pass filter to have a flat-response curve, its value of A_{CL} can be no greater than 1.586 (4 dB). This means that you cannot have a high-gain two-pole low-pass filter. The derivation of the value 1.586 involves calculus and is not covered here. In this case, we will simply accept its value as being valid.

Fulfilling the requirement of $A_{CL} < 1.586$ is no problem for the *unity-gain* filter in Figure 15.13a. This circuit, which is actually a voltage-follower, has a standard value of $A_{CL} = 1$. Note that R_f in the circuit is included for compensation, and its value is unimportant.

The circuit in Figure 15.13b can be designed for any value of A_{CL} between 1 and the upper limit of 1.586. Note that this circuit is normally designed according to the following guidelines:

1. $R_2 = R_1$
2. $C_2 = 2C_1$
3. $R_{f1} \leq 0.586 R_{f2}$

The following example illustrates these relationships, along with the process for determining the bandwidth of a two-pole low-pass filter.

═══════ **EXAMPLE 15.5** ═══════

Verify that the relationships between component values listed above apply to the circuit shown in Figure 15.14. Also, determine the bandwidth of the filter and draw its response curve.

Solution: First, let's verify the relationships listed above. R_1 and R_2 are equal in value, while C_2 is twice the value of C_1. Also note that the value of R_{f1} is approximately equal to $0.586R_{f2}$.

The cutoff frequency (and thus, the bandwidth) of the filter is found as

$$f_2 = \frac{1}{2\pi\sqrt{R_1 R_2 C_1 C_2}}$$

$$= \frac{1}{2\pi\sqrt{(10\ k\Omega)(10\ k\Omega)(0.015\ \mu F)(0.033\ \mu F)}}$$

$$= 715\ kHz$$

Finally, using the value of f_2 for the circuit, the response curve is drawn as shown in Figure 15.15.

FIGURE 15.14   **FIGURE 15.15**

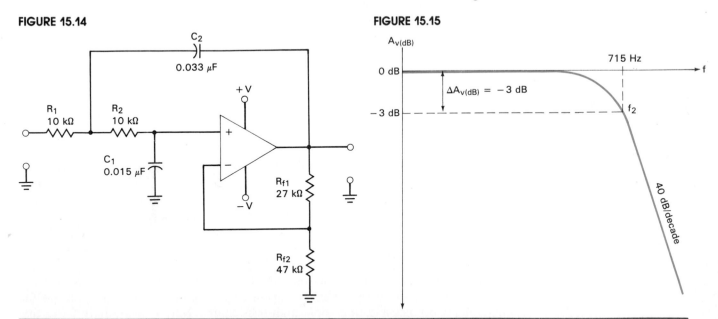

The Three-pole Low-pass Filter

In Chapter 13 you were shown that the roll-off rates of cascaded states *add* to form the total roll-off rate for the amplifier. It would follow that a three-pole filter, which has a roll-off rate of 60 dB/decade, could be formed by cascading a single-pole filter with a two-pole filter. This circuit would have the combined roll-off rate of 20 dB + 40 dB = 60 dB/decade. The three-pole low-pass filter is shown in Figure 15.16.

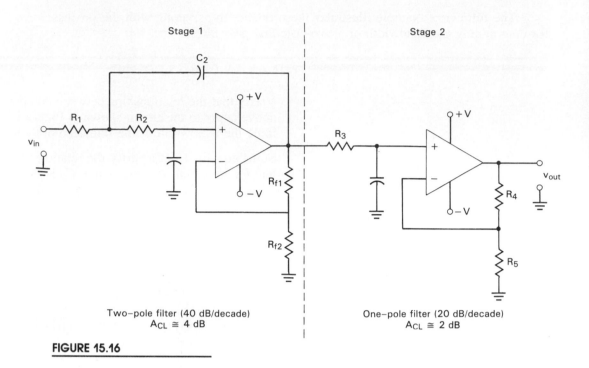

FIGURE 15.16

As you can see, we again have requirements for the values of A_{CL}. The two-pole circuit must have a close-loop gain of approximately 4 dB, while the single-pole filter must have a closed-loop gain of approximately 2 dB. As long as both of these requirements are fulfilled, the filter will have a Butterworth response curve. Also note that the two filters would be tuned to the same value of f_2. All circuit calculations for this amplifier are the same as those performed earlier.

How to analyze a multipole active filter.

Can active filters have more than three poles? Yes. When you see an active filter with more than three poles, you will have no problem in analyzing the circuit if you remember the following points:

1. The filter will have a 20-dB/decade roll-off *for each pole*. For example, a five-pole filter would have a roll-off rate of $5 \times 20 = 100$ dB/decade.
2. All stages will be tuned to the same cutoff frequency. Thus, to find the overall cutoff frequency, simply calculate the value of f_2 for any stage.

High-pass Filters

The difference between low-pass and high-pass active filters.

There is very little difference between the high-pass filter and the low-pass filters we have already discussed. This is illustrated in Figure 15.17, which shows the one-pole, two-pole, and three-pole high-pass filters. Note that the only thing that has changed is the positions of the capacitors and resistors. Since the capacitors are series elements, they will affect the low-frequency response of the circuits. The value of f_1 is found using the same equations we used to find f_2 for the low-pass filters.

As frequency decreases, the reactance of a given series capacitor will increase. This causes a larger portion of the input signal to be dropped across the capacitor.

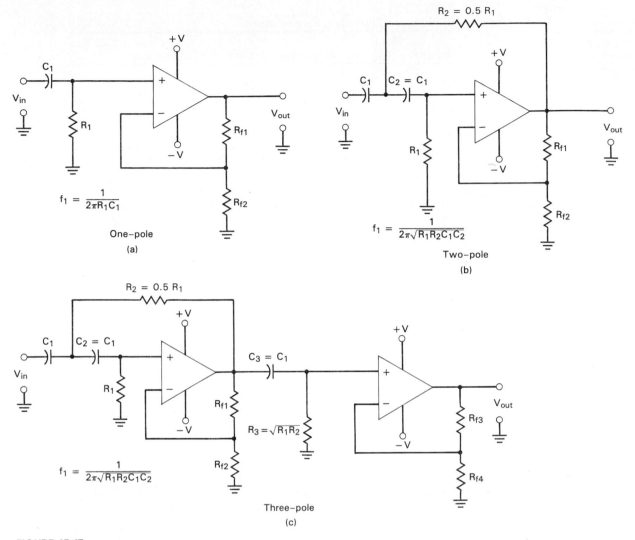

FIGURE 15.17

When the operating frequency has dropped to f_1, the series capacitors will have caused the gain of the amplifier to drop by 3 dB.

We are not going to spend any time on these circuits since their operating principles should be very familiar to you by now. Just remember, these circuits abide by all the rules established up to this point. The only difference is that they will pass frequencies *above* their cutoff frequencies.

Filter Gain Requirements

In this section, you have been provided with the gain requirements for several low-pass and high-pass active filters. Fulfilling these requirements provides a Butterworth response curve; that is, each filter has relatively constant gain until its cutoff frequency is reached. Then the gain rolls off at an approximate rate of 20 dB/ decade per pole.

The gain requirements for a variety of Butterworth active low-pass and high-pass filters are summarized in Table 15.1. The derivations of these gain requirements are way beyond the scope of this text. However, you should be aware that almost every type of low-pass or high-pass active filter has gain requirements that must be fulfilled if the circuit is to have a Butterworth response curve.

TABLE 15.1
Butterworth Filter Gain Requirements

No. of Poles	Maximum Overall dc Gain	Approximate Roll-off Rate (dB/Decade)
2	1.56 (4 dB)	40
3	2 (6 dB)	60
4	2.58 (8 dB)	80
5	3.29 (10 dB)	100
6	4.21 (12 dB)	120
7	5.37 (14 dB)	140

Decibel values are rounded off to nearest whole number.

Any of the multipole filters described in Table 15.1 could be constructed using the appropriate number of two-pole and one-pole cascaded states. For example, a five-pole circuit could be constructed by cascading two two-pole circuits and a one-pole circuit. To obtain the appropriate frequency-response characteristics, the following requirements would have to be met by the circuit:

1. Each of the two-pole circuits would require a closed-loop voltage gain of 4 dB and the one-pole circuit would require a closed-loop voltage gain of 2 dB. This would provide an overall voltage gain of 4 dB + 4 dB + 2 dB = 10 dB, the value given in Table 5.1 for a five-pole active filter.
2. The resistor and capacitor values would have to be chosen so that all three stages have identical cutoff frequencies.

FIGURE 15.18

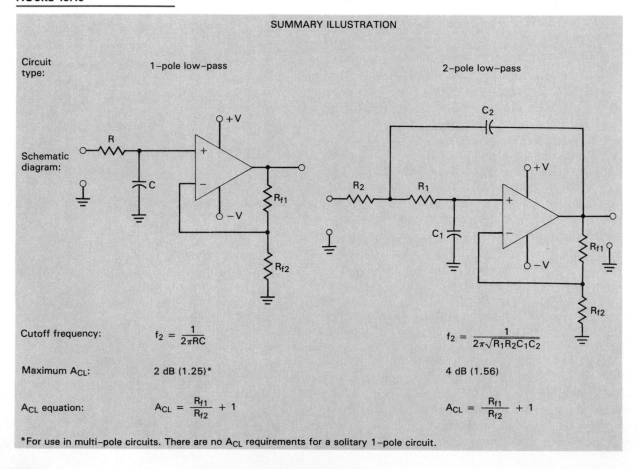

SUMMARY ILLUSTRATION

Circuit type: **1–pole low–pass**   **2–pole low–pass**

Schematic diagram:

Cutoff frequency: $f_2 = \dfrac{1}{2\pi RC}$ $f_2 = \dfrac{1}{2\pi\sqrt{R_1 R_2 C_1 C_2}}$

Maximum A_{CL}:   2 dB (1.25)*   4 dB (1.56)

A_{CL} equation: $A_{CL} = \dfrac{R_{f1}}{R_{f2}} + 1$ $A_{CL} = \dfrac{R_{f1}}{R_{f2}} + 1$

*For use in multi–pole circuits. There are no A_{CL} requirements for a solitary 1–pole circuit.

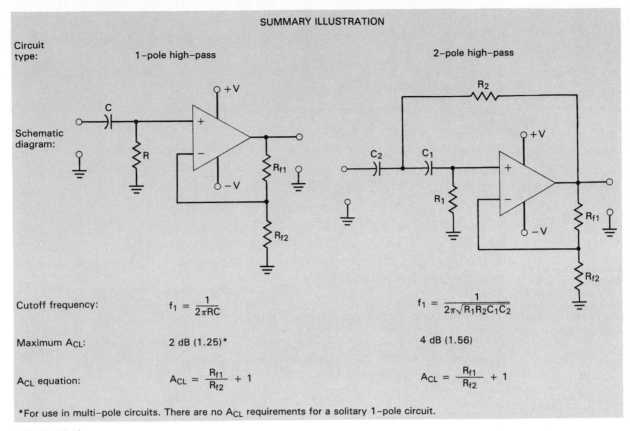

SUMMARY ILLUSTRATION

Circuit type:	1–pole high–pass	2–pole high–pass
Schematic diagram:		
Cutoff frequency:	$f_1 = \dfrac{1}{2\pi RC}$	$f_1 = \dfrac{1}{2\pi\sqrt{R_1 R_2 C_1 C_2}}$
Maximum A_{CL}:	2 dB (1.25)*	4 dB (1.56)
A_{CL} equation:	$A_{CL} = \dfrac{R_{f1}}{R_{f2}} + 1$	$A_{CL} = \dfrac{R_{f1}}{R_{f2}} + 1$

*For use in multi–pole circuits. There are no A_{CL} requirements for a solitary 1–pole circuit.

FIGURE 15.19

As long as these requirements are met, the circuit will operate as a Butterworth filter.

The component requirements for one-pole and two-pole Butterworth low-pass and high-pass active filters are summarized in Figures 15.18 and 15.19. Note that the values of R and C determine the cutoff frequencies, according to the proper form of the following equation:

$$f_c = \frac{1}{2\pi RC}$$

Also note that any of the circuits shown can be converted to unity-gain (0 dB) circuits by removing R_{f2}.

---

1. What is an *active filter*? What are the four types of active filters? [Objective 7]
2. Describe the frequency-response characteristics of each type of active filter. [Objective 7]
3. What is a *pole*? Why is the number of poles in an active filter important? [Objective 8]
4. Contrast the Butterworth response curve and the Chebyshev response curve. [Objective 9]

Sec. 15.3   Active Filters: Band-pass and Notch Circuits

5. What steps are involved in analyzing an active one-pole, two-pole, and three-pole low-pass filter? [Objective 10]

6. What is the difference between low-pass and high-pass active filters? [Objective 11]

7. Describe the dB gain and roll-off characteristics of low-pass and high-pass filters. [Objective 12]

15.3
Active Filters: Band-pass and Notch Circuits

A Practical Consideration. The order of the low-pass and high-pass filters is unimportant. We could have a high-pass first stage and a low-pass second stage, and the results would be the same.

You may recall that *band-pass filters* are designed to *pass* all frequencies within their bandwidths, while *notch filters* (*band-stop filters*) are designed to *block* all frequencies within their bandwidths. In this section, we will take a look at several band-pass and notch filters.

The Two-stage Band-pass Filter

It is common for a band-pass filter to be constructed by cascading a high-pass filter and a low-pass filter. Such a circuit is shown in Figure 15.20a. The first stage of the amplifier will pass all frequencies that are below its value of f_2. All

FIGURE 15.20

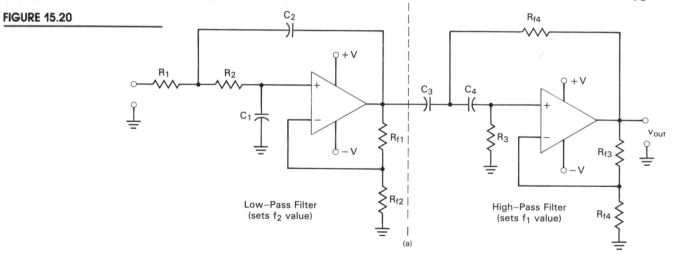

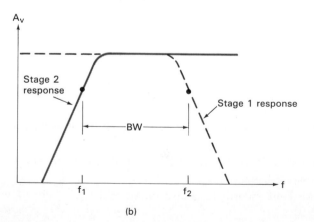

the frequencies passed by the first stage will head into the second stage, which will pass all frequencies above its value of f_1. The result of this circuit action is shown in Figure 15.20b. Note that the only frequencies that will pass through the amplifier are those that fall within the pass band of *both* amplifiers.

The frequency analysis of a circuit like the one in Figure 15.20a is relatively simple. The values of f_1 and f_2 are found as shown in Section 15.2. Once the values of f_1 and f_2 are known, the circuit values of bandwidth, center frequency, and Q are found as follows:

$$\begin{array}{|c|}
\hline
BW = f_2 - f_1 \\
f_0 = \sqrt{f_1 f_2} \\
Q = \dfrac{f_0}{BW} \\
\hline
\end{array}$$

These equations were used extensively in the first two sections of this chapter and need no further explanation. The following example illustrates the complete frequency analysis of the two-stage band-pass filter.

=== EXAMPLE 15.6 ===

Perform the complete frequency analysis of the amplifier in Figure 15.21.

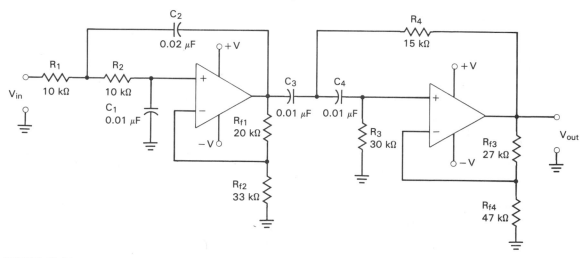

FIGURE 15.21

Solution: The value of f_2 is determined by the first stage of the circuit (the low-pass filter), as follows:

$$f_2 = \frac{1}{2\pi\sqrt{R_1 R_2 C_1 C_2}}$$

$$= \frac{1}{2\pi\sqrt{(10\ k\Omega)(10\ k\Omega)(0.01\ \mu F)(0.02\ \mu F)}}$$

$$= 1.13\ kHz$$

The value of f_1 is determined by the second stage of the circuit (the high-pass filter), as follows:

$$f_1 = \frac{1}{2\pi\sqrt{R_1 R_2 C_1 C_2}}$$

$$= \frac{1}{2\pi\sqrt{(30 \text{ k}\Omega)(15 \text{ k}\Omega)(0.01 \text{ }\mu\text{F})(0.01 \text{ }\mu\text{F})}}$$

$$= 750 \text{ Hz}$$

Thus, the circuit would pass all frequencies between 750 Hz and 1.13 kHz. All others would effectively be blocked by the band-pass filter.

Now that the values of f_1 and f_2 are known, we can solve for the amplifier bandwidth as follows:

$$\text{BW} = f_2 - f_1$$
$$= 1.13 \text{ kHz} - 750 \text{ Hz}$$
$$= 380 \text{ Hz}$$

The center frequency is found as

$$f_0 = \sqrt{f_1 f_2}$$
$$= \sqrt{(750 \text{ Hz})(1.13 \text{ kHz})}$$
$$= 921 \text{ Hz}$$

Finally, the value of Q is found as:

$$Q = \frac{f_0}{\text{BW}}$$
$$= \frac{921 \text{ Hz}}{380 \text{ Hz}}$$
$$= 2.42$$

PRACTICE PROBLEM 15–6

A filter like the one in Figure 15.21 has values of $R_1 = R_2 = 12 \text{ k}\Omega$, $C_1 = C_3 = C_4 = 0.01 \text{ }\mu\text{F}$, $C_2 = 0.02 \text{ }\mu\text{F}$, $R_3 = 39 \text{ k}\Omega$, and $R_4 = 20 \text{ k}\Omega$. Perform the frequency analysis of the filter.

The two-stage band-pass filter is the easiest of the band-pass filters to analyze. However, it has the disadvantage of requiring two op-amps and a relatively large number of resistors and capacitors. As you will see, the physical construction of the *multiple-feedback band-pass filter* is much simpler than that of the two-stage filter. At the same time, the frequency analysis of the multiple-feedback band-pass filter is more difficult than that of the two-stage filter.

Multiple-feedback Band-pass Filters

Multiple-feedback filter. A band-pass (or band-stop) filter that has a single op-amp and two feedback paths, one resistive and one capacitive.

The *multiple-feedback band-pass filter*, which is shown in Figure 15.22, derives its name from the fact that it has two feedback networks, one capacitive and one resistive. Note the presence of the input series capacitor (C_1) and the input shunt capacitor (C_2). The series capacitor limits the low-frequency response of the filter,

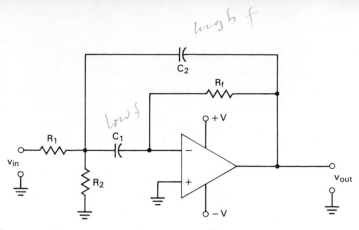

FIGURE 15.22

A bandpass active filter.

while the shunt capacitor limits its high-frequency response. This point is illustrated in Figure 15.23.

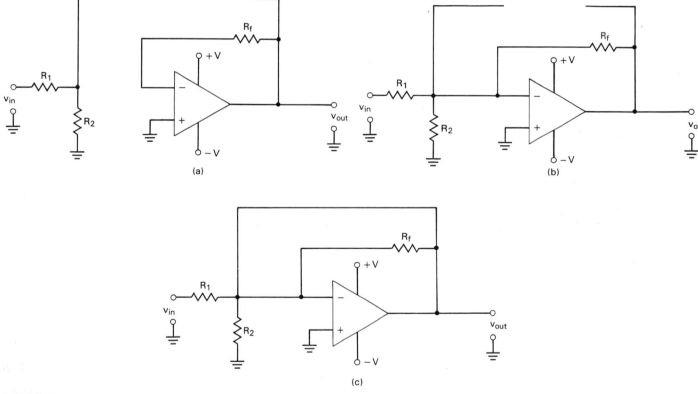

FIGURE 15.23

Bandpass active filter circuit operation.

Let's begin our discussion of the circuit operation by establishing some ground rules. To simplify our discussion, we will assume that:

1. $C_2 << C_1$.
2. A given capacitor will act as an open circuit until a *short-circuit frequency* is reached. At that point, the capacitor can be represented as a short circuit.

Sec. 15.3   Active Filters: Band-pass and Notch Circuits

Granted, the operation of the capacitors is more complicated than this, but we want to start by getting the overall picture of how the circuit works.

When the input frequency is below the short-circuit frequency for C_1, *both capacitors will act as open circuits*. Since $C_2 << C_1$, we know that the short-circuit frequency for C_1 is *lower* than that of C_2. Therefore, as long as C_1 is an open, C_2 is also an open. Having both capacitors acting as opens gives us the equivalent circuit in Figure 15.23a. As you can see, the input signal is completely isolated from the op-amp. Therefore, the circuit will have no output.

Now assume that the lower cutoff frequency for the filter (f_1) is equal to the short-circuit frequency for C_1. When f_1 is reached, C_1 becomes a short circuit, while C_2 remains open. This gives us the equivalent circuit in Figure 15.23b. For this circuit, v_{in} has no problem making it to the input of the op-amp, and the filter acts as an inverting amplifier. Thus, at frequencies above f_1, the filter has an output.

Assume now that f_2 is equal to the short-circuit frequency of C_2. When f_2 is reached, we have the equivalent circuit shown in Figure 15.23c. Now both capacitors are acting as short circuits. The input signal has no problem making it to the op-amp, but now the output is shorted back to the input. Since the gain of an inverting amplifier is found as

Note. The actual equation used to find A_{CL} for the multiple-feedback filter is a variation on the equation shown here. However, the equation shown here is valid for the point being made.

$$A_{CL} = \frac{R_f}{R_i}$$

and R_f is shorted, it is easy to see that the closed-loop voltage gain of the circuit is zero at frequencies above f_2. If we put these three equivalent circuits together, we have a circuit with zero voltage gain when operated at frequencies below f_1, relatively high voltage gain when operated at frequencies between f_1 and f_2, and zero voltage gain when operated at frequencies above f_2. This, by definition, is a band-pass filter.

Circuit Frequency Analysis

17

As you know, the cutoff frequency for a given low-pass or high-pass active filter is found using

$$f_c = \frac{1}{2\pi\sqrt{R_1 R_2 C_1 C_2}}$$

(15.6)

This equation can be modified to provide us with the following equation for the center frequency of a multiple-feedback band-pass filter:

$$f_0 = \frac{1}{2\pi C\sqrt{(R_1 \| R_2)R_f}}$$

(15.6')

(15.6)

The following example illustrates the use of this equation in determining the center frequency of a multiple-feedback filter.

━━━━━━━━━━ **EXAMPLE 15.7** ━━━━━━━━━━

Determine the value of f_0 for the filter in Figure 15.24.

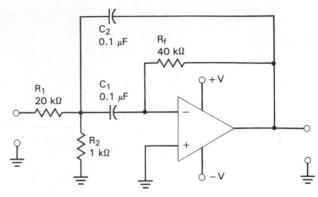

FIGURE 15.24

Solution: The center frequency of the circuit is found as

$$f_0 = \frac{1}{2\pi C \sqrt{(R_1 \| R_2)R_f}}$$

$$= \frac{1}{2\pi(0.1 \ \mu F)\sqrt{(20 \ k\Omega \| 1 \ k\Omega)(40 \ k\Omega)}}$$

$$= \frac{1}{2\pi(0.1 \ \mu F)(6.17 \ k\Omega)}$$

$$\cong 258 \ Hz$$

PRACTICE PROBLEM 15–7

A multiple-feedback filter has the following values: $R_1 = 30 \ k\Omega$, $R_2 = 820 \ \Omega$, $R_f = 68 \ k\Omega$, and $C_1 = C_2 = 0.1 \ \mu F$. Determine the center frequency of the circuit.

Once the center frequency of a multiple-feedback filter is known, we can calculate the value of Q for the circuit using the following equation:

$$\boxed{Q = \pi f_0 R_f C} \qquad \textbf{(15.7)}$$

The following example shows how equation (15.7) is used to determine the Q of a mulitple-feedback filter.

EXAMPLE 15.8

Determine the values of Q and bandwidth for the filter in Figure 15.24.

Solution: In Example 15.7, the center frequency of the filter was found to be 258 Hz. Using this value, the Q of the circuit is found as

$$Q = \pi f_0 R_f C$$
$$= \pi(258 \ Hz)(40 \ k\Omega)(0.1 \ \mu F)$$
$$= 3.24$$

Sec. 15.3   Active Filters: Band-pass and Notch Circuits

The bandwidth of the circuit is found as

$$BW = \frac{f_0}{Q}$$
$$= \frac{258 \text{ Hz}}{3.24}$$
$$= 80 \text{ Hz}$$

PRACTICE PROBLEM 15–8

Determine the values of Q and bandwidth for the filter described in Practice Problem 15–7.

Once the values of center frequency, Q, and bandwidth are known, the values of f_1 and f_2 for the multiple-feedback filter can be determined. As you will see, the equations used to determine the cutoff frequencies of the filter will vary with the Q of the circuit. That is, we will use one set of equations to find f_1 and f_2 when $Q > 2$ and another set to find f_1 and f_2 when $Q < 2$.

You may recall that the value of f_0 approaches the *algebraic* average (f_{ave}) of f_1 and f_2 when $Q > 2$, as was discussed in Section 15.1. Since f_{ave} is halfway between the cutoff frequencies, we can use the following equations to approximate the values of f_1 and f_2 when $Q > 2$:

$$\boxed{f_1 \cong f_0 - \frac{BW}{2} \qquad Q \geqslant 2} \qquad \textbf{(15.8)}$$

and

$$\boxed{f_2 \cong f_0 + \frac{BW}{2} \qquad Q \geqslant 2} \qquad \textbf{(15.9)}$$

Remember, the above equations may be used *only if Q is greater than or equal to 2*. If Q is less than 2, we have to use two equations that are a bit more complex than equations (15.8) and (15.9).

The actual relationship between f_{ave} and f_0 is given as

$$\boxed{f_{ave} = f_0 \sqrt{1 + \left(\frac{1}{2Q}\right)^2}} \qquad \textbf{(15.10)}$$

When the Q of a multiple-feedback filter is less than 2, equations (15.8) and (15.9) must be modified to take into account the difference between f_{ave} and f_0, as follows:

$$\boxed{f_1 = f_0 \sqrt{1 + \left(\frac{1}{2Q}\right)^2} - \frac{BW}{2} \qquad Q < 2} \qquad \textbf{(15.11)}$$

$$\boxed{f_2 = f_0 \sqrt{1 + \left(\frac{1}{2Q}\right)^2} + \frac{BW}{2} \qquad Q < 2} \qquad \textbf{(15.12)}$$

The following example illustrates the fact that equations (15.8) and (15.9) can be used to approximate the values of f_1 and f_2 when Q is greater than (or equal to) 2.

A multiple-feedback filter has values of $f_0 = 12$ kHz and BW $= 6$ kHz. Approximate the values of f_1 and f_2. Then determine the percent of error in the approximated values.

Solution: With the values of f_0 and BW given, we know that the Q of the circuit is 2. Thus, the values of f_1 and f_2 can be approximated as

$$f_1 = f_0 - \frac{BW}{2}$$
$$= 12 \text{ kHz} - 3 \text{ kHz}$$
$$= 9 \text{ kHz}$$

and

$$f_2 = f_0 + \frac{BW}{2}$$
$$= 12 \text{ kHz} + 3 \text{ kHz}$$
$$= 15 \text{ kHz}$$

Now we'll use equations (15.10) and (15.11) to validate the approximated values of f_1 and f_2, as follows:

$$f_1 = f_0 \sqrt{1 + \left(\frac{1}{2Q}\right)^2} - \frac{BW}{2}$$
$$= (12 \text{ kHz}) \sqrt{1 + \left(\frac{1}{4}\right)^2} - 3 \text{ kHz}$$
$$= (12 \text{ kHz})(1.031) - 3 \text{ kHz}$$
$$= 9.37 \text{ kHz}$$

and

$$f_2 = f_0 \sqrt{1 + \left(\frac{1}{2Q}\right)^2} + \frac{BW}{2}$$
$$= (12 \text{ kHz}) \sqrt{1 + \left(\frac{1}{4}\right)^2} + 3 \text{ kHz}$$
$$= (12 \text{ kHz})(1.031) + 3 \text{ kHz}$$
$$= 15.37 \text{ kHz}$$

Finally, the percents of error in the approximated values of f_1 and f_2 are found as

$$f_1: \quad \% \text{ of error} = \frac{|9.37 \text{ kHz} - 9 \text{ kHz}|}{9.37 \text{ kHz}} \times 100$$
$$= 4.1\%$$

and

$$f_2: \quad \% \text{ of error} = \frac{|15.37 \text{ kHz} - 15 \text{ kHz}|}{15.37 \text{ kHz}} \times 100$$
$$= 2.5\%$$

These percents of error are well below the acceptable value of 10%, and we can use the approximated values with confidence.

A multiple-feedback filter has values of $f_0 = 20$ kHz and BW = 5 kHz. Approximate the values of f_1 and f_2 for the circuit, and then determine the percents of error in those approximated values.

For the majority of active filters, the value of Q will be greater than 2. However, in those cases when $Q < 2$, you should be aware that the approximated values of f_1 and f_2 may be too inaccurate. This point is illustrated in the following example.

EXAMPLE 15.10

An active filter has values of $f_0 = 100$ Hz and $Q = 1.02$. Approximate the values of f_1 and f_2. Then determine the percents of error in those approximated values.

Solution: First, the bandwidth of the filter is found as

$$BW = \frac{f_0}{Q}$$
$$= \frac{100 \text{ Hz}}{1.02}$$
$$\cong 98 \text{ Hz}$$

The values of f_1 and f_2 are approximated as

$$f_1 \cong f_0 - \frac{BW}{2}$$
$$= 100 \text{ Hz} - 49 \text{ Hz}$$
$$= 51 \text{ Hz}$$

and

$$f_2 \cong f_0 + \frac{BW}{2}$$
$$= 100 \text{ Hz} + 49 \text{ Hz}$$
$$= 149 \text{ Hz}$$

Using equations (15.11) and (15.12), the cutoff frequencies for the circuit are found as

$$f_1 = f_0 \sqrt{1 + \left(\frac{1}{2Q}\right)^2} - \frac{BW}{2}$$
$$= (100 \text{ Hz}) \sqrt{1 + \left(\frac{1}{2.04}\right)^2} - 49 \text{ Hz}$$
$$= (100 \text{ Hz})(1.113) - 49 \text{ Hz}$$
$$= 62 \text{ Hz}$$

and

$$f_2 = f_0 \sqrt{1 + \left(\frac{1}{2Q}\right)^2} + \frac{BW}{2}$$

$$= (100 \text{ Hz}) \sqrt{1 + \left(\frac{1}{2.04}\right)^2} + 49 \text{ Hz}$$

$$= (100 \text{ Hz})(1.113) + 49 \text{ Hz}$$

$$= 160 \text{ Hz}$$

Finally, the percents of error are found as

$$f_1: \quad \% \text{ of error} = \frac{62 \text{ Hz} - 51 \text{ Hz}}{62 \text{ Hz}} \times 100$$

$$= 17.74\%$$

$$f_2: \quad \% \text{ of error} = \frac{160 \text{ Hz} - 149 \text{ Hz}}{160 \text{ Hz}} \times 100$$

$$= 6.88\%$$

While the percent of error in the approximated value of f_2 in the example is acceptable, the percent of error in the f_1 approximation is not. Therefore, we cannot approximate the cutoff frequencies for a filter with a Q that is less than 2.

Filter Gain

✔ 18

In most cases, the analysis of a multiple-feedback filter begins and ends with calculating the frequency response of the circuit. However, being able to predict the closed-loop voltage gain of the circuit provides us with a very valuable trouble-shooting tool, as you will see in our discussion on troubleshooting active filters.

The following equation is used to find the closed-loop voltage gain of a multiple-feedback filter:

$$A_{CL} = \frac{R_f}{2R_i} \qquad (15.13)$$

where R_i = the circuit series input resistor.

The following example illustrates the calculation of A_{CL} for a multiple-feedback filter.

━━━━━━━ **EXAMPLE 15.11** ━━━━━━━

Determine the closed-loop voltage gain for the filter in Figure 15.24.

Solution: For this circuit, R_1 is the circuit series input resistor. Using the values of R_f and R_1, the value of A_{CL} is found as

Note. The value of $A_{CL} = 1$ obtained in this example was a function of the circuit. It is not a constant.

$$A_{CL} = \frac{R_f}{2R_i}$$

$$= \frac{40 \text{ k}\Omega}{40 \text{ k}\Omega}$$

$$= 1$$

PRACTICE PROBLEM 15–11

Determine the closed-loop voltage gain for the filter described in Practice Problem 15–7. Assume that R_1 is the series input resistor for the circuit.

This completes the analysis of the multiple-feedback band-pass filter. At this point, we will move on to discuss the operation of two notch filters. The first will be a cascaded notch filter that is similar to the two-stage band-pass filter that we covered earlier. Then take a brief look at the multiple-feedback notch filter.

Notch Filters

Summing amplifier. A circuit that produces an output voltage that is proportional to the sum of its input voltages.

The notch filter is designed to block all frequencies that fall within its bandwidth. A multistage notch filter block diagram and frequency-response curve are shown in Figure 15.25. As the block diagram shows, the circuit is made up of a *high-pass filter*, a *low-pass filter*, and a *summing amplifier*. A summing amplifier is a circuit that produces an output voltage that is proportional to the sum of its input voltages. The operation of the summing amplifier is covered in detail in Chapter 16. For now, all you need to understand is that the summing amplifier in the notch filter will have an output that is equal to the sum of the filter output voltages.

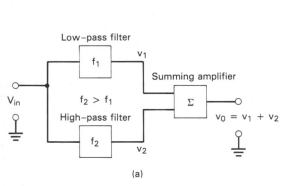

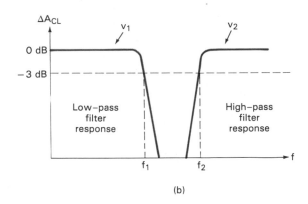

(a)   (b)

FIGURE 15.25

The circuit is designed so that f_1 (which is set by the high-pass filter) is lower in value than f_2 (which is set by the low-pass filter). The gap between the values of f_1 and f_2 is the bandwidth of the filter. This point is illustrated in the frequency-response curve in Figure 15.25b.

When the circuit input frequency is lower than f_1, the input signal will pass through the low-pass filter to the summing amplifier. Since the input frequency is below the cutoff frequency of the high-pass filter, v_2 will be zero. Thus, the output from the summing amplifier will equal the output from the low-pass filter.

When the circuit input frequency is higher than f_2, the input signal will pass through the high-pass filter to the summing amplifier. Since the input frequency is above the cutoff frequency of the low-pass filter, v_1 will be zero. Now the summing amplifier output will equal the output from the high-pass filter. Note that the frequencies below f_1 and above f_2 have been passed by the notch filter.

When the circuit input frequency is between f_1, and f_2, neither of the filters will produce an output (*ideally*). Thus, v_1 and v_2 will both be zero, and the output from the summing amplifier will also equal zero. In practice, of course, the exact output from the notch filter will depend on how close the input frequency is to either f_1 or f_2. But, in any case, the output from the notch filter will be greatly reduced when it is operated within its bandwidth.

The block diagram in Figure 15.25 is actually constructed as shown in Figure 15.26. While the circuit appears confusing at first, closer inspection shows that it is made up primarily of circuits that we have already discussed. The low-pass filter consists of the op-amp labeled IC1 and all the components with the L subscript. This low-pass filter, though drawn differently than you are used to seeing, is identical to the low-pass input stage of the band-pass filter in Figure 15.21. The high-pass filter consists of the op-amp labeled IC2 and all the components with the H subscript. This circuit is identical to the high-pass output stage of the band-pass filter in Figure 15.21. Finally, the summing amplifier is made up of the op-amp labeled IC3 and all the components with the S subscript. Despite the complex appearance of the circuit, it operates exactly as described in our discussion of the notch filter block diagram.

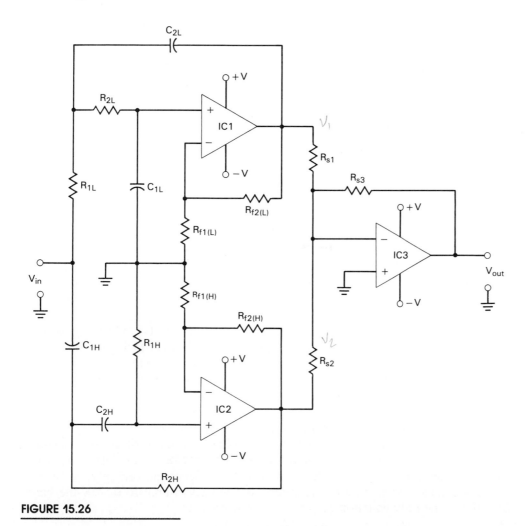

FIGURE 15.26

The frequency analysis of the notch filter shown in Figure 15.26 is identical to that of the two-stage band-pass filter. First, the cutoff frequencies of the low-pass and high-pass filters are determined. Then, using the calculated cutoff frequencies, the bandwidth, center frequency, and Q values for the circuit are determined.

The Multiple-feedback Notch Filter

The multiple-feedback notch filter, which is shown in Figure 15.27, is very similar to its band-pass counterpart.

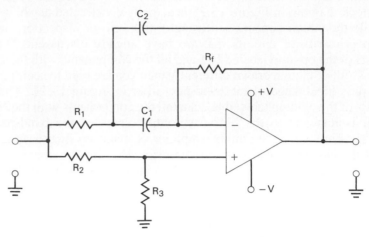

FIGURE 15.27

The capacitors in the notch filter (C_1 and C_2) react to a change in frequency exactly as described in our discussion on multiple-feedback band-pass filters. However, the connection of v_{in} to the noninverting input of the op-amp (via the R_2–R_3 voltage divider) radically alters the overall circuit response, as follows:

1. When $f_{in} < f_1$, C_1 prevents the input voltage from reaching the inverting input of the op-amp. However, the signal is still allowed to reach the noninverting input, and a noninverted output is produced by the circuit.

A Practical Consideration. C_1 will always produce a phase shift. Even though this phase shift may be extremely small, it will still prevent v_{in} from appearing at both inputs of the op-amp at the same time. Therefore, the circuit will still have a small output signal when operated within its bandwidth.

2. When $f_{in} > f_1$, the input signal is applied to both the inverting and noninverting inputs of the op-amp. In other words, v_{in} starts to be seen by the op-amp as a common-mode signal. As you recall, op-amps reject common-mode input signals. Therefore, there is little or no output from the filter when the input signal is within the bandwidth of the op-amp.

3. When $f_{in} > f_2$ both capacitors will act as short circuits. While C_2 partially shorts out the signal at the inverting input of the op-amp (reducing its amplitude), v_{in} is still being applied to the noninverting op-amp input. In this case, the op-amp will amplify the difference between the two input signals. Again, the circuit will have a measurable output.

In a summary, the circuit will act as a noninverting amplifier when the input frequency is lower than f_1 or higher than f_2. When the input frequency is within the bandwidth of the filter, the op-amp will, to one degree or another, see v_{in} as a pair of common-mode signals and will reject those signals accordingly.

The frequency analysis of the notch filter is very similar to that of the band-pass filter. The only equation modification is as follows:

$$f_0 = \frac{1}{2\pi C\sqrt{R_1 R_f}} \qquad (15.14)$$

The final difference lies in the A_{CL} characteristic of the circuit. When operated above f_2, C_2 acts as a short circuit, and the filter acts as a *voltage follower*. To have equal gain values on both sides of the bandwidth (that is, below f_1 and above f_2), the values of R_f and R_1 are normally selected to provide a low-frequency gain that is as close to unity (1) as possible. Thus, the closed-loop voltage gain can be approximated as

$$\boxed{A_{\text{CL}} \cong 1 \qquad (0 \text{ dB})} \qquad \textbf{(15.15)}$$

One Final Note

There are far too many types of active filters to cover in a textbook of this type. In this chapter, you have been introduced to several active filters. It should not surprise (or discourage) you to know that there are literally hundreds of active filter configurations. However, the most commonly used are still Butterworth filters.

Although we have not been able to completely cover the topic of active filters, you should now have the knowledge needed to pursue the study of active filters further.

SECTION REVIEW

1. Describe how cascaded low-pass and high-pass filters can form a band-pass filter. [Objective 13]

2. List, in order, the steps that are taken to perform the frequency analysis of a two-stage band-pass filter. [Objective 14]

3. In terms of circuit construction and analysis, contrast the two-stage band-pass filter to the multiple-feedback band-pass filter. [Objective 15]

4. Which capacitor in the multiple-feedback band-pass filter controls the high-frequency response of the circuit? Which one controls its low-frequency response? [Objective 16]

5. Briefly describe the operation of the multiple-feedback band-pass filter. [Objective 16]

6. List, in order, the steps that are taken to analyze the frequency response of a multiple-feedback band-pass filter. [Objective 17]

7. How does the analysis of a $Q < 2$ multiple-feedback filter differ from that of a $Q \geq 2$ filter? [Objective 17]

8. Describe the operation of the multistage notch filter shown in Figure 15.26. [Objective 19]

9. How would you analyze the frequency response of a notch filter like the one in Figure 15.26? [Objective 19]

10. Briefly describe the operation of the multiple-feedback notch filter. [Objective 20]

15.4
Active Filter Applications and Troubleshooting

It was stated at the beginning of this chapter that tuned circuits are used primarily in communications electronics. Audio and video circuits make extensive use of tuned circuits. At the same time, tuned circuits are also used in other areas of electronics, such as *biomedical electronics*, the area of electronics that deals with medical test and treatment equipment.

Biomedical electronics. The area of electronics that deals with medical test and treatment equipment.

In this section, we will take a look at several audio applications for active filters. We will also take a look at troubleshooting basic active filters.

An Audio Crossover Network

21

Crossover network. A circuit that is designed to separate high-frequency audio from low-frequency audio.

A *crossover network* is a circuit that is designed to split the audio signal from a stereo, television, or other communications system so that the high-frequency portion of the audio goes to a small high-frequency speaker (called a *tweeter*) and the low-frequency portion goes to a relatively large low-frequency speaker (called a *woofer*). A crossover network may be placed in an audio system as shown in Figure 15.28.

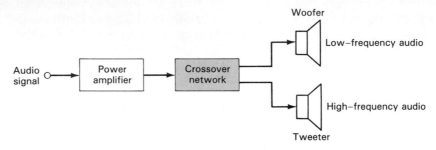

FIGURE 15.28

Let's assume that the block diagram represents the output of a stereo. In this stereo, the audio signal is applied to a power amplifier to boost the power level of the audio signal. Then the high-power signal is applied to the crossover network. This network splits the audio, sending the high-frequency portion of the audio to the tweeter and the low-frequency portion of the audio to the woofer.

The crossover network in Figure 15.28 would be a circuit that is similar to the one shown in Figure 15.29. This crossover network consists of an input buffer, followed by a two-pole low-pass filter and a two-pole high-pass filter.

FIGURE 15.29

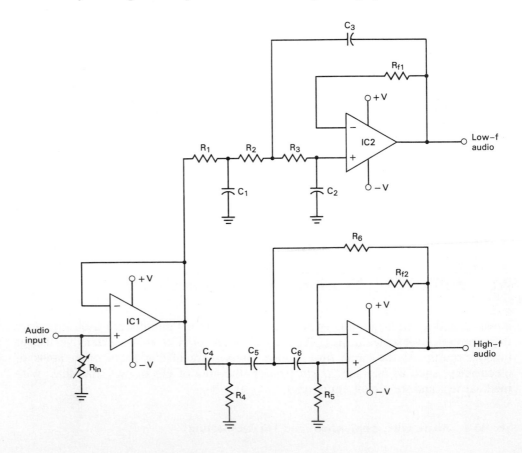

710

The input buffer is a voltage-follower that consists of IC1 and R_{in}. R_{in} is used to match the input impedance of the crossover network to the output impedance of the power amplifier.

The output signal from the buffer is applied to both of the active filters. IC2 and its associated circuitry pass the low-frequency audio to the woofer, while blocking the high-frequency audio. IC3 and its associated circuitry pass the high-frequency audio to the tweeter, while blocking the low-frequency audio.

A Simple Graphic Equalizer

A *graphic equalizer* is a circuit or system that is designed to allow you to control the amplitude of different audio-frequency ranges. A simplified graphic equalizer block diagram is shown in Figure 15.30. This graphic equalizer is made up of a series of low-Q active filters and a summing amplifier. The number of active filters used would increase with the sophistication of the system.

Graphic equalizer. A circuit or system designed to allow you to control the amplitude of different audio-frequency ranges.

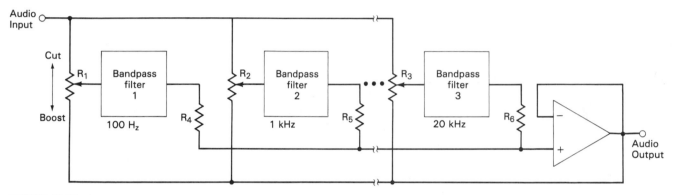

FIGURE 15.30

A simple graphic equalizer.

The summing amplifier is made up of the op-amp and the resistors labeled R_4, R_5, and R_6. The audio output from the summing amplifier is fed back to the band-pass filters via their input control potentiometers. When the potentiometer of a given filter is adjusted toward the audio output, the signal strength at the filter input *increases*. This has the effect of *boosting* (amplifying) that particular frequency. When the input potentiometer of a given filter is adjusted toward the audio input, the signals strength at the filter input *decreases*. This has the effect of *cutting* (attenuating) that particular frequency. By adjusting the input controls to the various active filters, you are able to boost the frequencies you want to hear while cutting others.

Some Other Active Filter Applications

Notch filters and high-pass filters can be used to eliminate the low-frequency noise that can be generated in many audio systems. For example, a notch filter can be used to eliminate the 30-Hz "rumble" that can be produced by a turntable. By tuning the notch filter to 30 Hz, this noise will be blocked by filter, while the high audio frequencies are allowed to pass.

Another application would be the use of a high-pass filter to eliminate both the 30-Hz rumble and the 60-Hz power line noise. By tuning the high-pass filter so that it has a lower cutoff frequency of 60 Hz, both the power line noise and the turntable noise are eliminated with a single circuit.

There are far more applications for active filters than could possibly be covered here. However, you should now have a good idea of how versatile these circuits are.

Active Filter Fault Symptoms

✔ 24

Active filter troubleshooting is relatively simple when you keep in mind the fact that some of the circuit components are used to determine its frequency-response characteristics while others are used to determine its gain characteristics. For example, consider the two-pole low-pass filter shown in Figure 15.31a. The frequency response of the circuit is controlled by the combination of R_1, R_2, C_1, and C_2. The gain of the circuit is controlled primarily by R_{f1} and R_{f2}.

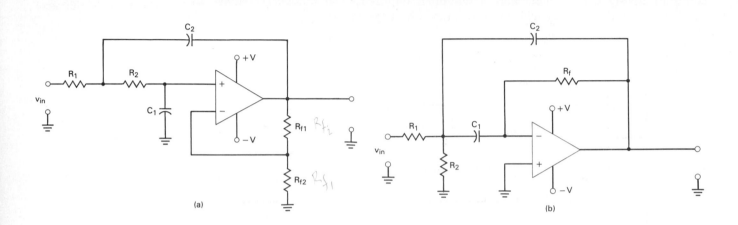

(a)　　　　　　　　　　　　(b)

FIGURE 15.31

If either of the gain components opens, the amplitude of the filter output will change. For example, if R_{f1} opens, the circuit configuration changes to that of a voltage follower, and the A_{CL} of the circuit will drop to 1 (0 dB). If R_{f2} opens, the value of A_{CL} will jump to the open-loop voltage gain of the op-amp, and the output signal will be clipped.

While R_1 and R_2 are a part of the frequency-response circuit, the output of the filter will drop to zero if either of these components opens. This is because the input signal will be isolated from the op-amp itself. The same thing will occur if C_1 is shorted.

If either C_1 opens or C_2 opens, there will be two results:

1. The roll-off rate of the circuit will decrease.
2. The cutoff frequency of the circuit will shift.

The first point is easy to see when you consider the case of C_2 opening. If this capacitor opens, we go from a two-pole filter to a single-pole filter. As you recall, the filter will have a roll-off rate of 20 dB/decade per pole. Thus, the roll-off rate for the filter will drop from 40 to 20 dB/decade (or from 12 to 6 dB/octave).

If either capacitor in the filter opens, the cutoff frequency for the circuit will *decrease*. You see, C_1 and C_2 are both shunt capacitors when viewed from the op-amp. Thus, from the op-amp point of view, they can be (loosely) considered to be parallel components. If either opens, the total shunt capacitance *increases*, causing a decrease in the cutoff frequency of the circuit.

TABLE 15.2
Low-pass Filter Fault Symptoms and Their Causes

Symptom	Possible Cause(s)
A_{CL} drops to 1	R_{f1} open, R_{f2} shorted[a]
$A_{CL} = A_{OL}$	R_{f2} open
$A_{CL} = 0$ (no output)	R_1 open, R_2 open, C_1 shorted, C_2 shorted, R_{f2} shorted[a]
Low f_c and roll-off rate	C_1 or C_2 open

[a] Highly unlikely.

Table 15.2 summarizes the faults that can occur in the low-pass two-pole filter and the symptoms of each. The fault symptoms that can develop in the two-pole high-pass filter are very similar to those listed for the low-pass filter, as can be seen by comparing Table 15.3 with Table 15.2.

A Practical Consideration. Don't forget that the circuit won't work if there is no input signal or if either of the power supply connections is faulty.

TABLE 15.3
High-pass Filter Fault Symptoms and Their Causes

A_{CL} drops to 1	R_{f1} open, R_{f2} shorted[a]
$A_{CL} = A_{OL}$	R_{f2} open
$A_{CL} = 0$ (no output)	R_1 shorted,[a] R_2 shorted,[a] C_1 open, C_2 open, R_f shorted[a]
Low f_c and roll-off rate	C_1 or C_2 shorted

[a] Highly unlikely.

Don't forget. A leaky capacitor will act (to some degree) like a shorted capacitor.

The two-stage band-pass filter (see Figure 15.21) can develop a fault in either of the two stages. A loss of gain in either stage will direct you to the appropriate components. If f_1 is low and the low-frequency roll-off rate is low, the fault is located in the high-pass filter stage. If f_2 is low and the high-frequency roll-off rate is low, the fault is located in the low-pass filter stage.

The multiple-feedback band-pass filter is relatively easy to troubleshoot if you keep in mind the equations for the circuit center frequency and gain. These equations were given earlier as

$$f_0 = \frac{1}{2\pi C \sqrt{(R_1 \| R_2)R_f}}$$

and

$$A_{CL} = \frac{R_f}{2R_i}$$

For the circuit in Figure 15.31b, R_i is the resistor labeled R_1. If this resistor opens, the output of the filter will drop to zero. If R_2 opens, the circuit will act as an inverting amplifier. You may recall that the gain of an inverting amplifier is found as

$$A_{CL} = \frac{R_f}{R_i}$$

Thus, if R_2 in the filter opens, the value of A_{CL} for the circuit will *double*. Also, the value of f_0 for the circuit will decrease, since R_2 will no longer be in parallel with R_1 (which increases the denominator value in the f_0 equation).

If R_f opens, the value of A_{CL} will increase to the value of A_{OL} for the op-amp. At the same time, the value of f_0 will dramatically decrease.

Sec. 15.4   Active Filter Applications and Troubleshooting

If C_1 shorts, the circuit will act as a single-pole low-pass filter. This is due to the fact that C_1 determines the lower cutoff frequency for the circuit. If C_2 shorts, the output of the op-amp will be shorted back to the input, and the circuit will not have an output. If C_1 opens, the input signal will be isolated from the filter, and the output will drop to zero. If C_2 opens, the circuit will act as a single-pole low-pass filter since C_2 controls the value of f_2. The fault symptoms for the multiple-feedback filter are listed in Table 15.4.

TABLE 15.4
Band-pass Filter Fault Symptoms and Their Causes

Symptom	Possible Causes
$A_{CL} = A_{OL}$	R_f open, R_1 shorted[a]
A_{CL} doubles	R_2 open
$A_{CL} = 0$ (no output)	R_1 open, R_2 shorted,[a] C_1 open, C_2 shorted, R_f shorted[a]
Circuit acts as a low-pass filter	C_1 shorted
Circuit acts as a high-pass filter	C_2 open

[a] Highly unlikely.

Active Filter Troubleshooting

✔ 25

Once you know the common fault symptoms for active filters, the troubleshooting procedure for these circuits is easy:

1. Determine the type of filter you are dealing with.
2. Verify that the filter has an input signal.
3. Verify that the load is not the cause of the problem by isolating the filter output from the load.
4. Verify that both supply inputs to the filter are working properly.
5. Using your observations and Tables 5.2 through 5.4, determine the source of the fault within the filter.
6. If none of the passive components is faulty, replace the op-amp.

SECTION REVIEW

1. What is a *crossover network*? [Objective 21]
2. Describe the operation of the crossover network in Figure 15.29. [Objective 21]
3. What is a *graphic equalizer*? [Objective 22]
4. Describe the operation of the graphic equalizer in Figure 15.30. [Objective 22]
5. Describe how low-pass and notch filters can be used to eliminate low-frequency noise in audio systems. [Objective 23]
6. List the fault symptoms that can occur in active filters (those described in the text) and the possible cause(s) of each. [Objective 24]
7. List the steps involved in troubleshooting an active filter. [Objective 25]

15.5
Discrete Tuned Amplifiers

✔ 26

Discrete component circuits are tuned using LC (inductive–capacitive) tank circuits in place of a collector (or drain) resistor. A typical circuit is shown in Figure 15.32. Note the presence of the parallel LC tank circuit in the collector circuit of the transistor. Since it is this circuit that determines the operating characteristics of the amplifier, we will take a moment to review some important characteristics of parallel LC circuits.

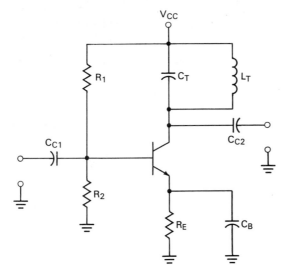

FIGURE 15.32

A discrete tuned amplifier.

✔ 27

Parallel LC Circuits

If we were to graph the *reactance versus frequency* characteristics of a capacitor and an inductor on the same chart, the plot would look like that shown in Figure 15.33. The *origin* of the graph represents 0-Ω impedance on the y-axis and 0-Hz

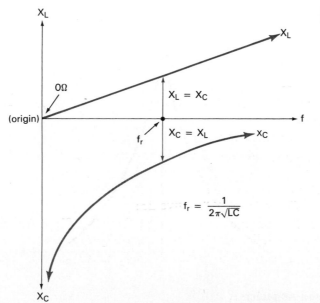

FIGURE 15.33

$$f_r = \frac{1}{2\pi\sqrt{LC}}$$

frequency on the x-axis. Note that at 0 Hz the impedance of the inductor is shown to be 0 Ω, while the value of X_c is shown to be ∞ Ω. As frequency increases, the value of X_L will *increase*, while the value of X_C will *decrease*. At some frequency, the values of X_C and X_L will be *equal* for a given capacitor–inductor pair. This frequency, called the *resonant frequency*, is found as

$$f_r = \frac{1}{2\pi\sqrt{LC}}$$

(15.16)

You should recall from your study of basic electronics that:

1. The current in a capacitive branch will *lead* the voltage across that branch by 90°.
2. The current in an inductive branch will *lag* the voltage across that branch by 90°.

In a parallel *LC* circuit, the voltage across the capacitor is equal to the voltage across the inductor. Combining this fact with the statements above, it is easy to see that I_C and I_L are 180° out of phase in a parallel *LC* tank. Thus

$$I_{net} = |I_C - I_L|$$

(15.17)

where I_{net} is the net current entering (and leaving) the tank circuit.

At resonance, $X_C = X_L$.

At resonance, the values of X_C and X_L are equal. Since the voltages across the parallel components are equal, I_C and I_L are equal. Thus, at resonance,

$$\begin{aligned} I_{net} &= |I_C - I_L| \\ &= 0 \text{ A} \end{aligned}$$

Below resonance, a tank circuit acts inductive. Above resonance, it acts capacitive.

Since the net current through the *LC* tank circuit is 0 A at resonance, the overall impedance of the tank circuit is infinite for all practical purposes. At frequencies below resonance, $X_L < X_C$. Thus $I_L > I_C$, and the circuit acts inductive. At frequencies above resonance, $X_C < X_L$. Thus $I_C > I_L$, and the circuit acts capacitive. These points are illustrated in Figure 15.34. Now let's apply these principles to the tuned amplifier shown in Figure 15.32.

FIGURE 15.34

Frequency response of a parallel LC tank circuit.

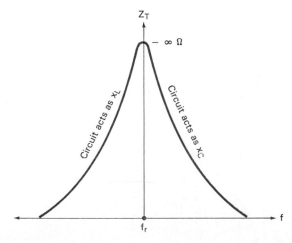

AC Circuit Conditions

The ac equivalent of Figure 15.32 is derived as usual, the only exception being the fact that the tank circuit components are not shorted. This ac equivalent circuit is shown in Figure 15.35. The load resistance is included to aid in our discussion.

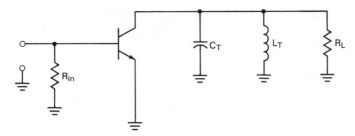

FIGURE 15.35

The ac equivalent circuit for a discrete tuned amplifier.

To completely understand the operation of this circuit, we need to take a look at three frequency conditions:

1. $f_{in} < f_r$
2. $f_{in} = f_r$
3. $f_{in} > f_r$

When $f_{in} < f_r$, the circuit will act *inductive*. This is not to say that there is no capacitive reactance present. To the contrary, X_C will be greater than X_L when the input frequency is below f_r. Since this is the case, most of the current through the tank circuit will be passing through the inductor. Because of this, we say that the circuit acts inductive. As frequency decreases, a point will be reached when $(X_C - X_L) = R_L$. When this happens, the voltage gain of the amplifier will drop by 3 dB, as we have discussed with *RC* circuits in the past. The only difference is that now we have an *LC* circuit causing the gain of the circuit to drop. Thus f_1 for the circuit occurs when $(X_C - X_L) = R_L$. When $f_{in} < f_r$.

When $f_{in} = f_r$, the tank circuit acts as an open, as was discussed earlier in this section. Since R_L represents the only path to ground in the collector circuit, all the ac collector current flows through R_L. Thus the efficiency of this amplifier is extremely high. When $f_{in} = f_r$.

When $f_{in} > f_r$, the reactance of the inductor starts to increase. However, now we see the reactance of the capacitor in the tank circuit decreasing rapidly. As f_{in} continues to increase, the point will be reached where $(X_L - X_C) = R_L$, and again the gain of the amplifier will drop by 3 dB. This is exactly the same thing that happened with the *RC* circuits in Chapter 13. Note that since the collector current now passes through R_L and the capacitor, the circuit is considered to be capacitive in nature. Also, f_2 occurs when $(X_L - X_C) = R_L$. When $f_{in} > f_r$.

As you may have figured out by now, the resonant frequency of the tank circuit is the center frequency of the tuned amplifier. By formula,

$$f_0 = f_r \qquad\qquad (15.18)$$

or

$$f_0 = \frac{1}{2\pi\sqrt{LC}} \qquad\qquad (15.19)$$

EXAMPLE 15.12

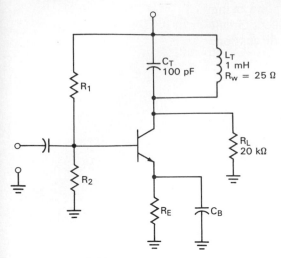

FIGURE 15.36

Determine the center frequency of the tuned amplifier shown in Figure 15.36.

Solution: The center frequency is equal to the resonant frequency of the *LC* tank circuit. Thus, using equation (15.19) and the circuit values shown, f_0 is found as

$$f_0 = \frac{1}{2\pi\sqrt{LC}}$$

$$= \frac{1}{2\pi\sqrt{(1\text{ mH})(100\text{ pF})}}$$

$$= \frac{1}{1.987 \times 10^{-6}}$$

$$= 503.29\text{ kHz}$$

PRACTICE PROBLEM 15–12

A tuned BJT amplifier has values of $L = 33$ mH and $C = 0.1$ μF. Calculate the value of f_0 for the circuit.

Once you have determined the value of f_0 for a tuned amplifier, the next step is to find the bandwidth. However, to determine bandwidth, we need to know the value of Q for the circuit. We take a look at the process for calculating Q now.

The Figure of Merit, Q

Earlier in this chapter, we stated that the Q of a tuned amplifier is equal to the ratio of center frequency to bandwidth. Although this statement helps to explain the use of Q in bandwidth calculations, it does not really tell you what Q is or where it comes from.

The Q of a tuned amplifier is a property of the tank circuit; more specifically, it is a property of the inductor. Strictly speaking, the Q of a tank circuit is *the ratio of energy stored in the circuit to energy lost in the circuit.* By formula,

Q (tank circuit). The ratio of energy stored in the circuit to energy lost by the circuit.

$$Q = \frac{\text{energy stored}}{\text{energy lost}}$$

Another way of writing this is

$$Q = \frac{\text{reactive power}}{\text{resistive power}}$$

Apparent power. Power stored in the field of a reactive component.

If you were to take the voltage across an inductor and multiply it by the current through the component, you would obtain an *apparent* power rating. The reason we say *apparent* is because the power value obtained is not really dissipated by the component, but, rather, is stored in the electromagnetic field that surrounds the component. This *apparent power* is actually *energy stored* by the component.

Another way of finding this *apparent power* (or *energy stored*) is

$$P_{APP} = (I_L^2)(X_L)$$

While most of the apparent power is stored in the electromagnetic field, *some* of the power is actually dissipated across the small amount of winding resistance in the coil. This *resistive power*, or *energy lost*, is found as

$$P_{Rw} = (I_L^2)(R_L)$$

Now the Q of the coil is found as the ratio of apparent (reactive) power to resistive power. By formula,

$$Q = \frac{P_{APP}}{P_{Rw}}$$
$$= \frac{(I_L^2)(X_L)}{(I_L^2)(R_w)}$$

or, simply,

$$\boxed{Q = \frac{X_L}{R_w}} \qquad \text{(15.20)}$$

where Q = the quality of the coil, measured at the center frequency
X_L = the reactance of the coil, measured at the center frequency
R_w = the resistance of the winding

The following example illustrates the process for determining the Q of a tank circuit.

EXAMPLE 15.13

Determine the Q of the tank circuit in Figure 15.36.

Solution: In Example 15.12, we determined the center frequency of the amplifier to be 503.29 kHz. Using this value, X_L for the circuit is determined as

$$X_L = 2\pi f L$$
$$= 2\pi(503.29 \text{ kHz})(1 \text{ mH})$$
$$= 3.16 \text{ k}\Omega$$

Using this value and the value of $R_w = 25 \ \Omega$ shown in the figure, the Q of the tank circuit is found as

$$Q = \frac{X_L}{R_w}$$
$$= \frac{3.16 \text{ k}\Omega}{25 \ \Omega}$$
$$= 126.5$$

PRACTICE PROBLEM 15–13

The inductor in Practice Problem 15–12 has a value of $R_w = 18 \ \Omega$. Determine the Q of the tank circuit.

The value of Q obtained in Example 15.13 is accurate for the tank circuit. However, before we can use this value for bandwidth calculations, we have to take into account the effects of *circuit loading*; that is, the effect that R_L will have on the Q of the tank. We will take a look at this *loaded-Q* now.

Loaded-Q

To see the effects of loading on circuit Q, we have to take another look at the ac equivalent of the collector circuit. The ac equivalent is shown in Figure 15.37a. In this figure, we have included the tank circuit capacitor, C_T. However, since it does not really weigh into this discussion, we will not refer to it.

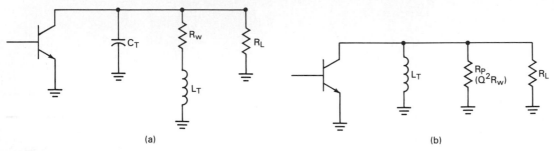

(a)   (b)

FIGURE 15.37

The effects of loading on circuit Q.

If you look closely at Figure 15.37a, you'll notice that R_w is in parallel with R_L. Thus, if we want an accurate value of Q, we must include the resistance of the load into our calculations. The first step in determining the loaded-Q of the circuit is to replace R_w with an equivalent parallel resistance, R_p. This resistance is shown in Figure 15.37b. The value of R_p was derived by determining the total series impedance of X_L and R_w and then solving for the equivalent parallel resistance value. If we were to do this here, we would eventually end up with the following relation:

$$R_P = Q^2 R_w$$

(15.21)

The derivation of equation (15.21) is shown in Appendix D. The total ac resistance of the collector circuit would now be found as

$$r_C = R_P \| R_L$$

(15.22)

and the *loaded-Q* would now be found as

$$Q_L = \frac{r_C}{X_L}$$

(15.23)

Equation (15.23) is also derived in Appendix D. The following example illustrates the determination of the value of loaded-Q for a tuned amplifier.

EXAMPLE 15.14

Determine the loaded-Q for the circuit shown in Figure 15.36.

Solution: The Q of the tank circuit was found in Example 15.13 to be 126.5. Using this value and the value of $R_w = 25 \ \Omega$, the value of R_P is found as

$$R_P = Q^2 R_w$$
$$= (126.5)^2 (25 \ \Omega)$$
$$= 400 \ k\Omega$$

Since $R_P >> R_L$, we can approximate the parallel combination of the two resistors (r_C) to be equal to R_L, 20 kΩ. Using this value in equation (15.23), along with the value of $X_L = 3.16$ kΩ (obtained in Example 15.13), we can find Q_L as follows:

$$Q_L = \frac{r_C}{X_L}$$
$$= \frac{20 \ k\Omega}{3.16 \ k\Omega}$$
$$= 6.33$$

PRACTICE PROBLEM 15–14

The circuit described in Practice Problems 15–12 and 15–13 has a 10-kΩ load. Determine the value of Q_L for the circuit.

As you can see, this loaded-Q value for the circuit is considerably lower than the original value of Q found for the tank circuit.

Now that we have determined the loaded-Q value for the circuit, the bandwidth of the amplifier can be found as shown in the following example.

=== **EXAMPLE 15.15** ===

Determine the bandwidth for the amplifier shown in Figure 15.36.

Solution: Using the values of $f_0 = 503.29$ kHz and $Q_L = 6.33$ found in previous examples, the bandwidth of the amplifier is found as

$$BW = \frac{f_0}{Q_L}$$
$$= \frac{503.29 \ kHz}{6.33}$$
$$= 79.5 \ kHz$$

PRACTICE PROBLEM 15–15

Determine the bandwidth of the amplifier in Practice Problem 15–14.

Once the values of f_0 and bandwidth are known, we can calculate the cutoff frequencies of the amplifier using the same guidelines that were established in our discussion on active filters. For circuits with $Q_L > 2$, the cutoff frequencies can be approximated as

$$f_1 \cong f_0 - \frac{BW}{2}$$

and

$$f_2 \cong f_0 + \frac{BW}{2}$$

When $Q_L < 2$, we need to use the more exact equations:

$$f_1 = f_0 \sqrt{1 + \left(\frac{1}{2Q_L}\right)^2} - \frac{BW}{2}$$

and

$$f_2 = f_0 \sqrt{1 + \left(\frac{1}{2Q_L}\right)^2} + \frac{BW}{2}$$

The only difference between these equations and those presented earlier in the text is the use of the *loaded-Q*, Q_L.

Now that you understand the operation of discrete tuned amplifiers, we will discuss some practical situations.

SECTION REVIEW

1. What is the circuit recognition feature of the discrete tuned amplifier? [Objective 26]
2. Why is the net current in a parallel resonant tank circuit approximately equal to zero? [Objective 27]
3. Describe the tuned discrete amplifier response to frequencies that are less than, equal to, and greater than the value of f_r. [Objective 28]
4. What is the Q of a tank circuit? [Objective 29]
5. How does the presence of a load affect the Q of the tank circuit? [Objective 29]

15.6
Discrete Tuned Amplifiers: Practical Considerations and Troubleshooting

 30

Why variable components are used.

The primary concern here is this: It is common for a technician to build a tuned amplifier and find that the center frequency is not what was expected. There are several reasons for this:

1. Capacitors and inductors usually have large tolerances. Thus, there can be a significant difference between the *rated* value of an inductor (or capacitor)

and its actual value. This can throw your calculations off by a signficant margin.

2. Amplifiers tend to have many "natural" capacitances that are not accounted for in the center–frequency equation. Values of stray capacitance and junction capacitance (for example) are always present in a discrete amplifier. If the capacitor used in the tank circuit is in the picofarad range, the other circuit capacitances can have a drastic effect on the actual center frequency of the circuit.

To overcome these problems, discrete tuned amplifiers are usually built using either a variable capacitor or a variable inductor. Both circuit types are shown in Figure 15.38. In the circuit shown in Figure 15.38a, a variable inductor is used for tuning the amplifier. If the value of L_T is *decreased*, the value of f_0 will *increase*, and vice versa. The same principle holds true for the circuit in Figure 15.38b. Of the two configurations, the variable-inductor circuit is used more commonly. This is due to the fact that variable inductors tend to be less expensive than variable capacitors.

Tuning the amplifier.

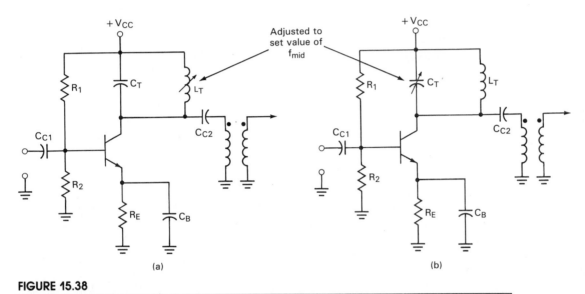

FIGURE 15.38

Amplifier tuning.

Another method of adjusting the tuning of a circuit is shown in Figure 15.39. This type of tuning is referred to as *electronic tuning*. This name comes from the fact that the tuning of the amplifier is *voltage controlled*. The component labeled D_1 in the circuit is a *varactor diode*. Varactor diodes are solid-state components that have relatively *high* junction capacitance values. By varying the amount of reverse bias on a varactor, the capacitance of the component junction is varied. This point is discussed in detail in Chapter 21. At this point we only need to establish the basic operating principle of the component.

When the biasing of D_1 is adjusted, its junction capacitance changes. This causes the total capacitance in the tuned circuit to change. Changing the total capacitance in the tuned circuit causes the resonant frequency of the circuit to change. Thus the tuning of the amplifier is changed by varying the reverse bias applied to the varactor.

Do not get caught up in trying to figure out all of Figure 15.39. This circuit, or one like it, will be covered in detail when you study communications electronics. It is introduced here solely to show you the concept of electronic tuning.

Electronic tuning. Using voltages to control the tuning of a circuit.

Varactor. A diode with high junction capacitance.

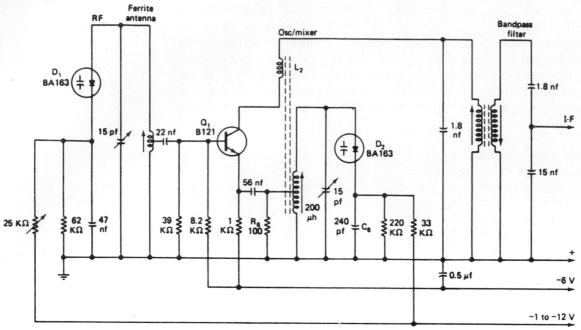

Broadcast band AM receiver front end with electronic tuning.

FIGURE 15.39

Electronic (varactor) tuning. (Courtesy of Prentice-Hall.)

Tuning a Circuit

 31

The next question is, how do you go about tuning an amplifier? What test procedure is used for adjusting the variable component to get the right center frequency? One method is to adjust the variable component using an *impedance meter*, or *Z-meter*. By connecting the meter across the variable component, you can adjust it for the proper value of L or C.

There are a couple of problems with using a Z-meter for tuning purposes:

1. The Z-meter cannot account for any other values of L and C in the circuit, and thus will usually only come close to giving you the needed value of L or C for the desired f_r.
2. Z-meters cost a *lot* of money. (This problem can tend to outweigh problem 1!)

A more practical approach to the problem is illustrated in Figure 15.40. Here a *dc current meter* is connected to the tuned amplifier between the tank circuit and V_{CC}. With no signal applied, this meter would read the dc value of I_C. However, when f_r is applied to the circuit, the meter reading will vary *slightly* around I_C.

The procedure for tuning the circuit is as follows:

1. Using an oscilloscope (or a frequency meter), adjust f_r to the desired value.
2. *Slowly* adjust L_T while keeping an eye on the meter.
3. When the meter shows a *rapid dip* in value, the circuit is properly tuned.

This test is based on the fact that the net current in the tank circuit is 0 A when it is operated at resonance. When it is operated either above or below f_r, a small amount of "alternating" current will start to flow through the tank. This

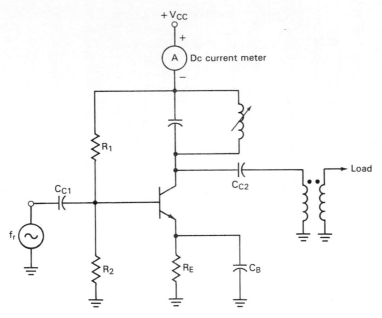

FIGURE 15.40

change in current will cause a *momentary* dip in the reading on the dc current meter.

By holding the input frequency at the desired value of f_r and adjusting L_T, the dip in the meter reading will indicate that the tank circuit has been tuned to the value of f_r.

Circuit Troubleshooting

The most common problem that develops in a tuned amplifier is frequency *drift*. Component aging can cause the tuning of an amplifier to change. When this happens, the amplifier may start to pass frequencies that may otherwise have been stopped. A simple cure for this problem is to retune the amplifier. If this does not solve the problem, one or both of the tank circuit components will probably have to be replaced. (*Do not neglect the possibility that the coupling transformer could cause the drift.*)

If either the capacitor or the inductor *fails* (opens or shorts), the effects will be much more drastic than those of a drift problem. Possible "fatal" faults that can occur in the tank circuit and their respective symptoms are listed in Table 15.5.

Drift. A change in tuning caused by component aging.

32

TABLE 15.5
Tuned Amplifier Troubleshooting

Fault	Symptoms
L_T open	V_{CC} is effectively removed from the transistor collector circuit (capacitors block dc voltages). Therefore, I_C and V_C both drop to zero.
L_T shorted	The ac output from the amplifier is developed across the tank circuit. Shorting L_T effectively removes the tank circuit from the picture. The circuit will now act as an emitter follower, and V_C will stay at V_{CC}.
C_T open	Capacitors are "open" circuits by nature. However, the capacitor *connections* can open, effectively removing the capacitor from the circuit. When this happens, the amplifier will act inductive, and all tuning is lost (the amplifier will pass *all* high frequencies).
C_T shorted	The symptoms for this condition are the same as those that occur when L_T shorts.

Don't forget. A leaky capacitor will act (to a degree) like a shorted component.

Sec. 15.6   Discrete Tuned Amplifiers: Practical Considerations and Troubleshooting

SUMMARY ILLUSTRATION

Discrete tuned amplifiers

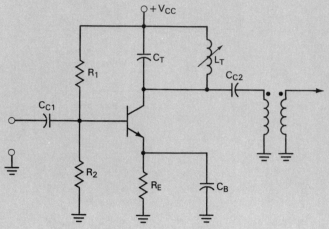

Range of Q values:	Normally greater than 2
Center frequency:	$f_O = \dfrac{1}{2\pi\sqrt{LC}}$

Practical considerations:

*Circuits are normally transformer-coupled to reduce Q-loading

*Variable components are normally included to allow adjustment of the circuit center frequency.

*Low C_T values can cause significant errors between calculated and measured center frequency values, due to effects of circuit stray and internal capacitances.

*Subject to center frequency drift. When this occurs, readjustment of the tuning is required.

*Many other circuit configurations are possible. The common factor is that all of them will contain at least one parallel LC circuit.

FIGURE 15.41

Since either component shorting will cause the same symptoms, further testing is required when V_C stays at V_{CC}. The simplest thing to do is measure the resistance of the capacitor. If the capacitor shows signs of charging, the inductor is the problem. If the capacitor resistance reading stays at 0 Ω, the capacitor is the cause of the problem.

As with any circuit troubleshooting, you should start by making sure that your tuned amplifier is the cause of the problem. Once you have narrowed the problem down to a given tuned amplifier, the table in this section should help you to establish whether or not the fault is in the tank circuit. If it is not, troubleshoot the rest of the circuit components using the procedures taught earlier in the book.

Summary

Discrete tuned amplifiers use *LC* circuits to provide narrowband tuning. The characteristics of the BJT tuned amplifier are summarized in Figure 15.41.

SECTION REVIEW

1. Why is amplifier tuning needed? [Objective 30]
2. What is *electronic tuning*? [Objective 30]

CHAP. 15 Tuned Amplifiers

3. Describe the methods that are normally used to adjust the tuning of an amplifier. [Objective 31]
4. What is *drift*? [Objective 32]
5. How is drift corrected? [Objective 32]
6. What are the symptoms that can occur when a major fault occurs in the tank circuit of a tuned amplifier? [Objective 32]

15.7
Class C Amplifiers

Earlier in the text, you were exposed to class A, class B, and class AB amplifiers. In this section we will take a brief look at class C amplifiers. Class C amplifiers have been neglected up to this point because they are tuned amplifiers.

Class C amplifiers are circuits containing transistors that conduct for *less than 180° of the input cycle*. A basic class C amplifier is shown in Figure 15.48.

Class C amplifiers. Circuits containing transistors that conduct for less than 180° of the input cycle.

DC Operation

The transistor in a class C amplifier is biased deeply into *cutoff*. This is often done by connecting the base resistor to a negative supply voltage. (for *npn* circuits) as shown in Figure 15.42.

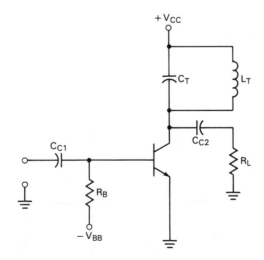

FIGURE 15.42

The class C amplifier.

Since a class C amplifier is biased in cutoff, the value of V_{CE} will be approximately equal to V_{CC}. This results in a dc load line like the one shown in Figure 15.43. Note that this dc load line can be somewhat misleading. You see, with the amplifier biased in cutoff, $I_C = 0$ A. The dc load line for the circuit would lead you to believe that the value of I_C can be varied, as long as $V_{CE} = V_{CC}$. This just is not the case. As long as there is no input signal to the transistor, $V_{CEQ} = V_{CC}$ and $I_{CQ} = 0$ A. Perhaps it would be more appropriate to have a dc

The dc load line for the class C amplifier is misleading.

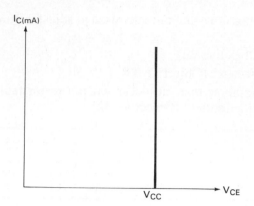

FIGURE 15.43

Class C amplifier dc load line.

load *point* for this circuit, rather than a dc load *line*. However, the dc operation of the circuit is traditionally represented by a load line, so. . . .

When a negative supply voltage is used, it is normally set at a value that fulfills the following relation:

$$\boxed{V_{in(pk)} + (-V_{BB}) = 1 \text{ V}}$$

or

$$\boxed{-V_{BB} = 1 \text{ V} - V_{in(pk)}} \qquad (15.24)$$

Thus, if the amplifier in Figure 15.48 had an input peak value of $+4$ V, the value of V_{BB} would be approximately -3 V. This would ensure that the transistor would turn on only at the positive peaks of the input cycle. The purpose served by this type of biasing will be discussed later in this section.

AC Operation

FIGURE 15.44

LC tank operation.

The ac operation of this circuit is based on the characteristics of the parallel resonant tank circuit. These characteristics are illustrated in Figure 15.44. In Figure 15.44a,

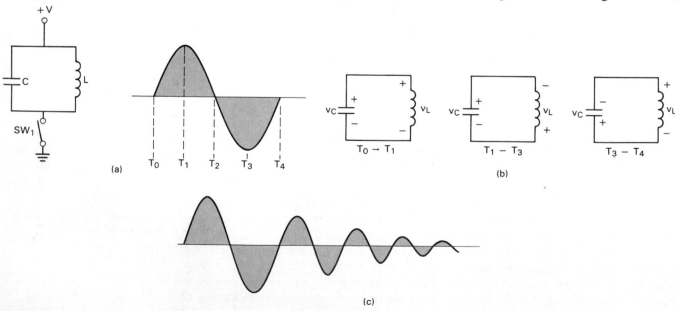

CHAP. 15 Tuned Amplifiers

we have a parallel tank circuit. The waveform shown on the right will be produced by the tank circuit if SW_1 is closed for an instant and then opened again. The source of the signal is easy to understand if we consider the circuit operation during each time interval shown (see Table 15.7).

TABLE 15.6
Circuit Operation

Time Interval	Circuit Action
$T_0–T_1$	During this time interval, the switch is closed. While the switch is closed, current flows through the circuit, charging both components to the polarities shown in Figure 15.50b. Note that at the end of T_1 the electromagnetic field around the inductor is at its *maximum* strength.
$T_1–T_3$	At the end of T_1, the switch is reopened. At this time, the field around the inductor starts to collapse. Recall that this collapsing field causes a *counter emf*, thus the reversed polarity of v_L shown in Figure 15.50b. With the polarities shown, the current in the tank circuit is reversed. The current will continue in this direction until the capacitor has charged fully and v_L has changed to the value represented at T_3 in Figure 15.50a.
$T_3–T_4$	At T_3, the capacitor has charged fully and v_L has again reversed polarity. Now the capacitor discharges through the inductor, causing the inductor voltage to go toward the value indicated at T_4.

If the process described were to continue, we would get a waveform like that shown in Figure 15.44c. Each cycle would be caused by the charge/discharge cycle set up by the capacitor and the inductor. Since the tank circuit would lose some power with each cycle, the waveform would eventually die out.

The charge/discharge cycle described here is known as the *flywheel effect*. Note that the waveform produced by the flywheel effect will have a frequency that is equal to the resonant frequency of the tank circuit.

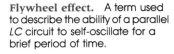

Flywheel effect. A term used to describe the ability of a parallel *LC* circuit to self-oscillate for a brief period of time.

If we wanted to keep the flywheel effect from dying out, we could do so by closing SW_1 for an instant on each positive transition from T_0 to T_1. This point is illustrated in Figure 15.45.

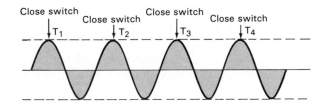

FIGURE 15.45

If we could close the switch in Figure 15.44a for just an instant during each cycle produced by the tank circuit, the waveform produced would not die out. This is due to the fact that we would be returning the power lost by the circuit on each cycle. For example, from T_1 to T_2, the tank circuit will lose a small amount of power. You should recall that this power lost is the result of current flowing through the R_w of the coil. At T_2, the closing of the switch would add enough power to the circuit to make up for the loss. If we could continue this cycle of closing and opening the switch, we could get the waveform produced by the tank circuit to continue indefinitely. With this in mind, let's take another look at the original class C amplifier circuit. For convenience, this circuit is shown again in Figure 15.46.

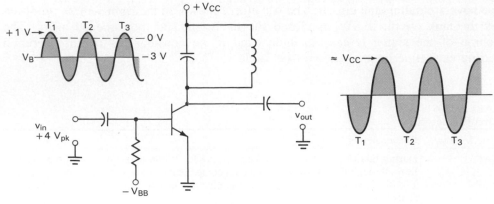

FIGURE 15.46

Class C circuit action.

Now, consider the transistor in this circuit to be the switch we talked about earlier. The positive peaks of the input signal would cause the transistor to saturate, while the transistor would be in cutoff the rest of the time. Each time the transistor saturates, it is like closing the switch in Figure 15.44. In other words, on each positive peak of the input, the transistor switch is closed, and the tank circuit gets the current pulse it requires to continue producing the output waveform.

Do not forget that the common-emitter transistor produces a 180° phase shift. Thus, the output signal is labeled on the negative peaks. If we had shown current waveforms, the input and output would have been shown to be in phase, as is the case with the common-emitter amplifier.

Another important point is this: For the class C amplifier to work properly, the tank circuit must be tuned to either the same frequency as v_{in} or to some *harmonic* of that frequency. A *harmonic* is a whole number multiple of a given frequency. For example, a 2 kHz signal would have harmonics of 4 kHz, 6 kHz, 8 kHz, and so on.

Harmonic. A whole number multiple of a given frequency.

When a class C amplifier is tuned to a harmonic of v_{in}, the output frequency of the circuit will be equal to that harmonic. For example, a class C amplifier that has an input frequency of 3 kHz and a tank circuit that is tuned to 6 kHz will produce a 6 kHz output. Thus, the class C amplifier can be used as a type of *frequency multiplier*.

Since the bandwidth, Q, and Q_L characteristics of the class C amplifier are the same as those described earlier in this chapter, we will not elaborate on them further.

SECTION REVIEW

1. What is the transistor conduction characteristic of the class C amplifier? [Objective 33]
2. Why is the dc load line of the class C amplifier misleading? [Objective 33]
3. Describe *flywheel effect*. [Objective 33]
4. How is the transistor used to maintain the flywheel effect in a class C amplifier? [Objective 33]
5. What is the input frequency requirement for proper class C operation? [Objective 33]

KEY TERMS

The following terms were introduced and defined in this chapter:

active filter	drift	passband
apparent power	electronic tuning	pole
band-pass filter	flywheel effect	Q (tank circuit)
band-stop (notch) filter	graphic equalizer	quality (Q)
biomedical electronics	high-pass filter	stop band
Butterworth filter	loaded-Q	summing amplifier
Chebyshev filter	low-pass filter	tuned amplifier
class C amplifiers	multiple-feedback filter	varactor
crossover network	notch (band-stop) filter	

PRACTICE PROBLEMS

§15.1

1. An amplifier has values of $f_0 = 14$ kHz and BW $= 2$ kHz. Calculate the value of Q for the circuit. [4a]

2. An amplifier has values of $f_0 = 1200$ kHz and BW $= 300$ kHz. Calculate the value of Q for the circuit. [4a]

3. An amplifier with a center frequency of 800 kHz has a Q of 6.2. Calculate the bandwidth of the circuit. [4b]

4. An amplifier with a center frequency of 1100 kHz has a Q of 25. Calculate the bandwidth of the circuit. [4b]

5. Complete the following table. [4]

f_0 (kHz)	BW	Q
740		2.4
388	40 kHz	
1050		5.6
920	600 kHz	

6. An amplifier has cutoff frequencies of 1180 and 1300 kHz. Show that $f_0 = f_{ave}$ for the circuit. Also, determine the Q of the circuit. [4b]

§15.2

7. Calculate the bandwidth and closed-loop voltage gain for the filter in Figure 15.47. [10]

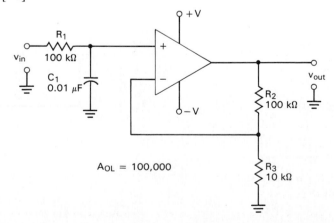

FIGURE 15.47

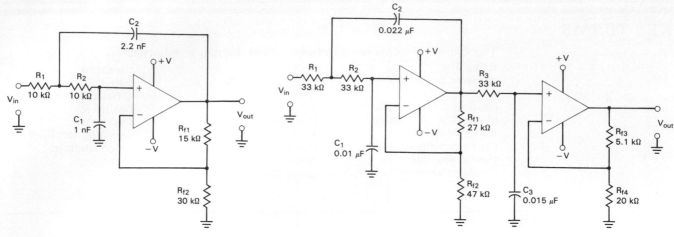

FIGURE 15.48

FIGURE 15.49

8. A filter like the one in Figure 15.47 has values of $R_1 = 82$ kΩ, $C_1 = 0.015\mu F$, $R_2 = 150$ kΩ, and $R_3 = 20$ kΩ. Calculate the bandwidth and closed-loop voltage gain of the circuit. [10]

✓ 9. Calculate the bandwidth and closed-loop voltage gain of the filter in Figure 15.48. [10]

10. A filter like the one in figure 15.14 has values of $R_1 = R_2 = 33$ kΩ, $C_1 = 100$ pF, $C_2 = 200$ pF, and $R_f = 220$ kΩ. Calculate the bandwidth and closed-loop voltage gain of the circuit. [10]

⌣ 11. Calculate the bandwidth and closed-loop voltage gain of the filter in Figure 15.49. [10]

12. Calculate the bandwidth and closed-loop voltage gain of the filter in Figure 15.50. [10]

✓ 13. Calculate the lower cutoff frequency and closed-loop voltage gain of the filter in Figure 15.51. [10]

FIGURE 15.50

FIGURE 15.51

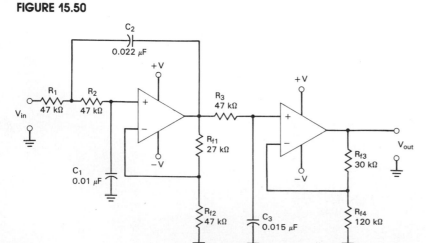

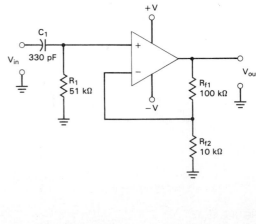

CHAP. 15 Tuned Amplifiers

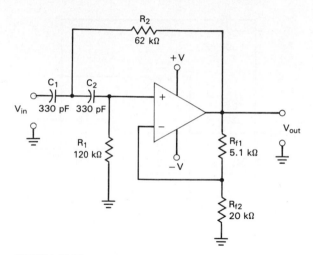

FIGURE 15.52

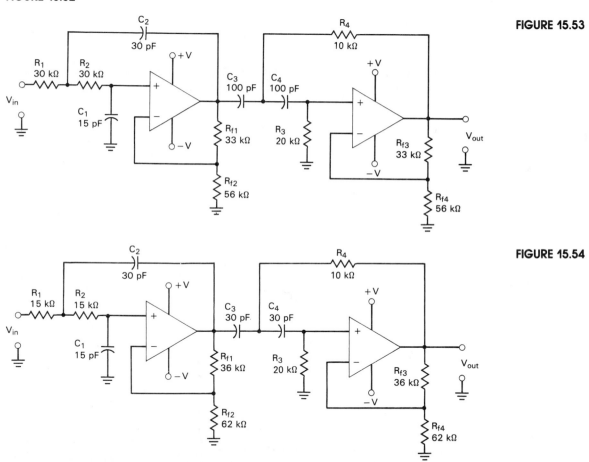

FIGURE 15.53

FIGURE 15.54

14. Calculate the lower cutoff frequency and closed-loop voltage gain of the filter in Figure 15.52. [10]

§15.3

15. Calculate the values of f_1, f_2, bandwidth, center frequency, and Q for the band-pass filter in Figure 15.53. [14]

16. Calculate the values of f_1, f_2, bandwidth, center frequency, and Q for the band-pass filter in Figure 15.54. [14]

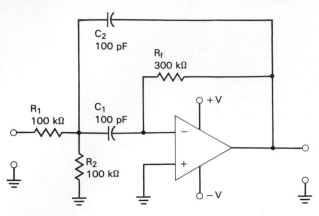

FIGURE 15.55

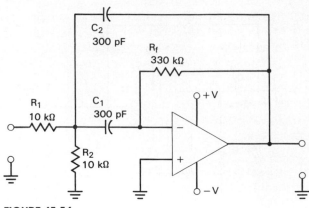

FIGURE 15.56

✓ 17. Calculate the values of f_1, f_2, bandwidth, center frequency, and Q for the band-pass filter in Figure 15.55. [17]

18. Calculate the values of f_1, f_2, bandwidth, center frequency, and Q for the band-pass filter in Figure 15.56. [17]

19. A multiple-feedback filter has values of $f_0 = 62$ kHz and BW $= 40$ kHz. Approximate the values of f_1 and f_2 for the circuit and determine the percent of error in those approximations. [17]

FIGURE 15.57

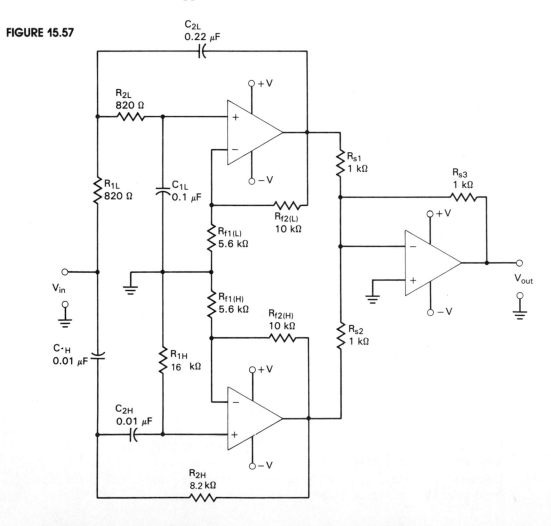

20. A multiple-feedback filter has values of $f_0 = 482$ kHz and BW $= 200$ kHz. Approximate the values of f_1 and f_2 for the circuit and determine the percent of error in those approximations. [17]

21. Calculate the value of A_{CL} for the filter in Figure 15.55. [18]

22. Calculate the value of A_{CL} for the filter in Figure 15.56. [18]

23. Calculate the values of f_1, f_2, bandwidth, center frequency, and Q for the notch filter in Figure 15.57. [19]

24. Calculate the values of f_1, f_2, bandwidth center frequency, and Q for the notch filter in Figure 15.58. [19]

25. Calculate the values of f_1, f_2, bandwidth, center frequency, and Q for the notch filter in Figure 15.59. [20]

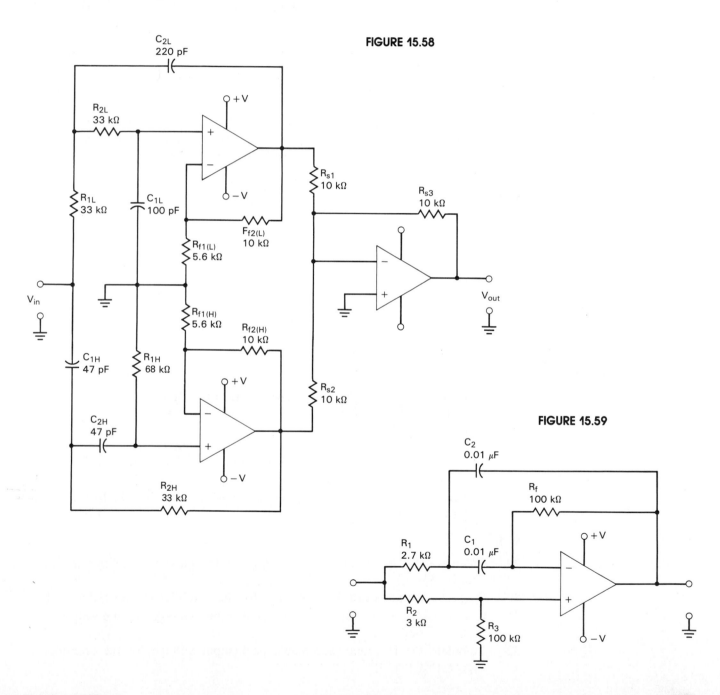

FIGURE 15.58

FIGURE 15.59

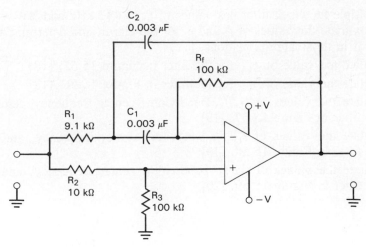

FIGURE 15.60

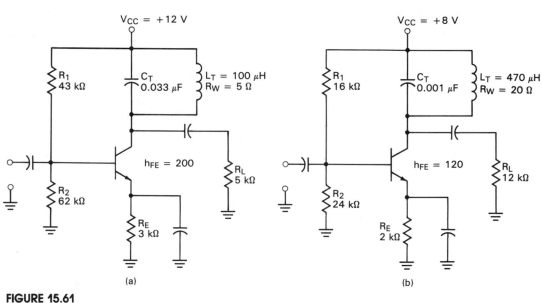

FIGURE 15.61

26. Calculate the values of f_1, f_2, bandwidth, center frequency, and Q for the notch filter in Figure 15.60. [20]

§15.5

√ 27. Calculate the values of Q, Q_L, center frequency, bandwidth, f_1, and f_2 for the circuit in Figure 15.61a. [28]

28. Calculate the values of Q, Q_L, center frequency, bandwidth, f_1, and f_2 for the circuit in Figure 15.61b. [28]

§15.7

√ 29. Calculate the values of V_{CEQ} and I_{CQ} for the class C amplifier in Figure 15.62. [33]

30. Calculate the values of V_{CEQ} and I_{CQ} for the amplifier in Figure 15.63. [33]

√ 31. Calculate the minimum acceptable peak input voltage for the amplifier in Figure 15.62. [33]

32. Calculate the minimum acceptable peak input voltage for the amplifier in Figure 15.63. [33]

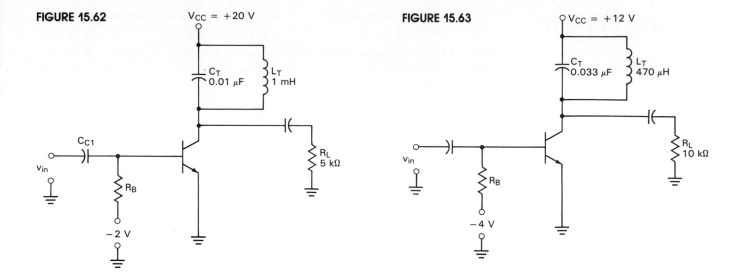

FIGURE 15.62

$V_{CC} = +20$ V

C_T 0.01 μF L_T 1 mH

C_{C1}

v_{in}

R_B

-2 V

R_L 5 kΩ

FIGURE 15.63

$V_{CC} = +12$ V

C_T 0.033 μF L_T 470 μH

v_{in}

R_B

-4 V

R_L 10 kΩ

TROUBLESHOOTING PRACTICE PROBLEMS

33. What fault is indicated by the readings in Figure 15.64? Explain your answer.
34. What fault is indicated by the readings in Figure 15.65? Explain your answer.

FIGURE 15.64

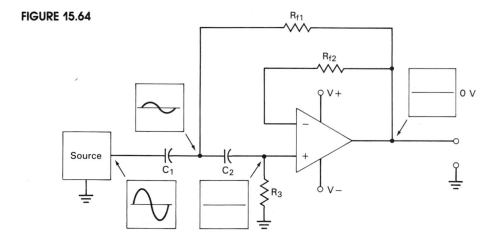

FIGURE 15.65

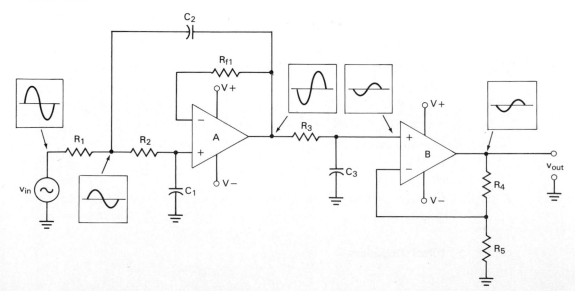

35. The inductor in Figure 15.66 is replaced by one with a winding resistance of 17 Ω. Calculate the shift that occurs in each of the cutoff frequencies as a result of this change.

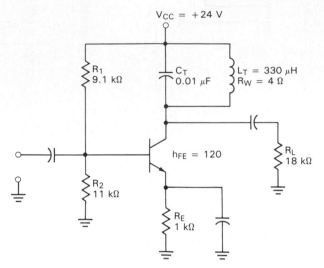

FIGURE 15.66

SUGGESTED COMPUTER APPLICATIONS PROBLEMS

36. Write a program that will determine the values of f_0, Q, and bandwidth for a multiple-feedback band-pass active filter.
37. Modify the program in problem 45 so that it will provide the correct values of f_1 and f_2 for any value of Q.
38. Write a program that will determine the values of f_0, Q, Q_L, and BW for a discrete tuned amplifier.

ANSWERS TO THE EXAMPLE PRACTICE PROBLEMS

15.1   60

15.2. 56 kHz

15.3. $f_0 = 1029$ kHz, $f_{ave} = 1030$ kHz, $Q = 10.29$

15.4. $f_2 = 102.6$ Hz, $A_{CL} = 2$

15.6. $f_1 = 570$ Hz, $f_2 = 938$ Hz, BW $= 368$ Hz, $f_0 = 731$ Hz, $Q = 1.99$

15.7. 216 Hz

15.8. $Q = 4.6$, BW $= 47$ Hz

15.9. % of error $(f_1) = 0.91\%$, % of error $(f_2) = 0.71\%$

15.11. 1.13

15.12. 2.77 kHz

15.13. 31.9

15.14. 11.3

15.15. 245 kHz

Additional Op-Amp Applications

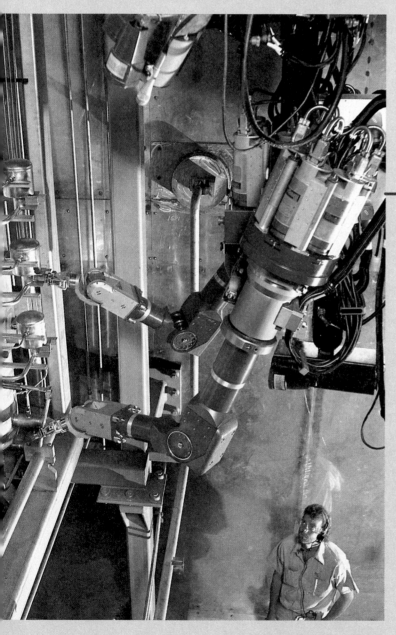

Industrial systems like the one shown require a wide variety of circuits to perform the operations for which they are designed. Op-amp circuits, like some of those covered in this chapter, are often used to aid in these operations.

OBJECTIVES

After studying the material in this chapter, you should be able to:

☐ 1. State the purpose served by a *comparator*. (§16.1)

☐ 2. Discuss the use of comparators in digital systems. (§16.1)

☐ 3. Explain the operation of the basic comparator. (§16.1)

☐ 4. Analyze the operation of a given comparator to determine its reference voltage (V_{ref}) and input/output phase relationship. (§16.1)

☐ 5. List the typical comparator fault symptoms and the possible causes of each. (§16.1)

☐ 6. State the purpose of the *integrator* and describe its operation. (§16.2)

☐ 7. Calculate the following for an integrator: its cutoff frequency and the frequency at which it starts to lose its linear characteristics. (§16.2)

☐ 8. List the common integrator faults and their symptoms. (§16.2)

☐ 9. State the purpose of a *differentiator* and describe its operation. (§16.2)

☐ 10. List the common differentiator faults and their symptoms. (§16.2)

☐ 11. Describe the *summing amplifier* and explain its operation. (§16.3)

☐ 12. Derive the general-class equation for a given summing amplifier and use the equation to calculate the output voltage for any set of input voltages. (§16.3)

☐ 13. Describe the use of the summing amplifier in a *digital-to-analog (D/A) converter*. (§16.3)

☐ 14. Describe the typical procedure for diagnosing and troubleshooting a faulty summing amplifier. (§16.3)

☐ 15. Describe the operation of the *averaging amplifier* and the *subtractor*. (§16.3)

☐ 16. Discuss the use of the op-amp in an audio amplifier. (§16.4)

☐ 17. Describe the operation of the op-amp high-impedance voltmeter. (§16.4)

☐ 18. Describe the operation of the voltage-controlled current source. (§16.4)

☐ 19. Describe the operation of the precision diode. (§16.4)

As you master each of the objectives listed, place a check mark (✔) in the appropriate box.

I n this chapter, we are going to take a look at some op-amp circuits that are widely used for a variety of applications. These circuits do not necessarily relate to each other, except for the fact that they are all constructed using one or more op-amps.

Most of the chapter will be spent dealing with four extremely common circuits: *comparators*, *integrators*, *differentiators*, and *summing amplifiers*. Each of these circuits will be given extensive coverage, including their operating principles, applications, and troubleshooting techniques. Then we will take a brief look at several other typical op-amp circuits. Remember as you go through the chapter that, other than the use of one or more op-amps, the circuits being covered are not necessarily related to each other. Each section should be approached as a separate entity.

16.1
Comparators

Comparator. A circuit used to compare two voltages.

The *comparator* is a relatively simple circuit that is used to compare two voltages and provide an output indicating the relationship between those two voltages. Generally, comparators are used to compare either:

1. Two changing voltages to each other, as in comparing two sine waves, or
2. A changing voltage to a set dc reference voltage

We will look at both of these circuit types extensively in this section. However, you may find it helpful if we take a moment to discuss the applications of comparators before going into any circuit detail.

Applications

Comparators are most commonly used in *digital* applications. Digital circuits are circuits that respond to *alternating dc voltage levels*. The signals that typically appear in a digital system will consist of alternating dc voltage levels, rather than sinusoidal waveforms. Typical digital signals are shown in Figure 16.1. You will spend a great deal of time studying digital systems, such as personal computers, in the future. For now, we need only establish that digital systems tend to respond only to alternating dc levels, or *square* waves.

◤ 2

Digial circuits. Circuits designed to respond to specific alternating dc voltage levels.

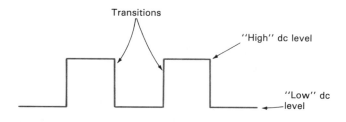

FIGURE 16.1
Digital waveform characteristics.

Now consider the problem illustrated in Figure 16.2. Here we see a digital system and a sine-wave source. Let's assume for a moment that the digital system is to perform some function whenever the sine wave reaches a peak value that is greater than 10 V. The nature of the function is not important at this point; only that the digital system must have a way of knowing whether or not V_{pk} of the sine wave is greater than 10 V. From the illustration, the problem is obvious. The variable source circuit is outputting a sine wave, and the digital system requires

FIGURE 16.2

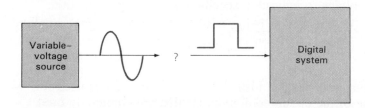

a square-wave input. The problem, then, is how to inform the digital system whether or not the output from the source is greater than 10 V in a manner to which the digital system can respond. A comparator would be used for this purpose. The solution to the problem is shown in Figure 16.3. In Figure 16.3a, a comparator is shown placed between the voltage source and the digital system. The inverting input of the comparator is connected to the reference voltage, +10 V. When the noninverting input is greater than this value, the output will go to the high-voltage level. It remains here until the noninverting input decreases below +10 V. At that time the output of the comparator goes back to the low-voltage level. This input/output action is illustrated in Figure 16.3b.

Since the digital system is designed to respond to high and low dc levels, we have successfully converted information about the sine wave into a form that the digital system can deal with. This is the primary function served by a comparator.

Several points can be made at this time:

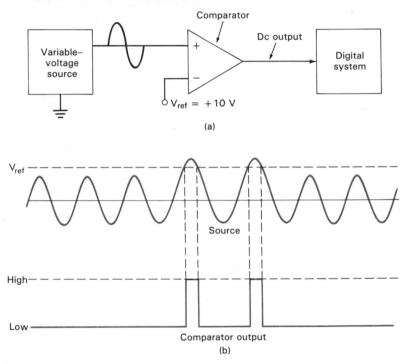

FIGURE 16.3

All illustration of basic comparator operation.

1. The output from a comparator is normally a dc voltage that indicates the polarity (or magnitude) relationship between the two input voltages.

2. A comparator does not normally convert a sine wave to a square wave. This function is normally performed by a circuit called a Schmitt trigger, which is discussed in Chapter 18. Instead, the comparator will merely change its dc output voltage to indicate a specific input voltage condition.

Level detector. Another name for a comparator being used to compare an input voltage to a fixed dc reference voltage.

3. When a comparator is being used to compare a signal amplitude to a fixed dc level (as was the case in Figure 16.3), the circuit is referred to as a *level detector*.

Now that you have an idea of what a comparator is used for, we will take a look at some of the circuit configurations that are commonly used.

CHAP. 16 Additional Op-amp Applications

Comparator Circuits

The most noticeable circuit recognition feature of the comparator is that there is *no feedback path* in the circuit. A typical comparator circuit is shown in Figure 16.4. Without a feedback path, the voltage gain of the circuit will be equal to the value of A_{OL} for the op-amp.

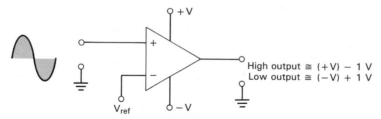

High output $\cong (+V) - 1$ V
Low output $\cong (-V) + 1$ V

FIGURE 16.4

When we discussed linear op-amp circuits in Chapter 12, you were shown that an open in the feedback path of an amplifier caused the output to clip at the positive and negative extremes. This clipping was caused by the high gain of the op-amp, A_{OL}, which is normally limited by the feedback path. In the case of the comparator, the clipping caused by the high gain of the op-amp is a desired feature. Since the gain of a comparator is equal to A_{OL}, virtually any difference voltage at the input will cause the output to go to one of the voltage extremes and stay there until the voltage difference is removed. The polarity of the input difference voltage will determine to which extreme the output of the comparator goes. This point is illustrated in the following example.

EXAMPLE 16.1

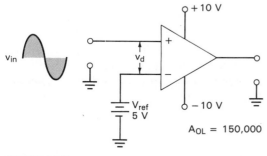

FIGURE 16.5

$A_{OL} = 150,000$

The inverting input of the op-amp in Figure 16.5 is shown to be connected to a +5-V reference. With the values of $+V$, $-V$, and A_{OL} shown, determine the output voltage from the circuit when the inverting input is at +4.9 and +5.1 V.

Solution: When the input voltage is at +5.1 V, the difference voltage is found as

$$v_d = v_{in} - V_{ref}$$
$$= 5.1 \text{ V} - 5 \text{ V}$$
$$= +0.1 \text{ V}$$

The output voltage is then found as

$$V_{out} = A_{OL}v_d$$
$$= (150,000)(0.1 \text{ V})$$
$$= 1500 \text{ V}$$

Since this is clearly beyond the output limits of the circuit, the output voltage is found as

$$V_{out} \cong V + - 1 \text{ V}$$
$$= +9 \text{ V}$$

When the input voltage is at +4.9 V, the difference voltage is found as

$$v_d = v_{in} - V_{ref}$$
$$= +4.9 \text{ V} - 5 \text{ V}$$
$$= -0.1 \text{ V}$$

Using this value, we obtain the following output voltage:

$$V_{out} = A_{OL}v_d$$
$$= (150,000)(-0.1 \text{ V})$$
$$= -1500 \text{ V}$$

Again, this output is clearly impossible for the circuit shown. Thus the output is found as

$$V_{out} \cong (V-) + 1 \text{ V}$$
$$= -9 \text{ V}$$

PRACTICE PROBLEM 16–1

A noninverting comparator has its inverting input connected to a $+2\text{-V}_{dc}$ reference. The op-amp is connected to $\pm 12\text{-V}$ supplies and has a value of $A_{OL} = 70,000$. Determine the output voltage for input voltages of +2.001 and +1.999 V_{dc}.

As Example 16.1 showed, the comparator output voltage will go to one of the extremes even when there is very little difference between the two inputs. In fact, if you divide the 9-V maximum output voltage by the value of A_{OL}, you will see that the circuit needs a difference voltage of only 60 μV to cause the output to clip at $+V$ or $-V$. Also, as you were shown, the polarity of the input will determine to which extreme the output will go.

Setting the Reference Level

The circuit shown in Figure 16.5 is not very practical, simply because most systems do not have a wide variety of power supply values available. In many cases, an electronic system will have one or two different supply voltages, and the reference voltage for the comparator would have to come from one of them. Most often, a voltage-divider circuit is used to set the reference voltage for a given-level detector. Such a circuit is shown in Figure 16.6. For the circuit shown, the reference voltage

FIGURE 16.6

A comparator with a reference-setting circuit.

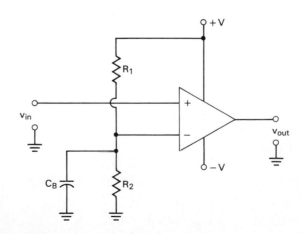

CHAP. 16   Additional Op-amp Applications

would be found using the standard voltage-divider formula, as follows:

$$V_{ref} = V \frac{R_2}{R_1 + R_2}$$ **(16.1)**

The following example illustrates the application of this formula in a practical analysis situation.

═══════════════════ **EXAMPLE 16.2** ═══════════════════

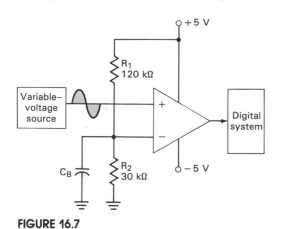

FIGURE 16.7

The digital system in Figure 16.7 is being used to perform some predetermined function when the variable-voltage source reaches a certain output level. What is this output level?

Solution: The reference voltage is being set by the voltage-divider circuit. Thus V_{ref} is found as

$$V_{ref} = V \frac{R_2}{R_1 + R_2}$$
$$= (+5 \text{ V})(0.2)$$
$$= +1 \text{ V}$$

Thus the digital system is "looking" for a change either above or below +1 V.

PRACTICE PROBLEM 16–2

A comparator like the one in Figure 16.7 has ±8-V supplies and resistor values of $R_1 = 3 \text{ k}\Omega$ and $R_2 = 1.8 \text{ k}\Omega$. Determine the input reference voltage for the circuit.

The circuits shown in both Figures 16.6 and 16.7 contain a *bypass capacitor* in the voltage-divider circuit. This bypass capacitor is included to prevent the variations in v_{in} from being coupled to the voltage-divider circuit through the op-amp. The result is that the output of the circuit will be much more reliable.

That is about all there is to this circuit. Next, we will take a look at some other comparator circuit configurations. As you will see, each comparator circuit configuration is used for a specific application. However, they all work according to the same basic principles.

Circuit Variations

Up to this point, we have assumed that the comparator is used to provide a *positive* output when the input voltage is more *positive* than some *positive* reference voltage. However, this is not always the case. In fact, you could substitute any combination of the words "positive" and "negative" into the following statement, and you would be correct:

■ A comparator is a circuit used to provide a _____ output when the input voltage is more _____ than some _____ reference voltage.

For example, take a look at the circuits shown in Figure 16.8. In Figure 16.8a, the circuit will have a *positive* output when the input voltage is more *negative* than some *positive* reference voltage. When v_{in} is more positive than this reference voltage, the output will be negative. The circuit in Figure 16.8b will have a *positive* output whenever the input voltage is more *positive* than a *negative* reference voltage. When v_{in} is more negative than this reference voltage, the output will be negative. The circuit in Figure 16.8c will have a *positive* output whenever the input voltage is *positive*, and a negative output otherwise. The circuit in Figure 16.8d will have a *positive* output whenever v_{in} is *negative*, and a negative output otherwise. If you apply your knowledge of the relationship between the inputs of an op-amp and its output, the statements above are easy to visualize.

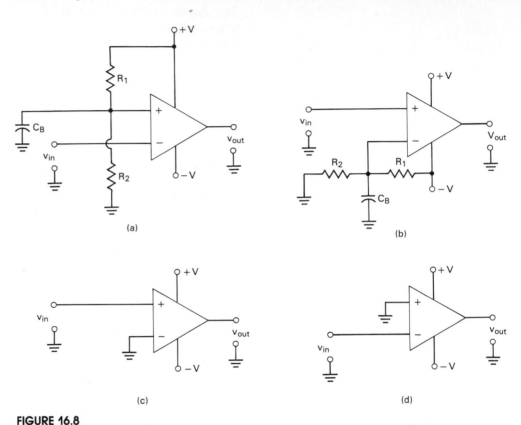

(a)

(b)

(c)

(d)

FIGURE 16.8

Comparator circuit variations.

Several of the problems at the end of the chapter involve analyzing the function performed by a specific comparator. You may want to go through these exercises before moving on.

Another circuit variation is the *variable* comparator. The variable comparator allows you to change the dc reference voltage. Such a circuit is shown in Figure 16.9. By adjusting the value of R_2 in the circuit, the reference voltage can be set to any value desired. Note that any of the comparators in Figure 16.8 can be made variable in the same way, with the exception of the ground reference comparators, of course.

Variable comparator. A comparator with an adjustable reference voltage.

Troubleshooting Comparators

One of the really nice things about comparators is that they're extremely easy to troubleshoot. For example, refer back to the circuit in Figure 16.9. There really

CHAP. 16   Additional Op-amp Applications

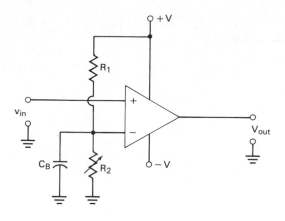

FIGURE 16.9

A comparator with a variable reference circuit.

isn't a whole lot here that can go wrong. In fact, there are only three common problems that could develop. These problems and their possible causes are as follows:

TABLE 16.1
Comparator Troubleshooting

Problem	Possible Causes
No output at all.	No input signal One or both supply voltages out A bad op-amp
The output changes at an input voltage that is too *high*.	R_2 is out of adjustment R_1 is shorted (not likely)
The output changes at an input voltage that is too *low*.	R_1 is open R_2 is out of adjustment C_B is shorted (not likely)

That is about it. Now, let's go through a troubleshooting problem or two so that you can see how to apply the table to some practical situations.

━━━━━━ **EXAMPLE 16.3** ━━━━━━

Determine the cause of the problem in the circuit shown in Figure 16.10.

Solution: The first step, of course, is to determine whether any problem exists at all. If you look at the voltage-divider circuit, you will notice that $R_1 = R_2$. Therefore, the reference voltage should be one-half of $+V$, or $+5$ V. Since the output is changing at the point when the input is at ground rather than $+5$ V, there *is* a problem.

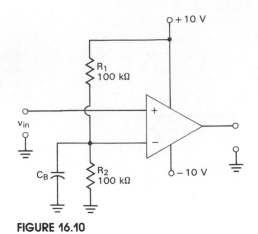

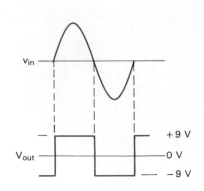

FIGURE 16.10

Table 16.1 indicates that this problem would be caused by R_1 open, R_2 shorted, or C_B shorted. Since the problem of R_1 open is by far the most likely, the resistance of this component is checked and found to be too high to measure. Replacing the component solves the problem.

Simple enough? Good. Now let's do one more, just for good measure.

EXAMPLE 16.4

Determine the cause of the problem in the circuit shown in Figure 16.11.

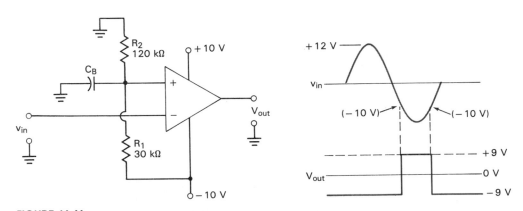

FIGURE 16.11

Solution: First, we need to determine what *should* be happening in the circuit. We know that the input signal is applied to the inverting input of the op-amp. Therefore, the output will go positive when v_{in} is more *negative* than the reference voltage. The reference voltage for this circuit is equal to the voltage across R_2 and is found as

$$V_{\text{ref}} = -V \frac{R_2}{R_1 + R_2}$$
$$= (-10 \text{ V})(0.8)$$
$$= -8 \text{ V}$$

Note that the value of $-V$ was used in place of $+V$ because the voltage-divider circuit is connected to the $-V$ supply.

The value of $V_{\text{ref}} = -8$ V tells us that there is a problem with the circuit, since it is triggering (changing output states) when v_{in} is equal to -10 V. This problem could be caused by R_1 shorted (not likely) or R_2 open. Checking the circuit shows that R_2 is open. Replacing the component solves the problem.

As you can see, the comparator is an easy circuit to analyze and troubleshoot. No matter which circuit configuration is used, you will have no problem dealing

FIGURE 16.12

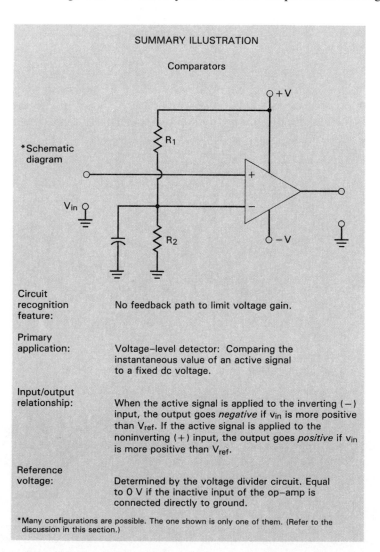

SUMMARY ILLUSTRATION

Comparators

*Schematic diagram

Circuit recognition feature:	No feedback path to limit voltage gain.
Primary application:	Voltage–level detector: Comparing the instantaneous value of an active signal to a fixed dc voltage.
Input/output relationship:	When the active signal is applied to the inverting ($-$) input, the output goes *negative* if v_{in} is more positive than V_{ref}. If the active signal is applied to the noninverting ($+$) input, the output goes *positive* if v_{in} is more positive than V_{ref}.
Reference voltage:	Determined by the voltage divider circuit. Equal to 0 V if the inactive input of the op–amp is connected directly to ground.

*Many configurations are possible. The one shown is only one of them. (Refer to the discussion in this section.)

with the circuit if you remember the basic operating principles of the comparator. As long as you know what the input/output relationship is supposed to be, you will have no problem determining the cause of any circuit fault. The characteristics of the comparator are summarized in Figure 16.12.

SECTION REVIEW

1. What is a *comparator*? [Objective 1]
2. Discuss the purpose served by comparators in digital systems. [Objective 2]
3. What is the circuit recognition feature of a comparator? [Objective 3]
4. What does the voltage gain of a comparator equal? [Objective 3]
5. How do you determine the reference voltage and the input/output phase relationship of a comparator? [Objective 4]
6. What is a *variable comparator*? [Objective 4]
7. List the common comparator faults and their causes. [Objective 5]

16.2
Integrators and Differentiators

In this section, we're going to take a look at the operation of op-amp integrators and differentiators. These two circuits are shown in Figure 16.13. As you can see, the two circuits are nearly identical in terms of their construction. Each contains

FIGURE 16.13

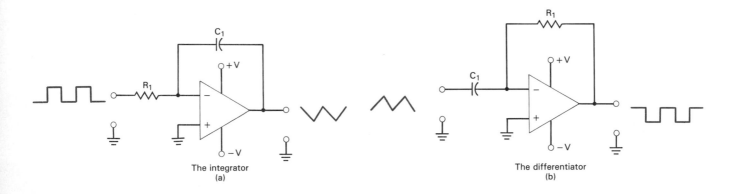

The integrator
(a)

The differentiator
(b)

a single op-amp and an *RC* circuit. However, the difference in resistor/capacitor placement in the two circuits causes them to have input/output relationships that are exact opposites. For example, the integrator will convert a square wave into a triangular wave, as is shown in Figure 16.13. At the same time, the differentiator will convert a triangular wave into a square wave.

Integrators

Integrator. A circuit whose output is proportional to the area of the input waveform.

Technically, the *integrator* is a circuit whose output is proportional to the *area* of its input waveform. The concept of waveform area is illustrated in Figure 16.14a.

CHAP. 16 Additional Op-amp Applications

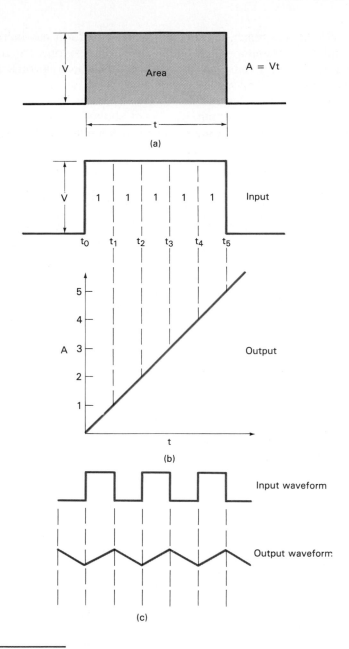

FIGURE 16.14

In geometry, we are shown that the area of a rectangle is equal to the product of its length and height. For the square wave in Figure 16.14a, the length would be the *pulse width* (measured in units of time) and the height would be its *amplitude* (measured in volts). Thus, for the waveform shown, area is found as

$$A = Vt$$

where A = the area of the waveform
V = the peak voltage of the waveform
t = the pulse width of the waveform

Figure 16.14b shows how the output of the integrator varies with the area of the input waveform. As you can see, the input square wave has been divided into five equal sections. For ease of discussion, each section has been assumed to have an area that is normalized to equal 1. From t_0 to t_1, we have one unit of area. In response to this input condition, the output waveform of the integrator goes to 1. When t_2 is reached, the area of the waveform has increased to 2 units of area, and the integrator output goes to 2, and so on. As you can see, the output of the integrator indicates the total number of area units at any point on the input waveform.

The integrator can be viewed as a square wave to triangular wave converter.

With continuing cycles of the input waveform, the integrator output will produce the corresponding output waveform shown in Figure 16.14c. Thus, the integrator can be viewed as a square wave to triangular wave converter.

Integrator Operation

The basic integrator is simply an *RC* circuit like the one shown in Figure 16.15. With the square-wave input shown, the *RC* integrator would *ideally* have a linear triangular output. This ideal output waveform is shown in the figure. The only problem is that the capacitor will not charge/discharge at a linear rate, but rather at an exponential rate. This produces the actual waveform shown in Figure 16.15. Note that the *actual* V_{out} waveform would be produced by the circuit when the time duration of the pulse is equal to five time constants.

In Chapter 3 you were shown that the capacitor in a given *RC* circuit requires five *time constants* to reach its full potential charge, where a time constant is the time period found as

$$\tau = RC$$

Thus the time required for the capacitor to reach full charge can be calculated as

$$T_C = 5RC$$

When the value of T_C is equal to the pulse width of the input, the *RC* integrator will produce the nonlinear output shown in Figure 16.15.

Miller integrator. Another name for the basic op-amp integrator.

Unfortunately, the nonlinear output of the *RC* integrator is not nearly close enough to the ideal integrator output for most applications. However, by adding an op-amp to the circuit, we can get a linear output from the integrator that is much closer to the ideal. The op-amp integrator, also known as the *Miller integrator*, was shown in Figure 16.13.

For you to understand the linear operation of this circuit, we have to consider the cause of the nonlinear output from a simple *RC* integrator. The nonlinear

FIGURE 16.15

The RC integrator.

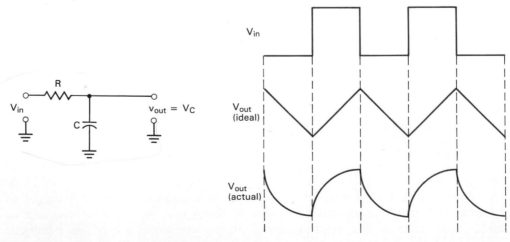

output from the circuit in Figure 16.15 is caused by the fact that *the available charging current decreases as the capacitor charges*. This is easy to see if we view the circuit as being a closed loop as shown in Figure 16.16. At the instant the switch is closed, the charge on the capacitor (V_C) is 0 V. The voltage across R is shown to be proportional to the difference between V_{in} and V_C. Thus, when V_C is zero, charging current is at its maximum value, as shown in the illustration. As the capacitor accepts a charge, V_C increases and the amount of available charging current decreases. Since the charging current is decreasing, the *rate* at which the capacitor charges is also decreasing. This gives us the *exponential* output waveform shown in Figure 16.15. (If you have trouble with this concept, review your basic electronics text material on *RC* time constants.)

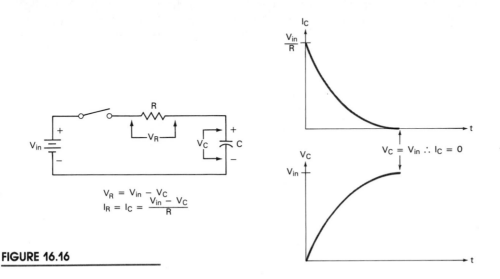

FIGURE 16.16

The key to eliminating the nonlinear output from the integrator is to provide a *constant-current* charge path for the capacitor. In other words, if we can make the capacitor charging current constant, the *rate* of charge for the component will become constant and the output will be linear. Now, keeping this in mind, let's take a look at the circuit in Figure 16.17. The *constant-current* characteristic of this circuit is based on two well-known points:

How the op-amp integrator produces a linear output.

1. The inverting input to the op-amp is held at *virtual ground* by the differential amplifier in the component's input circuit. This point was covered in Chapter 12.
2. The input impedance of the op-amp is so high that virtually all of I_1 will flow to the capacitor. This point was also covered in Chapter 12.

FIGURE 16.17

Op-amp integrator constant current characteristics.

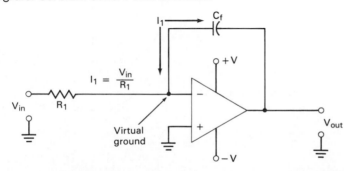

Since the inverting input is held at virtual ground, the value of input current (I_1) is found as

$$I_1 = \frac{V_{in}}{R_1}$$

Assuming that V_{in} is constant for a given period of time and R_1 is a fixed value, I_1 can be assumed to also be a constant value. Since virtually all of I_1 flows to the capacitor, C is being charged by a constant-current source. Thus, as long as V_{in} is constant, the capacitor will charge/discharge at a linear rate. This produces the linear *ramp* output shown in Figure 16.13.

Ramp. Another name for a voltage that changes at a constant rate.

Since the input to the integrator is applied to the inverting input, the output of the circuit will be 180° out of phase with the input. Thus, when the input goes positive, the output will be a negative ramp. When the input is negative, the output will be a positive ramp. This relationship is shown in Figure 16.18.

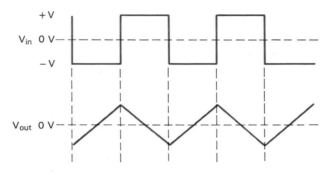

FIGURE 16.18

Op-amp integrator phase relationship.

The waveforms shown in Figure 16.18 lead us to a potential problem with the op-amp integrator. Note that the output is shown to be centered around 0 V. In practice, this may not be the case. If the op-amp has not been properly compensated for the input offset voltage, the output may not center itself around 0 V. Thus we must add a parallel resistor in the feedback path to ensure that any input offset voltage will be compensated for, and the output will be centered around 0 V. This circuit is shown in Figure 16.19. The addition of the feedback resistor has the benefit of centering the output around 0 V, but it also has a disadvantage.

The feedback resistor (Figure 16.19) eliminates any output offset voltage.

FIGURE 16.19

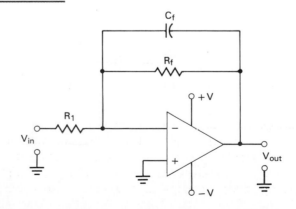

CHAP. 16   Additional Op-amp Applications

The *RC* feedback circuit will, like any *RC* circuit, have a cutoff frequency. In this case it is a *lower* cutoff frequency with which we are concerned. As frequency decreases, the reactance of the capacitor will increase. At some point, the reactance of the capacitor will be greater than the value of R_f, and the integrating action of the circuit will stop. As usual, the cutoff frequency is found using

$$f_1 = \frac{1}{2\pi R_f C_f} \tag{16.2}$$

The integrator will *start* to lose its linear output characteristics before the frequency found in equation (16.2) is reached. For an optimum output, the value of X_C should always be less than $0.1R_2$. Therefore, the circuit should not be operated below the frequency found by

$$f_{min} = \frac{10}{2\pi R_f C_f} \qquad \text{linear output} \tag{16.3}$$

EXAMPLE 16.5

7

0.01 μF

100 kΩ

+V

R_1

V_{in}

1 kΩ

$-V$

V_{out}

FIGURE 16.20

Determine the cutoff frequency for the circuit shown in Figure 16.20. Also, determine the frequency at which the output starts to lose its linear characteristics.

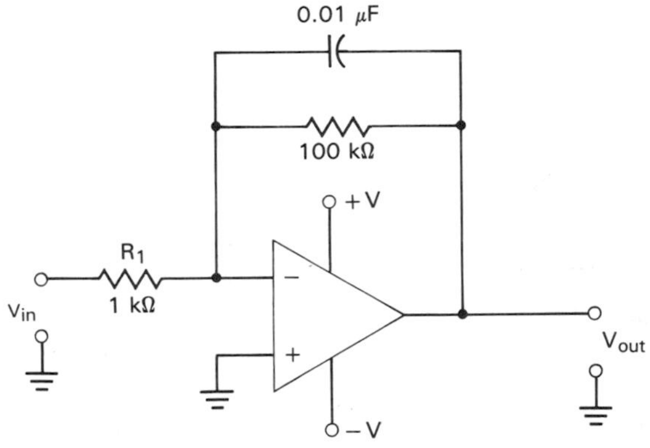

Solution: The cutoff frequency for the circuit is found as

$$f_1 = \frac{1}{2\pi R_f C_f} \qquad \text{nonlinear op'}$$
$$= \frac{1}{2\pi(100 \text{ k}\Omega)(0.01 \text{ }\mu\text{F})}$$
$$= 159 \text{ Hz}$$

If you take a close look at equations (16.2) and (16.3), you will see that the linearity will start to go when the operating frequency is 10 times the cutoff frequency. Therefore, we can save some time and trouble by using

$$f_{min} = 10f_c$$
$$= 1.59 \text{ kHz}$$

PRACTICE PROBLEM 16–5

An integrator has values of $C_f = 0.1$ μF and $R_f = 51$ kΩ. Determine the cutoff frequency for the circuit. Also determine the frequency at which it will start to lose its linear characteristics.

The name *Miller integrator* comes from the fact that the circuit contains a feedback capacitor. As you recall, a feedback capacitor may be broken into two capacitors, one at the input and one at the output. The Miller equivalent of the op-amp integrator is shown in Figure 16.21. Obviously, this circuit will have an upper cutoff frequency. This should make sense even without the Miller equivalent

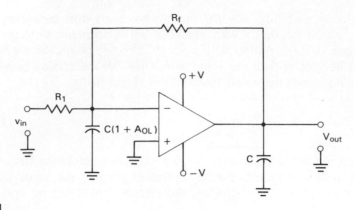

FIGURE 16.21
Miller ac equivalent circuit.

circuit. As operating frequency increases, the feedback capacitor will act more and more like a short circuit. Eventually, if the process were allowed to continue, the capacitor would short out the feedback resistor completely. In this case, the closed-loop gain of the circuit would drop to zero.

Integrator Troubleshooting

The integrator is another circuit that is relatively easy to troubleshoot. In this section we will take a look at the potential problems that may develop and the symptoms of each (Table 16.2). Any other problems that would develop with the amplifier would be diagnosed as discussed earlier. In other words, a problem with R_1, the op-amp, or the supply voltages would cause the same symptoms that we described earlier.

TABLE 16.2
Integrator Troubleshooting

Fault	Symptoms
R_f open	If R_f opens, the resistor is effectively removed from the circuit. Since this resistor is used to keep the output referenced around 0 V, the major symptom is the loss of this output reference. Also, the circuit cutoff frequency will drop to a much lower value.
R_f shorted	As you know, resistors do not usually short. However, if something were to short out R_f, integration would be lost, as would the gain of the circuit. In other words, the circuit would have little, if any, output signal.
C_f open	As you know, a capacitor is an open circuit by nature. However, if the component were to open in such a way as to prevent coupling, the circuit would start to act as a common inverting amplifier. The gain of the circuit would become $$A_{CL} = \frac{R_f}{R_i}$$ and the output would be a square wave.
C_f shorted	This would have the same effect as R_f shorting. In fact, if either of these components shorts, the result is that the other component will also be shorted. Therefore, you must test C_f and R_f individually to see which component is actually shorted.

The Differentiator

If we reverse the integrator capacitor and resistor (as shown in Figure 16.13b), we have a circuit called a *differentiator*. Technically, the differentiator is a circuit whose output is proportional to the rate of change of its input signal. The input/ output relationship of the differentiator is illustrated in Figure 16.22.

Differentiator. A circuit whose output is proportional to the rate of change of its input signal.

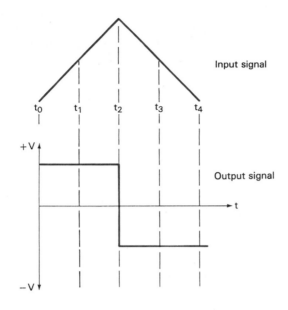

FIGURE 16.22

The input signal for the differentiator is shown to be a triangular waveform. Between times t_0 and t_2, the rate of change of the input is constant. Since the change is in a *positive* direction, the rate of change is a *positive constant*. Thus, the output of the circuit is constant at some positive voltage. Between times t_2 and t_4, the rate of change is in a negative direction. Thus, with the triangular input waveform, we get a square-wave output.

Equation (16.2) gave us a limit on the low-frequency operation of the integrator. The same equation can be used to provide us a limit on the *high-frequency* operation of the differentiator. Once again, the operating characteristics of the circuits are opposites.

Just as the integrator will start to lose its operating characteristics at 10 times the cutoff frequency, the differentiator will start to lose its operating characteristics at one-tenth the value of f_2. By formula,

$$f_{max} = \frac{1}{20\pi RC}$$ *Linear operation* **(16.4)**

If the input frequency to the differential exceeds the limit determined by equation (16.4), the output square wave will start to become distorted.

Differentiator Troubleshooting

Like the integrator, the differentiator is relatively easy to troubleshoot. The primary faults that can develop (other than a bad op-amp) are listed in Table 16.3, along with the symptoms of each.

TABLE 16.3
Differentiator Troubleshooting

Fault	Symptoms
R_1 open	The input will be removed from the op-amp, and the output of the circuit will remain at zero.
R_1 shorted	If something were to short out R_1, the gain of the circuit would increase drastically and the differentiating action of the circuit would be lost.
C_1 open	In this case, the gain of the circuit equals the value of A_{OL} for the op-amp. Also, differentiating action is lost.
C_1 shorted	The gain of the circuit will drop to zero, and there will be little (if any) output signal.

If the differentiator fails to operate and none of the symptoms in Table 16.3 appears, the op-amp is the source of the fault.

Summary

While they are also used for other applications, the integrator and differentiator are used primarily for waveform conversion. The characteristics of these two circuits are summarized for you in Figure 16.23.

FIGURE 16.23

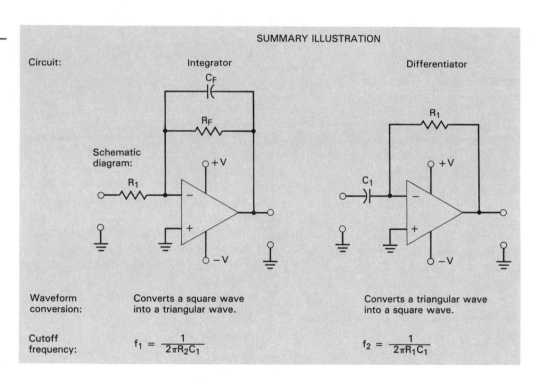

SUMMARY ILLUSTRATION

Circuit: Integrator Differentiator

Schematic diagram:

Waveform conversion: Converts a square wave into a triangular wave. | Converts a triangular wave into a square wave.

Cutoff frequency: $f_1 = \dfrac{1}{2\pi R_2 C_1}$ $f_2 = \dfrac{1}{2\pi R_1 C_1}$

SECTION REVIEW

1. What is an *integrator*? [Objective 6]
2. Describe the circuit operation of the Miller integrator. [Objective 6]
3. What is a *ramp*? [Objective 6]
4. List the common faults that occur in Miller integrators and the symptoms of each. [Objective 7]
5. What is a *differentiator*? [Objective 9]

CHAP. 16 Additional Op-amp Applications

6. Describe the circuit operation of the differentiator. [Objective 9]
7. List the common differentiator faults and the symptoms of each. [Objective 10]

16.3
Summing Amplifiers

The *summing amplifier* is an op-amp circuit that accepts several inputs and provides an output that is proportional to the *sum* of the inputs. The basic summing amplifier is shown in Figure 16.24. The key to understanding this circuit is to start by

Summing amplifier. Produces an output that is proportional to the sum of the input voltages.

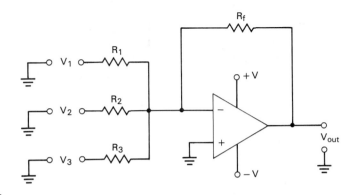

FIGURE 16.24
The summing amplifier.

considering each input as an individual circuit. If we had an input voltage only at V_1, the output would be found as

$$V_{out} = -V_1 \frac{R_f}{R_1}$$

Similarly, if V_2 were the only input, we would have

$$V_{out} = -V_2 \frac{R_f}{R_2}$$

And if V_3 were the only input, we would have

$$V_{out} = -V_3 \frac{R_f}{R_3}$$

If we have voltage inputs at all three of the terminals, the output will be the sum of the equations for each, as follows:

$$V_{out} = \frac{-V_1 R_f}{R_1} + \frac{-V_2 R_f}{R_2} + \frac{-V_3 R_f}{R_3}$$

Or, after factoring out the value of $-R_f$,

$$V_{out} = -R_f\left(\frac{V_1}{R_1} + \frac{V_2}{R_2} + \frac{V_3}{R_3}\right)$$ **(16.5)**

The following example illustrates the procedure for determining the output from a summing amplifier.

━━━ **EXAMPLE 16.6** ━━━

Determine the output voltage from the summing amplifier shown in Figure 16.25.

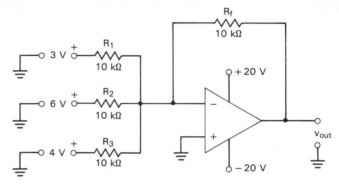

FIGURE 16.25

Solution: Using the values shown in the circuit and equation (16.5), the output voltage is found as

$$V_{out} = -R_f\left(\frac{V_1}{R_1} + \frac{V_2}{R_2} + \frac{V_3}{R_3}\right)$$
$$= -10\ k\Omega\left(\frac{+3\ V}{10\ k\Omega} + \frac{+6\ V}{10\ k\Omega} + \frac{+4\ V}{10\ k\Omega}\right)$$
$$= -13\ V$$

PRACTICE PROBLEM 16–6

A summing amplifier like the one in Figure 16.25 has the following values: $R_f = 2\ k\Omega$, $R_1 = R_2 = R_3 = 1\ k\Omega$, $V_1 = 1\ V$, $V_2 = 500\ mV$, and $V_3 = 1.5\ V$. The supply voltages for the circuit are $\pm 18\ V$. Determine the value of V_{out} for the circuit.

If you compare the result in Example 16.6 with the original input voltages, you will see that the amplifier output is equal to the sum of the input voltages.* Thus the term *summing amplifier* is appropriate for the circuit.

In practice, it is not always possible to have an output that is equal to the sum of the input voltages. For example, what if the inputs to Figure 16.25 had been $+10$ V, $+8$ V, and $+7$ V? Clearly, the sum of these voltages is $+25$ V. However, the maximum possible output voltage from the circuit is ± 19 V. Thus

*This is neglecting the negative polarity of the output, of course.

CHAP. 16   Additional Op-amp Applications

an output of 25 V would be impossible. To solve this problem, the value of R_f is usually set up so that the output voltage is *proportional* to the sum of the inputs. This point is illustrated in the following example.

EXAMPLE 16.7

Determine the output voltage from the circuit shown in Figure 16.26.

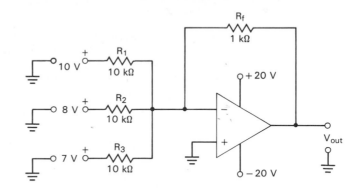

FIGURE 16.26

Solution: Using the values shown in the circuit and equation (16.5), the output voltage is found as

$$V_{out} = -R_f\left(\frac{V_1}{R_1} + \frac{V_2}{R_2} + \frac{V_3}{R_3}\right)$$
$$= -1\text{ k}\Omega\left(\frac{+10\text{ V}}{10\text{ k}\Omega} + \frac{+8\text{ V}}{10\text{ k}\Omega} + \frac{+7\text{ V}}{10\text{ k}\Omega}\right)$$
$$= -2.5\text{ V}$$

PRACTICE PROBLEM 16–7

A summing amplifier like the one in Figure 16.26 has the following values: $R_f = 1$ kΩ, $R_1 = R_2 = R_3 = 3$ kΩ, $V_1 = 2$ V, $V_2 = 1$ V, and $V_3 = 6$ V. The supply voltages for the circuit are ±12 V. Determine the value of V_{out} for the circuit.

In Example 16.7, we solved the problem of dealing with +10-, +8-, and +7-V inputs by *decreasing* the value of R_f. Now the output of the amplifier is not equal to the sum of the inputs, but rather is proportional to the sum of the inputs. In this case it is equal to one-tenth of the input sum. Picking random values of V_1, V_2, and V_3 and using them in equation (16.5) will show that the circuit always provides an output that is one-tenth of the sum of the inputs. This assumes, of course, that the value obtained by the equation does not exceed ±19 V.

At this time, there are several points that should be made regarding the summing amplifier:

1. There is no practical limit on the number of summing inputs. As long as the amplifier is capable of providing the correct output voltage, the circuit can

have any number of inputs. In this case, equation (16.5) would be expanded (or reduced) to include all the amplifier inputs.

2. The resistors used at the amplifier input do not necessarily have the same value. In other words, the inputs can be *weighted* so that certain inputs have more of an effect on the output.

The second point will be discussed in detail when we look at the most common application for the summing amplifier later in this section. First, we will look at how you can develop a general formula for any summing amplifier that will make the analysis of that circuit easier.

Circuit Analysis

☛ 12

Often, you need to be able to quickly predict the output from a summing amplifier under a variety of input combinations. This task is made easier if you derive a *general-class* output equation for the amplifier.

The general-class equation for a given summing amplifier is derived as follows:

1. Determine the $R_f : R$ ratio for each input branch.
2. Represent each branch as a product of its resistance ratio times its input voltage.
3. Add the products found in step 2.

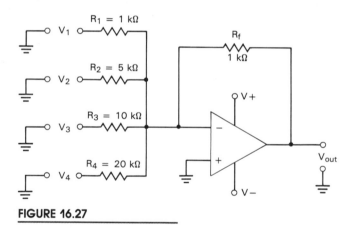

FIGURE 16.27

Let's apply this process to the circuit shown in Figure 16.27. For the first branch in the circuit, the resistance ratio is

$$\frac{R_f}{R_1} = 1$$

and the branch would be represented as this ratio times V_1, or simply V_1. The resistance ratio of the second branch would be

$$\frac{R_f}{R_2} = 0.2$$

and the branch would be represented as $0.2V_2$. Performing this procedure on the third and fourth branches would yield results of $0.1V_3$ and $0.05V_4$. The output of the circuit can now be found as

$$\boxed{V_{\text{out}} = V_1 + 0.2V_2 + 0.1V_3 + 0.05V_4}$$

CHAP. 16 Additional Op-amp Applications

Note that this formula would be the *general-class* equation for the circuit and could be used to determine the output voltage for any combination of input voltages. This is illustrated in the following example.

General-class equation. An equation derived for a summing amplifier that is used to predict the output voltage for any combination of input voltages.

EXAMPLE 16.8

Determine the output from the circuit shown in Figure 16.28 for each of the following input combinations:

V_1 (v)	V_2 (v)	V_3 (v)
+10	0	+10
0	+10	+10
+10	+10	+10

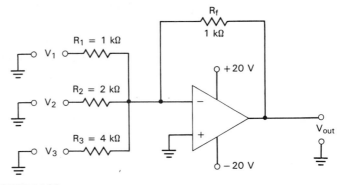

FIGURE 16.28

Solution: First, the circuit can be represented by its general-class equation. For this circuit the equation would be derived as

$$-V_{out} = \frac{R_f}{R_1} V_1 + \frac{R_f}{R_2} V_2 + \frac{R_f}{R_3} V_3$$
$$= V_1 + 0.5\, V_2 + 0.25 V_3$$

Now, using this equation, we can determine the output for the first set of inputs as

$$-V_{out} = 10\text{ V} + 0.5(0\text{ V}) + 0.25(10\text{ V})$$
$$= 12.5\text{ V}$$

For the second set of inputs,

$$-V_{out} = 0\text{ V} + 0.5(10\text{ V}) + 0.25(10\text{ V})$$
$$= 7.5\text{ V}$$

Finally, for the third set of inputs,

$$-V_{out} = 10\text{ V} + 0.5(10\text{ V}) + 0.25(10\text{ V})$$
$$= 17.5\text{ V}$$

PRACTICE PROBLEM 16–8

A summing amplifier like the one in Figure 16.28 has the following values: $R_f = 2$ kΩ, $R_1 = 200$ Ω, $R_2 = 400$ Ω, and $R_3 = 2$ kΩ. Derive the general-class equation for the circuit. Then use that equation to determine the value of V_{out} when $V_1 = V_2 = V_3 = 1$ V.

As you can see, deriving the general-class equation for a given summing amplifier makes the analysis of the circuit simpler, especially when you are trying to determine the output for several different input conditions.

Summing Amplifier Applications

Digital-to-analog (D/A) converter. A circuit that converts digital circuit outputs to equivalent voltages.

The most common use of the summing amplifier is as a *digital-to-analog converter*, or D/A converter. This application is illustrated in Figure 16.28.

First, a word or two about the *digital circuit*. This circuit has only two possible output values on any given line. For our circuit, we will assume these two voltages to be +5 and 0 V. Thus, each of the output lines, labeled 8, 4, 2, and 1, will always be at either +5 V or at 0 V. The exact value on a given line depends on the value that the circuit is presenting at the output. When a given line has +5 V on it, that value is part of the output. For example, if the digital circuit was putting out the equivalent of the decimal number 10, it would have the following output conditions:

$V_1(8)$	$V_2(4)$	$V_3(2)$	$V_4(1)$
+5 V	0 V	+5 V	0 V

With the +5 V outputs at (8) and (2), the output is representing the value $8 + 2 = 10$. If the circuit was outputting the value 6, the output voltages, from top to bottom, would be 0, +5, +5, and 0 V.

The D/A converter is used to convert the output value from the digital circuit into a single voltage level. This level is proportional to the numerical output of the circuit. To see how this works, we start by deriving the general-class formula for the circuit. This formula is

$$-V_{out} = V_1 + 0.5V_2 + 0.25V_3 + 0.125V_4$$

Using this equation, the output values in the following table were derived.

Decimal Value	$V_1(V)$	$V_2(V)$	$V_3(V)$	$V_4(V)$	$V_{out}(V)$
0	0	0	0	0	−0
1	0	0	0	+5	−0.625
2	0	0	+5	0	−1.25
3	0	0	+5	+5	−1.875
4	0	+5	0	0	−2.5
5	0	+5	0	+5	−3.125
.					
.					
.					
15	+5	+5	+5	+5	−9.375

It is not important for you to understand the exact workings of the digital circuit output. It is hoped that you have seen a pattern in the +5-V outputs from the circuit. However, if you have not, don't worry about it; you will have plenty of time to learn the output pattern when you formally study digital electronics. The main point is this: Digital circuits use two-voltage inputs and outputs to represent numerical values. Often, it is necessary to convert the output from a group of digital lines into a single voltage. This is accomplished using a process called D/A conversion. The conversion is performed by a summing amplifier whose inputs are weighted proportionally to the weights of the digital outputs. Thus, as the digital circuit output increases in value, the output voltage from the D/A converter increases.

Circuit Troubleshooting

The biggest problem with troubleshooting a summing amplifier is the fact that one of the input resistors can go bad and the circuit will still work for the other inputs. For example, look at the circuit shown in Figure 16.29. If the resistor on the output 4 line were to open, the rest of the input lines to the summing amplifier would still work, and the amplifier would still be providing an output. Granted, this output may be incorrect for a given set of digital input values, but this is not always easy to see with an oscilloscope.

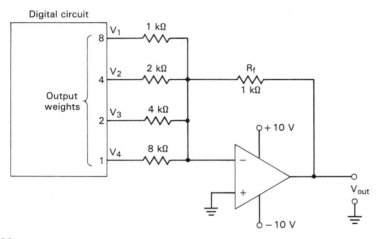

FIGURE 16.29

If you isolate a problem in a given circuit down to a summing amplifier, the best procedure is simply to measure the resistance of each component in the circuit. Since the only components contained in the circuit are resistors (other than the op-amp itself), a series of resistance checks is the fastest way to isolate a problem. If all the resistors prove to be good, replace the op-amp.

How do you tell when a given summing amplifier is faulty? To answer this question, let's take a look at the block diagram shown in Figure 16.30. Let's say that circuit B in the figure is not working properly. Tests indicate that it is not receiving the proper inputs. Testing circuit A shows that it *is* working properly.

FIGURE 16.30

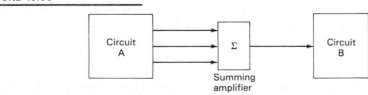

With these conditions, the problem must be the summing amplifier. In other words, in this case *the summing amplifier was shown to be faulty by proving that the problem wasn't anywhere else*. As depressing as this may sound, it is often the case since summing amplifiers tend to be used in relatively complex digital circuits.

Circuit Variations

Averaging amplifier. A summing amplifier that provides an output that is proportional to the *average* of the input voltage.

By using the proper input and feedback resistor values, a summing amplifier can be designed to provide an output that is equal to the *average* of any number of inputs. For example, consider the averaging amplifier shown in Figure 16.31.

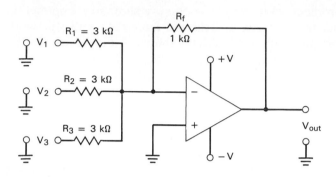

FIGURE 16.31

If we were to derive the general-class formula for the circuit in Figure 16.31, we would get

$$-V_{out} = \frac{V_1}{3} + \frac{V_2}{3} + \frac{V_3}{3}$$

or

$$-V_{out} = \frac{V_1 + V_2 + V_3}{3}$$

This, by definition, is the average of the three input voltages.

A summing amplifier will act as an averaging circuit whenever *both* of the following conditions are met:

1. All input resistors (R_1, R_2, and so on) are *equal in value*.
2. The ratio of any input resistor to the feedback resistor is equal to the number of circuit inputs.

For example, in Figure 16.31, all input resistors are equal in value (3 kΩ). If we take the ratio of any input resistor to the feedback resistor, we get 3 kΩ/1 kΩ = 3. This is equal to the number of inputs to the circuit.

Subtractor. A summing amplifier that provides an output that is proportional to the difference between the input voltages.

Another variation on the summing amplifier is shown in Figure 16.32. This circuit, called a *subtractor*, will provide an output that is equal to the difference between V_1 and V_2.

V_1 is applied to a standard inverting amplifier that has unity gain. Because of this, the output from the inverting amplifier will equal $-V_1$. This output is then applied to a summing amplifier (also having unity gain) along with V_2. Thus,

CHAP. 16   Additional Op-amp Applications

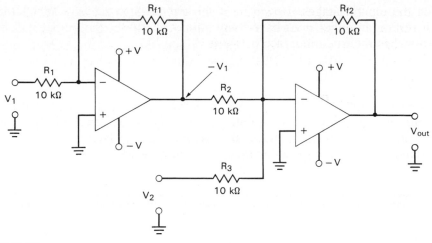

FIGURE 16.32

the output from the second op-amp is found as

$$-V_{out} = V_2 + (-V_1)$$
$$= V_2 - V_1$$

It should be noted that the gain of the second stage in the subtractor can be varied to provide an output that is proportional to (rather than equal to) the difference

FIGURE 16.33

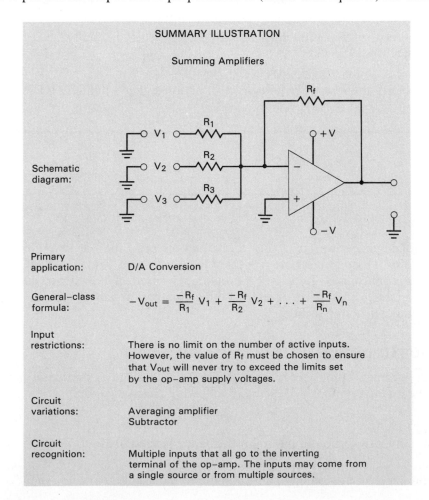

SUMMARY ILLUSTRATION

Summing Amplifiers

Schematic diagram:	(circuit diagram)
Primary application:	D/A Conversion
General–class formula:	$-V_{out} = \dfrac{-R_f}{R_1} V_1 + \dfrac{-R_f}{R_2} V_2 + \ldots + \dfrac{-R_f}{R_n} V_n$
Input restrictions:	There is no limit on the number of active inputs. However, the value of R_f must be chosen to ensure that V_{out} will never try to exceed the limits set by the op–amp supply voltages.
Circuit variations:	Averaging amplifier Subtractor
Circuit recognition:	Multiple inputs that all go to the inverting terminal of the op–amp. The inputs may come from a single source or from multiple sources.

between the input voltages. However, if the circuit is to act as a subtractor, the input inverting amplifier *must* have unity gain. Otherwise, the output will not be proportional to the true difference between V_1 and V_2.

Summary

The summing amplifier can be designed to provide an output voltage that is equal to the sum of any number of input voltages. Two variations on the basic summing amplifier are the averaging amplifier and the subtractor. The characteristics of the basic summing amplifier are summarized in Figure 16.33.

SECTION REVIEW

1. What is a *summing amplifier*? [Objective 11]
2. Why is the output from a summing amplifier often proportional (rather than equal to) the sum of its input voltages? [Objective 11]
3. Is there a limit on the number of inputs to a summing amplifier? Explain your answer. [Objective 11]
4. What is the *general-class equation* of a summing amplifier? [Objective 12]
5. List the steps used to derive the general-class equation for a summing amplifier. [Objective 12]
6. What is a *digital-to-analog (D/A) converter*? [Objective 13]
7. Describe the use of a summing amplifier as a D/A converter. [Objective 13]
8. Why are summing amplifiers difficult to troubleshoot? [Objective 14]
9. How is a summing amplifier usually determined to be faulty? [Objective 14]
10. What is an *averaging amplifier*? [Objective 15]
11. What are the resistor requirements of an averaging amplifier? [Objective 15]
12. What is a *subtractor*? [Objective 15]

16.4
Other Op-Amp Circuits

In this final section we are going to take a *very brief* look at several other op-amp circuits. These circuits are presented merely to give you an idea of the wide variety of applications for which the op-amp is used. We will not go into the details of circuit analysis or troubleshooting for the circuits in this section. It is strictly an applications section.

Audio Amplifier

In most communications receivers, the final output stage is the *audio amplifier*. The ideal audio amplifier will have the following characteristics:

Audio amplifier. The final audio stage in communications receivers; used to drive the speaker(s).

1. High gain
2. Minimum distortion in the audio-frequency range (20 Hz to 20 kHz)
3. High Z_{in}
4. Low Z_{out}, to provide optimum coupling to the speaker

An op-amp used in an audio amplifier will fulfill the requirements listed above very nicely. An op-amp audio amplifier is shown in Figure 16.34. The first thing that you will probably notice about this circuit is the fact that the $-V$ input to the op-amp is *grounded*. With the grounded $-V$ supply terminal, the output will be between the limits of $+V - 1$ V and $+1$ V (approximate), the $+1$ V being $-V + 1$ V $= 0$ V $+ 1$ V $= +1$ V. Note that the coupling capacitor between the op-amp and the speaker is necessary to reference the speaker signal around ground. Also, C_5 is included in the V_{CC} line to prevent any transient current caused by the operation of the op-amp from being coupled back to Q_1 through the power supply.

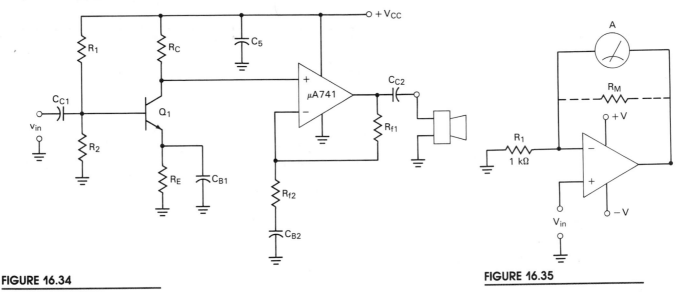

FIGURE 16.34

FIGURE 16.35

The high-gain requirement is accomplished by the combination of the two amplifier stages. The high Z_{in}/low Z_{out} of the audio amplifier is accomplished by the op-amp itself, as is the low-distortion characteristic.

High-impedance Voltmeter

✔ 17

The op-amp can be used in conjunction with a current meter to produce a relatively high Z_{in} voltmeter. Such a circuit is shown in Figure 16.35. In the circuit shown, the closed-loop gain depends on the internal resistance of the meter, R_M. The input voltage will be amplified, and the output voltage will cause a proportional current to flow through the meter. By adding a small series potentiometer in the feedback loop, the meter can be calibrated to provide a more accurate reading.

The high Z_{in} of the op-amp reduces the circuit loading that is caused by use of the meter. Although this type of circuit would cause *some* circuit loading, it would be much more accurate than a VOM with an input impedance of 20 kΩ/V.

Voltage-controlled Current Source

✔ 18

Earlier, you were shown how the constant-current capability of the feedback circuit can be used to produce a linear integrator. The voltage-controlled constant-current circuit works on the same principle. Such a circuit is shown in Figure 16.36.

To give you an idea of how the circuit works, we label the two sides of the zener diode A and B. Side B of the diode is common to the $+V$ side of R_2, so this point is also labeled as B. The lower side of the zener and the input to the op-amp are both labeled A, since these points are common to each other. The

Voltage-controlled current source. A circuit with a constant-current output that is controlled by the circuit input voltage.

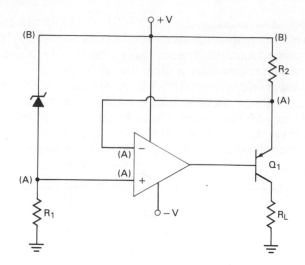

FIGURE 16.36

virtual-ground principle allows us to also label the inverting input to the op-amp as A. Continuing the process, the lower side of R_2 is also labeled as A.

Since the zener diode and R_2 have the same A–B labels, the voltages across these components must be equal. Since the zener voltage is constant, so is the voltage across R_2, and thus the current through the resistor. With I_{R2} being constant, so is the emitter current of the *pnp* transistor. Thus the load current is constant, regardless of the value of R_L (within reason). The maximum load resistance allowable for the circuit is found as

$$R_{L(max)} = \frac{+V - V_z}{I_E}$$

If this formula is exceeded, the transistor will saturate, and current regulation is lost.

The Precision Diode

The precision diode is a relatively simple circuit that can be used in any diode application. For example, consider the ~~negative~~ positive clipper shown in Figure 16.37a.

FIGURE 16.37

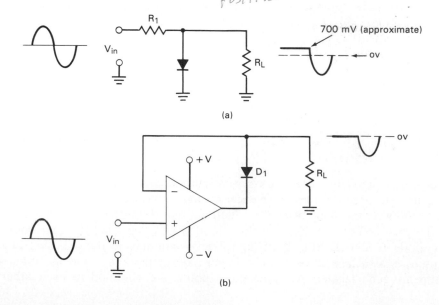

CHAP. 16 Additional Op-amp Applications

You may recall that this circuit is used to clip an input signal at approximately 0.7 V (700 mV). There's just one problem with this circuit. What if you want to clip a 500-mV$_{pk}$ signal at 0 V? The standard diode clipper cannot be used in this application because the 500 mV$_{pk}$ isn't sufficient to turn on the diode.

The precision diode (Figure 16.37b) can be used in this application. When V_{in} to the circuit is more positive than zero (by even a few millivolts), the op-amp output will go positive, cutting diode D_1 off. Depending on the sensitivity of the op-amp, clipping will occur by the time V_{in} reaches +5 mV or so. When the input goes negative, the output from the op-amp will also go negative, turning D_1 on. As always, the op-amp will adjust its output voltage until the voltage at the inverting input is approximately equal to the voltage at the noninverting input. Thus, the full negative peak value of V_{in} will be felt at the load. The 0.7-V drop across D_1 will be offset by the action of the op-amp.

Reversing the direction of D_1 will produce a *positive clipper*. The positive clipper works just like the negative clipper. The only difference is the portion of the input signal that is eliminated.

Precision diode. A circuit that consists of a diode and an op-amp. Characterized by the ability to conduct at extremely low forward voltages.

One Final Point

Obviously, there is an endless list of applications for the op-amp. We could not hope to cover them all in this chapter. In fact, we have barely scratched the surface.

In the following chapters you will see quite a few more op-amp circuits. In the meantime, if you wish to see more applications for op-amps, refer to the bibliography at the end of the book for a list of reference materials on op-amp circuits.

SECTION REVIEW

1. What is an audio amplifier? [Objective 16]
2. What characteristics of the op-amp make it ideal for audio-amplifier applications? [Objective 16]
3. What characteristics of the op-amp make it ideal for use as the input circuit for a voltmeter? [Objective 17]
4. Describe the operation of the voltage-controlled current source. [Objective 18]
5. Describe the operation of the precision diode. [Objective 19]

KEY TERMS

The following terms were introduced and defined in this chapter:

audio amplifier	general-class equation	ramp
averaging amplifier	integrator	subtractor
comparator	level detector	summing amplifier
differentiator	Miller integrator	variable comparator
digital circuits	precision diode	voltage-controlled current source
digital-to-analog (D/A) converter		

PRACTICE PROBLEMS

§16.1

1. Determine the value of V_{ref} for the comparator in Figure 16.38. [4]
2. Determine the value of V_{ref} for the comparator in Figure 16.39. [4]

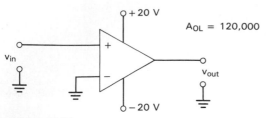

FIGURE 16.38

$A_{OL} = 120,000$

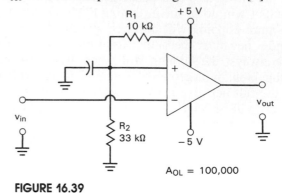

FIGURE 16.39

$A_{OL} = 100,000$

3. Determine the value of V_{ref} for the comparator in Figure 16.40. [4]
4. Determine the value of V_{ref} for the comparator in Figure 16.41. [4]

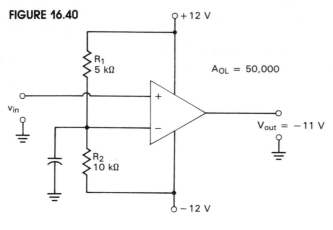

FIGURE 16.40

$A_{OL} = 50,000$

$V_{out} = -11$ V

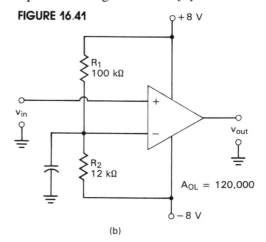

FIGURE 16.41

$A_{OL} = 120,000$

(b)

§16.2

5. Determine the cutoff frequency for the circuit in Figure 16.42. [7]
6. Determine the frequency at which the circuit in Figure 16.42 will start to lose its linear output characteristics. [7]

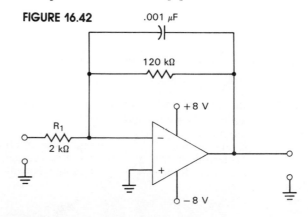

FIGURE 16.42

.001 µF

120 kΩ

R_1
2 kΩ

+8 V

−8 V

7. Determine the cutoff frequency of the circuit in Figure 16.43. [7]
8. Determine the frequency at which the circuit in Figure 16.43 will start to lose its linear output characteristics. [7]
9. Determine the cutoff frequency of the circuit in Figure 16.44. [7]
10. Determine the frequency at which the circuit in Figure 16.44 will start to lose its linear output characteristics. [7]
11. A differentiator has values of $R_1 = 10 \text{ k}\Omega$ and $C_1 = 0.01 \text{ μF}$. Determine its maximum linear operating frequency and its cutoff frequency. [9]
12. A differentiator has values of $R_1 = 2.2 \text{ k}\Omega$ and $C_1 = 0.15 \text{ μF}$. Determine its maximum linear operating frequency and its cutoff frequency. [9]

FIGURE 16.43

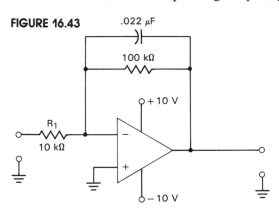

FIGURE 16.44

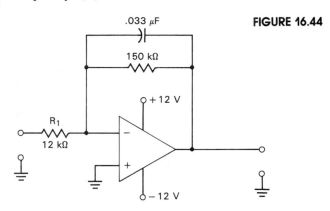

§16.3

13. Derive the general-class equation for the summing amplifier in Figure 16.45a. Then use the equation to determine the output voltage for the circuit. [12]
14. Derive the general-class equation for the summing amplifier in Figure 16.45b. Then use the equation to determine the output voltage for the circuit. [12]

FIGURE 16.45

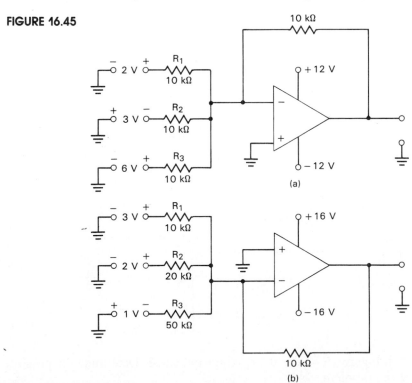

FIGURE 16.46

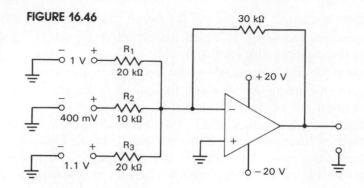

15. Derive the general-class equation for the summing amplifier in Figure 16.46. Then use the equation to determine the output voltage for the circuit. [12]

16. The feedback resistor in Figure 16.46 is changed to 24 kΩ. Derive the new general-class equation for the circuit and determine its new output voltage. [12]

17. A five-input summing amplifier has values of $R_1 = R_2 = R_3 = R_4 = R_5 = 15$ kΩ. What value of feedback resistor is required to produce an averaging amplifier? [15]

18. A four-input summing amplifier has values of $R_1 = R_2 = R_3 = R_4 = 20$ kΩ. What value of feedback resistor is required to produce an averaging amplifier? [15]

19. Refer back to Figure 16.32. Assume that $V_1 = 4$ V and $V_2 = 6$ V. Determine the output voltage from the circuit. [15]

20. Refer back to Figure 16.32. Assume that R_{f2} in the circuit has been changed to 30 kΩ. Also, assume that the input voltages are $V_1 = 1$ V and $V_2 = 3$ V. Determine the output voltage from the circuit. [15]

TROUBLESHOOTING PRACTICE PROBLEMS

21. The circuit in Figure 16.47 has the readings indicated. Determine the possible cause(s) of the problem.

FIGURE 16.47

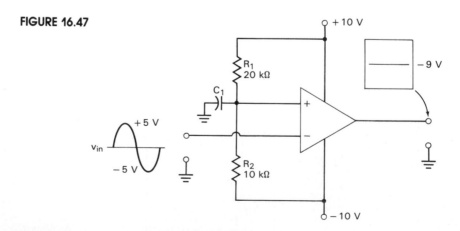

22. The circuit in Figure 16.48 has the readings indicated. Determine the possible cause(s) of the problem.

FIGURE 16.48

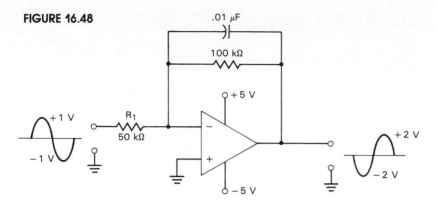

23. The circuit in Figure 16.49 has the readings indicated. Determine the possible cause(s) of the problem.

FIGURE 16.49

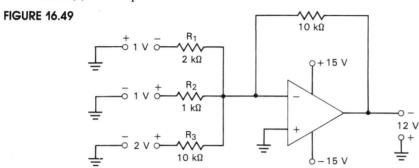

THE BRAIN DRAIN

Calculate the voltage gain of the audio amplifier in Figure 16.50.

25. The circuit in Figure 16.51 has the component values shown. Determine the value of V_{CE} for $R_L = 1\ k\Omega$ and $R_L = 3\ k\Omega$.

26. Design a circuit to solve the following equation:

$$-V_{out} = \frac{V_1 - V_2}{2} + \frac{V_3}{3}$$

The circuit is to contain only two op-amps.

FIGURE 16.50

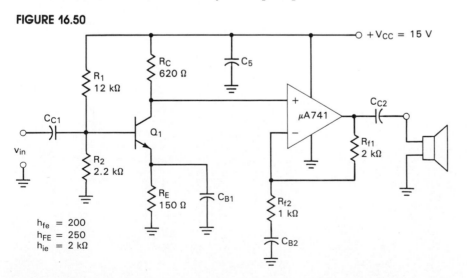

FIGURE 16.51

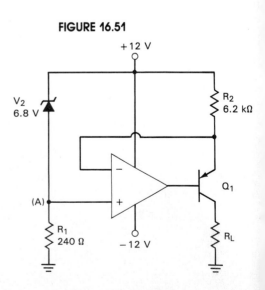

The Brain Drain

775

ANSWERS TO THE EXAMPLE PRACTICE PROBLEMS

16–1. $V_{out} = \pm 11$ V

16–2. $+3$ V

16–5. $f_1 = 31.21$ Hz, $f_{min} = 312.1$ Hz

16–6. -6 V

16–7. -3 V

16–8. $-V_{out} = 10V_1 + 5V_2 + V_3$, $V_{out} = -16$ V

17

Oscillators

An important quality of all electronic circuits, including oscillators, is reliability, especially when those circuits are intended for use in remote locations.

OBJECTIVES

After studying the material in this chapter, you should be able to:

☐ 1. State the function of the *oscillator*. (Introduction)

☐ 2. Define *positive feedback* and explain (in terms of phase shifts) how it is generally produced. (§17.1)

☐ 3. Describe how positive feedback maintains oscillations after an oscillator is triggered. (§17.1)

☐ 4. State the *Barkhousen Criterion* and explain what happens if it is not fulfilled. (§17.1)

☐ 5. List the three requirements for proper operation of an oscillator. (§17.1)

☐ 6. Describe the construction and operation of the *phase-shift oscillator*. (§17.2)

☐ 7. Explain why the phase-shift oscillator is rarely used. (§17.2)

☐ 8. Explain how a multistage discrete amplifier can go into oscillations and how this problem can be prevented. (§17.2)

☐ 9. Explain why $\alpha_v A_v$ must be slightly greater than 1 in a practical oscillator. (§17.2)

☐ 10. Explain how a trigger signal is produced in a practical oscillator. (§17.2)

☐ 11. Describe the construction of the *Wien-bridge oscillator*. (§17.3)

☐ 12. Explain the function and operation of the positive feedback network in the Wien-bridge oscillator. (§17.3)

☐ 13. Explain the function and operation of the negative feedback network in the Wien-bridge oscillator. (§17.3)

☐ 14. Define *propagation delay* and explain its effect on the operating frequency limits of Wien-bridge oscillators. (§17.3)

☐ 15. Describe the construction and operation of the *Colpitts oscillator*. (§17.4)

☐ 16. Calculate the feedback factor (α_v) for a Colpitts oscillator. (§17.4)

☐ 17. Calculate the value of A_{vL} for a Colpitts oscillator. (§17.4)

☐ 18. Explain why transformer coupling is preferred for connecting a load to a Colpitts oscillator. (§17.4)

☐ 19. Describe the construction and operation of the *Hartley* oscillator. (§17.5)

☐ 20. Describe the construction and operation of the *Clapp* oscillator. (§17.5)

☐ 21. Describe the construction and operation of the *Armstrong* oscillator. (§17.5)

☐ 22. List the circuit recognition for the various *LC* oscillators. (§17.5)

☐ 23. Define *crystal-controlled oscillator* and state the advantage that this type of oscillator has over other types of LC oscillators. (§17.6)

☐ 24. Discuss the electrical characteristics of quartz crystals. (§17.6)

☐ 25. Explain the stabilizing effects of a quartz crystal in an *LC* oscillator.

☐ 26. Discuss the procedure for oscillator troubleshooting. (§17.7)

As you master each of the objectives listed, place a check mark (✔) in the approximate box.

✔ 1

An *oscillator* is a circuit that produces an output waveform without any external signal source. The only input to an oscillator is the dc power supply. As such, the oscillator can be viewed as being a *signal generator*.

There are several types of oscillators, each classified according to the type of output waveform it produces. In this chapter, we will cover the operation of *sine-wave oscillators*, those that have sinusoidal outputs. Then, in Chapter 18, we will discuss (along with a variety of circuits) the operation of *square-wave oscillators*.

Oscillator. An ac signal generator.

17.1
Introduction

In Chapter 14 you were introduced to *positive feedback*. Positive feedback is the key to the operation of oscillators. As you recall, a positive feedback amplifier produces a feedback voltage, v_f, that is *in phase* with the original input signal. This is illustrated in Figure 17.1. In this circuit we show an input signal, v_{in}. This signal is applied to the amplifier, which introduces a 180° phase shift. The output signal is applied to the input of the feedback network, which introduces another 180° phase shift. The result is that the signal has been shifted 360° as it has traveled around the loop. As you know, shifting a signal 360° is the same as not shifting it at all. Therefore, the feedback signal is in phase with the original input signal.

Positive feedback. When a feedback signal is in phase with the original amplifier input signal.

Oscillators: The Basic Idea

Even though the circuit in Figure 17.1 is useful in explaining positive feedback, it has an input signal. This is inconsistent with our definition of an oscillator.

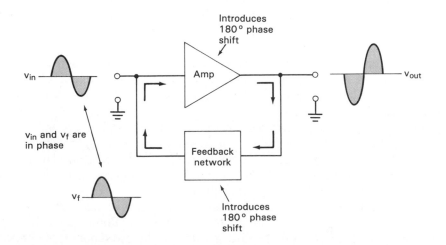

FIGURE 17.1

Positive feedback.

However, by modifying Figure 17.1, we can develop a circuit that is very useful for showing you the basic operating principle of the oscillator. This modified circuit is shown in Figure 17.2.

FIGURE 17.2

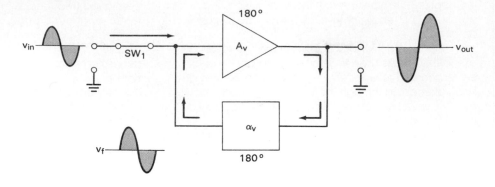

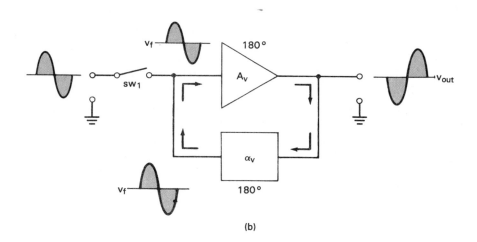

(b)

In Figure 17.2 we have added a switch in series with the amplifier input. When the switch is closed, the circuit waveforms are as shown in the figure. Now, assume that the switch is opened while the circuit is in operation. If this happens, v_{in} is removed from the circuit. However, v_f (which is in phase with the original input) is still applied to the amplifier input. The amplifier will respond to this signal in the same way that it did to v_{in}. In other words, v_f will be amplified and sent to the output. Since the feedback network sends a portion of the output back to the input, the amplifier receives another input cycle, and another output cycle is produced. As you can see, this process will continue as long as the amplifier is turned on, and the amplifier will produce a sinusoidal output with no external signal source.

Regenerative feedback is another name for positive feedback.

In any oscillator, the feedback network is used to *generate* an input to the amplifier, which, in turn, is used to *generate* an input to the feedback network. Since positive feedback produces this circuit action, it is often referred to as *regenerative feedback*. Regenerative feedback is the basis of operation for all oscillators. This point will be seen throughout this chapter.

Oscillators need a *trigger* signal.

It should be noted that an oscillator needs only a quick *trigger* signal to start the oscillating circuit action. In other words, anything that causes a slight signal variation at any point in the circuit will start the oscillator. It is not necessary for us to provide a complete input cycle from an external source. In fact, most oscillators will provide their own trigger signals. The sources of these trigger signals will be made clear in the next section. For now, just remember the basic requirements for an oscillator to work:

Basic oscillator circuit requirements.

1. The circuit must have a positive (regenerative) feedback network. This means that there must be a 180° phase shift produced by the feedback network.
2. The circuit must receive some trigger, either internally or externally generated, to start the oscillating action.

There is one other requirement that must be fulfilled for an oscillator to work. We will look at this requirement now.

The Barkhausen Criterion

In Chapter 14 you were shown that the active component in a feedback amplifier will have a voltage gain, while the feedback network itself will have a voltage loss. In other words, the A_v of the amplifier will be greater than 1, while the α_v of the feedback network will be less than 1.

For an oscillator to operate properly, the following relationship must be fulfilled:

$$\alpha_v A_v = 1 \qquad (17.1)$$

Barkhausen criterion. The relationship between the circuit feedback factor (α_v) and voltage gain (A_v) for proper oscillator operation.

This relationship is called the *Barkhausen criterion*. The results of *not* following this criterion are as follows:

1. If $\alpha_v A_v < 1$, the oscillations will fade out within a few cycles.
2. If $\alpha_v A_v > 1$, the oscillator will drive itself into saturation and cutoff clipping.

What happens if the Barkhausen criterion is *not* met.

These points are easy to understand if we apply them to a couple of circuits using several different gain–value combinations. As a reference, we will use the circuits shown in Figure 17.3. In Figure 17.3a, the output voltage is shown to be the product of the amplifier gain (A_v) and the input voltage (v_f). By formula,

$$v_{\text{out}} = A_v v_f$$

The value of v_f depends on the values of α_v and v_{out}, as follows:

$$v_f = \alpha_v v_{\text{out}}$$

Now we will use these equations to see what is happening in Figure 17.3a as the circuit progresses through several cycles of operation. We are going to assume that the initial input to the amplifier is a 0.1–V_{pk} signal. Starting with this value, and using $A_v = 100$ and $\alpha_v = 0.005$ (as shown), the circuit progresses as follows:

Cycle	v_{in}	v_{out}	v_f
1	0.1 V_{pk}	10 V_{pk}	0.05 V_{pk}
2	0.05 V_{pk}	5 V_{pk}	0.025 V_{pk}
3	0.025 V_{pk}	2.5 V_{pk}	0.0125 V_{pk}
4	0.0125 V_{pk}	1.25 V_{pk}	0.00625 V_{pk}

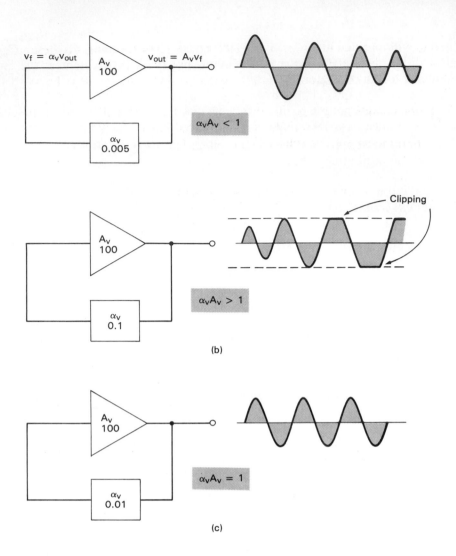

FIGURE 17.3

The effects of $\alpha_v A_v$ on oscillator operation.

Note how the value of v_f produced by each cycle is used as v_{in} for the next. This follows the basic operating principle of the oscillator. Now take a look at the progression of v_{out} values. As you can see, v_{out} decreases from each cycle to the next. If this progression were to continue, v_{out} would eventually reach 0 V for all practical purposes. This output deterioration is illustrated in Figure 17.3a. Note that this circuit has an $\alpha_v A_v$ product that is less than 1. As you can see, when $\alpha_v A_v < 1$, the oscillations will lose amplitude on each progressive cycle and will eventually fade out. This loss of signal is called *damping*.

Damping. The fading and loss of oscillations that occurs when $\alpha_v A_v < 1$, as shown in Figure 17.3a.

When $\alpha_v A_v$ is greater than 1, the opposite situation will occur. This situation is shown in Figure 17.3b. Using the same initial 0.1-V_{pk} input as before and the circuit values shown, the circuit in Figure 17.3b would progress as follows:

Cycle	v_{in}	v_{out}	v_f
1	0.1 V_{pk}	10 V_{pk}	1 V_{pk}
2	1 V_{pk}	100 V_{pk}	10 V_{pk}
3	10 V_{pk}	1000 V_{pk}	100 V_{pk}

For this circuit, it only took one or two cycles to start hitting some ridiculous values for v_{out}. The point is that v_{out} is increasing from each cycle to the next. Eventually, the output will start to experience saturation and cutoff clipping. Thus $\alpha_v A_v$ cannot be greater than 1.

The only condition that will provide a constant sinusoidal output from an oscillator is to have $\alpha_v A_v = 1$. This condition is shown in Figure 17.3c. Now, using the same 0.1-V_{pk} initial input, let's construct the cycle chart for Figure 17.3c.

Cycle	v_{in}	v_{out}	v_f
1	0.1 V_{pk}	10 V_{pk}	0.1 V_{pk}
2	0.1 V_{pk}	10 V_{pk}	0.1 V_{pk}
3	0.1 V_{pk}	10 V_{pk}	0.1 V_{pk}

It is obvious that the cycle is going to continue over and over. Thus, having a product of $\alpha_v A_v = 1$ will cause the oscillator to have a consistent sinusoidal output, which is the only acceptable situation.

Now we have established three requirements for an oscillator:

Oscillator circuit requirements.

1. Regenerative feedback
2. An initial input trigger to start the oscillations
3. $\alpha_v A_v = 1$ (the Barkhausen criterion)

As long as all these conditions are fulfilled, we have an oscillator. Now we will start to look at some circuits to see the various ways in which these requirements can be fulfilled.

SECTION REVIEW

1. What is an *oscillator*? [Objective 1]
2. What is *positive feedback*? [Objective 2]
3. In terms of phase shifts, how is positive feedback usually produced? [Objective 2]
4. How does positive feedback maintain the oscillations that are started by a trigger? [Objective 3]
5. What is the *Barkhausen criterion*? [Objective 4]
6. What happens when $\alpha_v A_v > 1$? [Objective 4]
7. What happens when $\alpha_v A_v < 1$? [Objective 4]
8. What is *damping*? [Objective 4]
9. List the three requirements for proper oscillator operation. [Objective 5]

17.2
Phase-shift Oscillators

Probably the easiest oscillator to understand is the *phase-shift oscillator*. This circuit contains *three RC* circuits in its feedback network, as shown in Figure 17.4.

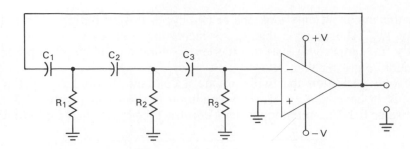

FIGURE 17.4

A basic phase shift oscillator.

Phase-shift oscillator. An oscillator that uses three *RC* circuits in its feedback network to produce a 180° phase shift.

But each network does not introduce 60° phase shift

A Circuit Variation. Figure 17.4 shows a phase-shift oscillator made up of series resistors and shunt capacitors. You can also construct a phase-shift oscillator using shunt resistors and series capacitors, that is, by reversing the capacitor and resistor locations.

In your study of basic electronics, you were introduced to the concept of the *phase shift* that is produced by an *RC* circuit at a given frequency. As you recall, the phase shift of a given *RC* circuit is found as

$$\theta = \tan^{-1} \frac{-X_C}{R} \qquad (17.2)$$

where θ = the phase angle of the circuit
$\tan^{-1}$ = the *inverse* tangent of the fraction

At this point, we are not interested in the exact value of θ. What we *are* interested in is the fact that θ changes with X_C, and thus with frequency. In other words, a given *RC* circuit can be designed to produce a specific phase shift at a given frequency.

Now assume that the phase-shift oscillator is designed so that the three *RC* circuits produce a combined phase shift of 180° at a resonant frequency, f_r. This will cause the circuit to oscillate at that frequency, provided that the Barkhausen criterion has been met. For example, let's assume that the circuit in Figure 17.4 has been designed to produce a 180° phase shift at 10 kHz and that at this frequency the Barkhausen criterion is met. Then the circuit will oscillate at 10 kHz, producing a sinusoidal output waveform.

You would think that each *RC* circuit in the phase-shift oscillator would be designed to produce a 60° phase shift, with the three 60° shifts combining to produce the 180° shift needed for regenerative feedback. However, this is not the case. Each *RC* circuit in the phase-shift oscillator acts as a *load* on the previous *RC* circuit. Just as the loaded *Q* of an amplifier was different than the unloaded *Q*, the phase shift of a loaded *RC* circuit changes from that of an unloaded *RC* circuit. Thus the exact phase shift of each *RC* circuit in the phase-shift oscillator will be different from the next. However, the overall phase shift of the three will still total 180°.

Practical Considerations

Oscillator stability. A measure of an oscillator's ability to maintain constant output amplitude and frequency.

Regulated power supplies prevent unwanted oscillations.

You should be aware that phase-shift oscillators are rarely used because they are extremely unstable. The *stability* of an oscillator is a measure of its ability to maintain an output that is constant in *frequency* and *amplitude*. Phase-shift oscillators are extremely difficult to stabilize in terms of *frequency*. Thus they cannot be used in any application where timing is critical.

Even though the phase-shift oscillator is rarely used, it is introduced here for two reasons. First, as was stated, the phase-shift oscillator is easy to understand. Thus it is a valuable learning tool. The second reason is that it is very easy to *accidentally* build a phase-shift oscillator. For example, consider the circuit shown in Figure 17.5. In this circuit we have three *RC* networks and a combined transistor phase shift of 180° (the first two 180° shifts cancel out). The feedback path is

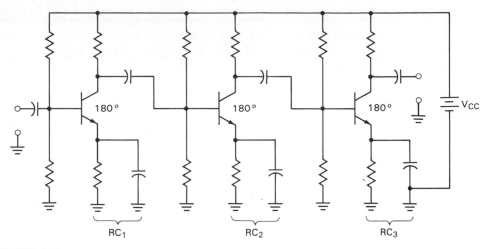

FIGURE 17.5

provided by the dc power supply, V_{CC}. Remember that V_{CC} is simply a dc source connected between the "high" side of the circuit and all the ground connections. If the internal impedance of the dc supply is high enough, any oscillations fed back can cause a significant amount of alternating current to be developed across the supply's internal impedance. This current, fed back to the first stage through its biasing circuit, can make a first-rate (although unintentional) oscillator out of the circuit. What's the solution? Using a *regulated* power supply. Regulated power supplies have extremely low internal impedance values. This low impedance keeps the ac current caused by the oscillations at a minimum. Thus, while the oscillating current *will* be present, it will be too small to cause any problem. It will simply appear as a small ripple current in the power supply.

Another practical consideration involves the Barkhausen criterion. The relationship $\alpha_v A_v = 1$ holds true for *ideal* circuits. However, practical circuits need an $\alpha_v A_v$ product that is *slightly greater than* 1. If you look back at Figure 17.4, you see that there are several resistors in the circuit. Each resistor will dissipate some small amount of power. This results in a power loss in the overall circuit. If $\alpha_v A_v$ is *exactly* equal to 1, the power lost in the resistors is not returned to the circuit by the feedback loop. Eventually, this would result in a loss of oscillations. By making $\alpha_v A_v$ slightly greater than 1, the power lost during each cycle will be returned to the circuit. How much greater than 1? Just enough to sustain oscillations. Remember, the higher $\alpha_v A_v$, the more likely the oscillator is to experience clipping.

In practice, $\alpha_v A_v$ must be *slightly greater than* 1.

Finally, we must address the question of how the oscillations start to begin with. Consider what happens when power is first applied to the circuit. Since the output is fed back to the *inverting* input, there must exist a 180° phase shift between the two ends of the feedback network. When the power is applied, the circuit *will* establish this phase shift in one of two ways:

How practical oscillators are triggered.

1. The input will remain stable and the output will change to the opposite polarity extreme, or
2. The output will remain stable, and the input offset voltage will change to the appropriate opposite-polarity voltage.

Which happens is of no consequence. The fact is that, either way, a *transition* occurs between the output and input. This transition is enough to trigger the oscillating process.

1. What is a *phase-shift oscillator*? [Objective 6]
2. How is regenerative feedback produced by the phase-shift oscillator? [Objective 6]
3. What is *oscillator stability*? [Objective 7]
4. Why are phase-shift oscillators rarely used? [Objective 7]
5. Explain how a three-stage BJT amplifier can become a phase-shift oscillator. [Objective 8]
6. Explain how the problem in question 5 is prevented. [Objective 8]
7. Why must $\alpha_v A_v$ be slightly greater than 1 in a practical oscillator? [Objective 9]
8. How is a trigger produced in a practical oscillator? [Objective 10]

17.3
The Wien-bridge Oscillator

Wien-bridge oscillator. An oscillator that achieves regenerative feedback by producing no phase shift at f_r.

Why there are *two* feedback paths.

The *Wien-bridge* oscillator is one of the more commonly used oscillators, up to frequencies of around 1 MHz. This circuit achieves regenerative feedback by producing *no phase shift* at the resonant frequency. In other words, neither the amplifier nor the feedback network will produce a phase shift. This has the same effect as using two 180° phase-shift circuits to produce oscillations. The basic Wien-bridge oscillator is shown in Figure 17.6.

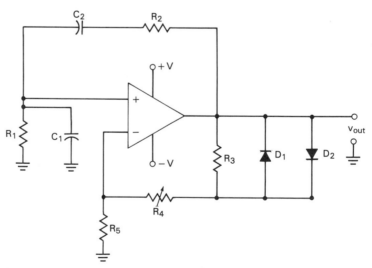

FIGURE 17.6
The Wien bridge oscillator.

The Wien-bridge oscillator has *two* feedback paths: a positive feedback path (to the non-inverting input) and a negative feedback path (to the inverting input). The positive feedback path is used to produce oscillations, while the negative feedback path is used to control the A_{CL} of the circuit.

The Positive Feedback Path

The positive feedback path consists of R_1, C_1, R_2, and C_2. R_1 and C_1 form a low-pass filter, while R_2 and C_2 form a high-pass filter. As you were shown in Chapter 15, the series combination of a low-pass filter and a high-pass filter forms a band-pass filter. The circuit will oscillate at the resonant frequency of this band-pass filter.

The positive feedback circuit is a band-pass filter.

In a typical Wien-bridge oscillator, $R_1 = R_2$ and $C_1 = C_2$. This means that the two circuits will have the same cutoff frequency. The result of this condition is illustrated in Figure 17.7. Note that the frequency-response curves cross at the f_c of each circuit. This frequency will be the frequency of oscillation for the circuit.

How the circuit works.

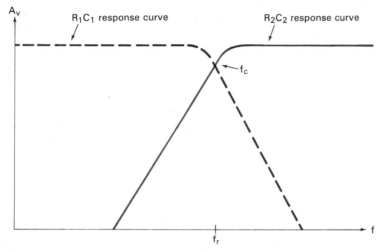

FIGURE 17.7

The frequency response of the positive feedback path.

As you know, a band-pass filter does not introduce a phase shift when operated at its resonant frequency. Also, there is no phase shift between the noninverting input of the op-amp and the component output terminal. Therefore, the output and input signals are in phase, and the feedback is regenerative.

One final note on this circuit. You will often see *trimmer potentiometers* added in series with R_1 and R_2. These trimmers allow the feedback circuit (and thus the resonant frequency) to be "fine-tuned." The circuit in Figure 17.8 shows the added trimmer pots. The pots are both adjusted to set the frequency of oscillations to the precise value desired.

Why trimmer pots are used.

FIGURE 17.8

Trimmer pots are used for oscillator tuning.

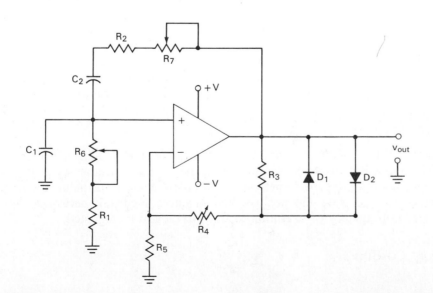

Why add two potentiometers? Why not just make R_1 and R_2 potentiometers and save the extra components? The answer to these questions is based on the fact that R_1 and R_2 are commonly selected to be in the neighborhood of 10 to 200 kΩ. For the circuit to be fine tuned, you want the potentiometers to have a range that is equal to approximately 10% of the total series resistance or less. For example, if $R_1 = R_2 = 100$ kΩ, you would want the potentiometers to have a maximum range of 10 kΩ or less. In all probability, you would want them to be as low as 1 kΩ. This gives you the ability to easily vary f_c around a relatively small frequency range.

The Negative Feedback Circuit

As stated earlier, the negative feedback circuit is used to set the closed-loop voltage gain (A_{CL}) of the circuit. You may recall from earlier discussions that voltage-divider circuits are commonly used for this purpose.

The only differences between this feedback path and the other negative feedback circuits covered up until now are:

1. The added potentiometer, R_4
2. The diodes in parallel with R_3

The potentiometer is added to the circuit to allow adjustment of the circuit A_{CL}. The diodes in the feedback network are used to limit the gain of the oscillator. If the oscillator output tries to exceed ($V_{R4} + V_{R5}$) by more than 0.7 V, one of the diodes will turn on, shorting out R_3. This results in an increase in the feedback current and a reduction of the overall gain of the op-amp. Thus, the output is prevented from exceeding a specified range. Note that the diodes are essentially acting as *clippers* in this circuit (and you thought you wouldn't have to worry about clippers anymore).

Frequency Limits

Propagation delay. The time required for a signal to pass through a component.

As you increase the operating frequency of an op-amp, you will eventually reach the point where a phase shift will start to occur between the noninverting input and the output terminal. This phase shift is caused by the *propagation delay* of the op-amp. Propagation delay is the time requried for a signal to pass through the component.

When the op-amp starts to introduce a phase shift, the oscillating action of the circuit will start to lose its stability. Remember, the operation of the circuit depends heavily on the phase relationship between the input and output. When you change this relationship, you affect the operation of the oscillator.

For most Wien-bridge oscillators, the upper frequency limit is in the range of 1 MHz. Above this frequency, the stability of the oscillator starts to drop. For any oscillator application, this is an unacceptable situation. Thus for high-frequency applications we must use *LC* oscillators. As you will see in the next two sections, *LC* oscillators are much better suited to high-frequency operation.

Summary

The Wien-bridge oscillator contains an op-amp and two feedback networks. The positive feedback network is used to control the operating frequency of the circuit, while the negative feedback network is used to control its gain. By including variable components in the positive and negative feedback networks, both the gain and the frequency of operation for the circuit can be adjusted.

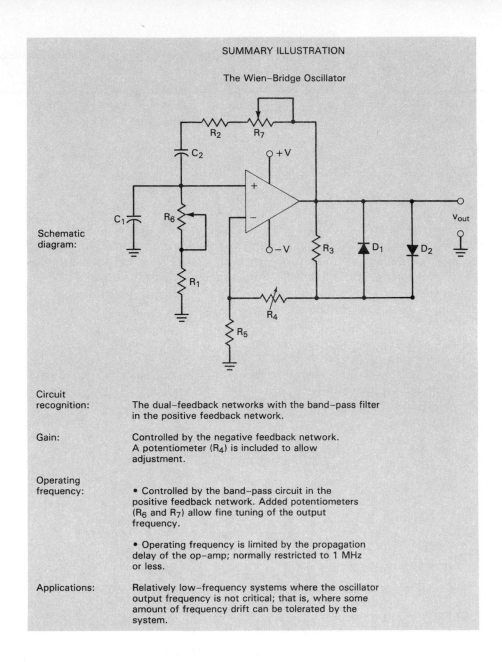

SUMMARY ILLUSTRATION

The Wien–Bridge Oscillator

Schematic
diagram:

Circuit
recognition: The dual–feedback networks with the band–pass filter
in the positive feedback network.

Gain: Controlled by the negative feedback network.
A potentiometer (R_4) is included to allow
adjustment.

Operating
frequency:
• Controlled by the band–pass circuit in the
positive feedback network. Added potentiometers
(R_6 and R_7) allow fine tuning of the output
frequency.

• Operating frequency is limited by the propagation
delay of the op–amp; normally restricted to 1 MHz
or less.

Applications: Relatively low–frequency systems where the oscillator
output frequency is not critical; that is, where some
amount of frequency drift can be tolerated by the
system.

A Practical Consideration. The
frequency stability of the Wien-
bridge oscillator is much higher
than that of the phase-shift
oscillator. However, it will still
experience some frequency drift
due to component tolerances and
heating. Thus, it is normally used
in low-frequency applications
where the *exact* operating
frequency isn't critical.

FIGURE 17.9

Because of the propagation delay of the op-amp, Wien-bridge oscillators are usually restricted to operating frequencies of 1 MHz or less. The Wien-bridge oscillator characteristics are summarized in Figure 17.9.

1. Describe the construction of the Wien-bridge oscillator. [Objective 11]
2. How does the positive feedback network of the Wien-bridge oscillator control its operating frequency? [Objective 12]
3. Explain why potentiometers are normally included in the positive feedback network of a Wien-bridge oscillator. [Objective 12]
4. Explain why a potentiometer is normally included in the negative feedback network of a Wien-bridge oscillator. [Objective 13]
5. What is *propagation delay*? [Objective 14]

SECTION REVIEW

6. How does propagation delay limit the operating frequency of a Wien-bridge oscillator? [Objective 14]

17.4
The Colpitts Oscillator

Colpitts oscillator. An oscillator that uses a pair of tapped capacitors and an inductor to produce the 180° phase shift in the feedback network.

The *Colpitts oscillator* is a discrete amplifier that uses a pair of *tapped capacitors* and an inductor to produce the regenerative feedback required for oscillations. The Colpitts oscillator is shown in Figure 17.10. As you can see, the transistor configuration is common emitter. Therefore, there is a 180° phase shift from the transistor base to the collector. This means that the feedback network must produce a 180° phase shift if the feedback is to be regenerative.

The Feedback Network

Obviously, the key to understanding the Colpitts oscillator is knowing how the feedback network achieves a 180° phase shift. The feedback network for the circuit in Figure 17.10 consists of C_1, C_2, and L_2.

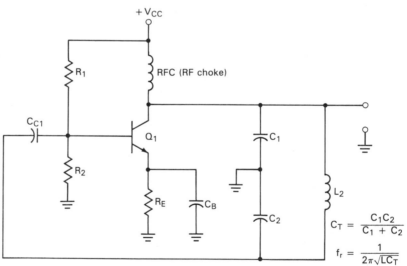

FIGURE 17.10

The Colpitts oscillator.

There are several key points that must be made clear if you are to get the picture of the overall circuit operation. These points are:

Circuit operation key points.

1. The output voltage from the amplifier is felt across C_1.
2. The feedback voltage is developed across C_2.
3. The voltage across C_2 is 180° out of phase with the voltage across C_1. Therefore, the feedback voltage is 180° out of phase with the output voltage.

Points 1 and 2 are easy to see if we simplify the circuit as shown in Figure 17.11. In Figure 17.11a, we have dropped the inductor L_2 from the circuit. Then,

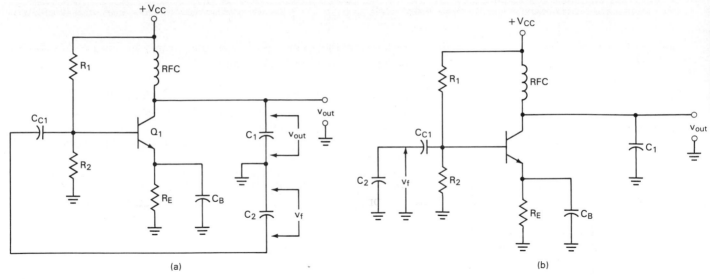

FIGURE 17.11

in Figure 17.11b, we split the capacitors at the ground connection. As you can see, the input voltage to the amplifier is developed across C_2. Therefore, C_2 must be the source of the feedback voltage. Also, C_1 is obviously across the output of the amplifier. Therefore, v_{out} is measured across this component. Now the trick is to see how the voltages across these two components are always 180° out of phase.

The phase relationship between v_{out} and v_f is caused by the *current* action in the feedback network. This point is easy to see if we redraw the feedback network as shown in Figure 17.12. By comparing the circuit in Figure 17.12 with the feedback network in Figure 17.10, you can see that we are dealing with exactly the same circuit in both illustrations. Now let's assume for a moment that l_2 has a voltage across it with the polarities shown. If we view L_2 as being the voltage source, it is easy to see that it will produce a current in the circuit. With this current in the circuit, voltages are developed across C_1 and C_2 with the polarities shown. As you can see, these voltages are 180° out of phase. If you reverse the voltage polarity across L_2, all polarity signs and current directions reverse, but the voltages across C_1 and C_2 are still 180° out of phase. When you look at the feedback network in Figure 17.10, it is easy to see that v_{out} and v_f are 180° out of phase when you keep in mind the circuit current action.

The amount of feedback voltage in the Colpitts oscillator depends on the α_v of the circuit. For this oscillator, α_v is the ratio of X_{C2} to X_{C1}. By formula,

How the feedback circuit produces a 180° phase shift.

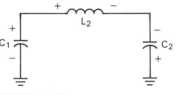

FIGURE 17.12

$$\alpha_v = \frac{X_{C2}}{X_{C1}} \qquad (17.3)$$

Since X_{C2} and X_{C1} are both inversely proportional to the values of C_2 and C_1 at a given frequency, equation (17.3) can be rewritten as

$$\alpha_v = \frac{C_1}{C_2} \qquad (17.4)$$

The validity of equation (17.4) is illustrated in the following example.

SEC. 17.4   The Colpitts Oscillator

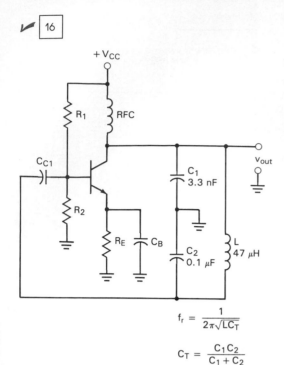

✔ 16

$f_r = \dfrac{1}{2\pi\sqrt{LC_T}}$

$C_T = \dfrac{C_1 C_2}{C_1 + C_2}$

FIGURE 17.13

Determine the value of α_v for the circuit in Figure 17.13 using both equations (17.3) and (17.4).

Solution: Before we can determine α_v with equation (17.3), we have to determine the reactance of the two capacitors. Before we can determine X_{C1} and X_{C2}, we have to find the operating frequency of the circuit. As always, the operating frequency is equal to the resonant frequency of the feedback network. To find f_r, we need to determine the total capacitance in the C_1–C_2–L_2 circuit.

Figure 17.11 demonstrated the fact that C_1 and C_2 are in *series*. Therefore,

$$C_T = \frac{C_1 C_2}{C_1 + C_2}$$
$$= 3.19 \text{ nF}$$

Now, we can use this value to find f_r as follows:

$$f_r = \frac{1}{2\pi\sqrt{LC}}$$
$$= 411 \text{ kHz}$$

Using the value of $f = 411$ kHz, we can now find the value of $X_{C1} = 117.34\ \Omega$ and $X_{C2} = 3.87\ \Omega$. Using these values in equations (17.3), we get an α_v of

$$\alpha_v = \frac{X_{C2}}{X_{C1}}$$
$$= \frac{3.87\ \Omega}{117.34\ \Omega}$$
$$= 0.03298 \cong 0.033$$

Now, if we determine the value of α_v using equation (17.4), we get

$$\alpha_v = \frac{C_1}{C_2}$$
$$= \frac{3.3 \text{ nF}}{0.1\ \mu\text{F}}$$
$$= 0.033$$

PRACTICE PROBLEM 17–1

A Colpitts oscillator like the one in Figure 17.13 has the following values: $C_1 = 10$ nF, $C_2 = 1.5\ \mu$F, and $L = 10\ \mu$H. Calculate the value of α_v for the circuit using both equations (17.3) and (17.4).

Given the fact that none of us enjoy doing things the hard way, equation (17.4) is definitely the way to go when you want to determine the value of α_v.

Circuit Gain

17

As with any other oscillator, the product of $\alpha_v A_v$ for the Colpitts oscillator must be slightly greater than 1. This prevents loss of oscillations due to power losses within the amplifier and, at the same time, prevents the circuit from driving itself into saturation and cutoff clipping.

Since the feedback network in the Colpitts oscillator is a parallel resonant tank circuit, it will draw very little current from the transistor collector circuit. Therefore, it can be ignored in determining the value of A_{vL} for the circuit. The value of A_{vL} is found as

$$A_{vL} = \frac{v_{out}}{v_f} \cong \frac{C_2}{C_1}$$

Amplifier Coupling

18

As with any parallel resonant tank circuit, the feedback network loses some efficiency when loaded down. To reduce the loading effects of R_L, Colpitts oscillators are commonly transformer coupled to the load. A transformer-coupled Colpitts oscillator is shown in Figure 17.14. For this circuit, the primary winding of T_1 is the feedback network inductance. Remember that the transformer reduces circuit loading because of the effects of the turns ratio on the reflected load impedance. This point was discussed in detail in Chapter 15.

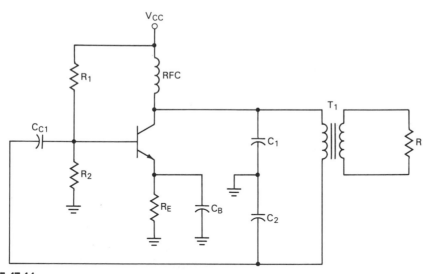

FIGURE 17.14

A transformer copuled Colpitts oscillator.

Also, it is acceptable to use capacitive coupling for the Colpitts oscillator provided that the following relationship is fulfilled:

$$C_C \ll C_T \qquad \textbf{(17.5)}$$

where C_T is the total series capacitance of the feedback network. By fulfilling this relationship, you ensure that the reactance of the coupling capacitor is much greater than that of the series combination of C_1 and C_2. This prevents circuit loading almost as well as using transformer coupling.

Summary

The Colpitts oscillator uses a pair of tapped capacitors and an inductor (all in its feedback network) to produce the 180° feedback phase shift required for oscillations. The frequency of operation for the circuit is approximately equal to the resonant frequency of the feedback circuit.

SECTION REVIEW

1. What is the *Colpitts oscillator*? [Objective 15]
2. Explain how the feedback network in the Colpitts oscillator produces a 180° voltage phase shift. [Objective 15]
3. List the key points to remember about the operation of the Colpitts oscillator. [Objective 15]
4. How do you calculate the value of A_{vL} for a Colpitts oscillator? [Objective 17]
5. Why is transformer coupling used in Colpitts oscillators? [Objective 18]

17.5
Other LC Oscillators

Although the Colpitts is the most commonly used *LC* oscillator, several others are worth mentioning here. In this section we are going to take a brief look at three other oscillator circuits starting with the Hartley oscillator.

Hartley Oscillators

The *Hartley oscillator* is almost identical to the Colpitts oscillator. The only difference is that the Hartley oscillator uses *tapped inductors* and a single capacitor. The Hartley oscillator is shown in Figure 17.15.

FIGURE 17.15

The Hartley oscillator.

A Practical Consideration. C_3 in Figure 17.15 is a *blocking capacitor*. It is included to prevent the RF choke (RFC) and L_2 from shorting V_{CC} to ground. The value of C_3 is generally too high to consider in any circuit frequency calculations.

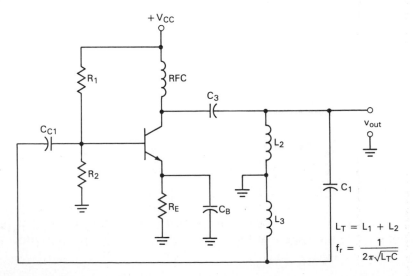

$$L_T = L_1 + L_2$$

$$f_r = \frac{1}{2\pi\sqrt{L_T C}}$$

For the Hartley oscillator, the feedback voltage is measured across L_3 and the output voltage is measured across L_2. Thus, the value of α_v is found as

Hartley Oscillator. An oscillator that uses a pair of tapped inductors and a parallel capacitor in its feedback network to produce the 180° voltage phase shift required for oscillation.

$$\alpha_v = \frac{X_{L3}}{X_{L2}} \qquad\qquad (17.6)$$

or

$$\alpha_v = \frac{L_3}{L_2} \qquad\qquad (17.7)$$

Since the inductors are in series in the Hartley oscillator, the total inductance is found as

$$L_T = L_2 + L_3 + 2L_M$$

where L_M = the *mutual* inductance between the two components

This value of L_T must be used whenever you are calculating the value of f_r for the circuit. Outside of these two differences, all calculations and circuit principles are the same for this circuit as for the Colpitts oscillator.

The Clapp Oscillator

The *Clapp oscillator* is a variation on the Colpitts, as is the Hartley oscillator. The Clapp oscillator was designed to eliminate the effects of *stray capacitance* on the operation of the basic Colpitts. This is done by adding an additional capacitor, as is shown in Figure 17.16. Just for a moment, ignore the presence of C_3 in the Clapp oscillator. As you can see, all you have without C_3 is a standard Colpitts oscillator.

Clapp oscillator. A Colpitts oscillator with an added capacitor (in series with the feedback inductor) that is used to reduce the effects of stray capacitance.

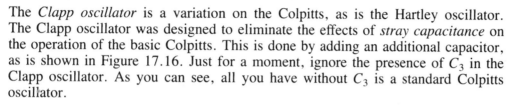

FIGURE 17.16

The Clapp oscillator.

If you refer back to Figure 17.11, you will see that C_1 and C_2 in the feedback network of the Colpitts oscillator are in parallel with the rest of the transistor circuit. As you have seen countless times, the transistor has junction capacitance values. These capacitance values are in parallel with C_1 and C_2 and thus affect the total amount of capacitance in the feedback network. This affects the operating frequency of the circuit.

The Clapp oscillator eliminates the effects of transistor capacitance by adding the capacitor C_3. This capacitor is normally much lower in value than C_1 or C_2. Because of this, it becomes the dominant component in all frequency calculations for the circuit. For the Clapp oscillator.

$$f_r = \frac{1}{2\pi \sqrt{LC_3}}$$

(17.8)

You may be wondering why C_1 and C_2 are even included in the Clapp oscillator if C_3 is the dominant component. The answer to this is simple. Even though the operating frequency of the circuit depends on the value of C_3, C_1 and C_2 are still needed to provide the 180° phase shift required for regenerative feedback. C_3 has not replaced C_1 and C_2; it has just eliminated their values from any frequency considerations. In fact, except for the f_r formula, all other calculations for the Clapp oscillator are exactly the same as those for the Colpitts oscillator.

The Armstrong Oscillator

21

Armstrong oscillator. An oscillator that uses a transformer in its feedback network to achieve the required 180° voltage phase shift.

The *Armstrong oscillator* is a simple but effective oscillator circuit. The basic circuit is shown in Figure 17.17. In the Armstrong oscillator, the required 180° phase shift in the feedback network is provided by the transformer, T_1. As the polarity dots indicate, the output of the amplifier will be inverted from the transformer primary to its secondary. This provides the 180° phase shift of v_f. Note that v_f is developed across the secondary of the oscillator.

FIGURE 17.17

The Armstron oscillator.

Note: C_2 is a blocking capacitor, used to prevent the RFC and the primary of T_1 from shorting out the dc supply.

22

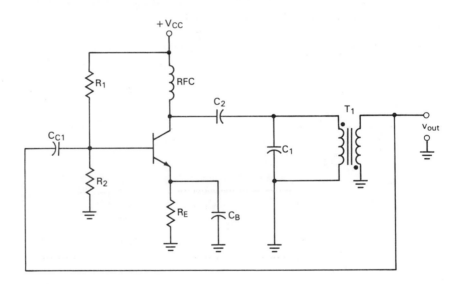

The capacitor in the output circuit (C_1) is there to provide the tuning of the oscillator. The resonant frequency of the circuit will be determined by the value of C_1 and the transformer primary.

One Final Note

You have been introduced to quite a few *LC* oscillators in the past two sections. The BJT oscillators make up only part of the entire picture. Any *LC* oscillator can be constructed using either a FET or an op-amp. For example, consider the circuits shown in Figure 17.18. Figure 17.18a shows a Colpitts oscillator using an op-amp. Figure 17.18b shows a FET Hartley oscillator. These oscillators work

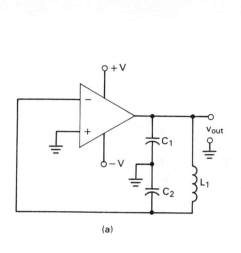

(a)

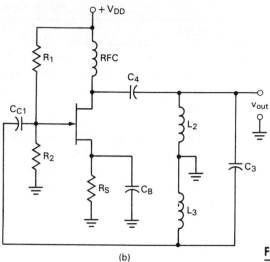

(b)

FIGURE 17.18

according to the same basic principles as their BJT counterparts, so do not let the different active components confuse you.

Another situation that you will commonly see is an oscillator that has the feedback network going from the output of one stage back to the emitter, source, or noninverting input of a previous stage. Again, the overall principles of such a circuit are the same as those discussed earlier.

The key to being able to deal with the various oscillators is to learn the circuit recognition features of the feedback networks. For example, the Colpitts oscillator has two tapped capacitors in the feedback network. Anytime that you see an oscillator with this configuration, it is a Colpitts. It really does not matter what the active component is or to which terminal the feedback network leads. If the circuit is an oscillator and has two tapped capacitors in the feedback network, it is a Colpitts. The recognition features of the other common *LC* oscillators are as follows:

Recognizing oscillator circuits

Oscillator Type	Recognition Feature(s)
Hartley	Tapped inductors or a center-tapped transformer with a single parallel capacitor
Clapp	Looks like a Colpitts, with a capacitor added in series with the inductor
Armstrong	A single transformer with a parallel capacitor

Once you have established that a given circuit is an oscillator, just remember the recognition features listed. They will tell you which type of circuit you are dealing with.

SECTION REVIEW

1. Where are the feedback and output voltages measured in the feedback network of a Hartley oscillator? [Objective 19]
2. How is the construction of the Clapp oscillator different from that of the Colpitts oscillator? What purpose does this difference serve? [Objective 20]
3. How is the required 180° voltage phase shift accomplished in the feedback network of an Armstrong oscillator? [Objective 21]
4. List the circuit recognition features of the *LC* oscillators covered in this section. [Objective 22]

Most communications applications require the use of oscillators that have *extremely* stable output signals. This can pose a problem for any "conventional" oscillator, since there are so many things that can cause the output from such an oscillator to change. For example, refer back to the circuit shown in Figure 17.10. The output from this circuit could change if any of the following were to occur:

1. The transistor is replaced. This could change r_e', causing A_v to change, which would change the $\alpha_v A_v$ product. As you know, that can cause you all sorts of grief.
2. Changing any component in the feedback network could change the resonant frequency.
3. The circuit will warm up. This can change resistance values, changing the load on the feedback network and so on.

Crystal-controlled oscillators. Oscillators that use quartz crystals to produce extremely stable output frequencies.

In any system where the stability of the oscillator is critical, any of the foregoing problems are intolerable. For these types of applications, a *crystal-controlled* oscillator is normally used. Crystal-controlled oscillators have a *quartz crystal* that is used to control the frequency of operation. To help you understand the importance of the crystal, we will take a brief look at what a crystal is and how it works.

Quartz Crystals

A quartz crystal is made of silicon dioxide, SiO_2. You may recall that this is the same compound that is used as the insulation layer in the gate of a MOSFET. Quartz crystals are very common in nature. They develop as six-sided compounds, as shown in Figure 17.19. When being used in an electronic component, a thin slice of crystal is placed between two conducting plates, like those of a capacitor.

Piezoelectric effect. The tendency of a crystal to vibrate at a fixed frequency when subjected to an electric field.

The key to the operation of the crystal is called the *piezoelectric effect*. This simply means that the crystal will vibrate at a constant rate when it is subjected to an electric field. The frequency of the vibrations depends on the physical dimensions of the crystal. Thus it is possible to produce crystals with very exact frequency ratings by simply cutting them to the right dimension.

FIGURE 17.19

A quartz crystal

The only factor that will normally alter the physical dimensions of a crystal (and thus its operating frequency) is *temperature*. However, by using cooling methods to hold the crystal temperature constant, the problem of frequency variation is eliminated.

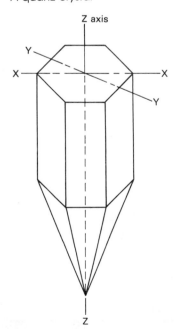

The electrical operation of the crystal is based on its mechanical properties. However, this does not prevent us from representing the crystal as an equivalent circuit as shown in Figure 17.20. Figure 17.20a shows the schematic symbol for the crystal. The equivalent circuit for the device is shown in Figure 17.20b. The components shown represent specific characteristics of the crystal, as follows:

C_C = the capacitance of the crystal itself

C_M = the *mounting capacitance* (this is the capacitance caused by the parallel conducting plates that hold the crystal)

L = the inductance of the crystal

R = the resistance of the crystal, determined by the amount of energy lost by the component when it is vibrating

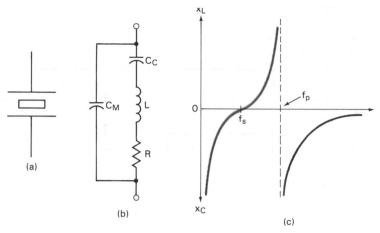

FIGURE 17.20

Crystal symbol, equivalent circuit, and frequency response.

That strange-looking graph in Figure 17.20c tells us quite a bit about the crystal. At very low frequencies, the reactance of the crystal is controlled by the extremely high values of X_{CM} and X_{CC}. As the frequency of operation increases, the circuit composed of C_C and L approaches its resonant frequency. At the frequency f_s, the crystal acts like a *series* resonant circuit. At this frequency, the reactances of C_C and L cancel, and the reactance of the component drops to zero. If the frequency continues to climb, the reactance of the parallel capacitor, C_M, starts to drop to a low enough value that it starts to affect the circuit. When f_p is reached, C_M and L resonate, and the circuit acts as a *parallel* resonant circuit. At frequencies above f_p, the continuing drop in C_M eventually causes the quartz to act as a short circuit. The bottom line of all this is as follows:

1. At f_s, the crystal will act as a series resonant circuit.
2. At f_p, the crystal will act as a parallel resonant circuit.

This means that we can use a crystal in place of a series LC circuit or in place of a parallel LC circuit. If we use it in place of a series LC circuit, the oscillator will operate at f_s. If we use it in place of a parallel LC circuit, the oscillator will operate at f_p.

CCO Circuits

The Colpitts oscillator can be modified to be a crystal-controlled oscillator (CCO) as shown in Figure 17.21. As you can see, the only difference here is the addition of the crystal (Y_1) in the feedback network. The crystal will act as a parallel-resonant circuit at f_p, allowing the maximum allowable energy transfer through the feedback network at that frequency.

    To understand the effectiveness of this oscillator, you must realize that the parallel resonant characteristics of the filter will exist only for an extremely limited band of frequencies. Even the smallest deviation from f_p will cause the crystal to act as an effective short. Therefore, the oscillator is capable of oscillating only at f_p. Thus we have an extremely stable oscillator. Note that placing the crystal in the same relative position in either a Hartley or Clapp oscillator will have the same effect.

◣ 25

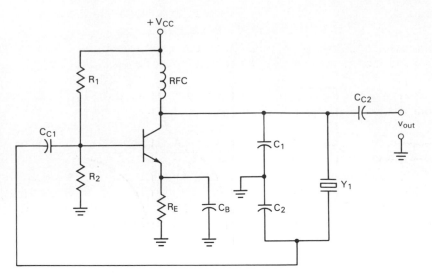

FIGURE 17.21

A crystal controlled Colpitts oscillator

SECTION REVIEW

1. What is a *crystal-controlled oscillator*? [Objective 23]
2. What is the *piezoelectric effect*? [Objective 24]
3. Explain how a crystal will act as a series resonant circuit at one frequency (f_s) and as a parallel resonant component at another (f_p). [Objective 24]
4. Explain how the addition of a crystal to an *LC* oscillator stabilizes the operating frequency of the circuit. [Objective 25]

17.7
Oscillator Troubleshooting

Of all the circuits that we have discussed in this text, oscillators are by far the most challenging when it comes to troubleshooting. They can also be the most time-consuming circuits to troubleshoot. The basic problem is this: Every component in the oscillator, with the exception of the biasing resistors, is involved in the production of the output signal. If any one of them goes bad, nothing will work. For example, refer back to Figure 17.14. Whether a failure of the circuit is caused by C_1, C_2, T_1, or C_{C1}, the results are exactly the same—the oscillator does not have an output.

As with any circuit, you start to troubleshoot an oscillator by making sure that it is the source of the problem. If the circuit does not have an output, check the power supply connections in the circuit. If there is no problem here, the oscillator has a fault. (You do not need to check the oscillator's signal source—it doesn't have one.)

When you have verified that an oscillator is bad, there are a few points to remember that may make the troubleshooting process go a little more smoothly:

1. Remember that for dc analysis purposes the oscillator is biased as an emitter,

source, or voltage follower. Thus your dc troubleshooting procedure should be the same as it is for these circuit types.

2. In the case of the oscillator, it is often faster (and therefore less costly) to go ahead and replace the reactive components in the feedback network.

If circumstances require you to test the components in the feedback network, use an ohmmeter to test them, remembering the following points:

1. A good capacitor will initially show a low resistance. Then, as the capacitor is charged by the ohmmeter, the resistance reading will rise.
2. An inductor will always have a low resistance reading. When a transformer is used, as in Figure 17.14, remember that you should measure a low resistance between the two primary connections and the two secondary connections. However, you should read a very high resistance when measuring from either primary connection to either secondary connection.

If you decide to go the component-replacement (or *swapping*) route, replace the reactive components one at a time. After you replace each component, apply power to the circuit and retest its operation. When the circuit starts to operate at the proper frequency, your task is complete.

A Practical Consideration. Most oscillators contain one or more variable components for output tuning. When you replace any component in the circuit, you must check the output frequency and, if necessary, retune the circuit.

SECTION REVIEW

1. Why are oscillators difficult to troubleshoot? [Objective 27]
2. What points should be remembered when troubleshooting an oscillator? [Objective 27]
3. When testing the reactive components in an oscillator, what points should be kept in mind? [Objective 27]

KEY TERMS

The following terms were introduced and defined in this chapter:

Armstrong oscillator	Hartley oscillator	positive feedback
Barkhausen criterion	oscillator	propagation delay
Clapp oscillator	oscillator stability	regenerative feedback
Colpitts oscillator	phase-shift oscillator	Wien-bridge oscillator
crystal-controlled oscillator	piezoelectric effect	

PRACTICE PROBLEMS

§17.3

1. Calculate the operating frequency of the circuit in Figure 17.22. [13]
2. Calculate the operating frequency of the circuit in Figure 17.23. [13]

§17.4

3. Calculate the operating frequency for the circuit in Figure 17.24. [16]
4. Calculate the value of α_v for the circuit in Figure 17.25. [16]

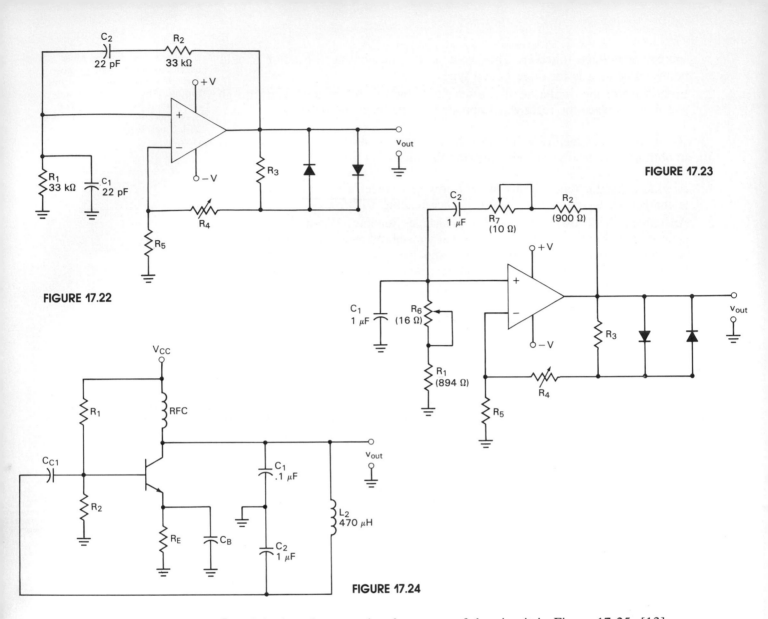

FIGURE 17.23

FIGURE 17.22

FIGURE 17.24

5. Calculate the operating frequency of the circuit in Figure 17.25. [13]
6. Calculate the value of α_v for the circuit in Figure 17.25. [13]
7. Calculate the value of A_{vL} for the circuit in Figure 17.24. [17]
8. Calculate the value of A_{vL} for the circuit in Figure 17.25. [17]

FIGURE 17.25

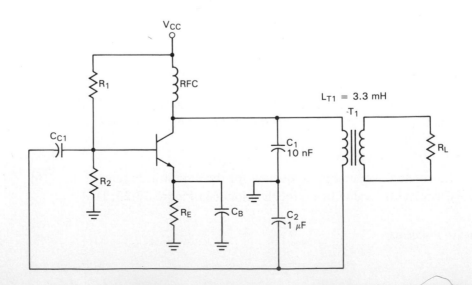

9. Calculate the operating frequency of the circuit in Figure 17.26. [19]

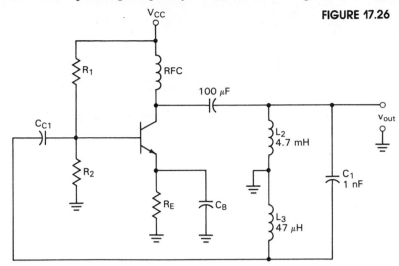

FIGURE 17.26

10. Calculate the operating frequency of the circuit in Figure 17.27. [19]

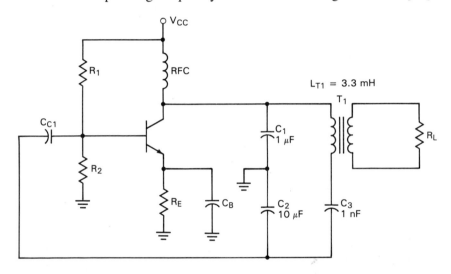

FIGURE 17.27

11. Calculate the operating frequency of the circuit in Figure 17.28. [20]

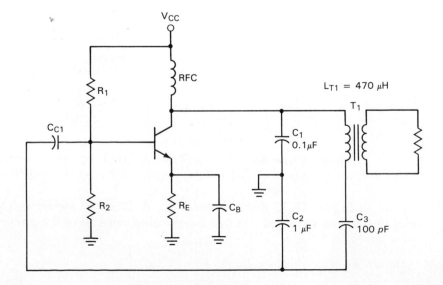

FIGURE 17.28

12. Calculate the operating frequency of the circuit in Figure 17.29. [20]

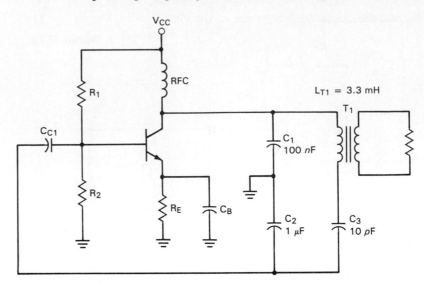

TROUBLESHOOTING PRACTICE PROBLEMS

13. The circuit in Figure 17.30 does not oscillate. A check of the dc voltages provides the readings indicated. Discuss the possible cause(s) of the problem.

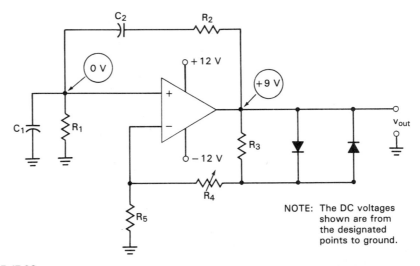

FIGURE 17.30

14. The circuit in Figure 17.31 does not oscillate. A check of the dc voltages provides the readings indicated. Discuss the possible cause(s) of the problem.

15. The circuit in Figure 17.32 does not oscillate. A check of the dc voltages provides the reading indicated. Discuss the possible cause(s) of the problem.

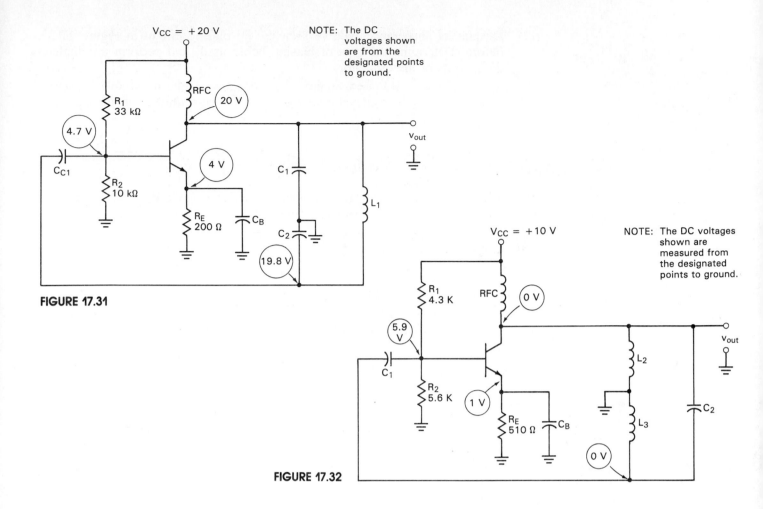

FIGURE 17.31

FIGURE 17.32

THE BRAIN DRAIN

16. Refer to Figure 17.27. Use circuit calculations to show that the circuit fulfills the Barkhausen criterion.

FIGURE 17.33

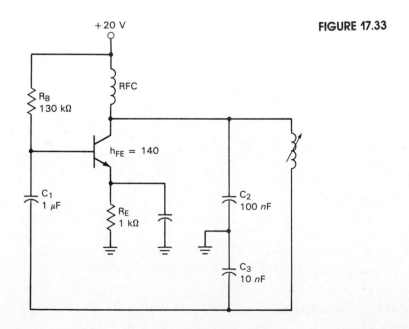

17. Despite its appearance, you've seen the dc biasing circuit in Figure 17.33 before. Determine the type of biasing being used, and perform a complete dc analysis of the oscillator.

18. The feedback inductor in Figure 17.33 is shown to be a variable inductor. Determine the adjusted value of L that would be required for an operating frequency of 22 kHz.

SUGGESTED COMPUTER APPLICATIONS PROBLEMS

19. Write a program to determine the operating frequency of a Wien-bridge oscillator, given the positive feedback values of R and C.

20. Write a program to determine the operating frequency and α_v values for a Colpitts oscillator, given the feedback network values of L and C.

ANSWERS TO THE EXAMPLE PRACTICE PROBLEMS

17.1. $\alpha_v = 0.0067$ (for both solutions)

18

Solid-State Switching Circuits

The study of discrete switching circuits will help you to prepare for the study of modern digital circuits, most of which are made up almost entirely of integrated circuits. The digital circuit shown here is an Erasable-Programmable Read-Only Memory (*EPROM*).

OBJECTIVES

After studying the material in this chapter, you should be able to:

1. Contrast *switching circuits* with *linear circuits*. (Introduction)

2. Describe and analyze the operation of the BJT switch. (§18.1)

3. Describe and analyze the operation of the JFET switch. (§18.1)

4. Describe and analyze the operation of the MOSFET switch. (§18.1)

5. Describe the function and operation of the op-amp switch. (§18.1)

6. Describe the function and operation of the *driver* circuit. (§18.1)

7. Define each of the following terms and discuss the relationship between them: *rectangular waveform, pulse width, space width, cycle time,* and *square wave.* (§18.2)

8. Describe the oscilloscope measurements of cycle time, pulse width, and space width. (§18.2)

9. Calculate the duty cycle of a given rectangular waveform. (§18.2)

10. Discuss *propagation delay* and its four sources. (§18.2)

11. Calculate the practical limit on the operating frequency of a basic switching circuit. (§18.2)

12. Discuss the causes of the various BJT switching times and the effects that a *speed-up capacitor* has on those times. (§18.2)

13. Describe FET switching time and the effect that a speed-up capacitor has on that time. (§18.2)

14. Contrast *switching transistors* with standard devices. (§18.2)

15. Discuss the factors that affect op-amp switching time. (§18.2)

16. Compare and contrast the *inverter* and the *buffer*. (§18.2)

17. Measure each of the following with an oscilloscope: *delay time* (t_d), *rise time* (t_r), *storage time* (t_s), and *fall time* (t_f). (§18.2)

18. State the purpose served by the *Schmitt trigger* and discuss its input/output relationship. (§18.3)

19. Describe and analyze the operation of the noninverting Schmitt trigger. (§18.3)

20. Describe and analyze the operation of the inverting Schmitt trigger. (§18.3)

21. Compare and contrast the Schmitt trigger and the comparator. (§18.3)

22. Contrast the output characteristics of the following circuits: the *astable multivibrator* (*free-running multivibrator*), the *monostable multivibrator* (*one-shot*), and the *bistable multivibrator* (*flip-flop*). (§18.4)

23. Describe the 555 timer and its primary application. (§18.4)

24. Explain the internal operation of the 555 timer. (§18.4)

25. Describe the operation of the monostable multivibrator (one-shot). (§18.4)

26. Calculate the pulse width for a one-shot. (§18.4)

27. List the circuit checks that are made when troubleshooting a one-shot. (§18.4)

28. Calculate the maximum allowable input trigger voltage for a one-shot. (§18.4)

29. Describe the operation of the astable (free-running) multivibrator. (§18.4)

30. Calculate the values of PRR, duty cycle, and pulse width for a free-running multivibrator. (§18.4)

31. Compare the procedure for troubleshooting a free-running multivibrator with that for the one-shot. (§18.4)

32. Describe the circuit construction and operation of the *voltage-controlled oscillator* (*VCO*). (§18.4)

As you master each of the objectives listed, place a check mark (✔) in the appropriate box.

Switching circuits. Circuits designed to respond to (or generate) nonlinear waveforms, such as square waves.

Throughout the text we have concentrated almost entirely on the *linear* operation of solid-state devices and circuits. Another important aspect of solid-state electronics is the way in which devices and circuits respond to *nonlinear* waveforms, such as *square waves*. Circuits designed to respond to nonlinear waveforms (or generate them) are referred to as *switching circuits*. In this chapter we will discuss the most basic types of switching circuits. As you will see, these circuits are viewed quite differently from linear circuits. They are also easier to understand and troubleshoot.

Solid-state switching circuits are the fundamental components of modern computer systems. It seems appropriate then that we should take a minute to look at the effects that developments in sold-state electronics have had on the digital computer.

The first general-purpose digital computer, called ENIAC, was developed in the 1930s. This computer was just about as large as a small building, weighed about 30 tons, and contained over 18,000 vacuum tubes. As the story is told, a couple of college students spent their days running around the computer shop with shopping carts full of vacuum tubes, since ENIAC burned out several tubes each hour. Amazingly, ENIAC was capable of doing less than your calculator.

When the first transistor computer circuits were developed, the *minicomputer* was born. Minicomputers were capable of performing more advanced functions and were a lot smaller than any vacuum tube computer.

The development of the solid-state *microprocessor*, a complete data-processing unit constructed on a single chip, in the 1970s led to the birth of the *microcomputer*. Since the birth of the microcomputer, computers have been accessible and affordable for almost everyone.

One interesting point: As computer chips became smaller and more powerful, they also became much less expensive. It has been said that if automobiles had gone through the same evolution as computers, you could now buy a Rolls Royce for less than $35.00. Of course, the car would be smaller than a pencil eraser!

18.1
Introductory Concepts

For you to understand some of the more complex switching circuits, you must be comfortable with the idea of using a BJT, FET, or op-amp as a switch. In this section we will discuss some of the basic principles involved in using a solid-state device as a switch.

The BJT as a Switch

A BJT can be used as a switch simply by driving the component back and forth between *saturation* and *cutoff*. A basic transistor switching circuit is shown in Figure 18.1a. As the figure shows, a square-wave input produces a square-wave

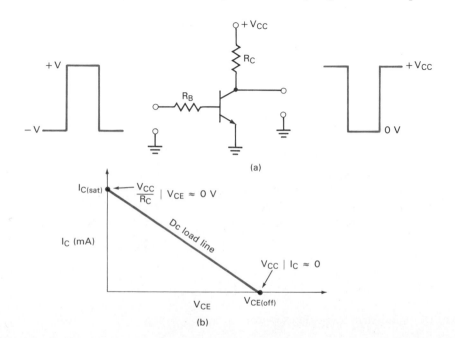

FIGURE 18.1

A basic transistor switch.

809

output. The way this is accomplished is easy to see if you refer to the dc load line shown in Figure 18.1b.

When the input to the transistor is at $-V$, the emitter–base junction of the transistor is biased off. When the transistor is biased off, the following conditions exist:

$$V_{CE} = V_{CC} \text{ (ideal)} \quad \text{and} \quad I_C = 0 \text{ (ideal)}$$

Ideal cutoff conditions are similar to those of an *open* switch.

As Figure 18.2 shows, these conditions are the same as those caused by an *open* switch.

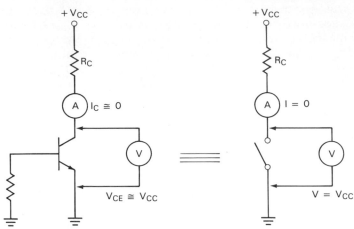

FIGURE 18.2

The "open" transistor switch.

When the input to the transistor is at $+V$, the transistor will saturate. This assumes, of course, that the input current is sufficient to cause saturation. When the transistor saturates, the following conditions exist:

$$V_{CE} = 0 \text{ V (ideal)} \quad \text{and} \quad I_C = \frac{V_{CC}}{R_C} \text{ (ideal)}$$

Ideal saturation conditions are similar to those of a *closed* switch.

As Figure 18.3 shows, these conditions are the same as those caused by a *closed* switch.

FIGURE 18.3

The "closed" transistor switch.

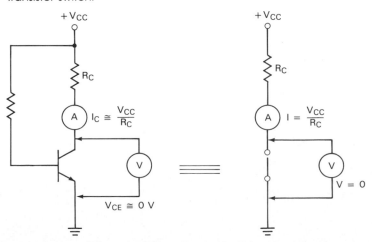

CHAP. 18   Solid-state Switching Circuits

For the circuit in Figure 18.1, we made several assumptions to simplify the discussion. These are:

1. $-V$ is low enough to prevent Q_1 from conducting.
2. $+V$ is high enough to produce enough base current to saturate Q_1.
3. The transistor is an ideal component.

In a practical transistor switching circuit, the value of $-V$ will usually be equal to the emitter supply connection. For example, in Figure 18.1, $-V$ would be 0 V, since the transistor emitter is returned to ground. Had the emitter been tied to a negative voltage, say -10 V, then $-V$ would have been -10 V. By making the low input voltage equal to V_{EE}, you ensure that the transistor will be in cutoff when the input equals $-V$.

The high input voltage, $+V$, must ensure that the transistor saturates. In a practical transistor switch, this is accomplished by two things:

1. $+V$ is usually equal to $+V_{CC}$.
2. R_B is made small enough so that the calculated value of $I_B \gg I_{C(sat)}/h_{FE}$.

Practical input signal characteristics

The result of fulfilling both of these requirements can be seen in Example 18.1. Here we use the relationships established in Chapter 6 to determine the minimum value of $+V$ required to saturate a transistor switch.

EXAMPLE 18.1

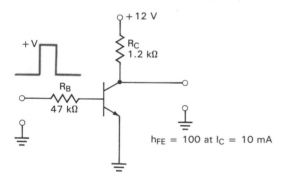

FIGURE 18.4

Determine the minimum high input voltage ($+V$) required to saturate the transistor switch shown in Figure 18.4.

Solution: Assuming that the transistor is an ideal component ($V_{CE} = 0$ V at saturation), the value of $I_{C(sat)}$ is found as

$$I_{C(sat)} = \frac{V_{CC}}{R_C}$$
$$= 10 \text{ mA}$$

As shown, the value of h_{FE} for the transistor is 100 when I_C is 10 mA. Using this value of h_{FE}, the value of base current required to cause saturation is found as

$$I_B = \frac{I_{C(sat)}}{h_{FE}}$$
$$= 100 \text{ } \mu A$$

Now recall that the base current for a base-biased circuit like the one shown is found as

$$I_B = \frac{+V - V_{BE}}{R_B}$$

This equation can be rewritten to give us an equation for $+V$, as follows:

$$+V = I_B R_B + V_{BE}$$

Using the values of $I_B = 100$ μA, $R_B = 47$ kΩ, and $V_{BE} = 0.7$ V, the minimum value of $+V$ required to saturate the transistor is found as

$$+V = I_B R_B + V_{BE}$$
$$= 4.7 \text{ V} + 0.7 \text{ V}$$
$$= 5.4 \text{ V}$$

Since we only need $+V = 5.4$ V to saturate the transistor, a value of $+V = V_{CC}$ will definitely do the trick.

PRACTICE PROBLEM 18–1

A BJT switch like the one in Figure 18.4 has the following values: $V_{CC} = +10$ V, $R_C = 100$ Ω, $R_B = 51$ kΩ, and $h_{FE} = 70$ at $I_C = 100$ mA. Determine the minimum high input voltage required to saturate the transistor.

Practical output values for a BJT switch.

Finally, in our discussion, we assumed that the transistor was an *ideal* component. That is, we assumed that $V_{CE} = V_{CC}$ when the transistor is in cutoff, and $V_{CE} = 0$ V when the transistor is saturated. In practice, the outputs from a transistor switch will fall in the following ranges:

Condition	Output Voltage (V_{CE}) Range
Cutoff	Within 1 V of V_{CC}
Saturation	Between 0.2 and 0.4 V

When a transistor is in cutoff, there is still some leakage current through the component. This leakage current will cause some voltage to be dropped across the collector resistor. Thus the output voltage may be a bit lower than V_{CC}. When the transistor is saturated, you still have some voltage being developed across the internal resistance of the component. Thus the output will never actually hit 0 V. However, the range of output voltages for a saturated transistor is so low that we can easily idealize the situation and say that the output is 0 V.

The JFET as a Switch

3

The differences between BJT and JFET switches.

The basic JFET switch differs from the BJT switch in several aspects:

1. The JFET switch has a much higher input impedance.
2. A *negative* input square wave (or *pulse*) is used to produce a *positive* output pulse.

Both of these points can be seen by looking at the basic JFET switch shown in Figure 18.5. The high input impedance of the JFET switch is caused by the relatively high value of R_G (typically in megohms) and the extremely high input impedance of the JFET gate. For the circuit shown in Figure 18.5, the input impedance of the switch would be equal to the value of R_G.

JFET switch input/output relationships.

To understand the input/output voltage relationship of the circuit, we need to refer to the transconductance curve shown in the insert in Figure 18.5. Recall that I_D is at its maximum value, I_{DSS}, when $V_{GS} = 0$ V. Thus when the input is

SEC. 18.1   Introductory Concepts

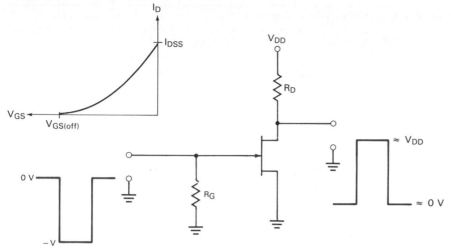

FIGURE 18.5

A JFET switch.

at 0 V, the JFET is saturated and the output is found as

$$V_{out} = V_{DD} - I_{DSS}R_D$$

Assuming that the value of R_D has been selected properly, the output from the JFET switch will be approximately 0 V when the input is at 0 V.

When the input goes to $-V$, the current through the JFET will drop nearly to zero, assuming that $-V$ is greater than or equal to $V_{GS(off)}$. With I_D being nearly zero, little voltage is dropped across R_D, and the output voltage will be close to V_{DD}. By formula,

$$V_{out} = V_{DD}$$

when V_{in} is more negative than $V_{GS(OFF)}$. Using these two relationships for V_{out}, it is easy to see that the input pulse shown in Figure 18.5 will produce the output pulse shown. The following example illustrates the analysis of a basic JFET switch.

EXAMPLE 18.2

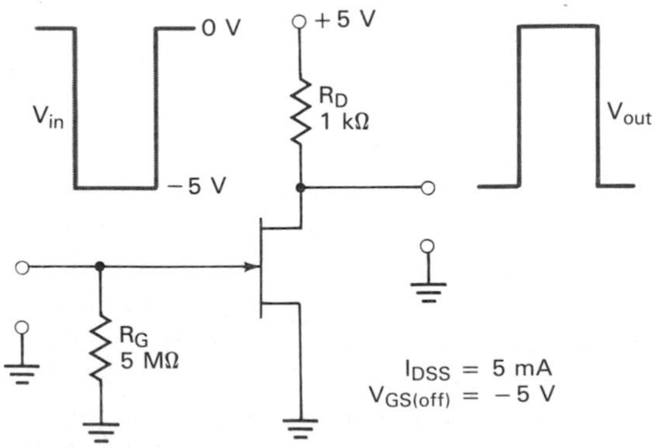

Determine the high and low output voltage values for the circuit shown in Figure 18.6.

Solution: When the input is at 0 V, the JFET has minimum resistance from source to drain, and $I_D = I_{DSS}$. Thus,

$$\begin{aligned} V_{out} &= V_{DD} - I_{DSS}R_D \\ &= 5\ V - 5\ V \\ &= 0\ V \end{aligned}$$

When the input is at -5 V, it is equal to $V_{GS(OFF)}$, and thus I_D is zero. Using this value in place of I_{DSS} in the above equation yields

$$\begin{aligned} V_{out} &= V_{DD} - I_D R_D \\ &= +5\ V - 0\ V \\ &= +5\ V \end{aligned}$$

FIGURE 18.6

R_D 1 kΩ

R_G 5 MΩ

$I_{DSS} = 5$ mA
$V_{GS(off)} = -5$ V

PRACTICE PROBLEM 18–2

A JFET switch like the one in Figure 18.6 has the following values: $V_{DD} = +10$ V, $R_D = 1.8$ kΩ, $I_{DSS} = 5$ mA, and $V_{GS(off)} = -3$ V. The input signal to the circuit is a square wave that has low and high output voltages of -4 and 0 V, respectively. Determine the high and low output voltage values for the circuit.

Practical output values for a JFET switch.

In Example 18.2 we have once again assumed that the active component is *ideal*. In practice, the output values would have been slightly off from the ideal values of 0 V and V_{DD} (+5 V). These discrepancies would be caused by the *drain cutoff current,* $I_{D(off)}$ and the *drain–source on-state voltage,* $V_{DS(on)}$, of the JFET.

The MOSFET as a Switch

The difference between MOSFET and JFET switches.

The MOSFET switch has the input impedance advantage of the JFET switch, while having the input/output polarity relationship of the BJT. The basic MOSFET switch is shown in Figure 18.7.

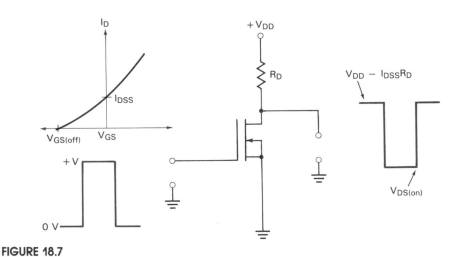

FIGURE 18.7

The MOSFET switch.

Since the depletion-type MOSFET can be operated with its gate–source junction forward biased, we are able to produce a positive output pulse with an input signal that is also positive. While there is still a 180° phase shift from input to output, both the input and output voltages from the MOSFET switch have a positive polarity. In contrast, the JFET switch uses a negative polarity input to produce a positive polarity output. This polarity relationship can be seen by referring back to Figure 18.5.

When the input to the circuit in Figure 18.7 is at 0 V, $I_D = I_{DSS}$, and

$$\boxed{V_{out} = V_{DD} = I_{DSS}R_D}$$

For the MOSFET switch shown, lower values of R_D will produce an output that is closer to the value of V_{DD}. This point is illustrated in the following example.

EXAMPLE 18.3

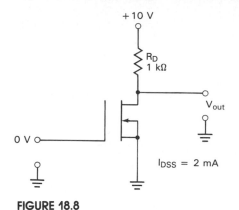

FIGURE 18.8

Determine the output voltage for the circuit shown in Figure 18.8 when the input is at 0 V. Repeat the procedure for the same circuit using $R_D = 100 \ \Omega$.

Solution: When the input is at 0 V, the current through the MOSFET is equal to I_{DSS}. Therefore, the output voltage is found as

$$V_{out} = V_{DD} - I_{DSS}R_D$$
$$= 10 \ V - 2 \ V$$
$$= 8 \ V$$

If the resistor in the circuit is replaced with a 100-Ω resistor, the output changes to

$$V_{out} = V_{DD} - I_{DSS}R_D$$
$$= 10 \ V - 200 \ mV$$
$$= 9.8 \ V$$

Thus, with a smaller value of R_D, the high output voltage is much closer to the ideal value, V_{DD}.

PRACTICE PROBLEM 18–3

A circuit like the one in Figure 18.8 has the following values: $V_{DD} = +8 \ V$, $R_D = 2 \ k\Omega$, and $I_{DSS} = 500 \ \mu A$. Determine the value of V_{out} for the circuit when $V_{in} = 0 \ V$.

A smaller R_D may help to produce a high output that is closer to V_{DD}, but it will also cause the low output voltage to be farther from ground. With a smaller value of R_D, the value of $I_D R_D$ will be lower for a given value of I_D. This means that when the MOSFET is conducting to its maximum capability R_D may not drop enough voltage to produce an output that is near 0 V. Thus we have a possible trade-off. If having an output close to V_{DD} is important, a low value of R_D *may* be required. This would depend on the current capability of the particular MOSFET.

The CMOS switch discussed in Chapter 11 eliminates the potential trade-off problem that can be caused by R_D. As a review, the CMOS switch is shown in Figure 18.9. Recall from our previous discussion that the two MOSFETs shown

The effect of R_D on the output voltage of a MOSFET switch.

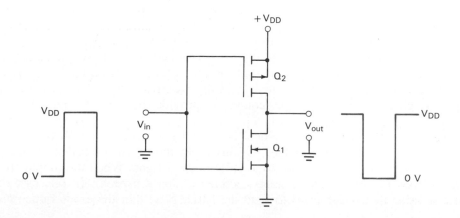

FIGURE 18.9

The CMOS switch.

in the switch are always in opposite operating states. Thus, when Q_1 is *on*, Q_2 is *off*, and vice versa. The following table is based on the relationship between the operating states of Q_1 and Q_2.

Q_1 State	Q_2 State	Q_2 Resistance
On	Off	High
Off	On	Low

When Q_1 is off, the resistance of Q_2 is very low. Thus the voltage dropped across Q_2 will be very low, and the output voltage will be very close to V_{DD}. When Q_1 is on, the resistance of Q_2 is very high. Therefore, the voltage drop across Q_2 will be very high, and the output voltage will be very close to 0 V. In this way, the CMOS switch eliminates the potential problem with the basic MOSFET switch. This is one of the reasons that CMOS switches are almost always used in place of other MOS switches.

The Op-Amp as A Switch

You should have no problem at all with this concept. An op-amp switch is nothing more than a *comparator*. You may recall that a comparator is an open-loop op-amp circuit that is used as a voltage-level detector. The basic comparator (op-amp switch) is shown in Figure 18.10. For the circuit shown, the high input voltage can be anything above a few millivolts. Since the gain of the comparator will be equal to the open-loop gain of the op-amp (A_{OL}), very little change is needed at the input to cause a change at the output. As usual, the op-amp circuit is by far the simplest to work with and understand.

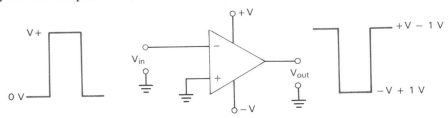

FIGURE 18.10

An op amp switch.

Basic Switching Applications

Driver. A circuit that is used to couple a low-current output to a relatively high current device.

A typical application for the simple BJT switch can be seen in Figure 18.11. Here we see the BJT switch being used as an LED *driver*. A driver is a circuit that is used to couple a low-current output to a relatively high current device. In Figure 18.11, the block labeled *circuit A* is assumed to be a switching circuit with a low-current output. Assuming that the output current from circuit A would be insufficient to drive the LED (which typically requires at least 1 mA of forward current to light), a driver would be required to provide the current needed to light the LED.

How does the circuit work? When the output from circuit A is at 0 V, Q_1 is in cutoff (acting as an open switch). In this case, the transistor collector current is approximately equal to zero, and the LED doesn't light. When the output from circuit A goes to +5 V, Q_1 saturates (acts as a closed switch). In this case, I_C rises to a value that is sufficient to light the LED. Note that the exact value of I_C

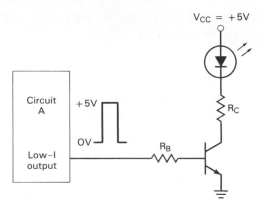

FIGURE 18.11

depends on the forward voltage drop of the LED and the value of R_C (which is acting as a series current-limiting resistor), as given in equation (2.11). In this case, the source voltage (V_S) is the +5-V value of V_{CC} for the transistor, and R_S is the collector resistor.

Equation (2.11):

$$I_F = \frac{V_S - V_F}{R_S}$$

Even though the driver shown in Figure 18.11 uses a BJT, drivers are commonly made with each of the active devices covered in this section. The device used in any driver application is determined by the system engineer, and no single active device seems to dominate this application.

Summary

BJT, JFET, MOSFET, and op-amp switches are commonly used as *drivers* in switching applications. A driver is a circuit used to couple a low-current circuit output to a relatively high current device, such as an LED.

The basic switching circuits are used to provide high and low dc outputs when driven by an input signal. A high output is one that is within 1 V of the circuit supply voltage (typically). A low voltage is one that is within 1 V of ground (or V_{EE} in the case of the op-amp).

SECTION REVIEW

1. What is a switching circuit? [Objective 1]
2. What is the primary difference between linear circuits and switching circuits? [Objective 1]
3. Describe the operation of the BJT switch. [Objective 2]
4. Describe the operation of the JFET switch. [Objective 3]
5. Describe the operation of the MOSFET switch. [Objective 4]
6. Describe the operation of the op-amp switch. [Objective 5]
7. What is a *driver*? [Objective 6]

18.2
Basic Switching Circuits: Practical Considerations

In this section, we're going to take a look at some of the factors that affect the operation and analysis of basic switching circuits. Most of these factors explain the differences between *ideal* and *practical* time and frequency measurements.

We will also look at two switching circuit classifications and several devices that are specially designed for switching applications.

Practical Measurements

✔ 7

Before we can discuss any practical time or frequency measurements, we must define some of the terms that are commonly used to describe square waves. We will define these terms with the help of Figure 18.12.

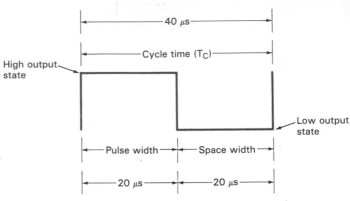

FIGURE 18.12

Rectangular waveform. A waveform made up of alternating (high and low) dc voltages.

Pulse width (PW). The time spent in the high dc voltage state.

Space width (SW). The time spent in the low dc voltage stage.

Cycle time (T_C). The sum of pulse width and space width.

A waveform that is made up of alternating (high and low) dc voltages is generally classified as a *rectangular waveform*. The time a rectangular waveform spends in the high dc voltage state is generally called the *pulse width* of the waveform. The time spent in the low dc voltage state is generally called the *space width* of the waveform. The sum of pulse width and space width gives you the *cycle time* (T_C) of the waveform. For example, if a rectangular waveform has a 20-μs pulse width and a 20-μs space width, the total cycle time is found as

$$T_C = PW + SW$$
$$= 20 \ \mu s + 20 \ \mu s$$
$$= 40 \ \mu s$$

where PW = pulse width
SW = space width

Square wave. A special-case rectangular waveform that has equal PW and SW values.

The *square wave* is a special-case rectangular waveform that has equal pulse width and space width values. Since PW = SW in Figure 18.12, the waveform shown would be called a square wave. When PW ≠ SW, the waveform is simply referred to as a rectangular waveform.

✔ 8

When discussing the *ideal* waveform, it is easy to determine the value of pulse width, space width, or cycle time. For example, take a look at the ideal waveform in Figure 18.13. The scale that is shown between the ideal and practical waveforms is used as a time reference. Assume (for the sake of discussion) that the scale is measuring time in μs (microseconds). For the ideal waveform, it is easy to see that the value of PW is

$$PW = 10 \ \mu s - 4 \ \mu s = 6 \ \mu s$$

Similarly, the value of SW can be found as

$$SW = 14 \ \mu s - 10 \ \mu s = 4 \ \mu s$$

CHAP. 18   Solid-state Switching Circuits

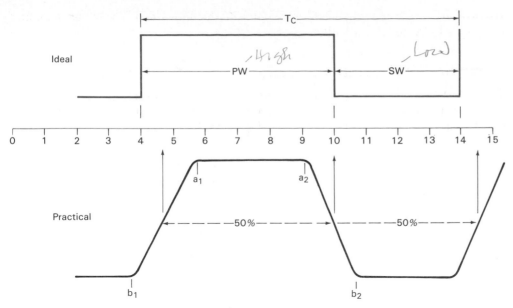

FIGURE 18.13

Measuring pulse width, space width, and cycle time.

and the cycle time can be found as

$$T_C = 14 \ \mu s - 4 \ \mu s = 10 \ \mu s$$

When we wish to make the time measurements of a practical rectangular waveform, we run into a bit of a problem. You see, the transitions from low to high and from high to low are not perfectly vertical.* If we measure the PW from a_1 to a_2, the value of PW is found to be approximately 3.5 μs. However, if we measure the PW from b_1 to b_2, the value of PW is found to be approximately 7 μs. The question here is: Where do we measure the values of PW, SW, and T_C so that everyone gets the same results?

To eliminate the problem of measurement variations, the values of PW, SW, and T_C are *always* measured at the 50% points on the waveform. That is, they are measured at the points where the voltage is halfway between the high and low values. For example, if the practical waveform in Figure 18.13 had peak values of 0 V (low) and +5 V (high), we would measure the values of PW, SW, and T_C at the +2.5-V points on the waveform. This point is illustrated further in the following example.

* The reasons for this are discussed later in this section.

A Practical Consideration. Of course, the value of T_C could have been found as $T_C = $ PW + SW.

Time measurements are always made at the 50% points on the waveform.

━━━━━━━━━━━━━━ **EXAMPLE 18.4** ━━━━━━━━━━━━━━

Figure 18.14a represents the output from an inverter as seen on an oscilloscope. If the oscilloscope is set to a horizontal calibration of 50 μs/div, what are the values of PW and T_C for the waveform?

Solution: The waveform has peak values of ±10 V. Therefore, we will measure PW and T_C and the 0-V points, since these are halfway between the peak values.

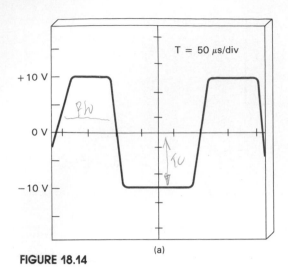

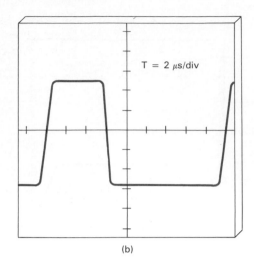

FIGURE 18.14

A Practical Consideration. In most circuit analyses, we are only interested in measuring the values of PW and T_C. If the value of SW is needed, it is then determined by subtracting PW from T_C.

Assuming that the PW is measured during the positive transition, we can see that there are approximately 2.3 major divisions between the two 0-V points. Since each major division represents 50 μs, the PW is found as

$$PW = 2.3 \times 50 \text{ μs}$$
$$= 115 \text{ μs}$$

Now, looking at the illustration, we can see that there are approximately 5.3 divisions from the 0-V point on one rising edge to the 0-V point on the next. Since these points correspond to one complete cycle, the cycle time is found as

$$T_C = 5.3 \times 50 \text{ μs}$$
$$= 265 \text{ μs}$$

PRACTICE PROBLEM 18–4

Determine the values of PW and T_C for the waveform in Figure 18.14b.

Duty cycle (dc). The ratio of pulse width (PW) to cycle time (T_c), measured as a percent.

Another time-relative measurement that is commonly made with switching circuits is *duty cycle*. Duty cycle is the ratio of pulse width to total cycle time, measured as a percent. By formula,

$$\boxed{\text{duty cycle} = \frac{\text{pulse width}}{\text{cycle time}} \times 100} \qquad (18.1)$$

The following example illustrates the use of this equation in determining the duty cycle of a given waveform.

=== **EXAMPLE 18.5** ===

Determine the duty cycle for the waveform in Figure 18.14a.

Solution: In Example 18.4 we determined the waveform to have values of PW = 115 μs and T_C = 265 μs. Using these values and equation (18.1), the value of duty cycle is found as

$$dc = \frac{PW}{T_C} \times 100$$
$$= \frac{115\ \mu s}{265\ \mu s} \times 100\%$$
$$= 43.4\%$$

PRACTICE PROBLEM 18–5

Determine the duty cycle of the waveform in Figure 18.4b.

The duty cycle of a waveform indicates the percentage of each cycle that is taken up in the pulse width. If a given waveform has a duty cycle of 35%, the pulse width takes up 35% of each cycle. An ideal square wave would always have a duty cycle of 50%, since, by definition, the pulsewidth is one-half of the cycle time in a square wave.

The duty cycle of a waveform can become a critical issue when you are dealing with certain IC switching circuits. Many digital (switching) ICs are rated at a maximum duty cycle. If a waveform is applied to such an IC, and the duty cycle of that waveform exceeds the maximum allowable for the IC, the result will be a circuit that will not work.

Now that you are familiar with the basic switching circuit measurements, we will examine the measurement characteristics of the BJT, FET, and op-amp individually.

BJT Switching Time

In an *ideal* BJT switch, the output would change at the *exact* instant that the input changes. The output would also have transitions that would be perfectly vertical. These characteristics of the ideal switch are illustrated in Figure 18.15.

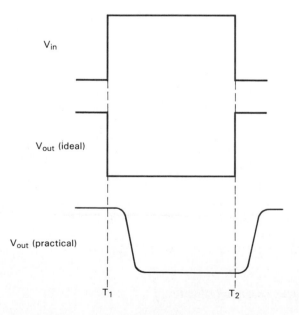

FIGURE 18.15

Propagation delay.

In this figure, we see the input to a BJT switch along with the ideal and practical output waveforms. As you can see, the ideal output is shown to be changing states *instantly* at T_1 and T_2, that is, at the same instant that the input is changing. Also, the output transitions at T_1 and T_2 are shown to be perfectly vertical, implying that the transition from one output state to the other would be instantaneous. In practice, the BJT would have a response that more closely resembles the *practical* output waveform. This waveform differs from the ideal in that:

1. There is a delay between each input transition and the time when the output transition starts.
2. The output transitions are not vertical, implying that the transitions require some measurable amount of time to occur.

The overall delay between the time that the input changes and the output changes (as measured at the 50% points on the two waveforms) is referred to as *propagation delay*.

There are actually *four* different sources of propagation delay for a BJT. Each of these sources is measured during a specific period, as shown in Figure 18.17. Here we see the input signal for a BJT, along with the output current (I_C)

Propagation delay. The time delay between input and output changes, as measured at the 50% point on the two waveforms. *A Practical Consideration.* FETs and op-amps also experience propagation delay. The sources of FET and op-amp propagation delay are discussed later in this section.

FIGURE 18.16

The causes of propagation delay.

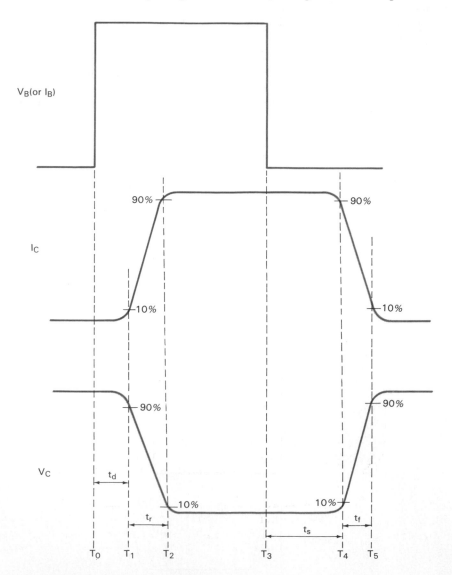

CHAP. 18   Solid-state Switching Circuits

and voltage (V_C) waveforms. Note that the input waveform could be used to represent the change in either the input voltage (V_B) or input current (I_B) since V_B and I_B are always in phase. Also, V_C is shown to be out of phase with I_C, as is always the case.

The first contributor to propagation delay is called *delay time* (t_d). This is the time required for the transistor to come out of cutoff. Delay time is measured between times T_0 and T_1 (Figure 18.16). In terms of I_C, delay time is *the time required for I_C to reach 10% of its maximum value*.

When you are viewing the output from a BJT switch on an oscilloscope, you are not actually looking at I_C. Thus, we need to define t_d in terms that are observed in practical situations. Since the oscilloscope is used to view voltages; we need to define t_d in terms of collector voltage. In terms of V_C, delay time is *the time required for V_C to drop to 90% of its maximum value*, in most cases, 90% of V_{CC}.

Rise time (t_r) is the time required for the BJT to make the transition from cutoff to saturation. It is measured between times T_1 and T_2 (Figure 18.17). In terms of I_C, rise time is *the time required for I_C to rise from 10% up to 90% of its maximum value*. In terms of V_C, it is *the time required for V_C to drop from 90% down to 10% of its maximum value*.

Delay time (t_d). The time required for a BJT to come out of cutoff. In terms of I_C, it is the time required for I_C to reach 10% of its maximum value.

In terms of oscilloscope measurements, t_d is the time required for V_C to drop to 90% of its maximum value.

Rise time (t_r). The time required for the BJT to go from cutoff to saturation. In terms of I_C, the time required for the 10% to 90% transition. In terms of V_C, the time required for the 90% to 10% transition.

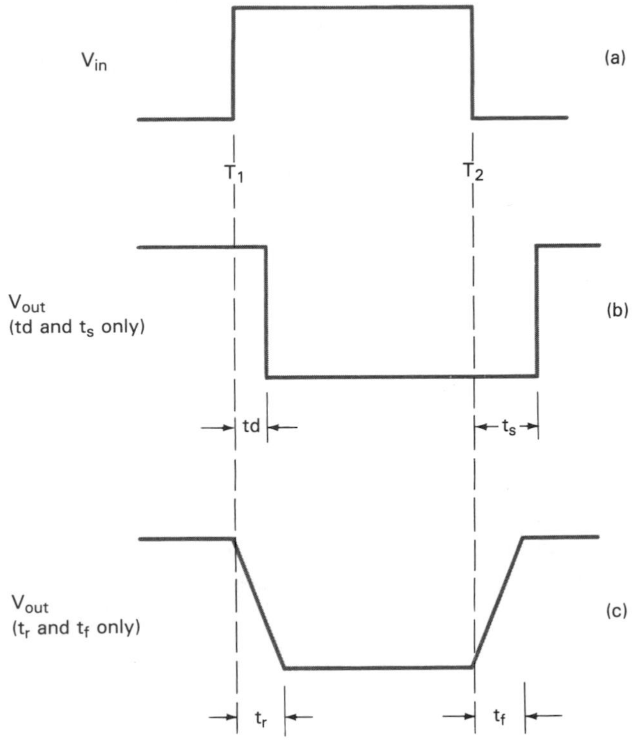

FIGURE 18.17

Storage time (t_s) is the time required for a BJT to come out of saturation. In terms of I_C, it is *the time required for I_C to drop from 100% down to 90% of its maximum value*, as can be seen between T_3 and T_4. In terms of V_C, it is the time required for V_C to rise to 10% of its maximum value. Note that storage time (like delay time) is shown to begin at the point when the input signal changes.

Finally, *fall time* (t_f) is the time required for the BJT to make the transition from saturation to cutoff, as shown between T_4 and T_5. In terms of I_C, it is the time required for I_C to drop from 90% to 10% of its maximum value. In terms of

Storage time (t_s). The time required for a BJT to come out of saturation In terms of I_C, the time required for I_C to drop to 90% of its maximum value. In terms of V_C, the time required for V_C to rise to 10% of its maximum value.

V_C, it is the time required for V_C to increase from 10% to 90% of its maximum value.

Fall time (t_f). The time required for the BJT to make the transition from saturation to cutoff. In terms of I_C, the time required for I_C to drop from 90% to 10% of its maximum value. In terms of V_C, the time required to increase from 10% to 90% of its maximum value.

There is an important point that needs to be made at this time. Delay time and storage time account for the delay between the input transition and *the start* of the output transition. Rise time and fall time account for the *slope* of the output transitions. These points are illustrated in Figure 18.17. If a BJT had only delay time and storage time, the output would resemble the waveform shown in Figure 18.17b. If a BJT had only rise time and fall time, the output would resemble the waveform shown in Figure 18.17c. As you will see later in this section, we can use external components to greatly reduce delay time and storage time. However, little can be done to change rise time and fall time. Thus, the practical output waveform for a BJT switch will most closely resemble the one in Figure 18.17c.

The values of t_d, t_r, t_s, and t_f are all provided on the specification sheet of a given BJT. For example, the 2N3904 spec sheet (page 241) lists the following maximum values under *switching characteristics*:

$$t_d = 35 \text{ ns}, \qquad t_r = 35 \text{ ns}, \quad t_s = 200 \text{ ns}, \qquad t_f = 50 \text{ ns}$$

If we add the values of delay time and rise time, we get the maximum time required for the 2N3904 to make the transition from a high output state to a low output state, 70 ns. If we add the values of storage time and fall time, we get the maximum time required for the 2N3904 to make the transition from a high output state to a low output state, 250 ns. As you can see, it takes more time for the device to go from high to low than it takes to go from low to high. The reasons for this is explained later in this section.

If we add all four of the listed switching times together, we get a maximum switching time of 320 ns for the 2N3904. This switching time can be used to determine the maximum switching frequency for the device, as follows:

$$f_{max} = \frac{1}{t_d + t_r + t_s + t_f}$$
$$= \frac{1}{320 \text{ ns}}$$
$$= 3.125 \text{ MHz}$$

☞ 11

An important point needs to be made: While a 2N3904 switching circuit is capable of operating at 3.125 MHz, the output from such a circuit would be an extremely distorted square wave. The reason for this lies in the relationship between the upper cutoff frequency (f_2) of the BJT and the frequency of its square-wave input.

In Appendix D, the following equation is derived for calculating the value of f_2 for a switching circuit:

$$f_2 = \frac{0.35}{t_r} \qquad \textbf{(18.2)}$$

where t_r = the *rise time* of the active device. Using this equation, the value of f_2 for a 2N3904 switching circuit would be found as

$$f_2 = \frac{0.35}{35 \text{ ns}} \qquad (t_r = 35 \text{ ns for the 2N3904})$$
$$= 10 \text{ MHz}$$

Thus, the 2N3904 has an upper cutoff frequency of 10 MHz.

To pass a square wave with minimum distortion, *the upper cutoff frequency of a switching circuit must be at least 10 times the frequency of the input square wave*. Thus, the practical limit on the frequency of a switching circuit input signal will be one-tenth the value of f_2 for the circuit. By formula,

$$f_{max} = \frac{0.35}{10t_r} \quad \text{(practical limit)} \tag{18.3}$$

In the case of the 2N3904 switch, this means that the practical limit on f_{in} for a circuit is 1 MHz.

Figure 18.18 shows what happens if the practical limit on the f_{in} for a circuit is exceeded. As you can see, the leading and trailing edges of the circuit output become more and more rounded as the frequency of the input waveform increases. While there is *some* rounding of the square wave corners when $f_2 = 10f_{in}$, this small amount of rounding is considered to be acceptable. The following example demonstrates the complete frequency analysis of a BJT switching circuit.

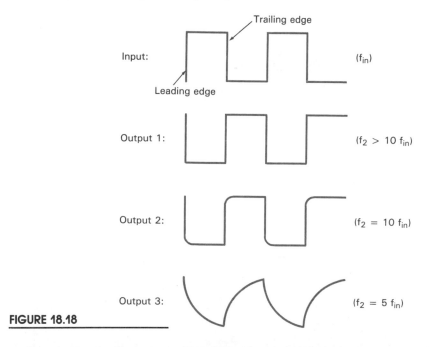

FIGURE 18.18

================ **EXAMPLE 18.6** ================

The Motorola MMBT2907 transistor has the following maximum values listed on its spec sheet: $t_d = 10$ ns, $t_r = 40$ ns, $t_s = 80$ ns, $t_f = 30$ ns. Determine the theoretical and practical limits on the operating frequency of the device.

Solution: The theoretical limit on f_{in} for the device is found using the same time-to-frequency conversion that we used earlier in this section, as follows:

$$f_{max} = \frac{1}{t_d + t_r + t_s + t_f} \quad \text{(maximum, theoretical)}$$
$$= \frac{1}{160 \text{ ns}}$$
$$= 6.25 \text{ MHz}$$

The value of f_2 for the device is now found as

$$f_2 = \frac{0.35}{t_r}$$
$$= \frac{0.35}{40 \text{ ns}}$$
$$= 8.75 \text{ MHz}$$

The practical limit on f_{in} is one-tenth the value of f_2 for the device, 875 kHz. If f_{in} exceeds 875 kHz, the output from the device will be distorted.

PRACTICE PROBLEM 18–6

A transistor has the following maximum values listed on its spec sheet: $t_d = 20$ ns, $t_r = 25$ ns, $t_s = 120$ ns, and $t_f = 20$ ns. Determine the theoretical and practical limits on f_{in} for the device.

Improving BJT Switching Time

✔ 12

Before you can understand the methods used to improve BJT switching time, we need to take a moment to discuss the causes of switching time. The various times are easily explained with the help of an illustration that appeared earlier in the text. Figure 5.7 is repeated as Figure 18.19.

FIGURE 18.19

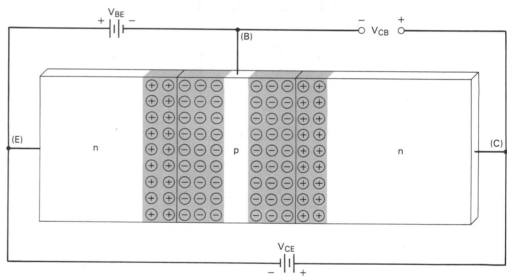

The cause of delay time.

When the BJT is in cutoff, the depletion layer is at its maximum width, and I_C is essentially at zero. When the input to the BJT goes positive, the depletion layer will start to "dissolve," allowing I_C to begin to flow. Delay time is the time required for the depletion layer to dissolve to the point where 10% of the maximum value of I_C is allowed to flow through the component. How long is this period of time? That depends on three things:

1. The physical characteristics of the particular BJT
2. The amount of reverse bias initially applied to the component
3. The amount of I_B that the input causes when it goes positive

We cannot do much about the physical characteristics of the BJT, but we can do several things to improve t_d. By keeping the initial reverse bias at a minimum, we can keep the depletion layer at a minimum width. The narrower the depletion layer, the less time it will take for it to dissolve. Also, you should recall that the BJT is a *current-controlled* device. That is, the width of the depletion layer is determined by the amount of I_B. Therefore, by providing a very high initial value of I_B, delay time is further reduced. Later in this section we discuss the method by which I_B is made initially to be very large, thus reducing t_d.

How to reduce delay time.

After delay time has passed, the depletion layer will continue to dissolve, causing I_C to continue to increase. Rise time is the time required for the depletion layer to dissolve to the point where I_C reaches 90% of its maximum value. Rise time is strictly a function of the physical characteristics of the BJT, and nothing can be done to improve it.

The biggest overall delay is *storage time*. Referring back to the specifications of the 2N3904, you can see that storage time is at least four times as long as any of the other switching times. Therefore, reducing t_s will have a major impact on the overall switching time of a given BJT.

When a transistor is in saturation, the base region is flooded with electrons. When the input voltage goes low, it takes a great deal of time for all these electrons to leave the base region and for a new depletion layer to begin to form. The time required for this to happen is storage time.

The cause of storage time.

The actual value of storage time depends on three things:

1. The physical characteristics of the BJT
2. The value of I_C
3. The initial value of reverse bias applied to the base

Again, we cannot do anything about the physical characteristics of the BJT, but we *can* do something about the other two variables. Storage time can be greatly reduced by keeping I_C low enough so that the BJT never really saturates. In other words, if we keep the BJT just *below* saturation, the number of electrons in the base region of the BJT is greatly reduced, and storage time is also reduced. Also, if we apply a *very large* initial value of reverse bias, the electrons will be forced out of the base region at a much faster rate, and again storage time is reduced.

How to reduce storage time.

Like rise time, fall time is a function of the physical characteristics of the BJT and cannot be reduced by any practical means. Fall time is the time required for the depletion layer to grow to the point where I_C drops to 10% of its maximum value.

Now, as a summary, let's review the steps we can take to reduce the overall switching time of a given BJT:

A Practical Consideration. We can reduce propagation delay as described in this section. However, since the practical frequency limit is determined by rise time, improving propagation delay will not improve the overall frequency response of the switch.

1. By applying a high *initial* value of I_B, delay time is reduced.
2. By using the *minimum* value of reverse bias required to hold the BJT in cutoff, delay time is further reduced.
3. By limiting I_B to a value lower than that required for the BJT to saturate, storage time is reduced.
4. By applying an *initial* reverse bias that is very large, storage time is further reduced.

Now, look at statements 1 and 3. These statements would indicate that we want I_B to *initially* be very high (to reduce delay time) and then settle down to some level below that required for saturation (to reduce storage time). Statements

2 and 4 can be combined in a similar fashion. We want a very high *initial* value of reverse bias (to reduce storage time) and then a minimum reverse bias (to reduce delay time).

Speed-up capacitor. A capacitor that is used to reduce propagation delay by reducing t_d and t_s.

The desired I_B and reverse-bias characteristics described above are both achieved by using what is called a *speed-up* capacitor. A basic BJT switch with an added speed-up capacitor is shown in Figure 18.20. The speed-up capacitor will perform all the required functions when its value is properly selected. The means by which it accomplishes its function are easy to understand if you take a look at the waveforms shown in Figure 18.21.

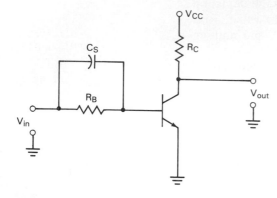

FIGURE 18.20

A speed-up capacitor (C_s) improves switching time.

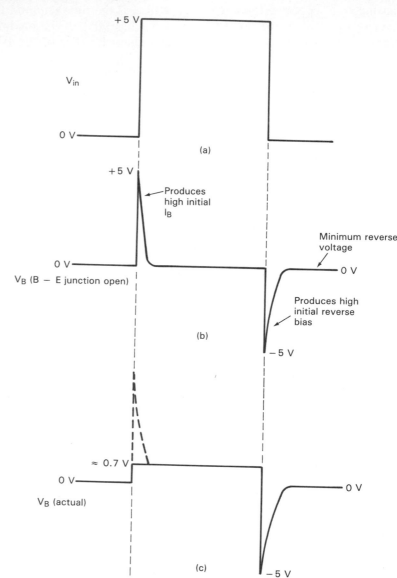

FIGURE 18.21

Waveforms caused by a speed-up capacitor.

In Figure 18.20, R_B and C_S form an *RC differentiator*. If the base–emitter junction of the transistor was open, the base waveform produced by the differentiator would look like the waveform shown in Figure 18.21b. As indicated, the positive spike produces a very high initial value of I_B. Then, as the spike returns to the 0-V level, I_B decreases to some value less than that required for saturation. Thus we have reduced the delay time while making sure that the value of I_B becomes low enough to prevent saturation.

When the input signal returns to 0 V, the output of the differentiator is driven to -5 V. This is the high initial value of reverse bias that is needed to reduce storage time. Since the transistor was prevented from saturating by I_B, we have now done all that can be done to reduce storage time. Also, since V_B returns to 0 V at the end of the negative spike, the final value of reverse bias is at a minimum. This, coupled with the high initial I_B spike, ensures that delay time is held to a minimum.

The value of C_S is normally chosen so that the time constant of the RC circuit is very short at the maximum operating frequency. This requirement is fulfilled by using the following equation to select C_S:

$$C_S < \frac{1}{20 R_B f_{max}}$$ **(18.4)**

The value of f_{max} used in equation (18.4) is the practical f_{max} of the circuit; found as $f_2/10$.

Equation (18.4) is based on the *short time constant* relationship

$$RC < \frac{\text{pulse width}}{10}$$

For a square wave, the pulse width is one-half of the cycle time. Thus

$$PW = \frac{1}{2 f_{max}}$$

can be used in place of the pulse-width value in the RC relationship to produce

$$RC < \frac{1}{20 f_{max}}$$

Finally, both sides of the equation are divided by R to produce equation (18.4). As long as the speed-up capacitor fulfills the requirement of equation (18.4), the switching times of the BJT will be greatly reduced.

In Figure 18.21b, we assumed the transistor had an open emitter–base junction to simplify things. The actual waveform that you would see under working conditions would be similar to the one shown in Figure 18.21c. Here we see the effects of the BJT on the base waveform. On the positive transition, the base–emitter junction will turn on and clip off a majority of the positive spike produced by the RC circuit. Also, since the transistor stays on during the entire positive half-cycle of the input, the output from the RC circuit will not actually return to 0 V. Rather, it will stay at the value of V_{BE} for the transistor, approximately 0.7 V.

JFET Switching Time

JFET spec sheets usually list values of t_d, t_r, t_s, and t_f. The causes of these times in the JFET are similar to those of the BJT. When the value of t_r is listed on the JFET spec sheet, the practical limit on f_{in} can be determined in the exact same fashion as it is for the BJT.

Some JFET spec sheets list only *turn-on time* and *turn-off time*. Turn-on time is the sum of delay time and rise time. Turn-off time is the sum of storage time and fall time. When these are the only values listed in the JFET spec sheet, it is difficult to determine the exact practical limit on f_{in} for the device. This is because you cannot determine the portion of turn-on time that is actually taken up by the rise time of the device. In other words, there is no conversion formula for calculating rise time when turn-on time is known.

Turn-on time (t_{on}). The sum of t_d and t_r.

Turn-off time (t_{off}). The sum of t_s and t_f.

When only t_{on} is given, use this value in place of t_r in equation (18.3) to determine the practical operating frequency limit. While the actual limit on f_{in} will be higher than your calculated value, you will still have a good approximation of this limit.

Since JFETs are voltage-controlled devices, speed-up capacitors will only improve the turn-off time of the component. By initially supplying a gate–source voltage that is greater than $V_{GS(off)}$, the speed-up capacitor will decrease the time required to turn off the JFET. However, the capacitor will do little to improve the turn-on time of the component.

Switching Devices

☑ 14

Switching transistors. Devices with extremely low switching times.

Types of BJTs and JFETs have been developed especially for switching applications. These devices, called *switching transistors*, are designed to have extremely short values of t_d, t_r, t_s, and t_f. For example, the 2N2369 has the following maximum values listed on its data sheet:

$$t_d = 5 \text{ ns}, \qquad t_r = 18 \text{ ns}, \qquad t_s = 13 \text{ ns}, \qquad t_f = 15 \text{ ns}$$

With the extremely small delay and storage time values, a 2N2369 switching circuit probably wouldn't need a speed-up capacitor. Also, with the given rise time of 18 ns, the practical limit on f_{in} for the device would be found as

$$f_{max} = \frac{0.35}{10 t_r}$$
$$= \frac{0.35}{180 \text{ ns}}$$
$$= 1.94 \text{ MHz}$$

This is almost twice the acceptable f_{in} limit of the 2N3904. Thus, the 2N2369 can be used at much higher switching frequencies than the 2N3904.

Op-amp Switching Time

☑ 15

The switching time of an op-amp is limited by its *slew rate*. This point was discussed in Chapter 12. Recall that the slew rate of an op-amp is the maximum rate at which it can change its output level. For example, the μA741 op-amp has a slew rate that is 0.5 V/μs.

Using the slew rate of an op-amp, its maximum switching frequency can be determined as

$$f_{max} = \frac{\text{slew rate}}{\pi \, V_{PP}} \tag{18.5}$$

where V_{PP} = the difference between the high and low output voltages. Remember, when you use the equation to find f_{max} for a given op-amp, the slew rate must be converted into a frequency value. This is illustrated in the following example.

═══ **EXAMPLE 18.7** ═══

A given op-amp has a slew rate of 0.6 V/μs. The supply voltages for the device are ± 5 V_{dc}. What is the practical frequency limit for the device?

Solution: The output voltage swing will be limited to ± 4 V, giving a peak-to-peak output of 8 V_{PP}. Thus,

$$f_{max} = \frac{\text{slew rate}}{V_{PP}}$$

$$= \frac{0.6 \text{ V/}\mu s}{8 \text{ V}}$$

$$= \frac{600 \text{ kHz}}{8} \quad \Bigg| \quad 600 \text{ kHz} = \frac{0.6 \text{ V}}{1 \text{ }\mu s}$$

$$= 75 \text{ kHz}$$

PRACTICE PROBLEM 18–7

An op-amp has values of slew rate $= 0.4$ V/μs, $+V = 10$ V, and $-V = -10$ V. Determine the practical frequency limit for the device.

In any practical application, there are only two things you can do to improve the switching time of an op-amp switch:

1. Reduce the output voltage swing. This, according to equation (18.5), will cause the value of f_{max} to increase.
2. Buy a faster op-amp.

Switching Circuit Classifications

✔ 16

Our discussion on basic switching circuits has been limited to a specific type of switching circuit called an *inverter*. An inverter has a 180° voltage phase shift from input to output. Looking back at the various illustrations of BJT, JFET, and op-amp switching circuits, you will see that every circuit had a high output voltage when the input voltage was low, and vice versa.

Inverter. A basic switching circuit that produces a 180° voltage phase shift.

Another basic switching circuit is the *buffer*. A buffer is a switching circuit that does not introduce a 180° voltage phase shift. The BJT, JFET, and op-amp buffers are shown in Figure 18.22. As you can see, these circuits are simply the emitter follower, source follower, and voltage follower. When a square wave is applied to any of these circuits, the output will be a square wave that is in phase

Buffer. A switching circuit that does not produce a voltage phase shift.

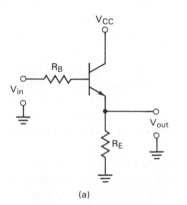

(a)

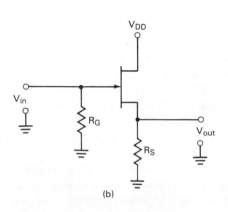

(b)

FIGURE 18.22

Some basic buffer circuits.

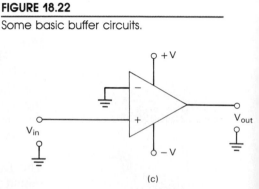

(c)

with the input signal. Because of this, we must redefine the V_C descriptions of t_d, t_r, t_s, and t_f. For the *buffer*:

Buffer time measurements.

1. t_d is the time required for V_C to reach 10% of its maximum value.
2. t_r is the time required for V_C to rise from 10% to 90% of its maximum value.
3. t_s is the time required for V_C to drop to 90% of its maximum value.
4. t_f is the time required for V_C to drop from 90% to 10% of its maximum value.

The changes in the V_C definitions for the buffer are necessary because it does not produce a 180° voltage phase shift, meaning that V_C and I_C will be in phase. Outside of this difference, the operation of the buffer and inverter circuits are nearly identical.

Summary

A waveform that is made up of alternating (high and low) dc voltages is classified as a *rectangular* waveform. The square wave is a special-wave rectangular waveform that has equal *pulse-width* (high) and *space-width* (low) values.

The ideal rectangular waveform would have instantaneous transitions from high to low and from low to high. However, these transitions are not perfectly vertical in practice. To ensure that pulse-width, space-width, and cycle time measurements are consistent, these values are always measured at the 50% points on rectangular waveforms, that is, at the points where the amplitude of the waveform is halfway between the high and low output voltage levels.

The duty cycle of a rectangular waveform indicates the percentage of the waveform that is taken up by the pulse width. Since pulse width and space width are always equal for a square wave, the duty cycle of a square wave will always be 50%.

The time between the input and output changes of a switching circuit is referred to as propagation delay. For discrete devices, there are four switching times that contribute to propagation delay. These times are as follows:

- Delay time (t_d): The time required for the device to come out of cutoff.
- Rise time (t_r): The time required for the device to make the transition from cutoff to saturation.
- Storage time (t_s): The time required for the device to come out of saturation.
- Fall time (t_f): The time required for the device to make the transition from saturation to cutoff.

The times listed are measured using an oscilloscope. Figure 18.23 summarizes these times and shows where they are measured on the output voltage waveforms for the inverter and the buffer. As you can see, the inverter has a 180° voltage phase shift from input to output, while the buffer does not.

The propagation delay of a discrete switch can be reduced by the use of a *speed-up capacitor*. The speed-up capacitor will reduce delay and storage time in the BJT and storage time (only) in a FET switch. The rise time of a device determines the maximum practical operating frequency of a switching circuit. Since a speed-up capacitor does not affect rise time, f_{max} is not increased when a speed-up capacitor is added.

A speed-up capacitor won't affect the propagation delay of an op-amp switch. The frequency characteristics of the op-amp switch depend on the peak-to-peak output transition and the slew rate of the op-amp. You can only improve the

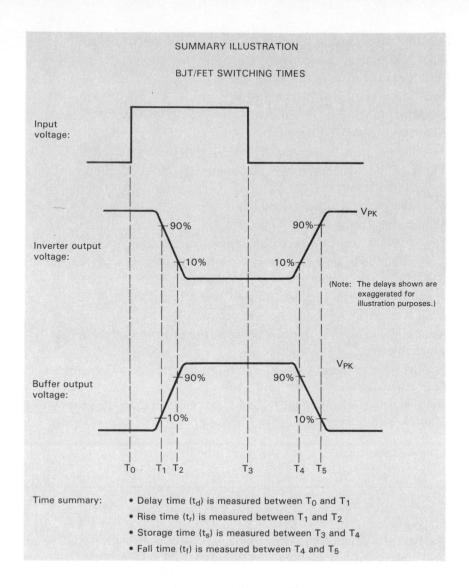

SUMMARY ILLUSTRATION

BJT/FET SWITCHING TIMES

Input voltage:

Inverter output voltage:

V_{PK}

90%

10%

90%

10%

(Note: The delays shown are exaggerated for illustration purposes.)

Buffer output voltage:

V_{PK}

90%

10%

90%

10%

T_0 T_1 T_2 T_3 T_4 T_5

Time summary:
- Delay time (t_d) is measured between T_0 and T_1
- Rise time (t_r) is measured between T_1 and T_2
- Storage time (t_s) is measured between T_3 and T_4
- Fall time (t_f) is measured between T_4 and T_5

FIGURE 18.23

frequency response of an op-amp switch by reducing the output transition or by getting a faster op-amp (one with a higher slew rate).

SECTION REVIEW

1. What is a *rectangular waveform*? [Objective 7]
2. Define each of the following terms: *pulse width*, *space width*, and *cycle time*. [Objective 7]
3. What is a *square wave*? [Objective 7]
4. Where are each of the values in question 2 measured on a practical rectangular waveform? [Objective 8]
5. Why do we need to have a standard for rectangular waveform time measurements? [Objective 8]
6. What is *duty cycle*? [Objective 9]
7. Why do all square waves have the same duty cycle value? What is this value? [Objective 9]
8. What is *propagation delay*? [Objective 10]
9. What is delay time? [Objective 10]

10. Describe delay time in terms of I_C and V_C. [Objective 10]
11. What is *rise time*? [Objective 10]
12. Describe rise time in terms of I_C and V_C. [Objective 10]
13. What is *storage time*? [Objective 10]
14. Describe storage time in terms of I_C and V_C. [Objective 10]
15. What is *fall time*? [Objective 10]
16. Describe fall time in terms of I_C and V_C. [Objective 10]
17. Which of the switching times determines the practical limit on f_{max} for the device? [Objective 11]
18. What causes delay time? [Objective 12]
19. What causes storage time? [Objective 12]
20. How does the use of a *speed-up capacitor* reduce t_d and t_s? [Objective 12]
21. What effect does a speed-up capacitor have on the practical frequency limit of a BJT? Explain your answer. [Objective 12]
22. Define *turn-on time* (t_{on}) and *turn-off time* (t_{off}). [Objective 13]
23. What is a *switching transistor*? [Objective 14]
24. What is the primary difference between the *inverter* and the *buffer*? [Objective 16]
25. List the points where t_d, t_r, t_s, and t_f are measured on the output voltage of an inverter. [Objective 17]
26. List the points where t_d, t_r, t_s, and t_f are measured on the output voltage of a buffer. [Objective 17]

18.3
Schmitt Triggers

Schmitt trigger. A voltage-level detector.

Now that we have covered the basic principles associated with switching circuits, it is time to start discussing some special-purpose switching circuits. As you will see, these more complicated circuits are used to perform functions that cannot be accomplished with the basic inverter or buffer. The first circuit we will cover is the *Schmitt trigger*.

The Schmitt trigger is a *voltage-level detector*. When the input to a Schmitt trigger reaches a specified level, the output will change states. When the input then goes below another specified level, the output will return to its original state. This input/output relationship is illustrated in Figure 18.24.

FIGURE 18.24

Schmitt trigger input and output signals.

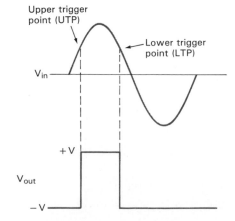

In Figure 18.24, the sine wave being applied to the Schmitt trigger passes the *upper trigger point* (UTP) on its positive-going transition. When the UTP is reached, the output from the Schmitt trigger goes to $+V$. As long as the sine-wave input stays above the UTP, the output from the Schmitt trigger will stay at $+V$. When the input goes negative again, it passes the *lower output point* (LTP). When this occurs, the output from the Schmitt trigger returns to $-V$. The output will now stay at $-V$ until the sine-wave input passes the UTP again.

Several points should be made at this time:

1. The UTP and LTP values are determined by the component values in the Schmitt trigger.
2. The Schmitt trigger can be designed so that the UTP and LTP values are not equal. However, the LTP can never be more positive than the UTP.
3. The Schmitt trigger can handle any type of input waveform.

The relationship between the LTP, UTP, and circuit component values will be seen when we start to discuss some specific circuits. The fact that the UTP and LTP values do not have to be equal to each other is illustrated in Figure 18.25. When the Schmitt trigger has different UTP and LTP values, the output will not go to $-V$ until the LTP is reached. Therefore, it is possible for the input to go below the UTP without the circuit output changing. In Figure 18.25, the sine wave passes the UTP at v_1 and again at v_2. At v_1, the output goes to $+V$, since the UTP has been reached on a *positive-going transition* of the input signal. When the input passes point v_2, the output does not change. Even though the input voltage is now below the UTP, the output is still at $+V$ and will remain there until the LTP is reached. This happens at v_3, and the output then returns to $-V$.

It is obvious from the waveforms in Figure 18.25 that the output from a Schmitt trigger will not be affected by any voltage between the UTP and the LTP. This is illustrated further in Figure 18.26. Even though the input is going

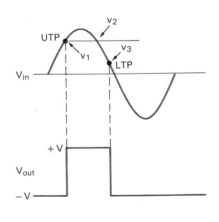

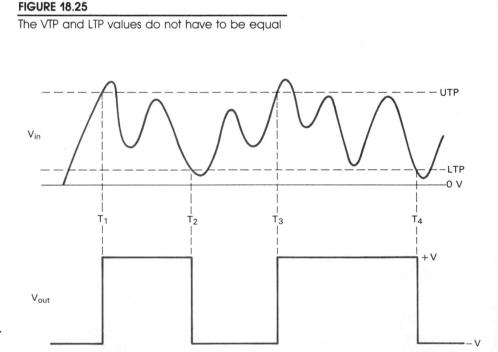

FIGURE 18.25

The VTP and LTP values do not have to be equal

FIGURE 18.26

Schmitt trigger hysteresis.

through some significant changes, the output only changes under the following conditions:

1. The UTP is reached by a positive-going transition.
2. The LTP is reached by a negative-going transition.

No input voltage between the UTP and the LTP will have any effect on the output of the circuit. Note that the range of voltages between the UTP and LTP that will not affect the output is (loosely) referred to as *hysteresis*. For example, if Figure 18.26 had values of UTP = +10 V and LTP = +2 V, the hysteresis of the circuit would be the input voltage range of $+2$ V $< V_{in} < +10$ V. No voltage in this range could ever cause the output of the circuit to change states.

Noninverting Schmitt Triggers

✔ 19

The noninverting Schmitt trigger uses a simple feedback resistor, as shown in Figure 18.27. Note the similarity between this circuit and the linear inverting amplifiers described earlier. The only difference is the reversal of the op-amp inputs.

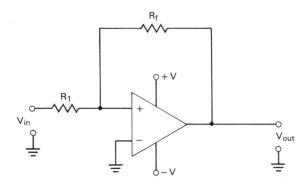

FIGURE 18.27

A noninverting Schmitt trigger.

If we add some component values to the circuit in Figure 18.27, it is easy to understand how the circuit works. The circuit is redrawn with component values added in Figure 18.28a. For this circuit, the values of R_1 and R_f are equal. Also, you know from earlier discussions that the current through the two resistors must be equal. Therefore, for the circuit shown in Figure 18.28a, the voltage across R_1 is always equal to the voltage across R_f.

FIGURE 18.28

Noninverting Schmitt trigger operation

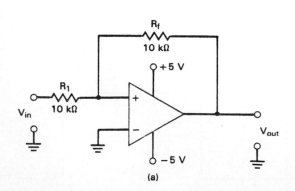

(a)

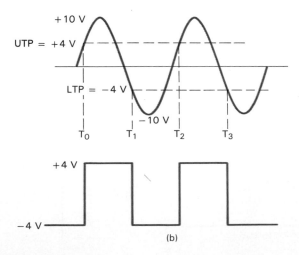

(b)

CHAP. 18   Solid-state Switching Circuits

Noninverting Schmitt trigger operation.

Now we will start to discuss the circuit operation by assuming that V_{in} is at +4 V. At this point, the input to the op-amp is at 0 V. As soon as the input starts to go more positive than +4 V, the noninverting input to the op-amp becomes more positive than the inverting input, and the output goes to +4 V (+V −1 V). This is illustrated in Figure 18.28b. At T_0, the input to the Schmitt trigger goes above +4 V, and the output responds by changing to +4 V. Remember, this happens because the noninverting input is now more positive than the inverting input.

The same circuit action would explain what happens on the negative transition of the input. The only difference is that the polarities are reversed. With the output of the circuit being at +4 V, the input must pass −4 V before the noniverting input goes to a value that is more negative than the inverting input. When this happens, the output returns to −4 V and the cycle repeats itself.

As it is drawn, the Schmitt trigger in Figure 18.28 has UTP and LTP values that are equal to (+V − 1 V) and (−V + 1 V). By changing the ratio of R_f to R_1, we can set the circuit up for other trigger point values. Generally, the UTP and LTP values can be determined using the following relationships:

$$\boxed{\text{UTP} = -\frac{R_1}{R_f}(-V + 1\text{ V})} \tag{18.6}$$

and

$$\boxed{\text{LTP} = -\frac{R_1}{R_f}(+V - 1\text{ V})} \tag{18.7}$$

The following example illustrates the use of these two equations.

EXAMPLE 18.8

Determine the UTP and LTP values for the circuit shown in Figure 18.29. Also, verify that the output will change states when the UTP and LTP values are exceeded.

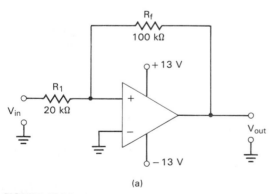

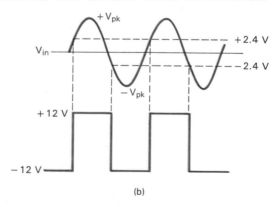

FIGURE 18.29

Solution: Using equation (18.6), the UTP is found as

$$\text{UTP} = -\frac{R_1}{R_f}(-V + 1\text{ V})$$
$$= -0.2(-12\text{ V})$$
$$= 2.4\text{ V}$$

The LTP is found using equation (18.7), as follows:

$$\text{LTP} = -\frac{R_1}{R_f}(+V - 1\text{ V})$$
$$= -0.2(+12\text{ V})$$
$$= -2.4\text{ V}$$

Now the UTP value can be verified as follows. When the input is at $+2.4$ V, the output is still at -12 V. Using the standard voltage-divider formula, we can obtain the voltage drop across R_1. This voltage drop is found as

$$V_1 = (V_{in} - V_{out})\frac{R_1}{R_1 + R_f}$$
$$= (14.4\text{ V})(0.167)$$
$$= 2.4\text{ V}$$

With 2.4 V being applied to the circuit and a 2.4-V drop across R_1, it is clear that the voltage at the op-amp input is 0 V. Therefore, if the input goes any higher than $+2.4$ V, the noninverting input to the op-amp will be more positive than the inverting input, and the output will go to $+12$ V. The same method can be used to validate the value of LTP $= -2.4$ V. This is left as an exercise for you to do on your own.

With the UTP and LTP values found, the input signals would appear as shown in Figure 18.29b

---

PRACTICE PROBLEM 18–8

A noninverting Schmitt trigger like the one in Figure 18.29 has values of $R_f = 33$ kΩ and $R_1 = 11$ kΩ. Determine the UTP and LTP values for the circuit.

The noninverting Schmitt trigger in Figure 18.29 is restricted in that the absolute values of the UTP and the LTP must be equal. For example, if the UTP is $+2.4$ V, the LTP must be -2.4 V. If the UTP were 4 V, the LTP would have to be -4 V. However, the noninverting Schmitt trigger can be modified so that the absolute values of the UTP and the LTP are not equal. A circuit that contains this modification is shown in Figure 18.30.

FIGURE 18.30

A noninverting Schmitt trigger designed for unequal VTP and LTP values.

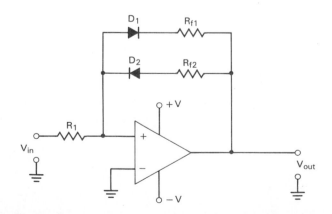

The key to the operation of this circuit is the fact that D_1 and D_2 will conduct on opposite transitions of the output. When the output from the Schmitt trigger is *negative*, D_1 will conduct and D_2 will be off. This means that R_{f2} is effectively removed from the circuit, and the UTP is found as

$$\boxed{\text{UTP} = -\frac{R_1}{R_{f1}}(-V + 1\text{ V})} \quad D_1 \qquad \textbf{(18.8)}$$

When the output is *positive*, D_1 will be off and D_2 will conduct. This means that R_{f1} is effectively removed from the circuit, and the LTP is found as

$$\boxed{\text{LTP} = -\frac{R_1}{R_{f2}}(+V - 1\text{ V})} \quad D_2 \qquad \textbf{(18.9)}$$

Since the UTP and LTP are determined by separate feedback resistors, these values are completely independent of each other. This point is illustrated in the following example.

EXAMPLE 18.9

Determine the UTP and LTP values for the noninverting Schmitt trigger shown in Figure 18.31a.

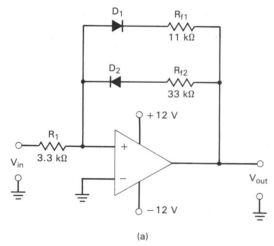

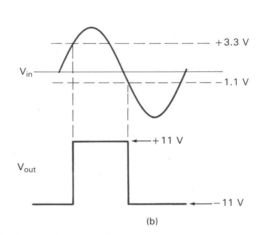

FIGURE 18.31

Solution: The UTP is found as

$$\text{UTP} = -\frac{R_1}{R_{f1}}(-V + 1\text{ V})$$
$$= -\frac{3.3\text{ k}\Omega}{11\text{ k}\Omega}(-11\text{ V})$$
$$= 3.3\text{ V}$$

The LTP is found as

$$\text{LTP} = -\frac{R_1}{R_{f2}}(+V - 1\text{ V})$$
$$= -\frac{3.3\text{ k}\Omega}{33\text{ k}\Omega}(11\text{ V})$$
$$= -1.1\text{ V}$$

With these UTP and LTP values, the circuit in Figure 18.31a would have the input/output relationship shown in Figure 18.31b.

PRACTICE PROBLEM 18–9

A noninverting Schmitt trigger like the one in Figure 18.31 has ±15-V supply voltages and the following resistor values: $R_1 = 2$ kΩ, $R_{f1} = 20$ kΩ, and $R_{f2} = 14$ kΩ. Determine the UTP and LTP values for the circuit.

The circuits in this section are all noninverting circuits; that is, their outputs go *high* when the UTP is passed by a *positive-going* input signal and negative when the LTP is passed by a *negative-going* input signal. It is possible to wire an op-amp as an *inverting Schmitt trigger* that will have the opposite input/output relationship from the one described here. That is, the output will go *negative* when the UTP is passed and *positive* when the LTP is passed. We will take a look at the operation of this circuit now.

Inverting Schmitt Triggers

[20]

The basic inverting Schmitt trigger is shown in Figure 18.32. At first, this circuit appears to be exactly the same as the linear amplifier we covered in Chapter 14. While the two circuits *are* very similar, there is one very important difference. The linear amplifier covered earlier had the input applied to the noninverting input and the feedback network connected to the inverting input of the op-amp. This is the exact opposite of the Schmitt trigger, as Figure 18.32 shows.

FIGURE 18.32

An inverting op-amp Schmitt trigger.

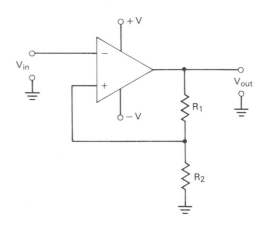

To understand the operation of the circuit, we must first establish the equations for the UTP and LTP. The UTP for the circuit in Figure 18.32 is found as

$$\text{UTP} = \frac{R_2}{R_1 + R_2}(+V - 1 \text{ V}) \tag{18.10}$$

and the LTP is found as

$$\text{LTP} = \frac{R_2}{R_1 + R_2}(-V + 1 \text{ V}) \tag{18.11}$$

The following example illustrates the procedure for determining the UTP and LTP values for a given inverting Schmitt trigger circuit.

================================ **EXAMPLE 18.10** ================================

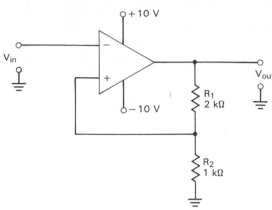

FIGURE 18.33

Determine the UTP and LTP values for the circuit shown in Figure 18.33.

Solution: The quickest approach is to determine the resistance ratio in the circuit first. This ratio is found to be

$$\frac{R_2}{R_1 + R_2} = 0.3333$$

Now the UTP is found as

$$UTP = 0.33(10 \text{ V} - 1 \text{ V})$$
$$= 3 \text{ V}$$

and the LTP is now found as

$$LTP = 0.33(-10 \text{ V} + 1 \text{ V})$$
$$= -3 \text{ V}$$

PRACTICE PROBLEM 18–10

An inverting Schmitt trigger has ±10-V supply voltages and values of $R_1 = 3$ kΩ and $R_2 = 1$ kΩ. Determine the UTP and LTP values for the circuit.

Next, we will use the values obtained in Example 18.10 to explain the operation of the circuit. First, remember that we are dealing with an *inverting* Schmitt trigger. Therefore, the output will go *low* when the UTP is passed and *high* when the LTP is passed. This point is illustrated in Figure 18.34. The waveforms shown here are the input and output waveforms for the circuit in Figure 18.33. As you can see, the input and output waveforms are out of phase by 180°.

The basic inverting Schmitt trigger has one limitation. If you look at Figure 18.33, it is not very difficult to figure out that the UTP and LTP values must

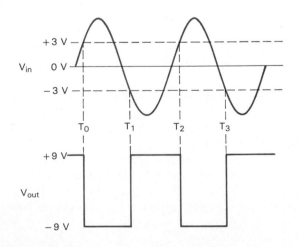

FIGURE 18.34

Input/output waveforms for an inverting Schmitt trigger

always be equal in magnitude for this circuit. This is due to the constant resistance ratio and the equal magnitudes of the two output voltages. In other words, you could not have an UTP of, say, +5 V and a LTP of −3 V for the circuit. The UTP and LTP values, while being opposite in polarity, would have to be equal in value.

The inverting Schmitt trigger can be modified to allow different values of UTP and LTP, as shown in Figure 18.35. In this circuit the UTP would be determined by R_1 and R_2, as before. The LTP would be determined by R_3, R_4, and D_1. By adjusting R_3 and R_4, the anode voltage of D_1 can be set to any value between $+V$ and $-V$. The LTP would then be 0.7 V more negative than the anode voltage. For example, let's say that R_3 and R_4 are set so that the anode voltage of D_1 is +2 V. The LTP for the circuit is then +1.3 V. If R_3 and R_4 are set to provide −4 V at the anode of D_1, the LTP is −4.7 V, and so on. Again, the values of R_1 and R_2 are adjusted to provide the desired UTP, and R_3 and R_4 are adjusted to provide the desired LTP.

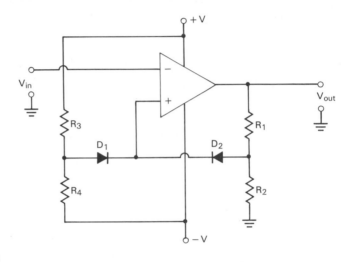

FIGURE 18.35

An inverting Schmitt trigger designed for unequal VTP and LTP values.

One Final Note

You have seen that the Schmitt trigger is a voltage-level detector that can be constructed in noninverting or inverting form. When the input signal to a Schmitt trigger makes a positive-going transition past the UTP, the output of the circuit goes to one dc output level. The output will stay at that level until the input signal makes a negative-going transition that passes the LTP. At that time, the output goes to the other dc output level. The dc output level for a specific input condition depends on whether the Schmitt trigger is an inverting or a noninverting circuit.

In Chapter 16, you were introduced to the *comparator*. This simple op-amp circuit is also a voltage-level detector that can be constructed in noninverting or inverting form. You may be wondering, then, what the difference is between the comparator and the Schmitt trigger.

The comparator is limited in that it has a single reference voltage (V_{ref}) that acts as both its UTP and its LTP. The Schmitt trigger, on the other hand, can be designed for totally independent UTP and LTP values, as you have seen in this section. Otherwise, the two circuits serve the same basic purpose.

1. What is a *Schmitt trigger*? [Objective 18]
2. What is the *upper trigger point (UTP)*? [Objective 18]
3. What is the *lower trigger point (LTP)*? [Objective 18]
4. What is *hystersis*? [Objective 18]
5. Explain the operation of the noninverting Schmitt trigger shown in Figure 18.28. [Objective 19]
6. Explain the operation of the noninverting Schmitt trigger shown in Figure 18.31. [Objective 19]
7. Explain the operation of the inverting Schmitt trigger in Figure 18.34. [Objective 20]
8. Explain the operation of the inverting Schmitt trigger in Figure 18.35. [Objective 20]
9. Compare and contrast the Schmitt trigger and the comparator. [Objective 21]

18.4
Multivibrators: The 555 Timer

Multivibrators are circuits that are designed to have zero, one, or two *stable* output states. The concept of *stable* output states is illustrated in Figure 18.36.

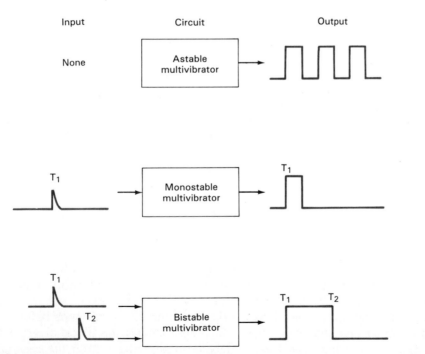

FIGURE 18.36

Multivibrator input/output relationships.

The *astable multivibrator* is a switching circuit that has *no stable output state*. In its own way, this circuit can be viewed as being a square-wave oscillator.

Multivibrators. Circuits designed to have zero, one, or two stable output states.

The circuit will produce a constant square-wave output with only a supply-voltage input. Because it continually produces this square-wave output, it is often referred to as a *free-running multivibrator*.

The *monostable multivibrator* has a single stable output state. For the circuit represented in Figure 18.36, the stable output would be $-V$. The circuit will stay at this output level as long as no input signal is applied. When the circuit receives an input, generally referred to as a *trigger*, it will produce a single output pulse. The duration of the output pulse depends on the component values used in the circuit. Since the monostable multivibrator produces a single output pulse for each input trigger, it is generally referred to as a *one-shot*.

The *bistable multivibrator* has two output states. When the proper input trigger is received, it will switch from one output state to the other. It will then stay at the new output state until another trigger is received, returning the output to its original output state. The bistable multivibrator is generally referred to as a *flip-flop*.

As you have probably figured out by now, the term *astable* means *not stable*, the term *monostable* means *one stable state*, and the term *bistable* means *two stable states*. If you look closely at Figure 18.36, you might notice something else that is appropriate about these names. The astable multivibrator has no active input. The one-shot generally has one active input. The flip-flop generally has two active inputs. Although there are plenty of exceptions to this, the input relationships shown generally hold true.

Flip-flops are covered extensively in any digital electronics course, and thus are not covered here. Although the astable and monostable multivibrators are also covered in digital electronics, they are discussed here to give you an introductory idea of how these circuits operate.

At one time, multivibrators were commonly constructed using discrete components. However, this is no longer the case. More often than any type of circuit we have discussed thus far, multivibrators are constructed using ICs. One of these ICs is the *555 timer*, which is an eight-pin (lead) IC that can be used for a variety of switching applications. Among its many applications are the astable and monostable multivibrator.

The advantage in using the 555 timer is the fact that all the active components required to produce a multivibrator are within a single chip. Simply by adding the proper passive components to the IC, it can be used as either a free-running multivibrator or a one-shot. In this section we are going to look at the 555 timer internal circuitry and its use as a free-running circuit or a one-shot.

The 555 Timer

The 555 timer is a switching circuit that comes in an eight-pin IC. The IC contains two comparators, a flip-flop, an inverter, and several resistors and transistors. The block diagram of the 555 timer is shown in Figure 18.37.

Since you may never have worked with an IC of this type before, there are a few points that should be made. In Figure 18.37, the block diagram is shown inside the IC casing. The small numbered blocks that border the case are the IC pins. These are used to connect the internal circuitry to any external components or supply voltages.

Looking at the left side of the case, you will see that there is a small indentation. This indentation is used to identify pin 1. When you hold a 555 timer, you hold it so that the indentation is on the left. Then pin 1 is the lower-left pin. The rest of the pins are set in counterclockwise order around the case.

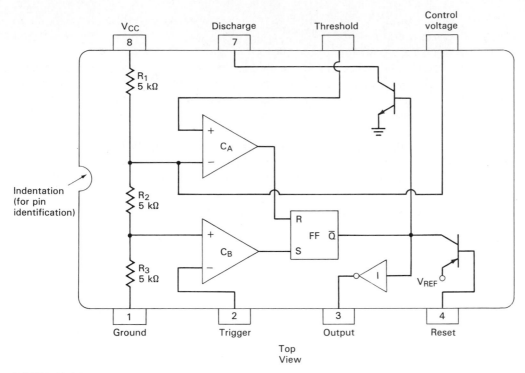

FIGURE 18.37

The 555 timer.

Inside the chip, comparator A (C_A), comparator B (C_B), the flip-flop (FF), and the inverter (I) all need V_{CC} and ground connections to operate. These connections are made automatically when you wire pin 8 to V_{CC} and pin 1 to ground. *Note that the IC will not work if the proper supply connections to pins 1 and 8 are not made.* Even though it is not shown in the block diagram, pins 1 and 8 do more than supply a complete current path for the resistive ladder (R_1, R_2, and R_3). They also provide the proper supply voltages for all the internal active components.

Next, we are going to discuss the internal operation of the 555 timer, starting with the operation of the comparator circuit. For the sake of discussion, we are going to start with several assumptions:

1. The 555 timer is connected to a +15-V supply voltage (pin 8).
2. All component outputs will be idealized as being either *high*, meaning +15 V, or *low*, meaning 0 V. The actual internal voltages would be near these levels, so nothing is lost in using ideal values.

The comparator circuit is shown in Figure 18.38.

First, a word or two about the resistive ladder, the circuit made up of R_1, R_2, and R_3. These three resistors are equal in value and therefore will drop equal voltages. In our circuit, each resistor drops 5 V. Thus the ($-$) input to C_A is at +10 V and the ($+$) input to C_B is at +5 V. Note that the comparator inputs from the resistive ladder will be equal to $\frac{1}{3}V_{CC}$ and $\frac{2}{3}V_{CC}$ as is shown. This is an important point to remember, since the operation of the entire 555 timer depends on these values.

The *comparators* are simply high-speed op-amps. Other than the fact that they are capable of switching at a faster rate, they act just as any op-amp does. With the ($-$) input to C_A at +10 V, the output from the comparators will be *high* if the ($+$) input is greater than +10 V and *low* otherwise. With the ($+$) input to

The supply connections must be made at pins 1 and 8 for proper circuit operation.

 24

Comparator operation.

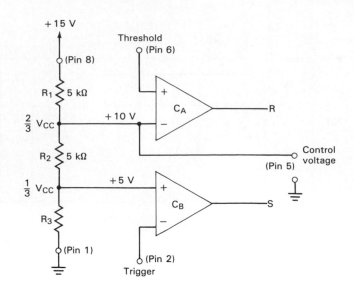

FIGURE 18.38

The 555 timer comparator circuit.

C_B at $+5$ V, the output from this comparator will be *high* when the $(-)$ input is less than $+5$ V, and low otherwise. These operating states are represented in Table 18.1.

TABLE 18.1
Comparator Input/Output Relationships

Comparator	Inverting Input $(-V)$	Noninverting Input $(+V)$	Output
A	$+10$	Threshold < 10	$R = $ low
	$+10$	Threshold > 10	$R = $ high
B	Trigger < 5	$+5$	$S = $ high
	Trigger > 5	$+5$	$S = $ low

Before we move on to include the flip-flop, one more point needs to be made about the comparator circuit. Looking at Figure 18.38, you will see that the $(-)$ input to C_A is also connected to pin 5, the *control voltage* input. This input allows you to set reference levels for the comparators that are independent of V_{CC}. For example, let's say that you want the reference voltages to be $V_{\text{ref}(A)} = 5$ V and $V_{\text{ref}(B)} = 2.5$ V. This could be done by connecting the control voltage input to $+5$ V, as shown in Figure 18.39. With the connections shown,

FIGURE 18.39

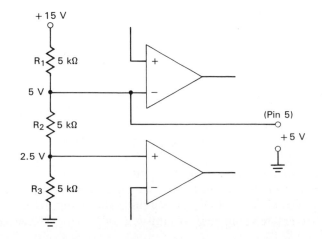

CHAP. 18   Solid-state Switching Circuits

V_{CC} is still $+15$ V. However, the $\frac{2}{3}V_{CC}$ and $\frac{1}{3}V_{CC}$ relationships no longer hold true. Since the $(-)$ input to C_A is tied directly to a $+5$-V supply, the lower side of R_1 is at $+5$ V. Thus R_1 is dropping 10 V, the difference between V_{CC} and the control voltage. Now, since $R_2 = R_3$, the $(+)$ input to C_B is one-half of the control voltage, or 2.5 V. Note that V_{CC} must still be connected in order for the circuit to work. Now, however, the comparators have lower reference voltages.

The comparator outputs are shown to be connected to a *flip-flop*, or bistable multivibrator. The output from the flip-flop depends on the voltages present at its *set* (S) and *reset* (R) inputs. The operation of the flip-flop is illustrated in Figure 18.40.

Flip-flop operation.

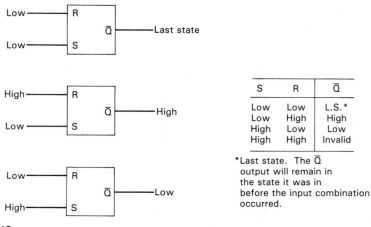

S	R	$\overline{Q}$
Low	Low	L.S.*
Low	High	High
High	Low	Low
High	High	Invalid

*Last state. The $\overline{Q}$ output will remain in the state it was in before the input combination occurred.

FIGURE 18.40

Flip-flop operation in the 555 timer.

When *both* inputs to the flip-flop are *low*, the output will remain in the state it was in before the two inputs went low. For example, if the flip-flop output was *high* when both inputs went low, the output would stay in the *high* state. If the output was *low*, it will remain *low*. If the S input goes *high* while the R input is *low*, the output will go *low*. If the S input is *low* and the R input goes *high*, the output will go *high*. Note that the output always equals the R input when the two inputs are in opposite states. Both inputs being high is an invalid input condition and will never occur during normal operation. This will be seen when we discuss some 555 timer circuits.

Flip-flops are somewhat complex circuits that are normally covered in a course on digital electronics. For this reason we are not going into a lengthy discussion of them here. At this point we are merely establishing the most basic rules of operation for the flip-flop shown in the 555 timer so that you can understand the overall operation of the IC. All you need to know about the flip-flop in the 555 timer is:

1. When the inputs to the flip-flop are *not* equal, the output will equal the R input.
2. When both inputs are *low*, the output will not change from its previous state.
3. Two high inputs will cause the circuit to malfunction.

Remembering these points will give you sufficient background to understand the 555 timer. Now let's tie the flip-flop and the comparator circuits together. The combination circuit is shown in Figure 18.41, which contains a chart that

SEC. 18.4 Multivibrators: The 555 Timer

847

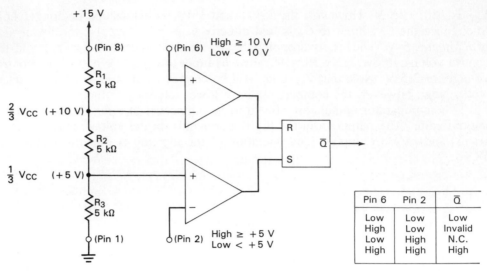

FIGURE 18.41

The combined comparator and flip-flop circuits.

Pin 6	Pin 2	$\overline{Q}$
Low	Low	Low
High	Low	Invalid
Low	High	N.C.
High	High	High

Note: N.C. = No change

The operation of the combined comparator and flip-flop circuits.

shows the value of $\overline{Q}$ for various combinations of pins 6 and 2 inputs. We'll go through this chart line by line to show how the combination circuit operates.

- *Line 1: pin 6 = low, pin 2 = low.* A *low* input at pin 6 means that this pin is at any potential that is less than +10 V. Therefore, the output from the comparator is *low*. When the input to pin 2 is *low*, meaning that it is less than +5 V, the output from this comparator is *high*. Since the inputs to the flip-flop are not equal, $\overline{Q} = R$, and the output is *low*.

- *Line 2: pin 6 = high, pin 2 = low.* When pin 6 is high, meaning greater than +10 V, the output from C_A is *high*. As was already shown, a low input to pin 2 will cause the output from C_B to be *high*. This causes an invalid input condition to the flip-flop. Therefore, *the 555 timer can never be operated with pin 6 high while pin 2 is low.*

- *Line 3: pin 6 = low, pin 2 = high.* As you have seen, a low input at pin 6 causes the output of C_A to be low. A *high* input to pin 2 will cause the output of C_B to be *low*. Thus the flip-flop has two low inputs, and the output will not change from its previous state.

- *Line 4: pin 6 = high, pin 2 = high.* With a high input to C_A, its output is high. With a high input to C_B, its output is low. This gives us the input condition of R = high and S = low. Since these inputs are unequal, $\overline{Q} = R$, and the output from the flip-flop is *high*.

  Now let's summarize the operation of the combination circuit as follows:

Summary of the comparator/flip-flop circuitry operation.

1. When pins 6 and 2 are in the same state, the output from the flip-flop is in the same state. If both are high, the output is high. If both are low, the output is low.

2. If pin 6 is low and pin 2 is high, the output will not change from its previous output state.

3. The input combination of pin 6 = high *and* pin 2 = low is invalid and will cause the circuit to malfunction.

Between the flip-flop and the 555 timer output (pin 3) is the commonly used block diagram symbol for the *inverter*. Don't let the strange symbol confuse you.

The symbol simply means that the 555 timer output will be opposite to the output of the flip-flop. Thus:

1. When pins 6 and 2 are in the same state, the output from the 555 timer (pin 3) will be in the *opposite* state. If both inputs are high, the output from the timer will be low, and vice versa.

2. Statements 2 and 3 on page 848 still hold true.

Timer input (pins 2 and 6)/output (pin 3) relationship.

Now we only need to clear up a few more points and we will be ready to look at some circuits. First, the flip-flop output is tied to the *discharge* output (pin 7) via an *npn* transistor. This transistor is merely acting as a current-controlled switch. When the output from the flip-flop is *high*, pin 7 is grounded through the transistor. When the output from the flip-flop is *low*, pin 7 will appear as an open to any external circuitry. The purpose served by this pin will be made clear when we discuss the basic 555 timer circuits. Incidentally, the inverter (pin 3) and the transistor (pin 7) both invert the flip-flop output. The reason that different symbols are used for the two circuits is that the inverter is being used to provide some gain for the flip-flop output, whereas the transistor is being used only as a switch.

The *reset* input (pin 4) is used to disable the 555 timer. Even though the timer will function (internally) as usual, the output (pin 3) will be held *low* whenever pin 4 is grounded. This input is used only in special applications and is usually held at an inactive level by being tied to V_{CC}. When pin 4 is tied to V_{CC}, the *pnp* transistor is biased *off* and the 555 timer output will operate normally. When the *reset* input is *low*, the *pnp* transistor is biased *on*, and the output from the flip-flop is shorted out before it reaches the inverter. In all our circuits, pin 4 will be tied to V_{CC}, thus disabling the reset circuit.

The purpose of the *reset input*.

When the reset input isn't being used, it should be connected to V_{CC}.

The Monostable Multivibrator

The 555 timer can be converted into a one-shot with the addition of a single resistor and a single capacitor. The 555 timer one-shot is shown in Figure 18.42.

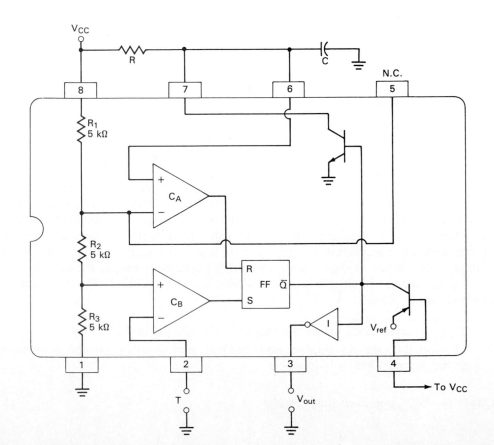

FIGURE 18.42

The 555 timer one-shot (monostable multivibrator).

Before we get into the operation of this circuit, a few practical observations should be made:

1. The *reset* input to the 555 timer (pin 4) is shown to be connected to V_{CC}. This disables the *reset* circuit.
2. Pins 8 and 1 are tied to V_{CC} and ground, respectively, as they must be for the IC to operate.
3. The *control voltage* input (pin 5) is shown as *not connected* (N.C.). This means that this pin is not connected to anything. This is acceptable when the control voltage input is not being used.
4. Pins 6 and 7 are tied together between R and C. Thus the voltage at *both* of these pins will always equal V_C.

Now let's look at the operation of the circuit. The appropriate waveforms are shown in Figure 18.43. To start with, we are going to have to make an assumption. We are going to assume that something caused the output of the flip-flop to be *high* before T_0. You will see what caused this condition in a minute. However, we need to assume this high flip-flop output to understand the waveforms at T_0.

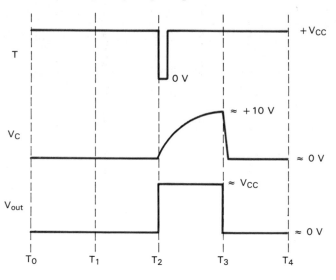

FIGURE 18.43

The 555 timer one-shot waveforms.

Between T_0 and T_2, the inactive circuit voltages are shown. Normally, the trigger input (T) will be approximately equal to V_{CC} when in its inactive state. This ensures that the output from C_B is low.

Assuming that the output from the flip-flop is high, we know that pins 3 and 7 are both tied *low*. The low output at pin 7 ensures that V_C is equal to 0 V, as shown in Figure 18.43. This low voltage is applied to C_A via pin 6. With a low input to C_A, its output is low. Since both inputs to the flip-flop are low, its output does not change from the *high* voltage that was already present. The circuit voltages will remain as shown until something happens to change them.

At T_2, the (T) input is driven *low*. When this occurs, the following sequence of events takes place:

1. The low input to C_B drives the output of the comparator *high*.
2. The inputs to the flip-flop are now unequal, and $\overline{Q}$ will go to the same level as the R input, which is *low*.

3. The low-flip-flop output is inverted and pin 3 goes *high*, as shown in Figure 18.43.

4. The low flip-flop output biases the *npn* transistor *off*. Thus the capacitor will now start to charge toward V_{CC}. The charge time of the capacitor depends on the values of R and C.

5. Eventually, the capacitor charges to the point where $V_C = +10$ V. At this point, the input to pin 6 is *high*.

6. The high input to C_A causes the comparators output to go *high*. Since the trigger signal is no longer present, the output from C_B has returned to a low.

7. Since the inputs to the flip-flop are unequal again, $\overline{Q}$ will switch to the level of R, which is *high*.

8. The *high* output from the flip-flop causes pin 3 to go low again (shown at T_3).

9. The high flip-flop output turns the *npn* transistor back on, which in turn discharges the capacitor. This returns all circuit potentials to their original values (shown at T_4).

As you can see, our initial assumption of $\overline{Q}$ = high is based on the fact that the flip-flop is always set to this value when it goes through a trigger cycle.

At this point, you may be thinking that this circuit is very complex. In a way it is, but in a far more practical sense, it is not. In this discussion, we have considered all the details of the internal 555 timer response to a trigger input. If you were working with this circuit in a practical situation, you would be concerned only with the input/output function of the timer. In fact, look at the diagram shown in Figure 18.44. This is what you would see if you were working with the circuit we just discussed. All that you have is one IC, one resistor, and one capacitor. If the circuit does not work, all you need to do is make a few basic measurements. If the passive components check out, you replace the 555 timer. How is *that* for simple? We will discuss the entire troubleshooting process later in this section.

FIGURE 18.44

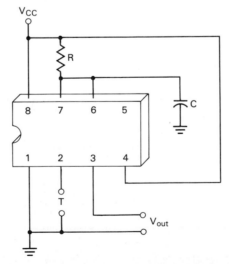

The pulse width of a 555 timer one-shot is determined by the values of R and C used in the circuit. The charge time of the capacitor in an RC circuit is found as

$$t = RC \left[\ln \frac{V_S - V_I}{V_S - v_C} \right] \qquad (18.12)$$

Note: The equation shown here is derived in a basic electronics course on RC circuits. In our discussion, we will accept its validity without going through its derivation.

SEC. 18.4   Multivibrators: The 555 Timer

where t = the time for the capacitor to charge to v_C
V_S = the source voltage that is being used to charge the capacitor
V_I = the initial charge on the capacitor, usually 0 V
v_C = the capacitor voltage of interest

Note that "ln" indicates that we are interested in the *natural log* of the fraction. For the 555 timer one-shot, the charging voltage (V_S) for the capacitor is V_{CC} and the initial charge (V_I) on the capacitor is 0 V. Since the pulse width is determined by the time required for the capacitor to charge to $\frac{2}{3}V_{CC}$, we use this value as v_C in the charge time equation. With these values, we can derive a useful equation that allows us to determine the pulse width of the one-shot for given values of R and C, as follows:

$$t = RC \left[\ln \frac{V_S - V_I}{V_S - v_C} \right]$$

$$= RC \left[\ln \frac{V_{CC} - 0 \text{ V}}{V_{CC} - \frac{2}{3}V_{CC}} \right]$$

$$= RC \left[\ln \frac{V_{CC}}{\frac{1}{3}V_{CC}} \right]$$

$$= RC \left[\ln(3) \right]$$

or

$$\boxed{PW = 1.1RC}$$

(18.13)

The following example demonstrates the use of this equation.

EXAMPLE 18.11

A one-shot like the one in Figure 18.44 has values of $R = 1.2 \text{ k}\Omega$ and $C = 0.1 \text{ μF}$. Determine the pulse width of the output.

Solution: The pulse width is found as

$$PW = 1.1RC$$
$$= 1.1(1.2 \text{ k}\Omega)(0.1 \text{ μF})$$
$$= 132 \text{ μs}$$

PRACTICE PROBLEM 18–11

A one-shot like the one in Figure 18.44 has values of $R = 18 \text{ k}\Omega$ and $C = 1.5$ μF. Determine the pulse width of the time output signal.

Circuit Troubleshooting

The one-shot in Figure 18.44 is extremely simple to troubleshoot. The process involves making the following checks:

Circuit checks.

■ Are the V_{CC} and ground connections good?
■ Is the *reset* input held at an inactive level?
■ Is the circuit receiving a *valid* trigger signal?

CHAP. 18 Solid-state Switching Circuits

- Is the resistor good?
- Is the capacitor good?

If the answer to all these questions is *yes*, replace the 555 timer. If any question has a *no*, you have found the problem.

What is a *valid* trigger signal?

The troubleshooting procedure shown involves checking for a *valid* trigger signal. The question at this point is this: just what *is* a *valid* trigger signal? A valid trigger signal is an input to pin 2 that is *low enough* to cause C_B to have a high output. The value of input voltage required to do this depends on whether or not the *control voltage* input is being used. When the control voltage input is *not* being used, the maximum input trigger value that will work can be found as

$$\boxed{V_{T(\text{max})} = \tfrac{1}{3}V_{CC}} \qquad \textbf{(18.14)}$$

This equation is based on the fact that the (+) input of C_B is at $\tfrac{1}{3}V_{CC}$ when the control voltage input is not being used. Therefore, the (T) input must go *below* this voltage level to cause the output of C_B to go high.

When the control voltage input is being used, it will be dropped equally across R_2 and R_3 in the resistive ladder. Therefore, the voltage at the (+) input of C_B will be equal to one-half of the control voltage. By formula,

$$\boxed{V_{T(\text{max})} = \tfrac{1}{2}V_{\text{con}}} \qquad \textbf{(18.15)}$$

where V_{con} is the voltage at the *control voltage* input. The following example illustrates the process for determining the maximum acceptable input trigger voltage for a given one-shot circuit.

═══════════════ **EXAMPLE 18.12** ═══════════════

Determine the value of $V_{T(\text{max})}$ for both of the one-shot circuits shown in Figure 18.45.

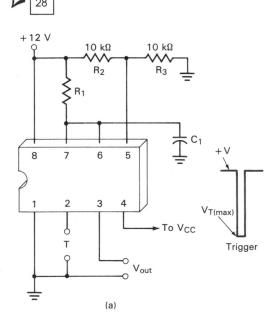

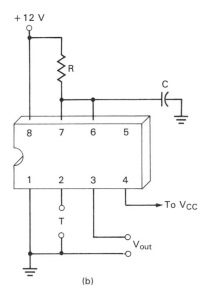

FIGURE 18.45

Solution: The circuit in Figure 18.45a has a voltage-divider circuit that is providing a voltage to the control voltage input. Using the standard voltage-divider formula, the control voltage

If a one-shot circuit has a frequency-related intermittent problem, the bypass capacitor will eliminate the problem by holding the supply current (I_{CC}) relatively constant during the switching time. Note that C_B will typically be around 0.1 μF.

How do you know when an intermittent problem is caused by f_{in}? Check the trigger signal. If the trigger signal is definitely valid, odds are that the problem is related to the input frequency of the circuit.

The Astable Multivibrator

✔ 29

Circuit recognition

The 555 timer can be wired for free-running operation as shown in Figure 18.49. Note that the circuit contains *two resistors and one capacitor*, and *does not have an input trigger from any other circuit*. The lack of a trigger signal from an external source is the circuit recognition feature of the free-running multivibrator.

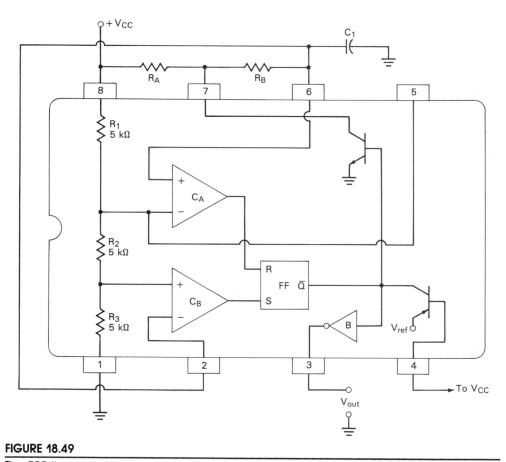

FIGURE 18.49

The 555 timer astable (free-running) multivibrator.

The key to the operation of this circuit is the fact that the capacitor is connected to *both* the trigger input (pin 2) and the threshold input (pin 6). Thus, if the capacitor is caused to charge and discharge between the threshold voltage ($\frac{2}{3}V_{CC}$) and the trigger voltage ($\frac{1}{3}V_{CC}$), the output will be a steady train of pulses. This point is illustrated in Figure 18.50. Since pins 6 and 2 are tied together, their voltage values will always be equal. When V_C charges to $\frac{2}{3}V_{CC}$, both inputs are high and the output from the timer goes *low*. When V_C discharges to $\frac{1}{3}V_{CC}$, both inputs are low and the output from the timer goes *high*.

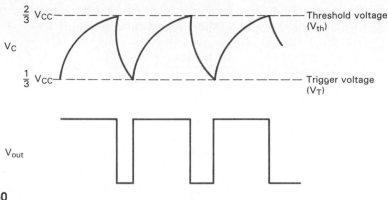

FIGURE 18.50

Free-running multivibrator waveforms.

The charge–discharge action of C_1 is caused by the 555 timer internal circuitry. To see how this is accomplished, we will start by making a few assumptions:

1. The capacitor is just starting to charge toward the threshold voltage, V_{th}.
2. The output from the flip-flop is *low*, causing pins 3 and 7 to be *open*.

With pin 7 being open, the capacitor will be charged by V_{CC} via R_A and R_B. As the capacitor charges, it eventually reaches V_{th}. At this point, both inputs to the 555 timer are *high*. This causes the output from the flip-flop to go high. The high voltage out of the flip-flop causes the output (pin 3) to go low and biases the *npn* transistor *on*. Thus pin 7 is now acting as a short to ground and will allow the capacitor to discharge. When it discharges down to V_T, both inputs will again be low, driving the output from the flip-flop low. This in turn causes the output to go high again and causes pin 7 to return to the *open* condition, and the cycle starts again.

Circuit operation

The output cycle time for the free-running multivibrator is found as

$$\boxed{T = 0.693[(R_A + 2R_B)C_1]} \tag{18.16}$$

Converting equation (18.16) into a frequency form yields the following equation for the operating frequency of the circuit:

$$f_0 = \frac{1}{0.693[(R_A + 2R_B)C_1]}$$

or

$$\boxed{f_0 = \frac{1.44}{(R_A + 2R_B)C_1}} \tag{18.17}$$

The duty cycle of the free-running multivibrator is the ratio of the pulse width to the cycle time. If you look at the waveforms in Figure 18.49, you'll see that the pulse width and cycle time of the circuit are determined by the charge and discharge times of C_1. We already know from equation (18.16) that the cycle time is equal to $0.693[(R_A + 2R_B)C_1]$. The pulse width of the circuit is found as

$$\boxed{PW = 0.693[(R_A + R_B)C_1]} \tag{18.18}$$

SEC. 18.4 Multivibrators: The 555 Timer

857

If we use the charge/discharge equations in place of PW and cycle time in the duty cycle formula, we obtain the following useful equation:

$$\text{duty cycle} = \frac{0.693[(R_A + R_B)C_1]}{0.693[(R_A + 2R_B)C_1]} \times 100$$

or

$$\boxed{\text{duty cycle} = \frac{R_A + R_B}{R_A + 2R_B} \times 100} \qquad \textbf{(18.19)}$$

The following example illustrates the basic analysis of a free-running multivibrator.

EXAMPLE 18.13

30

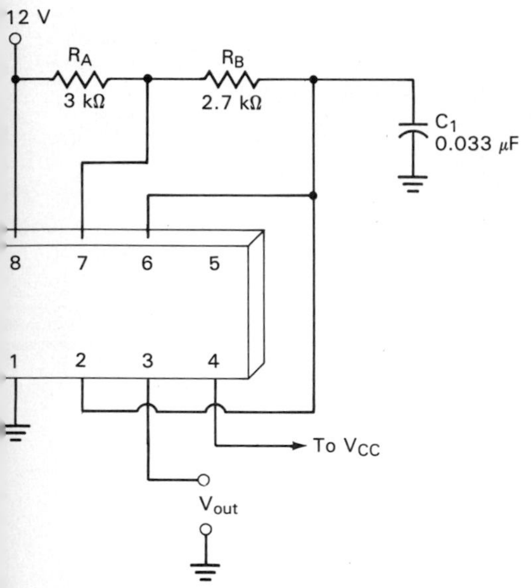

FIGURE 18.51

Determine the values of f_0 and duty cycle of the circuit shown in Figure 18.51. Also, determine the pulse width of the circuit.

Solution: The f_0 for the circuit is found as

$$\begin{aligned} f_0 &= \frac{1.44}{(R_A + 2R_B)C_1} \\ &= \frac{1.44}{(3\text{ k}\Omega + 5.4\text{ k}\Omega)(0.033\ \mu\text{F})} \\ &= 5.19\text{ kHz} \end{aligned}$$

The duty cycle of the circuit is found as

$$\begin{aligned} dc &= \frac{R_A + R_B}{R_A + 2R_B} \times 100 \\ &= 67.85\% \end{aligned}$$

Finally, the pulse width of the circuit is found as

$$\begin{aligned} PW &= 0.693[(R_A + R_B)C_1] \\ &= 0.693(5.7\text{ k}\Omega)(0.033\ \mu\text{F}) \\ &= 130\ \mu\text{s} \end{aligned}$$

PRACTICE PROBLEM 18–13

A circuit like the one in Figure 18.51 has the following values: $V_{CC} = +10$ V, $R_A = 5.1$ kΩ, $R_B = 2.2$ kΩ, and $C_1 = 0.022$ μF. Determine the values of f_0, duty cycle, and output pulse width for the circuit.

Why two resistors are used in the *RC* circuit.

At this point, you might be wondering why the free-running multivibrator requires two resistors in its *RC* network. After all, we only required a single resistor for the one-shot. To answer this question, let's take a look at what would happen if either R_A or R_B was shorted.

If we short R_A, we have a single resistor, R_B, between pins 6 and 7. This situation is unacceptable because the 555 timer will short circuit the power supply when pin 7 goes low. Recall that pin 7 goes low during every discharge cycle. If R_A is shorted, this low output is connected directly to V_{CC}. The result would be a "well-done" 555 timer. Thus, we must have a resistor between pin 7 and V_{CC}.

If R_B was shorted, there would be no RC circuit during the discharge cycle of C_1. As you have been shown, C_1 charges through $(R_A + R_B)$ and discharges through R_B. If R_B is replaced with a short circuit, the discharge cycle time will be 0, and the duty cycle will be 100%, an unacceptable situation. Thus, R_B is also required for proper operation.

Circuit Troubleshooting

The process for troubleshooting the free-running circuit is nearly the same as the procedure used on the one-shot. The only difference is that you have two resistors that must be verified as being good. All other tests are the same.

Voltage-controlled Oscillators

The free-running multivibrator can be modified to form a circuit called a *voltage-controlled oscillator*, or *VCO*. A VCO is a free-running circuit whose output frequency depends on a given input control voltage. The lower the input control voltage, the higher the output frequency, and vice versa. The 555 timer VCO is shown in Figure 18.52a. Note that this circuit is exactly the same as the one shown in Figure 18.51, with the exception of the potentiometer (R_1) added at the control voltage input of the 555 timer.

Voltage-controlled oscillator (VCO). A square-wave oscillator whose output frequency is controlled by a dc input voltage.

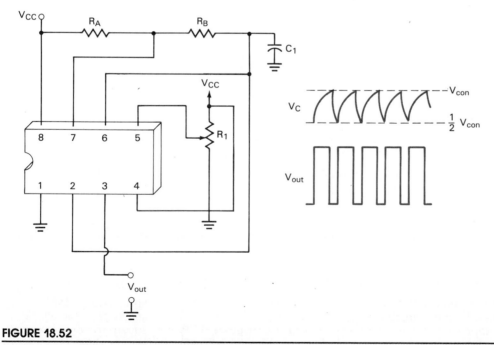

FIGURE 18.52

A voltage controlled oscillator (VCO).

When the control voltage input is reduced, the difference between V_{th} and V_T is also reduced. For example, when V_{con} is $+10$ V, $V_{th} = +10$ V, and $V_T = +5$ V, the difference being $+5$ V. When V_{con} is $+5$ V, $V_{th} = +5$ V, and $V_T =$

+2.5 V, the difference being +2.5 V. Thus, by reducing V_{con}, we reduce the difference between V_{th} and V_T.

When the difference between V_{th} and V_T is reduced, it takes less time for the capacitor to charge or discharge between the two values. Therefore, the cycle time is reduced and the operating frequency is increased. In this way we are able to control the operating frequency by varying V_{con}.

One Final Note

In this chapter we have barely scratched the surface of the entire subject of switching circuits. The study of digital electronics provides complete coverage of the subject. However, this chapter has provided you with the fundamental principles of switching circuits in such a way as to allow you to continue the study of digital electronics on your own.

SECTION REVIEW

1. What is a *multivibrator*? [Objective 22]
2. Describe the output characteristics of each of the following circuits: [Objective 22]
   a. *Astable multivibrator (free-running multivibrator)*
   b. *Monostable multivibrator (one-shot)*
   c. *Bistable multivibrator (flip-flop)*
3. What is the *555 timer*? [Objective 23]
4. Explain the operation of the 555 timer's internal comparator circuit. [Objective 24]
5. Explain the operation of the 555 timer's internal flip-flop. [Objective 24]
6. List the comparator/flip-flop circuitry input/output relationships. [Objective 24]
7. List the input/output relationships for the 555 timer. [Objective 24]
8. Briefly describe the operation of the monostable multivibrator (one-shot). [Objective 25]
9. What checks are made when troubleshooting a one-shot? [Objective 27]
10. Briefly describe the operation of the astable (free-running) multivibrator. [Objective 29]
11. Describe the operation of the voltage-controlled oscillator (VCO). [Objective 32]

KEY TERMS

The following terms were introduced and defined in this chapter:

astable multivibrator	hysteresis	space width (SW)
bistable multivibrator	inverter	speed-up capacitor
buffer	lower trigger point (LTP)	square wave
cycle time (T_C)	monostable multivibrator	storage time (t_s)
delay time (t_d)	multivibrator	switching circuits
driver	one-shot	switching transistors
duty cycle (dc)	propagation delay	turn-off time (t_{off})
fall time (t_f)	pulse width (PW)	turn-on time (t_{on})
555 Timer	rectangular waveform	upper trigger point (UTP)
flip-flop	rise time (t_r)	voltage-controlled
free-running multivibrator	Schmitt trigger	oscillator (VCO)

PRACTICE PROBLEMS

§18.1

1. Determine the minimum high voltage required to saturate the transistor in Figure 18.53. [2]

2. A switch like the one in Figure 18.53 has the following values: $V_{CC} = +12$ V, $h_{FE} = 150$, $R_C = 1.1$ kΩ, and $R_B = 47$ kΩ. Determine the minimum high input voltage required to saturate the transistor. [2]

3. Determine the minimum high input voltage required to saturate the transistor in Figure 18.54 [2]

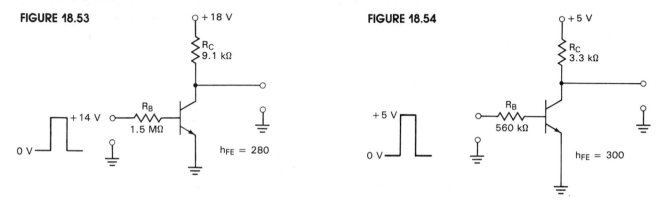

FIGURE 18.53

FIGURE 18.54

4. Determine the minimum high input voltage required to saturate the transistor in Figure 18.55. [2]

5. Determine the high and low output voltages for the circuit in Figure 18.56. [3]

6. Determine the high and low output voltages for the circuit in Figure 18.57. [3]

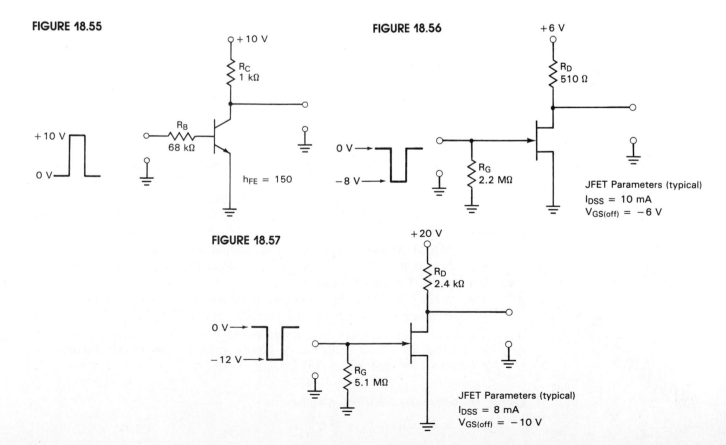

FIGURE 18.55

FIGURE 18.56

FIGURE 18.57

7. Determine the values of PW and T_C for the waveform shown in Figure 18.58. [8]

8. Determine the values of PW and T_C for the waveform shown in Figure 18.59. [8]

9. Calculate the duty cycle of the waveform in Figure 18.58. [9]

10. Calculate the duty cycle of the waveform in Figure 18.59. [9]

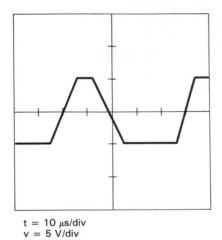

t = 10 μs/div
v = 5 V/div

FIGURE 18.58

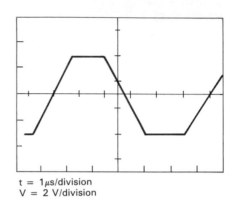

t = 1μs/division
V = 2 V/division

FIGURE 18.59

11. The 2N4264 BJT has the following parameters: $t_d = 8$ ns, $t_r = 15$ ns, $t_s = 20$ ns, and $t_f = 15$ ns. Determine the theoretical and practical limits on f_{in} for the device. [11]

12. The 2N4400 BJT has the following parameters: $t_d = 15$ ns, $t_r = 20$ ns, $t_s = 225$ ns, and $t_f = 30$ ns. Determine the theoretical and practical limits on f_0 for the device. [11]

13. The BCX587 BJT has the following parameters: $t_d = 16$ ns, $t_r = 29$ ns, $t_s = 475$ ns, and $t_f = 40$ ns. Determine the theoretical and practical limits on f_0 for the device. [11]

14. The MPS2222 BJT has the following parameters: $t_d = 10$ ns, $t_r = 25$ ns, $t_s = 225$ ns, and $t_f = 60$ ns. Determine the theoretical and practical limits on f_{in} for the device. [11]

15. The 2N3971 JFET has the following parameters: $t_d = 15$ ns, $t_r = 15$ ns, and $t_{off} = 60$ ns. Determine the theoretical and practical limits on f_{in} for the device. [13]

16. The 2N4093 JFET has the following parameters: $t_d = 20$ ns, $t_r = 40$ ns, and $t_{off} = 80$ ns. Determine the theoretical and practical limits on f_{in} for the device. [13]

17. The 2N4351 MOSFET has the following parameters: $t_d = 45$ ns, $t_r = 65$ ns, $t_s = 60$ ns, and $t_f = 100$ ns. Determine the theoretical and practical limits on f_{in} for the device. [13]

18. The 2N4393 JFET has the following parameters: $t_{on} = 15$ ns and $t_{off} = 50$ ns. Determine the theoretical and practical limits on f_{in} for the device. [13]

19. The LM318 op-amp has a slew rate of 70 V/μs. The device is being used in a switching circuit with ±14-V supply voltages. Determine the practical limit on f_{in} for the circuit. [15]

20. The TL071 op-amp has a slew rate of 13 V/μs. The device is being used in a switching circuit with ±15-V supply voltages. Determine the practical limit on f_{in} for the circuit. [15]

21. The LM741C op-amp has a slew rate of 0.5 V/μs. The device is being used in a switching circuit with ±5-V supply voltages. Determine the practical limit on f_{in} for of the circuit.

22. The NE531 op-amp has a slew rate of 35 V/μs. The device is being used in a switching circuit with ±10-V supply voltages. Determine the practical limit on f_{in} for the circuit. [15]

23. Determine the values of t_d, t_r, t_s, and t_f for the waveform in Figure 18.60. [17]

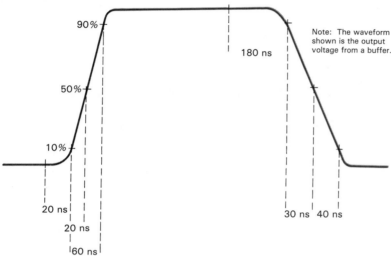

FIGURE 18.60

24. Determine the values of t_d, t_r, t_s, and t_f for the waveform in Figure 18.61. [17]

§18.3

25. Determine the UTP and LTP values for the Schmitt trigger in Figure 18.62. [19]

FIGURE 18.61

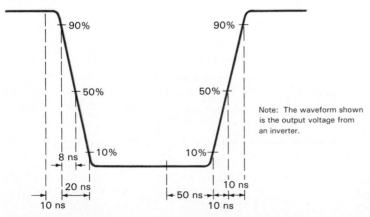

FIGURE 18.62

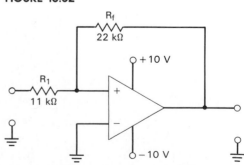

26. Determine the UTP and LTP values for the Schmitt trigger in Figure 18.63. [19]

27. Determine the UTP and LTP values for the Schmitt trigger in Figure 18.64. [19]

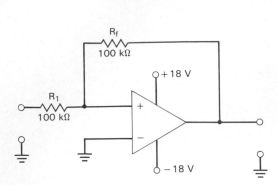

FIGURE 18.63

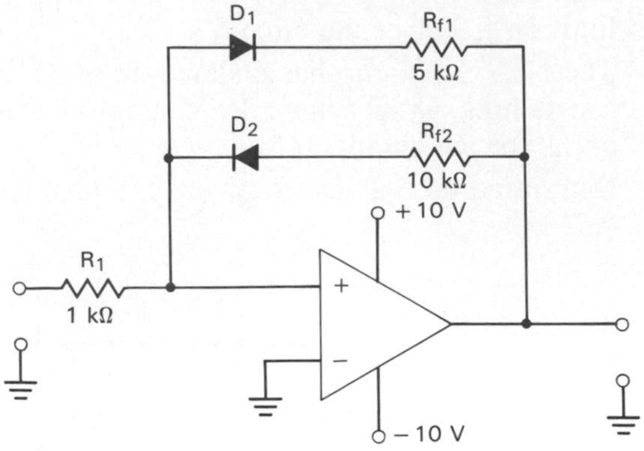

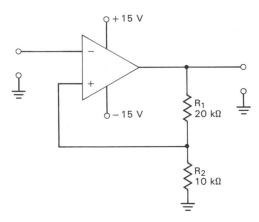

FIGURE 18.64

28. Determine the UTP and LTP values for the Schmitt trigger in Figure 18.65. [19]

29. Determine the UTP and LTP values for the Schmitt trigger in Figure 18.66. [20]

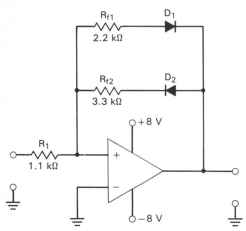

FIGURE 18.65

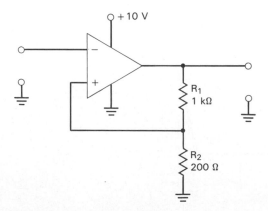

FIGURE 18.66

30. Determine the UTP and LTP values for the Schmitt trigger in Figure 18.67. [20]

FIGURE 18.67

CHAP. 18   Solid-state Switching Circuits

31. Determine the UTP and LTP values for the Schmitt trigger in Figure 18.68. [20]

32. Determine the UTP and LTP values for the Schmitt trigger in Figure 18.69. [20]

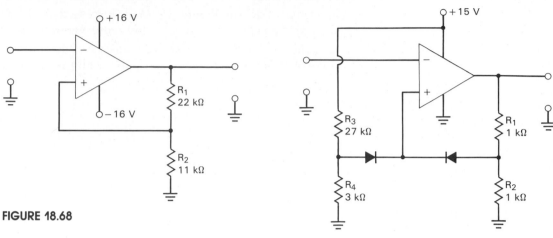

FIGURE 18.68

FIGURE 18.69

§18.4

33. Calculate the output pulse width of the one-shot in Figure 18.70. [26]
34. Calculate the output pulse width of the one-shot in Figure 18.71. [26]
35. Calculate the value of $V_{T(max)}$ for the switch in Figure 18.72. [28]
36. The resistor values in Figure 18.72 are changed to $R_1 = 33 \ k\Omega$ and $R_2 = 11 \ k\Omega$. Recalculate the value of $V_{T(max)}$ for the circuit. [28]

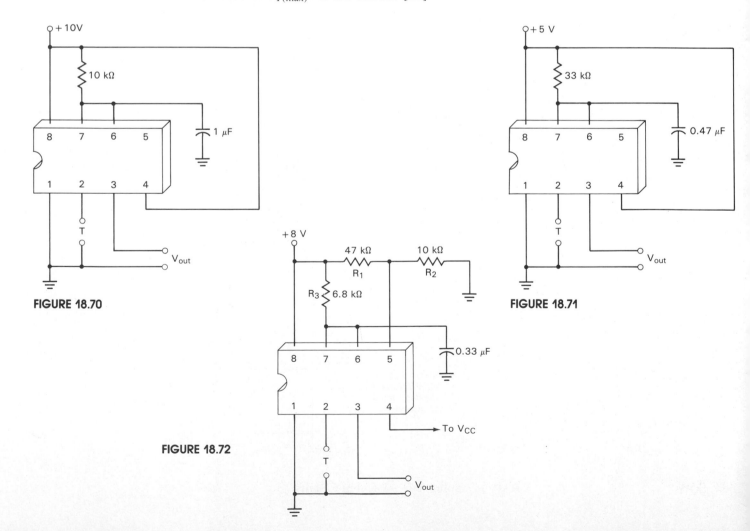

FIGURE 18.70

FIGURE 18.71

FIGURE 18.72

37. Calculate the operating frequency, duty cycle, and pulse-width values for the free-running multivibrator in Figure 18.73. [30]

38. Calculate the operating frequency, duty cycle, and pulse-width values for the free-running multivibrator in Figure 18.74. [30]

39. The value of R_B in Figure 18.73 is increased to 10 kΩ. Recalculate the operating frequency, duty cycle, and pulse-width values for the circuit. [30]

40. The value of C_1 in Figure 18.74 is increased to 0.033 μF. Recalculate the values of operating frequency and pulse width for the circuit. [30]

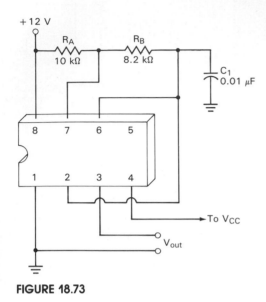

FIGURE 18.73

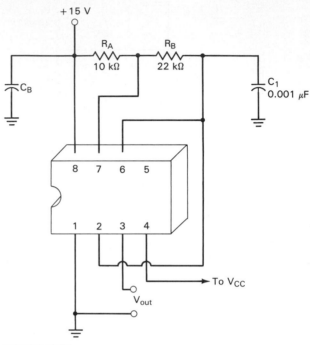

FIGURE 18.74

TROUBLESHOOTING PRACTICE PROBLEMS

41. The free-running multivibrator in Figure 18.75 has the dc readings indicated when power is applied. Discuss the possible cause(s) of the problem.

FIGURE 18.75

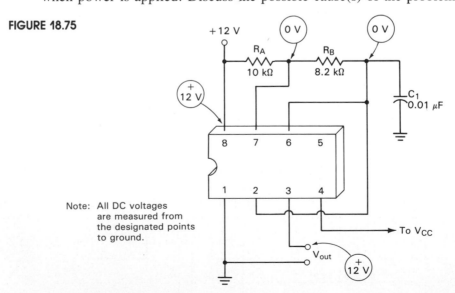

Note: All DC voltages are measured from the designated points to ground.

42. The free-running multivibrator in Figure 18.76 has the dc readings indicated when power is applied. Discuss the possible cause(s) of the problem.

FIGURE 18.76

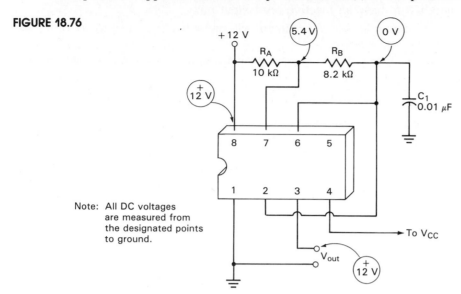

Note: All DC voltages are measured from the designated points to ground.

THE BRAIN DRAIN

43. The BJT in Figure 18.77a is faulty. Determine which (if any) of the transistors listed in the selector guide can be used as a substitute component. (Be careful, this one is more difficult than it appears!)

FIGURE 18.77

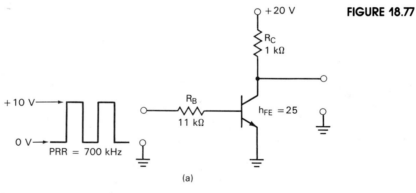

(a)

TABLE 2. Switching Transistors

Pinout: 1-Base, 2-Emitter, 3-Collector

Devices are listed in order of descending f_T.

| Device | Marking | Switching Time (ns) | | $V_{(BR)CEO}$ | h_{FE} | | | f_T |
		t_{on}	t_{off}		Min	Max	@ I_C (mA)	Min (MHz)
NPN								
MMBT2369	1J	12	18	15	20	—	100	—
BSV52	B2	12	18	12	40	120	10	400
MMBT2222	1B	35	385	30	30	—	500	250
MMBT2222A	1P	35	385	40	40	—	500	200
MMBT4401	2X	35	255	40	40	—	500	250
MMBT3903	1Y	70	225	40	15	—	100	250
MMBT3904	1A	70	250	40	30	—	100	200
PNP								
MMBT3638A	BN	75	170	25	20	—	300	—
MMBT3638	AM	75	170	25	20	—	300	—
MMBT3640	2J	25	35	12	20	—	50	500
MMBT4403	2T	35	225	40	90	180	1	150
MMBT2907	2B	45	100	40	30	—	500	200
MMBT2907A	2F	45	100	60	50	—	500	200
MMBT3906	2A	70	300	40	100	300	10	250

44. Using a 555 timer, design a free-running multivibrator that produces a 50-kHz output with a 60% duty cycle. The supply voltage for the timer is +10 V, and it is to have an inactive reset input pin.

45. Refer to Figure 18.78. The circuit was intended (by design) to fulfill the following requirements:
    a. The LED should light when a positive-going input signal passes an UTP value of +3 V.
    b. The LED should turn off when a negative-going input signal passes a LTP of −3 V. There is a flaw in the circuit design. Find the flaw and suggest a correction that will make the circuit work properly.

FIGURE 18.78

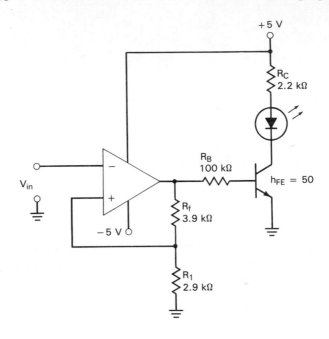

SUGGESTED COMPUTER APPLICATIONS PROBLEMS

46. Write a program that will determine the UTP and LTP values for both of the following:
    a. A noninverting Schmitt trigger like the one in Figure 18.30.
    b. An inverting Schmitt trigger like the one in Figure 18.32.

47. Write a program that will determine the PRR, duty cycle, and pulse-width values for a free-running multivibrator.

ANSWERS TO EXAMPLE PRACTICE PROBLEMS

18.1. 7.99 V

18.2. When $V_{GS} = 0$ V, $V_{out} = 1$ V. When $V_{GS} = -4$ V, $V_{out} = 10$ V.

18.3. 7 V

18.4. PW = 6 μs, $T_C = 18$ μs

18.5. 33.3%

18.6. Theoretical: 5.41 MHz,
Practical: 1.4 MHz

18.7. 22.2 kHz

18.8. UTP = +4 V, LTP = −4 V

18.9. UTP = +1.4 V, LTP = −2 V

18.10. UTP = +2.25 V, LTP = −2.25 V

18.11. 29.7 ms

18.12. 5.63 V

18.13. $f_0 = 6.9$ kHz, dc = 76.8%, PW = 111.3 μs

19
Thyristors and Optoelectronic Devices

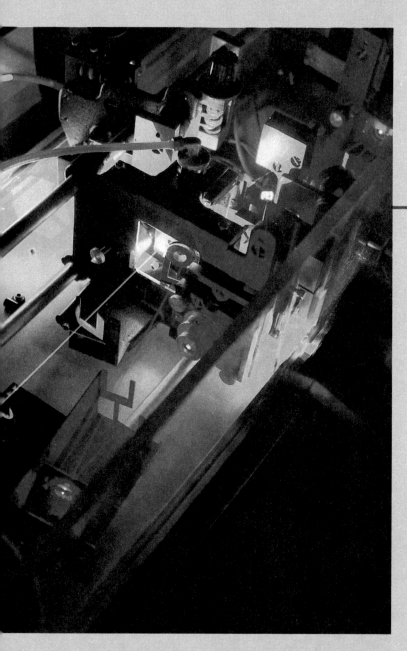

Industrial-optical systems, like the one shown, use both thyristors and optoelectronic devices to perform the functions for which they are designed.

OBJECTIVES

After studying the material in this chapter, you should be able to:

1. Contrast *thyristors* with other switching devices. (Introduction)
2. Define *optoelectronic devices*. (Introduction)
3. Describe the construction and operation of the *silicon unilateral switch* (*SUS*). (§19.1)
4. Describe the methods used to drive the SUS into cutoff. (§19.1)
5. Describe the various forward operating parameters of the SUS and explain how the device can be triggered by an increase in temperature. (§19.1)
6. Describe the construction and operation of the *silicon-controlled rectifier* (*SCR*). (§19.1)
7. List the methods that are used to trigger the SCR and drive it into cutoff. (§19.2)
8. Describe SCR triggering using the forward operating curve of the device. (§19.2)
9. List and define the primary SCR parameters and ratings. (§19.2)
10. Calculate the surge current limits for an SCR circuit, given the *circuit fusing* (I^2t) rating of the device. (§19.2)
11. Determine whether or not the rise time of an anode noise signal is sufficient to cause *false triggering*. (§19.2)
12. Discuss the methods used to prevent false triggering in SCR circuits. (§19.2)
13. Describe the function and operation of the *SCR crowbar*. (§19.2)
14. Describe the function and operation of the *SCR phase controller*. (§19.2)
15. Describe the function and operation of the *SCR motor-speed controller*. (§19.2)
16. Describe the procedures for testing SCRs. (§19.2)
17. Identify the most commonly used *bidirectional thyristors*. (§19.3)
18. Describe the symbol, construction, and operation of the *diac*. (§19.3)
19. State the most common application for the diac. (§19.3)
20. Identify the terminals of the *triac* on its schematic symbol. (§19.3)
21. Describe the operation and triggering of the triac. (§19.3)
22. Describe the methods that are commonly used to control triac triggering. (§19.3)
23. Describe the operation of the *triac phase controller*. (§19.3)

24. Describe the symptoms for the common triac faults. (§19.3)
25. Describe the construction and operation of the *unijunction transistor* (*UJT*). (§19.4)
26. Define *intrinsic standoff ratio* (η) and explain its significance. (§19.4)
27. Calculate the peak voltage (V_P) value for a UJT, given the values V_{BB} and η. (§19.4)
28. Define *negative resistance* and discuss the negative resistance region of UJT operation. (§19.4)
29. Describe the operation of the UJT relaxation oscillator. (§19.4)
30. Describe the operation of the *programmable unijunction transistor* (*PUT*) and its use as the active component in a relaxation oscillator. (§19.4)
31. Describe the construction, operation, and applications of the *silicon bilateral switch* (*SBS*). (§19.5)
32. Describe the construction and operation of the *gate turn-off* (*GTO*) device. (§19.5)
33. Discuss the primary limitation on the use of the GTO device and the means used to compensate for it. (§19.5)
34. Describe the operation and applications of the *SIDAC*. (§19.5)
35. Contrast light *emitters* and *detectors*. (§19.6)
36. Explain the concept of *wavelength* (λ). (§19.6)
37. Calculate the wavelength (in nanometers) of a signal with a specified frequency. (§19.6)
38. Explain why wavelength is an important concept in optoelectronics. (§19.6)
39. Define *light intensity*. (§19.6)
40. Explain the operation, parameters, and ratings of the *photodiode*. (§19.6)
41. Describe the operation of the *phototransistor*. (§19.6)
42. State the difference between the *photo-Darlington* and the standard phototransistor. (§19.6)
43. State the difference between the *LASCR* and the standard SCR. (§19.6)
44. List the primary optical ratings that must be considered when working with any photodetector. (§19.6)
45. Discuss the construction, operation, parameters, and ratings of the *optoisolator*. (§19.7)
46. Discuss the operation of the *solid-state relay* (*SSR*). (§19.7)
47. Compare and contrast the *optointerrupter* with the *optoisolator*. (§19.7)

As you master each of the objectives listed, place a check mark (ν) in the appropriate box.

I n this chapter, we're going to take a look at two independent groups of electronic devices. The first group of devices we will discuss are classified as *thyristors*. Thyristors are devices that are designed specifically for high-power switching applications. Unlike BJTs, FETs, and op-amps (which can also be used as switches), most thyristors most are not designed to be used as linear amplifying devices.

The second group of devices we will discuss are the *optoelectronic* devices. These devices are controlled by and/or emit (generate) light. In Chapter 2, we discussed the most basic of the optoelectronic devices, the *light-emitting diode*, or *LED*. In this chapter, we will discuss some of the other devices that belong to this group.

It should be noted that there is no central theme that ties thyristors and optoelectronic devices together. These two groups of components are being covered in the same chapter simply because it is convenient to do so.

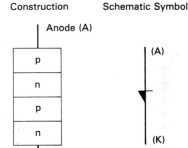

Thyristors. Devices designed specifically for high-power switching applications.

Optoelectronic devices. Devices that are controlled by and/or emit light.

19.1
Introduction To Thyristors: The Silicon Unilateral Switch (SUS)

The simplest of the thyristors is the *silicon unilateral switch* (*SUS*). The SUS is a two-terminal, four-layer device that can be triggered into conduction by applying a specified forward voltage across its terminals. The construction of the SUS (along with its schematic symbol) is shown in Figure 19.1. As you can see, the device contains two *n*-type regions and two *p*-type regions. Because of its construction, the SUS is often referred to as a *pnpn diode* or a *four-layer diode*. Note that the term *diode* is considered appropriate because the device has two terminals, called the *cathode* (*K*) and the *anode* (*A*).

Silicon unilateral switch (SUS). A two-terminal, four-layer device that can be triggered into conduction by applying a specified forward voltage across its terminals.

SUS Operation

The overall operation of the SUS is easy to understand if we discuss it in terms of a two-transistor equivalent circuit. The derivation of this equivalent circuit is illustrated in Figure 19.2. If we split the SUS as shown in Figure 19.2a, we can see that the device is effectively made up of two transistors: one *pnp* (Q_1) and one *npn* (Q_2). Note that the collector of each transistor is connected to the base of the other.

Construction   Schematic Symbol

FIGURE 19.1

SUS construction and schematic symbol.

FIGURE 19.2

SUS split construction and equivalent circuit.

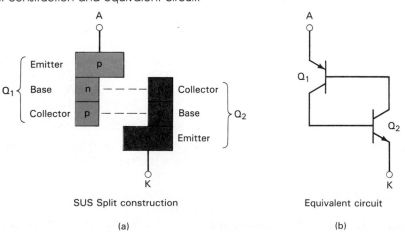

SUS Split construction   Equivalent circuit

(a)   (b)

The SUS is actually known by a variety of names. In addition to *SUS*, *pnpn diode*, and *four-layer diode*, the device is also referred to as a *Shockley diode* and a *reverse blocking diode thyristor* (the JEDEC designated name).

Figure 19.3a illustrates the response of an *off* (nonconducting) SUS to an increase in supply voltage. Assuming that both transistors are in cutoff, the total current through the device (I_F) is approximately equal to zero, as is the total current through the circuit. Since I_T is approximately equal to zero, there is no voltage dropped across the series resistor, and V_{AK} is approximately equal to V_S. Note that these circuit conditions will exist as long as V_{AK} is less than the *forward breakover voltage* [$V_{BR(F)}$] rating of the SUS. The $V_{BR(F)}$ rating of an SUS is the value of forward voltage that will force the device into conduction.

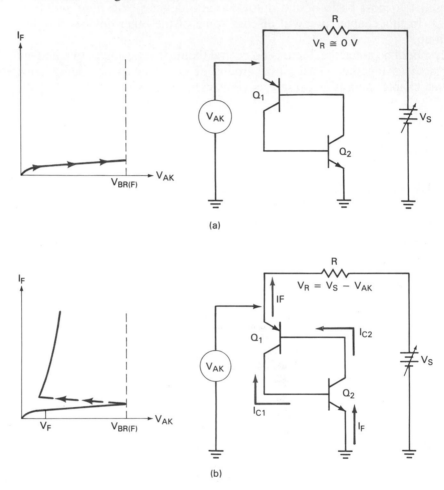

FIGURE 19.3

Figure 19.3b shows what happens when the value of V_{AK} reaches the forward breakover voltage of the device. When this occurs, the two transistors break down and begin to conduct. As soon as conduction starts, both transistors in the SUS equivalent circuit are driven into *saturation*. This causes the voltage across the SUS to drop as shown in the operating curve. Thus, when the SUS is forced into conduction, two things happen:

1. The device current (I_F) rapidly increases as the device is driven into saturation.
2. The value of V_{AK} rapidly decreases, due to the low resistance of the saturated device materials. (The relatively low value of V_{AK} is identified as V_F in the operating curve.)

As you can see, the SUS acts as an open circuit, dropping the applied voltage until $V_{BR(F)}$ is reached. At that time, the device breaks down and becomes a low-impedance conductor.

You may be wondering why the start of conduction causes both transistors in the SUS equivalent circuit to go into saturation. The answer is simple. If you look closely at the SUS equivalent circuit, you'll see that two conditions exist:

1. I_{C2} goes to the base of Q_1. Thus, $I_{C2} = I_{B1}$.
2. I_{B2} goes to the collector of Q_1. Thus, $I_{B2} = I_{C1}$.

Now, consider what happens when I_{B1} starts to flow. An increase in I_{B1} causes an increase in I_{C1}. This, in turn, causes an increase in I_{B2}, which causes an additional increase in I_{B1}. As this cycle continues, the device is rapidly forced into saturation.

The fact that the SUS forces itself into saturation is important. You may recall that the current through a standard device is controlled by the values of resistance and voltage that are external to the component. Thus, the forward current through the saturated SUS in Figure 19.3b is found as

$$I_F = \frac{V_S - V_F}{R_S} \qquad (19.1)$$

where V_F is the forward voltage drop across the saturated SUS. Equation (19.1) is important for several reasons. Obviously, it provides us with the means of determining the forward current through a saturated SUS. It also shows that the SUS requires the use of a *series current-limiting resistor*. Without such a resistor, there will be no limit on the device current, and the SUS will most likely be destroyed.

Driving the SUS into Cutoff

Once an SUS is driven into conduction, the device will continue to conduct as long as I_F is greater than a specified value, called the *holding current* (I_H). The holding current rating of an SUS indicates the minimum forward current required to maintain conduction. Once I_F drops below I_H, the device is driven into cutoff and will remain in cutoff until V_{AK} reaches $V_{BR(F)}$ again.

Several methods are used to drive I_F below the value of I_H. The first of these methods, called *anode current interruption*, is illustrated in Figure 19.4. In Figure 19.4a, a series switch is opened to interrupt the path for I_F, causing the device current to drop to zero. In Figure 19.4b, a shunt switch is closed to divert the circuit current (I_T) around the SUS, again causing the device current to drop to zero. In both cases, I_F has been forced to a value that is less than the I_H rating of the SUS, and the device is driven into cutoff.

Holding current (I_H). The minimum value of I_F required to maintain conduction.

Anode current interruption. A method of driving an SUS into cutoff by breaking the diode current path or shorting circuit current around the diode.

FIGURE 19.4

Anode current interruption.

Series interruption
(a)

Shunt interruption
(b)

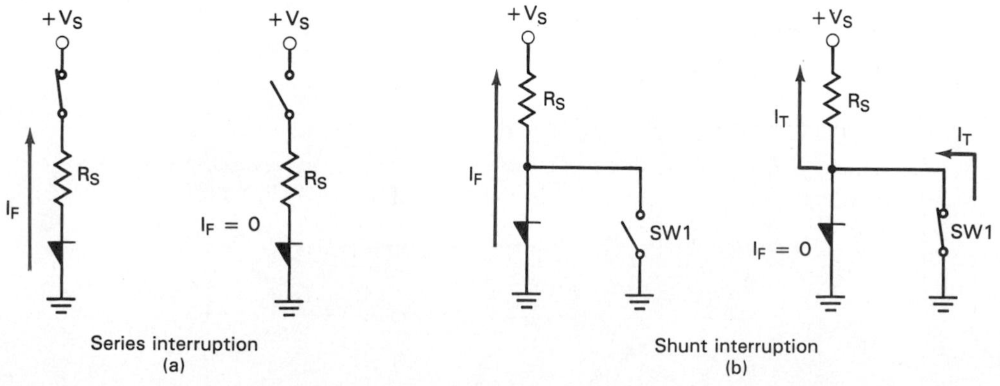

Forced commutation. Driving an SUS into cutoff by forcing a reverse current through the device. The SUS turns off when $I_F - I_R < I_H$.

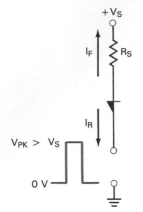

FIGURE 19.5

Forced commutation.

In practice, *series interruption* is generally caused by the *opening of a fuse*. As you'll see later in this chapter, thyristors can be used to protect voltage-sensitive loads from overvoltage conditions. When such a circuit is activated, the thyristor will stay on and protect the load just long enough for a fuse to blow. The blown fuse then acts as the open switch in Figure 19.4a. *Shunt interruption* is also used for load protection. However, the switch in the shunt interruption circuit would normally be a *reset* switch that is pressed when the overvoltage situation has passed. By closing the reset switch, the SUS is driven into cutoff, and normal circuit operation is restored.

Another method for driving an SUS into cutoff is called *forced commutation*. In this method (shown in Figure 19.5), a positive pulse is applied to the cathode of the device, forcing a *reverse current* (I_R) into the SUS. Since this current is in the opposite direction of I_F, the net current through the SUS is equal to the difference between I_F and I_R. The device turns off when $I_F - I_R < I_H$.

SUS Parameters

Figure 19.6a shows the operating curve of the SUS. As you can see, the reverse region of the curve is nearly identical to that of the *pn* junction diode and the zener diode. The part of the curve that falls between 0 V and the reverse breakdown voltage [$V_{BR(R)}$] is called the *reverse blocking region*. When the SUS is operating in this region, it will have the same characteristics as a reverse-biased *pn* junction diode; that is, the reverse current (I_R) will be in the low-microampere range (under normal circumstances), and the reverse voltage across the device (V_R) will be equal to the applied voltage. If the applied voltage reaches the reverse breakdown rating of the SUS, the device will break down and conduct in the reverse direction. As with the *pn* junction diode, this reverse conduction will usually destroy the component.

Forward blocking region. The *off-state* (nonconducting) region of operation.

The forward operating curve of the SUS is enlarged in Figure 19.6b. Forward operation is divided into two regions. The *forward blocking region* is the area of the curve that illustrates the *off-state* (nonconducting) operation of the device.

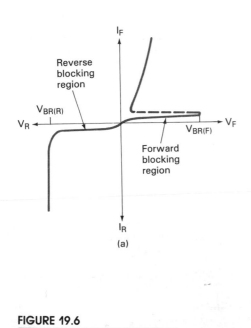

(a)

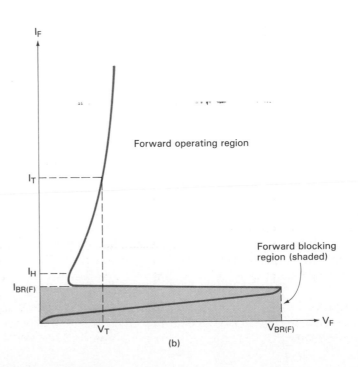

(b)

FIGURE 19.6

The SUS operating curve.

CHAP. 19   Thyristors and Optoelectronic Devices

The *forward operating region* is the area of the curve that illustrates the *on-state* (conducting) operation of the device.

The forward blocking region is defined by two SUS parameters: $V_{BR(F)}$ and $I_{BR(F)}$. As you know, $V_{BR(F)}$ (the forward breakover voltage) is the value of V_F that will cause the SUS to break down and conduct in the forward direction. The *forward breakover current* [$I_{BR(F)}$] rating of the SUS is the value of I_F at the point where breakover occurs. For example, the ECG6404 has the following parameters: $V_{BR(F)} = 10$ V (maximum) and $I_{BR(F)} = 500$ μA. This means that the value of I_F will be approximately 500 μA when V_F reaches the breakover rating of 10 V.

The importance of the $I_{BR(F)}$ rating is illustrated in Figure 19.7. Here we see the curve for a SUS that has a anode-to-cathode voltage (V_{AK}) that is less than the breakover rating of the device. As temperature increases, the leakage current through the device increases. When the temperature reaches 150°C, the leakage current reaches the value of $I_{BR(F)}$, and the devices switches into the on state. Thus, an SUS can be triggered into conduction by a significant increase in operating temperature, as well as having a forward voltage that reaches $V_{BR(F)}$. It should be noted that the temperature values shown in Figure 19.7 are for illustrative purposes and are not standard values.

Forward operating region. The on-state (conducting region of operation.

Forward breakover current, $I_{BR(F)}$. The value of I_F at the point where breakover occurs.

The effects of temperature on SUS breakover.

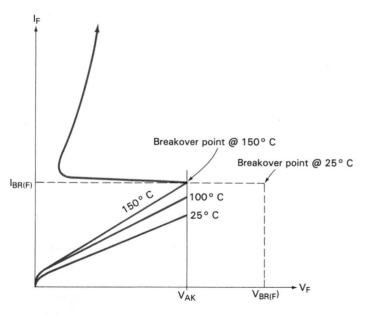

FIGURE 19.7

The effects of temperature on $V_{BR(F)}$.

A Practical Consideration. As temperature increases, the value of $V_{BR(F)}$ decreases. For the curve in Figure 19.7, the value of $V_{BR(F)}$ would fall on the V_{AK} value when $T = 150°$C. In other words, the intersection of the dashed $I_{BR(F)}$ and $V_{BR(F)}$ lines would have moved to the 150°C breakover point.

Once the SUS turns on, the device moves into the forward operating region. The holding current (I_H) is the minimum forward current required to maintain the on-state condition. When SUS current drops below I_H (which is typically in the neighborhood of 20 mA), the device returns to the off state, as was stated earlier in this section.

The *average on-state current* (I_T) is the maximum average forward current for the SUS. The series current-limiting resistor for an SUS is selected to ensure that the dc forward current through the device is held below this value. The *average on-state voltage* (V_T) rating is the value of V_F when the current through the SUS equals I_T.

Average on-state current (I_T). The maximum average forward current for the SUS.

Average on-state voltage (V_T). The value of V_F when $I_F = I_T$.

In addition to the parameters discussed here, the SUS has the standard maximum current, power dissipation, and voltage ratings. Since we have covered these parameters for other components, we will not cover them here.

Section Summary

The *silicon unilateral switch* (*SUS*) is a thyristor that is forced into conduction when the forward voltage across the device reaches a specified *forward breakover voltage*, $V_{BR(F)}$. Once triggered, the device becomes a low-impedance conductor. It remains in this *on state* (conducting state) until its forward current (I_F) drops below the *holding current* (I_H) rating of the device. At that time, the device returns to the *off state* (nonconducting state).

The SUS can be driven from the on state to the off state by one of two methods. *Anode current interruption* involves blocking or diverting the diode current so that I_F drops below the value of I_H. *Forced commutation* involves forcing a reverse current through the diode so that the net diode current ($I_F = I_R$) is less than I_H. In either case, the device returns to the off state and will remain there until triggered again.

One Final Note

You may be wondering why we haven't discussed SUS applications and troubleshooting. The fact is that the SUS is rarely used in modern circuit design. In fact, many device manufacturers have discontinued the production of silicon unilateral switches.

Despite its bleak future, the operation of the SUS was covered in detail for several good reasons. First, many of the SUS operating principles apply equally to many of the more commonly used thyristors. In fact, you will see in the next two sections that many of the commonly used thyristors are nothing more than variations on the basic SUS. Second, the fact that a component is being phased out does not mean that you will never run into one. In the event that you are faced with working with an SUS-based circuit, you will have the knowledge needed to handle the situation.

SECTION REVIEW

1. What are *thyristors*? [Objective 1]
2. What distinguishes the thyristors from other switching devices, such as BJTs, FETs, and op-amps? [Objective 1]
3. What are *optoelectronic* devices? [Objective 2]
4. What is the *silicon unilateral switch* (*SUS*)? [Objective 3]
5. Describe the construction of the SUS. [Objective 3]
6. What is *forward breakover voltage* [$V_{BR(F)}$] rating? [Objective 3]
7. Describe what happens when the voltage applied to an SUS reaches the $V_{BR(F)}$ rating of the device. [Objective 3]
8. Describe the two methods of *anode current interruption*. [Objective 4]
9. Explain how *forced commutation* drives an SUS into cutoff. [Objective 4]
10. What is the *forward blocking region* of an SUS? [Objective 5]
11. What is the *forward operating region* of an SUS? [Objective 5]
12. Explain how temperature can drive an SUS into the forward operating region. [Objective 5]

19.2
Silicon-Controlled Rectifier (SCR)

☞ 6

The *silicon-controlled rectifier*, or *SCR*, is a three-terminal device that is very similar in construction and operation to the SUS. The construction of the SCR is shown in Figure 19.8 along with its schematic symbol. As you can see, the construction of the SCR is identical to that of the SUS, outside of the addition of a third terminal called the *gate*. As you will see, the gate provides an additional means of triggering the device into the on state. Outside of that, the operation of the SCR is identical to that of the SUS.

Silicon-controlled rectifier (SCR). A three-terminal device that is very similar in construction and operation to the SUS. The third terminal, called the *gate*, provides another method for triggering the device.

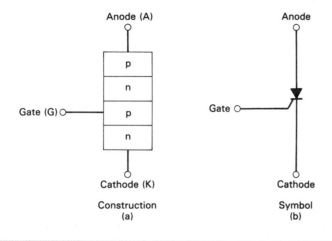

FIGURE 19.8

SCR construction and schematic symbol.

SCR Triggering

☞ 7

Figure 19.9a shows the two-transistor equivalent circuit for the SCR. Again, note the similarities between the SCR and the SUS. Like that SUS, the SCR can be triggered into conduction by applying an anode-to-cathode voltage (V_{AK}) that is equal to the $V_{BR(F)}$ rating of the device.

FIGURE 19.9

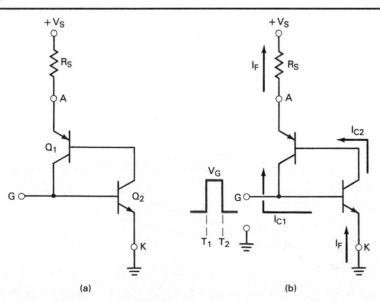

The addition of the gate terminal provides an additional means of triggering the device, as is shown in Figure 19.9b. Here we see a positive pulse being applied to the gate terminal. When the gate pulse goes positive (T_1), Q_2 is forced into conduction. At that point, the device takes itself into saturation, just like the SUS.

When the gate pulse is removed (T_2), the device continues its forward conduction. The forward conduction continues until the forward current (I_F) drops below the *holding current* (I_H) rating of the device. Note that the SCR is driven into cutoff (the off state) using the same two methods we use for the SUS: *anode current interruption* and *forced commutation*.

SCR Operating Curve

8

Figure 19.10a shows the operating curve of the SCR. As you can see, it is identical to that of the SUS. The enlarged forward operating curve (Figure 19.10b) shows the effects of gate current on forward conduction. As the value of I_G is increased, the device breaks over into the on state at decreasing values of V_F. For example, let's say that we have an applied SCR voltage that is equal to V_{AK} in Figure 19.10b. If $I_G = I_{G1}$, the device will be in the off state. However, if we increase I_G to the value of I_{G2}, the value of V_{AK} is more than high enough to drive the device into the on state. Again, when the device turns on, the forward current must be driven below the component's I_H rating to return it to the off state.

SCR Specifications

Many of the specifications for the SCR are identical to those for the SUS. For example, take a look at the spec sheet for the 2N682–92 series SCRs in Figure 19.11. Table 19.1 lists the specifications contained in this spec sheet that have

FIGURE 19.10

The SCR operating curve.

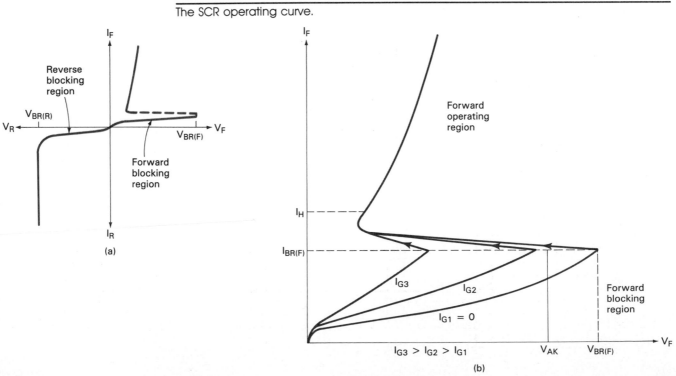

Silicon Controlled Rectifiers
Reverse Blocking Triode Thyristors

. . . designed primarily for half-wave ac control applications, such as motor controls, heating controls and power supplies; or wherever half-wave silicon gate-controlled, solid-state devices are needed.

- Glass Passivated Junctions and Center Gate Fire for Greater Parameter Uniformity and Stability
- Blocking Voltage to 800 Volts

2N682 thru 2N692

SCRs 25 AMPERES RMS 50 thru 800 VOLTS

CASE 263-04 STYLE 1

MAXIMUM RATINGS (T_J = 125°C unless otherwise noted.)

Rating	Symbol	Value	Unit
*Peak Repetitive Off-State Blocking Voltage, Note 1	V_{RRM} or V_{DRM}		Volts
2N682		50	
2N683		100	
2N685		200	
2N688		400	
2N690		600	
2N692		800	
*Peak Non-Repetitive Reverse Voltage	V_{RSM}		Volts
2N682		75	
2N683		150	
2N685		300	
2N688		500	
2N690		720	
2N692		960	
*RMS On-State Current (All Conduction Angles)	$I_{T(RMS)}$	25	Amps
*Average On-State Current (T_C = 65°C)	$I_{T(AV)}$	16	Amps
*Peak Non-Repetitive Surge Current (One cycle, 60 Hz, preceded and followed by rated current and voltage)	I_{TSM}	150	Amps
Circuit Fusing Considerations (t = 8.3 ms)	I^2t	93	A^2s
*Peak Gate Power	P_{GM}	5	Watts
*Average Gate Power	$P_{G(AV)}$	0.5	Watt

*Indicates JEDEC Registered Data   (cont.)

Note 1. V_{DRM} and V_{RRM} for all types can be applied on a continuous dc basis without incurring damage. Ratings apply for zero or negative gate voltage. Devices should not be tested for blocking capability in a manner such that the voltage supplied exceeds the rated blocking voltage.

FIGURE 19.11

The 2N682–92 SCR specification sheet.
(Courtesy of Motorola, Inc.)

MAXIMUM RATINGS — continued (T_J = 125°C unless otherwise noted.)

Rating	Symbol	Value	Unit
*Peak Forward Gate Current 2N682–2N688 2N690, 2N692	I_{GM}	2 1.2	Amps
*Peak Gate Voltage — Forward Reverse	V_{FGM} V_{RGM}	10 5	Volts
*Operating Junction Temperature Range	T_J	−65 to +125	°C
*Storage Temperature Range	T_{stg}	−65 to +150	°C
Stud Torque	—	30	in. lb.

THERMAL CHARACTERISTICS

Characteristic	Symbol	Max	Unit
Thermal Resistance, Junction to Case	$R_{\theta JC}$	2	°C/W

ELECTRICAL CHARACTERISTICS (T_J = 25°C unless otherwise noted.)

Characteristic	Symbol	Min	Typ	Max	Unit
*Average Forward or Reverse Blocking Current (Rated V_{DRM} or V_{RRM}, gate open, T_J = 125°C) 2N682–2N683 2N685 2N688 2N690 2N692	$I_{D(AV)}$, $I_{R(AV)}$	 — — — — —	 — — — — —	 6.5 6 4 2.5 2	mA
Peak Forward or Reverse Blocking Current (Rated V_{DRM} or V_{RRM}, gate open) T_J = 25°C T_J = 125°C	I_{DRM}, I_{RRM}	 — —	 — —	 10 20	 μA mA
*Peak On-State Voltage (I_{TM} = 50.3 A peak, Pulse Width ≤ 1 ms, Duty Cycle ≤ 2%)	V_{TM}	—	—	2	Volts
Gate Trigger Current (Continuous dc) (V_{AK} = 12 Vdc, R_L = 50 Ω) *(V_{AK} = 12 Vdc, R_L = 50 Ω, T_C = −65°C)	I_{GT}	 — —	 — —	 40 80	mA
Gate Trigger Voltage (Continuous dc) (V_{AK} = 12 Vdc, R_L = 50 Ω) *(V_{AK} = 12 Vdc, R_L = 50 Ω, T_J = −65°C)	V_{GT}	 — —	 0.65 —	 2 3	Volts
*Gate Non-Trigger Voltage (Rated V_{DRM}, R_L = 50 Ω, T_J = 125°C)	V_{GD}	0.25	—	—	Volts
Holding Current (V_{AK} = 12 Vdc, Gate Open)	I_H	—	7.3	50	mA
Critical Rate of Rise of Off-State Voltage (Rated V_{DRM}, Exponential Waveform, T_J = 125°C, Gate Open)	dv/dt	—	30	—	V/μs

*Indicates JEDEC Registered Data.

FIGURE 19.11
(continued)

already been discussed. One important point can be seen in Table 19.1. The forward and reverse blocking voltages and currents for an SCR will always be equal in magnitude. The same thing holds true for the SUS.

TABLE 19.1

Specification	*Symbol*	*Definition*
Peak repetitive off-state blocking voltage	V_{RRM} or V_{DRM}	The maximum forward or reverse blocking voltage. The same thing as $V_{BR(F)}$ and $V_{BR(R)}$.
Peak forward or reverse blocking current	I_{DRM} or I_{RRM}	The maximum forward or reverse blocking current. The same thing as $I_{BR(F)}$ and $I_{BR(R)}$.
Holding current	I_H	The minimum forward current required to maintain conduction.

In addition to the specifications listed in Table 19.1, several other parameters warrant discussion. The first of these is the *circuit fusing* (I^2t) rating. This rating indicates *the maximum forward surge current capability* of the device. For example, the I^2t rating of the 2N682–92 series SCRs is 93 A²s. If the product of the square of surge current times the duration (time) of the surge exceeds 93 A²s, the device will be destroyed by excessive power dissipation. The following example illustrates the determination of I^2t for a given circuit.

Circuit fusing (I^2t) rating. The maximum forward surge current capability of an SCR.

══ EXAMPLE 19.1 ══

In a given circuit, the 2N682 is subjected to a 50-A current surge that lasts for 10 ms. Determine whether or not this surge will destroy the device.

Solution: The I^2t value for the circuit is found as

$$I^2t = (50 \text{ A})^2(10 \text{ ms})$$
$$= (2500 \text{ A})(10 \text{ ms})$$
$$= 25 \text{ A}^2\text{s}$$

Since this value is well below the maximum rating of 93 A²s, the device will not be destroyed.

PRACTICE PROBLEM 19–1

The 2N6394 has a circuit fusing rating of 40 A²s. Determine whether or not the device will survive a 60-A surge that lasts for 15 ms.

When the circuit fusing rating for an SCR is known, we can determine the maximum allowable duration of a surge with a known current value, as follows:

$$\boxed{t_{\max} = \frac{I^2t \text{ (rated)}}{I_S^2}} \qquad\qquad \textbf{(19.2)}$$

where I_S = the known value of surge current.

The following example illustrates the use of equation (19.2) in determining the maximum allowable duration (time) of a surge.

══ EXAMPLE 19.2 ══

The 2N1843A has a circuit fusing rating of 60 A²s. The device is being used in a circuit where it could be subjected to a 100-A surge. Determine the limit on the duration of such a surge.

Solution: The maximum allowable duration of the surge is found as

$$t_{\max} = \frac{I^2t \text{ (rated)}}{I_S^2}$$
$$= \frac{60 \text{ A}^2\text{s}}{(100 \text{ A})^2}$$
$$= 6 \text{ ms}$$

Thus, if the 100-A surge is present for more than 6 ms, the 2N1843A will not survive the experience.

PRACTICE PROBLEM 19–2

The C35 series SCRs have a circuit fusing rating of 75 A²s. Determine the maximum allowable duration of a 150-A surge that passes through one of these devices.

We can also use a variation on equation (19.2) to determine the maximum allowable surge current value for a given period of time, as follows:

$$I_{S(max)} = \sqrt{\frac{I^2t \text{ (rated)}}{t_S}} \qquad (19.3)$$

where t = the time duration of the current surge.

The following example illustrates the use of equation (19.3) In determining the maximum allowable value of a current surge.

▬ EXAMPLE 19.3 ▬

Determine the highest surge current value that the 2N1843A can withstand for a period of 20 ms.

Note: The value of 60 A²s for the 2N1843A was given in Example 19.2.

Solution: The peak value for the surge is found as

$$I_{S(max)} = \sqrt{\frac{I^2t \text{ (rated)}}{t_S}}$$

$$= \sqrt{\frac{60 \text{ A}^2\text{s}}{20 \text{ ms}}}$$

$$= 54.77 \text{ A}$$

Thus, a 20-ms current surge would have to exceed 54.77 A to damage the 2N1843A.

PRACTICE PROBLEM 19–3

Determine the highest surge current value that a C35 series SCR can withstand for a duration of 50 ms.

Nonrepetitive surge current (I_{TSM}). The absolute limit on the forward surge current through an SCR.

Gate nontrigger voltage (V_{GD}). The maximum gate voltage that can be applied without triggering the SCR into conduction.

Later in this section, we're going to look at several SCR applications. In one of these applications, the SCR is being used as a surge protection device. When the SCR is being used in a surge protection application, the circuit fusing rating becomes critical.

The SCR has another maximum current rating called the *nonrepetitive surge current* (I_{TSM}) rating. This is the absolute limit on the surge current through the device. If this rated value is exceeded for any length of time, the device will be destroyed.

If you look under the *electrical characteristics* heading in Figure 19.11, you'll see the *gate nontrigger voltage* (V_{GD}) rating. This rating indicates the maximum gate voltage that can be applied without triggering the SCR into conduction. If V_G exceeds this rating, the SCR will be triggered into the on state.

The V_{GD} rating is important because it points out one of the potential causes of *false triggering*. False triggering is a situation where the SCR is accidentally triggered into conduction, usually by some type of noise. For example, the 2N682–92 series of SCRs has a V_{GD} rating of 250 mV. If a noise signal with a peak value greater than 250 mV appears at the gate, the device may be triggered into conduction.

Another common cause of false triggering is a rise in anode voltage that exceeds the *critical rise (dv/dt)* rating of the SCR. This rating indicates the *maximum rate of increase in anode-to-cathode voltage that the SCR can handle without false triggering occurring.* For example, the *dv/dt* rating of the 2N682–92 series SCRs is 30 V/μs. This means that a noise signal in V_{AK} with a rate of rise equal to 30 V/μs may cause false triggering.

Obviously, 30 V of noise would never occur under normal circumstances. However, even a relatively small amount of noise can have a sufficient *rate of increase* to cause false triggering. To determine the rate of increase of a given noise signal in V/μs, the following conversion formula can be used:

✔ 11

$$\Delta V = \frac{dv}{dt} \Delta t \qquad (19.4)$$

where $\dfrac{dv}{dt}$ = the critical rise rating of the component

Δt = the rise time of the increase in V_{AK}

ΔV = the amount of change in V_{AK}

As the following example illustrates, even a small amount of noise in V_{AK} can cause false triggering if the rise time of the noise signal is short enough.

EXAMPLE 19.4

A 2N682 SCR is being used in a circuit where 2-ns (rise time) noise signals randomly occur. Determine the noise amplitude required to cause false triggering.

Solution: With a critical rise rating of 30 V/μs, the amplitude that will cause false triggering is found as

$$\Delta V = \frac{dv}{dt} \Delta t$$
$$= (30 \text{ V/μs})(2 \text{ ns})$$
$$= 60 \text{ mV}$$

Thus, if V_{AK} increases by as little as 60 mV during the 2-ns duration of the noise signal, false triggering may occur.

PRACTICE PROBLEM 19–4

The C35D SCR has a critical rise rating of 25 V/μs. Determine the amplitude of noise needed to cause false triggering if the rise time of the noise signal is 100 ps.

As you can see, many of the critical SCR parameters deal with the current limits and false triggering limits of the device. At this point, we'll move on to look at the methods that are commonly used to prevent false triggering.

Preventing False Triggering

You have been shown that false triggering is usually caused by one of two conditions:

1. A noise signal at the gate terminal
2. A high-rise-time noise signal in V_{AK}

A Practical Consideration. **False triggering can also be caused by temperature, as is the case with the SUS.**

Noise from the gate signal source is generally reduced using one of the two methods shown in Figure 19.12. In Figure 19.12a, the gate is connected (via a gate resistor) to a negative dc power supply. For the sake of discussion, let's assume that the gate supply voltage is −2 V. If this is the case, the noise at the gate would have to be greater than $(2\ V + V_{GD})$ in order to cause false triggering. This is not likely to occur under any normal circumstances.

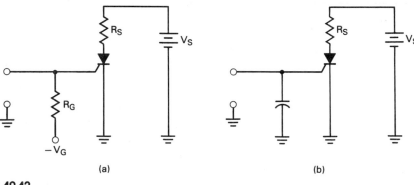

(a)   (b)

FIGURE 19.12

Preventing false triggering.

The circuit in Figure 19.12b uses a bypass capacitor that is connected between the gate terminal and the anode terminal (ground). This capacitor, which would short any gate noise signals to ground, is normally selected at a value between 0.1 μF and 0.01 μF.

Of the two methods shown in Figure 19.12, the bypass capacitor in Figure 19.12b is by far the most commonly used. This is because the capacitor provides a very inexpensive solution to the gate noise problem.

Snubber network. An *RC* circuit that is connected between the SCR anode and cathode to eliminate false triggering.

Noise in V_{AK} is normally eliminated by the use of a *snubber network*. A snubber is an *RC* circuit that is connected between the anode and cathode terminals of the SCR. A snubber network is shown in Figure 19.13. The capacitor (C_S) will effectively short any noise in V_{AK} to ground. The resistor (R_S) is included to limit the capacitor discharge current when the SCR turns on.

FIGURE 19.13

An RC snubber network.

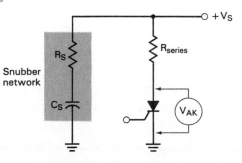

CHAP. 19   Thyristors and Optoelectronic Devices

It should be noted that noise in V_{AK} (Figure 19.13) would originate in the dc voltage source, V_S. Assuming that V_S comes from a standard dc power supply, any noise that is generated in the supply will be coupled to the SCR circuit. This noise will then be reduced by the snubber to a level that prevents it from causing false triggering.

Snubbers are used in most SCR circuits, since exceeding the critical rise (dv/dt) rating is the most common cause of false triggering. As you will see later in this chapter, snubbers are also used in a variety of circuits that contain some of the other thyristors. In other words, SCRs are not the only components that require the use of snubber networks.

SCR Applications

The SCR is the most commonly used of the thyristors. Most SCR applications are found in the area of *industrial electronics*, the area that deals with the devices, circuits, and systems that are used in manufacturing. Although we cannot cover all the possible applications for the SCR, we will touch briefly on several of them in this section.

Industrial electronics. The area of electronics that deals with the devices, circuits, and systems that are used in manufacturing.

The SCR Crowbar

One common use for the SCR is in a type of circuit called a *crowbar*. A crowbar is a circuit that is used to protect a voltage-sensitive load from excessive dc power supply output voltages. The crowbar circuit (shown in Figure 19.14) consists of a zener diode, a gate resistor (R_G), and an SCR. The circuit in Figure 19.14 also contains a snubber to prevent false triggering.

13

Crowbar. A circuit used to protect a voltage-sensitive load from excessive dc power supply output voltages.

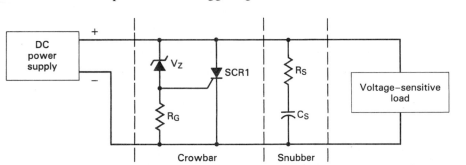

FIGURE 19.14
The SCR crowbar.

In the circuit shown, the zener diode and the SCR are normally off. With the zener diode being in cutoff, there is no path for current through R_G, and no voltage is developed across the resistor. With no voltage being developed across R_G, the gate of the SCR is at 0 V, holding the device in the off state. Thus, as long as the zener diode is off, the SCR is acting as an open and will not affect either the dc power supply or the load.

If the output voltage from the dc power supply suddenly *increases*, the zener diode breaks down and conducts. The current through the zener diode causes a voltage to be developed across R_G. This voltage then forces the SCR into conduction. When the SCR conducts, the output from the dc power supply is shorted through the device and the load is protected.

There are several points that should be made regarding this circuit:

1. The SCR would only be activated by some major fault within the dc power supply that causes its output voltage to increase drastically. Once activated, the SCR will continue to conduct until the power supply fuse opens.

The SCR in a crowbar is turned off by a blown primary fuse in the dc power supply. This is an example of *series anode current interruption*.

2. The *circuit fusing* (I^2t) rating of the SCR must be high enough to ensure that the SCR isn't destroyed before the power supply fuse has time to blow. If the SCR gives out before the power supply fuse, the power supply will continue to operate and the load will be damaged after all.

The I^2t rating of the SCR is especially critical when the dc power supply contains a slow-blow fuse.

3. The SCR in the crowbar is considered to be expendable. This is why there is no series current-limiting resistor in the circuit. The object of the circuit is to protect the load, not the SCR.

4. The SCR in a crowbar should be replaced after activation of the circuit as a precaution against future component failure from possible internal damage.

Circuit protection is far from being the only application for the SCR. A completely different type of application can be found in the SCR phase controller.

The SCR Phase Controller

Phase controller. A circuit used to control the conduction phase angle through a load, and thus the average load voltage.

Trigger-point voltage (V_{TP}). The voltage required at the anode of the gate diode to trigger the SCR.

The *phase controller* is a circuit that is used to control the conduction angle through a load, and thus the average load voltage. For example, by varying the setting of the potentiometer (R_2) in the phase controller in Figure 19.15a, we can vary the conduction angle (θ) through the load as shown in Figure 19.15b. Since the *average* (dc equivalent) load voltage varies with the conduction angle, adjusting the setting of R_2 also varies the average load voltage. This point is discussed later in this section.

Conduction through the load in Figure 19.15a is controlled by the SCR. When V_{in} causes the voltage at point A to reach a specified level, called the *trigger-point voltage* (V_{TP}), the SCR is triggered into conduction. Once triggered, the SCR allows V_{in} to pass through to the load. Note that the SCR then continues to conduct for the remainder of the positive alternation of V_{in}. The relationship between the voltage at point A (caused by V_{in}), the value of V_{TP}, and load conduction can be seen in the three waveforms in Figure 19.15b.

For the SCR to be triggered into conduction, the sine wave at point A must reach the level specified by the following equation:

$$V_{TP} = V_F + V_{GT}$$

where V_F = the forward voltage drop across the gate diode (D_1)
V_{GT} = the *gate trigger voltage* rating of the SCR

FIGURE 19.15

The SCR phase controller.

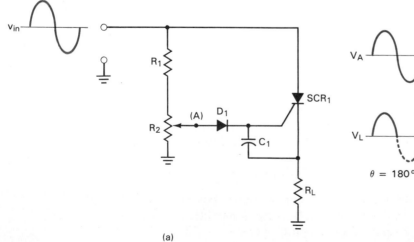
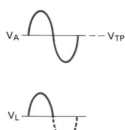
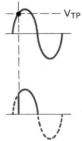
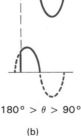
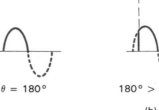
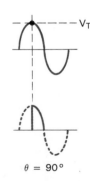

(a)

(b)

CHAP. 19 Thyristors and Optoelectronic Devices

The gate trigger voltage of the SCR is the value of V_{GK} that will cause I_G to become great enough to trigger the SCR into conduction.

When the sine wave at point A reaches the value of V_{TP}, the voltage is sufficient to overcome the forward voltage drop across D_1 and the value of V_{GT} for the SCR. This triggers the SCR into conduction for the remainder of the positive alternation of V_{in}. When V_{in} goes negative, the anode voltage of the SCR also goes negative. This causes the SCR current to drop below the holding current (I_H) value of the device, and it returns to the off state.

An important point needs to be made regarding the circuit and waveforms shown in Figure 19.15. The waveforms shown in the figure are somewhat misleading. You see, the value of V_{TP} for a given phase controller is a fixed value, while the amplitude of the sine wave at point A is variable. For example, let's say that the value of V_{GT} for the SCR in Figure 19.15a is 1.5 V. This gives the circuit a trigger-point voltage that is found as

$$V_{TP} = V_F + V_{GT}$$
$$= 0.7 \text{ V} + 1.5 \text{ V}$$
$$= 2.2 \text{ V}$$

This is the voltage required at point A to trigger the SCR into conduction. The voltage divider (R_1 and R_2) applies a sine wave to point A that is *proportional to* V_{in}. As the setting of R_2 is increased, the amplitude of the sine wave at point A is also increased. Thus, by increasing the setting of R_2, we decrease the value of V_{in} required to cause the voltage at point A to reach the value of V_{TP}. In the same fashion, decreasing the setting of R_2 increases the value of V_{in} required to cause the voltage at point A to reach the value of V_{TP}. From this discussion we can see that adjusting R_2 changes the amplitude of V_A, not the value of V_{TP} for the circuit.

A few more points should be made with regard to the circuit in Figure 19.15a:

1. The capacitor between the SCR gate and cathode terminals (C_1) is included to prevent false triggering.
2. SCRs tend to have relatively low gate breakdown voltage ratings, typically 10 V or less. For this reason, D_1 is used in the circuit to prevent the negative alternation of V_{in} from being applied to the SCR gate–cathode junction.
3. The load resistance is in series with the SCR and thus acts as the series current-limiting resistor for the device.
4. The mathematical relationship between the load conduction angle (θ) and the average (dc equivalent) load voltage (V_{AVE}) is defined using integral calculus and thus is not given here. However, it can be stated that the value of V_{AVE} for a waveform varies with *the area under the waveform* (shown as shaded areas in Figure 19.15b). As the area under the waveform decreases, so does its value of V_{AVE}. Therefore, by reducing the R_2 setting in Figure 19.15a, we reduce the load values of θ and V_{AVE}.

The fact that the value of V_{AVE} for a phase controller is variable makes the circuit very useful as a *motor-speed controller*.

 15

SCR Motor-speed Controller

A very practical variation on the phase controller is shown in Figure 19.16a. The circuit shown is a *motor-speed controller* for a *universal motor*. A universal motor is one whose rotation speed is directly proportional to the average motor voltage

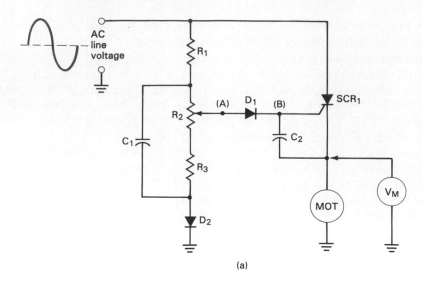

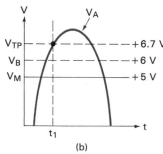

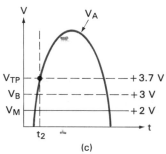

FIGURE 19.16

An SCR motor-speed
controller.

(V_M). When the value of V_M increases, so does the rotation speed of the motor, and vice versa. Note that there is a specific value of V_M that corresponds to any given rotation speed for a universal motor.

The trigger-point voltage for the motor-speed controller is found as

$$V_{TP} = V_F + V_{GT} + V_M$$

where V_M = the *average* motor voltage. This equation, along with the relationship between rotation speed and V_M, leads us to the primary difference between the phase controller (Figure 19.15) and the motor-speed controller: *The trigger-point voltage for the motor-speed controller varies with the speed of the motor.* As motor speed increases, so do V_M and V_{TP}. As motor speed decreases, so do V_M and V_{TP}.

Let's assume that the circuit in Figure 19.16a has values of $V_M = 5$ V and $V_{GT} = 1$ V. These values give us a trigger-point voltage of

$$\begin{aligned} V_{TP} &= V_F + V_{GT} + V_M \\ &= 0.7 \text{ V} + 1 \text{ V} + 5 \text{ V} \\ &= 6.7 \text{ V} \end{aligned}$$

The circuit conditions that correspond to the values of $V_M = 5$ V and $V_{TP} = 6.7$ V are represented in Figure 19.16b. Note that the voltage at point B is the sum of V_M and V_{GT}. When the positive alternation of V_{in} causes V_A to reach the value of V_{TP}, the SCR triggers and power is applied to the motor. As with the phase controller, the SCR will remain in the conducting state until V_{in} goes negative, causing forced commutation of the SCR. As long as the motor speed remains constant, this cycle repeats itself over and over again.

CHAP. 19   Thyristors and Optoelectronic Devices

Now let's take a look at how the motor-speed controller responds to a change in motor speed. Assume for a moment that some mechanical stress on the motor causes it to slow down, decreasing V_M to 2 V. We now have the situation represented in Figure 19.16c. The trigger-point voltage is now 3.7 V. Note that V_{GT} and V_F have not changed, only the value of V_M. Since a lower V_A is required to trigger the SCR, the device is triggered earlier in the cycle. This increases the dc average of the voltage applied to the motor, causing its rotation speed to increase to its original value. In other words, a decrease in motor speed results in a higher dc average voltage being applied to the motor, causing the motor to return to its original speed. Likewise, an increase in motor speed (caused by light mechanical stress) causes V_M to increase. The increase in V_M causes V_{TP} to increase, reducing the conduction angle and decreasing the dc average of the voltage applied to the motor. Again, after a few cycles of the input, the motor returns to its original speed.

The crowbar, phase controller, and motor-speed controller should give you a good idea of the wide variety of applications that exists for the SCR. At this point, we will move on to the fault tests that are normally performed on the SCR.

SCR Testing

Whenever possible, in-circuit testing on an SCR is performed. An open SCR will remain in the nonconducting state when triggered. A shorted SCR will usually conduct regardless of the polarity of the anode–cathode voltage (V_{AK}). In either case, the SCR fault can be diagnosed by checking the gate, anode, and cathode voltages of the device with a voltmeter or an oscilloscope.

Incidentally, there is one classic symptom of a shorted SCR in a motor-speed controller. If the SCR in Figure 19.16 is shorted, the complete cycle of V_{in} is applied to the motor. The constant change in the polarity of V_{in} causes the motor to *vibrate* rather than rotate in its normal direction.

In some circuits (like crowbars), it is impractical to perform any type of in-circuit tests on the SCR. In this case, the device must be tested using a circuit like the one shown in Figure 19.17. The circuit is initially set up so that V_{GK} is approximately 0 V. The value of R_L is selected to equal the value used in the V_{GT} listing on the spec sheet of the device. For example, the 2N682–92 series SCRs list a value of $R_L + 50\ \Omega$ for the testing of V_{GT} (see the spec sheet in Figure 19.11). Thus, when testing one of these SCRs, R_L would be a 51-Ω resistor. The value of $+V$ is set to equal the $V_{BR(F)}$ rating of the SCR under test.

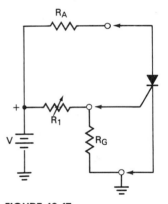

$V = V_{TM} + I_H R_A$

FIGURE 19.17

An SCR test circuit.

To test the SCR, the setting of R_1 is increased to the value of V_{GT}. At this point, the SCR should trigger, and V_{AK} should drop to a relatively low value. If the device fails to trigger, R_1 is varied across its entire range. If the device continues in the nonconducting state, it is open and must be replaced.

SCR Summary

The SCR is a three-terminal device that acts as a gated SUS. The device is normally triggered into conduction by applying a gate–cathode voltage (V_{GK}) that causes a specific level of gate current (I_G). This gate current triggers the device into conduction.

Like the SUS, the SCR is returned to its nonconducting state by either *anode current interruption* or *forced commutation*. When turned off, the device remains in its nonconducting state until another trigger is received.

Two critical parameters for the SCR are the *circuit fusing* and *critical rise* ratings. If the *circuit fusing* (I^2t) rating is exceeded, the device is destroyed by

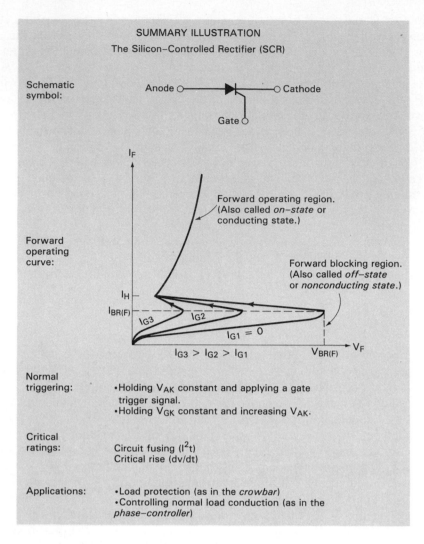

SUMMARY ILLUSTRATION
The Silicon–Controlled Rectifier (SCR)

Schematic symbol:

Anode ○———▷|———○ Cathode

Gate ○

Forward operating curve:

I_F

Forward operating region. (Also called *on–state* or conducting state.)

Forward blocking region. (Also called *off–state* or *nonconducting state*.)

I_H

$I_{BR(F)}$

I_{G3} I_{G2}

$I_{G1} = 0$

$I_{G3} > I_{G2} > I_{G1}$

$V_{BR(F)}$

V_F

Normal triggering:
- Holding V_{AK} constant and applying a gate trigger signal.
- Holding V_{GK} constant and increasing V_{AK}.

Critical ratings:
Circuit fusing (I^2t)
Critical rise (dv/dt)

Applications:
- Load protection (as in the *crowbar*)
- Controlling normal load conduction (as in the *phase–controller*)

FIGURE 19.18

excessive power dissipation. If the *critical rise (dv/dt)* rating is exceeded by any noise in the anode voltage, the device may experience *false triggering*. False triggering, a condition where the device is unintentionally triggered into conduction, can also be caused by excessive noise in the gate circuit biasing voltage. Note that false triggering is normally prevented by the use of *snubber networks* and *gate bypass capacitors*.

SCR applications generally fall into one of two categories: load overvoltage protection, and the limiting of the normal conduction through a load. The characteristics of the SCR are summarized in Figure 19.18.

SECTION REVIEW

1. Describe the construction of the SCR. [Objective 6]
2. List the methods that are used to trigger an SCR. [Objective 7]
3. List the methods used to force an SCR into cutoff. [Objective 7]
4. Using the operating curve in Figure 19.10, describe the forward operation of the SCR. [Objective 8a]
5. What is the *circuit fusing (I^2t)* rating of an SCR? Why is it important? [Objective 9a]
6. What is *false triggering*? [Objective 9b]

CHAP. 19   Thyristors and Optoelectronic Devices

7. Which SCR ratings are tied directly to false triggering problems? [Objective 9b]
8. How does a gate bypass capacitor prevent false triggering? [Objective 12]
9. What is a *snubber network*? What is it used for? [Objective 12]
10. What is the function of the SCR *crowbar*? [Objective 13]
11. Describe the operation of the crowbar. [Objective 13]
12. What is the function of the SCR *phase controller*? [Objective 14]
13. Describe the operation of the SCR phase controller. [Objective 14]
14. What is the function of the SCR *motor-speed controller*? [Objective 15]
15. Describe the operation of the motor-speed controller, including its response to a change in motor speed. [Objective 15]
16. What are the typical in-circuit symptoms of *open* and *shorted* SCRs? [Objective 16]

19.3
Diacs and Triacs

Diacs and *triacs* are components that are classified as *bidirectional thyristors*. The term *bidirectional* means that they are capable of conducting in two directions. For all practical purposes, the *diac* can be viewed as a bidirectional SUS, and the *triac* can be viewed as a bidirectional SCR. Since these components are extremely similar in operation to the thyristors we have discussed, the material in this section will seem more like a review than new material.

Bidirectional thyristor. A thyristor that is capable of conducting in two directions.

Diacs

The diac is a two-terminal, three-layer device whose forward and reverse operating characteristics are identical to the forward characteristics of the SUS. The diac symbol and construction are shown in Figure 19.19.

As you can see, the construction of the diac is extremely similar to that of the *npn* (or *pnp*) transistor. However, there are several crucial differences:

1. There is no base connection.
2. The three regions are nearly identical in size.
3. The three regions have nearly identical doping levels.

Diac. A two-terminal, three-layer device whose forward and reverse characteristics are identical to the forward characteristics of the SUS.

Note that diacs are often referred to as *bidirectional diodes*.

FIGURE 19.19

Diac construction and schematic symbol.

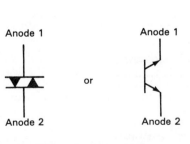

Commonly used symbols

(a)

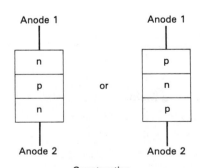

Construction

(b)

In contrast, the BJT has an extremely narrow and lightly doped base (center) region as you were shown in Chapter 5. Note that the diac can be constructed in either *npn* or *pnp* form. The operating principles and symbols are identical for both *npn* and *pnp* diacs.

The operating curve of the diac is shown in Figure 19.20. As you can see, the forward and reverse operating characteristics are identical to the forward characteristics of the SUS. In either direction, the device will act as an open until the breakover voltage (V_{BR}) is reached. At that time, the device is triggered and conducts in the appropriate direction. Conduction then continues until the device current drops below its holding current (I_H) value. Note that the breakover voltage and holding current values are identical for the forward and reverse regions of operation.

Both *anode current interruption* and *forced commutation* are used to drive a diac into cutoff.

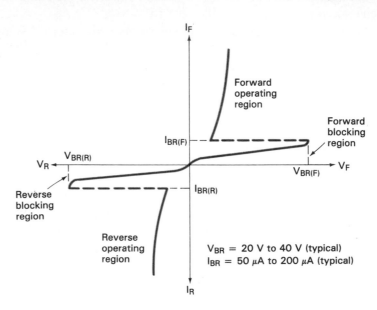

FIGURE 19.20

The diac operating curve.

Note that the diac can be triggered into conduction by an increase in temperature, as was the case with the SUS. The effects of temperature on both forward and reverse breakover can be seen in the SUS curve shown in Figure 19.7.

The most common application for the diac is as a triggering device for a triac in ac control circuits. This application will be discussed later in this section.

✔ 19

✔ 20

Triacs

The triac is a three-terminal, five-layer device whose forward and reverse characteristics are identical to the forward characteristics of the SCR. The triac symbol and construction are shown in Figure 19.21.

FIGURE 19.21

Triac construction and schematic symbol.

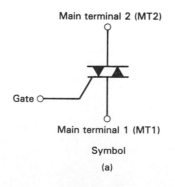

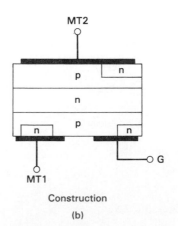

The primary conducting terminals are identified as *main terminal 1 (MT1)* and *main-terminal 2 (MT2)*. Note that MT1 is the terminal that is on the same side of the device as the gate.

If we split the triac as shown in Figure 19.22, it is easy to see that we have complementary SCRs that are connected in parallel. When MT2 is *positive* and MT1 is *negative*, a triggered triac will have the conduction path shown on the left in Figure 19.22b. When MT2 is *negative* and MT1 is *positive*, a triggered triac will have the conduction path shown on the right in Figure 19.22b. Note that the triggered triac will conduct in the appropriate direction as long as the current through the device is greater than its *holding current* (I_H) rating.

Triac. A bidirectional thyristor whose forward and reverse characteristics are identical to the forward characteristics of the SCR.

Triacs are also referred to as *triodes* and *bidirectional triode thyristors* (the JEDEC designated name).

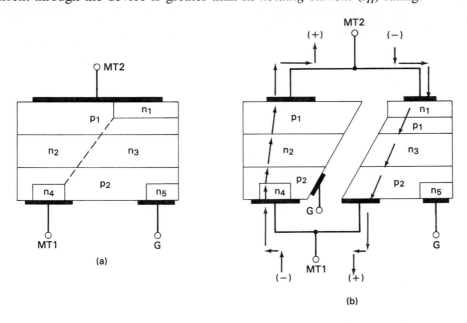

FIGURE 19.22

Triac split construction and currents.

Because of its unique structure, the operating curve of the triac is a bit more complicated than any we have seen thus far. The triac operating curve is shown in Figure 19.23. The graph in Figure 19.23 is divided into four *quadrants*, labeled *I*, *II*, *III*, and *IV*. Each quadrant has the MT2 and G (gate) polarities shown. For example, quadrant I indicates the operation of the device when MT2 is positive

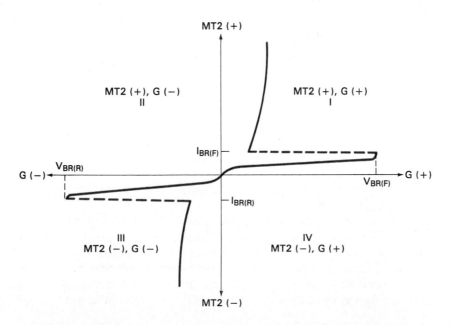

FIGURE 19.23

The triac operating curve.

(with respect to MT1) and a positive potential is applied to the gate. Quadrant II indicates the operation of the device when MT2 is positive (with respect to MT1) and a negative potential is applied to the gate, and so on.

Operation in quadrant I is identical to the operation of the SCR. All the SCR characteristics apply to the triac that is being operated in the first quadrant. For example:

1. The device will remain in the off state until $V_{BR(F)}$ is reached, or a positive gate trigger voltage (V_{GT}) causes a gate current (I_G) that is sufficient to trigger the device.
2. The device may be false triggered by increases in temperature, gate noise, or anode (MT2) noise.
3. Once triggered, the device continues to conduct until the forward current drops below its holding current (I_H) rating.

Note that, when operated in quadrant I, the current through the triac is in the direction shown on the left side of Figure 19.22b.

Quadrant III operation is simply a reflection of quadrant I operation. In other words, when MT2 is negative (with respect to MT1) and a negative gate trigger voltage is applied to the triac, it is triggered into reverse conduction, as shown on the right side of Figure 19.22b. Again, all the SCR characteristics apply to the triac; the direction of conduction has simply been reversed.

Quadrants II and IV illustrate the unique triggering combinations of the triac. The triac can be triggered into conduction by either of the following conditions:

1. Applying a negative gate trigger while MT2 is positive (quadrant II)
2. Applying a positive gate trigger while MT2 is negative (quadrant IV)

Quadrants I and II have the same MT2 polarities. The same is true for quadrants III and IV.

These triggering combinations are possible because of the fact that the gate is physically connected to both the p_2 and n_5 regions, as shown in Figure 19.22a. When quadrant II triggering occurs, the current through the device will be in the direction indicated on the left side of Figure 19.22b. When quadrant IV triggering occurs, the current through the device will be in the direction shown on the right side of Figure 19.22b. Since the polarity of MT2 determines the direction of current through the triac, quadrant II triggering will cause current in the same direction as quadrant I triggering. The same can be said for quadrant III triggering and quadrant IV triggering.

Confusing? Let's take a look at the V_{GT} parameters for a typical triac. The spec sheet for the 2N6157–65 series triacs (shown in Figure 19.24) lists the following values:

V_{GT} (Maximum Value)	Listed Conditions	Indicated Quadrant
2 V	MT2(+), G(+)	I
2.1 V	MT2(+), G(−)	II
2.1 V	MT2(−), G(−)	III
2.5 V	MT2(−), G(+)	IV

Now let's apply the values listed to the triacs shown in Figure 19.25. Figure 19.25a shows the quadrant I triggering scheme. MT2 is positive, and a +2-V gate input triggers the device into conduction. Quadrant II triggering (Figure 19.25b) occurs when MT2 is positive and a −2.1-V input is applied to the gate. Note that the triac conduction is the same for quadrants I and II triggering. Quadrant III triggering (Figure 19.25c) occurs when MT2 is negative and a −2.1-V input

CHAP. 19 Thyristors and Optoelectronic Devices

Triacs
Silicon Bidirectional Triode Thyristors

. . . designed primarily for industrial and military applications for the control of ac loads in applications such as light dimmers, power supplies, heating controls, motor controls, welding equipment and power switching systems; or wherever full-wave, silicon gate controlled solid-state devices are needed.

- Glass Passivated Junctions and Center Gate Fire
- Isolated Stud for Ease of Assembly
- Gate Triggering Guaranteed In All 4 Quadrants

2N6145–47
(See 2N5571)

2N6157
thru
2N6165

TRIACs
30 AMPERES RMS
200 thru 600 VOLTS

MAXIMUM RATINGS

Rating	Symbol	Value	Unit
*Peak Repetitive Off-State Voltage ($T_J = -65$ to $+125°C$) 1/2 Sine Wave 50 to 60 Hz, Gate Open	V_{DRM}		Volts
*Peak Principal Voltage 2N6157, 2N6160, 2N6163 2N6158, 2N6161, 2N6164 2N6159, 2N6162, 2N6165		200 400 600	
*Peak Gate Voltage	V_{GM}	10	Volts
*RMS On-State Current ($T_C = -65$ to $+85°C$) ($T_C = +100°C$) Full Sine Wave, 50 to 60 Hz	$I_{T(RMS)}$	30 20	Amps
*Peak Non-Repetitive Surge Current (One Full Cycle of surge current at 60 Hz, preceded and followed by a 30 A RMS current, $T_C = 85°C$)	I_{TSM}	250	Amps
Circuit Fusing Considerations (t = 8.3 ms)	I^2t	260	A^2s
*Peak Gate Power ($T_J = +80°C$, Pulse Width = 2 μs)	P_{GM}	20	Watts
*Average Gate Power ($T_J = +80°C$, t = 8.3 ms)	$P_{G(AV)}$	0.5	Watt
*Peak Gate Current	I_{GM}	2	Amps
*Operating Junction Temperature Range	T_J	-65 to $+125$	°C
*Storage Temperature Range	T_{stg}	-65 to $+150$	°C
*Stud Torque 2N6160 thru 2N6165	—	30	in. lb.

CASE 174-04
(TO-203AA)
STYLE 3
2N6157-59

CASE 263-04
STYLE 2
2N6160-62

CASE 311-02
STYLE 2
2N6163-65

THERMAL CHARACTERISTICS

Characteristic	Symbol	Max	Unit
*Thermal Resistance, Junction to Case	$R_{\theta JC}$	1	°C/W

*Indicates JEDEC Registered Data.

FIGURE 19.24

The 2N6157–65 triac specification sheet.
(Courtesy of Motorola, Inc.)

2N6157 thru 2N6165

ELECTRICAL CHARACTERISTICS (T_C = 25°C unless otherwise noted.)

Characteristic	Symbol	Min	Typ	Max	Unit
*Peak Forward or Reverse Blocking Current (Rated V_{DRM} or V_{RRM}) T_J = 25°C T_J = 125°C	I_{DRM}, I_{RRM}	— —	— —	10 2	μA mA
*Peak On-State Voltage (Either Direction) (I_{TM} = 42 A Peak, Pulse Width = 1 to 2 ms, Duty Cycle ≤ 2%)	V_{TM}	—	1.5	2	Volts
Gate Trigger Current (Continuous dc), Note 1 (Main Terminal Voltage = 12 Vdc, R_L = 50 Ohms) MT2(+), G(+) MT2(+), G(−) MT2(−), G(−) MT2(−), G(+) *MT2(+), G(+); MT2(−), G(−) T_C = −65°C *MT2(+), G(−); MT2(−), G(+) T_C = −65°C	I_{GT}	 — — — — — —	 15 20 20 30 — —	 60 70 70 100 200 250	mA
Gate Trigger Voltage (Continuous dc) (Main Terminal Voltage = 12 Vdc, R_L = 50 Ohms) MT2(+), G(+) MT2(+), G(−) MT2(−), G(−) MT2(−), G(+) *All Quadrants, T_C = −65°C *Main Terminal Voltage = Rated V_{DRM}, R_L = 10 k ohms, T_J = +125°C	V_{GT}	 — — — — — 0.2	 0.8 0.7 0.85 1.1 — —	 2 2.1 2.1 2.5 3.4 —	Volts
Holding Current (Main Terminal Voltage = 12 Vdc, Gate Open) (Initiating Current = 500 mA) MT2(+) MT2(−) *Either Direction, T_C = −65°C	I_H	 — — —	 8 10 —	 70 80 200	mA
*Turn-On Time (Main Terminal Voltage = Rated V_{DRM}, I_{TM} = 42 A, Gate Source Voltage = 12 V, R_S = 50 Ohms, Rise Time = 0.1 μs, Pulse Width = 2 μs)	t_{gt}	—	1	2	μs
Blocking Voltage Application Rate at Commutation, f = 60 Hz, T_C = 85°C On-State Conditions: (I_{TM} = 42 A, Pulse Width = 4 ms, di/dt = 17.5 A/ms) Off-State Conditions: (Main Terminal Voltage = Rated V_{DRM} (200 μs min), Gate Source Voltage = 0 V, R_S = 50 Ω)	dv/dt(c)	—	5	—	V/μs

*Indicates JEDEC Registered Data.
Note 1. All voltage polarities referenced to main terminal 1.

FIGURE 19.24

(continued)

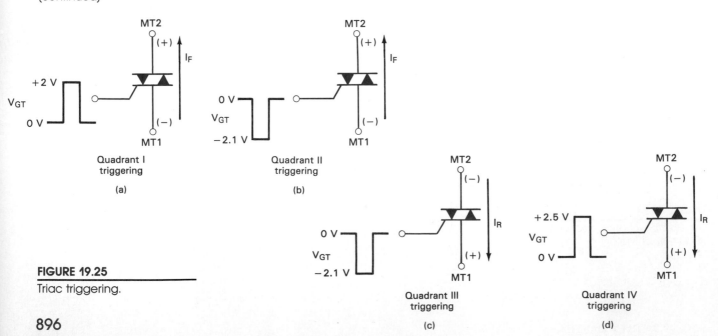

FIGURE 19.25
Triac triggering.

is applied to the gate. Quadrant IV triggering occurs when MT2 is negative, and a +2.5-V input is applied to the gate. Note that the triac conduction is in the same direction for quadrants III and IV triggering.

Several points should be made at this time:

1. Regardless of the triggering method used, triacs are forced into cutoff in the same fashion as the SCR. Either device current interruption or force commutation must be used to decrease the device current below its holding current (I_H) rating. At that point, the device enters its nonconducting state and remains in cutoff until another trigger is received.

2. Except for the quadrant triggering considerations, the spec sheet for a triac is nearly identical to that of an SCR. This can be seen by comparing the triac spec sheet in Figure 19.24 with the SCR spec sheet in Figure 19.11.

Controlling Triac Triggering

✔ 22

The low gate voltages required to trigger the triac can cause problems in many circuit applications. For example, consider the 2N6157 triac shown in Figure 19.26a. Here we see the approximate triggering voltages required for the device (quadrants I and II triggering). Three questions arise when you think about these gate triggering voltages:

1. What if we want the 2N6157 to be triggered *only* by a *positive* gate voltage?
2. What if we want the 2N6157 to be triggered *only* by a *negative* gate voltage?
3. What if we want positive *and* negative triggering, but at much higher gate voltage values?

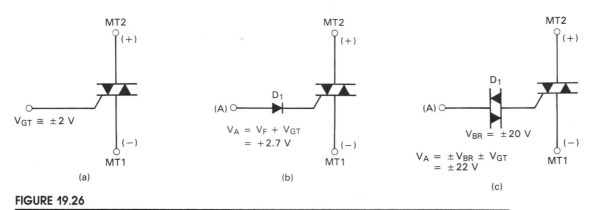

FIGURE 19.26

Controlling triac triggering.

The circuit in Figure 19.26b illustrates the method commonly used to allow *single-polarity* triggering. In the circuit shown, D_1 is forward biased when V_{GT} is *positive* and reverse biased when V_{GT} is *negative*. Since D_1 will conduct only when V_{GT} is positive, the triac can only be triggered by a positive gate signal. The potential required to trigger the device would be equal to the sum of V_F for D_1 and the required gate triggering voltage for the 2N6157, +2.7 V (approximate). By reversing the direction of the diode, we limit the triggering of the triac to *negative* gate voltages.

In many applications, the dual-polarity triggering characteristic of the triac is desirable, but the rated values of V_{GT} for the device are too low. When this is the case, the triggering levels can be raised by using a *diac* in the gate circuit, as shown in Figure 19.26c. With the diac present, the triac can still be triggered by both positive and negative gate voltages. However, to get conduction in the gate terminal (which is required for triggering), the gate trigger signal must overcome the V_{BR} values for the diac. For example, in Figure 19.26c, we have placed a diac with V_{BR} values of ± 20 V in the gate circuit of the triac. For the 2N6157 to be triggered into conduction, V_A will have to reach a value of

$$V_A = \pm V_{BR} \pm V_{GT}$$
$$= \pm 20 \text{ V} \pm 2 \text{ V}$$
$$= \pm 22 \text{ V}$$

We still have the dual-polarity triggering that we want, but now the gate trigger signal (V_A) must reach ± 22 V before the triac will be triggered into conduction. Note that thyristor triggering control is the primary application for the diac. It is also one of the applications for a device we will discuss in the next section, the *unijunction transistor (UJT)*.

Triac Applications

A typical triac application can be seen in the *phase controller* in Figure 19.27. The primary difference between the triac phase controller and the SCR phase controller is the triac's ability to conduct during both alternations of the input signal.

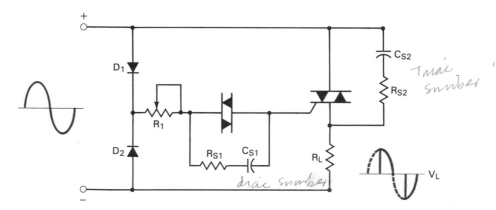

FIGURE 19.27

A triac phase controller.

With the input polarity indicated in Figure 19.27, D_1 is forward biased and D_2 is reverse biased. This puts a positive voltage at point A in the circuit. When V_A is high enough to overcome the V_{BR} rating of the diac and the V_{GT} rating of the triac, the triac is triggered into conduction, providing the positive output cycle shown. When the input polarity reverses, D_1 becomes reverse biased and D_2 becomes forward biased. In this case, point A is still positive, but MT2 is now negative. When the triac is triggered, the current direction through the load is reversed, and the negative output cycle is produced. Note that the triac in Figure 19.27 is quadrant I triggered during the positive alternation of the input and quadrant IV triggered during the negative alternation of the input. Also, note the presence of the diac and triac *snubber networks*. These are included because the diac and triac are subject to the same false triggering problems as the SCR.

Component Testing

Diacs and triacs are almost always in-circuit tested. An oscilloscope can be used to check the operation of each of these components. When shorted, both components will act as low-value resistors. When open, the components will not conduct, regardless of the presence of a triggering input.

CHAP. 19 Thyristors and Optoelectronic Devices

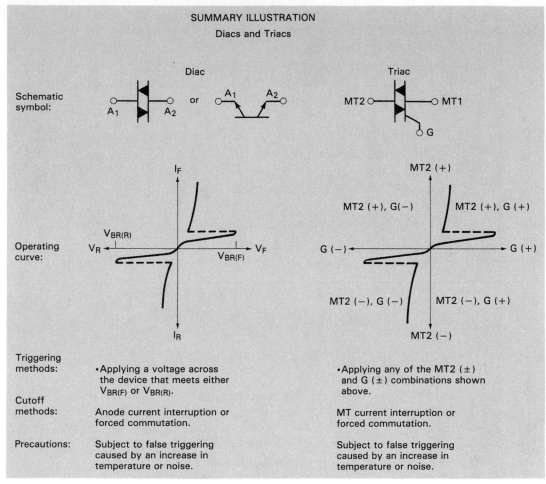

SUMMARY ILLUSTRATION
Diacs and Triacs

	Diac	Triac

Schematic symbol: — Diac: A_1 — A_2 or A_1 A_2; Triac: MT2 — MT1, G

Operating curve: (Diac) I_F, V_R, $V_{BR(R)}$, $V_{BR(F)}$, V_F, I_R; (Triac) MT2 (+), MT2 (+), G(−), MT2 (+), G (+), G (−), G (+), MT2 (−), G (−), MT2 (−), G (+), MT2 (−)

Triggering methods:
• (Diac) Applying a voltage across the device that meets either $V_{BR(F)}$ or $V_{BR(R)}$.
• (Triac) Applying any of the MT2 (±) and G (±) combinations shown above.

Cutoff methods:
(Diac) Anode current interruption or forced commutation.
(Triac) MT current interruption or forced commutation.

Precautions:
(Diac) Subject to false triggering caused by an increase in temperature or noise.
(Triac) Subject to false triggering caused by an increase in temperature or noise.

FIGURE 19.28

Section Summary

The *diac* acts as a bidirectional SUS. The device will conduct in either direction (*anode 1 to anode 2, or anode 2 to anode 1*) when the terminal voltage reaches the breakover potential of the device. The device then continues to conduct until its current is driven below the holding current (I_H) rating by either anode current interruption or forced commutation.

The *triac* acts as a bidirectional SCR. The triac can be triggered into conduction by any one of four biasing conditions. Because of the ease of triac triggering, external components are often required in triac circuits to restrict the conditions under which triggering will occur. Like all the devices we have covered thus far, the diac and triac are subject to false triggering from increases in temperature, gate noise, and anode (or main terminal) noise. The characteristics of the diac and triac are summarized in Figure 19.28.

1. What is a *bidirectional thyristor*? [Objective 17]
2. What is a *diac*? [Objective 18]
3. What are the two terminals of the diac called? [Objective 18]
4. How is a diac triggered into conduction? [Objective 18]

5. How is a diac driven into cutoff? [Objective 18]

6. What is the most common diac application? [Objective 18]

7. What is a *triac*? [Objective 19]

8. Draw the schematic symbol for a triac and identify the component terminals. [Objective 19]

9. Describe the four-quadrant triggering characteristics of the triac. [Objective 19]

10. What methods are used to drive a triac into cutoff? [Objective 21]

11. Describe the operation of each of the trigger control circuits in Figure 19.26. [Objective 22]

12. Describe the operation of the triac phase controller in Figure 19.27. [Objective 23]

13. What are symptoms of a *shorted* diac or triac? [Objective 24]

14. What are the symptoms of an *open* diac or triac? [Objective 24]

19.4
Unijunction Transistors (UJTs)

 25

Unijunction transistor (UJT). A three-terminal device whose trigger voltage is proportional to its applied biasing voltage.

Peak voltage (V_P). The value of V_{EB1} that triggers the UJT into conduction.

Peak current (I_P). The minimum value of I_E required to maintain emitter conduction.

The *unijunction transistor (UJT)* is a three-terminal switching device whose trigger voltage is proportional to its applied biasing voltage. The schematic symbol for the UJT is shown in Figure 19.29a. As you can see, the terminals of the UJT are called the *emitter*, *base 1 (B1)*, and *base 2 (B2)*. The switching action of the UJT is explained with the help of the circuit in Figure 19.29b. A biasing voltage (V_{BB}) is applied across the two base terminals. The emitter–base 1 junction acts as an open until the voltage across the terminals (V_{EB1}) reaches a specified value, called the *peak voltage (V_P)*. When $V_{EB1} = V_P$, the emitter–base 1 junction triggers into conduction and will continue to conduct until the emitter current drops below a specified value, called the *peak current (I_P)*. At that time, the device returns to its nonconducting state. Note that the peak voltage (emitter–base 1 triggering voltage) is proportional to the value of V_{BB}. Thus, when we change V_{BB}, we change the value of V_{EB1} needed to trigger the device.

UJT Construction and Operation

Figure 19.30 shows the construction and equivalent circuit of the UJT. As you can see, the UJT structure is very similar to that of the *n*-channel JFET. In fact,

FIGURE 19.29

UJT schematic symbol and biasing.

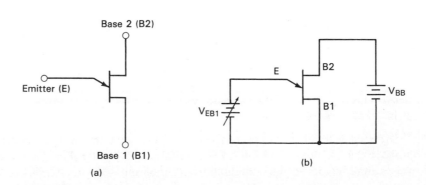

(a)

(b)

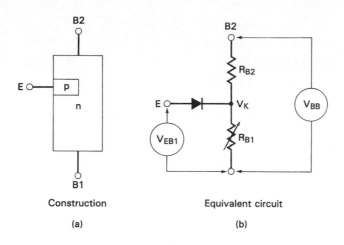

B2

B2

R_{B2}

E○─┤ p │

E○──▷├── V_K

n

V_{BB}

V_{EB1}

R_{B1}

B1

Construction

Equivalent circuit

(a)

(b)

FIGURE 19.30

UJT construction and equivalent circuit.

the only difference between the two components is the fact that the *p*-type (gate) material of the JFET *surrounds* the *n*-type (channel) material, as was shown in Chapter 10. The similarity in the construction of the UJT and the *n*-channel JFET is the reason that their schematic symbols are so similar.

Looking at the UJT equivalent circuit (Figure 19.29c) we see that the emitter terminal is connected (in effect) to a *pn* junction diode. The cathode of the diode is connected to a voltage-divider circuit that is made up of R_{B1} and R_{B2}. Note that these resistors represent the resistance of the *n*-type material that lies between the terminals and the *pn* junction.

The overall principle of operation is simple. For the emitter–base diode to conduct, its anode voltage must be approximately 0.7 V more positive than its cathode. The cathode potential of the diode (V_k) is determined by the combination of V_{BB}, R_{B1}, and R_{B2}. Using the standard voltage-divider equation, the value of V_k is found as

 26

$$V_k = V_{BB} \frac{R_{B1}}{R_{B1} + R_{B2}} \tag{19.5}$$

The resistance ratio in equation (19.5) is called the *intrinsic standoff ratio (η)* of the UJT. Since

Intrinsic standoff ratio (η). The ratio of emitter–base 1 resistance to the total *interbase resistance* in a UJT. The *interbase resistance* is the total resistance between base 1 and base 2 when the device is not conducting.

$$\eta = \frac{R_{B1}}{R_{B1} + R_{B2}}$$

equation (19.5) can be rewritten as

$$V_k = \eta V_{BB} \tag{19.6}$$

Adding 0.7 V to the value found in equation (19.6), we get an equation for the triggering voltage (V_P) for the UJT, as follows:

$$V_P = \eta V_{BB} + 0.7 \text{ V} \tag{19.7}$$

where η = the intrinsic standoff ratio of the UJT
V_P = the peak voltage of the device

The intrinsic standoff ratio for a given UJT is listed on its spec sheet. For example, take a look at the 2N5431 UJT spec sheet in Figure 19.31. Under the *electrical characteristics* heading, the spec sheet lists an intrinsic standoff ratio of

PN Unijunction Transistor
Silicon Annular Unijunction Transistor

2N5431

PN UJT

**CASE 22A-01
STYLE 1**

... characterized primarily for low interbase-voltage operation in sensing, pulse triggering, and timing circuits.

- Low R_{BB} Spread — 6 to 8.5 kΩ
- Low Peak-Point Current — I_P = 4 μA (Max) @ V_{B2B1} = 4 V
- Low Emitter Saturation Voltage — $V_{EB1(sat)}$ = 3 V (Max)
- Narrow Intrinsic Standoff Ratio — η = 0.72 to 0.80

MAXIMUM RATINGS (T_A = 25°C unless otherwise noted.)

Rating	Symbol	Value	Unit
RMS Power Dissipation, Note 1	P_D	360	mW
RMS Emitter Current	I_e	50	mA
Peak-Pulse Emitter Current, Note 2	i_e	1.5	Amp
Emitter Reverse Voltage	V_{B2E}	30	Volts
Interbase Voltage, Note 3	V_{B2B1}	35	Volts
Operating Junction Temperature Range	T_J	−65 to +125	°C
Storage Temperature Range	T_{stg}	−65 to +200	°C

Notes: 1. Derate 3 mW/°C increase in ambient temperature.
2. Duty Cycle ≤ 1%, PRR = 10 PPS (see Figure 5).
3. Based upon power dissipation at T_A = 25°C.

ELECTRICAL CHARACTERISTICS (T_A = 25°C unless otherwise noted.)

Characteristic	Fig. No.	Symbol	Min	Max	Unit
Intrinsic Standoff Ratio, Note 1 (V_{B2B1} = 10 V)	4	η	0.72	0.80	—
Interbase Resistance (V_{B2B1} = 3 V, I_E = 0)		R_{BB}	6	8.5	kΩ
Interbase Resistance Temperature Coefficient (V_{B2B1} = 3 V, I_E = 0, T_A = 0 to 100°C)		αR_{BB}	0.4	0.8	%/°C
Emitter Saturation Voltage, Note 2 (V_{B2B1} = 10 V, I_E = 50 mA)		$V_{EB1(sat)}$	—	3	Volts
Modulated Interbase Current (V_{B2B1} = 10 V, I_E = 50 mA)		$I_{B2(mod)}$	5	30	mA
Emitter Reverse Current (V_{B2E} = 30 V, I_{B1} = 0)		I_{EB2O}	—	10	nA
Peak-Point Emitter Current (V_{B2B1} = 25 V) (V_{B2B1} = 4 V)		I_P	— —	0.4 4	μA
Valley-Point Current (2) (V_{B2B1} = 20 V, R_{B2} = 100 ohms)		I_V	2	—	mA
Base-One Peak Pulse Voltage (V_{BB} = 4 Volts)	3	V_{OB1}	1	—	Volts

Notes: 1. η, Intrinsic standoff ratio, is defined in terms of the peak-point voltage, V_P, by means of the equation: $V_P = \eta V_{B2B1} + V_F$, where V_F is about 0.45 volt at 25°C @ I_F = 10 μA and decreases with temperature at about 2.5 mV/°C. The test circuit is shown in Figure 4. Components R_1, C_1, and the UJT form a relaxation oscillator; the remaining circuitry serves as a peak-voltage detector. The forward drop of Diode D_1 compensates for V_F. To use, the "cal" button is pushed, and R_3 is adjusted to make the current meter, M_1, read full scale. When the "cal" button is released, the value of η is read directly from the meter, if full scale on the meter reads 1.

2. Use pulse techniques: PW ≈ 300 μs, Duty Cycle ≤ 2% to avoid internal heating, which may result in erroneous readings.

FIGURE 19.31

The 2N5431 UJT specification sheet. (Courtesy of Motorola, Inc.)

$\eta = 0.8$ (maximum). The following example shows how this value is used to calculate the peak voltage for the 2N5431 with a given value of V_{BB}.

27

====== **EXAMPLE 19.5** ======

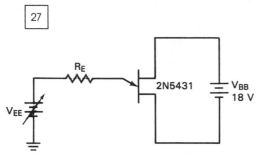

FIGURE 19.32

Determine the peak voltage (V_P) value for the 2N5431 in Figure 19.32.

Solution: With $\eta = 0.8$ (maximum) and $V_{BB} = +18$ V, the peak voltage of the circuit is found as

$$V_P = \eta V_{BB} + 0.7 \text{ V}$$
$$= (0.8)(+18 \text{ V}) + 0.7 \text{ V}$$
$$= 15.1 \text{ V}$$

PRACTICE PROBLEM 19–5

The 2N4870 UJT has a rating of $\eta = 0.75$ (maximum). Determine the maximum value of V_P for the device when it is being used in a circuit with $V_{BB} = +12$ V.

The peak voltage is applied across the emitter–base 1 terminals. Because of this, almost all of I_E is drawn through R_{B1} when the device is triggered, as is shown in Figure 19.33. Since I_E is being drawn through R_{B1}, this resistance value drops off as I_E increases. The value of R_{B2} is not affected very much by the presence of I_E. The results of this component operation are shown in the UJT characteristic curve in Figure 19.34. As you can see, V_{EB1} can be increased until the *peak point* is reached. When the peak point is reached, $V_{EB1} = V_P$, and the emitter diode begins to conduct. The initial value of I_E is referred to as the *peak emitter current* (I_P). This term can be confusing, since we tend to think of a peak current as being a maximum value. However, for the UJT, the peak current (I_P) is the *minimum* rating, as it is associated with the first point of conduction. Note that when $I_E < I_P$ the UJT is said to be in its *cutoff* region of operation.

When I_E increases above the value of I_P, the value of R_{B1} drops drastically. As a result, the value of V_{EB1} actually *decreases* as I_E increases. The decrease in V_{EB1} continues until I_E reaches the *valley current* (I_V) rating of the UJT. When I_E is increased beyond I_V, the device goes into *saturation*.

FIGURE 19.33

UJT currents.

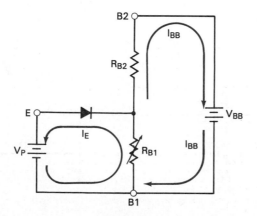

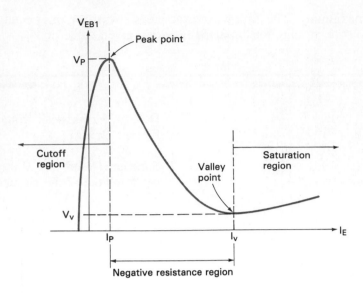

FIGURE 19.34

FIGURE 19.34

The UJT operating curve.

Negative resistance region. The region of operation that lies between the peak and valley points of the UJT curve.

Negative resistance. A term used to describe any device with current and voltage values that are *inversely proportional*.

 29

Relaxation oscillator. A circuit that uses the charge/discharge characteristics of a capacitor or inductor to produce a pulse output.

As the characteristic curve in Figure 19.34 shows, V_{EB1} and I_E are inversely proportional between the peak point and valley point of the curve. The region of operation between these two points is called the *negative resistance region*. Note that the term *negative resistance* is used to describe any device with current and voltage values that are *inversely proportional*.

Once the UJT is triggered, the device continues to conduct as long as I_E is greater than I_P. When I_E drops below I_P, the device returns to the cutoff region of operation and remains there until another trigger is received.

UJT Applications

UJTs are used almost exclusively as *thyristor triggering devices*. An example of this application can be seen in Figure 19.35. The circuit in Figure 19.35a is called a *relaxation oscillator*. A relaxation oscillator is a circuit that uses the charge/discharge characteristics of a capacitor or inductor to produce a pulse output. This pulse output is normally used as the triggering signal for a thyristor, such as an SCR or a triac.

FIGURE 19.35

A simple UJT relaxation oscillator.

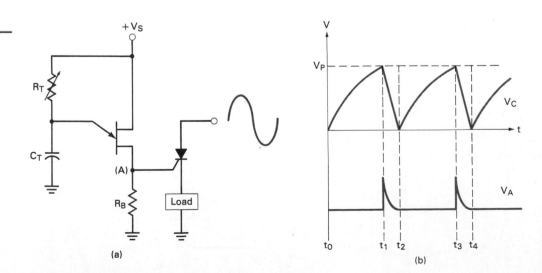

CHAP. 19   Thyristors and Optoelectronic Devices

When power is applied to the circuit in Figure 19.35a, the capacitor charges through R_T. This produces the capacitor charge curve shown between t_0 and t_1 in Figure 19.35b. At t_1, the charge on the capacitor reaches the V_P value for the UJT, and the device is triggered into conduction. The low-resistance emitter–base 1 junction of the UJT discharges the capacitor during the time interval between t_1 and t_2. The discharge current flows through the base 1 circuit of the UJT, and thus through R_B. This causes a voltage to be developed across R_B (shown as V_A in Figure 19.35b) that is applied to the gate of the SCR. This voltage triggers the SCR into conduction.

At t_2, the capacitor has discharged completely. At that point, I_E is too low to maintain the *on* condition of the UJT, and the component returns to the cutoff state. When this occurs, the capacitor starts to charge again through R_T, causing the charge curve shown between t_2 and t_3, and the cycle repeats itself.

If you look closely at the V_C waveform, you'll see that the capacitor requires a lot more time to charge than it does to discharge. This is because of the low value of R_{B1} when I_E is flowing through the component.

As shown, the UJT relaxation oscillator in Figure 19.35a isn't very practical. The problem is the fact that the capacitor charge/discharge cycle must be synchronized (timed) with the SCR anode voltage for proper phase control to occur. In a more practical relaxation oscillator, the dc supply voltage (V_S) is derived from the SCR anode voltage to control the timing of the oscillator. An example of this type of circuit is shown in Figure 19.36a.

Don't forget. The time required for a capacitor to charge or discharge is found as $T = 5RC$. Since R_{B1} (during conduction) is much less than R_T, the capacitor takes less time to discharge than it does to charge.

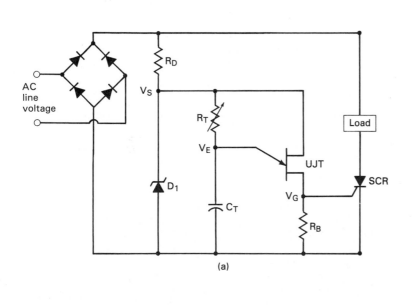

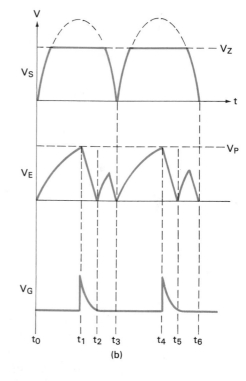

FIGURE 19.36

A UJT phase-control circuit.

The ac input to the circuit is applied to a bridge rectifier. The full-wave rectified signal is then applied to the rest of the circuit. The zener diode (D_1) clips the rectified signal, producing the V_S waveform shown in Figure 19.36b. Note that V_S is the biasing voltage for the UJT and its trigger circuit (R_T and C_T).

From t_0 to t_1, C_T charges through R_T. When the emitter voltage of the UJT (V_E) reaches V_P, the UJT fires, causing conduction through R_B. The current through R_B then produces the gate trigger pulse (V_G) required to trigger the SCR into conduction.

During the time period from t_1 to t_2, C_T is discharging. When the UJT returns to its cutoff state (at t_2), the capacitor begins to charge again through R_T. However, this time the capacitor charge doesn't make it to V_P. The reason for this is shown in Figure 19.36c. As V_E is starting to increase for the second time in the alternation, V_S is decreasing in value. Before V_E can reach the original value of V_P, the supply voltage is effectively removed from the UJT. Thus, the second charge cycle of C_T is cut short, producing the V_E waveform shown.

Even if C_T was able to completely recharge during a given alternation, it wouldn't have any effect on the load. The reason for this is simple: The SCR is triggered into conduction during the discharge cycle of C_T and continues to conduct for the remainder of the alternation. Since the SCR is already conducting when the second C_T charge cycle occurs, the production of another trigger signal by the UJT would have no effect on the SCR, and thus on the load.

Since the initial charge cycle of C_T is controlled by the cycles of V_S, the relaxation oscillator is synchronized to the ac load cycle. This method of syncronizing the oscillator to the ac load cycle is used extensively in practice.

One Final Note

The UJT is technically classified as a *thyristor-triggering device* rather than as a thyristor. Since triggering thyristors, such as SCRs and triacs, is the only common application for the UJT, the device is normally covered along with the thyristors. Also, you will normally find the spec sheets for UJTs in the same section of a parts manual as the thyristors.

Another thyristor trigger that is commonly used is the *programmable unijunction transistor (PUT)*. We will conclude this section with a brief introduction to this device.

The Programmable UJT (PUT)

The PUT is a four-layer device that is very similar to the SCR. In this case, however, the gate is used as a *reference* terminal rather than a triggering terminal. The schematic symbol for the PUT is shown, along with an application circuit, in Figure 19.37. Note the similarity between the PUT schematic symbol and the SCR schematic symbol. The similarity in schematic symbols between the two devices indicates the similarity in their operation.

The PUT blocks anode current until the value of V_{AK} reaches the value of V_{GK}. For example, if V_{GK} is +5 V, the device will not permit anode conduction until V_{AK} reaches +5 V. At that time, the device triggers and anode current increases rapidly. Once triggered, the PUT continues to conduct until the anode current drops below the I_P rating of the device. At that point, the device returns to cutoff and will block anode current until triggered again.

The circuit shown in Figure 19.37b demonstrates the use of the PUT in a relaxation oscillator. R_1 and R_2 are used to establish the value of V_{GK} for the device.

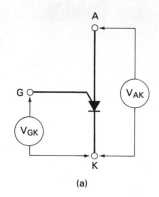

(a)

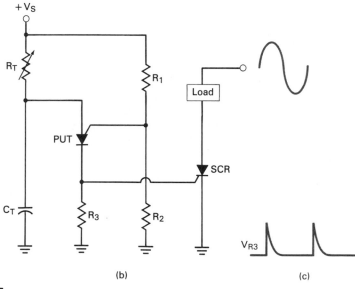

(b)   (c)

FIGURE 19.37

A simple PUT relaxation oscillator.

C_T is charged through R_T until V_{AK} reaches the value of V_{GK}. At that point, the PUT is triggered, causing conduction through R_3. The conduction through R_3 produces the voltage waveform shown in Figure 19.37c. This pulse then triggers the SCR into conduction.

When the PUT is triggered, C_T discharges through the device. When the current being supplied by the capacitor drops below the I_P rating of the PUT, the device returns to cutoff, and the charge cycle begins again.

It should be noted that the PUT relaxation oscillator would be synchronized to the SCR anode signal, as was the case with the UJT relaxation oscillator. Thus, the practical PUT oscillator is actually a bit more complex than the one shown in Figure 19.37b.

Section Summary

The UJT and PUT are devices that are not classified as thyristors, but rather as *thyristor triggers*. These devices are usually used to control the triggering of SCRs and triacs.

The most common UJT and PUT triggering circuits are *relaxation oscillators*. These oscillators are usually synchronized to the ac signal that is being applied to

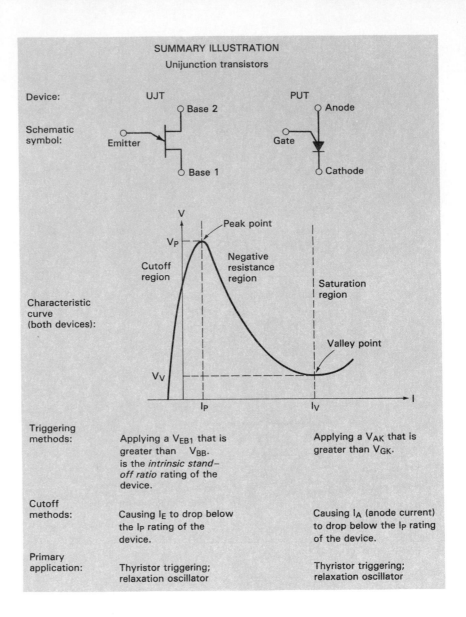

FIGURE 19.38

the circuit thyristor (SCR or triac) so that proper phase control is provided. The characteristics of the UJT and PUT are summarized in Figure 19.38.

SECTION REVIEW

1. What is the *unijunction transistor (UJT)*? [Objective 25]
2. What is the *peak voltage (V_P)* value of a UJT? [Objective 25]
3. What is the *peak current (I_P)* value of a UJT? [Objective 25]
4. What is the *intrinsic standoff ratio (η)* rating of a UJT? [Objective 26]
5. How does the value of η for a UJT relate to the value of V_P? [Objective 26]
6. What is *negative resistance*? [Objective 28]
7. What current and voltage values define the limits of the negative resistance region of UJT operation? [Objective 28]
8. What is the relationship between I_E and V_{EB1} in the negative resistance region of UJT operation? [Objective 28]

CHAP. 19 Thyristors and Optoelectronic Devices

9. Describe the operation of the *relaxation oscillator* in Figure 19.35. [Objective 29]

10. Why isn't the circuit in Figure 19.35 used in practice? [Objective 29]

11. Explain the operation of the relaxation oscillator in Figure 19.36. [Objective 29]

12. What is the primary difference between the PUT and the SCR? [Objective 30]

19.5
Other Thyristors and Thyristor Triggers

In this section, we are going to conclude our coverage of thyristors and thyristor triggers by taking a brief look at some of the other devices that are currently available. Most of the thyristors and triggers covered in this section are simply variations on one or more of the devices we have already discussed.

Silicon Bilateral Switch (SBS)

The *silicon bilateral switch* (*SBS*) is a thyristor trigger that is designed to provide high-current trigger pulses for power thyristors. These devices have very fast break-over switching times and relatively low breakover voltage (V_{BR}) ratings. The SBS is actually an integrated circuit, rather than a single component like the UJT or PUT. The schematic symbol and circuit construction for the SBS are shown in Figure 19.39.

The SBS circuit in Figure 19.39b shows that the device contains *complementary* silicon unilateral switches (SUSs); one a *pnpn* (made up of Q_1 and Q_2) and the

31

Silicon bilateral switch (SBS). A thyristor trigger that is designed to provide high-current pulses for triggering high-power thyristors.

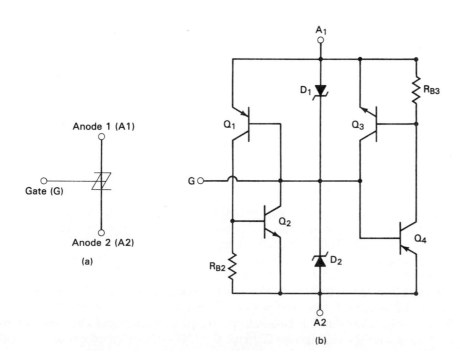

FIGURE 19.39
SBS construction and schematic symbol.

other an *npnp* (made up of Q_3 and Q_4). These circuits provide the *bilateral* (two-way) conduction ability of the device.

When A1 is more *positive* than A2 by an amount that is greater than the V_Z rating of D_2, the Q_1/Q_2 circuit will break over and conduct. When A1 is more *negative* than A2 by an amount that is greater than the V_Z rating of D_2, the Q_3/Q_4 circuit will break over and conduct. The zener ratings of D_1 and D_2 are always equal, so the breakover characteristics of the device are extremely symmetrical, meaning that the forward breakover characteristics are nearly identical to the reverse breakover characteristics.

The gate of the SBS is used to vary the breakover points (forward and reverse) in the same manner as the SCR gate. Once the device has started to conduct in either direction, conduction will continue until the device current drops below the *holding current* (I_H) rating.

The characteristic curve of the SBS is shown in Figure 19.40a. The typical breakover voltage (V_{BR}) ratings are 10 V or less, and the typical breakover current (I_{BR}) ratings are in the low to mid-microampere range. The on-state forward current ratings for the SBS are typically in the range of 100 mA and higher.

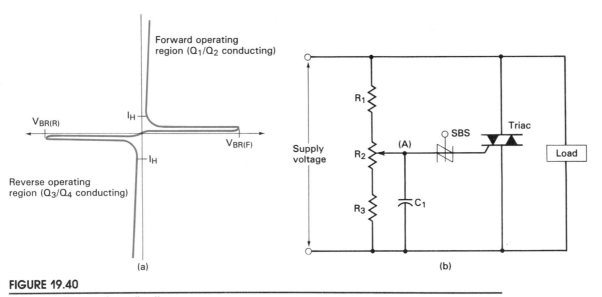

FIGURE 19.40

SBS operation and application.

A typical SBS application is shown in the crowbar circuit in Figure 19.40b. When the input voltage causes the voltage at point A to increase to the breakover rating of the SBS, the device turns on, triggering the triac. The triac then shorts out the supply voltage input, protecting the load. Note that the resistors in the circuit are used to set the trigger point for the SBS, while C_1 is included to prevent false (dv/dt) triggering.

Gate Turn-off (GTO) Devices

Gate turn-off (GTO) device. type of SCR that can be cut off by a negative gate pulse.

The JEDEC name for the GTO device is the **turn-off thyristor.**

The *gate turn-off* (*GTO*) *device* is a variation on the SCR. The primary difference between the GTO device and the SCR is the fact that a *negative gate pulse* can be used to drive the GTO device from its conducting state to its nonconducting state. The schematic symbol and equivalent circuit for the GTO device are shown in Figure 19.41.

As you can see, the GTO device has the same equivalent circuit as the SCR. However, the internal *n*-type and *p*-type regions are much narrower than those of the SCR. This means that the base regions of the two transistors are very narrow, as is the case with the real-life BJT. Since the GTO device more closely resembles

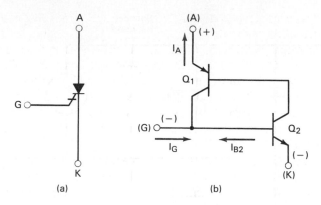

FIGURE 19.41

GTO device schematic symbol and equivalent circuit.

an actual two-transistor circuit (like the one shown in Figure 19.41b), we can turn the device off by causing the Q_2 base current to drop below the value needed to saturate the device.

For the circuit shown, the anode current (I_A) is found as

$$I_A = I_G + I_{B2}$$

where I_G = the gate current

I_{B2} = the base current of Q_2

If we rearrange this equation, we get

$$I_{B2} = I_A - I_G$$

Thus, if we apply a negative gate voltage to the GTO device, the value of I_G increases. If I_G becomes great enough, the value of I_{B2} will drop below the value required to maintain the saturated state of Q_2, and the device will go into cutoff. This causes the entire device to go into cutoff.

The effect of $-V_G$ on device conduction.

The narrow-base construction of the GTO device has a drawback. The reverse blocking voltage rating of the device tends to be very low; in the range of 15 to 20 V. If the reverse voltage that is being applied to the GTO device is greater than 15 V, the device must be protected using one of the diode circuits in Figure 19.42. In Figure 19.42a, a *reverse blocking diode* increases the overall reverse breakdown rating of the GTO device circuit. For example, if D_1 has a reverse breakdown voltage rating of 600 V and the GTO device has a reverse breakdown voltage rating of 20 V, V_R will have to reach 620 V to cause breakdown in the circuit. Thus, the GTO device is protected.

The *reverse conducting diode* in Figure 19.42b protects the GTO device by shorting any reverse voltage that is applied to the device. While this may seem to defeat the blocking function of the device, this is not the case. For example, consider the GTO device application shown in Figure 19.43.

FIGURE 19.42

GTO device protection.

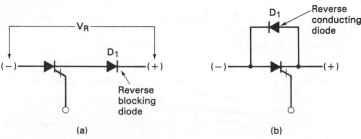

SEC. 19.5   Other Thyristors and Thyristor Triggers

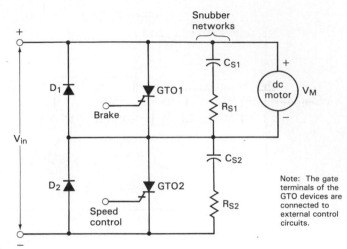

FIGURE 19.43

A GTO device application.

The circuit in Figure 19.43 is a dc motor controller that uses two GTO devices to provide braking and speed control. V_{in} to the circuit would be either a pure dc or a rectified ac signal that causes the motor voltage to have the polarities shown. When the motor is turned off, it produces a *counter emf*. As you know, this counter emf has a polarity that is the opposite of that shown in the figure. If this counter emf wasn't shorted out by D_1 and D_2, it would be applied across the GTO devices, destroying both of them. Thus, by conducting during the time that a reverse voltage is applied to the GTO devices, the diodes are providing the required reverse protection.

The fact that GTO2 in Figure 19.43 is being used as a *speed controller* points out another difference between the GTO device and the standard SCR. The physical construction of the GTO device gives it two distinct regions of operation, as is illustrated in Figure 19.44. When I_G is kept below a specified value, called the *latching current*, the device can be operated as a linear transistor. That is, the gate current can be varied to control the anode current of the device. If I_G exceeds the latching current rating of the device, it latches and enters the thyristor (saturated) region of operation. With this in mind, look again at the motor-speed controller in Figure 19.43. GTO2 in the circuit is being operated in the transistor (active) region. When its control circuit varies the value of I_G, the anode current through the device is also controlled. Since the total motor current is drawn through GTO2, controlling I_{G2} controls the total motor current and therefore the motor rotational

◢ 32b

FIGURE 19.44

The GTO device operating curves.

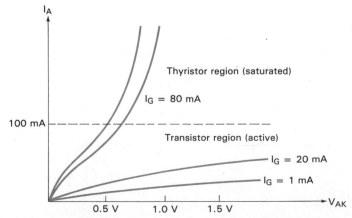

CHAP. 19   Thyristors and Optoelectronic Devices

speed. Note that GTO1 in the circuit is normally off. When activated by its control circuit, it short-circuits the motor, causing it to stop rotating.

The SIDAC

The SIDAC is a two-terminal, bidirectional device that has extremely high breakover voltage ratings; typically in the range of 100 to 280 V. *SIDAC* is an acronym for *SI*licon *D*iode–*AC*. The schematic symbol and characteristic curve for the SIDAC are shown in Figure 19.45.

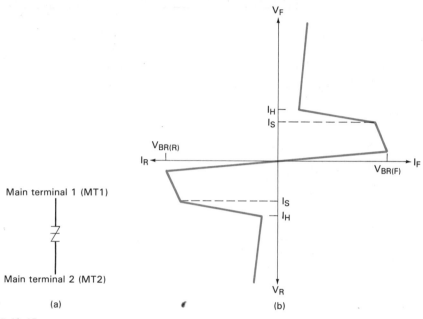

FIGURE 19.45

The SIDAC schematic symbol and operating curve.

The overall operation of the SIDAC is very similar to that of the SUS and the SBS. When the voltage across the component (in either direction) reaches the breakover rating of the device (V_{BR}), the SIDAC switches from its nonconducting state to its low resistance on state. The device continues to conduct as long as its current is greater than its holding current (I_H) rating.

The high breakover rating of the SIDAC makes it useful as a replacement for some of the phase-control circuits we discussed earlier. For example, refer back to Figure 19.27. The entire diac/triac phase-control circuit could be replaced by the SIDAC circuit shown in Figure 19.46. Assuming that the peak input voltage values are equal to the breakover ratings of the SIDAC, we have the phase-controlled output shown.

Two other SIDAC applications are illustrated in Figure 19.47. You may recall (from Chapter 4) that symmetrical zener diodes can be used as a *transient suppressor*,

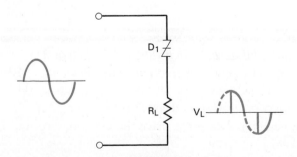

FIGURE 19.46

A SIDAC phase controller.

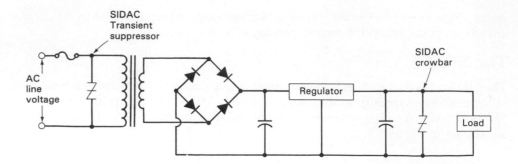

FIGURE 19.47

SIDAC transient suppressor
and crowbar

a circuit designed to break down and conduct when there is a surge in the ac line voltage that is applied to the primary of a power supply. In Figure 19.47, a SIDAC is used for this purpose. If a surge in the ac line voltage exceeds the breakover rating of the SIDAC, the device breaks down and conducts, shorting out the surge before it can be coupled to the transformer secondary. The SIDAC that is connected in parallel with the load is being used as a *crowbar*. If a fault in the power supply causes an increase in secondary voltage that exceeds the V_{BR} rating of the SIDAC, the device turns on and protects the load until the primary fuse in the power supply gives out.

Section Summary

The *silicon bilateral switch* (*SBS*) is a thyristor trigger that is closely related (in terms of operation) to the SCR. The SBS is an extremely fast triggering device that can handle relatively large currents and thus is well suited for triggering high-power thyristors. The gate of the SBS can be used to control the breakover point of the device in the same fashion as the SCR gate. However, the SBS is most commonly used as an open-gate breakover device. Thus, it is most commonly used to perform the same basic function as the diac.

The *gate turn-off* (*GTO*) *device* is a variation on the SCR. It is a unilateral (one-way conduction) device that can be driven into cutoff by applying a negative voltage to its gate terminal. The unique internal structure of the GTO device makes it the only thyristor that can also be used as a linear device. That is, a linear change in gate current can be used to produce a linear change in anode current, much like the base current in a BJT can be used to control its collector current. The GTO device will display these linear characteristics as long as gate current is kept below a specified value, called the *latching current* (L_L). If I_G reaches the value of I_L, the device switches to its on state and acts as a thyristor. That is, it becomes a low-impedance conductor until it is driven again into cutoff.

The SIDAC is a bilateral breakover thyristor that can be used in phase-control and load-protection circuits, as well as any other type of circuit that requires the use of a high-voltage breakover device.

SECTION REVIEW

1. What is a *silicon bilateral switch* (*SBS*)? [Objective 31]
2. In terms of its physical construction, how does the SBS differ from the other thyristor triggers we have discussed? [Objective 31]
3. What is a *gate turn-off* (*GTO*) device? [Objective 32a]
4. How does a negative gate pulse drive the GTO device into cutoff? [Objective 32a]

5. What is the primary limitation on the use of the GTO device? [Objective 33]

6. Describe the circuits used to overcome the limitation in question 5. [Objective 33]

7. In terms of operating regions, what distinguishes the GTO device from the other thyristors? [Objective 32b]

8. What is the SIDAC? [Objective 34]

9. Discuss the use of the SIDAC as a phase-control circuit. [Objective 34]

10. Discuss the use of the SIDAC as a transient suppressor. [Objective 34]

11. Discuss the use of the SIDAC as a crowbar. [Objective 34]

19.6
Discrete Photo-Detectors

In Chapter 2, we discussed the operation of the *light-emitting diode* (*LED*). The LED is classified as a *light emitter* because it gives off light when biased properly. In this section, we are going to take a look at the operation of several *light detectors*. Light detectors are components whose electrical output characteristics are controlled by the amount of light they receive. In other words, while *emitters* produce light, *detectors* respond to light.

When we discuss photodetectors, we need to consider a variety of parameters that are related to light. For this reason, we need to start this section with a brief discussion on light and its properties.

Light emitters. Optoelectronic devices that produce light.

Light detectors. Optoelectronic devices that respond to light.

Characteristics of Light

Light is *electromagnetic energy* that falls within a specific range of frequencies, as is shown in Figure 19.48. The entire light spectrum falls within the range of 30 THz to 3 PHz and is further broken down as shown in Table 19.2.

Light. Electromagnetic energy that falls within a specific range of frequencies.

FIGURE 19.48

The light spectrum.

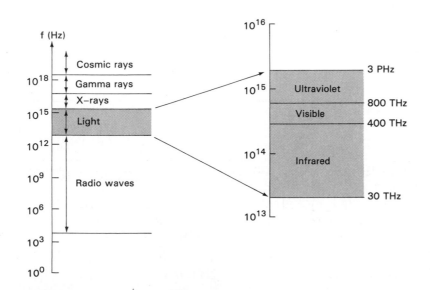

TABLE 19.2
Light Frequency Ranges

Type of Light	Frequency Range (approximate)
Infrared	30 THz to 400 THz
Visible	400 THz to 800 THz
Ultraviolet	800 THz to 3 PHz[a]

[a] P stands for *peta* (10^{15}), and T for *tera* (10^{12}).

Wavelength. The physical length of one cycle of a transmitted electromagnetic wave.

Two parameters are commonly used to describe light. The first of these is *wavelength* (λ). Wavelength is the physical length of one cycle of a transmitted

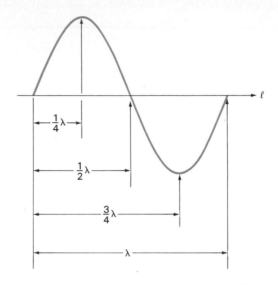

FIGURE 19.49

Wavelength.

electromagnetic wave. The concept of wavelength is illustrated in Figure 19.49. The wavelength for a signal with a given frequency is found as

$$\lambda = \frac{c}{f} \qquad (19.8)$$

A Practical Consideration. The wavelength parameters of many optoelectronic devices are given in *nanometers* (*nm*).

where λ = the wavelength of the signal (in *nanometers*)
c = the *speed of light*, given as 3×10^{17} nm/s
f = the frequency of the transmitted signal

The following example illustrates the calculation of wavelength for a given signal.

EXAMPLE 19.6

Determine the wavelength of a 150-THz light signal.

Solution: The wavelength of the signal is found as

$$\lambda = \frac{c}{f}$$
$$= \frac{3 \times 10^{17} \text{ nm/s}}{150 \times 10^{12} \text{ Hz}}$$
$$= 2000 \text{ nm}$$

Thus, the physical length of one cycle of a 150-THz signal is 2000 nm, or 2 μm.

PRACTICE PROBLEM 19–6

The infrared frequency spectrum falls between 30 THz and 400 THz. Determine the range of wavelengths for this range of frequencies.

Wavelength is an important concept because photoemitters and photodetectors are rated for specific wavelengths. You see, photoemitters are used to drive the various photodetectors. For a given photoemitter to be used with a given photodetector, the two devices must be rated for the same approximate wavelength. This point will be discussed further later in this section.

The second important light parameter is *intensity*. Light intensity is the amount of light per unit area that is received by a given photodetector. For example, consider the emitter/detector combination shown in Figure 19.50. The light intensity is a measure of the amount of light measured at the detector. Note that the intensity decreases as the distance between the emitter and detector (*d*) decreases. As you will see, the output characteristics of most photodetectors are controlled by the light intensity at their inputs, as well as the input wavelength.

Light intensity. The amount of light per unit area that is received by a given photodetector, also called **irradiance**.

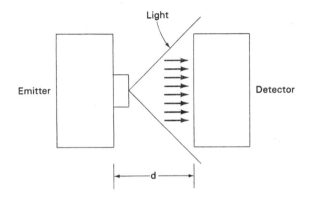

FIGURE 19.50

The Photodiode

The *photodiode* is a diode whose reverse conduction is light intensity controlled. When the light intensity at the optical input to a photodiode increases, the reverse current through the device also increases. This point is illustrated in Figure 19.51.

Photodiode. A diode whose reverse conduction is light intensity controlled.

FIGURE 19.51

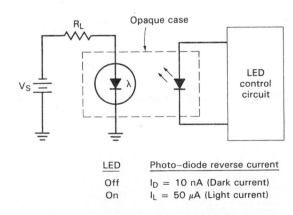

LED	Photo–diode reverse current
Off	I_D = 10 nA (Dark current)
On	I_L = 50 μA (Light current)

Opaque. A term used to describe anything that blocks light.

Light current. The reverse current through a photodiode with an active light input.

Dark current. The reverse current through a photodiode with no active light input.

In Figure 19.51, a photodiode is enclosed with an LED in an *opaque case*. The term *opaque* is used to describe anything that blocks light. Thus, when in the opaque case, the photodiode will only receive a light input when the LED is turned on by its control circuit. When the LED is off, the photodiode reverse current is shown to be 10 nA. When the LED lights, the photodiode reverse current increases to 50 μA. Note that the *light current* (I_L) is shown to be approximately 5000 times as great as the *dark current* (I_D). This ratio of light current to dark current is not at all uncommon.

FIGURE 19.52

The MRD500–510 photo detector specification sheet.
(Courtesy of Motorola, Inc.)

MOTOROLA
■ **SEMICONDUCTOR** ■
TECHNICAL DATA

Photo Detectors
Diode Output

| MRD500 |
| MRD510 |

**PHOTO DETECTORS
DIODE OUTPUT
PIN SILICON
250 MILLIWATTS
100 VOLTS**

. . . designed for application in laser detection, light demodulation, detection of visible and near infrared light-emitting diodes, shaft or position encoders, switching and logic circuits, or any design requiring radiation sensitivity, ultra high-speed, and stable characteristics.

- Ultra Fast Response — (<1 ns Typ)
- High Sensitivity — MRD500 (1.2 μA/mW/cm² Min)
  MRD510 (0.3 μA/mW/cm² Min)
- Available With Convex Lens (MRD500) or Flat Glass (MRD510) for Design Flexibility
- Popular TO-18 Type Package for Easy Handling and Mounting
- Sensitive Throughout Visible and Near Infrared Spectral Range for Wide Application
- Annular Passivated Structure for Stability and Reliability

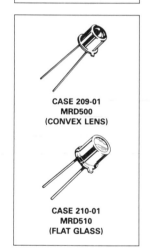

CASE 209-01
MRD500
(CONVEX LENS)

CASE 210-01
MRD510
(FLAT GLASS)

MAXIMUM RATINGS (T_A = 25°C unless otherwise noted)

Rating	Symbol	Value	Unit
Reverse Voltage	V_R	100	Volts
Total Power Dissipation @ T_A = 25°C Derate above 25°C	P_D	250 2.27	mW mW/°C
Operating Temperature Range	T_A	−55 to +125	°C
Storage Temperature Range	T_{stg}	−65 to +150	°C

STATIC ELECTRICAL CHARACTERISTICS (T_A = 25°C unless otherwise noted)

Characteristic	Fig. No.	Symbol	Min	Typ	Max	Unit
Dark Current (V_R = 20 V, R_L = 1 megohm) Note 2 T_A = 25°C T_A = 100°C	2 and 3	I_D	— —	— 14	2 —	nA
Reverse Breakdown Voltage (I_R = 10 μA)	—	$V_{(BR)R}$	100	200	—	Volts
Forward Voltage (I_F = 50 mA)	—	V_F	—	—	1.1	Volts
Series Resistance (I_F = 50 mA)	—	R_S	—	—	10	Ohms
Total Capacitance (V_R = 20 V, f = 1 MHz)	5	C_T	—	—	4	pF

OPTICAL CHARACTERISTICS (T_A = 25°C unless otherwise noted)

Characteristic		Fig. No.	Symbol	Min	Typ	Max	Unit
Light Current (V_R = 20 V) Note 1	MRD500 MRD510	1	I_L	6 1.5	9 2.1	— —	μA
Sensitivity at 0.8 μm (V_R = 20 V) Note 3	MRD500 MRD510	—	$S_{(\lambda = 0.8 \mu m)}$	— —	6.6 1.5	— —	μA/mW/ cm²
Response Time (V_R = 20 V, R_L = 50 Ohms)		—	$t_{(resp)}$	—	1	—	ns
Wavelength of Peak Spectral Response		5	λ_S	—	0.8	—	μm

NOTES: 1. Radiation Flux Density (H) equal to 5 mW/cm² emitted from a tungsten source at a color temperature of 2870 K.
2. Measured under dark conditions. (H ≈ 0).
3. Radiation Flux Density (H) equal to 0.5 mW/cm² at 0.8 μm.

918   CHAP. 19   Thyristors and Optoelectronic Devices

The spec sheet for a photodiode lists its values of I_D and I_L, as can be seen in Figure 19.52. Note that the value of I_D for the MRD500 photodiode is 2 nA (maximum), while the value of I_L for the device is 6 μA (minimum). These two values show the device to have a minimum ratio of light-to-dark current of 3000.

Several other photodiode ratings listed under the *optical characteristics* heading are of interest. The first of these is *wavelength of peak spectral response* (λ_S). This rating tells you the wavelength that will cause the strongest response in the photodiode. For the MRD500, the optimum input wavelength is 0.8 μm (800 nm). As the following example shows, this value can be used to determine the optimum operating frequency for the device.

Wavelength of peak spectral response (λ_S). A rating that indicates the wavelength that will cause the strongest response in a photodetector.

FIGURE 19.52

(continued)

TYPICAL CHARACTERISTICS

MRD500

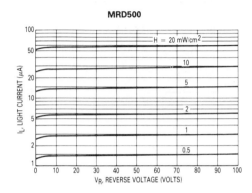

MRD510

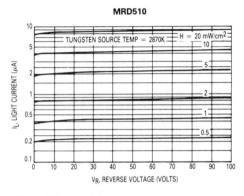

Figure 1. Irradiated Voltage — Current Characteristic

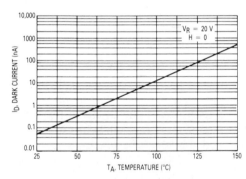

Figure 2. Dark Current versus Temperature

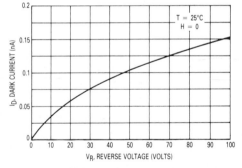

Figure 3. Dark Current versus Reverse Voltage

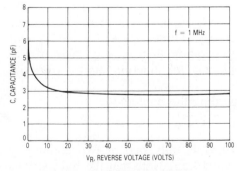

Figure 4. Capacitance versus Voltage

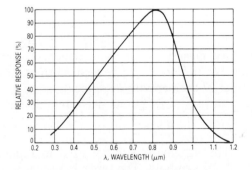

Figure 5. Relative Spectral Response

Determine the optimum operating frequency for the MRD500 photodiode.

Solution: Rearranging equation (19.8) gives us the following equation for frequency:

$$f = \frac{c}{\lambda}$$

Using the given optimum wavelength of 800 nm, the optimum operating frequency for the MRD 500 is found as

$$f = \frac{3 \times 10^{17} \text{ nm/s}}{800}$$
$$= 375 \text{ THz}$$

Since 375 THz is in the *infrared* frequency spectrum, the MRD500 is classified as an *infrared detector*.

PRACTICE PROBLEM 19–7

The Motorola MRD821 photodiode has an optimum wavelength of 940 nm. Determine the optimum operating frequency for the device.

Sensitivity. A rating that indicates the response of a photodetector to a specified light intensity.

The *sensitivity* rating of the photodiode indicates the response of the device to a specified light intensity. For the MRD500, the sensitivity rating (listed under *optical characteristics*) is given as 6.6 $\mu A/mW/cm^2$. This means that the reverse current of the device will increase by 6.6 μA for every 1 mW/cm ^{2} of light that is applied to the device. For example, refer back to Figure 19.51. If the light intensity at the photodiode is 1 mW/cm^2, the value of I_R for the device is approximately 6.6 μA. If the light intensity at the input increases to 2 mW/cm^2, I_R increases to approximately 13.2 μA, and so on.

Spectral response. A measure of a photodetector's response to a change in input wavelength. Note that response is measured in terms of the device *sensitivity*.

The *spectral response* of a photodiode is a measure of the device's response to a change in input wavelength. The spectral response of the MRD500 is shown in the *relative spectral response* curve in Figure 19.52b. As you can see, the response of the photodiode peaks at 800 nm, the value given on the spec sheet. If the input wavelength goes as low as 520 nm or as high as 950 nm, the *sensitivity* of the device drops to 50% of its rated value. Thus, at these two frequencies, the sensitivity of the device drops to approximately 3.3 $\mu A/mW/cm^2$.

The *irradiated voltage–current characteristic* curves in Figure 19.52b show that the light current (I_L) of the device is relatively independent of the amount of reverse bias across the diode. The dark current (I_D) curves show that the amount of dark current is far more dependent on temperature than on the value of V_R. This is similar to the reverse current through any *pn* junction diode, which is primarily temperature dependent.

Several final points need to be made regarding the photodiode:

1. The schematic symbol used in Figure 19.51 is only one of two commonly used symbols. The other is shown in Figure 19.53. Note that the light arrows in the symbol point *toward* the diode, rather than away from it.

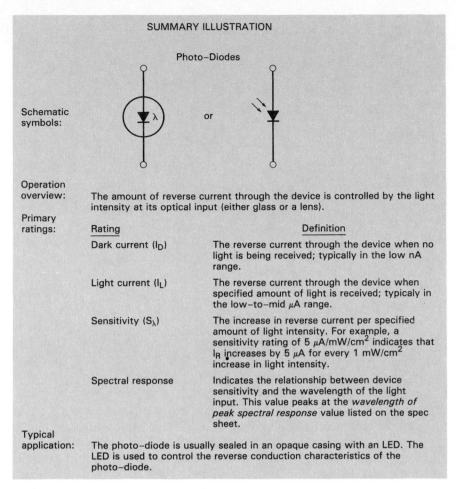

SUMMARY ILLUSTRATION

Photo–Diodes

Schematic symbols:	(schematic symbols) or (schematic symbols)
Operation overview:	The amount of reverse current through the device is controlled by the light intensity at its optical input (either glass or a lens).

Primary ratings:

Rating	Definition
Dark current (I_D)	The reverse current through the device when no light is being received; typically in the low nA range.
Light current (I_L)	The reverse current through the device when specified amount of light is received; typicaly in the low–to–mid μA range.
Sensitivity (S_λ)	The increase in reverse current per specified amount of light intensity. For example, a sensitivity rating of 5 μA/mW/cm^2 indicates that I_R increases by 5 μA for every 1 mW/cm^2 increase in light intensity.
Spectral response	Indicates the relationship between device sensitivity and the wavelength of the light input. This value peaks at the *wavelength of peak spectral response* value listed on the spec sheet.

Typical application:	The photo–diode is usually sealed in an opaque casing with an LED. The LED is used to control the reverse conduction characteristics of the photo–diode.

FIGURE 19.53

2. The photodiode is always operated in its reverse operating region. Since light affects the *reverse conduction* of the device, it wouldn't make any sense to operate it in its forward operating region.

3. Most of the photodiode parameters and ratings we have discussed apply to many of the other photodetectors. This point will become clear as our coverage of these devices continues.

The Phototransistor

The *phototransistor* is a three-terminal photodetector whose collector current is controlled by the intensity of the light at its optical input (base). Figure 19.54 shows one possible configuration for a phototransistor amplifier.

In Figure 19.54, the phototransistor is shown to be enclosed with an LED in an opaque casing. When the light from the LED is varied by its control circuit, the change in light causes a proportional change in base current in the phototransistor. This change in base current causes a change in emitter (and collector) current that causes a proportional voltage to be developed across R_E. This voltage is applied to the inverting amplifier, which responds accordingly.

The phototransistor is identical to the standard BJT in every respect, outside of the fact that its collector current is (ultimately) light intensity controlled. Thus, the phototransistor in Figure 19.54 could be a linear amplifier or a switch, depending on the output of the LED control circuit. The biasing circuit (R_1 and R_2) is used

Phototransistor. A three-terminal photodetector whose collector current is controlled by the intensity of the light at its optical input (base).

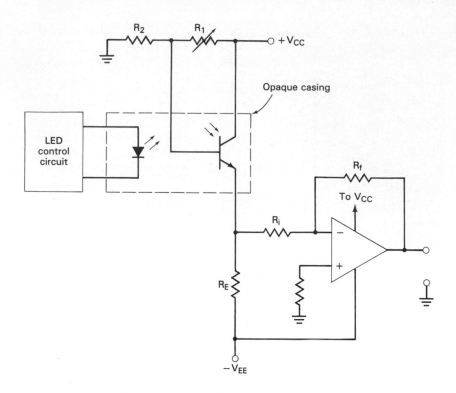

FIGURE 19.54

Optocoupling.

to adjust the dark (quiescent) operation of the phototransistor. When R_1 is varied, the dark emitter current of the phototransistor varies. Thus, by adjusting R_1, the input voltage to the op-amp is varied, allowing the output of the op-amp to be set to zero. Note that R_1 in this circuit would be called a *zero adjust* because it is used to set the final circuit output to zero volts.

The arrangement in Figure 19.54 is referred to as *optocoupling* because the output from the LED control circuit is coupled via light to the phototransistor/op-amp circuit. The advantage of optocoupling is that it provides complete dc isolation between the source (the LED control circuit) and the load (the phototransistor/op-amp circuit).

Several points should be made regarding the phototransistor: First, the spec sheet for the phototransistor contains the same basic parameters and ratings as found on the spec sheet for a photodiode. The phototransistor is subject to the same wavelength and sensitivity characteristics as the photodiode. The difference is the fact that the optical characteristics of the phototransistor relate the light input to *collector current* (rather than diode reverse current). Second, the base terminal of the phototransistor is normally returned to ground through some type of biasing circuit, as shown in Figure 19.54. The biasing circuit not only allows for adjusting the output from the device, but it also adds *temperature stability* to the circuit. You see, transistors have the same leakage current considerations as any other device. Without a return path from base to ground, any base leakage current is forced into the emitter terminal of the device. This can cause the phototransistor to be forced into conduction by a change in temperature. By providing a resistive path from the base terminal to ground, you prevent base leakage current from affecting the emitter circuit of the device.

The Photo-Darlington

The photo-Darlington is a phototransistor that is constructed in a Darlington configuration, as is shown in Figure 19.55. Outside of the increased output current capability

◤ 42

FIGURE 19.55

The photo-Darlington schematic symbol.

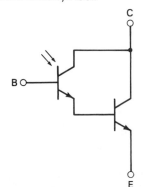

CHAP. 19 Thyristors and Optoelectronic Devices

(as is always the case with Darlingtons), the operation and characteristics of the device are identical to those of the standard phototransistor.

Photo-Darlington. A phototransistor with a Darlington pair output.

The LASCR

The *light-activated SCR (LASCR)*, or *photo-SCR*, is a three-terminal light-activated SCR. The schematic symbol for the LASCR is shown in Figure 19.56.

Again, all the properties of the photodiode can be found in the LASCR. In this case, however, the SCR is being *latched* into its conducting state by the light input to the device. Thus, the LASCR could be used in an optically coupled phase-control circuit.

Summary

Most discrete devices can be obtained in *photodetector* form. The photodetector devices work in a fashion that is very similar to their current-controlled counterparts. However, the photodetector devices are *light-controlled* devices.

When working with photodetectors, you need to remember the *wavelength* and *sensitivity* (intensity) characteristics of the devices. If a light emitter and photodetector are not matched in terms of wavelength and sensitivity, the optocoupling circuit will operate at reduced efficiency or may not work at all.

The wavelength and sensitivity concerns of discrete photodetector circuits are eliminated by the use of *optoisolators* and *optointerrupters*. These are integrated circuits that contain both a light emitter and a light detector. As you will see in the next section, these ICs are far easier to deal with than their discrete component counterparts.

Light-activated SCR (LASCR). A three-terminal light-activated SCR.

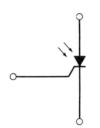

FIGURE 19.56
The LASCR schematic symbol.

SECTION REVIEW

1. What is the difference between a light *emitter* and a light *detector*? [Objective 35]
2. What is *wavelength*? [Objective 36]
3. Why is the concept of wavelength important when working with optoelectronic devices? [Objective 38]
4. What is *light intensity*? [Objective 39]
5. What is a *photodiode*? [Objective 40]
6. In terms of the photodiode, what is *dark current* (I_D)? What is *light current* (I_L)? [Objective 40]
7. What is the relationship between photodiode light current and dark current? [Objective 40]
8. What is the *wavelength of peak spectral response* for a photodiode? [Objective 40]
9. What is *sensitivity*? [Objective 40]
10. Explain the $\mu A/mW/cm^2$ unit of measure for sensitivity. [Objective 40]
11. What is *spectral response*? Which photodiode unit of measure is affected by spectral response? [Objective 40]
12. What is a *phototransistor*? [Objective 41]
13. What is the relationship between input light and collector current for a phototransistor? [Objective 41]
14. What is the difference between the optical ratings of the phototransistor and those of the photodiode? [Objective 41]

15. What is the difference between the *photo-Darlington* and the standard photo-transistor? [Objective 42]

16. What is the *LASCR*? How does it differ from the standard SCR? [Objective 43]

17. List the characteristics that must be considered whenever you are working with any discrete photodetector. [Objective 44]

19.7
Optoisolators and Optointerrupters

Optoisolators and *optointerrupters* are ICs that contain both a photoemitter and a photodetector. While the photoemitter is always an LED, the photodetector can be any one of the discrete photodetectors we discussed in the last section.

Optoisolator. An optocoupler. A device that uses light to couple a signal from its input (a photoemitter) to its output (a photodetector).

Optoisolators

The *optoisolator* is an optocoupler, that is, a device that uses light to couple a signal from its input (a photoemitter) to its output (a photodetector). For example, an optoisolator with a transistor output would contain the circuitry shown in the opaque casing in Figure 19.54.

The typical optoisolator comes in a six-pin DIP (*dual in-line package*), as is shown on the spec sheet in Figure 19.57. As the schematic diagram in the figure shows, this transistor-output optoisolator contains an LED and a phototransistor. Thus, the 4N35–7 series optoisolators could be used in the circuit in Figure 19.54.

Most of the optoisolator parameters and ratings are the standard LED and phototransistor specifications. However, there are several new ratings that warrant discussion.

Isolation source voltage (V_{ISO}). The voltage that, if applied across the input and output pins, will destroy an optoisolator.

The *isolation source voltage* rating indicates the input-to-output voltage that will cause the optoisolator to break down and conduct. In other words, it is the voltage that, if applied across the input and output pins, will destroy the device. For the 4N35–7 series chips, the value of V_{ISO} is 7500 V_{ac}.

Soldering temperature. The maximum amount of heat that can be applied to any pin of the IC without causing damage to the chip.

The *soldering temperature* is the maximum amount of heat that can be applied to any pin of the IC without causing internal damage to the chip. If the total heat applied to any pin of the 4N35 reaches 260°C for a duration of 10s, internal damage will result.

Isolation current (I_{ISO}). The amount of current that can be forced between the input and output at the rated voltage.

The *isolation current* (I_{ISO}) rating indicates the amount of current that can be forced between the input and output at the rated voltage. For example, a 3550 V_{pk} input that is applied across the 4N35 will cause 100 μA to flow between the input and output pins of the device. The 4N36 requires a 2500 V_{pk} input to cause 100 μA to flow, and so on.

Isolation resistance (R_{ISO}). The total resistance between the device input and output pins.

The *isolation resistance* (R_{ISO}) rating indicates the total resistance between the input pins and output pins of the device. For the 4N35–7 series, this rating is 10^{11} Ω (100 GΩ).

Isolation capacitance (C_{ISO}). The total capacitance between the device input and output pins.

Finally, the *isolation capacitance* (C_{ISO}) rating indicates the total capacitance between the device input and output pins. For the 4N35–7 series, this rating is 2 pF.

Outside of the ratings listed here, all of the parameters and ratings for the optoisolator are fairly standard. This can be seen by quickly looking through the spec sheet. As you glance through the spec sheet, you'll also notice two standard parameters that are missing: those for *wavelength* and *sensitivity*. Since the photoemit-

6-Pin DIP Optoisolators
Transistor Output

These devices consist of a gallium arsenide infrared emitting diode optically coupled to a monolithic silicon phototransistor detector.

- Convenient Plastic Dual-In-Line Package
- High Current Transfer Ratio — 100% Minimum at Spec Conditions
- Guaranteed Switching Speeds
- High Input-Output Isolation Guaranteed — 7500 Volts Peak
- UL Recognized. File Number E54915 ℞
- VDE approved per standard 0883/6.80 (Certificate number 41853), with additional approval to DIN IEC380/VDE0806, IEC435/VDE0805, IEC65/VDE0860, VDE0110b, covering all other standards with equal or less stringent requirements, including IEC204/VDE0113, VDE0160, VDE0832, VDE0833, etc. ⚠️ 883
- Meets or Exceeds All JEDEC Registered Specifications
- Special lead form available (add suffix "T" to part number) which satisfies VDE0883/6.80 requirement for 8 mm minimum creepage distance between input and output solder pads.
- Various lead form options available. Consult "Optoisolator Lead Form Options" data sheet for details.

4N35 **4N36** **4N37**	
6-PIN DIP OPTOISOLATORS TRANSISTOR OUTPUT	

CASE 730A-02 PLASTIC

MAXIMUM RATINGS (T_A = 25°C unless otherwise noted)

Rating	Symbol	Value	Unit
INPUT LED			
Reverse Voltage	V_R	6	Volts
Forward Current — Continuous	I_F	60	mA
LED Power Dissipation @ T_A = 25°C with Negligible Power in Output Detector	P_D	120	mW
Derate above 25°C		1.41	mW/°C
OUTPUT TRANSISTOR			
Collector-Emitter Voltage	V_{CEO}	30	Volts
Emitter-Base Voltage	V_{EBO}	7	Volts
Collector-Base Voltage	V_{CBO}	70	Volts
Collector Current — Continuous	I_C	150	mA
Detector Power Dissipation @ T_A = 25°C with Negligible Power in Input LED	P_D	150	mW
Derate above 25°C		1.76	mW/°C
TOTAL DEVICE			
Isolation Source Voltage (1) (Peak ac Voltage, 60 Hz, 1 sec Duration)	V_{ISO}	7500	Vac
Total Device Power Dissipation @ T_A = 25°C Derate above 25°C	P_D	250 2.94	mW mW/°C
Ambient Operating Temperature Range	T_A	−55 to +100	°C
Storage Temperature Range	T_{stg}	−55 to +150	°C
Soldering Temperature (10 seconds, 1/16" from case)	—	260	°C

(1) Isolation surge voltage is an internal device dielectric breakdown rating.
For this test, Pins 1 and 2 are common, and Pins 4, 5 and 6 are common.

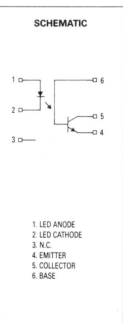

SCHEMATIC

1. LED ANODE
2. LED CATHODE
3. N.C.
4. EMITTER
5. COLLECTOR
6. BASE

FIGURE 19.57

The 4N35–7 Optoisolator specification sheet.
(Courtesy of Motorola, Inc.)

Characteristic		Symbol	Min	Typ	Max	Unit
INPUT LED						
Forward Voltage (I_F = 10 mA) T_A = 25°C T_A = −55°C T_A = 100°C		V_F	0.8 0.9 0.7	1.15 1.3 1.05	1.5 1.7 1.4	V
Reverse Leakage Current (V_R = 6 V)		I_R	—	—	10	μA
Capacitance (V = 0 V, f = 1 MHz)		C_J	—	18	—	pF
OUTPUT TRANSISTOR						
Collector-Emitter Dark Current (V_{CE} = 10 V, T_A = 25°C) (V_{CE} = 30 V, T_A = 100°C)		I_{CEO}	— —	1 —	50 500	nA μA
Collector-Base Dark Current (V_{CB} = 10 V) T_A = 25°C T_A = 100°C		I_{CBO}	— —	0.2 100	20 —	nA
Collector-Emitter Breakdown Voltage (I_C = 1 mA)		$V_{(BR)CEO}$	30	45	—	V
Collector-Base Breakdown Voltage (I_C = 100 μA)		$V_{(BR)CBO}$	70	100	—	V
Emitter-Base Breakdown Voltage (I_E = 100 μA)		$V_{(BR)EBO}$	7	7.8	—	V
DC Current Gain (I_C = 2 mA, V_{CE} = 5 V)		h_{FE}	—	400	—	—
Collector-Emitter Capacitance (f = 1 MHz, V_{CE} = 0)		C_{CE}	--	7	—	pF
Collector-Base Capacitance (f = 1 MHz, V_{CB} = 0)		C_{CB}	—	19	—	pF
Emitter-Base Capacitance (f = 1 MHz, V_{EB} = 0)		C_{EB}	—	9	—	pF
COUPLED						
Output Collector Current (I_F = 10 mA, V_{CE} = 10 V) T_A = 25°C T_A = −55°C T_A = 100°C		I_C	10 4 4	30 — —	— — —	mA
Collector-Emitter Saturation Voltage (I_C = 0.5 mA, I_F = 10 mA)		$V_{CE(sat)}$	—	0.14	0.3	V
Turn-On Time	(I_C = 2 mA, V_{CC} = 10 V, R_L = 100 Ω, Figure 11)	t_{on}	—	7.5	10	μs
Turn-Off Time		t_{off}	—	5.7	10	
Rise Time		t_r	—	3.2	—	
Fall Time		t_f	—	4.7	—	
Isolation Voltage (f = 60 Hz, t = 1 sec)		V_{ISO}	7500	—	—	Vac(pk)
Isolation Current (V_{I-O} = 3550 Vpk) 4N35 (V_{I-O} = 2500 Vpk) 4N36 (V_{I-O} = 1500 Vpk) 4N37		I_{ISO}	— — —	— — 8	100 100 100	μA
Isolation Resistance (V = 500 V)		R_{ISO}	10^{11}	—	—	Ω
Isolation Capacitance (V = 0 V, f = 1 MHz)		C_{ISO}	—	0.2	2	pF

FIGURE 19.57

(continued)

ter and photodetector are contained in the same chip, the wavelength and sensitivity values of the devices are unimportant.

As was stated earlier, optoisolators come with a variety of output devices. Among these are the transistor output, Darlington output, triac output, and SCR output. For each of these devices, the input/output parameters and ratings vary only to reflect the characteristics of the photodetector that is being used. Other than that, they operate exactly as described here and in Section 19.6.

An Optoisolator Application: The Solid-state Relay

Solid-state relay (SSR). A circuit that uses a dc input voltage to pass or block an ac signal.

A *solid-state relay* (SSR) is a circuit that uses a dc input voltage to pass or block an ac signal. The diagram in Figure 19.58a illustrates the input/output relationship for the SSR. When the dc input has the polarity shown, the ac waveform at the ac input (ac 1) is coupled to the ac output (ac 2), and thus to the load. As long as the dc potential remains, the ac signal will be coupled to the load. If the dc potential is removed, the output goes open circuit, and the ac signal is blocked from the load.

The schematic for the SSR is shown in Figure 19.58b. The dc circuit is coupled to the ac circuit via the triac-output optoisolator. As long as the dc input

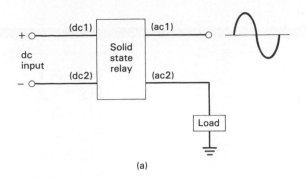

(a)

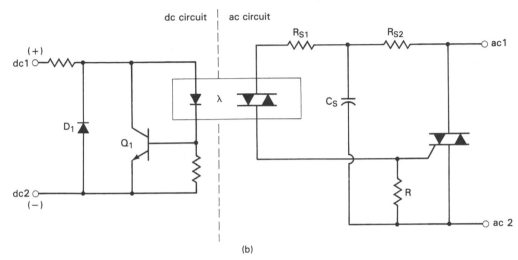

(b)

FIGURE 19.58

The Solid-state relay.

voltage has the polarity shown, the LED in the optoisolator is forward biased and emits light. This light keeps the triac photodetector conducting. As long as the triac photodetector is conducting, the triac in the ac circuit continues to conduct, passing the ac signal from the input terminal (ac 1) to the output terminal (ac 2).

If the dc input voltage is removed, the LED turns off, and the triac photodetector shuts off. This prevents a gate trigger voltage from being developed across R_3, and the output triac shuts off. When the output triac shuts off, the ac input is blocked from the ac output terminal.

The SSR is typically used to allow a digital circuit to control the operation (on or off) of a motor. This application is easy to understand if you picture the input to Figure 19.58a as coming from a digital system and the load as being a motor. Note that SSRs are also available in integrated form. While the exact circuitry of the integrated SSR varies from the circuitry shown in Figure 19.58b, the overall principle of operation is the same.

Optointerrupters

The *optointerrupter* is an IC optocoupler that is designed to allow an outside object to block the light path between the photoemitter and the photodetector. Two of the many common optointerrupter configurations are shown on the spec sheet in Figure 19.59. Note that the photoemitter is on one side of the slot, while the photodetector is on the other side.

The open gap between the photoemitter and the photodetector is the primary difference between the optointerrupter and the optoisolator. This gap allows the optointerrupter to be activated by some outside object, such as a piece of paper. For example, let's say that the optointerrupter in Figure 19.59 is being used in a photocopying machine, and that the device is placed so that the edge of a given

 47

Optointerrupter. An IC optocoupler that is designed to allow an outside object to block the light path between the photoemitter and the photodetector.

Slotted Optical Switches
Transistor Output

Each device consists of a gallium arsenide infrared emitting diode facing a silicon NPN phototransistor in a molded plastic housing. A slot in the housing between the emitter and the detector provides the means for mechanically interrupting the infrared beam. These devices are widely used as position sensors in a variety of applications.

- Single Unit for Easy PCB Mounting
- Non-Contact Electrical Switching
- Long-Life Liquid Phase Epi Emitter
- 1 mm Detector Aperture Width

H21A1
H21A2
H21A3
H22A1
H22A2
H22A3

SLOTTED
OPTICAL SWITCHES
TRANSISTOR OUTPUT

MAXIMUM RATINGS

Rating	Symbol	Value	Unit
INPUT LED			
Reverse Voltage	V_R	6	Volts
Forward Current — Continuous	I_F	60	mA
Input LED Power Dissipation @ T_A = 25°C Derate above 25°C	P_D	150 2	mW mW/°C
OUTPUT TRANSISTOR			
Collector-Emitter Voltage	V_{CEO}	30	Volts
Output Current — Continuous	I_C	100	mA
Output Transistor Power Dissipation @ T_A = 25°C Derate above 25°C	P_D	150 2	mW mW/°C
TOTAL DEVICE			
Ambient Operating Temperature Range	T_A	−55 to +100	°C
Storage Temperature	T_{stg}	−55 to +100	°C
Lead Soldering Temperature (5 seconds max)	—	260	°C
Total Device Power Dissipation @ T_A = 25°C Derate above 25°C	P_D	300 4	mW mW/°C

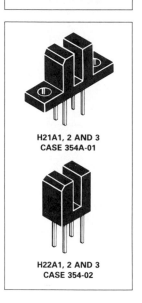

H21A1, 2 AND 3
CASE 354A-01

H22A1, 2 AND 3
CASE 354-02

FIGURE 19.59

The H21–2 series optointerruptors specification sheet.
(Courtesy of Motorola, Inc.)

copy passes through the slot as it (the paper) passes through the machine. When no paper is present in the optointerrupter gap, the light from the emitter reaches the phototransistor, causing it to saturate. Thus, the output from the phototransistor is *low*. When a piece of paper passes through the gap, the emitter light is blocked (by the paper) from the phototransistor. This causes the phototransistor to go into cutoff, causing its collector voltage to go *high*. The low and high output voltages from the phototransistor are used to tell the machine whether or not a piece of paper has passed the point where the optointerrupter is located.

The type of application discussed here is the primary application for the optointerruptor. Any other type of optical-coupling application is usually performed by a discrete optocoupler or an optoisolator.

SECTION REVIEW

1. What is an *optoisolator*? [Objective 45]
2. List and define the common *isolation* parameters and ratings of the optoisolator. [Objective 45]

3. Describe the operation of the *solid-state relay (SSR)*. [Objective 46]
4. Contrast the *optointerrupter* with the *optoisolator*. [Objective 47]

KEY TERMS

The following terms were introduced and defined in this chapter.

anode current interruption
average on-state current (I_T)
average on-state voltage (V_T)
bidirectional thyristor
circuit fusing rating (I^2t)
critical rise rating (dv/dt)
crowbar
dark current (I_D)
diac
false triggering
forced commutation
forward blocking region
forward breakover current [$I_{BR(F)}$]
forward breakover voltage [$V_{BR(F)}$]
forward operating region
gate nontrigger voltage (V_{GD})
gate trigger voltage (V_{GT})
gate turn-off (GTO) device
holding current (I_H)
industrial electronics
interbase resistance
intrinsic standoff ratio (η)
isolation capacitance (C_{ISO})

isolation current (I_{ISO})
isolation resistance (R_{ISO})
isolation source voltage (V_{ISO})
light
light-activated SCR (LASCR)
light current (I_L)
light detectors
light emitters
light intensity
negative resistance
negative resistance region
nonrepetitive surge current (I_{TSM})
opaque
optoelectronic devices
optointerruptor
optoisolator
peak current (I_P)
peak voltage (V_P)
phase controller
photo-Darlington
photodiode
phototransistor
programmable unijunction transistor (PUT)

relaxation oscillator
sensitivity
silicon bilateral switch (SBS)
silicon-controlled rectifier (SCR)
silicon diode–AC (SIDAC)
silicon unilateral switch (SUS)
snubber network
soldering temperature
solid-state relay (SSR)
spectral response
thyristors
triac
trigger-point voltage (V_{TP})
unijunction transistor (UJT)
universal motor
wavelength (λ)
wavelength of peak spectral response (λ_S)

PRACTICE PROBLEMS

§19.2

1. The 2N6237 SCR has a circuit fusing rating of 2.6 A^{2}s. Determine whether or not the component can withstand a 25-A surge that lasts for 120 ms. [10]

2. The 2N6342 SCR has a circuit fusing rating of 40 A^{2}s. Determine whether or not the device can withstand an 80-A surge that lasts for 8 ms. [10]

3. The 2N6237 SCR has a circuit fusing rating of 2.6 A^{2}s. Determine the maximum allowable duration of a 50-A surge through the device. [10]

4. The 2N6342 has a circuit fusing rating of 40 A^{2}s. Determine the maximum allowable duration of a 95-A surge through the device. [10]

5. The 2N6237 has a circuit fusing rating of 2.6 A^{2}s. Determine the maximum surge current that the device can withstand for 50 ms. [10]

6. The 2N6342 has a circuit fusing rating of 40 A^{2}s. Determine the maximum surge current that the device can withstand for 100 ms. [10]

7. The 2N6237 has a critical rise rating of $dv/dt = 10$ V/μs. Determine the anode noise amplitude at $t_r = 10$ ns required to cause false triggering. [11]

8. The 2N6342 has a critical rise rating of $dv/dt = 5$ V/μs. Determine the anode noise amplitude at $t_r = 25$ ns required to cause false triggering. [11]

§19.4

9. The 2N2646 has a range of $\eta = 0.56$ to 0.75. Determine the range of V_P values for the device when $V_{BB} = +14$ V. [27]
10. The 2N4871 UJT has a range of $\eta = 0.7$ to 0.85. Determine the range of V_P values for the device when $V_{BB} = +12$ V. [27]
11. The 2N4948 UJT has a range of $\eta = 0.55$ to 0.82. Determine the range of V_P values for the device when $V_{BB} = +16$ V. [27]
12. The 2N4949 UJT has a range of $\eta = 0.74$ to 0.86. Determine the range of V_P values for the device when $V_{BB} = +26$ V. [27]

§19.6

13. Determine the wavelength (in nm) for a 650-THz signal. [37]
14. Determine the wavelength (in nm) for a 220-THz signal. [37]
15. Determine the wavelength (in nm) for a 180-THz signal. [37]
16. The visible light spectrum includes all frequencies between approximately 400 THz and 800 THz. Determine the range of wavelength values for this band of frequencies. [27]
17. A photo-detector has a value of $\lambda_S = 0.94$ nm. Determine the optimum operating frequency for the device. [40]
18. A photo-detector has a rating of $\lambda_S = 0.84$ nm. Determine the optimum operating frequency for the device. [40]

THE BRAIN DRAIN

19. The spec sheet for the SCR in Figure 19.60 is located in Figure 19.11 (pages 879–80). The output short-circuit current for the dc power supply is 4 A, and will remain at that level until its primary fuse blows. For the circuit, determine:
    a. the maximum value of Vdc required to trigger the SCR.
    b. the maximum value of zener current at the instant that the SCR triggers.
    c. whether the fuse or the SCR will open first.
20. Refer to Figure 19.61. R_2 in the circuit is set to the value shown. The spec sheet for the 2N682 is located in Figure 19.11 (pages 879–80). Determine the conduction angle for the circuit load. (Hint: Start by finding the maximum value of V_{in} required to trigger the SCR.)
21. Refer to Figure 19.36 (page 905). The relaxation oscillator shown contains a 51 V zener diode. The UJT has a range of $\eta = 0.7$ to 0.82. Determine the minimum value of V_E required to trigger the UJT.

FIGURE 19.60

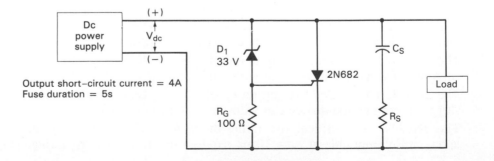

Output short–circuit current = 4A
Fuse duration = 5s

CHAP. 19 Thyristors and Optoelectronic Devices

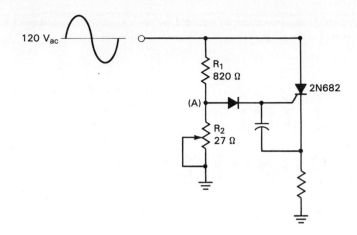

FIGURE 19.61

ANSWERS TO EXAMPLE PRACTICE PROBLEMS

19.1. For the application, $I^2t = 5.4$ A^{2}s.

19.2. 3.3 ms

19.3. 38.73 A

19.4. 2.5 mV

19.5. 9.7 V

19.6. 750 nm to 10,000 nm (10 μm)

19.7. 319.1 THz

20
Discrete and Integrated Voltage Regulators

As you know, a lot of research is now being done in the area of superconductivity. Someday, superconductors like the one shown here may completely change the entire field of electronics.

After studying the material in this chapter, you should be able to:

☐ 1. List the two purposes served by a voltage regulator. (Introduction)

☐ 2. Discuss the response of an *ideal* voltage regulator to a change in input voltage or load current demand. (§20.1)

☐ 3. Define *line regulation*. (§20.1)

☐ 4. Calculate the line regulation rating of a given voltage regulator. (§20.1)

☐ 5. State the relationship between the line regulation rating of a voltage regulator and the quality of the circuit. (§20.1)

☐ 6. List and define the commonly used line regulation ratings. (§20.1)

☐ 7. Define *load regulation*. (§20.1)

☐ 8. Calculate the load regulation rating of a given voltage regulator. (§20.1)

☐ 9. List and define the commonly used load regulation ratings. (§20.1)

☐ 10. State the relationship between the load regulation rating of a voltage regulator and the quality of the circuit. (§20.1)

☐ 11. Define and explain the single *regulation* rating of a voltage regulator. (§20.1)

☐ 12. Describe the basic difference between the *series regulator* and the *shunt regulator*. (§20.1)

☐ 13. Discuss the operation of the *pass-transistor regulator*. (§20.2)

☐ 14. Discuss the operation of the *Darlington pass-transistor regulator*. (§20.2)

☐ 15. Discuss the operation of the *series feedback regulator*. (§20.2)

☐ 16. State the primary drawback of series voltage regulators and the means by which it is overcome. (§20.2)

☐ 17. List the other disadvantages of using series voltage regulators. (§20.2)

☐ 18. Describe the construction and operation of the *shunt feedback regulator*. (§20.3)

☐ 19. Discuss the need for *overvoltage protection* in the shunt feedback regulator and the means by which it is provided. (§20.3)

☐ 20. Explain why the series voltage regulator is preferred over the shunt voltage regulator. (§20.3)

☐ 21. List and describe the four types of IC voltage regulators. (§20.4)

☐ 22. State the restriction on the input voltage polarity for a given IC voltage regulator. (§20.4)

☐ 23. List and define the commonly used IC voltage regulator parameters and ratings. (§20.4)

☐ 24. Calculate the maximum allowable input voltage for a given adjustable regulator. (§20.4)

☐ 25. Calculate the value of regulated output voltage (V_{dc}) for an adjustable regulator. (§20.4)

☐ 26. Discuss the use of a *pass transistor* for increasing the output current capability of an IC voltage regulator. (§20.4)

As you master each of the objectives listed, place a check mark (✔) in the appropriate box.

I n Chapter 3, you were introduced to the operation of the basic power supply. At that time, you were shown that a *rectifier* is used to convert the ac output of the transformer to *pulsating dc*. This pulsating dc is then applied to a *filter*, which reduces the varations in the rectifier output. Finally, the filtered dc is applied to a *voltage regulator*. This regulator serves two purposes:

1. It reduces the variations (ripple) in the filtered dc.
2. It maintains a constant output voltage regardless of changes in the load current demand.

The block diagram of the basic power supply is shown (along with its waveforms) in Figure 20.1.

The voltage regulator we used in Chapter 3 was simply a *zener diode* connnected in parallel with the load. In practice, this type of regulation (known as *brute-force regulation*) is rarely used. The primary problem with the simple zener regulator is the fact that the zener wastes a tremendous amount of power.

✔ 1

Introduction

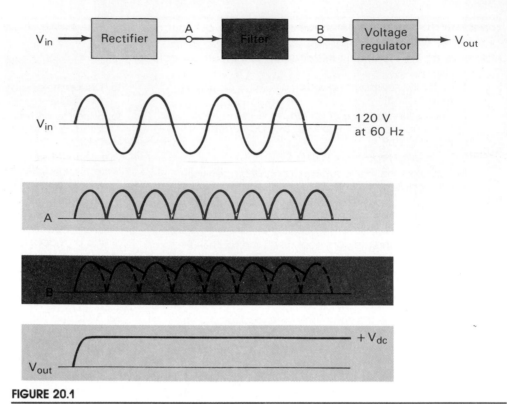

FIGURE 20.1

Basic power supply block diagram and waveforms.

Practical voltage regulators contain a number of discrete and/or integrated active devices. In this chapter, we will look at the operation of many of these practical voltage regulators.

20.1
Voltage Regulation: An Overview

▶ 2

The *ideal* voltage regulator will maintain a constant dc output voltage, regardless of changes in either its input voltage or its load current demand. For example, consider the +10 Vdc regulator shown in Figure 20.2. A change in the regulator

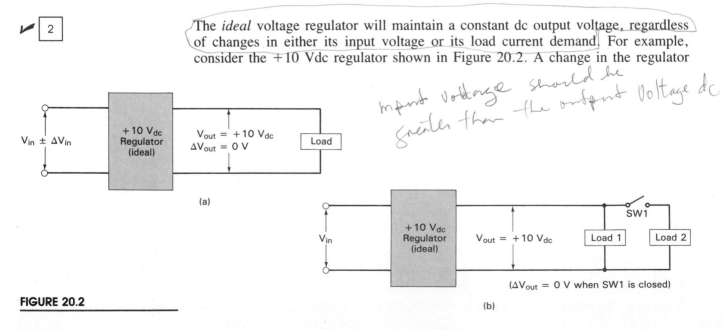

FIGURE 20.2

input voltage (shown as ΔV_{in} in Figure 20.2a) would not be coupled to the regulator output. Note that ΔV_{in} could be either a change in the steady-state (dc) value of V_{in} or could be some ripple voltage. In either case, the change in V_{out} for the ideal regulator would be 0 V. This assumes, of course, that the value of V_{in} does not decrease below the value required to maintain the operation of the regulator. (A regulator cannot have a 10 Vdc output if the input is less than 10 Vdc.)

Figure 20.2b shows how the ideal voltage regulator would respond to a change in load current demand. Assume that *load 1* and *load 2* have the same resistance. When SW1 is closed, the added load causes the current demand on the regulator to *double*. The ideal voltage regulator maintains a constant output voltage ($\Delta V_{out} = 0$) despite the change in load current demand.

Line Regulation

In practice, a change in the input voltage to a regulator *will* cause a change in its output voltage. The *line regulation* rating of a voltage regulator indicates the change in output voltage that will occur per unit change in the input voltage. The line regulator of a voltage regulator is found as

☑ 3

Line regulation. A rating that indicates the change in regulator output voltage that will occur per unit change in input voltage.

$$\text{line regulation} = \frac{\Delta V_{out}}{\Delta V_{in}} \qquad (20.1)$$

where ΔV_{out} = the change in output voltage, usually in microvolts or millivolts
ΔV_{in} = the change in input voltage, usually in volts

The following example illustrates the calculation of line regulation for a voltage regulator.

☑ 4

═══════════════ **EXAMPLE 20.1** ═══════════════

A voltage regulator experiences a 10-μV change in its output voltage when its input voltage changes by 5 V. Determine the value of line regulation for the circuit.

Solution: The line regulation of the circuit is found as

$$\text{line regulation} = \frac{\Delta V_{out}}{\Delta V_{in}}$$
$$= \frac{10 \ \mu V}{5 \ V}$$
$$= 2 \ \mu V/V$$

The 2 μV/V rating means that the output voltage will change by 2 μV for every 1-V change in the regulator's input voltage.

───

PRACTICE PROBLEM 20–1

The change in output voltage for a voltage regulator is measured at 100 μV when the input voltage changes by 4 V. Calculate the line regulation rating for the regulator.

───

5

As was stated earlier, the *ideal* voltage regulator would have a ΔV_{out} of 0 V when the input voltage changes. Thus, for the ideal voltage regulator,

$$\text{line regulation} = \frac{\Delta V_{out}}{\Delta V_{in}}$$
$$= \frac{0 \text{ V}}{\Delta V_{in}}$$
$$= 0$$

Based on the fact that the ideal line regulation rating is *zero*, we can make the following statement: *The lower the line regulation rating of a voltage regulator, the higher the quality of the circuit.*

Line regulation is commonly expressed in a variety of units. The $\mu V/V$ rating used in Example 20.1 is only one of them. The commonly used line regulation units and their meanings are listed in Table 20.1.

A Practical Consideration. Technically, line regulation should have no unit of measure, since it is the ratio of one voltage value to another. However, the units listed in Table 20.1 are commonly used on spec sheets, so you need to know what they are.

TABLE 20.1
Commonly Used Line Regulation Units

Unit	Meaning
$\mu V/V$	The change in output voltage (in microvolts) per 1 V change in input voltage.
ppm/V	*Parts-per-million per volt.* Another way of saying *microvolts per volt.*
%/V	The percent of change in the output voltage that can occur per 1-V change in the input voltage.
%	The total percent of change in output voltage that can occur over the rated range of input voltages.
mV (or μV)	The actual change in output voltage that can occur over the rated range of input voltages.

Load Regulation

7

Load regulation. A rating that indicates the change in regulator output voltage per unit change in load current.

The practical voltage regulator will also experience a slight change in output voltage when there is a change in load current demand. The *load regulation* rating of a voltage regulator indicates the change in output voltage that will occur per unit change in load current. The load regulation of a voltage regulator is found as

$$\boxed{\text{load regulation} = \frac{V_{NL} - V_{FL}}{\Delta I_L}} \qquad (20.2)$$

where V_{NL} = the no-load output voltage (the output voltage when the load is *open*)
V_{FL} = the full-load output voltage (the output voltage when the load current demand is at its maximum value)
ΔI_L = the *change in* load current demand

8

The following example illustrates the calculation of load regulation for a given voltage regulator.

EXAMPLE 20.2

A voltage regulator is rated for an output current of $I_L = 0$ to 20 mA. Under no-load conditions, the output voltage from the circuit is 5 V. Under full-load conditions, the output voltage from the circuit is 4.9998 V. Determine the value of load regulation for the circuit.

Solution: The load regulation of the regulator is found as

$$\text{load regulation} = \frac{V_{NL} - V_{FL}}{\Delta I_L}$$
$$= \frac{5\ \text{V} - 4.9998\ \text{V}}{20\ \text{mA}}$$
$$= 10\ \mu\text{V/mA}$$

The 10-μV/mA rating indicates that the output voltage changes by 10 μV for each 1-mA change in output current.

PRACTICE PROBLEM 20–2

A voltage regulator is rated for an output current of $I_L = 0$ to 40 mA. Under no-load conditions, the output voltage from the circuit is 8 V. Under full-load conditions, the output voltage from the circuit is 7.996 V. Determine the value of load regulation for the circuit.

Like the line regulation rating, load regulation is common expressed in a variety of units. The commonly used load regulation units and their meanings are listed in Table 20.2.

TABLE 20.2
Commonly Used Load Regulation Units

Unit	Meaning
μV/mA	The change in output voltage (in microvolts) per 1-mA change in load current.
%/mA	The percent of change in the output voltage that can occur per 1-mA change in load current.
%	The total percent of change that can occur in output voltage over the rated range of load current values.
mV (or μV)	The actual change in output voltage that can occur over the rated range of load current values.
Ω	V/mA, expressed as a resistance value. The rating times 1 mA gives you the V/mA rating of the regulator.

The ideal voltage regulator would not experience a change in output voltage when the load current demand increases. Thus, for the ideal regulator, $V_{NL} = V_{FL}$, and load regulation equals *zero*. Based on this fact, we can state that *the lower the load regulation rating of a voltage regulator, the higher the quality of the circuit*.

Line and Load Regulation: Some Practical Considerations

Some manufacturers combine the *line regulation* and *load regulation* ratings into a single *regulation* rating. This rating indicates the *maximum* change in output voltage that can occur when input voltage *and* load current are varied over their

Regulation. A rating that indicates the maximum change in regulator output voltage that can occur when input voltage *and* load current are varied over their entire rated ranges.

entire rated ranges. For example, a given voltage regulator has the following ratings:

$$V_{in} = 12 \text{ to } 24 \text{ V}_{dc}$$
$$I_L = 40 \text{ mA (maximum)}$$
$$\text{regulation} = 0.33 \%$$

These ratings indicate that the output voltage of the regulator will vary by no more than 0.33% as long as V_{in} remains between 12 and 24 Vdc and load current does not exceed 40 mA. A single regulation rating is always given as a percent, or in millivolts or microvolts.

There is one other important consideration we need to discuss. The unit of measure used for line regulation and/or load regulation is a good indicator of the quality of that regulator. For example, let's say that we are going to choose between two voltage regulators. These regulators have the following ratings:

	Dc Output Voltage	Line Regulation
Regulator A:	+10 V	0.12 %/V
Regulator B:	+10 V	40 μV/V

If the input voltage to regulator A increases by 5 V, the output voltage of the circuit will be

$$V_{out} = \text{Vdc} + (V_{dc})(0.12\%)(\Delta V_{in})$$
$$= 10 \text{ V} + 0.06 \text{ V}$$
$$= 10.06 \text{ V}$$

If the input voltage to regulator B increases by 5 V, the output voltage of the circuit will be

$$V_{out} = 10 \text{ V} + (40 \text{ μV})(\Delta V_{in})$$
$$= 10 \text{ V} + 200 \text{ μV}$$
$$= 10.0002 \text{ V}$$

As you can see, the regulator with the μV/V rating had a smaller change in output voltage than the regulator with the %/V rating. Since the ideal voltage regulator would have no change in output voltage when the input voltage changed, regulator B came much closer to the ideal regulator than regulator A.

◣ 12

Series regulator. A voltage regulator that is in series with the load.

Types of Regulators

There are two basic types of discrete voltage regulators: the *series regulator* and the *shunt regulator.* These two types of regulator are represented by the blocks in Figure 20.3. The *series regulator* is placed in *series* with the load, as shown in

FIGURE 20.3

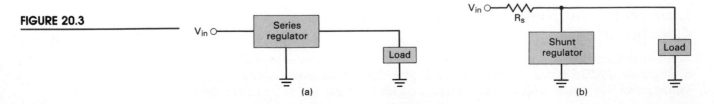

Figure 20.3a. The *shunt regulator* is placed in *parallel* with the load, as shown in Figure 20.3b. As you will see, both series and shunt regulators have their advantages and disadvantages.

Shunt regulator. A voltage regulator that is in parallel with the load.

SECTION REVIEW

1. What two purposes are served by a voltage regulator? [Objective 1]
2. How would the *ideal* voltage regulator respond to a change in input voltage? [Objective 2]
3. How would the *ideal* voltage regulator respond to a change in load current demand? [Objective 2]
4. What is *line regulation*? [Objective 3]
5. What is the line regulation value of an ideal voltage regulator? [Objective 5]
6. What is the relationship between the quality of a voltage regulator and its line regulation rating? [Objective 5]
7. List and define the commonly used line regulation ratings. [Objective 6]
8. What is *load regulation*? [Objective 7]
9. List and define the commonly used load regulation ratings. [Objective 9]
10. What is the load regulation value of an ideal voltage regulator? [Objective 10]
11. What is the relationship between the quality of a voltage regulator and its load regulation rating? [Objective 10]
12. Explain the single *regulation* rating. [Objective 11]
13. What is a *series regulator*? [Objective 12]
14. What is a *shunt regulator*? [Objective 12]

20.2 Series Voltage Regulators

Series voltage regulators can take a variety of forms. However, they all have one or more active devices that are placed in series with the load. In this section, we will take a look at three of the commonly used series voltage regulators.

Pass-transistor Regulator

13

The *pass-transistor regulator* uses a series transistor, called a *pass-transistor*, to regulate load voltage. The term *pass transistor* comes from the fact that the load current passes through the series transistor, Q_1 as shown in Figure 20.4.

Pass-transistor regulator. A regulator that uses a series transistor to maintain a constant load voltage.

The key to the operation of the pass-transistor regulator is the fact that the base voltage is held to the relatively constant voltage across the zener diode. For example, if a 9.1-V zener diode is being used, the base voltage of Q_1 is held at approximately 9.1 V. Since Q_1 is an *npn* transistor, V_L is found as

Don't forget: The emitter voltage of a conducting *npn* transistor is *less than* the base voltage.

$$\boxed{V_L = V_Z - V_{BE}} \tag{20.3}$$

Equation (20.3) is important because it provides the basis for explaining the response of the pass-transistor regulator to a change in load resistance. If load

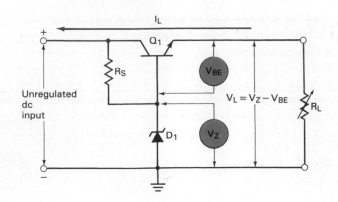

FIGURE 20.4

resistance increases, load voltage starts to increase. The increase in load voltage (and thus V_E) results in a decrease in V_{BE} (since V_B is constant). The decrease in V_{BE} reduces the conduction of the pass transistor, causing its value of V_{CE} to increase. The increase in V_{CE} offsets the initial increase in V_E, and a relatively constant load voltage is maintained.

In a similar fashion, a decrease in load resistance causes V_E to start to decrease. The decrease in V_E causes V_{BE} to increase, increasing conduction through the pass transistor. This increased conduction causes V_{CE} to decrease, offsetting the initial decrease in V_E. Again, a relatively constant load voltage is maintained by the regulator.

The pass-transistor regulator in Figure 20.4 has relatively good line and load regulation characteristics, but there is a problem with the circuit. If the input voltage or load current values increase, the zener diode may have to dissipate a relatively high amount of power. You see, increases in V_{in} or I_L result in an increase in zener conduction, since more current is drawn through R_S. The increased zener conduction results in higher power dissipation by the device. This problem, however, is reduced by the use of the *Darlington pass-transistor regulator*.

Darlington Pass-transistor Regulator

What purpose is served by R_S? R_S is used to bias the zener diode in its reverse operating region.

The *Darlington pass-transistor regulator* uses a Darlington pair, Q_1 and Q_2, in place of the single pass transistor, as shown in Figure 20.5. The load voltage for the Darlington circuit is found as

14

Darlington pass-transistor regulator. A series regulator that uses a Darlington pair in place of a single pass transistor.

$$\boxed{V_L = V_Z - 2V_{BE}} \qquad (20.4)$$

Since the total current gain of a Darlington pair is equal to $h_{FE1}h_{FE2}$, an increase in load current causes very little, if any, increase in zener current. Therefore, the

FIGURE 20.5

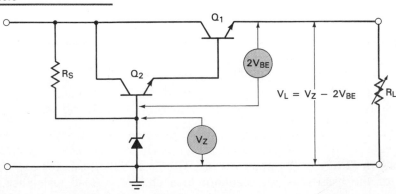

CHAP. 20   Discrete and Integrated Voltage Regulators

regulator in Figure 20.5 is not subject to the same power concerns as the regulator in Figure 20.4. However, since zener conduction *is* affected by temperature, and the zener current is applied to the high-current-gain Darlington pair, the load current in the Darlington pass-transistor regulator can be severely affected by significant increases in operating temperature. For this reason, the Darlington pass-transistor regulator must be kept relatively cool.

Series Feedback Regulator

The *series feedback regulator* uses an *error detector* to improve on the line and load regulation characteristics of the other pass-transistor regulators. The block diagram for the series feedback regulator is shown in Figure 20.6.

Series feedback regulator. A series regulator that uses an error-detection circuit to improve the line and load regulation characteristics of other pass-transistor regulators.

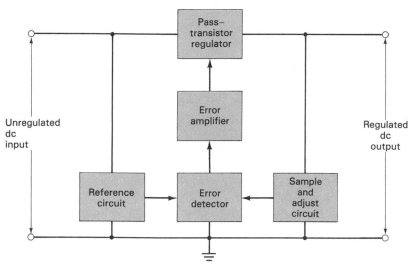

FIGURE 20.6

The error detector receives two inputs: a *reference voltage* that is derived from the unregulated dc input voltage and a *sample voltage* from the regulated output voltage. The error detector compares the reference and sample voltages and provides an output voltage that is proportional to the difference between the two. This output voltage is amplified and used to drive the pass-transistor regulator.

The series feedback regulator is capable of responding very quickly to differences between its sample and reference input voltages. This gives the circuit much better line and load regulation characteristics than the other two circuits we have discussed in this section.

The schematic diagram for the series feedback regulator is shown in Figure 20.7. The *sample and adjust circuit* is the voltage divider that consists of R_3, R_4, and R_5. The reference voltage is set by the zener diode. Q_2 detects and amplifies the difference between the reference and sample voltages and adjusts the conduction of the pass transistor accordingly.

If the load resistance of the feedback regulator increases, V_L starts to increase. This causes the voltage at the base of Q_2 to increase. Since the emitter voltage of Q_2 is clamped to the value of V_Z, the increase in $V_{B(Q2)}$ causes an increase in $V_{BE(Q2)}$. This increases the conduction through Q_2 and its collector resistor (R_1). The increased conduction causes $V_{C(Q2)}$ to decrease, which reduces the value of $V_{B(Q1)}$. The reduction in $V_{B(Q1)}$ reduces the conduction through the pass transistor, causing V_E to decrease. The decrease in V_E offsets the initial increase caused by the load.

How the circuit responds to an increase in load resistance.

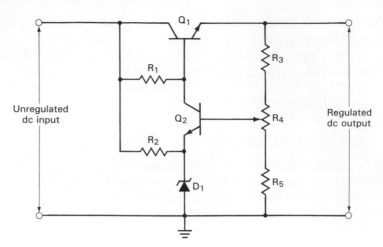

FIGURE 20.7

How the circuit responds to a
decrease in load resistance.

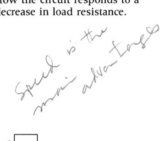

16

If the load resistance decreases, V_L starts to decrease. The decrease in V_L causes $V_{B(Q2)}$ to decrease, which reduces the conduction through the transistor. The reduced conduction through Q_2 causes $V_{C(Q2)}$ to increase, increasing $V_{B(Q1)}$. The increase in $V_{B(Q1)}$ causes the pass-transistor conduction to increase. This increases the value of V_E, offsetting the initial decrease in load voltage. Again, load regulation is maintained.

Short-circuit Protection

The primary drawback of any series regulator is the fact that the pass transistor can be destroyed by excessive load current if the load is shorted. To prevent a shorted load from destroying the pass transistor, a *current-limiting circuit* is usually added to the regulator as shown in Figure 20.8.

The current-limiting circuit consists of a transistor (Q_3) and a series resistor (R_S) that is connected between its base and emitter terminals. In order for Q_3 to

FIGURE 20.8

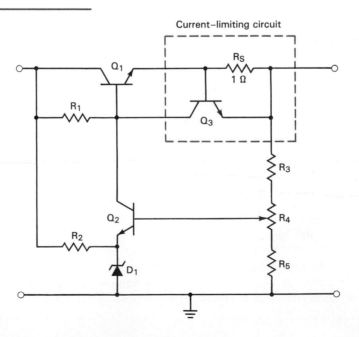

conduct, the voltage across R_S must reach approximately 0.7 V. This happens when

$$I_L = \frac{0.7 \text{ V}}{1 \text{ }\Omega}$$
$$= 700 \text{ mA}$$

Thus, if load current is less than 700 mA, Q_3 is in cutoff and the circuit acts exactly as described earlier. If the load current increases above 700 mA, Q_3 conducts, decreasing the voltage at the base of Q_1. The decreased $V_{B(Q1)}$ reduces the conduction of the pass transistor, preventing any further increases in load current. Thus, the load current for the circuit is limited to approximately 700 mA. In fact, for the type of current-limited regulator shown, the maximum load current can be found as

$$I_{L(\max)} = \frac{V_{BE(Q3)}}{R_S} \qquad (20.5)$$

Thus, the maximum allowable load current for the regulator can be set to any value by using the appropriate value of R_S.

You may be wondering why we don't simply use a single current-limiting resistor to protect the pass transistor. If a current-limiting resistor is used, it must have a fairly significant value in order to protect the pass transistor. The voltage drop across such a resistor would be much higher than that caused by a 1-Ω resistor. The current-limiting circuit is used to provide short-circuit protection without causing a significant drop in output voltage from the regulator.

One Final Note

The fact that a shorted load can damage a pass transistor is not the only drawback to the series regulator. For one thing, the fact that I_L passes through the pass transistor means that this transistor dissipates a fairly significant amount of power. Also, there is a voltage drop across the pass transistor. This voltage drop reduces the maximum possible output voltage.

Even with its power dissipation and short-circuit protection problems, the series regulator is used more commonly than the shunt regulator. As you will see in the next section, the shunt regulator has several problems that make the series regulator the more desirable of the two.

17

1. Describe the basic *pass-transistor regulator*. [Objective 13]
2. Describe how the pass-transistor regulator in Figure 20.4 responds to a change in load resistance. [Objective 13]
3. Explain how the *Darlington pass-transistor regulator* reduces the problem of excessive zener diode power dissipation. [Objective 14]
4. Describe the response of the *series feedback regulator* to a change in load resistance. [Objective 15]
5. Describe the operation of the current-limiting circuit in Figure 20.8. [Objective 16]
6. List the disadvantages of using series voltage regulators. [Objective 17]

SECTION REVIEW

Shunt Voltage Regulators

The most basic shunt regulator is the simple zener regulator that we discussed in Chapter 3. As you know, this regulator wastes far too much power for most applications and thus is rarely used. In this section, we will discuss the *shunt feedback regulator*. As you will see, this regulator is nearly identical to the series feedback regulator we covered in the last section.

Shunt Feedback Regulator

◄ 18

The *shunt feedback regulator* uses an error detector to control the conduction of a *shunt transistor*. This shunt transistor is shown as Q_1 in the shunt feedback regulator in Figure 20.9.

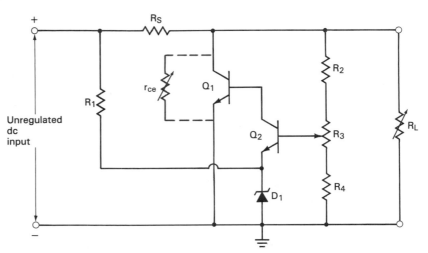

FIGURE 20.9

Shunt feedback regulator. A circuit that uses an error detector to control the conduction of a *shunt* transistor.

The *sample circuit* is (once again) a simple voltage-divider circuit. The *reference circuit* is made up of D_1 and R_1. The outputs from the sample and reference circuits are applied to the *error detector/amplifier*, Q_2. The output from Q_2 is then used to control the conduction of the shunt transistor, Q_1.

The operation of the shunt feedback regulator is easiest to understand if we view the shunt transistor (Q_1) as a variable resistor, r_{ce}. When Q_1 is not conducting, r_{ce} is at its maximum value. When Q_1 is saturated, Q_1 is at its minimum value. Thus, we can say that *the value of r_{ce} is inversely proportional to the conduction of Q_1.*

Under normal circumstances, R_3 is set so that Q_2 is conducting. Since $I_{B1} = I_{C2}$, Q_1 is biased somewhere in its active region of operation. The conduction of the two-transistor circuit is set so that r_{ce} is approximately midway between its two extremes in value.

How the circuit responds to a decrease in load resistance.

If the load resistance decreases, V_L also starts to decrease. This decrease causes $V_{B(Q2)}$ to decrease. Since $V_{E(Q2)}$ is set to a fixed value by the zener diode, $V_{BE(Q2)}$ decreases when $V_{B(Q2)}$ decreases. The reduction in $V_{BE(Q2)}$ reduces the conduction through the component, decreasing I_{B1}. This causes r_{ce} to increase, increasing the value of $V_{C(Q1)}$. Since $V_L = V_{C(Q1)}$, the increase in $V_{C(Q1)}$ offsets the initial decrease in V_L.

How the circuit responds to an increase in load resistance.

If R_L increases in value, V_L starts to increase. The increase in V_L increases $V_{B(Q2)}$, and thus $V_{BE(Q2)}$. This increases the conduction through Q_2, causing an increase in $I_{B(Q1)}$. The increase in $I_{B(Q1)}$ causes the shunt transistor's conduction to increase, reducing the value of r_{ce}. When r_{ce} decreases, $V_{C(Q1)}$ decreases, offsetting the initial increase in V_L.

The series resistor (R_S) is used to provide shorted-load protection for the power supply circuitry that is providing the unregulated dc input. Normally, this resistor must be a high-wattage resistor since all the regulator current and I_L must pass through it.

The purpose served by R_s.

Overvoltage Protection

Just as the series regulator must be protected against shorted-load conditions, the shunt regulator must be protected from input overvoltage conditions. If the unregulated dc input voltage to the regulator increases, the conduction of the shunt transistor increases to maintain the constant output voltage. Assuming that V_{CE} of Q_1 remains relatively constant, the increased conduction through the transistor will cause an increase in its power dissipation.

◢ 19

Several things can be done to ensure that an increase in unregulated dc input voltage will not destroy the shunt transistor. First, you can use a transistor whose $P_{D(max)}$ rating is far greater than the maximum power dissipation that would ever be required in the circuit. For example, assume that the regulator in Figure 20.9 has an input voltage that can go as high as 20 V and that the regulator is designed to provide a regulated 10-V_{dc} output. The worst-case value of transistor current could be approximated as

$$I_{C(max)} = \frac{V_{in} - V_{dc}}{R_C}$$

Assuming that a 100-Ω resistor is being used, $I_{C(max)}$ would be

$$I_{C(max)} = \frac{20\ V - 10\ V}{100}$$
$$= 100\ mA$$

Since the transistor is dropping 10 V across its collector–emitter terminals, the maximum power dissipation would be found as

$$P_D = V_{CE}I_{C(max)}$$
$$= (10\ V)(100\ mA)$$
$$= 1\ W$$

Thus, for the circuit described, any transistor with a $P_{D(max)}$ rating that is *greater than* 1 W could be used.

Of course, there are always circumstances where the unregulated dc input voltage can exceed its maximum *rated* value. To protect the circuit from such a circumstance, a *crowbar* circuit can be added to the input of the regulator. You may recall (from Chapter 19) that a *crowbar* is a circuit that uses an SCR to protect its load from an overvoltage condition. If a crowbar is added to the regulator input, the circuit is protected from any extreme input overvoltage condition that may arise.

Another practical consideration involves the potentiometer, R_3. This potentiometer is included in the circuit to provide an adjustment for the dc output voltage. Adjusting R_3 varies the conduction of Q_1, and thus the value of r_{ce}. Since this resistance forms a voltage divider with R_S, varying R_3 sets the value of the regulated dc output voltage.

One Final Note

20

Don't forget. High-voltage problems are far more common than shorted-component problems.

The series voltage regulator is preferred over the shunt regulator for several reasons. The primary problem with the shunt regulator is the fact that a fairly significant portion of the total current through R_S goes through the shunt transistor, rather than to the load. Another problem involves the voltage drop across R_S and the resulting power dissipation. Finally, realize that an input overvoltage condition is far more likely to occur than the shorted-load condition of the series regulator. In other words, the fault that can damage the shunt regulator is more likely to occur than the fault that can damage the series regulator. When you consider all these drawbacks, it is easy to understand why the series regulator is preferred over the shunt regulator.

SECTION REVIEW

1. Describe the response of the shunt feedback regulator to a decrease in load resistance. [Objective 18]
2. Describe the response of the shunt feedback regulator to an increase in load resistance. [Objective 18]
3. Explain how an input overvoltage condition can destroy the shunt transistor in a shunt feedback regulator. [Objective 19]
4. Discuss the means by which the overvoltage problem can be eliminated. [Objective 19]
5. Why are series regulators preferred over shunt regulators? [Objective 20]

20.4
IC Voltage Regulators

21

IC voltage regulator. A device that is used to hold the output voltage from a dc power supply relatively constant over a wide range of line and load voltages.

Fixed-positive regulator. A regulator with a predetermined $+V_{dc}$ output.

Fixed-negative regulator. A regulator with a predetermined $-V_{dc}$ output.

Adjustable regulator. A regulator whose output V_{dc} can be set to any value within specified limits.

The *IC voltage regulator* is a device that is used to hold the output voltage from a dc power supply relatively constant over a wide range of line and load variations. Most of the commonly used IC voltage regulators are three-terminal devices, though some types require more than three terminals. The schematic symbol for a three-terminal regulator is shown in Figure 20.10.

There are basically four types of IC voltage regulators: *fixed positive*, *fixed negative*, *adjustable*, and *dual tracking*. The *fixed-positive* and *fixed-negative* IC voltage regulators are designed to provide specific output voltages. For example, the LM309 (fixed positive) provides a $+5\text{-}V_{dc}$ output (as long as the regulator input voltages are within their specified ranges). The *adjustable regulator* can be

FIGURE 20.10

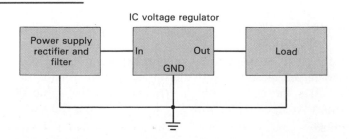

adjusted to provide any dc output voltage that is within its two specified limits. For example, the LM317 output can be adjusted to any value between its limits of $+1.2$ and $+32$ V_{dc}. Both positive and negative variable regulators are available. The *dual-tracking* regulator provides equal positive and negative output voltages. For example, the RC4195 provides outputs of $+15$ and -15 V_{dc}. Adjustable dual-tracking regulators are also available. These regulators have outputs that can be varied between their two rated limits. A single adjustment controls both outputs so that they are always equal in value. For example, if an adjustable dual-tracking regulator is adjusted for a positive output of $+2$ V_{dc}, the negative output will automatically be adjusted to -2 V_{dc}.

Dual-tracking regulator. A regulator that provides equal $+V_{dc}$ and $-V_{dc}$ outputs.

Regardless of the type of regulator used, the regulator input-voltage polarity must match the device's rated output polarity. In other words, positive regulators must have positive input voltages and negative regulators must have negative input voltages. Dual-tracking regulators require *both* positive and negative input voltages.

✔ 22

IC voltage regulators are *series regulators*. They contain internal pass transistors and transistor control components. Generally, the internal circuitry of an IC voltage regulator resembles that of the series feedback regulator.

IC Regulator Specifications

✔ 23a

For our discussion on common IC regulator specifications, we'll use the spec sheet for the LM317 voltage regulator. As was stated earlier, this device is an adjustable regulator whose output can be varied between $+1.2$ and $+32$ V_{dc}. The spec sheet for the LM317 is shown in Figure 20.11.

The *input/output voltage differential* rating (which is listed in the absolute maximum ratings section of the spec sheet) indicates the maximum difference between V_{in} and V_{out} that can occur without damaging the device. For the LM317, this rating is 40 V. The differential voltage rating can be used to determine the maximum allowable value of V_{in} as follows:

Input/output voltage differential. The maximum allowable difference between V_{in} and V_{out} for an adjustable regulator.

$$\boxed{V_{in(max)} = V_{out(adj)} + V_d} \qquad \text{(20.6)}$$

where $V_{in(max)}$ = the maximum allowable unrectified dc input voltage
$V_{out(adj)}$ = the *adjusted* output voltage of the regulator
V_d = the input/output voltage differential rating of the regulator

========== **EXAMPLE 20.3** ==========

24

The LM317 is adjusted to provide a $+8$-V_{dc} regulated output voltage. Determine the maximum allowable input voltage to the device.

Solution: With a V_d rating of 40 V, the maximum allowable value of V_{in} is found as

$$\begin{aligned} V_{in(max)} &= V_{out(adj)} + V_d \\ &= +8 \ V_{dc} + 40 \ V \\ &= +48 \ V \end{aligned}$$

PRACTICE PROBLEM 20–3

An adjustable IC voltage regulator has a V_d rating of 32 V. Determine the maximum allowable input voltage to the device when it is adjusted to $+6$ V_{dc}.

FIGURE 20.11

The LM317 specification sheet (Courtesy of Fairchild, a division of National Semiconductor)

Absolute Maximum Ratings

Power Dissipation	Internally limited
Input—Output Voltage Differential	40V
Operating Junction Temperature Range	
LM117	$-55°C$ to $+150°C$
LM217	$-25°C$ to $+150°C$
LM317	$0°C$ to $+125°C$
Storage Temperature	$-65°C$ to $+150°C$
Lead Temperature (Soldering, 10 seconds)	$300°C$

Preconditioning

Burn-In in Thermal Limit	**100% All Devices**

Electrical Characteristics (Note 1)

PARAMETER	CONDITIONS	LM117/217			LM317			UNITS
		MIN	TYP	MAX	MIN	TYP	MAX	
Line Regulation	$T_A = 25°C$, $3V \leq V_{IN} - V_{OUT} \leq 40V$ (Note 2)		0.01	0.02		0.01	0.04	%/V
Load Regulation	$T_A = 25°C$, $10\ mA \leq I_{OUT} \leq I_{MAX}$							
	$V_{OUT} \leq 5V$, (Note 2)		5	15		5	25	mV
	$V_{OUT} \leq 5V$, (Note 2)		0.1	0.3		0.1	0.5	%
Thermal Regulation	$T_Z = 25°C$, 20 ms Pulse		0.03	0.07		0.04	0.07	%/W
Adjustment Pin Current			50	100		50	100	μA
Adjustment Pin Current Change	$10\ mA \leq I_L \leq I_{MAX}$ $3V \leq (V_{IN} - V_{OUT}) \leq 40V$		0.2	5		0.2	5	μA
Reference Voltage	$3V \leq (V_{IN} - V_{OUT}) \leq 40V$, (Note 3) $10\ mA \leq I_{OUT} \leq I_{MAX}$, $P \leq P_{MAX}$	1.20	1.25	1.30	1.20	1.25	1.30	V
Line Regulation	$3V \leq V_{IN} - V_{OUT} \leq 40V$, (Note 2)		0.02	0.05		0.02	0.07	%/V
Load Regulation	$10\ mA \leq I_{OUT} \leq I_{MAX}$, (Note 2)							
	$V_{OUT} \leq 5V$		20	50		20	70	mV
	$V_{OUT} \leq 5V$		0.3	1		0.3	1.5	%
Temperature Stability	$T_{MIN} \leq T_j \leq T_{MAX}$		1			1		%
Minimum Load Current	$V_{IN} - V_{OUT} = 40V$		3.5	5		3.5	10	
Current Limit	$V_{IN} - V_{OUT} \leq 15V$							
	K and T Package	1.5	2.2		1.5	2.2		A
	H and P Package	0.5	0.8		0.5	0.8		A
	$V_{IN} - V_{OUT} = 40V$, $T_j = +25°C$							
	K and T Package	0.03	0.4		0.15	0.4		A
	H and P Package	0.15	0.07		0.075	0.07		A
RMS Output Noise, % of V_{OUT}	$T_A = 25°C$, $10\ Hz \leq f \leq 10\ kHz$		0.003			0.003		%
Ripple Rejection Ratio	$V_{OUT} = 10V$, $f = 120\ kHz$		65			65		dB
	$C_{ADJ} = 10μF$	66	80		66	80		dB
Long-Term Stability	$T_A = 125°C$		0.3	1		0.3	1	%
Thermal Resistance, Junction to Case	H Package		12	15		12	15	°C/W
	K Package		2.3	3		2.3	3	°C/W
	T Package					4		°C/W
	P Package					12		°C/W

Note 1: Unless otherwise specified, these specifications apply $-55°C \leq T_j \leq +150°C$ for the LM117, $-25°C \leq T_j \leq +150°C$ for the LM217, and $0°C \leq T_j \leq +125°C$ for the LM317; $V_{IN} - V_{OUT} = 5V$; and $I_{OUT} = 0.1A$ for the TO-39 and TO-202 packages and $I_{OUT} = 0.5A$ for the TO-3 and TO-220 packages. Although power dissipation is internally limited, these specifications are applicable for power dissipations of 2W for the TO-39 and TO-202, and 20W for the TO-3 and TO-220. I_{MAX} is 1.5A for the TO-3 and TO-220 packages and 0.5A for the TO-39 and TO-202 packages.

Note 2: Regulation is measured at constant junction temperature, using pulse testing with a low duty cycle. Changes in output voltage due to heating effects are covered under the specification for thermal regulation.

Note 3: Selected devices with tightened tolerance reference voltage available.

The *line regulation* rating of the LM317 is 0.04%/V (maximum). This rating is measured under the following conditions: $T_A = 25°C$ and $3\ V \leq V_{in} - V_{out} \leq 40\ V$. This means that the difference between the input and output voltages can be no less than 3 V and no greater than 40 V. For example, if we set the LM317 for a $+10\text{-}V_{dc}$ output, the input voltage to the device must be between $+13$ and $+50$ V. As long as these conditions are met, the output will vary by no more than $10\ V \times 0.04\% = 4\ mV$.

The *load regulation* rating for the LM317 depends on whether the device is being operated for a V_{out} that is less than or greater than $+5\ V_{dc}$. If V_{out} is less than $+5\ V_{dc}$, the output voltage will not vary by more than 25 mV when the load current is varied within the given range of values. If the output voltage is set to a value that is greater than or equal to $+5\ V_{dc}$, the same change in load current will cause a maximum output voltage change of 0.5%. Note that the rated range of output currents for the device is 10 mA to approximately 1.5 A.

Looking further down the spec sheet, you'll notice another set of line/load regulation ratings. This lower set of ratings indicates the maximum line/load regulation ratings of the device *over the entire range of allowable operating junction temperatures*. For the LM317, this range is 0° to 125°C, as is listed in the absolute maximum ratings section of the spec sheet. Note that the ratings we discussed earlier assumed an operating temperature of 25°C. Thus, the lower set of ratings indicates the *worst-case* operation of the device.

The *minimum load current* rating indicates the minimum allowable load current demand for the regulator. For the LM317, this rating is 10 mA. If the load current drops below 10 mA (such as when the load is open), regulation of the output voltage is lost.

Minimum load current. The value of I_L below which regulation is lost.

The *ripple rejection ratio* is the ability of the regulator to drop any ripple voltage at its input. For the LM317, any input ripple is reduced by 65 dB at the output. Using the decibel conversion process discussed in Chapter 13, we find that the input ripple will be reduced by a factor of approximately 1800 by the regulator.

Ripple rejection ratio. The ratio of regulator input ripple to maximum output ripple.

Reference:

Output Voltage Adjustment

The output voltage of the LM317 is adjusted using a voltage divider, as is shown in Figure 20.12. The *complete* LM317 spec sheet (not shown) contains the following equation for determining the dc output voltage of the circuit:

$$V_{dc} = 1.25 \left(\frac{R_2}{R_1} + 1 \right) \qquad (20.7)$$

where V_{dc} = the regulated dc output voltage of the regulator

FIGURE 20.12

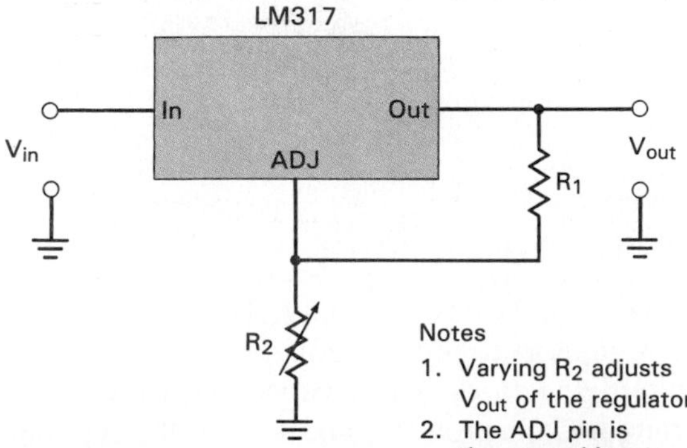

LM317

V_{in} In   Out V_{out}

ADJ R_1

R_2

Notes
1. Varying R_2 adjusts V_{out} of the regulator.
2. The ADJ pin is the same thing as the GND pin in Figure 20.10.

SEC. 20.3   Shunt Voltage Regulators

The following example shows how equation (20.7) is used to determine the regulated dc output voltage for an LM317 regulator circuit.

EXAMPLE 20.4

R_2 in Figure 20.13 is adjusted to 2.4 kΩ. Determine the regulated dc output voltage for the circuit.

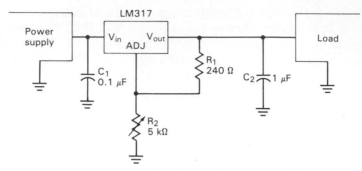

FIGURE 20.13

LM317 circuit connections

Solution: The regulated dc output voltage is found as

$$V_{dc} = 1.25 \left(\frac{R_2}{R_1} + 1 \right)$$
$$= 1.25 \left(\frac{2.4 \text{ k}\Omega}{240 \text{ }\Omega} + 1 \right)$$
$$= (1.25)(11)$$
$$= 13.75 \text{ V}$$

PRACTICE PROBLEM 20–4

R_2 in Figure 20.13 is adjusted to 1.68 kΩ. Determine the regulated dc output voltage for the LM317.

Note that the V_{dc} equation for a given adjustable regulator will always be given on its spec sheet.

Figure 20.13 shows two shunt capacitors that are connected to the regulator input and output pins. The input shunt capacitor is used to prevent the input ripple from driving the regulator into self-oscillations. The output shunt capacitor is used to improve the ripple reduction of the regulator.

IC Regulator Applications: A Complete Dual-polarity Power Supply

Figure 20.14 shows a complete dual-polarity power supply. The circuit uses *matched* fixed-positive and fixed-negative regulators to provide equal $+V_{dc}$ and $-V_{dc}$ outputs.

Capacitors C_1 and C_2 are *filter capacitors*. These capacitors would have values in the mid to high-microfarad range. Capacitors C_3 and C_4 are the regulator input

CHAP. 20 Discrete and Integrated Voltage Regulators

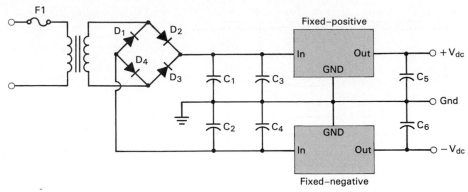

FIGURE 20.14

shunt capacitors and would be less than 1 μF in value. Capacitors C_5 and C_6 are output ripple reduction capacitors and would have values in the neighborhood of 1 μF.

Circuit Variations

Figure 20.15 shows how a *pass transistor* can be added in parallel with an IC voltage regulator to increase the maximum possible output current. The value of R_S in the circuit is selected to bias Q_1 on, as follows:

✔ 26

$$R_S = \frac{V_{BE(Q1)}}{I_{in}}$$

where I_{in} = the regulator input current. With Q_1 on, an increase in load current demand (above the capability of the regulator output) results in increased Q_1 conduction. This provides the needed load current.

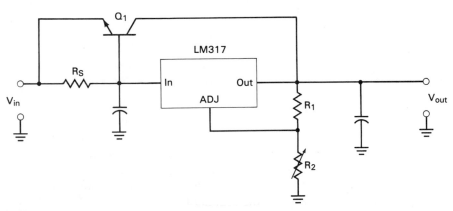

FIGURE 20.15

There is one serious drawback to the circuit in Figure 20.15. The input *ripple current* to the circuit is coupled (via Q_1) to the load. This problem, however, can be minimized by adding the *current-limiting circuit* (Q_2 and R_{S2}) shown in Figure 20.16. This current-limiting circuit acts just like the one we discussed for the series feedback regulator in Figure 20.8. When the value of R_{S2} is properly selected,

SEC. 20.3   Shunt Voltage Regulators

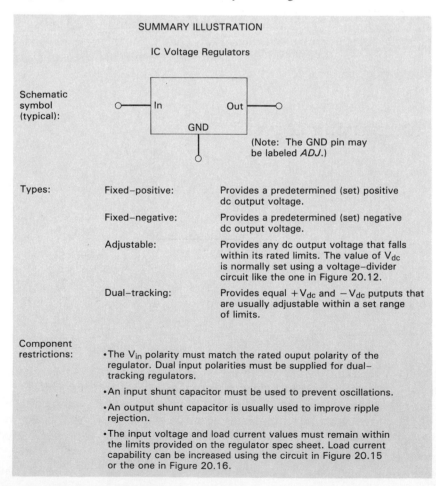

FIGURE 20.16

any excessive ripple current at the circuit input is prevented from reaching the load.

One Final Note

IC voltage regulators are extremely common. While we cannot possibly hope to cover the operation of every type of IC voltage regulator, you should have no problem in dealing with "everyday" IC voltage regulators. The characteristics of IC voltage regulators are summarized for you in Figure 20.17.

FIGURE 20.17

SUMMARY ILLUSTRATION

IC Voltage Regulators

Schematic symbol (typical):

In   Out

GND

(Note: The GND pin may be labeled *ADJ*.)

Types:

Fixed–positive: Provides a predetermined (set) positive dc output voltage.

Fixed–negative: Provides a predetermined (set) negative dc output voltage.

Adjustable: Provides any dc output voltage that falls within its rated limits. The value of V_{dc} is normally set using a voltage–divider circuit like the one in Figure 20.12.

Dual–tracking: Provides equal $+V_{dc}$ and $-V_{dc}$ putputs that are usually adjustable within a set range of limits.

Component restrictions:

- The V_{in} polarity must match the rated ouput polarity of the regulator. Dual input polarities must be supplied for dual-tracking regulators.

- An input shunt capacitor must be used to prevent oscillations.

- An output shunt capacitor is usually used to improve ripple rejection.

- The input voltage and load current values must remain within the limits provided on the regulator spec sheet. Load current capability can be increased using the circuit in Figure 20.15 or the one in Figure 20.16.

1. What is an *IC voltage regulator*? [Objective 21]
2. List and describe the four types of IC voltage regulators. [Objective 21]
3. What input polarity (or polarities) is/are required for each type of voltage regulator? [Objective 22]
4. What is the *input/output differential* rating? [Objective 23a]
5. What is the *minimum load current* rating? [Objective 23b]
6. What is the *ripple rejection ratio* rating? [Objective 23b]
7. What type of circuit is normally used to provide the adjustment of the dc output voltage from an adjustable regulator? [Objective 25]
8. Where can you find the V_{dc} equation for a given adjustable regulator? [Objective 25]
9. What purpose is served by the circuit in Figure 20.15? [Objective 26]
10. Why does the circuit in Figure 20.15 need to be modified as shown in Figure 20.16? [Objective 26]

KEY TERMS

The following terms were introduced and defined in this chapter:

adjustable regulator
Darlington pass-transistor
  regulator
dual-tracking regulator
fixed-negative regulator
fixed-positive regulator
IC voltage regulator

input/output voltage
  differential
line regulation
load regulation
minimum load current
pass-transistor regulator
regulation

ripple rejection ratio
series feedback regulator
series regulator
shunt feedback regulator
shunt regulator

PRACTICE PROBLEMS

§20.1

1. A voltage regulator experiences a 20-µV change in its output voltage when its input voltage changes by 4 V. Determine the line regulation of the circuit. [4]

2. A voltage regulator experiences a 14-µV change in output voltage when its input voltage changes by 10 V. Determine the line regulation rating of the circuit. [4]

3. A voltage regulator experiences a 15-µV change in its output voltage when its input voltage changes by 5 V. Determine the line regulation rating of the circuit. [4]

4. A voltage regulator experiences a 12-mV change in output voltage when its input voltage changes by 12 V. Determine the line regulation rating of the circuit. [4]

5. A voltage regulator is rated for an output current of 0 to 150 mA. Under no-load conditions, the output voltage of the circuit is 6 V. Under full-load conditions, the output from the circuit is 5.98 V. Determine the load regulation rating of the circuit. [8]

6. A voltage regulator experiences a 20-mV change in output voltage when the load current increases from 0 to 50 mA. Determine the load regulation rating of the circuit. [8]

7. A voltage regulator experiences a 1.5-mV change in output voltage when the load current increases from 0 to 20 mA. Determine the load regulation rating of the circuit. [8]

8. A voltage regulator experiences a 14-mV change in output voltage when the load current increases from 0 to 100mA. Determine the load regulation rating of the circuit. [8]

§20.2

9. Determine the approximate value of V_L for the circuit in Figure 20.18. [13]

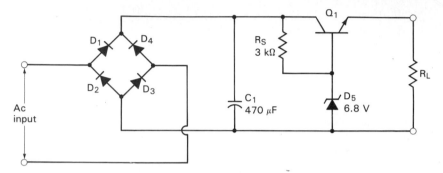

FIGURE 20.18

10. Determine the approximate value of V_L for the circuit in Figure 20.19. [14]

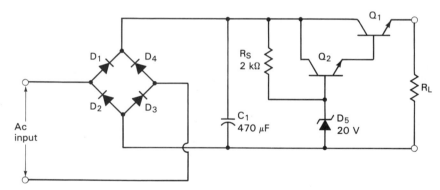

FIGURE 20.19

11. Determine the maximum possible value of load current for the circuit in Figure 20.20. [16]

FIGURE 20.20

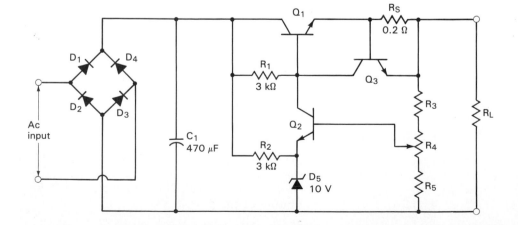

CHAP. 20 Discrete and Integrated Voltage Regulators

12. R_S in Figure 20.20 is changed to 1.2Ω. Determine the maximum possible load current for the new circuit. [16]

§20.4

13. An adjustable IC voltage regulator is set for a $+3$-V_{dc} output. The V_d rating for the device is 32 V. Determine the maximum allowable input voltage for the device. [24]

14. An adjustable IC voltage regulator is set for a $+6$-V_{dc} output. The V_d rating for the device is 24 V. Determine the maximum allowable input voltage for the device. [24]

15. The LM317 is being used in a circuit with adjustment resistance values of $R_1 = 330\ \Omega$ and $R_2 = 2.848\ k\Omega$ (adjusted potentiometer value). Determine the output voltage for the circuit. [25]

16. The LM317 is being used in a circuit with adjustment resistance values of $R_1 = 510\ \Omega$ and $R_2 = 6.834\ k\Omega$ (adjusted potentiometer value). Determine the output voltage for the circuit. [25]

THE BRAIN DRAIN

17. A $+12$-V_{dc} regulator has a line regulation rating of 420 ppm/V. Determine the output voltage for the circuit when V_{in} increases by 10 V.

18. A voltage regulator has the following measured values: $V_{out} = 12.002\ V_{dc}$ when $V_{in} = +20$ V (rated maximum allowable input), and $V_{out} = 12\ V_{dc}$ when $V_{in} = +10$ V (rated minimum allowable input). Determine the line regulation rating of the device in mV, %, and %/V.

19. A $+5$-V regulator has a load regulation rating of 2 Ω over a range of $I_L = 0$ to 100 mA (maximum). Express the load regulation of the circuit in V/mA, %, and %/mA.

20. A $+15$-V_{dc} regulator has a 0.02%/mA load regulation rating for a range of $I_L = 0$ to 50 mA (maximum). Assuming that the load current stays within its rated limits, determine the maximum load power that can be delivered by the regulator. (Assume that V_{out} increases as I_L increases.)

ANSWERS TO EXAMPLE PRACTICE PROBLEMS

20.1. 25 μV/V
20.2. 100 μV/mA
20.3. 38 V
20.4. 10 V

21
Special Applications Diodes

Several of the devices discussed in this chapter are used in ultra-high frequency communications applications.

OBJECTIVES

After studying the material in this chapter, you should be able to:

☐ 1. State the purpose served by a *varactor*. (§21.1)

☐ 2. Discuss the relationship between varactor bias and junction capacitance. (§21.1)

☐ 3. List, define, and explain the significance of the commonly used varactor parameters and specifications. (§21.1)

☐ 4. Describe the method by which a varactor can be used as the tuning component in a parallel (or series) *LC* circuit. (§21.1)

☐ 5. Given the values of L, C_t, and C_R for a varactor-tuned *LC* circuit, determine the range of resonant frequencies for the circuit. (§21.1)

☐ 6. Explain the relationship between the C_R rating of a varactor, *coarse tuning*, and *fine tuning*. (§21.1)

☐ 7. Describe the overall operation of *transient suppressors* and *constant-current diodes*. (§21.2)

☐ 8. Describe a *surge* and the danger it presents to an electronic system. (§21.2)

☐ 9. List the characteristics that every surge-protection circuit needs. (§21.2)

☐ 10. Compare and contrast *transient suppressors* and *standard zener diodes*. (§21.2)

☐ 11. List and define the commonly used transient suppressor ratings. (§21.2)

☐ 12. Calculate the value of P_{Pk} for a given transient suppressor at a given temperature using the power derating curve for the device. (§21.2)

☐ 13. Discuss the differences between standard transient suppressors and *back-to-back suppressors*. (§21.2)

☐ 14. Describe the differences between *constant-current diodes* and *pn* junction diodes. (§21.2)

☐ 15. Identify the schematic symbol for the constant-current diode. (§21.2)

☐ 16. List and define the commonly used constant-current diode ratings. (§21.2)

☐ 17. State the purpose served by the *series current regulator* and describe its operation. (§21.2)

☐ 18. State the purpose served by the *shunt current regulator* and describe its operation. (§21.2)

☐ 19. Describe the operation of the *tunnel diode*. (§21.3)

☐ 20. Describe the operation of the *tunnel diode oscillator*. (§21.3)

☐ 21. Describe the construction and operation of the Schottky diode. (§21.4)

☐ 22. Describe the construction and operation of the PIN diode. (§21.4)

☐ 23. Compare and contrast the forward operation of the PIN and *pn* junction diodes. (§21.4)

☐ 24. Describe the construction and operation of the step-recovery diode. (§21.4)

As you master each of the objectives listed, place a check mark (✓) in the appropriate box.

I n the first four chapters, we discussed the operation, common circuit applications, and troubleshooting of *pn*-junction diodes, zener diodes, and LEDs. In this chapter, we are going to take a relatively brief look at several other types of diodes.

The diodes covered in this chapter are called *special applications diodes* because they are used to perform functions other than those performed by the *pn* junction diode, zener diode, and LED. In other words, the diodes covered in this chapter would rarely be used for any purposes other than those we will be discussing.

The diodes we will be discussing have very little in common with each other. For this reason, each section in this chapter will read as a *stand-alone* section. This means that you do not have to cover the sections in this chapter in any particular order. You can choose those sections you wish to cover and the order you wish to cover them in.

Introduction

21.1
Varactor Diodes

Varactor. A diode that has high junction capacitance when reverse biased.

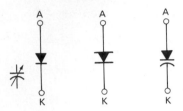

FIGURE 21.1

Varactor schematic symbols.

The relationship between bias and junction capacitance.

The varactor is a type of *pn*-junction diode that has relatively high junction capacitance when reverse biased. The amount of capacitance varies with the amount of reverse bias, and thus the varactor is most often used as a *voltage-controlled capacitor*. The varactor is also referred to as a *varicap*, *tuning diode*, and *epicap*. The commonly used schematic symbols for the varactor are shown in Figure 21.1.

The capacitance of the varactor junction is found as

$$C_T = \epsilon \frac{A}{W_d}$$

(21.1)

where C_T = the total junction capacitance
A = the cross-sectional area of the junction
ϵ = the permittivity of the semiconductor material
W_d = the *width* of the depletion layer

Equation 21.2 is really pretty useless for any practical applications. Let's face it: it would take a lot of intricate mathematical analyses to determine all the values required to solve the equation. However, the equation is introduced because it does serve to show you one important concept. That is the fact that the value of C_T is *inversely proportional* to the width of the depletion layer. Since the width of the depletion layer is directly proportional to the reverse voltage applied to the diode, we can draw the following conclusion: *The value of C_T is inversely proportional to the amount of reverse bias applied to the diode*. The relationship between C_T and V_R is illustrated in Figure 21.2. As the illustration shows, you increase or decrease the junction capacitance simply by changing the reverse bias on the diode. This makes the varactor ideal for use in circuits that require voltage-controlled tuning. Such an application is discussed later in this section and was briefly mentioned in Chapter 15.

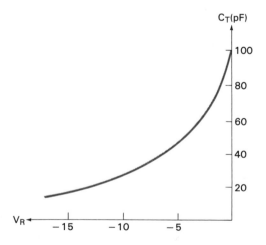

FIGURE 21.2

Varactor bias versus capacitance curve.

Varactor Specifications

The spec sheet of any varactor will provide you with the information you need for any circuit analyses. Figure 21.3 shows the spec sheet and operating curves

CHAP 21 Special Applications Diodes

for the Motorola MV209 series varactors. We will use this spec sheet for our discussion on the commonly used varactor parameters and characteristics.

The *maximum ratings* for the varactor are the same as those that are used for the *pn*-junction diode and the zener diode. The same holds true for the *reverse breakdown voltage* and *reverse leakage current* ratings that appear in the *electrical characteristics* section of the spec sheet. Since you are already familiar with these ratings, we will not discuss them here. If you need to review any of them, refer back to the appropriate discussions in Chapter 2.

The *diode capacitance temperature coefficient* (TC_C) rating of a varactor tells you how much the component's capacitance will change for each 1°C rise in temperature above 25°C. The TC_C rating for the MV209 series varactors is 300 ppm/°C. This means that the varactor capacitance will *increase* by 300 *parts per million* for each 1°C rise in temperature above 25°C.

What is meant by *parts per million*? It means that the capacitance will increase by *300-millionths of its normal value*. If a capacitor's normal value was 1 pF, the change would be found as

> Varactor maximum ratings are the same as those for *pn*-junction diodes and zener diodes.

> **Diode capacitance temperature coefficient.** The amount by which varactor capacitance changes when temperature changes.

> How to determine the ΔC for a specific ΔT.

$$\Delta C = (0.0003)(1 \text{ pF}) = 0.0003 \text{ pF}$$

FIGURE 21.3

(Courtesy of Motorola, Inc.)

SILICON EPICAP DIODE

. . . designed for general frequency control and tuning applications; providing solid-state reliability in replacement of mechanical tuning methods.

- High Q with Guaranteed Minimum Values at VHF Frequencies
- Controlled and Uniform Tuning Ratio
- Available in Surface Mount Package

MMBV109L MV209

CASE 182-02, STYLE 1 (TO-226AC)

2 o——|◀——o 1
Cathode   Anode

CASE 318-03, STYLE 8 SOT-23 (TO-236AB)

3 o——|◀——o 1
Cathode   Anode

26–32 pF VOLTAGE VARIABLE CAPACITANCE DIODES

MAXIMUM RATINGS

Rating	Symbol	MV209	MMBV209,L	Unit
		Value		
Reverse Voltage	V_R	30		Volts
Forward Current	I_F	200		mA
Forward Power Dissipation @ T_A = 25°C Derate above 25°C	P_D	280 2.8	200 2.0	mW mW/°C
Junction Temperature	T_J	+125		°C
Storage Temperature Range	T_{stg}	−55 to +150		°C

DEVICE MARKING

MMBV109L = 4A

ELECTRICAL CHARACTERISTICS (T_A = 25°C unless otherwise noted.)

Characteristic	Symbol	Min	Typ	Max	Unit
Reverse Breakdown Voltage (I_R = 10 μAdc)	$V_{(BR)R}$	30	—	—	Vdc
Reverse Voltage Leakage Current (V_R = 25 Vdc)	I_R	—	—	0.1	μAdc
Diode Capacitance Temperature Coefficient (V_R = 3.0 Vdc, f = 1.0 MHz)	TC_C	—	300	—	ppm/°C

Device	C_t, Diode Capacitance V_R = 3.0 Vdc, f = 1.0 MHz pF			Q, Figure of Merit V_R = 3.0 Vdc f = 50 MHz (Note 1)	C_R, Capacitance Ratio C_3/C_{25} f = 1.0 MHz (Note 2)	
	Min	Nom	Max	Min	Min	Max
MMBV109L, MV209	26	29	32	200	5.0	6.5

FIGURE 1 — DIODE CAPACITANCE

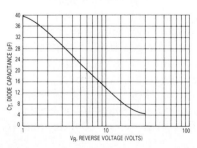

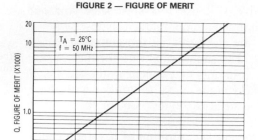

FIGURE 2 — FIGURE OF MERIT

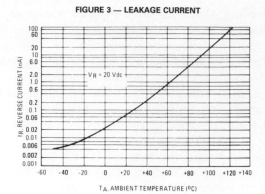

FIGURE 3 — LEAKAGE CURRENT

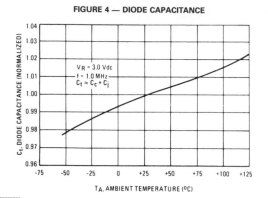

FIGURE 4 — DIODE CAPACITANCE

NOTES ON TESTING AND SPECIFICATIONS

1. Q is calculated by taking the G and C readings of an admittance bridge, such as Boonton Electronics Model 33AS8, at the specified frequency and substituting in the following equation:

$$Q = \frac{2\pi fC}{G}$$

2. C_R is the ratio of C_t measured at 3.0 Vdc divided by C_t measured at 25 Vdc.

FIGURE 21.3

Continued

If the capacitor's normal value was 20 pF, the change would be found as

$$\Delta C = (0.0003)(20 \text{ pF}) = 0.006 \text{ pF}$$

and so on. Note that the value (0.0003) was found by converting 300-millionths to decimal form. Since the normal capacitance rating of the MV209 is typically 29 pF, the increase in capacitance per 1°C rise in temperature would be found as

$$\Delta C = (0.0003)(29 \text{ pF}) = 0.0087 \text{ pF}$$

Temperature has little effect on the capacitance rating of most varactors.

Thus, the MV209 varactor would have to experience a temperature increase of nearly 115°C in order to experience an increase in capacitance of 1 pF. This leads us to the following conclusion: *The capacitance of a varactor is relatively independent of operating temperature.*

Diode capacitance (C_t). The rated value (or range) of C for a varactor at a specified value of V_R.

The *diode capacitance* (C_t) rating of a varactor is self-explanatory. For the MV209 series varactors, the C_t rating is 26 to 32 pF when V_R is 3 Vdc. Note that the *nominal* value (29 pF) is the *rated* value of the component. In other words, the MV209 would be rated as a 29-pF varactor diode, even though the actual value could fall anywhere within the specified range at $V_R = 3$ Vdc.

Capacitance ratio (C_R). The factor by which C changes from one specified value of V_R to another.

The *capacitance ratio* (C_R) rating of the varactor tells you how much the junction capacitance varies over the given range of voltages. For the MV209 series, the C_R rating ranges from 5.0 to 6.5 for $V_R = 3$ to 25 Vdc. This means that the capacitance at $V_R = 3$ Vdc will be 5.0 to 6.5 times as high as it is at $V_R = 25$ Vdc.

The *higher* the value of C_R, the wider the range of capacitance values for a given varactor. For example, let's say that we have two varactors, each having a C_t rating of 51 pF when $V_R = 5$ Vdc. Let's also assume that D_1 has a rating of $C_R = 1.5$ for $V_R = 5$ to 10 Vdc, and D_2 has a rating of $C_R = 4$ for $V_R = 5$ to 10

CHAP 21   Special Applications Diodes

Vdc. To find the capacitance of each diode at $V_R = 10$ Vdc, we would divide the value of C_t by the value of C_R. Thus,

$$C_{D1} = \frac{51 \text{ pF}}{1.5} = 34 \text{ pF} \qquad \text{(when } V_R = 10 \text{ V}_{dc}\text{)}$$

and

$$C_{D2} = \frac{51 \text{ pF}}{4} = 12.75 \text{ pF} \qquad \text{(when } V_R = 10 \text{ V}_{dc}\text{)}$$

The importance of the varactor C_R rating.

Thus, D_1 would have a range of $C = 34$ to 51 pF, while D_2 would have a range of $C = 12.75$ to 51 pF.

So, why is the C_R rating of a varactor important? A varactor with a *high C_R* rating could be used in a *coarse tuning* circuit, while a varactor with a *low C_R* rating could be used in a *fine tuning* circuit. Tuning was discussed in Chapter 15, and it will be discussed further when we take a look at varactor applications later in this section.

You can use C_R to find the range of C_t values. However, you must use the *capacitance versus reverse voltage* graph to find the value of C at various values of V_R.

While the C_R rating of a varactor can be used to determine its capacitance *range*, it is not used to determine the specific diode capacitance at a specific value of V_R. For example, we would not use the C_R rating of the MV209 to determine its capacitance at $V_R = 6$ Vdc. Instead, you need to use the *capacitance versus reverse voltage* curve that is shown in Figure 21.3. Using this curve, we could determine the value of capacitance at $V_R = 6$ Vdc to be approximately 23 pF.

The *figure of merit* rating is the Q of the junction capacitance. You may recall that the Q of a capacitor is the *ratio of energy stored in the capacitor to the energy lost through leakage current*. In other words, it is the ratio of power returned to the circuit by the capacitor to power dissipated by the capacitor. For the MV209, the Q rating is 200 (minimum). This means that the power returned by the capacitor to the circuit (when it discharges) will be at least 200 times the power that it dissipated.

Varactor Applications

Varactors are used almost exclusively in *tuned circuits*. A tuned *LC* circuit that contains a varactor is shown in Figure 21.4. Note that the capacitance of the varactor is in *parallel* with the inductor. Thus, the varactor and the inductor form a *parallel LC circuit*, or *LC tank circuit*.

Before we analyze the operation of the varactor in the circuit, a few points should be made. First, note the direction of the varactor. Since the varactor is pointing toward the positive source voltage, we know that it is reverse biased. Therefore, it is acting as a voltage-variable capacitance. *For normal operation, a varactor diode is almost always operated in its reverse operating region.* A forward-biased varactor diode would serve no useful purpose. Second, R_W in the circuit is the *winding resistance* of the inductor. This winding resistance is in series with the potentiometer, R_1. Thus, R_w and R_1 form a voltage divider that is used to determine the amount of reverse bias across D_1, and therefore its capacitance. By

For normal operation, the varactor is almost always reverse biased.

◢ 4

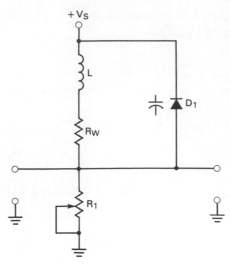

FIGURE 21.4

LC tank circuit.

adjusting the setting of R_1, we can vary the diode capacitance. This, in turn, varies the *resonant frequency* of the *LC* circuit.

The resonant frequency of the *LC* tank circuit is found using

$$f_r = \frac{1}{2\pi \sqrt{LC}}$$

(21.2)

If the amount of varactor reverse bias is *decreased*, the value of C for the component *increases*. The increase in C will cause the resonant frequency of the circuit to *decrease*. Thus, *a decrease in reverse bias causes a decrease in resonant frequency*. By the same token, an increase in varactor reverse bias causes an increase in the value of f_r. This point is illustrated in the following example.

════════════ **EXAMPLE 21.1** ════════════

☑ 5

The *LC* tank circuit in Figure 21.4 has a 1-mH inductor. The varactor has the following specifications: $C_t = 100\ pF$ when $V_R = 5\ V_{dc}$, and $C_R = 2.5$ for $V_R = 5$ to $10\ V_{dc}$. Determine the resonant frequency for the circuit at $V_R = 5$ V_{dc} and $V_R = 10\ V_{dc}$.

Solution: When the varactor reverse bias is 5 V_{dc}, the value of C is 100 pF. For this condition, the value of f_r is found as

$$f_r = \frac{1}{2\pi \sqrt{LC}}$$

$$= \frac{1}{2\pi \sqrt{(1\ \text{mH})(100\ \text{pF})}}$$

$$= 503.29\ \text{kHz}$$

The value of C at $V_R = 10\ V_{dc}$ is found by dividing the C_t rating by the value of C_R, as follows:

$$C = \frac{100\ \text{pF}}{2.5}$$

$$= 40\ \text{pF}$$

CHAP 21   Special Applications Diodes

Now, the value of f_r at $V_R = 10$ V$_{dc}$ is found as

$$f_r = \frac{1}{2\pi \sqrt{LC}}$$

$$= \frac{1}{2\pi \sqrt{(1 \text{ mH})(40 \text{ pF})}}$$

$$= 795.77 \text{ kHz}$$

Thus, when the varactor reverse bias increased from 5 to 10 V$_{dc}$, the value of f_r increased from 503.29 to 795.77 kHz.

PRACTICE PROBLEM 21–1

A circuit like the one shown in Figure 21.4 has a 3.3-mH inductor and a varactor with the following specifications: $C_t = 51$ pF at $V_R = 4$ V$_{dc}$ and $C_R = 1.8$ for $V_R = 4$ to 10 V$_{dc}$. Determine the value of f_r for $V_R = 4$ V$_{dc}$ and $V_R = 8$ V$_{dc}$.

You were told in our discussion on varactor parameters (page 961) that high-C_R varactors are used in *coarse tuning* circuits, while low-C_R varactors are used in *fine tuning* circuits. This point can be illustrated with the aid of the following example.

EXAMPLE 21.2

The diode in Figure 21.4 is replaced with one that has ratings of $C_t = 100$ pF when $V_R = 5$ V$_{dc}$ and $C_R = 1.02$ for $V_R = 5$ to 10 V$_{dc}$. Determine the frequency range of the circuit for $V_R = 5$ to 10 V$_{dc}$.

Solution: In Example 21.1, we determined the value of f_r at $V_R = 5$ V$_{dc}$ to be 503.29 kHz. This value has not changed. For the new circuit, the value of C at $V_R = 10$ V$_{dc}$ is found as

$$C = \frac{100 \text{ pF}}{1.02}$$

$$= 98 \text{ pF}$$

The value of f_r at $V_R = 10$ V$_{dc}$ is now found as

$$f_r = \frac{1}{2\pi \sqrt{LC}}$$

$$= \frac{1}{2\pi \sqrt{(1 \text{ mH})(98 \text{ pF})}}$$

$$= 508.4 \text{ kHz}$$

Table 21.1 summarizes the values found in Examples 21.1 and 21.2. As you can see, the range of frequencies for the circuit in Example 21.2 is much smaller than that for the circuit in Example 21.1.

TABLE 21.1
Results from Examples 21.1 and 21.2

Example	f_r at $V_R = 5 V_{dc}$	f_r at $V_R = 10 V_{dc}$	Δf_r
5.1	503.29 kHz	795.77 kHz	292.48 kHz
5.2	503.29 kHz	508.4 kHz	5.11 kHz

Coarse tuning versus *fine tuning.*

The circuit in Example 21.1 would be a *coarse tuning* circuit. That is, it would be used to vary the value of f_r over a wide range of values. The circuit in Example 21.2, on the other hand, would be a *fine tuning* circuit. It would be used to select a frequency within a smaller range. It would be used to center in on an *exact* frequency, while the coarse tuning circuit would be used to obtain a *ballpark* frequency value.

SECTION REVIEW

1. A varactor acts as what type of capacitance? [Objective 1]
2. What is the relationship between the amount of reverse bias applied to a varactor and its capacitance? Use equation (21.1) to explain your answer. [Objective 2]
3. What is the *diode capacitance temperature coefficient* rating? What is its unit of measure? [Objective 3]
4. What is the relationship between varactor capacitance and temperature? [Objective 3]
5. What is the *capacitance ratio* of a varactor? [Objective 3]
6. Why is the C_R rating of a varactor important? [Objective 3]
7. What is the Q of a capacitor? How does it relate to circuit bandwidth? [Objective 3]
8. Explain the operation of the circuit shown in Figure 21.4. [Objective 4]
9. What is the relationship between varactor reverse bias and the resonant frequency of its tuned circuit? [Objective 4]
10. What is the difference between *coarse tuning* and *fine tuning*? [Objective 6]
11. What is the relationship between the C_R rating of a varactor and the type of tuning provided? [Objective 6]

21.2
Transient Suppressors and Constant-current Diodes

Transient suppressors. Zener diodes that have extremely high surge-handling capabilities.

Constant-current diodes. Diodes that maintain a constant device current over a wide range of forward operating voltages.

In this section, we are going to discuss two diodes that are very similar in operation to the standard zener diode. The first is the *transient suppressor*. Transient suppressors are zener diodes that have extremely high *surge*-handling capabilities. These diodes are used to protect voltage-sensitive circuits from surges that can occur under a variety of circumstances. The second diode we are going to discuss is the *constant-current diode*. The constant-current diode is an extremely high impedance diode that will maintain a constant device current over a wide range of forward operating voltages. Thus, the constant-current diode can be viewed as a "current version" of the zener diode. Note that the constant-current diode is not actually a zener diode, despite the similarities between the two. Rather, it is a variation on the *pn* junction diode.

Transient Suppressors

In Chapter 4, you were introduced to the idea of using a *symmetical zener clipper* to protect a power supply from a voltage *surge*, or *transient*. Such a surge-protection circuit is shown in Figure 21.5.

You may recall that a surge is an abrupt high-voltage condition that lasts for a very brief period of time; usually in the microsecond or millisecond range. Left unchecked, a surge in the ac power lines can cause serious damage to the power supply of any electronic system, such as a television or personal computer. The circuit shown in Figure 21.5 would protect the power supply from such surges by shorting out any voltages greater than the V_Z ratings of the diodes.

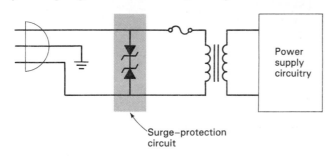

Surge–protection
circuit

FIGURE 21.5

For a surge-protection circuit to operate properly, it must have several character-istics:

9

1. The diodes used must have extremely high power dissipation ratings. This is due to the fact that most ac power line surges contain a relatively high amount of power, generally in the hundreds of watts or higher.

2. The diodes must be able to turn on very rapidly. If the diodes in the surge-protection circuit are too slow, the power supply could be damaged before they have a chance to turn on.

Requirements for surge protection circuits.

These requirements for the surge-protection circuit are easily fulfilled by using *transient suppressors*.

Transient suppressors have the same general operating characteristics as the standard zener diode. In fact, the schematic symbol for the transient suppressor is identical to that of a standard zener diode.

The main difference between the transient suppressor and the standard zener diode is the suppressor's *surge-handling capability*. The transient suppressor is designed to dissipate extremely high amounts of power for a very limited period of time. For example, the Motorola 1N5908 through 1N6389 series of transient suppressors can dissipate up to 1.5 kW for a period of slightly less than 10 ms. This amount of power, even for a short period of time, would destroy any standard zener diode.

The transient suppressor can dissipate power values in the kilowatt range for very brief periods of time.

You might think that the time limit on the power rating of a transient suppressor could be a problem, but it isn't. Remember, surges generally last only a few milliseconds. Thus, the transient suppressor can generally handle any surge that occurs.

Transient Suppressor Specifications

The *maximum ratings* and *electrical characteristics* shown in Figure 21.6 are those for the 1N5908 through 1N6389 series transient suppressors. We will use the values shown for our discussion on the typical suppressor specifications.

11a

MAXIMUM RATINGS

Rating	Symbol	Value	Units
Peak Power Dissipation (1) @ $T_L \leq 25^\circ C$	P_{PK}	1500	Watts
Steady State Power Dissipation @ $T_L \leq 75^\circ C$, Lead Length = 3/8'' Derated above $T_L = 75^\circ C$	P_D	5.0 50	Watts mW/$^\circ$C
Forward Surge Current (2) @ $T_A = 25^\circ C$	I_{FSM}	200	Amps
Operating and Storage Temperature Range	T_J, T_{stg}	–65 to +175	$^\circ$C

Lead Temperature not less than 1/16'' from the case for 10 seconds: 230°C

MECHANICAL CHARACTERISTICS

CASE: Void-free, transfer-molded, thermosetting plastic

FINISH: All external surfaces are corrosion resistant and leads are readily solderable and weldable

POLARITY: Cathode indicated by polarity band. When operated in zener mode, will be positive with respect to anode

MOUNTING POSITION: Any

ELECTRICAL CHARACTERISTIC (T_A = 25°C unless otherwise noted) V_F# = 3.5 V max, I_F** = 100 A) (C suffix denotes standard back to back versions. Test both polarities)

JEDEC Device	Device	Breakdown Voltage V_{BR} Volts Min	@ I_T (mA)	Maximum Reverse Stand-Off Voltage V_{RWM}*** (Volts)	Maximum Reverse Leakage @ V_{RWM} I_R (μA)	Maximum Reverse Surge Current I_{RSM}† (Amps)	Maximum Reverse Voltage @ I_{RSM}† (Clamping Voltage) V_{RSM}(Volts)	Clamping Voltage Peak Pulse Current @ I_{pp1}† = 1.0 A V_{C1} (Volts max)	Peak Pulse Current @ I_{pp2}† = 10 A V_{C2} (Volts max)
1N6373	ICTE-5/MPTE-5	6.0	1.0	5.0	300	160	9.4	7.1	7.5
—	ICTE-5C/MPTE-5C	6.0	1.0	5.0	300	160	9.4	8.1	8.3
1N6374	ICTE-8/MPTE-8	9.4	1.0	8.0	25	100	15.0	11.3	11.5
1N6382	ICTE-8C/MPTE-8C	9.4	1.0	8.0	25	100	15.0	11.4	11.6
1N6375	ICTE-10/MPTE-10	11.7	1.0	10	2.0	90	16.7	13.7	14.1
1N6383	ICTE-10C/MPTE-10C	11.7	1.0	10	2.0	90	16.7	14.1	14.5
1N6376	ICTE-12/MPTE-12	14.1	1.0	12	2.0	70	21.2	16.1	16.5
1N6384	ICTE-12C/MPTE-12C	14.1	1.0	12	2.0	70	21.2	16.7	17.1

FIGURE 21.6

Peak power dissipation rating. Indicates the amount of surge power that the suppressor can dissipate.

The maximum ratings are fairly standard, with the exception of the *peak power dissipation* rating. This rating indicates the surge-handling capability of the suppressors. Several notes on the specification sheet (not shown) indicate that this rating has some conditions:

1. The operating temperature is 25°C.
2. The surge has a duration of approximately 8 ms.
3. The maximum number of surges is four per minute.

P_{Pk} has *time, temperature,* and *frequency* restrictions.

These conditions indicate that the P_{Pk} rating of the suppressor is:

1. Temperature dependent
2. Time dependent
3. Frequency dependent

CHAP 21   Special Applications Diodes

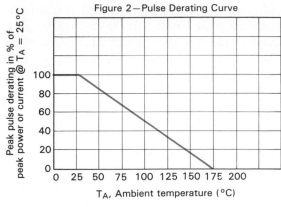

Figure 2—Pulse Derating Curve

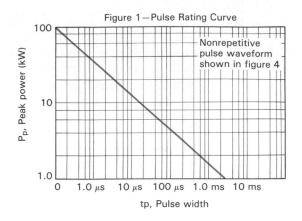

Figure 1—Pulse Rating Curve

FIGURE 21.7

The temperature dependency of the P_{pK} rating is illustrated in Figure 21.7a. As you can see, the P_{Pk} rating is derated (by percent) at temperatures above 25°C. The following example illustrates the use of this type of power derating curve.

EXAMPLE 21.3

A 1N5908 suppressor is being used in a circuit whose ambient temperature is 100°C. What is the value of P_{Pk} at this temperature?

Solution: The point where the curve intersects the $T = 100$°C line corresponds to a derating percentage of 50%. Thus, P_{Pk} will be 50% of its maximum value at this temperature, or 750 W.

PRACTICE PROBLEM 21–3

What is the P_{Pk} rating of the 1N5908 at $T = 150$°C?

You may be wondering why we bothered with the temperature curve when the spec sheet for the 1N5908 series suppressors lists a derating value. *The derating value listed in the Maximum Ratings table applies to the P_D rating of the components, not their P_{Pk} rating.* To derate the P_{Pk} rating, you must use the power versus temperature curve.

The derating value given on the suppressor spec sheet does *not* apply to the P_{Pk} rating.

The time dependency of the P_{Pk} rating is illustrated in Figure 21.7b. As you can see, P_{Pk} *and surge duration (pulse width) are inversely proportional.* As the duration of the surge increases, the power-dissipating capability of the suppressors decreases.

The final maximum rating is the *forward surge current* rating. As you can see, the forward surge current rating for the transient suppressor is extremely high. This is due to the combination of their high power dissipation ratings and the typically low value of V_F (approximately 0.7 V).

Two of the *electrical characteristics* should be very familiar to you by now. The *reverse breakdown voltage* (V_{BR}) and *reverse leakage current* (I_R) ratings are standard stuff. You've seen them both a dozen times or more already.

Maximum reverse stand-off voltage. The maximum allowable peak or dc reverse voltage that will not turn on a transient suppressor.

The *maximum reverse stand-off voltage* (V_{RWM}) rating is extremely important. This rating indicates the *maximum allowable peak or dc reverse voltage for any circuit using the component.* For example, refer back to Figure 21.5. Assuming that the circuit shown is being driven by a 115-V_{ac} line, the peak primary voltage would be approximately 163 V_{pk}. Thus, the diodes used in the circuit would need to have V_{RWM} ratings that are *greater than* 163 V. Otherwise, the diodes would turn on during the normal operating cycles of the power supply. *Note that V_{RWM} is the primary rating that is considered when attempting to substitute one suppressor for another.*

The I_{RSM} rating decreases as the V_{BR} rating increases.

The *maximum reverse surge current* (I_{RSM}) rating is self-explanatory. It is the maximum surge current that the suppressor can handle for the rated period of time (in this case, approximately 8 ms). Note that this rating *decreases* as the V_{BR} rating for the suppressors increases. This is due to the fact that the product of reverse voltage and reverse current must be *less than* the value of P_{Pk}. (Do you remember your basic power formulas?)

Clamping voltage. The maximum value of V_Z for a transient suppressor.

The *clamping voltage* (V_{RSM}) is the maximum rated value of V_Z. When turned on, the reverse voltage across the transient suppressor will be less than or equal to this value.

Maximum temperature coefficient. The percent of change in V_{BR} per 1°C rise in operating temperature.

The *maximum temperature coefficient* rating indicates the percentage by which V_{BR} will change per 1°C increase in temperature above 25°C. At this point, you shouldn't have any trouble with this type of rating.

Selector Guides

Selector guides for transient suppressors contain all the ratings that we have discussed in this section. The selector guides for a given series of transient suppressors can be obtained from the series manufacturer.

Back-to-Back Suppressors

Back-to-back suppressor. A single package containing two transient suppressors that are connected as shown in Figure 21.5.

If you look closely at the ratings for the 1N6374 and the 1N6382, you'll see that their ratings are identical. You may be wondering, then, what the difference is between these two suppressors.

The 1N6382 is a *back-to-back suppressor*. This type of suppressor, whose schematic symbol is shown in Figure 21.8, actually contains two transient suppressors that are connected *internally* like the two diodes shown in Figure 21.5. The obvious advantages to using this type of component are reduced circuit manufacturing costs and simpler circuitry. Note that the back-to-back suppressors have no V_F rating, since they are designed to break down at the rated value of V_{BR} in both directions.

(anode 1)——————(anode 2)

FIGURE 21.8

How can you tell whether a given suppressor is a single diode or a back-to-back suppressor? The component listings will contain some type of indicator that will be clearly identified in the *Notes* listing on the spec sheet. For the 1N5908 series sheet shown in Figure 21.6, the use of the suffix *C* in the ICTE number indicates that the device is a back-to-back suppressor.

Constant-current Diodes

Current regulator diode. Another name for the constant-current diode.

Constant-current diodes are drastically different form any of the diodes you have seen so far. This can be seen by taking a look at Figure 21.9, which shows the forward operating curves for the Motorola 1N5283 through 1N5314 series *current regulator diodes*.

When a *pn*-junction diode is forward biased, V_F will be approximately 0.7 V and I_F will be determined by the components that are in the diode circuit. This

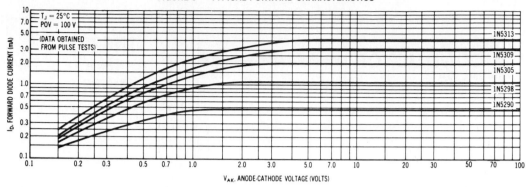

FIGURE 3 — TYPICAL FORWARD CHARACTERISTICS

FIGURE 21.9

does not hold true for the constant-current diode. As the forward curve in Figure 21.9 shows, the value of V_F for a constant-current diode can have a wide range of values, in this case anywhere from 0.1 to 100 V. At the same time, the value of I_F for the constant-current diode will be limited by the diode itself, not the components that are in the diode circuit. For example, the 1N5290 curve shows that the value of I_F for the device will increase as V_F increases from 0.15 V to approximately 1 V. At that point, the diode regulates the value of I_F to around 500 µA (0.5 mA). The value of I_F through the 1N5290 will be held at this value for any value of V_F between 1 and 100 V.

Because the operation of the constant-current diode is so radically different form any other diode, it has its own schematic symbol. This symbol is shown in Figure 21.10.

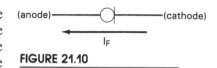

FIGURE 21.10

Constant-current Diode Specifications

Figure 21.11 shows the spec sheet for the 1N5283 through 1N5314 series constant-current diodes. The only new *maximum rating* is the *peak operating voltage (POV)* rating. The POV is the *maximum allowable value of V_F*. As you can see, the constant-current diode can have values of V_F other than 0.7 V, in this case anywhere up to 100 V maximum.

For reasons that you'll see in a moment, we're going to start our coverage of the *electrical characteristics* with the last rating given, the *maximum limiting voltage*, V_L. The V_L rating indicates *the voltage at which the diode begins to regulate current*. For the 1N5290, the value of V_L is 1.05 V. This is approximately equal to the value of V_L that we obtained from the current versus voltage graph in Figure 21.9.

The *regulator current (V_P)* rating is *the regulated value of forward current for values of V_F that are between V_L and POV*. As long as the forward voltage is kept between the rated values of V_L and POV, the current through the diode will be maintained at the value of I_P. Note that the value of I_P for the 1N5290 is shown to be 423 µA (0.423 mA). This value is approximately equal to the value we obtained from the current versus voltage graph in Figure 21.9.

The *minimum dynamic impedance (Z_T)* rating of the constant-current diode demonstrates another big difference between this diode and the *pn*-junction diode. You may recall that the bulk resistance of a *pn*-junction diode is typically very small, around 10 Ω or less. The impedance of the constant-current diode is extremely *high*, typically in the high kilohm to low megohm range. The *minimum knee impedance (Z_k)* is even higher!

Peak operating voltage (POV). The maximum allowable value of V_F.

Maximum limiting voltage (V_L). The voltage at which the diode starts to limit current.

Regulator current (I_P). The regulated value of forward current for forward voltages that are between V_L and POV.

SEC. 21.2 Transient Suppressors and Constant-current Diodes

FIGURE 21.11

(Courtesy of Motorola, Inc.)

MOTOROLA
■ **SEMICONDUCTOR** ■
TECHNICAL DATA

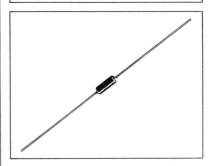

**1N5283
thru
1N5314**

**CURRENT
REGULATOR
DIODES**

CURRENT REGULATOR DIODES

Field-effect current regulator diodes are circuit elements that provide a current essentially independent of voltage. These diodes are especially designed for maximum impedance over the operating range. These devices may be used in parallel to obtain higher currents.

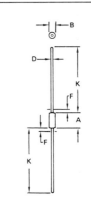

MAXIMUM RATINGS

Rating	Symbol	Value	Unit
Peak Operating Voltage ($T_J = -55°$ C to $+200°$C)	POV	100	Volts
Steady State Power Dissipation @ $T_L = 75°$C Derate above $T_L = 75°$C Lead Length $= 3/8''$ (Forward or Reverse Bias)	P_D	600 4.8	mW mW/°C
Operating and Storage Junction Temperature Range	T_J, T_{stg}	-55 to $+200$	°C

	MILLIMETERS		INCHES	
DIM	MIN	MAX	MIN	MAX
A	5.84	7.62	0.230	0.300
B	2.16	2.72	0.085	0.107
D	0.46	0.56	0.018	0.022
F	–	1.27	–	0.050
K	25.40	38.10	1.000	1.500

All JEDEC dimensions and notes apply

**CASE 51-02
DO-204AA
GLASS**

NOTES:
1. PACKAGE CONTOUR OPTIONAL WITHIN DIA B AND LENGTH A. HEAT SLUGS, IF ANY, SHALL BE INCLUDED WITHIN THIS CYLINDER, BUT SHALL NOT BE SUBJECT TO THE MIN LIMIT OF DIA B.
2. LEAD DIA NOT CONTROLLED IN ZONES F, TO ALLOW FOR FLASH, LEAD FINISH BUILDUP, AND MINOR IRREGULARITIES OTHER THAN HEAT SLUGS.

ELECTRICAL CHARACTERISTICS ($T_A = 25\,°C$ unless otherwise noted)

Type No.	Regulator Current I_P (mA) @ $V_T = 25$ V			Minimum Dynamic Impedance @ $V_T = 25$ V Z_T (MΩ)	Minimum Knee Impedance @ $V_K = 6.0$ V Z_K (MΩ)	Maximum Limiting Voltage @ $I_L = 0.8\ I_P$ (min) V_L (Volts)
	nom	min	max			
1N5283	0.22	0.198	0.242	25.0	2.75	1.00
1N5284	0.24	0.216	0.264	19.0	2.35	1.00
1N5285	0.27	0.243	0.297	14.0	1.95	1.00
1N5286	0.30	0.270	0.330	9.0	1.60	1.00
1N5287	0.33	0.297	0.363	6.6	1.35	1.00
1N5288	0.39	0.351	0.429	4.10	1.00	1.05
1N5289	0.43	0.387	0.473	3.30	0.870	1.05
1N5290	0.47	0.423	0.517	2.70	0.750	1.05
1N5291	0.56	0.504	0.616	1.90	0.560	1.10
1N5292	0.62	0.558	0.682	1.55	0.470	1.13
1N5293	0.68	0.612	0.748	1.35	0.400	1.15
1N5294	0.75	0.675	0.825	1.15	0.335	1.20
1N5295	0.82	0.738	0.902	1.00	0.290	1.25
1N5296	0.91	0.819	1.001	0.880	0.240	1.29
1N5297	1.00	0.900	1.100	0.800	0.205	1.35
1N5298	1.10	0.990	1.210	0.700	0.180	1.40
1N5299	1.20	1.08	1.32	0.640	0.155	1.45
1N5300	1.30	1.17	1.43	0.580	0.135	1.50
1N5301	1.40	1.26	1.54	0.540	0.115	1.55
1N5302	1.50	1.35	1.65	0.510	0.105	1.60
1N5303	1.60	1.44	1.76	0.475	0.092	1.65
1N5304	1.80	1.62	1.98	0.420	0.074	1.75
1N5305	2.00	1.80	2.20	0.395	0.061	1.85
1N5306	2.20	1.98	2.42	0.370	0.052	1.95
1N5307	2.40	2.16	2.64	0.345	0.044	2.00
1N5308	2.70	2.43	2.97	0.320	0.035	2.15
1N5309	3.00	2.70	3.30	0.300	0.029	2.25
1N5310	3.30	2.97	3.63	0.280	0.024	2.35
1N5311	3.60	3.24	3.96	0.265	0.020	2.50
1N5312	3.90	3.51	4.29	0.255	0.017	2.60
1N5313	4.30	3.87	4.73	0.245	0.014	2.75
1N5314	4.70	4.23	5.17	0.235	0.012	2.90

Constant-current Diode Applications

Most of the circuits that would use constant-current diodes are relatively complex and thus are not covered here. However, we *can* (at this point) cover some of the *types* of applications that these diodes would typically be used for.

The *series current regulator* is used to maintain a constant input current to a circuit. The basic configuration for a series current regulator is shown in Figure 21.12a. The block labeled A is the circuit whose input current must be regulated. With the constant-current diode being placed in series with the input to A, its input current will be maintained at the value of I_P for the diode.

The series current regulator maintains a constant circuit input current over a wide range of input voltages.

Two important points need to be made about the series current regulator shown:

1. The constant-current diode will maintain a constant value of I_P, thus eliminating the sinusoidal voltage that would otherwise appear at the input to circuit A.

FIGURE 21.12

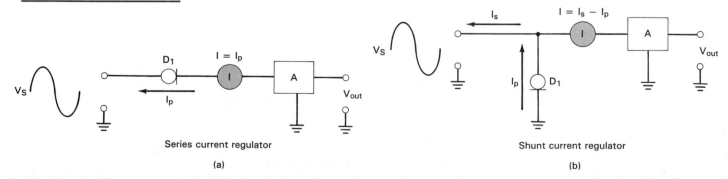

Series current regulator

(a)

Shunt current regulator

(b)

In other words, there will be a constant voltage at the input to A as long as the input resistance of A does not change.

2. For the circuit to operate properly, the voltage across the diode must fall between its V_L and POV ratings.

✔ 18

The shunt current regulator shown in Figure 21.12b provides circuit A with an input current that is equal to the *difference between* the source current (I_S) and the regulated current (I_P). Unlike the series current regulator, the voltage at the input of A would be approximately equal to V_S. Thus, the shunt current regulator would be used to provide circuit A with an input voltage that is equal to V_S and an input current that is less than I_S.

Figure 21.13 summarizes the operation of the series and shunt current regulators.

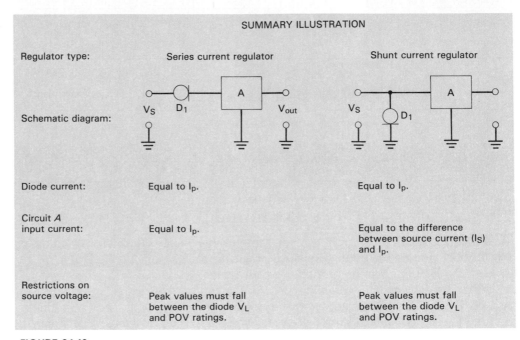

SUMMARY ILLUSTRATION

Regulator type:	Series current regulator	Shunt current regulator
Schematic diagram:		
Diode current:	Equal to I_p.	Equal to I_p.
Circuit A input current:	Equal to I_p.	Equal to the difference between source current (I_S) and I_p.
Restrictions on source voltage:	Peak values must fall between the diode V_L and POV ratings.	Peak values must fall between the diode V_L and POV ratings.

FIGURE 21.13

SECTION REVIEW

1. What is a *transient suppressor*? [Objective 7]
2. What is a *constant-current diode*? [Objective 7]
3. What is a *surge*? How are ac power line surges commonly generated? [Objective 8]
4. Why is a surge in the ac power line hazardous? [Objective 8]
5. What are the required characteristics for a surge-protection circuit? [Objective 9]
6. How does a *transient suppressor* differ from a standard zener diode? [Objective 10]
7. Define each of the following transient suppressor ratings: [Objective 11]
   a. Peak power dissipation (P_{Pk})
   b. Maximum reverse stand-off voltage (V_{RWM})
   c. Clamping voltage (V_{RSM})
   d. Maximum temperature coefficient

8. Why don't back-to-back suppressors have V_F ratings? [Objective 13]
9. What are the differences between constant-current diodes and *pn* junction diodes? [Objective 14]
10. Draw the schematic symbol for the constant-current diode and label its terminals. [Objective 15]
11. Define each of the following constant-current diode specifications:
    a. Peak operating voltage (POV)
    b. Maximum limiting voltage (V_L)
    c. Regulator current (I_P)
12. What purpose is served by the series current regulator? [Objective 17]
13. Describe the operation of the series current regulator. [Objective 17]
14. What purpose is served by the shunt current regulator. [Objective 18]
15. Describe the operation of the shunt current regulator. [Objective 18]

21.3
Tunnel Diodes

Tunnel diodes are components that are used in high-frequency communications applications, such as UHF and microwave communications systems. The schematic symbol and operating curve for the tunnel diode are shown in Figure 21.14. The operating curve is a result of the extremely heavy doping that is used in the manufacturing of the tunnel diode. In fact, the tunnel diode is doped approximately 1000 times as heavily as standard *pn*-junction diodes.

Tunnel diode. A heavily doped diode that is used in high-frequency communications circuits.

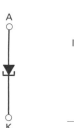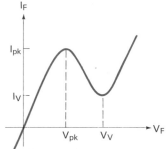

FIGURE 21.14

The tunnel diode symbol and characteristic curve.

In the forward operating region of the tunnel diode, we are interested in the area between the *peak voltage* (V_{Pk}) and the *valley voltage* (V_V). At $V_F = V_{Pk}$, forward current is called *peak current* (I_{Pk}). Note that I_{Pk} is the maximum value that I_F will reach under normal circumstances. As V_F is increased to the value of V_V, I_F *decreases* to its minimum value, called *valley current* (I_V). As you can see, *forward voltage and current are inversely proportional when the diode is operated between the values of V_{Pk} and V_V*. The term used to describe a device whose current and voltage are inversely proportional is *negative resistance*. Thus, the region of operation between V_{Pk} and V_V is called the *negative resistance region*.

Tunnel diodes are operated almost exclusively in the negative resistance region. When operated in this region, they work very well as the active device in an *oscillator*.

The **peak** and **valley** voltage and current values for a tunnel diode can be obtained from the component spec sheet.

When operated between V_{Pk} and V_V, tunnel diode forward voltage and current are inversely proportional.

Negative resistance. A term used to describe any device whose current and voltage are inversely proportional.

Tunnel Diode Oscillator

✔ 20

Negative resistance oscillator. Another name for the tunnel diode oscillator.

You may recall that an oscillator is *a circuit that is used to convert dc to ac*. The basic tunnel diode oscillator, or *negative resistance oscillator*, is shown in Figure 21.15. Note that the circuit is shown to have a dc input voltage and an ac output signal. The output frequency of the circuit will be approximately equal to the resonant frequency of the *LC* tank circuit.

The tank circuit in the negative resistance oscillator is formed by the primary of the transformer and C_2.

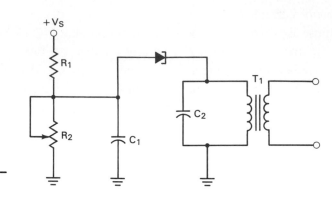

FIGURE 21.15

Tunnel diode oscillator.

The negative resistance oscillator in Figure 21.15 uses a tunnel diode to sustain oscillations. Note that R_1 and R_2 in the circuit are used to bias the tunnel diode. The biasing of the diode is set (by design) so that the following conditions are met:

1. When the waveform produced by the tank circuit is at its *maximum* (positive peak) value, the *difference* between this peak voltage and the biasing voltage is equal to the V_V rating of the tunnel diode. Thus, at the positive peak of the tank circuit output, the voltage across the tunnel diode is equal to V_V.
2. When the waveform produced by the tank circuit is at its *minimum* (negative peak) value, the difference between this peak voltage and the biasing voltage is equal to the V_{PK} rating of the tunnel diode. Thus, at the negative peak of the tank circuit output, the voltage across the tunnel diode is equal to V_{Pk}.

You may recall that the current through a tunnel diode is maximum when the voltage across the component is equal to V_{Pk}. Thus, when the tank circuit waveform is at its negative peak, the diode conduction is maximum, and power is returned to the tank circuit. When the tank circuit is at its positive peak, the voltage across the diode is equal to V_V, and the diode current is at its *minimum* value, I_V. Thus, the tunnel diode is effectively performing the same function as the active devices you saw in the oscillator circuits in Chapter 17.

You may be wondering how the oscillations start in the first place. When power is first applied to the circuit, the diode will momentarily conduct. The pulse of current that passes through the diode will produce the first half-cycle in the tank circuit. The rest, as they say, is history.,

One Final Note

Negative resistance oscillators cannot be used efficiently at low output frequencies.

The negative resistance oscillator has one major drawback. While the circuit works very well at extremely high frequencies (in the upper megahertz range and higher), it cannot be used efficiently at lower frequencies. Lower-frequency oscillators are generally made with *transistors*, as you saw in Chapter 17.

1. What is a tunnel diode? [Objective 19]
2. Describe the negative resistance operating region of the tunnel diode. Include the relationship between diode forward current and forward voltage. [Objective 19]
3. Why does the tunnel diode have the operating curve shown in Figure 21.14? [Objective 19]
4. What is an *oscillator*? [Objective 20]
5. What determines the output frequency of a negative resistance oscillator? [Objective 20]
6. How does the diode in Figure 21.15 keep the tank circuit oscillations from dying out? [Objective 20]
7. What does the diode forward voltage in Figure 21.15 equal when the tank circuit signal is at its positive peak value? Its negative peak value? [Objective 20]

21.4
Other Diodes

The diodes covered in this section are used less commonly than the other diodes we have discussed. Outside of this fact, they are not necessarily related to each other in characteristics or applications.

Schottky Diodes

The *Schottky diode* is a diode that has very little junction capacitance. Because of this, it can be operated at much higher frequencies than the typical *pn* junction diode. The reduced junction capacitance also results in a much faster *switching time*. For this reason, Schottky devices are being used more and more in digital switching applications.

The Schottky diode is often referred to by any of the following names: *Schottky barrier diode*, *hot-carrier diode*, and *surface-barrier diode*. The schematic symbol and characteristic curve for this component are shown in Figure 21.16. As the

Schottky diode. A high-speed diode with very little junction capacitance.

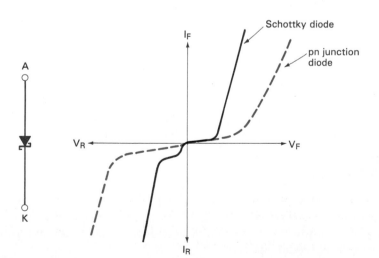

Schottky diode

pn junction diode

FIGURE 21.16

The Schottky diode symbol and characteristic curve.

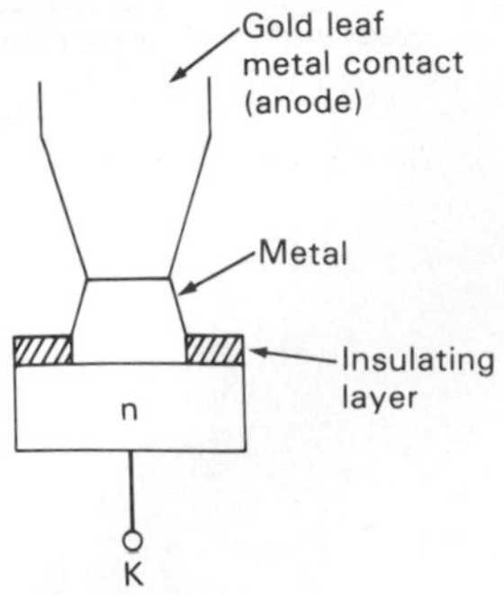

FIGURE 21.17

Schottky diode construction.

characteristic curve indicates, the Schottky diode has lower V_F and V_{BR} values than those of *pn* junction diode. Typically, the Schottky diode will have a V_F of approximately 0.3 V and a V_{BR} of less than -50 V. These are much lower than the typical *pn* junction ratings of $V_F = 0.7$ V and $V_{BR} = -150$ V.

The low junction capacitance and high-switching-speed capability of the Schottky diode are the result of the construction of the component. Schottky diodes have a junction that uses metal in place of the *p*-type material, as shown in Figure 21.17.

By forming a junction with a semiconductor and metal, you still have a junction, but now there is very little junction capacitance. With very little junction capacitance, the Schottky diode can be switched back and forth very rapidly between forward and reverse operation. While many components are capable of switching at this frequency rate, most of them are *low-current* devices. The Schottky diode is a relatively high current device, capable of switching rapidly while providing forward currents in the neighborhood of 50 A. In sinusoidal and *low-current* switching circuits, the Schottky diode is capable of operating at frequencies of 20 GHz and more.

PIN Diodes

22

The PIN diode is made up of *three* semiconductor materials. The construction of the PIN diode is illustrated in Figure 21.18. The center material is made up of *intrinsic* (pure) silicon. The *p*- and *n*-type materials are very heavily doped and therefore have very low resistances.

FIGURE 21.18

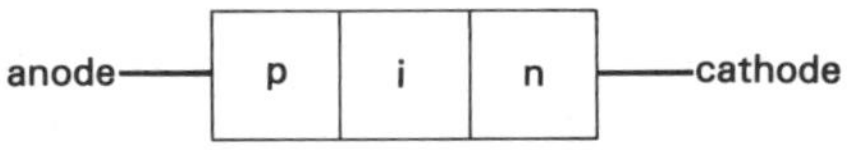

When reverse biased, the PIN diode acts as a capacitor.

When reverse biased, the PIN diode acts as a capacitor. The reason for this is illustrated in Figure 21.19. You may recall that an intrinsic semiconductor acts as an insulator. Thus, the intrinsic material in the PIN diode can be viewed as the dielectric of a capacitor.

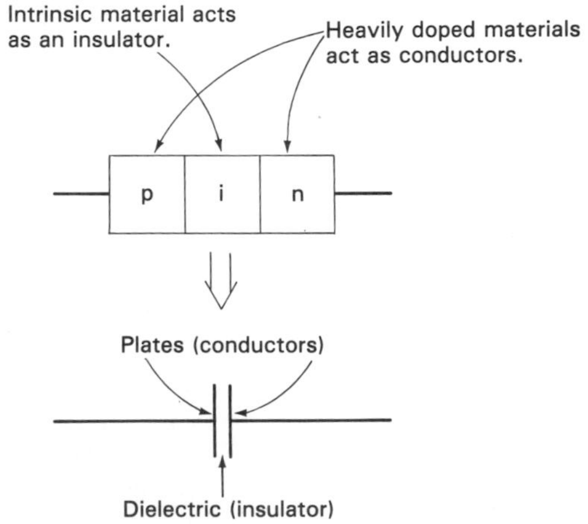

FIGURE 21.19

By comparison to the intrinsic material, the heavily doped *p*- and *n*-type materials can be viewed as being *conductors*. Therefore, we have a dielectric (the intrinsic material) sandwiched between two conductors (the *p*- and *n*-type materials). This forms the PIN diode capacitor.

The capacitance of a reverse-biased PIN diode will be relatively constant over a wide range of V_R values. Figure 21.20 shows the *capacitance versus V_R* curve for the Motorola MMBV3700 PIN diode. Note that the capacitance of the device remains at approximately 0.65 pF for values of V_R between -10 and -50 V. At values of V_R between 0 and -10 V, the MMBV3700 has the same capacitance characteristics as the varactor. This is typical for PIN diodes.

The capacitance of a reverse-biased PIN diode remains relatively constant over a wide range of reverse voltages.

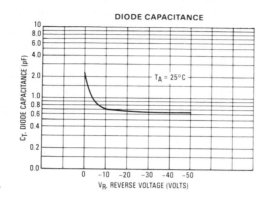

FIGURE 21.20

When forward biased, the intrinsic material is forced into conduction. As the number of free carriers in the intrinsic material increases, the resistance of the material decreases. Thus, when forward biased, the PIN diode acts as a *current-controlled resistance*. This point is illustrated in Figure 21.21. Note that the *series resistance* (diode resistance) *decreases* as forward current increases. This is due to the increase in the number of free carriers that are in the intrinsic material.

When forward biased, the PIN diode acts as a *current-controlled* resistor.

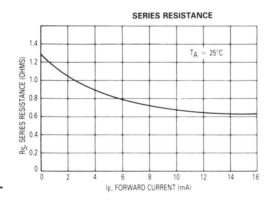

FIGURE 21.21

Figure 21.22 shows the forward operating curve for the PIN diode. As you can see, the current versus voltage curve gradually increases, starting at the

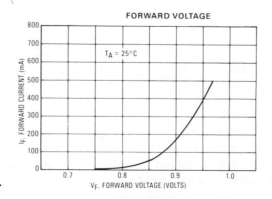

FIGURE 21.22

$V_F = 0.75$ V point. If you compare this forward operating curve to the *pn* junction diode curve, you'll see two major differences:

1. The *pn*-junction diode curve shows conduction to start at nearly 0 V, while the PIN diode *starts* conducting (in this case) at 0.75 V (750 mV).
2. The *pn*-junction diode has a definite turning point in the curve (called the *knee voltage*), while the PIN diode shows no definite knee voltage.

These two differences are caused by the construction of the PIN diode. For the PIN diode to conduct, V_F must overcome the resistance of the insulating intrinsic material. For the MMBV3700, V_F must be at least 750 mV before the intrinsic material will allow conduction.

The lack of a knee voltage, or turning point in the curve, is due to the fact that the PIN diode does not have a *pn* junction. Without a *pn* junction, the device does not have any sudden *turn-on* point. Rather, conduction increases at a gradual rate.

The PIN diode is used primarily in microwave communications. Its low reverse capacitance and current-controlled resistance characteristics make it useful in microwave communications circuits, most of which operate in the 1-GHz to infrared frequency range. Most commonly, they are used in *modulators*. Modulators are used to combine two signals of different frequencies into a single signal.

Step-recovery Diodes

The step-recovery diode is an ultrafast diode. Like the PIN diode, the step-recovery diode's characteristics are due to the unusual method of doping used. In the case of the step-recovery diode, the *p* and *n*-type materials are doped much more heavily at the ends of the component than they are at the junction. This point is illustrated by the graph in Figure 21.23, which shows the doping level increasing as the distance from the junction increases.

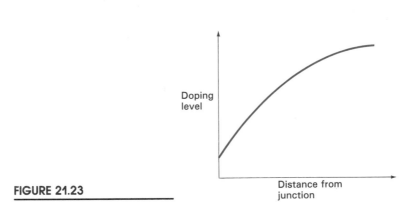

FIGURE 21.23

The unusual doping of the step-recovery diode affects the time required for the device to switch from *off* to *on*, and vice versa. The typical switching time for a step-recovery diode is in the low-picosecond range. This makes them ideal for ultrahigh-frequency switching applications.

One Final Note

There are many more types of diodes, too many to adequately cover in one chapter. However, the diodes covered in this chapter have been selected to give you a basis for understanding some of the other diode types.

PIN diode operation versus *pn* junction diode operation.

PIN diodes are used primarily in microwave communications.

Modulator. A circuit that combines two signals of different frequencies into a single signal.

 24

Step-recovery diode. A heavily doped diode with an ultrafast switching time.

CHAP 21 Special Applications Diodes

1. What are the commonly used names for the Schottky diode? [Objective 21]
2. Why can the Schottky diode be operated at much higher frequencies than the typical *pn*-junction diode? [Objective 21]
3. How does the construction of the Schottky diode differ from that of the *pn* junction diode? [Objective 21]
4. Describe the construction of the PIN diode. [Objective 22]
5. Explain the reverse characteristics of the PIN diode. [Objective 22]
6. Explain the forward characteristics of the PIN diode. [Objective 22]
7. Why is the forward operating curve of the PIN diode so different from that of a *pn*-junction diode? [Objective 23]
8. Describe the doping of the step-recovery diode. [Objective 24]
9. What is the result of the step-recovery diode doping scheme? [Objective 24]

KEY TERMS

The following terms were introduced and defined in this chapter:

back-to-back suppressor
capacitance ratio
clamping voltage
coarse tuning
constant-current diode
current-controlled
  resistance
correct regulator diode
diode capacitance
temperature coefficient
epicap
fine tuning
flywheel effect
hot-carrier diode

maximum limiting voltage
maximum reverse stand-
  off voltage
maximum reverse surge
  current
maximum temperature
  coefficient
modulator
negative resistance
peak operating voltage
peak power dissipation
  rating
PIN diode

regulator current
Schottky barrier diode
Schottky diode
series current regulator
shunt current regulator
step-recovery diode
surface-carrier diode
transient suppressor
tunnel diode oscillator
valley current
valley voltage
varactor
varicap

PRACTICE PROBLEMS

§21.1

1. A varactor has ratings of $C_t = 50$ pF and $TC_C = 500$ ppm/°C. Determine the change in capacitance for each 1°C rise in temperature. [3]
2. A varactor has ratings of $C_t = 48$ pF at 25°C and $TC_C = 800$ ppm/°C. Determine the change in capacitance for each 1°C rise in temperature. [3]
3. The varactor in Figure 21.24 has the following values: $C_t = 91$ pF at $V_R = 3$ V_{dc}, and $C_R = 4.8$ for $V_R = 3$ to 12 V_{dc}. Determine the resonant frequency for the circuit at $V_R = 3$ V_{dc} and $V_R = 12$ V_{dc}. [5]

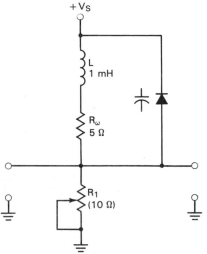

FIGURE 21.24

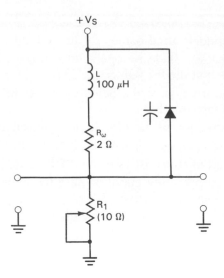

FIGURE 21.25

4. The varactor in Figure 21.25 has the following values: $C_t = 68$ pF at $V_R = 4$ V_{dc}, and $C_R = 1.12$ for $V_R = 4$ to 10 V_{dc}. Determine the resonant frequency for the circuit at $V_R = V_{dc}$ and $V_R = 10$ V_{dc}. [5]

§21.2

5. Figure 21.26 shows the surge power curves for the Motorola MPZ-16 series transient suppressors. How much power can these diodes handle if the surge duration is 8 ms and $T_C = 35°C$? [12]

6. Refer to Figure 21.26. Determine the maximum power dissipation for the MPZ-16 series diodes when the surge duration is 15 ms and $T_C = 35°C$. [12]

FIGURE 21.26

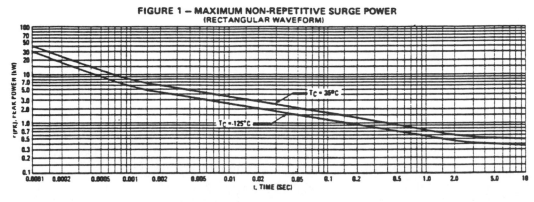

THE BRAIN DRAIN

7. The varactor in Figure 21.27a has the specifications and characteristics curve shown in Figure 21.27b. What value of V_R is required to set the resonant frequency of the tank circuit to 100 kHz?

FIGURE 21.27

(Courtesy of Motorola, Inc.)

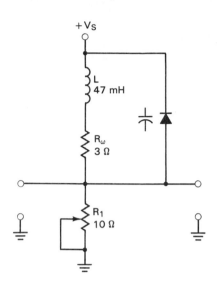

BB 139
VHF/FM VARACTOR DIODE
DIFFUSED SILICON PLANAR

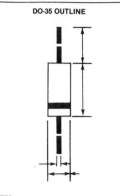

DO-35 OUTLINE

* C_3/C_{25} 5.0–6.5
* **MATCHED SETS** (Note 2)

ABSOLUTE MAXIMUM RATINGS (Note 1)

Temperatures
Storage Temperature Range   −55°C to +150°C
Maximum Junction Operating Temperature   +150°C
Lead Temperature   +260°C

Maximum Voltage
WIV   Working Inverse Voltage   30 V

NOTES:
Copper clad steel leads, tin plated
Gold plated leads available
Hermetically sealed glass package
Package weight is 0.14 gram

ELECTRICAL CHARACTERISTICS (25°C Ambient Temperature unless otherwise noted)

SYMBOL	CHARACTERISTIC	MIN	TYP	MAX	UNITS	TEST CONDITIONS
BV	Breakdown Voltage	30			V	$I_R = 100\ \mu A$
I_R	Reverse Current		10	50	nA	$V_R = 28$ V
			0.1	0.5	μA	$V_R = 28$ V, $T_A = 60°c$
C	Capacitance		29		pF	$V_R = 3.0$ V. 1 = 1 MHz
		4.3	5.1	6.0	pF	$V_R = 25$ V, 1 = 1 MHz
C_3/C_{25}	Capacitance Ratio	5.0	5.7	6.5		$V_R = 3$ V / 25 V, 1 = 1 MHz
Q	Figure of Merit		150			$V_R = 3.0$ V, 1 = 100 MHz
R_S	Series Resistance		0.35		Ω	C510 pF, f = 600 MHz
L_S	Series Inductance		2.5		nH	1.5 mm from case
f_o	Series Resonant Frequency		1.4		GHz	$V_R = 25$ V

NOTES
1 These ratings are limiting values above which the serviceability of the diode may be impaired.
2 The capacitance difference between any two diodes in one set is less than 3% over the reverse voltage range of 0.6 V to 28 V.

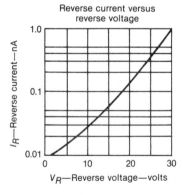

Reverse current versus
reverse voltage

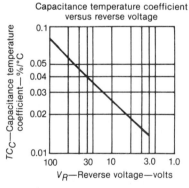

Capacitance temperature coefficient
versus reverse voltage

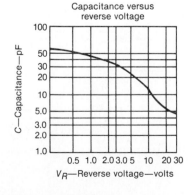

Capacitance versus
reverse voltage

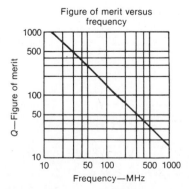

Figure of merit versus
frequency

FIGURE 21.28

(Courtesy of Motorola, Inc.)

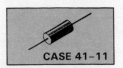

CASE 41–11

PEAK POWER DISSIPATION @ 1.0 ms = 1500 WATTS

Breakdown Voltage		Device Type		I$_{RSM}$ Maximum Reverse Surge Current Amp	V$_{RSM}$ Maximum Reverse Voltage @ I$_{RSM}$ Volts	Case
V(BR) Volts Nom	@I$_T$ mA					
6.0	1.0	1N5908		120	8.5	41–11
6.8	10	1N6267	1.5KE6.8	139	10.8	
7.5	10	1N6268	1.5KE7.5	128	11.7	
8.2	10	1N6269	1.5KE8.2	120	12.5	
9.1	1.0	1N6270	1.5KE9.1	109	13.8	
10	1.0	1N6271	1.5KE10	100	15.0	
11	1.0	1N6272	1.5KE11	93	16.2	
12	1.0	1N6273	1.5KE12	87	17.3	
13	1.0	1N6274	1.5KE13	79	19.0	
15	1.0	1N6275	1.5KE15	68	22.0	
16	1.0	1N6276	1.5KE16	64	23.5	
18	1.0	1N6277	1.5KE18	56.5	26.5	
20	1.0	1N6278	1.5KE20	51.5	29.1	
22	1.0	1N6279	1.5KE22	47.0	31.9	
24	1.0	1N6280	1.5KE24	43.0	34.7	
27	1.0	1N6281	1.5KE27	38.5	39.1	
30	1.0	1N6282	1.5KE30	34.5	43.5	
33	1.0	1N6283	1.5KE33	31.5	47.7	
36	1.0	1N6284	1.5KE36	29.0	52	
39	1.0	1N6285	1.5KE39	26.5	56.4	
43	1.0	1N6286	1.5KE43	24	61.9	
47	1.0	1N6287	1.5KE47	22.2	67.8	
51	1.0	1N6288	1.5KE51	20.4	73.5	
56	1.0	1N6289	1.5KE56	18.6	80.5	
62	1.0	1N6290	1.5KE62	16.9	89	
68	1.0	1N6291	1.5KE68	15.3	98	
75	1.0	1N6292	1.5KE75	13.9	108	
82	1.0	1N6293	1.5KE82	12.7	118	
91	1.0	1N6294	1.5KE91	11.4	131	
100	1.0	1N6295	1.5KE100	10.4	144	
110	1.0	1N6296	1.5KE110	9.5	158	
120	1.0	1N6297	1.5KE120	8.7	173	
130	1.0	1N6298	1.5KE130	8.0	187	
150	1.0	1N6299	1.5KE150	7.0	215	
160	1.0	1N6300	1.5KE160	6.5	230	
170	1.0	1N6301	1.5KE170	6.2	244	
180	1.0	1N6302	1.5KE180	5.8	258	
200	1.0	1N6303	1.5KE200	5.2	287	
220	1.0		1.5KE220	4.3	344	
250	1.0		1.5KE250	5.0	360	

Breakdown Voltage for Standard is ± 10% tolerance; ± 5% version is available by adding "A",
i.e., 1N6267A. 1.5KE6.8A. Clipper (back to back) versions are available by ordering the 1.5KE
series with a "C" or "CA" suffix, i.e., 1.5KE6.8C or 1.5KE6.8CA.

8. The 1N6303 transient suppressor has the characteristics shown in the selector guide in Figure 21.28. Determine the dynamic zener impedance (Z_Z) of this suppressor. (Hint: Review the method used to determine Z_Z that was shown in Chapter 2.)

9. The circuit shown in Figure 21.29 cannot tolerate an input voltage surge that is greater than 30% above the rated peak value of V_{in}. Which of the suppressors listed in Figure 21.31 would be best suited for use in this circuit?

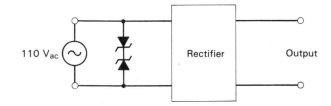

FIGURE 21.29

ANSWERS TO EXAMPLE PRACTICE PROBLEMS

21–1. 388 kHz at $V_r = 4$ V$_{dc}$, 289 kHz at $V_r = 10$ V$_{dc}$

21–3. $P_{Pk} = 225$ W at $T = 150°C$

APPENDIX A

Device Listing

The following device specification sheets and/or operating curves are located on the pages indicated:

Part Number	Type of Device	Code*	Page
BB139	Varactor		981
BFW43	High-voltage pnp transistor	MR	207
H21–22	Optical switches		928
LM317	IC Voltage regulator		948
MDA2500	Intergrated bridge rectifier		102
MLL755	Zener diode		74
MMBV3700	PIN diode	OC	977
MUR 5005	Power rectifier		101
MRD500	Photodiode		918
MV209	Varactor diode		959
μA741	Operational amplifier	OC	538
μA741	Operational amplifier	EC	540
1N746–59	Zener diodes		57
1N4001–7	Rectifier diodes		48
1N5283–314	Current-regulator diodes	OC	969
1N5391–9	Rectifier diodes		71
1N5908–6389	Transient suppressors		966
2N682–92	Silicon-controlled recifiers		879
2N2222	Npn transistors	OC	311
2N3251	Npn transistors	OC	313
2N3902	High-power npn transistor		209
2N3903–4	Npn transistor		203, 241
2N4237	High-current pnp transistor	MR	208
2N4400	Npn transistor	EC	298
2N4400	Npn transistor	OC	299
2N5431	Unijunction transistor		902
2N5484–6	N-channel JFET		469
2N6157–65	Triac		985
3N169–71	E-MOSFET		494
4N35–7	Optoisolator		925
–	Example rectifier diode selector guide		50
–	Example zener diode selector guide		60

*MR = Maximum ratings only
 EC = Electrical characteristics listing only
 OC = Operating curves only

Standard Resistor Device Listing

(See page 989.)

MAXIMUM RATINGS

Rating	Symbol	Value	Unit
Collector-Emitter Voltage	V_{CEO}	40	Vdc
Collector-Base Voltage	V_{CBO}	40	Vdc
Emitter-Base Voltage	V_{EBO}	5.0	Vdc
Collector Current — Continuous	I_C	200	mAdc
Total Device Dissipation @ $T_A = 25°C$ Derate above 25°C	P_D	625 5.0	mW mW/°C
Total Power Dissipation @ $T_A = 60°C$	PD	250	mW
Total Device Dissipation @ $T_C = 25°C$ Derate above 25°C	P_D	1.5 12	Watts mW/°C
Operating and Storage Junction Temperature Range	T_J, T_{stg}	−550 to +150	°C

THERMAL CHARACTERISTICS

Characteristic	Symbol	Max	Unit
Thermal Resistance, Junction to Case	$R_{\theta JC}$	83.3	°C/W
Thermal Resistance, Junction to Ambient	$R_{\theta JA}$	200	°C/W

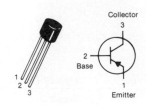

2N3905
2N3906

CASE 29-04, STYLE 1
TO-92 (TO-226AA)

GENERAL PURPOSE TRANSISTOR

NPN SILICON

ELECTRICAL CHARACTERISTICS ($T_A = 25°C$ unless otherwise noted.)

Characteristic		Symbol	Min	Max	Unit
OFF CHARACTERISTICS					
Collector-Emitter Breakdown Voltage(1) (I_C 1.0 mAdc, $I_B = 0$)		$V_{(BR)CEO}$	40	—	Vdc
Collector-Base Breakdown Voltage (I_C 10 µAdc, $I_E = 0$)		$V_{(BR)CBO}$	40	—	Vdc
Emitter-Base Breakdown Voltage ($I_E = 10$ µAdc, $I_C = 0$)		$V_{(BR)EBO}$	5.0	—	Vdc
Base Cutoff Current ($V_{CE} = 30$ Vdc, $V_{BE} = 3.0$ Vdc)		I_{BL}	—	50	nAdc
Collector Cutoff Current ($V_{CE} = 30$ Vdc, $V_{BE} = 3.0$ Vdc)		I_{CEX}	—	50	nAdc
ON CHARACTERISTICS(1)					
DC Current Gain ($I_C = 0.1$ mAdc, $V_{CE} = 1.0$ Vdc)	2N3905 2N3906	h_{FE}	30 60	— —	—
($I_C \doteq 1.0$ mAdc, $V_{CE} = 1.0$ Vdc)	2N3905 2N3906		40 80	— —	
($I_C = 10$ mAdc, $V_{CE} = 1.0$ Vdc)	2N3905 2N3906		50 100	150 300	
($I_C = 50$ mAdc, $V_{CE} = 1.0$ Vdc)	2N3905 2N3906		30 60	— —	
($I_C = 1]]$ mAdc, $V_{CE} = 1.0$ Vdc)	2N3905 2N3906		15 30	— —	
Collector-Emitter Saturation Voltage ($I_C = 10$ mAdc, $I_B = 1.0$ mAdc) ($I_C = 50$ mAdc, $I_B = 5.0$ mAdc)		$V_{CE(sat)}$	— —	0.25 0.4	Vdc
Base-Emitter Saturation Voltage ($I_C = 10$ mAdc, $I_B = 1.0$ mAdc) ($I_C = 50$ mAdc, $I_B = 5.0$ mAdc)		$V_{BE(sat)}$	0.65 —	0.85 0.95	Vdc
SMALL-SIGNAL CHARACTERISTICS					
Current-Gain — Bandwidth Product ($I_C = 10$ mAdc, $V_{CE} = 20$ Vdc, f = 100 MHz)	2N3905 2N3906	f_T	200 250	— —	MHz
Output Capacitance ($V_{CB} = 5.0$ Vdc, $I_E = 0$, f = 100 kHz)		C_{obo}	—	4.5	pF

ELECTRICAL CHARACTERISTICS (continued) (T_A = 25°C unless otherwise noted.)

Characteristic		Symbol	Min	Max	Unit
Input Capacitance (V_{CB} = 0.5 Vdc, I_C = 0, f = 100 kHz)		C_{ibo}	—	10.0	pF
Input Impedance (I_C = 1.0 mAdc, V_{CE} = 10 Vdc, f = 1.0 kHz)	2N3905 2N3906	h_{ie}	0.5 2.0	8.0 12	k ohms
Voltage Feedback Ratio (I_C = 1.0 mAdc, V_{CE} = 10 Vdc, f = 1.0 kHz)	2N3905 2N3906	h_{re}	0.1 0.1	5.0 10	X 10^{-4}
Small-Signal Current Gain (I_C = 1.0 mAdc, V_{CE} = 10 Vdc, f = 1.0 kHz)	2N3905 2N3906	h_{fe}	50 100	200 400	—
Output Admittance (I_C = 1.0 mAdc, V_{CE} = 10 Vdc, f = 1.0 kHz)	2N3905 2N3906	h_{oe}	1.0 3.0	40 60	μmhos
Noise Figure (I_C = 100 μAdc, V_{CE} = 5.0 Vdc, R_S = 1.0 k ohms, f = 10 Hz to 15.7 kHz)	2N3905 2N3906	NF	— —	5.0 4.0	dB

SWITCHING CHARACTERISTICS

Delay Time	(V_{CC} = 3.0 Vdc, V_{BE} = 0.5 Vdc,		t_d	—	35	ns
Rise Time	I_C = 10 mAdc, I_{B1} = 1.0 mAdc)		t_r	—	35	ns
Storage Time		2N3905 2N3906	t_s	— —	200 225	ns
Fall Time	(V_{CC} = 3.0 Vdc, I_C = 10 mAdc, I_{B1} = I_{B2} = 1.0 mAdc)	2N3905 2N3904	t_f	— —	60 75	ns

(1) Pulse Test: Pulse Width $\leq$ 300 μs, Duty Cycle $\leq$ 2.0%.

FIGURE 1 — DELAY AND RISE TIME
EQUIVALENT TEST CIRCUIT

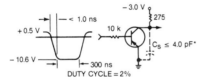

FIGURE 2 — STORAGE AND FALL TIME
EQUIVALENT TEST CIRCUIT

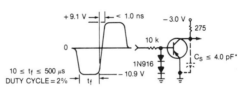

*Total shunt capacitance of test jig and connectors

2N4013
2N4014

CASE 22-03, STYLE 1
TO-18 (TO-206AA)

SWITCHING TRANSISTORS

NPN SILICON

MAXIMUM RATINGS

Rating	Symbol	2N4013	2N4014	Unit
Collector-Emitter Voltage	V_{CEO}	30	50	Vdc
Collector-Base Voltage	V_{CBO}	50	80	Vdc
Emitter-Base Voltage	V_{EBO}	6.0		Vdc
Collector Current — Continuous — Peak	I_C	1.0 2.0		Adc
Total Device Dissipation @ $T_A = 25°C$ Derate above 25°C	P_D	0.5 28.6		Watt mW/°C
Total Device Dissipation @ $T_C = 25°C$ Derate above 25°C	P_D	1.4 6.8		Watts mW/°C
Operating and Storage Junction Temperature Range	T_J, T_{stg}	-65 to $+200$		°C

ELECTRICAL CHARACTERISTICS ($T_A = 25°C$ unless otherwise noted.)

Characteristic		Symbol	Min	Typ	Max	Unit
OFF CHARACTERISTICS						
Collector-Emitter Breakdown Voltage(1) ($I_C = 10$ mAdc, $I_B = 0$)	2N4014 2N4013	$V_{(BR)CEO}$	50 30	— —	— —	Vdc
Collector-Emitter Breakdown Voltage ($I_C = 10$ μAdc, $V_{BE} = 0$)	2N4014 2N4013	$V_{(BR)CES}$	80 50	— —	— —	Vdc
Collector-Base Breakdown Voltage ($I_C = 10$ μAdc, $I_E = 0$)	2N4014 2N4013	$V_{(BR)CBO}$	80 50	— —	— —	Vdc
Emitter-Base Breakdown Voltage ($I_E = 10$ μAdc, $I_C = 0$)		$V_{(BR)EBO}$	6.0	—	—	Vdc
Collector Cutoff Current ($V_{CB} = 60$ Vdc, $I_E = 0$) ($V_{CB} = 40$ Vdc, $I_E = 0$) ($V_{CB} = 60$ Vdc, $I_E = 0$, $T_A = 100°C$) ($V_{CB} = 40$ Vdc, $I_E = 0$, $T_A = 100°C$)	2N4014 2N4013 2N4014 2N4013	I_{CBO}	— — — —	0.12 0.12 — —	1.7 1.7 120 120	μAdc
Collector Cutoff Current ($V_{CE} = 80$ Vdc, $V_{EB} = 0$) ($V_{CE} = 50$ Vdc, $V_{EB} = 0$)	2N4014 2N4013	I_{CES}	— —	0.15 0.15	10 10	μAdc
ON CHARACTERISTICS(1)						
DC Current Gain ($I_C = 10$ mAdc, $V_{CE} = 1.0$ Vdc) ($I_C = 100$ mAdc, $V_{CE} = 1.0$ Vdc) ($I_C = 100$ mAdc, $V_{CE} = 1.0$ Vdc, $T_A = -55°C$) ($I_C = 300$ mAdc, $V_{CE} = 1.0$ Vdc) ($I_C = 500$ mAdc, $V_{CE} = 1.0$ Vdc) ($I_C = 500$ mAdc, $V_{CE} = 1.0$ Vdc, $T_A = -55°C$) ($I_C = 800$ mAdc, $V_{CE} = 2.0$ Vdc) ($I_C = 1.0$ Adc, $V_{CE} = 5.0$ Vdc)	 2N4014 2N4013 2N4014 2N4013	h_{FE}	30 60 30 40 35 20 20 25 25 30	— — — — — — —	— 150 — — — — —	—

MAXIMUM RATINGS

Rating	Symbol	Value	Unit
Drain-Source Voltage	V_{DS}	25	Vdc
Drain-Gate Voltage	V_{DG}	25	Vdc
Reverse Gate-Source Voltage	V_{GSR}	−25	Vdc
Gate Current	I_G	10	mAdc
Total Device Dissipation @ T_A = 25°C Derate above 25°C	P_D	310 2.82	mW mW/°C
Junction Temperature Range	T_J	125	°C
Storage Channel Temperature Range	T_{stg}	−65 to +150	°C

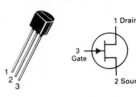

2N5457
thru
2N5459

CASE 29-04, STYLE 5
TO-92 (TO-226AA)

1 Drain
3 Gate
2 Source

JFETs
GENERAL PURPOSE

N-CHANNEL — DEPLETION

Refer to 2N4220 for graphs.

ELECTRICAL CHARACTERISTICS (T_A = 25°C unless otherwise noted.)

Characteristic		Symbol	Min	Typ	Max	Unit
OFF CHARACTERISTICS						
Gate-Source Breakdown Voltage (I_G = −10 μAdc, V_{DS} = 0)		$V_{(BR)GSS}$	−25	—	—	Vdc
Gate Reverse Current (V_{GS} = −15 Vdc, V_{DS} = 0) (V_{GS} = −15 Vdc, V_{DS} = 0, T_A = 100°C)		I_{GSS}	 — —	 — —	 −1.0 −200	nAdc
Gate Source Cutoff Voltage (V_{DS} = 15 Vdc, I_D = 10 nAdc)	2N5457 2N5458 2N5459	$V_{GS(off)}$	−0.5 −1.0 −2.0	— — —	−6.0 −7.0 −8.0	Vdc
Gate Source Voltage (V_{DS} = 15 Vdc, I_D = 100 μAdc) (V_{DS} = 15 Vdc, I_D = 200 μAdc) (V_{DS} = 15 Vdc, I_D = 400 μAdc)	2N5457 2N5458 2N5459	V_{GS}	— — —	−2.5 −3.5 −4.5	— — —	Vdc
ON CHARACTERISTICS						
Zero-Gate-Voltage Drain Current* (V_{DS} = 15 Vdc, V_{GS} = 0)	2N5457 2N5458 2N5459	I_{DSS}	1.0 2.0 4.0	3.0 6.0 9.0	5.0 9.0 16	mAdc
SMALL-SIGNAL CHARACTERISTICS						
Forward Transfer Admittance Common Source* (V_{DS} = 15 Vdc, V_{GS} = 0, f = 1.0 kHz)	2N5457 2N5458 2N5459	$|y_{fs}|$	1000 1500 2000	— — —	5000 5500 6000	μmhos
Output Admittance Common Source* (V_{DS} = 15 Vdc, V_{GS} = 0, f = 1.0 kHz)		$|y_{os}|$	—	10	50	μmhos
Input Capacitance (V_{DS} = 15 Vdc, V_{GS} = 0, f = 1.0 MHz)		C_{iss}	—	4.5	7.0	pF
Reverse Transfer Capacitance (V_{DS} = 15 Vdc, V_{GS} = 0, f = 1.0 MHz)		C_{rss}	—	1.5	3.0	pF

*Pulse Test: Pulse Width ≤ 630 ms; Duty Cycle ≤ 10%.

MAXIMUM RATINGS

Rating	Symbol	Value	Unit
Collector-Emitter Voltage	V_{CEO}	40	Vdc
Collector-Base Voltage	V_{CBO}	40	Vdc
Emitter-Base Voltage	V_{EBO}	12	Vdc
Collector Current — Continuous	I_C	500	mAdc
Total Device Dissipation @ $T_A = 25°C$ Derate above 25°C	P_D	625 5.0	mW mW/°C
Total Device Dissipation @ $T_C = 25°C$ Derate above 25°C	P_D	1.5 12	Watts mW/°C
Operating and Storage Junction Temperature Range	T_J, T_{stg}	−55 to +150	°C

THERMAL CHARACTERISTICS

Characteristic	Symbol	Max	Unit
Thermal Resistance, Junction to Case	$R_{\theta JC}$	83.3	°C/W
Thermal Resistance, Junction to Ambient	$R_{\theta JA}$(1)	200	°C/W

(1) $R_{\theta JA}$ is measured with the device soldered into a typical printed circuit board.

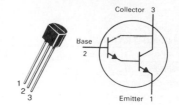

2N6426
2N6427

CASE 29-04, STYLE 1
TO-92 (TO-226AA)

DARLINGTON TRANSISTORS

NPN SILICON

ELECTRICAL CHARACTERISTICS ($T_A = 25°C$ unless otherwise noted.)

Characteristic	Symbol	Min	Typ	Max	Unit
OFF CHARACTERISTICS					
Collector-Emitter Breakdown Voltage(2) ($I_C = 10$ mAdc, $V_{BE} = 0$)	$V_{(BR)CES}$	40	—	—	Vdc
Collector-Base Breakdown Voltage ($I_C = 100$ μAdc, $I_E = 0$)	$V_{(BR)CBO}$	40	—	—	Vdc
Emitter-Base Breakdown Voltage ($I_E = 10$ μAdc, $I_C = 0$)	$V_{(BR)EBO}$	12	—	—	Vdc
Collector Cutoff Current ($V_{CE} = 25$ Vdc, $I_B = 0$)	I_{CEO}	—	—	1.0	μAdc
Collector Cutoff Current ($V_{CB} = 30$ Vdc, $I_E = 0$)	I_{CBO}	—	—	50	nAdc
Emitter Cutoff Current ($V_{BE} = 10$ Vdc, $I_C = 0$)	I_{EBO}	—	—	50	nAdc
ON CHARACTERISTICS					
DC Current Gain(2) ($I_C = 10$ mAdc, $V_{CE} = 5.0$ Vdc)　　　　　2N6426 2N6427	h_{FE}	20,000 10,000	— —	200,000 100,000	—
($I_C = 100$ mAdc, $V_{CE} = 5.0$ Vdc)　　　　2N6426 2N6427		30,000 20,000	— —	300,000 200,000	
($I_C = 500$ mAdc, $V_{CE} = 5.0$ Vdc)　　　　2N6426 2N6427		20,000 14,000	— —	200,000 140,000	
Collector-Emitter Saturation Voltage ($I_C = 50$ mAdc, $I_B = 0.5$ mAdc) ($I_C = 500$ mAdc, $I_B = 0.5$ mAdc)	$V_{CE(sat)}$	— —	0.71 0.9	1.2 1.5	Vdc
Base-Emitter Saturation Voltage ($I_C = 500$ mAdc, $I_B = 0.5$ mAdc)	$V_{BE(sat)}$	—	1.52	2.0	Vdc
Base-Emitter On Voltage ($I_C = 50$ mAdc, $V_{CE} = 5.0$ Vdc)	$V_{BE(on)}$	—	1.24	1.75	Vdc
SMALL-SIGNAL CHARACTERISTICS					
Output Capacitance ($V_{CB} = 10$ Vdc, $I_E = 0$, $f = 1.0$ MHz)	C_{obo}	—	5.4	7.0	pF
Input Capacitance ($V_{BE} = 1.0$ Vdc, $I_C = 0$, $f = 1.0$ MHz)	C_{ibo}	—	10	15	pF

ELECTRICAL CHARACTERISTICS (continued) (T_A = 25°C unless otherwise noted.)

Characteristic		Symbol	Min	Typ	Max	Unit
Input Impedance (I_C = 10 mAdc, V_{CE} = 5.0 Vdc, f = 1.0 kHz) 2N6426 2N6427		h_{ie}	100 50	— —	2000 1000	k Ω
Small-Signal Current Gain (I_C = 10 mAdc, V_{CE} = 5.0 Vdc, f = 1.0 kHz) 2N6426 2N6427		h_{fe}	20,000 10,000	— —	— —	—
Current Gain — High Frequency (I_C = 10 mAdc, V_{CE} = 5.0 Vdc, f = 100 MHz) 2N6426 2N6427		$|h_{fe}|$	1.5 1.3	2.4 2.4	— —	—
Output Admittance (I_C = 10 mAdc, V_{CE} = 5.0 Vdc, f = 1.0 kHz)		h_{oe}	—	—	1000	μmhos
Noise Figure (I_C = 1.0 mAdc, V_{CE} = 5.0 Vdc, R_S = 100 kΩ, f = 1.0 kHz)		NF	—	3.0	10	dB

(2) Pulse Test: Pulse Width ≤ 300 μs, Duty Cycle ≤ 2.0%.

Standard Resistor Values

10% Tolerance

Ω				kΩ			MΩ	
0.10	1.0	10	100	1.0	10	100	1.0	10.0
0.12	1.2	12	120	1.2	12	120	1.2	12.0
0.15	1.5	15	150	1.5	15	150	1.2	15.0
0.18	1.8	18	180	1.8	18	180	1.8	18.0
0.22	2.2	22	220	2.2	22	220	2.2	22.0
0.27	2.7	27	270	2.7	27	270	2.7	
0.33	3.3	33	330	3.3	33	330	3.3	
0.39	3.9	39	390	3.9	39	390	3.9	
0.47	4.7	47	470	4.7	47	470	4.7	
0.56	5.6	56	560	5.6	56	560	5.6	
0.68	6.8	68	680	6.8	68	680	6.8	
0.82	8.2	82	820	8.2	82	820	8.2	

In addition to the above values, multiples of the following values are available with 5% tolerance only:

1.1	2.4	5.1
1.3	3.0	6.2
1.6	3.6	7.5
2.0	4.3	9.1

APPENDIX B
Approximating Circuit Values

The analysis of a given circuit can often be simplified by approximating many of the resistance and current values in the circuit. In this appendix, we're going to look at the methods by which you can approximate circuit values and the circumstances that make circuit approximations valid.

What Is Meant By *Approximating Circuit Values*?

When you approximate circuit values, you ignore any resistance and/or current values that will have little impact on the analysis of a given circuit. For example, consider the simple series circuit shown in Figure B.1a. What is total resistance in this circuit? Obviously, it is $10,000,010\Omega$. However, *for all practical purposes*, couldn't we just say that we have *approximately* 10 MΩ? Of course we can. In this circuit, the value of R_1 will have very little, if any, visible effect on the values of R_T, I_T, or V_{R2}. In other words, these three values would be very nearly equal whether calculated using the value of R_1 or assuming that R_1 is not in the circuit. Therefore, the value of R_1 can be dropped from the circuit with very little loss in the accuracy of our calculations.

Now, take a look at the circuit shown in Figure B.1b. In terms of circuit *current*, which component could be dropped from the circuit? In this case, it would be valid to ignore the value of R_2, since very little current will flow through this branch. In fact, if we assume the voltage source to be 10 V, we could use Ohm's law to calculate the current values of $I_1 = 1$ A and $I_2 = 1$ μA. In this case, the value of I_2 has virtually no effect on the value of total circuit current, and thus, can be ignored in circuit calculations.

Later, we'll establish the guidelines for approximating circuit values. At this point, we'll take a look at *when* it is valid to use circuit approximations.

When Can You Use Circuit Approximations?

Any time that you are troubleshooting a given circuit, you can assume that it is allright to use approximated values. Generally, when you are troubleshooting, you are only interested in whether or not a voltage or resistance is *close to* its rated value. For example, consider the circuit shown in Figure B.2. Assume that you are troubleshooting this circuit. As the Figure shows, V_B should be a 1 mV$_{PP}$ sine wave, V_E should be a 1 V$_{PP}$ sine wave, V_C should be an 8 V$_{PP}$ sine wave that is $180°$ out of phase with the other two voltages. When checking this circuit with an oscilloscope, you would not be concerned with whether or not a given value of V_{PP} was off by 0.001V. You would only be concerned with whether or not the value of V_{PP} was *close* to the given value, and whether or not the phase relationships were correct. Therefore, any approximations of circuit V_{PP} values would be completely acceptable.

There are two circumstances when you would not use the circuit approximations covered here. The first is when you are performing an exact circuit analysis involving *h*-parameters. *H*-parameter equations, such as those introduced in Appendix D, lead to very exact values for transistor gain and input/output impedence. It wouldn't make much sense to use these exact equations and then approximate all of the values external to the transistor.

The other time you wouldn't use circuit approximations is when you are designing a circuit for a specific application.

As a summary, the circuit approximation techniques that we are about to discuss can usually be used for general circuit analysis and troubleshooting. They may not be used when an exact analysis is required (such as those involving *h*-parameters) or when designing a circuit.

Approximating Circuit Resistance Values

Generally, you can ignore the value of a resistor in a series circuit *when the value of that component is less than 10% of the value of the next smallest resistance value*. For example, take a look at the voltage-dividers shown in Figure B.3.

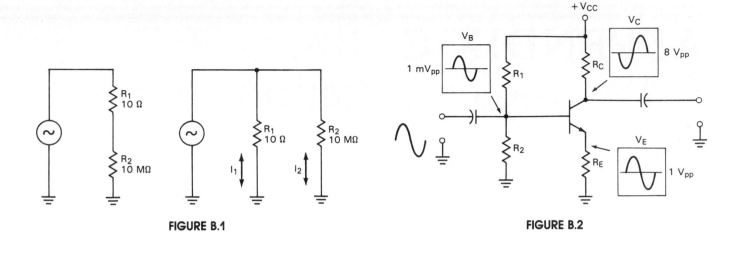

FIGURE B.1

FIGURE B.2

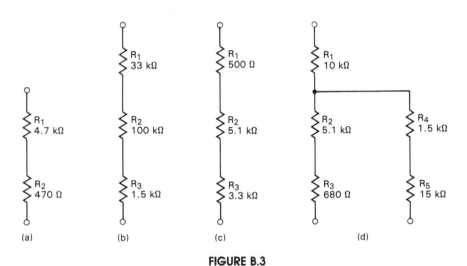

FIGURE B.3

In Figure B.3a, the value of R_2 is equal to $0.1R_1$. Therefore, we can approximate the total resistance is the circuit as being equal to the value of R_1, 4.7 kΩ. In Figure B.3B, the value of R_3 can be ignored because it is less than $0.1R_1$, and R_1 is the next lowest resistance value in the circuit. Note that the value of R_2 is not considered because it is greater than the value of R_1. In Figure B.3c, none of the resistors can be dropped from the circuit. Why? Even though R_1 is less than $0.1R_2$, it is not less than $0.1R_3$. Since it is not less than (or equal to) 10% of the *next lowest individual resistance value*, it must be kept in the circuit.

Which resistor (if any) can be dropped from Figure B.3d? If you picked R_4, you were correct. The value of R_5 is 10% of the value of R_t, so it can be dropped from the circuit. Why doesn't the value of R_1 (10 kΩ) prevent us from dropping R_4? The circuit in Figure B.3d is a *series-parallel* circuit. While R_1 is in series with the *combination* of R_2, R_3, R_4, and R_5, it is not considered to be in series with R_4 directly. This is due to the fact that the current through R_1 is not necessarily equal to the current through R_4. By definition, two components are in series only when the current through the two components are equal.

For parallel circuits, the 10% rule still holds. In this case, however, *the larger resistance values is the one that can be dropped*. The reason for this can be seen by referring back to Figure B.1b. For the circuit shown, the larger resistor provided less than 10% of the total circuit current. Thus, its value could easily be ignored.

APPENDIX C
H-Parameter Equations & Derivations

In Chapter 7, you were introduced to the four transistor *h*-parameters and their use in basic circuit analysis. In this appendix, we'll take a more in-depth look at *h*-parameter equations and the transistor hybrid equivalent circuit.

The Transistor Hybrid Equivalent Circuit

Using the *h*-parameter values described in Chapter 7, the *hybrid equivalent circuit* of a transistor is developed. To understand this AC equivalent circuit, the transistor must be viewed as a 4-terminal device. This representation of the transistor is shown in Figure C.1. Since the emitter terminal of the transistor is common to both the input and output circuits in a common emitter amplifier, it is represented as two separate terminals. Realize however, that both of these terminals represent the single transistor emitter. The input and output voltages are represented as v_1 and v_2 respectively.

Input Circuit

The AC equivalent of the input circuit is derived using Thevenin's theorem. Thevenin's theorem states that any circuit can be represented as a single voltage source in series with a single resistance, as is reviewed in Appendix B. The hybrid equivalent circuit for the transistor input is derived as illustrated in Figure C.2.

The thevenin voltage for the input is equal to the value of v_1 with the base circuit open. As Figure C.2 shows, this voltage is found as

$$v_1 \cong h_{re}v_2 \tag{C.1}$$

This equation is simply another form of equation (7.10). Since $v_{be} = v_1$, and $v_{ce} = v_2$, equation (7.10) can be rewritten as above.

The Thèvenin resistance of the transistor input is equal to h_{ie}. This value is found as shown in Chapter 7 (page 222).

So, why is the input represented as a Thevenin equivalent circuit? Since the voltage across the base-emitter junction remains fairly constant, it is convenient to define the operation of the input circuit in terms of this voltage. Any changes in i_b or v_b do not change the voltage drop across the emitter-base junction inside of the transistor. Therefore, it is a good reference for discussing the AC operation of the input circuit.

Output Circuit

The hybrid equivalent of the output circuit is derived using Norton's theorem. Norton's theorem represents a circuit as a current source in parallel with a single resistance. The AC equivalent of the output is represented in Figure C.3.

The norton current source is represented as the current gain of the transistor (h_{fe}) times the base input current (i_l). For formula,

$$i_2 = h_{fe}i_1 \tag{C.2}$$

This equation is simply another form of equation (7.8). Since $i_b = i_1$, and $i_c = i_2$, equation (7.8) can be rewritten as above.

The norton resistance is equal to the collector voltage (v_2) divided by the collector current (i_1). By formula,

$$r = \frac{v_c}{i_c} = \frac{v_2}{i_2}$$

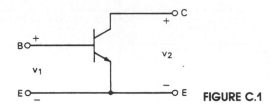

FIGURE C.1

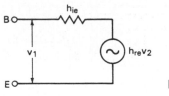

FIGURE C.2

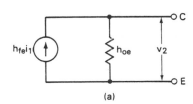

(a)

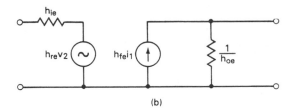

(b)

FIGURE C.3

Since h_{oe} is the reciprocal of the above relationship, the norton equivalent resistance is found as

$$\frac{1}{h_{oe}} = \frac{v_c}{i_c} = \frac{v_2}{i_2} \qquad \text{(C.3)}$$

The output circuit is represented as a norton equivalent circuit because of the fact that i_c is dependent upon factors outside of the collector-emitter circuit under normal circumstances. For example, changing the value of R_C will not cause i_c to change significantly. However, changing i_b, *will*. Since the transistor output current is relatively independent of the external component values, it is the reference value for output calculations.

The complete hybrid equivalent circuit for the transistor is shown in Figure C.3b. This circuit will be used throughout our discussions on *h*-parameter calculations.

Calculations Involving *H*-Parameters

Very exact values of current gain (A_i), voltage gain (A_v), transistor input impedance (Z_{in}), and transistor output impedance (Z_{out}) can be obtained using *h*-parameters. *Note that these calculations give you exact values of operation at specific values of* I_c. If I_C changes, the values of A_v, A_i, Z_{in}, and Z_{out} obtained with the *h*-parameters will all change. For this reason, *h*-parameters are not usually used for everyday circuit calculations.

Current Gain (A_i)

The current gain of a transistor is defined as the ratio of output current to input current. By formula,

$$A_i = \frac{i_2}{i_1} \qquad \text{(C.4)}$$

Now, let's apply this equation to the circuit shown in Figure C.4. This circuit is an amplifier AC equivalent circuit. The only difference between this circuit and ones discussed earlier is that the transistor has been replaced by its hybrid equivalent circuit. The boundaries of the transistor are represented by the dotted line.

The output current (i_2) is equal to the sum of $h_{fe}i_1$ and the current produced by the application of v_2 across the norton resistance. By formula,

$$i_2 = h_{fe}i_1 + v_2 h_{oe} \qquad \text{(C.5)}$$

By substituting this value in place of i_2 in equation (7.14), we get

$$A_i = \frac{h_{fe}i_1 + v_2 h_{oe}}{i_1}$$

Which can be rewritten as

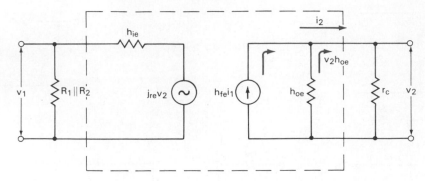

FIGURE C.4

$$A_i = \frac{h_{fe}i_1}{i_1} + \frac{v_2 h_{oe}}{i_1}$$

or

$$A_i = h_{fe} + \frac{v_2 h_{oe}}{i_1}$$

Since $v_2 = i_2 r_C$,

$$A_i = h_{fe} - \frac{i_2 r_C h_{oe}}{i_1}$$

which can be rewritten as

$$A_i = h_{fe} - \left(\frac{i_2}{i_1}\right) r_C h_{oe}$$

This is now rewritten as

$$A_i = h_{fe} - A_i r_C h_{oe}$$

With further algebraic manipulation, this becomes

$$A_i = \frac{h_{fe}}{1 + r_C h_{oe}} \qquad \textbf{(C.6)}$$

Equation (C.6) is our final goal; a formula which defines A_i strictly in terms of the transistor h-parameters and the AC resistance of the collector circuit. The following example shows how equation (C.6) is used.

EXAMPLE C.1

Determine the value of A_i for the circuit shown in Figure C.5. Assume that the h-parameter values for the transistor are as follows:

$$h_{ie} = 1 \text{ k}\Omega$$
$$h_{re} = 2.5 \times 10^{-4}$$
$$h_{fe} = 50$$
$$h_{oe} = 25 \text{ }\mu\text{S}$$

Solution: To determine the value of A_i, we need to determine r_C. For the circuit shown.

$$r_C = R_C \| R_L$$
$$= 3.2 \text{ k}\Omega$$

Now, A_1 is determined as

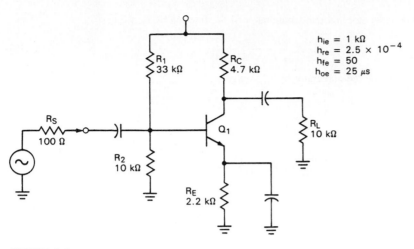

FIGURE C.5

$$h_{ie} = 1 \text{ k}\Omega$$
$$h_{re} = 2.5 \times 10^{-4}$$
$$h_{fe} = 50$$
$$h_{oe} = 25 \text{ μs}$$

$$A_i = \frac{h_{fe}}{1 + r_C h_{oe}}$$

$$= \frac{50}{1 + (3.2 \text{ k}\Omega)(25 \text{ μS})}$$

$$= 46.3$$

If you take a look at the final result in Example C.1, you'll notice that it was very close to the original value of h_{fe}. The reason for this is the fact that $(r_c h_{oe}) << 1$. Since this is normally the case, equation (C.6) can be approximated to

$$A_i \cong h_{fe} \tag{C.7}$$

This is the value we have been using for A_i in all of our circuit calculations.

Voltage Gain (A_v)

The voltage gain of an amplifier is defined as the ratio of output voltage to input voltage. By formula,

$$A_v = \frac{v_2}{v_1} \tag{C.8}$$

Now, take a look at the circuit shown in Figure C.6. As is shown, v_1 is the sum of $h_{re}v_2$ and the voltage developed across h_{ie}. The voltage across h_{ie}. The voltage across h_{ie} is found as $h_{ie}i_1$, therefore,

$$v_1 = h_{re}v_2 + h_{ie}i_1 \tag{C.9}$$

It was already shown that $v_2 = i_2 r_C$, and, substituting these two equations into equation (C.8) we get

$$A_v = \frac{v_2}{v_1} = \frac{i_2 r_C}{h_{re}v_2 + h_{ie}i_1}$$

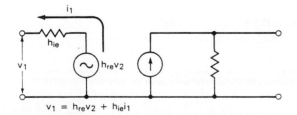

$$v_1 = h_{re}v_2 + h_{ie}i_1$$

FIGURE C.6

If we divide both the numerator and the denominator of the fraction by i_2, we get

$$A_v = \frac{r_C}{h_{re}r_C + h_{ie}/A_i} \qquad \text{(C.10)}$$

Confused? When you divide the numerator by i_2, you get

$$\frac{i_2 r_C}{i_2} = r_C$$

When you divide $h_{re}v_2$ by i_2, you get

$$\frac{h_{re}v_2}{i_2} = h_{re}\left(\frac{v_2}{i_2}\right) = h_{re}r_C$$

Finally, when you divide $h_{ie}i_1$ by i_2, you get

$$\frac{h_{ie}i_1}{i_2} = h_{ie}\left(\frac{i_1}{i_2}\right) = h_{ie}\left(\frac{1}{A_i}\right) = \frac{h_{ie}}{A_i}$$

These values were substituted into equation (C.10).

The final exact equation for A_v is derived by substituting equation (C.6) for A_i in equation (C.10). This gives us

$$A_v = \frac{h_{fe}r_C}{h_{ie} + (h_{ie}h_{oe} - h_{re}h_{fe})r_C} \qquad \text{(C.11)}$$

Again, we have defined a gain value strictly in terms of the transistor h-parameters and the AC collector resistance. The following example shows the use of this equation.

EXAMPLE C.2

Determine the voltage gain for the circuit shown in Figure C.5. (Example C.1).

Solution: Using the h-parameter values listed in example C.1, we get

$$
\begin{aligned}
A_v &= \frac{h_{fe}r_C}{h_{ie} + (h_{ie}h_{oe} - h_{re}h_{fe})r_C} \\[2mm]
&= \frac{(50)(3.2 \text{ k}\Omega)}{1 \text{ k}\Omega + [(1 \text{ k}\Omega)(25 \text{ }\mu\text{S}) - (2.5 \times 10^{-4})(50)]3.2 \text{ k}\Omega} \\[2mm]
&= \frac{1.6 \times 10^5}{1.04 \text{ k}\Omega} \\[2mm]
&= 153.85
\end{aligned}
$$

Anyone care for an aspirin? Don't give up yet . . . it gets better. You see, the input impedance of a transistor can be approximated as being equal to h_{ie}. When we determined $Z_{in(\text{base})}$, we defined it as being equal to $h_{fe}r'_e$. Now, based on these two relationships, we'll define r'_e as follows:

$$r'_e = \frac{h_{ie}}{h_{fe}} \qquad \text{(C.12)}$$

Now, let's try attacking the problem in example D.2 from another angle.

EXAMPLE C.3

Determine the voltage gain for the circuit shown in Figure C.5. (Example C.1).

Solution: First, using the *h*-parameters listed in example C.1, we'll determine the value of r'_e as

$$r'_e = \frac{1 \text{ k}\Omega}{50}$$

$$= 20 \ \Omega$$

The voltage gain can now be calculated as

$$A_v = \frac{r_C}{r'_e}$$

$$= \frac{3.2 \text{ k}\Omega}{20 \ \Omega}$$

$$= 160$$

You will find that the method of determining A_v used in example C.3 will always give you a result that is well within 10% of the exact value. For virtually every analysis and design problem, this value of A_v will work just fine.

Input Impedance (Z_{in})

The input impedance to a transistor is defined as the ratio of input voltage (v_1) to input current (i_1). By formula,

$$Z_{in} = \frac{v_1}{i_1} \tag{C.13}$$

If you refer back to Figure C.6 you'll recall that $v_1 = h_{re}v_2 + h_{ie}i_1$. Substituting this value of v_1 into equation (C.13) gives us

$$Z_{in} = \frac{h_{re}v_2 + h_{ie}i_1}{i_1}$$

Which simplifies to

$$Z_{in} = h_{ie} + \frac{h_{re}v_2}{i_i}$$

This equation can be rewritten as

$$Z_{in} = h_{ie} + h_{re}\left(\frac{v_2}{i_1}\right)$$

And

$$Z_{in} = h_{ie} + h_{re}\left(\frac{i_2 r_C}{i_1}\right)$$

Since $\frac{i_2}{i_1} = A_1$, the above equation can be rewritten as

$$Z_{in} = h_{ie} + h_{re}A_i r_C$$

Finally, substituting equation (C.6) into the above equation gives us

$$Z_{in} = h_{ie} + \frac{h_{re}h_{fe}r_C}{1 + r_C h_{oe}} \tag{C.14}$$

Remember earlier when we assumed that $Z_{in} \cong h_{ie}$? The following example will show this assumption to be valid.

EXAMPLE C.4

Determine the exact value of Z_{in} for the amplifier shown in Example C.1.

Solution: Using the h-parameter values from example C.1, the exact value of Z_{in} is found as

$$Z_{in} = h_{ie} + \frac{h_{re}h_{fe}r_C}{1 + r_C h_{oe}}$$

$$= 1 \text{ k}\Omega + \frac{(2.5 \times 10^{-4})(50)(3.2 \text{ k}\Omega)}{1 + (3.2 \text{ k}\Omega)(25 \text{ } \mu\text{S})}$$

$$= 1 \text{ k}\Omega + \frac{40}{1.08} \text{ } \Omega$$

$$= 1037 \text{ } \Omega$$

As you can see, the result of the calculation in example C.4 was only 37 Ω away from the value of h_{ie}.

Output Impedance

The output impedance of a transistor is defined as the ratio of output voltage (v_2) to output current (i_2). By formula,

$$Z_{out} = \frac{v_2}{i_2} \qquad \textbf{(C.15)}$$

Substituting equation (C.5) in place of i_2, gives us

$$Z_{out} = \frac{v_2}{h_{fe}i_1 + v_2 h_{oe}} \qquad \textbf{(C.16)}$$

Now, take a look at the circuit shown in Figure C.7.

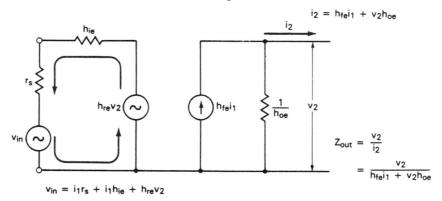

$$v_{in} = i_1 r_s + i_1 h_{ie} + h_{re}v_2$$

FIGURE C.7

The Kirchoff's voltage equation for the input is

$$v_{in} = i_1 r_s + i_1 h_{ie} + h_{re}v_2$$

If we short out the voltage source (a normal step in thevenizing a circuit), we get

$$0 = i_1 r_s + i_1 h_e + h_{re}v_2$$

$$= i_1(r_s + h_{ie}) + h_{re}v_2$$

Subtracting ($h_{re}v_2$) from both sides of the equation gives us

$$i_1(r_s + h_{ie}) = -h_{re}v_2$$

And finally,

$$i_1 = \frac{-h_{re}v_2}{r_s + h_{ie}}$$ (C.17)

Now, if you substitute equation (C.17) for the value of i_1 in equation (C.16), you get

$$Z_{out} = \frac{r_s - h_{ie}}{(r_s + h_{ie})h_{oe} - h_{re}h_{fe}}$$ (C.18)

or

$$Z_{out} = \frac{1}{h_{oe} - \left(\dfrac{h_{fe}h_{re}}{h_{ie} + r_s}\right)}$$ (C.19)

=========================== EXAMPLE C.5 ===========================

Determine the value of output impedance for the transistor used in example C.1.

Solution: The source resistance is shown to be 100 Ω. Using this value, and the parameters given in example C.1, z_{out} is found as

$$Z_{out} = \frac{1}{h_{oe} - \left(\dfrac{h_{fe}h_{re}}{h_{ie} + r_s}\right)}$$

$$= \frac{1}{(25 \ \mu S) - \dfrac{(50)(2.5 \times 10^{-4})}{(1 \ k\Omega) + (100 \ \Omega)}}$$

$$= 73.3 \ k\Omega$$

H-Parameter Approximations

Most of the *h*-parameter formulas covered in this section can be approximated to a form that is easier to handle. While these formula approximations will not produce results that are as accurate as the original equations, they may be used for most applications.

Equation (C.6) shows current gain (A_i) to be found as

$$A_i = \frac{h_{fe}}{1 + r_C h_{oe}}$$

Since $r_C h_{oe} \ll 1$, the denominator of the equation can be approximated as being equal to one. This leaves us with

$$A_i \cong h_{fe}$$ (C.20)

The validity of this approximation is demonstrated in example C.1. In this example, the exact value of A_i was determined to be 46.3. The value of h_{fe} for the transistor was given as 50. The difference between these two gain values is 3.7; a difference of 7.4%.

The voltage gain (A_v) of an amplifier was found in equation (C.11) as

$$A_v = \frac{h_{fe}r_C}{h_{ie} + (h_{ie}h_{oe} - h_{re}h_{fe})r_C}$$

The approximation of this equation is based on several factors. First, we originally defined A_v as

$$A_v = \frac{r_C}{r'_e}$$

This could also be written as

$$A_v = r_C \left(\frac{1}{r'_e} \right)$$

Now, equation (C.12) defined $r'e$ as

$$r'_e = \frac{h_{ie}}{h_{fe}}$$

Therefore,

$$\frac{1}{r'_e} = \frac{h_{fe}}{h_{ie}}$$

Thus, A_v can be approximated as

$$A_v \cong \frac{h_{fe} r_e}{h_{ie}} \qquad \textbf{(C.21)}$$

The validity of this approximation was demonstrated in examples C.2 and C.3. Example C.2 obtained an exact value of A_v equal to 153.85. Example C.3 approximated the value of A_v for the same circuit as 160. The % of error for the approximated value was about 4%.

Example C.4 calculated a Z_{in} for an amplifier transistor of 1037 Ω. The value of h_{ie} for the transistor was 1 kΩ. Thus, we can approximate the value of Z_{in} as

$$Z_{in} \cong h_{ie} \qquad \textbf{(C.22)}$$

Unfortunately, there is no quick approximation for the output impedance of a transistor. Whenever you need to calculate this value, you need to use equation (C.19). However, there is some good news here. In most cases, there is no reason to calculate the output impedance of a transistor. This is due to the fact that it rarely weighs into any common emitter circuit calculations. When you do need to get a general idea of how large Z_{out} is for a given amplifier, you can use

$$Z_{out} > \frac{1}{h_{oe}} \qquad \textbf{(C.23)}$$

While equation (C.23) will not tell you exactly what the value of Z_{out} is, it will give you an idea of how large the value is.

Other Transistor Configurations

H-parameters apply not only to common emitter circuits, but to common base and common collector circuits as well. However, you must use several conversion factors to obtain the values needed to calculate the A_i, A_v, Z_{in}, and Z_{out} values for these circuit configurations. In this section, you will be introduced to the h-parameter conversions, as well as the common base and common collector formulas used for circuit calculations. We will not analyze these formulas and conversions. They are simply listed for future reference.

The common base h-parameters are identified by the use of the subscript letter b in place of e. The following chart illustrates this point.

Parameter	Common Emitter	Common Base
Current Gain	h_{fe}	h_{fb}
V Feedback Ratio	h_{re}	h_{rb}
Input Impedance	h_{ie}	h_{ib}
Output Admittance	h_{oe}	h_{ob}

To convert the common emitter parameters to common base parameters, you use the following conversion formulas:

$$h_{fb} = \frac{h_{fe}(1 - h_{re}) - h_{ie}h_{oe}}{(1 + h_{fe})(1 - h_{re}) + h_{ie}h_{oe}} \qquad \textbf{(C.24)}$$

$$h_{rb} = \frac{h_{ie}h_{oe} - h_{re}(1 + h_{oe})}{(1 + h_{fe})(1 - h_{re}) + h_{ie}h_{oe}} \qquad \textbf{(C.25)}$$

$$h_{ib} = \frac{h_{ie}}{(1 + h_{fe})(1 - h_{re}) + h_{ie}h_{oe}} \tag{C.26}$$

$$h_{ob} = \frac{h_{oe}}{(1 + h_{fe})(1 - h_{re}) + h_{ie}h_{oe}} \tag{C.27}$$

The circuit gain and impedance values are now found as

$$A_i = \frac{h_{fb}}{1 + h_{ob}r_C} \tag{C.28}$$

$$A_v = \frac{h_{fb}r_C}{h_{ib} + (h_{ib}h_{ob} - h_{fb}h_{rb})r_C} \tag{C.29}$$

$$Z_{in} = h_{ib} - \frac{h_{rb}h_{fb}r_C}{1 + h_{ob}r_C} \tag{C.30}$$

$$Z_{out} = \frac{1}{h_{ob} - \dfrac{h_{fb}h_{rb}}{h_{ib} + r_s}} \tag{C.31}$$

The *common collector h*-parameters are identified by the letter c in the subscript in place of e. The conversions from common emitter parameters to common collector parameters are as follows:

$$h_{fc} = h_{fe} + 1 \tag{C.32}$$

$$h_{rc} = 1 - h_{re} \tag{C.33}$$

$$h_{ic} = h_{ie} \tag{C.34}$$

$$h_{oc} = h_{oe} \tag{C.35}$$

The gains and impedances of the common collector amplifier are now found as

$$A_i = \frac{h_{fc}}{1 + h_{oc}r_E^*} \tag{C.36}$$

$$A_v = \frac{h_{fc}r_E}{h_{ic} + (h_{ic}h_{oc} - h_{fc}h_{rc})r_E} \tag{C.37}$$

$$Z_{in} = h_{ic} - \frac{h_{rc}h_{fc}r_E}{1 + h_{oc}r_E} \tag{C.38}$$

$$Z_{out} = \frac{1}{h_{oe} + \dfrac{h_{fc}h_{rc}}{h_{ic} + r_s}} \tag{C.39}$$

The actual gain and impedance equations shown in this section are exactly the same as those used for the common emitter circuit. However, before they can be used, the common emitter *h*-parameters must be converted to the needed form.

---

* For the common collector circuit, the output is taken from the emitter. Therefore, the output AC resistance is the AC resistance of the emitter circuit, r_E.

APPENDIX D
Selected Equation Derivations

Equation 7.2

Schockley's equation for the total current through a pn junction is

$$I_T = I_S(\epsilon^{Vq/kT} - 1) \qquad \textbf{(D.1)}$$

where

I_S = the reverse saturation current through the diode

ϵ = the *exponential constant*; approximately 2.71828

V = the voltage across the depletion layer

q = the charge on an electron; approximately 1.6×10^{-19} V

k = Boltzmann's constant; approximately 1.38×10^{-23} J/°K

T = the temperature of the device, in degrees Kelvin (°K = °C + 273)

We can solve for the value of q/kT at room temperature (approximately 21°C) as

$$q/kT = \frac{1.6 \times 10^{-19}\, V}{(1.38 \times 10^{-23})(294°K)}$$
$$= 40$$

Using this value, the equation for I_T is rewritten as

$$I_T = I_S(\epsilon^{40V} - 1) \qquad \textbf{(D.2)}$$

Now, differentiating the above equation gives us

$$\frac{dI}{dV} = 40 I_S \epsilon^{40V} \qquad \textbf{(D.3)}$$

If we rearrange equation (D.2) as

$$I_S \epsilon^{40V} = (I_T + I_S)$$

we can use this equation to rewrite equation (D.3) as

$$\frac{dI}{dV} = 40(I_T + I_S) \qquad \textbf{(D.4)}$$

Now, if we take the reciprocal of equation (D.4), we will have an equation for the AC resistance of the junction, r'_e, as follows:

$$\frac{dV}{dI} = \frac{1\,V}{40(I_T + I_S)}$$
$$= \frac{1\,V}{40}\frac{1}{I_T + I_S}$$
$$= 25\,mV\frac{1}{I_T + I_S}$$
$$= \frac{25\,mV}{I_T + I_S}$$

In this case, $(I_T + I_S)$ is the current through the emitter-base junction of the transistor, I_E. Therefore, we can rewrite the above equation as

$$r'_e = \frac{25 \; mV}{I_E}$$

Equation 8.29

Equation (8.15) gives the value of base input impedance as

$$Z_{in(base)} = h_{fc}(r'_e + r_E)$$

where r_E is the parallel combination of R_E and R_L. Multiplying on the right side of the equation, we obtain

$$Z_{in(base)} = h_{fc}r'_e + h_{fc}r_E$$

Now, since $h_{ic} = h_{fc}r'_e$, we can rewrite the above equation as

$$Z_{in(base)} = h_{ic} + h_{fc}r_E$$

Thus, for the Darlington amplifier shown in Figure D–1, the value of $Z_{in(base)}$ for Q_2 is

$$Z_{in(base)2} = h_{ic2} + h_{fc2}r_{E2}$$

And, the value of $Z_{in(base)}$ for Q_1 is

$$Z_{in(base)1} = h_{ic1} + h_{fc1}r_{E1}$$

As Figure D–1 shows, Q_2 is the ac load on Q_1. Thus, $r_{E1} = Z_{in(base)2}$. Thus,

$$Z_{in(base)1} = h_{ic1} + h_{fc1}r_{E1}$$
$$= h_{ic1} + h_{fc1}(h_{ic2} + h_{fc2}r_{E2})$$

or

$$Z_{in(base)} = h_{ic1} + h_{fc1}(h_{ic2} + h_{fc2}r_E)$$

where

$$Z_{in(base)} = \text{the input impedance of } Q_1$$
$$r_E = \text{the parallel combination of } R_E \text{ and } R_L$$

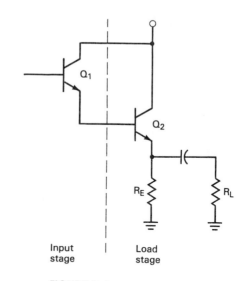

Input | Load
stage | stage

FIGURE D.1

RC-Coupled Class A Efficiency

The ideal class A amplifier will have the following characteristics:

$$1. \quad V_{CEQ} = \frac{V_{CC}}{2}$$

$$2. \quad V_{PP} = V_{CC}$$

3. $I_{CQ} = \dfrac{I_{C(sat)}}{2}$

4. $I_{PP} = I_{C(sat)}$

These ideal characteristics will be used in our derivation of the maximum ideal efficiency rating of 25%.

Equation (9.11), page 367, gives us the following equation for load power:

$$P_L = \frac{V_{PP}^2}{8R_L}$$

Since $R_L = V_{PP}/I_{PP}$, we can rewrite the above equation as

$$P_L = \left(\frac{V_{PP}^2}{8}\right)\left(\frac{1}{R_L}\right)$$
$$= \left(\frac{V_{PP}^2}{8}\right)\left(\frac{I_{PP}}{V_{PP}}\right) \qquad\qquad \textbf{(D.5)}$$
$$= \frac{V_{PP}I_{PP}}{8}$$

Now, we will use the characteristics above and equation (D.5) to determine the ideal maximum load power as follows:

$$P_L = \frac{V_{PP}I_{PP}}{8}$$
$$= \frac{V_{CC}I_{C(sat)}}{8}$$
$$= \frac{(2V_{CEQ})(2I_{CQ})}{8}$$
$$= \frac{4(V_{CEQ}I_{CQ})}{8}$$
$$= \frac{V_{CEQ}I_{CQ}}{2} \qquad (\textit{maximum, ideal})$$

Now, recall that the power drawn from the supply of an amplifier is found as

$$P_S = V_{CC}I_{CC}$$

where

$$I_{CC} = I_{CQ} + I_1$$

Since I_{CQ} is normally *much greater than* I_1, we can approximate the equation above to

$$P_S \cong V_{CC}I_{CQ}$$

Note that this equation is valid for most amplifier power analyses. We can rewrite the for the above the ideal amplifier as follows:

$$P_S = 2V_{CEQ}I_{CQ}$$

We can now use the derived values of P_L and P_S to detremine the *maximum ideal value* of η, as follows:

$$\eta = \frac{P_L}{P_S} \times 100\%$$
$$= P_L \frac{1}{P_S} \times 100\%$$
$$= \frac{V_{CEQ}I_{CQ}}{2} \frac{1}{2V_{CEQ}I_{CQ}} \times 100\%$$
$$= \frac{V_{CEQ}I_{CQ}}{4V_{CEQ}I_{CQ}} \times 100\%$$

$$= \frac{1}{4} \times 100\%$$

$$= 25\%$$

Transformer-Coupled Class A Efficiency

Assume for a minute that we are dealing with an *ideal* transformer-coupled amplifier. This amplifier will have characteristics of $V_{CEQ} = V_{CC}$ and $V_{CE(\max)} = 2V_{CC}$. These characteristics ignore the fact that there will be some voltage dropped across the emitter resistor of the amplifier. Now, recall that the maximum power that can be delivered to a load from a class A amplifier is found as

$$P_{L(\max)} = \frac{V_{CEQ}I_{CQ}}{2}$$

This equation was derived in our initial discussion on class A amplifiers. Since the ideal transformer-coupled amplifier has a value of $V_{CEQ} = V_{CC}$, the equation above can be rewritten as

$$P_{L(\max)} = \frac{V_{CC}I_{CQ}}{2} \quad (maximum,\ ideal)$$

You may also recall that the power drawn from the supply can be approximated

$$P_S = V_{CC}I_{CQ}$$

We can use these two equations to calculate the maximum *ideal* efficiency of the transformer-coupled amplifier, as follows:

$$\eta = \frac{P_L}{P_S} \times 100\%$$

$$= P_L \frac{1}{P_S} \times 100\%$$

$$= \frac{V_{CC}I_{CQ}}{2} \frac{1}{V_{CC}I_{CQ}} \times 100\%$$

$$= \frac{1}{2} \times 100\%$$

$$= 50\%$$

Class B Amplifier Efficiency

Equation (9.29) defines $I_{C1(\text{ave})}$ as

$$I_{C1(\text{ave})} = \frac{0.159 V_{CC}}{R_L}$$

This equation is based on the relationship

$$I_{C1(\text{ave})} \cong 0.318 I_{Pk}$$

The value *0.318* is the "round-off" equivalent of $1/\pi$. A closer approximation of the multiplier used is 0.3183099. Using this value, equation (9.22) becomes

$$I_{C1(\text{ave})} = \frac{0.1591549 V_{CC}}{R_L}$$

The multiplier shown *is* more accurate, but its length makes it impractical for everyday analysis purposes. However, we will need it in proving that the maximum *ideal* efficiency of a class B amplifier is approximately 78.5%.

In the ideal class B amplifier, $I_{C1(\text{ave})} \gg I_1$. Therefore, we can assume that $I_{CC} = I_{C1(\text{ave})}$ in the ideal class B amplifier. Based on this point, the determination of the ideal maximum efficiency for the class B amplifier proceeds as follows:

$$\eta = \frac{P_L}{P_S} \times 100\%$$

$$= P_L \left(\frac{1}{P_S}\right) \times 100\%$$

$$= \left(\frac{V_{CC}^2}{8R_L}\right) \left(\frac{1}{V_{CC}(I_{C1,\text{ave}})}\right) \times 100\%$$

$$= \left(\frac{V_{CC}^2}{8R_L}\right) \left(\frac{R_L}{0.1591549V_{CC}^2}\right) \times 100\%$$

$$= \left(\frac{1}{1.2732495}\right) \times 100\%$$

$$= 78.54\%$$

The final result can, of course, be assumed to be equal to 78.5%. Remember that this maximum efficiency rating is ideal. Any practical value of efficiency for a class B amplifier would have to be less than this value.

Equation 9.34

The instantaneous power dissipation in a Class B amplifier is found as

$$p = V_{CEQ}(1 - k \sin \Theta)I_{C(Sat)}(k \sin \Theta) \qquad \textbf{(D.6)}$$

where

V_{CEQ} = the quiescent value of V_{CE} for the amplifier
$I_{C(Sat)}$ = the saturation current for the amplifier, as determined by the AC load line for the circuit.
Θ = the phase angle of the output at the instant that p is being measured
k = a constant factor, representing the percent of the load line that is actually being used by the circuit. K is represented as a decimal value between 0 and 1.

The average power dissipation can be found by integrating equation (D.6) for one half-cycle. Note that a half-cycle is represented as being between 0 and π *radians* in the integration:

$$p_{ave} = \frac{1}{2\pi} \int_0^\pi p \, d\Theta \qquad \textbf{(D.7)}$$

Performing the integration yields:

$$p_{ave} = \left(\frac{V_{CEQ}I_{C(Sat)}}{2\pi}\right) \left(2k - \frac{\pi k^2}{2}\right) \qquad \textbf{(D.8)}$$

The next step is to find the maximum value of k in equation (D.8). The first step in finding this maximum value is to differentiate p_{ave} with respect to k, as follows:

$$\frac{dp_{ave}}{dk} = \frac{V_{CEQ}I_{C(Sat)}}{2} (2 - \pi k) \qquad \textbf{(D.9)}$$

Next, we set the right hand side of equation (D.9) equal to zero. This provides us with the following equations:

$$\frac{V_{CEQ}I_{C(Sat)}}{2} = 0 \qquad \text{and} \qquad 2 - \pi k = 0$$

Now, we can solve the equation on the right to find the maximum value:

$$k = \frac{2}{\pi} \qquad \textbf{(D.10)}$$

At this point, we want to take the value of k obtained in equation (D.10) and plug it into equation (D.7). Reducing this new equation gives us

$$p_{ave} = 0.1V_{CEQ}I_{C(Sat)} \qquad \textbf{(D.11)}$$

Finally, we replace the values of V_{CEQ} and $I_{C(Sat)}$ in equation (D.11) using the following relationships:

$$V_{CEQ} = \frac{PP}{2} \quad \text{and} \quad I_{C(Sat)} = \frac{V_{CEQ}}{R_L}$$

This gives us

$$P_D = 0.1 \left(\frac{V_{CEQ}PP}{2R_L} \right)$$

or

$$P_D = \left(\frac{V_{CEQ}PP}{20R_L} \right)$$

Now, since V_{CEQ} is equal to $\frac{PP}{2}$, we can rewrite the above equation as

$$P_D = (V_{CEQ}) \left(\frac{PP}{20R_L} \right)$$
$$= \left(\frac{PP}{2} \right) \left(\frac{PP}{20R_L} \right)$$

or

$$P_D = \frac{PP^2}{40R_L}$$

This is another form of equation 9.34.

Equation 12.6:

The equation for the instantaneous voltage at any point on a sine wave is given as

$$v = V_{Pk}(\sin \omega t)$$

where

V_{Pk} = the peak input voltage

$\omega = 2\pi f$

t = the time from the start of the cycle to the instant that v occurs

If we differentiate v with respect to t, we get

$$\frac{dv}{dt} = \omega V_{Pk}(\cos \omega t)$$

Since slew rate is a maximum rating of a given op-amp, and the rate of change in a sine wave varies, we need to determine the point at which the rate of change in the sine wave is at its maximum value. This point occurs when the sine wave passes the reference voltage, or at $t = 0$. The value of $\frac{dv}{dt}$ at $t = 0$ determines the maximum operating frequency of the op-amp, as follows:

$$\frac{dv}{dt}_{MAX} = \omega_{MAX}V_{Pk}$$

or

$$\text{Slew Rate} = 2\pi f_{MAX}V_{Pk}$$

Finally, this equation is rearranged into the form of equation (12.6):

$$f_{MAX} = \frac{\text{Slew Rate}}{2\pi V_{Pk}}$$

Equation 13.25

The proof of this equation starts by taking a look at the RC circuit shown in Figure D.2. Since the current through a series circuit is constant, the ratio of v_{out} to v_{in} is equal to the ratio of X_C (the output shunt component) to the total impedance of the circuit. By formula,

$$\frac{v_{out}}{v_{in}} = \frac{X_C}{Z_T} \tag{D.12}$$

or

$$\frac{v_{out}}{v_{in}} = \frac{\dfrac{1}{j\omega C}}{\dfrac{1}{j\omega C} + R}$$

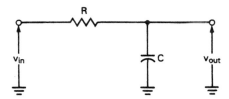

FIGURE D.2

Dividing both the numerator and the denominator in equation (D.12) by $\dfrac{1}{j\omega C}$, we get

$$\frac{v_{out}}{v_{in}} = \frac{1}{1 + j\omega RC} \tag{D.13}$$

Now, since $f_c = \dfrac{1}{2\pi RC}$ and $\omega = 2\pi f$, equation (D.13) can be rewritten as

$$\frac{v_{out}}{v_{in}} = \frac{1}{1 + j\dfrac{1}{f_c}} \tag{D.14}$$

Now, the ratio of X_C to Z_T varies with the ratio of operating frequency of f_c. Thus, equation (D.14) may be rewritten as

$$\frac{v_{out}}{v_{in}} = \frac{1}{1 + j(f/f_c)} \tag{D.15}$$

Finally, we use the relationships

$$A_v = \frac{v_{out}}{v_{in}}$$

and

$$1 + jX = \sqrt{1_1 + X^2}$$

(where X = any quantity) to rewrite equation (E.14) as

$$\Delta A_v = \frac{1}{\sqrt{1 + (f/f_c)^2}} \tag{D.16}$$

Rewriting equation (D.16) in dB form, we obtain

$$\Delta A_{v(dB)} = 20 \log \left(\frac{1}{\sqrt{1 + (f/f_c)^2}} \right)$$

Equations 13.53 and 13.54

In this derivation, we will concentrate on the *low frequency* responses of a multistage amplifier. We will then expand the derivation to include the *high frequency response* of a given multistage amplifier.

In order to minimize any confusion, we now define two variables that will be used extensively in this derivation:

$$A_{v(\text{mid})} = \text{the \textit{midband} gain of an amplifier}$$
$$A_v = \text{the instantaneous gain of an amplifier.}$$

Note that $A_v = A_{v(\text{mid})}$ when the amplifier is operated above f_1, and $A_v < A_{v(\text{mid})}$ when the amplifier is operated below f_1.

Earlier in the text, you were shown that the ratio of amplifier gain to its midband gain (A_v') is found as

$$A_v' = \frac{A_v}{A_{v(\text{mid})}} = \frac{1}{\sqrt{1 + (f_1/f)^2}} \tag{D.17}$$

It follows that two amplifier stages with identical values of f_1 and operated at the same value of f will have values of A_v' that are equal. By formula,

$$A_{v1}' = A_{v2}'$$

when

$$f_{c1} = f_{c2} \text{ and the two circuits are at the same operating frequency}$$

Now, recall that the total gain of n cascaded amplifier stages is equal to the product of the individual stage gain values. By formula,

$$A_{vT} = (A_{v1})(A_{v2}) \cdots (A_{vn})$$

The same relationship holds true for the gain ratios of the individual stages. By formula,

$$A_{vT}' = (A_{v1}')(A_{v2}') \cdots (A_{vn}') \tag{D.18}$$

Now, assuming that we have n stages with equal values of A_v', the value of A_{vT}' is found as

$$A_{vT}' = (A_v')^n \tag{D.19}$$

where

$$A_v' = \text{the gain ratio of any individual stage}$$

Substituting equation (D.17) into equation (D.18) gives us

$$A_{vT}' = (A_v')^n = \left(\frac{1}{\sqrt{1 + (f_1/f)^2}} \right)^n \tag{D.20}$$

$$A_{vT}' = \left(\frac{1}{\sqrt{1 + (f_1/f)^2}} \right)^n \tag{D.21}$$

Now, when $f = f_1$ (the lower 3 dB point), equation (E.16) simplifies as follows:

$$\left(\frac{1}{\sqrt{1 + (f_1/f_1)^2}} \right) = \left(\frac{1}{\sqrt{1 + 1}} \right) = \left(\frac{1}{\sqrt{2}} \right) = (2)^{1/2}$$

If we set A_{vT}' to the value of $(2)^{1/2}$, equation (E.19) can be rewritten as

$$(2)^{1/2} = \left\{ \left[1 + (f_1/f)^2 \right]^{1/2} \right\}^n$$

Since the exponent value of $(1/2)$ appears on both sides of the equation, it can be dropped completely, leaving

$$2 = \left[1 = (f_1/f)^2 \right]^n$$

or

$$(2)^{1/n} = 1 + (f_1/f)^2 \tag{D.22}$$

Finally, solving equation (D.22) for f yields

$$f = \frac{f_1}{\sqrt{2^{1/n} - 1}}$$

Or, in another form,

$$f_{1T} = \frac{f_1}{\sqrt{2^{1/n} - 1}} \tag{D.23}$$

This is equation (13.53). The same process is then used to derive equation (13.54).

Equation 14.5

Equation (14.1) defines v_{in} of an amplifier with feedback as

$$v_{in} = v_s - v_f$$

According to equation (14.3),

$$v_f = \alpha_v v_{out}$$

Substituting this equation for v_f in equation (14.1), we get

$$v_{in} = v_s - \alpha_v v_{out} \tag{D.24}$$

Now, equation (D.24) can be rewritten to solve for v_s as

$$v_s = v_{in} + \alpha_v v_{out} \tag{D.25}$$

Since $v_{out} = A_v v_{in}$, we can rewrite equation (D.25) as

$$v_s = v_{in} + \alpha_v A_v v_{in}$$

or

$$v_s = v_{in}(1 + \alpha_v A_v) \tag{D.26}$$

Now, A_v is the total gain of the feedback amplifier from source to output. By formula,

$$A_{vf} = \frac{v_{out}}{v_s} \tag{D.27}$$

Substituting equation (D.26) for v_s in (D.27) gives us

$$A_{vf} = \frac{v_{out}}{v_{in}(1 + \alpha_v A_v)}$$

or

$$A_{vf} = \frac{A_v}{1 + \alpha_v A_v}$$

Equations 15.21 and 15.23

We start these proofs with an assumption that can be seen intuitively; that is, for any given series reactive-resistive circuit there exists a parallel equivalent reactive-resistive circuit. These two circuits are represented in Figure D.3.

All this assumption does is let us assume that it *is* possible to derive a parallel equivalent circuit for a series reactive-resistive circuit.

For the series circuit shown in Figure D.3,

$$Q = \frac{X_S}{R_S} \tag{D.28}$$

and

$$Z_S = R_S \pm jX_S \tag{D.29}$$

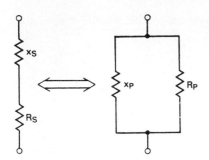

FIGURE D.3

For the parallel circuit shown in Figure D.3, the total impedance (Z_p) is found as

$$Z_P = \frac{(R_P)(\pm jX_P)}{R_P \pm jX_P} \tag{D.30}$$

If we multiply the right hand side of equation (D.30) by *1* in the form of

$$\frac{R_P \pm jX_P}{R_P \pm jX_P}$$

we get

$$Z_P = \frac{R_P X_P^2}{R_P^2 + X_P^2} \pm j\frac{X_P R_P^2}{R_P^2 + X_P^2} \tag{D.31}$$

Now, refer back to our original assumption regarding the two circuits in Figure D.3. If these two circuits are equivalent circuits, then the total impedance of the two circuits must be equal. By formula,

$$Z_S = Z_P$$

Substituting this relationship into equation (D.31) give us

$$Z_S = \frac{R_P X_P^2}{R_P^2 + X_P^2} \pm j\frac{X_P R_P^2}{R_P^2 + X_P^2} \tag{D.32}$$

Now, if we substitute equation (D.29) for Z_S in equation (E.31) we get

$$R_S \pm jX_s = \frac{R_P X_P^2}{R_P^2 + X_P^2} \pm j\frac{X_P R_P^2}{R_P^2 + X_P^2} \tag{D.33}$$

For the two sides of equation (D.33) to be equal, the real components on each side must be equal, and the imaginary components on each side must also be equal. Therefore,

$$R_S = \frac{R_P X_P^2}{R_P^2 + X_P^2} \tag{D.34}$$

and

$$X_S = \frac{X_P R_P^2}{R_P^2 + X_P^2} \tag{D.35}$$

Using these two equations, we can rewrite equation (D.28) as

$$Q = \frac{\dfrac{X_P R_P^2}{R_P^2 + X_P^2}}{\dfrac{R_P X_P^2}{R_P^2 + X_P^2}}$$

$$= \frac{X_P R_P^2}{R_P X_P^2}$$

and

$$Q = \frac{R_P}{X_P} \tag{D.36}$$

Equation (D.36) will be used to prove equation (15.23) at the end of this proof. For now, we'll concentrate on equation (D.35). Using the relationship

$$Q = \frac{R_P}{X_P}$$

we can rewrite equation (D.35) as

$$X_S = \frac{X_P}{1 + (1/Q)^2} \tag{D.37}$$

Since tuned amplifiers are *high-Q* circuits, it is safe to assume that $(1/Q)^2 \ll 1$. Based on this assumption, equation (D.37) simplifies to

$$X_S \cong X_P \tag{D.38}$$

Now, we can rewrite equation (D.36) as

$$R_P = QX_P$$

Or, using equation (D.38),

$$R_P = QX_S$$

Since $X_S = QR_S$ (from equation D.28) we can solve the above equation as

$$R_P = QR_S$$
$$= Q(QR_S)$$

or

$$R_P = Q^2 R_S \tag{D.39}$$

Now, let's relate equations (D.36) and (D.39) to the *LC* tuned circuit. In an *LC* tuned circuit, the series reactance is X_L and the series resistance R_W. Replacing R_S in equation (D.39) with R_W gives us

$$R_P = Q^2 R_W$$

Finally, replacing X_P in equation (D.36) with X_L gives us

$$Q = \frac{R_P}{X_L}$$

or

$$Q = \frac{R_P \parallel R_L}{X_L}$$

when a load is connected to the equivalent parallel circuit. Of course, this equation is the equivalent of equation (15.23).

Equation 18.2

The time required for a capacitor to charge to a given values is found as

$$t = RC \left[\ln \frac{V - V_0}{V - v_c} \right] \tag{D.40}$$

where

  V = the charging voltage
  V_0 = the initial charge on the capacitor
  v_c = the capacitor charge of interest
  $\ln$ = an operator, indicating that we are to take the *natural log* of the fraction

We can use equation (D.40) to find the time required for v_c to reach 0.1 V (10% of V) as follows:

$$t = RC \left[\ln \frac{V - 0\,V}{V - 0.1\,V} \right]$$

$$= RC \left[\ln \frac{V}{0.9\,V} \right]$$

$$= RC\,[\ln{(1.11)}]$$

$$= 0.1RC$$

Now, we can use equation (D.40) to find the time required for v_c to reach 0.9 V (90% of V) as follows:

$$t = RC \left[\ln \frac{V - 0\,V}{V - 0.9\,V} \right]$$

$$= RC \left[\ln \frac{V}{0.1\,V} \right]$$

$$= RC\,[\ln{(10)}]$$

$$= 2.3RC$$

Since rise time is defined as the time for v_c to fall from 90% of its maximum value to 10% of its maximum value, it can be found as

$$t_r = t_{(90\%)} - t_{(10\%)}$$

$$= 2.3RC - 0.1RC \qquad\qquad \textbf{(D.41)}$$

$$= 2.2RC$$

Now, the upper cutoff frequency of a given RC circuit is found as

$$f_2 = \frac{1}{2\pi RC}$$

Rewriting the f_2 equation, we get

$$RC = \frac{1}{2\pi f_2}$$

Substituting this equation in equation (D.41), we get

$$t_r = 2.2RC$$

$$= 2.2 \left(\frac{1}{2\pi f_2} \right)$$

$$= \frac{2.2}{2\pi f_2}$$

$$= \frac{0.35}{f_2}$$

or

$$f_2 = \frac{0.35}{t_r}$$

APPENDIX E
Glossary

AC beta (β_{ac}). The ratio of ac collector current to ac base current, listed on specification sheets as h_{fe}.

AC emitter resistance. The dynamic resistance of the transistor emitter, used in voltage gain and impedance calculations.

AC equivalent circuit. A representation of a circuit that shows how the circuit appears to an ac signal source.

AC load line. A graph that represents all possible combinations of i_c and v_{ce} for a given BJT amplifier.

Acceptor atoms. Another name for trivalent atoms.

Active region. The BJT operating region between saturation (maximum I_C) and cutoff (minimum I_C).

Adjustable regulator. A regulator whose output dc voltage can be set to any value within specified limits.

Alphanumeric displays. LED (and other) circuits used to display letters, numbers, and punctuation marks.

Amplification. Increasing the power of an ac signal.

Amplifier. A circuit used to provide current, voltage, and/or power gain.

Anode. The p-type material of a diode.

Anode current interruption. A method of driving an SUS into cutoff by breaking the device current path or shorting current around the device.

Apparent power. Power stored in the field of a reactive component.

Armstrong oscillator. An oscillator that uses a transformer in its feedback network to achieve the required 180° voltage phase shift.

Astable multivibrator. A switching circuit that has no stable output state. A square wave oscillator. Also known as a *free-running* multivibrator.

Attenuation factor (α_v). The ratio of feedback voltage to output voltage. The value of α_v is always less than one.

Audio amplifier. The final audio stage in communications receivers, used to drive the speaker(s).

Avalanche current. The current that occurs when a *pn*-junction diode goes into reverse breakdown.

Average forward current (I_0). The maximum allowable value of dc forward current for a *pn*-junction diode.

Average load current. The dc equivalent current of a rectified ac signal.

Average load voltage. The dc equivalent voltage of a rectified ac signal.

Average on-state current (I_T). The maximum average (dc) forward current for an SUS.

Average on-state voltage (V_T). The value of V_F for an SUS when $I_F = I_T$ (the average on-state current rating of the device).

Averaging amplifier. A summing amplifier that provides an output which is proportional to the average of the input voltages.

Back-to-back suppressor. A single package containing two transient suppressors that are connected cathode-to-cathode or anode-to-anode.

Band. Another name for an orbial shell (in atomic theory) or a range of frequencies (in frequency analysis).

Band-pass filter. A filter designed to pass all frequencies that fall between f_1 and f_2.

Band-stop ("notch") filter. A filter designed to block all frequencies that fall between f_1 and f_2.

Bandwidth. The range of frequencies over which gain is relatively constant.

Barkhausen criterion. The relationship between the feedback factor (α_v) and voltage gain (A_v) for proper oscillator operation.

Barrier potential. The natural potential across a *pn* junction.

Base bias. A BJT biasing circuit that consists of a single base resistor between the base terminal and V_{CC} and no emitter resistor. Also known as *fixed bias*.

Base curve. A curve illustrating the relationship between I_B and V_{BE}.

Base cutoff current. The maximum amount of current through a reverse biased emitter-base junction.

Base-emitter saturation voltage. The maximum value of V_{BE} when a transistor is in saturation.

Beta curve. A curve that shows the relationship between beta and temperature and/or collector current.

Beta-dependent circuit. A BJT biasing circuit whose Q-point values are affected by changes in the dc beta (h_{FE}) of the transistor.

Beta-independent circuit. A BJT biasing circuit whose Q-point values are relatively independent of changes in the dc beta (h_{FE}) of the transistor.

Bias. A potential used to establish depletion layer width.

Biased clamper. A clamper that allows a waveform to be shifted above or below a dc reference other than 0 V.

Biased clipper. A shunt clipper that uses a dc voltage source to set the voltage at which the input voltage is clipped.

Bidirectional thyristor. A thyristor that is capable of conducting in two directions.

Biomedical electronics. The area of electronics that deals with medical test and treatment equipment.

Bistable multivibrator. A switching circuit with two stable output states. Also called a *flip-flop*.

Bode plot. A frequency response graph that assumes ΔA_p is zero until the cutoff frequency is reached.

Buffer. An emitter follower that is commonly used for impedance matching. The buffer has high input impedance, low output impedance, and no voltage phase shift from input to output.

Bulk resistance. The natural resistance of a forward-biased diode.

Butterworth filter. An active filter characterized by flat pass-band response and a 20 dB/decade roll-off rate.

Bypass capacitor. A capacitor that is used to establish an ac ground at a specific point in a circuit.

Capacitance ratio (CR). The factor by which the capacitance of a varactor varies from one value of V_R to another.

Cascaded. A term meaning *connected in series*.

Cascode amplifier. A low C_{in} amplifier used in high-frequency applications.

Cathode. The n-type terminal of a diode.

Center frequency. Another name for midpoint frequency.

Channel. The material that connects the source and drain terminals of an FET.

Chebyshev filter. An active filter characterized by irregular pass-band response and 40 dB/decade roll-off rates.

Circuit fusing (I^2t) rating. The maximum forward surge current capability of an SCR.

Clamper. Circuits used to change the dc reference of an input signal. Also known as *dc restorers*.

Clamping voltage. The maximum value of V_Z for a transient suppressor.

Clapp oscillator. A Colpitts oscillator with an added capacitor (in series with the feedback inductor) that is used to reduce the effects of stray capacitance.

Class A amplifier. An amplifier containing a transistor that conducts during the entire cycle of the ac input.

Class AB operation. When the transistors in an amplifier conduct for a portion of the input cycle that is between 180° and 360°.

Class B amplifier. An amplifier containing two transistors that each conduct for approximately 180° of the ac input cycle.

Class C amplifier. An amplifier containing a transistor that conducts for less than 180° of the ac input cycle.

Clipper. Circuits used to eliminate a portion of an ac signal. Also called a *limiter*.

Collector-base junction. One of two junctions in a BJT. The other is the emitter-base junction.

Collector bypass capacitor. A capacitor connected between V_{CC} and ground, parallel with the dc power supply. Often referred to as a *decoupling capacitor*.

Collector curve. A curve that relates the values of I_C, I_B, and V_{CE} for a given transistor.

Collector cutoff current (I_{CEX}). The maximum value of I_C through a cutoff transistor.

Collector-emitter saturation voltage. The maximum value of V_{CE} when a transistor is in saturation.

Collector-feedback bias. A BJT biasing circuit constructed so that V_C directly affects V_B.

Colpitts oscillator. An oscillator that uses a pair of tapped capacitors and an inductor to produce the 180° phase shift in the feedback network.

Common-anode display. A multisegment display in which all LED anodes are tied to a single pin.

Common-base amplifier. An amplifier configuration in which the base is common to all supply voltages. The common-base amplifier has high voltage gain, low current gain, and no input/output voltage phase shift.

Common-collector amplifier. An amplifier configuration in which the collector is common to all supply voltages. The common-collector amplifier (also called an *emitter-follower*) has low voltage gain, high current gain, and no input/output voltage phase shift.

Common-cathode display. A multisegment display in which all LED cathodes are tied to a single pin.

Common-drain amplifier. The FET equivalent of a common-collector amplifier. Also referred to as a *source follower*.

Common-emitter amplifier. An amplifier configuration in which the emitter is common to all supply voltages. The common-emitter amplifier has high voltage and current gain, and a 180° voltage phase shift from input to output.

Common-gate amplifier. The FET equivalent of a common-base amplifier.

Common-source amplifier. The FET equivalent of a common-emitter amplifier.

Comparator. A circuit whose dc output is determined by the polarity relationship of its inputs.

Complementary MOS (CMOS). A logic family that uses a combination of p-channel and n-channel MOSFETs.

Complementary-symmetry amplifier. A class B (or class AB) circuit configuration that uses complementary transistors. Also called a *push-pull emitter follower*.

Complementary transistors. A pair, one npn and one pnp, with matched characteristics and parameters.

Compliance. The maximum V_{PP} output of an amplifier.

Conduction band. The band outside of the valence shell.

Constant-current diodes. Diodes that maintain a constant device current over a wide range of forward operating voltages.

Counter emf. A voltage produced by a transformer that has the opposite polarity of the voltage that caused it. Often referred to as *kick emf*.

Coupling capacitor. A capacitor connected between amplifier stages to provide dc isolation between the stages while allowing the ac signal to pass with little or no distortion.

Covalent bonding. One of the means by which atoms are held together. In a covalent bond, shared electrons hold the atoms together.

Critical rise (dv/dt) rating. The maximum rate of increase in V_{AK} for an SCR that will not cause false triggering.

Crossover distortion. Distortion caused by class B transistor biasing. It occurs when both transistors are in cutoff.

Crossover network. A circuit that is designed to separate high-frequency audio from low-frequency audio.

Crowbar. A circuit used to protect a voltage-sensitive load from excessive dc power supply output voltages.

Crystal-controlled oscillator. An oscillator that uses a quartz crystal to produce an extremely stable output frequency.

Current feedback. A method of controlling amplifier characteristics by feeding (sending) a portion of the amplifier output current back to the input.

Current gain. The factor by which current increases from the base of a transistor (the amplifier input) to the collector of the transistor (the amplifier output), represented as h_{fe}, A_i, and/or β_{ac}.

Current-gain bandwidth product. The frequency at which BJT current gain drops to unity. Also referred to as *beta-cutoff-frequency*.

Current limiting resistor. A resistor placed in series with a device in order to limit the amount of current that can be drawn through that device.

Current regulator diode. Another name for a constant-current diode.

Cutoff. A BJT operating state where I_C is near zero.

Cutoff clipping. A type of distortion caused by driving a transistor into cutoff.

Cutoff frequencies. The frequencies at which the power gain (A_p) of an amplifier drops to approximately 50% of its midband value ($A_{p,mid}$).

Cycle time. The sum of pulse width and space width.

Damping. The fading and loss of oscillations that occurs when $\alpha_v A_v < 1$. (See *Barkhausen criterion*).

Dark current. The reverse current through a photo-detector with no active light input.

Darlington amplifier. A special-case emitter follower that uses a two-transistor configuration (called a *Darlington Pair*) to increase the overall amplifier current gain and input impedance.

Darlington pair. Two transistors, connected so that their collectors are tied together and the emitter of the first is tied to the base of the second.

Darlington pass-transistor regulator. A series regulator that uses a Darlington pair in place of a single pass transistor.

dBm reference. A dB value that reverences output power to 1 mW.

DC alpha (α). The ratio of dc collector current to dc emitter current.

DC beta (β or h_{FE}). The ratio of dc collector current to dc base current.

DC load line. A graph of all possible combinations of I_C and V_{CE} for a given amplifier.

DC power dissipation rating. The maximum allowable value of P_D for a given device.

DC restorer. See *Clamper*.

Decade. A multiplier of 10.

Decibel (dB). A representation of gain, given as 10 times the common logarithm of the gain ratio.

Decoupling capacitor. See *Collector-bypass capacitor*.

Delay time (t_d). The time required for a BJT to come out of cutoff when provided with a pulse input.

Depletion layer. The area around a *pn* junction that is depleted of free carriers. Also called the *depletion region*.

Depletion-mode operation. Using an input voltage to reduce the channel size of a FET.

Depletion MOSFET (D-MOSFET). A MOSFET that can be operated in both the depletion mode and the enhancement mode.

Diac. A two-terminal device whose forward and reverse characteristics are identical to the forward characteristics of the silicon unilateral switch (SUS).

Differentiator. A circuit whose output is proportional to the rate of change of its input signal.

Diffusion current. The I_F through a pn junction when V_F is less than the knee voltage.

Digital circuits. Circuits designed to respond to specific alternating dc voltage levels.

Digital communications. A method of transmitting and receiving information in digital form.

Digital-to-analog converter. A circuit that converts digital circuit outputs to equivalent analog voltages.

Diode. A two-terminal device designed to conduct under specific conditions.

Diode bias. A biasing circuit that uses two diodes in place of the resistor(s) used in class B biasing.

Diode capacitance (C_T). The rated value (or range) of capacitance for a varactor at a specified value of V_R.

Diode capacitance temperature coefficient. A multiplier used to determine the amount by which varactor capacitance changes when temperature changes.

Direct-coupled amplifier. A multistage amplifier where the output from the first stage is connected directly to the input of the next.

Discrete. A term used to describe devices that are packaged in individual casings.

Distortion. An unwanted change in the shape of an ac signal.

Donor atoms. Another name for pentavalent atoms.

Doping. Adding impurity elements to intrinsic semiconductors to improve their conductivity.

Drain. The FET equivalent of the BJT's collector.

Drain-feedback bias. The MOSFET equivalent of collector-feedback bias.

Drift. A change in tuning caused by component aging.

Driver. A circuit that is used to couple a low-current output to a relatively high current device.

Dual-gate MOSFET. A MOSFET constructed with two gates to reduce the overall gate input capacitance.

Dual-polarity power supply. A dc supply that provides both positive and negative dc output voltages.

Dual-tracking regulator. A regulator that provides equal +dc and −dc output voltages.

Duty cycle. The ratio of pulse width (PW) to cycle time (T_C), given as a percent.

Efficiency. The ratio of amplifier ac output voltage to dc input voltage, given as a percent.

Electrical characteristics. The guaranteed operating characteristics of a device, listed on the device specification sheet.

Electron-hole pair. An electron and its matching valence-band hole.

Electronic tuning. Using voltages to control the tuning of a circuit.

Emitter. One of three input terminals to a BJT.

Emitter-base junction. One of the two junctions in a BJT. The other is the collector-base junction.

Emitter bias. A BJT biasing circuit that consists of a dual-polarity power supply and a grounded base resistor.

Emitter-feedback bias. A bias circuit constructed so that V_E has a direct effect on V_B.

Emitter follower. See *common-collector amplifier*.

Energy gap. The space between two orbital shells.

Enhancement-mode operation. Using an input voltage to increase the channel size of a MOSFET.

Enhancement MOSFET (E-MOSFET). MOSFETs that are restricted to enhancement-mode operation.

Extrinsic. Another word for *impure*.

Fall time (t$_f$). The time required for a BJT to make the transition from saturation to cutoff.

False triggering. When a noise signal triggers a thyristor into conduction.

Feedback amplifier. An amplifier that provides a signal path from its output back to its input.

Feedback bias. A term used to describe a circuit in which the output values of the circuit directly affect its biasing.

Feedback factor. A value that appears in almost all the equations for a given negative feedback amplifier.

Feedback network. The feedback signal path.

Field effect transistor. A three-terminal, voltage-controlled device used in amplification and switching applications.

Filter. A circuit that reduces the variations in the output from the rectifier in a dc power supply.

Fixed bias. See *base bias*.

Fixed-negative regulator. A regulator with a predetermined negative dc output voltage.

Fixed-positive regulator. A regulator with a predetermined positive dc output voltage.

Flip-flop. Another name for the *bistable multivibrator*.

Flywheel effect. A term used to describe the ability of a parallel LC circuit to self-oscillate for a brief period of time.

Forced commutation. Driving a thyristor into cutoff by forcing a reverse current through the device.

Forward bias. A potential used to reduce the resistance of a pn junction.

Forward blocking region. The off-state (nonconducting) region of thyristor operation.

Forward breakover current. The value of I$_F$ at the point where breakover occurs for a given thyristor.

Forward breakover voltage. The value of forward voltage that forces a breakover thyristor into conduction.

Forward operating region. The on-state (conducting) region of device operation.

Forward power dissipation. A rating that indicates the maximum allowable power dissipation of a forward-biased device.

Forward voltage. The voltage across a forward biased pn junction.

Free-running multivibrator. Another name for the *astable multivibrator*.

Frequency-response curve. A curve showing the relationship between amplifier gain and operating frequency.

Full load. A term used to describe a minimum load resistance which draws maximum load current.

Gain. A multiplier that exists between the input and output of an amplifier.

Gain-bandwidth product. A constant, equal to the unity-gain frequency of an op-amp. The product of A$_{CL}$ and bandwidth will always be approximately equal to this constant.

Gain-stabilized amplifier. A term that is often used to describe a *swamped amplifier*.

Gate. The JFET equivalent of the BJT base.

Gate bias. The JFET equivalent of base bias.

Gate current. The maximum amount of current that can be drawn through the gate of a JFET without damaging the device.

Gate nontrigger voltage. The maximum gate voltage that can be applied to an SCR without triggering the device into conduction.

Gate reverse current. The maximum amount of gate current that will occur in a JFET when the gate-source junction is reverse biased.

Gate-source breakdown voltage. The value of V$_{GS}$ that will cause the gate-source junction to break down.

Gate-source cutoff voltage. The value of V$_{GS}$ that causes FET drain current to drop to approximately zero.

Gate trigger voltage. The value of V$_{GK}$ that will cause I$_G$ to become great enough to trigger an SCR into conduction.

Gate turn-off device. A type of SCR that can be driven into cutoff by applying a negative pulse to its gate.

General-class equation. An equation derived for a summing amplifier that is used to predict the output voltage for any combination of input voltages.

Graphic equalizer. A circuit or system designed to allow you to control the amplitudes of different audio frequency ranges.

h-parameters. Transistor specifications that describe the operation of the device under full-load or no-load conditions.

Half-wave rectifier. A circuit that produces pulsating dc by eliminating with the positive alternations or the negative alternations of its ac input.

Half-wave voltage doubler. A half-wave circuit which provides a dc output voltage that is approximately twice its ac peak input voltage.

Hartley oscillator. An oscillator that uses a pair of tapped inductors and a parallel capacitor in its feedback network to produce the 180° phase shift required for oscillations.

Heat sink. A large metallic object that helps to cool devices by increasing their effective surface area.

Heat-sink compound. A chemical used to aid in the transfer of heat from a device to its heat sink.

High-current transistors. BJTs with high maximum I$_C$ ratings.

High-pass filter. A filter designed to pass all frequencies above its cutoff frequency.

High-power transistors. BJTs with high forward power dissipation ratings.

High-voltage transistors. BJTs with high reverse breakdown voltage ratings.

Holding current. The minimum value of I$_F$ required to maintain thyristor forward conduction.

Hole. A gap in a covalent bond.

Hysteresis. A term that is (loosely) used to describe the range of voltages between the LTP and UTP values of a Schmitt trigger.

IC voltage regulator. A device that is used to hold the output voltage from a dc power supply relatively constant over a range of line and load demands.

Impedance matching. A circuit design technique that ensures maximum power transfer between a source and its load.

Industrial electronics. The area of electronics that deals with the devices, circuits, and systems that are used in manufacturing.

Input/output voltage differential. The maximum allowable difference between V$_{in}$ and V$_{out}$ for an IC voltage regulator.

Integrated circuit (IC). An entire circuit that is constructed on a single piece of semiconductor material.

Integrator. A circuit whose output is proportional to the area of its input waveform.

Interbase resistance. The total resistance between the two bases of a UJT, measured when the emitter of the device is not conducting.

Intrinsic. Another term for *pure*.

Intrinsic standoff ratio. The ratio of emitter-base 1 resistance to the interbase resistance for a given UJT.

Inverter. A basic switching circuit that produces a 180° voltage phase shift. A logic-level converter.

Isolation capacitance. The total capacitance between the input and output pins of an optoisolator.

Isolation current. The amount of current that can be forced between the input and output pins of an optoisolator at the rated voltage.

Isolation resistance. The total resistance between the input and output pins of an optoisolator.

Isolation source voltage. The voltage that, if applied across the input and output pins of an optoisolator, will destroy the device.

Kick emf. See *counter emf*.

Knee voltage. The voltage at which device current suddenly increases or decreases.

Leaky. A term used to describe a component that is partially shorted.

Level detector. Another name for a comparator that is being used to compare an input signal to a fixed dc reference voltage.

Lifetime. The time between electron-hole pair generation and recombination.

Light. Electromagnetic energy that falls with a specific range of frequencies.

Light-activated SCR. LASCR. A three-terminal SCR that is optically activated.

Light current. The reverse current through a photodiode with an active light input.

Light emitter. An optoelectronic device that generates light.

Light detector. An optoelectronic device that responds to an active light input.

Light intensity. The amount of light per unit area that is received by a given photodetector.

Limiter. Circuits used to eliminate a portion of an ac input signal. Also called a *clipper*.

Line regulation. A rating that indicates the change in regulator output voltage that will occur per unit change in input voltage.

Load regulation. A rating that indicates the change in regulator output voltage that will occur per unit change in load current.

Loaded-Q. The Q of a tank circuit when a load is present.

Logic family. A group of digital circuits with nearly identical input and output characteristics.

Logic levels. The dc signal voltages for a digital circuit.

Low-pass filter. A filter designed to pass all frequencies below a given cutoff frequency.

Low trigger point (LTP). A reference voltage for a Schmitt trigger. When a negative-going input passes the LTP, the output of the circuit goes to a specified dc level.

Majority carriers. The primary charge carriers in doped semiconductors. In n-type materials, conduction-band electrons. In p-type materials, valence-band holes.

Maximum limiting voltage. The voltage at which a constant-current diode starts to limit current.

Maximum ratings. Device parameters that must not be exceeded under any circumstances.

Maximum reverse stand-off voltage. The maximum peak voltage that will not turn on a transient suppressor.

Maximum temperature coefficient. The percent of change in V_{BR} per 1°C rise in operating temperature.

Midpoint bias. Having a Q-point that is centered on the circuit load line.

Midpoint frequency. The frequency that equals the geometric average of the cutoff frequencies (f_1 and f_2) for a given amplifier.

Miller integrator. Another name for the basic op-amp integrator.

Miller's theorem. A theorem that derives equivalent input and output capacitances for a feedback capacitor in an inverting amplifier.

Minimum load current. The value of I_L below which regulation is lost.

Minority carriers. The secondary charge carriers in doped semiconductors. In n-type materials, valence-band holes. In p-type materials, conduction-band electrons.

Model. A representation of a component or circuit.

Modulator. A circuit that combines two signals of different frequencies into a single composite signal.

Monostable multivibrator. A switching circuit with one stable output state. Also called a *one-shot*.

Multicolor LED. An LED that emits one color when forward biased and another color when reverse biased.

Multiple-feedback filter. A band-pass (or band-stop) filter that has a single op-amp and two feedback paths, one resistive and one capacitive.

Multisegment display. An LED circuit used to display alphanumeric symbols.

Multistage amplifier. An amplifier made up of a series of cascaded stages.

Multivibrators. Switching circuits designed to have zero, one, or two stable output states.

N-type material. Silicon or germanium that is doped with pentavalent impurities.

Negative feedback. A type of feedback in which the feedback signal is 180° out of phase with the amplifier input signal.

Negative resistance. A term used to describe a region of operation where current and voltage are inversely proportional.

Negative resistance oscillator. An oscillator whose operation is based on a negative resistance device, such as a tunnel diode or a unijunction transistor.

Noise. Any unwanted interference with a transmitted signal.

Nominal zener voltage. The rated value of V_Z for a given zener diode.

Nonlinear distortion. A type of distortion that is caused by driving the base-emitter junction of a BJT into its nonlinear operating region.

Nonrepetitive surge current (I_{TSM}). An SCR rating that indicates the absolute limit on any forward surge current through the device.

Npn transistor. A BJT with n-type emitter and collector materials and a p-type base material.

Octave. A frequency multiplier of 2.

Off characteristics. The characteristics of a discrete device that is in cutoff.

On characteristics. The characteristics of a discrete device that is in either the active or saturation regions of operation.

One-shot. Another name for the monostable multivibrator.

Opaque. A term used to describe anything that blocks light.

Optocoupler. A device that uses light to couple a signal from its input (a photoemitter) to its output (a photodetector).

Optoelectronic devices. Devices that are controlled by and/or emit light.

Optointerruptor. An optocoupler designed to allow an outside object block the path between its photoemitter and its photodetector.

Optoisolator. Another name for an optocoupler.

Oscillations. Output cycles produced by an oscillator.

Oscillator. A circuit designed to produce a sinusoidal, triangular, or rectangular output. A circuit that converts dc to ac.

Oscillator stability. A measure of an oscillator's ability to maintain constant output amplitude and frequency.

P-type material. Silicon or germanium that has trivalent impurities.

Parameter. A limit.

Passband. The range of frequencies that is passed by an amplifier.

Pass-transistor regulator. A regulator that uses a series transistor to maintain a constant load voltage.

Peak inverse voltage (PIV). The maximum reverse bias that a diode will be exposed to in a circuit.

Peak repetitive reverse voltage (V_{RRM}). The maximum allowable value of reverse bias for a given pn junction diode.

Pentavalent elements. Elements with five valence shell electrons.

Phase controller. A circuit used to control the conduction phase angle through a load, and thus, the average load voltage.

Phase-shift oscillator. An oscillator that uses three RC circuits in its feedback network to produce the 180° phase shift required for oscillations.

Photo-Darlington. A phototransistor with a Darlington-pair output.

Photodiode. A diode whose reverse conduction is light intensity controlled.

Phototransistor. A transistor whose collector current is controlled by the light intensity at its optical input (base).

Piezoelectric effect. The tendency of a crystal to vibrate at a fixed frequency when subjected to an electric field.

Pinchoff voltage. With V_{GS} constant, the value of drain-source voltage that causes maximum drain current in a FET.

Pnp transistor. A BJT with p-type collector and emitter terminals and an n-type base material.

Positive feedback. A type of feedback in which the feedback signal is in phase with the amplifier input signal.

Power gain. The factor by which ac signal power increases from the input of an amplifier to its output.

Power rectifiers. Diodes with extremely high forward current and/or power dissipation ratings.

Power supply. A group of circuits used to convert ac to dc.

Precision diode. A circuit that consists of a diode and an op-amp. Characterized by the ability to conduct at extremely low forward voltages.

Propagation delay. The time required for a signal to pass through a component or circuit, measured at the 50% points on the input and output waveforms.

Pulse width. The time spent in the high voltage state.

Push-pull emitter follower. See Complementary-symmetry amplifier.

Q (tank circuit). The ratio of energy stored in a circuit to energy lost by the circuit.

Q-point. A point on a load line that corresponds to the circuit dc output current and voltage values.

Q-point shift. A condition where a change in operating temperature indirectly causes the Q-point values of current and voltage to change.

Quiescent. At rest.

Ramp. Another name for a voltage that changes at a constant rate.

Recombination. When a free electron returns to the valence shell.

Rectangular waveform. A waveform made up of alternating (high and low) dc voltages.

Rectifier. A circuit that converts ac to pulsating dc.

Regenerative feedback. Another name for positive feedback.

Regulated dc power supply. A dc power supply with extremely low internal resistance.

Regulation. A rating that indicates the maximum change in regulator output voltage that can occur when input voltage and load current are varied over their entire rated ranges.

Relaxation oscillator. A circuit that uses the charge/discharge characteristics of a capacitor or inductor to produce a pulse output.

Reverse bias. A potential that causes a pn junction to have extremely high resistance.

Reverse breakdown voltage. The minimum reverse voltage large enough to cause a pn junction to break down and conduct.

Reverse current. The minority carrier current through a reverse biased pn junction.

Reverse saturation current (I_S). A current that is caused by thermal activity in a reverse biased pn junction. I_S is temperature dependent.

Reverse voltage. The voltage across a reverse biased pn junction.

Reverse voltage feedback ratio (h_{re}). The ratio of v_{ce} to v_{be} for a given transistor, measured under no-load conditions.

RF amplifier. Radio-frequency amplifier. The first circuit that the input signal to a radio receiver is applied to.

Ripple rejection ratio. The ratio of regulator input ripple voltage to maximum regulator output ripple voltage.

Ripple voltage. The remaining variations in the output voltage from a filtered rectifier.

Rise time. The time required for a BJT to go from cutoff to saturation; measured as the time required for I_C to rise from 10% to 90% of its maximum value.

Roll-off rate. The rate of gain reduction for a circuit when operated beyond its cutoff frequencies.

Saturation. A BJT operating region where I_C reaches the maximum value that the circuit will allow.

Saturation clipping. A type of distortion caused by driving a BJT into saturation.

Schmitt trigger. A voltage level detection circuit.

Schottky diode. An ultra-fast diode with very little junction capacitance.

Self-bias. A JFET biasing circuit that uses a source resistor to establish a negative value of V_{GS}.

Semiconductor. An element that falls halfway between an insulator and a conductor.

Sensitivity. A rating that indicates the response of a photodetector to a specified light intensity.

Series current regulator. A circuit that uses a constant-current diode to maintain a constant circuit input current over a wide range of input voltages.

Series feedback regulator. A series voltage regulator that uses an error-detection circuit to improve the line and load regulation characteristics of other pass-transistor regulators.

Series regulator. A voltage regulator in which the primary active component is in series with the load.

Seven-segment display. A display made up of seven LEDs shaped in a figure-8 that is used (primarily) to display numbers.

Shorted-gate drain current (I_{DSS}). The value of I_D for an FET when $V_{GS} = 0$ V. The maximum value of I_D for a JFET.

Shunt clipper. A clipper with the diode placed in parallel with the load.

Shunt feedback regulator. A voltage regulator that uses an error detection circuit to control the conduction of a shunt regulating transistor.

Shunt regulator. A voltage regulator in which the primary active device is in parallel with the load.

Silicon bilateral switch (SBS). A thyristor trigger that is designed to provide high-current pulses for triggering high-power thyristors.

Silicon-controlled rectifier (SCR). A three terminal device that is very similar in construction and operation to the SUS. A third terminal, called the gate, provides another means of triggering the device.

Silicon unilateral swtich (SUS). A two-terminal, four-layer device that can be triggered into conduction by applying a specified forward voltage across its terminals.

Snubber network. An RC circuit that is connected between the SCR anode and cathode to eliminate false triggering.

Solid-state relay (SSR). A circuit that uses a dc input voltage to pass or block an ac signal.

Source. The FET equivalent of the BJT's emitter terminal.

Source follower. The common-drain amplifier. The FET equivalent of the emitter follower.

Space width. The time spent in the low voltage state. The time between pulses in switching circuits.

Specification sheet. A listing of all the important parameters and operating characteristics of a device or circuit.

Spectral response. A measure of a photodetector's response to a change in input wavelength.

Speed-up capacitor. A capacitor that is used to reduce propagation delay in a BJT switching circuit by reducing t_s and t_d.

Square wave. A special-case rectangular waveform that has equal pulse width and space width values.

Stage. A single amplifier in a cascaded group of amplifiers.

Static reverse current (I_R). The reverse current through a diode when $V_R < V_{BR}$.

Step-recovery diode. A heavily doped diode with an ultra-fast switching time.

Stop band. The range of frequencies that is outside of an amplifier's pass band.

Storage time (t_s). The time required for a BJT to come out of saturation.

Substrate. The foundation material of a MOSFET.

Subtractor. A summing amplifier that provides an output which is proportional to the difference between the input voltages.

Summing amplifier. A circuit that produces an output which is proportional to the sum of its input voltages.

Surface leakage current (I_{SL}). A current that passes along the surface of a reverse biased pn junction. I_{SL} is dependent on the value of reverse voltage.

Surge current. The high initial current in a power supply. Any current caused by a non-repeating high-voltage condition.

Swamped amplifier. An amplifier that uses a partially bypassed emitter resistance to reduce variations in gain caused by changes in ac emitter resistance (r'_e).

Switching circuits. Circuits designed to respond to (or generate) nonlinear waveforms, such as square waves.

Switching time constants. A term used to describe a condition where a capacitor has different charge and discharge times.

Switching transistor. A device with extremely low switching times.

Symmetrical zener shunt clipper. A shunt clipper that uses two back-to-back zener diodes.

Thermal contact. Placing two or more components in physical contact with each other so that their operating temperatures will be equal.

Threshold voltage ($V_{GS,th}$). The value of V_{GS} that establishes a channel in an E-MOSFET.

Thyristors. Devices designed specifically for high-power switching applications.

Transconductance (g_m). A ratio of a change in drain current to a change in V_{GS}, measured in microsiemens (μS) or micromhos (μmhos).

Transconductance curve. A plot of all possible combinations of V_{GS} and I_D.

Transient. An abrupt current or voltage spike. A surge.

Transient suppressor. A zener diode that has extremely high surge-handling capabilities.

Transistor. A three terminal device whose output current is controlled (under normal circumstances) by its input current.

Triac. A bidirectional thyristor whose forward and reverse characteristics are identical to the forward characteristics of the SCR.

Trigger-point voltage (V_{TP}). The voltage required at the gate of an SCR to trigger the device into conduction.

Trivalent elements. Elements with three valence electrons.

Troubleshooting. The process of locating faults in electronic equipment.

Tuned amplifier. An amplifier designed to have a specific bandwidth.

Tunnel diode. A heavily doped diode that is used in high-frequency communications systems.

Turn-off time (t_{off}). The combination of storage time (t_s) and fall time (t_f).

Turn-on time (t_{on}). The combination of delay time (t_d) and rise time (t_r).

Unijunction transistor (UJT). A three-terminal device whose trigger voltage is proportional to its applied biasing voltage.

Universal motor. A motor whose rotational speed is directly proportional to the average motor voltage.

Upper trigger point (UTP). A reference voltage. When a positive-going input reaches the UTP, the Schmitt trigger output switches states.

Valence shell. The outermost shell of an atom. The number of electrons in the valence shell determines the conductivity of the atom.

Varactor. A diode that has high junction capacitance when reverse biased.

Variable comparator. A comparator with an adjustable reference voltage.

Vertical MOS (VMOS). An E-MOSFET that is designed to handle high values of drain current.

Voltage-controlled current source. A circuit with a constant-current output that is controlled by the input voltage.

Voltage-controlled oscillator (VCO). A square wave oscillator whose output frequency is controlled by a dc input voltage.

Voltage-divider bias. A biasing circuit that contains a voltage-divider in the input circuit.

Voltage feedback. A circuit configuration in which a portion of the output voltage is fed back to the circuit input.

Voltage gain (A_v). The factor by which ac voltage increases from the input of an amplifier to its output.

Voltage multiplier. A circuit that produces a dc output voltage which is some multiple of its peak input voltage.

Voltage regulator. A circuit designed to maintain a constant dc load voltage despite changes in line voltage or load current demand.

Wavelength. The physical length of one cycle of a transmitted electromagnetic wave.

Wavelength of peak spectral response. A rating that indicates the wavelength that will cause the strongest response in a photodetector.

Wien-bridge oscillator. An oscillator that achieves regenerative feedback using a series RC circuit that is in series with a parallel RC circuit.

Worst-case analysis. A mathematical analysis that gives you the absolute limits of normal circuit operation.

Zener breakdown. A type of reverse breakdown that occurs at very low values of V_R.

Zener clamper. A clamper that uses a zener diode to set the reference voltage.

Zener shunt clipper. A clipper that uses a zener diode that is connected in parallel with the load.

Zener diode. A diode that is designed to work in its reverse breakdown region of operation.

Zener impedance (Z_Z). A zener diode's opposition to a change in zener current.

Zener voltage (V_Z). The approximate voltage across a zener diode when it is operated in its reverse breakdown region of operation.

Zero bias. A term used to describe the biasing of a pn junction at room temperature with no potential applied. A D-MOSFET biasing circuit.

APPENDIX F
Answers To Selected Odd-Numbered Problems

Chapter 2:

7. V_{D1} = 10 V, V_{D2} = 0 V **9.** V_{D1} = 0.7 V, V_{R1} = 0.3 V, I_T = 3 mA **11.** V_{D1} = 10 V, V_{R1} = 0 V, I_1 = 0 A, V_{D2} = 0.7 V, V_{R2} = 9.3 V, I_2 = 5.17 mA **13.** V_{D1} = V_{D2} = 0.7 V, I_T = 25.3 mA, V_{R1} = 2.53 V, V_{R2} = 5.06 V **15.** 14.8%, not acceptable **17.** 5.51% **19.** V_{RRM} > 100 V **21.** V_{RRM} = 200 V **23.** 6.77 mW **25.** 1.07 A **27.** 0.698 V (698 mV) **29.** 100 mV **31.** 2.16 V **33.** 300 μA **35.** MR2504 **37.** MR756 **39.** Figures a, c, d, and e. **41.** 147 mA **43.** 4.44 W **45.** 1N5362A **47.** 1N5349A **49.** 1.6 kΩ (5% tolerance) or 1.8 kΩ (10% tolerance) **51.** good **53.** open

Chapter 3:

1. 29.3 V_{pk} **3.** 16.27 V_{pk} **5.** 9.33 V_{dc} **7.** 5.18 V_{dc} **9.** −15.57 V_{pk}, −4.96 V_{dc}, 972.55 μA **11.** 16.27 V_{pk} **13.** 20.52 V_{pk}, 13.06 V_{dc}, 1.31 mA **15.** 20.52 V_{pk}, 13.06 V_{dc}, 2.11 mA **17.** 33.95 V_{pk} **19.** −20.52 V_{pk}, −13.06 V_{dc}, 1.3 mA **21.** 41.03 V_{pk}, 4.1 mA, 26.12 V_{dc}, 2.61 mA **23.** 22.63 V **25.** 2.02 A **27.** 13.42 V_{dc}, 44.84 mV_{PP} **29.** 520 mV_{PP}, 23.8 V_{dc}, 29.34 mA **31.** 33.95 V_{pk} **33.** 13.33 mA **35.** 10.33 mA **37.** 487Ω **39.** 52.53 mA **41.** 25 mA, 20 mA, 5 mA **43.** 9.1 V_{dc}, 4.55 mA **45.** 33 V, 33 mA, 139.29 mV_{PP} **47.** 33 V, 33 mA, 4.71 V_{PP} **49.** One of the diodes is open. **51.** Yes **53.** There is a problem.

Chapter 4:

1. 18.02 V_{pk} **3.** −11.25 V_{pk} **5.** −0.7 V, 3.98 V_{pk} **7.** −0.7V, +10 V_{pk} **9.** 0.7 V, −12 V_{pk} **11.** ±10.7 V **13.** −8.7 V, 12 V_{pk} **15.** The signal is not clipped. **19.** 0 V_{pk} and 14 V_{pk} **21.** 70 μs, 7.5 ms **23.** 3 V_{pk} and 12 V_{pk} **25.** 6 V_{pk} and −12 V_{pk} **27.** 15 V_{dc}, 30 V_{dc} **29.** 24 V_{dc}, 24 V_{dc}, 48 V_{dc} **31.** 56.58 V_{dc} **33.** 20 V_{dc}, 40 V_{dc}, 20V_{dc}, 60 V_{dc} **35.** 18 V_{dc}, 36 V_{dc}, 18 V_{dc}, 36 V_{dc}, 72 V_{dc} **37.** ±67.89 V_{dc} **39.** D_1 is open. **41.** D_1 is open. **43.** D_1 is shorted. **45.** V_{C4} is low. **51.** 1000 V_{dc}

Chapter 5:

1. 3.84 mA **3.** 256.54 mA **5.** 1.12 mA **7.** 3.75 mA, 20, 24 mA, 61.54 μA **11.** 60 mA, 60.15 mA **13.** 27.3 μA, 12.03 mA **15.** 170.6 μA, 64.83 mA **17.** 27.944 mA, 56 μA **19.** 13.02 mA, 1.02 mA **21.** 0.9977 **23.** 1 mA **25.** 34 V **27.** 2 V

Chapter 6:

7. 4 mA, 5 V **9.** 4.71 mA, −7.29 V **11.** Midpoint biased **13.** Not midpoint biased **15.** 6 mA, 2.5 V **17.** 1.88 mA, 6.36 V **19.** 2.67 mA, 24 V **21.** 4.77 mA, 7.37 V, 6.25 V, 30 V **23.** 660 kΩ **25.** 2.03 mA, 17.8 V, 13.44 μA **27.** 11.75 mA, −5.9 V, 64.92 μA **29.** 75 **31.** 5 mA, 30 V, not midpoint biased **33.** 16.67 mA, −20 V, not midpoint biased **35.** Not midpoint biased. **37.** 1.83 mA, 4.34 V **39.** 1.52 mA, −5.44 V **41.** 1 mA, 30 V **43.** 11.27 mA, −6.32 V **45.** Open transistor (C-E) **47.** R_C is open. **49.** R_1 is open. **51.** 10 μA

Chapter 7:

1. 2.08 Ω **3.** 10.33 Ω **5.** 12.02 Ω **7.** 32.34 Ω **13.** 33.33 **15.** 96.49 **17.** 108 **19.** 185.53 **21.** 319.72 **23.** 3.84 Vac **25.** 27.8 × 10^3 **27.** 47.96 × 10^3 **29.** 44 W **31.** 110.22 **33.** 63.13 **35.** 4.85 kΩ, 1.34 kΩ **37.** 3.83 kΩ, 2.71 kΩ **39.** 150.8 **41.** 67.6 **43.** 8 × 10^3, 2.31 × 10^4, 1.85 × 10^8 **45.** 6.24 × 10^5, 1.2 × 10^6, 7.49 × 10^{11} **47.** 21.49 **49.** 61.5 kΩ, 1.8 kΩ **51.** 100, 5 kΩ, 50 Ω, 76 **53.** 132, 3.35 kΩ, 97.7 **55.** 176, 968 Ω, 159.6

Chapter 8:

1. 6 V, 5.3 V, 2.65 mA **5.** 0.9862, 100, 98.62 **7.** 4.83 kΩ **9.** 21 Ω **11.** A_{vL} = 0.9639, A_i = 80, A_p = 77.1, Z_{in} = 3.25 kΩ **13.** 85 kΩ **15.** 190.3 **17.** 17.5Ω, 5 kΩ **19.** 10.13 mA, 6.3 MΩ **21.** 6.4 × 10^3 **23.** 19.89 kΩ, 18.9 kΩ **25.** A_i = 6.4 × 10^3, A_v ≅ 1, Z_{in} ≅ 19.89 kΩ, Z_{out} ≅ 18.75 Ω **27.** R_1 is open or Q_1 is open. **29.** R_6 is open. **31.** R_1 is open or Q_1 is open. **33.** Q_1 is leaky.

Chapter 9:

1. 9.06 V_{PP} **3.** 6.8 V_{PP}, saturation **5.** 53.6 mW **7.** 187.5 μW **9.** 855 μW **11.** 1.6% **13.** 131.64 mW, 1.93 mW, 1.47% **15.** 2.83% **19.** 250 mW **21.** 37.4% **25.** 13.5 mW **27.** 23.14 W **29.** 18 W **31.** 77.8% **33.** 4 V, 4.7 V, 3.3 V **35.** 68.99% **37.** 78% **39.** 34 mW **41.** 468.3 mW **43.** 133.3 mW **45.** 312.5 mW **51.** 7.41 mW **53.** Yes, cutoff clipping

Chapter 10:

1. 2 mA **3.** 14 mA **5.** 4.32 mA **13.** 1.5 mA, 5.5 V **15.** 2 mA to 3.3 mA **17.** 0 μA to 640 μA **19.** V_{GS} = −150 mV to −1.4 V, I_D = 250 μA to 1.4 mA, V_{DS} = 5.98 V to 10.93 V **21.** V_{GS} = −500 mV to −2.8 V, I_D = 500 μA to 3 mA, V_{DS} = 7 V to 14.5 V **23.** 1 mA to 1.5 mA, 13.35 V to 18.9 V **25.** 2.2 mA to 3.2 mA, 11.2 V to 15.2 V **27.** 1000 μS to 5600 μS **29.** 4800 μS, 3000 μS, 1200 μS **31.** 6.25 to 7.82 **33.** 1.37 to 2.09 **35.** 2.54 to 3.34 **37.** 667 kΩ **39.** 0.3333, 2.55 MΩ, 500Ω **41.** 0.5327, 3.1 MΩ, 216 Ω **43.** 3.82, 192Ω, 714Ω **45.** The JFET gate is shorted. **47.** C_{C2} is shorted.

Chapter 11:

5. 1000 μS, 2000 μS, 3000 μS **7.** 3 mA **9.** 0 mA, 8 mA, 14.22 mA **11.** 8 mA, 16 V **13.** 10 mA, 7.2 V **15.** 20 mA, 6 V **17.** Q_1 is open. **19.** 2.6 to 3.2 **21.** 0 mA to 1.8 mA **23.** 200 Ω

Chapter 12:

1. (+), (−), (+) **3.** 16 V_{PP} **5.** 133.3 mV_{PP} **7.** 150 mV_{PP} **9.** 12 mV **11.** 143.3 kHz **13.** 35.4 kHz **15.** A_{CL} = 120, Z_{in} ≅ 1 kΩ, Z_{out} < 50 Ω, CMRR = 6000, f_{max} = 79.6 kHz **17.** A_{CL} = 1, Z_{in} ≅ 100 kΩ, Z_{out} ≥ 75 Ω, CMRR = 200, f_{max} = 159 kHz **19.** A_{CL} = 101, Z_{in} > 1 MΩ, Z_{out} < 40 Ω, CMRR = 50,500, f_{max} = 78.3 kHz **21.** A_{CL} = 41, Z_{in} > 2 MΩ, Z_{out} < 100 Ω, CMRR = 1242, f_{max} = 291.3 kHz **23.** A_{CL} = 1, Z_{in} = 5 MΩ, Z_{out} = 40 Ω, CMRR = 1000, f_{max} = 318.5 kHz **25.** R_f is open. **31.** 500,000

Chapter 13:

1. 638 kHz, 27.7 kHz **3.** 484.4 kHz, 50.5 kHz **5.** Ratio: 13.64 **7.** 9.5 kHz, 538.5 kHz **9.** 1.152 MHz, 1.15 MHz **11.** 26.99 dB, 10 dB, 35.03 dB, −3.01 dB **13.** 232.98 mW, 25 W, 792.45 mW, 477.73 mW **15.** 1.79 × 10⁸ (82.53 dB) **17.** 39.8 mW **19.** 80 dB, 102.5 dB, −6 dB, −0.24 dB **21.** 75.7 mV, 17.8 mV, 70.8 mV, 6.35 V **23.** 520 Hz **25.** 169 Hz **27.** 203 Hz **29.** 28.4 dB, 26.5 dB, 19.7 dB, 13.9 dB **31.** 520 Hz **33.** 318.3 pF **35.** 7 pF **37.** 533.7 kHz **39.** 13.45 MHz **41.** 80 Hz, 583.6 kHz **43.** 79 Hz **45.** 241 Hz **47.** 2.1 MHz, 7.96 MHz **49.** 9 pF, 5 pF, 1 pF **51.** 796 Hz, 1.3 MHz **53.** 30 kHz **55.** 198.6 kHz **57.** It can be used. **59.** 12 MHz **61.** 25 MHz **63.** 102 kHz **65.** 4.66 kHz **67.** 539.7 kHz **69.** 165.8 kHz

Chapter 14:

1. 4.52 **3.** 76.27 **5.** 10.6 MΩ, 0.995 Ω **7.** 19.6, 1.02 MΩ, 9.8 Ω **9.** 3.9, 255 kΩ, 15.7 Ω **11.** 39.2 Hz, 42.1 MHz, 42.1 MHz **13.** 19.6 Hz, 23.6 MHz, 23.5 MHz **15.** 23.2 **17.** 1.99 **19.** 24 GΩ, 0.005 Ω **21.** 2 MHz **23.** 100, 1.8 kΩ, 100Ω, 10 kHz **25.** 7.57 **27.** 15.47 MHz **29.** 1.14 kΩ, 1.27 kΩ **31.** 37.68, 4.17 kΩ, 14.15 kΩ, 6.56 MHz **33.** 26.3, 27.4 GΩ, 18 MΩ, 45.63 kHz **35.** 2, 240 GΩ, 12 MΩ, 1.5 MHz **37.** 5, 250Ω, 900 MΩ, 1 MHz **39.** R₁ is open. **43.** 14.1, 4.45Ω, 1.2 MΩ

Chapter 15:

1. 7 **3.** 129 kHz **5.** 308.3 kHz, 9.7, 187.5 kHz, 1.53 **7.** 159.2 kHz, 11 **9.** 10.7 kHz, 1.5 **11.** 325.2 Hz, 1.98 **13.** 9.46 kHz, 11 **15.** 112.5 kHz, 250.1 kHz, 137.6 kHz, 167.7 kHz, 1.22 **17.** 8.74 kHz, 19.34 kHz, 10.6 kHz, 13.0 kHz, 1.23 **19.** f₁: 6.87%, f₂: 3.64%. **21.** 1.5 **23.** 1.31 kHz, 1.39 kHz, 80 Hz, 1.35 kHz, 16.88 **25.** 809.3 Hz, 1127.9 Hz, 318.6 Hz, 968.6 Hz, 3.04 **29.** 20 V, 0 A **31.** +3 V **33.** C₂ is open **35.** Δf₁ = Δf₂ = 3.1 kHz

Chapter 16:

1. 0 V **3.** +4 V **5.** 1.33 kHz **7.** 72.34 Hz **9.** 32.15 Hz **11.** 159 Hz, 1.59 kHz **13.** −V_{out} = V₁ + V₂ + V₃, −5 V **15.** −V_{out} = 1.5V₁ + 3V₂ + 1.5V₃, −4.35 V **17.** 3 kΩ **19.** −2 V **25.** 4.1 V, 1.9 V

Chapter 17:

1. 219.2 kHz **3.** 24.35 kHz **5.** 27.84 kHz **7.** 10 **9.** 22.13 kHz **11.** 734.1 kHz **15.** RFC is open.

Chapter 18:

1. 11.3 V **3.** 3.54 V **5.** 0.9 V, 6 V **7.** 17.5 μs, 50 μs **9.** 35% **11.** 17.2 MHz, 2.33 MHz **13.** 1.79 MHz, 1.21 MHz **15.** 11.1 MHz, 2.33 MHz **17.** 3.70 MHz, 538.5 kHz **19.** 2.69 MHz **21.** 62.5 kHz **23.** 20 ns, 60 ns, 180 ns, 70 ns **25.** 4.5 V, −4.5 V **27.** 1.8 V, −0.9 V **29.** 4.62 V, −4.62 V **31.** 5 V, −5 V **33.** 11 ms **35.** 0.7 V **37.** 5.45 kHz, 68.9%, 126 μs **39.** 4.8 kHz, 66.7%, 138.6 μs **41.** R_A is open. **43.** MMBT2222, MMBT2222A, MMBT4401

Chapter 19:

1. No **3.** 1.04 ms **5.** 7.21 A **7.** 100 mV **9.** 8.54 V to 11.20 V **11.** 9.5 V to 13.8 V **13.** 461.5 nm **15.** 1666.7 nm **17.** 319 PHz (P = 10¹⁵) **21.** 34.6 V

Chapter 20:

1. 5 μV/V **3.** 3 μV/V **5.** 133.3 μV/mA **7.** 75 μV/mA **9.** 6.1 V **11.** 3.5 A **13.** 35 V **15.** 12.0 V **17.** 4.2 mV **19.** 2 mV/mA, 4.17%, 0.0417%/mA

Chapter 21:

1. 25 × 10⁻¹⁵ F/°C **3.** 726.4 kHz, 331.6 kHz **5.** 4 kW **7.** 0.4 Vdc **9.** 1N6298

Index